Biotechnology of Lactic Acid Bacteria

Novel Applications

Biotechnology of Lactic Acid Bacteria

Novel Applications

Second Edition

Edited by

Fernanda Mozzi, Raúl R. Raya, and Graciela M. Vignolo

WILEY Blackwell

This edition first published 2016

Registered Office
John Wiley & Sons, Ltd, The Atrium, Southern Gate, Chichester, West Sussex, PO19 8SQ, UK

Editorial Offices
9600 Garsington Road, Oxford, OX4 2DQ, UK
The Atrium, Southern Gate, Chichester, West Sussex, PO19 8SQ, UK
111 River Street, Hoboken, NJ 07030-5774, USA

For details of our global editorial offices, for customer services and for information about how to apply for permission to reuse the copyright material in this book please see our website at www.wiley.com/wiley-blackwell.

Library of Congress Cataloging-in-Publication data applied for.

A catalogue record for this book is available from the British Library.

Wiley also publishes its books in a variety of electronic formats. Some content that appears in print may not be available in electronic books.

Set in 9.5/11.5pt Times by SPi Global, Pondicherry, India

1 2016

Contents

List of Contributors

Editors

Fernanda Mozzi. Centro de Referencia para Lactobacilos (CERELA)-CONICET, Argentina

Raúl R. Raya. Centro de Referencia para Lactobacilos (CERELA)-CONICET, Argentina

Graciela M. Vignolo. Centro de Referencia para Lactobacilos (CERELA)-CONICET, Argentina

Contributors

Tamara Aleksandrzak-Piekarczyk. Institute of Biochemistry and Biophysics, Polish Academy of Sciences, Poland

Thibault Allain. INRA, Commensal and Probiotics-Host Interactions Laboratory, UMR 1319 Micalis, France
AgroParisTech, UMR1319, France

Camille Aubry. INRA, Commensal and Probiotics-Host Interactions Laboratory, UMR 1319 Micalis, France
AgroParisTech, UMR1319, France

M. Andrea Azcarate-Peril. Department of Cell Biology and Physiology, and Microbiome Core Facility, University of North Carolina at Chapel Hill, Chapel Hill, USA

Luis G. Bermúdez-Humarán. INRA, Commensal and Probiotics-Host Interactions Laboratory, UMR 1319 Micalis, France
AgroParisTech, UMR1319, France

Ana Binetti. Instituto de Lactología Industrial (INLAIN), Universidad Nacional del Litoral-CONICET, Argentina

Damien Bouchard. INRA, AGROCAMPUS OUEST UMR1253 Science et Technologie du Lait et de l'Œuf, France

Dag A. Brede. Department of Chemistry, Biotechnology and Life Science, Norwegian University of Life Sciences, Norway

Romain Briandet. INRA and AgroParisTech, UMR1319 Micalis, France

Mariángeles Briggiler Marcó. Instituto de Lactología Industrial (INLAIN), Universidad Nacional del Litoral-CONICET, Argentina

Domenico Carminati. Consiglio per la Ricerca e la Sperimentazione in Agricoltura, Centro di Ricerca per le Produzioni Foraggere e Lattiero-Casearie (CRA-FLC), Italy

Marie-Christine Champomier-Vergès. INRA, UMR 1319 Micalis, France
AgroParisTech, UMR Micalis, France

Jean-Marc Chatel. INRA, Commensal and Probiotics-Host Interactions Laboratory, UMR 1319 Micalis, France
AgroParisTech, UMR1319, France

Alejandra de Moreno de LeBlanc. Laboratorio de Inmunología, Centro de Referencia para Lactobacilos (CERELA)-CONICET, Argentina

Willem M. de Vos. Department of Veterinary Biosciences, University of Helsinki, Finland
Laboratory of Microbiology, Wageningen University, The Netherlands

Luc De Vuyst. Research Group of Industrial Microbiology and Food Biotechnology (IMDO), Department of Bioengineering Sciences, Faculty of Sciences and Bioengineering Sciences, Vrije Universiteit Brussel, Belgium

Raffaella Di Cagno. Department of Soil, Plant and Food Sciences, University of Bari Aldo Moro, Italy

Dzung B. Diep. Department of Chemistry, Biotechnology and Life Science, Norwegian University of Life Sciences, Norway

Grace L. Douglas. Human Health & Performance Directorate, NASA Johnson Space Center, Houston, USA

François P. Douillard. Department of Veterinary Biosciences, University of Helsinki, Finland

Sergine Even. INRA, AGROCAMPUS OUEST UMR1253 Science et Technologie du Lait et de l'Œuf, France

Giovanna E. Felis. Department of Biotechnology, University of Verona, Italy

María Fernández. Departamento de Microbiología y Bioquímica, Instituto de Productos Lácteos de Asturias (IPLA-CSIC), Spain

Pasquale Filannino. Department of Soil, Plant and Food Sciences, University of Bari Aldo Moro, Italy

Christina Gabrielsen. Department of Chemistry, Biotechnology and Life Science, Norwegian University of Life Sciences, Norway

Pilar García. DairySafe Group, Instituto de Productos Lácteos de Asturias (IPLA-CSIC), Spain

Giorgio Giraffa. Consiglio per la Ricerca e la Sperimentazione in Agricoltura, Centro di Ricerca per le Produzioni Foraggere e Lattiero-Casearie (CRA-FLC), Italy

Marco Gobbetti. Department of Soil, Plant and Food Sciences, University of Bari Aldo Moro, Italy

Daniela Guglielmotti. Instituto de Lactología Industrial (INLAIN), Universidad Nacional del Litoral-CONICET, Argentina

Elvira M. Hebert. Centro de Referencia para Lactobacilos (CERELA)-CONICET, Argentina

Marianela Juárez del Valle. Centro de Referencia para Lactobacilos (CERELA)-CONICET, Argentina

Todd R. Klaenhammer. Department of Food, Bioprocessing, and Nutrition Sciences, and Southeast Dairy Foods Research Center, North Carolina State University, Raleigh, USA

Magdalena Kowalczyk. Institute of Biochemistry and Biophysics, Polish Academy of Sciences, Poland

Jonathan Emiliano Laiño. Centro de Referencia para Lactobacilos (CERELA)-CONICET, Argentina

Philippe Langella. INRA, Commensal and Probiotics-Host Interactions Laboratory, UMR 1319 Micalis, France
AgroParisTech, UMR1319, France

Yves Le Loir. INRA, AGROCAMPUS OUEST UMR1253 Science et Technologie du Lait et de l'Œuf, France

Jean Guy LeBlanc. Centro de Referencia para Lactobacilos (CERELA)-CONICET, Argentina

Aline Lonvaud-Funel. University of Bordeaux, ISVV, France

Graciela L. Lorca. Department of Microbiology and Cell Science, Genetics Institute and Institute of Food and Agricultural Sciences, University of Florida, Gainesville, USA

Sylvie Lortal. INRA, AGROCAMPUS OUEST UMR1253 Science et Technologie du Lait et de l'OEuf, France

Carolina Maldonado Galdeano. Laboratorio de Inmunología, Centro de Referencia para Lactobacilos (CERELA)-CONICET, Argentina
Catedra de Inmunologia, Instituto de Microbiologia, Facultad de Bioquimica, Quimica y Farmacia, Universidad Nacional de Tucuman, Argentina

Beatriz Martínez. DairySafe Group, Instituto de Productos Lácteos de Asturias (IPLA-CSIC), Spain

Baltasar Mayo. Departamento de Microbiología y Bioquímica, Instituto de Productos Lácteos de Asturias (IPLA-CSIC), Spain

Maria Fiorella Mazzeo. Centro di Spettrometria di Massa Proteomica e Biomolecolare, Istituto di Scienze dell'Alimentazione, CNR, Italy

Fernanda Mozzi. Centro de Referencia para Lactobacilos (CERELA)-CONICET, Argentina

Jane M. Natividad. INRA, Commensal and Probiotics-Host Interactions Laboratory, UMR 1319 Micalis, France
AgroParisTech, UMR1319, France

Ingolf F. Nes. Department of Chemistry, Biotechnology and Life Science, Norwegian University of Life Sciences, Norway

Ivanna Novotny Nuñez. Laboratorio de Inmunología, Centro de Referencia para Lactobacilos (CERELA)-CONICET, Argentina

Martin Manuel Palomar. Laboratorio de Inmunología, Centro de Referencia para Lactobacilos (CERELA)-CONICET, Argentina

Gabriela Perdigón. Laboratorio de Inmunología, Centro de Referencia para Lactobacilos (CERELA)-CONICET, Argentina
Catedra de Inmunologia, Instituto de Microbiologia, Facultad de Bioquimica, Quimica y Farmacia, Universidad Nacional de Tucuman, Argentina

Jean-Christophe Piard. INRA and AgroParisTech, UMR1319 Micalis, France

Mariana Piuri. Departamento de Química Biológica, FCEyN, Universidad de Buenos Aires, Argentina

Tomislav Pogačić. Department of Dairy Science, Faculty of Agriculture University of Zagreb, Croatia

Raúl R. Raya. Centro de Referencia para Lactobacilos (CERELA)-CONICET, Argentina

Jorge Reinheimer. Instituto de Lactología Industrial (INLAIN), Universidad Nacional del Litoral-CONICET, Argentina

Ana Rodríguez González. DairySafe Group, Instituto de Productos Lácteos de Asturias (IPLA-CSIC), Spain

Milton H. Saier, Jr. Division of Biological Sciences, University of California at San Diego, San Diego, USA

Elisa Salvetti. Department of Biotechnology, University of Verona, Italy

Graciela Savoy de Giori. Centro de Referencia para Lactobacilos (CERELA)-CONICET, Argentina

Fernando Sesma. Centro de Referencia para Lactobacilos (CERELA)-CONICET, Argentina

Rosa Anna Siciliano. Centro di Spettrometria di Massa Proteomica e Biomolecolare, Istituto di Scienze dell'Alimentazione, CNR, Italy

María Pía Taranto. Centro de Referencia para Lactobacilos (CERELA)-CONICET, Argentina

Anne Thierry. INRA, AGROCAMPUS OUEST UMR1253 Science et Technologie du Lait et de l'OEuf, France

Sandra Torriani. Department of Biotechnology, University of Verona, Italy

Taylor A. Twiddy. Department of Microbiology and Cell Science, Genetics Institute and Institute of Food and Agricultural Sciences, University of Florida, Gainesville, USA

Magalie Weber. INRA, AGROCAMPUS OUES, UMR1253 Science et Technologie du Lait et de l'OEuf, France

Stefan Weckx. Research Group of Industrial Microbiology and Food Biotechnology (IMDO), Department of Bioengineering Sciences, Faculty of Sciences and Bioengineering Sciences, Belgium

Miriam Zago. Consiglio per la Ricerca e la Sperimentazione in Agricoltura, Centro di Ricerca per le Produzioni Foraggere e Lattiero-Casearie (CRA-FLC), Italy

Monique Zagorec. INRA, UMR 1014 Secalim, France
LUNAM Université, Oniris, France

Preface

We have witnessed an explosion of new findings in the field of Lactic Acid Bacteria (LAB) over the last five years that have passed since the previous edition of our book. Rapid advances in the "omics" (genomics, proteomics, pangenomics, metagenomics) sciences and next generation sequencing technologies have revolutionized the characterization of LAB; the discovery of new LAB species has been achieved and complete genomes of all major groups are now available. The insights arisen from these basic studies have been translated into innovation leading to the use of LAB in applications beyond classic food fermentations. Thus, LAB have been engineered or manipulated for their use as live vaccines or as microbial factories for the production of food ingredients, nutraceuticals, commodity chemicals, and other high-value metabolites. LAB microbiota involved in the fermentation of vegetables, fruits, cocoa, wine, meat, and dairy products, as well as their contribution to sensory and safety, have been updated and/or included as new fields. Results from comparative and functional genomics have been used to understand the response of LAB to their environment, leading to better understanding of their adaption and safety in traditional/industrial foods and their interactions with the human host. Thus, chapters in this new edition of our book have been updated or present for the first time a wide range of topics including basic issues (metabolism, biodiversity, biofilms and transport systems), comprehensive information on new advanced approaches (comparative and functional genomics), human-health LAB-related aspects, LAB safety, as well as traditional and novel biotechnological applications. We hope this book finds its audience among microbiologists, food scientists, nutritionists, clinical and advanced students. The editors greatly appreciate the hard work and the time dedicated of many well-known leading scientists who contributed to this book. We also thank to CERELA, CONICET, and FONCyT from Argentina.

This book is devoted to the memory of our very good friend and colleague, Fernando Sesma, who passed away in July 2014.

Chapter 1

Updates on Metabolism in Lactic Acid Bacteria in Light of "Omic" Technologies

Magdalena Kowalczyk[1], Baltasar Mayo[2], María Fernández[2], and Tamara Aleksandrzak-Piekarczyk[1,*]

[1] *Institute of Biochemistry and Biophysics, Polish Academy of Sciences, Poland*
[2] *Departamento de Microbiología y Bioquímica, Instituto de Productos Lácteos de Asturias (IPLA-CSIC), Spain*

1.1. Sugar Metabolism

Sugars are the primary carbon and energy source for LAB that are grown for fermented food and feed production as well as in laboratory media. Many different transport systems are involved in LAB carbohydrate uptake, including phosphotransferase systems (PTS), ATP-binding cassettes (ABC), and glycoside–pentoside–hexuronide transporters. In *Lactococcus lactis*, glucose is imported by either a mannose or cellobiose PTS or one or more non-PTS permease(s) (Castro et al. 2009). LAB prefer glucose but can also metabolize several common hexoses; however, the ability to ferment other sugars is strain dependent. Dairy LAB can use milk's most abundant sugar, lactose, as a carbon source, whereas plant-associated bacteria utilize a large variety of other carbohydrates, including β-glucosides (Aleksandrzak-Piekarczyk 2013). Furthermore, phenotypic and genotypic analyses of twenty *L. lactis* subsp. *lactis* and *cremoris* genotypes showed strain-to-strain variations (Fernandez et al. 2011). These two groups had distinctive carbohydrate fermentation and enzyme activity profiles with *cremoris* genotypes exhibiting broader profiles.

Monosaccharides incorporated by the cell or liberated in the cytoplasm by disaccharide hydrolysis enter glycolysis at the glucose-6P (G6P) level or are processed by the Leloir pathway (Figure 1.1). In *L. lactis*, lactose that is transported by the PTS system is hydrolyzed to galactose-6P, transformed by the tagatose pathway (Tag6P), and then enters glycolysis at the triose phosphate level. In some LAB, only the glucose moiety of lactose is fermented, while the galactose moiety is excreted; this leads to galactose accumulation in the medium, which yields poor-quality dairy products (Neves et al. 2010; Aleksandrzak-Piekarczyk, 2013). Galactose can be imported by the non-PTS permease GalP and metabolized via the Leloir pathway (*galMKTE*). Alternatively, galactose can be imported by PTS^{Lac} (*lacFE*) and further metabolized to triose phosphates by the Tag6P pathway (*lacABCD*). Recently, an alternative uptake route was discovered that consists of galactose translocation via the

* Corresponding author: Institute of Biochemistry and Biophysics, Polish Academy of Sciences, Pawinskiego 5a, 02-106 Warsaw, Poland. tamara@ibb.waw.pl

Biotechnology of Lactic Acid Bacteria: Novel Applications, Second Edition. Edited by Fernanda Mozzi, Raúl R. Raya, and Graciela M. Vignolo.

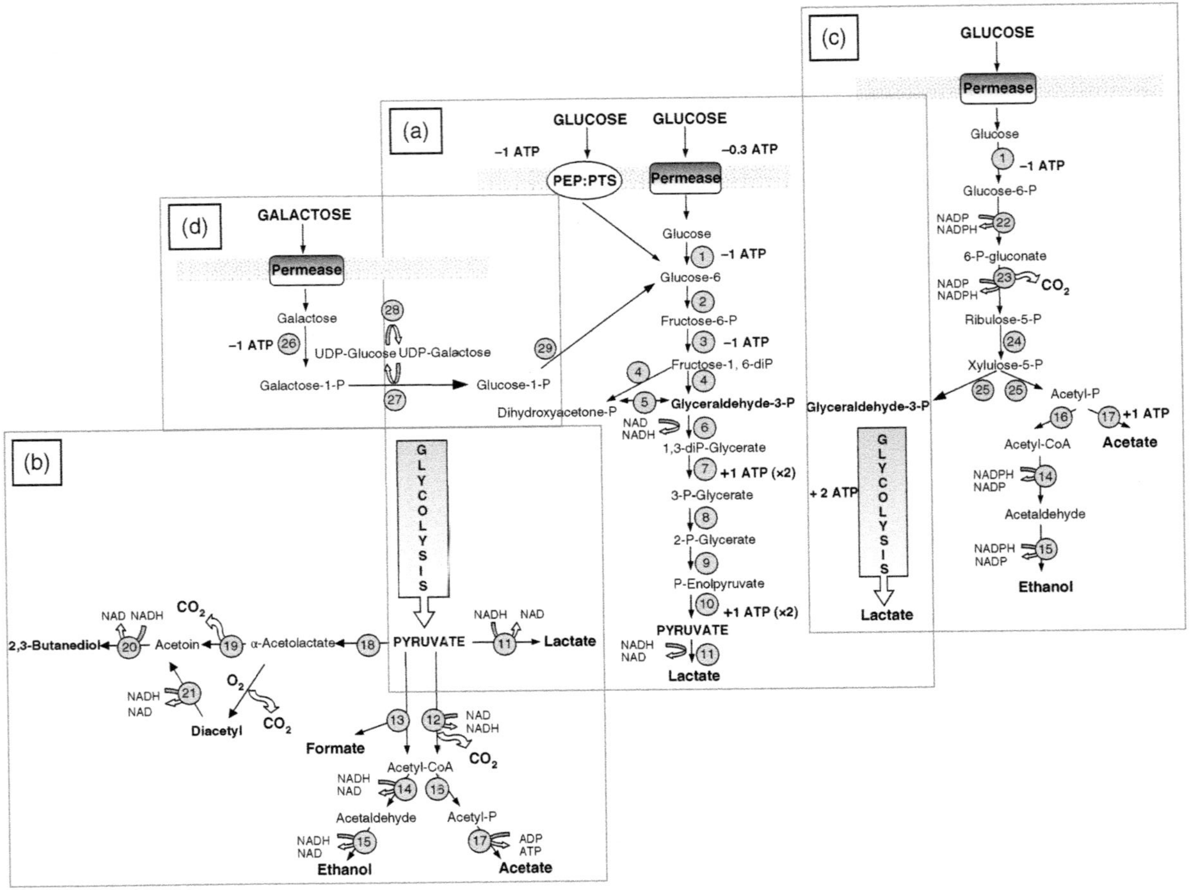

Figure 1.1. Pathways of glucose metabolism. (a) Homofermentative pathway. (b) Mixed-acid metabolism. (c) Heterofermentative pathway. (d) Leloir pathway. Reactions are catalyzed by the following enzymes: 1, glucokinase (GLK); 2, glucose-phosphate isomerase (GPI); 3, phosphofructokinase (PFK); 4, fructose-bisphosphate aldolase (FBPA); 5, triose-phosphate isomerase (TPI); 6, glyceraldehyde-phosphate dehydrogenase (GAPDH); 7, phosphoglycerate kinase (PGK); 8, phosphoglycerate mutase (PMG); 9, enolase (ENO); 10, pyruvate kinase (PK); 11, lactate dehydrogenase (LDH); 12, pyruvate dehydrogenase (PDH); 13, pyruvate formate lyase (PFL); 14, acetaldehyde dehydrogenase (ACDH); 15, alcohol dehydrogenase (ADHE); 16, phosphotransacetylase (PTA); 17, acetate kinase (ACK); 18, α-acetolactate synthase (ALS); 19, α-acetolactate decarboxylase (ALD); 20, 2,3-butanediol dehydrogenase (BDH); 21, diacetyl reductase (DR); 22, glucose-6-P dehydrogenase (G6PDH); 23, 6-P-gluconate dehydrogenase (6PGDH); 24, ribulose-5-P-3-epimerase (RPPE); 25, D-xylulose-5P phosphoketolase (XPK); 26, galactokinase (GK); 27, galactose-1-P-uridylyltransferase (GPUDLT); 28, UDP-galactose-1-epimerase (UDPE); 29, phosphoglucomutase (PGM).

galactose PTS, followed by Gal6P dephosphorylation to galactose, which is further metabolized via the Leloir pathway (Neves et al. 2010). This knowledge has been used to genetically engineer strains that enhance galactose consumption rates by up to 50% (Neves et al. 2010).

Sugar fermentation leads to the formation of lactic acid alone or in combination with other organic acids and ethanol. Variations in the metabolic products of LAB have yielded three categories of fermentation: homofermentation, mixed-acid metabolism, and heterofermentation (Figure 1.1). These three types of fermentation and their regulatory mechanisms have been comprehensively reviewed in the first edition of this book (Mayo et al. 2010). Currently, the control and regulation of glycolytic flux in LAB are not fully understood (Martinussen et al. 2013). The control of glycolytic flux in *L. lactis* is not due to the actions of a single enzyme, sugar transport, or ATP-dependent mechanism. However, a combination of these mechanisms cannot be ruled out as a possible explanation (Martinussen et al. 2013).

In recent years, studies have shifted from digestible disaccharides to indigestible higher oligosaccharides as interest in intestinal microbial ecology and the commercial use of prebiotic oligosaccharides has emerged. Mono- and disaccharide metabolism is well understood; however, few data are available on the metabolism of higher oligosaccharides, which are abundant in cereals, milk, fruits, and the upper intestinal tract of animals. The metabolism of four major oligosaccharide groups have been examined in detail: (i) starch, maltodextrins, and isomalto-oligosaccharides (IMO); (ii) fructo-oligosaccharides (FOS); (iii) β-galacto-oligosaccharides (βGOS); and (iv) raffinose-family oligosaccharides and α-galacto-oligosaccharides (ROF and αGOS, respectively) (Gänzle and Follador 2012).

In addition to metabolizing sugar, LAB are able to direct sugar towards exopolysaccharide (EPS) biosynthesis. These long-chain saccharides are loosely attached to the cell surface to form capsule-like structures or are secreted into the environment. EPS production by lactobacilli has been previously discussed in a comprehensive review (Badel et al. 2011). Several studies of biopolymer diversity in LAB from cereal (Bounaix et al. 2009, 2010; Palomba et al. 2012) and in intestinal LAB (Salazar et al. 2009; Górska-Frączek et al. 2011, 2013; Sims et al. 2011) have also been published since the first edition of this book.

EPSs come in many different structures, sizes, and sugar compositions and are classified into two groups: homopolysaccharides (HoPS), which consist of one type of monosacharide (α-D-glucan, β-D-glucan, fructan, or a polygalactan); and heteropolysaccharides (HePS), which consist of different types of monosaccharides (D-glucose, D-galactose, L-rhamnose, and their derivatives). HoPS are synthesized extracellularly by highly specific glycosyltransferase enzymes as well as glucan- or fructan-sucrases. HoPS synthesis specifically requires sucrose as a substrate and the energy generated by its hydrolysis. The crystal structures, reaction and product specificities of glucansucrases as well as structural analyses of α-glucan polymers have been recently reviewed (Leemhuis et al. 2013).

HePS are synthesized from glucose, galactose, or other monosaccharides by the combined actions of several types of glycosyltransferases. HePS biosynthesis involves four major consecutive steps: (i) sugar transport into the cytoplasm, (ii) sugar-1P synthesis, (iii) polymerization of repeating unit precursors, and (iv) EPS export outside the cell. The synthesis of two EPSs in *Lactobacillus johnsonii* FI9785 is dependent on the 14-kb *eps* gene cluster; however, the precise regulation of EPS biosynthesis has yet to be identified (Dertli et al. 2013). It is thought that EPS production can be regulated at each of these four steps. Genome sequencing of EPS-related genes and their organization (Koryszewska-Baginska et al. 2014) may provide additional insight into whether this is the case.

1.1.1. Practical Aspects of Sugar Catabolism

The ability of LAB to ferment sugars has been widely utilized in various foods' production. However, LAB also have the potential for increasing the production value of biofuels and biochemical products due to their robustness and tolerance for ethanol, low pH, and high temperatures (Martinussen et al. 2013). Focus has increased on optimizing lactate production from natural substrates such as starchy or lignocellulosic materials from agricultural, agro-industrial, and forestry sources due to their abundance, low price, high polysaccharide content, and renewability (Okano et al. 2009; Abdel-Rahman et al. 2011; Castillo Martinez et al. 2013). In addition, EPS from LAB can play an important role in the food industry as an emulsifier, thickener, viscosifier, and stabilizer. EPS has been used in the rheology and texture of fermented milks (Ramchandran and Shah 2009) and other fermented products, such as sourdough (Katina et al. 2009; Galle et al. 2010) and cereal-based beverages (Zannini et al. 2013). It may also improve the quality, safety, and acceptability of gluten-free bread (Moroni et al. 2009) and replace hydrocolloids in sorghum sourdough (Galle et al. 2011). EPSs from LAB are also of great interest to agro-food industries since their vast structural diversity may lead to innovative applications. However, the majority of LAB only produce low levels of polysaccharides; therefore, optimized methodologies for increased EPS production and recovery are still required (Notararigo et al. 2013).

EPS from LAB also has beneficial physiological properties for humans (Patel et al. 2011). Cell-bound EPSs from *Lactobacillus acidophilus* 606 (Kim et al. 2010) and *Lactobacillus plantarum* 70810 (Wang et al. 2014) have been shown to have antitumor properties. EPSs have also been shown to exhibit immunomodulatory activity for macrophages (Liu et al. 2011; Ciszek-Lenda et al. 2011) as well as intestinal epithelial cells (Patten et al. 2014). In addition, EPSs produced by *Lactobacillus reuteri* can inhibit enterotoxigenic *Escherichia coli*-induced hemagglutination of porcine erythrocytes, which further indicates that EPS has therapeutic potential (Wang et al. 2010). EPS from *Lact. plantarum* 70810 has a metal binding capacity and could be used as a potential biosorbent for lead removal from the environment (Feng et al. 2012). LAB can also produce a variety of functional oligosaccharides that can be used as prebiotics (Pepe et al. 2013), nutraceuticals, sweeteners, humectants, drugs against colon cancer, and immune stimulators (Patel et al. 2011). Some probiotic LAB can also utilize prebiotic compounds, including non-digestible FOSs, inulin-type fructans, or β-glucans (Russo et al. 2012), which stimulate the growth of beneficial commensals in the gastrointestinal tract.

1.2. Citrate Metabolism and Formation of Aroma Compounds

In addition to sugars, several LAB species can metabolize citrate. Citrate fermentation in LAB leads to the production of volatile compounds. In fermented dairy products, these compounds are C_4 compounds, such as diacetyl, acetoin, and butanediol, which are responsible for the typical aroma of many fermented dairy products. Therefore, citrate metabolizing LAB, such as *L. lactis* subsp. *lactis* biovar. diacetylactis (*L. diacetylactis*) and some *Leuconostoc* and *Weissella* species, are currently used as starter and adjunct cultures for the production of these C_4 compounds. However, in other fermented products, such as wine, beer, and sausages, the volatile compounds produced from the fermentation of citrate by LAB are considered off-flavors, and their presence should be avoided. Citrate utilization by LAB has been previously described in detail (Quintans et al. 2008) and summarized in the first edition of this book (Mayo et al. 2010). Therefore, in this chapter, we are presenting only citrate metabolism in LAB in the context of recent achievements.

1.2.1. Citrate Transport

Citrate transport is a limiting step for citrate utilization and is performed by a variety of membrane-associated permeases. In contrast, volatile compounds formed in the cytoplasm are secreted without requiring specific transporters. Most LAB species internalize citrate using a 2-hydroxycarboxylate (2-HCT) transporter, which can transport dicarboxylic and tricarboxylic acids. The 2-HCT family of transporters includes CitP from *Lactococcus*, *Leuconostoc*, and *Weissella* (Pudlik and Lolkema 2010). CitP is an antiport transporter that exchanges H-citrate^{2-} and lactate^{1-} to generate a membrane potential (Figure 1.2). In *L. diacetylactis*, CitP is encoded by the *citQRP* operon located on the "citrate

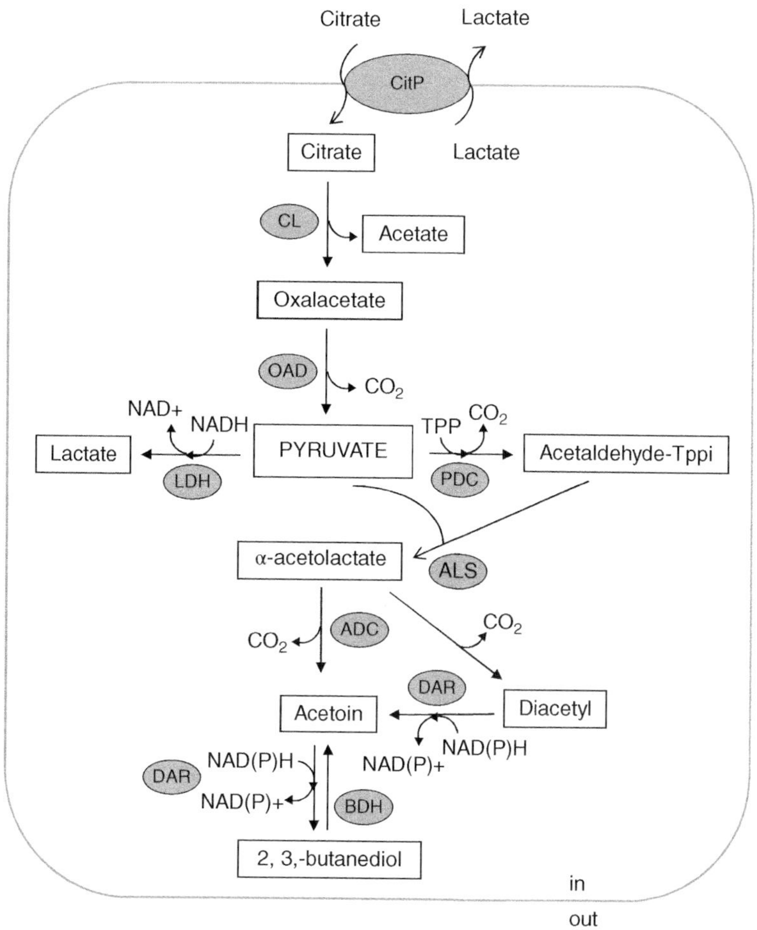

Figure 1.2. Citrate metabolism in *Lactococcus*, *Leuconostoc*, and *Weissella* species. Key for the enzymes: CL, citrate lyase; OAD, oxaloacetate decarboxylase; LDH, lactate dehydrogenase; PDC, pyruvate decarboxylase; ALS, α-acetolactate synthase; ADC, α-acetolactate decarboxylase; DAR, diacetyl acetoin reductase; BDH, 2,3-butanediol dehydrogenase; Tppi, thiamine pyrophosphate.

plasmid" (Drider et al. 2004; Kelly et al. 2010). In *L. diacetylactis*, transcription of the promoters that control *cit* operons are specifically activated by low pH environments as an adaptive response to acid stress. This has been confirmed by transcriptomics analysis of *L. diacetylactis* in milk (Raynaud et al. 2005) and cheese (Cretenet et al. 2010). In *Weissella paramesenteroides* and *Leuconostoc mesenteroides* subsp. *cremoris*, the *citP* genes (*citMCDEFGRP*) are located in a plasmidic or chromosomal *citI* cluster, respectively (Martín et al. 1999, 2000; Bekal et al. 1998). Enlarged citrate plasmids (15 to 23 kbp) have also been observed in wild (non-starter) strains of *L. diacetylactis* (Drici et al. 2010; Kelly et al. 2010).

In *Enterococcus faecalis* and *Lactobacillus casei*, citrate is transported by CitM and CitH transporters, respectively, both belonging to the family of citrate-metal symporters (CitMHS) (for a review see Lensbouer and Doyle 2010). CitMHS transporters transport citrate in cation (Ca^{2+}, Mn^{2+}, or Fe^{3+}) complexes. Recently, CitH in *Lact. casei* has been shown to be a proton symporter that transports a Ca^{2+}-citrate complex and uses Ca^{2+} as a substrate (Mortera et al. 2013). Other types of citrate transporters have also been identified in *Lact. plantarum*, *Oenococcus oeni* (for a review see Mayo et al. 2010), and the atypical citrate-fermenting wild *L. diacetylactis* strain (Passerini et al. 2013b).

1.2.2. Conversion of Citrate into Pyruvate and Production of Aroma Compounds

Once inside the cell, citrate is converted into acetate and oxaloacetate in a reaction catalyzed by the citrate lyase (CL) enzyme complex (Figure 1.2). Next, oxaloacetate is decarboxylated by oxaloacetate decarboxylase (OAD), which generates pyruvate and CO_2 (Figure 1.2). Analysis of various LAB genomes have identified genes encoding the α-, β-, and δ-subunits of OAD (Makarova et al. 2006). However, the physiological role of OAD remains poorly understood in LAB species; only enzymes from *E. faecalis* (Repizo et al. 2013) and *Lact. casei* (Mortera et al. 2013) have been recently investigated. In other LAB species, including *L. diacetylactis*, *W. paramesenteroides*, *Leuc. mesenteroides*, *Lact. plantarum*, and *O. oeni*, oxaloacetate is decarboxylated by the soluble and cytoplasmic malic enzyme (ME) (Sender et al. 2004). Surprisingly, the *cit* locus of *E. faecalis* has been recently shown to contain genes that encode both OAD and ME (Espariz et al. 2011).

Pyruvate metabolism in LAB can yield different end products, including lactate, formate, acetate, and ethanol as well as the important aroma compounds diacetyl, acetoin, and butanediol (Neves et al. 2005).

1.2.3. Conversion of Citrate into Succinate

Some LAB species cannot truly convert citrate into pyruvate. Instead, the CitT transporter generates succinate via malate and fumarate. Furthermore, the complete tricarboxylic acids (TCA) pathway has recently been identified in the *Lact. casei* genome using *in silico* analysis (Díaz-Muñiz et al. 2006). In this LAB species, the dominant end-products of citrate metabolism were acetic acid and L-lactic acid at both excess and limiting amounts of carbohydrates. Trace amounts of D-lactic acid, acetoin, formic acid, ethanol, and diacetyl confirm OAD activity; however, succinic acid, malic acid, and butanendiol were not observed (Díaz-Muñiz et al. 2006; Mortera el al. 2013).

1.2.4. Bioenergetics of Citrate Metabolism

The co-metabolism of glucose and citrate produces different physiological effects in homofermentative and heterofermentative LAB. In homofermentative LAB, citrate utilization has a protective effect against acid stress. In heterofermentative LAB, "citrolactic" fermentation generates one extra mol of ATP per mol of citrate. In milk, *L. diacetylactis* metabolizes lactose and produces lactic acid, which is exchanged by the antiporter CitP for citrate during excretion. *L. diacetylactis* is homofermentative and converts glucose into lactate, generating two moles of ATP per mol of glucose. The NAD^+ consumed in the first steps of this pathway is regenerated during the transformation of pyruvate into lactate, thereby maintaining the redox potential. In the presence of glucose and citrate, each mol of citrate produces one mol of pyruvate without generating NADH. This excess of pyruvate is diverted to α-acetolactate synthesis and the subsequent production of aroma compounds. Similar to *L. diacetylactis*, the higher biomass of *Lact. casei* during Ca^{2+}-citrate and carbohydrate co-metabolism has been attributed to counteracting the growth inhibition of carbohydrate metabolism that is induced by acidification in its final stages (Mortera et al. 2013).

1.3. The Proteolytic System of Lactic Acid Bacteria

L. lactis strains isolated from dairy products are characterized by a high number of amino acid auxotrophies and the ability to utilize milk proteins as an amino acid source. These properties are common to dairy strains even from distant geographic locations in Asia, Europe, North America, and

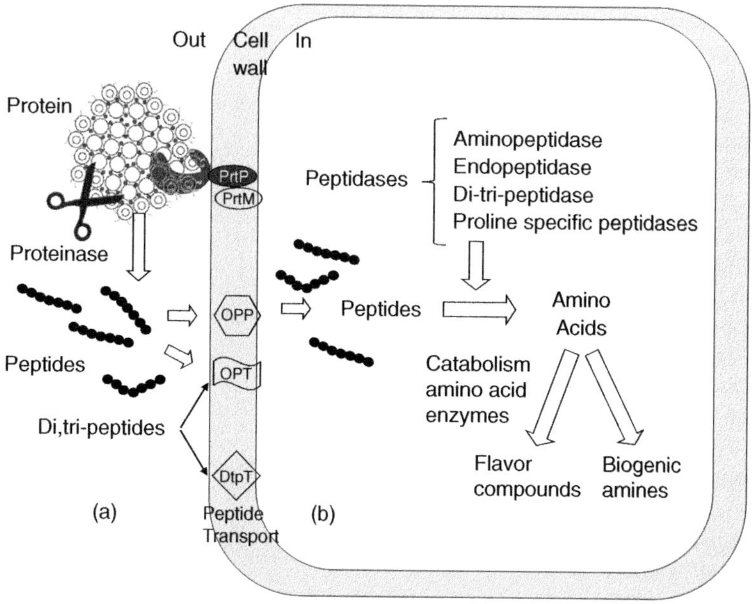

Figure 1.3. Diagram of the proteolytic systems of lactic acid bacteria. (a) Extracellular components: PrtP, cell-envelope proteinase; PrtM, proteinase maduration protein; Opp, oligopetide permease; DtpT, the ion linked trasnsporter for di-and tripeptides; and Opt, the ABC transporter for peptides. (b) Intracelullar components: pool of about 20–25 peptidases, including general (PepN, PepC) and specific (PepX, PepQ) peptidases, and amino acid catabolic enzymes (carboxylases, aminotransferases, etc.).

New Zealand (Rademaker et al. 2007; Kelly et al. 2010). The use of environmental proteins as an amino acid source in LAB explains the importance of their proteolytic system. LAB depend on this system to obtain the essential amino acids that are then used as precursors for not only peptides and proteins but also for many other biomolecules. Amino acids are the precursors of aromatic compounds, which are important to the final flavor of food products. Proteolytic activity also generates other molecules, such as bioactive peptides, that have functions related to the probiotic properties of LAB.

The proteolytic system in LAB can be divided into several steps: protein degradation, peptide transport, peptide degradation, and amino acid catabolism (Figure 1.3).

1.3.1. Protein Degradation

Studies of protein degradation in LAB initially focused on casein degradation using *L. lactis* as a model organism. Casein hydrolysis in LAB is initiated by a cell-envelope proteinase (CEP), which degrades proteins into oligopeptides, and gene deletion studies have shown that LAB are unable to grow in milk in the absence of a functional CEP. However, since CEP is extracellular, peptides produced by CEP can also be consumed by protease-negative variants, allowing them to survive in culture (Bachmann et al. 2012).

Lactocepins are a diverse group of CEPs that belong to the subtilisin protein family of serine proteases. CEPs are anchored to the cell wall via sortase A (SrtA) (Dandoy et al. 2011). LAB lactocepins are encoded by *prtP*, *prtB*, *prtS*, and/or *prtH*, which differ in their number of functional domains. CEP distribution varies widely among strains. Overall, the most abundant paralog among LAB is *prtH3*, which is carried by over 80% of strains tested, followed by paralogs of *prtH* and *prtH4* (Broadbent et al. 2011). Most LAB possess only one CEP. However, four CEP genes (*prtH*,

prtH2, *prtH3*, and *prtH4*) have been identified in *Lactobacillus helveticus* CNRZ32 (Jensen et al. 2009). The presence of several protease genes in *Lact. helveticus* could explain its high proteolytic efficiency. Only *prtH2* is common to all characterized *Lact. helveticus* strains (Genay et al. 2009). However, analysis of the proteinase in BGRA43 showed that the only active gene was *prtH*. CEP activation requires the maturation proteinase PrtM. Two PrtMs (PrtM1 and PrtM2) have been identified in *Lact. helveticus* CNRZ32 (Savijoki et al. 2006). Additional studies (Genay et al. 2009; Broadbent et al. 2011) have reported that PrtM1 is required for PrtH activation, while PrtM2 plays a role in the activation of other CEP paralogs.

As mentioned above, CEP activity was initially evaluated using casein as a substrate. However, LAB strains isolated from non-milk or non-fermentation environments have also exhibited casein hydrolysis. For instance, *Lact. helveticus* BGRA43, which is isolated from human feces, has strong proteolytic activity and is able to completely hydrolyze α_{s1}-, β-, and κ-caseins (Strahinic et al. 2013). In addition, the lactocepin secreted by *Lactobacillus paracasei* VSL#3 can selectively degrade cell-associated and tissue-distributed IP-10 and other proinflammatory chemokines *in vitro* (von Schillde et al. 2012). These findings indicate that lactocepin, which is encoded by *prtP*, is highly selective despite targeting a broad spectrum of cleavage sites. Therefore, additional protein characteristics, such as surface charge and/or three-dimensional structure, determine whether a protein is cleaved or not. The selective degradation of pro-inflammatory chemokines could be also related to differences in the regulation of *prtP* expression.

In *Streptococcus thermophilus*, the cell wall associated proteinase PrtS is highly conserved (95% identity) with the PrtS of *Streptococcus suis*. Although recent studies have suggested that PrtS contributes to the virulence of *Strep. suis* (Bonifait el al. 2010), the primary role of PrtS in *Strep. thermophilus* strains is cleaving casein to oligopeptides. This function is clearly related to the adaptation of *Strep. thermophilus* to dairy environments; analyses of *prtS* distribution in *Strep. thermophilus* found that it occurs infrequently in historical specimens but frequently in more recent industrial ones. Furthermore, this "ecological" island confers an important metabolic trait for milk adaptation and appears to be laterally transferred in *Strep. thermophilus*. Together, these data suggest that *Strep. thermophilus* evolved via gene acquisition and selection as the result of the environmental pressures of food production (Delorme et al. 2010).

The second stage of protein degradation is the transport of di-, tri-, and oligo-peptides into the cell. Three oligo-, di-, and tri-peptide transport systems (Opp, Dpp and DtpT, respectively) have been described in LAB. *Lact. acidophilus*, *Lactobacillus brevis*, *Lact. casei*, *Lactobacillus rhamnosus*, and *L. lactis*, possess all three of these peptide transport systems. Some *Lact. helveticus* strains, such as DPC4571, also have three peptide transport systems, while others, such as H10, only have two (Opp and DtpT). These results indicate that the proteolytic systems differ between different strains of even the same species. Finally, *Lact. reuteri* only has one functional peptide transport system (DtpT) (Liu et al. 2010).

1.3.2. Peptidases

Once casein-derived peptides are taken up by LAB cells, they are degraded via the concerted action of peptidases with differing and partially overlapping specificities. Peptidases are a very important part of the proteolytic system in LAB and are involved in peptide hydrolysis and the release of essential amino acids. Peptidases can be divided into two main groups: endopeptidases and exopeptidases. Endopeptidases hydrolyze the internal peptide bonds of oligopeptides to generate peptides that can be further degraded by exopeptidases. Exopeptidases act on the terminal ends of oligopeptides to generate smaller peptides or amino acids.

The main LAB endopeptidases that have been characterized are PepO, PepF, PepG, and PepE, all of which act on NH_2-X_n↓X_n-COOH substrates. PepO is encoded by three paralogous genes (*pepO*,

pepO2, and *pepO3*). In *Lact. helveticus* strains, these genes are identical; however, strain heterogeneity may be observed due to the loss of gene function or sequence polymorphisms, which may affect the specificity or relative activity of the individual peptidase enzymes (Broadbent et al. 2011).

Three paralogous genes also encode PepF (*pepF*, *pepF1*, and *pepF2*). In *L. lactis*, *pepF1* is located on the chromosome, while *pepF2* is located on a plasmid. The location of *pepF2* explains the variability in its absence or presence among different *Lactococcus* strains (Liu et al. 2010).

The PepE and PepG proteins are absent in lactococci and streptococci. In *Lactobacillus*, two paralogous genes have been described for PepE (*pepE* and *pepE2*). In *Lact. helveticus*, these genes are almost universally conserved among strains from different origins, including cheese, whey, whiskey malt, and commercial cultures. PepE also plays an important role in the debitterness function of *Lact. helveticus* (Broadbent et al. 2011).

Exopeptidases are traditionally classified by their specificity, and four exopeptidase groups have been identified in LAB: aminopeptidases, dipeptidases, tripeptidases, and proline-specific peptidases. Aminopeptidases hydrolyze a single amino acid from the *N*-terminal oligopeptide NH_2–X↓Xn–COOH. They can be divided into either general or specific aminopeptidases. General aminopeptidases (PepN and PepC) are present in all genomes, usually as one gene per genome. Some LAB genomes have two peptidase homologs that potentially have the same function; for example, there are two PepC homologs in *Lact. johnsonii* (Liu et al. 2010). Specific aminopeptidases are categorized according to the type of residue they hydrolyze. PepS, which has only been described in *Strep. thermophilus*, is specific for aromatic residues. PepA is specific for Glu and Asp residues; it is found in streptococci and some *Lactobacillus* and *Lactococcus* strains but is absent in *Pediococcus* and *Oenococcus*. PepM is specific for methionine residues and is present in *Lactobacillus*, *Lactococcus*, *Streptococcus*, *Pediococcus*, and *Oenococcus* strains (Liu et al. 2010).

Dipeptidases are exopeptidases specific for NH2–X↓X–COOH dipeptides. The PepD dipeptidase family has a broad specificity for various dipeptides. The *pepD* genes are distributed heterogeneously in LAB genomes and can vary in number from 0 to 6 paralogs. PepV is also encoded by multiple paralogous genes. It is present in all LAB and also has a broad specificity.

Tripeptidases release the amino acid from the *N*-terminal position of NH2–X↓X–X–COOH tripeptides. They have a broad specificity but preferentially target hydrophobic peptides and do not hydrolyze proline residues. The only tripeptidase currently identified in LAB is PepT. The *pepT* gene is present in all LAB genomes and can occur as two paralogous genes in some LAB (*Lact. acidophilus*, *Lact. johnsonii*, and *Lactobacillus gasseri*).

Proline-specific peptidases can hydrolyze proline residues from the *N*-terminal position of peptides. Proline iminopeptidase (PepI) has aminopeptidase activity towards *N*-terminal proline peptides and prefers tri-peptides (NH2–Pro↓Xn–COOH). The prolinase PepR has a broad specificity for dipeptides (NH2–Pro↓X–COOH). These proline-specific peptidases are absent from all *L. lactis* strains. The activity of cell extract from *Lact. helveticus* and *Lact. rhamnosus* for several proline dipeptides was significantly reduced in PepR-deletion mutant. Those observations suggest that PepI and PepR may contribute to the specific proteolytic capacity for breaking down peptides containing proline in *Lactobacillus* strains (Liu et al. 2010).

PepP is a proline peptidase that can cleave any *N*-terminal amino acid linked to proline in an oligopeptide (NH2–X↓Pro–Xn–COOH). PepQ is also a proline peptidase but is specific for proline dipeptides (NH2–X↓Pro–COOH). One *pepP* gene is found in each LAB genome, except in *Lactobacillus sakei* and *Pediococcus pentosaceus*. The absence of *pepP* in both these genomes is likely due to gene loss. PepQ genes are equally distributed in all LAB genomes as one copy per genome. However, the *Lactobacillus delbrueckii* subsp. *bulgaricus* (*Lact. bulgaricus*) strains have two *pepQ* paralogs; one paralog is clustered with other orthologs, while the second is located in a separate cluster. This may be the result of an ancient duplication or horizontal gene transfer (HGT) event. PepX is a proline-specific endopeptidase (NH2–X–Pro↓Xn–COOH) that is present in all LAB genomes as one gene per genome. However, some LAB genomes have two peptidase homologs, and these are normally the genomes of strains from dairy environments. PepX2 is a PepX homolog of *L. lactis* subsp. *lactis* IL1403. This putative *pepX2* gene was originally annotated as a hypothetical protein named *ymgC*. The only members of the PepX2 (YmgC) group are from *L. lactis* subsp. *lactis* IL1403 and *Pediococcus* (Liu et al. 2010).

1.3.3. Technological Applications of the Proteolytic System

The proteolytic system provides LAB with the amino acids they require to grow. For instance, *Lact. bulgaricus* and *Lact. helveticus* have a very extensive set of proteolytic enzymes, which is consistent with previous findings that *Lact. bulgaricus*, rather than *Strep. thermophilus*, serves as the proteolytic organism in yogurt. *Lact. helveticus* is a proteolytic adjunct culture of cheese that is used to degrade bitter peptides. Interestingly, *Lact. bulgaricus* encodes the Dpp system, which prefers the uptake of hydrophobic di/tripeptides. This complements the presence of *Strep. thermophilus*, which encodes the general di/tripeptide transporter DtpT and suggests that more peptides can be utilized by both bacteria when they are grown together. LAB from plants, such as *Lact. plantarum*, *O. oeni*, and *Leuc. mesenteroides*, encode fewer proteolytic enzymes, which is consistent with their fiber-rich ecological niche that contains less proteins. Milk proteins have also been used as raw material for obtaining bioactive peptides. Various peptides with immunostimulating, antimicrobial, opioid, anticancer, mineral binding, and antihypertensive properties have been isolated as proteolytic products in LAB (Griffiths and Tellez, 2013).

1.3.4. Amino Acid Catabolism

The catabolism of amino acids has quality (flavor compound formation) and safety (biogenic amine synthesis) implications for fermented foods. Amino acid catabolism is also thought to have an important role in the abilities of LAB to obtain energy in nutrient-limited conditions. Furthermore, amino acid catabolism has also been identified as a mechanism of pH control.

Amino acid degradation is important for the synthesis of volatile compounds and the transamination of some amino acids: methionine, branched-chain, and aromatic amino acids. Transamination is the main degradation pathway that leads to the formation of α-keto acids, which are then degraded into various aromatic compounds. The conversion of amino acids to keto- and hydroxyl acids is initiated by lactobacilli, and *Lactococcus* strains further convert these products to carboxylic acid. This cooperation between LAB and non-starter LAB can enhance cheese flavor.

Leucine, valine, and isoleucine catabolism can be divided into two parts. The first comprises the main degradation pathway; an aminotransferase reaction converts the amino acid to α-keto acid or, in the case of leucine, α-keto isocaproate (KICA). The resulting compounds are then converted to aldehyde, alcohol, or carboxylic acid via three different pathways: α-keto acid decarboxylation, oxidative decarboxylation, or an alternative dehydrogenation route that generates α-hydroxy-isocaproate (HICA) (Smit et al. 2005). Aromatic amino acids are also degraded by the amino transferase AraT. Transamination generates indole pyruvate, phenyl pyruvate, and p-hydroxy-phenyl pyruvate from tryptophan, phenylalanine, and tyrosine, respectively. Alpha-ketoacids generated by aromatic amino acid transamination are further degraded to various compounds by enzymatic (dehydrogenation, decarboxylation, or oxidative decarboxylation) or chemical reactions.

Sulfur compounds are generated by sulfur amino acid catabolism and are potent odorants that contribute flavor to many fermented foods. Methionine catabolism produces various volatile sulfur compounds (VSCs) such as H_2S, methanethiol, dimethyl sulfide (DMS), dimethyl disulfide (DMDS), and dimethyl trisulfide (DMTS) (Fernández et al. 2000). The enzymes in LAB strains from raw goats' milk cheeses crucial for VSC formation from L-methionine have very diverse enzyme capabilities.

Lactococci had higher C-S lyase and aminotransferase activities specific for sulfur-containing compounds than *Lactobacillus* and *Leuconostoc* (Hanniffy et al. 2009).

Amino acid catabolism also generates biogenic amines (BA). BA are basic nitrogenous organic compounds that are primarily formed by amino acid decarboxylation. BA are present in a wide range of foods, including dairy products, and can accumulate to elevated concentrations. Consuming foods with high amounts of these amines has toxicological effects. Tyrosine, histidine, lysine, ornithine, and arginine can all be decarboxylated into tyramine, histamine, cadaverine, putrescine, and agmatine, respectively. Agmatine can be further deaminated to produce putrescine. Amino acid decarboxylation pathways involve the transport of the amino acid into the cell, decarboxylation, and transport of the resulting BA out of the cell. The amino acid/BA interchange is performed by a transporter protein, and the tyramine, histamine, and putrescine biosynthesis pathways have been characterized in LAB. The genes encoding the decarboxylase and transporter proteins are located on the chromosomes for most LAB strains. Although BA production in LAB has been proposed as strain specific, genomic analysis of *E. faecalis*, *Enterococcus durans*, and *Enterococcus faecium* have shown that tyramine production is species specific (Ladero et al. 2012). Tyramine biosynthesis requires tyrosine, the decarboxylase (TDC), and a transporter protein (TyrP). The genes that encode TDC and TyrP form a chromosomal cluster, and a third open reading frame that encodes a protein homologous to tyrosyl tRNA synthase is always present in this cluster, upstream of the *tdc* gene (Linares et al. 2012). In cheese, *Lactobacillus* and *Enterococcus* genera are the main producers of tyramine.

In dairy products, the agmatine pathway (AgDI) has been proposed as the main route for putrescine synthesis since the ornithine decarboxylase pathway is not present in dairy LAB. These bacteria make putrescine from agmatine in three steps that are catalyzed by agmatine deiminase (AgDI), putrescine transcarbamylase (PTC), and carbamate kinase (CK). An antiporter then exchanges putrescine for agmatine. In addition to putrescine, ATP and NH_3 are also products of this set of reactions. The agmatine pathway has been identified in *E. faecium*, *E. faecalis*, *Lact. brevis*, and *Lactobacillus curvatus*. In the last two species, both the tyramine and putrescine pathways are linked on the chromosome. Some *Lactococcus* strains are also able to produce putrescine from agmatine (Ladero et al. 2012). In *Lactococcus*, some strains carry the gene cluster and some have an insertion element that renders the cluster non-functional. These combinations suggest that these strains putatively evolved from an ancestral strain that carried the gene cluster; the gene cluster could then accumulate mutations or even be lost if agmatine was not present in the strain's environment.

The agmatine pathway is similar to the arginine deiminase pathway (ADI), which is comprised of three reactions catalyzed by arginine deiminase, ornithine carbamoyl-transferase, and CK. The ADI pathway is widely distributed among LAB and has been described in strains from the *Enterococcus*, *Lactobacillus*, *Lactococcus*, *Leuconostoc*, *Oenococcus*, *Streptococcus*, and *Weissella* genera.

Histamine is another BA that occurs frequently in fermented products. It is produced from histidine in a reaction catalyzed by histidine decarboxylase (HdcA). Histamine synthesis also requires an antiporter protein that interchanges histidine and histamine. The histamine gene cluster has a gene encoding for the HdcB protein, which is cotranscribed with *hdcA* and is necessary for HdcA maturation (Trip et al. 2011). A gene encoding a protein homologous to histidyl tRNA synthetase is also found in this cluster, although its position varies from one LAB species to another (Calles-Enriquez et al. 2010). The HDC cluster has been characterized in different species of *Lactobacillus*, *O. oeni*, *Strep. thermophilus*, *Tetragenococcus muriaticus*, and *Tetragenococcus halophilus* and is located on the chromosome, except for *Lactobacillus hilgardii* (Lucas et al. 2005).

BA synthesis always requires the presence of amino acids; however, other parameters such as pH, carbon source, or temperature differ from one BA to another (Linares et al. 2012).

1.4. LAB Metabolism in Light of Genomics, Comparative Genomics, and Metagenomics

The use of next generation sequencing technology revolutionized microbial science by rapidly increasing the number of publically available genomes. The first LAB genome was published in 2001 (Bolotin et al. 2001) and, since then, nearly 100 complete genomes from typical LAB species (excluding enterococci) have been deposited at NCBI, and more than 600 whole genome sequencing projects with a permanent draft or incomplete status have been reported, according to GOLD (as of 20 May 2015). Among draft genome sequences, a dozen *Lactobacillus* strains have been generated as part of the Human Microbiome Project (The Human Microbiome Jumpstart Reference Strains Consortium et al. 2010). Furthermore, the technological advances and cost reductions of sequencing have allowed the resequencing of the *L. lactis* MG1363 genome and its comparison to the genome of its derivative *L. lactis* NZ9000 (Linares et al. 2010). This comparison identified six differences that were not attributable to known errors in the published sequence. Two specific mutations were found in the *ptcC* promoter, which plays a key role in regulating cellobiose and glucose uptake. A comparative analysis of the genomes of six *Lactobacillus* and several other LAB has also identified nine niche-specific genes (O'Sullivan et al. 2009). A comparison of 20 complete *Lactobacillus* genomes showed that their bacterial pan-genome consists of approximately 14,000 genes, including a core genome of 383 orthologous genes (Kant et al. 2011).

The analysis of mixed cultures, which is very important in industrial fermentations, has also become possible using metagenomic approaches. Novel molecular technologies such as metagenomics can be used to analyze the metabolic potentials of LAB. In one study, LAB isolates from African pearl millet slurries and the metagenomes of amylaceous fermented foods were screened for genes involved in probiotic functions and starch metabolism (Turpin et al. 2011). Metagenomic approaches were used to monitor changes in bacterial populations, metabolic potential, and the overall genetic features of the microbial community during a 29-day fermentation process of the traditional Korean food kimchi (Jung et al. 2011). The resulting metagenomic sequences were categorized by their functions. Sequences in the carbohydrate category indicated that the kimchi microbiome was enriched with genes associated with mono-, di-, and oligosaccharide fermentation. Similarly, sequences in the fermentation category were enriched for various genes involved in lactate fermentation and acetoin and butanediol metabolism. The number of metabolic genes involved in carbohydrate metabolism and fermentation generally increased as kimchi fermentation progressed.

In addition to fermented food products, the human gastrointestinal tract and its complex microbial community have also garnered considerable interest. More than three million bacterial genes in the human gastrointestinal tract have been characterized (Qin et al. 2010; Methé et al. 2012).

1.5. Novel Aspects of Metabolism Regulation in the Post-genomic Age

Transcriptional regulation is essential for microorganism adaptations to changes in external or intracellular conditions. In bacteria, transcription can be activated or inhibited by various transcription factors (TFs) that recognize specific *cis*-regulatory DNA elements and TF-binding sites (TFBSs) in gene promoter regions. The set of genes or operons under direct control of the same TF is known as a regulon, and all the regulons in an organism form a transcriptional regulatory network.

In recent years, the number of dairy and non-dairy LAB genomic sequences has dramatically increased. A study of 30 LAB representatives found that 3.5% (*Strep. thermophillus, Lact. delbrueckii, Lact. helveticus*) to 7.5% (*Lact. plantarum*) of the proteome helped regulate transcription (Figure 1.4; Ravcheev et al. 2013). In addition, the total amount of TFs varied significantly between species, ranging from 63–64 (*Lact. helveticus* and *Strep. thermophilus*, respectively) to 240 (*Lact. plantarum*). Among 30 *Lactobacillales* genomes, putative TFs were identified in 49 protein families (average of 36 TFs per genome), and approximately 90% belonged to 24 major families (at least two TFs per genome) (Ravcheev et al. 2013). The most represented TF family was the Xre family (298 TFs in total), and a dozen or even several dozen Xre members were identified in each genome (average of 19 per genome). Other TF families were considerably less represented; those that had at least four representatives per genome included TetR, GntR, MarR, OmpR, LacI, LysR, MerR, and AraC (Figure 1.4) (Ravcheev et al. 2013).

Other systematic genome-wide analyses of transcriptional regulatory networks have been performed for single LAB species, including *Lact. plantarum* (Wels et al. 2011) and *L. lactis* (de Jong et al. 2012, 2013). These sequence- and motif-based reconstructions of gene regulatory networks have led to the development of several analysis tools, such as PEPPER, RegTransBase, PRODORIC, RegPrecise, FITBAR, RegAnalyst, and MGcV (Overmars et al. 2013). Tools for discovering new motifs have also been developed, including MEME (Bailey et al. 2009), Tmod (Sun et al. 2009), GIMSAN (Ng and Keich 2008), and EXTREME (Quang and Xie 2014).

(a)

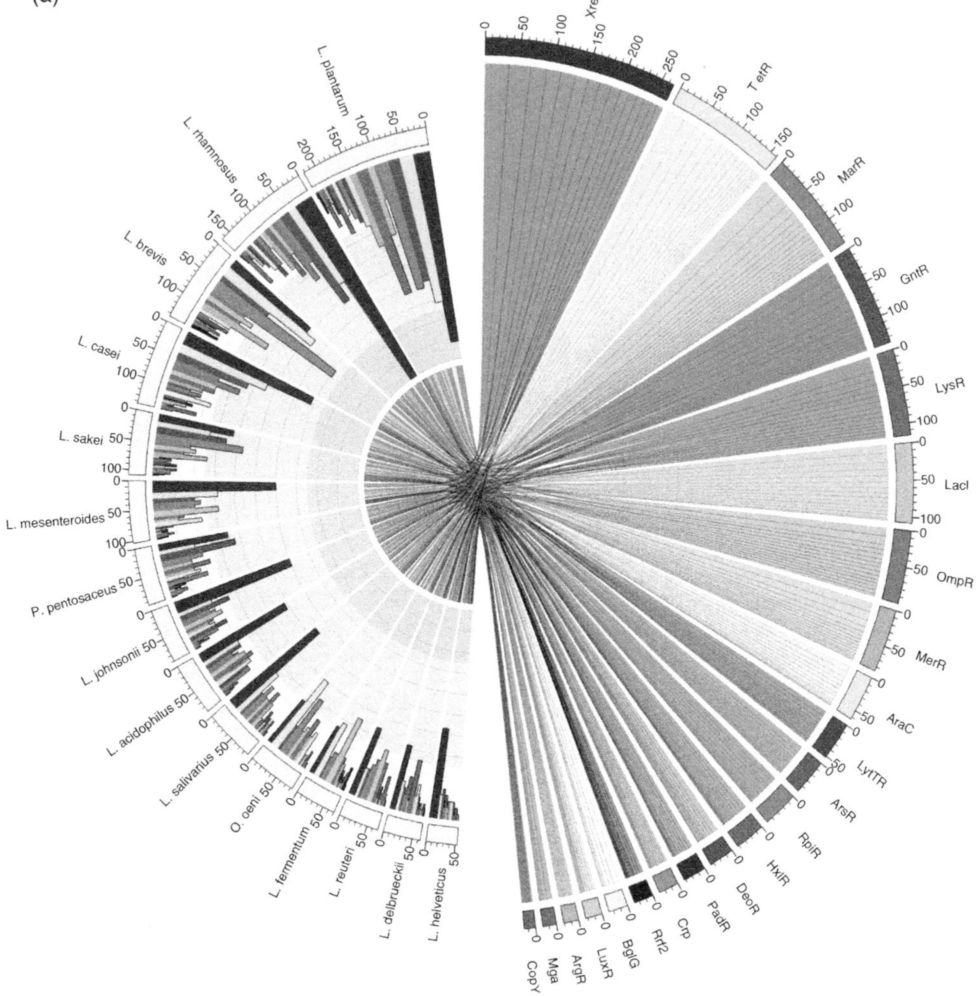

Figure 1.4. *(Continued)*

(b)

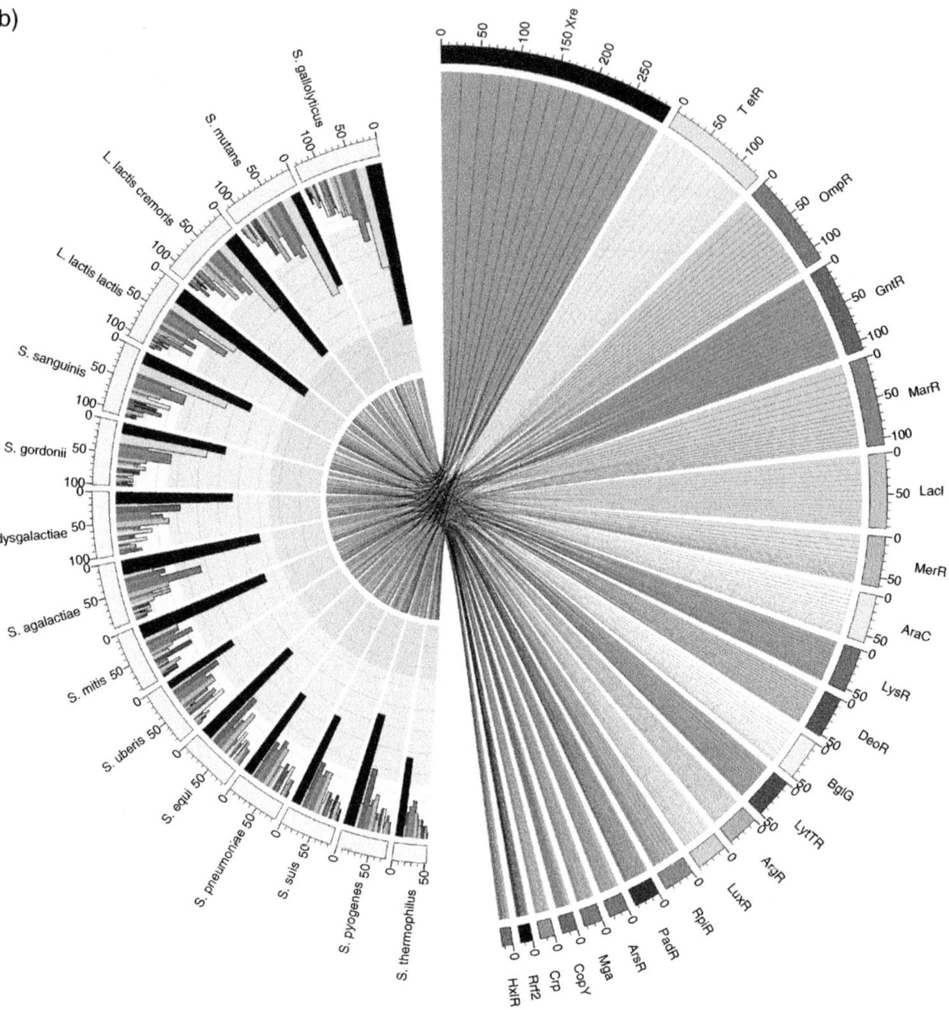

Figure 1.4. Distribution of predicted transcription factors (TFs) in selected Lactobacillaceae (a) and Streptococcaceae (b) genomes. The least-represented TFs families (AbrB, AsnC, BirA, CcpN, CodY, ComK, CtsR, DtxR, Fis, Fur, GlnL, GutM, HrcA, HTH_11, IclR, LexA, NiaR, NrdR, NrtR, PF04394, PurR, Rex, ROK, SdaR, SfsA, SorC, YobV) are not presented on the ideograms. Row data derive from (Ravcheev et al. 2013). Graphical presentation of these data has been developed with Circos (v0.64) (Krzywinski et al. 2009). Strains of presented *Lactobacillales* species: *L. lactis cremoris* (*Lactococcus lactis* subsp. *cremoris* SK11), *L. lactis lactis* (*Lactococcus lactis* subsp. *lactis* IL1403), *S. thermophilus* (*Streptococcus thermophilus* CNRZ1066), *S. agalactiae* (*Streptococcus agalactiae* 2603V/R) *S. uberis* (*Streptococcus uberis* 0140J), *S. equi* (*Streptococcus equi* MGCS10565), *S. dysgalactiae* (*Streptococcus dysgalactiae* GGS_124), *S. pyogenes* (*Streptococcus pyogenes* M1 GAS), *S. gallolyticus* (*Streptococcus gallolyticus* UCN34), *S. mutans* (*Streptococcus mutans* UA159), *S. suis* (*Streptococcus suis* 05ZYH33), *S. mitis* (*Streptococcus mitis* B6), *S. pneumoniae* (*Streptococcus pneumoniae* TIGR4), *S. gordonii* (*Streptococcus gordonii* CH1), *S. sanguinis* (*Streptococcus sanguinis* SK36), *L. sakei* (*Lactobacillus sakei* 23K), *L. casei* (*Lactobacillus casei* ATCC 334), *L. rhamnosus* (*Lactobacillus rhamnosus* GG), *L. delbrueckii* (*Lactobacillus delbrueckii* ATCC BAA-365), *L. acidophilus* (*Lactobacillus acidophilus* NCFM), *L. helveticus* (*Lactobacillus helveticus* DPC 4571), *L. johnsonii* (*Lactobacillus johnsonii* NCC 533), *P. pentosaceus* (*Pediococcus pentosaceus* ATCC 25745), *L. brevis* (*Lactobacillus brevis* ATCC 367), *L. plantarum* (*Lactobacillus plantarum* WCFS1), *L. fermentum* (*Lactobacillus fermentum* IFO 3956), *L. reuteri* (*Lactobacillus reuteri* JCM 1112), *O. oeni* (*Oenococcus oeni* PSU-1), *L. mesenteroides* (*Leuconostoc mesenteroides* ATCC 8293), *L. salivarius* (*Lactobacillus salivarius* UCC118).

Despite extensive analyses of the available genomes, comprehensive, experimental studies of LAB regulatory networks remain lacking. Most gene regulation studies in LAB focus on individual regulons in primarily model species, such as *L. lactis* subsp. *cremoris* MG1363 or *L. lactis* subsp. *lactis* IL1403. These studies have identified both global (CcpA – catabolite control protein A) and specific (ClaR, FruR, MalR, and XylR from the RpiR, DeoR, LacI, and AraC families, respectively) TFs for different α- or β-glucosides, fructose, galactose, lactose, maltose, sorbose, or xylose assimilation genes (reviewed by Aleksandrzak-Piekarczyk 2013; de Jong et al. 2012). CcpA is a component of carbon catabolite repression (CCR), which also includes HPr, HPr kinase, and the glycolytic enzymes fructose 1,6-bisphosphate and glucose-6-phosphate (Deutscher 2008; Görke and Stülke 2008). Large-scale approaches, such as transcriptomics and proteomics, have shown that CcpA is also involved in the regulation of tens of genes associated with membrane transport, nucleotide and nitrogen metabolism, protein biosynthesis, and folding (Mazzeo et al. 2012; Zomer et al. 2006). In addition, CcpA-dependent CCR has been shown to impede the expression of two operons involved in citrate metabolism in *E. faecalis* (Suárez et al. 2011).

Many genes undergo diverse regulation by CcpA or other TFs. In milk, transcriptomics has identified several regulatory networks in *L. lactis* (de Jong et al. 2013). A comparative systems biology approach revealed a crucial role for phosphate in regulating central metabolism and glucose uptake in *L. lactis* and *Streptococcus pyogenes* (Levering et al. 2012). Comparing kinetics models in both strains revealed variations that can be explained by differences in phosphate levels in their natural environments.

Recent studies have sought to determine the scope of proteolytic systems and amino acid biosynthesis regulation in *L. lactis* cells. Previous studies identified a significant role for CodY, a global transcriptional regulator, in negatively regulating components of the proteolytic system, which are stimulated by the intracellular pool of branched chain amino acids (BCAAs) (den Hengst et al. 2005). Subsequent studies have reported on the CodY-dependant regulation of several proteolytic system components in other LAB, including *Strep. thermophilus* (Liu et al. 2009), *Streptococcus pneumoniae* (Hendriksen et al. 2008), *Streptococcus mutans* (Lemos et al. 2008), and *O. oeni* (Ritt et al. 2009) but not in lactobacilli. The latest studies have indicated that a proteolysis regulator separate from CodY exists in lactobacillaceae, enterococcaceae, leuconostocaceae, carnobacteriaceae, listeriaceae, exiguobacteria, and bacillaceae. This has been thoroughly examined in *Lact. helveticus*, and a novel BCARR protein has been identified that represses proteolysis gene expression in response to BCAAs (Wakai and Yamamoto 2013). No BCARR orthologs are present in streptococcaceae, including lactococci, which have a CodY homolog.

Amino acid assimilation is crucial for LAB that are generally auxotrophic for amino acids. In addition to RNA structural switches, regulating sulfur amino acid metabolism is also dependent on LysR-family transcriptional regulators, including CmbR in *L. lactis* and CysR, HomR, and MetR in *Strep. mutans* (Liu et al. 2012 and references within). Global ArgR and AhrC regulons have been established by transcriptome analyses, which indicate that both regulators are dedicated to controlling arginine metabolism in *L. lactis* (Larsen et al. 2008). Interestingly, pneumococcal ArgR and AhrC are not involved in regulating arginine biosynthesis and breakdown, as is the case in other bacteria, and instead control the expression of genes associated with arginine and peptide uptake (Kloosterman and Kuipers 2011). AhrC has also recently been reported to have a role in *Strep. pneumoniae* (Kloosterman and Kuipers 2011) and *E. faecalis* (Frank et al. 2013) virulence. Very few global studies of modifications in LAB cell metabolism and the various regulatory mechanisms involved with amino acid shortages have been performed. One study analyzed the global transcriptome and proteomic response of *L. lactis* during progressive isoleucine starvation. CodY appeared to be specifically dedicated to regulating the supply of isoleucine, while the other regulators identified were associated with growth rate and stringent response (Dressaire et al. 2011).

1.6. Functional Genomics and Metabolism

1.6.1. Transcriptomics, Proteomics, and Metabolomics

One of the greatest achievements of recent molecular biology is the development of high-throughput functional genomics approaches, such as transcriptomics, proteomics, or metabolomics. Transcriptomics have been used to analyze the LAB response to different growth conditions, culturing regimens, and various stresses (de Vos 2011). A variety of studies, including attempts at functional genomics, have contributed to the discovery and further characterization of respiratory metabolism in many LAB species (for recent reviews see Lechardeur et al. 2011a; Pedersen et al. 2012). Respiration was found to be activated by cofactors such as exogenous heme or heme and menaquinone. Some LAB species were also found to use nitrate as a terminal electron acceptor (Brooijmans et al. 2009). Furthermore, activating electron transport chains leads to higher biomass production and increased robustness, which is beneficial for industrial applications (Pedersen et al. 2012).

Transcriptional analysis identified genes involved in oxygen respiration. For example, the *ygfCBA* operon from *L. lactis*, which encodes a putative transport system and regulator, is strongly induced by heme and is involved in heme tolerance and homeostasis (Pedersen et al. 2008). YgfC (or HrtR4) is a heme-responsive repressor that regulates *hrtRBA* (formerly *ygfCBA*) operon expression in response to free heme molecules (Lechardeur et al. 2011b). Recently, the crystal structures of HrtR have confirmed that it senses and binds a heme molecule to regulate expression of the heme-efflux system that is responsible for heme homeostasis in *L. lactis* (Sawai et al. 2012). A community transcriptomic approach revealed huge differences in gene expression profiles between *Lact. plantarum* from the human intestine and that from laboratory media (Marco et al. 2010). Moreover, similar expression profiles were observed in mice and humans, which suggests that there are some common intestinal adaptation mechanisms. The metabolic capacity of *Lact. plantarum* is specifically adapted in the intestine for carbohydrate acquisition and the expression of exopolysaccharide and proteinaceous cell surface compounds. The transcriptome response in complex food ecosystems has also been analyzed in bacteria in yogurt fermentation. In milk, *Strep. thermophilus* LMD-9 in the presence of *Lact. bulgaricus* overexpresses genes involved in amino acid transport and metabolism as well as DNA replication (Goh et al. 2011). The metatranscriptomic approach in yogurt showed that specific compounds and metabolic pathways are involved in interactions between *Strep. thermophilus* and *Lact. bulgaricus* strains (Sieuwerts et al. 2010). Similar approaches have been applied to complex communities in fermented foods, such as kimchi (Nam et al. 2009) or sourdough (Weckx et al. 2011).

Proteomic tools can also be used to improve the understanding of metabolism in food microorganisms. Recently, the proteomic approach was used to quantitatively analyze the bacterial enzymes released in Swiss-type cheese during ripening (Jardin et al. 2012). This approach has also been used to investigate stress responses in LAB to acidic growth conditions (Nezhad 2010) or low temperatures (Garnier et al. 2010). Recent modeling approaches of protein and mRNA stability based on transcriptome and proteome *L. lactis* data have also increased the quantitative abilities of functional genomics (Dressaire et al. 2009; Picard et al. 2009).

Global metabolomics studies of LAB have not been frequently reported (de Vos 2011). One study of folate-overproducing *Lact. plantarum* cells (Wegkamp et al. 2010) found that reductions in growth rate occurred when the folate gene cluster was overexpressed. In another study, the response of *L. lactis* to acid stress was analyzed at the metabolite level and complemented by performing a genome-wide transcriptome analysis (Carvalho et al. 2013). The H+/lactate stoichiometry of lactic acid export was determined, and the study findings were used to propose a metabolic model that accounted for the molecular mechanisms of the *L. lactis* response to acid stress.

1.6.2. Global Phenotypic Characterization of Microbial Cells

Phenotype MicroArray™ (PM) technology (Biolog Inc.) is a system of cellular assays designed for the simultaneous and high-throughput screening of a large variety of phenotypes. This technology can be used to complement traditional genomic, transcriptomic, and proteomic approaches. PM can test for nearly 2000 microbial phenotypes (C, N, P, and S metabolism; pH growth range; pH control; sensitivity to NaCl and other ions; and sensitivity to chemical agents) to gain a comprehensive overview of pathway functions in a single experiment (Bochner 2009). In recent years, several studies have presented the results of PM analysis in LAB. PM is most often used to analyze gene function by comparing LAB cell lines that differ by a single gene mutation. For example, the role of the lactate dehydrogenase enzyme (LDH) in *L. lactis*, *E. faecalis*, and *Strep. pyogenes* was identified by screening a large variety of carbon sources using PM arrays. The *ldh* deletion mutants were not able to utilize all carbon sources as efficiently as their corresponding wild-types, which explained why their growth was impaired in a rich but not chemically-defined medium (Fiedler et al. 2010). In another study, the *liaS* gene in *Strep. mutants* was found to encode a cell-envelope stress-sensing histidine kinase, which is thought to be important for the expression of virulence factors (Zhang and Biswas 2009). The *liaS* mutant was more tolerant to various inhibitors that targeted protein synthesis, DNA synthesis, and cell-wall biosynthesis.

PM can also be used to analyze naturally occurring microbial strains, the biological properties of these strains, the effects of chemicals on their cells, and as a tool in bioprocess optimization. For instance, PM was used to understand the phenotypic manifestations of bacterial environmental adaptation (Di Cagno et al. 2010). In this study, the metabolisms of 72 *Lact. plantarum* isolates from raw vegetables and fruits were compared on a variety of carbon sources. Eight clusters of *Lact. plantarum* isolates were categorized based on their metabolism on 27 carbon sources, which interestingly, was not associated with their original habitats (Di Cagno et al. 2010). In another study, combined genomic, transcriptomic, and PM analyses were used to understand the survival capacity of microbes in sourdough (Passerini et al. 2013a). Bacteria that could effectively utilize five-carbon (C5) and six-carbon (C6) sugars from lignocellulose would significantly improve the economic conversion of lignocellulose to biofuels and bioproducts. PM analyses indicated that the *Lactobacillus buchneri* strain can metabolize a broad spectrum of carbon sources, including various C5 and C6 monosaccharides, disaccharides, and oligosaccharides. Furthermore, *Lact. buchneri* had better metabolic rates under anaerobic conditions, high tolerance to ethanol and other stresses, and therefore was an attractive candidate for biomass conversion to various bioproducts (Liu et al. 2007).

1.7. Systems Biology of LAB

Different modeling approaches have been successfully applied to LAB monocultures (Santos et al. 2013). Kinetics models are used to integrate experimental data from biochemical studies of transport, enzyme kinetics, as well as flux and metabolites measurements (Teusink et al. 2011). In *L. lactis*, kinetics modeling has been primarily used to model glycolysis (Hoefnagel et al. 2002a, 2002b; Voit et al. 2006; Andersen et al. 2009; Oh 2011; Levering et al. 2012). Genome sequencing, genome analysis, and omic techniques (proteomics, transcriptomics, metabolomics, etc.) have been used to construct genome-scale metabolic models (Teusink et al. 2006; Pastink et al. 2009; Santos et al. 2009; Flahaut et al. 2013), which can help elucidate many cellular biochemical pathways. Coupling genomic data with metabolic fingerprinting using high-throughput techniques can further facilitate the extraction of relevant biological information. Such techniques have been used to characterize citrate utilization and the subsequent generation of flavor compounds (Bachmann et al. 2009; de Bok et al. 2011). The latest developments in and future of modeling lie in mixed culture and complex ecosystem applications (for a review, see Santos et al. 2013). These analyses have been applied to food consortia, such as *Strep. thermophilus* and *Lact. bulgaricus* in yogurt (Sieuwerts 2009) as well as the gut microbiota of the human gastrointestinal tract (de Graaf et al. 2010). Unravelling the physiology of multi-species microbial communities is a great challenge and will require the combination of phylogenetic, metagenomic, and meta-functional genomics data (Martins dos Santos et al. 2010).

Furthermore, systems biology is another powerful tool that can be used to decipher metabolism in LAB. Systems biology studies microorganisms as integrated and interacting networks of genes, proteins, and biochemical reactions (Teusink et al. 2011; de Vos 2011).

Acknowledgments

Work in the authors' laboratories has been supported by the Applied Research Program (PBS) of the National Centre for Research and Development (NCBR) funds (PBS2/B8/12/2014) and the National Science Centre (NCN) (grant UMO-2011/01/B/NZ2/05377). Projects were also funded by the Spanish Ministry of Economy and Competitiveness and Innovation (Ref. AGL2011-24300) and INIA (Ref. RM2010-00017-00-00 and RM2011-00005-00-00).

References

Abdel-Rahman, M.A., Tashiro, Y., and Sonomoto, K. (2011) Lactic acid production from lignocellulose-derived sugars using lactic acid bacteria: Overview and limits. *J Biotechnol* **156**, 286–301.

Aleksandrzak-Piekarczyk, T. (2013) Lactose and β-glucosides metabolism and its regulation in *Lactococcus lactis*: A review. In: *Lactic Acid Bacteria - R& D for Food, Health and Livestock Purposes*. ed Marcelino Kongo, J. InTech. Available at http://cdn.intechopen.com/pdfs-wm/42338.pdf [accessed 30 January 2013].

Andersen, A.Z., Carvalho, A.L., Neves, A.R., et al. (2009) The metabolic pH response in *Lactococcus lactis*: An integrative experimental and modelling approach. *Comput Biol Chem* **33**, 71–83.

Bachmann, H., Kruijswijk, Z., Molenaar, D., et al. (2009) A high-throughput cheese manufacturing model for effective cheese starter culture screening. *J Dairy Sci* **92**, 5868–5882.

Bachmann, H., Starrenburg, M.J.C., Molenaar, D., et al. J.E.T. (2012) Microbial domestication signatures of *Lactococcus lactis* can be reproduced by experimental evolution. *Genome Res* **22**, 115–124.

Badel, S., Bernardi, T., and Michaud, P. (2011) New perspectives for Lactobacilli exopolysaccharides. *Biotechnol Adv* **29**, 54–66.

Bailey, T.L., Boden, M., Buske, F.A., et al. (2009) MEME SUITE: tools for motif discovery and searching. *Nucleic Acids Res* **37** (Web Server), W202–W208.

Bekal, S., Van Beeumen, J., Samyn, B., et al. (1998) Purification of *Leuconostoc mesenteroides* citrate lyase and cloning and characterization of the *citCDEFG* gene cluster. *J Bacteriol* **180**, 647–654.

Bochner, B.R. (2009) Global phenotypic characterization of bacteria. *FEMS Microbiol Rev* **33**, 191–205.

De Bok, F.A.M., Janssen, P.W.M., Bayjanov, J.R., et al. (2011) Volatile compound fingerprinting of mixed-culture fermentations. *Appl Environ Microbiol* **77**, 6233–6239.

Bolotin, A., Wincker, P., Mauger, S., et al. (2001) The complete genome sequence of the lactic acid bacterium *Lactococcus lactis* ssp. *lactis* IL1403. *Genome Res* **11**, 731–753.

Bonifait, L., de la Cruz Dominguez-Punaro, M., Vaillancourt, K., et al. (2010) The cell envelope subtilisin-like proteinase is a virulence determinant for *Streptococcus suis*. *BMC Microbiol* **10**, 42.

Bounaix, M.-S., Gabriel, V., Morel, S., et al. (2009) Biodiversity of exopolysaccharides produced from sucrose by sourdough Lactic Acid Bacteria. *J Agric Food Chem* **57**, 10889–10897.

Bounaix, M.-S., Gabriel, V., Robert, H., et al. (2010) Characterization of glucan-producing *Leuconostoc* strains isolated from sourdough. *Int J Food Microbiol* **144**, 1–9.

Broadbent, J.R., Cai, H., Larsen, R.L., et al. (2011) Genetic diversity in proteolytic enzymes and amino acid metabolism among *Lactobacillus helveticus* strains. *J Dairy Sci* **94**, 4313–4328.

Brooijmans, R., de Vos, W.M., and Hugenholtz, J. (2009) Electron transport chains of lactic acid bacteria-walking on crutches is part of their lifestyle. *F1000 Biol Rep* **1**. Available at http://F1000.com/Reports/Biology/content/1/34 [accessed 29 April 2009].

Di Cagno, R., Minervini, G., Sgarbi, E., et al. (2010) Comparison of phenotypic (Biolog System) and genotypic (random amplified polymorphic DNA-polymerase chain reaction, RAPD-PCR, and amplified fragment length polymorphism, AFLP) methods for typing *Lactobacillus plantarum* isolates from raw vegetables and fruits. *Int J Food Microbiol* **143**, 246–253.

Calles-Enriquez, M., Eriksen, B.H., Andersen, P.S., et al. (2010) Sequencing and transcriptional analysis of the *Streptococcus thermophilus* histamine biosynthesis gene cluster: factors that affect differential *hdcA* expression. *Appl Environ Microbiol* **76**, 6231–6238.

Carvalho, A.L., Turner, D.L., Fonseca, L.L., et al. (2013) Metabolic and transcriptional analysis of acid stress in *Lactococcus lactis*, with a focus on the kinetics of lactic acid pools. *PLoS One* **8**, e68470.

Castillo Martinez, F.A., Balciunas, E.M., Salgado, J.M., et al. (2013) Lactic acid properties, applications and production: A review. *Trends Food Sci Technol* **30**, 70–83.

Castro, R., Neves, A.R., Fonseca, L.L., et al. (2009) Characterization of the individual glucose uptake systems of *Lactococcus lactis*: mannose-PTS, cellobiose-PTS and the novel GlcU permease. *Mol Microbiol* **71**, 795–806.

Ciszek-Lenda, M., Nowak, B., Śróttek, M., et al. (2011) Immunoregulatory potential of exopolysaccharide from *Lactobacillus rhamnosus* KL37. Effects on the production of inflammatory mediators by mouse macrophages: *L. rhamnosus* exopolysaccharide and inflammation. *Int J Exp Pathol* **92**, 382–391.

Cretenet, M., Laroute, V., Ulve, V., et al. (2010) Dynamic analysis of the *Lactococcus lactis* transcriptome in cheeses made from milk concentrated by ultrafiltration reveals multiple strategies of adaptation to stresses. *Appl Environ Microbiol* **77**, 247–257.

Dandoy, D., Fremaux, C., de Frahan, M.H., et al. (2011) The fast milk acidifying phenotype of *Streptococcus thermophilus* can be acquired by natural transformation of the genomic island encoding the cell-envelope proteinase PrtS. *Microb Cell Fact* **10** (Suppl 1), S21.

Delorme, C., Bartholini, C., Bolotine, A., et al. (2010) Emergence of a cell wall protease in the *Streptococcus thermophilus* population. *Appl Environ Microbiol* **76**, 451–460.

Dertli, E., Colquhoun, I.J., Gunning, A.P., et al. (2013) Structure and biosynthesis of two exopolysaccharides produced by *Lactobacillus johnsonii* FI9785. *J Biol Chem* **288**, 31938–31951.

Deutscher, J. (2008) The mechanisms of carbon catabolite repression in bacteria. *Curr Opin Microbiol* **11**, 87–93.

Díaz-Muñiz, I., Banavara, D.S., Budinich, M.F., et al. (2006) *Lactobacillus casei* metabolic potential to utilize citrate as an energy source in ripening cheese: a bioinformatics approach. *J Appl Microbiol* **101**, 872–882.

Dressaire, C., Gitton, C., Loubière, P., et al. (2009) Transcriptome and proteome exploration to model translation efficiency and protein stability in *Lactococcus lactis*. *PLoS Comput Biol* **5**, e1000606.

Dressaire, C., Redon, E., Gitton, C., et al. (2011) Investigation of the adaptation of *Lactococcus lactis* to isoleucine starvation integrating dynamic transcriptome and proteome information. *Microb Cell Fact* **10** (Suppl 1), S18.

Drici, H., Gilbert, C., Kihal, M., and Atlan, D. (2010) Atypical citrate-fermenting *Lactococcus lactis* strains isolated from dromedary's milk. *J Appl Microbiol* **108**, 647–657.

Drider, D., Bekal, S., and Prévost, H. (2004) Genetic organization and expression of citrate permease in lactic acid bacteria. *Genet Mol Res* **3**, 271–281.

Espariz, M., Repizo, G., Blancato, V., et al. (2011) Identification of malic and soluble oxaloacetate decarboxylase enzymes in *Enterococcus faecalis*: *E. faecalis* malic enzyme family proteins. *FEBS J* **278**, 2140–2151.

Feng, M., Chen, X., Li, C., et al. (2012) Isolation and identification of an exopolysaccharide-producing lactic acid bacterium strain from chinese paocai and biosorption of Pb(II) by its exopolysaccharide. *J Food Sci* **77**, T111–T117.

Fernandez, E., Alegria, A., Delgado, S., et al. (2011) Comparative phenotypic and molecular genetic profiling of wild *Lactococcus lactis* subsp. *lactis* strains of the *L. lactis* subsp. *lactis* and *L. lactis* subsp. *cremoris* genotypes, isolated from starter-free cheeses made of raw milk. *Appl Environ Microbiol* **77**, 5324–5335.

Fernández, M., van Doesburg, W., Rutten, G.A., et al. (2000) Molecular and functional analyses of the *metC* gene of *Lactococcus lactis*, encoding cystathionine β-lyase. *Appl Environ Microbiol* **66**, 42–48.

Fiedler, T., Bekker, M., Jonsson, M., et al. (2010) Characterization of three lactic acid bacteria and their isogenic *ldh* deletion mutants shows optimization for YATP (cell mass produced per mole of ATP) at their physiological pHs. *Appl Environ Microbiol* **77**, 612–617.

Flahaut, N.A.L., Wiersma, A., van de Bunt, B., et al. (2013) Genome-scale metabolic model for *Lactococcus lactis* MG1363 and its application to the analysis of flavor formation - Springer. *Appl Microbiol Biotechnol* **97**, 8729–8739.

Frank, K.L., Guiton, P.S., Barnes, A.M.T., et al. (2013) AhrC and Eep are biofilm infection-associated virulence factors in *Enterococcus faecalis*. *Infect Immun* **81**, 1696–1708.

Galle, S., Schwab, C., Arendt, E., and Gänzle, M. (2010) Exopolysaccharide-forming *Weissella* strains as starter cultures for sorghum and wheat sourdoughs. *J Agric Food Chem* **58**, 5834–5841.

Galle, S., Schwab, C., Arendt, E.K., and Gänzle, M.G. (2011) Structural and rheological characterisation of heteropolysaccharides produced by lactic acid bacteria in wheat and sorghum sourdough. *Food Microbiol* **28**, 547–553.

Gänzle, M.G. and Follador, R. (2012) Metabolism of oligosaccharides and starch in lactobacilli: A review. *Front Microbiol* **3**. Available at http://www.ncbi.nlm.nih.gov/pmc/articles/PMC3458588/pdf/fmicb-03-00340.pdf [accessed 26 September 2012].

Garnier, M., Matamoros, S., Chevret, D., et al. (2010) Adaptation to cold and proteomic responses of the psychrotrophic biopreservative *Lactococcus piscium* strain CNCM I-4031. *Appl Environ Microbiol* **76**, 8011–8018.

Genay, M., Sadat, L., Gagnaire, V., and Lortal, S. (2009) *PrtH2*, not *prtH*, is the ubiquitous cell wall proteinase gene in *Lactobacillus helveticus*. *Appl Environ Microbiol* **75**, 3238–3249.

Goh, Y.J., Goin, C., O'Flaherty, S., et al. (2011) Specialized adaptation of a lactic acid bacterium to the milk environment: the comparative genomics of *Streptococcus thermophilus* LMD-9. *Microb Cell Fact* **10** (Suppl 1), S22.

Görke, B. and Stülke, J. (2008) Carbon catabolite repression in bacteria: many ways to make the most out of nutrients. *Nat Rev Microbiol* **6**, 613–624.

Górska-Frączek, S., Sandström, C., Kenne, L., et al. (2013) The structure and immunoreactivity of exopolysaccharide isolated from *Lactobacillus johnsonii* strain 151. *Carbohydr Res* **378**, 148–153.

Górska-Frączek, S., Sandström, C., Kenne, L., et al. (2011) Structural studies of the exopolysaccharide consisting of a nonasaccharide repeating unit isolated from *Lactobacillus rhamnosus* KL37B. *Carbohydr Res* **346**, 2926–2932.

De Graaf, A.A., Maathuis, A., de Waard, P., et al. (2010) Profiling human gut bacterial metabolism and its kinetics using [U-13C] glucose and NMR. *NMR Biomed* **23**, 2–12.

Griffiths, M.W. and Tellez, A.M. (2013) *Lactobacillus helveticus*: the proteolytic system. *Front Microbiol* **4**. Available at http://www.ncbi.nlm.nih.gov/pmc/articles/PMC3587842/pdf/fmicb-04-00030.pdf [accessed 5 March 2013].

Hanniffy, S.B., Pelaez, C., Martinez-Bartolome, M.A., et al. (2009) Key enzymes involved in methionine catabolism by cheese lactic acid bacteria. *Int J Food Microbiol* **135**, 223–230.

Hendriksen, W.T., Bootsma, H.J., Estevao, S., et al. (2008) CodY of *Streptococcus pneumoniae*: link between nutritional gene regulation and colonization. *J Bacteriol* **190**, 590–601.

Hoefnagel, M.H., van der Burgt, A., Martens, D.E., et al. (2002a) Time dependent responses of glycolytic intermediates in a detailed glycolytic model of *Lactococcus lactis* during glucose run-out experiments. *Mol Biol Rep* **29**, 157–161.

Hoefnagel, M.H., Starrenburg, M.J., Martens, D.E., et al. (2002b) Metabolic engineering of lactic acid bacteria, the combined approach: kinetic modelling, metabolic control and experimental analysis. *Microbiology* **148**, 1003–1013.

Jardin, J., Mollé, D., Piot, M., et al. (2012) Quantitative proteomic analysis of bacterial enzymes released in cheese during ripening. *Int J Food Microbiol* **155**, 19–28.

Jensen, M.P., Ardö, Y., and Vogensen, F.K. (2009) Isolation of cultivable thermophilic lactic acid bacteria from cheeses made with mesophilic starter and molecular comparison with dairy-related *Lactobacillus helveticus* strains. *Lett Appl Microbiol* **49**, 396–402.

De Jong, A., Hansen, M.E., Kuipers, O.P., et al. (2013) The transcriptional and gene regulatory network of *Lactococcus lactis* MG1363 during Growth in Milk. *PLoS One* **8**, e53085.

De Jong, A., Pietersma, H., Cordes, M., et al. (2012) PePPER: a webserver for prediction of prokaryote promoter elements and regulons. *BMC Genomics* **13**, 299.

den Hengst, C.D., van Hijum, S.A., Geurts, J.M., et al. (2005) The Lactococcus lactis CodY regulon: identification of a conserved cis-regulatory element. *J Biol Chem* **7**, 343323–34342.

Jung, J.Y., Lee, S.H., Kim, J.M., et al. (2011) Metagenomic analysis of kimchi, a traditional korean fermented food. *Appl Environ Microbiol* **77**, 2264–2274.

Kant, R., Blom, J., Palva, A., Siezen, R.J., and de Vos, W.M. (2011) Comparative genomics of *Lactobacillus*: Comparative analysis of 20 complete *Lactobacillus* genomes. *Microb Biotechnol* **4**, 323–332.

Katina, K., Maina, N.H., Juvonen, R., et al. (2009) *In situ* production and analysis of *Weissella confusa* dextran in wheat sourdough. *Food Microbiol* **26**, 734–743.

Kelly, W.J., Ward, L.J.H., and Leahy, S.C. (2010) Chromosomal diversity in *Lactococcus lactis* and the origin of dairy starter cultures. *Genome Biol Evol* **2**, 729-744.

Kim, Y., Oh, S., Yun, H.S., Oh, S., and Kim, S.H. (2010) Cell-bound exopolysaccharide from probiotic bacteria induces autophagic cell death of tumour cells. *Lett Appl Microbiol* **51**, 123-130.

Kloosterman, T.G. and Kuipers, O.P. (2011) Regulation of arginine acquisition and virulence gene expression in the human pathogen *Streptococcus pneumoniae* by transcription regulators ArgR1 and AhrC. *J Biol Chem* **286**, 44594–44605.

Koryszewska-Baginska, A., Bardowski, J., and Aleksandrzak-Piekarczyk, T. (2014) Genome sequence of the probiotic strain *Lactobacillus rhamnosus* (Formerly *Lactobacillus casei*) LOCK908. *Genome Announc* **2**, e00120-14–e00120-14.

Krzywinski, M., Schein, J., Birol, I., et al. (2009) Circos: An information aesthetic for comparative genomics. *Genome Res* **19**, 1639–1645.

Ladero, V., Fernández, M., Calles-Enríquez, M., et al. (2012) Is the production of the biogenic amines tyramine and putrescine a species-level trait in enterococci? *Food Microbiol* **30**, 132–138.

Larsen, R., van Hijum, S.A.F.T., Martinussen, J., et al. (2008) Transcriptome analysis of the *Lactococcus lactis* ArgR and AhrC regulons. *Appl Environ Microbiol* **74**, 4768–4771.

Lechardeur, D., Cesselin, B., Fernandez, A., et al. (2011a) Using heme as an energy boost for lactic acid bacteria. *Curr Opin Biotechnol* **22**, 143–149.

Lechardeur, D., Cesselin, B., Liebl, U., et al. (2011b) Discovery of intracellular heme-binding protein HrtR, which controls heme efflux by the conserved HrtB-HrtA transporter in *Lactococcus lactis*. *J Biol Chem* **287**, 4752–4758.

Leemhuis, H., Pijning, T., Dobruchowska, J.M., et al. (2013) Glucansucrases: Three-dimensional structures, reactions, mechanism, α-glucan analysis and their implications in biotechnology and food applications. *J Biotechnol* **163**, 250–272.

Lemos, J.A., Nascimento, M.M., Lin, V.K., et al. (2008) Global regulation by (p)ppGpp and CodY in *Streptococcus mutans*. *J Bacteriol* **190**, 5291–5299.

Lensbouer, J.J. and Doyle, R.P. (2010) Secondary transport of metal–citrate complexes: the CitMHS family. *Crit Rev Biochem Mol Biol* **45**, 453–462.

Levering, J., Musters, M.W.J.M., Bekker, M., et al. (2012) Role of phosphate in the central metabolism of two lactic acid bacteria - a comparative systems biology approach: Modelling glycolysis of lactic acid bacteria. *FEBS J* **279**, 1274–1290.

Linares, D.M., Kok, J., and Poolman, B. (2010) Genome sequences of *Lactococcus lactis* MG1363 (Revised) and NZ9000 and comparative physiological studies. *J Bacteriol* **192**, 5806–5812.

Linares, D.M., del Río, B., Ladero, V., et al. (2012) Factors influencing biogenic amines accumulation in dairy products. *Front Microbiol* **3**. Available at http://www.ncbi.nlm.nih.gov/pmc/articles/PMC3390585/pdf/fmicb-03-00180.pdf [accessed 28 May 2012].

Liu, C.-F., Tseng, K.-C., Chiang, S.-S., et al. (2011) Immunomodulatory and antioxidant potential of *Lactobacillus* exopolysaccharides. *J Sci Food Agric* **91**, 2284-2291.

Liu, F., Du, L., Du, P., and Huo, G. (2009) Possible promoter regions within the proteolytic system in *Streptococcus thermophilus* and their interaction with the CodY homolog. *FEMS Microbiol Lett* **297**, 164–172.

Liu, M., Bayjanov, J.R., Renckens, B., et al. (2010) The proteolytic system of lactic acid bacteria revisited: a genomic comparison. *BMC Genomics* **11**, 36.

Liu, M., Prakash, C., Nauta, A., et al. (2012) Computational analysis of cysteine and methionine metabolism and its regulation in dairy starter and related bacteria. *J Bacteriol* **194**, 3522–3533.

Liu, S., Skinner-Nemec, K.A., and Leathers, T.D. (2007) *Lactobacillus buchneri* strain NRRL B-30929 converts a concentrated mixture of xylose and glucose into ethanol and other products. *J Ind Microbiol Biotechnol* **35**, 75–81.

Lucas, P.M., Wolken, W.A.M., Claisse, O., et al. (2005) Histamine-producing pathway encoded on an unstable plasmid in *Lactobacillus hilgardii* 0006. *Appl Environ Microbiol* **71**, 1417–1424.

Makarova, K., Slesarev, A., Wolf, Y., et al. (2006) Comparative genomics of the lactic acid bacteria. *Proc Natl Acad Sci U S A* **103**, 15611–15616.

Marco, M.L., de Vries, M.C., Wels, M., et al. (2010) Convergence in probiotic *Lactobacillus* gut-adaptive responses in humans and mice. *ISME J* **4**, 1481–1484.

Martín, M., Corrales, M.A., Mendoza, D., et al. (1999) Cloning and molecular characterization of the citrate utilization *citMCDEFGRP* cluster of *Leuconostoc paramesenteroides*. *FEMS Microbiol Lett* **174**, 231–238.

Martín, M., Magni, C., López, P., and de Mendoza, D. (2000) Transcriptional control of the citrate-inducible *citMCDEFGRP* operon, encoding genes involved in citrate fermentation in *Leuconostoc paramesenteroides*. *J Bacteriol* **182**, 3904–3912.

Martins dos Santos, V.A., Müller, M., and De Vos, W.M. (2010) Systems biology of the gut: the interplay of food, microbiota and host at the mucosal interface. *Curr Opin Biotechnol* **21**, 539–550.

Martinussen, J., Solem, C., Holm, A.K., and Jensen, P.R. (2013) Engineering strategies aimed at control of acidification rate of lactic acid bacteria. *Curr Opin Biotechnol* **24**, 124–129.

Mayo, B., Aleksandrzak-Piekarczyk, T., Fernández, M., et al. (2010) Updates in the metabolism of lactic acid bacteria. In: *Biotechnology of Lactic Acid Bacteria* ((eds) Mozzi, F., Raya, R.R., Vignolo, G.M.) pp. 3–33. Wiley-Blackwell. Available at http://dx.doi.org/10.1002/9780813820866.ch1 [accessed 25 March 2010].

Mazzeo, M.F., Cacace, G., Peluso, A., et al. (2012) Effect of inactivation of *ccpA* and aerobic growth in *Lactobacillus plantarum*: A proteomic perspective. *J Proteomics* **75**, 4050–4061.

Methé, B.A., Nelson, K.E., Pop, M., et al. (2012) A framework for human microbiome research. *Nature* **486**, 215–221.
Moroni, A.V., Dal Bello, F., and Arendt, E.K. (2009) Sourdough in gluten-free bread-making: An ancient technology to solve a novel issue? *Food Microbiol* **26**, 676–684.
Mortera, P., Pudlik, A., Magni, C., et al. (2013) Ca2+-citrate uptake and metabolism in *Lactobacillus casei* ATCC 334. *Appl Environ Microbiol* **79**, 4603–4612.
Nam, Y.D., Chang, H.W., Kim, K.H., et al. (2009) Metatranscriptome analysis of lactic acid bacteria during kimchi fermentation with genome probing microarrays. *Int J Food Microbiol* **130**, 140–146.
Neves, A., Pool, W., Kok, J., et al. (2005) Overview on sugar metabolism and its control in – The input from *in vivo* NMR. *FEMS Microbiol Rev* **29**, 531–554.
Neves, A.R., Pool, W.A., Solopova, A., et al. (2010) Towards enhanced galactose utilization by *Lactococcus lactis*. *Appl Environ Microbiol* **76**, 7048–7060.
Nezhad, Mh. (2010) Effect of growth at low pH on the cell surface properties of a typical strain of *Lactobacillus casei* group. *Iran J Microbiol* **2**. Available at http://www.ncbi.nlm.nih.gov/pmc/articles/PMC3279780/pdf/IJM-2-147.pdf [accessed July 2010].
Ng, P. and Keich, U. (2008) GIMSAN: a Gibbs motif finder with significance analysis. *Bioinformatics* **24**, 2256–2257.
Notararigo, S., Nácher-Vázquez, M., Ibarburu, I., et al. (2013) Comparative analysis of production and purification of homo- and hetero-polysaccharides produced by lactic acid bacteria. *Carbohydr Polym* **93**, 57–64.
O'Sullivan, O., O'Callaghan, J., Sangrador-Vegas, A., et al. (2009) Comparative genomics of lactic acid bacteria reveals a niche-specific gene set. *BMC Microbiol* **9**, 50.
Oh, E. (2011) Dynamic modeling of lactic acid fermentation metabolism with *Lactococcus lactis*. *J Microbiol Biotechnol* **21**, 162–169.
Okano, K., Tanaka, T., Ogino, C., et al. (2009) Biotechnological production of enantiomeric pure lactic acid from renewable resources: recent achievements, perspectives, and limits. *Appl Microbiol Biotechnol* **85**, 413–423.
Overmars, L., Kerkhoven, R., Siezen, R.J., and Francke, C. (2013) MGcV: the microbial genomic context viewer for comparative genome analysis. *BMC Genomics* **14**, 209.
Palomba, S., Cavella, S., Torrieri, E., et al. (2012) Polyphasic screening, homopolysaccharide composition, and viscoelastic behavior of wheat sourdough from a *Leuconostoc lactis* and *Lactobacillus curvatus* exopolysaccharide-producing starter culture. *Appl Environ Microbiol* **78**, 2737–2747.
Passerini, D., Coddeville, M., Le Bourgeois, P., et al. (2013a) The carbohydrate metabolism signature of *Lactococcus lactis* strain A12 reveals its sourdough ecosystem origin. *Appl Environ Microbiol* **79**, 5844–5852.
Passerini, D., Laroute, V., Coddeville, M., et al. (2013b) New insights into *Lactococcus lactis* diacetyl- and acetoin-producing strains isolated from diverse origins. *Int J Food Microbiol* **160**, 329–336.
Pastink, M.I., Teusink, B., Hols, P., et al. (2009) Genome-scale model of *Streptococcus thermophilus* LMG18311 for metabolic comparison of lactic acid bacteria. *Appl Environ Microbiol* **75**, 3627–3633.
Patel, S., Majumder, A., and Goyal, A. (2011) Potentials of Exopolysaccharides from lactic acid bacteria. *Indian J Microbiol* **52**, 3–12.
Patten, D.A., Leivers, S., Chadha, M.J., et al. (2014) The structure and immunomodulatory activity on intestinal epithelial cells of the EPSs isolated from *Lactobacillus helveticus* sp. *Rosyjski* and *Lactobacillus acidophilus* sp. 5e2. *Carbohydr Res* **384**, 119–127.
Pedersen, M.B., Garrigues, C., Tuphile, K., et al. (2008) Impact of aeration and heme-activated respiration on *Lactococcus lactis* gene expression: identification of a heme-responsive operon. *J Bacteriol* **190**, 4903–4911.
Pedersen, M.B., Gaudu, P., Lechardeur, D., et al. (2012) Aerobic respiration metabolism in lactic acid bacteria and uses in biotechnology. *Annu Rev Food Sci Technol* **3**, 37–358.
Pepe, O., Ventorino, V., Cavella, S., et al. (2013) Prebiotic content of bread prepared with flour from immature wheat grain and selected dextran-producing lactic acid bacteria. *Appl Environ Microbiol* **79**, 3779–3785.
Picard, F., Dressaire, C., Girbal, L., and Cocaign-Bousquet, M. (2009) Examination of post-transcriptional regulations in prokaryotes by integrative biology. *C R Biol* **332**, 958–973.
Pudlik, A.M. and Lolkema, J.S. (2010) Citrate uptake in exchange with intermediates in the citrate metabolic pathway in *Lactococcus lactis* IL1403. *J Bacteriol* **193**, 706–714.
Qin, J., Li, R., Raes, J., Arumugam, M., et al. (2010) A human gut microbial gene catalogue established by metagenomic sequencing. *Nature* **464**, 59–65.
Quang, D. and Xie, X. (2014) EXTREME: An online EM algorithm for motif discovery. *Bioinformatics* btu093.
Quintans, G., Nieves, Blancato, V., Repizo, G., et al. (2008) Citrate metabolism and aroma compound production in lactic acid bacteria. In: *Molecular Aspects of Lactic Acid Bacteria for Traditional and New Applications* ((eds) Mayo, B. López, P. and Pérez-Martínez, G.) pp. 65–88. Kerala: Research Signpost. Available at http://www.cib.csic.es/repositorio_bd/publicacion/1773/urls_documento/chapter_3_publicado.pdf.

Rademaker, J.L.W., Herbet, H., Starrenburg, M.J.C., et al. (2007) Diversity analysis of dairy and nondairy *Lactococcus lactis* isolates, using a novel multilocus sequence analysis scheme and (GTG)5-PCR fingerprinting. *Appl Environ Microbiol* **73**, 7128–7137.

Ramchandran, L. and Shah, N.P. (2009) Effect of exopolysaccharides on the proteolytic and angiotensin-I converting enzyme-inhibitory activities and textural and rheological properties of low-fat yogurt during refrigerated storage. *J Dairy Sci* **92**, 895–906.

Ravcheev, D.A., Best, A.A., Sernova, N.V., et al. (2013) Genomic reconstruction of transcriptional regulatory networks in lactic acid bacteria. *BMC Genomics* **14**, 94.

Raynaud, S., Perrin, R., Cocaign-Bousquet, M., and Loubiere, P. (2005) Metabolic and transcriptomic adaptation of *Lactococcus lactis* subsp. *lactis* Biovar *diacetylactis* in response to autoacidification and temperature downshift in skim milk. *Appl Environ Microbiol* **71**, 8016–8023.

Repizo, G.D., Blancato, V.S., Mortera, P., et al. (2013) Biochemical and genetic characterization of the *Enterococcus faecalis* oxaloacetate decarboxylase complex. *Appl Environ Microbiol* **79**, 2882–2890.

Ritt, J.-F., Remize, F., Grandvalet, C., et al. (2009) Peptidases specific for proline-containing peptides and their unusual peptide-dependent regulation in *Oenococcus oeni*. *J Appl Microbiol* **106**, 801–813.

Russo, P., López, P., Capozzi, V., et al. (2012) Beta-glucans improve growth, viability and colonization of probiotic microorganisms. *Int J Mol Sci* **13**, 6026–6039.

Salazar, N., Prieto, A., Leal, J.A., et al. (2009) Production of exopolysaccharides by *Lactobacillus* and *Bifidobacterium* strains of human origin, and metabolic activity of the producing bacteria in milk. *J Dairy Sci* **92**, 4158–4168.

Santos, F., Teusink, B., Molenaar, D., et al. (2009) Effect of amino acid availability on vitamin B12 production in *Lactobacillus reuteri*. *Appl Environ Microbiol* **75**, 3930–3936.

Santos, F., de Vos, W.M., and Teusink, B. (2013) Towards metagenome-scale models for industrial applications - the case of Lactic Acid Bacteria. *Curr Opin Biotechnol* **24**, 200–206.

Savijoki, K., Ingmer, H., and Varmanen, P. (2006) Proteolytic systems of lactic acid bacteria. *Appl Microbiol Biotechnol* **71**, 394–406.

Sawai, H., Yamanaka, M., Sugimoto, H., et al. (2012) Structural basis for the transcriptional regulation of heme homeostasis in *Lactococcus lactis*. *J Biol Chem* **287**, 30755–30768.

Von Schillde, M.-A., Hörmannsperger, G., Weiher, M., et al. (2012) Lactocepin secreted by *Lactobacillus* exerts anti-inflammatory effects by selectively degrading proinflammatory chemokines. *Cell Host Microbe* **11**, 387–396.

Sender, P.D., Martín, M.G., Peirú, S., and Magni, C. (2004) Characterization of an oxaloacetate decarboxylase that belongs to the malic enzyme family. *FEBS Lett* **570**, 217–222.

Sieuwerts, S. (2009) Analysis of molecular interactions between yoghurt bacteria by an integrated genomics approach. PhD thesis Wageningen University, the Netherlands.

Sieuwerts, S., Molenaar, D., van Hijum, S.A.F.T., et al. (2010) Mixed-culture transcriptome analysis reveals the molecular basis of mixed-culture growth in *Streptococcus thermophilus* and *Lactobacillus bulgaricus*. *Appl Environ Microbiol* **76**, 7775–7784.

Sims, I.M., Frese, S.A., Walter, J., et al. (2011) Structure and functions of exopolysaccharide produced by gut commensal *Lactobacillus reuteri* 100-23. *ISME J* **5**, 1115–1124.

Smit, G., Smit, B., and Engels, W. (2005) Flavour formation by lactic acid bacteria and biochemical flavour profiling of cheese products. *FEMS Microbiol Rev* **29**, 591–610.

Strahinic, I., Lozo, J., Terzic-Vidojevic, A., et al. (2013) Technological and probiotic potential of BGRA43 a natural isolate of *Lactobacillus helveticus*. *Front Microbiol* **4**. Available at http://www.ncbi.nlm.nih.gov/pmc/articles/PMC3552238/pdf/fmicb-04-00002.pdf [accessed 23 January 2013].

Suárez, C.A., Blancato, V.S., Poncet, S., et al. (2011) CcpA represses the expression of the divergent cit operons of *Enterococcus faecalis* through multiple *cre* sites. *BMC Microbiol* **11**, 227.

Sun, H., Yuan, Y., Wu, Y., et al. (2009) Tmod: toolbox of motif discovery. *Bioinformatics* **26**, 405–407.

Teusink, B., Bachmann, H., and Molenaar, D. (2011) Systems biology of lactic acid bacteria: a critical review. *Microb Cell Fact* **10** (suppl 1): S11. Available at http://www.microbialcellfactories.com/content/10/S1/S11 [accessed 30 August 2011].

Teusink, B., Wiersma, A., Molenaar, D., et al. (2006) Analysis of growth of *Lactobacillus plantarum* WCFS1 on a complex medium using a genome-scale metabolic model. *J Biol Chem* **281**, 40041–40048.

The Human Microbiome Jumpstart Reference Strains Consortium, Nelson, K.E., Weinstock, G.M., et al. (2010) A catalog of reference genomes from the human microbiome. *Science* **328**, 994–999.

Trip, H., Mulder, N.L., Rattray, F.P., and Lolkema, J.S. (2011) HdcB, a novel enzyme catalysing maturation of pyruvoyl-dependent histidine decarboxylase: HdcB catalyses maturation of histidine decarboxylase. *Mol Microbiol* **79**, 861–871.

Turpin, W., Humblot, C., and Guyot, J.-P. (2011) Genetic screening of functional properties of lactic acid bacteria in a fermented pearl millet slurry and in the metagenome of fermented starchy foods. *Appl Environ Microbiol* **77**, 8722–8734.

Voit, E.O., Neves, A.R., and Santos, H. (2006) The intricate side of systems biology. *Proceedings of the National Academy of Sciences* **103**, 9452–9457.

De Vos, W.M. (2011) Systems solutions by lactic acid bacteria: from paradigms to practice. *Microb Cell Fact* **10** (Suppl 1), S2. Available at http://www.microbialcellfactories.com/content/10/S1/S2 [accessed 30 August 2011].

Wakai, T. and Yamamoto, N. (2013) A novel branched chain amino acids responsive transcriptional regulator, BCARR, negatively acts on the proteolytic system in *Lactobacillus helveticus*. *PLoS One* **8**, e75976.

Wang, K., Li, W., Rui, X., et al. (2014) Characterization of a novel exopolysaccharide with antitumor activity from *Lactobacillus plantarum* 70810. *Int J Biol Macromol* **63**, 133–139.

Wang, Y., Ganzle, M.G., and Schwab, C. (2010) Exopolysaccharide synthesized by *Lactobacillus reuteri* decreases the ability of enterotoxigenic *Escherichia coli* to bind to porcine erythrocytes. *Appl Environ Microbiol* **76**, 4863–4866.

Weckx, S., Allemeersch, J., Van der Meulen, R., et al. (2011) Metatranscriptome analysis for insight into whole-ecosystem gene expression during spontaneous wheat and spelt sourdough fermentations. *Appl Environ Microbiol* **77**, 618–626.

Wegkamp, A., Mars, A.E., Faijes, M., et al. (2010) Physiological responses to folate overproduction in *Lactobacillus plantarum* WCFS1. *Microb Cell Fact* **9**, 100–106.

Wels, M., Overmars, L., Francke, C., et al. (2011) Reconstruction of the regulatory network of *Lactobacillus plantarum* WCFS1 on basis of correlated gene expression and conserved regulatory motifs: Reconstruction of the *L. plantarum* regulatory network. *Microb Biotechnol* **4**, 333–344.

Zannini, E., Mauch, A., Galle, S., et al. (2013) Barley malt wort fermentation by exopolysaccharide-forming *Weissella cibaria* MG1 for the production of a novel beverage. *J Appl Microbiol* **115**, 1379–1387.

Zhang, J. and Biswas, I. (2009) A phenotypic microarray analysis of a *Streptococcus mutans liaS* mutant. *Microbiology* **155**, 61–68.

Zomer, A.L., Buist, G., Larsen, R., et al. (2006) Time-resolved determination of the CcpA regulon of *Lactococcus lactis* subsp. *cremoris* MG1363. *J Bacteriol* **189**, 1366–1381.

Chapter 2

Systematics of Lactic Acid Bacteria: Current Status

Giovanna E. Felis, Elisa Salvetti, and Sandra Torriani*

Department of Biotechnology, University of Verona, Italy

2.1. Families and Genera of Lactic Acid Bacteria

It is widely recognized that the ability to produce lactic acid from carbohydrates is a characteristic shared among a set of Gram (+) bacteria and has no phylogenetic meaning. On the other side, phylogenetic trees constitute the backbone of modern bacterial systematics so, taxonomically speaking, lactic acid-forming bacteria are distributed in different lineages: family Bacillaceae (genus *Bacillus*, Zhou et al. 2013), family Sporolactobacillaceae (genus *Sporolactobacillus*, Fritze and Claus 1995), family Bifidobacteriaceae (genera *Aeriscardovia*, *Alloscardovia*, *Bifidobacterium*, *Gardnerella*, *Metascardovia*, *Parascardovia*, and *Scardovia*, Mattarelli et al. 2014) and order Lactobacillales, which are traditionally considered the genuine lactic acid bacteria (LAB) (Schleifer and Ludwig 1995).

This phylogenetic group is composed of about 500 validly described species, belonging to six Families of low-GC content Gram-staining (+) bacteria, in particular:

1. Aerococcaceae, with genera *Abiotrophia*, *Aerococcus*, *Dolosicoccus*, *Eremococcus*, *Facklamia*, *Globicatella*, and *Ignavigranum*.
2. Carnobacteriaceae, with genera *Alkalibacterium*, *Allofustis*, *Alloiococcus*, *Atopobacter*, *Atopococcus*, *Atopostipes*, *Carnobacterium*, *Desemzia*, *Dolosigranulum*, *Granulicatella*, *Isobaculum*, *Lacticigenium*, *Marinilactibacillus*, *Pisciglobus*, and *Trichococcus*.
3. Enterococcaceae, with genera *Bavariicoccus*, *Catellicoccus*, *Enterococcus*, *Melissococcus*, *Pilibacter*, *Tetragenococcus*, and *Vagococcus*.
4. Lactobacillaceae, with genera *Lactobacillus* and *Pediococcus*.
5. Leuconostocaceae, with genera *Leuconostoc*, *Fructobacillus*, *Oenococcus*, and *Weissella*.
6. Streptococcaceae, with genera *Lactococcus*, *Lactovum*, and *Streptococcus*.

Interestingly, genera *Agitococcus* and *Sharpea* are included in families Carnobacteriaceae and Lactobacillaceae, respectively, in the List of Prokaryotic Names with Standing in Nomenclature (LPNS, www.bacterio.net) but not according to Mattarelli et al. (2014). Moreover, over the last few years, the continuous description of novel species has deeply modified the phylogenetic structure of the order, and in particular, families Carnobacteriaceae and Enterococcaceae seem to be paraphyletic with respect to other families of the order (Figure 2.1).

* Corresponding author: Department of Biotechnology, University of Verona, Strada Le Grazie 15, 37134 Verona, Italy. sandra.torriani@univr.it

Biotechnology of Lactic Acid Bacteria: Novel Applications, Second Edition. Edited by Fernanda Mozzi, Raúl R. Raya, and Graciela M. Vignolo.

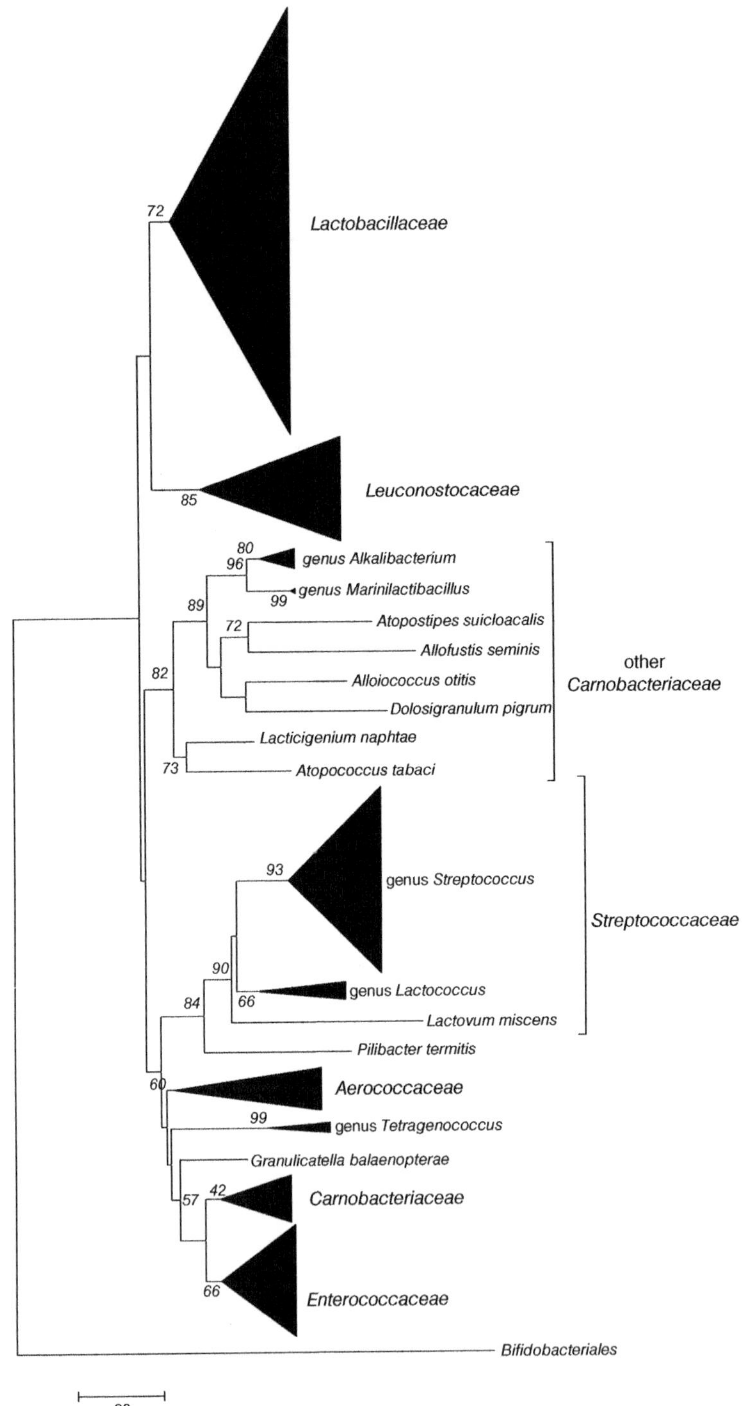

Figure 2.1. Phylogenetic tree depicting the relationship between families of order Lactobacillales based on 16S rRNA gene sequence. The tree was calculated using Neighbor-Joining and the number of differences as a method. Bootstrap values (1000 replicates) are reported in percentages at nodes. The scale bar represents the number of substitutions per site.

As general characteristics, Aerococcaceae include generally catalase-negative, non-motile, facultatively anaerobic cocci associated with human and animal infections. Phylogenetically, two paraphyletic groups are recognized (Mattarelli et al. 2014).

Family Carnobacteriaceae include usually catalase-negative cocci or rods, generally facultatively anaerobic or microaerophilic, with variable motility, and peculiar chemotaxonomic characteristics (Ludwig et al. 2009). Also, interestingly, these bacteria can perform respiration under aerobic conditions with involvement of cytochromes b and d (Mattarelli et al. 2014).

Family Enterococcaceae comprise catalase-negative cocci, non-spore formers, facultatively anaerobic, anaerobic or microaerophilic, of major concern due to possible role in spreading of antibiotic resistance (Hollenbeck and Rice 2012).

Lactobacillaceae is the most numerous and heterogeneous family, as about 180 species are recognized (www.bacterio.net/lactobacillus), although novel species are described on an almost monthly basis. These are non-spore-forming rods or coccobacilli (able to divide in tetrads in the genus *Pediococcus*), which can be motile, generally oxygen-tolerant, often acidophilic, and obligately saccharoclastic, with different end-products of fermentation (see below) but also with the ability of respiration, at least in some species (Brooijmans et al. 2009). All other characteristics usually considered in taxonomic studies, such as temperature and pH ranges for growth, isomers of lactic acid produced, type of peptidoglycan in the cell wall, are very variable (Salvetti et al. 2012).

Leuconostocaceae are non-endospore-forming and facultatively anaerobic rods or cocci, usually characterized by the production of equimolar amounts of lactic acid, ethanol, and carbon dioxide from glucose (obligately heterofermentative metabolism, see below). Exceptions are members of the genus *Fructobacillus*, which produce acetic acid instead of ethanol (Endo and Okada 2008). Also, some species of the genus *Leuconostoc* possess heme-dependent respiratory capability. Genus *Oenococcus*, in particular species *O. oeni*, is considered important thanks to the ability to convert L-malic acid to L-lactic acid in wine (Henick-Kling 1993).

Streptoccocaceae consist of facultatively anaerobic, non-sporing, homofermentative spherical or ovoid cells, typically arranged in chains or pairs. The majority of species belonging to genus *Streptococcus* are parasitic in man and other animals and the association between some streptococci and a variety of human and animal diseases is established (Hardie and Whiley 1995). On the other hand, members of genus *Lactococcus* are widely used in the dairy industry due to their ability to produce L(+)-lactic acid from lactose in spontaneously fermented raw milk at 20–30 °C (Teuber 1995).

From an applied viewpoint, families Lactobacillaceae and Streptococcaceae represent the two most important lineages since they encompass the highest number of commercialized LAB (lactobacilli, *Streptococcus thermophilus* and *Lactococcus lactis*).

2.2. A Focus on the Family Lactobacillaceae

Lactobacillaceae includes the highest number of GRAS species (Generally Recognized As Safe) and many strains are among the most important bacteria in food microbiology and human nutrition due to their contribution to fermented food production or their use as probiotics. The massive description of novel species over the past 20 years has led to a progressive revision of the genus with the recognition of a growing number of variable phylogenetic groups (Dellaglio and Felis 2005; Felis and Dellaglio 2007). After the last taxonomic update (Salvetti et al. 2012), a number of novel species has been described, namely *Lact. backii* (Tohno et al. 2013a), *Lact. curieae* (Lei et al. 2013), *Lact. faecis* (Endo et al. 2013), *Lact. ginsenosidimutans* (Jung et al. 2013), *Lact. heilongjiangensis* (Gu et al. 2013a), *Lact. iwatensis* (Tohno et al. 2013a), *Lact. mudanjangensis* (Gu et al. 2013b), *Lact. nenjiangensis* (Gu et al. 2013b), *Lact. okkaidonensis* (Tohno et al. 2013b), *Lact. oryzae* (Tohno et al. 2013c),

Lact. silagei (Tohno et al. 2013d), *Lact. songhuajiangensis* (Gu et al. 2013b), *Lact. yonginensis* (Yi et al. 2013), *Lact. apis* (Killer et al. 2014a), and *Lact. rodentium* (Killer et al. 2014b). These species were ascribed to different phylogenetic groups and further confound the already complex phylogenetic structure of the family. Nevertheless, some phylogenetic groups are recognizable and are foreseen to become, in the future, novel genera when agreement on criteria for their delineation is reached in the scientific community. Among the different groups, the *Lact. delbrueckii*-group, enriched with *Lact. rodentium* and *Lact. apis* with respect to Salvetti et al. (2012), would keep the genus name in case of reclassification.

At every taxonomic level, that is order Lactobacillales, family Lactobacillaceae, genus *Lactobacillus*, and *Lact. delbrueckii*-group (*Lactobacillus sensu stricto*), only a little correlation exists between the traditional classification based on metabolic properties and phylogenetic relatedness (Claesson et al. 2008).

Knowledge of LAB metabolism mostly derives from genus *Lactobacillus*, and it is usually defined as follows: species performing homolactic fermentation (namely, homofermentative metabolism) produce more than 85% lactic acid from hexoses, using the glycolytic pathway (EMPP). Heterofermenters can utilize the phosphogluconate pathway (PP pathway) to degrade not only hexoses but also pentoses yielding, besides lactic acid, other side-products, such as ethanol, acetic acid, and carbon dioxide (Hammes and Vogel 1995). Moreover, heterofermentative metabolism can be either facultative or obligate: the former determines degradation of hexoses and pentoses by the EMP and PP pathways, respectively, while in the latter, pentoses and hexoses are degraded by the phosphogluconate pathway (the first phase of the PP pathway) with the concurrent production of CO_2 (Hammes and Vogel 1995). It must be emphasized that this information derives from studies on a very limited number of species with respect to the actual diversity and, as already mentioned, metabolic groupings do not correspond to phylogenetic clusters (Vandamme et al. 1996).

Recently, the availability of genome sequences allowed analysis of the distribution and organization of 42 genes belonging to the glycolysis (EMPP) and pentose phosphate (PP) pathways (Salvetti et al. 2013) in the order Lactobacillales. The analysis revealed the occurrence of lineage-specific trends of gene loss/gain within the two metabolic pathways examined. In the genus *Lactobacillus*, the *Lact. delbrueckii*-group was the LAB cluster lacking the highest number of central EMPP genes. Also, all 11 strains analyzed lacked operon coding for the pyruvate dehydrogenase complex (PDHc), constituted by four genes (*pdhA*, *pdhB*, *pdhC*, and *pdhD*), which could be the key that connects the genotype to their phenotype: they are homofermentative (Hammes and Vogel 1995) and, as such, glucose is converted exclusively into lactic acid via lactate dehydrogenase (Ldh). On the other side, Leuconostocaceae and the obligately heterofermentative *Lactobacillus* strains (members of the *Lact. reuteri*-group, *Lact. brevis*, and *Lact. buchneri*) are missing phosphofructokinase (*pfkA*), a critical priming ATP-dependent reaction at the beginning of the EMPP (Ronimus and Morgan 2001). Again, the absence of this gene can be linked strictly to their metabolism, as glucose fermentation is driven through the PP pathway, leading to the observed production of CO_2 in obligate heterofermenters (Salvetti et al. 2013).

In the *Lact. delbrueckii*-group, the absence of a complete PDHc operon and presence of *pfkA* gene was observed, thus indicating a coherent genetic makeup despite the phenotypic diversity (Salvetti et al. 2013). In fact, facultatively and obligately heterofermentative species such as *Lact. acetotolerans*, *Lact. fornicalis*, *Lact. hamsteri*, *Lact. intestinalis*, *Lact. jensenii*, and *Lact. psittaci* revealed a genotypic background comparable with homofermentative species, therefore their metabolic phenotype is probably due to alternative pathways (unpublished data). Interestingly, in homofermentative species *Lact. iners* LMG18914^T and *Lact. amylotrophicus* LMG11400^T, PCR products were obtained for *pdhA*- and *pdhB*-targeting primers. Remarkably, the sequences obtained revealed a high similarity with *Staphylococcus warneri* L37603 (98%) and *Streptococcus haemolyticus* JCSC1435 (91%), suggesting that horizontal gene transfer could have occurred as they share the same ecological niches as human stomach, vagina, and clinical samples (Falsen et al. 1999; Coton et al. 2009; Pereira et al. 2010). This hypothesis is reinforced by previous observations on a more limited number of genomes (Makarova and Koonin 2007), which revealed that the enolase gene (*eno*) was present in Lactobacillales in more than one copy, one the ancestral version of that in Gram-(+) bacteria, and the other probably acquired from a different bacterial lineage, most likely Actinobacteria (Makarova and Koonin 2007).

Those observations suggest that genomic data are of utmost importance to clarify the correct relationships between genotype and phenotype, and could make it possible to tri-dimensionally describe taxa, harmonizing genotypic, phylogenetic, and phenotypic data. Moreover, these predictions on the evolution of LAB might have important consequences also in the development of a "natural" (i.e., reflecting evolutionary relationships) taxonomic scheme of genera and families (Makarova et al. 2006; Salvetti et al. 2013).

Genome sequence-based approaches can be a buffer against the distorting evolutionary events at a single locus; the investigation of several sequences can lead to a further comprehension of the events, which occurred during the evolution of an organism (Konstantinidis and Tiedje 2005).

2.3. Taxonomic Tools in the Genomic Era

Taxonomic procedures have always evolved dependently of technology advances, as the introduction of innovative methods has always provided new opportunities for prokaryotic systematics (Stackebrandt et al. 2002). In this context, an active debate is ongoing in the scientific community of taxonomists in order to find the best way to integrate genomic data in taxonomic procedures (Konstantinidis and Tiedje 2005; Rossello-Mora 2012; Ramasamy et al. 2014). Thus, the second meeting of Bergey's International Society for Microbial Systematics (BISMiS), held in Edinburgh, Scotland (7–10 April, 2014) was entitled "Defining Microbial Diversity in the Genomic Era" (bismis.org).

Genome sequence data can be retrieved from a number of databases, among which the most important is probably GOLD (Genomes OnLine Database www.genomesonline.org), which allows comprehensive access to information regarding genome and metagenome sequencing projects and their associated data around the world.

Some other online databases and tools, which can be employed to gain taxonomic data either from entire genomes or single sequences, are CMR (Comprehensive Microbial Resource, http://cmr.jcvi.org/tigr-scripts/CMR/CmrHomePage.cgi), DNA Bank Network (http://www.dnabank-network.org), Global Genome Biodiversity Network (http://www.ggbn.org/) SILVA (Comprehensive Ribosomal RNA Databases, http://www.arb-silva.de/), RDP (Ribosomal Database Project, http://rdp.cme.msu.edu/), and RibAlign (http://www.megx.net/ribalign/index.html) and software available at the Nucleic Acid Research website (http://www.oxfordjournals.org/nar/database/subcat/5/19).

An ongoing huge sequencing effort, named GEBA (Genomic Encyclopedia of Bacteria and Archaea, http://genome.jgi.doe.gov/programs/bacteria-archaea/GEBA.jsf), aspires to sequence thousands of bacterial and archaeal genomes from diverse branches of the Tree of Life and it represents the first systematic attempt to use the Tree of Life itself as a guide to sequencing target selection (Klenk and Göker 2010). Extraction of reliable information requires a standardized description of genomes and analytical procedures to exchange and integrate genomic data, which is the goal of the Genomic Standards Consortium (GSC, http://gensc.org/) and its related open access journal Standards in Genomic Sciences (http://www.standardsingenomics.org/index.php/sigen).

It is expected that in the near future species' description will include genome sequencing of the type strains, thanks also to the decrease in sequencing costs, and that digital DNA-DNA Hybridization (DDH) will replace wet lab DDH (http://ggdc.dsmz.de/) as well as GC determination (Meier-Kolthoff et al. 2013). At that point, it is expected that the subcommittee on the taxonomy of *Bifidobacterium*, *Lactobacillus*, and related organisms, which very recently published the "Recommended minimal standards for description of new taxa of the genera *Bifidobacterium*, *Lactobacillus* and related genera" (Mattarelli et al. 2014), will give indications for the integration of genome data in LAB taxonomy too.

References

Brooijmans, R.J.W., de Vos, W.H., and Hugenholtz, J. (2009) The electron transport chains of *Lactobacillus plantarum* WCFS1. *Appl Environ Microbiol* **75**, 3580–3585.

Claesson, M.J., van Sinderen, D., and O'Toole, P.W. (2008) *Lactobacillus* phylogenomics – towards a reclassification of the genus. *Int J Syst Evol Microbiol* **58**, 2945–2954.

Coton, E., Desmonts, M.H., Leroy, S., et al. (2009) Biodiversity of coagulase-negative staphylococci in French cheeses, dry fermented sausages, processing environments and clinical samples. *Int J Food Microbiol* **28**, 221–229.

Dellaglio, F. and Felis G.E. (2005) Taxonomy of lactobacilli and bifidobacteria. In: *Probiotics and Prebiotics: Scientific Aspects*. ((ed.) G.W. Tannock), 1st edn, pp. 25–50. Caister Academic Press, Norfolk.

Endo, A. and Okada, S. (2008) Reclassification of the genus *Leuconostoc* and proposals of *Fructobacillus fructosus* gen. nov., comb. nov., *Fructobacillus durionis* comb. nov., *Fructobacillus ficulneus* comb. nov. and *Fructobacillus pseudoficulneus* comb. nov. *Int J Syst Evol Microbiol* **58**, 2195–2205.

Endo, A., Irisawa, T., Futagawa-Endo, Y., et al. (2013) *Lactobacillus faecis* sp. nov., isolated from animal faeces. *Int J Syst Evol Microbiol* **63**, 4502–4507.

Falsen, E., Pascual, C., Sjodén, B., et al. (1999) Phenotypic and phylogenetic characterization of a novel *Lactobacillus* species from human sources: description of *Lactobacillus iners* sp. nov. *Int J Syst Evol Microbiol* **49**, 217–221.

Felis, G.E. and Dellaglio F. (2007) Taxonomy of lactobacilli and bifidobacteria. *Curr Issues Intest Microbiol* **8**, 44–61.

Fritze, D. and Claus, D. (1995) Spore-forming, lactic acid producing bacteria of the genera *Bacillus* and *Sporolactobacillus*. In: *The Genera of Lactic Acid Bacteria* ((eds) B.J.B. Wood and W.H. Holzapfel), 1st edn, pp. 368–387. Blackie Academic & Professional, Glasgow, UK.

Gu, C.T., Li, C.Y., Yang, L.J., and Huo G.C. (2013a) *Lactobacillus heilongjiangensis* sp. nov., isolated from Chinese pickle. *Int J Syst Evol Microbiol* **63**, 4094–4099.

Gu, C.T., Li, C.Y., Yang, L.J., and Huo, G.C. (2013b) *Lactobacillus mudanjiangensis* sp. nov., *Lactobacillus songhuajiangensis* sp. nov. and *Lactobacillus nenjiangensis* sp. nov., isolated from Chinese traditional pickle and sourdough. *Int J Syst Evol Microbiol* **63**, 4698–4706.

Hammes, W.P. and Vogel R.F. (1995) The genus *Lactobacillus*. In: *The Genera of Lactic Acid Bacteria* ((eds) B.J.B. Wood and W.H. Holzapfel), 1st edn, pp. 19–55. Blackie Academic & Professional, Glasgow, UK.

Hardie, J.M and Whiley, R.A. (1995) The genus *Streptococcus*. In: *The Genera of Lactic Acid Bacteria* ((eds) B.J.B. Wood and W.H. Holzapfel), 1st edn, pp. 55–115. Blackie Academic & Professional, Glasgow, UK.

Henick-Kling, T. (1993) Malolactic fermentation. In: *Wine Microbiology and Biotechnology* ((ed.) G.H. Fleet), 1st edn, pp. 289–326. Harwood Academic Publisher, Amsterdam.

Hollenbeck B.L. and Rice L.B. (2012) Intrinsic and acquired resistance mechanisms in *Enterococcus*. *Virulence* **3**, 421–433.

Jung, H.M., Liu, Q.M., Kim, J.K., et al. (2013) *Lactobacillus ginsenosidimutans* sp. nov., isolated from kimchi with the ability to transform ginsenosides. *Antonie van Leeuwenhoek* **103**, 867–876.

Killer, J., Dubná, S., Sedláček, I., and Švec P. (2014a) *Lactobacillus apis* sp. nov., from the stomach of honeybees (*Apis mellifera*), having an in vitro inhibitory effect on the causative agents of American and European foulbrood. *Int J Syst Evol Microbiol* **64**, 152–157.

Killer, J., Havlik, J., Vlková, E., et al. (2014b) *Lactobacillus rodentium* sp. nov., from the digestive tract of wild rodents. *Int J Syst Evol Microbiol* **64**, 1526–1533. Klenk, H.P. and Göker, M. (2010) En route to a genome-based classification of Archaea and Bacteria? *Syst Appl Microbiol* **33**, 175–182.

Konstantinidis, K.T. and Tiedje, J.M. (2005) Towards a genome-based taxonomy for prokaryotes. *J Bacteriol* **187**, 6258–6264.

Lei, X., Sun, G., Xie, J., and Wei D. (2013) *Lactobacillus curieae* sp. nov., isolated from stinky tofu brine. *Int J Syst Evol Microbiol* **63**, 2501–2505.

Ludwig, W., Schleifer, K.H., and Whitman, W.B. (2009) Revised road map to the phylum Firmicutes. In: *Bergey's Manual of Systematic Bacteriology* ((eds) P. De Vos, G.M. Garrity, D. Jones, N.R. Krieg, W. Ludwig, F.A. Rainey, K.-H. Schleifer & W.B. Whitman) 2nd edn, vol **3**, pp. 1–13. Springer, New York.

Makarova, K., Slesarev, A., Wolf, Y., et al. (2006) Comparative genomics of the lactic acid bacteria. *Proc Natl Acad Sci USA* **103**, 15611–15616.

Makarova, K.S. and Koonin, E.V. (2007) Evolutionary genomics of lactic acid bacteria. *J Bacteriol* **189**, 1199–1208.

Mattarelli, P., Holzapfel, W., Franz, M.A.P.C., et al. (2014) Recommended minimal standards for description of new taxa of the genera *Bifidobacterium*, *Lactobacillus* and related genera. *Int J Syst Evol Microbiol* **64**, 1434–1451.

Meier-Kolthoff J.P., Göker, M., Spröer, C., and Klenk H.P. (2013) When should a DDH experiment be mandatory in microbial taxonomy? *Arch Microbiol* **195**, 413–418.

Pereira, E.M., Shuenck, R.P., Malvar, K.L., et al. (2010) *Staphylococcus aureus*, *Staphylococcus epidermidis* and *Staphylococcus haemolyticus*: methicillin-resistant isolates are detected directly in blood cultures by multiplex PCR. *Microbiol Res* **165**, 243–249.

Ramasamy, D., Mishra, A.K., Lagier, J.C., et al. (2014) A polyphasic strategy incorporating genomic data for the taxonomic description of novel bacterial species. *Int J Syst Evol Microbiol* **64**, 384–391.

Ronimus, R.S. and Morgan, H.W. (2001) The biochemical properties and phylogenies of phosphofructokinases from extremophiles. *Extremophiles* **5**, 357–373.

Rosselló-Móra, R. (2012) Towards a taxonomy of Bacteria and Archaea based on interactive and cumulative data repositories. *Environ Microbiol* **14**, 318–334.

Salvetti, E., Torriani, S., and Felis, G.E. (2012) The genus *Lactobacillus*: a taxonomic update. *Probiotics Antimicr Prot* **4**, 217–226.

Salvetti, S., Fondi, M., Fani, R., et al. (2013) Evolution of lactic acid bacteria in the order Lactobacillales as depicted by analysis of glycolysis and pentose phosphate pathways. *Syst Appl Microbiol* **36**, 291–305.

Schleifer, K.H. and Ludwig, W. (1995) Phylogenetic relationships of lactic acid bacteria. In: *The Genera of Lactic Acid Bacteria* ((eds) B.J.B. Wood and W.H. Holzapfel), 1st edn, pp. 7–17. Blackie Academic & Professional, Glasgow, UK.

Stackebrandt, E., Frederiksen, W., Garrity, G.M., et al. (2002) Report of the *ad hoc* committee for the re-evaluation of the species definition in bacteriology. *Int J Syst Evol Microbiol* **52**, 1043–1047.

Teuber M. (1995) The genus *Lactococcus*. In: *The Genera of Lactic Acid Bacteria* ((eds) B.J.B. Wood and W.H. Holzapfel), 1st edn, pp. 173–175. Blackie Academic & Professional, Glasgow, UK.

Tohno, M., Kitahara, M., Irisawa, T., et al. (2013a) Description of *Lactobacillus iwatensis* sp. nov., isolated from orchardgrass (*Dactylis glomerata* L.) silage, and *Lactobacillus backii* sp. nov. *Int J Syst Evol Microbiol* **63**, 3854–3860.

Tohno, M., Kitahara, M., Uegaki, R., et al. (2013b) *Lactobacillus hokkaidonensis* sp. nov., isolated from subarctic timothy grass (*Phleum pratense* L.) silage. *Int J Syst Evol Microbiol* **63**, 2526–2631.

Tohno, M., Kitahara, M., Irisawa, T., et al. (2013c) *Lactobacillus oryzae* sp. nov., isolated from fermented rice grain (*Oryza sativa* L. subsp. *japonica*). *Int J Syst Evol Microbiol* **63**, 2957–2962.

Tohno, M., Kitahara, M., Irisawa, T., et al. (2013d) *Lactobacillus silagei* sp. nov., isolated from orchardgrass silage. *Int J Syst Evol Microbiol* **63**, 4613–4618.

Vandamme, P., Pot, B., Gillis, M., et al. (1996) Polyphasic taxonomy, a consensus approach to bacterial systematics. *Microbiol Rev* **60**, 407–438.

Yi, E.J., Yang, J.E., Lee, J.M., et al. (2013) *Lactobacillus yonginensis* sp. nov., a lactic acid bacterium with ginsenoside converting activity isolated from Kimchi. *Int J Syst Evol Microbiol* **63**, 3854–3860.

Zhou, X., Ye, L., and Wu, J.C. (2013) Efficient production of L-lactic acid by newly isolated thermophilic *Bacillus coagulans* WCP10-4 with high glucose tolerance. *Appl Microbiol Biotechnol* **97**, 4309–4314.

Chapter 3

Genomic Evolution of Lactic Acid Bacteria: From Single Gene Function to the Pan-genome

Grace L. Douglas[1], M. Andrea Azcarate-Peril[2], and Todd R. Klaenhammer[3,*]

[1]*Human Health & Performance Directorate, NASA Johnson Space Center, Houston, USA*
[2]*Department of Cell Biology and Physiology, and Microbiome Core Facility, University of North Carolina at Chapel Hill, Chapel Hill, USA*
[3]*Department of Food, Bioprocessing, and Nutrition Sciences, and Southeast Dairy Foods Research Center, North Carolina State University, Raleigh, USA*

3.1. The Genomics Revolution

Twenty years have passed from the completion of the first bacterial sequenced genome, *Haemophilus influenzae* Rd (Fleischmann et al. 1995), and 14 since the public release of the first lactic acid bacteria (LAB) genome, sequenced in 2001. In fact, by 2001 four genomes belonging to LAB were released, three of them pathogens (*Streptoccocus pyogenes* M1 GAS (Ferretti et al. 2001), *Streptococcus pneumoniae* TIGR4 (Tettelin et al. 2001), and *S. pneumoniae* R6 (Hoskins et al. 2001)). The fourth one was the food-grade microbe, *Lactococcus lactis* subsp. *lactis* IL1403 (Bolotin et al. 2001), a strain extensively used in laboratory research.

The genera currently included in the LAB group are: *Lactobacillus*, *Weissella*, *Carnobacterium*, *Streptococcus*, *Enterococcus*, *Lactococcus*, *Vagococcus*, *Leuconostoc*, *Oenococcus*, *Pediococcus*, and *Tetragenococcus* (Holzapfel et al. 2001). Over the past several decades, a series of genetic techniques have been developed and employed to compare and characterize the properties of LAB. In recent years, rapid, low-cost sequencing methods have led to an explosion in genomic sequencing of LAB. At the time of this analysis (May 2015) there were 3202 (complete and partial) publically available at the National Center for Biotechnology Information (http://www.ncbi.nlm.nih.gov/; Table 3.1), compared to only 128 just seven years ago. The availability of increasing numbers of sequences has enabled genomic comparisons leading to improved identification of evolutionary trends, interactions, and molecular mechanisms, and most recently to the foundational development and analyses of LAB pan-genomes. A pan-genome contains the entire set of genes within a defined group, including core genes shared between all members of the group, variable genes shared between at least two but not all members in a group, and unique genes encoded by only one member of a group (Lukjancenko et al. 2012). Comparative analyses on a pan-genome level are beginning to reveal the genetic variation of genera, species, and strains to provide insight into adaptation

* Corresponding author: Department of Food, Bioprocessing, and Nutrition Sciences, and Southeast Dairy Foods Research Center, North Carolina State University, Raleigh, NC, USA. trk@unity.ncsu.edu

Biotechnology of Lactic Acid Bacteria: Novel Applications, Second Edition. Edited by Fernanda Mozzi, Raúl R. Raya, and Graciela M. Vignolo.

Table 3.1. Lactic acid bacteria genomes, number of complete or partially complete sequences per genus. Summarized from http://www.ncbi.nlm.nih.gov/genome/browse/.

Genus	Complete Sequences	Partial Sequences
Lactobacillus	70	332
Weissella	4	11
Carnobacterium	3	16
Streptococcus	156	1737
Enterococcus	15	691
Lactococcus	17	41
Vagococcus	0	1
Leuconostoc	9	22
Oenococcus	2	59
Pediococcus	3	9
Tetragenococcus	1	3

(Frese et al. 2011; Kant et al. 2011; Lukjancenko et al. 2012; Douillard et al. 2013b). This chapter will focus on the genomics of the generally recognized as safe (GRAS), non-pathogenic LAB. Furthermore, although traditionally included in the LAB group, the genus *Bifidobacterium* will not be included in this analyses because it is only weakly phylogenetically related to the group and uses different metabolic pathways for carbohydrate utilization (Felis and Dellaglio 2007). Further development of the LAB pan-genome and subsequent comparative and functional genomics studies are key to characterizing advantageous genetic capabilities and interactions, which are now enabling improvements in application-directed strain selection (van Hijum et al. 2013).

3.2. Genomic Adaptations of LAB to the Environment

Traditionally, LAB have been associated with dairy products (Cogan et al. 2007), plants (Di Cagno et al. 2008), meats (Hugas et al. 2003; Rantsiou and Cocolin 2006; Tanasupawat et al. 2008), alcoholic beverages (Mills et al. 2005; Moreno-Arribas and Polo 2005), and living organisms, including humans (Reuter 2001; Azcarate-Peril et al. 2008; Busconi et al. 2008; Laukova et al. 2008; Nazef et al. 2008; Walter, 2008). Subsequent sections will review the genomic characteristics that permit LAB strains to adapt to diverse environments.

3.2.1. LAB Evolution in the Dairy Environment

One of the most prominent examples of the adaptation of a microorganism to its environment is illustrated by the genome evolution of some LAB in response to dairy. Dairy LAB have a high number of pseudogenes in metabolic and transport pathways that are not essential in a nutritious environment that provides lactose as the main carbohydrate source, casein as the main source of amino acids, and most vitamins and minerals (Douglas and Klaenhammer 2010). As an example, the most important characteristics of the genome sequence of *Lactobacillus helveticus* DPC 4571 (Callanan et al. 2008), a Swiss cheese isolate commonly used as starter and adjunct culture in cheese manufacture, are: a predicted dependency on external supplies of amino acids and cofactors similar to that described for closely related gastrointestinal (GI) tract isolates, *Lact. acidophilus* (Altermann et al. 2005) and *Lact. johnsonii* (Pridmore et al. 2004), a high peptidolytic activity, and the ability to lyse rapidly in the cheese matrix, features that play a critical part in cheese ripening. Also, an unusually high number of insertion sequences (IS) suggest that horizontal gene transfer played a very important role in the origin of LAB that have specialized for growth in milk (Bolotin et al. 2004; Makarova

Table 3.2. Components of the *Lactobacillus helveticus* CNRZ32 proteolytic enzyme system: a look before and after genome sequence determination (Cogan et al., 2007). Reproduced with permission.

Genes identified and characterized before sequencing project	Genes identified upon completion of the draft genome sequence
Proteinases	
*prt*H	*prt*H2 plus 9 additional proteases
Endopeptidases	
*pep*E, *pep*O, *pep*O2 endopeptidases	*pep*E2, *pep*F, *pep*O3, plus 2 glycoproteins
Aminopeptidases	
*pep*C, *pep*N, *pep*X	*pep*C2 plus 7 additional aminopeptidases
Di-Tripeptidases	
*pep*D, *pep*I, *pep*Q, *pep*R	*pep*D2, *pep*D3, *pep*D4, *pep*Q2, *pep*T1, and *pep*T2
Other	
	Oligo- and di-tripeptide transport systems: *opp*A, *opp*A2, *opp*B-D, *opp*F, and *dtp*A, *dtp*A2, and *dtp*T
	Multiple amino acid transporters

et al. 2006). The sequence of another strain of *Lact. helveticus* (CNRZ32) characterized more essential components of the proteolytic enzyme system that confirmed and expanded the knowledge of this system (Table 3.2) (Cogan et al. 2007). Transcriptome profiling tools helped define the relationship between LAB and the milk environment. Microarray technology confirmed the overexpression of several members of the proteolytic system in cultures of *Lact. helveticus* CNRZ32 growing in milk versus a complex medium, including previously characterized genes (*pep*E, *pep*N, *pep*R, *pep*O2, *pep*O, *pep*X) and genes identified by comparative genomics (*prt*H2 and the *opp* operon) (Smeianov et al. 2007). In *Lact. acidophilus* NCFM, transcriptome analysis during growth in milk identified expression of similar members of the proteolytic system as well as a two-component regulatory system involved in the regulation of oligopeptide transport systems in this probiotic bacteria (Azcarate-Peril et al. 2005, 2009).

Comparative genomic analysis has begun to identify a variety of traits encoded within dairy fermentation microbes and elucidate the impact of fermentation and metabolism properties on product quality. Comparative analysis of 27 *Streptococcus thermophilus* strains distinguished 15 different *eps* gene clusters and 67 glycosyltransferases, resulting in the production of unique repeating unit heteropolysaccharides (De Vuyst et al. 2011). Characterization of these clusters under different fermentation conditions is likely to facilitate strain screening and selection of desirable textural properties in *Strep. thermophilus* strains used in fermented yogurt products (De Vuyst et al. 2011; Garrigues et al. 2013).

3.2.2. LAB Evolution in Vegetable and Meat Fermentations

The microbiota of fresh vegetables normally consists of Gram (−) aerobic bacteria, yeasts, and a low number of LAB (Harris 1998). However, conditions of anaerobiosis, moisture levels, concentration of salt and temperature can result in changes in the population balance and select for spontaneous fermentation by LAB. The same process occurs in meat products where LAB are initially present at low numbers (10^2–10^3 colony forming units, CFU/g), but rapidly dominate the fermentation due to favorable conditions (Rantsiou and Cocolin 2006). In addition to their ability to produce lactic acid and reduce pH, LAB are competitive in some vegetable and meat fermentations because of their ability to produce bacteriocins. Bacteriocins are small peptides (30–60 aa) with antimicrobial properties against bacteria usually of the same or closely related species (narrow spectrum), and occasionally against a broader spectrum of species (Klaenhammer 1988). Bacteriocins produced by LAB have been consumed by humans in fermented food products for millennia. Hence, they are considered safe and are of great interest in food biopreservation. A number of excellent reviews deal with the uses and applications of bacteriocins produced by LAB in food preservation (Rantsiou et al. 2006; Galvez et al. 2008; Settanni and Corsetti 2008; O'Shea et al. 2013). The approach taken traditionally to study bacteriocin production has dramatically changed with the advent of genome sequencing. One example is the comparative-genomics approach used to identify bacteriocin-like genes in the Lactobacillales (Makarova and Koonin 2007). Since the gene sequences of the bacteriocin determinants are highly divergent, Makarova and Koonin (2007) identified genes that are commonly found clustered and associated with bacteriocin-coding sequences. Using this approach they were able to identify such clusters in seven *Lactobacillus* genomes. The candidate bacteriocins were homologs of pediocin (Fimland et al. 2005) and divercin V41, a class IIa bacteriocin naturally produced by *Carnobacterium divergens* V41 and similar to pediocins and enterocins (Rihakova et al. 2008).

3.2.3. Fast-evolving LAB

Wines that undergo desirable malolactic fermentation are often described as "fuller," more complex and less acidic, sometimes with "buttery" overtones. It is well known that LAB are responsible for the reduction in the acidity of wine by the conversion of malic acid into lactic acid and CO_2. The LAB isolated from the wine environment mostly belong to the genera *Pediococcus*, *Lactobacillus*, *Leuconostoc*, and *Oenococcus* (Pilone et al. 1966; du Plessis et al. 2004; Bae et al. 2006). In an interesting study, Marcobal and collaborators (2008) analyzed the rate of spontaneous mutations for rifampin and erithromycin resistance and demonstrated that *O. oeni* PSU-1 (Mills et al. 2005) and the other species in the genus *O. kitaharae* NRIC0645 exhibited a higher spontaneous mutation rate compared to their phylogenetic neighbors, *Leuconostoc mesenteroides* ATCC 8293 and *Pediococcus pentosaceus* ATCC 25745. The main difference between the strains of *Oenococcus* and the other strains was that the latter contain functional *mut*S and *mut*L genes (Makarova et al. 2006), which encode two key enzymes involved in DNA mismatch repair. It is necessary for MutS to bind to a DNA mismatch, and MutL to join the MutS–DNA complex to activate MutH, which is an endonuclease that cuts the non-methylated strand within an hemimethylated GATC sequence and enables the mismatched nucleotide to be replaced (Stanislawska-Sachadyn and Sachadyn 2005). The consequences of absence of *mut*S and *mut*L in *O. oeni* is of technological relevance since this is the most rapidly evolving species of the LAB and the organoleptic properties of wine that has undergone malolactic fermentation are usually highly strain specific (de Las Rivas et al. 2004).

3.2.4. LAB in the GI Tract

The presence of LAB in the gastrointestinal (GI) tract of animals and humans is widely recognized and has been reviewed both before (Salminen and Deighton 1992) and after (Walter 2008; Stolaki et al. 2011) the explosion of genome sequencing projects. In the GI tract system, determination of the individual components is a complex effort as the GI tract provides an anaerobic environment rich in enzymatic activities, bile, and extreme pH conditions (Sanderson 1999). As a consequence, the human microbiota is composed of a vast number of non-cultivable bacterial species compared to only a fraction that can be cultured and manipulated in a laboratory. The Human Microbiome Project (HMP: https://commonfund.nih.gov/hmp/) initiative has begun to elucidate the fascinating symbiotic relationship between microbial commensal organisms and their human hosts.

The coevolution of our microbiota with us has resulted in development of mechanisms to harvest otherwise inaccessible nutrients as an intrinsic part of our metabolism, and to modulate immune function through interaction with host epithelial cells (Backhed et al. 2005; O'Flaherty and Klaenhammer 2010). The human body maintains a relationship that varies from parasitic to beneficial with bacterial numbers ranging from 10^1 to 10^3 CFU/ml in the stomach, 10^4 to 10^7 CFU/ml in the distal small intestine, and a maximum of 10^{11} to 10^{12} CFU/ml in the colon (O'Hara and Shanahan 2006). Human microbial communities are overwhelmingly dominated by four phyla: Firmicutes, Bacteroidetes, Actinobacteria, and Proteobacteria. Representation of *Lactobacillus* sequences in molecular-phylogenetic analysis of the human gut microbiota varies from 0.03 to 47% depending on the type of methodology and analysis used, the number of sequences, and the sampling area (Walter 2008; Booijink et al. 2010). At the present time, a number of probiotic bacteria of human origin are exploited commercially, including *Lact. casei* Shirota (Morishita et al. 1974), *Lact. acidophilus* NCFM (Gilliland et al. 1975), *Lact. rhamnosus* GG (Silva et al. 1987), and *Lact. acidophilus* LA-1 (Bernet et al. 1994).

Probiotics, defined as "live microorganisms which when administered in adequate amounts confer a health benefit on the host" (FAO/WHO 2001), include LAB strains that promote benefits such as alleviation of GI issues, reduction of cold and flu-like symptoms, strengthening of the intestinal epithelial barrier, and protection against infection (Corr et al. 2007; Leyer et al. 2009; Mennigen et al. 2009; Miele et al. 2009; Ohland and Macnaughton 2010). With the characterization of the human microbiome under way, identification of these important taxonomic groups in existing human samples will be a key contribution to the HMP. Furthermore, the identification of a potential relationship between genetics, lifestyle, and the human probiome, defined as commensal intestinal bacteria considered to have a beneficial influence on human health (Azcarate-Peril et al. 2008), will begin to unravel the complex host–microbiome equilibrium.

One of the main objectives of the HMP is to link members of the human microbiome to states of health or disease, such as obesity-related diabetes, through comparative analyses. Although strain analyses through the HMP supports pan-genome development (Nelson et al. 2010), most obesity and health related microbiome studies to date only reveal order-level comparisons and gene expression profiles that provide little insight into the impact of individual LAB (Turnbaugh et al. 2009; Larsen et al. 2010; Brown et al. 2011; Greenblum et al. 2012; Faith et al. 2013). It is expected that as microbiome studies become more developed, they will begin to provide insight into strain specific evolution and beneficial mechanisms for targeted approaches to improve human health (Preidis and Versalovic 2009; Ventura et al. 2009).

3.3. "Probiotic Islands"?

Considerable discussion followed genomic studies of organisms considered true autochthonous species of the human intestinal probiome (Azcarate-Peril et al. 2008). According to Reuter (2001) *Lact. gasseri* and *Lact. reuteri* are the *Lactobacillus* species native to the human GI tract. True transient lactobacilli include *Lact. salivarius*, *Lact. plantarum*, *Lact. casei*, *Lact. buchneri*, *Lact. brevis*, and *Lact. fermentum* (Reuter 2001). Recent reviews suggest that most *Lactobacillus* species found in the mammalian GI tract are not true intestinal inhabitants and that it is difficult to differentiate autochthonous species from those that originate from proximal or exogenous sources (Walter 2008; Stolaki et al. 2011). Interestingly, a number of factors allow probiotic lactobacilli to persist in the gut, and some of those factors have also been reported as contributors to bacterial "virulence" indicating, not surprisingly, that both probiotic and pathogenic bacteria use similar strategies to survive in the GI tract (Hill 2012). The concept of a "probiotic island" as a genomic region encompassing genes involved in probiotic functionalities arises as a counterpart of the "pathogenicity island" described in virulent species such as *Ent. faecalis* (Coburn et al. 2007) and *Salmonella enterica* (Halici et al. 2008).

One example of such probiotic islands originated from the analysis of the complete genome sequences of the reuterin-producing *Lact. reuteri* JCM 1112^T and its closely related species *Lact. fermentum* IFO 3956 (Morita et al. 2008). Comparative genome analysis revealed that JCM 1112^T has a unique cluster of 58 genes for the biosynthesis of reuterin and cobalamin (Taranto et al. 2003), which has a lower GC content and is apparently inserted into a conserved region resulting from a horizontal gene transfer (HGT) event (Morita et al. 2008). Production of 3-hydroxypropionaldehyde (3-HPA) from glycerol by *Lact. reuteri* was first reported by Talarico and Dobrogosz (1989). The authors characterized this substance, termed reuterin, and showed that it inhibited growth of Gram (+) and Gram (–) bacteria as well as yeasts, fungi, and protozoa. Synthesis of reuterin is mediated by glycerol dehydratase (E.C. 4.2.1.30). According to Morita et al. (2008) the genome of JCM 1112^T contains three genes (LAR_1633–1635) with dehydratase subunit motifs (Pfam PF02286–02288) in the propanediol utilizing operon (LAR_1616–1640). The genes have been designated *gup*CDE (glycerol utilization gene candidates in the *pdu* operon) based on homology with other bacterial strains. The encoded amino acid sequences exhibited the highest identity to their homologues from *Lact. brevis* ATCC 367 (Makarova et al. 2006), showing 81, 66, and 57% identities, respectively. Additionally, a cluster responsible for cobalamin biosynthesis (*cbi*, *cob*, and *hem*) has been identified and located adjacent to the *pdu* operon. Notably, the organization of this genomic island represents a close association of the *pdu* and *cbi–cob* operons reflecting the cobalamin requirement for glycerol dehydratase activity. A comparison of *pdu* and *cbi–cob* operons from *Lact. reuteri* with *Lact. fermentum* IFO 3956 (Morita et al. 2008) and *Lact. plantarum* WCFS1 (Kleerebezem et al. 2003) revealed not only that the *pdu–cbi–cob–hem* gene cluster comprising 58 ORFs (LAR_1583–1640) was absent in these LAB but also that it was inserted in a locus that was common to all three bacteria (Morita et al. 2008). Interestingly, a comparison with the draft genome sequence of *Lact. reuteri* 100-23 (NCBI bioprojects PRJNA54165 and PRJNA13431), a rodent-specific strain, showed that the *pdu–cbi–cob–hem* gene cluster is absent.

Recently, the SpaCBA pilus was discovered encoded on a genomic island in *Lact. rhamnosus* GG, a strain known for its ability to adhere and persist in the GI tract, through comparative analysis with *Lact. rhamnosus* Lc705, a strain with poor GI persistence (Kankainen et al. 2009). The SpaCBA pilus was shown to be involved in both intestinal epithelial cell adhesion and immunomodulation (Kankainen et al. 2009; Lebeer et al. 2012). In a comparative analysis of 100 *Lact. rhamnosus* strains, 31 out of the 34 strains encoding a functional SpaCBA pilus were human isolates, indicating the importance of pili for persistence in this niche (Douillard et al. 2013b). The *spaCBA* gene cluster was likely acquired by HGT, indicated by IS elements, enabling these strains to compete with similarly equipped enteric microbes and persist in the human GI tract (Kankainen et al. 2009; Douillard et al. 2013b). The *spaCBA* gene cluster was found to be conserved in *Lact. casei* BL23 as well as in two *Lact. casei* and two *Lact. rhamnosus* isolates from probiotic products. While all of the *Lact. rhamnosus* strains expressed functional pili, *Lact. casei* strains did not, likely due to the absence of an upstream IS element that established an active promoter in the *Lact. rhamnosus* strains. The pili were only one of several probiotic differences identified in the two species, supporting the importance of application-directed strain selection (Douillard et al. 2013a).

While beneficial/commensal and pathogenic microbes can share the same intestinal environment and they all may have similar stress responses, survival, and adherence mechanisms, they are genetically distinct, even within LAB strains. A comparative analysis of 81 genomes encompassing six genera (*Lactobacillus*, *Lactococcus*, *Leuconostoc*, *Streptococcus*, *Enterococcus*, and *Bifidobacterium*) revealed a similar distribution between Clusters of Orthologous Groups (COG) in both pathogenic and nonpathogenic pan-genomes, but varied core genomes (Lukjancenko et al. 2012). The pathogenic core genome was enriched for metabolism and cell wall/membrane biogenesis genes while the non-pathogenic was enriched for translation, ribosomal structure, and biogenesis (Figure 3.1), suggesting a contribution of genes in these COGs to virulence or to

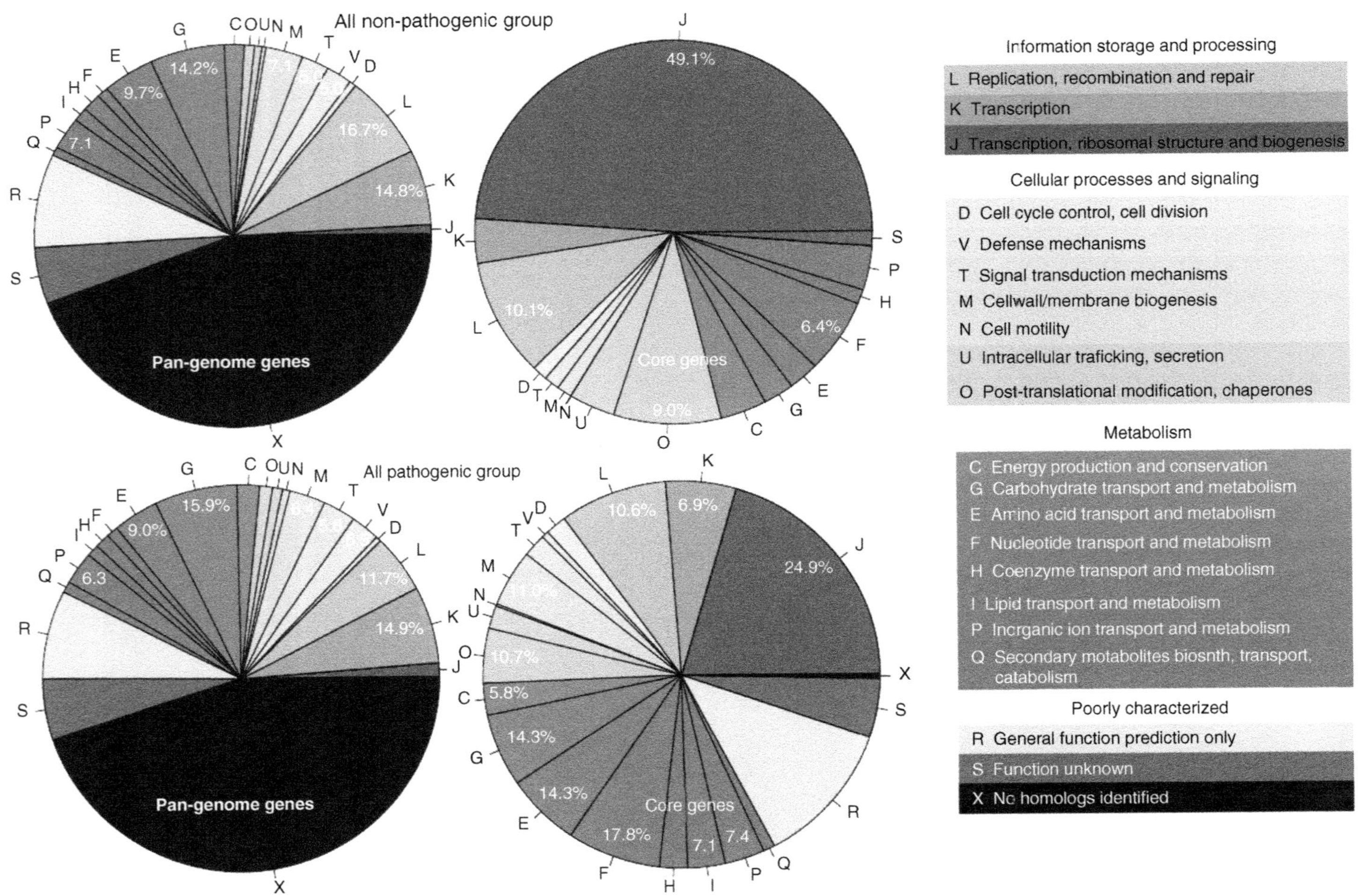

Figure 3.1. COG statistics for the pan- and core genomes of nonpathogenic and pathogenic isolates from the genera *Lactobacillus*, *Lactococcus*, *Leuconostoc*, *Enterococcus*, *Streptococcus*, and *Bifidobacterium*. Modified and reproduced with permission from Lukjancenko et al. (2012).

probiotic/fermentative traits, respectively (Lukjancenko et al. 2012), which may be revealed by functional studies.

3.4. Stress Resistance and Quorum Sensing Mechanisms

A number of survival features have been identified in LAB, although most of them are not as manifestly organized as the *pdu–cbi–cob–hem* island from *Lact. reuteri*. The majority of the probiotic features are related to strain survival in the GI tract. The intestinal environment results from three main factors: dietary intake, bacterial ecology, and host physiology, including factors such as peristalsis and glandular secretions. Several factors restrict bacterial cell growth including gastric acidity, oxidative stress, digestive enzymes, bile salts, peristalsis, mucus, the resident commensal microbiota, exfoliation of enterocytes during epithelial renewal, epithelial translocation of secretory IgA, $CD8^+$ intraepithelial T lymphocytes, and innate host defense mechanisms mediated by gene-encoded antimicrobial peptides (Sanderson 1999; Ouellette 2004). LAB are typically regarded as aerotolerant anaerobes (Axelsson 1998) that can grow in the presence of oxygen and generate partially reduced reactive oxygen species (ROS) including the superoxide radical (O_2·–), hydrogen peroxide (H_2O_2), and the hydroxyl radical (HO·). ROS are both cytotoxic and mutagenic and, in the GI tract, ROS are central players in inflammatory bowel diseases (IBD) (Sartor 2004). To atone for the harmful effects of ROS, both bacterial and intestinal cells have evolved protective mechanisms that utilize antioxidant enzymes such as superoxide dismutases (SODs) and hydroperoxidases to prevent the formation of HO·. Since SOD levels in IBD patients are frequently depleted (Lih-Brody et al. 1996), there is potential for elevated levels of this enzyme to function therapeutically for IBD. LAB encoding genes involved with both oxidative stress resistance and regulatory mechanisms have been identified (Miyoshi et al. 2003; El-Sharoud 2005) and oxidative resistance genes, encoding SOD and catalase, have been expressed in probiotic strains such as *Lact. gasseri* (Bruno-Barcena et al. 2004; Carroll et al. 2007) and technologically relevant organisms like *Lact. delbrueckii* subsp. *bulgaricus* (Rochat et al. 2006).

One important attribute of probiotic LAB is their ability to resist and thrive in acidic environments (Gueimonde and Salminen 2006), like the stomach of mammals and fermented low pH foods. Several mechanisms of acid tolerance have been described in LAB. The most important and universally present in the group, the multisubunit F_1F_0 ATPase, which links the production of ATP molecules to the transmembrane proton motive force (PMF), can either generate ATP at the expense of PMF or produce PMF consuming ATP. The PMF facilitates the extrusion of protons from the cytoplasm. The F_0 complex has proton translocating activity, while the peripherally bound F_1 complex has ATPase activity (Nath 2002). Another mechanism of acid resistance is the Arginine Deiminase (ADI, E.C. 3.5.3.6) pathway. Present only in a limited number of members of the LAB group (Arena et al. 2002; Cotter and Hill 2003), this enzyme also exhibits anti-carcinogenic activities (Kim et al. 2007b). The anti-inflammatory activities of the purified recombinant ADI from *L. lactis* subsp. *lactis* ATCC7962 (Kim et al. 2007b) showed that ADI strongly inhibits the generation of nitric oxide (NO) and prostaglandin E2 (PGE2) in LPS-treated macrophages by inhibiting expression of the inducible NO synthase (iNOS) and cyclooxygenase (COX-2) proteins. The purified enzyme also inhibited the production of IL-6 in LPS-treated RAW 264.7 macrophages. The enzyme responsible for the over-production of NO, iNOS, is often observed during inflammation and tumor development.

Other acid resistance mechanisms include urease, a nickel-containing oligomeric enzyme that catalyzes the hydrolysis of urea to two molecules of ammonia and one molecule of carbon dioxide (Mobley et al. 1995). The enzyme has maximal activity at pH 7 but loses activity at more extreme acidic pHs, being essentially inactive at pH 4.3. Ureases have been mostly associated with pathogens with the most studied system encoded by the highly urolytic bacteria *Helicobacter pylori* (Bartnik

2008). However, urease homologs have been identified in *Lact. fermentum* (Kakimoto et al. 1990), *Lact. brevis* ATCC 367 (LVIS_1854), and *Leuconostoc mesenteroides* subsp. *mesenteroides* ATCC 8293 (LEUM_2023). Interestingly, *Strep. thermophilus* is the only member of the food grade LAB to encode a functional urease (Makarova et al. 2006).

Amino acid decarboxylases are also involved in acid resistance. These enzymes process the bond holding the carboxylic (-COOH) group to the rest of the amino acid. As a result, the end product is a basic chemical that causes the pH to increase. The glutamate decarboxylase (Gad) system of acid resistance has been extensively characterized in Gram (−) and Gram (+) bacteria (Sanders et al. 1998; Waterman and Small 2003; Tramonti et al. 2006). In *L. lactis*, expression of *gad*CB in presence of chloride increases in modified M17 medium, in which β-glycerophosphate was omitted to lower the buffering capacity of the medium, when the culture pH decreases. Expression of *gad*CB is also induced by glutamate (Sanders et al. 1998). Glutamate decarboxylase homologs were identified in a limited number of LAB besides *L. lactis*, including *Lact. plantarum* WCFS (lp_3420) and *Lact. brevis* ATCC 367 (LVIS_0079, LVIS_1847, and LVIS_2213). However, the transport component of the Gad system (GadC) appears to be widely distributed in LAB. In *Lact. acidophilus* NCFM, GadC is one of the mechanisms responsible for the intrinsic acid resistance of the strain (Azcarate-Peril et al. 2004).

Another mechanism that has received attention involves the density-dependent recognition of an autoinducer molecule by bacterial cells with consequent changes in gene expression. Several studies have reviewed the role of this mechanism, termed *quorum sensing*, in human health (Jayaraman and Wood 2008; Sifri 2008; Willcox et al. 2008), food fermentation (Fleet 2003; Sieuwerts et al. 2008), food spoilage (Ammor et al. 2008), and production of antimicrobial molecules (van der Ploeg 2005; Petersen et al. 2006; Straume et al. 2007; Navarro et al. 2008). One important *quorum sensing* system used to communicate among and between species is based on a furanosyl borate diester called autoinducer-2 (AI-2), produced in four enzymatic steps from methionine. The gene encoding the AI-2 synthase, *lux*S, has been identified in many different Gram (+) and Gram (−) species (Vendeville et al. 2005; Lebeer et al. 2007; Buck et al. 2009). AI-2 regulates the expression of various phenotypes including virulence factors, DNA processing, cell morphology, motility, biofilm formation, toxin production, light production, cell division, and adhesion (Xavier and Bassler, 2003; Lebeer et al., 2007; Sztajer et al., 2007 Buck et al., 2009). Highly conserved homologs of *luxS* are present in over 55 species of Gram (-) and Gram (+) bacteria (Kaper and Sperandio 2005). Figure 3.2 shows a comparison of the region containing *lux*S in several LAB genomes that exhibit over 70% similarity levels in this gene. As previously reported (Buck et al. 2009), it is notable that, despite the high degree of similarity of *lux*S, different species show markedly different organization in the surrounding genomic regions. Furthermore, *Lact. delbrueckii* subsp. *bulgaricus* ATCC BAA-365 (Makarova et al. 2006) encodes two copies of *lux*S in distant loci (LBUL_0088 and LBUL_1231) with no apparent gene synteny.

3.5. The Impact of Genome Sequencing on Characterization, Taxonomy, and Pan-genome Development of Lactic Acid Bacteria

The explosion of available genomic information has impacted our phylogenetic view of the traditional concept of "species." The most accepted definition of species today is the phylo-phenetic concept: "a species is considered a monophyletic and genomically coherent cluster of individual organisms that shows a high degree of overall similarity with respect to many independent characteristics, and is diagnosable by a discriminative phenotypic property" (Rossello-Mora and Amann 2001; Felis and Dellaglio 2007). Prior to the genome sequencing boom, the criteria used to define "species," which is the essential entity of biological diversity, included phenotypic (such as morphological and physiological

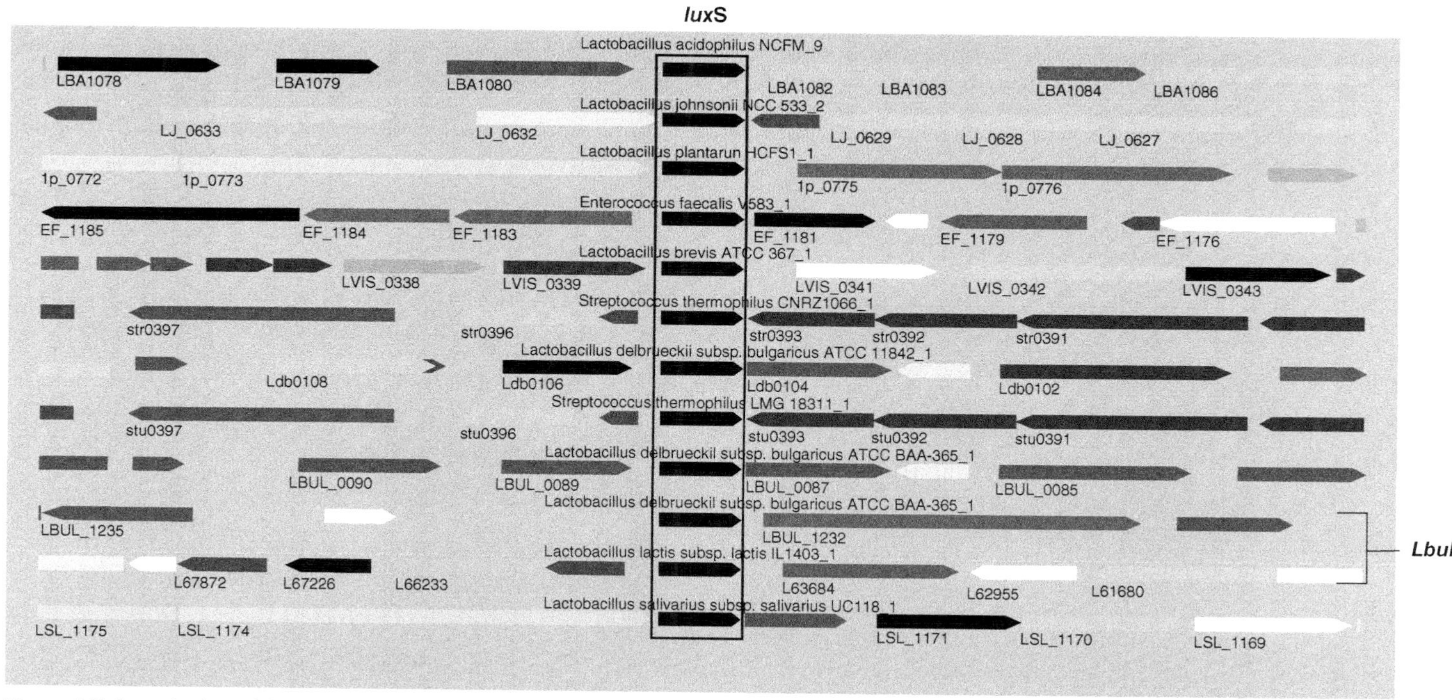

Figure 3.2 Organization of the genome region of selected LAB containing the autoinducer-2 (AI-2) producer gene *luxS*. The comparison was performed using the Multi-Genome Region Comparison tool in the Comprehensive Microbial Resource website at JCVI (http://www.jcvi.org/). The cutoff selection for protein matches was 40% minimum similarity. The two chromosomal regions of *Lact. delbrueckii* subsp. *bulgaricus* ATCC BAA-365 containing *luxS* are indicated.

characteristics, cell wall composition, protein fingerprinting, electrophoretic mobility of enzymes, and fatty acid methylester analysis) as well as genotypic analysis (such as GC content analysis, DNA-DNA and DNA-RNA hybridization, ribotyping, DGGE, RAPD-PCR, and multiple locus sequence typing (MLST)) (Ludwig and Schleifer 1994; Vandamme et al. 1996; Klein et al. 1998; Carr et al. 2002; Felis and Dellaglio 2007). A study by Laursen et al. (2005) characterized over 100 *Carnobacterium* strains (50 from beef and pork products, and 52 from cod, halibut, salmon, shrimps, and roe products) using numerical taxonomy based on classical biochemical reactions, carbohydrate fermentation, and inhibition tests (temperature, salt, pH, chemical preservatives, antibiotics, bacteriocins), SDS-PAGE of whole cell proteins, plasmid profiling, intergenic spacer region (ISR) analysis, and examination of amplified-fragment length polymorphism (AFLP). The approach divided the *Carnobacterium* strains into 24 groups that shared less than 89% similarity. These groups were identified mainly as *C. divergens* and *C. maltaromaticum* and both species exhibited high phenotypic intra-species variability, although reliable identification was obtained by SDS-PAGE of whole cell proteins and by 16S-23S rDNA ISR polymorphism. Presently, no comparative analyses have been possible for *Carnobacterium* isolated from animal products, as no strains of *C. divergens* and only one of *C. maltaromaticum* have been sequenced. More recently, pulsed field gel electrophoresis and MLST identified 22 pulsotypes and 19 sequence types, respectively, from 24 *Lactobacillus sanfranciscensis* sourdough strains, with strain evolution mainly influenced by processing conditions and environment (Picozzi et al. 2010). Currently, only the sequence for one *Lact. sanfranciscensis* strain is available at NCBI. As accurate strain discrimination is necessary to achieve expected product characteristics, the authors suggest using MLST to clarify and discriminate strains used in commercial starter cultures (Picozzi et al. 2010).

Comparative genome hybridization (CGH) is commonly employed when whole genome sequences of multiple strains from the same species are not available. CGH was used to compare strains of *L. lactis* (Kok et al. 2005). DNA microarray hybridizations of *L. lactis* IL1403 with a mixture of randomly Cy3- or Cy5-labeled DNA fragments from the *L. lactis* IL1403 and *L. lactis* MG1363 showed a positive correlation between gene similarity and the obtained ratio for genes with a similarity index of > 75%. The study also highlighted genes present in IL1403 but not in MG1363 (*glp*D, *glpK*, *dex*B, *yab*E, *rma*F, *lcn*C, *umu*C, and pi251). Also, analysis of the genome sequence of MG1363 (Wegmann et al., 2007) revealed approximately 85% DNA sequence identity between coding sequences (CDS) present in both *L. lactis* MG1363 and *L. lactis* IL1403 and 97.7% between those in the two subspecies *cremoris* strains, *L. lactis* MG1363 and *L. lactis* SK11 (Makarova et al. 2006). Additionally, a core genome containing approximately 58% of CDS was defined for *Strep. thermophilus* (Rasmussen et al. 2008). The authors selected 2200 relevant CDS from available *Strep. thermophilus* genomes (Bolotin et al. 2004; Makarova et al. 2006) as input to design 65- to 75-mer oligonucleotides for a microarray platform. CGH of 47 dairy *Strep. thermophilus* strains identified 1271 core genes and confirmed the absence of virulence genes.

Recent CGH studies indicate the evolution of niche specific genes in LAB. CGH of 18 *Lact. sakei* strains revealed conservation in metabolism and stress response genes relevant to their meat niche (Nyquist et al. 2011). Host-specific gene content was revealed in a CGH study of 57 *Lact. reuteri* strains isolated from six different vertebrate hosts. The human strains were characterized by a high percentage of mobile elements and pseudogenes. Comparative analysis between the human strain F275 and the mouse strain 100-23 revealed reductive evolution in F275, especially in surface proteins, possibly enabling immune tolerance. Genes specific to mice strains included a two-component regulatory system, TCS2, predicted to be involved in quorum sensing, and a SecA2 protein secretion system. Mutants in the TCS2 and SecA2 pathways significantly reduced colonization of the 100-23 rodent isolate, indicating the importance of these pathways in the rodent GI niche. Further indication of specificity was demonstrated by the unsuccessful mouse colonization of strains originating from other hosts (Frese et al. 2011).

With the beginning of the genome sequencing era, comparative genomic studies confirmed the notion that bacterial species share a core genome. As the pan-genome increases in size with the addition of more strains, species, or genera, the core genome is reduced as long as consistent selection criteria are used for inclusion (Lukjancenko et al. 2012).

In a comparison of five *Lactobacillus* strains, the DNA level was found to be too diverse for phylogenetic alignment, but proteome comparison based on FASTA identified 593 shared orthologs (Canchaya et al. 2006). The largest proportion of core genes were grouped into COGs involved in housekeeping functions, including metabolism, replication, transcription, and translation, but the second largest proportion had no predicted function. A greater number of unique transcription and signal transcription proteins in the larger genomes was suggested to be reflective of the increasing regulatory complexity required by the more intricate metabolic and cellular pathways in these genomes. When more stringent selection criteria were applied, including a requirement for identification by both FASTA and BLAST, the core genome for the five *Lactobacillus* strains was reduced to 354 proteins (Canchaya et al. 2006). Seven more *Lactobacillus* genome sequences became available in the next two years, and all twelve, plus six additional Firmicutes sequences, were compared using similarly stringent selection criteria, and both FASTA and BLAST (Canchaya et al. 2006; Claesson et al. 2008). The analysis identified only 141 core proteins, with the largest COGs in translation and replication. The diversity found between strains resulted in varied phylogenetic tree associations and a range of stability based on the tree-building approach, and led the authors to suggest GroEL as a more valid phylogenetic marker than 16S rRNA. The analysis identified only four niche specific proteins, predicted to be involved in purine metabolism in meat associated strains (Claesson et al. 2008).

Another comparative study with less stringent selection criteria identified 567 shared genes in the core genome of 12 sequenced LAB from 6 genera within the order *Lactobacillales*. The majority of core genes were involved in replication, transcription, and translation, but approximately 100 genes were poorly characterized (Makarova and Koonin 2007), similar to the distribution of the core identified from the comparison of only five *Lactobacillus* genomes (Canchaya et al. 2006). The analyses revealed only two core genes with no detectable orthologs outside of *Lactobacillales*.

In a comparison of eleven LAB genomes, of which three were from dairy, five from the GI tract, and three from multi-niche environments, nine genes were identified as a niche specific barcode differentiating dairy and GI associated strains (O'Sullivan et al. 2009). The six dairy specific genes were part of the proteolytic system or the R/M system, while the three gut specific genes were involved in bile salt hydrolysis or sugar metabolism. The authors inferred that adaptation of dairy-related strains to the dairy environment through lateral gene transfer of metabolism genes, and decay of GI related genes, supports the development of niche-related genetic barcodes. Although the barcode was verified against all sequenced LAB at the time, a later study that included 20 *Lactobacillus* genomes determined that all 9 niche specific genes were found to be present in strains from other niches (O'Sullivan et al. 2009; Kant et al. 2011). The results indicate that, although it is tempting to look for unique gene barcodes to profile a bacterial niche, in the current state of continuously increasing sequence availability, these identifiers are unreliable as they may turn up in an unexpected niche. It is suggested that as genome sequences become commonly available for all bacterial species, more reliable barcodes may be conclusively identified.

The first LAB pan-genome was identified from the comparison of 20 *Lactobacillus* genomes, exposing the diversity in this group. The pan-genome was calculated using EDGAR, a high-throughput comparative genomics platform based on BLAST score ratios, revealing 14,000 protein coding genes (Blom et al. 2009; Kant et al. 2011). The core genome included 383 genes, which is significantly higher than the core of 141 genes identified previously for 18 Firmicutes, likely due to differences in both the stringency of the selection criteria and the diversity of genera included (Claesson et al. 2008;

Table 3.3. Pan- and core genomes, based on gene families, for specific strains from five genera of LAB bacteria. Reproduced with permission (Lukjancenko et al., 2012).

Genus	Number of genomes	Number of species	Average genome size (kbp)	Average % CG	Average number of genes (min–max)	Average number of gene families (min–max)	Pan-genome	Core genome
Lactobacillus	*21*	*14*	*2369*	*42.4*	*2235 (1562–3059)*	*2071 (1437–2873)*	*13,069*	*363*
Lactococcus	*4*	*1*	*2532*	*35.4*	*2465 (2266–2504)*	*2238 (2118–2341)*	*3389*	*1522*
Leuconostoc	*3*	*3*	*2025*	*37.9*	*1986 (1820–2130)*	*1896 (1724–2050)*	*2927*	*1164*
Enterococcus	*11*	*4*	*3041*	*36.6*	*3078 (2573–2515)*	*2707 (2439–3114)*	*7519*	*1092*
Streptococcus	*23*	*12*	*1981*	*38.9*	*2018 (1696–2270)*	*1923 (1643–2180)*	*9785*	*638*

Kant et al. 2011). Although the analysis did not reveal any potential niche specific genes, a larger secretome and greater number of LPXTG-cell wall anchor sortases were identified in probiotic strains than in dairy strains (Kant et al. 2011). Also, in a comparative analysis between secretomes from 26 LAB strains, the largest secretomes were typically found in probiotic or GI-related strains (Zhou et al. 2010). This is not surprising, as probiotic strains are expected to secrete surface proteins for survival and interaction in the GI tract (Kleerebezem et al. 2010).

In a comparison of 62 genomes that included pathogens, probiotics, and fermentative strains from the genera *Lactobacillus*, *Lactococcus*, *Leuconostoc*, *Streptococcus*, and *Enterococcus*, genes were grouped into families based on similarity, and these families were compared between genomes. The overall pan-genome included 30,615 gene families with a core genome of 243 gene families. In separate genus analyses, 21 *Lactobacillus* strains ranged in proteome homology from 11.5 to 93.3%, reflected by a significantly larger pan-genome and smaller core genome than the other five genera analyzed (Table 3.3) (Lukjancenko et al. 2012). The *Lactobacillus* taxonomic group has often been considered to be too diverse, and it has been suggested that the genus be reclassified into smaller, more defined groups (Canchaya et al. 2006; Felis and Dellaglio 2007; Claesson et al. 2008). As fully sequenced genomes become available, attempts to cluster genomes and identify signature group genes have shown that the most closely related group includes *Lact. acidophilus* NCFM (Claesson et al. 2008; Kant et al. 2011). However, analyses of other single genus pan-genomes in recent years have demonstrated similar diversity, providing support for keeping the *Lactobacillus* genus intact (Ussery et al. 2009; Lukjancenko et al. 2010, 2012; Vesth et al. 2010).

A second source of variability among genomes, beyond the divergence of the core genome, is due to the rapid gain and loss of large DNA portions, especially within species (Deloger et al. 2008). From this concept, the authors identify different approaches to estimate genome distances. The "Average Nucleotide Identity" (ANI) is "the average nucleotide identity of the total genomic sequence shared between two strains" (Konstantinidis et al. 2006). This method defines the list of orthologs and derives the overall divergence of the core genome by averaging the percentage of identity at the nucleotide level of all orthologs found. Other methods approximate distances based on estimating the proportion of common genes or DNA. The approach proposed by Deloger et al. (2008) intends to capture both dimensions of genome variability at inter and intra-species levels based on the amount of Maximal Unique and Exact Matches (MUMs) of a given minimal length shared by the compared genomes. A MUM index (MUMi) analysis of 67 genera for which at least two genomes are available indicated relative homogeneity within species, and also showed a correlation between MUMi, ANI, and the more classical method, MLST analysis. The MUMi for *L. lactis* subspecies *lactis* versus *L. lactis* subspecies *cremoris* was 0.74 indicating high diversity (the MUMi varies between 0 for very similar, to 1 for very distant genomes) at the subspecies level. This analysis correlates well with other observations of high intra-species variability in LAB. For example, phenotypic analyis of nine strains of

Lact. gasseri of intestinal origin revealed high intra-species variability. This level of variability highlights the importance of strain sequencing and in depth studies of strain-specific genetic systems (Azcarate-Peril et al. 2008).

3.6. Functional Genomic Studies to Unveil Novel LAB Utilities

The presence within bacterial genomes of a large number of "ORFans," which are defined as annotated genes that are restricted to a particular genome and that possess no known homologs (Daubin and Ochman 2004), has been observed widely in a number of genomes and comparative genome papers. A MUMmer plot of the *Strep. thermophilus* CNRZ1066 chromosome versus that of *Strep. thermophilus* LMG 18311 (Figure 3.3) shows very high similarity levels; however, several genes can be identified as unique for each one of the analyzed strains (represented as red and green dots). The uniqueness of ORFans within a genome has not allowed the use of classic comparative approaches to determine their function and evolution. While other bioinformatics methods can provide a theoretical guide for the functionality of ORFan genes, only functional genomics approaches can confirm that function to potentially uncover novel LAB features.

Numerous studies have demonstrated the importance of functional genomics in LAB research. In 2002, Jansen et al. (2002) identified a novel family of repetitive DNA sequences, present among both domains of the prokaryotes (Archaea and Bacteria), but absent from eukaryotes or viruses.

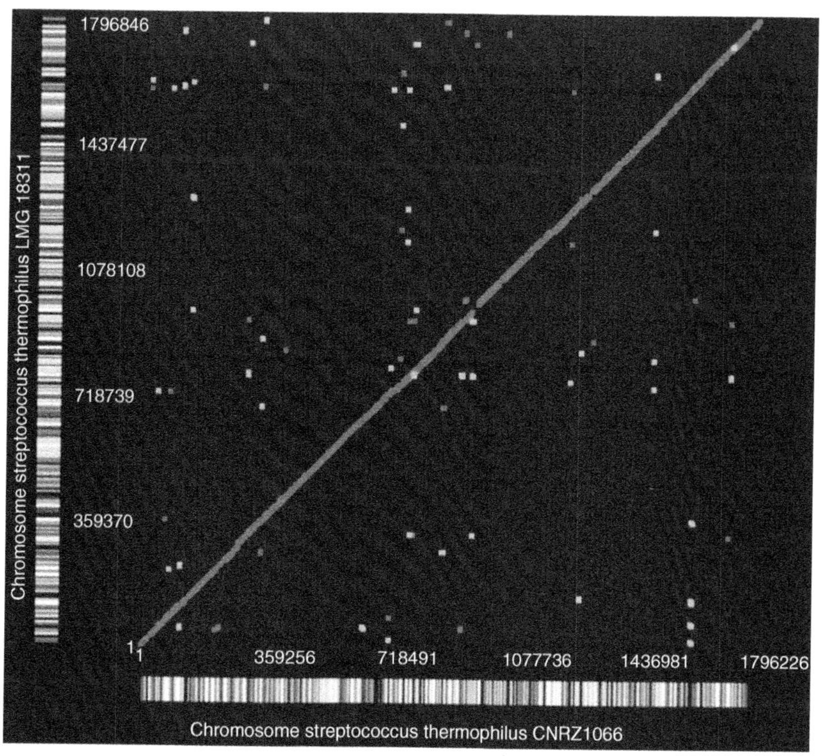

Figure 3.3. Whole-genome alignment of *Strep. thermophilus* CNRZ1066 versus *Strep. thermophilus* LMG 18311 using MUMmer (Maximal Unique Match) (Kurtz et al. 2004). Minimum Match Length was set at 150. Unique regions in each chromosome are depicted as green (unique to CNRZ1066) and red dots (unique to LMG18311).

The family, identified by *in silico* analysis, was characterized by direct repeats from 21 to 37 bp, interspaced by similarly sized non-repetitive sequences. This family was termed Clustered Regularly Interspaced Short Palindromic Repeats (CRISPR) (Jansen et al. 2002). As genome sequencing projects progressed, CRISPRs were identified in an increasing number of bacterial genomes; however, their function in phage resistance was only elucidated in 2007 (Barrangou et al. 2007) by altering the CRISPR1 locus of a dairy strain of *Strep. thermophilus* DGCC7710. Through adding and deleting spacers, derived isogenic strains were shown to be sensitive or resistant to two virulent bacteriophages isolated from industrial yogurt samples, phage 858 and phage 2972 (Levesque et al. 2005). The discovery that CRISPR loci included environmentally-driven acquired immunity provided insight into LAB evolution. In an analysis of 102 LAB genomes, 47 were found to contain CRISPR loci. Comparative analysis between these sequences revealed eight CRISPR repeat families, with variability in leader sequences and *cas* gene content. The CRISPR loci contained 104 different spacers that showed 100% identity to either phage, prophage, plasmid, or chromosomal sequences, indicating the impact of environment on the evolution of each LAB (Horvath et al. 2009).

The development of molecular tools like vectors and systems to insertionally inactivate, delete, or overexpress specific loci in the genome has been invaluable in the definition of gene function and creation of recombinant LAB. Genetic manipulations systems were developed earlier for *L. lactis*, natural gene transfer systems (transduction, conjugation, transposition) were routinely employed over 25 years ago (Fitzgerald and Gasson 1988). A few years later, a gene replacement vector based on the *Escherichia coli* vector pMTL23p with a Cm^r gene from pC194 (Iordanescu and Surdeanu 1980) and a *nis*A gene from pFI172 (Dodd et al. 1990) inactivated by insertion of an Em^r gene from pE194 was constructed for insertional inactivation of the nisin production gene *nis*A (Dodd et al. 1992). Although some early attempts of genetic manipulation in lactobacilli were successful (O'Sullivan and Klaenhammer 1993; Bhowmik and Steele 1994; Skaugen and Nes 1994), only a restricted number of strains were susceptible to such manipulation and the developed tools achieved limited success. More recently, systems for insertional inactivation or deletion, chromosomal insertion and expression, and heterologous expression of interesting genes were developed based on available genomic information to support functional genomic studies (Bruno-Barcena et al. 2005; Klaenhammer et al. 2005; Lambert et al. 2007; Goh et al. 2009; Douglas et al. 2011; Russell and Klaenhammer 2001). The available expression and secretion systems for heterologous protein secretion in *L. lactis*, including promoters, signal peptides, and mutant host strains have been reviewed by (Morello et al. 2008). Earlier, functional studies that allowed the identification of gene function based on genomic information in lactobacilli were reviewed by Claesson et al. (2007) and Klaenhammer et al. (2005). Comparative genomics will be useful in identifying strains with advantageous traits that are susceptible to genetic manipulation strategies.

One of the most exciting promises of LAB is the prospect of using GRAS members of this group as delivery vehicles for vaccines, antimicrobial compounds, and therapeutics. An early study demonstrated that intragastric administration of a recombinant *L. lactis* strain, secreting murine IL-10, prevented onset of colitis in IL-10 knockout mice, and caused a 50% reduction of the inflammation in dextran sulfate sodium-induced chronic colitis (Steidler et al. 2000). More recently, the potential for orally delivered LAB vaccines was demonstrated in *Lact. acidophilus* and *Lact. gasseri* strains expressing a dendritic cell-targeted protective antigen for *Bacillus anthracis*, which conferred protection to mice against a lethal dose of anthrax (Mohamadzadeh et al. 2009; Mohamadzadeh 2010). In 2006, the first pilot trial with a recombinant LAB in humans was published (Braat et al. 2006). The authors replaced the *thy*A gene of *L. lactis* with a synthetic sequence encoding mature human IL-10, fused at its *N*-terminus to a lactococcal secretion signal. They chose this method because it provided an adequate containment for the recombinant bacteria since the modified strain of *L. lactis* was unable to survive in the environment without thymidine or thymine. The results of this study were very promising as clinical benefit was observed in eight of ten patients. Comparative techniques will enable identification of strains that may improve or augment these targeted delivery methods. Wells and Mercenier (2008) reviewed the accumulated scientific data existent on the use of LAB as mucosal delivery vehicles of vaccine antigens, microbicides, and therapeutics (for another version, see this book, *Chapter 11). Current applications of LAB delivery include the production and delivery of single-chain variable fragments, DNA, peptides, antigens, cytokines, enzymes, and allergens.

3.7. Conclusions

With the rapidly increasing availability of complete microbial genomes at NCBI, comparative genomics studies are becoming more informative, revealing complex diversity and functional capability within specific bacterial phylogenetic groups and also answering essential evolutionary questions. In particular, LAB genomics has contributed significantly to our understanding of the relationships between genotype and phenotype, which will have important ramifications to control biotechnologically relevant processes and to develop novel vaccines and improved diagnostics. Furthermore, strain sequencing is becoming routine with the widespread implementation of next generation, massively parallel ultra high-throughput sequencing platforms like HiSeq and MiSeq (Illumina), SOLiD (Applied Biosystems), 454 (Roche), PacBio RS II (Pacific Biosciences), PGM (Ion Torrent), HeliScope (Helicos BioSciences Corp.), and Polonator G.007 (Danaher Motion and George Church). These platforms support a wide range of genetic analysis applications from reference assembly of genomes, resequencing, *de novo* assembly, SNP detection, digital gene expression, metagenomics, genomics, and transcriptomics analysis (Shendure and Ji 2008; Metzker 2010; Loman et al. 2012; Mardis 2013), and downstream applications like polymerase colony (polony) multiplex analysis of gene expression (PMAGE) (Kim et al. 2007a). Given the vast amounts of data generated by these methods, High Performance Computing (HPC) systems to analyze, visualize, cluster, compare, and store data have driven development of increasing numbers of bioinformatics tools and databases (Altermann and Klaenhammer 2005; Teusink et al. 2005; Pons et al. 2008; Waagmeester et al. 2008; Kiefer et al. 2009; Li and Homer 2010). Bioinformatics may remain the bottleneck in the elucidation of important pathways in LAB. While comparative analyses may improve identification and application-directed strain selection for specific traits, functional genomics studies, in a high-throughput fashion, will be the pillar to unveil novel functionalities in these microbes that are so beneficial in food preservation, bioprocessing, and health.

References

Altermann, E., and Klaenhammer, T.R. (2005). PathwayVoyager: pathway mapping using the Kyoto Encyclopedia of Genes and Genomes (KEGG) database. *BMC Genomics* **6**, 60.

Altermann, E., Russell, W.M., Azcarate-Peril, M.A., et al. (2005). Complete genome sequence of the probiotic lactic acid bacterium Lactobacillus *acidophilus* NCFM. *Proc Natl Acad Sci USA* **102**, 3906–3912.

Ammor, M. S., Michaelidis, C., and Nychas, G.J. (2008). Insights into the role of quorum sensing in food spoilage. *J Food Prot* **71**, 1510–1525.

Arena, M. E., Manca de Nadra, M.C., and Munoz, R. (2002). The arginine deiminase pathway in the wine lactic acid bacterium *Lactobacillus hilgardii* X1B: structural and functional study of the arcABC genes. *Gene* **301**, 61–66.

Axelsson, L. (1998). Lactic acid bacteria: classification and physiology. In: *Lactic Acid Bacteria: Microbiology and Functional Aspects* ((eds) S. Salminen and A. von Wright), pp. 1–12. Marcel Dekker, Inc., New York.

Azcarate-Peril, M.A., Altermann, E., Goh, Y.J., et al. (2008). Analysis of the genome sequence of *Lactobacillus gasseri* ATCC 33323 reveals the molecular basis of an autochthonous intestinal organism. *Appl Environ Microbiol* **74**, 4610–4625.

Azcarate-Peril, M.A., Altermann, E., Hoover-Fitzula, R.L., et al. (2004). Identification and inactivation of genetic loci involved with *Lactobacillus acidophilus* acid tolerance. *Appl Environ Microbiol* **70**, 5315–5322.

Azcarate-Peril, M.A., McAuliffe, O., Altermann, E., et al. (2005). Microarray analysis of a two-component regulatory system involved in acid resistance and proteolytic activity in *Lactobacillus acidophilus*. *Appl Environ Microbiol* **71**, 5794–5804.

Azcarate Peril, M.A., Tallon, R., and Klaenhammer, T.R. (2009). Temporal gene expression and probiotic attributes of *Lactobacillus acidophilus* during growth in milk. *J Dairy Sci* **92**, 870–886.

Backhed, F. and Ley, R.E., Sonnenburg, J.L., et al. (2005). Host-bacterial mutualism in the human intestine. *Science* **307**, 1915–1920.

Bae, S., Fleet, G.H., and Heard, G.M. (2006). Lactic acid bacteria associated with wine grapes from several Australian vineyards. *J Appl Microbiol* **100**, 712–727.

Barrangou, R., Fremaux, C., Deveau, H.,et al. (2007). CRISPR provides acquired resistance against viruses in prokaryotes. *Science* **315**, 1709–1712.

Bartnik, W. (2008). Clinical aspects of *Helicobacter pylori* infection. *Pol Arch Med Wewn* **118**, 426–430.

Bernet, M.F., Brassart, D., Neeser, J.R., and Servin, A. L. (1994). *Lactobacillus acidophilus* LA 1 binds to cultured human intestinal cell lines and inhibits cell attachment and cell invasion by enterovirulent bacteria. *Gut* **35**, 483–489.

Bhowmik, T. and Steele, J.L. (1994). Cloning, characterization and insertional inactivation of the *Lactobacillus helveticus* D(-) lactate dehydrogenase gene. *Appl Microbiol Biotechnol* **41**, 432–439.

Blom, J., Albaum, S.P., Doppmeier, D., et al. (2009). EDGAR: a software framework for the comparative analysis of prokaryotic genomes. *BMC Bioinformatics* **10**, 154.

Bolotin, A., Quinquis, B., Renault, P., et al. (2004). Complete sequence and comparative genome analysis of the dairy bacterium *Streptococcus thermophilus*. *Nat Biotechnol* **22**, 1554–1558.

Bolotin, A., Wincker, P., Mauger, S., et al. (2001). The complete genome sequence of the lactic acid bacterium *Lactococcus lactis* ssp. *lactis* IL1403. *Genome Res* **11**, 731–753.

Booijink, C.C., El-Aidy, S., Rajilic-Stojanovic, M., et al. (2010). High temporal and inter-individual variation detected in the human ileal microbiota. *Environ Microbiol* **12**, 3213–3227.

Braat, H., Rottiers, P., Hommes, D.W., et al. (2006). A phase I trial with transgenic bacteria expressing interleukin-10 in Crohn's disease. *Clin Gastroenterol Hepatol* **4**, 754–759.

Brown, C.T., Davis-Richardson, A.G., Giongo, A., et al. (2011). Gut microbiome metagenomics analysis suggests a functional model for the development of autoimmunity for type 1 diabetes. *PLoS One* **6**, e25792.

Bruno-Barcena, J.M., Andrus, J.M., Libby, S.L., et al. (2004). Expression of a heterologous manganese superoxide dismutase gene in intestinal lactobacilli provides protection against hydrogen peroxide toxicity. *Appl Environ Microbiol* **70**, 4702–4710.

Bruno-Barcena, J.M., Azcarate-Peril, M.A., Klaenhammer, T.R., and Hassan, H.M. (2005). Marker-free chromosomal integration of the manganese superoxide dismutase gene (sodA) from *Streptococcus thermophilus* into *Lactobacillus gasseri*. *FEMS Microbiol Lett* **246**, 91–101.

Buck, B.L., Azcarate-Peril, M.A., and Klaenhammer, T.R. (2009). Role of autoinducer-2 on the adhesion ability of *Lactobacillus acidophilus*. *J Appl Microbiol* **107**, 269–279.

Busconi, M., Reggi, S., and Fogher, C. (2008). Evaluation of biodiversity of lactic acid bacteria microbiota in the calf intestinal tracts. *Antonie Van Leeuwenhoek* **94**, 145–155.

Callanan, M., Kaleta, P., O'Callaghan, J., et al. (2008). Genome sequence of *Lactobacillus helveticus*, an organism distinguished by selective gene loss and insertion sequence element expansion. *J Bacteriol* **190**, 727–735.

Canchaya, C., Claesson, M.J., Fitzgerald, G.F., et al. (2006). Diversity of the genus Lactobacillus revealed by comparative genomics of five species. *Microbiology* **152**, 3185–3196.

Carr, F.J., Chill, D., and Maida, N. (2002). The lactic acid bacteria: a literature survey. *Crit Rev Microbiol* **28**, 281–370.

Carroll, I.M., Andrus, J.M., Bruno-Barcena, J.M., et al. (2007). Anti-inflammatory properties of *Lactobacillus gasseri* expressing manganese superoxide dismutase using the interleukin 10-deficient mouse model of colitis. *Am J Physiol Gastrointest Liver Physiol* **293**, G729–738.

Claesson, M.J., van Sinderen, D., and O'Toole, P.W. (2007). The genus *Lactobacillus* – a genomic basis for understanding its diversity. *FEMS Microbiol Lett* **269**, 22–28.

Claesson, M.J., van Sinderen, D., and O'Toole, P.W. (2008). *Lactobacillus* phylogenomics--towards a reclassification of the genus. *Int J Syst Evol Microbiol* **58**, 2945–2954.

Coburn, P.S., Baghdayan, A.S., Dolan, G.T., and Shankar, N. (2007). Horizontal transfer of virulence genes encoded on the *Enterococcus faecalis* pathogenicity island. *Mol Microbiol* **63**, 530–544.

Cogan, T.M., Beresford, T.P., Steele, J., et al. (2007). Invited review: Advances in starter cultures and cultured foods. *J Dairy Sci* **90**, 4005–4021.

Corr, S.C., Li, Y., Riedel, C.U., et al. (2007). Bacteriocin production as a mechanism for the antiinfective activity of *Lactobacillus salivarius* UCC118. *Proc Natl Acad Sci USA* **104**, 7617–7621.

Cotter, P.D., and Hill, C. (2003). Surviving the acid test: responses of gram-positive bacteria to low pH. *Microbiol Mol Biol Rev* **67**, 429–453, table of contents.

Daubin, V. and Ochman, H. (2004). Bacterial genomes as new gene homes: the genealogy of ORFans in E. coli. *Genome Res* **14**, 1036–1042.

de Las Rivas, B., Marcobal, A., and Munoz, R. (2004). Allelic diversity and population structure in *Oenococcus oeni* as determined from sequence analysis of housekeeping genes. *Appl Environ Microbiol* **70**, 7210–7219.

De Vuyst, L., Weckx, S., Ravyts, F., et al. (2011). New insights into the exopolysaccharide production of *Streptococcus thermophilus*. *International Dairy Journal* **21**, 586–591.

Deloger, M., El Karoui, M., and Petit, M.A. (2008). A genomic distance based on MUM indicates discontinuity between most Bacterial Species and Genera. *J Bacteriol.* **191**, 91–99.

Di Cagno, R., Surico, R.F., Siragusa, S.,et al. (2008). Selection and use of autochthonous mixed starter for lactic acid fermentation of carrots, French beans or marrows. *Int J Food Microbiol* **127**, 220–228.

Dodd, H.M., Horn, N., and Gasson, M.J. (1990). Analysis of the genetic determinant for production of the peptide antibiotic nisin. *J Gen Microbiol* **136**, 555–566.

Dodd, H.M., Horn, N., Hao, Z., and Gasson, M.J. (1992). A lactococcal expression system for engineered nisins. *Appl Environ Microbiol* **58**, 3683–3693.

Douglas, G.L., Goh, Y.J., and Klaenhammer, T.R. (2011). Integrative food grade expression system for lactic acid bacteria. *Methods Mol Biol* **765**, 373–387.

Douglas, G.L. and Klaenhammer, T.R. (2010). Genomic evolution of domesticated microorganisms. *Annu Rev Food Sci Technol* **1**, 397–414.

Douillard, F.P., Ribbera, A., Jarvinen, H.M., et al. (2013a). Comparative genomic and functional analysis of *Lactobacillus casei* and *Lactobacillus rhamnosus* strains marketed as probiotics. *Appl Environ Microbiol* **79**, 1923–1933.

Douillard, F. P., Ribbera, A., Kant, R., et al. (2013b). Comparative genomic and functional analysis of 100 *Lactobacillus rhamnosus* strains and their comparison with strain GG. *PLoS Genet* **9**, e1003683.

du Plessis, H.W., Dicks, L.M., Pretorius, I.S., et al. (2004). Identification of lactic acid bacteria isolated from South African brandy base wines. *Int J Food Microbiol* **91**, 19–29.

El-Sharoud, W.M. (2005). Two-component signal transduction systems as key players in stress responses of lactic acid bacteria. *Sci Prog* **88**, 203—228.

Faith, J.J., Guruge, J.L., Charbonneau, M., et al. (2013). The long-term stability of the human gut microbiota. *Science* **341**, 1237439.

FAO/WHO (2001). Health and nutritional properites of probiotics in food including powder milk with live lactic acid bcteria. FAO/WHO, Cordoba, Argentina.

Felis, G.E. and Dellaglio, F. (2007). Taxonomy of Lactobacilli and Bifidobacteria. *Curr Issues Intest Microbiol* **8**, 44–61.

Ferretti, J.J., McShan, W.M., Ajdic, D., et al. (2001). Complete genome sequence of an M1 strain of *Streptococcus pyogenes*. *Proc Natl Acad Sci USA* **98**, 4658–4663.

Fimland, G., Johnsen, L., Dalhus, B., and Nissen-Meyer, J. (2005). Pediocin-like antimicrobial peptides (class IIa bacteriocins) and their immunity proteins: biosynthesis, structure, and mode of action. *J Pept Sci* **11**, 688–696.

Fitzgerald, G.F., and Gasson, M.J. (1988). In vivo gene transfer systems and transposons. *Biochimie* **70**, 489–502.

Fleet, G.H. (2003). Yeast interactions and wine flavour. *Int J Food Microbiol* **86**, 11–22.

Fleischmann, R.D., Adams, M.D., White, O., et al. (1995). Whole-genome random sequencing and assembly of *Haemophilus influenzae* Rd. *Science* **269**, 496–512.

Frese, S.A., Benson, A.K., Tannock, G.W., et al.(2011). The evolution of host specialization in the vertebrate gut symbiont *Lactobacillus reuteri*. *PLoS Genet* **7**, e1001314.

Galvez, A., Lopez, R.L., Abriouel, H., et al. (2008). Application of bacteriocins in the control of foodborne pathogenic and spoilage bacteria. *Crit Rev Biotechnol* **28**, 125–152.

Garrigues, C., Johansen, E., and Crittenden, R. (2013). Pangenomics--an avenue to improved industrial starter cultures and probiotics. *Curr Opin Biotechnol* **24**, 187–191.

Gilliland, S.E., Speck, M.L., and Morgan, C.G. (1975). Detection of *Lactobacillus acidophilus* in feces of humans, pigs, and chickens. *Appl Microbiol* **30**, 541–545.

Goh, Y.J., Azcarate-Peril, M.A., O'Flaherty, S., et al. (2009). Development and application of a upp-based counterselective gene replacement system for the study of the S-layer protein SlpX of *Lactobacillus acidophilus* NCFM. *Appl Environ Microbiol* **75**, 3093–3105.

Greenblum, S., Turnbaugh, P.J., and Borenstein, E. (2012). Metagenomic systems biology of the human gut microbiome reveals topological shifts associated with obesity and inflammatory bowel disease. *Proc Natl Acad Sci USA* **109**, 594–599.

Gueimonde, M. and Salminen, S. (2006). New methods for selecting and evaluating probiotics. *Dig Liver Dis* **38** Suppl 2, S242-7.

Halici, S., Zenk, S.F., Jantsch, J., and Hensel, M. (2008). Functional analysis of the *Salmonella* pathogenicity island 2-mediated inhibition of antigen presentation in dendritic cells. *Infect Immun* **76**, 4924–4933.

Harris, L. (1998). The microbiology of vegetable fermentations. In: *Microbiology of Fermented Foods*" ((ed.) B. J. B. Wood), 2nd edn, pp. 46–72. Blackie Academic and Professional, London.

Hill, C. (2012). Virulence or niche factors: what's in a name? *J Bacteriol* **194**, 5725-7.

Holzapfel, W.H., Haberer, P., Geisen, R., et al. (2001). Taxonomy and important features of probiotic microorganisms in food and nutrition. *Am J Clin Nutr* **73**, 365S–373S.

Horvath, P., Coute-Monvoisin, A.C., Romero, D.A., et al. (2009). Comparative analysis of CRISPR loci in lactic acid bacteria genomes. *Int J Food Microbiol* **131**, 62–70.

Hoskins, J., Alborn, W.E., Jr., Arnold, J., et al. (2001). Genome of the bacterium *Streptococcus pneumoniae* strain R6. *J Bacteriol* **183**, 5709–5717.

Hugas, M., Garriga, M., and Aymerich, M.T. (2003). Functionality of enterococci in meat products. *Int J Food Microbiol* **88**, 223–233.

Iordanescu, S. and Surdeanu, M. (1980). New incompatibility groups of *Staphylococcus aureus* plasmids. *Plasmid* **4**, 256–260.

Jansen, R., Embden, J.D., Gaastra, W., and Schouls, L.M. (2002). Identification of genes that are associated with DNA repeats in prokaryotes. *Mol Microbiol* **43**, 1565–1575.

Jayaraman, A. and Wood, T.K. (2008). Bacterial quorum sensing: signals, circuits, and implications for biofilms and disease. *Annu Rev Biomed Eng* **10**, 145–167.

Kakimoto, S., Sumino, Y., Kawahara, K., et al. (1990). Purification and characterization of acid urease from *Lactobacillus fermentum*. *Appl Microbiol Biotechnol* **32**, 538–543.

Kankainen, M., Paulin, L., Tynkkynen, S., et al. (2009). Comparative genomic analysis of *Lactobacillus rhamnosus* GG reveals pili containing a human- mucus binding protein. *Proc Natl Acad Sci USA* **106**, 17193–17198.

Kant, R., Blom, J., Palva, A., et al. (2011). Comparative genomics of *Lactobacillus*. *Microb Biotechnol* **4**, 323–332.

Kaper, J.B. and Sperandio, V. (2005). Bacterial cell-to-cell signaling in the gastrointestinal tract. *Infect Immun* **73**, 3197–3209.

Kiefer, F., Arnold, K., Kunzli, M., et al. (2009). The SWISS-MODEL Repository and associated resources. *Nucleic Acids Res* **37**, D387–392.

Kim, J.B., Porreca, G.J., Song, L., et al. (2007a). Polony multiplex analysis of gene expression (PMAGE) in mouse hypertrophic cardiomyopathy. *Science* **316**, 1481–1484.

Kim, J.E., Hur, H.J., Lee, K.W., and Lee, H.J. (2007b). Anti-inflammatory effects of recombinant arginine deiminase originating from *Lactococcus lactis* ssp. *lactis* ATCC 7962. *J Microbiol Biotechnol* **17**, 1491–1497.

Klaenhammer, T.R. (1988). Bacteriocins of lactic acid bacteria. *Biochimie* **70**, 337–349.

Klaenhammer, T.R., Barrangou, R., Buck, B.L., et al. (2005). Genomic features of lactic acid bacteria effecting bioprocessing and health. *FEMS Microbiol Rev* **29**, 393–409.

Kleerebezem, M., Boekhorst, J., van Kranenburg, R., et al. (2003). Complete genome sequence of *Lactobacillus plantarum* WCFS1. *Proc Natl Acad Sci USA* **100**, 1990–1995.

Kleerebezem, M., Hols, P., Bernard, E.,et al. (2010). The extracellular biology of the lactobacilli. *FEMS Microbiol Rev* **34**, 199–230.

Klein, G., Pack, A., Bonaparte, C., and Reuter, G. (1998). Taxonomy and physiology of probiotic lactic acid bacteria. *Int J Food Microbiol* **41**, 103–125.

Kok, J., Buist, G., Zomer, A.L., et al. (2005). Comparative and functional genomics of lactococci. *FEMS Microbiol Rev* **29**, 411–433.

Konstantinidis, K.T., Ramette, A., and Tiedje, J.M. (2006). The bacterial species definition in the genomic era. *Philos Trans R Soc Lond B Biol Sci* **361**, 1929–1940.

Kurtz, S., Phillippy, A., Delcher, A.L., et al. (2004). Versatile and open software for comparing large genomes. *Genome Biol* **5**, R12.

Lambert, J.M., Bongers, R.S., and Kleerebezem, M. (2007). Cre-lox-based system for multiple gene deletions and selectable-marker removal in *Lactobacillus plantarum*. *Appl Environ Microbiol* **73**, 1126–1135.

Larsen, N., Vogensen, F.K., van den Berg, F.W., et al. (2010). Gut microbiota in human adults with type 2 diabetes differs from non-diabetic adults. *PLoS One* **5**, e9085.

Laukova, A., Strompfova, V., Simonova, M., et al. (2008). Species of *Enterococcus faecalis* associated with free-living rodents. *Folia Microbiol (Praha)* **53**, 173–178.
Laursen, B.G., Bay, L., Cleenwerck, I., et al. (2005). *Carnobacterium divergens* and *Carnobacterium maltaromaticum* as spoilers or protective cultures in meat and seafood: phenotypic and genotypic characterization. *Syst Appl Microbiol* **28**, 151–164.
Lebeer, S., Claes, I., Tytgat, H.L., et al. (2012). Functional analysis of *Lactobacillus rhamnosus* GG pili in relation to adhesion and immunomodulatory interactions with intestinal epithelial cells. *Appl Environ Microbiol* **78**, 185–193.
Lebeer, S., De Keersmaecker, S.C., Verhoeven, T.L., et al. (2007). Functional analysis of luxS in the probiotic strain *Lactobacillus rhamnosus* GG reveals a central metabolic role important for growth and biofilm formation. *J Bacteriol* **189**, 860–871.
Levesque, C., Duplessis, M., Labonte, J., et al. (2005). Genomic organization and molecular analysis of virulent bacteriophage 2972 infecting an exopolysaccharide-producing *Streptococcus thermophilus* strain. *Appl Environ Microbiol* **71**, 4057–4068.
Leyer, G.J., Li, S., Mubasher, M.E., et al. (2009). Probiotic effects on cold and influenza-like symptom incidence and duration in children. *Pediatrics* **124**, e172–179.
Li, H. and Homer, N. (2010). A survey of sequence alignment algorithms for next-generation sequencing. *Brief Bioinform* **11**, 473–483.
Lih-Brody, L., Powell, S.R., Collier, K.P., et al. (1996). Increased oxidative stress and decreased antioxidant defenses in mucosa of inflammatory bowel disease. *Dig Dis Sci* **41**, 2078–2086.
Loman, N.J., Misra, R.V., Dallman, T.J., et al. (2012). Performance comparison of benchtop high-throughput sequencing platforms. *Nat Biotechnol* **30**, 434–439.
Ludwig, W. and Schleifer, K. H. (1994). Bacterial phylogeny based on 16S and 23S rRNA sequence analysis. *FEMS Microbiol Rev* **15**, 155–173.
Lukjancenko, O., Ussery, D.W., and Wassenaar, T.M. (2012). Comparative genomics of *Bifidobacterium*, *Lactobacillus* and related probiotic genera. *Microb Ecol* **63**, 651–673.
Lukjancenko, O., Wassenaar, T.M., and Ussery, D.W. (2010). Comparison of 61 sequenced *Escherichia coli* genomes. *Microb Ecol* **60**, 708–720.
Makarova, K., Slesarev, A., Wolf, Y., et al. (2006). Comparative genomics of the lactic acid bacteria. *Proc Natl Acad Sci USA* **103**, 15611–15616.
Makarova, K.S. and Koonin, E.V. (2007). Evolutionary genomics of lactic acid bacteria. *J Bacteriol* **189**, 1199–1208.
Mardis, E.R. (2013). Next-generation sequencing platforms. *Annu Rev Anal Chem (Palo Alto Calif)* **6**, 287–303.
Mennigen, R., Nolte, K., Rijcken, E., et al. (2009). Probiotic mixture VSL#3 protects the epithelial barrier by maintaining tight junction protein expression and preventing apoptosis in a murine model of colitis. *Am J Physiol Gastrointest Liver Physiol* **296**, G1140–1149.
Metzker, M.L. (2010). Sequencing technologies - the next generation. *Nat Rev Genet* **11**, 31–46.
Miele, E., Pascarella, F., Giannetti, E., Quaglietta, L., et al. (2009). Effect of a probiotic preparation (VSL#3) on induction and maintenance of remission in children with ulcerative colitis. *Am J Gastroenterol* **104**, 437–443.
Mills, D.A., Rawsthorne, H., Parker, C., et al. (2005). Genomic analysis of *Oenococcus oeni* PSU-1 and its relevance to winemaking. *FEMS Microbiol Rev* **29**, 465–475.
Miyoshi, A., Rochat, T., Gratadoux, J.J., et al. (2003). Oxidative stress in *Lactococcus lactis*. *Genet Mol Res* **2**, 348–359.
Mobley, H.L., Island, M.D., and Hausinger, R.P. (1995). Molecular biology of microbial ureases. *Microbiol Rev* **59**, 451–480.
Mohamadzadeh, M., Duong, T., Sandwick, S.J., et al. (2009). Dendritic cell targeting of *Bacillus anthracis* protective antigen expressed by *Lactobacillus acidophilus* protects mice from lethal challenge. *Proc Natl Acad Sci USA* **106**, 4331–4336.
Mohamadzadeh, M., Durmaz, E., Zadeh, M., et al. (2010). Targeted expression of anthrax protective antigen by *Lactobacillus gasseri* as an anthrax vaccine. *Future Microbiol* **5**, 1289–1296.
Morello, E., Bermudez-Humaran, L.G., Llull, D., et al. (2008). *Lactococcus lactis*, an efficient cell factory for recombinant protein production and secretion. *J Mol Microbiol Biotechnol* **14**, 48–58.
Moreno-Arribas, M.V. and Polo, M. C. (2005). Winemaking biochemistry and microbiology: current knowledge and future trends. *Crit Rev Food Sci Nutr* **45**, 265–286.
Morishita, T., Fukada, T., Shirota, M., and Yura, T. (1974). Genetic basis of nutritional requirements in *Lactobacillus casei*. *J Bacteriol* **120**, 1078–1084.
Morita, H., Toh, H., Fukuda, S., et al. (2008). Comparative genome analysis of *Lactobacillus reuteri* and *Lactobacillus fermentum* reveal a genomic island for reuterin and cobalamin production. *DNA Res* **15**, 151–161.

Nath, S. (2002). The molecular mechanism of ATP synthesis by F1F0-ATP synthase: a scrutiny of the major possibilities. *Adv Biochem Eng Biotechnol* **74**, 65–98.

Navarro, L., Rojo-Bezares, B., Saenz, Y., et al. (2008). Comparative study of the pln locus of the quorum-sensing regulated bacteriocin-producing *L. plantarum* J51 strain. *Int J Food Microbiol* **128**, 390–394.

Nazef, L., Belguesmia, Y., Tani, A., et al. (2008). Identification of lactic acid bacteria from poultry feces: evidence on anti-campylobacter and anti-listeria activities. *Poult Sci* **87**, 329–334.

Nelson, K.E., Weinstock, G.M., Highlander, S.K., et al. (2010). A catalog of reference genomes from the human microbiome. *Science* **328**, 994–9999.

Nyquist, O.L., McLeod, A., Brede, D.A., et al. (2011). Comparative genomics of *Lactobacillus sakei* with emphasis on strains from meat. *Mol Genet Genomics* **285**, 297–311.

O'Flaherty, S. and Klaenhammer, T. (2010). The role and potential of probiotic bacteria in the gut, and the communication between gut microflora and gut/host. *International Dairy Journal* **20**, 262–268.

O'Hara, A.M., and Shanahan, F. (2006). The gut flora as a forgotten organ. *EMBO Rep* **7**, 688–693.

O'Shea, E. F., Cotter, P.D., Ross, R.P., and Hill, C. (2013). Strategies to improve the bacteriocin protection provided by lactic acid bacteria. *Curr Op Biotech* **24**, 130–134.

O'Sullivan, D.J. and Klaenhammer, T.R. (1993). High- and low-copy-number *Lactococcus* shuttle cloning vectors with features for clone screening. *Gene* **137**, 227–231.

O'Sullivan, O., O'Callaghan, J., Sangrador-Vegas, A., et al. (2009). Comparative genomics of lactic acid bacteria reveals a niche-specific gene set. *BMC Microbiol* **9**, 50.

Ohland, C.L. and Macnaughton, W.K. (2010). Probiotic bacteria and intestinal epithelial barrier function. *Am J Physiol Gastrointest Liver Physiol* **298**, G807–819.

Ouellette, A.J. (2004). Defensin-mediated innate immunity in the small intestine. *Best Pract Res Clin Gastroenterol* **18**, 405–419.

Petersen, F.C., Fimland, G., and Scheie, A.A. (2006). Purification and functional studies of a potent modified quorum-sensing peptide and a two-peptide bacteriocin in *Streptococcus mutans*. *Mol Microbiol* **61**, 1322–1334.

Picozzi, C., Bonacina, G., Vigentini, I., and Foschino, R. (2010). Genetic diversity in Italian *Lactobacillus sanfranciscensis* strains assessed by multilocus sequence typing and pulsed-field gel electrophoresis analyses. *Microbiology* **156**, 2035–2045.

Pilone, G.J., Kunkee, R.E., and Webb, A.D. (1966). Chemical characterization of wines fermented with various malo-lactic bacteria. *Appl Microbiol* **14**, 608–615.

Pons, N., Batto, J.M., Ehrlich, S.D., and Renault, P. (2008). Development of software facilities to characterize regulatory binding motifs and application to streptococcaceae. *J Mol Microbiol Biotechnol* **14**, 67–73.

Preidis, G.A., and Versalovic, J. (2009). Targeting the human microbiome with antibiotics, probiotics, and prebiotics: gastroenterology enters the metagenomics era. *Gastroenterology* **136**, 2015–2031.

Pridmore, R.D., Berger, B., Desiere, F., et al. (2004). The genome sequence of the probiotic intestinal bacterium *Lactobacillus johnsonii* NCC 533. *Proc Natl Acad Sci USA* **101**, 2512–2517.

Rantsiou, K. and Cocolin, L. (2006). New developments in the study of the microbiota of naturally fermented sausages as determined by molecular methods: a review. *Int J Food Microbiol* **108**, 255–267.

Rantsiou, K., Drosinos, E.H., Gialitaki, M., et al. (2006). Use of molecular tools to characterize *Lactobacillus* spp. isolated from Greek traditional fermented sausages. *Int J Food Microbiol* **112**, 215–222.

Rasmussen, T.B., Danielsen, M., Valina, O., et al. (2008). *Streptococcus thermophilus* core genome: comparative genome hybridization study of 47 strains. *Appl Environ Microbiol* **74**, 4703-10.

Reuter, G. (2001). The *Lactobacillus* and *Bifidobacterium* microflora of the human intestine: composition and succession. *Curr Issues Intest Microbiol* **2**, 43–53.

Rihakova, J., Belguesmia, Y., Petit, V.W., et al. (2008). Divercin V41 from gene characterization to food applications: 1998-2008, a decade of solved and unsolved questions. *Lett Appl Microbiol* **48**, 1–7.

Rochat, T., Gratadoux, J.J., Gruss, A.,et al. (2006). Production of a heterologous nonheme catalase by *Lactobacillus casei*: an efficient tool for removal of H2O2 and protection of *Lactobacillus bulgaricus* from oxidative stress in milk. *Appl Environ Microbiol* **72**, 5143–5149.

Rossello-Mora, R. and Amann, R. (2001). The species concept for prokaryotes. *FEMS Microbiol Rev* **25**, 39–67.

Russell, W.M. and Klaenhammer, T.R. (2001). Efficient system for directed integration into the *Lactobacillus acidophilus* and *Lactobacillus gasseri* chromosomes via homologous recombination. *Appl Environ Microbiol* **67**, 4361–4364.

Salminen, S. and Deighton, M. (1992). Lactic acid bacteria in the gut in normal and disordered states. *Dig Dis* **10**, 227–238.

Sanders, J.W., Leenhouts, K., Burghoorn, J., et al. (1998). A chloride-inducible acid resistance mechanism in *Lactococcus lactis* and its regulation. *Mol Microbiol* **27**, 299–310.

Sanderson, I.R. (1999). The physicochemical environment of the neonatal intestine. *Am J Clin Nutr* **69**, 1028S–1034S.

Sartor, R.B. (2004). Therapeutic manipulation of the enteric microflora in inflammatory bowel diseases: antibiotics, probiotics, and prebiotics. *Gastroenterology* **126**, 1620–1633.

Settanni, L. and Corsetti, A. (2008). Application of bacteriocins in vegetable food biopreservation. *Int J Food Microbiol* **121**, 123–138.

Shendure, J. and Ji, H. (2008). Next-generation DNA sequencing. *Nat Biotechnol* **26**, 1135–1145.

Sieuwerts, S., de Bok, F.A., Hugenholtz, J., and van Hylckama Vlieg, J.E. (2008). Unraveling microbial interactions in food fermentations: from classical to genomics approaches. *Appl Environ Microbiol* **74**, 4997–5007.

Sifri, C.D. (2008). Healthcare epidemiology: quorum sensing: bacteria talk sense. *Clin Infect Dis* **47**, 1070–1076.

Silva, M., Jacobus, N. V., Deneke, C., and Gorbach, S. L. (1987). Antimicrobial substance from a human *Lactobacillus* strain. *Antimicrob Agents Chemother* **31**, 1231–1233.

Skaugen, M. and Nes, I. F. (1994). Transposition in *Lactobacillus sake* and its abolition of lactocin S production by insertion of IS1163, a new member of the IS3 family. *Appl Environ Microbiol* **60**, 2818–2825.

Smeianov, V.V., Wechter, P., Broadbent, J.R., et al. (2007). Comparative high-density microarray analysis of gene expression during growth of *Lactobacillus helveticus* in milk versus rich culture medium. *Appl Environ Microbiol* **73**, 2661–2672.

Stanislawska-Sachadyn, A. and Sachadyn, P. (2005). MutS as a tool for mutation detection. *Acta Biochim Pol* **52**, 575–583.

Steidler, L., Hans, W., Schotte, L., et al. (2000). Treatment of murine colitis by *Lactococcus lactis* secreting interleukin-10. *Science* **289**, 1352–1355.

Stolaki, M., De Vos, W.M., Kleerebezem, M., and Zoetendal, E.G. (2011). Lactic acid bacteria in the gut. In: *Lactic Acid Bacteria: Microbiological and Functional Aspects* ((eds) S. Lahtinen, A. C. Ouwehand, S. Salminen and A. von Wright), 4the edn, CRC Press, Boca Raton, FL.

Straume, D., Kjos, M., Nes, I.F., and Diep, D.B. (2007). Quorum-sensing based bacteriocin production is down-regulated by N-terminally truncated species of gene activators. *Mol Genet Genomics* **278**, 283–293.

Sztajer, H., Lemme, A., Vilchez, R., et al. (2007). Autoinducer-2 regulated genes in *Streptococcus mutans* UA159 and global metabolic effect of the luxS mutation. *J Bacteriol* **190**, 401–415.

Talarico, T.L. and Dobrogosz, W J. (1989). Chemical characterization of an antimicrobial substance produced by *Lactobacillus reuteri*. *Antimicrob Agents Chemother* **33**, 674–69.

Tanasupawat, S., Sukontasing, S., and Lee, J.S. (2008). *Enterococcus thailandicus* sp. nov., isolated from fermented sausage ('mum') in Thailand. *Int J Syst Evol Microbiol* **58**, 1630–1634.

Taranto, M.P., Vera, J.L., Hugenholtz, J., et al. (2003). *Lactobacillus reuteri* CRL1098 produces cobalamin. *J Bacteriol* **185**, 5643–5647.

Tettelin, H., Nelson, K.E., Paulsen, I.T., et al. (2001). Complete genome sequence of a virulent isolate of *Streptococcus pneumoniae*. *Science* **293**, 498–506.

Teusink, B., van Enckevort, F.H., Francke, C., et al. (2005). In silico reconstruction of the metabolic pathways of *Lactobacillus plantarum*: comparing predictions of nutrient requirements with those from growth experiments. *Appl Environ Microbiol* **71**, 7253–7262.

Tramonti, A., De Canio, M., Delany, I., et al. (2006). Mechanisms of transcription activation exerted by GadX and GadW at the gadA and gadBC gene promoters of the glutamate-based acid resistance system in *Escherichia coli*. *J Bacteriol* **188**, 8118–8127.

Turnbaugh, P.J., Hamady, M., Yatsunenko, T., et al. (2009). A core gut microbiome in obese and lean twins. *Nature* **457**, 480–484.

Ussery, D.W., Kiil, K., Lagesen, K., et al. (2009). The genus burkholderia: analysis of 56 genomic sequences. *Genome Dyn* **6**, 140–157.

van der Ploeg, J.R. (2005). Regulation of bacteriocin production in *Streptococcus mutans* by the quorum-sensing system required for development of genetic competence. *J Bacteriol* **187**, 3980–3989.

van Hijum, S.A., Vaughan, E.E., and Vogel, R. F. (2013). Application of state-of-art sequencing technologies to indigenous food fermentations. *Curr Opin Biotechnol* **24**, 178–186.

Vandamme, P., Pot, B., Gillis, M., et al. (1996). Polyphasic taxonomy, a consensus approach to bacterial systematics. *Microbiol Rev* **60**, 407–438.

Vendeville, A., Winzer, K., Heurlier, K., et al. (2005). Making 'sense' of metabolism: autoinducer-2, LuxS and pathogenic bacteria. *Nat Rev Microbiol* **3**, 383–396.

Ventura, M., O'Flaherty, S., Claesson, M.J., et al. (2009). Genome-scale analyses of health-promoting bacteria: probiogenomics. *Nat Rev Microbiol* **7**, 61–71.

Vesth, T., Wassenaar, T.M., Hallin, P.F., et al. (2010). On the origins of a *Vibrio* species. *Microb Ecol* **59**, 1–13.

Waagmeester, A.S., Kelder, T., and Evelo, C.T. (2008). The role of bioinformatics in pathway curation. *Genes Nutr* **3**, 139–142.

Walter, J. (2008). Ecological role of lactobacilli in the gastrointestinal tract: implications for fundamental and biomedical research. *Appl Environ Microbiol* **74**, 4985–4996.

Waterman, S.R., and Small, P.L. (2003). Identification of the promoter regions and sigma(s)-dependent regulation of the gadA and gadBC genes associated with glutamate-dependent acid resistance in *Shigella flexneri*. *FEMS Microbiol Lett* **225**, 155–160.

Wegmann, U., O'Connell-Motherway, M., Zomer, A., et al. (2007). Complete genome sequence of the prototype lactic acid bacterium *Lactococcus lactis* subsp. *cremoris* MG1363. *J Bacteriol* **189**, 3256–3270.

Wells, J.M., and Mercenier, A. (2008). Mucosal delivery of therapeutic and prophylactic molecules using lactic acid bacteria. *Nat Rev Microbiol* **6**, 349–362.

Willcox, M.D., Zhu, H., Conibear, T.C., et al. (2008). Role of quorum sensing by *Pseudomonas aeruginosa* in microbial keratitis and cystic fibrosis. *Microbiology* **154**, 2184–2194.

Xavier, K.B. and Bassler, B.L. (2003). LuxS quorum sensing: more than just a numbers game. *Curr Opin Microbiol* **6**, 191–197.

Zhou, M., Theunissen, D., Wels, M., and Siezen, R.J. (2010). LAB-Secretome: a genome-scale comparative analysis of the predicted extracellular and surface-associated proteins of lactic acid bacteria. *BMC Genomics* **11**, 651.

Chapter 4

Lactic Acid Bacteria: Comparative Genomic Analyses of Transport Systems

Graciela L. Lorca[1,*], Taylor A. Twiddy[1], and Milton H. Saier, Jr.[2]

[1] *Department of Microbiology and Cell Science, Genetics Institute and Institute of Food and Agricultural Sciences, University of Florida, Gainesville, USA*
[2] *Division of Biological Sciences, University of California at San Diego, San Diego, USA*

4.1. Introduction

Molecular phylogeny has revealed the existence of three domains of life: bacteria, archaea, and eukaryotes (Olsen et al. 1994; Rosselló-Mora 2005). The former two domains include all prokaryotic life forms with small cells and a lack of eukaryotic organelles. But prokaryotic organisms display a huge diversity of cellular activities including many metabolic capabilities that cannot be found in eukaryotes (Guerrero and Berlanga 2006). There are far more bacterial kingdoms than archaeal or eukaryotic kingdoms, probably reflecting their role as the primordial, ancestral cell type that gave rise to the other two.

One bacterial kingdom consists of the firmicutes or low G+C Gram (+) bacteria. They include spore-forming *Bacillus* and *Clostridium* species, disease-causing *Staphylococcus*, *Listeria*, and *Streptococcus* species and a group of organisms collectively known as the lactic acid bacteria (LAB). These last mentioned organisms are uniquely of industrial importance because of their roles in the fermentation of fruits, vegetables, and meats as well as their use in the dairy industry for cheese and yogurt production (Hayes et al. 2007; Sanz et al. 2007; Reddy et al. 2008). However, these bacteria also function as producers (but also cause spoilage) of wines and beer (Coton et al. 1998; Lonvaud-Funel 1999; Rouse and van Sinderen 2008; Justé et al. 2014). *Lactobacillus* is one of the main bacteria used as probiotics for humans and animals because of its prophylactic or therapeutic properties (Hayes et al. 2007; Sanz et al. 2007; Shavakhi et al. 2014). They provide sources of biopolymers and other chemicals (i.e., bioactive peptides, short chain acids, vitamins) which have prebiotic activity by promoting the growth of other beneficial commensal gut bacteria (i.e., *Bifidobacterium*) (Hayes et al. 2007; de Vrese and Schrezenmeir 2008; van Zanten et al. 2014).

LAB all have remarkable genetically encoded synthetic capabilities including the production of excreted bacteriocins together with immunity devices that render them insensitive to their own peptide-toxins. The secretion of these antimicrobial peptides is used by lactic acid bacteria for biological warfare and cell communication (Eijsink et al. 2002).

* Corresponding author: Department of Microbiology and Cell Science, Genetics Institute and Institute of Food and Agricultural Sciences, University of Florida, Gainesville, FL 32608, USA. glorca@ufl.edu

Biotechnology of Lactic Acid Bacteria: Novel Applications, Second Edition. Edited by Fernanda Mozzi, Raúl R. Raya, and Graciela M. Vignolo.

The LAB group is remarkably diverse in their habitats, which are well reflected by their metabolic and transport capabilities. These organisms include the closely related *Lactobacillus casei* and *Lact. brevis*, *Pediococcus pentosaceus*, as well as the more distant *Lact. gasseri* and *Lact. delbrueckii* which are often found in plants and their products as well as natural inhabitants of the human gut (Vaughan et al. 2002; Makarova et al. 2006; Lorca et al. 2007). These organisms are used by man for the fermentation of plant, dairy, and meat products. Two other LAB, *Lactococcus lactis* subspecies *cremoris* and *Streptococcus thermophilus*, distantly related to the other LAB but closely related to each other, are both found in milk and are useful for dairy product fermentations. Finally, two additional LAB, closely related to each other but more distantly related to the others, are *Leuconostoc mesenteroides*, useful for food fermentation purposes, and *Oenococcus oeni*, important in wine production (Hastings et al. 1991; Mills et al. 2005).

Certain other bacteria, including high G+C Gram (+) actinobacteria, can provide similar useful functions. *Bifidobacterium longum* is found in the intestinal tracts of animals including humans, and is an important probiotic organism that, together with other bifidobacteria, dominates the intestinal tracts of infants that are breast fed but not of those that are fed formula (cow milk). Bifidobacterial species are believed to be responsible in part for the greater health of breast-fed children (Parracho et al. 2007; Sela and Mills 2014). *Brevibacterium linens* is another actinobacterium used for cheese ripening as well as vitamin and carotenoid production (Rattray and Fox 1999).

Transmembrane transport is essential to virtually all of the useful functions provided by LAB, and the presence of these transporters must reflect the evolutionary histories of the organisms in which they are found (Ren and Paulsen 2005). In earlier work we summarized some of the unusual complement of transporters encoded within LAB genomes, sequenced by the LAB genome consortium (Markarova et al. 2006; Lorca et al. 2007, 2010). In this chapter we report analyses of the distributions of the various transporters found in 11 LAB, their associations with the different transporter families, and their functional significance. We show that LAB have most unusual distributions of transporters, compared to other bacterial types, including actinobacteria, Gram (–) bacteria, and archaea, and we provide potential evolutionary explanations.

4.2. Channel-forming Proteins

As summarized in Lorca et al. (2007), LAB have a small percentage of their transport proteins as channel proteins (1.7–4.1%). Here, we highlight new findings regarding their functional role.

The MIP (major intrinsic protein; TC# 1.A.8 in the Transporter Classification Database (TCDB; www.tcdb.org)) family of aquaporins and glycerol facilitators were found in the 11 bacteria originally studied. Proteins in this family facilitate the passage of several neutral molecules such as water, glycerol, dihydroxyacetone, CO_2, urea, and ammonia (Jensen et al. 2002; Carbrey et al. 2003; Uehlein et al. 2003). All the microorganisms analyzed have at least one member of this family while some have up to five members (Table 4.1).

The first member to be functionally characterized was Gla_Llac from *L. lactis* and it was found to facilitate permeability to water and glycerol (Froger et al. 2001). Lactic acid bacteria in general cannot utilize glycerol as a sole carbon source (Alvarez et al. 2004) but some species are able to co-metabolize glucose and glycerol (Lüthi-Peng et al. 2002). These phenotypic data suggest that these channels may be involved in facilitating the diffusion or permeability to a broader variety of compounds than originally demonstrated in *in vitro* experiments.

A recent functional study aimed at identifying substrates for members of the MIP was performed using *Lact. plantarum* where six paralogs were identified in its genome (Bienert et al. 2013). A combination of *in silico* predictions and *in vivo* testing was used to identify substrates for four out of six members present (summarized in Table 4.1). They were found to facilitate the movement of water

Table 4.1. Summary of functionally characterized aquaporins and glycerol facilitators in LAB.

Protein	Microorganism	Substrate	Reference
GlaLlac	*L. lactis*	Permeability to water and glycerol	Froger et al. 2001
LpGlpF1	*Lac. plantarum*	Permeability to: H_2O_2 Facilitate urea and lactic acid diffusion	Bienert et al. 2013
LpGlpF2	*Lact. plantarum*	Permeability to: Water, glycerol, dihydroxyacetone, H_2O_2	Bienert et al. 2013
LpGlpF3	*Lact. plantarum*	Permeability to: Water, glycerol, dihydroxyacetone, H_2O_2	Bienert et al. 2013
LpGlpF4	*Lact. plantarum*	Permeability to: Water, glycerol, dihydroxyacetone, H_2O_2, and lactic acid Facilitate urea and lactic acid diffusion	Bienert et al. 2013
LpGlpF5	*Lact. plantarum*	ND	Bienert et al. 2013
LpGlpF6	*Lact. plantarum*	ND	Bienert et al. 2013
LsGlpF1	*Lact. sakei*	Permeability to: Water, glycerol, and lactic acid	Bienert et al. 2013
PpGlpF1	*P. pentosaceus*	Permeability to: Water, glycerol, and lactic acid	Bienert et al. 2013

ND, not determined; not involved in permeability of diffusion of water, glycerol, dihydroxyacetone, lactic acid or H_2O_2

(LpGlpF2, LpGlpF3, LpGlpF4), glycerol (LpGlpF2, LpGlpF3, LpGlpF4), dihydroxyacetone (LpGlpF2, LpGlpF3, LpGlpF4), and urea (LpGlpF1, LpGlpF4). Interestingly, they were involved in H_2O_2 (LpGlpF1, LpGlpF2, LpGlpF3), and lactic acid movement (LpGlpF1, LpGlpF4). The involvement of aquaporins in lactic acid metabolism was the first description of this function in prokaryotes. MIP channels that diffuse lactic acid were found previously in humans (AQP9), *Arabidopsis* (NIP2), and trematodes (SmAQP) (Tsukaguchi et al. 1999; Choi and Roberts 2007; Faghiri et al. 2010). The involvement of aquaporins in the movement of lactic acid is most relevant to these group bacteria since acid stress induced by lactic acid is the main factor associated with the transition into stationary phase of growth under excess of carbon source.

Pore-Forming Toxins (Proteins and Peptides; TC# 1.C.) represent a group of proteins that have considerably expanded over the last few years within the lactic acid bacteria group (Table 4.2). The production of bacteriocins is a very desirable trait with biotechnological applications. Bacteriocins are bacterially produced peptide antibiotics with the ability to kill a limited range of bacteria, usually but not always those that are closely related to the producer bacterium.

They are usually synthesized as small precursor proteins or peptides that are processed with proteolytic elimination of their *N*-terminal leader sequences, and the resultant mature peptides form one, two, or more putative amphipathic transmembrane α-helical spanners (TMSs). Many bacteriocins are encoded in operons that also encode an immunity protein and an ABC transport system (TC #3.A.1) with a protease domain at the *N*-terminus. The ABC systems export the bacteriocins while the protease domains cleave the *N*-terminal leader sequence. A few bacteriocins are exported by the type II general secretory pathway rather than by ABC-type export systems. In some cases, expression of the bacteriocin-encoding operon is induced by a bacteriocin-like peptide which acts in conjunction with a two component sensor kinase-response regulator.

Peptide bacteriocins produced by lactic acid bacteria are categorized into two different classes according to their biochemical and genetic properties (Drider et al. 2006; Nes et al. 2007). Class I peptides are the lantibiotics, which are small, post-translationally modified peptides that contain unusual amino acids such as lanthionine (TC# 1.C.20). Class II includes unmodified bacteriocins that are subdivided into three subclasses, namely, class IIa (pediocin-like bacteriocins), class IIb (two-peptide bacteriocins), and IIc (other (i.e., non-pediocin-like), one-peptide bacteriocins).

Table 4.2. Summary of functionally characterized Pore-Forming Toxins in LAB.

TCID	Name	Source	Reference
1.C.20.1.1	Class I lantibiotic bacteriocin Nisin precursor	*L. lactis*	Buchman et al. 1988
1.C.21.1.1	Class I lantibiotic bacteriocin Lacticin 481	*L. lactis*	Piard et al. 1993
1.C.22.1.1	Class I lantibiotic bacteriocin Lactococcin A	*L. lactis*	Stoddard et al. 1992
1.C.22.1.2	Thiol-activated peptide Lactococcin B	*L. lactis*	van Belkum et al. 1992
1.C.22.1.5	Curvaticin FS47	*Lact. curvatus*	Garver and Muriana, 1994
1.C.22.1.6	Plantaricin A precursor	*Lact. plantarum*	Nissen-Meyer et al. 1993
1.C.22.1.8	Amylovorin L471 (Lactobin A)	*Lact. amylvorus*	Callewaert et al. 1999
1.C.23.1.1	Class I lantibiotic bacteriocin Lactocin S	*Lact. sake* L45	Mørtvedt et al. 1991
1.C.24.1.1	Class IIa bacteriocin Pediocin PA-1 precursor	*P. acidilactici*	Henderson et al. 1992
1.C.24.1.2	Class IIa bacteriocin Sakacin P precursor	*Lact. sake*	Tichaczek et al. 1994
1.C.24.1.5	Class IIa bacteriocin Mesentericin Y105 precursor	*Leuc. mesenteroides*	Héchard et al. 1992
1.C.24.1.6	Class IIa bacteriocin Leucocin A precursor	*Leuc. gelidum*	Hastings et al. 1991
1.C.24.1.8	Class IIa bacteriocin Sakacin A precursor / Curvacin A	*Lact. sakei, Lact. curvatus*	Holck et al. 1992
1.C.24.1.9	Class IIa bacteriocin Bavaricin MN precursor	*Lact. sakei*	Kaiser and Montville, 1996
1.C.24.1.14	Listeria-active class IIa peptide, Lactococcin MMFII	*L. lactis*	Ferchichi et al. 2001
1.C.25.1.1	Class IIb two peptide bacteriocin Lactococcin G	*L. lactis*	Nissen-Meyer et al. 1992
1.C.26.1.1	Class IIb two peptide bacteriocin Lactacin F (LafA)-Lactacin X (LafX)	*Lact. johnsonii*	Muriana and Klaenhammer, 1991
1.C.29.1.1	Cation-selective class IIb two peptide bacteriocin, plantaricin EF	*Lact. Plantarum*	Oppegard et al. 2007
1.C.30.1.1	Anion-selective classIIb two peptide bacteriocin, plantaricin J, K	*Lact. plantarum*	Oppegard et al. 2007
1.C.30.1.2	Plantaricins Sb, Sa precursors	*Lact. plantarum*	Stephens et al. 1998
1.C.30.1.3	Thermophilin 1, 2, precursors ThmA, B	*Strep. thermophilus*	Marciset et al. 1991
1.C.37.1.1	Class IIc bacteriocin, Lactococcin 972	*L. lactis*	
1.C.53.1.1	Cyclic bacteriocin, Group I, Lactocyclicin Q	*L.* sp-strain QU12	Sawa et al. 2009
1.C.83.1.1	Cyclic bacteriocin, Group II, Gassericin A	*Lact. gasseri*	Kawai et al. 1998
1.C.90.1.2	Cyclic bacteriocin, Group I, Garvicin ML	*Lact. garvieae*	Borrero et al. 2011
1.C.93.1.1	Lacticin Q	*L.*	Fujita et al. 2007

The best-studied example from Class I lantibiotics is nisin (TC# 1.C.20.1.1). Nisin apparently forms channels in bacterial membranes using Lipid II, the prenyl chain-linked donor of the peptidoglycan building block, both as a receptor and as an intrinsic component of the pore (Breukink et al. 2003; Wiedemann et al. 2004).

Class II non lanthionine-containing heat-stable bacteriocins are small membrane active peptides of less than 10 kDa characterized by a Gly-Gly-1 Xaa+1 processing site in the bacteriocin precursor, processed by the protease domain linked to the ABC-type bacteriocin export permease (e.g., TC #3.A.1.42.2). Subgroups in Class II bacteriocins include class IIa, Listeria-active peptides with a consensus sequence in the *N*-terminus of Y-G-N-G-V-X-C. Pediocin PA-1, and Sakacin P are members of this family (TC #1.C.24.1.1 and 1.C.24.1.2, respectively). While Pediocin PA-1 has a *C*-terminal disulfide bridge, Sakacin P does not. Introducing such a bridge in Sakacin P broadened its target cell specificity and rendered it 10–20 times more potent against many bacterial strains but not others. The disulfide bridge also increased its heat stability.

4.3. The Major Facilitator Superfamily

The largest superfamily of secondary carriers present in nature is the Major Facilitator Superfamily (MFS), having the transporter classification (TC) number in the TC database (TCDB; www/tcdb.org) of 2.A.1 (Pao et al. 1998; Saier et al. 1999; Reddy et al. 2012). It currently consists of 90 families and over 100,000 sequenced members that can transport virtually any type of small substrate of biological importance. Members can be solute importers, catalyzing the active uptake of solutes against concentration gradients using a proton symport mechanism with the proton motive force (pmf) serving as the energy source, but they can also catalyze solute efflux against a concentration gradient using a proton antiport mechanism, thus extruding solutes out of the cell. Finally, a few MFS carriers are solute equilibrators, catalyzing facilitated diffusion via a solute uniport mechanism, and several – both symporters and uniporters – can also catalyze solute : solute exchange. The type of vectorial reaction catalyzed is determined by the carrier.

Almost all of the families of the currently recognized MFS consist of members that catalyze either uptake or efflux, but not both. However, there are six potential exceptions (SP, DHA1, DHA2, OPA, NNP, and OCT), based on reports from different laboratories (see TCDB). The sugar porter (SP) family (TC# 2.A.1.1), includes members that usually utilize sugar : H^+ symport but can also use equilibrative sugar uniport. Uniporters of the SP family occur primarily in animals that maintain constant high sugar levels in the extracellular plasma due to strict homeostatic mechanisms. However they can also be found in yeast and occasionally in bacteria that grow in environments containing high sugar concentrations. Some SP carriers have evolved receptor functions, thereby gaining regulatory sensor activities (Kruckeberg et al. 1998). However, no member of this huge family within the MFS has been shown to catalyze substrate : H^+ antiport.

Other reported examples of facilitated diffusion include a fructose uniporter in the yeast *Zygosaccharomyres* (2.A.1.2.23), which is a member of the DHA1 family of drug : H^+ antiporters, and the putative bacterial bile acid uptake porter (2.A.1.3.13) in the DHA2 family of drug exporters. The SP and DHA1 families are among the largest families in the MFS. It seems that in the MFS and certain other families, the loss of cation symport activity to generate substrate uniporters has occurred relatively frequently during evolution, that symporters and uniporters frequently can catalyze substrate : substrate exchange, but that interconversion of antiporters and symporters has been relatively rare during evolutionary history (see examples in TCDB).

In the Organophosphate : Phosphate Antiporter (OPA) family (2.A.1.4), most members preferentially catalyze substrate : substrate antiport, but they can also catalyze substrate : H^+ symport, and some *preferentially* seem to use a symport rather than an antiport mechanism. Also, in the Nitrate/Nitrite Porter (NNP) family (2.A.1.8), members of similar sequence can catalyze either uptake, efflux, or NO^-_3 : NO^-_2 exchange. Finally, members of the Organocation Transporter (OCT) family in animals (2.A.1.19) appear to be mechanistically promiscuous, catalyzing substrate uptake, export, exchange, and/or uniport. Sometimes the mechanism and substrate : H^+ stoichiometry depends on

Table 4.3. Functional types of Major Facilitator Superfamily (MFS) carriers and their relative occurrences in five different prokaryotic groups of organisms.

Group	LAB	Non-LAB Firmicutes	Actinobacteria	Gram (–) bacteria	Archaea
Drugs	63%	62%	48%	41%	59%
Anions	11%	16%	36%	32%	19%
Sugars	8%	8%	8%	9%	10%
Amino Acids	0%	0%	0%	2%	0%
Peptides	10%	8%	5%	11%	9%
Nucleosides	9%	7%	3%	5%	3%
# Systems	**291**	**298**	**138**	**990**	**142**
#Organisms	**11**	**13**	**5**	**41**	**16**
# Systems/Organism	**26**	**23**	**28**	**24**	**9**

the conditions and the specific substrate being transported. A few MFS carriers, but not many, can use Na^+ instead of H^+ as the co-transported cation. Na^+-dependent antiport appears to be rare.

MFS carriers are usually about 400 amino acyl residues (aa) long and have 12 transmembrane α-helical spanners (TMSs) with two homologous repeat units, each of six TMSs. They are found ubiquitously in bacteria, archaea, and eukaryotes. Over 20,000 sequenced MFS members are available for sequence analysis, and x-ray structures have been reported for several of them (Law et al. 2008). Some evidence suggests that the basic 6-TMS repeat unit arose by duplication of a primordial three TMS encoding genetic element and that MFS carriers arose from simple two TMS ion channels (Hvorup and Saier 2002; Vastermark et al. 2014).

As presented in Table 4.3, LAB and non-LAB firmicutes have a disproportionately high representation of MFS multidrug efflux pumps (62–63%), while the organic and inorganic anion uptake porters occur in low numbers (11% for LAB and 16% for the other firmicutes) compared to other bacterial groups (actinobacteria and Gram (–) bacteria with 36 and 32% representation, respectively). The high percentages of drug exporters in LAB reflects the apparent obsession of these organisms with biological warfare, while the low numbers of anion uptake systems may reflect their fermentative (sugar) modes of metabolism rather than respiratory (organic acid) modes of metabolism (Table 4.4). LAB also have 2–3-fold more nucleoside/nucleobase transporters than most other prokaryotes examined (Table 4.3). Table 4.4 summarizes the functionally characterized members of the MFS superfamily in LAB.

4.4. Other Large Superfamilies of Secondary Carriers

Excluding the MFS, the three largest superfamilies of secondary transporters include the Resistance/Nodulation/Division (RND) exporters (exclusively export carriers; Tseng et al. 1999; Delmar et al. 2014), the Drug/Metabolite Transporters (DMT) (both uptake and efflux systems; Jack et al. 2001; Yen et al. 2010), and the Multidrug/Oligosaccharide/Polysaccharide (MOP) porters (exclusively efflux pumps; Hvorup et al. 2003b). These three superfamilies will be examined sequentially in this section.

The RND superfamily (TC# 2.A.6) is essentially absent in the LAB characterized in this study, with only one exception: a single lipid-exporting HAE2 family member (Table 4.5; Lorca et al. 2007). The HME heavy metal exporters are found only in Gram (–) bacteria. HAE1 MDR pumps are present only in Gram (–) bacteria and non-LAB firmicutes with lower representation in the latter compared to the former. HAE3 homologues are present only in archaea and Gram (–) bacteria but not in the Gram (+) bacteria surveyed. Finally, the general protein secretory (Sec) auxiliary proteins,

Table 4.4. Summary of functionally characterized secondary carriers in LAB.

TCID	Name	Name	Microorganism	Reference
2.A.1.1.41	The D-xylose : H^+ symporter	XylT	*Lact. brevis*	Chaillou et al. 1998
2.A.1.2.5	Multidrug : H^+ antiporter	LmrP	*L. lactis*	Mazurkiewicz et al. 2004
2.A.1.3.37	The uridine/deoxyuridine/ 5-fluorouridine uptake transporter	UriP	*L. lactis*	Martinussen et al. 2010
2.A.1.17.2	Glucose transporter	OEOE_0819	*O. oeni*	Kim et al. 2011
2.A.1.23.1	Conjugated bile salt : H^+ symporter	CbsT1	*Lact. johnsonii* 100-100	Elkins et al. 2001
2.A.1.23.2	Taurocholate : cholate antiporter	CbsT2	*Lact. johnsonii* 100-100	Elkins and Savage 2003
2.A.1.1.41	The D-xylose : H^+ symporter	XylT	*Lact. brevis*	Chaillou et al. 1998a,b
2.A.1.68.1	The glucose transporter	OEOE_1574	*O. onei*	Kim et al. 2011
2.A.2.1.2	Probable fucosyl-α-1,6-*N*-acetylglucosamine uptake porter	AlfD	*Lact. casei*	Rodríguez-Díaz et al. 2012
2.A.2.2.1	Lactose permease, LacS	LacS	*Strep. thermophilus*	Andersen et al, 2011
2.A.2.2.3	Galactose permease	GalP	*Lact. lactis*	Grossiord et al. 2003
2.A.2.3.3	Isoprimeverose permease	XylP	*Lact. pentosus*	Heuberger et al. 2001
2.A.3.7.1	Glutamate : γ-aminobutyrate antiporter	GadC	*L. lactis*	De Biase and Pennacchietti 2012
2.A.7.5.2	Probable ribose transporter	RbsU	*Lact. sakei*	Stentz and Zagorec 1999
2.A.7.5.3	Glucose : H^+ symporter	GlcU (YxfA)	*L. lactis*	Castro et al. 2009
2.A.7.21.4	The orotate transporter	OroP	*L. lactis*	Defoor et al. 2007
2.A.17.1.6	POT famiy di- and tri-peptide porter	DtpT	*Strep. thermophilus*	Solcan et al. 2012
2.A.21.3.20	Sialic acid transporter	NanT	*Lact. sakei*	Anba-Mondoloni et al. 2013
2.A.24.2.2	Malate : lactate antiporter	MaeP (MleP)	*L. lactis*	Bandell et al. 1997
2.A.24.3.1	Electrogenic citrate : L-lactate exchanger	CitN	*L. lactis*	Pudlik and Lolkema 2012
2.A.24.3.2	Citrate:lactate antiporter	CitP	*Leuc. mesenteroides*	Bandell et al. 1997
2.A.26.1.3	Branched chain amino acid : H^+ symporter	BrnQ	*Lact. delbrueckii*	Schick et al. 1999
2.A.40.1.2	High affinity uracil permease	PyrP	*L. lactis*	Martinussen et al. 2001
2.A.66.2.16	Capsular polysaccharide exporter	CpsU	*Strep. thermophilus*	Cefalo et al. 2011
2.A.69.4.5	Malate permease	MleP	*O. oeni*	Labarre et al. 1996a,b
2.A.77.1.3	The Cadmium Resistance (CadD) Family	CadD	*Lact. fermentum*	Halttunen et al. 2007
2.A.87.1.1	The riboflavin uptake transporter	RibU	*L. lactis*	Burgess et al. 2006
2.A.88.3.2	Thiamin transporter (Thiamin ECF transporter S component)	ThiT	*L. lactis*	Majsnerowska et al. 2013
2.A.88.3.3	Thiamin transporter (Thiamin ECF transporter S component)	ThiT	*Lact. casei*	Rodionov et al. 2002, 2009
2.A.88.6.1	Putative queosine precursor uptake transporter	QrtT	*Lact. sakei*	Rodionov et al. 2009

Table 4.4. *(Continued)*

TCID	Name	Name	Microorganism	Reference
2.A.87.2.1	Pantothenate transporter (Pantothenic acid ECF transporter S component)	PanT	*Leuc. mesenteroides*	Rodionov et al. 2009
2.A.88.5.2	Putative niacin uptake transporter (niacin ECF transporter S component)	NiaX	*L. lactis* subsp. *cremoris*	Rodionov et al. 2009
2.A.88.9.1	Predicted queuosine precursor transporter (queuosine ECF transporter S component)	QueT	*Lact. lactis*	Rodionov et al. 2009
2.A.88.9.2	Predicted queuosine precursor transporter (queuosine ECF transporter S component)	QueT	*Leuc. gasicomitatum*	Rodionov et al. 2009
2.A.115.2.4	Novobiocin exporter	NbcE	*Lact. brevis*	Pagliai et al. 2010
2.A.118.1.7	The probable citruline/ornithine antiporter	ArcD	*Lact. sakei*	Rimaux et al. 2013
2.A.3.2.9	Histidine/histamine antiporter	HdcP	*Strep. thermophilus*	Trip et al. 2012
2.A.3.1.19	His, Arg, Lys transporter	HisP	*L. lactis* subsp *lactis and* subsp. *cremoris*	Trip et al. 2013
2.A.3.1.18	Lysine transporter	LysP	*L. lactis* subsp *lactis* and subsp. *cremoris*	Trip et al. 2013
2.A.3.1.21	Serine transporter	SerP1	*L. lactis* subsp *lactis and* subsp. *cremoris*	Trip et al. 2013
2.A.3.1.20	Serine transporter	SerP2	*L. lactis* subsp *lactis* and subsp. *cremoris*	Trip et al. 2013
2.A.3.1.22	Phenylalanine/tyrosine transporter	FywP	*L. lactis* subsp *lactis* and subsp. *cremoris*	Trip et al. 2013
2.A.3.2.11	Arginine/Ornithine transporter	ArcD1	*L. lactis* subsp *lactis* and subsp. *cremoris*	Trip et al. 2013
3.A.3.2.10	Arginine/Ornithine transporter	ArcD2	*L. lactis* subsp *lactis* and subsp. *cremoris*	Trip et al. 2013
2.A.3.7.6	Aspartate/Glutamate transporter	AcaP	*L. lactis* subsp *lactis* and subsp. *cremoris*	Trip et al. 2013
2.A.3.7.7	Putrescine transporter	AguD	*L. lactis* subsp *lactis* and subsp. *cremoris*	Trip et al. 2013
2.A.26.1.9	Leucine/Isoleucine/valine transporter	BrnQ	*L. lactis* subsp *lactis* and subsp. *cremoris*	Trip et al. 2013
2.A.3.3.23	Branched-chain amino acid transporter	CtrA (BcaP)	*L. lactis*	denHengst et al., 2006; PMID 16621821

Table 4.5. Distribution of RND superfamily members in five groups of prokaryotes.

	% Representation					# Systems/ Organism
Family Group	HME	HAE1	SecDF	HAE2	HAE3	(ratio)
LAB	0%	0%	0%	100%	0%	1/11 (0.1)
Non LAB Firmicutes	0%	28%	44%	28%	0%	18/13 (1.4)
Actinobacteria	0%	0%	25%	75%	0%	32/5 (6.4)
Gram (–) Bacteria	12%	58%	21%	0%	8%	264/41 (6.4)
Archaea	0%	0%	59%	0%	41%	34/16 (2.1)

Table 4.6. Distribution of DMT superfamily members in five groups of prokaryotes.

TC Family #: TC Family Name:	1 SMR	3 DME	5 GRP	7 RarD	21 BAT2	23 Trp-E	24 TPPT	# Proteins/Organism
Non LAB Firmicutes	21	33	13	5	7	6	15	85/11 (7.7)
Actinobacteria	17	50	0	25	8	0	0	12/5 (2.4)
Gram (−) bacteria	16	60	0	6	5	6	8	283/33 (8.6)
Archaea	9	64	0	0	0	7	20	44/14 (3.1)

Table 4.7. Distribution of MOP superfamily members in five groups of prokaryotes.

Family:	1 MATE	2 PST	4 MVF	7 U-MOP1	12 U-MOP4	# Proteins/Organism
LAB	34	47	2	15	0	47/11 (4.3)
Non LAB Firmicutes	53	33	4	8	1	76/11 (6.9)
Actinobacteria	50	8	42	0	0	12/5 (2.4)
Gram (−) bacteria	41	28	21	7	3	155/32 (4.8)
Archaea	43	41	1	15	0	87/14 (6.2)

SecDF, which together comprise the equivalent of a full length RND pump that is thought to facilitate integral cytoplasmic membrane protein insertion (Xie et al. 2006), are present in all prokaryotic types examined except the LAB. In *E. coli*, loss of these proteins decreases protein secretion efficiency and gives rise to a temperature-sensitive growth phenotype (Xie et al. 2006). SecDF is a proton conduit that functions as a membrane chaperone, powered by the pmf to achieve ATP-independent protein translocation (Tsukazaki et al. 2011). It is surprising that SecDF homologues are lacking in LAB that can grow at rapid rates. This may be because these glycolytic organisms generate enough ATP so that pmf-translocation is not required.

The Drug/Metabolite Transporter (DMT) superfamily (TC# 2.A.7; Jack et al. 2001) shows decreased proportions of members of (i) the small cationic multidrug resistance (SMR) efflux pumps, and (ii) the Drug/Metabolite Exporter (DME) family in LAB and other firmicutes relative to other bacterial types. DME porters characterized so far are mostly metabolite exporters. Archaea show the smallest percentages of SMR family members but the largest percentages of DME, Trp-E, and TPPT family members of all prokaryotic groups examined (Table 4.6). The RarD and BAT2 families were not found in archaea. Glucose/Ribose Porter (GRP) family members, which take up sugars, are present only in firmicutes with LAB showing more than twice as many family members as non-LAB firmicutes (see Table 4.6).

In LAB, the MOP superfamily (TC# 2.A.66; Table 4.7; Hvorup et al. 2003a) shows low percentages of MATE family MDR pumps and Mouse Virulence Factor (MVF) family cell wall lipid flippases compared to other bacteria examined (Mohammad and Valvano 2014). U-MOP family members (of unknown function) are poorly represented in all groups of prokaryotes. However, LAB exhibit increased proportions of polysaccharide exporters of the PST family, and of U-MOP1 family members relative to other types of bacteria (Table 4.7). It should be noted that in all of the superfamilies mentioned, several families of unknown specificity are present. Some of these are represented in LAB while others are not. This fact may be understood when the functions of these transporters are revealed. The underrepresentation of both DMT and MOP superfamilies relative to other firmicutes is also worthy of note.

4.5. ABC Transporters

Members of the ATP-binding Cassette (ABC) Superfamily (Davidson and Maloney 2007) belong to 92 currently described families, including 33 families of uptake systems and 59 families of exporters (see TCDB). Each family is, in general, specific for a different type of substrate, although exceptions to this generalization are known. They can be specific for all kinds of small molecules, and members of several families function in the export of macromolecules (proteins, lipids, and complex carbohydrates). They share the characteristic of being driven by ABC protein-dependent ATP hydrolysis (Davidson and Maloney 2007).

ABC primary active transporters form a unified group of proteins only in the sense that they use homologous energy-coupling ABC proteins. These are superimposed on the transmembrane porters that derive from at least three distinct families of independently evolving transporters (Wang et al. 2009; Zheng et al. 2013). One of these families (ABC1) exhibits six TMSs per polypeptide chain, having arisen from a primordial 2-TMS hairpin peptide as a result of two intragenic triplication events ($2 \times 3 = 6$) (Figure 4.1a). The second family (ABC2) arose from a 3-TMS precursor by a single intragenic duplication event ($3 \times 2 = 6$), giving rise to a set of independently evolving 6-TMS proteins showing no sequence similarity with ABC1 proteins (Figure 4.1b). A third family (ABC3) has members with four, eight, or ten TMS per polypeptide chain (Figure 4.1c). ABC3 proteins have a basic 4-TMS topology, but in the 8- and 10-TMS homologues, this unit has been duplicated, and in the 10-TMS proteins, duplication resulted in the creation of two extra nonhomologous TMSs separating the two repeat units (Figure 4.1c). These three types of systems are all present in the three domains of life, but in varying numbers, depending on the organismal type.

Over 100,000 sequenced members of the ABC superfamily are collectively available for analysis. They occur in relative numbers of 1>2>3 when all living organisms are considered, but the distributions of these porter types are organism specific. Thus, LAB and actinobacteria display relative proportions of 1>>3>2, archaea have 2>>3>1, and Gram (–) bacteria and eukaryotes exhibit proportions of 1>2>3 (Table 4.8 and unpublished data). Surprisingly, the distributions of the three topological

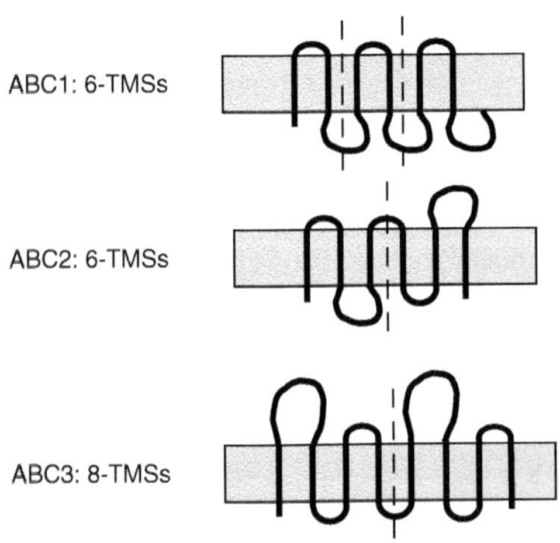

Figure 4.1. Schematic depiction of the structures of the three families of integral membrane ABC exporters that catalyze substrate efflux using ATP hydrolysis for energy coupling. (a) ABC1 proteins are 6-TMS permeases that arose by intragenic triplication of a 2-TMS hairpin-encoding gene. (b) ABC2 proteins are 6-TMS permeases that arose by intragenic duplication of a gene encoding a 3-TMS primordial structure giving a 6-TMS protein with the two repeat units having opposite orientation in the membrane. (c) ABC3 proteins are 4-, 8-, and 10-TMS permeases where duplication of the primordial 4-TMS element gave rise to the 8- and 10-TMS proteins. In the 10-TMS proteins, the extra two TMSs separate the two 4-TMS repeat units. (Wang et al., 2009; Zheng et al., 2013).

Table 4.8. Organism-type distribution of the ABC1, ABC2, and ABC3-type exporters of the ABC functional superfamily.

Type/Group	ABC1 (%)	ABC2 (%)	ABC3 (%)	Proportions	# Proteins/Organism
LAB	56	19	25	1>>3>2	305/11 (28)
Non LAB Firmicutes	43	29	28	1>2=3	335/11 (31)
Actinobacteria	55	17	28	1>>3>2	97/5 (19)
Gram (−) bacteria	37	34	29	1>2>3	641/33 (19)
Archaea	23	52	25	2>>3=1	200/14 (14)
Total	**41**	**32**	**27**	**1>2>3**	**1578/73 (22)**

types occurring in the five organismal groups examined are distinctive, with only a few exceptions. Thus, Gram (−) bacteria and archaea have fewer ABC1 and more ABC2 porters than the three types of Gram (+) bacteria. Archaea are most unusual as ABC2 porters predominate over ABC1 porters.

With respect to the relative proportions of the major functional types of ABC exporters, MDR pumps predominate in all groups of living organisms (Table 4.10). In LAB and other firmicutes, 64–70% of all ABC exporters probably serve protective functions of drug/hydrophobic compound efflus (see Table 4.10; Table 4.9 for experimentally characterized porters in LAB). LAB have three times as many ABC-type peptide exporters as do other firmicutes, more than G (−) bacteria and archaea and about the same proportions of protein exporters. This fact correlates with the use of signaling peptides in Gram (+) bacteria to a much greater extent than in Gram (−) bacteria. However, these organisms have fewer ABC lipid, and complex carbohydrate exporters than other prokaryotic phyla (Table 4.10).

ABC uptake permeases often exhibit five TMS per polypeptide chain although some have been internally duplicated to give ten TMS. Our recent work has led to the probability that the basic five TMS unit also evolved by intragenic duplication. The statistical analyses suggest that TMS 1–2 are homologous to TMS 4–5 (Zheng et al. 2013). Two possible routes can be considered. First, a 2-TMS hairpin polypeptide chain might have duplicated so that a 5-TMS protein resulted with the *N*-terminal hairpin facing inwards and the *C*-terminal hairpin facing outwards. In this case, the central TMS (TMS3) would have been generated *de novo* during the duplication event. Second, a 3-TMS precursor could have duplicated to give a 6-TMS protein, and then the first or last TMS would have been lost as a result of a small *N*-terminal or *C*-terminal deletion mutation early during the evolution of these ABC porters. We favor this second interpretation because these uptake porters show significant sequence similarity with ABC2 exporters (Zheng et al. 2013).

Some of these uptake porters, particularly those specific for vitamins and trace metals, are capable of functioning either as low affinity, high efficiency, secondary carriers (when the ABC protein is absent) or as high affinity, low efficiency, primary active transporters (when the ABC protein is superimposed on the transmembrane porter). This has been demonstrated for the *E. coli* biotin uptake system, BioY (Hebbeln et al. 2007). However, genomic evidence suggests that other homologous transport proteins (prokaryotic ThiW, TrpP, and phosphate : H^+ or Na^+ symporters of the PNaS family, see TCDB) may also be capable of catalyzing primary active uptake with appropriate ATP hydrolyzing subunits (E.I. Sun and M.H. Saier, unpublished observations). Moreover, recently published data suggests that this family of porters is much larger than previously thought and that many are specific for vitamins or vitamin precursors (Rodionov et al. 2009). These suggestions need to be substantiated by wet lab experimentation.

4.6. Heavy Metal Transporters

Heavy metal ions are both essential for life and toxic at high concentrations. Eight families of secondary active transporters are known to mediate divalent cation uptake and/or efflux (Nramp, VIT, ZIP, and NiCoT for uptake and CaCA, CDF, CadD, and RND/HME for efflux). In LAB, two of

Table 4.9. Summary of functionally characterized ABC transporters in LAB.

TCID	Name	Name	Microorganism	Reference
3.A.1.1.20	The fructooligosaccharide porter	MsmEFGK	*Lact. acidophilus*	Barrangou et al. 2003
3.A.1.2.17	General nucleoside uptake porter	NupABC/BmpA	*L. lactis*	Martinussen et al. 2010
3.A.1.3.25	Glutamine transporter, GlnQP.	GlnPQ	*L. lactis*	Fulyani et al. 2013
3.A.1.5.10	Oligopeptide porter	OppABCDF	*L. lactis*	Doeven et al. 2004
3.A.1.5.25	The ABC peptide/signalling peptide transporter	OptASBCDF	*L. lactis*	Lamarque et al. 2011
3.A.1.5.32	ABC α-galactoside uptake porter	MsmEFGK	*Lact. acidophilus*	Andersen et al. 2012
3.A.1.12.5	Uptake system for glycine-betaine (high affinity) and proline (low affinity)	BusAA-ABC	*L. lactis*	Biemans-Oldehinkel et al. 2006
3.A.1.25.4	The biotin uptake system	BioMNY	*L. lactis*	Berntsson et al. 2012
3.A.1.25.6	Riboflavin ECF transport system	EcfAA'T/RibU	*Strep. thermophilus*	Karpowich and Wang 2013
3.A.1.26.9	The folate transporter	FolT/EcfAA"T	*Lact. brevis*	Xu et al. 2013
3.A.1.28.2	The folate transporter	EcfAA'ST	*Lact. brevis*	Xu et al. 2013
3.A.1.29.1	The putative methionine precursor/uptake transporter	MtsTUV	*Lact. johnsonii*	Rodionov et al. 2009
3.A.1.111.3	Nisin exporter	NisT	*L. lactis*	Kuipers et al. 1998
3.A.1.111.5	The lacticin Q exporter	LcnDR3	*L. lactis*	Yoneyama et al. 2009
3.A.1.112.2	Pediocin PA-1 exporter	PedD	*P. acidilactici*	Venema et al. 1995
3.A.1.112.3	Bacteriocin (lactococcin) exporter	LcnC	*L. lactis*	Stoddard et al. 1992
3.A.1.117.1	The multidrug exporter, LmrA	LmrA	*L. lactis*	Balakrishnan et al. 2004
3.A.1.117.2	Hop resistance protein	HorA	*Lact. brevis*	Sakamoto et al. 2001
3.A.1.124.1	The 3-component nisin immunity exporter	NisFEG	*L. lactis*	Okuda et al. 2010
3.A.1.135.1	The heterodimeric multidrug exporter, YdaG/YbdA	YdaG/YbdA	*L. lactis*	Lubelski et al. 2004, 2006

the four families of heavy metal uptake systems (Nramp and VIT) are represented in much larger numbers than the other two families (ZIP and NiCoT) (Table 4.11). This situation is similar in non-LAB firmicutes, except that, most surprisingly, VIT family members are reduced by over 10-fold. Actinobacteria and Gram (−) bacteria exhibit more equal numbers of the four family members, but in archaea, the ZIP family is dominant over the Nramp and VIT families, which are in turn better represented than the NiCoT family (Table 4.11).

Of the heavy metal efflux systems, the CDF family predominates in all prokaryotic phyla examined, but the CadD family is well represented in Gram (+) bacteria (Table 4.11). CadD family members are not present in Gram (−) bacteria. On the other hand, within the RND superfamily, Heavy Metal Exporters (HME family) are well represented in Gram (−) but not in Gram (+) bacteria. Neither family is represented in archaea. Thus, the CadD and HME families appear to occur in restricted prokaryotic phyla. Only one, but not both, is present in any one type of organisms (Table 4.11).

The Ca^{2+} : H^+ or Na^+ cation antiporters (CaCA family) (Lytton 2007) are similarly represented in actinobacteria, Gram (−) bacteria, and archaea, but non-LAB firmicutes have substantially reduced numbers, and LAB lack these carriers altogether. These porters function primarily to exclude cytoplasmic Ca^{2+}, but some of them can also export other divalent ions. In the next section we shall

Table 4.10. Functional types of ABC exporter, expressed in percent, identified in different types of prokaryotes. The percentages are in bold print, while the relative proportions of ABC1, ABC2 and ABC3 types are provided below the percentages. CHO, complex carbohydrates; AAs, amino acids.

Substrate Type/ Organismal type	Drugs	CHOs	Lipids	Proteins	Peptides	AAs	Unk	Total #
LAB	**64**	**4**	**1**	**10**	**12**	**3**	**6**	**305**
	1>>2>3	2>3	3	1=3	1	1	3>>2,1	1>>3>2
Non LAB	**70**	**5**	**1**	**10**	**4**	**3**	**7**	**335**
Fermicutes	1>>2,3	2>>3	2	1=3>2	2>3>1	1	3>2>1	1>2=3
Actinobacteria	**42**	**6**	**1**	**10**	**14**	**8**	**19**	**97**
	1>2>3	3>>2	1	3	1	1	3>1	1>>3>2
Gram (–) bacteria	**39**	**10**	**10**	**16**	**7**	**4**	**14**	**641**
	2>1>3	2>>3>1	3>1>2	1=3	3>2>1	1	1=3>2	1>2>3
Archaea	**51**	**8**	**3**	**8**	**9**	**5**	**17**	**200**
	3>2>1	2	1>2	3>2>>1	1>2	2>>1	3>2>>1	2>>3=1
Total	**52**	**7**	**5**	**12**	**8**	**4**	**12**	**1578**
	1>2>3	2>>3>1	3>2>>1	3>1>>2	1>2>3	1>>2	3>2>1	1>2>3

Table 4.11. Distribution of eight families of secondary carriers specific for inorganic divalent cations in five groups of organisms. All values are expressed as the average numbers of transporters in each family divided by the number of genomes examined.

	# Transporter 5/#Organisms				
Organismal Type/Family	LAB	Non LAB (Firmicutes)	Actino-bacteria	Gram (–) Bacteria	Archaea
Nramp (Mn^{2+}, Fe^{2+}, Zn^{2-}, Cu^{2+}, Cd^{2+}, Ni^{2+}, Co^{2-} uptake)	1.7	1.2	1	0.4	0.4
VIT (Fe^{2+}, Mn^{2+} uptake)	1.3	0.1	0.5	0.2	0.4
ZIP (Zn^{2+}, Fe^{2+}, Co^{2+}, Mn^{2+}, Ni^{2+} uptake)	0.1	0.3	0.5	0.5	0.9
NiCoT (Ni^{2+}, Co^{2+} uptake or efflux)	0.2	0.1	0.2	0.6	0.2
CaCA (Ca^{2+}, Mn^{2+}, Zn^{2+}, Mg^{2+} efflux)	0	0.4	0.8	0.8	0.8
CDF (Co^{2+}, Ni^{2+}, Cd^{2+}, Zn^{2+}, Cu^{2+} efflux)	1.5	2	0.8	1.3	1.8
CadD (Cd^{2+} efflux)	0.4	0.2	0.4	0	0
RND/HME (Ni^{2+}, Co^{2+}, Zn^{2+}, Ag^{+}, Pb^{2+}, Hg^{2+} efflux)	0	0	0	0.6	0

see that the representation of P-type Ca^{2+}-ATPases is inversely proportional to the numbers of CaCA family members in the different bacterial types. Thus, it seems that Ca^{2+} export is essential, but that either CaCA family members or Ca^{2+}-ATPases can perform this function. Whether one or the other is used must depend on environmental or physiological conditions. For example, when the primary source of energy is ATP, resulting from fermentative metabolism, ATP-dependent primary active transporters tend to predominate over pmf-driven secondary carriers. However, when respiration provides a pmf as the primary energy source, and ATP is secondarily derived from the pmf via an F-type ATPase, secondary carriers predominate (Paulsen et al. 1998, 2000).

4.7. P-type ATPases in Prokaryotes

P-type ATPases control cation homeostasis in many organismal types (Fagan and Saier 1994; Palmgren and Axelson 1998; De Hertogh et al. 2004; Haupt et al. 2005; Kuhlbrandt 2004; Chan et al. 2010; Thever and Saier 2009). As noted above, firmicutes, which have low numbers of CaCA family members, have 2–3 times as many Ca^{2+}-ATPases as are found in actinobacteria and Gram (−) bacteria (Table 4.12). By contrast, archaea, which have good representation of CaCA carriers, also have excellent representation of the Ca^{2+}-ATPases.

Of great interest is the fact that LAB encode P-type ATPases (TC# 3.A.3) in their genomes that normally occur in eukaryotes but not in most other bacteria. These include members of the families of Na^+, K^+-exchange ATPases of animals (family 1 in TCDB), H^+/Mn^{2+}-ATPases of plants and fungi (family 3), and Na^+ or K^+ extruding ATPases of fungi and unicellular eukaryotes (family 9). Na^+, K^+-type ATPases are found in LAB and archaea, H^+/Mn^{2+}-ATPase are present in LAB and, in lower numbers, in Gram (−) bacteria and archaea, and Na^+ or K^+ exporters are present only in LAB (Table 4.12). The presence of eukaryotic K^+-ATPase types correlates with the absence of prokaryotic Kdp-type ATPases (Bramkamp et al. 2007) in LAB. These observations may reflect the close associations of various LAB with animals, plants, and fungi (Gobbetti et al. 1994; Vaughan et al. 2002; Reddy et al. 2008) although the possibility of vertical descent should not be excluded. It is also worth noting that copper and heavy metal exporters are present in firmicutes (especially LAB) in lower proportions compared to all other prokaryotic types examined (Table 4.12). Interestingly, three copper ATPase genes that mediate copper homeostasis in *Lact. bulgaricus* are induced upon acidification of the culture medium (Penaud et al. 2006).

Our studies have identified P-type ATPases that do not fall into one of the ten functionally well-characterized families of these enzyme/transporters (TC# 3.A.3.1-10; Chan et al. 2010; Thever and Saier, 2009). Altogether, we have identified about two dozen families, each represented only in eukaryotes or in prokaryotes, not in both. Some of the largest of these families, well represented in prokaryotes, are presented in Table 4.12. Molecular genetic, biochemical, and physiological experimentation will be required to determine the transport substrates of these enzymes.

4.8. The Prokaryote-specific Phosphotransferase System (PTS)

The PTS catalyzes group translocation, the coupling of sugar transport to sugar phosphorylation. Phosphoenolpyruvate is the initial phosphoryl donor. It phosphorylates Enzyme I, and then the phosphoryl group is sequentially passed to HPr, Enzyme IIA, Enzyme IIB, and finally to an incoming sugar

Table 4.12. P-type ATPase superfamily representation (%) in five types of organisms. The values are for the individual families found in LAB.

	1	2	3	4	5	6	7	9	23	25	32	
Family/Group	Na, K^+ (Out/ In)	Ca^{2+} (Out)	H^+, Mn^{2+} (Out)	Mg^{2+} (In)	Cu^+, Ag^+ or Cu^{2+} (Out/In)	HM (Out)	K^+ (In)	$Na^+; K^+$ (efflux)	Unk	Unk	Unk	# proteins/ organism
LAB	6	31	6	7	18	13	0	2	9	9	0	89/11 (8.1)
Non LAB Firmicutes	0	32	0	9	23	17	10	0	5	2	2	57/11 (5.2)
Actinobacteria	0	11	0	0	33	24	3	0	5	5	5	37/5 (7.4)
Gram (−) Proteobacteria	0	11	2	9	32	15	12	0	0	2	2	121/33 (3.7)
Archaea	5	35	3	2	31	19	0	0	0	2	3	58/14 (4.1)
Average %	2	24	3	6	27	17	6	0.3	3	4	2	362/73 (5.0)

in a reaction dependent on the transmembrane porter, Enzyme IIC. The cytoplasmic, general energy coupling proteins are Enzymes I and HPr while the Enzyme II complexes consist of IIA, IIB, and IIC subunits, sometimes fused to each other. In the case of the mannose systems, there are also IID subunits. These systems predominate in bacteria but can also be found in some archaea (Pickl et al. 2012).

As discussed previously (Hvorup et al. 2003a; Barabote and Saier 2005; Saier et al. 2005), there exist four independently evolving families of PTS group translocators (see Figure 4.2). These permeases consist of IIA, IIB, IIC, and in the Mannose (Man) family, IID protein domains. Family 1, the Glucose/Fructose/Lactose (Glc/Fru/Lac) superfamily, consists of transporters that fall into seven phylogenetic clusters, each specific for a particular sugar or sugar type as follows: (i) glucose, (ii) glucosides, (iii) fructose, (iv) mannitol, (v) lactose, (vi) diacetylchitobiose, and (vii) glucitol (Nguyen et al. 2006). All of these systems except the glucitol-type systems probably have ten TMS per polypeptide chain (Nguyen et al. 2006; Cao et al., 2011, PMID 21471968). In the glucitol systems, the ten TMS are probably split between two proteins, each with 5 TMS. Glucitol porters exhibit sufficient sequence similarity to establish that they are homologous to other members of the Glc/Fru/Lac family (Figure 4.2a) although these porters comprise the most sequence-divergent group of proteins in this superfamily (Nguyen et al. 2006).

The mannose (Man) family, family 2, is the only family to have a IID constituent (Saier et al. 2005). Only this family includes systems that each can exhibit broad specificity for hexoses (glucose, mannose, fructose, glucosamine, *N*-acetyl glucosamine, etc.). These 6-TMS proteins evolved independently of the 10 TMS permeases of the Glc/Fru/Lac family (Figure 4.2b). All constituents of the Man family permeases (IIA, IIB, IIC, and IID) are unique to this family, although the IIA proteins have been incorporated into the late evolving Dha PTS (see below).

Two distantly related families, the ascorbate (Asc) group translocators and the galactitol (Gat)-specific systems (Figure 4.2c), have 10–12 TMS per polypeptide chain and together comprise family 3 (Hvorup et al. 2003b). They are related by common descent, have a basic 5-TMS repeat unit, duplicated to give the 10-TMS topology plus 1 or 2 additional TMS (Luo et al, 2015, PMID# 25686089). They may have evolved from secondary carriers (Figure 4.2c). Surprisingly, it appears that members of the galactitol family, but not of the ascorbate family, may still be able to function as secondary carriers (Saier et al. 2005).

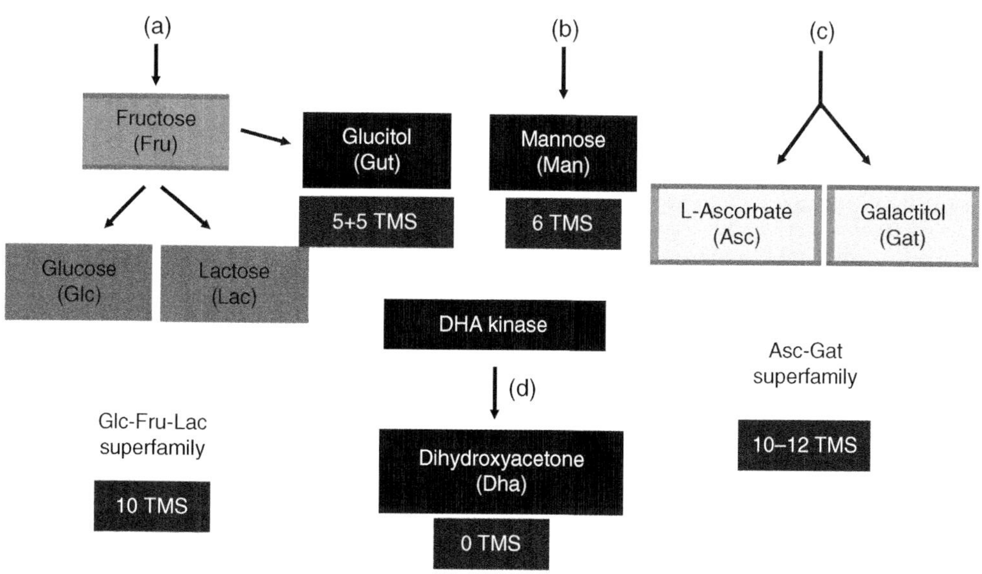

Figure 4.2. Four independently evolving families within the PTS functional superfamily. This schematic figure illustrates independent evolutionary origins for the four currently recognized families. (a, left) The Glc/Fru/Lac superfamily, (b, center) the Man family, (c, right) the Asc/Gat superfamily, and (d, bottom center) the Dha family.

Finally, the dihydroxyacetone (DHA) PTS systems (the Dha systems; Figure 4.2d) are not transporters at all. They phosphorylate DHA in the cytoplasm using phosphoenolpyruvate as the phosphoryl donor (Garcia-Alles et al. 2004). However, unlike all PTS permeases, the protein complex exhibits 0 TMS, is a multi-subunit cytoplasmic enzyme complex, and arose from soluble ATP-dependent DHA kinases. These unique kinases are found in bacteria, including some that also have the DHA-PTS (Barabote and Saier 2005).

Dha systems of the PTS have three protein constituents, DhaM (IIA), DhaL (IIB), and DhaK (IIC) (Siebold et al. 2003). DhaL corresponds in sequence to the *N*-termini of DHA kinases while DhaK corresponds to the *C*-termini of these kinases. DhaM contains a domain that is distantly related to IIA proteins of the mannose (Man) systems (family 2, Figure 4.2b) and can be fused to other PTS domains. Like the GAT systems, these systems are not full-fledged PTS systems because (i) DhaL contains tightly bound ADP which is phosphorylated rather than a histidyl or cysteyl residue in the proteins, and (ii) DhaK binds DHA covalently via a histidyl residue to provide specificity (Siebold et al. 2003; Garcia-Alles et al. 2004). Covalent bond formation between enzymes and the sugar substrate is not observed for any other PTS enzyme.

Numbers of PTS proteins encoded within the genome of an organism do not correlate with genome size (Barabote and Saier 2005). For example, both large genome organisms such as most actinobacteria and small genome organisms such as *Mycoplasma* species usually have few PTS permeases. Firmicutes, including LAB, have the most PTS permeases in spite of their reduced genome sizes (Table 4.13; Lorca et al. 2007). Yet not all LAB and other firmicutes have large numbers of PTS porters; these organisms differ in over a 20-fold range with respect to their PTS permease representation. Those firmicutes with highest representation have over 3% of their genomes encoding PTS proteins, up to about two dozen PTS porters (Barabote and Saier 2005).

Strain differences within a single species show surprising differences in PTS content. For example, various strains of *Strep. pyogenes* may have or lack a dihydroxyacetone (DHA)-PTS, while various strains of *Strep. agalactiae* differ with respect to the presence or absence of an extra lactose (Lac) permease. Finally, *Strep. pneumoniae* strains differ with respect to the presence of a fructose (Fru) system and whether or not IIA Fru is fused to a BglG-like transcriptional regulator (Barabote and Saier 2005).

Further studies revealed that, over evolutionary time, fusion and splicing of PTS proteins to give proteins of varying sizes and domain compositions have occurred frequently. Moreover,

Table 4.13. Summary of functionally characterized PTS systems in LAB.

TCID	Name	Example	Reference
4.A.3.1.2	Lactose PTS group translocator #1	*Lact. gasseri*	Francl et al. 2012
4.A.3.1.3	Lactose PTS group translocator #2	*Lact. gasseri*	Francl et al. 2012
4.A.3.2.4	The cellobiose-specific (PtcA-PtcB-CelB) porter	*L. lactis*	Kowalczyk et al. 2008
4.A.6.1.6	Glucose porter, ManLMN	*Lact. casei*	Yebra et al. 2006
4.A.6.1.7	The glucose/mannose/2-deoxyglucose/ fructose phosphotransferase systems, ManLMN	*Strep. thermophilus*	Cochu et al. 2003
4.A.6.1.10	The hexose (glucose and fructose demonstrated) PTS uptake system	*O. oeni*	Kim et al. 2011
4.A.6.1.11	Mannose enzyme II complex, IIAB, IIC, IID	*L. lactis*	Kjos et al. 2011
4.A.6.1.13	Fucosyl-α-1,3-*N*-acetylglucosamine PTS uptake porter (AlfEFG)	*Lact. casei*	Rodríguez-Díaz et al. 2012

horizontal transfer of PTS genes accounts for the varied PTS protein compositions of closely related organisms. It appears that transporter gene transfer between bacteria has occurred with frequencies that are roughly inversely proportional to phylogenetic distance, and horizontal transfer between bacterial phyla has been surprisingly rare. Moreover, transfer between the three domains of life has seldom occurred. For example, genes encoding PTS proteins could not be identified in the eukaryotic domain, and archaeal PTS proteins are only found in some of these organisms (Barabote and Saier 2005; Pickl et al. 2012). By contrast, genes encoding members of the Mitochondrial Carrier (MC) family are largely lacking in prokaryotes (unpublished observations). It is assumed that these two superfamilies of transporters evolved after the great divide between bacteria, archaea, and eukaryotes, and the presence of PTS proteins in some archaea could have resulted from horizontal gene transfer from bacteria. Finally, fusions of PTS domains to non-PTS enzymes and transport systems suggest novel PTS regulatory functions not yet recognized (Barabote and Saier 2005).

4.9. Multidrug Resistance Pumps

The occurrence of MFS-type MDR efflux pumps in LAB is much higher than in most other bacteria as noted in Section 4.3. For example, LAB have over 60% of their MFS carriers functioning as MDR/hydrophobic compound exporters compared to 40–50% for other bacteria (Table 4.3; Lorca et al. 2007), thus showing an unexpected trend. Moreover, in most LAB, MFS porters predominate: MFS (55%) > ABC (32%) > MOP (7%) > RND (4%) > DMT (2%). However, in *Strep. thermophilus*, the order is MOP (47%) > ABC (37%) > MFS (16%) > RND or DMT, and in *Lact. delbruekii* and *Lact. gasseri* these ratios are ABC (51%) > MFS (44%) > RND (<5%) > MOP or DMT (0%). It is therefore clear that while LAB exhibit unusual trends relative to other bacteria, they sometimes exhibit considerable variation among themselves.

4.10. Nutrient Transport in LAB

Three mechanisms account for sugar uptake in bacteria. In LAB, PTS permeases (52%) predominate over secondary carriers (34%) which predominate over primary active ABC transporters (15%). Gram (+) firmicute pathogens show similar proportions. However, in actinobacteria, the order is secondary carriers (50%) > ABC transporters (41%) > PTS group translocators (8%), and in select Gram (−) pathogens, the order is ABC (70%) > PTS (22%) > secondary carriers (8%). These relative numbers may reflect the modes of energy generation in the various groups of organisms under study (Paulsen et al. 1998, 2000; Lorca et al. 2007). It is also interesting to note that in contrast to most other groups of bacteria, LAB prefer mammal- or plant-derived glycosides and oligosaccharides to simple sugars (Lorca et al. 2007).

In virtually all groups of prokaryotes studied, ABC uptake porters of the PepT family (TC# 3.A.1.5) predominate (71–86% for bacteria and nearly 100% for archaea) relative to all other types of peptide uptake systems. Secondary active transport of peptides is relatively rare. In LAB, the order of secondary carrier family representation is POT (TC# 2.A.17; 16%) > OPT (TC# 2.A.67; 10%) > AbgT (TC# 2.A.68) > PAT of the MFS (TC# 2.A.1.25) or of PUP (TC# 9.A.18) (nearly 0% representation). This same trend is valid for other bacterial types although the percentages of these secondary carriers, relative to ABC transporter, is usually greater for LAB. However, Gram (−) bacteria exhibit appreciable representation of PAT and PUP family members, which are largely absent from Gram (+) bacteria. Since ABC systems usually transport their substrates with higher affinities than secondary active transporters, these observations may reflect peptide concentrations in the environments where these organisms evolved.

4.11. Conclusions and Perspectives

Transport proteins in LAB constitute 13–19% of the total genome-encoded genes, in spite of the fact that the average value for most organisms is close to 10% (Lorca et al. 2007). About 5% of these transporters are low specificity, bidirectional, channel proteins involved in adaptation to stress conditions. About 55% are for nutrient uptake while ~40% are for efflux. Specificities for uptake systems are amino acids > sugars > cations = anions > peptides. For exporters, the preferences are drugs >> peptides > macromolecules.

Most LAB have more secondary carriers than ATP-dependent, primary active transporters. However, in some LAB (e.g. *Lact. delbrueckii* and *Strep. thermophilus*), the situation is the reverse (Lorca et al. 2007). These observations must reflect the environmental niches and evolutionary histories that these organisms have experienced. They probably also reflect the environments these organisms prefer to inhabit today.

LAB have tremendous industrial importance for fermentation of food products, for production and spoilage of wine and beer, and for the purpose of promoting mammalian health (see Section 4.1). Based on our studies, presented and discussed in this chapter, we have found that these organisms share some unusual but nearly universal characteristics. They all have mechanosensitive channels for osmotic adaptation, and they have disproportionately large numbers of protective drug efflux pumps. They also have a preponderance of oligosaccharide and glycoside uptake porters relative to transporters that take up free sugars. These organisms largely lack electron carriers although a very limited array of them have been discovered (Vido et al. 2004).

LAB characteristically have substantial numbers of peptide uptake and efflux systems, used for nutrition, signaling, regulation, and biological warfare. All LAB secrete proteins via Sec/Oxa1 systems, but they lack the Sec auxillary proteins, SecDF. Because of their rapid growth rates, they may use ATP to drive secretion rather than the proton motive force (pmf) (Tsukazaki et al. 2011). Moreover, all LAB seem to have competence-related and septal DNA translocation proteins, although competencies for DNA uptake are not a demonstrated characteristic of these organisms. Finally, all LAB examined have at least one (and sometimes more) peptidoglycan (murein) precursor exporters.

Other strikingly unusual characteristics of LAB include:

1. the virtual absence of RND-type exporters;
2. the poor representation of DMT and MOP-type carriers, especially for drugs and metabolites;
3. the preponderance of MFS-type drug exporters (>60% of all MFS carriers);
4. the unusual distribution of the three independently arising ABC-types of integral membrane proteins, (ABC1 : ABC2 : ABC3 = 1>>3>2 for LAB as compared with most other bacteria (1>2>3>) and archaea (2>>1>3);
5. the preponderance of PTS group translocators over primary and secondary active transporters for sugars, although all types are represented;
6. the tremendous variability in PTS transporter representation in the different LAB examined;
7. the unusual representation of divalent cation carriers with the NRAMP and VIT families being used for uptake and the CDF family largely responsible for export; and
8. the surprising representation of eukaryotic-type Na^+ and H^+-ATPases (TC families 3.A.3.1, 3, and 9) absent in most other bacteria including other fermicutes.

Why do LAB have proportions of transporters so different from those of other bacteria, and why do they have a rich supply of transporters, some of which are normally found only in eukaryotes? Perhaps their complement of transporters reflects their unusual life-styles in association with plants and animals. But why do they have distributions of transporter types that differ so much from other microbes, and sometimes from each other? The obvious generalized answer has to be EVOLUTION, but what specifically? There are probably multiple answers: (i) their reduced genome sizes may have allowed the retention of only the most important transporters for these organisms, coping in their individualistic niches. (ii) Their unique habitats, reflecting unusual types of nutrient availability, certainly played a role. (iii) Their apparent obsession with competition and protection, and their preoccupation with biological warfare against other microbes rather than eukaryotes, must have been important. (iv) Possibly a "desire" or need for communal life, dependent on intra- and interspecies communication, played a role. And their close mutually beneficial associations with eukaryotes (animals and plants, for example) was probably important, particularly to facilitate horizontal gene transfer between the eukaryotic and bacterial domains.

Note

This chapter in part reviews aspects of published work (Lorca et al. 2007, 2010) but also reports previously unpublished work delivered in a plenary talk presented at the 9th Symposium on Lactic Acid Bacteria held August 31–September 4, 2008 at Egmond Aan Zee in the Netherlands.

Acknowledgments

We thank the NIH (GM077402 and RO3AI078001) for financial support. We wish to acknowledge the important contributions of Lakshmi Reddy, Anphong Nguyen, Eric I. Sun, John Tseng, Ming-Ren Yen, Cindy Yiu, Soma Patel, Dorjee G. Tamang, Bin Wang, and Foon Hoe Wong. Their efforts were instrumental in the generation of data included in this chapter.

References

Alvarez, M. de F., Medina, R., Pasteris, S.E., et al. (2004) Glycerol metabolism of *Lactobacillus rhamnosus* ATCC 7469: cloning and expression of two glycerol kinase genes. *J Mol Microbiol Biotechnol* **7**, 170–181.

Anba-Mondoloni, J., Chaillou, S., Zagorec, M., and Champomier-Vergès, M.C. (2013) Catabolism of N-acetylneuraminic acid, a fitness function of the food-borne lactic acid bacterium *Lactobacillus sakei*, involves two newly characterized proteins. *Appl Environ Microbiol* **79**, 2012–2018.

Andersen, J.M., Barrangou, R., Abou Hachem, M., et al. (2011) Transcriptional and functional analysis of galactooligosaccharide uptake by *lacS* in *Lactobacillus acidophilus*. *Proc Natl Acad Sci USA* **108**, 17785–17790.

Andersen, J.M., Barrangou, R., Hachem, M.A., et al. (2012) Transcriptional analysis of prebiotic uptake and catabolism by *Lactobacillus acidophilus* NCFM. *PLoS One* **7**, e44409.

Balakrishnan, L., Venter, H., Shilling, R.A., and van Veen, H.W. (2004) Reversible transport by the ATP-binding cassette multidrug export pump LmrA: ATP synthesis at the expense of downhill ethidium uptake. *J Biol Chem* **279**,11273–11280.

Bandell, M., Ansanay, V., Rachidi, N., et al. (1997) Membrane potential-generating malate (MleP) and citrate (CitP) transporters of lactic acid bacteria are homologous proteins. Substrate specificity of the 2-hydroxycarboxylate transporter family. *J Biol Chem* **272**, 18140–18146.

Barabote, R.D. and Saier, M.H., Jr. (2005) Comparative genomic analyses of the bacterial phosphotransferase system. *Microbiol Mol Biol Rev* **69**, 608–634.

Barrangou, R., E. Altermann, R. Hutkins, et al. (2003) Functional and comparative genomic analyses of an operon involved in fructooligosaccharide utilization by *Lactobacillus acidophilus*. *Proc Natl Acad Sci USA* **100**, 8957–8962.

Berntsson, R.P., ter Beek, J., Majsnerowska, M., et al. (2012) Structural divergence of paralogous S components from ECF-type ABC transporters. *Proc Natl Acad Sci USA*. **109**, 13990–13995.

Biemans-Oldehinkel, E., Mahmood, N.A., and Poolman, B. (2006) A sensor for intracellular ionic strength. *Proc Natl Acad Sci USA* **103**, 10624–10629.

Bienert, G.P., Desguin, B., Chaumont, F., and Hols, P. (2013) Channel-mediated lactic acid transport: a novel function for aquaglyceroporins in bacteria. *Biochem J* **454**, 559–570.

Borrero, J., Brede, D.A., Skaugen, M., et al. (2011) Characterization of garvicin ML, a novel circular bacteriocin produced by *Lactococcus garvieae* DCC43, isolated from mallard ducks (Anas platyrhynchos). *Appl Environ Microbiol* **77**, 369–373.

Bramkamp, M., Altendorf, K., and Greie, J.C. (2007) Common patterns and unique features of P-type ATPases: a comparative view on the KdpFABC complex from *Escherichia coli. Mol Membr Biol* **24**, 375–386.

Breukink, E., van Heusden, H.E., Vollmerhaus, P.J., et al. (2003) Lipid II is an intrinsic component of the pore induced by nisin in bacterial membranes. *J Biol Chem* **278**, 19898–19903.

Buchman, G.W., Banerjee, S., and Hansen, J.N. (1988) Structure, expression, and evolution of a gene encoding the precursor of nisin, a small protein antibiotic. *J Biol Chem* **263**, 16260–16266.

Burgess, C.M., Slotboom, D.J., Geertsma, E.R., et al. (2006) The riboflavin transporter RibU in *Lactococcus lactis*: molecular characterization of gene expression and the transport mechanism. *J Bacteriol* **188**, 2752–2760.

Callewaert, R., Holo, H., Devreese, B., et al. (1999) Characterization and production of amylovorin L471, a bacteriocin purified from *Lactobacillus amylovorus* DCE 471 by a novel three-step method. *Microbiology* **145**, 2559–2568.

Carbrey, J.M., Gorelick-Feldman, D.A., Kozono, D., et al. (2003) Aquaglyceroporin AQP9: solute permeation and metabolic control of expression in liver. *Proc Natl Acad Sci USA* **100**, 2945–2950.

Castro, R., Neves, A.R., Fonseca, L.L., et al. (2009) Characterization of the individual glucose uptake systems of *Lactococcus lactis*: mannose-PTS, cellobiose-PTS and the novel GlcU permease. *Mol Microbiol* **71**, 795–806.

Cao, Y., Jin, X., Levin, E.J., Huang, H., Zong, Y., Quick, M., Weng, J., Pan, Y., Love, J., Punta, M., Rost, B., Hendrickson, W.A., Javitch, J.A., Rajashankar, K.R., Zhou, M. (2011) Crystal structure of a phosphorylation-coupled saccharide transporter. *Nature* **473**, 50–54.

Cefalo, A.D., Broadbent, J.R., and Welker D.L. (2011) Protein-protein interactions among the components of the biosynthetic machinery responsible for exopolysaccharide production in Streptococcus thermophilus MR-1C. *J Appl Microbiol* **110**(3):801–812.

Chaillou, S., Bor, Y.C., Batt, C.A., et al. (1998) Molecular cloning and functional expression in *Lactobacillus plantarum* 80 of *xylT*, encoding the D-xylose-H$^+$ symporter of *Lactobacillus brevis. Appl Environ Microbiol* **64**, 4720–4728.

Chaillou, S., Postma, P.W., and Pouwels, P.H. (1998) Functional expression in *Lactobacillus plantarum* of *xylP* encoding the isoprimeverose transporter of *Lactobacillus pentosus. J Bacteriol* **180**, 4011–4014.

Chan, H., Babayan, V., Blyumin, E., et al. (2010) The p-type ATPase superfamily. *J Mol Microbiol Biotechnol* **19**, 5–104.

Choi, W. G., and Roberts, D.M. 2007. Arabidopsis NIP2;1, a major intrinsic protein transporter of lactic acid induced by anoxic stress. *J Biol Chem* **282**, 24209–24218.

Cochu, A., Vadeboncoeur, C., Moineau, S., and Frenette, M. (2003) Genetic and biochemical characterization of the phosphoenolpyruvate:glucose/mannose phosphotransferase system of *Streptococcus thermophilus. Appl Environ Microbiol* **69**, 5423–5432.

Costantini, A., Vaudano, E., Rantsiou, K., et al. (2011) Quantitative expression analysis of *mleP* gene and two genes involved in the ABC transport system in *Oenococcus oeni* during rehydration. *Appl Microbiol Biotechnol* **91**, 1601–1609.

Coton, E., Rollan, G.C., Bertrand, A., and Lonvaud-Funel, A. (1998) Histamine-producing lactic acid bacteria in wines: early detection, frequency and distribution. *Am J Enol Vitic* **49**, 199–203.

Davidson, A.L. and Maloney, P.C. (2007) ABC transporters: how small machines do a big job. *Trends Microbiol* **15**, 448–455.

De Biase, D. and Pennacchietti, E. (2012) Glutamate decarboxylase-dependent acid resistance in orally acquired bacteria: function, distribution and biomedical implications of the *gadBC* operon. *Mol Microbiol* **86**, 770–786.

Defoor, E., Kryger, M.B., and Martinussen, J. (2007) The orotate transporter encoded by *oroP* from *Lactococcus lactis* is required for orotate utilization and has utility as a food-grade selectable marker. *Microbiology* **153**, 3645–3659.

De Hertogh, B., Lantin, A.C., Baret, P.V., and Goffeau, A. (2004) The archaeal P-type ATPases. *J Bioenerg Biomembr* **36**, 135–142.

Delmar, J.A., Su, C.C., and Yu, E.W. (2014) Bacterial multidrug efflux transporters. *Annu Rev Biophys* **43**, 93–117.

de Vrese, M. and Schrezenmeir, J. (2008) Probiotics, prebiotics, and synbiotics. *Adv Biochem Eng Biotechnol* **111**, 1–66.

Doeven, M.K., Abele, R., Tampé, R., and Poolman, B. (2004) The binding specificity of OppA determines the selectivity of the oligopeptide ATP-binding cassette transporter. *J Biol Chem* **279**, 32301–32307.

Drider, D., Fimland, G., Héchard, Y., et al. (2006) The continuing story of class IIa bacteriocins. *Microbiol Mol Biol Rev* **70**, 564–582.

Eijsink, V.G., Axelsson, L., Diep, D.B., et al. (2002) Production of class II bacteriocins by lactic acid bacteria; an example of biological warfare and communication. *Antonie Van Leeuwenhoek* **81**, 639–654.

Elkins, C.A., Moser, S.A., and Savage, D.C. (2001) Genes encoding bile salt hydrolases and conjugated bile salt transporters in *Lactobacillus johnsonii* 100-100 and other *Lactobacillus* species. *Microbiology* **147**, 3403–3412.

Elkins, C.A. and Savage, D.C. (2003) CbsT2 from *Lactobacillus johnsonii* 100-100 is a transport protein of the major facilitator superfamily that facilitates bile acid antiport. *J Mol Microbiol Biotechnol* **6**, 76–87.

Fagan, M.J. and Saier, M.H., Jr. (1994) P-type ATPases of eukaryotes and bacteria: sequence analyses and construction of phylogenetic trees. *J Mol Evol* **38**, 57–99.

Faghiri, Z., Camargo, S.M., Huggel, K., et al. (2010) The tegument of the human parasitic worm *Schistosoma mansoni* as an excretory organ: the surface aquaporin SmAQP is a lactate transporter. *PLoS One* **5**, e10451.

Ferchichi, M., Frère, J., Mabrouk, K., and Manai, M. (2001) Lactococcin MMFII, a novel class IIa bacteriocin produced by *Lactococcus lactis* MMFII, isolated from a Tunisian dairy product. *FEMS Microbiol Lett* **205**, 49–55.

Francl, A.L., Hoeflinger, J.L., and Miller, M.J. (2012) Identification of lactose phosphotransferase systems in *Lactobacillus gasseri* ATCC 33323 required for lactose utilization. *Microbiology* **158**, 944–952.

Fremaux, C., Héchard, Y., and Cenatiempo, Y. (1995) Mesentericin Y105 gene clusters in *Leuconostoc mesenteroides* Y105. *Microbiology* **141**, 1637–1645.

Froger, A., Rolland, J.-P., Bron, P., et al. (2001) Functional characterization of a microbial aquaglyceroporin. *Microbiology* **147**, 1129–1135.

Fujita, K., Ichimasa, S., Zendo, T., et al. (2007) Structural analysis and characterization of lacticin Q, a novel bacteriocin belonging to a new family of unmodified bacteriocins of gram-positive bacteria. *Appl Environ Microbiol* **73**, 2871–2877.

Fulyani, F., Schuurman-Wolters, G.K., Zagar, A.V., et al. (2013) Functional diversity of tandem substrate-binding domains in ABC transporters from pathogenic bacteria. *Structure* **21**, 1879–1888.

Garcia-Alles, L.F., Siebold, C., Nyffeler, T.L., et al. (2004) Phosphoenolpyruvate and ATP-dependent dihydroxyacetone kinases: covalent substrate-binding and kinetic mechanism. *Biochemistry* **43**, 13037–13045.

Garver, K.I. and Muriana, P.M. (1994) Purification and partial amino acid sequence of curvaticin FS47, a heat-stable bacteriocin produced by *Lactobacillus curvatus* FS47. *Appl Environ Microbiol* **60**, 2191–2195.

Gobbetti, M., Corsetti, A., and Rossi, J. (1994) The sourdough microflora. Interactions between lactic acid bacteria and yeasts: metabolism of amino acids. *World J Microbiol Biotechnol* **10**, 275–279.

Grossiord, B.P., Luesink, E.J., Vaughan, E.E., et al.(2003) Characterization, expression, and mutation of the *Lactococcus lactis galPMKTE* genes, involved in galactose utilization via the Leloir pathway. *J Bacteriol* **185**, 870–878.

Guerrero, R. and Berlanga, M. (2006) Life's unity and flexibility: the ecological link. *Int Microbiol* **9**, 225–235.

Halttunen, T., Salminen, S., and Tahvonen, R. (2007) Rapid removal of lead and cadmium from water by specific lactic acid bacteria. *Int J Food Microbiol* **114**, 30–35.

Hao, Z., Chen, S., and Wilson, D.B. (1999) Cloning, expression, and characterization of cadmium and manganese uptake genes from *Lactobacillus plantarum*. *Appl Environ Microbiol* **65**, 4746–4752.

Hastings, J.W., Sailer, M., Johnson, K., et al. (1991) Characterization of leucocin A-UAL 187 and cloning of the bacteriocin gene from *Leuconostoc gelidum*. *J Bacteriol* **173**, 7491–7500.

Hastings, J.W., Stiles, M.E., and von Holy, A. (1994) Bacteriocins of leuconostocs isolated from meat. *Int J Food Microbiol* **24**, 75–81.

Haupt, M., Bramkamp, M., Coles, M., et al. (2005) Prokaryotic Kdp-ATPase: recent insights into the structure and function of KdpB. *J Mol Microbiol Biotechnol* **10**, 120–131.

Hayes, M., Ross, R.P., Fitzgerald, G.F., and Stanton, C. (2007) Putting microbes to work: dairy fermentation, cell factories and bioactive peptides. Part 1:overview. *Biotechnol J* **2**, 426–434.

Hebbeln, P., Rodionov, D.A., Alfandega, A., and Eitinger, T. (2007) Biotin uptake in prokaryotes by solute transporters with an optional ATP-binding cassette-containing module. *Proc Natl Acad Sci USA* **104**, 2909–2914.

Héchard, Y., Dérijard, B., Letellier, F., and Cenatiempo, Y. (1992) Characterization and purification of mesentericin Y105, an anti-Listeria bacteriocin from *Leuconostoc mesenteroides*. *J Gen Microbiol* **138**, 2725–2731.

Henderson, J.T., Chopko, A.L., and van Wassenaar, P.D. (1992) Purification and primary structure of pediocin PA-1 produced by *Pediococcus acidilactici* PAC-1.0. *Arch Biochem Biophys* **295**, 5–12.

Heuberger, E.H., Smits, E., and Poolman, B. (2001) Xyloside transport by XylP, a member of the galactoside-pentoside-hexuronide family. *J Biol Chem* **276**, 34465–34472.

Holck, A., Axelsson, L., Birkeland, S.E., et al. Purification and amino acid sequence of sakacin A, a bacteriocin from Lactobacillus sake Lb706. *J Gen Microbiol* **138**, 2715–2720.

Hvorup, R.N. and Saier, M.H., Jr. (2002) Sequence similarity between the channel-forming domains of voltage-gated ion channel proteins and the C-terminal domains of secondary carriers of the major facilitator superfamily. *Microbiology* **148**, 3760–3762.

Hvorup, R.N., Winnen, B., Chang, A.B., et al. (2003a) The multidrug/oligosaccharidyl-lipid/polysaccharide (MOP) exporter superfamily. *Eur J Biochem* **270**, 799–813.

Hvorup, R.N., Chang, A.B., and Saier, M.H., Jr. (2003b) Bioinformatic analyses of the bacterial L-ascorbate phosphotransferase system permease family. *J Mol Microbiol Biotechnol* **6**, 191–205.

Jack, D.L., Yang, N.M., and Saier, M.H., Jr. (2001) The drug/metabolite transporter superfamily. *Eur J Biochem* **268**, 3620–3639.

Jensen, M.O., Park, S., Tajkhorshid, E., and Schulten, K. (2002) Energetics of glycerol conduction through aquaglyceroporin GlpF. *Proc Natl Acad Sci USA* **99**, 6731–6736.

Justé, A., Malfliet, S., Waud, M., et al. (2014) Bacterial community dynamics during industrial malting, with an emphasis on lactic acid bacteria. *Food Microbiol* **39**, 39–46.

Kaiser, A.L. and Montville, T.J. (1996) Purification of the bacteriocin bavaricin MN and characterization of its mode of action against *Listeria monocytogenes* Scott A cells and lipid vesicles. *Appl Environ Microbiol* **62**, 4529–4535.

Karpowich, N.K. and Wang, D.N. (2013) Assembly and mechanism of a group II ECF transporter. *Proc Natl Acad Sci USA* **110**, 2534–2539.

Kawai, Y., Saito, T., Suzuki, M., and Itoh, T. (1998) Sequence analysis by cloning of the structural gene of gassericin A, a hydrophobic bacteriocin produced by *Lactobacillus gasseri* LA39. *Biosci Biotechnol Biochem* **62**, 887–892.

Kim, O.B., Richter, H., Zaunmüller, T., et al. (2011) Role of secondary transporters and phosphotransferase systems in glucose transport by *Oenococcus oeni*. *J Bacteriol* **193**, 6902–6911.

Kjos, M., Nes, I.F., and Diep, D.B. (2011) Mechanisms of resistance to bacteriocins targeting the mannose phosphotransferase system. *Appl Environ Microbiol* **77**, 3335–3342.

Kowalczyk, M., Cocaign-Bousquet, M., Loubiere, P., and Bardowski, J. (2008) Identification and functional characterisation of cellobiose and lactose transport systems in *Lactococcus lactis IL1403*. *Arch Microbiol* **189**, 187–196.

Kruckeberg, A.L., Walsh, M.C., and Van Dam, K. (1998) How do yeast cells sense glucose? *Bioessays* **20**, 972–976.

Kuhlbrandt, W. (2004) Biology, structure and mechanism of P-type ATPases. *Nat Rev Mol Cell Biol* **5**, 282–295.

Kuipers, O.P., Beerthuyzen, M.M., Siezen, R.J., and De Vos, W.M. (1993) Characterization of the nisin gene cluster *nisABTCIPR* of *Lactococcus lactis*. Requirement of expression of the *nisA* and *nisI* genes for development of immunity. *Eur J Biochem* **216**, 281–291.

Labarre, C., Divies, C., and Guzzo, J. (1996a) Genetic organization of the *mle* locus and identification of a *mleR*-like gene from *Leuconostoc oenos*. *Appl Env Microbiol* **62**, 4493–4498.

Labarre, C., Guzzo, J., Cavin, J.F., and Divies, C. (1996b) Cloning and characterization of the genes encoding the malolactic enzyme and the malate permease of *Leuconostoc oenos*. *Appl Environ Microbiol* **62**, 1274–1282.

Lamarque, M., Aubel, D., Piard, J.C., et al. (2011) The peptide transport system Opt is involved in both nutrition and environmental sensing during growth of *Lactococcus lactis* in milk. *Microbiology* **157**, 1612–1619.

Law, C.J., Maloney, P.C., and Wang, D.N. (2008) Ins and outs of major facilitator superfamily antiporters. *Annu Rev Microbiol* **62**, 289–305.

Lonvaud-Funel, A., (1999) Lactic acid bacteria in the quality improvement and depreciation of wine. *Antonie Van Leeuwenhoek* **76**, 317–331.

Lorca, G.L., Barabote, R.D., Zlotopolski, V., et al. (2007) Transport capabilities of eleven Gram-positive bacteria: comparative genomic analyses. *BBA- Biomembranes* **1768**, 1342–1366.

Lorca, G., Reddy, L., Nguyen, A., et al. (2010) Lactic acid bacteria: comparative genomic analyses of transport systems, in *Biotechnology of Lactic Acid Bacteria: Novel Applications* ((eds) F. Mozzi, R. R. Raya, and G. M. Vignolo), Wiley-Blackwell, Oxford, UK.

Lubelski, J., Mazurkiewicz, P., van Merkerk, R., et al. (2004) *ydaG* and *ydbA* of *Lactococcus lactis* encode a heterodimeric ATP-binding cassette-type multidrug transporter. *J Biol Chem* **279**, 34449–34455.

Lubelski, J., de Jong, A., van Merkerk, R., et al. (2006) LmrCD is a major multidrug resistance transporter in *Lactococcus lactis*. *Mol Microbiol* **61**, 771–781.

Lüthi-Peng, Q., Dileme, F.B., and Puhan, Z. (2002) Effect of glucose on glycerol bioconversion by *Lactobacillus reuteri*. *Appl Microbiol Biotechnol* **59**, 289–296.

Luo, P., Yu, X., Wang, W., Fan, S., Li, X., Wang J. (2015) Crystal structure of a phosphorylation-coupled vitamin C transporter. *Nat Struct Mol Biol*. **22**, 238–241.

Lytton, J. (2007) Na^+/Ca^+ exchangers: three mammalian gene families control Ca^{2+} transport. *Biochem J* **406**, 365–382.

Majsnerowska, M., Hänelt, I., Wunnicke, D., et al. (2013) Substrate-induced conformational changes in the S-component ThiT from an energy coupling factor transporter. *Structure* **21**, 861–867.

Makarova, K., Slesarev, A., Wolf, Y., et al. (2006) Comparative genomics of the lactic acid bacteria. *Proc Natl Acad Sci USA*. **103**, 15611–15616.

Martinussen, J., Schallert, J., Andersen, B., and Hammer, K. (2001) The pyrimidine operon *pyr*RPB-*car*A from *Lactococcus lactis*. *J Bacteriol* **183**, 2785–2794.

Martinussen, J., Sørensen, C., Jendresen, C.B., and Kilstrup, M. (2010) Two nucleoside transporters in *Lactococcus lactis* with different substrate specificities. *Microbiology* **156**, 3148–3157.

Marciset, O., Jeronimus-Stratingh, M.C., Mollet, B., and Poolman, B. (1997) Thermophilin 13, a nontypical antilisterial poration complex bacteriocin, that functions without a receptor. *J Biol Chem* **272**, 14277–14284.

Mazurkiewicz, P., Driessen, A.J., and Konings, W.N. (2004) Energetics of wild-type and mutant multidrug resistance secondary transporter LmrP of *Lactococcus lactis*. *Biochim Biophys Acta* **1658**, 252–261.

Mills, D.A., Rawsthorne, H., Parker, C., et al. (2005) Genomic analysis of *Oenococcus ceni* PSU-1 and its relevance to winemaking. *FEMS Microbiol Rev* **29**, 465–475.

Mohamed, Y. F. and Valvano, M. A. (2014) A *Burkholderia cenocepacia* MurJ (MviN) homolog is essential for cell wall peptidoglycan synthesis and bacterial viability. *Glycobiology* **24**, 564–576.

Mørtvedt, C.I., Nissen-Meyer, J., Sletten, K., and Nes, I.F. (1991) Purification and amino acid sequence of lactocin S, a bacteriocin produced by *Lactobacillus sake* L45. *Appl Environ Microbiol* **57**, 1829–1834.

Muriana, P.M. and Klaenhammer, T.R. (1991) Purification and partial characterization of lactacin F, a bacteriocin produced by *Lactobacillus acidophilus* 11088. *Appl Environ Microbiol* **57**, 114–121.

Nes, I.F., Diep, D.B., and Holo, H. (2007) Bacteriocin diversity in *Streptococcus* and *Enterococcus*. *J Bacteriol* **189**, 1189–1198.

Nguyen, T.X., Yen, M.R., Barabote, R.D., and Saier, M.H., Jr. (2006) Topological predictions for integral membrane permeases of the phosphoenolpyruvate:sugar phosphotransferase system. *J Mol Microbiol Biotechnol* **11**, 345–360.

Nissen-Meyer, J., Larsen, A.G., Sletten, K., et al. (1993) Purification and characterization of plantaricin A, a *Lactobacillus plantarum* bacteriocin whose activity depends on the action of two peptides. *J Gen Microbiol* **139**, 1973–1978.

Nissen-Meyer, J., Holo, H., Håvarstein, L.S.,et al. (1992) A novel lactococcal bacteriocin whose activity depends on the complementary action of two peptides. *J Bacteriol* **174**, 5686–5692.

Okuda, K., Yanagihara, S., Sugayama, T., et al. (2010) Functional significance of the E loop, a novel motif conserved in the lantibiotic immunity ATP-binding cassette transport systems. *J Bacteriol* **192**, 2801–2808.

Olsen, G.J., Woese, C.R., and Overbeek, R. (1994) The winds of (evolutionary) change: breathing new life into microbiology. *J Bacteriol* **176**, 1–6.

Oppegård, C., Rogne, P., Emanuelsen, L., et al. (2007). The two-peptide class II bacteriocins: structure, production, and mode of action. *J. Mol Microbiol Biotechnol* **13**, 210–219.

Pagliai, F.A., Gardner, C.L., Pande, S.G., and Lorca, G.L. (2010) LVIS553 transcriptional regulator specifically recognizes novobiocin as an effector molecule. *J Biol Chem* **285**, 16921–16930.

Palmgren, M.G. and Axelsen, K.B. (1998) Evolution of P-type ATPases. *Biochim Biophys Acta* **1365**, 37–45.

Pao, S.S., Paulsen, I.T., and Saier, M.H., Jr. (1998) The major facilitator superfamily. *Microbiol Mol Biol Rev* **62**, 1–32.

Parracho, H., McCartney, A.L., and Gibson, G.R. (2007) Probiotics and prebiotics in infant nutrition. *Proc Nutr Soc* **66**, 405–411.

Paulsen, I.T., Nguyen, L., Sliwinski, M.K., et al. (2000) Microbial genome analyses: comparative transport capabilities in eighteen prokaryotes. *J Mol Biol* **301**, 75–100.

Paulsen, I.T., Sliwinski, M.K., and Saier, M.H., Jr. (1998) Microbial genome analyses: global comparisons of transport capabilities based on phylogenies, bioenergetics and substrate specificities. *J Mol Biol* **277**, 573–592.

Penaud S., Fernandez, A., Boudebbouze, S., et al. (2006) Induction of heavy-metal-transporting CPX-type ATPases during acid adaptation in *Lactobacillus bulgaricus*. *Appl Environ Microbiol* **72**, 7445–7454.

Piard, J.C., Kuipers, O.P., Rollema, H.S., et al. (1993) Structure, organization, and expression of the *lct* gene for lacticin 481, a novel lantibiotic produced by *Lactococcus lactis*. *J Biol Chem* **268**, 16361–16368.

Pickl, A., Johnsen, U., and Schönheit, P. (2012) Fructose degradation in the haloarchaeon *Haloferax volcanii* involves a bacterial type phosphoenolpyruvate-dependent phosphotransferase system, fructose-1-phosphate kinase, and class II fructose-1,6-bisphosphate aldolase. *J Bacteriol* **194**, 3088–3097.

Poolman, B., Smid, E.J., and Konings, W.N. (1987) Kinetic properties of a phosphate-bond-driven glutamate-glutamine transport system in *Streptococcus lactis* and *Streptococcus cremoris*. *J Bacteriol* **169**, 2755–2761.

Pudlik, A.M. and J.S. Lolkema. (2012) Substrate specificity of the citrate transporter CitP of *Lactococcus lactis*. *J Bacteriol* **194**, 3627–3635.

Rattray, F.P. and Fox, P.F. (1999) Aspects of enzymology and biochemical properties of *Brevibacterium linens* relevant to cheese ripening: a review. *J Dairy Sci* **82**, 891–909.

Reddy, G., Altaf, M., Naveena, B.J., et al. (2008) Amylolytic bacterial lactic acid fermentation- a review. *Biotechnol Adv* **26**, 22–34.

Reddy, V.S., Shlykov, M.A., Castillo, R., et al. (2012) The major facilitator superfamily (MFS) revisited. *FEBS J* **279**, 2022–2035. Erratum in: *FEBS J* (2013) **280**, 3975.

Ren, Q. and Paulsen, I.T. (2005) Comparative analyses of fundamental differences in membrane transport capabilities in prokaryotes and eukaryotes. *PloS Comput Biol* **1**, e27.
Revilla-Guarinos, A., Gebhard, S., Alcántara, C., et al. (2013) Characterization of a regulatory network of peptide antibiotic detoxification modules in *Lactobacillus casei* BL23. *Appl Environ Microbiol* **79**, 3160–3170.
Rimaux, T., Rivière, A., Hebert, E.M., et al. (2013) A putative transport protein is involved in citrulline excretion and re-uptake during arginine deiminase pathway activity by *Lactobacillus sakei*. *Res Microbiol* **164**, 216–225.
Rodionov, D.A., Hebbeln, P., Eudes A., et al. (2009) A novel class of modular transporters for vitamins in prokaryotes. *J Bacteriol* **191**, 42–51.
Rodionov, D.A., Vitreschak, A.G., Mironov, A.A., and Gelfand, M.S. (2002) Comparative genomics of thiamin biosynthesis in procaryotes. New genes and regulatory mechanisms. *J Biol Chem* **277**, 48949–48959.
Rodríguez-Díaz, J., Rubio-del-Campo, A., and Yebra, M.J. (2012) *Lactobacillus casei* ferments the N-Acetylglucosamine moiety of fucosyl-α-1,3-N-acetylglucosamine and excretes L-fucose. *Appl Environ Microbiol* **78**, 4613–4619.
Rosselló-Mora, R. (2005) Updating prokaryotic taxonomy. *J Bacteriol* **187**, 6255–6257.
Rouse, S. and van Sinderen, D. (2008) Bioprotective potential of lactic acid bacteria in malting and brewing. *J Food Prot.* **71**, 1724–1733.
Saier, M.H., Jr., Beatty, J.T., Goffeau, A., et al. (1999) The major facilitator superfamily. *J Mol Microbiol Biotechnol* **1**, 257–279.
Saier, M.H., Jr., Hvorup, R.N., and Barabote, R.D. (2005) Evolution of the bacterial phosphotransferase system: from carriers and enzymes to group translocators. *Biochem Soc Trans* **33**, 220–224.
Sakamoto, K., Margolles, A., van Veen H.W., and Konings, W.N. (2001) Hop resistance in the beer spoilage bacterium *Lactobacillus brevis* is mediated by the ATP-binding cassette multidrug transporter HorA. *J Bacteriol* **183**, 5371–5375.
Sanz, Y., Nadal, I., and Sanchez, E. (2007) Probiotics as drugs against human gastrointestinal infections. *Recent Patents Anti-Infect Drug Disc* **2**, 148–156.
Sawa, N., Zendo, T., Kiyofuji, J.ct al. (2009). Identification and characterization of lactocyclicin Q, a novel cyclic bacteriocin produced by *Lactococcus* sp. strain QU 12. *Appl Environ Microbiol* **75**, 1552–1558.
Schick, J., Weber, B., Klein, J.R., and Henrich, B. (1999) PepR1, a CcpA-like transcription regulator of *Lactobacillus delbrueckii* subsp. *lactis*. *Microbiology* **145**, 3147–3154.
Sela, D.A. and Mills, D.A. (2014) The marriage of nutrigenomics with the microbiome: the case of infant-associated bifidobacteria and milk. *Am J Clin Nutr* **99**, 697S–703S.
Shavakhi, A., Minakari, M., Farzamnia, S., et al. (2014) The effects of multi-strain probiotic compound on symptoms and quality-of-life in patients with irritable bowel syndrome: A randomized placebo-controlled trial. *Adv Biomed Res* **3**, 140.
Siebold, C., Garcia-Alles, L.F., Erni, B., and Baumann, U. (2003) A mechanism of covalent substrate binding in the X-ray structure of subunit K of the *Escherichia coli* dihydroxyacetone kinase. *Proc Natl Acad Sci USA* **100**, 8188–1892.
Solcan, N., Kwok, J., Fowler, P.W.,et al. (2012) Alternating access mechanism in the POT family of oligopeptide transporters. *EMBO J* **31**, 3411–3421
Stentz, R. and Zagorec, M. (1999) Ribose utilization in *Lactobacillus sakei*: analysis of the regulation of the rbs operon and putative involvement of a new transporter. *J Mol Microbiol Biotechnol* **1**, 165–173.
Stephens, S.K., Floriano, B., Cathcart, D.P., et al. (1998) Molecular analysis of the locus responsible for production of plantaricin S, a two-peptide bacteriocin produced by *Lactobacillus plantarum* LPCO10. *Appl Environ Microbiol* **64**, 1871–1877.
Stoddard, G.W., Petzel, J.P., van Belkum, M.J., et al. (1992) Molecular analyses of the lactococcin A gene cluster from *Lactococcus lactis* subsp. *lactis* biovar diacetylactis WM4. *Appl Environ Microbiol* **58**, 1952–1961.
Thever, M.D. and Saier, M.H. Jr. (2009) Bioinformatic characterization of p-type ATPases encoded within the fully sequenced genomes of 26 eukaryotes. *J Membr Biol* **229**, 115–130.
Tichaczek, P.S., Vogel, R.F., and Hammes, W.P. (1994) Cloning and sequencing of *sakP* encoding sakacin P, the bacteriocin produced by *Lactobacillus sake* LTH 673. *Microbiology* **140**, 361–367.
Trip, H., Mulder, N.L., and Lolkema, J.S. (2012) Improved acid stress survival of *Lactococcus lactis* expressing the histidine decarboxylation pathway of *Streptococcus thermophilus* CHCC1524. *J Biol Chem* **287**, 11195–11204.
Trip, H., Mulder, N.L., and Lolkema, J.S. (2013) Cloning, expression, and functional characterization of secondary amino acid transporters of *Lactococcus lactis*. *J Bacteriol* **195**, 340–350.
Tseng, T.T., Gratwick, K.S., Kollman, J., et al. (1999) The RND permease superfamily: an ancient, ubiquitous and diverse family that includes human disease and development proteins. *J Mol Biotechnol* **1**, 107–125.
Tsukaguchi, H., Weremowicz, S., Morton, C.C., and Hediger, M.A. (1999) Functional and molecular characterization of the human neutral solute channel aquaporin-9. *Am J Physiol* **277**, F685–696.

Tsukazaki, T., Mori, H., Echizen, Y., et al. (2011) Structure and function of a membrane component SecDF that enhances protein export. *Nature* **474**, 235–238.

Uehlein, N., Lovisolo, C., Siefritz, F., and Kaldenhoff, R. (2003) The tobacco aquaporins NtAQP1 is a membrane CO2 pore with physiological functions. *Nature* **425**, 734–737.

van Belkum, M.J., Kok, J., and Venema, G. (1992) Cloning, sequencing, and expression in *Escherichia coli* of lcnB, a third bacteriocin determinant from the lactococcal bacteriocin plasmid p9B4-6. *Appl Environ Microbiol* **58**, 572–577.

van Zanten, G.C., Krych, L., Röytiö, H., et al. (2014) Synbiotic *Lactobacillus acidophilus* NCFM and cellobiose does not affect human gut bacterial diversity but increases abundance of lactobacilli, bifidobacteria and branched-chain fatty acids: a randomized, double-blinded cross-over trial. *FEMS Microbiol Ecol.* **XX,** 1–12.

Västermark, A., Lunt, B., and Saier, M. (2014) Major facilitator superfamily porters, LacY, FucP and XylE of *Escherichia coli* appear to have evolved positionally dissimilar catalytic residues without rearrangement of 3-TMS repeat units. *J Mol Microbiol Biotechnol* **24**, 82–90.

Vaughan, E.E., de Vries, M.C., Zoetendal, E.G., et al. (2002) The intestinal LAB. *Antonie Van Leeuwenhoek* **82,** 341–352.

Venema, K., Kok, J., Marugg, J.D., et al. (1995) Functional analysis of the pediocin operon of *Pediococcus acidilactici* PAC1.0: PedB is the immunity protein and PedD is the precursor processing enzyme. *Mol Microbiol* **17**, 515–522.

Vido, K., le Bars, D., Mistou, M-Y., et al. (2004) Proteome analyses of heme-dependent respiration in *Lactococcus lactis*: involvement of the proteolytic system. *J Bacteriol* **186**, 1648–1657.

Wang, B., Dukarevich, M., Sun, E.I., et al. (2009) Membrane porters of ATP-binding cassette transport systems are polyphyletic. *J Membr Biol* **231**, 1–10.

Wiedemann, I., Benz, R., and Sahl, H.G. (2004) Lipid II-mediated pore formation by the peptide antibiotic nisin: a black lipid membrane study. *J Bacteriol* **186**, 3259–3261.

Xie, K., Kiefer, G., Nagler, R.E., et al. (2006) Different regions of the nonconserved large periplasmic domain of *Escherichia coli* YidC are involved in the SecF interaction and membrane insertase activity. *Biochemistry* **45**, 13401–13408.

Xu, K., Zhang, M., Zhao, Q., et al. (2013) Crystal structure of a folate energy-coupling factor transporter from *Lactobacillus brevis*. *Nature.* **497**, 268–271.

Yen, M.R., Chen, J.S., Marquez, J.L., et al. (2010) Multidrug resistance: phylogenetic characterization of superfamilies of secondary carriers that include drug exporters. *Methods Mol Biol* **637**, 47–64.

Yebra, M.J., Monedero, V., Zuniga, M., et al. (2006) Molecular analysis of the glucose-specific phosphoenolpyruvate:sugar phosphotransferase system from *Lactobacillus casei* and its links with the control of sugar metabolism. *Microbiology* **152**, 95–104.

Yoneyama, F., Imura, Y., Ohno, K., et al. (2009). Peptide-lipid huge toroidal pore, a new antimicrobial mechanism mediated by a lactococcal bacteriocin, lacticin Q. *Antimicrob Agents Chemother* **53**, 3211–3217.

Zheng, W.H., Västermark, Å., Shlykov, M.A., et al. (2013) Evolutionary relationships of ATP-Binding Cassette (ABC) uptake porters. *BMC Microbiol* **13**, 13–98.

Chapter 5

Novel Developments in Bacteriocins from Lactic Acid Bacteria

Ingolf F. Nes[1,*], Christina Gabrielsen[2], Dag A. Brede[1], and Dzung B. Diep[1]

[1] *Department of Chemistry, Biotechnology and Life Science, Norwegian University of Life Sciences, Norway*

[2] *Norwegian University of Science and Technology, Norway*

5.1. Introduction

Production of antimicrobial peptides is an ancient evolutionary conserved antagonistic trait that is found in all domains of life, from the host defense peptides of the innate immune system in plants, animals, and humans, to the archaeocins produced by archaea and the bacteriocins produced by bacteria (Nissen-Meyer and Nes 1997; O'Connor and Shand 2002; Zasloff 2002). Bacteriocins are defined as ribosomally synthesized antimicrobial peptides produced by Gram (+) and Gram (−) bacteria (Nes et al. 1996; Riley and Wertz 2002; Tagg et al. 1976), which presumably give the producer strain an advantage over closely related bacteria in the competition for nutrition in an ecological niche (Eijsink et al. 2002). Bacteriocin production generally commences at high cell densities when resources become scarce, in response to quorum sensing mechanisms, and as such also serve as a means of intraspecies bacterial communication (Nes and Eijsink 1999; Nes et al. 2006). The overwhelming majority of bacteriocins discovered so far come from the Gram (+) lactic acid bacteria (LAB). Because LAB are ubiquitous in the microbiota of plants and animals, and are used in food preservation and production as starter cultures for a wide range of dairy and fermented foods, humans have likely benefited from their effects as inhibitors of common food-borne pathogens and food-spoiling bacteria.

5.2. Characteristics and Classification of Bacteriocins

LAB bacteriocins constitute a large and heterogeneous group, from which bacteriocins have been classified according to features such as size, homology, structure, and post-translational modifications and so on. The classification of LAB bacteriocins is frequently revised as the diversity increases and more bacteriocins with novel features are characterized (Nes et al. 1996; Cotter et al. 2005b; Nissen-Meyer et al. 2009b; Rea et al. 2011a), but the main distinction is made between the Class Ia lantibiotics, which undergo post-translational modifications, and the Class II non-modified peptides. Previously, a third class consisting of large heat-labile lytic proteins (bacteriolysins) was

* Corresponding author: Department of Chemistry, Biotechnology and Life Science, Norwegian University of Life Sciences, Box 5003, N1432 Aas, Norway. ingolf.nes@nmbu.no

Biotechnology of Lactic Acid Bacteria: Novel Applications, Second Edition. Edited by Fernanda Mozzi, Raúl R. Raya, and Graciela M. Vignolo.

included in this classification scheme; however, according to Cotter et al. (2005b) and some subsequent classifications, these are no longer regarded as bacteriocins.

5.2.1. Class Ia: Lantibiotics

The Class I bacteriocins are mainly comprised of a large group known as the lantibiotics, which constitute the antimicrobials of the groups of lanthipeptides (Arnison et al. 2013). The lanthipeptides constitute all lanthionine and methyllanthionine containing peptides of which lantibiotics form a large subgroup. It should be emphasized that the lanthipeptides also include peptides with different biological activities, not only antimicrobials. Arnison et al. (2013) define four different classes of lanthipeptides according to the four different types of biosynthetic enzymes that install the characteristic thioether crosslinks (Arnison et al. 2013).

The lantibiotics, which include the prototype bacteriocin nisin, are small peptides (approx. 19–38 residues) defined by the thioether-containing amino acids lanthionine (Lan) and β-methyllanthionine (MeLan). These residues are introduced by a two-step post-translational process catalyzed by dedicated dehydratase, cyclase, or synthetase enzymes. The Lan and MeLan residues form crosslinks between them, resulting in characteristic intramolecular cyclic structures (Bierbaum et al. 1996). Class I lantibiotics may also contain other post-translationally modified amino acids, for example didehydroalanine (Dha) and didehydrobutyrine (Dhb), D-alanine and others. As seen among Class IIb bacteriocins, some Class I bacteriocins need two peptide moieties for optimal activity. The lantibiotics have been subdivided further into 11 groups based on features such as mode of action, structure, and, more recently, also on sequence similarity of the peptides (Cotter et al. 2005a). Recently, two new subclasses have been included into the post-translationally modified bacteriocins (Rea et al. 2011a): The first new subclass (designated as Class Ib) consists of labyrinthopeptins that contain the modified amino acid labionin. Labionins are carbacyclic amino acid residues that form thioether and methylene bridges, resulting in the complex cyclic structures of these peptides (Meindl et al. 2010). Labyrinthopeptin A1 possesses antiviral activity which is beyond the scope of this review and therefore will not be discussed further and is not included in Figure 5.1 (Ferir et al. 2013). Members of the second new subclass (Class Ic) are known as the sactibiotics, and are characterized by several α-carbon to sulfur crosslinks (see Figure 5.1) (Murphy et al. 2011). The first characterized bacteriocin containing these unusual bridges was the head-to-tail circularized bacteriocin subtilosin A produced by *Bacillus subtilis* (Kawulka et al. 2004), while another well-known example is thuricin CD, a linear two-peptide bacteriocin produced by *B. thuringiensis* (Rea et al. 2010).

5.2.2. Class II: The Non-lantibiotics

Class II bacteriocins are a large and heterogeneous group of small (<10 kDa), heat stable peptides defined by their non-modified nature (Nes and Holo 2000; Cotter et al. 2005b). The common designation of these peptides as "unmodified" is, however, somewhat misleading: some Class II bacteriocins in fact contain modifications like disulfide bonds (Fimland et al. 2000), formylated *N*-methionine residues (Fujita et al. 2007; Liu et al. 2011), or have circular backbones (Maqueda et al. 2008; Martin-Visscher et al. 2009). This class of bacteriocins is divided into several subgroups based on structural and functional characteristics (Figure 5.1).

Subclass IIa are pediocin-like bacteriocins with a conserved *N*-terminal sequence motif known as the "pediocin box" and one or two intrachain disulfide bonds (Lozano et al. 1992; Nes and Holo 2000). The peptides form a very characteristic structure, consisting of a conserved *N*-terminal antiparallel β-sheet connected by a flexible hinge to the more variable *C*-terminal hairpin-like domain

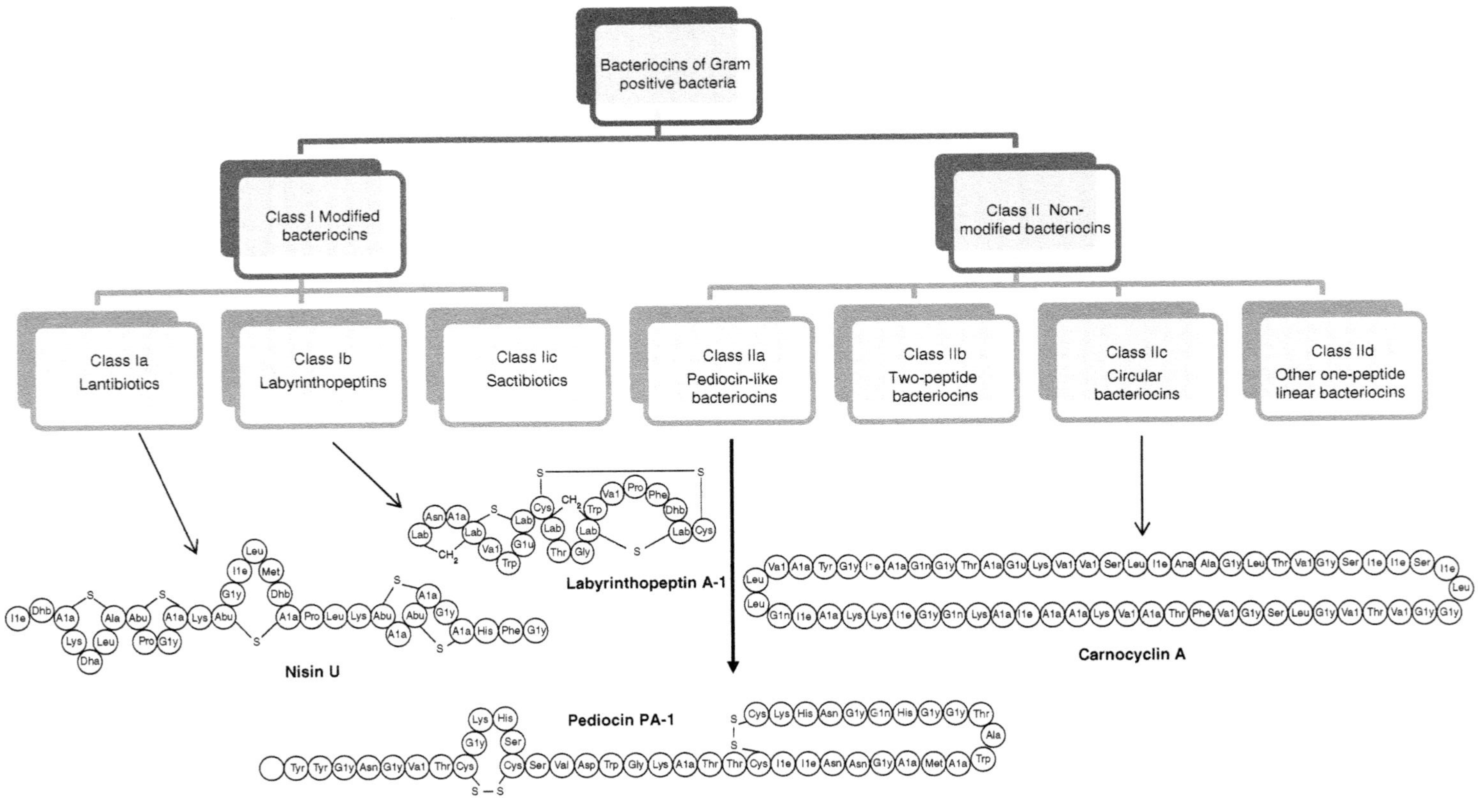

Figure 5.1. Classification scheme of the bacteriocins of Gram (+) bacteria, revised from Cotter et al. (2005b). The main distinction is made between the Class I lantibiotics, which undergo extensive post-translational modifications, and the Class II non-modified bacteriocins. The amino acid sequence and structural features of one example bacteriocin of each different class are shown.

shaped in membrane-mimicking environments (Nissen-Meyer et al. 2009b). Class IIa bacteriocins are synthesized as precursors with an *N*-terminal leader sequence of the type referred to as a double-glycine type leader (Havarstein et al. 1994). This leader is cleaved *C*-terminally to the double-glycine motif upon export out of the cell, which is processed by a dedicated ABC transporter containing an *N*-terminal proteolytic domain and the help of an accessory protein. Such leaders and dedicated ABC transporters are also found amongst some lantibiotics (Havarstein et al. 1995). The mature peptides generally consist of 37–48 amino acid residues and are cationic, usually with a pI between 8 and 10 (Nes et al. 1996). The subclass IIa bacteriocins have narrow inhibitory spectra, but they are known for their anti-listerial and anti-enterococcal activity. With approximately 50 characterized members to date (Cui et al. 2012), they constitute the largest subclass of Class II bacteriocins, and have been extensively studied with respect to genetics, structure, and mode of action (Nes et al. 2007; Oppegard et al. 2007; Diep et al. 2009; Nissen-Meyer et al. 2010; Nissen-Meyer et al. 2009a; Rogne et al. 2009). Subclass IIb consists of two-peptide bacteriocins whose optimal activity is dependent on the combined action of two non-identical peptide moieties (Nissen-Meyer et al. 1992). The individual peptides exert no or very low antimicrobial activity and for each two-peptide bacteriocin only one cognate immunity protein is required to protect the producer from being killed by its own bacteriocin, as seen with the Class II one-peptide bacteriocins. Two-peptide bacteriocins are generally cationic, but contain hydrophilic and/or amphiphilic regions, and are synthesized with a double-glycine leader like most Class IIa bacteriocins. The structures of two-peptide bacteriocins are in some cases characterized by a central α-helix with flexible *N*- and *C*-terminal regions, and the two peptides are thought to interact via helix–helix interactions within the target membrane (Nissen-Meyer et al. 2009b).

Class IIc bacteriocins comprise the circular bacteriocins, characterized by a head-to-tail covalent bond forming a circular structure (Figure 5.1) (van Belkum et al. 2011; Gabrielsen et al. 2014). These bacteriocins are known for a relatively broad inhibitory spectrum and resistance to heat, extreme pH, and even many proteases.

Lastly, Class IId is generally viewed as a collection of the remaining non-modified bacteriocins, which includes all other non-pediocin-like linear one-peptide bacteriocins. This class can be further divided based on leader sequences: (i) The *sec*-dependent bacteriocins contain a leader sequence that directs peptides for export by the general secretory pathway (GSP), (ii) double-glycine leaders are found most frequently, whereas (iii) the leaderless bacteriocins, which, as the name indicates, lack a leader sequence, are often secreted by dedicated ABC transporters (Kastin 2006). Numerous leaderless bacteriocins have been identified and they share the common characteristics observed for other bacteriocins in being small, heat-stable, and cationic with a high pI. Some of the leaderless bacteriocins have been shown to retain the formyl group of the *N*-terminal methionine residue (Liu et al. 2011). The leaderless bacteriocins have been found as one peptide, but more frequently at least two or even more homologous peptides are produced by the same bacterium. In 1998 the two homologous peptides enterocin A and B were isolated from an *Enterococcus faecium* strain. Both peptides showed antibacterial activity but together they acted strongly synergistically (Cintas et al. 1998). Several leaderless bacteriocins are found as one peptide, such as enterocin Q and LsbB produced by *Ent. faecium* and *L. lactis*, respectively (Cintas et al. 2000; Gajic et al. 2003). In *Staphylococcus aureus* up to four homologous leaderless bacteriocins have been identified and they act synergistically (Netz et al. 2001). All four peptides are encoded by genes located in the same operon and they are produced, expressed, and excreted without post-translational modifications. While the peptides that comprise the two-peptide bacteriocins (subclass IId) share no sequence homology, the peptides comprise a leaderless bacteriocin cluster sharing sequence homology. This might indicate that the two-peptide bacteriocins originate from the merging of two separate bacteriocins, while the leaderless bacteriocins have developed through a gene duplication mechanism.

5.3. Mode of Action

Generally, Class Ia bacteriocins have broad antimicrobial spectra and display antimicrobial activity towards Gram (+) bacteria. The latter may be due to restricted access to targets including the cytoplasmic membrane in Gram (–) bacteria, as compromising the Gram (–) outer membrane in some cases has been shown to sensitize cells to lantibiotics (Stevens et al. 1991; Schved et al. 1994). Although mechanistic studies have been performed for only few bacteriocins of this class, it has become evident that there are common features in the mode of action. First and foremost, the requirement for a target-, receptor- or docking molecule has been demonstrated for some lantibiotics to be the cell wall precursor lipid II (Brotz et al. 1998a; Wiedemann et al. 2001, 2006b). Different sites on the lipid II molecule might be involved in target recognition depending on the type of lantibiotics (Wiedemann et al. 2006a,b; Martinez et al. 2008; Bottiger et al. 2009). The model lantibiotic nisin, for example, binds to the pyrophosphate moiety of the lipid II molecule (Hsu et al. 2004). The binding of nisin to lipid II in turn facilitates a dual effect on target cells: (i) the inhibition of peptidoglycan biosynthesis by interaction with and sequestration of lipid II and (ii) the formation of pores in the cytoplasmic membrane (Wiedemann et al. 2001). While the docking (target)-independent membrane permeabilization activity of nisin occurs at micromolar concentrations *in vitro* (Brotz et al. 1998b), the target-specific interaction with lipid II has demonstrated an increase in antimicrobial activity by 1000-fold (i.e., at nanomolar range) (Breukink et al. 1999). This dual mode of action thus explains the high potency of many lantibiotics, which often display lipid II-mediated antimicrobial activity in the nanomolar range *in vivo* (Breukink et al. 1999). Nisin has a modular structure, where the two *N*-terminal ring structures (known as the "pyrophosphate cage") are essential for binding to the pyrophosphate moiety of lipid II, whereas the *C*-terminal region is able to insert into the cytoplasmic membrane, oligomerize, and form a pore consisting of eight nisin molecules and four lipid II molecules (Hasper et al. 2004). The hinge region separating these modules is important for movement of the modules relative each other, and thus also vital for pore formation (Yuan et al. 2004). Similar structural modules are believed to also form functional units in other lantibiotics; the pyrophosphate cage is, for instance, conserved in the nisin and epidermin group (Sahl and Bierbaum 1998). Variations in these structures may thus account for observed differences in antimicrobial activity and the spectrum of different types of lantibiotics. However, not all lantibiotics share the pore-forming mechanism with nisin, but instead target enzymatic reactions: Mersacidin and nukacin ISK-1 both bind lipid II and inhibit peptidoglycan biosynthesis in sensitive cells (Brotz et al. 1998a; Asaduzzaman et al. 2009), while lantibiotics of the cinnamycin group have been shown to target phosphatidylethanolamine in the cell membrane and function by inhibiting the activity of phospholipase A2 (Marki et al. 1991). The two-peptide lantibiotic (Smb) produced by a *Streptococcus mutans* strain, which displays inhibitory activity against a broad spectrum of bacteria, has been shown to need a receptor/target for its activity (Biswas and Biswas 2014). It was shown that Smb required a receptor-like protein on the susceptible *Strep. pyogenes* strain and this receptor-like protein was identified as a protease-like protein (named LsrS) that belongs to the CAAX protease family. Surprisingly, nisin also appeared to require LsrS for optimum inhibitory activity, while the closely related Smb lantibiotics, haloduracin and galolacticin, did not.

Class II bacteriocins generally have well-defined and narrow antimicrobial spectra. From early on, it was established that certain Class II bacteriocins kill target cells by permeabilization of the membrane of targeted bacteria or by interfering with the integrity of the cytoplasmic membrane (van Belkum et al. 1991; Chikindas et al. 1993; Moll et al. 1996, 1999a, b). It is now established that several bacteriocins target specific proteins (receptors), but for most Class II bacteriocins, however, the detailed molecular mechanisms and potential target molecule(s) are as yet unknown. The exception is class IIa bacteriocins and lactococcin A (Class IId), which have been demonstrated to specifically target the transmembrane IIC and IID proteins of the mannose phosphotransferase system (Man-PTS), a broad specificity glucose uptake system, of the sensitive cells (Ramnath et al. 2004; Diep et al. 2007). Furthermore, it has been shown that an *N*-terminal extracellular loop-containing region in the IIC protein is the major determinant responsible for species-specificity, potentially serving as an interaction site for the Class IIa bacteriocins (Kjos et al. 2010). Lactococcin A, which is a *Lactococcus*-specific bacteriocin, targets only Man-PTS of *L. lactis* but both the transmembrane proteins (IIC and IID) are involved in bacteriocin recognition (Diep et al. 2007). The interaction of the bacteriocins with the Man-PTS components in turn is believed to cause structural changes in the complex which render the permease irreversibly open as a pore leading to leakage of solutes, disruption of cell membrane integrity, and eventually cell death (Kjos et al. 2011a). The finding that the target molecule of a bacteriocin is in fact a permease thus opens up the possibility that, instead of polymerizing into pores by themselves, bacteriocins can actually target and disrupt the cell's own uptake systems to produce the same lethal effect.

Despite all bacteria containing at least one such Man-PTS in their genome, it is surprising to observe that the Man-PTS-dependent bacteriocins display a narrow inhibitory spectrum, targeting predominantly bacteria closely related to the producers. In order to shed more light on this aspect of specific-target recognition, 86 members of the Man-PTS protein family from a wide range of bacteria were analyzed with respect to phylogenetic relation and bacteriocin receptor/target (Kjos et al. 2009). The phylogenetic analysis grouped the 86 members of the Man-PTS protein family into three main clusters (group I–III) (Kjos et al. 2009). Fourteen Man-PTS members distributed across the phylogenetic tree (all three groups) were selected for experimental analysis if they could serve as receptors for Class IIa bacteriocins, and only the selected members from group I Man-PTS were prone to bacteriocin sensitivity (Kjos et al. 2009). It is also interesting to note that the interaction between the Man-PTS receptors and the bacteriocin is very specific since, of the four putative Man-PTSs of *Listeria monocytogenes*, only one (belonging to group I) serves as a receptor/target for the Class IIa bacteriocins (Kjos et al. 2009). It has also been shown that the targets/receptors of Class IIa bacteriocins and lactococcin A are not exchangeable, although they both belong to group I Man-PTS receptors (Kjos et al. 2010). By combining classical genetics (making mutants) with modern technologies such as genomics and proteomics, an efficient protocol for bacteriocin receptor identification was established (Kjos et al. 2014).

A *L. lactis* strain, known to produce a small (30 amino acid residues) and leaderless bacteriocin termed LsbB, was shown to need a receptor on target cells (Uzelac et al. 2013; Ovchinnikov et al. 2014). The receptor of LsbB was identified by use of whole-genome sequencing of the resistant mutants derived from the original susceptible strain *L. lactis* IL1403, and all resistant mutants were found to harbor mutations in the gene *yvjB*. The requirement of *yvjB*, probably as a receptor, was confirmed by gene disruption, which rendered sensitive strains resistant to LsbB. The most compelling evidence that *yvjB* was unconditionally required for LsbB activity was heterologous expression of *yvjB* in resistant strains of other species, such as *Lactobacillus paracasei* and *Ent. faecalis* that rendered them sensitive. The *yvjB* encodes a zinc-dependent membrane-bound protease (belonging to the M50 protease family) (Ovchinnikov et al. 2014).

The fact that several circular bacteriocins have been reported to act on lipid bilayers/liposomes may support the notion that such unspecific surface interactions are sufficient for antimicrobial activity (Abee et al. 1995; Galvez et al. 1991; van Belkum et al. 2011). However, several of these studies were performed on artificial liposomes/lipid bilayers and with bacteriocins at concentrations significantly above the levels required for antimicrobial activity *in vivo*. It was demonstrated that the maltose ABC transporter facilitates high-level sensitivity to the cyclic garvicin ML (Class IIc) (Gabrielsen et al. 2012). However, consistent with studies on other circular bacteriocins, receptor-independent (maltose ABC transporter independent) killing at higher concentrations of GarML was also perceived (Gabrielsen et al. 2012). The results suggest that this class of bacteriocins may indeed require a specific interaction with a target receptor/mediator for antimicrobial activity at low concentrations. Such a dual concentration-dependent mode of action has also been demonstrated previously for nisin and is supported by data from previous studies with the circular sactibiotic subtilosin A (Thennarasu et al. 2005; Shelburne et al. 2007).

It should also be mentioned that some Class II bacteriocins, namely lactococcin 972 (Class IId) and *garvicin* A, can act by inhibiting septum formation (Martinez et al. 2000, 2008; Maldonado-Barragan et al. 2013), but whether or not these activities are receptor-mediated has not been conclusively confirmed. But it is interesting to note that the activity of lactococcin 972 is lipid II-dependent, since such lipid II dependence is usually associated with lantibiotics.

The diverse bacteriocin receptors disclosed so far share a common feature by being membrane-located proteins, and this is in line with the general view that bacteriocins are membrane-active molecules, but also through binding/interacting to specific membrane-located receptor-like molecules thereby abolishing the transport/membrane integrity, causing uncontrolled cellular leakage and consequently killing the target cell. Though identification of several receptors has been accomplished, the molecular details behind target recognition and the succeeding steps leading to cellular leakage and membrane disruption, and particularly the nature of the pores, are not understood. It is of great importance to seek more knowledge of these aspects which are essential for developing and/or rational design of novel bacteriocins for improved and safe applications. A comprehensive study of the structures of and intermolecular interactions between the bacteriocins and their receptors seems to be needed in order to develop such compounds into useful and sustainable antimicrobials.

5.4. Bacteriocin Resistance

In order to effectively exploit the potential of bacteriocin applications we have to learn more about their mode of action since it becomes increasingly evident that the presumption that bacteriocins act by directly forming pores in the cytoplasmic membrane is not always true. The mechanisms by which these antimicrobial peptides act appear to be nearly as diverse as their structures and physicochemical properties.

Similar to the situation for antibiotics, both innate and acquired resistance to bacteriocins have been identified. Different from traditional antibiotics, the bacteriocin producer encodes a cognate immunity protein that protects the producers from self-destruction (Nes et al. 1996). At least for the Class IIa bacteriocins, we have good perception of the mode of action of the immunity protein and how it protects the bacteriocin producer (Diep et al. 2007). The self-destruction activity of the bacteriocin is prevented by the formation of a tripartite complex between the receptor (Man-PTS), immunity protein, and the bacteriocin (Diep et al. 2007). The sequestering reaction locks the bacteriocin in the tripartite complex and thereby prevents it from permeabilizing the bacterial membrane, but it leads also to a dramatic metabolic change by not allowing the bacterium to grow efficiently on glucose. The growth inhibitory effect is due to inactivation of the glucose transporter but is not affected when consuming other carbohydrates that employ different uptake systems such as galactose permease (Kjos et al. 2011b).

Some bacterial species and isolates are completely resistant to certain bacteriocins, while other species show mixed susceptibility to certain bacteriocins. Some bacteriocins seem to have high species/genus specificity, as seen with lactococcin A that specifically kills only lactococci, but the MIC (minimum inhibitory concentration) value can vary greatly between different isolates (Holo et al. 1991). Among the most studied bacteriocins with respect to target bacteria are the strong antilisterial Class IIa bacteriocins. In a previous study, approximately 200 natural *L. monocytogenes* isolates were tested against three Class IIa bacteriocins (pediocin PA1 and sakacin A and P) and all isolates were susceptible to the bacteriocins but the MIC values varied 60–70-fold for the two most potent bacteriocins and approximately 300-fold for the least efficient bacteriocin (sakacin A) (Katla et al. 2003). Nisin A was also included in this study and showed the same variability but was not as effective as the three Class IIa bacteriocins. In other words, even natural bacterial isolates of the same species show high variability in susceptibility to bacteriocins. A number of works have also shown that when a susceptible bacterium is exposed to a bacteriocin it will eventually develop some resistance but with different frequency and level depending on the bacteriocin and the target bacterium. It has been reported that the resistance incident to Class IIa bacteriocins varies greatly, ranging from 10^{-2} to 10^{-7} (Gravesen et al. 2002b; Vadyvaloo et al. 2002a).

Molecular mechanisms implied in bacteriocin resistance are many and in most instances are not well defined. Generally they involve changes in the bacterial cell envelope, which result in (i) lack of target recognition, (ii) physico-chemical changes in membrane and cell wall composition, (iii) stress responses, (iv) bacteriocin sequestering, (v) acquired new genes increasing the bacteriocin tolerance, and (vi) proteolytic degradation, among many. Strategies to overcome resistance have to be developed in order to make bacteriocin applications more attractive.

Bacteria can also develop resistance in a susceptible strain upon long exposure to bacteriocins. However, the frequency of resistance development varies greatly depending on both the type of bacteriocins and the sensitive bacteria, which likely reflects their different modes of action (Gravesen et al. 2002a). There are also examples of intrinsic resistance mechanisms (Collins et al. 2010; McBride and Sonenshein 2011).

In bacteriocins with receptor-mediated antimicrobial activity, loss or reduced expression of the receptor has been shown to result in increased resistance to the corresponding bacteriocin (Ramnath et al. 2000). This is the case for Class IIa bacteriocins and lactococcin A, where lost or reduced expression of the Man-PTSs caused resistance in *L. monocytogenes* and *L. lactis*, respectively (Kjos et al. 2011b). The Class II bacteriocin lactococcin 972 (which targets the cell wall precursor lipid II) resistance has been linked to peptidoglycan remodeling (Roces et al. 2012). Resistance in *L. monocytogenes* against the Class I lantibiotic nisin has also been correlated with changes in both altered fatty acid and phospholipid composition, which is believed to adversely affect how nisin interacts with the membrane (Verheul et al. 1997; Crandall and Montville 1998). Also, in Class IIa-resistant cells of *L. monocytogenes*, which express normal levels of the receptor Man-PTS, resistance has been associated with changes in membrane phospholipid composition (Vadyvaloo et al. 2002b) and cell surface charge (Vadyvaloo et al. 2004). A third known resistance mechanism is known as immune mimicry. Immune mimicry entails the expression of functional immunity homologs in non-producing strains, that is strains without a cognate bacteriocin gene cluster (Fimland et al. 2002; Draper et al. 2009). Such "immunity" genes are thought to be remnants of inactivated or deleted bacteriocin operons, which nevertheless provide the non-producing strain with a competitive advantage. Immunity-like determinants providing resistance towards the two-peptide lantibiotic lacticin 3147 are, for example, produced not only by lacticin 3147-producing strains, but also by *B. licheniformis* and *Ent. faecium* (Draper et al. 2009). Enzymatic activity that degrades peptide bacteriocins, consequently leading to loss or lowered antimicrobial activity, is another mechanism by which bacteria can acquire resistance towards bacteriocins. Extracellular proteases or peptidases of different specificities are likely candidates for such non-specific activity. Even so, there are not many reports of bacteriocin-resistance caused by enzymatic degradation. One known example is the extracellular gelatinase in *Ent. faecalis*, which has been shown to degrade Class IIa bacteriocins (Sedgley et al. 2009). For the Class I lantibiotic nisin, there are two known enzymes, nisin resistance protein (NRS) (Sun et al. 2009) and nisinase (Jarvis 1967), that specifically degrade or modify the *C*-terminal end of the peptide and consequently reduce the enzyme producers their sensitivity to nisin.

It is well known that transporters play a role in bacteriocin producer self-immunity for different classes of bacteriocins, including lantibiotics (Draper et al. 2008) and circular bacteriocins (Martinez-Bueno et al. 1998; Kemperman et al. 2003). Different transporter complexes have also been shown to contribute to intrinsic resistance of *L. monocytogenes* to nisin, and of *Clostridium difficile* to nisin and gallidermin (Collins et al. 2010; McBride and Sonenshein 2011).

It has become evident that some bacteriocin resistance traits can provide cross-resistance to other bacteriocins or antimicrobial agents (Rekhif et al. 1994; Crandall and Montville 1998; Mehla and Sood 2011), and that sequential exposure to a bacteriocin in some cases can provide increased or more stable resistance compared to single exposure (Ming and Daeschel 1993; Gravesen et al. 2002a). Bacteriocin resistance mechanisms are therefore not always specific, clear-cut, on/off mechanisms, but can result from an interplay between several contributing factors, most of which are not yet well defined. This is an area of bacteriocin research that requires more attention, because it is crucial to minimize resistance in practical applications of bacteriocins, especially if bacteriocins are ever to be considered as viable options to antibiotics in medical therapeutics.

5.5. Applications

5.5.1. Opportunities and Hurdles in Application of Bacteriocins

The bacteriocins are among our most potent antibacterial compounds available with many potential applications. The Gram (+) bacteriocins in general, and more specifically LAB bacteriocins, are shown to eradicate a diverse group of Gram (+) bacteria. But the antimicrobial activity is partly unpredictable since the concentration of bacteriocins required to kill bacteria within the same species can vary a lot. With regard to the killing/growth inhibition of different isolates of *L. monocytogenes*, the MIC values of pediocin PA-1 and sakacin P vary up to 50–70-fold and about 300-fold for sakacin A. Nevertheless the bacteriocin concentration needed to kill susceptible bacteria is usually very low, being 100–1000-fold more potent compared to defensins produced by eukaryotic organisms (Nissen-Meyer and Nes 1997). Some bacteriocins such as lactococcin A can kill *L. lactis* strains at extremely low concentration, almost equal to a few bacteriocin molecules per target cell (Holo et al. 1991)

Due to their strong antibacterial activity, being produced by the harmless LAB and having not shown any toxic effects on mammalian cells, they seem ideal for use in fighting harmful and pathogenic bacteria. Over the years bacteriocins have been assessed in food applications in order to inhibit growth of or kill pathogens and food-detrimental bacteria (Cleveland et al. 2001; De Vuyst and Leroy 2007; Balciunas et al. 2013). It has been shown that such bacteriocins reduce and prevent growth of unwanted bacteria in foods, but most often not as effectively as expected and consequently their uses have been limited. Presently only nisin and a few Class IIa bacteriocins such as pediocin PA-1 and in commercial starter cultures sakacin A (curvacin A) and P have been commercially exploited.

There are several reasons for the limited success in commercial application of bacteriocins. Their physiochemical properties seem to be an obstacle when applied in a chemically complex environment such as food and feed. Being positively charged and hydrophobic/amphiphilic they will attach to negatively charged and/or lipophilic surfaces/molecules that will actually make the bacteriocins less available to reach target bacteria. There have been a great number of excellent works and reviews in recent years that directly deal with applications of bacteriocins in foods, therefore this field will not be treated in this review. We will focus on medical-related applications of bacteriocins and the potential of bacteriocin-producing bacteria as probiotics.

5.5.2. Application of Bacteriocins in Medical-related and Personal Hygiene Products

The era of traditional broad spectrum antibiotics might seem to be at an end since an increasing number of pathogens have become resistant and even multi-resistant to the available antibiotics of today and there is no sign of new ones entering the market. The increasing numbers of hospital-acquired infections (e.g., urinary tract infections, pneumonia, bloodstream infections, etc.) caused by antibiotic resistant bacteria worldwide represent a progressive threat to global human health. Bacteriocins from Gram (+) bacteria and in particular LAB are extremely potent in their ability to kill other Gram (+) bacteria, including many pathogens and antibiotic resistant ones, and many should have great potential to be developed into efficient means to kill targeted pathogens. However, this has not been successfully achieved in human medicine yet, although many bacteriocins with strong killing preference for several pathogens are available.

There are a few examples where bacteriocins and bacteriocin-producing bacteria are presently in use. Mastitis is a bacterial infection of the mammary gland and udder tissue of dairy cattle and is usually treated with antibiotics. For years nisin A has been used in prophylactic treatment of and protection against mastitis. Several commercial nisin-based products are presently on the market and these include products under the labels MastPro™, Mast Out®, Wipe Out®, Consept®, Nisin Z Silver-Elephant®, and others.

In recent years several natural nisin-like lantibiotics (nisin Z (de Vos et al. 1993), nisin Q (Yoneyama et al. 2009), nisin F (de Kwaadsteniet et al. 2008), nisin U and U2 (Wirawan et al. 2006), and salivaricin D (Birri et al. 2012a)) have been characterized. The new natural variants have not been tested rigorously and might represent new and attractive antimicrobials for commercial applications. In addition some bioengineered nisin A derivatives have shown enhanced antimicrobial activity against many bacteria and further investigation on engineered nisin-variants might lead to new and improved antimicrobials for various uses (Healy et al. 2013).

The two-peptide lantibiotic lacticin 3148 has also proved a good replacement for antibiotics for mastitis prevention/treatment in both lactating and dry cows. The same bacteriocin has also been successfully tested in food and seems to be a promising candidate for combat of food pathogens (Galvez et al. 2007).

A novel Class I two-peptide bacteriocin thuricin CD (a sactibiotic) was isolated from a human fecal *B. thuringiensis* isolate. Thuricin CD differs from other two-peptide bacteriocins by the post-translational modifications, in that three specific cysteines are linked to the alpha-carbons of the modified residues thereby forming three sulfur to alpha-carbon bridges (Fig 5.1). Even more notably, thuricin CD was found to inhibit a narrow spectrum of spore forming Gram (+) bacteria including the clinically important *C. difficile* known to be responsible for the development of antibiotic-associated diarrhea and colitis (Rea et al. 2011b). Recent *ex vivo* mouse model studies of the distal colon showed that thuricin CD compared favorably with metronidazole, the first antibiotic of choice for the treatment of *C. difficile* associated diseases. Follow-up studies showed that rectal administration of thuricin CD to *C. difficile*-infected mice reduced its presence in feces by 95% compared to the non-treated animals (Rea et al. 2014). It is also important to note that thuricin CD has a strong preference for killing *C. difficile* and does not alter much the common gut microbiota, while lacticin 3846, a wide-spectrum bacteriocin, greatly alters the gastrointestinal microbiota, the latter being considered negative for the host.

The literature reports that although many bacteriocins are very efficient in *in vitro* tests, they often fail to confer such a strong antimicrobial activity in animal model systems (Eijsink et al. 1998; Dabour et al. 2009; Dobson et al. 2011). The lack of success in applying bacteriocins in treatment of infections might be due to *in situ* delivery problems. They might not reach the targets efficiently because of unspecific adsorbance to neighboring surfaces, tissues or/and molecules not related to the targeted pathogens or being degraded by proteinases. This is considered a major obstacle in successfully applying such compounds. To solve this problem one has to find measures to efficiently deliver the peptides to the bacterial targets and it seems that this aspect of bacteriocin application needs high priority with the involvement of multidisplinary sectors of academia, hospitals, and pharmaceutical companies. Presently, some publications have addressed the aspect of bacteriocin delivery. In other applications, such as in eukaryotic antimicrobial peptides and enzymes, the technologies of immobilization on various surfaces including plastic films have been investigated, while bacteriocins have not received the same focus yet (Ercolini et al. 2006; Liu et al. 2009; Soman et al. 2009; Costa et al. 2011a).

The use of new technologies in the implementation of bacteriocins in the treatment of topical infections embodies very promising opportunities. Implanted biomaterials are common and costly and infection is frequently associated with such materials. Infections can be extremely resistant to antibiotics and host defense systems and there is a need to develop surfaces that prevent bacterial adhesion. By coating such implant surfaces with bacteriocins it should be possible to inhibit bacterial colonization. Such technologies are applied in immobilization enzyme and antimicrobial peptides from eukaryotes on various surfaces (Chen et al. 2009; Costa et al. 2011b; Bali et al. 2014). It has been reported that carbohydrate nanoparticle-stabilized emulsion prolonged the efficacy of nisin against food *L. monocytogenes* (Bi et al. 2011). Nanofibers have been assessed to serve as a carrier matrix for the bacteriocins plantaricin 423 and ST4SA. This could represent a new and promising delivery system for various applications of bacteriocins (Heunis et al. 2011).

In a recent report it was shown that nisin-containing nanofibers significantly reduced the viability of *S. aureus* in a murine excisional skin infection model. Wounds treated with nisin-containing nanofibers reduced *S. aureus* to 4.3×10^2 CFU/wound, whereas the control wound (only nanofibers) had 2.2×10^7 CFU/wound after one week (Heunis et al. 2013).

An attractive delivery system for bacteriocins is the use of living bacteria that produce bacteriocins. This can be used both in food and feed fermenting industries to prevent growth of pathogens as well as food spoilage bacteria. It has also been applied in protective cultures but with limited success. In addition to potential problems outlined above in this review, another important issue is the status of bacteriocin production *in situ*. The delivery of bacteriocin-producing bacteria to human or animal is mostly considered within the concept of probiotics that will be discussed in the next section.

Over the years many applicative ideas involving bacteriocins have been suggested and some have also been investigated further. *Acne vulgaris* is the most common skin disease worldwide and the bacterium *Propionibacterium acnes* is known to play a pivotal role in such skin infection. Several studies have shown that bacteriocins might be an efficient remedy to treat such infections and a few products have been commercialized (http://www.probiotic-lab.com/acnehomepage.html) (Chung and Seo 2010). Skin lotions containing various ingredients including bacteriocin are also available today (http://www.nikelcosmetics.com/probiotic-cream-p19). Bacteriocin-containing toothpaste and mouthwash have been patented and today several products related to this use exist. Inventions related to cosmetic compositions and methods for treating body odors, in particular axillary odors, have recently been patented (Gavillon and Simonnet 2011). It is claimed that the use of a Class II bacteriocin should inhibit the development of the skin flora responsible for axillary odors.

Despite patents being filed and several products being launched with respect to applications of bacteriocins in cosmetic and health products, we do not have sufficient experience with and information on the products to justify conclusively if they work according to their claims.

5.5.3. Bacteriocin-producing Probiotics

Probiotic cultures have garnered great interest in recent years due to their positive effects on human health. In 2001 WHO defined probiotics as "live micro-organisms which, when administered in adequate amounts, confer a health benefit on the host." Among the many health-bringing claims are that probiotics and prebiotics exert a range of effects on mucosal barrier function and the responses of the underlying gut associated lymphoid tissue (Hardy et al. 2013). It has also been shown that probiotic bifidobacteria are able to reduce the severity of weanling diarrhea caused by rotavirus and *E. coli* in piglets (Shu et al. 2001). But it is still debatable which and how probiotic cultures convey these positive health effects. A quite recent review asked the question whether bacteriocin production is a probiotic trait (Dobson et al. 2012). In our view there is overwhelming support for such a notion. A large body of evidence demonstrates the presence of bacteriocin-producing LAB in gut contents such as feces. It has been shown that in infants bacteriocin-producing bacteria are established among the LAB that emerge shortly after birth (Birri et al. 2010).

In a recent study (Birri et al. 2013), 150 LAB were obtained from feces of 28 healthy Ethiopian infants. Bacteriocin-producing LAB species were isolated from eight of the infants. Many different bacteriocins were identified, including one new bacteriocin called salivaricin D (a nisin-like bacteriocin) from *Strep. salivarius* (Birri et al. 2012b), one Class IIa bacteriocin (avicin A) from *Ent. avium*, one Class IIa bacteriocin from *Ent. faecalis* strains, one unknown bacteriocin from *Ent. faecalis*, two unknown bacteriocins from *Lact. fermentum* strains, and the two-peptide gassericin T from a *Lact. gasseri* isolate (Birri et al. 2013). A previous study also showed that fecal bacteria obtained from Norwegian infants had the ability to produce bacteriocins and it should also be added that the novel bacteriocin avicin A producing *Ent. avium* is not limited to specific regions, as producer isolates of this bacteriocin were found in infants from both Norway and Ethiopia (Forberg 2005; Birri et al. 2010).

From the ileal–caecal region of the human GI tract, a probiotic *Lact. salivarius* subsp. *salivarius* was isolated and found to produce the two-peptide bacteriocin ABP-118, which is discussed at the end of this section (Flynn et al. 2002).

In a study with fecal samples from piglets, a *Lact. reuteri* isolate was shown to produce a bacteriocin-like substance and it was shown that the same type of bacteria were found in porcine milk obtained from the same population of pigs (Bohle et al. 2010). Probiotic bacteria are shown to effectively promote the health status of pigs (Cho 2011), and recently a patent for probiotic supplement for the diet of pigs was filed and one of the strains included in this patent does indeed produce a bacteriocin (Stanton et al. 2012).

The oral cavity is also an environment for numerous bacteria including pathogens, and the use of bacteriocin-producing bacteria to improve oral cavity hygiene seems feasible and has been shown to be successful (Wescombe et al. 2009; Burton et al. 2013). The *Strep. salivarius* strain M18 has been shown to confer oral health benefits as it produces bacteriocins targeting the cariogenic species *Streptococcus mutans*. In a randomized double-blind, placebo-controlled study of 100 dental caries-active children, it was demonstrated that treatment with this strain reduced the plaque score and gingival and soft-tissue health of the treated children (Burton et al. 2013). The human vagina is inhabited by a healthy microbiota and during a disease-associated imbalance of the bacterial biota this part of the body is vulnerable to infections (Larsen and Monif 2001). Supplement of probiotics to prevent vaginal infection is a promising therapeutic approach that has been investigated and proven to be experimentally successful, but further investigation is needed to assess efficacy and safety (Cribby et al. 2008). Bacteriocin-producing vaginal LAB have been characterized and one probiotic feature of these microorganisms is probably bacteriocin production (Dezwaan 2007; Pingitore et al. 2009).

An important and remaining question still is: Do bacteriocinogenic bacteria in the gut actually produce bacteriocin *in situ*? Indeed, we have a wealth of strong evidence that bacteriocin production takes place in the gut. The *Lact. salivarius* UCC118 strain producing Abp-118 bacteriocin can impede *L. monocytogenes* infection in mice, while the isogenic non-Abp-118 producer did not (Corr et al. 2007). The bacteriocin effect was further demonstrated when the pathogenic *L. monocytogenes* was protected from Abp-118 by expressing the cloned gene encoding the dedicated immunity protein of Abp-118 (Corr et al. 2007). The probiotic effect of *Lact. salivarius* UCC118 on the composition of the intestinal microbiota in mice and pigs was recently published (Riboulet-Bisson et al. 2012). When mice and pigs were administered either with the bacteriocin-producing *Lact. salivarius* UCC118 or the bac- mutant cells, a significant but differently affected microbiota was observed.

It can be concluded that there are a myriad of possibilities to apply bacteriocins into new and alternative means to fight infectious and damaging bacteria, but still further studies are needed to fully understand the mechanisms underlying their probiotic effects.

5.6. Future Perspectives

With the increasing development of resistance to classical antibiotics in recent years, there is an ever increasing need for novel antimicrobial agents, for medical, food, and feed industries. However, the search for new antibiotics is slow, as there are apparently fewer classes of antibiotics left to be discovered and cumbersome laws and regulations and large costs associated with research and development. Bacteriocins have many favorable characteristics that make them some of the most promising candidates for novel antimicrobials. First, most bacteriocins have relatively narrow inhibitory spectra that make it possible to target specific pathogens or unwanted bacteria. Consequently, the toxicity of bacteriocins to eukaryotic cells is generally low if present at all, especially at the concentrations required for antimicrobial activity. In contrast to most traditional antibiotics, which often act in micromolar concentrations, bacteriocins are commonly active in the nanomolar range, that is, displaying extremely high specific activity. Furthermore, it is possible that the activity of bacteriocins can be increased through combination with other antimicrobials or membrane-active substances. Some bacteriocins have broad antimicrobial spectra, and generally those produced by Gram (+) bacteria are broader than the Gram (+) producers. Broad spectrum antimicrobial agents can be useful for targeting infections of unknown origin. However, it has been established that use of broad spectrum antibiotics can have a negative impact on the commensal and beneficial human microbiota, and so the prospect of targeting specifically a pathogenic genus/species without adversely affecting the commensal population is highly appreciated. Importantly, the antimicrobial spectra of bacteriocins commonly include pathogenic and antibiotic resistant strains of *Streptococcus*, *Staphylococcus*, *Enterococcus*, *Propioonibacterium*, *Listeria*, and *Clostridium*, rendering possible numerous applications of such bacteriocins. Staphylococci and enterococci are responsible for a large number of nosocomial infections and instances where acquired resistance to antibiotics present a serious problem in medical treatment worldwide. The LAB bacteriocins have a special status, in that many of the bacteriocins as well as the producing strains are generally regarded as safe (GRAS) or Qualified Presumption of Safety (QPS), because they are produced by bacteria commonly found in fermented foods and in the human microbiota. Especially many species of *Lactobacillus*, *Leuconostoc*, and *Pediococcus* are approved and widely used as starter cultures. As a consequence, these bacteriocin producers are not only particularly suited for food applications and use as probiotics but also pave the way for their further development into other non-food applications. Indeed, several commercial probiotic strains used today produce bacteriocins and thus may provide added health benefits, but most of these bacteriocins have not been characterized in detail.

The physicochemical characteristics of bacteriocins also constitute an advantage: many bacteriocins are heat stable, tolerant to low pH, and in a few cases are also resistant to a range of proteolytic enzymes. Such characteristics make these bacteriocins or bacteriocin producers well suited to industrial applications, as probiotics or for therapeutic use through oral administration. Lastly and importantly, due to their gene-encoded nature bacteriocins present excellent targets for bioengineering. Peptides with a range of improved functional properties, such as increased thermal and pH stability, resistance to proteases, solubility, potency, and modified target specificity, could be bioengineered. Indeed, several of these traits have already been engineered for derivatives of the lantibiotic nisin (Rollema et al. 1995; Molloy et al. 2013), most notably the enhanced activity against both Gram (+) and Gram (−) pathogens (Healy et al. 2013), which demonstrates the real potential of bioengineering in creating "tailor-made" bacteriocins for specific applications.

The potential applications of bacteriocins in food are diverse: bacteriocins can be added to foods as concentrated preparations (as preservatives, additives, or ingredients), or bacterial starter cultures that produce bacteriocin *in situ*. Nisin produced by *L. lactis* ssp. is the only bacteriocin now widely in use as a food preservative due to its efficacy and food-grade status. Nisin was approved for use as a preservative by the UN's Food and Agriculture Organization (FAO) as early as in 1969, and is now approved in more than 80 countries worldwide, where it is mainly used for the inhibition of *L. monocytogenes* in cheeses and *C. botulinum* in meat products. Pediocin PA-1, which is produced by different species of pediococci, is yet another promising bacteriocin for prevention of common pathogens, for example *L. monocytogenes*, in meats, cheeses, and fermented vegetables. Pediocin PA-1 is not yet approved for use as a food additive, but is currently approved as a food ingredient or shelf-life extender.

With respect to clinical applications of bacteriocins, there have been some promising recent advances, and several bacteriocins are now being developed for human applications. The *in vivo* activity of bacteriocins against pathogens has been demonstrated for several bacteriocins of both Class I and Class II in animal models. Hopefully, the development of such bacteriocins will move into clinical trials, to be approved for use in human therapeutics in the near future.

As mentioned above, resistance development is a serious challenge for antibiotics and this might also be a future obstacle in the application of bacteriocins as antimicrobial agents. Some efforts have been made to lower resistance development by using bacteriocins in combination with other antimicrobial agents with different target/mechanisms of action (Brumfitt et al. 2002; Luders et al. 2003; Naghmouchi et al. 2012). With a detailed understanding of the mode of action and the bacteriocin resistance mechanism, together with the possibility of bioengineering and combination use with other antimicrobials or synergistic compounds, the use of bacteriocins will likely pave the way to many successful applications in the near future, both in the food and medicine sectors. The target specificity is the strength of bacteriocins and makes them extremely powerful as well as specific, which leaves the beneficial microbiota mostly unaffected. Development of efficient delivery systems will greatly minimize many of the abovementioned obstacles in bacteriocin applications.

References

Abee, T., Krockel, L., and Hill, C. (1995). Bacteriocins: modes of action and potentials in food preservation and control of food poisoning. *Int J Food Microbiol* **28**(2), 169–185.

Arnison, P.G., Bibb, M.J., Bierbaum, G., et al. (2013). Ribosomally synthesized and post-translationally modified peptide natural products: overview and recommendations for a universal nomenclature. *Natural Product Rep* **30**, 108–160.

Asaduzzaman, S.M., Nagao, J., Iida, H., et al. (2009). Nukacin ISK-1, a bacteriostatic lantibiotic. *Antimicrob Agents Chemother* **53**, 3595–3598.

Balciunas, E.M., Castillo Martinez, F.A., Todorov, S.D., et al. (2013). Novel biotechnological applications of bacteriocins: A review. *Food Control* **32**, 134–142.

Bali, V., Panesar, P.S., and Bera, M.B. (2014). Potential of Immobilization technology in bacteriocin production and antimicrobial packaging. *Food Rev Int* **30**, 244–263.

Bi, L., Yang, L., Bhunia, A.K., and Yao, Y. (2011). Carbohydrate nanoparticle-mediated colloidal assembly for prolonged efficacy of bacteriocin against food pathogen. *Biotechnol Bioeng* **108**, 1529–1536.

Bierbaum, G., Gotz, F., Peschel, A., et al. (1996). The biosynthesis of the lantibiotics epidermin, gallidermin, Pep5 and epilancin K7. *Antonie Van Leeuwenhoek* **69**, 119–127.

Birri, D.J., Brede, D.A., Forberg, T., et al. (2010). Molecular and genetic characterization of a novel bacteriocin locus in Enterococcus avium isolates from infants. *Appl Environ Microbiol* **76**, 483–492.

Birri, D.J., Brede, D.A., and Nes, I.F. (2012a). Salivaricin D, a novel intrinsically trypsin-resistant lantibiotic from Streptococcus salivarius 5M6c isolated from a healthy infant. *Appl Environ Microbiol* **78**, 402–410.

Birri, D.J., Brede, D.A., and Nes, I.F. (2012b). Salivaricin D, a novel intrinsically trypsin-resistant lantibiotic from Streptococcus salivarius 5M6c isolated from a healthy infant. *Appl Environ Microbiol* **78**, 402–410.

Birri, D.J., Brede, D.A., Tessema, G.T., and Nes, I.F. (2013). Bacteriocin production, antibiotic susceptibility and prevalence of haemolytic and gelatinase activity in faecal lactic acid bacteria isolated from healthy Ethiopian infants. *Microb Ecol* **65**, 504–516.

Biswas, S. and Biswas, I. (2014). A conserved streptococcal membrane protein, LsrS, exhibits a receptor-like function for lantibiotics. *J Bacteriol* **196**, 1578–1587.

Bohle, L.A., Brede, D.A., Diep, D.B., et al. (2010). Specific degradation of the mucus adhesion-promoting protein (MapA) of Lactobacillus reuteri to an antimicrobial peptide. *Appl Environ Microbiol* **76**, 7306–7309.

Bottiger, T., Schneider, T., Martinez, B., et al. (2009). Influence of Ca(2+) ions on the activity of lantibiotics containing a mersacidin-like lipid II binding motif. *Appl Environ Microbiol* **75**, 4427–4434.

Breukink, E., Wiedemann, I., van Kraaij, C., et al. (1999). Use of the cell wall precursor lipid II by a pore-forming peptide antibiotic. *Science* **286**, 2361–2364.

Brotz, H., Bierbaum, G., Leopold, K., et al. (1998a). The lantibiotic mersacidin inhibits peptidoglycan synthesis by targeting lipid II. *Antimicrob Agents Chemother* **42**, 154–160.

Brotz, H., Josten, M., Wiedemann, I., et al. (1998b). Role of lipid-bound peptidoglycan precursors in the formation of pores by nisin, epidermin and other lantibiotics. *Mol Microbiol* **30**, 317–327.

Brumfitt, W., Salton, M.R.J., and Hamilton-Miller, J.M.T. (2002). Nisin alone and combined with peptidoglycan-moduling antibiotics: activity against methicillin-resistant *Staphylococcus aureus* and vancomycin-resistant enterococci. *J Anticrobiol Chemother* **50**, 731–734.

Burton, J.P., Drummond, B.K., Chilcott, C.N., et al. (2013). Influence of the probiotic *Streptococcus salivarius* strain M18 on indices of dental health in children: a randomized double-blind, placebo-controlled trial. *J Med Microbiol* **62**, 875–884.

Chen, R.X., Cole, N., Willcox, M.D.P., et al. (2009). Synthesis, characterization and in vitro activity of a surface-attached antimicrobial cationic peptide. *Biofouling* **25**, 517–524.

Chikindas, M.L., Garcia-Garcera, M.J., Driessen, A.J., et al. (1993). Pediocin PA-1, a bacteriocin from Pediococcus acidilactici PAC1.0, forms hydrophilic pores in the cytoplasmic membrane of target cells. *Appl Environ Microbiol* **59**, 3577–3584.

Cho, J.H., Zhaou, P.Y., and Kim, I.H. (2011). Probiotics as a dietary additive for pigs: a review. *J Anim Vet Adv* **10**, 2127–2134.

Chung, M.J. and Seo, J.G. (2010). Novel use of bacteriocin derived from Enterococcus faecalis sl-5 (Google Patents).

Cintas, L.M., Casaus, P., Herranz, C., et al. (2000). Biochemical and genetic evidence that *Enterococcus faecium* L50 produces enterocins L50A and L50B, the sec-dependent enterocin P, and a novel bacteriocin secreted without an N-terminal extension termed enterocin Q. *J Bacteriol* **182**, 6806–6814.

Cintas, L.M., Casaus, P., Holo, H., et al. (1998). Enterocins L50A and L50B, two novel bacteriocins from *Enterococcus faecium* L50, are related to staphylococcal hemolysins. *J Bacteriol* **180**, 1988–1994.

Cleveland, J., Montville, T.J., Nes, I.F., and Chikindas, M.L. (2001). Bacteriocins: safe, natural antimicrobials for food preservation. *Int J Food Microbiol* **71**, 1–20.

Collins, B., Curtis, N., Cotter, P.D., et al. (2010). The ABC transporter AnrAB contributes to the innate resistance of *Listeria monocytogenes* to nisin, bacitracin, and various beta-lactam antibiotics. *Antimicrob Agents Chemother* **54**, 4416–4423.

Corr, S.C., Li, Y., Riedel, C.U., et al. (2007). Bacteriocin production as a mechanism for the antiinfective activity of *Lactobacillus salivarius* UCC118. *Proc Natl Acad Sci USA* **104**, 7617–7621.

Costa, F., Carvalho, I.F., Montelaro, R.C., et al. (2011a). Covalent immobilization of antimicrobial peptides (AMPs) onto biomaterial surfaces. *Acta Biomater* **7**, 1431–1440.

Costa, F., Carvalho, I.F., Montelaro, R.C., et al. (2011b). Covalent immobilization of antimicrobial peptides (AMPs) onto biomaterial surfaces. *Acta Biomaterialia* **7**, 1431–1440.

Cotter, P.D., Hill, C., and Ross, R.P. (2005a). Bacteriocins: developing innate immunity for food. *Nat Rev Microbiol* **3**, 777–788.

Cotter, P.D., Hill, C., and Ross, R.P. (2005b). Bacteriocins: developing innate immunity for food. *Nat Rev Microbiol* **3**, 777–788.

Crandall, A.D. and Montville, T.J. (1998). Nisin resistance in Listeria monocytogenes ATCC 700302 is a complex phenotype. *Appl Environ Microbiol* **64**, 231–237.

Cribby, S., Taylor, M. and Reid, G. (2008). Vaginal microbiota and the use of probiotics. *Interdisciplinary perspectives on infectious diseases* **9**, 256490.

Cui, Y., Zhang, C., Wang, Y., et al. (2012). Class IIa bacteriocins: diversity and new developments. *Int J Mol Sci* **13**, 16668–16707.

Dabour, N., Zihler, A., Kheadr, E., et al. (2009). *In vivo* study on the effectiveness of pediocin PA-1 and *Pediococcus acidilactici* UL5 at inhibiting Listeria monocytogenes. *Int J Food Microbiol* **133**, 225–233.

de Kwaadsteniet, M., Ten Doeschate, K., and Dicks, L.M. (2008). Characterization of the structural gene encoding nisin F, a new lantibiotic produced by a *Lactococcus lactis* subsp. lactis isolate from freshwater catfish (*Clarias gariepinus*). *Appl Environ Microbiol* **74**, 547–549.

de Vos, W.M., Mulders, J.W., Siezen, R.J., et al. (1993). Properties of nisin Z and distribution of its gene, nisZ, *in Lactococcus lactis*. *Appl Environ Microbiol* **59**, 213–218.

De Vuyst, L. and Leroy, F. (2007). Bacteriocins from lactic acid bacteria: production, purification, and food applications. *J Mol Microbiol Biotechnol* **13**, 194–199.

Dezwaan D.C., Littell, J.S., Allen, J.P., et al. (2007). Purification and characterization of enterocin 62-6, a two-peptide bacteriocin produced by a vaginal strain of *Enterococcus faecium*: Potential significance in bacterial vaginosis. *Microb Ecol Health Dis* **19**, 241–250.

Diep, D.B., Skaugen, M., Salehian, Z., et al., 2007. Common mechanisms of target cell recognition and immunity for class II bacteriocins. *Proc Natl Acad Sci USA* **104**, 2384–2389.

Diep, D.B., Straume, D., Kjos, M., et al. (2009). An overview of the mosaic bacteriocin pln loci from *Lactobacillus plantarum*. *Peptides* **30**, 1562–1574.

Dobson, A., Cotter, P.D., Ross, R.P. and Hill, C. (2012). Bacteriocin production: a probiotic trait? *Appl Environ Microbiol* **78**, 1–6.

Dobson, A., Crispie, F., Rea, M.C., et al. (2011). Fate and efficacy of Lacticin 3147-producing *Lactococcus lactis* in the Mammalian gastrointestinal tract. *FEMS Microbiol Ecol* **76**, 602–614.

Draper, L.A., Grainger, K., Deegan, L.H., et al. (2009). Cross-immunity and immune mimicry as mechanisms of resistance to the lantibiotic lacticin 3147. *Mol Microbiol* **71**, 1043–1054.

Draper, L.A., Ross, R.P., Hill, C., and Cotter, P.D. (2008). Lantibiotic immunity. *Curr Protein Pept Sci* **9**, 39–49.

Eijsink, V.G., Axelsson, L., Diep, D.B., et al. (2002). Production of class II bacteriocins by lactic acid bacteria; an example of biological warfare and communication. *Antonie Van Leeuwenhoek* **81**, 639–654.

Eijsink, V.G.H., Skeie, M., Middelhoven, P.H., et al. (1998). Comparative studies of class IIa bacteriocins of lactic acid bacteria. *Appl Environ Microbiol* **64**, 3275–3281.

Ercolini, D., Storia, A., Villani, F., and Mauriello, G. (2006). Effect of a bacteriocin-activated polythene film on *Listeria monocytogenes* as evaluated by viable staining and epifluorescence microscopy. *J Appl Microbiol* **100**, 765–772.

Ferir, G., Petrova, M.I., Andrei, G., et al. (2013). The lantibiotic peptide labyrinthopeptin A1 demonstrates broad anti-HIV and anti-HSV activity with potential for microbicidal applications. *PLoS One* **8**, e64010.

Fimland, G., Eijsink, V.G., and Nissen-Meyer, J. (2002). Comparative studies of immunity proteins of pediocin-like bacteriocins. *Microbiology* **148**, 3661–3670.

Fimland, G., Johnsen, L., Axelsson, L., et al. (2000). A C-terminal disulfide bridge in pediocin-like bacteriocins renders bacteriocin activity less temperature dependent and is a major determinant of the antimicrobial spectrum. *J Bacteriol* **182**, 2643–2648.

Flynn, S., van Sinderen, D., Thornton, G.M., et al. (2002). Characterization of the genetic locus responsible for the production of ABP-118, a novel bacteriocin produced by the probiotic bacterium *Lactobacillus salivarius* subsp. *salivarius* UCC118. *Microbiology* **148**, 973–984.

Forberg, T. (2005). *Lactic acid bacteria of different origin : production of antimicrobial substances and distribution of bacteriocin genes (In Norwegian)*. Master. Norwegian University of Life Sciences, Aas.

Fujita, K., Ichimasa, S., Zendo, T., et al. (2007). Structural analysis and characterization of lacticin Q: a novel bacteriocin belonging to a new family of unmodified bacteriocins of Gram-Positive bacteria. *Appl Environ Microbiol* **73**, 2871–2877.

Gabrielsen, C., Brede, D.A., Hernández, P.E., et al. (2012). The maltose ABC transporter in *Lactococcus lactis* facilitates high-level sensitivity to the circular bacteriocin garvicin ML. *Antimicrob Agents Chemother* **56**, 2908–2915.

Gabrielsen, C., Brede, D.A., Nes, I.F., and Diep, D.B. (2014). Circular bacteriocins: biosynthesis and mode of action. *Appl Environ Microbiol* **80**, 6854–6862.

Gajic, O., Buist, G., Kojic, M., et al. (2003). Novel mechanism of bacteriocin secretion and immunity carried out by lactococcal multidrug resistance proteins. *J Biol Chem* **278**, 34291–34298.

Galvez, A., Abriouel, H., Lopez, R.L., and Ben Omar, N. (2007). Bacteriocin-based strategies for food biopreservation. *Int J Food Microbiol* **120**, 51–70.

Galvez, A., Maqueda, M., Martinez-Bueno, M., and Valdivia, E. (1991). Permeation of bacterial cells, permeation of cytoplasmic and artificial membrane vesicles, and channel formation on lipid bilayers by peptide antibiotic AS-48. *J Bacteriol* **173**, 886–892.

Gavillon, R. and Simonnet, J.T. (2011). Cosmetic method for treating body odours using a bacteriocin based composition (Google Patents).

Gravesen, A., Jydegaard Axelsen, A.M., Mendes da Silva, J., et al. (2002a). Frequency of bacteriocin resistance development and associated fitness costs in Listeria monocytogenes. *Appl Environ Microbiol* **68**, 756–764.

Gravesen, A., Ramnath, M., Rechinger, K.B., et al. (2002b). High-level resistance to class IIa bacteriocins is associated with one general mechanism in *Listeria monocytogenes*. *Microbiology* **148**, 2361–2369.

Hardy, H., Harris, J., Lyon, E., et al. (2013). Probiotics, prebiotics and immunomodulation of gut mucosal defences: homeostasis and immunopathology. *Nutrients* **5**, 1869–1912.

Hasper, H.E., de Kruijff, B., and Breukink, E. (2004). Assembly and stability of nisin-Lipid II pores. *Biochemistry* **43**, 11567–11575.

Havarstein, L.S., Diep, D.B., and Nes, I.F. (1995). A family of bacteriocin ABC transporters carry out proteolytic processing of their substrates concomitant with export. *Mol Microbiol* **16**, 229–240.

Havarstein, L.S., Holo, H., and Nes, I.F. (1994). The leader peptide of colicin V shares consensus sequences with leader peptides that are common among peptide bacteriocins produced by gram-positive bacteria. *Microbiology* **140** (Pt 9), 2383–2389.

Healy, B., Field, D., O'Connor, P.M., et al. (2013). Intensive mutagenesis of the nisin hinge leads to the rational design of enhanced derivatives. *PLoS One* **8**, e79563.

Heunis, T., Bshena, O., Klumperman, B., and Dicks, L. (2011). Release of bacteriocins from nanofibers prepared with combinations of poly(d,l-lactide) (PDLLA) and poly(Ethylene Oxide) (PEO). *Int J Mol Sci* **12**, 2158–2173.

Heunis, T.D., Smith, C., and Dicks, L.M. (2013). Evaluation of a nisin-eluting nanofiber scaffold to treat *Staphylococcus aureus*-induced skin infections in mice. *Antimicrob Agents Chemother* **57**, 3928–3935.

Holo, H., Nilssen, O., and Nes, I.F. (1991). Lactococcin A, a new bacteriocin from Lactococcus lactis subsp. *cremoris*: isolation and characterization of the protein and its gene. *J Bacteriol* **173**, 3879–3887.

Hsu, S.T.D., Breukink, E., Tischenko, E.,et al. (2004). The nisin-lipid II complex reveals a pyrophosphate cage that provides a blueprint for novel antibiotics. *Nat Struct Mol Biol* **11**, 963–967.

Jarvis, B. (1967). Resistance to nisin and production of nisin-inactivating enzymes by several *Bacillus* species. *J Gen Microbiol* **47**, 33–48.

Kastin, A.J. (2006). *Handbook of Biologically Active Peptides*. Academic Press, Amsterdam, Boston, xliii, 1595 p., 1540 p. of col. plates pp.

Katla, T., Naterstad, K., Vancanneyt, M., et al. (2003). Differences in susceptibility of *Listeria monocytogenes* strains to sakacin P, sakacin A, pediocin PA-1, and nisin. *Appl Environ Microbiol* **69**, 4431–4437.

Kawulka, K.E., Sprules, T., Diaper, C.M., et al. (2004). Structure of subtilosin A, a cyclic antimicrobial peptide from *Bacillus subtilis* with unusual sulfur to alpha-carbon cross-links: formation and reduction of alpha-thio-alpha-amino acid derivatives. *Biochemistry* **43**, 3385–3395.

Kemperman, R., Jonker, M., Nauta, A., et al. (2003). Functional analysis of the gene cluster involved in production of the bacteriocin circularin A by *Clostridium beijerinckii* ATCC 25752. *Appl Environ Microbiol* **69**, 5839–5848.

Kjos, M., Borrero, J., Opsata, M., et al. (2011a). Target recognition, resistance, immunity and genome mining of class II bacteriocins from Gram-positive bacteria. *Microbiology* **157**, 3256–3267.

Kjos, M., Nes, I.F., and Diep, D.B., 2009. Class II one-peptide bacteriocins target a phylogenetically defined subgroup of mannose phosphotransferase systems on sensitive cells. *Microbiology* **155**, 2949–2961.

Kjos, M., Nes, I.F., and Diep, D.B. (2011b). Mechanisms of resistance to bacteriocins targeting the mannose phosphotransferase system. *Appl Environ Microbiol* **77**, 3335–3342.

Kjos, M., Oppegard, C., Diep, D.B., et al. (2014). Sensitivity to the two-peptide bacteriocin lactococcin G is dependent on UppP, an enzyme involved in cell-wall synthesis. *Mol Microbiol* **92**, 1177–1187.

Kjos, M., Salehian, Z., Nes, I.F., and Diep, D.B. (2010). An extracellular loop of the mannose phosphotransferase system component IIC is responsible for specific targeting by class IIa bacteriocins. *J Bacteriol* **192**, 5906–5913.

Larsen, B. and Monif, G.R.G. (2001). Understanding the bacterial flora of the female genital tract. *Clin Infect Dis* **32**, e69–e77.

Liu, L.H., Xu, K.J., Wang, H.Y., et al. (2009). Self-assembled cationic peptide nanoparticles as an efficient antimicrobial agent. *Nature Nanotechnol* **4**, 457–463.

Liu, X., Vederas, J.C., Whittal, R.M., et al. (2011). Identification of an N-terminal formylated, two-peptide bacteriocin from *Enterococcus faecalis* 710C. *J Agric Food Chem* **59**, 5602–5608.

Lozano, J.C.N., Meyer, J.N., Sletten, K., et al. (1992). Purification and amino acid sequence of a bacteriocin produced by *Pediococcus acidilactici*. *J Gen Microbiol* **138**, 1985–1990.

Luders, T., Birkemo, G.A., Fimland, G., et al. (2003). Strong synergy between a eukaryotic antimicrobial peptide and bacteriocins from lactic acid bacteria. *Appl Environ Microbiol* **69**, 1797–1799.

Maldonado-Barragan, A., Cardenas, N., Martinez, B., et al. (2013). Garvicin A, a novel class IId bacteriocin from *Lactococcus garvieae* that inhibits septum formation in *L. garvieae* strains. *Appl Environ Microbiol* **79**, 4336–4346.

Maqueda, M., Sanchez-Hidalgo, M., Fernandez, M., et al. (2008). Genetic features of circular bacteriocins produced by Gram-positive bacteria. *FEMS Microbiol Rev* **32**, 2–22.

Marki, F., Hanni, E., Fredenhagen, A., and van Oostrum, J. (1991). Mode of action of the lanthionine-containing peptide antibiotics duramycin, duramycin B and C, and cinnamycin as indirect inhibitors of phospholipase A2. *Biochem Pharmacol* **42**, 2027–2035.

Martínez, B., Rodríguez, A., and Suárez, J.E. (2000). Lactococcin 972, a bacteriocin that inhibits septum formation in lactococci. *Microbiology* **146**, 949–955.

Martin-Visscher, L.A., Gong, X., Duszyk, M., and Vederas, J.C. (2009). The three-dimensional structure of carnocyclin A reveals that many circular bacteriocins share a common structural motif. *J Biol Chem* **284**, 28674–28681.

Martinez-Bueno, M., Valdivia, E., Galvez, A., et al. (1998). Analysis of the gene cluster involved in production and immunity of the peptide antibiotic AS-48 in *Enterococcus faecalis*. *Mol Microbiol* **27**, 347–358.

Martinez, B., Bottiger, T., Schneider, T., et al. (2008). Specific interaction of the unmodified bacteriocin Lactococcin 972 with the cell wall precursor lipid II. *Appl Environ Microbiol* **74**, 4666–4670.

McBride, S.M. and Sonenshein, A.L. (2011). Identification of a genetic locus responsible for antimicrobial peptide resistance in *Clostridium difficile*. *Infect Immun* **79**, 167–176.

Mehla, J. and Sood, S.K. (2011). Substantiation in *Enterococcus faecalis* of dose-dependent resistance and cross-resistance to pore-forming antimicrobial peptides by use of a polydiacetylene-based colorimetric assay. *Appl Environ Microbiol* **77**, 786–793.

Meindl, K., Schmiederer, T., Schneider, K., et al. (2010). Labyrinthopeptins: a new class of carbacyclic lantibiotics. *Angew Chem Int Ed Engl* **49**, 1151–1154.

Ming, X.T. and Daeschel, M.A. (1993). Nisin Resistance of Foodborne bacteria and the specific resistance responses of *Listeria monocytogenes* Scott-A. *J Food Prot* **56**, 944–948.

Moll, G., Ubbink-Kok, T., Hildeng-Hauge, H., et al. (1996). Lactococcin G is a potassium ion-conducting, two-component bacteriocin. *J Bacteriol* **178**, 600–605.

Moll, G.N., Konings, W.N., and Driessen, A.J. (1999a). Bacteriocins: mechanism of membrane insertion and pore formation. *Antonie Van Leeuwenhoek* **76**, 185–198.

Moll, G.N., van den Akker, E., Hauge, H.H., et al. (1999b). Complementary and overlapping selectivity of the two-peptide bacteriocins plantaricin EF and JK. *J Bacteriol* **181**, 4848–4852.

Molloy, E.M., Field, D., O'Connor, P.M., et al. (2013). Saturation mutagenesis of lysine 12 leads to the identification of derivatives of nisin A with enhanced antimicrobial activity. *PLoS One* **8**, e58530.

Murphy, K., O'Sullivan, O., Rea, M.C., et al. (2011). Genome mining for radical SAM protein determinants reveals multiple sactibiotic-like gene clusters. *PLoS One* **6**, e20852.

Naghmouchi, K., Le Lay, C., Baah, J., and Drider, D., (2012). Antibiotic and antimicrobial peptide combinations: synergistic inhibition of *Pseudomonas fluorescens* and antibiotic-resistant variants. *Res Microbiol* **163**, 101–108.

Nes, I.F., Brede, D.A., and Holo, H. (2006). The non-lantibiotic heat-stable bacteriocins in Gram positive bacteria. In: *Handbook of Biological Active Peptides* ((ed.) A.J. Kastin), Academic Press, San Diego, Ca, USA.

Nes, I.F., Diep, D.B., Havarstein, L.S., et al. (1996). Biosynthesis of bacteriocins in lactic acid bacteria. *Antonie Van Leeuwenhoek* **70**, 113–128.

Nes, I.F. and Eijsink, V.G.H. (1999). Regulation of group II peptide bacteriocin synthesis by quorum-sensing mechansims. In: *Cell-Cell Signaling in Bacteria* ((eds) G.M. Dunny and S.C. Winans) ASM Press, Washington, D.C., 175–192.

Nes, I.F. and Holo, H. (2000). Class II antimicrobial peptides from lactic acid bacteria. *Biopolymers* **55**, 50–61.

Nes, I.F., Yoon, S.-S., and Diep, B.D. (2007). Ribosomally synthesiszed antimicrobial peptides (bacteriocins) in lactic acid bacteria: A Review. *Food Sci Biotechnol* **16**, 675–690.

Netz, D.J., Sahl, H.G., Marcelino, R., et al. (2001). Molecular characterisation of aureocin A70, a multi-peptide bacteriocin isolated from *Staphylococcus aureus*. *J Mol Biol* **311**, 939–949.

Nissen-Meyer, J., Holo, H., Havarstein, L.S., et al. (1992). A novel lactococcal bacteriocin whose activity depends on the complementary action of two peptides. *J Bacteriol* **174**, 5686–5692.

Nissen-Meyer, J. and Nes, I.F. (1997). Ribosomally synthesized antimicrobial peptides: their function, structure, biogenesis, and mechanism of action. *Arch Microbiol* **167**(2–3), 67–77.

Nissen-Meyer, J., Oppegard, C., Rogne, P., et al. (2010). Structure and mode-of-action of the two-peptide (Class-IIb) bacteriocins. *Probiot Antimicrob Prot* **2**, 52–60.

Nissen-Meyer, J., Rogne, P., Oppegard, C., et al. (2009a). Structure-function relationships of the non-lanthionine-containing peptide (class II) bacteriocins produced by gram-positive bacteria. *Curr Pharm Biotechnol* **10**, 19–37.

Nissen-Meyer, J., Rogne, P., Oppegard, C., et al. (2009b). Structure-function relationships of the non-lanthionine-containing peptide (class II) bacteriocins produced by gram-positive bacteria. *Curr Pharm Biotechnol* **10**, 19–37.

O'Connor, E.M. and Shand, R.F. (2002). Halocins and sulfolobicins: The emerging story of archaeal protein and peptide antibiotics. *J Ind Microbiol Biotechnol* **28**, 23–31.

Oppegard, C., Rogne, P., Emanuelsen, L., et al. J. (2007). The two-peptide class II bacteriocins: structure, production, and mode of action. *J Mol Microbiol Biotechnol* **13**, 210–219.

Ovchinnikov, K.V., Kristiansen, P.E., Uzelac, G., et al. (2014). Defining the structure and receptor binding domain of the leaderless bacteriocin LsbB. *J Biol Chem* **289**, 23838–23845.

Pingitore, E.V., Hebert, E.M., Nader-Macias, M.E., and Sesma, F. (2009). Characterization of salivaricin CRL 1328, a two-peptide bacteriocin produced by *Lactobacillus salivarius* CRL 1328 isolated from the human vagina. *Res Microbiol* **160**, 401–408.

Ramnath, M., Arous, S., Gravesen, A., et al. (2004). Expression of mptC of *Listeria monocytogenes* induces sensitivity to class IIa bacteriocins in *Lactococcus lactis*. *Microbiology* **150**, 2663–2668.

Ramnath, M., Beukes, M., Tamura, K., and Hastings, J.W. (2000). Absence of a putative mannose-specific phosphotransferase system enzyme IIAB component in a leucocin A-resistant strain of Listeria monocytogenes, as shown by two-dimensional sodium dodecyl sulfate-polyacrylamide gel electrophoresis. *Appl Environ Microbiol* **66**, 3098–3101.

Rea, M., Ross, R.P., Cotter, P., and Hill, C. (2011a). Classification of Bacteriocins from gram-positive bacteria. In: *Prokaryotic Antimicrobial Peptides* ((eds) D.Drider and S. Rebuffat) Springer, New York, 29–53.

Rea, M.C., Alemayehu, D., Casey, P.G., et al. (2014). Bioavailability of the anti-clostridial bacteriocin thuricin CD in gastrointestinal tract. *Microbiology* **160**, 439–445.

Rea, M.C., Dobson, A., O'Sullivan, O., et al. (2011b). Effect of broad- and narrow-spectrum antimicrobials on *Clostridium difficile* and microbial diversity in a model of the distal colon. *Proc Natl Acad Sci USA* **108** Suppl 1, 4639–4644.

Rea, M.C., Sit, C.S., Clayton, E., et al. (2010). Thuricin CD, a posttranslationally modified bacteriocin with a narrow spectrum of activity against *Clostridium difficile*. *Proc Natl Acad Sci USA* **107**, 9352–9357.

Rekhif, N., Atrih, A., and Lefebvre, G. (1994). Selection and Properties of spontaneous mutants of *Listeria monocytogenes* ATCC 15313 resistant to different bacteriocins produced by lactic acid bacteria strains. *Curr Microbiol* **28**, 237–241.

Riboulet-Bisson, E., Sturme, M.H., Jeffery, I.B., et al. (2012). Effect of *Lactobacillus salivarius* bacteriocin Abp118 on the mouse and pig intestinal microbiota. *PLoS One* **7**, e31113.

Riley, M.A. and Wertz, J.E. (2002). Bacteriocins: evolution, ecology, and application. *Annu Rev Microbiol* **56**, 117–137.

Roces, C., Courtin, P., Kulakauskas, S., et al. (2012). Isolation of *Lactococcus lactis* mutants simultaneously resistant to the cell wall-active bacteriocin Lcn972, lysozyme, nisin, and bacteriophage c2. *Appl Environ Microbiol* **78**, 4157–4163.

Rogne, P., Haugen, C., Fimland, G., et al. (2009). Three-dimensional structure of the two-peptide bacteriocin plantaricin JK. *Peptides* **30**, 1613–1621.

Rollema, H.S., Kuipers, O.P., Both, P., et al. (1995). Improvement of solubility and stability of the antimicrobial peptide nisin by protein engineering. *Appl Environ Microbiol* **61**, 2873–2878.

Sahl, H.G. and Bierbaum, G., 1998. Lantibiotics: biosynthesis and biological activities of uniquely modified peptides from gram-positive bacteria. *Annu Rev Microbiol* **52**, 41–79.

Schved, F., Henis, Y., and Juven, B.J. (1994). Response of spheroplasts and chelator-permeabilized cells of gram-negative bacteria to the action of the bacteriocins pediocin SJ-1 and nisin. *Int J Food Microbiol* **21**, 305–314.

Sedgley, C.M., Clewell, D.B., and Flannagan, S.E. (2009). Plasmid pAMS1-encoded, bacteriocin-related "Siblicide" in *Enterococcus faecalis*. *J Bacteriol* **191**, 3183–3188.

Shelburne, C.E., An, F.Y., Dholpe, V., et al. (2007). The spectrum of antimicrobial activity of the bacteriocin subtilosin A. *J Antimicrob Chemother* **59**, 297–300.

Shu, Q., Qu, F., Gill, H.S., and Access, F. (2001). Probiotic treatment using *Bifidobacterium lactis* HN019 reduces weanling diarrhea associated with rotavirus and *Escherichia coli* infection in a piglet model. *J Ped Gastroenterol Nutr* **33**, 171–177.

Soman, N.R., Baldwin, S.L., Hu, G., et al. (2009). Molecularly targeted nanocarriers deliver the cytolytic peptide melittin specifically to tumor cells in mice, reducing tumor growth. *J Clin Inv* **119**, 2830–2842.

Stanton, C., Hill, C.D.M., Fitzgerald, G.D.M., and Ross, P. (2012). Probiotic composition suitable for animals (Google Patents).

Stevens, K.A., Sheldon, B.W., Klapes, N.A., and Klaenhammer, T.R. (1991). Nisin treatment for inactivation of *Salmonella* species and other gram-negative bacteria. *Appl Environ Microbiol* **57**, 3613–3615.

Sun, Z., Zhong, J., Liang, X., et al. (2009). Novel mechanism for nisin resistance via proteolytic degradation of nisin by the nisin resistance protein NSR. *Antimicrob Agents Chemother* **53**, 1964–1973.

Tagg, J.R., Dajani, A.S., and Wannamaker, L.W. (1976). Bacteriocins of gram-positive bacteria. *Bacteriol Rev* **40**(3), 722–756.

Thennarasu, S., Lee, D.K., Poon, A., et al. (2005). Membrane permeabilization, orientation, and antimicrobial mechanism of subtilosin A. *Chem Phys Lipids* **137**, 38–51.

Uzelac, G., Kojic, M., Lozo, J.,et al. (2013). A Zn-dependent metallopeptidase is responsible for sensitivity to LsbB, a class II leaderless bacteriocin of *Lactococcus lactis* subsp. *lactis* BGMN1-5. *J Bacteriol* **195**(24), 5614–5621.

Vadyvaloo, V., Arous, S., Gravesen, A., et al. (2004). Cell-surface alterations in class IIa bacteriocin-resistant *Listeria monocytogenes* strains. *Microbiology* **150**, 3025–3033.

Vadyvaloo, V., Hastings, J.W., van der Merwe, M.J., and Rautenbach, M. (2002a). Membranes of class IIa bacteriocin-resistant *Listeria monocytogenes* cells contain increased levels of desaturated and short-acyl-chain phosphatidylglycerols. *Appl Environ Microbiol* **68**, 5223–5230.

Vadyvaloo, V., Hastings, J.W., van der Merwe, M.J., and Rautenbach, M. (2002b). Membranes of class IIa bacteriocin-resistant *Listeria monocytogenes* cells contain increased levels of desaturated and short-acyl-chain phosphatidylglycerols. *Appl Environ Microbiol* **68**, 5223–5230.

van Belkum, M.J., Kok, J., Venema, G., et al. (1991). The bacteriocin lactococcin A specifically increases permeability of lactococcal cytoplasmic membranes in a voltage-independent, protein-mediated manner. *J Bacteriol* **173**, 7934–7941.

van Belkum, M.J., Martin-Visscher, L.A., and Vederas, J.C. (2011). Structure and genetics of circular bacteriocins. *Trends Microbiol* **19**, 411–418.

Verheul, A., Russell, N.J., VantHof, R., et al. (1997). Modifications of membrane phospholipid composition in nisin-resistant *Listeria monocytogenes* Scott A. *Appl Environ Microbiol* **63**, 3451–3457.

Wescombe, P.A., Heng, N.C., Burton, J.P., et al. (2009). Streptococcal bacteriocins and the case for *Streptococcus salivarius* as model oral probiotics. *Future Microbiol* **4**, 819–835.

Wiedemann, I., Bottiger, T., Bonelli, R.R., et al. (2006a). Lipid II-based antimicrobial activity of the lantibiotic plantaricin C. *Appl Environ Microbiol* **72**, 2809–2814.

Wiedemann, I., Bottiger, T., Bonelli, R.R., et al. (2006b). The mode of action of the lantibiotic lacticin 3147 a complex mechanism involving specific interaction of two peptides and the cell wall precursor lipid II. *Mol Microbiol* **61**, 285–296.

Wiedemann, I., Breukink, E., van Kraaij, C., et al. (2001). Specific binding of nisin to the peptidoglycan precursor lipid II combines pore formation and inhibition of cell wall biosynthesis for potent antibiotic activity. *J Biol Chem* **276**, 1772–1779.

Wirawan, R.E., Klesse, N.A., Jack, R.W., and Tagg, J.R. (2006). Molecular and genetic characterization of a novel nisin variant produced by *Streptococcus uberis*. *Appl Environ Microbiol* **72**, 1148–1156.

Yoneyama, F., Imura, Y., Ichimasa, S., et al. (2009). Lacticin Q, a lactococcal bacteriocin, causes high-level membrane permeability in the absence of specific receptors. *Appl Environ Microbiol* **75**, 538–541.

Yuan, J., Zhang, Z.Z., Chen, X.Z., et al. (2004). Site-directed mutagenesis of the hinge region of nisinZ and properties of nisinZ mutants. *Appl Microbiol Biotechnol* **64**, 806–815.

Zasloff, M. (2002). Antimicrobial peptides of multicellular organisms. *Nature* **415**, 389–395.

Chapter 6

Bacteriophages of Lactic Acid Bacteria and Biotechnological Tools

Beatriz Martínez[1], Pilar García[1], Ana Rodríguez Gonzalez[1], Mariana Piuri[2], and Raúl R. Raya[3,*]

[1] *DairySafe Group, Instituto de Productos Lácteos de Asturias (IPLA-CSIC), Spain*
[2] *Departamento de Química Biológica, FCEyN, Universidad de Buenos Aires, Argentina*
[3] *Centro de Referencia para Lactobacilos (CERELA)-CONICET, Argentina*

6.1. Introduction

Lactic acid bacteria (LAB) have been used in the preservation and production of fermented foods of plant and animal origin for centuries. One of the most critical problems in these processes is the contamination of the starter cultures by bacteriophages (phages) that affect starter activity or cause bacterial lysis. The significance of phage infections in dairy fermentations was first highlighted by Whitehead and Cox (1935). To control phage contamination, the scientific community has focused its efforts on developing strategies for their elimination from the dairy environment. These strategies (Table 6.1) have been based on strain selection and rotation systems, phage-insensitive mutants, and anti-phage resistance mechanisms. Nevertheless, phage remains the major single cause of fermentation failure. Since *Lactococcus lactis* and *Streptococcus thermophilus* strains suffer the highest number of phage infections, the current knowledge on phage biology mainly relies on phages infecting these species. Phages infecting lactobacilli species have been less studied because they hardly cause any problems in dairy processes. However, the increasing role of lactobacilli as probiotics has fueled research into their phages. In fact, a number of dairy fermentation batches in which *Lactobacillus casei*, *Lact. acidophilus*, *Lact. delbrueckii* subsp. *lactis* and *Lact. helveticus* strains have suffered phage contamination have been reported.

The study of LAB phages has not only brought information about the types and characteristics of lytic phages present in the fermentation industry but, in particular, streptococcal comparative phage genomics has also significantly contributed to the field of phage taxonomy and to the characterization of the CRISPR (clustered regularly interspaced short palindromic repeats) system, an antiphage mechanism acquired by bacteria after phage infection. Some LAB phages have been used as transducing vehicles for genetic characterization as well as manipulation of LAB strains. Furthermore, phage genetic elements have been the DNA source in developing antiphage systems (i.e., using phage promoter sequences or repressor function) and integration vectors (i.e., using

* Corresponding author: Centro de Referencia para Lactobacilos (CERELA)-CONICET, Chacabuco 145, 4000 San Miguel de Tucumán, Tucumán, Argentina. rraya@cerela.org.ar

Biotechnology of Lactic Acid Bacteria: Novel Applications, Second Edition. Edited by Fernanda Mozzi, Raúl R. Raya, and Graciela M. Vignolo.

Table 6.1. Strategies applied in phage control

Survey: the reservoir and distribution of phages in the dairy environment, milk, and starters.
Control phage proliferation: includes, among others, direct inoculation of milk with concentrated starter cultures; use of antiphage starter medium; use of separate rooms for starter production and cheese manufacture; regular cleaning and disinfections (chlorinate) of vats, equipments, rooms (laboratory and factory), clothes, and air (positive filtration); rotation of starter strains.
Strain selection and starter rotation system: use mixture of well-characterized strains, starter rotation system with non-overlapping phage-insensitive patterns.
Bacteriophage-insensitive bacterial mutants (BIM)
Design of phage-resistant starter strains: conjugation of natural phage resistance systems (i.e., restriction/modification systems).
Antiphage systems: use of genetic phage elements that interfere with the phage replication cycle including: phage repressor; superinfection exclusion proteins; origin of replication; and antimessenger of phage genes.

integrative functions) to deliver and stabilize genes in the LAB genome. Excellent reviews on the field of LAB phages have been published recently (Brøndsted and Hammer 2006; Brüssow and Desiere 2006; Brüssow and Suárez 2006; Emond and Moineau 2007; Garneau and Moineau 2011; Mills et al. 2012; Mahony and van Sinderen 2014).

6.2. Bacteriophages of Lactic Acid Bacteria

LAB phages have been isolated from diverse natural sources (e.g., raw milk, yogurt, cheese, sauerkraut, pickles, cereals, wine, coffee, sewage, saliva, vagina) and upon induction of lysogens. The transition from lysogeny to lytic development of LAB phages, or prophage induction, occurs in response to DNA damage and has often been induced after treatment with mitomycin C. As in other phage systems, the frequency of lysogenization depends on the cell growth conditions and on the multiplicity of infection; starvation of cells prior to phage adsorption and high multiplicity of infection enhance lysogeny. Most of the isolated LAB phages showed very restricted host ranges, limited to one or a few starter strains of the same species; for most lysogens suitable indicator strains or conditions for lytic propagation are not easily defined and evidence of these phage-like particles is limited to their visualization under the transmission electron microscope. The narrow host range that most temperate phages exhibit suggests the presence of homoimmune or restriction modification systems. Plasmid DNA transduction has been used as an alternative to lytic plaque assays to define the host range of the *Lactobacillus* phage phi-adh; high-frequency transducing particles of phage phi-adh transduced plasmid DNA to several *Lact. gasseri* cells which, otherwise, did not support lytic growth of this phage (Raya et al. 1989).

The presence of prophage sequences can also be pointed to in the completed genome sequences of LAB strains by using phage proteins as markers, such as integrases, portal proteins, terminases, and tail tape measure proteins (Canchaya et al. 2003). Comparative analysis of complete bacterial genomes has shown that most LAB prophage genomes share the same genetic organization; involving two clusters of genes that control related functions and are transcribed divergently. One of these clusters comprises genes participating in integration and lysogeny maintenance; the other includes genes involved in lytic development (Canchaya et al. 2003, 2004). Also, prophage sequences are located at both sides of the terminus of the bacterial replication; the lytic gene cluster being transcribed in the same direction as the surrounding bacterial genes (Canchaya et al. 2004).

Phages of LAB known so far are tailed and belong to the order *Caudovirales* that comprises over 95% of all known phages (Forde and Fitzgerald 1999b). These phages possess double-stranded linear DNA and their G + C content is similar to that of their hosts, ranging from 33% in *Lact. delbrueckii* subsp. *lactis* phages to 48% in *Lact. casei* phages. The genome size ranges from 14 to 141 Kb.

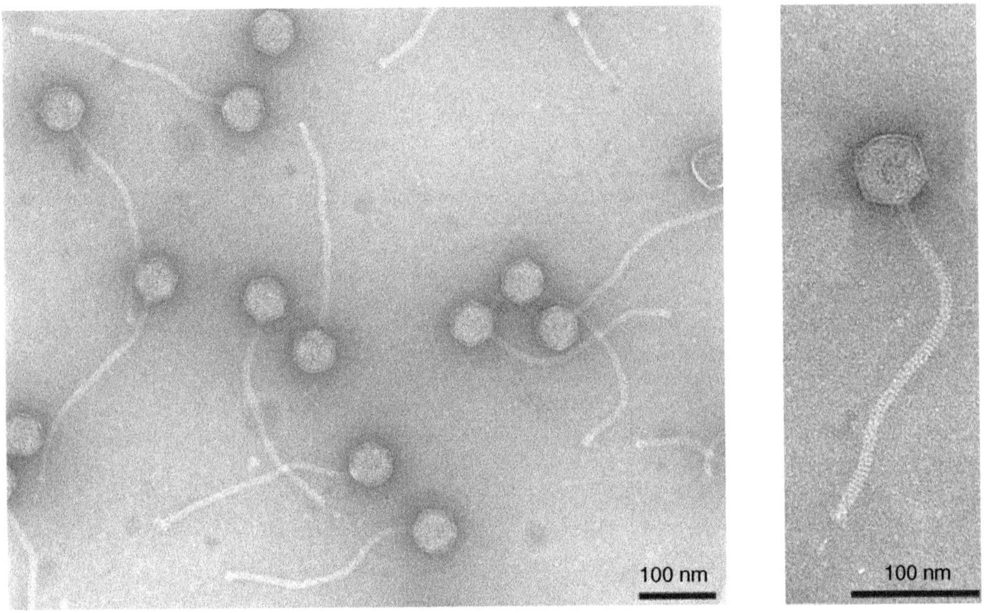

Figure 6.1. Electron micrograph of negatively stained bacteriophage A2 that infects strains of *Lact. casei* and *Lact. paracasei.*

Most lactococcal phages belong to the *Siphoviridae* family (long, non-contractile tails) while a few belong to the *Podoviridae* family (short tails). *Strep. thermophilus* phages belong to the *Siphoviridae* family. Virulent and temperate streptococcal phages are genetically related and form two distinct siphovirus groups according to the number of major structural proteins and the packaging mechanism of their double-stranded DNA: cos-like (group Sfi21) and pac-like (group Sfi11). Sfi11-like phages differ from the Sfi11 phages by having two distinct major head proteins instead of only one, by using a scaffold protein to aid assembly, and by not processing their major head proteins (Proux et al. 2002). Finally, *Lactobacillus* phages are mainly siphovirus with just a few myoviruses.

Lysogeny is rare in *Strep. thermophilus* (less than 10%), but a common phenomenon in *L. lactis* and *Lactobacillus* (Brøndsted and Hammer 2006). The close relationship observed between temperate and virulent phages of, for example, *Strep. thermophilus*, *Lact. delbrueckii* subsp. *bulgaricus* or *Lact. helveticus*, suggested that temperate phages could also be the source of phage outbreaks found in commercial fermentations, although it has not yet been proven unequivocally. However, virulent forms originating from FSW, a temperate phage of *Lact. casei* S-1 strain (ATCC 27139), were isolated from abnormal fermentations with the S-1 strain. Phage FSW was not inducible with mitomycin C or ultraviolet light, although spontaneous induction was observed (Shimizu-Kadota et al. 1983). DNA–DNA hybridization and heteroduplex analyses showed that the DNA of FSW and its virulent derivatives were identical, except that an extra 1.3 kb piece of DNA was present in the virulent forms. Sequence analysis of this extra fragment (1256 bp), showed characteristics of an insertion IS element named ISL1, the first element described in *Lactobacillus* (Shimizu-Kadota et al. 1985). The demonstration that prophage FSW was the source of virulent phages represented the first formal evidence that temperate bacteriophages can be a source of virulent variants in dairy fermentations.

At present, complete genome sequences of 75 phages (70 siphovirus, 3 myovirus, and 2 podovirus) infecting LAB, as well as many prophage sequences inserted in the in sequenced LAB genomes are available in databases (http://www.ncbi.nlm.nih.gov/genomes/GenomesGroup.cgi?taxid=10239&sort=taxonomy) (Shimizu-Kadota et al. 2000; Garcia et al. 2003; Lo et al. 2005).

6.2.1. Classification of Lactococcal Phages

L. lactis phages are commonly found in the dairy environment, representing a permanent threat in the manufacture of fermented dairy products. To better control phage infections, phages were isolated and characterized, which led to the discovery of natural phage defense systems present in LAB (among them, the CRISPR system) and to the identification of counter-defense mechanisms. Hitherto, these phages are the best characterized LAB phages.

An important characteristic of lactococcal phages is their great diversity. Deveau et al. (2006) differentiated them into 10 groups by DNA–DNA hybridization studies, electron microscopic observations, and comparative sequence analyses. Ninety-eight percent of known lactococcal phages responsible for most dairy fermentation breakdowns belong to the groups 936, c2, and P335. Possibly the systematic use of selected starters in the dairy industry have "filtered" these phages, facilitating their propagation as these phages are more often isolated. These phage groups are (i) shipovirures, (ii) the lactococcal phages that belong to the groups 949, P087, and 1706 (phages with longer tails), and (iii) the Q54 and 1358 phages. Conversely, phages of the groups P034 and KSY1 belong to the *Podoviridae* family. Based on this study, an improved multiplex PCR test has been designed and applied to rapidly identify members of the three predominant groups that are responsible for fermentation breakdowns.

6.3. Antiphage Strategies

Starter cultures are continuously exposed to phage attack in food fermentations. Dairy fermentations are particularly susceptible to phage attack due to the rather limited starter culture repertoire and the use of non-sterile substrates such as milk. Moreover, in many dairy plants starter cultures are cultivated in large volumes and in rich media, which are ideal conditions for phage propagation. Therefore, strategies to avoid the presence and survival of phages in dairy factories have always been a matter of concern.

Two streamlined approaches have been followed to minimize the negative impact of phages in dairy. One of them relies on the implementation of technological hurdles to reduce phage contamination. These control measurements imply a correct plant design to avoid aerosols and cross-contamination routes, strict cleaning and disinfection procedures, use of phage inhibitory growth media for starter propagation and starter rotation, among others (Sturino and Klanehammer 2004). The development of Phage Analysis and Critical Control Points (PACCP) tools has also been suggested (Samson and Moineau 2013). In this context, efficient and rapid detection methods, such as those based on the polymerase chain reaction (PCR) and other high-throughput sequencing technologies, have had a positive impact on monitoring the sources of phage contamination within the plant and in raw materials. The other approach to counteract phage contamination is based on a better understanding of natural resistance mechanisms and phage biology to pave the way for the development of specialized biotechnological tools to inhibit phage propagation. As described by Samson and Moineau (2013), bacteria and phages are engaged in a "continuous arms race" that results in the selection of sophisticated resistance mechanisms in bacteria that is translated into a strong selective pressure on phages that results in evolved phages able to surpass bacterial defenses. Due to their economical relevance, phage-resistance mechanisms have been extensively studied in LAB. However, there are comprehensive reviews in the literature that cover unique resistant mechanisms in other bacteria (Labrie et al. 2010; Samson et al. 2013a).

6.3.1. Natural Mechanisms of Phage Resistance

Phage resistant strains or strains with distinct phage susceptibility patterns that retain their technological traits are in great demand to facilitate starter rotation schemes. This has been one of the main driving forces promoting the study of phage–host interactions that has contributed to the discovery of natural resistance mechanisms in many LAB strains (Sturino and Klaenhammer 2006; Mahony

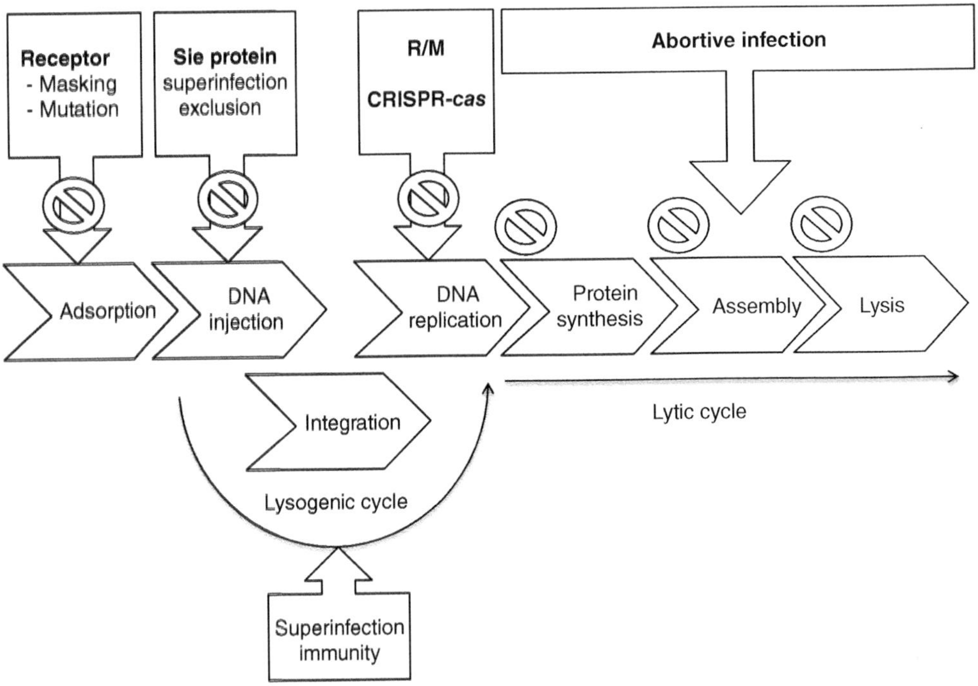

Figure 6.2. Natural phage resistance mechanisms described in lactic acid bacteria. R/M: restriction/modification systems; CRISPR-*cas*: clustered regularly interspaced short palindromic repeats (CRISPR) loci and *cas* (CRISPR-associated) genes.

et al. 2012). Natural resistance mechanisms have been categorized into five major groups according to the step inhibited during phage infection (Figure 6.2).

Prevention of adsorption (Figure 6.2) is an effective means of phage resistance. The first step of phage infection implies recognition of the host that occurs at the bacterial cell envelope and is mediated by structural phage tail proteins. Although the exact mechanisms of target recognition and phage receptors remain elusive, a common theme arises which is the involvement of (poly-)saccharide moieties, at least during the first step of reversible adsorption. Structural studies have shown the affinity and binding of the receptor binding proteins of the lactococcal phages p2 and TP901-1 to cell wall polysaccharides (CWP) (Spinelli et al. 2014). A distinct decoration of other CWP as lipoteichoic acids (LTA) has been shown to be required for efficient adsorption of the *Lact. delbrueckii* LL-H phage (Räisänen et al. 2007). Proteinaceous receptors have also been described including the phage infection protein Pip used by lactococcal phages of the c2 family and the S-layer protein in *Lact. helveticus* phages (Geller et al. 1993; Mahony et al. 2012). Phage resistance can be gained by altering the structure and composition of these putative receptors or by masking them by the production of exopolysaccharides (Forde and Fitzgerald 1999a; Räisänen et al. 2007). Abolition of phage receptor biosynthesis is only possible when their function in the host bacteria is not essential. This is the case of several bacteriophage insensitive *L. lactis* mutants that lack the protein Pip (Valyasevi et al. 1991).

Delivery of viral DNA into the bacterial host (Figure 6.2) can be inhibited by superinfection exclusion (Sie) proteins, which are encoded by some *L. lactis* and *Strep. thermophilus* temperate phages within their lysogeny module (Mahony et al. 2008; Ali et al. 2014). These membrane proteins are produced by the resident prophages and seem to block the insertion of the tape measure protein of the incoming phage into the bacterial membrane, inhibiting viral DNA ejection (Bebeacua et al. 2013).

Once the viral DNA has entered the host, it can be digested by site-specific restriction enzymes while the host DNA is protected by a cognate methylase. Three different groups of restriction-modification systems have been described in LAB, many of them belonging to type II (LlaDCHI, LlaAI, ScrFI, LlaBI) and often plasmid encoded. The activity of restriction/modification systems reduces the efficiency of plating (EOP) as a function of the number of recognition sites in the phage genome and several mechanisms have been developed by the phages to overcome DNA digestion, including acquisition of their own methylases (Samson et al. 2013b).

The presence of CRISPR loci and *cas* (CRISPR-associated) genes also interferes with phage replication by degrading phage or other invading DNA in a sequence-specific manner (Labrie et al. 2010). CRISPR loci contain series of short direct repeats (21–48 bp), separated by unrelated spacer sequences of similar size. CRISPR loci are transcribed and processed by Cas proteins. When these short RNAs in conjunction with Cas proteins find a complementary DNA sequence, this targeted DNA is cleaved. Upon new phage infections or plasmid uptake, new specific spacer sequences are incorporated into the CRISPR locus, providing the host and its progeny with acquired resistance. CRISPR/*cas* loci are widespread in LAB (Barrangou and Horvath 2012). CRISPR/*cas* loci have been firstly identified in several *Lactobacillus*, *Streptococcus*, and *Enterococcus* strains, and are thought to be absent in other industrially relevant LAB genera. However, as more sequencing projects are undertaken, the number of identified CRISPR/*cas* loci are currently increasing and have already been found in *Lactococcus*, *Leuconostoc*, *Oenococcus*, and *Pediococcus* strains.

Finally, up to 23 different abortive infection mechanisms (Abi) have been described that interfere with further steps during phage propagation (Chopin et al. 2005; Labrie et al. 2010). Most of them have been found in *L. lactis*. Contrary to the other resistance mechanisms, Abi systems lead to the premature death of the host cell, reducing burst sizes. Abi systems are often host-encoded proteins that target essential steps of phage DNA replication (A, F, K, P, T) or interfere with RNA transcription (B, G, U), protein translation (C, V), or induce premature lysis (Z) (Figure 6.2). Specific toxin/antitoxin (TA) pairs, which are often linked to postsegregational killing systems in plasmids, have also been shown to contribute to resistance to specific phages (Samson et al. 2013b).

6.3.2. Genetically Engineered Antiphage Systems

Many of the natural resistance mechanisms described above are plasmid-borne. Therefore, conjugation is a straightforward approach to transfer more than one phage resistance mechanism to industrially relevant starter bacteria in a food-grade manner (Durmaz and Klaenhammer 1995). The selection of bacteriophage insensitive mutants (BIM) may also help starter rotation schemes. However, BIMs may not fulfill the technologically relevant phenotypes as the wild-type strain does, and they quickly revert to the susceptible phage phenotype. Therefore, several genetically engineered antiphage systems have been elegantly designed, mostly based on the fundamental knowledge of phage biology and on the natural phage resistance mechanisms described in bacteria. These strategies have been reviewed elsewhere (Coffey and Ross 2002; Sturino and Klaenhammer 2006; Samson and Moineau 2013a) and some examples are outlined here.

Starter bacteria have been engineered to overproduce phage components that interfere with phage propagation. One example is cloning the origin of phage DNA replication in a multicopy plasmid. In this way, replication proteins are titrated away, slowing down the synthesis of new copies of phage DNA and, consequently, reducing the number of new phage particles (Moscoso and Suárez 2000). It is also possible to interfere with the phage cycle by overproducing mutated versions of specific proteins which work as multimeric enzymes, referred to as "subunit poisoning" (Sturino and Klaenhammer 2007). The incorporation of the mutated version prevents protein oligomerization and abolishes enzyme activity. Synthesis of the repressor of the lytic cycle, required by temperate phages to maintain the lysogenic state, has been shown to hamper phage infection by closely related phages. This approach has been shown to work against *Lactococcus* and *Lactobacillus* phages (Álvarez et al. 1999; Durmaz et al. 2002).

Antisense RNA strategies have also been explored against lactococcal and *Strep. thermopilus* phages. They basically consist of cloning genes or gene fragments in reverse orientation to promote the formation of inactive and short-lived double-stranded RNA molecules when the target phage RNA is synthesized. Therefore, antisense RNA strategies are highly specific and their efficacy to inhibit phage propagation also depends on the targeted gene and expression levels (Walker and Klaenhammer 2000). At the protein level, virion proteins can be targeted by specific antibodies providing means to inactivate them. Nanobodies (antibody fragments) and other protein-binding proteins known as DARPins (designed ankyrin repeat proteins), which are able to specifically recognize and bind to the baseplate of the lactococcal phage TP901-1, were shown to block phage adsorption and, subsequently, phage propagation (Desmyter et al. 2013).

Triggering bacterial death upon phage infection has been engineered by cloning the lactococcal restriction cassette LlaIR under the control of a phage promoter that is activated only when the bacterial host is infected (Djordjevic et al. 1997).

The abovementioned genetic approaches, although proven to overcome phage infection, have suffered from being rather specific since they rely on the degree of conservation and homology among the target molecules on which they have been designed. However, it is likely that increasing negative awareness of the use of genetically modified microorganisms (GMO) in food is the reason why these strategies have not been translated into improved starter cultures as yet. Therefore, it is important to seek other non-GMO approaches to engineer phage-resistant starters. Interestingly, *L. lactis* resistant to phage c2 could be isolated after several rounds of growth in the presence of the cell wall active bacteriocin Lcn972, which induced genetic rearrangements within the *L. lactis* chromosome including the partial deletion of *pip* (Roces et al. 2012). Of note, phages able to defeat bacterial defenses are continuously evolving (Samson et al. 2013a,b) and the "zero risk" of phage fermentation failures is unlike to be reached in food settings.

6.4. Phage-Based Molecular Tools

6.4.1. Phage Integrases and Integration Vectors

Temperate phages that integrate their genomes into the host chromosome encode an integrase enzyme that mediates recombination between a short sequence of phage DNA (the phage attachment site, *attP*) and a short sequence of bacterial DNA (the bacterial attachment site, *attB*). After integration, the integrated phage genome (or prophage) is flanked by two hybrid sites, *attL* (attachment left) and *attR* (attachment right), each consisting of half the *attP* sequence and half the *attB* sequence. These new sites are substrates for excisive recombination, generally mediated by the integrase in the presence of an additional phage-encoded protein called excisionase. Extra host factors might also be necessary for this reaction to occur. Environmental factors (usually stress effectors) can induce this excisive reaction, so the released phage can start a lytic cycle. The unidirectional reaction mediated by phage integrases and the requirement of just two short sequences of DNA for recombination makes them excellent candidates for genetic engineering purposes. Integrases are included in the category of site-specific recombinases; some require encoded host factors to mediate recombination and some are autonomous with no co-factor requirements (Groth and Calos 2004). Site-specific recombinases can be classified into two major families based on amino acid sequence homology and catalytic residues of either tyrosine (e.g lambda integrase) or serine (Phage C31, R4 and TP901-1) (Esposito and Scocca 1997; Smith and Thorpe 2002). No evidence exists that any of the phage integrases of the serine recombinase family require co-factors.

Relevant technological properties of LAB (e.g., sugar fermentation, citrate transport, protease synthesis, phage resistance mechanism, etc.) are plasmid-encoded. Plasmid instability is a major source of disruption of fermentative processes. Thus, integrative plasmids carrying a phage integrase gene and the correspondent *attP* site can be a powerful tool to clone any gene of interest for subsequent stable integration into the host chromosome. Also, integrative vectors can be very useful during genetic studies to insert a second copy of a gene under an inducible promoter while attempting to construct mutants of essential genes. Several vector systems based on the integration apparatus of temperate phages of *Lactobacillus* (Álvarez et al. 1998; Dupont et al. 1995; Raya et al. 1992) and *Lactococcus* (Christiansen et al. 1994; Lillehaug et al. 1997) have been constructed. In general, these systems allow integration at the *attB* site on the phage host genome but in some cases, if the *attB* sequence is present, integration can occur in other related (van de Guchte et al. 1994; Auvray et al. 1997; Lin et al. 2013) or even unrelated bacteria (Álvarez et al. 1998).

Integrative vectors can be extremely useful for the construction of industrially-relevant LAB strains, but the entire plasmid (including usually an antibiotic selection marker and an origin of replication in *Escherichia coli* for amplification of the vector) is left on the chromosome after integration so the generated strains lose their category of food grade or GRAS (Generally Recognized as Safe). To avoid this issue, Martín et al. (2000) developed a strategy combining a delivery and depuration system. First, they constructed a vector that does not replicate in Gram (+) bacteria containing an *E. coli* origin of replication (*ori*), the *Beta*-lactamase (*Ap*) and erythromycin (*Em*) resistance genes flanked by two directly oriented *six* sites, the *int* gene and *attP* of phage A2. To validate the use of the system they cloned in the same vector a DNA segment containing the *cI* gene, which encodes for the A2 phage repressor. After electroporation of *Lact. casei* ATCC 393 with this plasmid, Erm^R cells were selected and tested for correct integration at the bacterial *attB* site. These clones were then transformed with a shuttle *E. coli*-LAB plasmid carrying a chloramphenicol resistance marker and expressing *Beta*-recombinase, that mediates recombination between the two *six* sites. A high percentage of Cm^R cells were also Erm^S, implying that *Beta* recombinase-mediated deletion of the undesired DNA (*ori-Ap-Em*) took place. The plasmid carrying the *Beta* recombinase was subsequently cured from the cells to render a cI-expressing strain containing only food grade DNA in the host chromosome. As expected, this strain was resistant to phage A2 infection.

More recently, Petersen et al. (2013) described a repetitive, marker free method that allows multiple chromosomal integration of DNA in *Lactococcus* using the integration system of phage TP901-1. The first gene of interest to be integrated in the chromosome is cloned in a vector that carries *attP* and a minimal *attB* sequence ($attB_{min}$). This plasmid also contains an *E. coli* origin of replication, antibiotic resistance markers, and the *oroP* gene (codes for an orotate transporter that can be used as a counter-selection marker) flanked by the *lox66* and *lox71* sites inserted as directed repeats. The advantage of using these mutated *lox*P sites is that after Cre-mediated recombination a *lox72* site is generated that is a poor substrate for Cre. This is a necessary condition when the system is going to be reused since the presence of multiple *loxP* sites could result in undesirable recombination events.

The TP901-1 integrase expression is then carried from a circularized PCR fragment containing the *int* gene under control of a strong promoter that is co-transformed with the plasmid. Transient expression was sufficient to promote integration and did not affect the stability of the integrated DNA, a side effect observed when TP901-1 integrase was continuously expressed. Once the plasmid carrying the first gene is integrated in the genome, cells are transformed with a plasmid that is not able to replicate in *Lactococcus* and express Cre recombinase transiently. Cells are then plated in minimal media containing 5-FO. Only those clones where recombination between *lox66* and *lox71* has occurred and the orotate transporter has been lost (and also ori- antibiotic resistance markers) can grow in the presence of this toxic compound. This strategy can be repeated multiple times since integration of the new vector will take place at the $attB_{min}$ site provided by the previous integrated plasmid.

Another interesting application of phage integrases, particularly those that do not require host factors for activity, is that they can be used in foreign environments such as mammalian cells (Yamaguchi et al. 2011). It has been shown that *Streptomyces* phage C31 integrase works in human and mouse cells, mediating intramolecular integration in transfected plasmid DNA and intermolecular integration into the chromosomes (Groth et al. 2000; Thyagarajan et al. 2001). Phage C31 integrase can efficiently catalyze a site-specific cassette exchange reaction using introduced wild-type *att* sites in mouse ES cells (Belketi et al. 2003) and can recognize and insert *attB*-containing plasmids into pseudo *attP* sites in the murine genome (Olivares et al. 2002). The related integrase of *Streptomyces* phage R4 also works in human cells (Olivares et al. 2001). In the context of LAB phages, the TP901-1 serine integrase was able to carry out intramolecular integration on a transfected plasmid substrate (Stoll et al. 2002) and intrachromosomal deletions in human cells (Hollis et al. 2003). Also, the mycobacteriophage bxb1 integrase was successfully used for site-specific integration in *Plasmodium falciparum* chromosomes (Nkrumah et al. 2006) and in Drosophila (Huang et al. 2011). More recently, Yamaguchi et al. (2011) designed a human artificial chromosome (HAC) vector containing five recombination sites (phiC31 *attP*, R4 *attP*, TP901-1 *attP*, Bxb1 *attP* and FRT) to compare the activity of mammalian codon-optimized integrases in these cell types. This multi-integrase HAC vector was used to produce transgene-expressing cells and create a platform cell line for gene expression. The ability of phage integrases to specifically and efficiently recombine short DNA sequences in a genomic context in living cells makes them a potential tool in a variety of genetic engineering applications.

6.4.2. CRISPR Applications

Clustered regularly interspaced short palindromic repeats (CRISPR) are a family of DNA repeats that were described for the first time in 1987 in an intergenic region in the *E. coli* K12 genome (Ishino et al. 1987). Jansen *et al.* (2002) reported that these DNA repeats were often accompanied by CRISPR-associated proteins (Cas). Later on, several studies established that part of the repeated sequences (spacers) showed homology to foreign genetic elements such as viruses and plasmids (Bolotin et al. 2005; Mojica et al. 2005; Pourcel et al. 2005). Barrangou et al. (2007) provided the first experimental data about the function of CRISPR and Cas in adaptive phage resistance in *Strep. termophilus*. Since then, a large number of research articles have described the differences found between systems and organisms (Marraffini and Sontheimer 2010; Terns and Terns 2011; Richter et al. 2012; Westra et al. 2012; Wiedenheft et al. 2012). Using the CRISPRdb database (Grissa et al. 2007), it has been shown that about 45% of bacterial and 90% of archaeal genomes carry CRISPR-*cas* systems.

CRISPR and CRISPR-associated proteins provide adaptive immunity against viruses and plasmids. In general, small fragments from the foreign genetic element (spacers) are incorporated and arranged between direct repeats (DR) to originate the CRISPR arrays. This array is transcribed into a precursor RNA (pre-cRNA) that during a maturation process is diced into smaller CRISPR RNAs (crRNAs). These interfering RNAs use the Cas-encoded apparatus to specifically target and cleave nucleic acids complementary to the spacer sequence. The primary applications of CRISPR systems include crRNA-mediated interference for phage resistance in industrially relevant strains and the use of these arrays' hypervariability for genotyping purposes, especially for pathogenic strains (Shariat and Dudley 2014).

Subtyping or strain identification using CRISPRs has been used in diverse Gram (+) and Gram (−) bacteria. CRISPR applications existed long before their function was elucidated. Spacer-oligonucleotide typing, or "spoligotyping," has been used for decades to type *Mycobacterium* species (Groenen et al. 1993). For spoligotyping, the CRISPR array is amplified by PCR with labeled primers that recognize the direct repeats. The PCR products are hybridized to a membrane that contains probes corresponding to the spacer sequences. Since the spacer content is strain specific, differential hybridization patterns enable separation of the strains. With several improvements to the classical spoligotyping (including automation and use of microbeads instead of a membrane), this technique remains the gold standard for subtyping of members of the *M. tuberculosis* complex (MTBC) (Streicher et al. 2007). Next generation spoligotyping has also been applied to *Salmonella*. A 100% concordance between this method and CRISPR data from sequenced strains was found (Fabre et al. 2012). With upcoming DNA sequencing techniques, sequence-based analysis of CRISPR arrays has been increasingly used for microbial subtyping. The majority of sequenced-based CRISPR typing has been performed in *Salmonella* (Fabre et al. 2012) and for *Campylobacter jejuni*, for which it was shown that results from this technique are comparable to other subtyping methods (Schouls et al. 2003).

Delannoy et al. (2012) developed a real-time PCR assay with primers designed to anneal specific spacers present in different serotypes of *Shiga*-toxin-producing *E. coli* (STEC) that represent the most clinically predominant STECs within a particular O serogroup in the United States and Europe. For this approach, when a unique spacer is present in a pathogenic or predominant strain, the isolate can be identified using real-time PCR with fluorescent probes specific to that particular spacer. Also, differences in CRISPR array length as a consequence of variations in space numbers can be exploited to rapidly screen isolates by PCR and electrophoretic analysis (Le Fleche et al. 2001).

Eight distinct families of CRISPR loci based on CRISPR repeat sequences and *cas* gene content, organization, and sequences were identified in LAB genomes. These eight families do not cluster according to phylogenetic relationships for LAB (Horvath et al. 2009). The diversity of CRISPR loci in LAB likely reflects their lateral origin and their rapid evolution due to environmental pressure, notably linked to phage predation (Tyson and Banfield 2008). This polymorphism has been proposed as a potentially valuable tool for typing and comparative analyses of strains and microbial populations. Towards this application and evolutionary studies CRISPR systems have been characterized in several bacterial groups of LAB (Diancourt et al. 2007; Horvath et al. 2008; Guinane et al. 2011) but actual reports using CRISPR for typing of LAB strains are still missing. Finally, because spacers are characteristically sequentially inserted at the leader end of the locus, they can be used to track the historical path of a strain and establish phylogenetic relationships as well as common ancestry between strains (Barrangou 2013).

Based on the knowledge that crRNA-mediated interference results in cleavage of a specific target DNA, CRISPR-*cas* systems became a potential tool for genome editing. Three types of CRISPR-*cas* systems have been characterized (Makarova et al. 2011). In type I and type III systems, the pre-crRNAs are processed by specialized Cas endonucleases. Once mature, each crRNA assembles into a multi-Cas protein complex capable of recognizing and cleaving nucleic acids complementary to the crRNA. In type II, an additional transactivating cRNA (tracrRNA), necessary for pre-crRNAs maturation, exists. TracrRNA basepairs with each CRISPR repeat sequence within the pre-crRNA to form a double-stranded RNA template that is specifically cleaved by RNAse III in the presence of Cas9 (Deltcheva et al. 2011). After this processing, only the product of Cas9 is responsible for crRNA guided silencing of the target DNA (Barrangou et al. 2007; Garneau et al. 2010; Gottesman 2011).

The potential of type II system programmed DNA cleavage for genome editing was more evident when Jinek et al. (2012) designed a chimeric RNA containing a target recognition sequence at the 5′end followed by a hairpin structure retaining the base-pairing interactions that occur between the tracrRNA and the crRNA (guide RNA). This single transcript was effective to guide site-specific DNA cleavage by Cas9 *in vitro*. DNA double-stranded breaks (DSB) can be repaired by two different pathways: non-homologous end-joining (NHEJ) and homology directed repair (HDR). NHEJ can lead to the introduction of insertion/deletion mutations of different lengths. HDR can be used to introduce specific point mutations or insert a desired sequence through recombination if a DNA "donor template" is exogenously supplied. Using targeted nuclease-induced DSBs the frequency of the mutants is greater than 1%, sometimes even over 50% so the recombinants can be identified directly by screening without the need to use a marker selection (Sander and Joung 2014). After publication of Jinek's studies (2012), several groups focused on the adaptation of the described strategy to target genes in bacteria, human cell lines, and human pluripotent stem cells in culture (Cong et al. 2013; Jiang et al. 2013; Mali et al. 2013).

Mali et al. (2013) used a human codon-optimized Cas9 from *Strep. pyogenes* bearing a SV40 nuclear localization signal in combination with a guide RNA (gRNA). This gRNA was designed to target a specific 20-bp DNA sequence in the cell genome, followed by the trinucleotide (5′-NGG-3′) protospacer adjacent motif (PAM), which is recognized by Cas9 itself and is essential for cleavage. They targeted the AAVS1 locus in human 293T cells, K562 cells, and induced pluripotent stem cells. The targeting frequencies were compared with other editing technologies for targeting DSB-inducing nucleases to specific genomic sites like zinc finger nucleases (ZFNs) (Rebar and Pabo 1994; Kim et al. 1996) and transcription activator nucleases (TALENs) (Boch et al. 2009; Moscou and Bogdanove 2009). Mali et al. (2013) also proposed that this versatile method could be adapted to modify other genomic sites by simply modifying the sequence of the gRNA expression vector to match a sequence in the locus of interest. To depict the feasibility of this strategy, and keeping in mind that the target sequence has to be followed by a 5′-NGG-3′ PAM, they generated ~190,000 specific gRNA-targetable sequences, corresponding to ~40.5% exons of genes in the human genome, by using bioinformatics tools.

Simultaneously, Cong et al. (2013) reconstituted the *Strep. pyogenes* and *Strep. termophilus* CRISPR systems in mammalian cells to facilitate efficient genome editing. A single CRISPR array containing two different spacers to target two separate genes, proving that different sites in the genome can be edited simultaneously, was used.

Marraffini's group (2013) reported the use of CRISPR-*cas* systems for targeted genome editing in bacteria (Jiang et al. 2013). In particular, the strategy employed is based on the co-transformation of a targeting construct and an editing template. The editing template introduces the desired mutations, avoids cleavage by RNA : Cas9 complex and in consequence, wild-type cells are killed and recombinants survive. Editing was performed in *Strep. pneumoniae* and *E. coli*, and although editing was facilitated by a co-selection of transformable cells and a small induction of recombination at the target site by Cas9 RNA guided cleavage, the major advantage was the selection against non-edited cells. Although promising, this technique is only applicable to transformable bacteria. Using a similar strategy but a different type of CRISPR system, Kiro et al. (2014) could engineer *E. coli* bacteriophage T7 showing CRISPR-*cas* editing for modification of bacteriophage genomes. Additionally, the CRISPR-Cas Type II-A system was successfully used to engineered the genome of the virulent Streptococcus *thermophilus* virulent phage 2972 (Martel and Moineau, 2014).

More recently, Cas9 RNA-guided endonuclease was used to modify genes in several organisms including yeast, tobacco, thale cress, rice wheat, sorghum, mice, rats, rabbits, frogs, fruit flies, silkworms, and roundworms. A list of the published reports of cell types and organisms modified by Cas9 is available in Sander and Joung (2014).

The great progress made over the past few years in the development of Cas9-based technologies for genome editing has opened many questions concerning further applications of this valuable technique.

6.4.3. Recombineering

Recombineering refers to the engineering of recombinant DNA by homologous recombination (Court et al. 2002). Recombineering employs recombination systems encoded by bacteriophages to enhance the frequency of homologous recombination. This technique has the advantage of using DNA substrates with short regions of homology to the target sequences, allowing construction of chromosomal gene knock outs, deletions, insertions, point mutations, *in vivo* cloning mutagenesis of bacterial chromosomes, plasmids, genomic libraries, and phages (Muyrers et al. 1999; Datsenko and Wanner 2000; Yu et al. 2000; Copeland et al. 2001; Ellis et al. 2001; Swaminathan et al. 2001; Piuri and Hatfull 2006; Sarov et al. 2006).

The increasing number of sequenced genomes available has facilitated the use of recombineering as it is necessary to know the sequence of the target gene for its application.

The first systems described employed recombination functions encoded by bacteriophage lambda and the Rac prophage; these systems were originally used to modify *E. coli* genomes. In lambda, three proteins are involved in recombineering: Exo, Beta, and Gam. The Rac prophage only codes for RecE and RecT, which are equivalent to lambda Exo and Beta. Exo is an exonuclease that degrades one strand of dsDNA (double-stranded DNA) to generate a ssDNA (single-stranded DNA) substrate. Beta is the DNA pairing enzyme that anneals the recombineering substrate to its target. Gam inhibits the *E. coli* RecBCD and SbcD enzymes, preventing degradation of the dsDNA substrate (Kulkarni and Sthal 1989; Murphy 2007). The latest model to explain the mechanism of dsDNA recombination mediated by lambda phage proposes that Exo entirely degrades one strand, leaving the other strand intact as ssDNA. This ssDNA intermediate recombines via Beta recombinase catalyzed annealing at the replication fork (Maresca et al. 2010; Mosberg et al. 2010).

Beta belongs to a family of ssDNA annealing proteins and only this protein is necessary for single-stranded oligonucleotide recombination for the construction of point mutations and small deletions (Ellis et al. 2001). Higher frequencies of recombinants are obtained when an oligonucleotide that anneals to the DNA strand undergoing discontinuous synthesis is used. As result of the replication process, transient regions of ssDNA may be accessible for Beta-mediated annealing of the oligo. The increased recombination efficiency when using "lagging strand" oligos may reflect the increased availability of single-stranded regions during lagging vs. leading strand synthesis (Ellis et al. 2001; Zhang et al. 2003).

The *E. coli* system was successfully employed in other Gram (-) bacteria and further adapted for application in more divergent bacteria (Ranallo et al. 2006; Katashkina et al. 2009; Binder et al. 2013; Swingle et al. 2010b; Pontes and Dale 2011). The efficiency of recombineering in *E. coli* led to the search for proteins with RecE and RecT similarities in other phages of bacteria, particularly because adapting the lambda/Rac system for its use in other organisms has not always proved to give optimal results (van Kessel and Hatfull 2007; Datta et al. 2008; Swingle et al. 2010a).

Product expression of genes *60* and *61* of Mycobacteriophage Che9c-homologs to ReC and RecT, respectively, enhanced mycobacterial recombination frequencies. This new recombineering system facilitates the construction of gene knock outs and point mutations in the pathogenic bacterium *Mycobacterium tuberculosis* and also in *M. smegmatis* (van Kessel and Hatfull 2008a, b). Using a combination of bioinformatics tools, Swingle et al. (2010a) identified proteins homologous to RecE/RecT in *Pseudomonas syringae*. These genes were located in a region that corresponds to a prophage or prophage remnant containing the recombination functions. The requirements of the system were further studied in more detail to improve the efficiency of recombineering in this bacterium (Bao et al. 2012).

In 2012, van Pijkeren and Britton described the successful development and application of ssDNA recombineering in LAB. Using *E. coli recT* as a query, they searched the draft genome sequence of *Lact. reuteri* ATCC PTA 6475 and identified two *recT* homologs, *recT1* and *recT2*. Analysis of the genomic context of these genes suggests that they are probably part of a prophage or remnant prophage (MP's observation). *recT1* was cloned under control of an inducible promoter and, after induction, cells were transformed with an oligonucleotide containing a mutation that targets *RpoB*. When incorporated, this mutation causes an amino acid change that confers resistance to rifampicin. To increase the frequency of recombineering, an oligonucleotide targeting the lagging strand was used. In comparison to *E. coli*, much larger amounts of oligos had to be used; up to 100 micrograms of oligo were necessary to obtain a 1000-fold increase in recombineering efficiency over background levels, while in *E. coli* usually recombineering levels are saturated when using 115 nanograms of a

70-base oligonucleotide (Sawitzke et al. 2011). In order to avoid the MMR (methyl-directed mismatch repair system), and based on the results previously obtained in *E. coli* (Costantino and Court 2003), van Pijkeren and Britton (2012) assessed an oligo containing five adjacent mismatches. Using this targeting substrate a 100,000-fold increase in recombineering levels compared to the background was observed. The same values were obtained when assayed in MutS$^-$ or MutL$^-$ strains indicating that MMR can be completely avoided using this strategy in *Lact. reuteri*. Noteworthy, the level of recombineering frequency achieved was sufficient to generate mutations without the need of selection. To verify that non-selected recombineering could be done, after transformation using this type of oligo, 100 colonies from antibiotic free plates were replica-plated on plates containing rifampicin. Five percent of the colonies had incorporated the mutation in their genome. Genes located in different positions in the chromosome were also targeted and the recombineering rates ranged from 0.4 to 19%. In the same work, it was shown that the system could be successfully employed to generate point mutations in *L. lactis* MG1363, and two human isolates of *Lactobacillus* whose genomes are available: *Lact. plantarum* BAA 793 and *Lact. gasseri* ATCC 33323. Later on, in order to optimize parameters for ssDNA recombineering, the same group tested the efficiency of several RecT homologs derived from other bacteria van Pijkeren et al. (2012). From the tested proteins, only RecT from *Enterococcus faecalis* yielded a similar number of recombinants compared to the native *Lact. reuteri* RecT. Another strategy used in *E. coli* to increase the frequency of ssDNA recombineering is the use of phosphorothiotate oligonucleotides (PTOs) to avoid host nucleases (Sawitzke et al. 2011). Using this type of oligos at very high concentration, recombineering efficiency was increased 10-fold in *L. lactis* but not in *Lact. reuteri*, emphasizing that a panel of optimizations are required for different strains to improve recombineering; thus, results from *E. coli* cannot be directly extrapolated to other bacteria.

ssDNA recombineering was successfully employed to convert *Lact. reuteri* from vancomycin resistant to sensitive by mutating the active site of Ddl (D-Ala-D-Ala ligase) (van Pikjeren and Britton 2012) and to increase the production of the antimicrobial reuterin by modification of six bases in a transcriptional regulator element region of the *pdu*- operon van Pijkeren et al. (2012).

Recombineering can also be effectively applied to genetic modification of bacteriophage genomes, in both their prophage and lytically growing states (Marinelli et al. 2012). This technique has been especially useful for the modification of lytically growing phages, because high frequency events are required in the absence of a selectable system.

Oppenheim et al. (2004) originally described a method for engineering lambda using Red expressed from a defective prophage. In general, cells are first infected with phage and recombination functions are expressed; electrocompetent cells are prepared and transformed with target dsDNA or ssDNA. The phage progeny is then screened for the desired mutation. In 2008, Marinelli et al. described a new method termed Bacteriophage Recombineering of Electroporated DNA (BRED) used to manipulate mycobacteriophages. In this procedure, a *M. smegmatis* strain carrying an inducible plasmid coding for gp60 and 61 of Mycobacteriophage Che9c was induced and electrocompetent cells were prepared. Electroporation was performed with both the genomic phage DNA and a PCR-generated or synthetic DNA substrate that targeted the phage genome. After a recovery period, the reactions were mixed with plating cells of *M. smegmatis* and plated as in an infectious center assay. All of these primary plaques will contain wild-type phage DNA and a subset (5–50%) of the mutant allele, which can be identified by PCR screening. If the mutation did not ablate a region essential for phage growth, mutant phages can be isolated by plating dilutions of the primary plaques, followed by purification and PCR analysis of the resulting secondary plaques.

To date, BRED has been successfully used on several mycobacteriophages (Payne et al. 2009; Catalao et al. 2010, 2011; Dedrick et al. 2013) and was also adapted to make deletions in phage P1vir using an *E. coli* strain expressing lambda- Red recombinases (Feher et al. 2012).

Recombineering has been used for mutagenesis of the Tal_{2009} protein of *Lactococcus* phage TP901-1 (Stockdale et al. 2013). To achieve mutagenesis, they constructed a derivative of *L. lactis* NZ9000 (carrying the recombineering functions) containing TP901-1*erm* prophage (a derivative of TP901-1 with an erythromycin marker). After induction of recombineering functions, cells were transformed with oligos to introduce three different mutations in Tal_{2009}. Since the phage genome is integrated, until this point this protocol is essentially the same used for modification of the bacterial chromosome. After screening for the mutations, lysogens were induced by addition of mitomycin C and the recovered mutant phages were used for further experiments.

Although BRED sounds a promising tool for modification of lytically growing phages of LAB, the extremely large amounts of oligos that have to be used in recombineering experiments (about 200 µg in *Lact. reuteri* and up to 500 µg in *L. lactis*), in addition to a low efficiency of electroporation of phage DNA to recover infectious particles (Chassy and Flickinger, 1987; and our observations), suggests that several parameters will have to be optimized to adapt this technique for genome modification of this type of phage.

6.5. LAB Phages as Biocontrol Tools

The negative consequences of LAB infection by phages in food settings may be technologically profitable too. Right after their discovery, phages were already envisaged as powerful tools to treat human infections, but they were neglected due to the use of antibiotics, mostly in the Western countries (Sulakvelidze et al. 2001). Nowadays, due to the threat of antibiotic resistant bacteria, interest in phages as antimicrobials is returning in clinical settings and, moreover, is currently extended to other fields such as food biopreservation (García et al. 2010).

In this context, LAB phages can be regarded as a very valuable biocontrol tool to prevent growth of LAB in those biotechnological processes in which LAB are undesirable. For example, *Leuconostoc* species are responsible for the ropy slime formation causing the spoilage of meat products. The growth of *Lc. gelidum* in meat could be partially inhibited during cold storage by the presence of the phage ggg (Greer et al. 2007). Similarly, a virulent phage could be used to reduce beer spoilage by postponing post-acidification by *Lact. brevis* (Deasy et al. 2011).

Controlling bacterial populations in food by phages is a rather recent preservation strategy but is being rapidly adopted. Indeed, several phage-based food preservatives are already available in the US market. These products target relevant pathogenic bacteria such as *Listeria monocytogenes*, *E. coli*, and *Salmonella*. Considering the wealth of knowledge on LAB phages, it is expected that a similar approach will be easily followed to inhibit spoilage LAB by phages.

6.6. Conclusions

The last two decades have brought an enormous increase of knowledge about the genomics and molecular biology of LAB phages. A deep knowledge of LAB phages will certainly contribute to the design of new strategies to control industrially relevant starter cultures from phage infection, and to the study LAB diversity and evolution. Furthermore, phage-derived components may be exploited for food safety and biotechnological applications.

References

Ali, Y., Koberg, S., Hessner, S., Sun, X., et al. (2014) Temperate *Streptococcus thermophilus* phages expressing superinfection exclusion proteins of the Ltp type. *Front Microbiol* **5**, 98.

Álvarez, M.A., Herrero, M., and Suárez, J.E. (1998) The site-specific recombination system of the *Lactobacillus* species bacteriophage A2 integrates in Gram-positive and Gram-negative bacteria. *Virology* **250**, 185–193.

Álvarez, M.A., Rodríguez, A., and Suárez, J.E. (1999). Stable expression of the *Lactobacillus casei* bacteriophage A2 repressor blocks phage propagation during milk fermentation. *J Appl Microbiol* **86**, 812–816.

Auvray, F., Coddeville, M., Ritzenthaler, P., and Dupont, L. (1997) Plasmid integration in a wide range of bacteria mediated by the integrase of *Lactobacillus delbrueckii* bacteriophage mv4. *J Bacteriol* **179**, 1837–1845.

Bao, Z., Cartinhour, S., and Swingle, B. (2012) Substrate and target sequence length influence RecTE(Psy) recombineering efficiency in *Pseudomonas syringae*. *PloS One* **7**, e50617.

Barrangou, R. (2013) CRISPR-Cas systems and RNA-guided interference. *WIREs RNA* **4**, 267–278.

Barrangou, R., Fremaux, C., Deveau, H., et al. (2007) CRISPR provides acquired resistance against viruses in prokaryotes. *Science* **315**, 1709–1712.

Barrangou, R. and Horvath, P. (2012) CRISPR: new horizons in phage resistance and strain identification. *Annu Rev Food Sci Technol* **3**, 143–162.

Bebeacua, C., Lorenzo Fajardo, J.C., Blangy, S., et al. (2013) X-ray structure of a superinfection exclusion lipoprotein from phage TP-J34 and identification of the tape measure protein as its target. *Mol Microbiol* **89**, 152–165.

Belteki, G., Gertsenstein, M., Ow, D.W., and Nagy, A. (2003) Site-specific cassette exchange and germline transmission with mouse ES cells expressing phiC31 integrase. *Nat Biotechnol* **21**, 321–324.

Binder, S., Siedler, S., Marienhagen, J., et al. (2013) Recombineering in *Corynebacterium glutamicum* combined with optical nanosensors: a general strategy for fast producer strain generation. *Nucleic Acids Res* **41**, 6360–6369.

Boch, J., Scholze, H., Schornack, S., et al. (2009) Breaking the code of DNA binding specificity of TAL-type III effectors. *Science* **326**, 1509–1512.

Bolotin, A., Quinquis, B., Sorokin, A., and Ehrlich, S.D. (2005) Clustered regularly interspaced short palindrome repeats (CRISPRs) have spacers of extrachromosomal origin. *Microbiology* **151**, 2551–2561.

Brøndsted, L. and Hammer, K. (2006) Phages of *Lactococcus lactis*. In: *The Bacteriophages* ((ed.)R. Calendar) Oxford University Press, Inc., New York, USA, pp. 572–592.

Brüssow, H. and Desiere, F. (2006) Evolution of tailed phages. Insights from comparative phage genomics. In: *The Bacteriophages* ((ed.) R. Calendar) Oxford University Press, Inc., New York, USA, pp. 26–36.

Brüssow, H. and Suárez, J.E. (2006) *Lactobacillus* phages. In: *The Bacteriophages* ((ed.) R. Calendar) Oxford University Press, Inc., New York, USA, pp. 653–666.

Canchaya, C., Fournous, G., and Brüssow, H. (2004) The impact of prophages on bacterial chromosomes. *Mol Microbiol* **53**, 9–18.

Canchaya, C., Proux, C., Fournous, G., et al. (2003) Prophage genomics. *Microbiol Mol Biol Rev* **67**, 238–276.

Catalao, M.J., Gil, F., Moniz-Pereira, J., and Pimentel, M. (2010) The mycobacteriophage Ms6 encodes a chaperone-like protein involved in the endolysin delivery to the peptidoglycan. *Mol Microbiol* **77**, 672–686.

Catalao, M.J., Milho, C., Gil, F., et al. (2011) A second endolysin gene is fully embedded in-frame with the lysA gene of mycobacteriophage Ms6. *PloS One* **6**, e20515.

Chassy, B.M. and Flickinger J.L. (1987) Transformation of Lactobacillus casei by elctroporation. *FEMS Microbiol Lett* **44**, 173–177.

Chopin, M.C., Chopin A., and Bidnenko, E. (2005) Phage abortive infection in lactococci: variations on a theme. *Curr Opin Microbiol* **8**, 473–479.

Christiansen, B., Johnsen, M.G., Stenby, E., et al. (1994) Characterization of the lactococcal temperate phage TP901-1 and its site-specific integration. *J Bacteriol* **176**, 1069–1076.

Coffey, A. and Ross, R.P. (2002) Bacteriophage-resistance systems in dairy starter strains: molecular analysis to application. *Antonie van Leeuwenhoek* **82**, 303–321.

Cong, L., Ran, F.A., Cox, D., et al. (2013) Multiplex genome engineering using CRISPR/Cas systems. *Science* **339**, 819–823.

Copeland, N.G., Jenkins, N.A., and Court, D.L. (2001) Recombineering: a powerful new tool for mouse functional genomics. *Nat Rev Genet* **2**, 769–779.

Costantino, N. and Court, D.L. (2003) Enhanced levels of lambda Red-mediated recombinants in mismatch repair mutants. *PNAS (USA)* **100**, 15748–15753.

Court, D.L., Sawitzke, J.A., and Thomason, L.C. (2002) Genetic engineering using homologous recombination. *Ann Rev Genet* **36**, 361–388.

Datsenko, K.A. and Wanner, B.L. (2000) One-step inactivation of chromosomal genes in *Escherichia coli* K-12 using PCR products. *PNAS (USA)* **97**, 6640–6645.

Datta, S., Costantino, N., Zhou, X., and Court, D.L. (2008) Identification and analysis of recombineering functions from Gram-negative and Gram-positive bacteria and their phages. *PNAS (USA)* **105**, 1626–1631.

Deasy, T., Mahony, J., Neve, H., et al. (2011) Isolation of a virulent *Lactobacillus brevis* phage and its application in the control of beer spoilage. *J Food Prot* **74**, 2157–2161.

Dedrick, R.M., Marinelli, L.J., Newton, G.L., et al. (2013) Functional requirements for bacteriophage growth: gene essentiality and expression in mycobacteriophage Giles. *Mol Microbiol* **88**, 577–589.

Delannoy, S., Beutin, L., and Fach, P. (2012) Use of clustered regularly interspaced short palindromic repeat sequence polymorphisms for specific detection of enterohemorrhagic *Escherichia coli* strains of serotypes O26:H11, O45:H2, O103:H2, O111:H8, O121:H19, O145:H28, and O157:H7 by real-time PCR. *J Clin Microbiol* **50**, 4035–4040.

Deltcheva, E., Chylinski, K., Sharma, C.M., et al. (2011) CRISPR RNA maturation by trans-encoded small RNA and host factor RNase III. *Nature* **471**, 602–607.

Desmyter, A., Farenc, C., Mahony, J., et al. (2013) Viral infection modulation and neutralization by camelid nanobodies. *PNAS (USA)* **110**, 1371–1379.

Deveau, H., Labrie, S.J., Chopin, M.C., and Moineau, S. (2006) Biodiversity and classification of lactococcal phages. *Appl Environ Microbiol* **72**, 4338–4346.

Diancourt, L., Passet, V., Chervaux, C., et al. (2007) Multilocus sequence typing of *Lactobacillus casei* reveals a clonal population structure with low levels of homologous recombination. *Appl Environ Microbiol* **73**, 6601–6611.

Djordjevic, G.M., O'Sullivan, D.J., Walker, S.A., et al. (1997) A triggered-suicide system designed as a defense against bacteriophages. *J Bacteriol* **179**, 6741–6748.

Dupont, L., Boizet-Bonhoure, B., Coddeville, M., et al. (1995). Characterization of genetic elements required for site-specific integration of *Lactobacillus delbrueckii* subsp. *bulgaricus* bacteriophage mv4 and construction of an integration-proficient vector for *Lactobacillus plantarum*. *J Bacteriol* **177**, 586–595.

Durmaz, E. and Klaenhammer, T.R. (1995) A starter culture rotation strategy incorporating paired restriction/ modification and abortive infection bacteriophage defenses in a single *Lactococcus lactis* strain. *Appl Environ Microbiol* **61**, 1266–1273.

Durmaz, E., Madsen, S.M., Israelsen, H., and Klaenhammer, T.R. (2002) *Lactococcus lactis* lytic bacteriophages of the P335 group are inhibited by overexpression of a truncated CI repressor. *J Bacteriol* **184**, 6532–6544.

Ellis, H.M., Yu, D., DiTizio, T., and Court, D.L. (2001) High efficiency mutagenesis, repair, and engineering of chromosomal DNA using single-stranded oligonucleotides. *PNAS (USA)* **98**, 6742–6746.

Emond, E. and Moineau, S. (2007) Bacteriophages in food fermentations. In: *Bacteriophages. Genetics and Molecular Biology* ((eds) S. McGrath and D. van Sinderen) Cyster Academic Press, Norfolk, United Kingdom, pp. 93–124.

Esposito, D. and Scocca, J.J. (1997) The integrase family of tyrosine recombinases: evolution of a conserved active site domain. *Nucleic Acids Res* **25**, 3605–3614.

Fabre, L., Zhang, J., Guigon, G., et al. (2012) CRISPR typing and subtyping for improved laboratory surveillance of *Salmonella* infections. *PloS One* **7**, e36995.

Feher, T., Karcagi, I., Blattner, F.R., and Posfai, G. (2012) Bacteriophage recombineering in the lytic state using the lambda red recombinases. *Microb Biotechnol* **5**, 466–476.

Forde, A. and Fitzgerald, G.F. (1999a) Analysis of exopolysaccharide (EPS) production mediated by the bacteriophage adsorption blocking plasmid, pCI658, isolated from *Lactococcus lactis* ssp. *cremoris* HO2. *Int Dairy J* **9**, 465–472.

Forde, A. and Fitzgerald, G.F. (1999b) Bacteriophage defence systems in lactic acid bacteria. *Antonie van Leeuwenhoek* **76**, 89–113.

García, P., Ladero, V. and Suárez, J.E. (2003) Analysis of the morphogenetic cluster and genome of the temperate *Lactobacillus casei* bacteriophage A2. *Arch Virol* **148**, 1051–1070.

García, P., Rodríguez, L., Rodríguez, A., and Martínez, B. (2010) Food biopreservation: promising strategies using bacteriocins, bacteriophages and endolysins. *Trends Food Sci Technol* **21**, 373–382.

Garneau, J.E., Dupuis, M.E., Villion, M., et al. (2010) The CRISPR/Cas bacterial immune system cleaves bacteriophage and plasmid DNA. *Nature* **468**, 67–71.

Garneau, J.E. and Moineau, S. (2011) Bacteriophages of lactic acid bacteria and their impact on milk fermentations. *Microb Cell Fact* **10(Suppl 1)**, S20.

Geller, B.L., Ivey, R.G., Trempy, J.E., and Hettinger-Smith, B. (1993) Cloning of a chromosomal gene required for phage infection of *Lactococcus lactis* subsp. *lactis* C2. *J Bacteriol* **175**, 5510–5519.

Gottesman, S. (2011). Microbiology: Dicing defence in bacteria. *Nature* **471**, 588–589.

Greer, G.G., Dilts, B.D., and Ackermann, H.W. (2007) Characterization of a *Leuconostoc gelidum* bacteriophage from pork. *Int J Food Microbiol* **114**, 370–375.

Grissa, I., Vergnaud, G., and Pourcel, C. (2007) The CRISPRdb database and tools to display CRISPRs and to generate dictionaries of spacers and repeats. *BMC Bioinformatics* **8**, 172.

Groenen, P.M., Bunschoten, A.E., van Soolingen, D., and van Embden, J.D. (1993). Nature of DNA polymorphism in the direct repeat cluster of *Mycobacterium tuberculosis*; application for strain differentiation by a novel typing method. *Mol Microbiol* **10**, 1057–1065.

Groth, A.C. and Calos, M.P. (2004) Phage integrases: biology and applications. *J Mol Biol* **335**, 667–678.

Groth, A.C., Olivares, E.C., Thyagarajan, B., and Calos, M.P. (2000) A phage integrase directs efficient site-specific integration in human cells. *PNAS (USA)* **97**, 5995–6000.

Guinane, C.M., Kent, R.M., Norberg, S., et al. (2011) Host specific diversity in *Lactobacillus johnsonii* as evidenced by a major chromosomal inversion and phage resistance mechanisms. *PloS One* **6**, e18740.

Hollis, R.P., Stoll, S.M., Sclimenti, C.R., et al. (2003) Phage integrases for the construction and manipulation of transgenic mammals. *Reprod Biool Endocrinol* **1**, 79.

Horvath, P., Coûté-Monvoisin, A.-C., Romero, D.A., et al. (2009) Comparative analysis of CRISPR loci in lactic acid bacteria genomes. *Int J Food Microbiol* **131**, 62–70.

Horvath, P., Romero, D.A., Coute-Monvoisin, A.C., et al. (2008) Diversity, activity, and evolution of CRISPR loci in *Streptococcus thermophilus*. *J Bacteriol* **190**, 1401–1412.

Huang, J., Ghosh, P., Hatfull, G.F., and Hong, Y. (2011) Successive and targeted DNA integrations in the *Drosophila* genome by Bxb1 and phiC31 integrases. *Genetics* **189**, 391–395.

Ishino, Y., Shinagawa, H., Makino, K., et al. (1987) Nucleotide sequence of the *iap* gene, responsible for alkaline phosphatase isozyme conversion in *Escherichia coli*, and identification of the gene product. *J Bacteriol* **169**, 5429.

Jansen, R., Embden, J.D., Gaastra, W., and Schouls, L.M. (2002) Identification of genes that are associated with DNA repeats in prokaryotes. *Mol Microbiol* **43**, 1565–1575.

Jiang, W., Bikard, D., Cox, D., et al. (2013) RNA-guided editing of bacterial genomes using CRISPR-Cas systems. *Nat Biotechnol* **31**, 233–239.

Jinek, M., Chylinski, K., Fonfara, I., et al. (2012). A programmable dual-RNA-guided DNA endonuclease in adaptive bacterial immunity. *Science* **337**, 816–821.

Katashkina, J.I., Hara, Y., Golubeva, L.I., et al. (2009) Use of the lambda Red-recombineering method for genetic engineering of *Pantoea ananatis*. *BMC Mol Biol* **10**, 34.

Kim, Y.G., Cha, J., and Chandrasegaran, S. (1996) Hybrid restriction enzymes: zinc finger fusions to FokI cleavage domain. *PNAS (USA)* **93**, 1156–1160.

Kiro, R., Shitrit, D., and Qimron, U. (2014) Efficient engineering of a bacteriophage genome using the type I-E CRISPR-Cas system. *RNA Biology* **11**, 42–44.

Kulkarni, S.K. and Stahl, F.W. (1989) Interaction between the *sbcC* gene of Escherichia coli and the *gam* gene of phage lambda. *Genetics* **123**, 249–253.

Labrie, S.J., Samson, J.E., and Moineau, S. (2010) Bacteriophage resistance mechanisms. *Nat Rev Microbiol* **8**, 317–327.

Le Fleche, P., Hauck, Y., Onteniente, L., et al. (2001) A tandem repeats database for bacterial genomes: application to the genotyping of *Yersinia pestis* and *Bacillus anthracis*. *BMC Microbiol* **1**, 2.

Lillehaug, D., Nes, I.F., and Birkeland, N.K. (1997) A highly efficient and stable system for site-specific integration of genes and plasmids into the phage phiLC3 attachment site (attB) of the *Lactococcus lactis* chromosome. *Gene* **188**, 129–136.

Lin, C.F., Lo, T.C., Kuo, Y.C., and Lin, T.H. (2013) Stable integration and expression of heterologous genes in several lactobacilli using an integration vector constructed from the integrase and attP sequences of phage phiAT3 isolated from *Lactobacillus casei* ATCC 393. *Appl Microbiol Biotechnol* **97**, 3499–3507.

Lo, T.C., Shih, T.C., Lin, C.F., et al. (2005) Complete genomic sequence of the temperate bacteriophage phiAT3 isolated from *Lactobacillus casei* ATCC 393. *Virology* **339**, 42–55.

Mahony, J., Ainsworth, S., Stockdale, S., and van Sinderen, D. (2012) Phages of lactic acid bacteria: the role of genetics in understanding phage-host interactions and their co-evolutionary processes. *Virology* **434**, 143–150.

Mahony, J., McGrath, S., Fitzgerald, G.F., and van Sinderen, D. (2008) Identification and characterization of lactococcal-prophage-carried superinfection exclusion genes. *Appl Environ Microbiol* **74**, 6206–6215.

Mahony, J. and van Sinderen, D. (2014) Current taxonomy of phages infecting lactic acid bacteria. *Front Microbiol* **5**, 7 doi: 10.3389/fmicb.2014.00007.

Makarova, K.S., Haft, D.H., Barrangou, R., et al. (2011) Evolution and classification of the CRISPR-Cas systems. *Nat Rev Microbiol* **9**, 467–477.

Mali, P., Yang, L., Esvelt, K.M., et al. (2013) RNA-guided human genome engineering via Cas9. *Science* **339**, 823–826.

Maresca, M., Erler, A., Fu, J., et al. (2010) Single-stranded heteroduplex intermediates in lambda Red homologous recombination. *BMC Mol Biol* **11**, 54.

Marinelli, L.J., Hatfull, G.F., and Piuri, M. (2012) Recombineering: A powerful tool for modification of bacteriophage genomes. *Bacteriophage* **2**, 5–14.

Marinelli, L.J., Piuri, M., Swigonova, Z., et al. (2008) BRED: a simple and powerful tool for constructing mutant and recombinant bacteriophage genomes. *PloS One* **3**, e3957.

Marraffini, L.A. and Sontheimer, E.J. (2010) CRISPR interference: RNA-directed adaptive immunity in bacteria and archaea. *Nat Rev Genetics* **11**, 181–190.

Martín, M.C., Alonso, J.C., Suárez, J.E., and Álvarez, M.A. (2000) Generation of food-grade recombinant lactic acid bacterium strains by site-specific recombination. *Appl Environ Microbiol* **66**, 2599–2604.

Martel, B. and Moineau, S. (2014) CRISPR-Cas: an efficient tool for genome engineering of virulent bacteriophages. *Nucleic Acids Res* **42**, 9504–9513.

Mills, S., Ross, R.P., Neve, H., and Coffey, A. (2012) Bacteriophage and anti-phage mechanisms in lactic acid bacteria. In: *Lactic Acid Bacteria. Microbiological and Functional Aspects*, (ed. S. Lahtinen, A. Ouwehand, S. Salminen, and A. von Wright), 4th edition, CRC Press, London, United Kingdom, pp 165–186.

Mojica, F.J., Diez-Villasenor, C., Garcia-Martinez, J., and Soria, E. (2005) Intervening sequences of regularly spaced prokaryotic repeats derive from foreign genetic elements. *J Mol Evol* **60**, 174–182.

Mosberg, J.A., Lajoie, M.J., and Church, G.M. (2010) Lambda red recombineering in *Escherichia coli* occurs through a fully single-stranded intermediate. *Genetics* **186**, 791–799.

Moscoso, M. and Suárez, J.E. (2000) Characterization of the DNA replication module of bacteriophage A2 and use of its origin of replication as a defense against infection during milk fermentation by *Lactobacillus casei*. *Virology* **273**, 101–111.

Moscou, M.J. and Bogdanove, A.J. (2009) A simple cipher governs DNA recognition by TAL effectors. *Science* **326**, 1501.

Murphy, K.C. (2007) The lambda Gam protein inhibits RecBCD binding to dsDNA ends. *J Mol Biol* **371**, 19–24.

Muyrers, J.P., Zhang, Y., Testa, G., and Stewart, A.F. (1999) Rapid modification of bacterial artificial chromosomes by ET-recombination. *Nucleic Acids Res* **27**, 1555–1557.

Nkrumah, L.J., Muhle, R.A., Moura, P.A., et al. (2006) Efficient site-specific integration in *Plasmodium falciparum* chromosomes mediated by mycobacteriophage Bxb1 integrase. *Nat Methods* **3**, 615–621.

Olivares, E.C., Hollis, R.P., and Calos, M.P. (2001) Phage R4 integrase mediates site-specific integration in human cells. *Gene* **278**, 167–176.

Olivares, E.C., Hollis, R.P., Chalberg, T.W., et al. (2002) Site-specific genomic integration produces therapeutic Factor IX levels in mice. *Nat Biotechnol* **20**, 1124–1128.

Oppenheim, A.B., Rattray, A.J., Bubunenko, M., et al. (2004) In vivo recombineering of bacteriophage lambda by PCR fragments and single-strand oligonucleotides. *Virology* **319**, 185–189.

Payne, K., Sun, Q., Sacchettini, J., and Hatfull, G.F. (2009) Mycobacteriophage lysin B is a novel mycolylarabinogalactan esterase. *Mol Microbiol* **73**, 367–381.

Petersen, K.V., Martinussen, J., Jensen, P.R., and Solem, C. (2013) Repetitive, marker-free, site-specific integration as a novel tool for multiple chromosomal integration of DNA. *Appl Environ Microbiol* **79**, 3563–3569.

Piuri, M. and Hatfull, G.F. (2006) A peptidoglycan hydrolase motif within the mycobacteriophage TM4 tape measure protein promotes efficient infection of stationary phase cells. *Mol Microbiol* **62**, 1569–1585.

Pontes, M.H. and Dale, C. (2011) Lambda red-mediated genetic modification of the insect endosymbiont *Sodalis glossinidius*. *Appl Environ Microbiol* **77**, 1918–1920.

Pourcel, C., Salvignol, G., and Vergnaud, G. (2005) CRISPR elements in *Yersinia pestis* acquire new repeats by preferential uptake of bacteriophage DNA, and provide additional tools for evolutionary studies. *Microbiology* **151**, 653–663.

Proux, C., van Sinderen, D., Suarez, J., et al. (2002) The dilemma of phage taxonomy illustrated by comparative genomics of Sfi21-like *Siphoviridae* in lactic acid bacteria. *J Bacteriol* **184**, 6026–6036.

Räisänen, L., Draing, C., Pfitzenmaier, M., et al. (2007) Molecular interaction between lipoteichoic acids and *Lactobacillus delbrueckii* phages depends on D-alanyl and alpha-glucose substitution of poly(glycerophosphate) backbones. *J Bacteriol* **189**, 4135–4140.

Ranallo, R.T., Barnoy, S., Thakkar, S., et al. (2006) Developing live *Shigella* vaccines using lambda Red recombineering. *FEMS Immunol Med Microbiol* **47**, 462–469.

Raya, R.R., Fremaux, C., De Antoni, G.L., and Klaenhammer, T.R. (1992) Site-specific integration of the temperate bacteriophage phi adh into the *Lactobacillus gasseri* chromosome and molecular characterization of the phage (attP) and bacterial (attB) attachment sites. *J Bacteriol* **174**, 5584–5592.

Raya, R.R., Kleeman, E.G., Luchansky, J.B., and Klaenhammer, T.R. (1989) Characterization of the temperate bacteriophage phi-adh and plasmid transduction in *Lactobacillus acidophilus* ADH. *Appl Environ Microbiol* **55**, 2206–2213.

Rebar, E.J. and Pabo, C.O. (1994). Zinc finger phage: affinity selection of fingers with new DNA-binding specificities. *Science* **263**, 671–673.

Richter, C., Chang, J.T., and Fineran, P.C. (2012) Function and regulation of clustered regularly interspaced short palindromic repeats (CRISPR) / CRISPR associated (Cas) systems. *Viruses* **4**, 2291–2311.

Roces, C., Courtin, P., Kulakauskas, S., et al. (2012) Isolation of *Lactococcus lactis* mutants simultaneously resistant to the cell wall-active bacteriocin Lcn972, lysozyme, nisin, and bacteriophage c2. *Appl Environ Microbiol* **78**, 4157–4163.

Samson, J. and Moineau, S. (2013) Bacteriophages in food fermentations: new frontiers in a continuous arms race. *Ann Rev Food Sci Technol* **4**, 347–368.

Samson, J., Magadán, A.H., Sabri, M., and Moineau, S. (2013a) Revenge of the phages: defeating bacterial defences. *Nat Rev Microbiol* **11**, 675–687.

Samson, J.E., Spinelli, S., Cambillau, C., and Moineau, S. (2013b) Structure and activity of AbiQ, a lactococcal endoribonuclease belonging to the type III toxin-antitoxin system. *Mol Microbiol* **87**, 756–768.

Sander, J.D. and Joung, J.K. (2014) CRISPR-Cas systems for editing, regulating and targeting genomes. *Nat Biotechnol* **32**, 347–355.

Sarov, M., Schneider, S., Pozniakovski, A., et al. (2006) A recombineering pipeline for functional genomics applied to *Caenorhabditis elegans*. *Nat Methods* **3**, 839–844.

Sawitzke, J.A., Costantino, N., Li, X.T., et al. (2011) Probing cellular processes with oligo-mediated recombination and using the knowledge gained to optimize recombineering. *J Mol Biol* **407**, 45–59.

Schouls, L.M., Reulen, S., Duim, B., et al. (2003) Comparative genotyping of *Campylobacter jejuni* by amplified fragment length polymorphism, multilocus sequence typing, and short repeat sequencing: strain diversity, host range, and recombination. *J Clin Microbiol* **41**, 15–26.

Shariat, N. and Dudley, E.G. (2014) CRISPRs: molecular signatures used for pathogen subtyping. *Appl Environ Microbiol* **80**, 430–439.

Shimizu-Kadota, M., Kiwaki, M., Hirokawa, H., and Tsuchida, N. (1985) ISL1: a new transposable element in *Lactobacillus casei*. *Mol Gen Genet* **200**, 193–198.

Shimizu-Kadota, M., Kiwaki, M., Sawaki, S., et al. (2000) Insertion of bacteriophage phiFSW into the chromosome of *Lactobacillus casei* strain Shirota (S-1): characterization of the attachment sites and the integrase gene. *Gene* **249**, 127–134.

Shimizu-Kadota, M., Sakurai, T., and Tsuchida, N. (1983) Prophage origin of a virulent phage appearing on fermentations of *Lactobacillus casei* S-1. *Appl Environ Microbiol* **45**, 669–674.

Smith, M.C. and Thorpe, H.M. (2002) Diversity in the serine recombinases. *Mol Microbiol* **44**, 299–307.

Spinelli, S., Veesler, D., Bebeacua, C., and Cambillau, C. (2014) Structures and host-adhesion mechanisms of lactococcal siphophages. *Front Microbiol* **5**, 3.

Stockdale, S.R., Mahony, J., Courtin, P., et al. (2013) The lactococcal phages Tuc2009 and TP901-1 incorporate two alternate forms of their tail fiber into their virions for infection specialization. *J Biol Chem* **288**, 5581–5590.

Stoll, S.M., Ginsburg, D.S., and Calos, M.P. (2002) Phage TP901-1 site-specific integrase functions in human cells. *J Bacteriol* **184**, 3657–3663.

Streicher, E.M., Victor, T.C., van der Spuy, G., et al. (2007) Spoligotype signatures in the *Mycobacterium tuberculosis* complex. *J Clin Microbiol* **45**, 237–240.

Sturino, J.M. and Klaenhammer T.R. (2004) Bacteriophage defense systems and strategies for lactic acid bacteria. *Adv Appl Microbiol* **56**, 331–378.

Sturino, J.M. and Klaenhammer, T.R. (2006) Engineered bacteriophage-defence systems in bioprocessing. *Nat Rev Microbiol* **4**, 395–404.

Sturino, J.M. and Klaenhammer, T.R. (2007) Inhibition of bacteriophage replication in *Streptococcus thermophilus* by subunit poisoning of primase. *Microbiology* **153**, 3295–3302.

Sulakvelidze, A., Alavidze, Z., and Morris, J.G. Jr. (2001) Bacteriophage therapy. *Antimicrob Agents Chemother* **45**, 649–659.

Swaminathan, S., Ellis, H.M., Waters, L.S., et al. (2001) Rapid engineering of bacterial artificial chromosomes using oligonucleotides. *Genesis* **29**, 14–21.

Swingle, B., Bao, Z., Markel, E., et al. (2010a) Recombineering using RecTE from *Pseudomonas syringae*. *Appl Environ Microbiol* **76**, 4960–4968.

Swingle, B., Markel, E., Costantino, N., et al. (2010b) Oligonucleotide recombination in Gram-negative bacteria. *Mol Microbiol* **75**, 138–148.

Terns, M.P. and Terns, R.M. (2011) CRISPR-based adaptive immune systems. *Curr Opin Microbiol* **14**, 321–327.

Thyagarajan, B., Olivares, E.C., Hollis, R.P., et al. (2001) Site-specific genomic integration in mammalian cells mediated by phage phiC31 integrase. *Mol Cel Biol* **21**, 3926–3934.

Tyson, G.W. and Banfield, J.F. (2008) Rapidly evolving CRISPRs implicated in acquired resistance of microorganisms to viruses. *Environ Microbiol* **10**, 200–207.

Valyasevi, R., Sandine, W.E., and Geller, B.L. (1991) A membrane protein is required for bacteriophage c2 infection of *Lactococcus lactis* subsp. *lactis* C2. *J Bacteriol* **173**, 6095–6100.

van de Guchte, M., Daly, C., Fitzgerald, G.F., and Arendt, E.K. (1994) Identification of int and attP on the genome of lactococcal bacteriophage Tuc2009 and their use for site-specific plasmid integration in the chromosome of Tuc2009-resistant *Lactococcus lactis* MG1363. *Appl Environ Microbiol* **60**, 2324–2329.

van Kessel, J.C. and Hatfull, G.F. (2007) Recombineering in *Mycobacterium tuberculosis*. *Nat Methods* **4**, 147–152.

van Kessel, J.C. and Hatfull, G.F. (2008a) Efficient point mutagenesis in mycobacteria using single-stranded DNA recombineering: characterization of antimycobacterial drug targets. *Mol Microbiol* **67**, 1094–1107.

van Kessel, J.C. and Hatfull, G.F. (2008b) Mycobacterial recombineering. *Methods Mol Biol* **435**, 203–215.

van Pijkeren, J.P. and Britton, R.A. (2012) High efficiency recombineering in lactic acid bacteria. *Nucleic Acids Res* **40**, e76.

Van Pijkeren, J.P., Neoh, K.M., Sirias, D., et al. (2012) Exploring optimization parameters to increase ssDNA recombineering in *Lactococcus lactis* and *Lactobacillus reuteri*. *Bioengineered* **3**, 209–217.

Walker, S.A. and Klaenhammer, T.R. (2000) An explosive antisense RNA strategy for inhibition of a lactococcal bacteriophage. *App Environ Microbiol* **66**, 310–319.

Westra, E.R., Swarts, D.C., Staals, R.H., et al. (2012) The CRISPRs, they are a-changin': how prokaryotes generate adaptive immunity. *Ann Rev Genetics* **46**, 311–339.

Whitehead, H.R. and Cox, G.A. (1935) The occurrence of bacteriophages in starter cultures of lactic streptococci. *N Z J Dairy Sci Technol* **16**, 319–320.

Wiedenheft, B., Sternberg, S.H., and Doudna, J.A. (2012) RNA-guided genetic silencing systems in bacteria and archaea. *Nature* **482**, 331–338.

Yamaguchi, S., Kazuki, Y., Nakayama, Y., et al. (2011) A method for producing transgenic cells using a multi-integrase system on a human artificial chromosome vector. *PloS One* **6**, e17267.

Yu, D., Ellis, H.M., Lee, E.C., et al. (2000) An efficient recombination system for chromosome engineering in Escherichia coli. *PNAS (USA)* **97**, 5978–5983.

Zhang, Y., Muyrers, J.P., Rientjes, J., and Stewart, A.F. (2003) Phage annealing proteins promote oligonucleotide-directed mutagenesis in *Escherichia coli* and mouse ES cells. *BMC Mol Biol* **4**, 1.

Chapter 7

Lactic Acid Bacteria and the Human Intestinal Microbiome

François P. Douillard[1] and Willem M. de Vos[1,2,*]

[1] *Department of Veterinary Biosciences, University of Helsinki, Finland*
[2] *Laboratory of Microbiology, Wageningen University, The Netherlands*

7.1. Introduction

Since we are born virtually sterile, we are colonized by developing communities of microbes (collectively termed microbiota). These are most abundant and complex in the intestinal tract, where the microbiota outnumber our body cells by a factor of ten, while their collective genome (termed the microbiome) contains several millions of genes (Qin et al. 2010). The functions of the intestinal microbiota are manifold and vary from contributing to digesting our food to producing vitamins and other important compounds. Until the last decade, the intestinal microbial communities present were poorly characterized and comprehended, due to the limitation of analytical tools available to do so (Rajilić-Stojanović et al. 2009). Microbial cultivation from fecal samples and clinical biopsies were found to be not representative of the true richness and complexity of the intestinal microbiota, since the cultivable species constitute only a small fraction of the microbial communities. Various new developments, including high-throughput culturing approaches, have addressed this and now over 1000 species from the human intestinal tract have been cultured (Rajilić-Stojanović & de Vos, 2014). However, this is still far from a complete description of the complex intestinal microbiota. The development of high-throughput molecular analytic tools, such as next-generation DNA sequencing and phylogenetic and other microarray platforms have changed our way of approaching the complexity of the human intestinal microbiota (Vaughan et al. 1999; Rajilić-Stojanović et al. 2007; Clemente et al. 2012; Le Chatelier et al. 2013). While our understanding of the intestinal tract is in its early infancy, we are clearly at the dawn of major findings on its inhabitants and their intricate relationships with the host. Comparative and functional metagenomic studies already started to uncover the pivotal role and impact of intestinal microbiota on human health and dozens of diseases have been correlated with specific microbial signatures (de Vos and de Vos 2012). In some cases, bacterial therapies have been shown to be more effective than pharmaceutical treatments, indicating the causal relation between the intestinal microbiota and health (van Nood et al. 2013). As the majority of

* Corresponding author: Department of Veterinary Biosciences, University of Helsinki, Agnes Sjöberginkatu 2, 00790 Helsinki, Finland. Laboratory of Microbiology, Wageningen University, Dreijenplein 10, 6703 HB Wageningen, The Netherlands. willem.devos@wur.nl

Biotechnology of Lactic Acid Bacteria: Novel Applications, Second Edition. Edited by Fernanda Mozzi, Raúl R. Raya, and Graciela M. Vignolo.

intestinal microbes are anaerobic Gram (+) bacteria, a renewed interest in implementing these species in future therapeutic applications is emerging (de Vos and Nieuwdorp 2013).

Lactic acid bacteria (LAB) are related anaerobic Gram (+) bacteria that convert sugars predominantly into lactic acid. A variety of LAB have been isolated from the human intestine (Rajilić-Stojanović et al. 2007). However, their number is several dimensions smaller than previously estimated based on culturing studies. Due to their interactions with the human intestine and their associated health benefits, lactic acid bacteria are increasingly used in food products and preservatives. In this chapter, we will first discuss the LAB associated with the human intestinal microbiota and their potential roles in health and diseases. We will then address recent metagenomic studies that challenge the established belief that the human intestine is colonized by many lactobacilli and try to offer insights into one highly relevant question: are LAB originally occurring (autochthonous) and/or just transiently passing (allochthonous) in the human intestinal tract? To further illustrate this, we will describe one particular case study: the well-characterized *Lactobacillus rhamnosus* species, which has been extensively studied at both genomic and phenotypic levels. It is expected that these studies will provide a basis or model for future research work addressing the intestinal function of autochthonous and allochthonous LAB.

7.2. Ecology of the Human Intestinal Tract

7.2.1. The Human Microbiome in the Upper and Lower Intestinal Tract

The human intestinal tract is the most densely colonized ecosystem of our body consisting of over 100 trillion (10^{14}) of microbes, mainly Bacteria but also Archaea (around 1%) or Eukarya (<0.1%) (Zoetendal et al. 2008; Rajilić-Stojanović and de Vos 2014 unpublished results). Following the stomach that functions as a sterilization tank, mainly due to its low pH, the intestinal tract is progressively colonized with densities reaching 10^3–10^6 cells per ml of content in the upper intestine that is around 3–5 m in length. Here, the food as well as the microbiota encounter a large surface of intestinal cells; functional metagenomic approaches have shown that microbes compete with the host here for food derived sugars (Zoetendal et al. 2012). The small intestinal microbiota consists mainly of *Streptococcus* spp. and *Veillonella* spp. that form a trophic chain. The sugars are rapidly taken up by phosphotransferase systems in the streptococci that convert them mainly into lactic acid. The streptococci include *Strep. bovis*, *Strep. intermedius*, and *Strep. mitis* but also new streptococci have been identified (van den Bogert et al. 2013). Subsequently, *Veillonella* spp. convert the produced lactic acid into propionic and acetic acids that fuel the enterocytes. *Lactobacillus* spp. have often been isolated from the small intestine (Tannock et al. 2000) suggesting they are dominant at this site, but culture-independent molecular studies have shown that most lactobacilli are in a minority, representing only 0.001–0.1% of the total microbiota (Booijink et al. 2010). Similarly, molecular inventories based on 16S rRNA amplicon sequencing have not detected *Lactobacillus* spp. as an important group of inhabitants of the upper intestinal tract (van den Bogert et al. 2011). However, the upper intestinal tract is the most important reservoir of lactobacilli as the lower intestinal tract contains an even a lower amount despite the fact that it has the largest microbial densities of up to 10^{10}–10^{11} in the colon where food transit slows down due to its architecture. In a molecular survey of *Lactobacillus* rRNA sequences, it was observed that the upper and the lower intestinal tracts contain a variety of *Lactobacillus* spp. (Heilig et al. 2002). These LAB were obviously at low levels but due to amplification with *Lactobacillus*-specific primers they could be detected and further analysed. One remarkable finding in the analysis of *Lactobacillus*-like 16S rRNA sequences from both the upper and lower intestinal tracts is that a significant fraction of the recovered 16S rRNA sequences have no nearest neighbor, indicating that they are new species that have not yet been cultured.

This testifies to the fact that a great part of the intestinal tract microbiota has not yet been grown in the laboratory in pure culture.

7.2.2. Lactic Acid Bacteria Associated with the Human Intestine

Lactic acid bacteria are intimately associated with humans and have been found in a large variety of human cavities or sites, that is the skin, oral cavity, stomach, intestinal tract, or vagina (Walter 2005). The ecological versatility within the LAB group is high, ranging from specialists to generalists. For example, *Lact. iners* is a dominant member of the vaginal microbiota and is usually not detected in ecological niches other than the vagina, showing the specific adaptation of the species to one particular niche (Mendes-Soares et al. 2014). In contrast, members of the *Lact. rhamnosus* species have been isolated in cheese, the oral cavity, gastrointestinal tract, and vagina, where some *Lact. rhamnosus* isolates display specific metabolic, genomic or phenotypic adaptation traits (Douillard et al. 2013). In the human fecal microbiota, LAB represent a minor proportion of the bacterial community (Walter 2008; Qin et al. 2010), ruling out the initial idea that LAB constitute an abundant community in the gastrointestinal (GI) tract (see above). Based on specific qPCR approaches, it could be estimated that in fecal samples of adults, the abundance of *Lactobacillus/Enterococcus* and *Bifidobacterium* spp. is 0.01–1.8% and 2.5–4.9%, respectively (Louis et al. 2007). It is however likely that the enterococci are much more abundant in the GI tract than lactobacilli. Hence the overall proportion of lactobacilli is low in comparison to other bacterial communities present in the intestine, although their presence varies throughout the intestinal tract: $<10^4$ CFU/ml in the small intestine and 10^6 CFU/g in the fecal samples (Harmsen et al. 2002; Bello et al. 2003; Walter 2008). However, it should be realized that "abundance" is not the same as "importance," and hence the low level of lactobacilli does not imply a limited role within the human intestinal microbiota.

When defining a list of intestinal LAB, one has first to acknowledge that the isolation site of lactobacilli does not necessarily correspond to their true ecological habitat(s), as they may be simply transient. Thus, the distinction between autochtonous and allochtonous bacterial species is an important aspect that relates to their respective roles and effects directly to the ecology of the intestine. Based on rRNA sequences, for example, it remains difficult to conclude whether these lactobacilli are colonizers (autochtonous) or passengers (allochtonous). We will further discuss this in Section 7.3.4, illustrating our argumentation with the example of the *Lact. rhamnosus* species. Among LAB and other related bacteria, four genera are usually detected in fecal or intestinal samples: *Lactobacillus* (25 species), *Leuconostoc* (2 species), *Pediococcus* (1 species), and *Bifidobacterium* (8 species) (Table 7.1). For some of the LAB species detected in the intestine, it remains difficult to classify them as autochtonous or allochtonous, as there is no clear experimental evidence. Over the last year, it has been found in humans that a number of *Lact. gasseri*, *Lact. salivarius* subsp. *salivarius*, *Lact. ruminis*, *Lact. reuteri*, and *Lact. crispatus* strains were persistent in the human intestine, whereas some of the non-persistent (transient) LAB strains belonged to *Lact. acidophilus*, *Lact. crispatus*, and *Lact. gasseri* species, for example (Mitsuoka 1992; Tannock et al. 2000; Reuter 2001). This suggests that the remarkable intraspecies diversity (genotypic and phenotypic) observed within LAB species may derive from the grouping between gut-persistent (autochtonous) strains and transient (allochtonous) strains, reflecting ecological fitness and adaptation to specific habitats. Such an observation was recently described for the *Lact. rhamnosus* species (Douillard et al. 2013). Analysis of the LAB members residing in the human intestinal microbiota still remains challenging. Thus, LAB species that are commonly used in the food and dairy industry, such as *Lact. plantarum*, *Lact. casei*, or *Lact. rhamnosus* (Tannock 2004; Saxelin et al. 2005; Bernardeau et al. 2006), are found in the oral cavity (Reuter 2001; Walter 2008). Therefore, it cannot be excluded that species detected in fecal samples may derive from the oral cavity or from food products, as many are known to survive the gastric passage and can be recovered from the duodenum, such as *Lact. plantarum* (Vesa et al. 2000). Similar observations have been made with specific interventions where *Lact. rhamnosus* GG have been detected in fecal samples using molecular and culturing approaches (Lahti et al. 2013). In conclusion, only detailed

Table 7.1. List of LAB and bifidobacteria detected in the human intestinal tract as previously described (Molin et al. 1993; Ahrne et al. 1998; Heilig et al. 2002; Vaughan et al. 2002; Bello et al. 2003; Walter 2008).

***Lactobacillus*, *Pediococcus*, and *Leuconostoc* genera**		
Lactobacilli		
Lact. acidophilus	*Lact. fructivorans*	*Lact. pontis*
Lact. amylovorus	*Lact. gasseri*	*Lact. reuteri*
Lact. buchneri	*Lact. helveticus*	*Lact. rhamnosus*
Lact. brevis	*Lact. johnsonii*	*Lact. ruminis*
Lact. casei	*Lact. jensenii*	*Lact. salivarius*
Lact. crispatus	*Lact. mucosae*	*Lact. sakei*
Lact. curvatus	*Lact. oris*	*Lact. vaginalis*
Lact. delbrueckii	*Lact. plantarum*	
Lact. fermentum	*Lact. paracasei*	
Pediococcus		
Pediococcus pentosaceus		
Leuconostoc		
Leuconostoc argentinum	*Leuconostoc mesenteroides*	
***Bifidobacterium* genus**		
B. adolescentis	*B. breve*	*B. infantis*
B. angulatum	*B. catenulatun*	*B. longum*
B. bifidus	*B. gallicum*	*B. pseudocatenulatum*

genotypic and phenotypic characterization of isolates recovered from the intestinal tract may offer the possibility to identify adaptation traits and bring insights into their true ecological habitat.

7.2.3. Metagenomic Studies of the Intestine in Relation to LAB

Metagenomic studies have been intensively conducted to further comprehend the role of the microbiota in various pathologies, infections, and metabolic disorders. In that respect, molecular techniques have brought a wealth of information on the beneficial effects of LAB within the human intestine, which proves to be particularly relevant with regard to the inclusion of LAB in probiotic products. As we are born virtually sterile, there has been considerable interest in the early-life colonization process. Recent phylogenetic analysis has shown the meconium to contain a dozen different *Lactobacillus* spp. that disappear quickly in follow-up fecal samples (Moles et al. 2013). Based on the 16S rRNA sequences, only *Lact. reuteri* and in one case *Lact. vaginalis* appeared to stay at high levels, but culturing studies also revealed *Lact. fermentum, Lact. gasseri*, and *Lact. salivarius*; this fact highlighting the differences between 16S rRNA abundance and culturing efficiency.

In succession studies, mainly *Lact. rhamnosus*, *Lact. casei*, and *Lact. salivarius* were found in one month-old or older healthy infants (Heilig et al. 2002). In contrast, in the first metagenomic study of a new-born child followed for over a year, the specific detection of *Lactobacillus* spp. was not described (Koenig et al. 2011). The number of metagenomic reads in this study was quite small, suggesting that *Lactobacillus* spp. may have been missed. In a recent analysis of deep metagenomic reads of several 100 day-old babies, we observed a variety of *Lactobacillus* species, notably that of *Lact. gasseri*, *Lact. johnsonii*, and *Lact. rhamnosus*, providing an unbiased and PCR-independent view of the presence of these bacteria and their genomes in the early-life intestinal tract (de Been et al. unpublished results). Similarly, in a comparative metagenomic study of infants in Africa and America, genes derived from *Lact. casei*, *Lact. reuteri*, and *Lact. delbrueckii* were found to be present (Yatsunenko et al. 2012). A recent metagenomic study included a feeding trial in monozygotic twins and gnotobiotic mice with a product containing two strains of *Lact. delbrueckii* and some other related bacteria, including *Bifidobacterium animalis*, *Strep. thermophilus*, and *Lactococcus lactis*

(McNulty et al. 2011). In line with other phylogenetic studies, no major effect of this feeding trial on the intestinal microbiota was observed. Remarkably, the metagenomic approach was not sufficiently sensitive to reproducibly detect lactobacilli or *Strep. thermophilus* in feces of the human trial or in gnotobiotic mice fed with the same product. This finding contrasts to other human trials where *Lact. rhamnosus* GG could clearly be recovered from fecal samples using phylogenetic or culturing approaches (Kankainen et al. 2009; Lahti et al. 2013). Possible explanations include sensitivity of the used metagenomic methods as well as lysis of the *Lact. delbrueckii* during the intestinal tract passage. The *L. lactis* and *B. animalis* strains could be recovered from both humans and mice. The recovery of *L. lactis* from a human feeding trial has previously been described and indicates that the level of intake determines the number of cells in fecal samples (Klijn et al. 1995). In monozygotic twins, the *B. animalis* strain appeared the most prominent and active, expressing a variety of genes involved in sugar degradation. The comparison between human and mice suggested that similar responses were detected (McNulty et al. 2011). This is in accordance with an earlier feeding trial in humans where the expression of *Lact. plantarum* was studied and compared to that in mice (Marco et al. 2010). In that study, expression profiling in both systems allowed definition of a core intestinal transcriptome of *Lact. plantarum* and identified niche-specific adaptations to the intestinal ecosystem.

In conclusion, a number of metagenomic and other profiling studies have addressed the expression of LAB in the intestinal tract of humans and mice. However, the level of activity of autochthonous LAB is too low to get meaningful data and most studies have addressed the expression of transiently colonizing LAB. This fact has been shown to be a fruitful approach and also allows addressing the response of the host, as shown in model studies with *Lact. plantarum* that revealed upregulation of NFκB-mediated immune response in healthy volunteers (van Baarlen et al. 2009).

7.3. A Case Study: The *Lactobacillus rhamnosus* Species

7.3.1. *Genomic Diversity of* Lact. rhamnosus *and Intestinal Adaptation*

The *Lact. rhamnosus* species is highly versatile and its members have been isolated from a large range of ecological habitats including the human GI tract and the vaginal cavity. The adaptation of one bacterial isolate to a particular niche is reflected in its genome organization and content, as shown in LAB, that is *Lact. rhamnosus* (Douillard et al. 2013) but also in *Lact. casei* (Cai et al. 2009; Broadbent et al. 2012; Toh et al. 2013) and *Lact. paracasei* (Smokvina et al. 2013). Ecological adaptation patterns were observed among isolates from the *Lact. rhamnosus*, *Lact. casei*, and *Lact. paracasei* species. Our recent comparative genomic analysis of 100 *L. rhamnosus* isolates from different ecological habitats showed an association between niche adaptation, phenotype, and genome re-arrangements (Douillard et al. 2013). Typically, the genomic diversity observed among isolates was due to the acquisition, loss or decay of genes that were mostly located in 17 discrete chromosomal regions (Figure 7.1). Some of these regions, termed lifestyle islands, consist of genes encoding proteins that provide beneficial traits for persisting in the human intestine. When comparing *Lact. rhamnosus* isolates to the intestinal strain *Lact. rhamnosus* GG, the most variable regions were related to carbohydrate metabolism genes (PTS and ABC transporters), host interaction and signaling (mucus-binding pili), immune system (CRISPR-*cas* locus), and biofilm formation (exopolysaccharides) (Douillard et al. 2013).

One remarkable example is the *spa*CBA-*srtC1* gene cluster that encodes mucus-binding pili in *Lact. rhamnosus* GG (Kankainen et al. 2009). Cell surface pilus structures are commonly found in Gram (+) bacterial species (Ton-That and Schneewind 2004; Telford et al. 2006) and play an important role in colonization, signaling, and interaction with the host factor (Nallapareddy et al. 2006; Lebeer et al. 2007). Within the *Lact. rhamnosus* species, functional genes encoding mucus-binding pili were more prevalent in human isolates (40%) than in dairy isolates (13%). The percentage increases further when looking at only intestinal isolates (56%) (Douillard et al. 2013) indicating that such a trait is maintained in isolates that are found in mucosal niches, as it allows the bacterial cells to bind the intestinal mucosa.

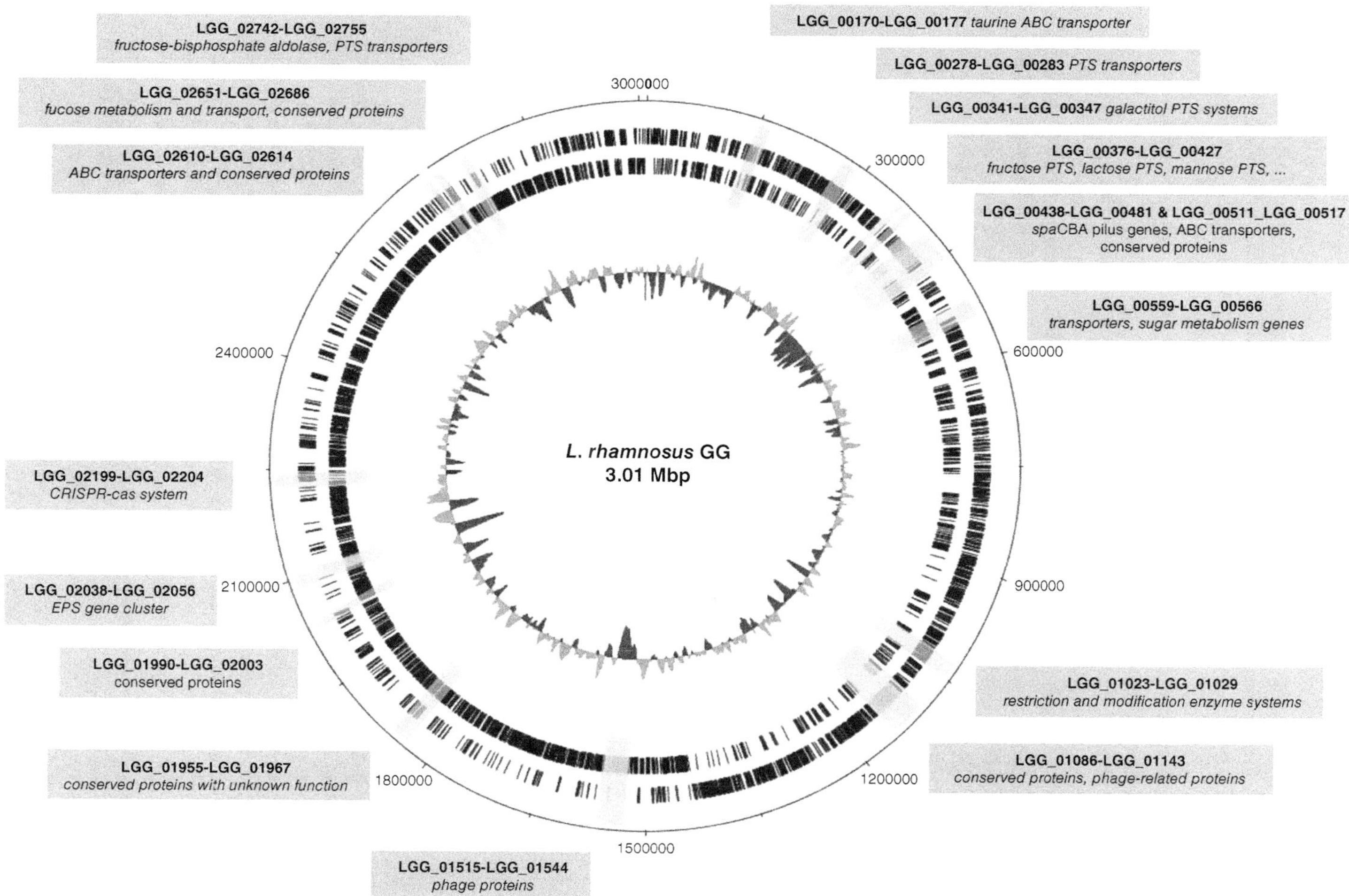

Figure 7.1. Hypervariable regions identified in *Lact. rhamnosus* strain GG chromosome when compared to other *Lact. rhamnosus* isolates. These regions, also termed lifestyle islands, mostly relate to carbohydrate metabolism, host interaction and signaling, DNA restriction/modification systems, CRISPR-*cas* locus and prophages.

Nutrient utilization is also another important trait that relates directly to niche specificity. In this regard, the remarkable specialization of dairy isolates illustrates how colonizing a stable lactose-rich niche resulted in the conservation of functional lactose metabolism, as observed in the dairy strain *Lact. rhamnosus* Lc705, and the loss of non-essential metabolic genes, such as genes involved in the fucose metabolic pathway (Kankainen et al. 2009; Douillard et al. 2013). In contrast, some *Lact. rhamnosus* isolates from the human GI tract have a larger repertoire of biological functions that promote a better fitness and versatility to compete with other bacterial communities, endure harsh physico-chemical environments (i.e. bile exposure), and utilize a distinct range of nutrients in a dynamic and changing environment (human intestine).

The CRISPR- (clustered regularly interspaced short palindromic repeats) based immunity system is commonly found in LAB. It provides protection and immunity against bacteriophages and other foreign mobile DNA elements, such as plasmids (Barrangou et al. 2007; Deveau et al. 2008; Horvath et al. 2008; Marraffini and Sontheimer 2008). Over the last few years, intense work has focused on understanding the CRISPR system, its role, and its mechanism. Aside from its protective function for bacteria, the CRISPR system has proved to constitute a marker of choice to differentiate strains in epidemiological studies, since they contain information relative to previous exposures to foreign entities potentially characteristic of a particular niche (Westra et al. 2012). Their evolution and dynamics are higher than the rest of the genome. Typically, the CRISPR system contains a leader sequence, an alternated array of CRISPRs and spacers, and a *cas* gene cluster (Marraffini and Sontheimer 2010). Evolving rapidly in response to the environment, the CRISPR-*cas* locus, especially the spacers, can reveal important data on each strain. *Lact. rhamnosus* GG possesses one Type II-A CRISPR-*cas* locus with an array of 24 spacers followed by 4 *cas* genes (Kankainen et al. 2009). However, not all *Lact. rhamnosus* strains harbor the CRISPR-*cas* locus. It was dominantly present in strains related to *L. rhamnosus* GG and consisted of a larger repertoire of spacers (Douillard et al. 2013), whereas dairy strains appeared to have a reduced set of spacers and were even in some cases lacking the whole CRISPR-*cas* locus. Dairy isolates had mostly around 6–8 spacers illustrating that bacteriophages and other mobile genetic elements present in each ecological habitat may have impacted differently on the CRISPR-*cas* system, that is the intestinal tract vs. dairy products (Douillard et al. 2013). Also, a higher number of spacers is indicative of a possible exposure to a larger variety of phages or transposable elements. In this regard, the complexity and richness of the intestine is more likely to solicit more such protection systems than in the milk environment, where bacterial cells may encounter a less diverse population of bacteriophages.

7.3.2. Lactobacillus rhamnosus *Metabolism and Adaptation to the Intestine*

LAB found in the human intestinal tract have undergone some metabolic changes reflecting their ecological adaptation to the intestine. Typically, these bacteria have the ability to metabolize complex sugars (Ventura et al. 2009), whereas LAB isolated from milk-derived products utilize a narrower range of carbohydrates. This feature illustrates their adaptation to an environment where the variety and accessibility to carbohydrates greatly differ from the intestine, that is lactose constitutes the main carbon source in milk products. Such niche-driven metabolic changes have also been reported for isolates from other habitats, such as *Lact. iners* from the vaginal cavity or *Lact. delbrueckii* subsp. *bulgaricus* (Macklaim et al. 2011; Mendes-Soares et al. 2014). In the *Lact. rhamnosus* species, the fermentative profiles of 100 strains isolated from various niches were analyzed and clear adaptation patterns could be identified (Douillard et al. 2013). Most dairy strains and their genetically related isolates were able to ferment D-lactose and D-maltose, in contrast to the intestinal strain *Lact. rhamnosus* GG and related strains. Inversely, the latter were more prone to utilize L-fucose (abundant in the intestine) and D-arabinose (Douillard et al. 2013). The distinctive fermentative profiles could be associated to some extent to the loss, decay, or mutations of carbohydrate-associated genes, located in discrete chromosomal regions as mentioned above. When considering amino acid transport and biosynthesis pathways, no significant differences could be reported between the dairy isolates *Lact. rhamnosus* Lc705 and the intestinal isolate *Lact. rhamnosus* GG. One notable difference was the lack of *cysE* and *cysK* genes (bioconversion serine-cysteine) in GG compared to Lc705 (Kankainen et al. 2009). Both *Lact. rhamnosus* strains also possess an equivalent number of peptidases (Kankainen et al. 2009), which is in line with other lactobacilli as seen in *Lact. casei* (Cai et al. 2009) or *Lact. johnsonii* (Pridmore et al. 2004). The presence of such genes encoding proteases and peptidases is beneficial in order to utilize proteins as a source of nutrients in their respective niches.

7.3.3. Host Interaction Factors in Lact. rhamnosus

Components located on the cell surface of LAB play a central role in interacting with host tissues (Sengupta et al. 2013) and protecting the bacterium integrity upon environmental stresses (Guerzoni et al. 2001). However, the cell envelope varies between LAB species and strains in terms of composition and architecture, implying that they have the ability to establish different interactions with the human host (Jacobsen et al. 1999). Over the recent years, the genome sequencing of a number of LAB offered a wealth of information to further characterize bacterial secretomes from a functional and ecological perspective. One of the first large-scale genome-wide analysis of LAB, performed by Zhou and colleagues, associated specific secretome proteins to niche-driven adaptation traits (Zhou et al. 2010), providing a basis for future functional studies on targeted host interaction candidates. Specifically, the secretome of the 26 LAB genomes analyzed consisted of 6–10% of the total genomes (Zhou et al. 2010). Although half of the predicted secretome proteins could be associated to cell-wall metabolism, transporters and proteins with adhesive properties were also identified. In *Lact. rhamnosus*, the main interaction factors beneficial for residing in the intestine relate to adherence to host tissues, immune modulation, and intestinal cell homeostasis (Figure 7.2 and Table 7.2). As mentioned above, one of the most studied features of *Lact. rhamnosus* GG is the presence of sortase-dependent pilus structures that have the ability to bind to mucus (Lebeer et al. 2007; Kankainen et al. 2009; Lebeer et al. 2012; Tripathi et al. 2013; von Ossowski et al. 2013). Within the *Lact. rhamnosus* species itself, the prevalence of pili in intestinal isolates indicates that it constitutes a trait associated with the intestinal mucosa (Douillard et al. 2013). In *Lact. rhamnosus* GG, two additional proteins MabA and MBF were found to display adhesion and biofilm formation properties (Velez et al. 2010; von Ossowski et al. 2011). However, SpaCBA pili remains the most important player to adhere to mucus, as the mucus binding properties of a number of *Lact. rhamnosus* isolates could be abolished upon addition of SpaC antiserum (Kankainen et al. 2009; Douillard et al. 2013). Other host factors were shown to be involved in immuno-signaling or modulation (Table 7.2). Despite these observations, it seems that *Lact. rhamnosus* GG has not yet revealed all its secrets. Recently, Gagic and colleagues identified a gene cluster encoding a secreted-docking protein pair in *Lact. rhamnosus* strain HN001 which is also present in strain GG (Gagic et al. 2013). Displayed on the bacterium cell surface, the secreted-docking protein pair function remains unknown, but it is tempting to associate it to potential novel sensing and signaling systems.

7.3.4. The Lact. rhamnosus *Species: Autochthonous or Allochthonous in the Human Intestine?*

The recent development of next-generation sequencing techniques has offered a deeper and more comprehensive view on the composition of the human microbiota and has also challenged some pre-conceived ideas about LAB residing in the human intestine. Both culturing techniques and molecular approaches indicated that lactobacilli account for a small percentage of the overall human intestinal microbiota (Walter 2008). However, it remains unclear whether these LAB are permanent

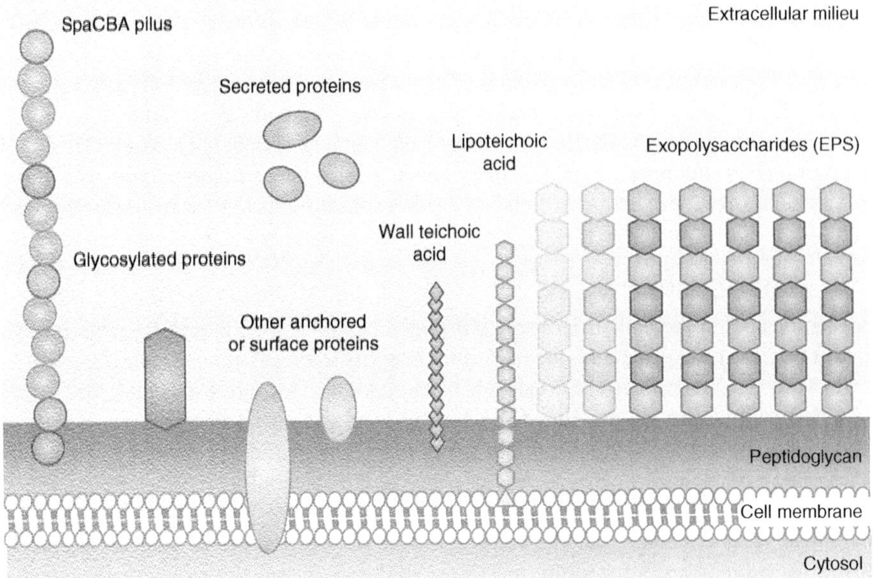

Figure 7.2. Architecture and composition of *Lact. rhamnosus* GG cell envelope associated with different cell-wall components. The cell membrane is typically covered with peptidoglycans that are decorated with mucus-binding pili, LPXTG proteins, glycoproteins, lipoteichoic acid (LTA), wall teichoic acids (WTA). A layer of exopolysaccharides (EPS) is juxtaposed onto the peptidoglycans.

Table 7.2. Host–microbe interaction factors in *Lact. rhamnosus* strain GG promoting ecological fitness in the human gastrointestinal tract. Representative and non-exhaustive list of reference(s) are shown for each factor.

Host–Microbe Factors	Biological Functions	References
Mucus-binding SpaCBA pili	*Adherence, immunomodulation, biofilm formation, aggregation*	(Lebeer et al. 2007; Kankainen et al. 2009; Lebeer et al. 2012; Tripathi et al. 2013; von Ossowski et al. 2013)
Mucus-binding factor (MBF)	*Mucus adherence*	(von Ossowski et al. 2011)
P40	*Peptidoglycan hydrolase, intestinal epithelium homeostasis*	(Yan et al. 2013)
P75	*Peptidoglycan hydrolase, intestinal cell homeostasis*	(Yan et al. 2007)
Exopolysaccharides	*Protection against innate immune factors*	(Lebeer et al. 2011)
Lipoteichoic acid	*Microbe-associated molecular pattern (MAMP) with pro-inflammatory activities*	(Claes et al. 2012)
MabA	*Biofilm formation and adhesion*	(Velez et al. 2010)
Spc proteins	*Secreted-docking protein pair*	(Gagic et al. 2013)

residents (autochthonous) or simply transients (allochthonous) in the human intestine. The question is particularly relevant since they are increasingly consumed as dairy products and exert potential beneficial properties on the consumer's health. In an effort to demonstrate the ability of LAB to persist in the human intestine, *in vitro* or animal models have been employed but have proved to be still too different from the human intestine due to technical limitations and simplifications of the intestine environment, and physiological and anatomical differences, respectively (Lebeer et al. 2008; Kleerebezem and Vaughan 2009). There are no ambiguities about the ability of a number of LAB species, such as *Lact. salivarius* or *Lact. reuteri*, to colonize their dedicated hosts (Guan et al. 2003; Tannock 2004; 2005; O'Callaghan and O'Toole 2013). Further human trials would certainly contribute to elucidate the question; these are particularly relevant for the selection of novel probiotic strains/species, since the isolation of a true autochthonous strain/species from the human intestine would possess the most adequate phenotypic traits to persist and stably colonize the human intestine upon consumption. For most other LAB species, it is generally acknowledged that they are allochthonous to the human intestine, such as *Lact. acidophilus*, *Lact. casei*, *Lact. paracasei*, *Lact. plantarum*, and *Lact. rhamnosus* (Mitsuoka 1992; Tannock et al. 2000; Reuter 2001). This last species should be further discussed in the light of different questions. The remarkable intraspecies diversity in *Lact. rhamnosus* suggests that rather than looking at the species, one should examine the question "transient or residents?" for each single strain. In that regard, stamping of *Lact. rhamnosus* as allochtonous or autochtonous to the human intestine may be related to the presence of adaptation traits to their respective ecological niche within the same species. In our recent work on *Lact. rhamnosus*, we showed that the *L. rhamnosus* species could be classified into two main pheno-genotypes (Douillard et al. 2013). The GG-like pheno-genotype consisted of strains harboring genomic and phenotypic traits that were clearly beneficial not only for surviving in the intestine but also for interacting and colonizing the human intestinal tract. Specifically, the GG-like group was shown to be bile resistant, to have higher antimicrobial activity and larger CRISPR systems, and to display mucus-binding pili on its cell surface (Douillard et al. 2013). Through the fastidious metabolism of the GG-like group, as most LAB, the use of by-products from other bacteria or the colonization of localized niches within the intestine may allow this group to maintain stable communities in defined and specific intestinal niches. Also, *Lact. rhamnosus* harbors metabolic genes that are clearly beneficial for residing in the human intestine. Thus, the use of L-fucose is an example of intestinal adaptation, since L-fucose is abundant in the intestinal tract and can be utilized as a nutrient source by these GG-like strains (Douillard et al. 2013). To conclude this section, there are a number of data supporting the hypothesis that some *Lact. rhamnosus* strains may be true inhabitants of the human intestine, since they have traits allowing them to survive, grow, compete, interact, and signal within the intestinal ecosystem. Future human feeding trials using *Lact. rhamnosus* strains may further support our view on this. The increasing use of molecular techniques combined with traditional culturing methods, phenotypic characterization, and full-genome sequencing of fecal LAB isolates would offer the possibility of identifying them as allochtonous or autochtonous and predicting their role within the intestinal ecosystem. This would offer the opportunity to select novel autochtonous LAB with high probiotic potential.

7.4. Concluding Perspectives and Future Directions

The great interest in the human microbiome (Brown et al. 2013) has revived attention paid to LAB presence in the human intestine. A large number of high-throughput metagenomic approaches has been developed in parallel to genome sequencing of numerous LAB (Zoetendal et al. 2008; de Vos 2011). One of the major findings has been that the level of LAB, particularly lactobacilli, is rather low in adults. This fact contrasts with their presence in early life and hence considerable attention is given to the early-life ecosystem where developmental aberrations of LAB microbiota may have profound effects (de Weerth et al. 2013). Moreover, a variety of approaches have been developed and applied to transiently passing LAB, notably *Lactobacillus* spp. marketed as probiotics. While this may support further strain development, detailed analysis of the expression of autochthonous LAB is needed to further advance our insight into microbial–host interactions that affect human health.

This touches upon the methods used to differentiate between autochthonous or allochthonous strains of LAB in the human intestinal tract. The recent comparative genome analysis of 100 strains of *Lact. rhamnosus* provides a number of molecular avenues to differentiate between these extremes (Douillard et al. 2013). This is highly relevant as the differentiation is based on molecular methods, including adaptation traits and CRISPR sequences that form a proxy of the past. It is expected that these and other molecular approaches, notably those based on ultra-deep functional metagenomics, will provide new insight into the adaptation, strain diversity, and host impact of LAB in the human intestinal tract.

Acknowledgments

FPD was financially supported by a postdoctoral research fellowship (Academy of Finland, grant number 252123). WMdV was supported by the European Research Council (ERC grant 250172 - Microbes Inside) and the Academy of Finland (grant number 141130) and the University of Helsinki. We apologize for not citing all relevant research work due to space limitations.

References

Ahrne, S., Nobaek, S., Jeppsson, B., et al. (1998) The normal *Lactobacillus* flora of healthy human rectal and oral mucosa. *J Appl Microbiol* **85**, 88–94.

Barrangou, R., Fremaux, C., Deveau, H., et al. (2007) CRISPR provides acquired resistance against viruses in prokaryotes. *Science* **315**, 1709–1712.

Bello, F.D., Walter, J., Hammes, W.P., and Hertel, C. (2003) Increased complexity of the species composition of lactic acid bacteria in human feces revealed by alternative incubation condition. *Microb Ecol* **45**, 455–463.

Bernardeau, M., Guguen, M., and Vernoux, J.P. (2006) Beneficial lactobacilli in food and feed: long-term use, biodiversity and proposals for specific and realistic safety assessments. *FEMS Microbiol Rev* **30**, 487–513.

Booijink, C.C.G.M., El-Aidy, S., Rajilić-Stojanović, M., et al. (2010) High temporal and inter-individual variation detected in the human ileal microbiota. *Environ Microbiol* **12**, 3213–3227.

Broadbent, J.R., Neeno-Eckwall, E., Stahl, B., et al. (2012) Analysis of the *Lactobacillus casei* supragenome and its influence in species evolution and lifestyle adaptation. *BMC Genomics* **13**, 533.

Brown, J., de Vos, W.M., DiStefano, P.S., et al. (2013) Translating the human microbiome. *Nat Biotechnol* **31**, 304–308.

Cai, H., Thompson, R., Budinich, M.F., et al. (2009) Genome sequence and comparative genome analysis of *Lactobacillus casei*: insights into their niche-associated evolution. *Genome Biol Evol* **1**, 239–257.

Claes, I.J.J., Segers, M.E., Verhoeven, T.L.A., et al. (2012) Lipoteichoic acid is an important microbe-associated molecular pattern of *Lactobacillus rhamnosus* GG. *Microb Cell Fact* **11**, 161.

Clemente, J.C., Ursell, L.K., Parfrey, L.W., and Knight, R. (2012) The impact of the gut microbiota on human health: an integrative view. *Cell* **148**, 1258–1270.

de Vos, W.M. (2011) Systems solutions by lactic acid bacteria: from paradigms to practice. *Microb Cell Fact* **10**, S2

de Vos, W.M. and de Vos, E.A. (2012) Role of the intestinal microbiome in health and disease: from correlation to causation. *Nutr Rev* **70**, S45–S46.

de Vos, W.M. and Nieuwdorp, M. (2013) Genomics: A gut prediction. *Nature* **498,** 48–49.

de Weerth, C., Fuentes, S., and de Vos, W.M. (2013) Crying in infants: On the possible role of intestinal microbiota in the development of colic. *Gut Microbes* **9**, 416–421.

Deveau, H., Barrangou, R., Garneau, J.E., et al. (2008) Phage response to CRISPR-encoded resistance in *Streptococcus thermophilus*. *J Bacteriol* **190**, 1390–1400.

Douillard, F.P., Ribbera, A., Kant, R.,et al. (2013) Comparative genomic and functional analysis of 100 *Lactobacillus rhamnosus* strains and their comparison with strain GG. *PLoS Genet* **9**, e1003683.

Gagic, D., Wen, W., Collett, M.A., and Rakonjac, J. (2013) Unique secreted-surface protein complex of *Lactobacillus rhamnosus*, identified by phage display. *Microbiology open* **2**, 1–17.

Guan, L.L., Hagen, K.E., Tannock, G.W., et al. (2003) Detection and identification of *Lactobacillus* species in crops of broilers of different ages by using PCR-denaturing gradient gel electrophoresis and amplified ribosomal DNA restriction analysis. *Appl Environ Microbiol* **69**, 6750–6757.

Guerzoni, M.E., Lanciotti, R., and Cocconcelli, P.S. (2001) Alteration in cellular fatty acid composition as a response to salt, acid, oxidative and thermal stresses in *Lactobacillus helveticus*. *Microbiology* **147**, 2255–2264.

Harmsen, H.J.M., Raangs, G.C., He, T., et al. (2002) Extensive set of 16S rRNA-based probes for detection of bacteria in human feces. *Appl Environ Microbiol* **68**, 2982–2990.

Heilig, H.G.H.J., Zoetendal, E.G., Vaughan, E.E., et al. (2002) Molecular diversity of *Lactobacillus* spp. and other lactic acid bacteria in the human intestine as determined by specific amplification of 16S ribosomal DNA. *Appl Environ Microbiol* **68**, 114–123.

Horvath, P., Romero, D.A., Coûté-Monvoisin, A.-C., et al. (2008) Diversity, activity, and evolution of CRISPR loci in *Streptococcus thermophilus*. *J Bacteriol* **190**, 1401–1412.

Jacobsen, C.N., Rosenfeldt Nielsen, V., Hayford, A.E., et al. (1999) Screening of probiotic activities of forty-seven strains of *Lactobacillus* spp. by *in vitro* techniques and evaluation of the colonization ability of five selected strains in humans. *Appl Environ Microbiol* **65**, 4949–4956.

Kankainen, M., Paulin, L., Tynkkynen, S., et al. (2009) Comparative genomic analysis of *Lactobacillus rhamnosus* GG reveals pili containing a human- mucus binding protein. *Proc Natl Acad Sci USA* **106**, 17193–17198.

Kleerebezem, M. and Vaughan, E.E. (2009) Probiotic and gut lactobacilli and bifidobacteria: molecular approaches to study diversity and activity. *Annu Rev Microbiol* **63**, 269–290.

Klijn, N., Weerkamp, A.H. and de Vos, W.M. (1995) Genetic marking of *Lactococcus lactis* shows its survival in the human gastrointestinal tract. *Appl Environ Microbiol*, **61**, 2771–2774.

Koenig, J.E., Spor, A., Scalfone, N., et al. (2011) Succession of microbial consortia in the developing infant gut microbiome. *Proc Natl Acad Sci* **108**, 4578-4585.

Lahti, L., Salonen, A., Kekkonen, R.A., et al. (2013) Associations between the human intestinal microbiota, *Lactobacillus rhamnosus* GG and serum lipids indicated by integrated analysis of high-throughput profiling data. *PeerJ* **1**, e32.

Le Chatelier, E., Nielsen, T., Qin, J., et al. (2013) Richness of human gut microbiome correlates with metabolic markers. *Nature* **500**, 541–546.

Lebeer, S., Claes, I.J., Tytgat, H.L.P., Verhoeven, T.L.A., et al. (2012) Functional analysis of *Lactobacillus rhamnosus* GG pili in relation to adhesion and immunomodulatory interactions with intestinal epithelial cells. *Appl Environ Microbiol* **78**, 185-193.

Lebeer, S., Claes, I.J., Verhoeven, T.L.A., et al. (2011) Exopolysaccharides of *Lactobacillus rhamnosus* GG form a protective shield against innate immune factors in the intestine. *Microb Biotechnol* **4**, 368–374.

Lebeer, S., Vanderleyden, J., and De Keersmaecker, S.C.J. (2008) Genes and molecules of lactobacilli supporting probiotic action. *Microbiol Mol Biol Rev* **72**, 728–764.

Lebeer, S., Verhoeven, T.L.A., Perea Vélez, M., et al. (2007) Impact of environmental and genetic factors on biofilm formation by the probiotic strain *Lactobacillus rhamnosus* GG. *Appl Environ Microbiol* **73**, 6768–6775.

Louis, P., Scott, K.P., Duncan, S.H., and Flint, H.J. (2007) Understanding the effects of diet on bacterial metabolism in the large intestine. *J Appl Microbiol* **102**, 1197–1208.

Macklaim, J.M., Gloor, G.B., Anukam, K.C., et al. (2011) At the crossroads of vaginal health and disease, the genome sequence of *Lactobacillus iners* AB-1. *Proc Natl Acad Sci USA* **1**, 4688–4695.

Marco, M.L., de Vries, M.C., Wels, M., et al. (2010) Convergence in probiotic *Lactobacillus* gut-adaptive responses in humans and mice. *ISME J* **4**, 1481–1484.

Marraffini, L.A. and Sontheimer, E.J. (2008) CRISPR interference limits horizontal gene transfer in *Staphylococci* by targeting DNA. *Science* **322**, 1843–1845.

Marraffini, L.A. and Sontheimer, E.J. (2010) CRISPR interference: RNA-directed adaptive immunity in bacteria and archaea. *Nat Rev Genet* **11**, 181–190.

McNulty, N.P., Yatsunenko, T., Hsiao, A., et al. (2011) The impact of a consortium of fermented milk strains on the gut microbiome of gnotobiotic mice and monozygotic twins. *Sci Transl Med* **3**, 106ra106.

Mendes-Soares, H., Suzuki, H., Hickey, R.J., and Forney, L.J. (2014) Comparative functional genomics of *Lactobacillus* spp. reveals possible mechanisms for specialization of vaginal lactobacilli to their environment. *J Bacteriol* **196**, 1458–1470.

Mitsuoka, T. (1992) The human gastrointestinal tract. In: *The lactic acid bacteria in health and disease* ((ed.) B.J.B. Wood), Elsevier Applied Science, London, pp. 69–114. Elsevier Applied Science, London.

Moles, L., Gómez, M., Heilig, H., et al. (2013) Bacterial diversity in meconium of preterm neonates and evolution of their fecal microbiota during the first month of life. *PLoS ONE* **8**, e66986.

Molin, G., Jeppsson, B., Johansson, M.L., et al. (1993) Numerical taxonomy of *Lactobacillus* spp. associated with healthy and diseased mucosa of the human intestines. *J Appl Bacteriol* **74**, 314–323.

Nallapareddy, S.R., Singh, K.V., Sillanpää, J., et al. (2006) Endocarditis and biofilm-associated pili of *Enterococcus faecalis*. *J Clin Investig* **116**, 2799–2807.

O'Callaghan, J. and O'Toole, P.W. (2013) *Lactobacillus*: Host–Microbe Relationships. In: *Between Pathogenicity and Commensalism* ((eds) U. Dobrindt, J.H. Hacker, and C. Svanborg), Springer, Berlin Heidelberg, pp.119–154.

Pridmore, R.D., Berger, B., Desiere, F., et al. (2004) The genome sequence of the probiotic intestinal bacterium *Lactobacillus johnsonii* NCC 533. *Proc Natl Acad Sci USA* **101**, 2512–2517.

Qin, J., Li, R., Raes, J., Arumugam, M., et al. (2010) A human gut microbial gene catalogue established by metagenomic sequencing. *Nature* **464**, 59–65.

Rajilić-Stojanović, M., de Vos, W.M. (2014). The first 1000 cultured species of the human gastrointestinal microbiota. *FEMS Microbiol Rev*. **38**, 996–1047.

Rajilić-Stojanović, M., Heilig, H.G.H.J., Molenaar, D., et al. (2009) Development and application of the human intestinal tract chip, a phylogenetic microarray: analysis of universally conserved phylotypes in the abundant microbiota of young and elderly adults. *Environ Microbiol* **11**, 1736–1751.

Rajilić-Stojanović, M., Smidt, H., and de Vos, W.M. (2007) Diversity of the human gastrointestinal tract microbiota revisited. *Environ Microbiol* **9**, 2125–2136.

Reuter, G. (2001) The *Lactobacillus* and *Bifidobacterium* microflora of the human intestine: composition and succession. *Curr Issues Intest Microbiol* **2**, 43–53.

Saxelin, M., Tynkkynen, S., Mattila-Sandholm, T., and de Vos, W.M. (2005) Probiotic and other functional microbes: from markets to mechanisms. *Curr Opin Biotechnol* **16**, 204–211.

Sengupta, R., Altermann, E., Anderson, R.C., et al. (2013) The role of cell surface architecture of lactobacilli in host-microbe interactions in the gastrointestinal tract. *Mediators Inflamm* **2013**, 237921.

Smokvina, T., Wels, M., Polka, J., et al. (2013) *Lactobacillus paracasei* comparative genomics: towards species pan-genome definition and exploitation of diversity. *PLoS One* **8**, e68731.

Tannock, G.W. (2004) A special fondness for lactobacilli. *Appl Environ Microbiol* **70**, 3189–3194.

Tannock, G.W. (2005) New perceptions of the gut microbiota: implications for future research. *Gastroenterol Clin North Am* **34**, 361–382.

Tannock, G.W., Munro, K., Harmsen, H.J.M., et al. (2000) Analysis of the fecal microflora of human subjects consuming a probiotic product containing *Lactobacillus rhamnosus* DR20. *Appl Env Microbiol* **66**, 2578–2588.

Telford, J.L., Barocchi, M., Margarit, I., et al. (2006) Pili in Gram-positive pathogens. *Nat Rev Microbiol* **4**, 509–519.

Toh, H., Oshima, K., Nakano, A., et al. (2013) Genomic adaptation of the *Lactobacillus casei* group. *PLoS One* **8**, e75073.

Ton-That, H. and Schneewind, O. (2004) Assembly of pili in Gram-positive bacteria. *Trends Microbiol* **12**, 228–234.

Tripathi, P., Beaussart, A., Alsteens, D., et al. (2013) Adhesion and nanomechanics of pili from the probiotic *Lactobacillus rhamnosus* GG. *ACS Nano* **7**, 3685–3697.

Walter, J. (2005) The microecology of lactobacilli in the gastrointestinal tract. In: *Prebiotics and probiotics: Scientific Aspects* ((ed.) G.W. Tannock), Caister Academic Press, Norfolk, pp.51–82.

Walter, J. (2008) Ecological role of lactobacilli in the gastrointestinal tract: implications for fundamental and biomedical research. *Appl Environ Microbiol* **74**, 4985–4996.

van den Bogert, B., de Vos, W.M., Zoetendal, E.G., and Kleerebezem, M. (2011) Microarray analysis and barcoded pyrosequencing provide consistent microbial profiles depending on the source of human intestinal samples. *Appl Environ Microbiol* **77**, 2071–2080.

van Baarlen, P., Troost, F.J., van Hemert, S., et al. (2009) Differential NF-kappaB pathways induction by *Lactobacillus plantarum* in the duodenum of healthy humans correlating with immune tolerance. *Proc Natl Acad Sci USA*, **106**, 2371–2376.

van den Bogert, B., Erkus, O., Boekhorst, J., et al. (2013) Diversity of human small intestinal *Streptococcus* and *Veillonella* populations. *FEMS Microbiol Ecol* **85**, 376–388.

van Nood, E., Keller, J.J., Kuijper, E.J., and Speelman, P. (2013) [New treatment options for infections with Clostridium difficile]. *Ned Tijdschr Geneeskd* **157**, A6880.

Vaughan, E.E., de Vries, M.C., Zoetendal, E.G., et al. (2002) The intestinal LABs. *Antonie Van Leeuwenhoek* **82**, 341–352.

Vaughan, E.E., Mollet, B., and deVos, W.M. (1999) Functionality of probiotics and intestinal lactobacilli: light in the intestinal tract tunnel. *Curr Opin Biotechnol* **10**, 505–510.

Vélez, M.P., Petrova, M.I., Lebeer, S., et al. (2010) Characterization of MabA, a modulator of *Lactobacillus rhamnosus* GG adhesion and biofilm formation. *FEMS Immunol Med Microbiol* **59**, 386–398.

Ventura, M., O'Flaherty, S., Claesson, M.J., et al. (2009) Genome-scale analyses of health-promoting bacteria: probiogenomics. *Nat Rev Microbiol* **7**, 61–71.

Vesa, T., Pochart, P., and Marteau, P. (2000) Pharmacokinetics of *Lactobacillus plantarum* NCIMB 8826, *Lactobacillus fermentum* KLD, and *Lactococcus lactis* MG 1363 in the human gastrointestinal tract. *Aliment Pharmacol Ther* **14**, 823–828.

Westra, E.R., Swarts, D.C., Staals, R.H.J., et al. (2012) The CRISPRs, they are a-changin': how prokaryotes generate adaptive immunity. *Annu Rev Genet* **46**, 311–339.

von Ossowski, I., Pietilä, T.E., Rintahaka, J., et al. (2013) Using recombinant lactococci as an approach to dissect the immunomodulating capacity of surface piliation in probiotic *Lactobacillus rhamnosus* GG. *PloS One* **8**, e64416.

von Ossowski, I., Satokari, R., Reunanen, J., et al. (2011) Functional characterization of a mucus-specific LPXTG surface adhesin from probiotic *Lactobacillus rhamnosus* GG. *Appl Environ Microbiol* **77**, 4465–4472.

Yan, F., Cao, H., Cover, T.L., et al. (2007) Soluble proteins produced by probiotic bacteria regulate intestinal epithelial cell survival and growth. *Gastroenterology* **132**, 562–575.

Yan, F., Liu, L., Dempsey, P.J., et al. (2013) A *Lactobacillus rhamnosus* GG-derived soluble protein, p40, stimulates ligand release from intestinal epithelial cells to transactivate epidermal growth factor receptor. *J Biol Chem* **288**, 30742–30751.

Yatsunenko, T., Rey, F.E., Manary, M.J., et al. (2012) Human gut microbiome viewed across age and geography. *Nature* **486**, 222–227.

Zhou, M., Theunissen, D., Wels, M., and Siezen, R.J. (2010) LAB-Secretome: a genome-scale comparative analysis of the predicted extracellular and surface-associated proteins of Lactic Acid Bacteria. *BMC Genomics* **11**, 651.

Zoetendal, E.G., Rajilić-Stojanović M., and de Vos, W.M. (2008) High-throughput diversity and functionality analysis of the gastrointestinal tract microbiota. *Gut* **11**, 1605–1615

Zoetendal, E.G., Raes, J., van den Bogert, B., et al. (2012) The human small intestinal microbiota is driven by rapid uptake and conversion of simple carbohydrates. *ISME J* **6**, 1415–1426.

Chapter 8

Probiotics and Functional Foods in Immunosupressed Hosts

Ivanna Novotny Nuñez[1], Martin Manuel Palomar[1], Alejandra de Moreno de LeBlanc[1], Carolina Maldonado Galdeano[1,2], and Gabriela Perdigón[1,2,*]

[1] *Laboratorio de Inmunología, Centro de Referencia para Lactobacilos (CERELA)-CONICET, Argentina*
[2] *Cátedra de Inmunología, Instituto de Microbiología, Facultad de Bioquímica, Química y Farmacia, Universidad Nacional de Tucumán, Argentina*

8.1. Introduction

The intestinal ecosystem is a complex microenvironment constituted by eukaryotic and prokaryotic cells where a complex network of signals that maintain the intestinal homeostasis is established. The gut microbiota is an important component of this ecosystem; commensal bacteria are key players participating in the digestion process, vitamin synthesis, and other metabolite production (bile acids, lipids, amino acids, and short-chain fatty acids), which are important for maintaining health in mammals (Brestoff and Artis 2013). However, it is known that the commensal microbiota is not in direct contact with the epithelial cells due to the presence of the mucus layer (Johansson et al. 2011).

In addition, there are specialized immune cells distributed in the gut that interact with antigens that reach the intestinal lumen, such as M cells and dendritic cells (DC) (Rescigno et al. 2001). The most important function of these cells is to discriminate between pathogenic and non-pathogenic microorganisms to protect the host from external agents. The immune response against foreign antigen entry involves a wide variety of cells, which trigger a number of mechanisms like cytokine production, cell migration and activation, and differential gene expression. These processes are demanding of energy, so nutrient requirements are highly relevant (Pearce and Pearce 2013). Recent studies have demonstrated the critical role of the commensal microbiota and their metabolic products in regulating the development, homeostasis, and function of the intestinal immune cells (Hill and Artis 2010; Tremaroli and Bächked 2012).

In healthy patients, the intestinal microbiota is characterized by its stability and diversity, but during metabolic and inflammatory diseases, the composition of the gut microbiota suffers alterations

* Corresponding author: Laboratorio de Inmunología, Centro de Referencia para Lactobacilos (CERELA-CONICET), Chacabuco 145, 4000 San Miguel de Tucumán, Tucumán, Argentina. Cátedra de Inmunología, Instituto de Microbiología, Facultad de Bioquímica, Química y Farmacia, Universidad Nacional de Tucumán, Ayacucho 491, 4000 San Miguel de Tucumán, Tucumán, Argentina. perdigon@cerela.org.ar

Biotechnology of Lactic Acid Bacteria: Novel Applications, Second Edition. Edited by Fernanda Mozzi, Raúl R. Raya, and Graciela M. Vignolo.

causing an imbalance with consequent changes in the homeostasis of the intestine, the intestinal barrier, and visceral perception (Verdu et al. 2004; Flint et al. 2012). Chronic stress in animal models has been also reported to increase the permeability of the intestinal barrier resulting in a visceral hypersensitivity (Ait-Belgnaoui et al. 2005). The composition of the intestinal microbiota is influenced by the microorganisms that enter with the diet, with lactic acid bacteria (LAB) one of the most studied due to their beneficial (probiotic) properties. The term "probiotics" was defined by FAO/WHO as "live microorganisms which, when administered in adequate amounts confer a health benefit on the host" (FAO/WHO, 2001). Previous studies have demonstrated that probiotics could be useful to reconstitute the intestinal mucosa (Ohland and Macnaughton 2010), to enhance the IgA-producing cells in the lamina propria of the small intestine, and to stimulate the systemic immune response (Ohashi and Ushida 2009; Maldonado Galdeano et al. 2012). These microorganisms can also improve the non-specific intestinal barrier by changing the composition and activity of the gut microbiota (Arora et al. 2012). It had been demonstrated that orally administered probiotic bacteria induce a mucosal immunity response in sites distant to the intestine, such as the respiratory tract and mammary glands (de Moreno de LeBlanc et al. 2005).

In this chapter we focus on the usefulness of a probiotic fermented milk in immunodeficient hosts using under-nourished or stressed experimental models.

8.2. Probiotic Fermented Milk in a Malnutrition Model

Malnutrition caused by deficient intake of food is a scourge in developing countries due to limited economic resources and compromises nutritional status. Protein-energy malnutrition (PEM) is classified according to weight loss as: slight (10–25%), moderate (25–39%), and severe (>40%). Its consequences include growth delay, often associated with diverse clinical syndromes such as the alteration of bowel epithelial mucosa and gut microbiota, which in turn affect the production of intestinal enzymes, carbohydrate digestion, and lipid absorption. Mineral, vitamin, or immune deficiencies are also consequences of malnutrition that may accompany weight loss. PEM can be reversed with appropriate re-nutrition, which can restore both gastrointestinal and immune functions, regenerate the intestinal mucosa, and improve the intestinal microbiota. In severely malnourished hosts, probiotic administration as supplement of a re-nutrition diet could restore the intestinal mucosa, reducing its permeability and improving the local (increase in IgA-secreting cells) and systemic immune responses (Gauffin Cano et al. 2002). Similarly, administration of *Lactobacillus pentosus* ONRIC b0240 to mice in a PEM model enhanced the IgA production and the humoral immune response (Shimosato et al. 2011). The effect of probiotic fermented milk (PFM) containing *Lact. casei* DN114001, currently named *Lact. paracasei* CNCMI-1518, or the bacterial free supernatant (BFS, obtained from this PFM), as diet supplements in a non-severe mouse PEM model was described by Maldonado Galdeano et al. (2011). PFM reversed the gut histological alterations induced in the PEM model, such as the decrease in the numbers and lengths of the intestinal villi. Mice given milk or BFS during the re-nutrition period showed a similar effect, although to a lesser extent.

In addition, the growth deficit of colon wall strata as well as the alteration of goblet cells/enterocytes ratio induced in malnourished rats was prevented by the probiotics *Bifidumbacterium bifidus* and *Lact. acidophilus* (Lima et al. 2012).

Malnourished mice showed a decrease in the number of the IgA-producing cells, T lymphocyte population (CD4+ and CD8+ cells) as well as antigen presenting cells (APCs), such as macrophages and dendritic cells (DC), in the small intestine as reported by Maldonado Galdeano et al. (2011). Re-nutrition with PFM significantly increased the IgA-producing cells and APCs; however, a re-nutrition diet using BFS restored only APCs while administration of whole milk induced macrophage proliferation (Table 8.1). In contrast, T cells were not restored by any supplied diets.

Table 8.1. Immune cells in the small intestine of malnourished and re-nourished mice.

Samples	Groups	IgA+	CD8+	CD4+	F4/80+	DC
5 days of re-nutrition	WC	143±26[a,b]	40±7[a]	47±9[a]	122±24[a,d]	40±9[a,c]
	MC	106±33[b]	24±7[b]	30±7[b,c]	83±11[b]	16±6[b]
	Milk	100±32[b]	26±10[a,b]	53±17[a,c]	114±19[a,c,d]	26±11[b,c]
	PFM	163±31[a,b]	20±5[b]	36±9[a,b]	116±24[a,d]	57±19[a,d]
	BFS	124±39[a,b]	24±8[b]	28±8[a,b]	101±17[c,d]	80±19[d]

The number of IgA+ cells, CD4+ and CD8+ T lymphocytes were determined by direct immunofluorescence; macrophages (F4/80+ cells) and dendritic cells (DC) were determined by indirect immunofluorescence on slides prepared with the small intestine tissue of mice from different experimental groups: well-nourished control (WC), malnourished control (MC), and mice re-nourished with whole milk (Milk), probiotic fermented milk (PFM) or bacterial free supernatant (BFS).

Results are expressed as the number of positive cells per ten fields of vision (1000X). Values are means for n = 9 mice ± SD.

[a,b,c,d] Means for each cell population without a common letter differ significantly ($P < 0.05$). Values are those obtained from Maldonado Galdeano et al. (2011).

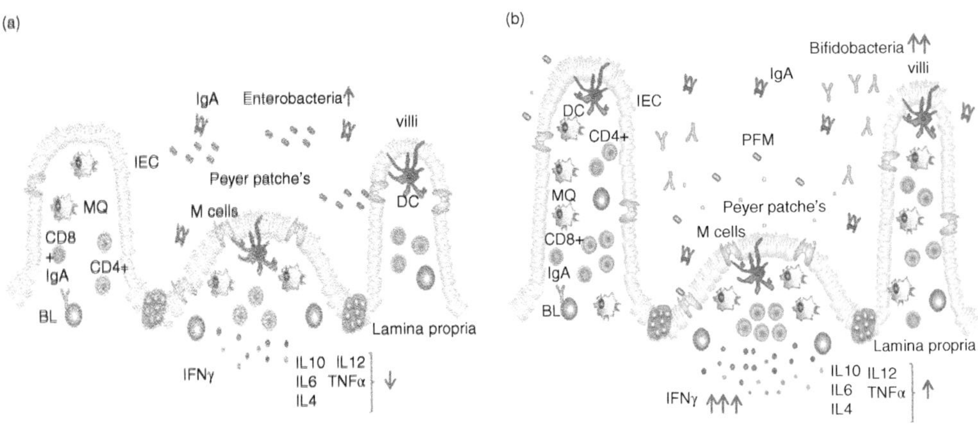

Figure 8.1. Improvement of the intestinal mucosal and the immune system by a PFM as dietary supplement after a malnutrition period. (a) Impairment of the intestinal mucosa by malnutrition. Malnutrition induces histological changes in the intestinal mucosa, with length diminution of the villi. The number of IgA producing cells diminished in the small intestine of malnourished mice. Also a decrease in the number of immune cells, such as MQ, DC, B and T lymphocytes (CD4 and CD8), was observed, as in cytokines producing cells (IL-10, IL-4, IL-6, IL-12, and TNFα). The intestinal microbiota showed a significant increase in the enterobacteria population. (b) Improvement of the intestinal mucosa of malnourished animals with PFM. The re-nutrition with PFM as a dietary supplement restores the length of the intestinal villi Significant increases in the number of IgA producing cells in the lamina propria of the intestinal villi were observed. The number of MQ, DC, BL, and T lymphocytes (CD4 and CD8) also improved. The cytokine producing cells (IL-10, IL-4, IL-6, IL-12 and TNF-α) increased mainly IFN-γ. The PFM induced increases in the bifidobacteria population accompanied by a diminution of enterobacteria.IEC: Intestinal epithelial cell; GC: goblet cell; PP: Peyer's patch; LP lamina propria; MQ: macrophages; DC: dendritic cells; BL: B lymphocytes; S-IgA: secretory IgA.

The different re-nutrition diets not only were able to reestablish the number of immune cells in the lamina propria of the small intestine, but also the functionality of them. Malnutrition induces impairment in the cytokine production by the intestinal immune cells, which was reverted when the re-nutrition diets were supplemented with PFM or BFS, reaching values near to those observed in the well-nourished control group (Figure 8.1). This finding is very important considering the role of cytokines in the enhancement of the immune response during a vaccination process (Rizza et al. 2002).

It is also important to consider that malnutrition is one of the most common causes that induce secondary immunodeficiency in children and is associated with increased risk for infections. Studies conducted by Taylor et al. (2013) showed that PEM increased susceptibility to influenza infection in mice. In addition, Cheung et al. (2012) observed that probiotic administration improved the general status of patients with liver disease and prevented complications, such as ascites and inflammation. Other studies performed in humans have demonstrated that children with PEM given a probiotic-like curd accelerated the immune recovery, improving both cellular and humoral immune responses (Dewan et al. 2007).

The improvement in the gut mucosal immunity induced by PFM was also effective against *Salmonella* infection (Maldonado Galdeano et al. 2011). Similar results were obtained when the probiotics strains *Lact. casei* CRL 431 and *Lact. rhamnosus* CRL 1505 were administered by oral or nasal routes respectively, against *Streptococcus pneumoniae* infections (Villena et al. 2006; Barbieri et al. 2013). Moreover, protection against *Candida albicans* infection in PEM mice was demonstrated by administering the probiotic strain *L. casei* CRL 431 (Villena et al. 2011).

Malnutrition affects not only the intestinal immune response but also the systemic immune response, decreasing the level of serum specific antibodies after immunization. It has been demonstrated that mice receiving re-nutrition diets supplemented with PFM or BFS before and after OVA immunization increased the titers of anti-OVA IgG in serum, titers being higher in mice given PFM (Maldonado Galdeano et al. 2011). These increases in the humoral immune response would be related with the phagocytic activity of macrophages from spleen and peritoneum. The ability of probiotics to increase the immunity against infections has been also demonstrated by Salva et al. (2012). In this study, administration of *Lact. rhamnosus* CRL 1505 induced the recovery of B cells, suggesting its beneficial effect to accelerate the recovery of lymphopoiesis in immunocompromised mice.

One of the most affected organs during malnutrition is the thymus (Savino 2002). Previous studies have demonstrated that the histological alterations produced by malnutrition induce histological atrophy and changes in the lymphoid compartment of the thymus (Lynch et al. 2009). Malnourished mice showed a loss of corticomedular differentiation in the thymus with lobules decreasing compared to well-nourished animals. These alterations were reversible in mice given PFM as a re-nutrition diet. An improvement in the thymus histology was also observed in mice that received milk or BFS; however, the changes observed were lower than those in the animals given PFM (Nuñez et al. 2012). Malnutrition decreases significantly the monocyte proliferation and the number of apoptotic cells in thymus (Ortiz et al. 2009). PFM administration recovered the number of apoptotic cells in the thymus, but the diet supplementation with BFS or whole milk did not (Nuñez et al. 2012).

Malnutrition produced an increase of the immature double-positive thymocytes (CD4+CD8+) and a decrease in the mature single-positive thymocytes (CD4+CD8−), which was reverted with the re-nutrition diets, mainly in the mice that received PFM, these values being similar to the well-nourished mice. No significant differences were observed in the percentage of single-positive (CD8+CD4−) cells among all the groups. These results agree with those obtained by Savino and Dardenne (2010), who showed that in the malnutrition period there is a severe thymus atrophy resulting from massive thymocyte apoptosis (particularly affecting the immature CD4+CD8+ cell subset) and a decrease in the cell proliferation.

The production of pro-inflammatory cytokines, such as IFNγ and TNFα, was increased in the thymus from malnourished mice, whereas the production of IL-10 and IL-6 was decreased. Re-nutrition with PFM diminished the number of TNFα-secreting cells and increased the number of IL-6-secreting cells, reaching similar values with respect to the well-nourished control group. PFM administration also maintained increased numbers of IFNγ and IL-10 producing cells compared to the well-nourished mice. The increase in the production of IL-10 is important to regulate the immune response and to maintain homeostasis in the thymus microenvironment (Oberholzer et al. 2001).

In summary, the administration of fermented milk containing *Lact. casei* DN 114001 was the most effective re-nutrition diet to improve the alterations induced by malnutrition in the intestinal and thymus histology of mice, compared to the BFS and whole milk. These improvements were accompanied by a beneficial balance in the intestinal microbiota, influencing the gut immune response, especially through the innate immune response, and maintaining intestinal homeostasis. Figure 8.1 shows occurring changes in the intestine of malnourished mice and those restorative effects observed in mice fed PFM. As other probiotic microorganisms (Finamore et al. 2012), PFM administration was able to stimulate the systemic immune response suggesting its use as an adjuvant of the immune system in a malnourished host. The beneficial effects observed in the mucosal and systemic immune responses by PFM administration will be related to the improvement in the thymus function. Thus, PFM administration could be useful to improve the gut and systemic immune response in immunosuppressed hosts (by malnutrition) or in risk groups, such as infants or older people. PFM could be used as a biological immunomodulator to improve the gut mucosal immune system, and also to beneficially affect distant mucosal sites and systemic immunity.

8.3. Probiotic Administration in Stress Process

Stress can be defined as the physiological reaction of an organism to external or internal stimuli in which the body initiates various defense mechanisms to maintain homeostasis (Ramsey 1982; Cruz et al. 2012). The stress response involves a complex network of mechanisms essential for survival, mediated by neurotransmitters, peptidic hormones, and endocrine hormones from the enteric nervous system (ENS). These molecules modulate the humoral and cellular components of the intestinal immune system (de Jonge 2013).

The microbiota can influence gut homeostasis by the regulation of bowel motility and modulation of intestinal pain, immune responses, and nutrient processing (Husebye et al. 2001; Ait-Belgnaoui et al. 2006). Furthermore it is known that the gut microbiota has a great influence on the health status of an individual and also has an impact on behavior, so the concept of a microbiota–gut–brain axis has emerged (Cryan and O'Mahony 2011). Recent neurobiological insights into this gut–brain crosstalk have revealed a complex, bidirectional communication: the microbiota influences the Central Nervous System (CNS) function and in turn the CNS influences the microbiota composition (Mayer 2011). Disturbances of this system have been implicated in a wide range of disorders, including functional and inflammatory gastrointestinal disorders and obesity and eating disorders, making the individual more susceptible to disease (Bartrop et al. 1977; Mason 1991; Maunder 2005; Cohen et al. 2012). The damage induced by stress varies depending on the time at which the body is exposed to the stressor agent (Dhabhar 2003).

Several studies have demonstrated that stress enhances the permeability of the intestinal epithelium, with activation of mast cells, overproduction of INFγ, and changes in the morphology of the colonic epithelium via reduced expression of tight junction protein 2 (zone occludes 2) and by occlusion of important components of the intestinal tight junction (Demaude et al. 2006). These changes in the gut microbiota are associated with decreased immunological parameters. In this sense, it is known that hormones released during stress situations (glucocorticoids) alter immune functions such as antigen presentation, leukocyte trafficking and proliferation, antibody secretion, and cytokine release (Dhabhar et al. 1995).

Several reports indicate that probiotic bacteria have a wide range of beneficial effects in healthy individuals, for example they can modulate the immune system by increasing the mucosal and systemic immune responses (Maldonado Galdeano and Perdigón 2004), reinforcing the epithelial barrier by reducing its permeability and enhancing the local immune response, and systemic immunity can be affected as well (Maldonado Galdeano and Perdigon 2006; Maldonado Galdeano et al. 2007, 2011). Probiotic microorganisms can improve the immune response against infection (Maragkoudakis et al. 2010; Castillo et al. 2011), modulate the inflammatory response, and influence the composition and activity of intestinal microbiota (de Moreno de LeBlanc et al. 2008; Chaves et al. 2011). Moreover, there are some clinical reports that indicate the role of probiotics in reducing anxiety, decreasing stress response, and improving the mood in individuals with chronic fatigue (Logan and Katzman 2005). Considering these previous results, the administration of the probiotic bacterium *Lact. casei* CRL 431 as an adjuvant on the intestinal immune system during a chronic stress process was studied (Palomar et al. 2014). In an experimental bifactorial stress protocol of immobilized (cylindrical plexiglass container) mice model the effectiveness of the probiotic strain administration to reverse the immunosuppressed state induced by stress was shown (Palomar et al. 2014).

Stressed mice that received a suspension of *Lact. casei* CRL 431 in drinking water throughout the experimental period exhibited diminished levels of serum corticosterone, which increased significantly in the stressed group compared to control mice (Palomar et al. 2014). A decrease in the length of the villi and a reduction on the crypt depth in the intestines were observed in stressed mice. Probiotic administration recovered the normal architecture of the small intestine from stressed mice, this effect being important to normalize nutrient absorption associated to villi length (Xu et al. 2003). The reduction in crypt depth, induced by stress, is due to an increase in the apoptosis of epithelial cells. Other studies demonstrated that glucocorticoids potently induced apoptosis in spleenocytes (Yin et al. 2000). In previous work, it was demonstrated that *Lact. casei* CRL 431 increased the expression of the Bcl2 protein that establishes a balance between mitosis and cellular apoptosis; this being a possible explanation of the positive effect of the probiotic administration (Perdigón et al. 2002).

Stress influences gut mucosal immune response producing a diminution in the number of IgA+ B-lymphocytes in the lamina propria of the small intestine and in the secretion of S-IgA into the intestinal fluid of stressed animals. The probiotic strain *Lact. casei* CRL 431 increased these parameters in agreement with another study in a mouse stress model (Jarillo-Luna et al. 2007). The decrease in IgA+ cell numbers in stressed mice could be due to high levels of glucocorticoids that induce apoptosis of B cells in the lamina propria of the gut (Fukuzuka et al. 2000; Brunner et al. 2001; Ruiz-Santana et al. 2001). Another possibility could be the low number of CD4+ T cells. These last cells produce cytokines such as IFNγ, IL-6, IL-5, IL-4, and IL-10, which are involved in the expansion and differentiation of IgA+ B cells (Iwakabe et al. 1998; Elenkov 2004; Brandtzaeg and Johansen 2005). These cell populations increased in stressed animals consuming probiotics.

In previous studies the level of IFN-γ, an important cytokine in the synthesis of secretory IgA by intestinal epithelial cells, was analyzed in healthy mice and it was demonstrated that *Lact. casei* CRL431 increased this cytokine production in the small intestine of the animals (Maldonado Galdeano et al. 2010). In contrast, the significant increase of IFN-γ observed in mice from the stressed group was reduced by the consumption of this probiotic. In the stressed model, probiotic administration restored the production of this cytokine to the levels of control mice, restoring the physiological status and regulating the IgA secretion.

Macrophages and DC are very important in innate and adaptive immunity, and CD11b and CD11c are used respectively as markers to identify them. In the gut, these cells are involved in the bacterial clearance to avoid invasion of translocating bacteria from the intestinal lumen to deep tissues and to maintain intestinal homeostasis. In previous work, we demonstrated that the main mechanism by which the probiotic strain *Lact. casei* CRL 431 influences the gut mucosal immunity is through activation of the innate immune response (Maldonado Galdeano and Perdigón 2006). For this reason we analyzed whether the stress induced changes in the expression of both markers in the immune cells present in the lamina propria of the small intestine of mice. The number of CD11b+ cells

Table 8.2. Secretory IgA in the intestinal fluid and immune cells in the small intestine of stressed mice.

Groups	CD4+	CD8+	CD11b+	CD11c+	IgA+	IgA pg/ml
NC	27±4[a]	33±5[a]	14±1[a]	7±2[a]	100±17[a]	40±7[a]
NC+P	21±4[ab]	36±5[a]	ND	ND	195±22[b]	78±7[b]
SC	15±5[b]	21±6[b]	6±1[b]	22±5[b]	43±12[c]	9.73±1.4[c]
SC+P	32±11[a]	24±7[ab]	20±1[c]	9±4[a]	76±10[a]	20.4±1.6[d]

The number of IgA+, CD11b+, CD11c+, CD4+ and CD8+ T cells were determined by direct immunofluorescence on the small intestine tissue slides of mice from different experimental groups: normal control (NC), normal control with *Lact. casei* CRL 431 (NC+P), stressed control (SC) and stressed control with *Lact. casei* CRL 431.

Results are expressed as the number of positive cells per 10 fields of vision (1000X). Values are means for $n = 9$ mice ± SD. Secretory IgA was measured by ELISA in the intestinal fluid of the small intestine of mice from different experimental groups. Data correspond to the mean ± SD of results of $n = 9$ animals from three separate experiments

[a,b,c,d] Means for each cell population without a common letter differ significantly ($P < 0.05$). Values are those obtained from Palomar et al. (2014).

decreased significantly while an increase in the CD11c+ cells was observed in stressed mice as compared to control animals. Administration of the probiotic bacterium normalized the number of CD11b+ cells in stressed animals (Table 8.2) (Palomar et al. 2014).

The intestinal microbiota, as was pointed out above, plays an important role in maintaining intestinal homeostasis and is affected by different stressor agents such as malnutrition (Zak-Gołąb et al. 2014) or pathologies such as inflammatory bowel disease (Hold et al. 2014). Under stress conditions, different populations, such as strict anaerobes, lactobacilli, and enterobacteria, can be affected (Wang and Kasper 2013). In this regard, studies performed in mice exposed to a chronic restraint stress showed that the composition of the intestinal microbiota differs from the non-stressed control mice principally with a decrease in *Bacteroides* spp. and an increase in *Clostridium* spp. (Bailey et al. 2011). In our studies, no changes in the composition of anaerobes, lactobacilli, and enterobacteria populations were observed in mice exposed to stress stimuli. However, the lactobacilli population significantly increased when the probiotic bacterium *Lact. casei* CRL 431 was administered to mice. These results agree with others obtained in healthy mice that received a probiotic fermented milk where probiotic administration was associated with decreases in enterobacteria and increases in lactobacilli (de Moreno de LeBlanc et al. 2008).

The balance and composition of microbiota is extremely important to modulate the immune system and to influence the development and physiology of the host (Brestoff and Artis 2013). During stress, permeability and the intestinal barrier function are altered. Probiotics are able to reinforce the epithelial barrier, as was demonstrated by different studies (Zareie et al. 2006; Ait-Belgnaoui et al. 2012). In addition, we demonstrated that the probiotic *Lact. casei* CRL 431 improves immune parameters related to the intestinal barrier, such as the S-IgA and the number of macrophages and DC in the lamina propria of the small intestine (Palomar et al. 2014).

8.4. Conclusions

The role of microbiota composition has been shown to be critical for normal human physiology, modulating the immune, endocrine, and neural systems, which altogether modulate behavioral response. Strategies to change the balance of the intestinal microbiota are of great interest and are being sought to influence the complicated network of signals among the gut immune cells, microbiota, and brain. The well-known beneficial effect of probiotics on the intestinal microbiota represents a potential strategy to improve the gut immune system and the intestinal barrier function in immunosupressed hosts as a consequence of stress and malnutrition.

Acknowledgments

This work had been financially supported by grants from (Consejo Nacional de Investigaciones Científicas y Técnicas) CIUNT 26/D 442 (Universidad Nacional de Tucumán), and CONICET PIP 0652 and PIP 1071, Argentina and DANONE-Argentina-CONICET

References

Ait-Belgnaoui, A., Bradesi, S., Fioramonti, J., et al. (2005) Acute stress-induced hypersensitivity to colonic. *Pain* **113**, 141–147.

Ait- Belgnaoui, A., Durand, H., Cartier, C., et al. (2012) Prevention of the gut leakiness by a probiotic treatment leads to attenuated HPA response to an acute psychological stress in rats. *Psychoneuroendocrinology* **37**, 1885–1895.

Ait-Belgnaoui, A., Han, W., Lamine, F., et al. (2006) Lactobacillus farciminis treatment suppresses stress induced visceral hypersensitivity: a possible action through interaction with epithelial cell cytoskeleton contraction. *Gut* **55**, 1090–1094.

Arora, T., Singh, S., and Sharma, R.K. (2013) Probiotics: Interaction with gut microbiome and antiobesity potential. *Nutrition* **29**, 591–596.

Bailey, M.T. Dowd, S.E., Galley, J.D., et al. (2011) Exposure to social stressor alters the structure of the intestinal microbiota: implications for stressor-induced immunomodulation. *Brain Behav Immun* **25**, 397–407.

Barbieri, N., Villena, J., Herrera, M., et al. (2013) Nasally administered *Lactobacillus rhamnosus* accelerate the recovery of humoral immunity in B lymphocyte-deficient malnourished mice. *J Nutr* **143**, 227–235.

Bartrop, R.W., Luckhurst, E., Lazarus, L., et al. (1977) Depressed lymphocyte function after bereavement. *Lancet* **1**, 834–836.

Brandtzaeg, P. and Johansen, F.E. (2005) Mucosal B cells: phenotypic characteristics, transcriptional regulation, and homing properties. *Immunol Rev* **206**, 32–36.

Brestoff, J.R. and Artis, D. (2013) Commensal bacteria at the interface of host metabolism and the immune system. *Nat Immunol* **14**, 676–684.

Brunner, T., Arnold, D., Wasem, C., et al. (2001) Regulation of cell death and survival in intestinal intraepithelial lymphocytes. *Cell Death Differ* **8**, 706–714.

Castillo, N.A., Perdigón, G., and de Moreno de Leblanc, A. (2011) Oral administration of a probiotic Lactobacillus modulates cytokine production and TLR expression improving the immune response against *Salmonella enterica* serovar Typhimurium infection in mice. *BMC Microbiol* **11**, 177.

Chaves, S., Perdigón, G., and de Moreno de LeBlanc, A. (2011) Yoghurt consumption regulates the immune cells implicated in acute intestinal inflammation and prevents the recurrence of the inflammatory process in a mouse model. *J Food Prot* **74**, 801–811.

Cheung, K., Lee, S.S., and Raman, M. (2012) Prevalence and mechanisms of malnutrition in patients with advanced liver disease, and nutrition management strategies. *Clinical Gastroenterol Hepatol* **10**, 117–125.

Cohen, S., Janicki-Deverts, D., Doyle, W.J., et al. (2012) Chronic stress, glucocorticoid receptor resistance, inflammation, and disease risk. *Proc Natl Acad Sci USA* **109**, 5995–5999.

Cruz, F.C., Marin, M.T., Leãoa, R.M., and Planeta, C.S. (2012) Behavioral and neuroendocrine effects of the exposure to chronic restraint or variable stress in early adolescent rats. *Int J Dev Neurosci* **30**, 19–23.

Cryan, J.F. and O´Mahony, S. M. (2011) The microbiome-gut-brain axis: from bowel to behavior. *Neurogastroenterology* **136**, 2003–2014.

de Jonge, W.J. (2013) The gut's little brain in control of intestinal immunity. *ISRN Gastroenterol* 630159. doi: 10.1155/2013/630159.

Demaude, J., Salvador-Cartier, C., Fioramonti, J., et al. (2006) Phenotypic changes in colonocytes following acute stress or activation of mast cells in mice: implications for delayed epithelial barrier dysfunction. *Gut* **55**, 655–661.

de Moreno de LeBlanc, A. Dogi C.A., Galdeano C.M., et al. (2008) Effect of the administration of a fermented milk containing *Lactobacillus casei* DN-114001 on intestinal microbiota and gut associated immune cells of nursing mice and after weaning until immune maturity. *BMC Immunol* **9**, 27–38.

de Moreno de LeBlanc, A., Maldonado Galdeano, C., Chaves, S., and Perdigón, G. (2005) Oral administration of *Lactobacillus casei* CRL 431 increases immunity in bronchus and mammary glands. *Eur J Inflamm* **3**, 23–28.

Dewan, P., Kaur, I., Chattopadhya, D.A., et al. (2007) A pilot study on the effects of curd (dahi) and leaf protein concentrate in children with protein energy malnutrition (PEM). *Indian J Med Res* **126**, 199–203.

Dhabhar, F.S. (2003) Stress, leukocyte trafficking, and the augmentation of skin immune function. *Ann N Y Acad Sci* **992**, 205–217.

Dhabhar, F.S., Miller, A.H., McEwen, B.S., and Spencer, R.L. (1995) Effects of stress on immune cell distribution. Dynamics and hormonal mechanisms. *J Immunol* **154**, 5511–5527.

Elenkov, I.J. (2004) Glucocorticoids and the Th1/Th2 balance. *Ann N Y Acad Sci* **1024**, 138–146.

FAO/WHO. (2001) Evaluation of health and nutritional properties of powder milk and live lactic acid bacteria. Food and Agriculture Organization of the United Nations and World Health Organization Expert Consultation Report. Available at: ftp://ftp.fao.org/docrep/fao/009/a0512e/a0512e00.pdf.

Finamore, A., Roselli, M., Britti, M.S., et al. (2012). *Lactobacillus rhamnosus* GG and *Bifidobacterium animalis* MB5 induce intestinal but not systemic antigen-specific hyporesponsiveness in ovalbumin-immunized rats. *J Nutr* **142**, 375–381.

Flint, H.J., Scott, K.P., Louis, P., and Duncan, S.H. (2012) The role of the gut microbiota in nutrition and health. *Nat Rev Gastroenterol Hepatol* **9**, 577–589.

Fukuzuka, K., Edwards C.K., and Clare-Salzer, M., et al. (2000) Glucocorticoid and Fas ligand induced mucosal lymphocyte apoptosis after burn injury. *J Trauma* **49**, 710–716.

Gauffin Cano, P., Aguero, G., and Perdigon, G. (2002) Adjuvant effects of *Lactobacillus casei* added to a renutrition diet in a malnourished mouse model. *Biocell* **26**, 35–48.

Hill, D.A. and Artis, D. (2010) Intestinal bacteria and the regulation of immune cell homeostasis. *Annu Rev Immunol* **28**, 623–667.

Hold, G. L., Smith, M., Grange, C., et al. (2014) Role of the gut microbiota in inflammatory bowel disease pathogenesis: What have we learnt in the past 10 years? *World J Gastroenterol* **7**, 1192–1210.

Husebye, E., Hellstrom, P.M., Sundler, F., et al. (2001) Influence of microbial species on small intestinal myoelectric activity and transit in germ-free rats. *Am. J Physiol Gastrointest Liver Physiol* **280**, G368–G380.

Iwakabe, K., Shimada, M., Ohta, A., et al. (1998) The restraint stress drives a shift in Th1/Th2 balance toward Th2-dominant immunity in mice. *Immunol Let* **62**, 39–43.

Jarillo-Luna, A., Rivera-Aguilar, V., Garfias, H.R., et al. (2007) Effect of repeated restraint stress on the levels of intestinal IgA in mice. *Psychoneuroendocrinology* **32**, 681–692.

Johansson, M.E., Larsson, J.M., and Hansson, G.C. (2011) The two mucus layers of colon are organized by the MUC2 mucin, whereas the outer layer is a legislator of host–microbial interactions. *Proc Nat Acad Sci USA* **108**, 4659–4665.

Lima, D.P., Azevedo, J.F., Hermes-Uliana, C., et al. (2012) Probiotics prevent growth deficit of colon wall strata of malnourished rats post-lactation. *Ann Acad Bras Cienc* **84**, 727–736.

Logan, A.C. and Katzman, M. (2005) Mayor depressive disorder: probiotics may be an adyuvant therapy. *Med Hypotheses* **64**, 533–538.

Lynch, H.E., Goldberg, G.L., Chidgey, A., et al. (2009) Thymic involution and immune reconstitution. *Trends Immunol* **30**, 366–373.

Maldonado Galdeano, C.M. and Perdigón, G. (2004) Role of viability of probiotic strains in their persistence in the gut and in mucosal immune stimulation. *J Appl Microbiol* **97**, 673–681.

Maldonado Galdeano, C., and Perdigón, G. (2006) The probiotic bacterium *Lactobacillus casei* induces activation of the gut mucosal immune system through the innate immunity. *Clin Vacc Immunol* **13**, 219–226.

Maldonado Galdeano, C., de Moreno de Leblanc, A., Dogi, C.A., and Perdigón, G. (2010) Lactic acid bacteria as immunomodulators of the gut associated immune system. In: *Biotechnology of Lactic Acid Bacteria: Novel Applications* ((ed.) F. Mozzi, R. R. Raya, G. Vignolo) Wiley- Blackwell, USA, pp. 125–140.

Maldonado Galdeano, C., de Moreno de LeBlanc, A, Vinderola, G., et al. (2007) Proposed model: mechanisms of immunomodulation induced by probiotic bacteria. *Clin Vac Immunol* **14**, 485–492

Maldonado Galdeano, C. Dogi, C.A., Bibas Bonet, M.E., et al. (2012) Probiotic bacteria as mucosal immune system adjuvants. In: *Bioactive Food as Dietary Interventions for Liver and Gastrointestinal Disease*, ((eds) V. Preedy and R. R. Watson), Elsevier, Oxford, UK, pp. 61–84.

Maldonado Galdeano, C., Novotny Núñez, I., de Moreno de LeBlanc, A., et al. (2011) Impact of a probiotic fermented milk in the gut ecosystem and in the systemic immunity using a non-sever protein-energy-malnutrition model in mice. *BMC Gastroenterol* **11**, 64.

Maragkoudakis, P.A., Chingwaru, W., Gradisnik, L., et al. (2010) Lactic acid bacteria efficiently protect human and animal intestinal epithelial and immune cells from enteric virus infection. *Int J Food Microbiol* **31**, S91–97.

Mason, D. (1991) Genetic variation in the stress response: susceptibility to experimental allergic encephalomyelitis and implications for human inflammatory disease. *Immunol Today* **12**, 57–60.

Maunder, R.G. (2005) Evidence that stress contributes to inflammatory bowel disease: evaluation, synthesis, and future directions. *Inflamm Bowel Dis* **11**, 600–608.

Mayer, E.A. (2011) Gut feelings: the emerging biology of gut-brain communication. *Nature Rev Neurosci* **12**, 453–466.

Nuñez, I.N., Maldonado Galdeano, C.M., Carmuega, E., et al. (2012) Effect of a probiotic fermented milk on the thymus in BALB/c mice under non-severe protein-energy malnutrition. *British J Nutr* **110**, 500–508.

Oberholzer, C., Oberholzer, A., Bahjat, F.R., et al. (2001) Targeted adenovirus-induced expression of IL-10 decreases thymic apoptosis and improves survival in murine sepsis. *Proc Natl Acad Sci USA* **98**, 11503–11508.

Ohashi, Y. and Ushida, K. (2009). Health-beneficial effects of probiotics: Its mode of action. *Anim Sci J* **80**, 361–371.

Ohland, C.L. and Macnaughton, W.K. (2010) Probiotic bacteria and intestinal epithelial barrier function. *Am J Physiol Gastrointest Liver Physiol* **298**, G807–819.

Ortiz, R., Cortés, L., Cortés, E., and Medina, H. (2009) Malnutrition alters the rates of apoptosis in splenocytes and thymocyte sub-populations of rats. *Clin Exp Immunol* **155**, 96–106.

Palomar, M.M., Maldonado Galdeano, C., and Perdigón, G. (2014) Influence of a lactobacillus probiotic strain on the intestinal ecosystem in a stress model in mouse. *Brain Behav Immun* **35**, 77–85.

Pearce, E.L. and Pearce, E.J. (2013) Metabolic pathways in immune cell activation and quiescence. *Immunity* **38**, 633–643.

Perdigón, G., Maldonado Galdeano, C., Valdez, J.C., and Medici, M. (2002) Interaction of lactic acid bacteria with the gut Immune system. *Eur J Clin Nutr* **56**, S21–S26.

Ramsey, J.M. (1982) Modern stress and the disease process. In: *Basic Physiology*. Addison-Wesley, California, pp. 177–179.

Rescigno, M., Urbano, M., Valzasina, B., et al. (2001) Dendritic cells express tight junction proteins and penetrate gut epithelial monolayers to sample bacteria. *Nat Immunol* **2**, 361–367.

Rizza, P., Ferrantini, M., Capone, I., and Belardelli, F. (2002) Cytokines as natural adjuvants for vaccines: where are we now? *Trends Immunol* **23**, 381–383.

Ruiz-Santana, S., Lopez, A., Torres, S., et al. (2001) Prevention of dexamethasone-induced lymphocytic apoptosis in the intestine and in Peyer patches by enteral nutrition. *J Parenter Enteral Nutr* **25**, 338–345.

Salva, S., Merino, M.C., Agüero, G., et al. (2012) Dietary supplementation with probiotics improves hematopoiesis in malnourished mice. *PLoS One* **7**, e31171.

Savino, W. (2002) The thymus gland is a target in malnutrition. *Eur J Clin Nutr* **56**, S46–S49.

Savino, W. and Dardenne, M. (2010) Nutritional imbalances and infections affect the thymus: consequences on T-cell-mediated immune responses. *Proc Nutr Soc* **69**, 636–643.

Shimosato, T., Tomida, K., and Otani, H. (2011) Effect of *Lactobacillus pentosus* ONRIC b0240 on intestinal IgA production in mice fed differing levels of protein. *J Agric Food Chem* **23**, 59 (6), 2646–2651.

Taylor, A.K., Cao, W., Vora, K.P., et al. (2013) Protein energy malnutrition decreases immunity and increases susceptibility to influenza infection in mice. *J Infect Dis* **207**, 501–510.

Tremaroli, V. and Bächked, F. (2012) Functional interactions between the gut microbiota and host metabolism. *Nature* **489**, 242–249.

Verdu, E.F., Bercik, P., Bergonzelli, G.E., et al. (2004) *Lactobacillus paracasei* normalizes muscle hypercontractility in a murine model of postinfective gut dysfunction. *Gastroenterology* **127**, 826–837.

Villena, J., Racedo, S., Agüero, G., and Alvarez, S. (2006) Yoghurt accelerates the recovery of defence mechanisms against *Streptococcus pneumoniae* in protein-malnourished mice. *Br J Nutr* **95**, 591–602.

Villena, J., Salva, S., Agüero, G., and Alvarez, S. (2011) Immunomodulatory and protective effect of probiotic *Lactobacillus casei* against *Candida albicans* infection in malnourished mice. *Microbiol Immunol* **55**, 434–45.

Wang, Y. and Kasper, L. H. (2013) The role of microbiome in central nervous system disorders. *Brain Behav Immun* **38**, 1–12. doi: 10.1016/j.bbi.2013.12.015.

Xu, Z.R., Hu, C.H., Xia, M.S., et al. (2003) Effects of dietary fructooligosaccharide on digestive enzyme activities, intestinal microflora and morphology of male broilers. *Poult Sci* **82**, 1030–1036.

Yin, D., Tuthill, D., Mufson, R.A., and Shi, Y. (2000) Chronic restraint stress promotes lymphocyte apoptosis by modulating CD95 expression. *J Exp Med* **17**, 1423–1428.

Zak-Gołąb, A., Olszanecka-Glinianowicz, M., Kocełak, P., and Chudek, J. (2014) The role of gut microbiota in the pathogenesis of obesity. *Postepy Hig Med Dosw* **68**, 84–90.

Zareie, M., Johnson-Henry, K., Jury, J., et al. (2006) Probiotics prevent bacterial traslocation and improve intestinal barrier function in rats following chronic psychological stress. *Gut* **55**, 1553–1560.

Chapter 9

Lactic Acid Bacteria in Animal Production and Health

Damien Bouchard, Sergine Even, and Yves Le Loir*

INRA, AGROCAMPUS OUEST UMR1253 Science et Technologie du Lait et de l'Œuf, France

9.1. Introduction

Probiotics are used in several animal production sectors, which include livestock productions (poultry, pigs, and ruminants) for meat, egg, and or milk production, and aquaculture for fish and shellfish production. Of course, each of these sectors presents peculiar characteristics with regards to the animal species (mammals, birds, fish, or molluscs), the environment (sea water or fresh water, farm settings …), the type of production (meat, egg, milk), and the major pathogens and the main diseases encountered in each species. Hence, although some objectives are common to all production sectors (e.g., enhancement of production yields, improvement of animal health and welfare), probiotics selected and used may vary for each sector. Therefore bacteria and yeasts can be found in probiotic formulae. Depending on the animal species, and the type of production (e.g., meat or egg for poultry) and period of life (early life, weaning, adulthood in mammals), specific microorganisms are chosen, for example yeasts in adult ruminants and lactobacilli in pigs and poultry.

Probiotics were initially used as feed additives to optimize the zootechnical performances of farming activities. The interest in probiotic use in animal farming recently increased to face new challenges such as the ban of antibiotic growth promoters (AGPs) (since 2006 in the European Union). Moreover, several outbreaks of food-borne diseases have urged the farming sectors to reduce bacterial pathogens in products of animal origin (Gaggia et al. 2010). Enteric and most often asymptomatic carriage of pathogens is indeed a direct source of carcass- and food-contamination during the process.

These new applications of probiotics are in line with a huge societal demand on the reduction of inputs (including phytosanitary and antibiotic treatments) for a more sustainable agriculture. This tendency has induced research efforts aimed at reducing the use of antibiotics while maintaining good sanitary standards and good production yields in livestock. The approaches adopted include the implementation of improved farming practices in terms of zootechnical and hygienic measures to circumvent infectious diseases and to improve animal welfare. Besides, the development and the use

*Corresponding author: INRA, AGROCAMPUS OUEST UMR1253 Science et Technologie du Lait et de l'Œuf, 65 rue de Saint Brieuc, F-35042 Rennes, France. yves.leloir@rennes.inra.fr

Biotechnology of Lactic Acid Bacteria: Novel Applications, Second Edition. Edited by Fernanda Mozzi, Raúl R. Raya, and Graciela M. Vignolo.

of vaccines and the improvement of animal genetic selection have also occurred in several sectors. In addition to these strategies, the concept of biological control of infectious diseases has been investigated. This is based on the use of bacterial strains as a biological barrier to prevent or even to treat infections. The use of lactic acid bacteria (LAB) in agricultural livestock gave encouraging results in various contexts and was shown to significantly improve overall animal health and production yields. Beneficial effects of probiotic LAB are reportedly due to their capacity to inhibit or eliminate pathogens through the production of organic acids, bacteriocins, and other antimicrobial compounds (Reid 1999; Espeche et al. 2012). LAB were also shown to inhibit colonization and to modulate and impair virulence expression in some pathogens. Last but not least, LAB can also exert their beneficial effect on host health through an immune stimulation of host defense allowing a clearance of the pathogens or the prevention of infections. Most of the scientific literature on the health benefits of LAB is focused on intestinal (digestive tract) and vaginal (urogenital tract) ecosystems. However, thanks to the development of molecular biology tools and new sequencing technologies, other ecosystems are now being described and investigated to understand their role and impact on organ health status (e.g., mammary ecosystem, rumen ecosystem). LAB could also play a beneficial role therein.

The impact of these ecosystems and of LAB on host health and the mechanisms that govern their relationship and interaction with the animal hosts are still poorly documented and understood. The molecular bases of the probiotic effects are difficult to investigate *in vivo*, but can be approached through animal models, cell cultures, and *in vitro* assays. Such studies improve our knowledge in the field and help in setting better selection criteria for new probiotic candidates. In this chapter, we will review recent data upon the potential of LAB to improve animal productions and animal health and well-being.

9.2. Lactic Acid Bacteria and Probiotics

LAB are a heterogeneous group of bacteria, which include *Lactobacillus*, *Lactococcus*, *Enterococcus*, *Streptococcus*, *Leuconostoc*, and *Oenococcus* genera as the better known members. Their common characteristic is their ability to produce lactic acid as the major end-product of fermentation. They have been used for centuries for their technological and functional properties in food fermentations. Most LAB species have a "Generally Recognized as Safe" status (delivered by the FDA, in the US) and a "Qualified Presumption of Safety" status (in Europe).

Their use in animal productions has gained interest in parallel to their use as probiotics in human consumption. LAB are indeed used as probiotics to improve animal health and to prevent infections. The most important expected benefits of probiotic LAB are growth-promoting effects (better growth rate and enhanced body weight gain, BWG) through better feed utilization and digestion, and biological control of pathogen colonization. They are also used to enhance milk and egg production.

The definition of probiotics has evolved. Probiotics were firstly defined as "A live microbial feed supplement which beneficially affects the host animal by improving its intestinal microbial balance" (Füller 1989). The new definition by the World Health Organization (FAO/WHO 2001), "live micro-organisms which, when administered in adequate amounts, confer a health benefit on the host," opens new fields of application of probiotics beyond feed supplementation. The use of LAB as vaginal- and mammary- probiotics is indeed now being investigated.

Beside LAB, many probiotic microorganisms are investigated and or used in animal feed (Table 9.1). Commercial preparations do not always use valid taxonomic designations although a clear indication is mandatory on the label of the product. Among the non-LAB species, *Bacillus* and *Saccharomyces* are the most used probiotics in livestock and poultry.

Table 9.1. Types of microorganisms used as probiotics in livestock and animal productions.

Genus	Species
LAB	
Lactobacillus	*Lact. acidophilus*, *Lact. amylovorus*, *Lact. brevis*, *Lact. casei* subsp. *casei* (*Lact. casei*), *Lact. crispatus*, *Lact. farmicinis*, *Lact. fermentum*, *Lact. murinus*, *Lact. plantarum* subsp. *plantarum* (*Lact. plantarum*), *Lact. reuteri*, *Lact. rhamnosus*, *Lact. salivarius*, *Lact. Amylovorus*/*Lact. sobrius*
Lactococcus	*L. lactis* subsp. *cremoris* (*Streptococcus cremoris*), *L. lactis* subsp. *lactis*
Leuconostoc	*L. citreum*, *L. lactis*, *L. mesenteroides*
Pediococcus	*P. acidilactici*, *P. pentosaceus* subsp. *pentosaceous*
Streptococcus	*Strep. infantarius*, *Strep. salivarius* subsp. *salivarius*, *Strep. thermophilus* (*Strep. salivarius* subsp. *thermophilus*)
Enterococcus	*Ent. faecalis* (*Streptococcus faecalis*), *Ent. faecium* (*Streptococcus faecium*)
High G+C content Gram positive bacteria	
Bifidobacterium	*B animalis* subsp. *animalis* (*B. animalis*), *B. lactis* subsp. *lactis* (*B. lactis*), *B. longum* subsp. *longum* (*B. longum*), *B. pseudolongum* subsp. *pseudolongum* (*B. pseudolongum*), *B. thermophilum*
Propionibacterium	*P. freudenreichii*
Sporulating Gram positive bacteria	
Bacillus	*B. cereus* (*B. cereus* var. *toyoi*), *B. licheniformis*, *B. subtilis*
Gram negative bacteria	
Prevotella	*P. bryantii*
Yeasts	
Saccharomyces	*S. cerevisiae* (*S. boulardii*), *S. pastorianus* (*S. carlsbergensis*)
Kluyveromyces	*K. fragilis*, *K. marxianus*
Aspergillus	*A. orizae*, *A. niger*
Candida	*C. pintolopessi*

9.3. Classifications and Regulatory Criteria of Probiotics in Animal Health

Probiotics of the first generation were identified as such mostly on the basis of their history of use and their observed (but rarely demonstrated) beneficial effects. Probiotics of second generations are submitted due to solid scientific criteria and subjected to strong regulations, which require a demonstration of the harmlessness of the proposed microorganisms as well as a proof of their efficiency prior to acceptance of any health allegation. According to their functions and properties, additives for animal feed are classified in one or several categories including technological, sensorial, nutritional, or zootechnical additives. Probiotics that target host health are most often used in animal feedstuff. Most of them address health benefits for the gastrointestinal tract and are affiliated to the category "zootechnical additives." In Europe, the authorization of microorganisms used as zootechnical additives for feed is subjected to strict regulation (1831/2003/EC; EFSA, 2003). Zootechnical additives are defined as "any additive used to affect favorably the performance of animals in good health or used to affect favorably the environment." Applications for approval for a strain or a mix of several strains must include (i) the molecular characterization and identification at species and strain levels and (ii) the demonstration of zootechnical performance (for at least one of the characteristics given in the definition). Of note, authorization is restricted to a given target animal species and for a category of age within this species (EFSA 2012). Besides these uses as feed additives, new applications are now explored in animal health, to prevent or to cure uro-genital and mammary infections.

9.4. Probiotic LAB and Animal Production Sectors

In the following paragraphs, we will address each of the main animal production sectors, their specificities in terms of the production types (meat, milk, egg), the ecosystems to be considered (gut, vaginal, and/or mammary) and presence of LAB therein, the main diseases encountered, and the use of probiotics to prevent these diseases or enhance productivity.

9.4.1. Probiotics in Ruminants

Ruminants rearing, that is cattle, buffaloes, goats, and sheep, represent a huge animal production sector for meat and milk production. The main targets and types of probiotics in ruminants depend on the type of productions or on the stage of development of the animals. In young ruminants, these targets include an optimal maturation of the rumen microbiota, a reduced risk of pathogen colonization, and an increased digestive safety at weaning. In beef cattle, the use of probiotics aims at promoting BWG, increasing feed efficiency, reducing acidosis, and limiting healthy carriage of pathogens and thus shedding of human pathogens. In dairy cattle, probiotics are used to increase milk yield and quality and feed efficiency, and to promote health by limiting acidosis (Caucheyras-Durand and Durand 2010).

Ruminant Microbiota Rumen microbes are essential for ruminant productions. Rumen microbiota is thus now intensively investigated through New Generation Sequencing (NGS) technologies in cattle, goats, and sheep (Lee et al. 2012; Omoniyi et al. 2014). Initial results show that the rumen microbiota is far more complex and diverse than the human cecum (Morgavi et al. 2013). The dynamics of rumen microbiota was recently investigated using 454 sequencing of 16S RNA and subsequent analysis of operational taxonomy units (OTU) at five periods of time between 1 day and 2 years, within the first two years of life in cattle. Rumen microbiota showed huge changes from early life to adulthood, with an abundance of Firmicutes (streptococci) during the three first days of life and the replacement of the aerobic and facultative anaerobic genera by anaerobic ones after only three days of life (Jami et al. 2013). Besides the abundance of streptococci, the presence of LAB was rarely reported. Of note, in most studies the methods used were not discriminative enough (identification at the phylum or, at the best, genus level) to precisely identify LAB species. Similarly, the presence of LAB was not reported in the rumen microbiota of pre-ruminant calves (Li et al. 2012).

Since the rumen is essential for feed digestion in ruminants, a possible link between its composition and physiological parameters is now investigated. A link between microbial composition of rumen and milk composition was thus recently shown, with a clear correlation between the milk-fat yield and the ratio of the phyla Firmicutes to Bacteroidetes (Jami et al. 2014). Besides, rumen composition is also an important parameter (but not the only one) in the production of greenhouse gas methane in ruminants (Morgavi et al. 2010). A better knowledge of this complex microbiota will open the way for strategies (which include the use of probiotics) promoting more robust, efficient, and less-polluting ruminants.

Two other ecosystems, namely the genital tract- and the mammary-ecosystems, are now being explored in ruminants. Their host organ (uterus and vagina, and udder) can indeed be sites of infections that are at the origin of serious peripartum diseases like metritis, endometritis, and mastitis. The endogenous microbiota could serve here as a barrier against pathogens and might also play a role in the immunomodulation of the organ (Nader-Macias et al. 2008).

The vaginal microbiota of cows comprises several LAB species including *Lactobacillus* spp., *Pediococcus* spp., *Leuconostoc* spp., and *Weissella* spp., some of which were isolated and tested for their probiotic capacities against urogenital infections (Otero et al. 2000; Wang et al. 2013a).

Pioneering works isolated several *Streptococcus* sp. and *Lactobacillus* sp. strains from the vagina of healthy cows (Otero et al. 2006) and reported promising results using *Lactobacillus gasseri* isolates, which showed inhibitory activity against *Staphylococcus aureus in vitro* (Otero and Nader-Macias 2006).

Uterine microbiota of cows was also recently investigated and showed a huge diversity in healthy cows compared to metritic and endometritic ones. Regardless of the health status, the uterine microbiota is mainly composed of members of the phyla Bacteroidetes, Fusobacteria, Firmicutes, Proteobacteria, and Tenericutes (Santos and Bicalho 2012).

The mammary microbiota is still poorly documented in ruminants, although there is increasing interest within the scientific community. Recent works showed a high diversity in the mammary microbiota isolated from the teat apex, using culture independent techniques such as denaturing gradient gel electrophoresis (DGGE on PCR-amplified fragments of 16S RNA and other genes) (Braem et al. 2012, 2013). Such approaches allowed the identification of *Lactococcus* sp., *Bifidobacterium* sp., and *Propionibacterium* sp. in the teat apex samples of lactating cows. Noteworthy, *Lactobacillus* sp. was not reported (Braem et al. 2012). Pyrosequencing of the 16S rRNA was recently used to describe the microbial diversity of bovine milk samples derived from unaffected quarters of cows, and revealed the presence of lactobacilli (*Lact. acidophilus*, *Lact. reuteri*, and *Lact. johnsonii*) at a low prevalence (Oikonomou et al. 2014).

Probiotics and Zootechnical Parameters in Ruminants In cattle, the most common microbial species used as probiotics are yeasts. Since lactic acid overproduction in the gut can lead to dramatic disorders, LAB are, by definition, poorly used as feed additives in this sector. In dairy herds and flocks, probiotic yeasts (e.g., *Saccharomyces cerevisiae*) improve performance with, among other beneficial effects, an increase in dry matter intake and milk production (Jouany 2006). Apart from yeasts, the use of *Bacillus* sp. strains (*B. subtilis* and *B. licheniformis*) was also shown to improve milk yields and composition in sheep (Kritas et al. 2006). In beef cattle, daily feed supplementation with live yeasts was shown to improve growth parameters (average daily gain, final weight, intake, BWG) (Lesmeister et al. 2004). Some of these benefits are correlated to a better balance in ruminal microbiota (Silberberg et al. 2013).

Probiotics and Disease Prevention in Ruminants

Diarrhea Before weaning, dairy calves are susceptible to pathogens such as *Salmonella* spp. and *Escherichia coli*, the most frequent bacterial etiologic agents in calves during the first week of life (Barrington et al. 2002). The resulting diarrhea can severely affect their zootechnical performances. Probiotic LAB have been tested to prevent the establishment of opportunistic bacterial pathogens and to maintain the microbial balance. Although a consensus has not been reached yet on the real efficiency of probiotics in reducing the prevalence of gastrointestinal diseases in calves (Barrington et al. 2002; Ewaschuk et al. 2006), several works report that the use of probiotics as feed additives in calves' diet reduces the incidence of diarrhea. Probiotics may thus be considered as alternatives to antibiotics in calves rearing (Signorini et al. 2012).

Acidosis Subacute ruminal acidosis (SARA) is a digestive disorder characterized by the instability of the microbial ecosystem and by fermentations that lead to lactate, butyrate, and/or propionate rather than acetate (i.e., lactic-, butyric-, and propionic- SARA). It is a major concern in cattle health and welfare. In addition to, or besides the use of, chemical buffers and ionophores, the use of probiotics as feed additives has been investigated as a strategy to balance the ruminal microbiota and fermentation, to stabilize the ruminal pH, and therefore to prevent SARA. Most of the probiotics used in this context contain yeasts (mainly *S. cerevisiae*) as the main active microorganism (Desnoyers et al. 2009). *Prevotella bryantii*, a Gram negative bacterial species found in the rumen microbiota, was also investigated as a probiotic but it was shown ineffective in preventing SARA in dairy cows (Chiquette et al. 2012).

Scientific literature on bacterial probiotics containing LAB and propionic acid bacteria (PAB) reports variable and sometimes contradictory results, which can be attributed to the instability of the microbiota. A preliminary determination of the acidosis type helps in choosing the most relevant probiotic microorganisms and ensures better prevention. It was thus shown that a combination of *Propionibacterium* sp. and *Lact. plantarum* or *Lact. rhamnosus* were inefficient against lactic acidosis whereas it showed promising results in minimizing butyric and propionic SARA (Lettat et al. 2012).

Postpartum Infections: Mastitis and Metritis Mastitis and metritis are inflammatory diseases commonly due to bacterial infections and are the major complications postpartum. Despite great progress in cattle management in terms of hygiene procedures, infectious diseases remain a huge economic problem and a cause of massive use of antibiotics. Mastitis (udder infection) and metritis (uterus infection) are among the most common diseases in dairy cattle, and affect animal welfare as well as productivity parameters and fertility. There are currently no efficient vaccines for these diseases and curative strategies are mainly based on antibiotherapy. This contributes to the risk of antibiotic resistance dissemination, which can further impact both animal and human health, and prompts the need for the development of alternative and sustainable strategies.

Dairy cows have a propensity for bacterial infections of the endometrium, which causes metritis in up to 40% of animals within 10 days of parturition. Chronic endometritis persists more than three weeks postpartum in 20% of cows, and subclinical endometritis affects an additional 15% of the animals. Although several pathogens are involved in metritis, *E. coli* is the first and most common pathogen isolated from the postpartum uterus, preceding infection with other bacteria. It also stimulates a robust innate immune response and inflammation of the endometrium.

Lactobacilli are present in the vaginal microbiota of healthy cows. Some works have explored their capacity to prevent pathogen colonization by mechanisms such as the production of antagonistic substances as lactic acid, H_2O_2, or bacteriocins (Otero et al. 2006; Otero and Nader-Macias 2006; Nader-Macias et al. 2008).

Mastitis affects over 40% of lactating dairy cows. *E. coli*, *Streptococcus uberis*, and *S. aureus* are among the most prevalent pathogens in ruminant mastitis. LAB were tested *in vivo* to assay the efficiency of an infusion of probiotic candidates in the mammary gland of infected cows. A first trial of infusion of *Lactobacillus* in the mammary gland of cows with high Somatic Cell Counts (SCC) resulted in a low cure rate and a rise in SCC (Greene et al. 1991). In another trial, cows with chronic subclinical or clinical mastitis were intrammamary infused with *Lactococcus lactis* DPC3147, which was as effective as an antibiotic treatment, but it did not succeed in curing mastitis (Crispie et al. 2008). The response of healthy cows to an infusion with *L. lactis* DPC3147 was further investigated and was shown to result in high Polymorphonuclear neutrophils (PMN) recruitment, a rise in acute phase protein, milk amyloid A (Klostermann et al. 2008), the formation of clots in milk, and a rise in IL-1β and IL-8 (Beecher et al. 2009). Taken together, these works suggest that *L. lactis* can exert a preventive and curative effect against bovine mastitis.

LAB are also found in the endogenous ecosystem of raw foremilk and in the teat canal. Some research works report their isolation, identification, and characterization in terms of their inhibitory potential against *S. aureus*, a major mastitis pathogen. In the mid 2000s, several pioneering works showed that a few LAB strains (only 1 out of 102 isolates) showed inhibitory properties against this pathogen, which correlated with the production of a bacteriocin-like compound (Espeche et al. 2009). Whether such a strain has potential in mastitis prevention similar to that of *L. lactis* DPC3147 has to be determined by *in vivo* experiments (Espeche et al. 2012; Gill et al. 2006). Recently, our team investigated the mechanisms underlying the preventive potential of *Lact. casei* strains *in vitro*, using a tripartite interaction model between *Lact. casei*, *S. aureus*, and bovine mammary epithelial cells (bMEC). We thus showed that three *Lact. casei* strains were able to inhibit, in a strain-dependent fashion, adhesion and internalization of *S. aureus* into bMEC (Bouchard et al. 2013). These results are encouraging since internalization is suspected to be involved in the persistence and chronicity of *S. aureus* mastitis.

9.4.2. Probiotics in Pigs

Pigs are bred essentially for meat production, most often in intensive systems where diseases such as diarrhea and pneumonia cause loss of growth and increased mortality, particularly in post-weaning piglets. Feed efficiency is a crucial parameter in pig farming to provide optimal growth. AGPs were widely used until they were recently banned and gave way to the development of alternative strategies including probiotics. Besides, as monogastric animals, pigs are often used as models for studies on the human digestive tract. Their gut microbiota has thus been investigated and there are numerous reports on the use and the effects of probiotics on pigs' health and zootechnical efficiency. Studies on probiotics in pigs have been reported in the literature since the mid 1990s and really took off in the early 2000s with reports on the probiotic use on both zootechnical parameters and infection prevention (Rychen and Nunes 1995).

Like ruminants, the main targets and the types of probiotics in pig rearing depend on the type of productions or on the stage of development of the animals. In gestating sows, probiotics are used to improve diet digestibility, to limit constipation, and to decrease stress. During lactation and in piglets, probiotics are recommended to increase litter size and viability, to improve colostrum quality, milk quality and quantity, and to increase piglet weight. In fattening pigs, the use of probiotics aims at increasing feed efficiency, improving meat quality, and reducing the risk of diarrhea.

Pig Microbiota The gut microbiota of pigs has recently been investigated through pyrosequencing of variable regions of the 16S rRNA and it was revealed that LAB are present at significant levels therein. In the ilea of young pigs (26 days of age), the genera *Clostridium*, *Lactobacillus*, *Streptococcus*, *Helicobacter*, *Ruminococcus*, *Veillonella*, *Candidatus*, *Actinobacillus*, *Bacillus*, and *Turicibacter* were the 10 most frequently encountered (in order of frequency) (Dowd et al. 2008). Some studies investigated the susceptibility of this microbiota to stresses (e.g., weaning) and or pathogens (e.g., *Salmonella* sp.). A culture-dependent approach showed that weaning indeed correlated with changes in the microbiota of the small intestines (Janczyk et al. 2007). Upon weaning, populations of *Lactobacillus* increase (with *Lact. sobrius*/*Lact. amylovorus* as the dominant species) and those of *Bifidobacterium* and *E. coli* decrease. These findings were confirmed using 16S rRNA gene amplification and DGGE (Dowd et al. 2007). Bacterial diversity in pig feces varies with age, as shown by a longitudinal investigation of fecal samples in pigs between 10 and 22 weeks of age. However, these variations were mostly due to less abundant bacterial components of the feces and lactobacilli remained the dominant species in pig fecal samples (Kim et al. 2011). Resistance to infection is one of the important roles ascribed to the gut microbiota, with a normal microbiota excluding pathogens either by occupying the niche or interfering with pathogens. On the other hand, intestinal microbiota can be affected by infections as measured by 16S rRNA gene sequencing in pigs experimentally challenged with *Salmonella enterica* serovar Typhimurium (Bearson et al. 2013). The use of AGPs such as tylosin was recently shown to induce population shifts in both abundant and less abundant species of the gut microbiota (Kim et al. 2012). An oral inoculum prepared from feces of a boar from a high-health herd was also shown to affect respiratory and gut microbial communities and to positively modulate the systemic immune response upon challenge with *Mycoplasma hyopneumoniae* (Schachtschneider et al. 2013).

Probiotics and Zootechnical Parameters in Pigs Enterococci were particularly investigated for their probiotic effects in pigs. *Ent. faecium* (strain NHRD IHARA) isolated from a healthy pig intestine was tested in oral administration to post-weaning piglets and was shown to improve growth performance and to stimulate the production of immunoglobulin A (IgA) in the gut and serum (Ihara et al. 2013). Similar results were recently reported when using heat-killed cells of this strain (Sukegawa et al. 2014). Since the *Lactobacillus* genus is an important component of gut microbiota in pigs, the use of *Lactobacillus* strains as probiotics has been given increasing attention in recent years. Of note, an abundance of somewhat pig-specific lactobacilli species in pig feces makes them good markers of pig manure. Recent works propose using quantification of *L. sobrius/amylovorus* using real-time PCR for identifying pig fecal contamination in surface waters (Marti et al. 2010).

Probiotics and *Disease Prevention in Pigs* Control of infectious diseases in pig farming is a key element in food production both from the animal health and from the public health point of view. *Lact. plantarum* used as diet supplementation of neonatal piglets showed promising results by increasing the total lactobacilli populations in the gut of weaned pigs (Takahashi et al. 2007). When administered together with maltodextrin and/or fructooligosaccharides (FOS), *Lact. plantarum* decreased the amount of *E. coli* O8:K88 in the intestines of piglets (Nemcova et al. 2007). Similarly a combination of FOS and *Lact. paracasei* significantly increased counts of *Bifidobacterium* spp., *Lactobacillus* spp., total anaerobes and aerobes in weaning piglets compared to a non-treated group. It also significantly decreased *Clostridium* and *Enterobacterium* counts in the feces of the piglets (Bomba et al. 2002).

Other *Lactobacillus* species exert beneficial effects on piglet health. *Lact. sobrius* (synonym of *Lact. amylovorus*, now considered a single species) (Jakava-Viljanen et al. 2008) was shown to significantly reduce the amount of ETEC in the ileum when administered to piglets after weaning. In parallel, daily weight gain was improved in the piglets receiving *Lact. sobrius* compared to the control (Konstantinov et al. 2008). A treatment with *Lact. rhamnosus* GG (LGG) reduced diarrhea induced by *E. coli* K88 in post-weaning piglets. Stimulation of intestinal defenses, modulation of gut microbiota, and regulation of inflammatory cytokines are among the probable mechanisms involved in these effects (Zhang and Zhang 2010).

Bifidobacteria, when administered together with lactobacilli immediately after birth, were shown to promote the colonization of neonatal piglet gut with a beneficial commensal microbiota. This treatment limited pathogen load in premature neonatal piglets, and thus lowered colonization by *Clostridium perfringens* and reduced the incidence and severity of necrotizing enterocolitis (Siggers et al. 2008).

Although the underlying mechanisms of such probiotic effects are not fully elucidated, it is admitted that reducing pathogen adhesion to mucosa reduces symptom severity. *B. lactis* and *Lact. rhamnosus* reduced pathogens' adherence to the intestinal mucosa as shown for *Salmonella*, *E. coli*, and *Clostridium* spp. in swine, being more effective when both species were combined (Collado et al. 2007). In a piglet model, post-weaning diarrhea associated with *E. coli* and rotavirus infections was reduced by a probiotic treatment with *B. lactis* HN019 (Shu et al. 2001) while in weaning piglets, *B. animalis* subsp. *lactis* was also shown to positively affect the growth performance and the bifidobacteria : *E. coli* ratio in the intestines (Modesto et al. 2009). *In vitro* experiments showed that *Ent. faecium* decreases by four log units the multiplication of swine influenza virus (H1N1 and H3N2 strains) in a porcine macrophage cell line. Experimental evidences showed that this probiotic effect occurred through the direct adsorptive trapping of influenza virus and by strengthening the innate defense at the cellular level (Wang et al. 2013b). Despite the interesting results obtained with *Ent. faecium* in *in vitro* models of viral infections, it appears that *Ent. faecium in vivo* effects on disease prevention are less promising in challenging tests with common bacterial pathogens. For instance, *Ent. faecium* NCIMB 10415 showed no beneficial effects on the performance of weaned piglets upon challenge with *Salmonella* Typhimurium in terms of pathogen shedding and specific (anti-*Salmonella* IgG) and nonspecific (haptoglobin) humoral immune responses (Kreuzer et al. 2012). Some works even report that in piglets that received *Ent. faecium* NCIMB 10415 and were challenged with *Salmonella* serovar Typhimurium DT104, fecal excretion and colonization of *Salmonella* in organs were significantly greater in the enterococci fed group than in the control group. Besides, the humoral immune response (IgM and IgA) against *Salmonella* was greater in the probiotic group animals than in control animals. This study actually suggests that *Ent. faecium* treatment enhanced the course of *Salmonella* infection in weaning piglets (Szabo et al. 2009). Together with the controversies on the GRAS status of *Enterococcus* sp., these latter results question the use of enterococci as probiotics in feed supplementation.

9.4.3. Probiotics in Poultry

Benefits of feed supplementation with live microorganisms (bacteria or yeasts) are reported in poultry farming on animals' performance and health. Notably, resistance of chickens to bacterial infections (*C. perfringens*, *E. coli*, or *Salmonella*), feed efficiency, productivity of laying hens, and egg quality have been reported (Chaucheyras-Durand and Durand 2010).

Poultry Microbiota The chicken gut microbiota is highly diverse and 16S RNA sequencing showed that it comprises over 900 and 460 species-equivalent OTUs in chicken and turkey sequence collections, respectively (Wei et al. 2013). All along the gastrointestinal tract (GIT) of chickens (gizzard, duodenum, ileum, cecum), lactobacilli are the predominant species (Lu et al. 2003; Guardia et al. 2011; Sekelja et al. 2012; Stanley et al. 2013). They represent over 68% of the ileum and duodenum microbiota in chicken. Such predominance is observed early after hatching, in the first weeks of life, and persists in the mature microbiota (Lu et al. 2003).

In contrast to mammals, in birds, the cecum is an important site of fermentation, it is a major site of water and nutrient absorption and its function affects animal health and feed performance. Cecal microbiota digests foods rich in starch, cellulose, and other polysaccharides present in grains and cereals. Culture-independent identification methods showed an abundance of lactobacilli but also a high abundance of OTUs with very low sequence similarity to known culturable isolates (Gong et al. 2002; Lu et al. 2003). These studies also highlighted qualitative and quantitative differences among the cecal microbiota of individual birds reared under very similar conditions (Gong et al. 2002).

In addition to its great importance in chicken health, the chicken cecum is considered to be a major pathogen reservoir. Chicken fecal samples were analyzed using 16S rRNA sequencing and compared to other GIT sections (Sekelja et al. 2012). This investigation showed that fecal microbiota varies greatly over time, suggesting that temporal shifts in fecal microbiota are a consequence of periodic emptying of different GIT sections. Such variation induced by discharges is a phenomenon generally overlooked in mammalian studies limited to feces (Stanley et al. 2014).

Probiotics and Zootechnical Parameters in Poultry In egg production, probiotics are used as feed additives to improve nutrient digestibility (Nahashon et al. 1994b, 1996) and hen overall performance (feed efficiency, egg production, and egg quality) (Tortuero 1973; Nahashon et al. 1994a, 1994b; Mohan et al. 1995; Abdulrahim et al. 1996; Haddadin et al. 1996; Balevi et al. 2001; Davis and Anderson 2002; Panda et al. 2003). They are also used to positively modulate the intestinal microbiota (Koenen et al. 2004; Yu et al. 2008) and gut mucosal immunity (Balevi et al. 2001; Koenen et al. 2004). Probiotics are also investigated for their possible role in altering the lipid metabolism of chickens, such as reduction of the cholesterol content in egg yolk (Mohan et al. 1995; Abdulrahim et al. 1996; Haddadin et al. 1996; Li et al. 2006) and in serum (Kalavathy et al. 2003). For example, feed supplementation with *P. acidilactici* MA18/5M in laying hens was shown to increase egg weight, eggshell thickness, and relative weight, and to improve feed efficiency ratio per kilogram of eggs. It also significantly decreases the number of broken eggs and drastically reduces the number of downgraded eggs (Mikulski et al. 2012). Similarly, supplementation with *Lact. salivarius* and *B. subtilis* probiotic strains to the diet of layer hens resulted in significant increases in egg production, daily egg yield, and damaged egg ratio. Interestingly, these positive effects were obtained with heat-inactivated bacteria. Another interesting effect observed in this study was the increase of antibody titer against the avian influenza virus in most treated groups. In some of the combinations tested, the titer of antibody against the Newcastle disease virus was also higher. Feed supplementation with probiotics can thus positively affect the performance, egg quality, and immune response of layer hens (Zhang et al. 2012). Of note, some of these combinations also included spore-forming bacteria (like *C. butyricum*, *B. subtilis*). The use of such spore-formers in probiotics preparation should take into account the risks of dissemination and development of spores and spoilage of egg products during their process and storage. The risk of development of spoiling bacteria (mostly spore-formers) is indeed a serious concern in egg alteration (Techer et al. 2014). Thus, LAB appear as the most promising and safer candidates in probiotic uses. Indeed, several *Lactobacillus* spp. were shown to increase performance parameters (growth rate, feed conversion, and meat quality). Feed supplementation of broiler diets with selected *Lactobacillus* strains indeed improved the feed conversion rate, BWG, and reduced abdominal fat deposition (Kalavathy et al. 2003). The gut of healthy chickens can also be a source of probiotic candidates, as shown for strains of *Lact. reuteri*, *Lact. salivarius*, *Ent. faecium*, *B. animalis*, and *P. acidilactici*, which were tested for their efficacy on BWG, feed intake, and feed conversion ratio in broilers. Such probiotic formula, when administered with water and feed, displayed a growth promoting effect comparable to that of an antibiotic treatment (Mountzouris et al. 2007). When compared to formulae containing probiotics from human origin, those containing probiotics isolated from chicken gut microbiota present a better efficiency in growth promotion (Timmerman et al. 2006).

The mechanisms underlying such probiotic effects are still unknown. Moreover, the literature in this specific field presents contradictory results concerning the efficacy of probiotics on poultry growth performance. It is important to note that the means and timing of the administration of probiotic formulae can dramatically affect their efficacy, as shown for lactobacilli preparations (Timmerman et al. 2006).

Probiotics and Disease Prevention in Poultry *Salmonella* spp. is one of the most studied pathogens in the poultry health sector because hens and chickens can be carriers or infected, and consequently this increases the risk of contamination along the food chain. Besides, *Campylobacter jejuni*, *Listeria monocytogenes*, pathogenic *E. coli*, *Yersinia enterocolitica*, and *C. perfringens* are also considered a threat for both poultry and human health. Probiotics are considered a relevant strategy to control pathogen shedding and maintain a balanced and healthy gut microbiota. Most probiotic applications in poultry are linked to the concept of competitive exclusion (CE) (Schneitz 2005). Feed supplementation with lactobacilli was indeed shown to significantly increase the recovery of neonatal broiler chicks after challenge with *Salmonella* enteritidis (Higgins et al. 2008).

In addition to LAB, some PAB species used in dairy fermentation are considered good probiotic candidates in human health (Cousin et al. 2012). Members of the PAB group were identified in the poultry GIT microbiota by culture-dependent (Salanitro et al. 1978) and culture-independent techniques (Apajalahti et al. 1998). Isolates of *P. acidipropionici* and *P. avidum* were recently tested for their probiotic potential in poultry. These PAB were shown to produce short chain fatty acids (SCFA) in an *in vitro* cecal model at concentrations high enough to inhibit *S. enterica* serovar Enteritidis. Moreover, *P. acidipropionici* strain LET 105 was also able to compete with *Salmonella* for adhesion sites on the intestinal mucosa in *ex vivo* assays (Arganaraz-Martinez et al. 2013).

Besides the "classical" feed supplementation, alternative means of *in ovo* inoculation probiotics have been explored since the late 1990s. Gut colonization of birds is indeed considered to take place after hatching. However, some works provide scientific evidence that the chick intestine can contain small numbers of live bacteria before hatch (Deeming 2005). Since a balanced and abundant gut microbiota is considered a good barrier against pathogens, it was thus tempting to establish it early through *in ovo* inoculation of probiotic strains. The first attempts failed due to high embryonic mortality. Besides, compared with oral administration of probiotics, they showed lower level of protection against *Salmonella* (Meijerhof and Hulet 1997). Advances in *in ovo* inoculation techniques now enable the inoculation of the digestive tract of the poultry embryo with external bacteria before hatching. *In ovo* administration of two lactococcal strains (*L. lactis* ssp. *lactis* SL1 and *L. lactis* ssp. *cremoris* IBB SC1) was thus shown to exert a slight but significantly positive effect on poultry meat (better feed conversion and greater pectoral muscle) compared to the non-treated group (Maiorano et al. 2012). In addition, *in ovo* administration of *Ent. faecium* resulted in a significant reduction in the number of *Salmonella* Enteritidis positive chicks after challenge at 4 days of age (de Oliveira et al. 2014).

9.5. Conclusions

Societal demand for a more sustainable agriculture providing the population with safer and better quality animal products has considerably increased over the last decade. This has led to strict sanitary rules and a restricted use of antibiotics in veterinary medicine to avoid the emergence of antibiotic resistance in livestock and in consumers. In livestock, probiotics appear as an interesting alternative to antibiotics as a feed supplement. Probiotics are used to enhance zootechnical performances and mostly to prevent various digestive disorders. Microorganisms used in livestock, sometimes used empirically, are very diverse and include yeasts, Gram (–) and Gram (+) bacteria including spore-formers. Even though their beneficial effects are observed on livestock in terms of production performances and/or disease prevention in animals, their fate along the production chain must be questioned and secured. To use spore-formers of the *B. cereus* group, for example, can be beneficial on farms but may have deleterious effects on the animal end-products. Similarly, the use of enterococci, which reportedly acquire antibiotic resistance determinants and are involved in nosocomial infections in humans, might be considered with precautions. Like in human health, probiotic LAB are of special interest because of their GRAS status and because they showed promising results and/or demonstrated benefits on animal health. Efforts should now be devoted to the elucidation of the underlying mechanisms of such probiotic effects.

References

Abdulrahim, S.M., Haddadin, S.Y., Hashlamoun, E.A., and Robinson, R.K. (1996) The influence of *Lactobacillus acidophilus* and bacitracin on layer performance of chickens and cholesterol content of plasma and egg yolk. *Brit Poultry Sci* **37**, 341–346.

Apajalahti, J.H., Sarkilahti, L.K., Maki, B.R., et al. (1998) Effective recovery of bacterial DNA and percent-guanine-plus-cytosine-based analysis of community structure in the gastrointestinal tract of broiler chickens. *Appl Environ Microbiol* **64**, 4084–4088.

Arganaraz-Martinez, E., Babot, J.D., Apella, M.C., and Perez, C.A. (2013) Physiological and functional characteristics of Propionibacterium strains of the poultry microbiota and relevance for the development of probiotic products. *Anaerobe* **23**, 27–37.

Balevi, T., Ucan, U.S., Coskun, B., et al. (2001) Effect of dietary probiotic on performance and humoral immune response in layer hens. *Brit Poultry Sci* **42**, 456–461.

Barrington, G.M., Gay, J.M., and Evermann, J.F. (2002) Biosecurity for neonatal gastrointestinal diseases. *Veterinary Clinics of North America: Food Anim Practice* **18**, 7–34.

Bearson, S.M., Allen, H.K., Bearson, B.L., et al. (2013) Profiling the gastrointestinal microbiota in response to Salmonella: low versus high *Salmonella* shedding in the natural porcine host. *Infec Gen Evol* **16**, 330–340.

Beecher, C., Daly, M., Berry, D.P., et al. (2009) Administration of a live culture of *Lactococcus lactis* DPC 3147 into the bovine mammary gland stimulates the local host immune response, particularly IL-1beta and IL-8 gene expression. *J Dairy Res* **76**, 340–348.

Bomba, A., Nemcova, R., Gancarcikova, S., et al. (2002) Improvement of the probiotic effect of microorganisms by their combination with maltodextrins, fructo-oligosaccharides and polyunsaturated fatty acids. *Brit J Nutr* **88** Suppl 1, S95–S99.

Bouchard, D.S., Rault, L., Berkova, N., et al. (2013) Inhibition of *Staphylococcus aureus* invasion into bovine mammary epithelial cells by contact with live *Lactobacillus casei*. *Appl Environ Microbiol* **79**, 877–885.

Braem, G., De Vliegher S., Verbist, B., et al. (2012) Culture-independent exploration of the teat apex microbiota of dairy cows reveals a wide bacterial species diversity. *Vet Microbiol* **157**, 383–390.

Braem, G., De Vliegher S., Verbist, B., et al. (2013) Unraveling the microbiota of teat apices of clinically healthy lactating dairy cows, with special emphasis on coagulase-negative staphylococci. *J Dairy Sci* **96**, 1499–1510.

Chaucheyras-Durand, F. and Durand, H. (2010) Probiotics in animal nutrition and health. *Benef Microb* **1**, 3–9.

Chiquette, J., Allison, M.J., and Rasmussen, M. (2012) Use of *Prevotella bryantii* 25A and a commercial probiotic during subacute acidosis challenge in mid lactation dairy cows. *J Dairy Sci* **95**, 5985–5995.

Collado, M.C., Grzeskowiak, L., and Salminen, S. (2007) Probiotic strains and their combination inhibit in vitro adhesion of pathogens to pig intestinal mucosa. *Curr Microbiol* **55**, 260–265.

Cousin, F.J., Foligne, B., Deutsch, S.M., et al. (2012) Assessment of the probiotic potential of a dairy product fermented by *Propionibacterium freudenreichii* in piglets. *J Agric Food Chem* **60**, 7917–7927.

Crispie, F., Alonso-Gomez, M., O'Loughlin, C., et al. (2008) Intramammary infusion of a live culture for treatment of bovine mastitis: effect of live lactococci on the mammary immune response. *J Dairy Res* **75**, 374–384.

Davis, G.S. and Anderson, K.E. (2002) The effects of feeding the direct-fed microbial, primalac, on growth parameters and egg production in single comb white leghorn hens. *Poultry Sci* **81**, 755–759.

de Oliveira, J.E., van der Hoeven-Hangoor, E., van de Linde, I.B., et al. (2014) *In ovo* inoculation of chicken embryos with probiotic bacteria and its effect on posthatch Salmonella susceptibility. *Poultry Sci* **93**, 818–829.

Deeming, D.C. (2005) Yolk sac, body dimensions and hatchling quality of ducklings, chicks and poults. *Brit Poultry Sci* **46**, 560–564.

Desnoyers, M., Giger-Reverdin, S., Bertin, G., et al. (2009) Meta-analysis of the influence of *Saccharomyces cerevisiae* supplementation on ruminal parameters and milk production of ruminants. *J Dairy Sci* **92**, 1620–1632.

Dowd, S.E., Callaway, T.R. and Morrow-Tesch, J. (2007) Handling may cause increased shedding of *Escherichia coli* and total coliforms in pigs. *Foodborne Path Dis* **4**, 99–102.

Dowd, S.E., Callaway, T.R., Wolcott, R.D., et al. (2008) Evaluation of the bacterial diversity in the feces of cattle using 16S rDNA bacterial tag-encoded FLX amplicon pyrosequencing (bTEFAP). *BMC Microbiol* **8**, 125.

EFSA. Regulation (EC) No 1831/2003 of the European Parliament and of the council of 22 September 2003 on additives for use in animal nutrition. *Official J Eur Union* (2003) **268**, 29–43.

EFSA. Guidance for the preparation of dossiers for zootechnical additives. (2012) EFSA Panel on Additives and Products or Substances used in Animal Feed (FEEDAP). European Food Safety Authority (EFSA), Parma, Italy. *EFSA J* **10**, 2536.

Espeche, M. C., Otero, M. C., Sesma, F., and Nader-Macias, M. E. (2009) Screening of surface properties and antagonistic substances production by lactic acid bacteria isolated from the mammary gland of healthy and mastitic cows. *Vet Microbiol* **135**, 346–357.

Espeche, M.C., Pellegrino, M., Frola, I., et al. (2012) Lactic acid bacteria from raw milk as potentially beneficial strains to prevent bovine mastitis. *Anaerobe* **18**, 103–109.

Ewaschuk, J.B., Zello, G.A., and Naylor, J.M. (2006) *Lactobacillus* GG does not affect D-lactic acidosis in diarrheic calves, in a clinical setting. *J Vet Int Med* **20**, 614–619.

FAO/WHO EXPERT CONSULTATION:" Health and Nutritional Properties of Probiotics in Food including Powder Milk with Live Lactic Acid Bacteria" Cordoba (Argentina) 2001.

Füller, R. (1989) Probiotics in man and animals. *J Appl Bacteriol* **66**, 365–378.

Gaggia, F., Mattarelli, P., and Biavati, B. (2010) Probiotics and prebiotics in animal feeding for safe food production. *Int J Food Microbiol* **141** Suppl 1, S15–S28.

Gill, J.J., Sabour, P.M., Gong, J., et al. (2006) Characterization of bacterial populations recovered from the teat canals of lactating dairy and beef cattle by 16S rRNA gene sequence analysis. *FEMS Microbiol Ecol* **56**, 471–481.

Gong, J., Forster, R.J., Yu, H., et al. (2002) Molecular analysis of bacterial populations in the ileum of broiler chickens and comparison with bacteria in the cecum. *FEMS Microbiol Ecol* **41**, 171–179.

Greene, W.A., Gano, A.M., Smith, K.L., et al. (1991) Comparison of probiotic and antibiotic intramammary therapy of cattle with elevated somatic cell counts. *J Dairy Sci* **74**, 2976–2981.

Guardia, S., Konsak, B., Combes, S., et al. (2011) Effects of stocking density on the growth performance and digestive microbiota of broiler chickens. *Poultry Sci* **90**, 1878–1889.
Haddadin, M.S., Abdulrahim, S.M., Hashlamoun, E.A., and Robinson, R.K. (1996) The effect of *Lactobacillus acidophilus* on the production and chemical composition of hen's eggs. *Poultry Sci* **75**, 491–494.
Higgins, S.E., Higgins, J.P., Wolfenden, A.D., et al. (2008) Evaluation of a *Lactobacillus*-based probiotic culture for the reduction of *Salmonella enteritidis* in neonatal broiler chicks. *Poultry Sci* **87**, 27–31.
Ihara, Y., Hyodo, H., Sukegawa, S., et al. (2013) Isolation, characterization, and effect of administration in vivo, a novel probiotic strain from pig feces. *Anim Sci J* **84**, 434–441.
Jakava-Viljanen, M., Murros, A., Palva, A., and Bjorkroth, K.J. (2008) *Lactobacillus sobrius* is a later synonym of *Lactobacillus amylovorus*. *Int J Syst Evol* **58**, 910–913.
Jami, E., Israel, A., Kotser, A., and Mizrahi, I. (2013) Exploring the bovine rumen bacterial community from birth to adulthood. *Int Syst Microbiol Evol J* **7**, 1069–1079.
Jami, E., White, B.A., and Mizrahi, I. (2014) Potential role of the bovine rumen microbiome in modulating milk composition and feed efficiency. *PLoS One* **9**, e85423.
Janczyk, P., Pieper, R., Smidt, H., and Souffrant, W. B. (2007) Changes in the diversity of pig ileal lactobacilli around weaning determined by means of 16S rRNA gene amplification and denaturing gradient gel electrophoresis. *FEMS Microbiol Ecol* **61**, 132–140.
Jouany, J.P. (2006) Optimizing rumen functions in the close-up transition period and early lactation to drive dry matter intake and energy balance in cows. *Anim Reprod Sci* **96**, 250–264.
Kalavathy, R., Abdullah, N., Jalaludin, S., and Ho, Y.W. (2003) Effects of *Lactobacillus* cultures on growth performance, abdominal fat deposition, serum lipids and weight of organs of broiler chickens. *Brit Poultry Sci* **44**, 139–144.
Kim, H.B., Borewicz, K., White, B.A., et al. (2011) Longitudinal investigation of the age-related bacterial diversity in the feces of commercial pigs. *Vet Microbiol* **153**, 124–133.
Kim, K.A., Gu, W., Lee, I.A., et al. (2012) High fat diet-induced gut microbiota exacerbates inflammation and obesity in mice via the TLR4 signaling pathway. *PLoS One* **7**, e47713.
Klostermann, K., Crispie, F., Flynn, J., et al. (2008) Intramammary infusion of a live culture of *Lactococcus lactis* for treatment of bovine mastitis: comparison with antibiotic treatment in field trials. *J Dairy Res* **75**, 365–373.
Koenen, M.E., Kramer, J., van der Hulst R., et al. (2004) Immunomodulation by probiotic lactobacilli in layer- and meat-type chickens. *Brit Poultry Sci* **45**, 355–366.
Konstantinov, S.R., Smidt, H., Akkermans, A.D., et al. (2008) Feeding of *Lactobacillus sobrius* reduces *Escherichia coli* F4 levels in the gut and promotes growth of infected piglets. *FEMS Microbiol Ecol* **66**, 599–607.
Kreuzer, S., Janczyk, P., Assmus, J., et al. (2012) No beneficial effects evident for *Enterococcus faecium* NCIMB 10415 in weaned pigs infected with *Salmonella enterica* serovar Typhimurium DT104. *Appl Environ Microbiol* **78**, 4816–4825.
Kritas, S. K., Govaris, A., Christodoulopoulos, G., and Burriel, A.R. (2006) Effect of *Bacillus licheniformis* and *Bacillus subtilis* supplementation of ewe's feed on sheep milk production and young lamb mortality. *J Vet Med A* **53**, 170–173.
Lee, H.J., Jung, J.Y., Oh, Y.K., et al. (2012) Comparative survey of rumen microbial communities and metabolites across one caprine and three bovine groups, using bar-coded pyrosequencing and (1)H nuclear magnetic resonance spectroscopy. *Appl Environ Microbiol* **78**, 5983–5993.
Lesmeister, K.E., Heinrichs, A.J., and Gabler, M.T. (2004) Effects of supplemental yeast (*Saccharomyces cerevisiae*) culture on rumen development, growth characteristics, and blood parameters in neonatal dairy calves. *J Dairy Sci* **87**, 1832–1839.
Lettat, A., Noziere, P., Silberberg, M., et al. (2012) Rumen microbial and fermentation characteristics are affected differently by bacterial probiotic supplementation during induced lactic and subacute acidosis in sheep. *BMC Microbiol* **12**, 142.
Li, L., Xu, C.L., Ji, C., et al. (2006) Effects of a dried *Bacillus subtilis* culture on egg quality. *Poultry Sci* **85**, 364–368.
Li, R.W., Connor, E.E., Li, C., et al. (2012) Characterization of the rumen microbiota of pre-ruminant calves using metagenomic tools. *Environ Microbiol* **14**, 129–139.
Lu, J., Idris, U., Harmon, B., et al. (2003) Diversity and succession of the intestinal bacterial community of the maturing broiler chicken. *Appl Environ Microbiol* **69**, 6816–6824.
Maiorano, G., Sobolewska, A., Cianciullo, D., et al. (2012) Influence of *in ovo* prebiotic and synbiotic administration on meat quality of broiler chickens. *Poultry Sci* **91**, 2963–2969.
Marti, R., Dabert, P., Ziebal, C., and Pourcher, A. M. (2010) Evaluation of *Lactobacillus sobrius/L. amylovorus* as a new microbial marker of pig manure. *Appl Environ Microbiol* **76**, 1456–1461.

Meijerhof, R. and Hulet R.M. (1997) *In ovo* injection of competitive exclusion culture in broiler hatching eggs. *J Appl Poultry Res* **6**, 260–266.

Mikulski, D., Jankowski, J., Naczmanski, J., et al. (2012) Effects of dietary probiotic (*Pediococcus acidilactici*) supplementation on performance, nutrient digestibility, egg traits, egg yolk cholesterol, and fatty acid profile in laying hens. *Poultry Sci* **91**, 2691–2700.

Modesto, M., D'Aimmo, M.R., Stefanini, I., et al. (2009) A novel strategy to select *Bifidobacterium* strains and prebiotics as natural growth promoters in newly weaned pigs. *Livestock Sci* **122**, 248–258.

Mohan, B., Kadirvel, R., Bhaskaran, M., and Natarajan, A. (1995) Effect of probiotic supplementation on serum/yolk cholesterol and on egg shell thickness in layers. *Brit Poultry Sci* **36**, 799-803.

Morgavi, D.P., Forano, E., Martin, C. and Newbold, C.J. (2010) Microbial ecosystem and methanogenesis in ruminants. *Animal* **4**, 1024–1036.

Morgavi, D.P., Kelly, W.J., Janssen, P.H. and Attwood, G.T. (2013) Rumen microbial (meta)genomics and its application to ruminant production. *Animal* **7** Suppl 1, 184–201.

Mountzouris, K.C., Tsirtsikos, P., Kalamara, E., et al. (2007) Evaluation of the efficacy of a probiotic containing *Lactobacillus, Bifidobacterium, Enterococcus*, and *Pediococcus* strains in promoting broiler performance and modulating cecal microflora composition and metabolic activities. *Poultry Sci* **86**, 309–317.

Nader-Macias, M.E., Otero, M.C., Espeche, M.C., and Maldonado, N. C. (2008) Advances in the design of probiotic products for the prevention of major diseases in dairy cattle. *J Ind Microbiol Biotechnol* **35**, 1387–1395.

Nahashon, S.N., Nakaue, H.S., and Mirosh, L. W. (1994a) Production variables and nutrient retention in single comb White Leghorn laying pullets fed diets supplemented with direct-fed microbials. *Poultry Sci* **73**, 1699–1711.

Nahashon, S.N., Nakaue, H.S., Snyder, S.P., and Mirosh, L.W. (1994b) Performance of single comb White Leghorn layers fed corn-soybean meal and barley-corn-soybean meal diets supplemented with a direct-fed microbial. *Poultry Sci* **73**, 1712-1723.

Nahashon, S.N., Nakaue, H.S., and Mirosh, L.W. (1996) Performance of single comb white leghorn fed a diet supplemented with a live microbial during the growth and egg laying phases. *Anim Feed Sci Technol* **57**, 25–38.

Nemcova, R., Bomba, A., Gancarcikova, S., et al. (2007) Effects of the administration of lactobacilli, maltodextrins and fructooligosaccharides upon the adhesion of *E. coli* O8:K88 to the intestinal mucosa and organic acid levels in the gut contents of piglets. *Vet Res Communic* **31**, 791–800.

Oikonomou, G., Bicalho, M.L., Meira, E., et al. (2014) Microbiota of cow's milk; distinguishing healthy, sub-clinically and clinically diseased quarters. *PLoS One.* **9**, e85904.

Omoniyi, L.A., Jewell, K.A., Isah, O. A., et al. (2014) An analysis of the ruminal bacterial microbiota in West African Dwarf sheep fed grass- and tree-based diets. *J Appl Microbiol* **116**, 1094–1105.

Otero, C., Saavedra, L., Silva de, R.C., et al. (2000) Vaginal bacterial microflora modifications during the growth of healthy cows. *Lett Appl Microbiol* **31**, 251–254.

Otero, M.C., Morelli, L., and Nader-Macias, M.E. (2006) Probiotic properties of vaginal lactic acid bacteria to prevent metritis in cattle. *Let Appl Microbiol* **43**, 91–97.

Otero, M.C. and Nader-Macias, M.E. (2006) Inhibition of *Staphylococcus aureus* by H2O2-producing *Lactobacillus gasseri* isolated from the vaginal tract of cattle. *Anim Reprod Sci* **96**, 35–46.

Panda, A.K., Reddy, M.R., Rama Rao, S.V., and Praharaj, N.K. (2003) Production performance, serum/yolk cholesterol and immune competence of white leghorn layers as influenced by dietary supplementation with probiotic. *Trop Anim Health Produc* **35**, 85–94.

Reid, G. (1999) The scientific basis for probiotic strains of *Lactobacillus*. *Appl Environ Microbiol* **65**, 3763–3766.

Rychen, G. and Nunes, C.S. (1995) Effects of three microbial probiotics on postprandial porto-arterial concentration differences of glucose, galactose and amino-nitrogen in the young pig. *Brit J Nutr* **74**, 19–26.

Salanitro, J.P., Blake, I.G., Muirehead, P.A., et al. (1978) Bacteria isolated from the duodenum, ileum, and cecum of young chicks. *Appl Environ Microbiol* **35**, 782–790.

Santos, T.M. and Bicalho, R.C. (2012) Diversity and succession of bacterial communities in the uterine fluid of postpartum metritic, endometritic and healthy dairy cows. *PLoS One* **7**, e53048.

Schachtschneider, K.M., Yeoman, C.J., Isaacson, R.E., et al. (2013) Modulation of systemic immune responses through commensal gastrointestinal microbiota. *PLoS One* **8**, e53969.

Schneitz C. (2005) Competitive exclusion in poultry—30 years of research. *Food Control* **16**, 657–667

Sekelja, M., Rud, I., Knutsen, S.H., et al. (2012) Abrupt temporal fluctuations in the chicken fecal microbiota are explained by its gastrointestinal origin. *Appl Environ Microbiol* **78**, 2941–2948.

Shu, Q., Qu, F., and Gill, H.S. (2001) Probiotic treatment using *Bifidobacterium lactis* HN019 reduces weanling diarrhea associated with rotavirus and *Escherichia coli* infection in a piglet model. *J Pediatr Gastroenterol Nutr* **33**, 171–177.

Siggers, R.H., Siggers, J., Boye, M., et al. (2008) Early administration of probiotics alters bacterial colonization and limits diet-induced gut dysfunction and severity of necrotizing enterocolitis in preterm pigs. *J Nutr* **138**, 1437–1444.

Signorini, M.L., Soto, L.P., Zbrun, M.V., et al. (2012) Impact of probiotic administration on the health and fecal microbiota of young calves: a meta-analysis of randomized controlled trials of lactic acid bacteria. *Res Vet Sci* **93**, 250–258.

Silberberg, M., Chaucheyras-Durand, F., Commun, L., et al. (2013) Repeated acidosis challenges and live yeast supplementation shape rumen microbiota and fermentations and modulate inflammatory status in sheep. *Animal* **7**, 1910–1920.

Stanley, D., Geier, M.S., Denman, S.E., et al. (2013) Identification of chicken intestinal microbiota correlated with the efficiency of energy extraction from feed. *Vet Microbiol* **164**, 85–92.

Stanley, D., Hughes, R.J., and Moore, R.J. (2014) Microbiota of the chicken gastrointestinal tract: influence on health, productivity and disease. *Appl Microbiol Biotechnol* **98**, 4301–4310.

Sukegawa, S., Ihara, Y., Yuge, K., et al. (2014) Effects of oral administration of heat-killed *Enterococcus faecium* strain NHRD IHARA in post-weaning piglets. *Anim Sci J* **85**, 454–460.

Szabo, I., Wieler, L.H., Tedin, K., et al. (2009) Influence of a probiotic strain of *Enterococcus faecium* on *Salmonella enterica* serovar Typhimurium DT104 infection in a porcine animal infection model. *Appl Environ Microbiol* **75**, 2621–2628.

Takahashi, S., Egawa, Y., Simojo, N., et al. (2007) Oral administration of *Lactobacillus plantarum* strain Lq80 to weaning piglets stimulates the growth of indigenous lactobacilli to modify the lactobacillal population. *J Gen Appl Microbiol* **53**, 325–332.

Techer, C., Baron, F., Delbrassinne, L., et al. (2014) Global overview of the risk linked to the *Bacillus cereus* group in the egg product industry: identification of food safety and food spoilage markers. *J Appl Microbiol* **116**, 1344–1358.

Timmerman, H.M., Veldman, A., van den Elsen E., et al. (2006) Mortality and growth performance of broilers given drinking water supplemented with chicken-specific probiotics. *Poultry Sci* **85**, 1383–1388.

Tortuero, F. (1973) Influence of the implantation of *Lactobacillus acidophilus* in chicks on the growth, feed conversion, malabsorption of fats syndrome and intestinal flora. *Poultry Sci* **52**, 197–203.

Wang, Y., Ametaj, B.N., Ambrose, D.J., and Ganzle, M.G. (2013a) Characterisation of the bacterial microbiota of the vagina of dairy cows and isolation of pediocin-producing *Pediococcus acidilactici*. *BMC Microbiol* **13**, 19.

Wang, Z., Chai, W., Burwinkel, M., et al. (2013b) Inhibitory influence of *Enterococcus faecium* on the propagation of swine influenza A virus in vitro. *PLoS One* **8**, e53043.

Wei, S., Morrison, M., and Yu, Z. (2013) Bacterial census of poultry intestinal microbiome. *Poultry Sci* **92**, 671–683.

Yu, M.H., Yu, X.L., Chen, C.L., et al. (2008) The change of intestinal microecology in rats after orthotopic liver transplantation. *Zhonghua Wai Ke Za Zhi* **46**, 1139–1142.

Zhang, C. and Zhang, W. (2010) *Escherichia coli* K88ac fimbriae expressing heat-labile and heat-stable (STa) toxin epitopes elicit antibodies that neutralize cholera toxin and STa toxin and inhibit adherence of K88ac fimbrial *E. coli*. *Clin Vaccine Immunol* **17**, 1859–1867.

Zhang, J.L., Xie, Q.M., Ji, J., et al. (2012) Different combinations of probiotics improve the production performance, egg quality, and immune response of layer hens. *Poultry Sci* **91**, 2755–2760.

Chapter 10

Proteomics for Studying Probiotic Traits

Rosa Anna Siciliano* and Maria Fiorella Mazzeo

Centro di Spettrometria di Massa Proteomica e Biomolecolare, Istituto di Scienze dell'Alimentazione, CNR, Italy

10.1. Introduction

The availability of complete genomic sequences of probiotic strains belonging to the *Lactobacillus* and *Bifidobacterium* genera has paved the way to proteomic studies aimed at investigating probiotic functionalities. In fact, proteins directly mediate probiotic adaption to the harsh physical-chemical environment of the gastrointestinal tract (GIT), adhesion to host epithelial cells and intestinal mucosa, as well as probiotic health-promoting effects.

Proteome, namely the entire set of proteins encoded by the genome of organisms (Wilkins et al. 1996), is an evolving entity, which quantitatively and qualitatively changes due to environmental/pathological conditions or cellular compartments. Furthermore, several isoforms of the same protein could exist due to post-translational modifications (PTMs, such as glycosylation and phosphorylation), alternative splicing or protease action. Therefore, proteomics, through the identification of differentially expressed proteins, could depict a global, dynamic, and integrated view of cell physiology.

Microbial proteomics widely contributed to elucidate metabolic and molecular processes underlying cell response to environmental stimuli (growth under stress conditions, starvation, etc.), to investigate the role of PTMs in the regulation of protein functions and biological processes, and to study virulence factors and surface proteins involved in host–probiotic/pathogen interactions (De Angelis and Gobbetti 2004; Schmidt and Völker 2011; Chao and Hansmeier 2012). More recently, proteomics has also facilitated insights into microbial communities' lifestyles in complex ecosystems including biofilms and GIT (metaproteomics) (Hettich et al. 2012).

Proteome complexity, also due to different abundance levels of proteins in cells (spanning 10 orders of magnitude), required the introduction of analytical strategies based on innovative and sensitive mass spectrometric instruments. In fact, two-dimensional electrophoresis (2-DE) (Klose 1975; O'Farrell 1975) allowed the separation and visualization of proteome components, but could not lead to protein identification. Improvements in mass spectrometric technologies and the development of computational bioinformatics have provided the appropriate tools for large-scale protein identification and quantification, thus opening the way to high-throughput proteomics.

* Corresponding author: Centro di Spettrometria di Massa Proteomica e Biomolecolare, Istituto di Scienze dell'Alimentazione, CNR, via Roma 64, 83100 Avellino, Italy. rsiciliano@isa.cnr.it

Biotechnology of Lactic Acid Bacteria: Novel Applications, Second Edition. Edited by Fernanda Mozzi, Raúl R. Raya, and Graciela M. Vignolo.

10.2. Mass Spectrometric Methodologies in Proteomics

10.2.1. The Classical Approach: 2-DE Separation and Protein Identification by Mass Spectrometry

Complex protein mixtures extracted from biological systems are separated and visualized at high resolution through 2-DE. Protein separation is carried out in the first dimension by isoelectric focusing (based on protein isoelectric point) and, orthogonally, in the second dimension by sodium dodecyl sulfate-polyacrylamide gel electrophoresis (SDS-PAGE) (based on protein molecular mass) in order to produce 2-DE proteomic maps. Specific image analysis software allows the comparison of 2-DE proteomic maps obtained from a biological system under different physiological and/or pathological conditions, thus detecting spots with quantitative differences in their intensities, likely originated from differentially expressed proteins. Proteins contained in the spots of interest are in-gel digested (commonly using trypsin) and identified by analyzing the obtained peptide mixtures by mass spectrometry (MS). Matrix assisted laser desorption ionization-time of flight-mass spectrometry (MALDI-TOF-MS) and liquid chromatography coupled to tandem mass spectrometry (LC-MS/MS) are commonly applied in proteomics (Henzel et al.1993; Pappin et al. 1993). Protein identification is achieved by using MS data for querying specific protein sequence databases (NCBInr, Uniprot, etc.) with search engines such as Mascot (http://www.matrixscience.com) or ProteinProspector (http://prospector.ucsf.edu) (Figure 10.1).

10.2.2. Gel-Free Proteomic Approaches

In gel-free proteomic approaches, also named shotgun proteomics, the entire proteome of a biological system is enzymatically digested and the very complex peptide mixture thus obtained, containing thousands of peptides, is directly analyzed by two-or multi-dimensional LC-MS/MS

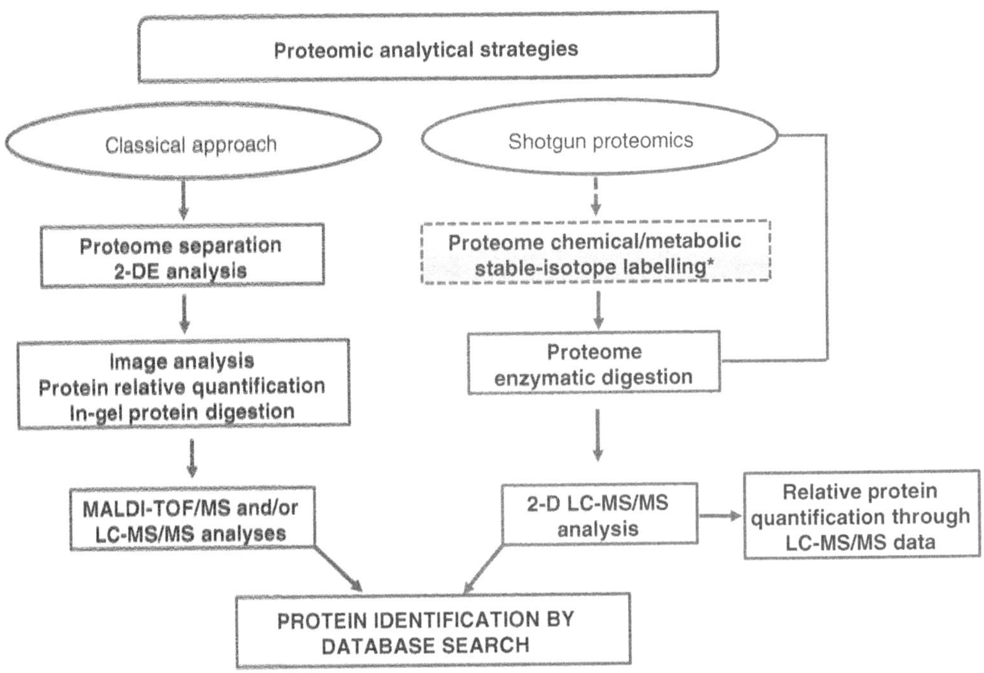

Figure 10.1. Workflow generally applied in proteomic studies.
*Label-free quantification methods do not include this step.

(Washburn et al. 2001; Fournier et al. 2007). High-resolution mass spectrometric instruments have been developed for shotgun proteomic experiments (Domon and Aebersold 2006, 2010; Nilsson et al. 2010). The MS dataset is used to perform database searches using dedicated bioinformatics tools in order to achieve the identification of proteome components (Figure 10.1).

In addition, quantitative MS-based methodologies have been developed to specifically identify proteins differentially expressed under different conditions (Domon and Aebersold 2010; Cox and Mann 2011). These include metabolic protein stable-isotope labeling methods (such as stable isotope labeling with amino acids in cell culture (SILAC) (Ong et al. 2002)), chemical protein/peptide stable-isotope labeling methods (such as isotope-coded affinity tag (ICAT) (Gygi et al. 1999) and isobaric tags for relative and absolute quantitation (iTRAQ) (Ross et al. 2004)) and label-free quantification methods such as spectral counting methods (Neilson et al. 2011). To better investigate PTMs, techniques based on affinity chromatography have been introduced for the selective enrichment of proteins or peptides carrying specific functional groups (such as immobilized metal affinity chromatography (IMAC) for phosphoproteome analysis (Zhao and Jensen 2009)).

Theoretical and technical aspects of mass spectrometry and proteomics have been extensively reviewed (Aebersold and Mann 2003; Walther and Mann 2010; Angel et al. 2012).

10.3. Proteomics for Studying Molecular Mechanisms of Probiotic Action

10.3.1. Adaptation Mechanisms to the GIT Environment

Exposure to acid pH (≈2) in the stomach and to bile in the small intestine (from 1 to 40 mM) are the major challenges that probiotics face in the GIT. The lowering of intracellular pH damages proteins and DNA and modifies the transmembrane ΔpH, thus affecting the proton motive force that acts as an energy source for different transmembrane transport processes. Bile, composed of bile acids (weak organic acids) and bile salts (taurine or glycine conjugates), mainly impairs the cell membrane, altering its architecture, and cell wall functionalities. Moreover, deconjugation of bile salts by bile salt hydrolases (BSHs) in the bacterial cytoplasm causes intracellular acidification. Therefore, probiotics have developed response mechanisms to acid and bile that share several features and allow them to adapt and survive in GIT (Sánchez et al. 2008a; Ruiz et al. 2013).

Proteomics clearly demonstrate the activation of general stress response mechanisms leading to an accumulation of molecular chaperones (Clp protease family, GroES, GroEL, DnaK, and DnaJ) to counteract protein misfolding, denaturation, and aggregation. In addition, proteins responsible for maintaining redox balance, such as thioredoxin, were overexpressed to withstand oxidative stress. Moreover, expression of proteins involved in transcription and translation (such as RNA polymerases, elongation factors, and ribosomal proteins), nucleotide and amino acid biosynthesis, and of those related to DNA-repair was affected. Harsh GIT conditions also caused a metabolic reorganization, mainly leading to the activation of glycolysis and bifid shunt, a unique hexose metabolism of *Bifidobacterium*, as an increased energy production is crucial to counteract acid pH or bile salts exposure. In fact, these pathways assure a greater production of energetic intermediates (ATP) to sustain ATP dependent processes, such as folding machinery, transport and detoxifying systems, and the activity of proton translocating ATPase (F_0-F_1-ATPase). This enzyme, overexpressed under both acid and bile exposure, has a key role in maintaining the cytoplasmic pH homeostasis in anaerobic microorganisms facilitating proton extrusion (Sánchez et al. 2007a,b; Lee et al. 2008a,b; Koponen et al. 2012) (Figure 10.2).

The activation of the arginine deiminase (ADI) pathway also contributes to the restoration of optimal cytoplasm pH in *Lactobacillus* (De Angelis and Gobbetti 2004; Lee et al. 2008b; Lee and Pi 2010). As this metabolic pathway is absent in *Bifidobacterium*, the synthesis of higher amounts of enzymes involved in the biosynthesis of branched-chain amino acids (BCAA) and glutamine

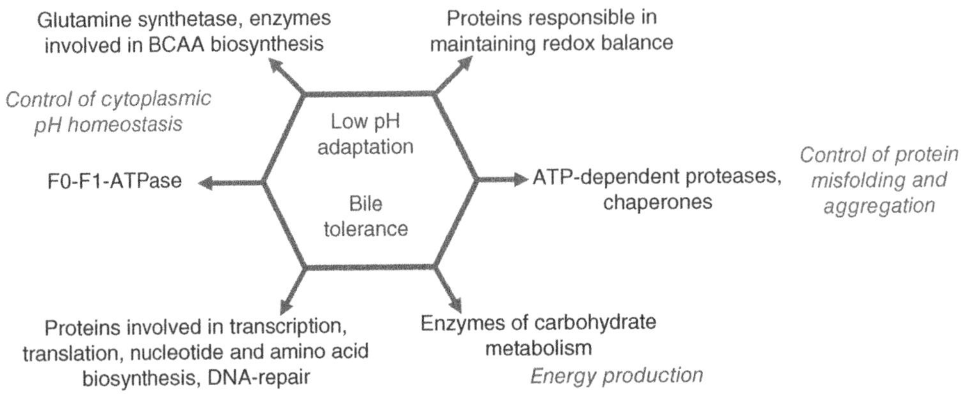

Figure 10.2. Response mechanisms triggered by both acid and bile exposure in lactobacilli and bifidobacteria.

synthetase, that leads to ammonia production, has been also associated to the intracellular pH control in these probiotics (Sánchez et al. 2007b; Ruiz et al. 2013).

Due to its lipophilic character, bile exposure could modify cell-envelope architecture and affect cell wall lipid composition, in a species- and strain-dependent fashion. The increased synthesis of exopolysaccharides (EPS) was induced in bifidobacteria by bile, aimed at creating a protective coating (Ruiz et al. 2013). Proteomics also demonstrated that acquisition of bile-resistance was correlated to a significant overproduction of enzymes involved in EPS biosynthesis in *Lact. delbrueckii* (Burns et al. 2010). Conversely, a decreased production of enzymes involved in EPS biosynthesis in a bile containing environment has been associated in *Lact. rhamnosus* GG to a reduction of EPS layer thickness in order to promote bacterial adhesion to the gut (Koskenniemi et al. 2011). Alterations in the synthesis of proteins involved in fatty acids metabolism also suggested rearrangements aimed at modifying cell wall composition (Sánchez et al. 2007a; Wu et al. 2010; Koskenniemi et al. 2011; Alcántara and Zúñiga 2012; Ruiz et al. 2013).

In addition, bile exposure induced the overexpression of ABC-type multidrug transporters and the activity of bile efflux pumps for bile salts extrusion (Ctr and BetA in *Bifidobacterium*) (Sánchez et al. 2008a; Ruiz et al. 2009; Hamon et al. 2011, 2012; Koskenniemi et al. 2011).

Intriguingly, the role of BSHs in conferring resistance to bile remains unclear. These proteins, responsible for bile salts deconjugation, were found at higher levels in a bile-adapted *B. animalis* strain, thus suggesting that overexpression of these proteins made this strain more tolerant (Sánchez et al. 2007a; Ruiz et al. 2013). BSH expression in *Lactobacillus* seemed to be species-specific: in fact, it was induced immediately upon bile exposure in *Lact. rhamnosus* GG (Koskenniemi et al. 2011) and in *Lact. johnsonii* PF01 (three BSHs with different bile salts specificity) (Lee et al. 2013) while it was not modified in *Lact. casei* Zhang (Wu et al. 2010) and decreased in *Lact. plantarum* (Hamon et al. 2011) (Figure 10.3).

10.3.2. Adhesion Mechanisms to the Host Mucosa

Adhesion of microorganisms to the intestinal mucosa is a key feature in colonization and is related to the ability of probiotics to interact with the host in order to exert beneficial health effects. Species- and strain-specific colonization strategies involve proteinaceous compounds as well as teichoic and lipoteicoic acids, peptidoglycans, and EPS. Extracellular proteins, that include cell-envelope and secreted proteins, represent the first-line interaction with the mucus layer of the GIT

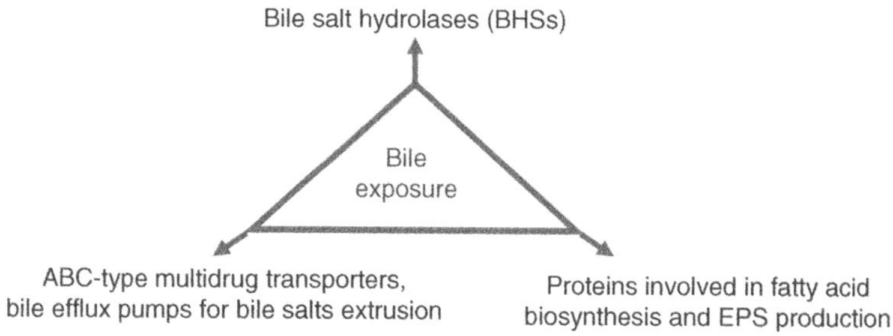

Figure 10.3. Response mechanisms specifically triggered by bile exposure in lactobacilli and bifidobacteria.

and/or extracellular matrix proteins (fibronectin, collagen, and laminin) and plasminogen (Sánchez et al. 2010). As to functional properties, these proteins are the leading actors of probiotic–host molecular crosstalk and mainly mediate interactions with the host immune system. Proteomics, properly integrated by *in vitro* tests, has been applied to the characterization of extracellular proteins, also prompted by the development of specific strategies for membrane and cell-wall proteomics (Solis and Cordwell 2011). In this frame, proteomics contributes to identify cytoplasmic proteins in the extracellular proteome acting as adhesion promoting factors for probiotics. Such proteins, defined as moonlighting proteins and lacking any extracytoplasmic sorting sequence, display different, seemingly unrelated, functions in different cell locations (Jeffery 2003). Proteins belonging to the glycolytic pathway (such as enolase, glyceraldehyde-3-phosphate dehydrogenase, phosphoglycerate mutase, triose-phosphate isomerase), ribosomal proteins, molecular chaperones (such as DnaK and GroEL), and translational factors (such as elongation factor-Tu (EF-Tu)) were included in the large and constantly growing list of moonlighting proteins (Sánchez et al. 2008b, 2010; Beck et al. 2011). These proteins also showed fibronectin, mucin or plasminogen binding activity (Granato et al. 2004; Bergonzelli et al. 2006; Antikainen et al. 2007; Sánchez et al. 2010; Candela et al. 2010; Vastano et al. 2013). As these proteins are also involved in the adhesion of pathogens to the GIT, probiotics competitively exclude enteropathogens from binding to intestinal cells and mucus (Lebeer et al. 2010).

Several factors involved in bacteria–host interactions were identified in a study aimed at achieving a detailed characterization of the secretome and membrane proteome of *B. animalis* subsp. *lactis* BB-12. In particular, plasminogen, collagen, fibronectin or mucin binding proteins were identified as well as proteins involved in fimbriae formation (Gilad et al. 2011, 2012).

Proteomics also proved to have a role in deciphering the molecular basis of the strain-specificity of probiotic features. The in-depth proteome analysis of two *Lact. rhamnosus* strains (the GG strain and the dairy strain Lc705) led to a comprehensive characterization of the predicted extracellular proteins, including lipoproteins, integral membrane proteins, proteins anchored to peptidoglycans, and secreted proteins. A large portion of these surface proteins (about 25%) was strain-specific, thus providing molecular evidence of the phenotypic and functional differences between these strains. In particular, the sortase enzyme Srt2 and the pilin elements SpaC and SpaA, involved in the assembly of a pilus structure and intestinal mucus binding, were specifically expressed by *Lact. rhamnosus* GG (Savijoki et al. 2011).

Interestingly, phosphorylated forms of proteins participating in central cellular pathways (mainly glycolysis) and moonlighting proteins were characterized in *Lact. rhamnosus* GG by phosphoproteomics. This study indicated that phosphorylation was induced under acidic growth conditions (Koponen et al. 2012). As phosphorylation is also implicated in infection and adhesion of pathogens (Jers et al. 2008), this PTM could have a crucial role in modulating the multifunctional biological properties and subcellular localization of moonlighting proteins.

Proteomic studies clearly highlighted that several proteins involved in adhesion mechanisms were also implicated in acid pH and bile adaptation, thus suggesting that GIT stress conditions would represent a key signal triggering the colonization process and more, in general, the probiotic–host crosstalk.

10.3.3. Molecular Mechanisms of Probiotic Immunomodulatory Effects

Beneficial effects of probiotic consumption in the prevention and treatment of several gastrointestinal diseases, such as inflammatory bowel diseases, antibiotic associated-diarrhoea, neonatal necrotizing enterocolitis, irritable bowel syndrome, *Helicobacter pylori* infection, as well as food allergies and intolerances, have been clearly assessed (Sanders et al. 2013). Furthermore, probiotics are effective in reducing cholesterol levels and lowering blood pressure (Kumar et al. 2012). However, molecular mechanisms underlying strain-specific probiotic action and the identity of effector molecules (peptidoglycan, teichoic acid, cell surface polysaccharides, extracellular proteins) still remain to be fully elucidated.

Proteomics integrated by *in vitro* and *in vivo* analyses played a pivotal role in the identification of extracellular proteins involved in probiotic–host immune system interplay. In *Lact. johnsonii* NCC533, extracellular proteins EF-Tu and GroEL elicited a CD14-dependent proinflammatory response stimulating interleukin-8 (IL-8) release in macrophages and HT29 cells (Granato et al. 2004; Bergonzelli et al. 2006). Overexpression of HSPs, GroES and GroEL, induced in *Lact. acidophilus* by a pH decrease in culture media, affected the secretion of cytokines (IL-12, IFN-γ and IL-10) from splenocytes (Kuwana and Yamamoto 2012). This evidence led to count immunomodulation among the multitasking functions of moonlighting proteins.

As to surface proteins, the S-layer protein SlpA of *Lact. acidophilus* was involved in the modulation of dendritic cells (DCs) and T-cell functions, as it directly interacted with a specific DCs receptor. In fact, the deletion of *slpA* in mutant strains led to a reduced DCs binding capacity and to an enhanced proinflammatory cytokine production (Konstantinov et al. 2008). Moreover, the amount of SlpA in different *Lact. acidophilus* strains was related to adhesion to Caco-2 cells as well as to the release of IL-12 from DCs (Ashida et al. 2011).

Proteomics provided a global picture of *Lact. acidophilus* NCFM surface proteins, also leading to the identification of a putative S-layer protein (LBA1029) that contributed to a proinflammatory TNF-α response from murine DCs (Johnson et al. 2013). Six proteins with potential immunogenic abilities were identified in the extracellular proteome of *B. animalis* BB12. In particular, two proteins showed sequence homology with the surface antigen protein of *Lact. rhamnosus* GG p40, and with cell wall-associated glycoside hydrolase p75 (Gilad et al. 2011). The proteins p40 and p75 were able to stimulate the activation of Akt (protein kinase B protein family) in intestinal epithelial cells and to inhibit TNF-induced apoptosis of intestinal epithelial cells (Yan et al. 2007).

Of note, it has been recently reported that specific soluble factors produced by probiotics, called postbiotics, could be sufficient to elicit host immunonomodulatory response. In fact, a peptide (STp) originated by the cleavage of an extracellular protein of *Lact. plantarum* by means of intestinal proteases was capable of modulating phenotype and function of human blood-enriched DCs (Bernardo et al. 2012). These findings set the basis for the development of novel nutraceuticals or functional foods that could become a safer alternative for specific clinical applications, such as chronic inflammatory conditions, where administration of probiotics seems to be less effective (Tsilingiri and Rescigno 2013).

10.3.4. Probiotics and Prebiotics

"A prebiotic is a selectively fermented ingredient that allows specific changes, both in the composition and/or activity in the gastrointestinal microflora that confers benefits upon host wellbeing and health" is the updated definition of prebiotics (Gibson et al. 2004). Currently, only three dietary ingredients have achieved prebiotic status in the European Union: fructooligosaccharides (FOS and inulin), galactooligosaccharides (GOS), and lactulose (Gibson et al. 2004). These oligosaccharides are neither absorbed by the host nor degraded by bile and digestive enzymes and move to the small intestine and colon, where they are metabolized by lactobacilli and bifidobacteria. The metabolic endpoint of bacterial carbohydrate fermentation is the production of short-chain fatty acids (SCFAs), predominantly acetate, propionate, and butyrate that host cells use as an additional energy source (Topping et al. 2001; Gibson and Glenn 2004). In addition, prebiotics exert a trophic effect on the intestinal epithelium, increasing cell growth and controlling proliferation and differentiation (Guarner and Malagelada 2003).

Recently, synbiotics, that are suitable combinations of prebiotics and probiotics, have been developed with the aim of further promoting well-being and health (Kolida and Gibson 2011).

In the last decade, omics sciences, in particular transcriptomics and proteomics, have contributed to reveal oligosaccharide metabolism in lactobacilli and bifidobacteria, showing that bacteria employ several strategies, sometime strain-specific, for utilizing different carbon sources. These strategies involve intra/extracellular or membrane associated enzymes with activity towards oligo- and polysaccharides (such as glycoside hydrolases (GHs)), assisted by complex sugar transport systems ((mainly ATP-binding cassette (ABC)-type systems) nested in membranes (Gänzle and Follador 2012; Abou Hachem et al. 2013).

Metabolic enzymes responsible for oligosaccharide degradation in lactobacilli, so far characterized, are intracellular, except amylase and levansucrase. Therefore, transport systems play a pivotal role in the prebiotic metabolism. In fact, the expression of the *lac* operon, encoding for lactose permease LacS, enzymes involved in galactose metabolism, two β-galactosidases (GH2 and GH42), and two transcriptional regulators, was induced in *Lact. acidophilus* NCFM grown on GOS. On the other hand, mutant strains obtained from the deletion of *lacS* were not able to grow on substrates containing GOS, lactose, and lactitol as a carbon source, thus confirming the cytoplasmic localization of these enzymes (Andersen et al. 2011). Increased amounts of LacS and GH2 were also observed at proteomic level in the same strain grown on lactitol (Majumender et al. 2011). Interestingly the *lac* operon was overexpressed also in response to bile acids (Pfeiler et al. 2007), suggesting a transversal role of proteins encoded by this operon in *Lact. acidophilus* adaptation to the gut niche. Similarly, FOS metabolism in *Lact. acidophilus* greatly relies on ABC transport systems due to the intracellular localization of fructo-furanosidases (Barrangou et al. 2006; Gänzle and Follador 2012).

Metabolism of FOS and GOS in bifidobacteria proceeds through a multistep process that includes binding and internalization of oligosaccharide molecules by ABC-type transport systems and their degradation by means of specific intracellular glycoside hydrolases. The produced intermediates are subsequently metabolized in the Leloir and glycolytic pathways as well as in the bifid shunt. Of note, β-galactosidases also exhibit transglycosylic activity that allows the synthesis of prebiotic substances from lactose (Pokusaeva et al. 2011). Proteomics highlights the overexpression of specific GHs in *B. longum* subsp. *infantis* grown on different carbon sources (GOS, FOS, human milk oligosaccharides), while only four GHs (out of the sixteen identified) were induced in all the experimental conditions (Kim et al. 2013). Similar strategies were also implicated in the metabolic utilization of emerging prebiotic candidates, such as β-glucans and xylo-oligosaccharides (Gilad et al. 2010; Zhao and Cheung 2013).

10.4. Concluding Remarks and Future Directions

Proteomics has significantly contributed to elucidate the molecular mechanisms underlying probiotic actions, revealing that some proteins actually mediate both adaptation and adhesion to the GIT as well as immunomodulation. Of note, many of these proteins are also involved in response mechanisms to industrial stresses. As it is well known that strains adapted to a particular stress usually display cross-resistance to other stress factors, specific technological strategies might be designed in order to select strains that are more tolerant and with enhanced probiotic functional features (Sánchez et al. 2012). In addition, proteomics could support the definition of novel probiotic markers, useful for the rapid selection of strains with predictable and improved functionality (Sánchez *et al.* 2012) and prompt the identification and development of new postbiotics (Tsilingiri and Rescigno 2013).

References

Abou Hachem, M., Andersen, J.M., Barrangou, R., Møller, M.S., Fredslund, F., Majumder, A., Ejby, M., Lahtinen, S.J., et al. (2013) Recent insight into oligosaccharide uptake and metabolism in probiotic bacteria. *Biocatal Biotransfor* **31**, 226–235.

Aebersold, R. and Mann, M. (2003) Mass spectrometry-based proteomics. *Nature* **422**, 198–207.

Alcántara, C. and Zúñiga, M. (2012) Proteomic and transcriptomic analysis of the response to bile stress of *Lactobacillus casei* BL23. *Microbiology* **158**, 1206–1218.

Andersen, J.M., Barrangou, R., Abou Hachem, M., Lahtinen, S., Goh, Y.J., Svensson, B. and Klaenhammer, T.R. (2011) Transcriptional and functional analysis of galactooligosaccharide uptake by lacS in *Lactobacillus acidophilus*. *Proc Natl Acad Sci U S A* **108**, 17785–17790.

Angel, T.E., Aryal, U.K., Hengel, S.M., Baker, E.S., Kelly, R.T., Robinson, E.W. and Smith, R.D. (2012) Mass spectrometry-based proteomics: existing capabilities and future directions. *Chem Soc Rev* **41**, 3912–3928.

Antikainen, J., Kuparinen, V., Lähteenmäki, K. and Korhonen, K.T. (2007) pH-dependent association of enolase and glyceraldehyde-3-phosphate dehydrogenase of *Lactobacillus crispatus* with the cell wall and lipoteichoic acids. *J Bacteriol* **189**, 4539–4543.

Ashida, N., Yanagihara, S., Shinoda, T. and Yamamoto, N. (2011) Characterization of adhesive molecule with affinity to Caco-2 cells in *Lactobacillus acidophilus* by proteome analysis. *J Biosci Bioeng* **112**, 333–337.

Barrangou, R., Azcarate-Peril, M.A., Duong, T., Conners, S.B., Kelly, R.M. and Klaenhammer, T.R. (2006) Global analysis of carbohydrate utilization by *Lactobacillus acidophilus* using cDNA microarrays. *Proc Natl Acad Sci U S A* **103**, 3816–3821.

Beck, H.C., Feddersen, S. and Petersen, J. (2011) Application of probiotic proteomics in enteric cytoprotection. In J.J. *Probiotic Bacteria and Enteric Infections*. eds. Malago, J.F.J.G. Koninkx and Marinsek-Logar, R. pp. 155–168. Springer Science + Business Media B.V., Dordrecht, Netherlands.

Bergonzelli, G.E. Granato, D., Pridmore, R.D., Marvin-Guy, L.F., Donnicola, D. and Corthésy-Theulaz, I.E. (2006) GroEL of *Lactobacillus johnsonii* La1 (NCC 533) is cell surface associated: potential role in interactions with the host and the gastric pathogen *Helicobacter pylori*. *Infect Immun* **74**, 425–434.

Bernardo, D., Sánchez, B., Al-Hassi, H.O., Mann, E.R., Urdaci, M.C., Knight, S.C. and Margolles, A. (2012) Microbiota/host crosstalk biomarkers: regulatory response of human intestinal dendritic cells exposed to *Lactobacillus* extracellular encrypted peptide. *PLoS ONE* **7**, e36262.

Burns, P., Sánchez, B., Vinderola, G., Ruas-Madiedo, P., Ruiz, L., Margolles, A., Reinheimer, J. and de los Reyes-Gavilán, C.G. (2010) Inside the adaptation process of *Lactobacillus delbrueckii* subsp. *lactis* to bile. *Int J Food Microbiol* **142**, 132–141.

Candela, M., Centanni, M., Fiori, J., Biagi, E., Turroni, S., Orrico, C., Bergmann, S., Hammerschmidt, S. and Brigidi, P. (2010) DnaK from *Bifidobacterium animalis* subsp. *lactis* is a surface-exposed human plasminogen receptor upregulated in response to bile salts. *Microbiology* **156**, 1609–1618.

Chao, T.C. and Hansmeier, N. (2012) The current state of microbial proteomics: where we are and where we want to go. *Proteomics* **12**, 638–650.

Cox, J. and Mann, M. (2011) Quantitative, high-resolution proteomics for data-driven systems biology. *Annu Rev Biochem* **80**, 273–299.

De Angelis, M. and Gobbetti, M. (2004) Environmental stress responses in *Lactobacillus*: a review. *Proteomics* **4**, 106–122.

Domon, B. and Aebersold, R. (2006) Mass spectrometry and protein analysis. *Science* **312**, 212–217.

Domon, B. and Aebersold, R. (2010) Options and considerations when selecting a quantitative proteomics strategy. *Nat Biotechnol* **28**, 710–721.

Fournier, M.L. Gilmore, J.M., Martin-Brown, S.A. and Washburn, M.P. (2007) Multidimensional separations-based shotgun proteomics. *Chem Rev* **107**, 3654–3686.

Gänzle, M.G. and Follador, R. (2012) Metabolism of oligosaccharides and starch in lactobacilli: A review. *Front Microbiol* **3**, 340.

Gibson, G.R. and Glenn, R. (2004) Fibre and effects on probiotics (the prebiotic concept). *Clin Nutr Suppl* **1**, 25–31.

Gibson, G.R., Probert, H.M., Loo, J.V., Rastall, R.A. and Roberfroid, M.B. (2004) Dietary modulation of the human colonic microbiota: updating the concept of prebiotics. *Nutr Res Rev* **17**, 259–275.

Gilad, O., Jacobsen, S., Stuer-Lauridsen, B., Pedersen, M.B., Garrigues, C. and Svensson, B. (2010) Combined transcriptome and proteome analysis of *Bifidobacterium animalis* subsp. *lactis* BB-12 grown on xylo-oligosaccharides and a model of their utilization. *Appl Environ Microbiol* **76**, 7285–7291.

Gilad, O., Svensson, B., Viborg, A.H., Stuer-Lauridsen, B. and Jacobsen, S. (2011) The extracellular proteome of *Bifidobacterium animalis* subsp. *lactis* BB-12 reveals proteins with putative roles in probiotic effects. *Proteomics* **11**, 2503–2514.

Gilad, O., Hjernø, K., Osterlund, E.C., Margolles, A., Svensson, B., Stuer-Lauridsen, B., Møller, A.L. and Jacobsen, S. (2012) Insights into physiological traits of *Bifidobacterium animalis* subsp. *lactis* BB-12 through membrane proteome analysis. *J Proteomics* **75**, 1190–1200.

Granato, D., Bergonzelli, G.E., Pridmore, R.D., Marvin, L., Rouvet, M. and Corthésy-Theulaz, I.E. (2004) Cell surface-associated elongation factor Tu mediates the attachment of *Lactobacillus johnsonii* NCC533 (La1) to human intestinal cells and mucins. *Infect Immun* **72**, 2160–2169.

Guarner, F. and Malagelada, J. (2003) Gut flora in health and disease. *Lancet* **361**, 512–519.

Gygi, S.P., Rist, B., Gerber, S.A., Turecek, F., Gelb, M.H. and Aebersold, R. (1999) Quantitative analysis of complex protein mixtures using isotope-coded affinity tags. *Nat Biotechnol* **17**, 994–999.

Hamon, E., Horvatovich, P., Izquierdo, E., Bringel, F., Marchioni, E., Aoudé-Werner, D. and Ennahar, S. (2011) Comparative proteomic analysis of *Lactobacillus plantarum* for the identification of key proteins in bile tolerance. *BMC Microbiol* **11**, 63.

Hamon, E., Horvatovich, P., Bisch, M., Bringel, F., Marchioni, E., Aoudé-Werner, D. and Ennahar, S. (2012) Investigation of biomarkers of bile tolerance in *Lactobacillus casei* using comparative proteomics. *J Proteome Res* **11**, 109–118.

Henzel, W.J., Billeci, T.M., Stults, J.T., Wong, S.C., Grimley, C. and Watanabe, C. (1993) Identifying proteins from two-dimensional gels by molecular mass searching of peptide fragments in protein sequence databases. *Proc Natl Acad Sci U S A* **90**, 5011–5015.

Hettich, R.L., Sharma, R., Chourey, K. and Giannone, R.J. (2012) Microbial metaproteomics: identifying the repertoire of proteins that microorganisms use to compete and cooperate in complex environmental communities. *Curr Opin Microbiol* **15**, 373–380.

Jeffery, C.J. (2003) Moonlighting proteins: old proteins learning new tricks. *Trends Genet* **19**, 415–417.

Jers, C., Soufi, B., Grangeasse, C., Deutscher, J. and Mijakovic, I. (2008) Phosphoproteomics in bacteria: towards a systemic understanding of bacterial phosphorylation networks. *Expert Rev Proteomics* **5**, 619–627.

Johnson, B., Selle, K., O'Flaherty, S., Goh, Y.J. and Klaenhammer, T. (2013) Identification of extracellular surface-layer associated proteins in *Lactobacillus acidophilus* NCFM. *Microbiology* **159**, 2269–2282.

Kim, J.H., An, H.J., Garrido, D., German, J.B., Lebrilla, C.B. and Mills, D.A. (2013) Proteomic analysis of *Bifidobacterium longum* subsp. *infantis* reveals the metabolic insight on consumption of prebiotics and host glycan. *PLoS ONE* **8**, e57535.

Klose, J. (1975) Protein mapping by combined isoelectric focusing and electrophoresis of mouse tissues. A novel approach to testing for induced point mutations in mammals. *Humangenetik* **26**, 231–243.

Kolida, S. and Gibson, G.R. (2011) Synbiotics in health and disease. *Annu Rev Food Sci Technol* **2**, 373–393.

Konstantinov, S.R., Smidt, H., de Vos, W.M., Bruijns, S.C., Singh, S.K., Valence, F., Molle, D., Lortal, S., et al. (2008) S layer protein A of *Lactobacillus acidophilus* NCFM regulates immature dendritic cell and T cell functions. *Proc Natl Acad Sci U S A* **105**, 19474–19479.

Koponen, J., Laakso, K., Koskenniemi, K., Kankainen, M., Savijoki, K., Nyman, T.A., de Vos, W.M., Tynkkynen, S., et al. (2012) Effect of acid stress on protein expression and phosphorylation in *Lactobacillus rhamnosus* GG. *J Proteomics* **75**, 1357–1374.

Koskenniemi, K., Laakso, K., Koponen, J., Kankainen, M., Greco, D., Auvinen, P., Savijoki, K., Nyman, T.A., et al. (2011) Proteomics and transcriptomics characterization of bile stress response in probiotic *Lactobacillus rhamnosus* GG. *Mol Cell Proteomics* **10**, M110.002741.

Kumar, M., Nagpal, R., Kumar, R., Hemalatha, R., Verma, V., Kumar, A., Chakraborty, C., Singh, B., et al. (2012) Cholesterol-lowering probiotics as potential biotherapeutics for metabolic diseases. *Exp Diabetes Res* **2012**, 902917.

Kuwana, R. and Yamamoto, N. (2012) Increases in GroES and GroEL from *Lactobacillus acidophilus* L-92 in response to a decrease in medium pH, and changes in cytokine release from splenocytes: Transcriptome and proteome analyses. *J Biosci Bioeng* **114**, 9–16.

Lebeer, S., Vanderleyden, J. and De Keersmaecker, S.C. (2010) Host interactions of probiotic bacterial surface molecules: comparison with commensals and pathogens. *Nat Rev Microbiol* **8**, 171–184.

Lee, J.Y., Pajarillo, E.A., Kim, M.J., Chae, J.P. and Kang, D.K. (2013) Proteomic and transcriptional analysis of *Lactobacillus johnsonii* PF01 during bile salt exposure by iTRAQ shotgun proteomics and quantitative RT-PCR. *J Proteome Res* **12**, 432–443.

Lee, K., Lee, H.G. and Choi, Y.J. (2008a) Proteomic analysis of the effect of bile salts on the intestinal and probiotic bacterium *Lactobacillus reuteri*. *J Biotechnol* **137**, 14–19.

Lee, K., Lee, H.G., Pi, K. and Choi, Y.J. (2008b) The effect of low pH on protein expression by the probiotic bacterium *Lactobacillus reuteri*. *Proteomics* **8**, 1624–1630.

Lee, K. and Pi, K. (2010) Effect of transient acid stress on the proteome of intestinal probiotic *Lactobacillus reuteri*. *Biochemistry (Mosc)* **75**, 460–465.

Majumder, A., Sultan, A., Jersie-Christensen, R.R., Ejby, M., Schmidt, B.G., Lahtinen, S.J., Jacobsen, S. and Svensson, B. (2011) Proteome reference map of *Lactobacillus acidophilus* NCFM and quantitative proteomics towards understanding the prebiotic action of lactitol. *Proteomics* **11**, 3470–3481.

Neilson, K.A., Ali, N.A., Muralidharan, S., Mirzaei, M., Mariani, M., Assadourian, G., Lee, A., van Sluyter, S.C. and Haynes, P.A. (2011) Less label, more free: approaches in label-free quantitative mass spectrometry. *Proteomics* **11**, 535–553.

Nilsson, T., Mann, M., Aebersold, R., Yates, J.R. III, Bairock, A. and Bergeron, J.J.M. (2010) Mass spectrometry in high-throughput proteomics: ready for the big time. *Nat Methods* **7**, 681–685.

O'Farrell, P.H. (1975) High resolution two-dimensional electrophoresis of proteins. *J Biol Chem* **250**, 4007–4021.

Ong, S.E., Blagoev, B., Kratchmarova, I., Kristensen, D.B., Steen, H., Pandey, A. and Mann, M. (2002) Stable isotope labeling by amino acids in cell culture, SILAC, as a simple and accurate approach to expression proteomics. *Mol Cell Proteomics* **1**, 376–386.

Pappin, D.J., Hojrup, P. and Bleasby, A.J. (1993) Rapid identification of proteins by peptide-mass fingerprinting. *Curr Biol* **3**, 327–332.

Pfeiler, E.A., Azcarate-Peril, M.A. and Klaenhammer, T.R. (2007) Characterization of a two component regulatory system implicated in the bile tolerance of *Lactobacillus acidophilus* NCFM. *J Bacteriol* **189**, 4624–4634.

Pokusaeva, K., Fitzgerald, G.F. and van Sinderen, D. (2011) Carbohydrate metabolism in *Bifidobacteria*. *Genes Nutr* **6**, 285–306.

Ross, P.L., Huang, Y.N., Marchese, J.N., Williamson, B., Parker, K., Hattan, S., Khainovski, N., Pillai, S., et al. (2004) Multiplexed protein quantitation in *Saccharomyces cerevisiae* using amine-reactive isobaric tagging reagents. *Mol Cell Proteomics* **3**, 1154–1169.

Ruiz, L., Couté, Y., Sánchez, B., de los Reyes-Gavilán, C.G., Sanchez, J.C. and Margolles, A. (2009). The cell-envelope proteome of *Bifidobacterium longum* in an *in vitro* bile environment. *Microbiology* **155**, 957–967.

Ruiz, L., Margolles, A. and Sánchez, B. (2013) Bile resistance mechanisms in *Lactobacillus* and *Bifidobacterium*. *Front Microbiol* **4**, 396.

Sánchez, B., Champomier-Vergès, M.C., Stuer-Lauridsen, B., Ruas-Madiedo, P., Anglade, P., Baraige, F., de los Reyes-Gavilán, C.G., Johansen, E., et al. (2007a) Adaptation and response of *Bifidobacterium animalis* subsp. *lactis* to bile: a proteomic and physiological approach. *Appl Environ Microbiol* **73**, 6757–6767.

Sánchez, B., Champomier-Vergès, M.C., Collado, C.M., Anglade, P., Baraige, F., Sanz, Y., de los Reyes-Gavilán, C.G., Margolles, A. and Zagorec, M. (2007b) Low-pH adaptation and the acid tolerance response of *Bifidobacterium longum* biotype *longum*. *Appl Environ Microbiol* **73**, 6450–6459.

Sanchez, B., Ruiz, L., de los Reyes-Gavilan, C.G. and Margolles, A. (2008a) Proteomics of stress response in *Bifidobacterium*. *Front Biosci* **13**, 6905–6919.

Sánchez, B., Champomier-Vergès, M.C., Anglade, P., Baraige, F., de los Reyes-Gavilan, C.G., Margolles, A. and Zagorec, M. (2008b) A preliminary analysis of *Bifidobacterium longum* exported proteins by two-dimensional electrophoresis. *J Mol Microbiol Biotechnol* **14**, 74–79.

Sánchez, B., Urdaci, M. and Margolles, A. (2010) Extracellular proteins secreted by probiotic bacteria as mediators of effects that promote mucosal-bacteria interactions. *Microbiology* **156**, 3232–3242.

Sánchez, B., Ruiz, L., Gueimonde M., Ruas-Madiedo P. and Margolles, A. (2012) Toward improving technological and functional properties of probiotics in foods. *Trends Food Sci Technol* **26**, 56–63.

Sanders, M.E., Guarner, F., Guerrant, R., Holt, P.R., Quigley, E.M., Sartor, R.B., Sherman, P.M. and Mayer, E.A. (2013) An update on the use and investigation of probiotics in health and disease. *Gut* **62**, 787–796.

Savijoki, K., Lietzén, N., Kankainen, M., Alatossava, T., Koskenniemi, K., Varmanen, P. and Nyman, T.A. (2011) Comparative proteome cataloging of *Lactobacillus rhamnosus* strains GG and Lc705. *J Proteome Res* **10**, 3460–3473.

Schmidt, F. and Völker, U. (2011) Proteome analysis of host-pathogen interactions: investigation of pathogen responses to the host cell environment. *Proteomics* **11**, 3203–3211.

Solis, N. and Cordwell, S.J. (2011) Current methodologies for proteomics of bacterial surface-exposed and cell envelope proteins. *Proteomics* **11**, 3169–3189.

Topping, D.L. and Clifton, P.M. (2001) Short-chain fatty acids and human colonic function: roles of resistant starch and nonstarch polysaccharides. *Physiol Rev* **81**, 1031–1064.

Tsilingiri, K. and Rescigno, M. (2013) Postbiotics: what else?. *Benef Microbes* **4**, 101–107.

Vastano, V., Capri, U., Candela, M., Siciliano, R.A., Russo, L., Renda, M. and Sacco, M. (2013) Identification of binding sites of *Lactobacillus plantarum* enolase involved in the interaction with human plasminogen. *Microbiol Res* **168**, 65–72.

Walther, T.C. and Mann, M. (2010) Mass spectrometry-based proteomics in cell biology. *J Cell Biol* **190**, 491–500.

Washburn, M.P., Wolters, D. and Yates, J.R. 3rd. (2001) Large-scale analysis of the yeast proteome by multidimensional protein identification technology. *Nat Biotechnol* **19**, 242–247.

Wilkins, M.R., Pasquali, C., Appel, R.D., Ou, K., Golaz, O., Sanchez JC, Yan JX, Gooley AA, et al. (1996) From proteins to proteomes: large scale protein identification by two-dimensional electrophoresis and amino acid analysis. *Biotechnology (N Y)* **14**, 61–65.

Wu, R., Sun, Z., Wu, J., Meng, H. and Zhang, H. (2010) Effect of bile salts stress on protein synthesis of *Lactobacillus casei* Zhang revealed by 2-dimensional gel electrophoresis. *J Dairy Sci* **93**, 3858–3868.

Yan, F., Cao, H., Cover, T.L., Whitehead, R., Washington, M.K. and Polk, D.B. (2007) Soluble proteins produced by probiotic bacteria regulate intestinal epithelial cell survival and growth. *Gastroenterology* **132**, 562–575.

Zhao, J. and Cheung, P.C. (2013) Comparative proteome analysis of *Bifidobacterium longum* subsp. *infantis* grown on β-glucans from different sources and a model for their utilization. *J Agric Food Chem* **61**, 4360–4370.

Zhao, Y. and Jensen, O.N. (2009) Modification-specific proteomics: strategies for characterization of post-translational modifications using enrichment techniques. *Proteomics* **9**, 4632–4641.

Chapter 11

Engineering Lactic Acid Bacteria and Bifidobacteria for Mucosal Delivery of Health Molecules

Thibault Allain, Camille Aubry, Jane M. Natividad, Jean-Marc Chatel, Philippe Langella*, and Luis G. Bermúdez-Humarán*

INRA, Commensal and Probiotics-Host Interactions Laboratory, UMR 1319 Micalis, France
AgroParisTech, UMR1319 Micalis, France

11.1. Introduction

For decades, attenuated pathogenic microorganisms have been used as either vaccine delivery vehicles (Roland and Brenneman 2013) or live vectors due to their ability to interact with the mucosal surfaces eliciting an immune response (Wells and Mercenier 2008; Bermudez-Humaran et al. 2011). However, one of the major drawbacks of this strategy is the capacity of these microorganisms to recover their pathogenic potential, rendering them not entirely safe for human use. For this reason, scientists have explored the ability to engineer food-grade lactic acid bacteria (LAB) to be used as live mucosal vectors for the production and *in situ* delivery of therapeutic molecules such as cytokines, antigens, and enzymes.

LAB constitute a heterogeneous group of acid-tolerant Gram-positive bacteria with diverse characteristics. This clade comprises many genera including *Lactococcus*, *Lactobacillus*, *Enterococcus*, and *Streptococcus*. Although not considered part of LAB, species belonging to the bifidobacteria genus are commonly associated with LAB due to their probiotic properties and their ability to produce lactic acid. Both LAB and bifidobacteria can survive the stress conditions (i.e., low pH of the stomach and bile efflux) of the gastrointestinal tract (GIT): *Lactococcus lactis* can survive in the intestine for up to 48 hours, while lactobacilli and bifidobacteria persist for 4 and 7 days, respectively. For many centuries, these non-pathogenic microorganisms have been commonly used for fermentation, food production, and preservation; as such, the United States Food and Drug Administration (FDA) consider them as Generally Recognized As Safe (GRAS). Over the past 20 years, more than 100 peer-reviewed publications have demonstrated the potential of these bacteria as live delivery vectors. This chapter focuses on the delivery vector potential of bacterial species belonging to the genera of *Lactococcus*, *Lactobacillus*, and *Bifidobacterium* as a novel strategy to prevent and treat a myriad of human diseases.

* Corresponding authors: INRA, Commensal and Probiotics-Host Interactions Laboratory, UMR 1319 Micalis, Jouy-en-Josas, France. AgroParisTech, UMR1319 Micalis, F-78350 Jouy-en-Josas, France.

Biotechnology of Lactic Acid Bacteria: Novel Applications, Second Edition. Edited by Fernanda Mozzi, Raúl R. Raya, and Graciela M. Vignolo.

11.2. *Lactococcus lactis*: A Pioneer Bacterium

Lactococcus lactis is a well-known food-grade LAB species commonly used in the food-processing industry, such as in the production of fermented milk products. This bacterium is fully characterized in terms of taxonomy and physiology. Furthermore, the genome of several strains has been successfully sequenced (Bolotin et al. 2001; Wegmann et al. 2007). In addition to its safety profile (i.e., GRAS FDA's status), this bacterium does not colonize the GIT and thus has been classified as a non-invasive microorganism (Steidler et al. 1998; Chatel et al. 2001; Nga 2005).

L. lactis is considered a model LAB for mucosal delivery of therapeutic molecules and production of heterologous proteins (Morello et al. 2008). Indeed, this bacterium possesses a number of advantages that make it ideal as a live vector. First, *L. lactis* exhibits a high transformation efficiency by electroporation (Gerber and Solioz 2007). This phenotype is in contrast to other kinds of Gram-positive bacteria, which are usually hard to transform. To improve the ability to produce and deliver molecules, several genetic tools including plasmids and expression systems (i.e., promoters, peptide signals) and transformation strategies have been developed. Second, this bacterium possesses simple secretory machinery and secretes only one protein, Usp45, which can be detected quantitatively (Nouaille et al. 2003); this fact is particularly important as it facilitates heterologous protein purification processes. In addition, laboratory strains have been developed to inactivate HtrA, the unique protease present on the extracellular surface of *L. lactis*, to avoid any extracellular degradation of recombinant proteins (Miyoshi et al. 2002). These *L. lactis* unique features make it a suitable candidate as a mucosal delivery vector.

11.3. *Lactobacillus* spp. as a Delivery Vector

Although *L. lactis* represents a promising candidate for a live mucosal vector delivery system, some laboratories have further explored the capacity of other bacteria with better intrinsic properties such as lactobacilli. Specific strains of lactobacilli have been characterized to possess probiotic properties (microorganisms that provide health benefits when consumed). Furthermore, some lactobacilli species such as *Lact. casei*, *Lact. acidophilus*, *Lact. gasseri*, *Lact. plantarum*, and *Lact. fermentum* are able to persist for longer periods than *L. lactis* in the GIT. Additionally, some species of lactobacilli are part of the microbiota that colonized the mammalian small intestine. The interaction of lactobacilli with the intestinal mucosa is an important factor because this enhances mucosal immune responses. Since the end of the 1990s, researchers have been exploiting these inherent properties of lactobacilli to make them an attractive alternative and safer delivery system for molecules and compounds with health benefit (Rush et al. 1995; Pouwels et al. 1996).

Similar to *L. lactis*, many genetic tools have been improved over the last decade for lactobacilli to produce heterologous proteins. Identically to *L. lactis*, lactobacilli have a thick cell wall, and there is huge molecular and biochemical interspecies variation among this genus. Despite the significant improvements made on the delivery of more than 50 molecules, lactobacilli are still difficult to transform and genetically manipulate (Pouwels et al. 1998; LeBlanc et al. 2013).

11.4. Bifidobacteria as a New Live Delivery Vehicle

As previously mentioned, Bifidobacterium does not belong to the LAB group (Actinobacteria phylum); however, this genus shares many properties with lactobacilli, such as their probiotic characteristic and their ability to persist in the GIT for a long period of time. Bifidobacteria are also known to be a major component of mammals' microbiota. Thus, these bacteria are excellent alternatives to be used as mucosal delivery vectors of heterologous proteins and to enhance innate and adaptive immune responses. However, they lack a well-characterized expression system. Furthermore, they remain hard to transform, and transformation protocols differ from one strain to another due to interspecies variation (Serafini et al. 2012).

Over the last decades, the capacity of LAB to produce and deliver molecules, such as proteins and DNA of health interest, has been widely explored. As described previously, *L. lactis* currently remains the best candidate among LAB for mucosal approaches due to the ease and availability of genetics tools to transform this bacteria. Nevertheless, unlike *Escherichia coli*, which is a Gram-negative bacterium, LAB are still unique as they do not possess LipoPolySaccharides (LPS). LPS have a local inflammatory effect and are known to be involved in several diseases such as Inflammatory Bowel Disease (IBD) and obesity. For this reason, LAB are better candidates as *in vivo* vehicles of recombinant proteins than *E. coli*. The equivalent of LPS in LAB, the LipoTeichoic Acid (LTA), has been shown to be only toxic at high concentrations (Lebeer et al. 2012). Finally, LAB are very good live vehicles to *in situ* produce and deliver therapeutic molecules to the GIT.

11.5. Engineering Genetic Tools for Protein and DNA Delivery

The delivery of heterologous proteins and DNA by LAB requires efficient and reliable genetic tools such as cloning vectors and various expression systems. A cloning vector plasmid is defined as a small circular extra-chromosomal self-replicating DNA molecule that can be present in low or high-copy number inside the bacterial cells. This plasmid carries an expression cassette containing either a constitutive or an inducible promoter upstream of the gene of interest to control heterologous protein production. In principle, production of heterologous antigens by recombinant LAB can be performed in three different cellular locations: (i) intracellular, which protects the recombinant protein from the drastic environmental conditions (such as gastric juices, bile salts. and low stomach pH). However, the release would require cellular lysis for protein delivery; (ii) extracellular, which allows a direct interaction of the recombinant protein with the environment (food product or the digestive tract); and (iii) cell wall-attached, a cellular location that combines the advantages of the two former cellular locations, that is, interactions between the cell wall-anchored protein and the environment limiting proteolysis degradation (Bermudez-Humaran et al. 2004).

Herein, we review the evolution and latest scientific advances of these genetic tools developed for LAB, more specifically for *L. lactis* and lactobacilli.

11.5.1. Cloning Vectors

For heterologous protein expression, a cloning vector must contain at least a functional origin of replication (*ori*) suitable for the host microorganism, one or more selection markers such as the antibiotic resistance gene or the wild-type gene for auxotrophic bacterial strains, and a multiple cloning site.

The main replication mechanisms for bacterial circular plasmids are theta-type, strand displacement, and rolling circle. Different elements are involved in this replication process; this includes the *ori* site and the Rep protein gene which participate in the initiation of replication (del Solar et al. 1998).

Selection markers are essential to allow the selection of transformant cells harboring the plasmid of interest. For decades, antibiotic resistance genes have been the most commonly used markers, including ampicillin, chloramphenicol, erythromycin, and tetracycline. Nowadays, in order to comply with the requirements of the food and pharmaceutical industries, laboratories are developing food-grade vectors without antibiotics. Among these, new alternative selection markers consist of using a plasmid expressing the wild-type gene marker (Bron et al. 2002) to complement growth of an auxotrophic bacteria.

11.5.2. Expression Systems

Several expression systems were first developed in *L. lactis* in order to express and control heterologous protein production (Bermudez-Humaran 2009). To do so, several constitutive promoters were initially isolated from a genomic library (van der Vossen et al. 1987) with varying protein expression levels (de Vos et al. 1999). However, expression of constitutive protein may lead to a cytoplasmic accumulation.

Inducible promoters were developed in *L. lactis* and later adapted to other LAB strains to control the gene expression of interleukins, antigens, or enzymes. This strategy regulates the expression and increases the production efficiency of therapeutic molecules, a pivotal advantage in mucosal targeting (Morello et al. 2008; Benbouziane et al. 2013). The most commonly used regulation systems in *L. lactis* and lactobacilli and future trends of this technology are the following (Figure 11.1):

1. **Nisin-controlled gene expression (NICE)**: The NICE system is the most commonly used inducible expression system for heterologous proteins production in LAB. The synthesis,

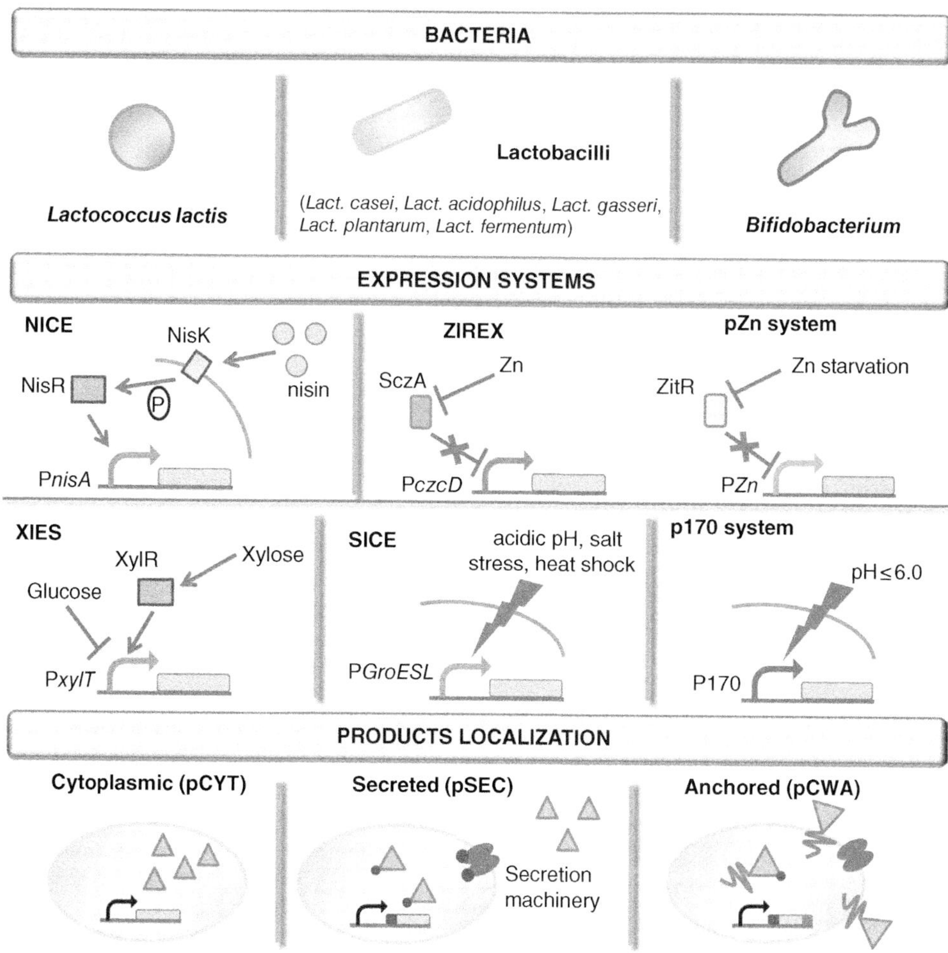

Figure 11.1. Schematic representation of broad choice of bacteria, expression systems, and product localizations mainly used in mucosal delivery.

post-translational modifications, and secretion of nisin, which is an antimicrobial peptide produced by some *L. lactis* strains, are regulated by the *nis* operon (11 chromosomal genes *nisABTCIPRKFEG*) (De Vos and Gasson 1989; Wells et al. 1993b; Eichenbaum et al. 1998). The NICE system consists of the nisA (PnisA), the nisF (PnisF), and the nisRK promoters. Upon detection of nisin (which is encoded by the *nisA* gene) by the nisRK regulator system complex, P*nisA* and P*nisF* promoters are induced and activated; this subsequently leads to gene transcription (Kuipers et al. 1993).

The laboratory strain *L. lactis* NZ9000 is a derivative of the *L. lactis* MG1363 strain carrying the *nisRK* genes into its chromosome. When the gene of interest is located downstream of P*nisA*, its expression can be triggered by adding minimal concentrations of nisin (0.1 to 10 ng/ml). This system is still considered the most efficient one for heterologous expression in *L. lactis* (Mierau and Kleerebezem 2005) and other LAB, and allows higher induction of protein quantities (Mierau et al. 2005).

2. **Zinc-inducible expression systems (P_{ZN}ZitR-driven Heterologous Expression and Zirex)**: In 2004, Llull and Poquet designed a new expression system based on the promoter from the *L. lactis zit* operon, which is involved in zinc regulation, the PznZitR-driven Heterologous Expression (Llull and Poquet 2004). In the absence of zinc, such as during starvation induced by EDTA or when using a chemically defined medium, ZitR, the transcriptional repressor of P*zn*, becomes inactive and the RNA polymerase can bind to the P*zn* promoter. The authors have shown that this expression system is highly inducible but the protein quantity obtained is lower compared to the NICE system.

   More recently, a new zinc-inducible expression system, called Zirex, was developed for *L. lactis* (Mu et al. 2013). When triggered by a non-toxic concentration of zinc, the streptococcal promoter PczcD (regulated by the SczA repressor) is activated and leads to elevated expression levels. Moreover, Mu and colleagues have successfully combined both the NICE system and Zirex to produce different proteins at different times in the same microorganism.
3. **Xylose-inducible expression system (XIES)**: XIES is a sugar-dependent system engineered to produce either cytoplasmic or secreted proteins, through the utilization of a xylose-inducible promoter, P*xylT*. This expression system was first described for heterologous protein delivery in *L. lactis* NCDO2118 (Miyoshi et al. 2004) using the *Staphylococcus aureus* nuclease gene, with or without protein targeting (using the Usp45 signal peptide). The system exhibits high amounts of protein production when induced by sugars such as xylose or glucose. This system constitutes a good alternative to the NICE system as its induction is strongly regulated (Miyoshi et al. 2004) and no protein degradation has been observed when secreted (de Azevedo et al. 2012).
4. **pSIP system**: Sorvig and colleagues (2003) developed another system exclusively suitable for inducible expression of lactobacilli genes (initially for two lactobacilli species *Lact. sakei* and *Lact. plantarum*). This system (SIP) is involved in the production of bacteriocins (sakacin A and P) in *Lact. sakei*. Upon stimulation by a peptide pheromone (IP), a protein kinase (cognate histidine protein kinase) is expressed; this then in turn activates a response regulator, leading to the bacteriocin production. Lactobacilli genomes were further screened to enhance expression yields and secretion efficiency (Karlskas et al. 2014). This system, which has been successfully used to produce several heterologous proteins in more than 10 Lactobacillus species (Sorvig et al. 2003; Halbmayr et al. 2008; Nguyen et al. 2011), can be currently applied to cytoplasmic expression, cell-wall anchoring, and secretion.
5. **Self-regulated promoters by environmental conditions: A new generation of inducible promoters has been developed** to dispense with the use of inducers. Indeed, for *in vivo* experiments, LAB harboring expression systems such as NICE, XIES, or SIP need to be induced beforehand

to produce heterologous protein. Here, the "Stress-Induced Controlled Expression" (SICE) and P170, both regulated by environmental conditions found in the GIT, are described.

Stress-Induced Controlled Expression (SICE): The SICE system has been recently described in lactococci, which has been helpful in designing alternative transformation strategies (Benbouziane et al. 2013). This "regulatory genes-free" expression system, controlled by the pGroESL promoter, is activated only under certain stress conditions such as the presence of bile salts and NaCl, temperature, and acidic pH, albeit a low constitutive expression may occur. Moreover, this system represents good potential in the production and delivery of therapeutic proteins *in situ* at mucosal surfaces in several human diseases such as IBD and human papillomavirus type-16 (HPV-16) (Benbouziane et al. 2013).

P170 system: Another system regulated by environmental conditions has been developed using the properties of LAB to acidify their environment. Indeed, the pH-inducible promoter P170, which is induced when a pH value of 6.0 or lower is reached, has been identified in *L. lactis* (Madsen et al. 1999). Therefore, when used for heterologous protein production, the P170 is self-activated at low pH and leads to high levels of recombinant protein production.

Signal peptide Engineering bacterial vectors for mucosal delivery of therapeutic molecules requires mastering their targeting mode. Certainly, protein locations may have a strong influence on efficiency of the treatment and immune response modulation (Chatel et al. 2001; Chatel et al. 2003; Bermudez-Humaran et al. 2005; Karlskas et al. 2014). Moreover, not all cellular location systems are suitable for some compounds such as insoluble proteins. The production of antigens, interleukins, or enzymes can occur in three different cellular locations as indicated previously (Figure 11.1). The three main expression vectors that have been developed depending on the desired location: pCYT (intracellular), pSEC (secreted form) and pCWA (cell-wall anchored) are presented:

pCYT: When using the pCYT system, the protein of interest is produced and remains inside the bacterial cell in the absence of a signal sequence. Therefore, this strategy relies on cellular lysis (pH, gastric juices, bile) to deliver accumulated recombinant protein to the extracellular environment after oral administration.

pSEC: This strategy consists in releasing the protein into the extracellular environment via a signal peptide (SP). The SP (*N*-terminal part), which is hydrophobic, is recognized by the secretion machinery (signal peptidase) leading the protein to be secreted (Bermudez-Humaran et al. 2011). For instance, for lactococci strains, the Usp45 SP has been widely used for interleukin secretion showing good yields. However, with this system, delivered proteins are instantly subject to proteolytic degradation encountered in the GIT. This secretion system can be optimized by adding a synthetic propeptide (LEISSTCDA) to enhance the production-secretion (Le Loir et al. 1998). Indeed, many studies exhibit an increase of production and secretion (4- to 10-fold higher than normal secretion) leading to a better immune response.

pCWA: In some cases, an alternative system is required when neither cytoplasmic nor secreted forms exhibit good immune response. To circumvent this problem, anchoring the product to the cell wall is recommended. To do so, cell-wall anchoring (CWA) combines a *N*-terminus signal sequence to translocate the protein outside the cell, and a *C*-terminus motif (LPXTG) allowing its anchoring to the cell surface (glycan peptide) by covalent bounds (Robinson et al. 1997; Robinson et al. 2004; Bermudez-Humaran et al. 2005; Dieye et al. 2010). Interestingly, several studies showed that for mucosal vaccination, cell-wall anchored antigens exposed to the surface of LAB result in a stronger immune response compared to intra- or extracellular strategies (Bermudez-Humaran et al. 2004; Wells and Mercenier 2008).

11.6. Therapeutic Applications

11.6.1. Inflammatory Bowel Disease (IBD)

The term inflammatory bowel disease (IBD) refers to a group of gut disorders characterized by an uncontrolled and chronic inflammation of the GIT. Epidemiological evidences suggest that the incidence of IBD is increasing dramatically worldwide. The two major forms of IBD are Crohn's disease (CD) and ulcerative colitis (UC). Although the main symptoms, which include chronic abdominal pain and cramping, bloody diarrhea, weight loss, ulcers, fever, and intense tiredness, are similar for both CD and UC, certain disease features are different between the two conditions (Kellermayer 2012). For instance, the inflammation for CD varies from mucosal to transmural (different layers of the intestinal epithelium), whereas in UC, the inflammation is non-transmural. Additionally, CD can affect any part of the GIT while UC is mainly restricted to the colon (Baumgart and Sandborn 2007).

These diseases are the result of the pertubation of the complex balance among commensal microbiota, intestinal epithelial cells, and the immune system. When this interaction is disturbed, an abnormal and significant production of inflammatory mediators, such as proteases, reactive oxygen species (ROS), and pro-inflammatory cytokines, might occur (Segui et al. 2004; LeBlanc et al. 2013).

Nowadays, therapies for IBD mostly target the symptoms and there are no efficient treatments (Levine and Lukawski-Trubish 1995). In this context, there is a need to find alternative therapies for IBD. Recent findings highlight that the microbiota of IBD patients, characterized by a significant decrease of *Faecalibacterium prausnitzii*, lactobacilli, and bifidobacteria, are different compared to healthy individuals (Ott et al. 2004; Mylonaki et al. 2005; Conte et al. 2006; Sokol et al. 2008). For this reason, there is an increased interest in studies looking at the ability of commensal microorganisms, including probiotic strains of lactobacilli and bifidobacteria, to prevent or ameliorate inflammation. Indeed, it has been shown that probiotic bacteria stimulate the microbiota, influence the intestinal barrier function such as intestinal permeability (Isolauri et al. 2004), modulate host immune responses, and reduce oxidative stress by producing antioxidant enzymes (Clifford et al. 1989; Tamboli et al. 2003; Marteau et al. 2004).

As we have previously shown, the ability of LAB to be genetically engineered combined with their inherent anti-inflammatory and immunomodulatory properties make them ideal candidates for the prevention and treatment of IBD. Here, the use of recombinant lactococci and lactobacilli producing and/or secreting anti-proteases (elafin), antioxidant enzymes (SOD), as well as cytokines to prevent and treat intestinal disorders is reviewed.

11.6.2. Anti-protease Enzyme-producing LAB: The Tole of Elafin

Recent discoveries have shown that proteases and endogenous anti-proteases play an important role in IBD. Specifically, we observe an increase in proteolytic activity in the gut of IBD patients (Cenac et al. 2007; Schmid et al. 2007), due to a protease upregulated expression and/or decrease expression of their endogenous inhibitors, such as Elafin (Sallenave 2010; Shaw and Wiedow 2011). Motta and colleagues showed that transgenic mice producing Elafin are protected from experimentally induced colitis that mimics IBD in mice (Motta et al. 2011). These authors further explored the anti-inflammatory properties of this protease inhibitor using recombinant LAB as a delivery system. Specifically, *L. lactis* and *L. casei* have been genetically engineered to produce and secrete Elafin. Oral administration of this recombinant LAB to mice subjected to experimental colitis reduced local inflammation and epithelial damage, restored colon intestinal homeostasis, and normalized elastolytic activity *in vivo* (Motta et al. 2012). Furthermore, *in vitro* experiments using human intestinal epithelial cells (IEC) from inflamed intestine showed that these recombinant LAB are efficient in normalizing the intestinal permeability of these cells and preventing the increased production of pro-inflammatory cytokines and chemokines. Overall, these results suggest that oral delivery of LAB secreting Elafin may be useful for treating IBD in humans.

11.6.3. Antioxidant Enzyme-producing Lactococci and Lactobacilli

It is well known that IBD is associated with oxidative stress; IBD patients have increased concentrations of reactive oxygen species (ROS) in the GIT and impaired antioxidant defenses (Weitzman and Gordon 1990). Under normal conditions, ROS, such as hydroxyl radicals, superoxide anions, and hydrogen peroxide, take part in cell signaling and regulatory pathways. Additionally, ROS are involved in the course of anti-infective processes by inducing cell damage (Berlett and Stadtman 1997). However, in some diseases such as IBD, there is a high concentration of ROS, causing major oxidative damage in the intestine (Farr and Kogoma 1991). To counteract this phenomenon, the normal intestinal mucosa secretes antioxidant enzymes to regulate and maintain the ROS level (catalase, glutathione peroxidase, glutathione reductase, glutathione-S-transferase, and superoxide dismutase), but the levels of these enzymes are significantly lower in IBD patients (Lih-Brody et al. 1996; Kruidenier et al. 2003).

Recent studies have shown that some LAB probiotic strains may play a protective role in IBD by expressing antioxidant enzymes (Kruidenier et al. 2003; Gosselink et al. 2004; LeBlanc et al. 2008). Therapeutic uses of recombinant LAB to locally deliver bioavailable antioxidant enzymes, such as superoxide dismutase (SOD) and catalase (CAT), have been explored to decrease ROS levels in the GIT in order to cure IBD.

LAB expressing SOD and CAT enzymes SOD is a metalloenzyme that converts ROS (superoxide anion) into H_2O_2 (Wispe et al. 1989; Koc et al. 2003). Previous studies displayed successful treatment of experimental colitis with SOD, by reducing peroxidation reactions in the gut as well as decreasing oxidative stress (Segui et al. 2004). SOD, which has very short circulatory half-life and exists in three different forms (manganese, copper–zinc, and iron forms), is not inherently present in *Lactobacillus* spp.

Ten years ago, Bruno-Barcena et al. reported on the protective role of lactobacilli strains expressing recombinant SOD manganese from *Strep. thermophilus* (*sodA*) against hydrogen peroxide *in vitro* (Bruno-Barcena et al. 2004). Then, Han and colleagues tested and confirmed the ability of genetically engineered *Lact. plantarum* and *L. lactis* strains *in situ* producing and releasing heterologous SOD to protect mice from TNBS-induced colitis, suggesting the anti-inflammatory role of SOD *in vivo* (Han et al. 2006).

Furthermore, recent evidence has shown that recombinant *Lact. gasseri* and *Lact. plantarum* producing manganese-specific SOD (MnSOD) are able to prevent and cure colitis in IL-10-deficient mice (Carroll et al. 2007) and in TNBS-induced colitis (Han et al. 2006). Both studies revealed that there is significant anti-inflammatory activity as well as a reduction of the infiltration of neutrophils and macrophages compared to untreated controls. Following on from these promising results, Watterlot and colleagues enhanced this strategy by cloning MnSOD from *L. lactis* into *Lact. casei* BL23, and compared the efficiency of this strategy in combination with *Lact. casei* BL23 expressing and producing CAT, another antioxidant enzyme (Watterlot et al. 2010). Tested on a murine model with dextran sodium sulfate (DSS) induced colitis, they observed histologically that the *Lact. casei* SOD and *Lact. casei* CAT/*Lact. casei SOD* combination reduced cecal and colonic inflammation. Similar findings have been reported by Leblanc et al. (LeBlanc et al. 2011).

More recently, a pioneer study explored the use of bifidobacteria as mucosal vector of antioxidant enzymes (Zuo et al. 2014). *Bifidobacterium longum* was genetically engineered to express the SOD gene from *Strep. thermophilus* or the heme-dependent catalase (KAT) gene from *Lact. plantarum* as well as a combination of both genes. By evaluating their oxidative-stress resistance, these recombinant strains were able to degrade ROS (H_2O_2 and peroxides) and cell viability was improved in both *B. longum* SOD and *B. longum* KAT compared to controls. These encouraging results open up great vistas in the use of genetically engineered lactobacilli and bifidobacteria producing antioxidant enzymes to prevent or cure IBD caused by oxidative stress.

LAB as cytokine delivery vector It is now recognized that regulatory T cells (Treg) play a central role in gut homeostasis by maintaining immune tolerance and controlling inflammation. Indeed, a hyper-activation of CD4(+) T helper cells leads to an excessive and uncontrolled immune response, such as in the case of CD and UC, which have been putatively described to possess the Th1/Th17 and Th2 phenotype respectively (Ruddy et al. 2004; Leon et al. 2006). Several studies have shown that certain profiles of cytokines, which are produced by immune cells and involved in T cell activation and differentiation, have been directly associated with IBD. In particular, cytokines such as IFN- γ, IL-17, IL-22, and IL-23 are dysregulated in IBD (Sarra et al. 2010).

On the other hand, increasing evidences relate the immunomodulatory properties of probiotic LAB and their beneficial role in mucosal inflammation. For instance, some LAB strains can increase anti-inflammatory cytokine levels such as IL-10 and downregulate the ongoing inflammation. These crucial observations prompted scientists to use genetically modified LAB expressing cytokines as an alternative treatment for IBD.

In a pioneering study, Steidler and colleagues (2000) showed that recombinant *L. lactis* strain secreting murine IL-10 reduced inflammation in DSS induced colitis and prevented colitis in IL-10 knockout mice (Schotte et al. 2000). Furthermore, this team scaled up to clinical trials with *L. lactis* expressing mature human IL-10, and observed decreased disease activity in Crohn's disease patients (Braat et al. 2006). In terms of safety concerns, this strain had been genetically modified by replacing the thymidylate synthase gene (*thyA*) with the synthetic expression cassette for human IL-10. This specific strain, which is unable to survive in the GIT environment without supplementation, is a feasible option and raised substantial expectations for the treatment of diseases using safer, live genetically modified bacteria.

In a recent study, the ability of LAB to transfer eukaryotic DNA to epithelial cells was explored to deliver murine IL-10 to the intestinal mucosa (Zurita-Turk et al. 2014). A successful transfer of the plasmid containing the mIL-10 cDNA cassette to host epithelial cells was achieved when an invasive *L. lactis* laboratory strain harboring the pValac:IL-10 was administered to mice subjected to DSS-induced inflammation. Delivery was associated with significantly diminished intestinal inflammation characterized by an increase of anti-inflammatory cytokines. Plasmid transfer is an attractive alternative to deliver active molecules to cure IBD.

Recently, the use of recombinant *L. lactis* expressing the immunomodulatory interleukin-27 (IL-27) as a novel approach for the treatment of IBD was reported. IL-27 plays an important role in both innate and adaptive immunity by mediating pro- and anti-inflammatory effects. As a matter of fact, IL-27 can initiate a Th1 response and have a suppressive role on Th2 and Th17 differentiation and proliferation (Yoshimura et al. 2006). Moreover, this cytokine can induce the production of anti-inflammatory cytokines such as IL-10.

Hanson and colleagues (2014) studied the protective role of orally administered recombinant IL-27 expressing *L. lactis* using two types of mice model of inflammation: T cell transfer-induced enterocolitis or DSS-induced colitis (Hanson et al. 2014). In both experiments, mucosal delivery of *L. lactis* IL-27 significantly reduced colitis by lowering inflammatory cytokines levels such as IL-17 and increasing the production of IL-10. In addition, *L. lactis* IL-27 was more effective than *L. lactis* IL-10 in reducing induced colitis, implying that this strategy is a better treatment for IBD.

11.7. Allergy

11.7.1. Use of LAB in Food Allergy

Cow's milk allergy is one of the most common food allergies in young infants, with an incidence of 2–3% in the first years of life, characterized by gastrointestinal and respiratory symptoms (Exl and Fritsche 2001; Crittenden and Bennett 2005; Vandenplas et al. 2014). Allergy to cow's milk is mainly due to a strong immunologic response to milk protein, such as the bovine β-lactoglobulin (BLG). In order to evaluate the immune response induced by BLG, Chatel and colleagues (2001) administrated recombinant *L. lactis* BLG in different forms (cytoplasmic and extracellular) in mice, using the NICE system (Chatel et al. 2001). After oral administration of *L. lactis* expressing and secreting BLG, BLG-specific fecal IgA was detected and mice were protected from further sensitization. In addition, intranasal co-administration of *L. lactis* BLG and *L. lactis* producing IL-12 induced BLG-specific IgG2a and interferon gamma (IFN-γ), leading to a protective Th1 response (Cortes-Perez et al. 2007).

The main drawback of this strategy is that BLG is mostly expressed and secreted in denatured form in *L. lactis*. However, *L. lactis* strains have been engineered to deliver plasmids containing the BLG expression cassette directly into epithelial cells *in vitro* (Guimaraes et al. 2006) and *in vivo* in order to produce *in situ* the native conformation, and thus enhance specific immune response. This strategy, previously explained in this chapter, consists of transferring an expression cassette encoding BLG cDNA under the control of the eukaryotic cytomegalovirus promoter (Chatel et al. 2008). The production and secretion of native BLG was demonstrated through co-incubation with Caco-2 epithelial cells. Moreover, after oral administration, the plasmid containing the BLG expression cassette and the native BLG were detected in the mucosa. BLG specific IgG2a were detected in the sera indicating a pro-inflammatory Th1 immune response. After sensitization, mice treated with *L. lactis* delivering plasmids showed lower levels of IgE in comparison with untreated mice.

Lately, Pontes and colleagues (2012) developed invasive recombinant *L. lactis* strains to increase plasmid transfer and host–bacteria crosstalks. However, by expressing invasins such as *Staphylococcus aureus* Fibronectin Binding Protein A (FnBPA) and *Listeria monocytogenes* Internalin A (mInlA) at their surface (Innocentin et al. 2009), neither *L. lactis* FnBPA nor *L. lactis* mInLA enhanced plasmid transfer *in vivo* (Pontes et al. 2012; Pontes et al. 2014). Moreover, these recombinant strains induced an anti-inflammatory Th2 immune response, which might not be ideal for allergy treatments.

In another study, *L. lactis* secreting murine IL-10 was developed to prevent food-induced IgE sensitization (Frossard et al. 2007). A significant decrease in allergic reactions such as anaphylaxis was observed in treated mice, with a significant inhibition of BLG-specific IgE and IgG1 production. Overall, the evidence confirms the potential of recombinant LAB as a novel approach to treat or understand the pathogenesis of food allergy.

11.7.2. Allergic Airways Diseases

Asthma is a chronic inflammatory disease strongly correlated with airway inflammation driven by Th2 cytokines (i.e., IL-4 and IL-5) secreted by allergen-specific CD4(+) T cells. Recent evidence suggests that asthmatic patients have increased Th2 responses with reduced Th1 responses and consequently smaller amounts of secreted IFN-γ (a pro-inflammatory cytokine). This cytokine as well as IL-12, a Th1-promoting cytokine, have been successfully used as regulators of allergic asthma. Wu and colleagues (2006) evaluated the immunomodulatory effects of a recombinant strain of *L. lactis* secreting biologically active IL-12 in a murine model of OVA-induced asthma (Wu et al. 2006). Intranasal mouse administration with this strain resulted in a switch from Th2 to Th1 with increased IFN-γ and decreased IL-4 levels. In addition, significant decreases in airway hyper-responsiveness and pulmonary inflammation were also observed in mice administered the recombinant *L. lactis*. The ability of this strain to decrease airway hyper-responsiveness and eosinophilia suggests the potential of our approach as a novel therapy for the treatment of chronic asthma.

Also, the use of a recombinant *L. lactis* strain producing murine IL-10 has been explored for its putative immunomodulatory effects in a murine model of OVA-induced asthma (Marinho et al. 2010). A significant decrease in IgE and IgG1 levels as well as a reduction of pulmonary inflammation were observed in treated mice compared with the asthmatic group. These encouraging results suggest that live vectors are an efficient therapeutic tool to ameliorate allergic airway inflammation.

11.8. Autoimmune Diseases

11.8.1. Type 1 Diabetes Mellitus

Type 1 diabetes mellitus (T1DM) is an autoimmune disease characterized by the destruction of pancreatic beta cells, which results in insulin deficiency and hyperglycemia. Presently, there is no efficient cure against T1DM, even though insulin administration remains the best treatment.

Many different strategies have been investigated to cure T1DM, including recombinant LAB strains secreting auto-antigenic peptides. For instance, the human insulin gene has been successfully expressed in *L. lactis* and *Lact. casei* (Chen et al. 2007). Insulin specific antibodies were detected in the sera of non-obese diabetic mice (NOD), and immune tolerance was induced. Numerous studies have demonstrated the role of regulatory T cells in this disease. To induce a suppression of T cell response and restore tolerance, Takiisih and colleagues (2012) combined orally administered *L. lactis* secreting both pro-insulin antigen and IL-10, with a low dose of systemic anti-CD3 (Takiishi et al. 2012). This combination was associated with remission of diabetes in NOD mice, and long-term tolerance was induced in 59% of treated mice. Recently, new immunotherapies based on the administration of recombinant LAB producing self-antigens, which are involved in diabetes, have been developed. Mucosal delivery of heterologous antigens such as glutamic acid decarboxylase (GAD)-65 (Robert et al. 2014), HSP65, and P277 (Ma et al. 2014) showed a reduction in the incidence of DM1 in NOD mice. Therefore, the administration of LAB-based vaccines could be an effective treatment strategy for type 1 diabetes in humans.

11.8.2. Celiac Disease

Celiac disease (CD) is a life-long autoimmune disease characterized by an insufficient tolerance towards gliadin, one of the main components of gluten, in genetically susceptible individuals. There is currently no cure for CD except a strict gluten-free diet (GFD) treatment, which is both expensive and restrictive. Human leukocyte antigens HLA-DQ2 and/or HLA-DQ8-restricted T cell responses mediate the loss of gliadin tolerance in CD (Stepniak and Koning 2006). A genetically modified *L. lactis* secreting the deamidated DQ8 (eDQ8d)-specific gliadin epitope have been administered in NOD AB^0 DQ8 transgenic mice as a means of inducing antigen (Ag)-specific tolerance and to treat CD (Huibregtse et al. 2009); this leads to a suppression of local and systemic DQ8-restricted T cell responses through an upregulation of Foxp3(+) T-cells. Thus, these findings suggest that recombinant LAB holds great promise in restoring Ag-specific tolerance.

More recently, scientists have explored the role of Elafin, an endogenous protease inhibitor, in CD. It is well known that the transglutaminase (TG-2) plays a key role in gliadin peptide deamidation, which increases the binding affinity of gliadin peptides to HLA-DQ2/DQ8 molecules (Kristiansen et al. 1998). Based on the fact that Elafin has been recently identified as a substrate for TG-2 (Baranger et al. 2011), Galipeau et al. orally administered a recombinant *L. lactis* strain expressing Elafin to gluten-sensitive mice to target and inhibit TG-2 activity (Galipeau et al. 2014). Small intestinal barrier functions and inflammation of treated mice were restored. These beneficial effects suggest that mucosal delivery of Elafin in CD patients may be a novel effective therapeutic approach for gluten-related disorders.

11.9. Infectious Diseases

11.9.1. Mucosal Delivery of Bacterial Antigens

Surface protein antigen of Streptococcus mutans In a pioneer study, Iwaki *et al.* (1990) constructed an inactivated recombinant *L. lactis* strain producing a surface protein antigen (PAc) of *Strep. mutans*, to develop an oral vaccine against dental caries (Iwaki et al. 1990). Mice orally immunized presented significant salivary IgA and serum IgG-specific responses to PAc in comparison with non-treated mice. These original findings described the use of *L. lactis* for mucosal delivery of antigens of interest and paved the way for alternative strategies in vaccination and immunization.

Tetanus Toxin (TTFC) The tetanus toxin is a neurotoxin produced by *Clostridium tetani* under certain conditions. The use of *L. lactis* as a live protein delivery vector was first published by Wells et al. with the fragment C of tetanus toxin (TTFC) to provide a proof-of-concept (Wells et al. 1993a). Mice immunized with subcutaneous injection of *L. lactis* strains producing TTFC were successfully protected against lethal challenge with tetanus toxin. Both oral (Robinson et al. 1997) and nasal (Norton et al. 1997) administration strategies induced similar systemic antibody responses at protective levels, albeit oral immunization produced lower serum IgG and mucosal IgA antibodies compared to nasal immunization.

Listeriolysin O-secreted Lactococcus lactis The pore-forming toxin listeriolysin O (LLO) is the major virulence factor of *Listeria monocytogenes*. This bacterium is the etiologic agent of listeriosis, a food-borne disease, which causes meningitis and septicemia in immunocompromised individuals, meningitis in newborns, as well as miscarriage in pregnant women. Henceforth, Bahey-El-Din et al. (2008) constructed several plasmids to allow constitutive or nisin-inducible expression of secreted or non-secreted LLO in *L. lactis* to develop a vaccine against *L. monocytogenes* (Bahey-El-Din et al. 2008). Mice immunized with one of the *L. lactis* strains were protected compared to the non-immunized mice; however, only the intraperitoneal (i.p.) and not the oral route of immunization seemed to be efficient. The greatest decrease of liver or spleen bacterial burden was observed when mice were challenged with the *L. lactis* strain secreting LLO after nisin induction. Furthermore, CD8+ T lymphocytes specific for LLO epitope were detected after an i.p. route challenge but not when *L. lactis* strains were given by oral route.

11.9.2. Mucosal Delivery of Viral Antigens

Human papillomavirus type-16 (HPV-16) E7 antigen Human papillomavirus type-16 (HPV-16) is the main causative agent of cervical cancer, the second most important cause of cancer-related deaths in women (Whitehead et al. 2014). Previous studies showed that the HPV-16 E7 oncoprotein, constitutively and exclusively expressed in cervical carcinomas, causes the immortalization of primary keratinocytes and the deregulation of cell transformations (Munger et al. 1989; Zwerschke and Jansen-Durr 2000). For these reasons, the HPV-16 E7 oncoprotein has become a major therapeutic target; which prompted Bermudez and colleagues to engineer recombinant *L. lactis* strains expressing the HPV-16 E7 protein (Bermudez-Humaran et al. 2002). As the protein location may have an influence on immune response modulation and on the treatment efficiency, different delivery systems, including cytoplasmic, secreted or anchored to the cell wall, were explored. After intranasal administration, E7-specific humoral and cellular responses were observed in both secreted and cell-wall anchored forms. However, E7-specific cellular responses, such as secretion of IL-2 and IFN-γ, were higher in mice treated with E7-HPV protein anchored to the cell surface.

In order to enhance humoral and immune responses, the same group co-administered *L. lactis* expressing E7-anchored protein and *L. lactis* secreting IL-12 in a TC-1 tumor mouse model (Bermudez-Humaran et al. 2005). In therapeutic immunization, they observed that the combination of antigen and adjuvant reduces the progression of tumors, and 50% of pre-treated mice exhibited complete prevention of HPV-related tumors. More recently, the co-delivery of antigens and cytokines has been extended to DNA vaccination strategy. A *L. lactis* strain containing the expression DNA cassette of both HPV-16 E7 and murine IL-12 was intranasally administered to a TC-1 tumor animal model (Li et al. 2014). *L. lactis* recombinant strains induced a strong E7-specific immune response and protected mice against TC-1-induced tumors. Moreover, the percentage of survival rate (over 80%) and tumor-free mice was higher in *L. lactis* carrying both the HPV-16 E7 and IL-12 expression cassette than in single recombinant *L. lactis* carrying either HPV-16 E7 or IL-12 DNA. One the other hand, the use of recombinant lactobacilli displaying E7-HPV antigen as a mucosal live vaccine has been explored. Engineered with a new cell-wall anchor from lactobacilli, E7 antigen-producing recombinant *Lact. plantarum* (Cortes-Perez et al. 2005) showed encouraging results in an animal model.

In a recent publication, a novel HPV-targeting immunotherapy for cervical cancer has been developed and tested in clinical trials in patients with HPV16-associated cervical intraepithelial neoplasia grade 3 (CIN3), which is a mucosal precancerous lesion (Kawana et al. 2014). Oral administration of E7-expressing *Lact. casei* succeeded in inducing an E7-specific mucosal immune response, but not a systemic cell-mediated immune response. This study was the first to report a link between antigen-specific mucosal immunity and the clinical response of an immunotherapy in human mucosal neoplasia. Moreover, this treatment was associated with the reduction of pathological severity from CIN3 to CIN2 in 70% of patients, which corresponds to a moderate dysplasia.

Rotavirus and infectious diarrhea Rotavirus is the most common cause of severe diarrhea in infants and young children worldwide. This virus, which is mainly transmitted through fecal–oral contamination, infects and damages mature enterocytes in the small intestine, and causes watery diarrhea, fever, and vomiting (Parashar et al. 2003).

In 2001, the lack of specific treatments and efficient vaccines for rotavirus infections led researchers to develop a mucosal vaccine for the production and delivery of rotavirus antigens (Enouf et al. 2001). Previous studies showed that rotavirus nonstructural protein 4 (NSP4), contributes to viral pathogenesis and induces diarrhea in mice (Tian et al. 1996). Therefore, Enouf and colleagues (2001) explored the antigenic and immunogenic properties of NSP4 by immunizing rabbits with a recombinant *L. lactis* strain secreting NSP4 (with the NICE system). They observed that a NSP4-specific immune and humoral response was elicited in treated rabbits. This study was the first to show the production of an entire viral antigen using *L. lactis*.

Since 2006, new vaccines (Rotarix™ and RotaTeq™) based on the use of attenuated live viruses have been licensed worldwide against rotavirus infection. However, mucosal live vaccine delivery remains a good alternative to overcome the risks associated with attenuated vaccines, such as genetic re-assortment with wild-type strain (Ward 2008). In this context, the ability of recombinant *L. lactis* strains to produce and deliver rotavirus capsid proteins has been investigated. Perez et al. showed that oral administration of *L. lactis* secreting VP7 induced IgG-specific antibody response against rotavirus in mice (Perez et al. 2005). Moreover, the secreted form of rVP7 induced a stronger anti-VP7 immune response than cytoplasmic and cell-wall anchored forms. Additional experiments demonstrated that *L. lactis* strains expressing the cell-wall anchored VP8 antigen, which plays a key role in viral infectivity, induced anti-VP8 intestinal and systemic antibody responses in a murine model (Marelli et al. 2011). Even though this strategy needs to be improved in order to obtain higher level responses with lower doses, its use offers a promising outlook on the prophylaxis of rotavirus infections.

Avian influenza viruses The known etiologic agents of influenza are influenza viruses, which are RNA viruses that belong to the Orthomyxoviridae family. Avian influenza viruses, such as H5N1 and H9N2, are considered a serious public health threat worldwide due to their ability to jump species barriers and cause human infections. For instance, the pathogenic avian influenza H5N1 virus, which emerged at the end of the 1990s, has a mortality rate of approximately 60% in humans. A recent study reported the engineering of a nisin-induced *L. lactis* expressing the influenza H5N1 hemagglutinin (HA) antigen as a mucosal vaccine (Lei et al. 2010). Mice treated with recombinant *L. lactis* secreting H5N1-HA exhibited high levels of HA-specific serum IgG and fecal IgA antibody compared to controls. This live vaccine has been subsequently packaged in enteric-coated mini-capsules to further aid in mass distribution of the product. Mice that were orally fed with the polymer mini-capsule system showed complete protection against lethal challenges of H5N1 virus due to induced strong humoral and systemic immune responses.

In a recent study, Shi and colleagues developed a recombinant *L. plantarum* producing the hemagglutinin of H9N2 (using pSIP system), an avian influenza virus affecting land-based birds, pigs, and humans (Shi et al. 2014). A cytotoxic T cell (CD8+) response and production of H9N2-HA specific IgA and IgG antibodies were observed in orally immunized BALB/c mice. With an urgent need to control the occurrence of H5N1 and H9N2 influenza, both of these studies indicate that hemagglutinin may be a good target for the development of oral vaccines.

11.9.3. Parasitic Diseases

Malaria Malaria is a mosquito-borne disease (transmitted by *Anopheles mosquito*) caused by intracellular parasites belonging to the genus *Plasmodium*. Among the five species that affect humans, *Plasmodium falciparum* is the most lethal, and poses a grave threat to public health (Gardner et al. 2002).

In the early 2000s, a recombinant LAB-based vaccine antigen was described for the first time against *P. falciparum* infection (Theisen et al. 2004). A fusion protein composed of the *P. falciparum* Merozoite (hepatic stage) Surface Protein 3 (MSP3) and the *P. falciparum* Glutamate-rich protein (GLURP) was successfully expressed and secreted by *L. lactis*. Mice immunized with GLURP-MSP3 fusion protein produced high levels of both GLURP- and MSP3-specific antibodies. In contrast to conventional methods, *L. lactis* was used as an expression system, not as a live vaccine delivery system.

In 2006, Ramasamy and colleagues reported the protective role of recombinant *L. lactis* strains expressing the *P. falciparum* Merozoite Surface Ag2 (MSA2) in a rabbit model (Ramasamy et al. 2006). This antigen was administered by two different methods at the surface of *L. lactis*: anchored to the cell wall, and non-covalently bound to non-genetically modified *L. lactis* with Gram-positive enhancer matrix (GEM) particles (van Roosmalen et al. 2006). When administered orally, both of these forms elicited the production of MSA2-IgG antibodies in rabbits, demonstrating a mucosal immunization and activation of MSA2-specific T cells. Moreover, this study demonstrates that the GEM system is equally efficient as recombinant *L. lactis* in inducing immune response, and illustrates its potential as a novel innocuous vaccine delivery system.

Complementary experiments explored this approach with lactobacilli using a murine model of disease. Interestingly, *Lact. .reuteri* and *Lact. .salivarius* with surface-bound MSA2 elicited significant IgA antibodies production in the gut, whereas IgG-specific antibodies levels were significantly lower compared to *L. lactis* (Moorthy and Ramasamy 2007).

Giardiasis *Giardia lamblia* is the protozoan parasite responsible for giardiasis, the most common intestinal parasitic disease in the world, characterized by intestinal malabsorption, diarrhea, and abdominal pain in humans (Ankarklev et al. 2010). Giardiasis is a water-borne infection characterized by the ingestion of cysts, that evolve into the trophozoite form in the small intestine after exposure to gastric acid and proteases in the stomach host. The trophozoite form adheres to the epithelial cells of the duodenum and is released into cyst form after the infectious stage. Since the infection disseminates in cyst stage, researchers explored cyst wall proteins as possible targets for immunization.

Lee et al. engineered a *L. lactis* expressing the *G. lamblia* Cyst Wall Protein 2 (CWP2), a major component of the cyst wall structure, at three different locations (cytoplasmic, secreted or cell-wall anchored form) using the NICE system (Lee and Faubert 2006). In this study, recombinant *L. lactis* CWP2 strains were successful at inducing CWP2-specific IgA antibodies, despite the subcellular location. When challenged with this recombinant lactococci, treated mice exhibited a significant (63%) reduction in cyst output compared to non-immunized control mice. This study has been the second report of the expression of a parasitic protein in LAB.

In complementary experiments, the authors compared the ability of two LAB species, *L. lactis* and *Strep. gordonii*, to deliver CWP2 to intestinal mucosal sites and stimulate immune responses (Lee et al. 2009). Oral immunization of mice with either recombinant *L. lactis* or *Strep. gordonii* expressing *G. lamblia*-CWP2 antigen induced CWP2-specific intestinal IgA antibodies production, and a reduction of the cyst output in mice challenged with *Giardia muris* (respectively 71 and 90%) was observed. Surprisingly, *Strep. gordonii* induced a higher level of IgA than *L. lactis* and was the only one to induce CWP2-specific intestinal IgG production. These recent results suggest that *Strep. gordonii* might be a better vaccine delivery vehicle in the prophylaxis of *G. lamblia* infections.

Cryptosporidiosis Cryptosporidiosis is a water-borne disease mainly caused by *Cryptosporidium parvum*, an intestinal parasite. It is characterized by diarrhea in immune-competent hosts and can be life-threatening in immunosuppressed individuals, such as AIDS patients. Regarding the lack of efficient strategies for controlling this disease, a live recombinant *Lact. casei* has been engineered for mucosal vaccination against *C. parvum* sporozoites (host stage). The immunogenicity of *Lact. casei* expressing the *C. parvum* P23 surface adhesion protein was evaluated in a mouse model (Geriletu et al. 2011) and showed that oral administration of *Lact. casei* secreting P23 elicited serum IgG and mucosal IgA-specific production. Moreover, an expression of IL-4, IL-6, and IFN-γ was detected in splenocytes after stimulation.

This pioneering study showed that recombinant *Lact. casei* induced specific antibody responses to *C. parvum* P23 antigen, and thus can be employed as a potentially safe oral vaccine for cryptosporidiosis.

References

Ankarklev, J., Jerlstrom-Hultqvist, J., Ringqvist, E., et al. (2010) Behind the smile: cell biology and disease mechanisms of *Giardia* species. *Nat Rev Microbiol* **8**, 413–422.

Bahey-El-Din, M., Griffin, B.T., and Gahan, C.G. (2008) Nisin inducible production of listeriolysin O in *Lactococcus lactis* NZ9000. *Microb Cell Fact* **7**, 24.

Baranger, K., Zani, M.L., Labas, V., et al. (2011) Secretory leukocyte protease inhibitor (SLPI) is, like its homologue trappin-2 (pre-elafin), a transglutaminase substrate. *PLoS One* **6**, e20976.

Baumgart, D.C. and Sandborn, W.J. (2007) Inflammatory bowel disease: clinical aspects and established and evolving therapies. *Lancet* **369**, 1641–1657.

Benbouziane, B., Ribelles, P., Aubry, C., et al. (2013) Development of a stress-inducible controlled expression (SICE) system in *Lactococcus lactis* for the production and delivery of therapeutic molecules at mucosal surfaces. *J Biotechnol* **168**, 120–129.

Berlett, B.S. and Stadtman, E.R. (1997) Protein oxidation in aging, disease, and oxidative stress. *J Biol Chem* **272**, 20313–20316.

Bermudez-Humaran, L.G. (2009) *Lactococcus lactis* as a live vector for mucosal delivery of therapeutic proteins. *Human vaccines* **5**, 264–267.

Bermudez-Humaran, L.G., Cortes-Perez, N.G., Le Loir, Y., et al. (2004) An inducible surface presentation system improves cellular immunity against human papillomavirus type 16 E7 antigen in mice after nasal administration with recombinant lactococci. *J Med Microbiol* **53**, 427–433.

Bermudez-Humaran, L.G., Cortes-Perez, N.G., Lefevre, F., et al. (2005) A novel mucosal vaccine based on live lactococci expressing E7 antigen and IL-12 induces systemic and mucosal immune responses and protects mice against human papillomavirus type 16-induced tumors. *J Immunol* **175**, 7297–7302.

Bermudez-Humaran, L.G., Kharrat, P., Chatel, J.M., and Langella, P. (2011) Lactococci and lactobacilli as mucosal delivery vectors for therapeutic proteins and DNA vaccines. *Microb Cell Fact* **10 Suppl 1**, S4.

Bermudez-Humaran, L.G., Langella, P., Miyoshi, A., et al. (2002) Production of human papillomavirus type 16 E7 protein in Lactococcus lactis. *Appl Environ Microbiol* **68**, 917–922.

Bolotin, A., Wincker, P., Mauger, S., et al. (2001) The complete genome sequence of the lactic acid bacterium *Lactococcus lactis* ssp. lactis IL1403. *Genome Res* **11**, 731–753.

Braat, H., Rottiers, P., Hommes, D.W., et al. (2006) A phase I trial with transgenic bacteria expressing interleukin-10 in Crohn's disease. *Clin Gastroenterol Hepatol* **4**, 754–759.

Bron, P.A., Benchimol, M.G., Lambert, J., et al. (2002) Use of the alr gene as a food-grade selection marker in lactic acid bacteria. *Appl Environ Microbiol* **68**, 5663–5670.

Bruno-Barcena, J.M., Andrus, J.M., Libby, S.L., et al. (2004) Expression of a heterologous manganese superoxide dismutase gene in intestinal lactobacilli provides protection against hydrogen peroxide toxicity. *Appl Environ Microbiol* **70**, 4702–4710.

Carroll, I.M., Andrus, J.M., Bruno-Barcena, J.M., et al. (2007) Anti-inflammatory properties of *Lactobacillus gasseri* expressing manganese superoxide dismutase using the interleukin 10-deficient mouse model of colitis. *Amer J Physiol Gastroint Liver Physiol* **293**, G729–738.

Cenac, N., Andrews, C.N., Holzhausen, M., et al. (2007) Role for protease activity in visceral pain in irritable bowel syndrome. *J Clin Invest* **117**, 636–647.

Chatel, J.M., Langella, P., Adel-Patient, K., et al. (2001) Induction of mucosal immune response after intranasal or oral inoculation of mice with *Lactococcus lactis* producing bovine beta-lactoglobulin. *Clin Diagn Lab Immunol* **8**, 545–551.

Chatel, J.M., Nouaille, S., Adel-Patient, K., et al. (2003) Characterization of a *Lactococcus lactis* strain that secretes a major epitope of bovine beta-lactoglobulin and evaluation of its immunogenicity in mice. *Appl Environ Microbiol* **69**, 6620–6627.

Chatel, J.M., Pothelune, L., Ah-Leung, S., et al. (2008) In vivo transfer of plasmid from food-grade transiting lactococci to murine epithelial cells. *Gene Therapy* **15**, 1184–1190.

Chen, S.W., Zhong, J., and Huan, L.D. (2007) Expression of human insulin in lactic acid bacteria and its oral administration in non-obese diabetic mice. *Wei Sheng Wu Xue Bao* **47**, 987–991.

Clifford, A.J., Wilson, D.S. and Bills, N.D. (1989) Repletion of folate-depleted rats with an amino acid-based diet supplemented with folic acid. *J Nutr* **119**, 1956–1961.

Conte, M.P., Schippa, S., Zamboni, I., et al. (2006) Gut-associated bacterial microbiota in paediatric patients with inflammatory bowel disease. *Gut* **55**, 1760–1767.

Cortes-Perez, N.G., Ah-Leung, S., Bermudez-Humaran, L.G., et al. (2007) Intranasal coadministration of live lactococci producing interleukin-12 and a major cow's milk allergen inhibits allergic reaction in mice. *Clin Vaccine Immunol* **14**, 226–233.

Cortes-Perez, N.G., Azevedo, V., Alcocer-Gonzalez, J.M., et al. (2005) Cell-surface display of E7 antigen from human papillomavirus type-16 in *Lactococcus lactis* and in *Lactobacillus plantarum* using a new cell-wall anchor from lactobacilli. *J Drug Targ* **13**, 89–98.

Crittenden, R.G. and Bennett, L.E. (2005) Cow's milk allergy: a complex disorder. *J Amer Col Nutr* **24**, 582S–591S.

de Azevedo, M., Karczewski, J., Lefevre, F., et al. (2012) In vitro and in vivo characterization of DNA delivery using recombinant *Lactococcus lactis* expressing a mutated form of *L. monocytogenes* Internalin A. *BMC Microbiol* **12**, 299.

de Vos, D., Clements, P., Pyke, S.M., et al. (1999) Characterisation and in vitro cytotoxicity of triorganophosphinegold(i) 2-mercaptobenzoate complexes. *Metal-based Drugs* **6**, 31–40.

De Vos, W.M. and Gasson, M.J. (1989) Structure and Expression of the *Lactococcus lactis* Gene for Phospho-{beta}-galactosidase (lacG) in *Escherichia coli* and *L. lactis*. *Microbiology* **135**, 1833.

del Solar, G., Giraldo, R., Ruiz-Echevarria, M.J., et al. (1998) Replication and control of circular bacterial plasmids. *Microbiol Mol Biol Rev* **62**, 434–464.

Dieye, Y., Oxaran, V., Ledue-Clier, F., et al. (2010) Functionality of sortase A in *Lactococcus lactis*. *Appl Environ Microbiol* **76**, 7332–7337.

Eichenbaum, Z., Federle, M.J., Marra, D., et al. (1998) Use of the lactococcal nisA promoter to regulate gene expression in gram-positive bacteria: comparison of induction level and promoter strength. *Appl Environ Microbiol* **64**, 2763.

Enouf, V., Langella, P., Commissaire, J., et al. (2001) Bovine rotavirus nonstructural protein 4 produced by *Lactococcus lactis* is antigenic and immunogenic. *Appl Environ Microbiol* **67**, 1423–1428.

Exl, B.M. and Fritsche, R. (2001) Cow's milk protein allergy and possible means for its prevention. *Nutrition* **17**, 642–651.

Farr, S.B. and Kogoma, T. (1991) Oxidative stress responses in Escherichia coli and Salmonella typhimurium. *Microbiol Rev* **55**, 561–585.

Frossard, C.P., Steidler, L., and Eigenmann, P.A. (2007) Oral administration of an IL-10-secreting *Lactococcus lactis* strain prevents food-induced IgE sensitization. *J Allergy Clin Immunol* **119**, 952–959.

Galipeau, H.J., Wiepjes, M., Motta, J.P., et al. (2014) Novel role of the serine protease inhibitor elafin in gluten-related disorders. *Amer J Gastroenterol* **109**, 748–756.

Gerber, S.D. and Solioz, M. (2007) Efficient transformation of *Lactococcus lactis* IL1403 and generation of knock-out mutants by homologous recombination. *J Basic Microbiol* **47**, 281–286.

Geriletu, Xu, R., Jia, H., et al. (2011) Immunogenicity of orally administrated recombinant *Lactobacillus casei* Zhang expressing *Cryptosporidium parvum* surface adhesion protein P23 in mice. *Curr Microbiol* **62**, 1573–1580.

Gosselink, M.P., Schouten, W.R., van Lieshout, L.M., et al. (2004) Delay of the first onset of pouchitis by oral intake of the probiotic strain *Lactobacillus rhamnosus* GG. *Dis Colon Rectum* **47**, 876–884.

Guimaraes, V.D., Innocentin, S., Lefevre, F., et al. (2006) Use of native lactococci as vehicles for delivery of DNA into mammalian epithelial cells. *Appl Environ Microbiol* **72**, 7091–7097.

Halbmayr, E., Mathiesen, G., Nguyen, T.H., et al. (2008) High-level expression of recombinant beta-galactosidases in *Lactobacillus plantarum* and *Lactobacillus sakei* using a Sakacin P-based expression system. *J Agric Food Chem* **56**, 4710–4719.

Han, W., Mercenier, A., Ait-Belgnaoui, A., et al. (2006) Improvement of an experimental colitis in rats by lactic acid bacteria producing superoxide dismutase. *Inflamm Bowel Dis* **12**, 1044–1052.

Hanson, M.L., Hixon, J.A., Li, W., et al. (2014) Oral delivery of IL-27 recombinant bacteria attenuates immune colitis in mice. *Gastroenterology* **146**, 210–221 e213.

Huibregtse, I.L., Marietta, E.V., Rashtak, S., et al. (2009) Induction of antigen-specific tolerance by oral administration of *Lactococcus lactis* delivered immunodominant DQ8-restricted gliadin peptide in sensitized nonobese diabetic Abo Dq8 transgenic mice. *J Immunol* **183**, 2390–2396.

Innocentin, S., Guimaraes, V., Miyoshi, A., et al. (2009) *Lactococcus lactis* expressing either *Staphylococcus aureus* fibronectin-binding protein A or Listeria monocytogenes internalin A can efficiently internalize and deliver DNA in human epithelial cells. *Appl Environ Microbiol* **75**, 4870–4878.

Isolauri, E., Salminen, S., and Ouwehand, A.C. (2004) Microbial-gut interactions in health and disease. Probiotics. *Best Pract Res Clin Gastroenterol* **18**, 299–313.

Iwaki, M., Okahashi, N., Takahashi, I., et al. (1990) Oral immunization with recombinant *Streptococcus lactis* carrying the *Streptococcus mutans* surface protein antigen gene. *Infect Immun* **58**, 2929–2934.

Karlskas, I.L., Maudal, K., Axelsson, L., et al. (2014) Heterologous protein secretion in lactobacilli with modified pSIP vectors. *PLoS One* **9**, e91125.

Kawana, K., Adachi, K., Kojima, S., et al. (2014) Oral vaccination against HPV E7 for treatment of cervical intraepithelial neoplasia grade 3 (CIN3) elicits E7-specific mucosal immunity in the cervix of CIN3 patients. *Vaccine* **2**, 6233–6239.

Kellermayer, R. (2012) Epigenetics and the developmental origins of inflammatory bowel diseases. *Can J Gastroenterol* **26**, 909–915.

Koc, M., Taysi, S., Emin Buyukokuroglu, M., and Bakan, N. (2003) The effect of melatonin against oxidative damage during total-body irradiation in rats. *Radiat Res* **160**, 251–255.

Kristiansen, O.P., Zamani, M., Johannesen, J., et al. (1998) Linkage and association between a CD4 gene polymorphism and IDDM in Danish IDDM patients. The Danish IDDM Epidemiology and Genetics Group, and The Danish Study Group of Diabetes in Childhood. *Diabetes* **47**, 281–283.

Kruidenier, L., van Meeteren, M.E., Kuiper, I., et al. (2003) Attenuated mild colonic inflammation and improved survival from severe DSS-colitis of transgenic Cu/Zn-SOD mice. *Free Rad Biol Med* **34**, 753–765.

Kuipers, O.P., Beerthuyzen, M.M., Siezen, R.J., and De Vos, W.M. (1993) Characterization of the nisin gene cluster nisABTCIPR of *Lactococcus lactis*. Requirement of expression of the nisA and nisI genes for development of immunity. *Eur J Biochem/FEBS* **216**, 281–291.

Le Loir, Y., Gruss, A., Ehrlich, S.D., and Langella, P. (1998) A nine-residue synthetic propeptide enhances secretion efficiency of heterologous proteins in *Lactococcus lactis*. *J Bacteriol* **180**, 1895–1903.

Lebeer, S., Claes, I.J., and Vanderleyden, J. (2012) Anti-inflammatory potential of probiotics: lipoteichoic acid makes a difference. *Trends Microbiol* **20**, 5–10.

LeBlanc, J.G., Aubry, C., Cortes-Perez, N.G., et al. (2013) Mucosal targeting of therapeutic molecules using genetically modified lactic acid bacteria: an update. *FEMS Microbiol Lett* **344**, 1–9.

LeBlanc, J.G., del Carmen, S., Miyoshi, A., et al. (2011) Use of superoxide dismutase and catalase producing lactic acid bacteria in TNBS induced Crohn's disease in mice. *J Biotechnol* **151**, 287–293.

LeBlanc, J.G., Ledue-Clier, F., Bensaada, M., et al. (2008) Ability of *Lactobacillus fermentum* to overcome host alpha-galactosidase deficiency, as evidenced by reduction of hydrogen excretion in rats consuming soya alpha-galacto-oligosaccharides. *BMC Microbiol* **8**, 22.

Lee, P., Abdul-Wahid, A., and Faubert, G.M. (2009) Comparison of the local immune response against *Giardia lamblia* cyst wall protein 2 induced by recombinant *Lactococcus lactis* and *Streptococcus gordonii*. *Microbes Infec/Institut Pasteur* **11**, 20–28.

Lee, P. and Faubert, G.M. (2006) Expression of the *Giardia lamblia* cyst wall protein 2 in *Lactococcus lactis*. *Microbiology* **152**, 1981–1990.

Lei, H., Xu, Y., Chen, J., et al. (2010) Immunoprotection against influenza H5N1 virus by oral administration of enteric-coated recombinant *Lactococcus lactis* mini-capsules. *Virology* **407**, 319–324.

Leon, F., Smythies, L.E., Smith, P.D., and Kelsall, B.L. (2006) Involvement of dendritic cells in the pathogenesis of inflammatory bowel disease. *Adv Exper Med Biol* **579**, 117–132.

Levine, J.B. and Lukawski-Trubish, D. (1995) Extraintestinal considerations in inflammatory bowel disease. *Gastroenterol Clin North America* **24**, 633–646.

Li, Y., Li, X., Liu, H., et al. (2014) Intranasal immunization with recombinant lactococci carrying human papillomavirus E7 protein and mouse interleukin-12 DNA induces E7-specific antitumor effects in C57BL/6 mice. *Oncol Lett* **7**, 576–582.

Lih-Brody, L., Powell, S.R., Collier, K.P., et al. (1996) Increased oxidative stress and decreased antioxidant defenses in mucosa of inflammatory bowel disease. *Dig Dis Sci* **41**, 2078–2086.

Llull, D. and Poquet, I. (2004) New expression system tightly controlled by zinc availability in *Lactococcus lactis*. *Appl Environ Microbiol* **70**, 5398–5406.

Ma, Y., Liu, J., Hou, J., et al. (2014) Oral Administration of Recombinant *Lactococcus lactis* Expressing HSP65 and Tandemly Repeated P277 Reduces the Incidence of Type I Diabetes in Non-Obese Diabetic Mice. *PLoS One* **9**, e105701.

Madsen, S.M., Arnau, J., Vrang, A., et al. (1999) Molecular characterization of the pH-inducible and growth phase-dependent promoter P170 of *Lactococcus lactis*. *Mol Microbiol* **32**, 75–87.

Marelli, B., Perez, A.R., Banchio, C., et al. (2011) Oral immunization with live *Lactococcus lactis* expressing rotavirus VP8 subunit induces specific immune response in mice. *J Virol Methods* **175**, 28–37.

Marinho, F.A., Pacifico, L.G., Miyoshi, A., et al. (2010) An intranasal administration of *Lactococcus lactis* strains expressing recombinant interleukin-10 modulates acute allergic airway inflammation in a murine model. *Clin Exper Allergy* **40**, 1541–1551.

Marteau, P., Lepage, P., Mangin, I., et al. (2004) Review article: gut flora and inflammatory bowel disease. *Alim Pharmacol Therapeut* **20 Suppl 4**, 18–23.

Mierau, I. and Kleerebezem, M. (2005) 10 years of the nisin-controlled gene expression system (NICE) in *Lactococcus lactis*. *Appl Microbiol Biotechnol* **68**, 705–717.

Mierau, I., Leij, P., van Swam, I.,et al. (2005) Industrial-scale production and purification of a heterologous protein in *Lactococcus lactis* using the nisin-controlled gene expression system NICE: the case of lysostaphin. *Microb Cell Fact* **4**, 15.

Miyoshi, A., Jamet, E., Commissaire, J., et al. (2004) A xylose-inducible expression system for *Lactococcus lactis*. *FEMS Microbiol Lett* **239**, 205–212.

Miyoshi, A., Poquet, I., Azevedo, V., et al. (2002) Controlled production of stable heterologous proteins in *Lactococcus lactis*. *Appl Environ Microbiol* **68**, 3141–3146.

Moorthy, G. and Ramasamy, R. (2007) Mucosal immunisation of mice with malaria protein on lactic acid bacterial cell walls. *Vaccine* **25**, 3636–3645.

Morello, E., Bermudez-Humaran, L.G., Llull, D., et al. and Poquet, I. (2008) *Lactococcus lactis*, an efficient cell factory for recombinant protein production and secretion. *J Mol Microbiol Biotechnol* **14**, 48–58.

Motta, J.P., Bermudez-Humaran, L.G., Deraison, C., et al. (2012) Food-grade bacteria expressing elafin protect against inflammation and restore colon homeostasis. *Sci Translat Med* **4**, 158ra144.

Motta, J.P., Magne, L., Descamps, D., et al. (2011) Modifying the protease, antiprotease pattern by elafin overexpression protects mice from colitis. *Gastroenterology* **140**, 1272–1282.

Mu, D., Montalban-Lopez, M., Masuda, Y., and Kuipers, O.P. (2013) Zirex: a novel zinc-regulated expression system for *Lactococcus lactis*. *Appl Environ Microbiol* **79**, 4503–4508.

Munger, K., Werness, B.A., Dyson, N., et al. (1989) Complex formation of human papillomavirus E7 proteins with the retinoblastoma tumor suppressor gene product. *EMBO J* **8**, 4099–4105.

Mylonaki, M., Rayment, N.B., Rampton, D.S., et al. (2005) Molecular characterization of rectal mucosa-associated bacterial flora in inflammatory bowel disease. *Inflam Bowel Dis* **11**, 481–487.

Nga, B.H. (2005) Genome analysis of lactic acid bacteria in food fermentations and biotechnological applications. *Curr Opin Microbiol* **8**, 307–312.

Nguyen, T.T., Nguyen, T.H., Maischberger, T., et al. (2011) Quantitative transcript analysis of the inducible expression system pSIP: comparison of the overexpression of *Lactobacillus* spp. beta-galactosidases in *Lactobacillus plantarum*. *Microb Cell Fact* **10**, 46.

Norton, P.M., Wells, J.M., Brown, H.W., et al. (1997) Protection against tetanus toxin in mice nasally immunized with recombinant *Lactococcus lactis* expressing tetanus toxin fragment C. *Vaccine* **15**, 616–619.

Nouaille, S., Ribeiro, L.A., Miyoshi, A., et al. (2003) Heterologous protein production and delivery systems for *Lactococcus lactis*. *Genet Mol Res* **2**, 102–111.

Ott, S.J., Musfeldt, M., Wenderoth, D.F., et al. (2004) Reduction in diversity of the colonic mucosa associated bacterial microflora in patients with active inflammatory bowel disease. *Gut* **53**, 685–693.

Parashar, U.D., Hummelman, E.G., Bresee, J.S., et al. (2003) Global illness and deaths caused by rotavirus disease in children. *Emerg Infect Dis* **9**, 565–572.

Perez, C.A., Eichwald, C., Burrone, O., and Mendoza, D. (2005) Rotavirus vp7 antigen produced by *Lactococcus lactis* induces neutralizing antibodies in mice. *J Appl Microbiol* **99**, 1158–1164.

Pontes, D., Azevedo, M., Innocentin, S., et al. (2014) Immune response elicited by DNA vaccination using *Lactococcus lactis* is modified by the production of surface exposed pathogenic protein. *PLoS One* **9**, e84509.

Pontes, D., Innocentin, S., Del Carmen, S., et al. (2012) Production of Fibronectin Binding Protein A at the surface of *Lactococcus lactis* increases plasmid transfer in vitro and in vivo. *PLoS One* **7**, e44892.

Pouwels, P.H., Leer, R.J., and Boersma, W.J. (1996) The potential of *Lactobacillus* as a carrier for oral immunization: development and preliminary characterization of vector systems for targeted delivery of antigens. *J Biotechnol* **44**, 183–192.

Pouwels, P.H., Leer, R.J., Shaw, M., et al. (1998) Lactic acid bacteria as antigen delivery vehicles for oral immunization purposes. *Int J Food Microbiol* **41**, 155–167.

Ramasamy, R., Yasawardena, S., Zomer, A., et al. (2006) Immunogenicity of a malaria parasite antigen displayed by *Lactococcus lactis* in oral immunisations. *Vaccine* **24**, 3900–3908.

Robert, S., Gysemans, C., Takiishi, T., et al. (2014) Oral delivery of glutamic acid decarboxylase (GAD)-65 and IL10 by *Lactococcus lactis* reverses diabetes in recent-onset NOD mice. *Diabetes* **63**, 2876–2887.

Robinson, K., Chamberlain, L.M., Lopez, M.C., et al. (2004) Mucosal and cellular immune responses elicited by recombinant *Lactococcus lactis* strains expressing tetanus toxin fragment C. *Infect Immun* **72**, 2753–2761.

Robinson, K., Chamberlain, L.M., Schofield, K.M., et al. (1997) Oral vaccination of mice against tetanus with recombinant *Lactococcus lactis*. *Nature Biotechnol* **15**, 653–657.

Roland, K.L. and Brenneman, K.E. (2013) Salmonella as a vaccine delivery vehicle. *Expert review of vaccines* **12**, 1033–1045.

Ruddy, M.J., Wong, G.C., Liu, X.K., et al. (2004) Functional cooperation between interleukin-17 and tumor necrosis factor-alpha is mediated by CCAAT/enhancer-binding protein family members. *J Biol Chem* **279**, 2559–2567.

Rush, C.M., Hafner, L.M., and Timms, P. (1995) Lactobacilli: vehicles for antigen delivery to the female urogenital tract. *Advan Experim Med Biol* **371B**, 1547–1552.

Sallenave, J.M. (2010) Secretory leukocyte protease inhibitor and elafin/trappin-2: versatile mucosal antimicrobials and regulators of immunity. *Amer J Respir Cell Mol Biol* **42**, 635–643.

Sarra, M., Pallone, F., Macdonald, T.T., and Monteleone, G. (2010) IL-23/IL-17 axis in IBD. *Inflamm Bowel Dis* **16**, 1808–1813.

Schmid, M., Fellermann, K., Fritz, P., et al. (2007) Attenuated induction of epithelial and leukocyte serine antiproteases elafin and secretory leukocyte protease inhibitor in Crohn's disease. *J Leukocyte Biol* **81**, 907–915.

Schotte, L., Steidler, L., Vandekerckhove, J., and Remaut, E. (2000) Secretion of biologically active murine interleukin-10 by *Lactococcus lactis*. *Enz Microbial Technol* **27**, 761–765.

Segui, J., Gironella, M., Sans, M., et al. (2004) Superoxide dismutase ameliorates TNBS-induced colitis by reducing oxidative stress, adhesion molecule expression, and leukocyte recruitment into the inflamed intestine. *J Leukocyte Biol* **76**, 537–544.

Serafini, F., Turroni, F., Guglielmetti, S., et al. (2012) An efficient and reproducible method for transformation of genetically recalcitrant bifidobacteria. *FEMS Microbiol Lett* **333**, 146–152.

Shaw, L. and Wiedow, O. (2011) Therapeutic potential of human elafin. *Biochem Soc Trans* **39**, 1450–1454.

Shi, S.H., Yang, W.T., Yang, G.L., et al. (2014) Immunoprotection against influenza virus H9N2 by the oral administration of recombinant *Lactobacillus plantarum* NC8 expressing hemagglutinin in BALB/c mice. *Virology* **464–465**, 166–176.

Sokol, H., Pigneur, B., Watterlot, L., et al. (2008) *Faecalibacterium prausnitzii* is an anti-inflammatory commensal bacterium identified by gut microbiota analysis of Crohn disease patients. *Proc Natl Acad Sci U S A* **105**, 16731–16736.

Sorvig, E., Gronqvist, S., Naterstad, K., et al. (2003) Construction of vectors for inducible gene expression in *Lactobacillus sakei* and *Lactobacillus plantarum*. *FEMS Microbiol Lett* **229**, 119–126.

Steidler, L., Robinson, K., Chamberlain, L., et al. (1998) Mucosal delivery of murine interleukin-2 (IL-2) and IL-6 by recombinant strains of *Lactococcus lactis* coexpressing antigen and cytokine. *Infect Immun* **66**, 3183–3189.

Stepniak, D. and Koning, F. (2006) Celiac disease--sandwiched between innate and adaptive immunity. *Human Immunol* **67**, 460–468.

Takiishi, T., Korf, H., Van Belle, T.L., et al. (2012) Reversal of autoimmune diabetes by restoration of antigen-specific tolerance using genetically modified *Lactococcus lactis* in mice. *J Clin Invest* **122**, 1717–1725.

Tamboli, C.P., Caucheteux, C., Cortot, A., et al. (2003) Probiotics in inflammatory bowel disease: a critical review. *Best Pract Res Clin Gastroenterol* **17**, 805–820.

Theisen, M., Soe, S., Brunstedt, K., et al. (2004) A Plasmodium falciparum GLURP-MSP3 chimeric protein; expression in *Lactococcus lactis*, immunogenicity and induction of biologically active antibodies. *Vaccine* **22**, 1188–1198.

Tian, P., Ball, J.M., Zeng, C.Q., and Estes, M.K. (1996) The rotavirus nonstructural glycoprotein NSP4 possesses membrane destabilization activity. *J Virol* **70**, 6973–6981.

van der Vossen, J.M., van der, Lelie, D. and Venema, G. (1987) Isolation and characterization of Streptococcus cremoris Wg2-specific promoters. *Appl Environ Microbiol* **53**, 2452–2457.

van Roosmalen, M.L., Kanninga, R., El Khattabi, M., et al. (2006) Mucosal vaccine delivery of antigens tightly bound to an adjuvant particle made from food-grade bacteria. *Methods* **38**, 144–149.

Vandenplas, Y., De Greef, E., and Devreker, T. (2014) Treatment of Cow's Milk Protein Allergy. *Ped Gastroenterol Hepatol Nutr* **17**, 1–5.

Ward, R.L. (2008) Rotavirus vaccines: how they work or don't work. *Exp Rev Mol Med* **10**, e5.

Watterlot, L., Rochat, T., Sokol, H., et al. (2010) Intragastric administration of a superoxide dismutase-producing recombinant *Lactobacillus casei* BL23 strain attenuates DSS colitis in mice. *Int J Food Microbiol* **144**, 35–41.

Wegmann, U., O'Connell-Motherway, M., Zomer, A., et al. (2007) Complete genome sequence of the prototype lactic acid bacterium *Lactococcus lactis* subsp. cremoris MG1363. *J Bacteriol* **189**, 3256–3270.

Weitzman, S.A. and Gordon, L.I. (1990) Inflammation and cancer: role of phagocyte-generated oxidants in carcinogenesis. *Blood* **76**, 655–663.

Wells, J.M. and Mercenier, A. (2008) Mucosal delivery of therapeutic and prophylactic molecules using lactic acid bacteria. *Nat Rev Microbiol* **6**, 349–362.

Wells, J.M., Wilson, P.W., Norton, P.M., et al. (1993a) *Lactococcus lactis*: high-level expression of tetanus toxin fragment C and protection against lethal challenge. *Mol Microbiol* **8**, 1155–1162.

Wells, J.M., Wilson, P.W., Norton, P.M., et al. (1993b) *Lactococcus lactis*: high-level expression of tetanus toxin fragment C and protection against lethal challenge. *Mol Microbiol* **8**, 1155–1162.

Whitehead, M., Ohlschlager, P., Almajhdi, F.N., et al. (2014) Human papillomavirus (HPV) type 16 E7 protein bodies cause tumour regression in mice. *BMC Cancer* **14**, 367.

Wispe, J.R., Clark, J.C., Burhans, M.S., et al. (1989) Synthesis and processing of the precursor for human mangano-superoxide dismutase. *Bioch Biophys Acta* **994**, 30–36.

Wu, C., Yang, G., Bermudez-Humaran, L.G., et al. (2006) Immunomodulatory effects of IL-12 secreted by *Lactococcus lactis* on Th1/Th2 balance in ovalbumin (OVA)-induced asthma model mice. *Int Immunopharmacol* **6**, 610–615.

Yoshimura, T., Takeda, A., Hamano, S., et al. (2006) Two-sided roles of IL-27: induction of Th1 differentiation on naive CD4+ T cells versus suppression of proinflammatory cytokine production including IL-23-induced IL-17 on activated CD4+ T cells partially through STAT3-dependent mechanism. *J Immunol* **177**, 5377–5385.

Zuo, F., Yu, R., Feng, X., et al. (2014) Combination of heterogeneous catalase and superoxide dismutase protects *Bifidobacterium longum strain* NCC2705 from oxidative stress. *Appl Microbiol Biotechnol* **98**, 7523–7534.

Zurita-Turk, M., Del Carmen, S., Santos, A.C., et al. (2014) *Lactococcus lactis* carrying the pValac DNA expression vector coding for IL-10 reduces inflammation in a murine model of experimental colitis. *BMC Biotechnol* **14**, 73.

Zwerschke, W. and Jansen-Durr, P. (2000) Cell transformation by the E7 oncoprotein of human papillomavirus type 16: interactions with nuclear and cytoplasmic target proteins. *Adv Cancer Res* **78**, 1–29.

Chapter 12

Lactic Acid Bacteria for Dairy Fermentations: Specialized Starter Cultures to Improve Dairy Products

Domenico Carminati[1], Giorgio Giraffa[1], Miriam Zago[1], Mariángeles Briggiler Marcó[2], Daniela Guglielmotti[2], Ana Binetti[2], and Jorge Reinheimer[2,*]

[1] *Consiglio per la Ricerca e la Sperimentazione in Agricoltura, Centro di Ricerca per le Produzioni Foraggere e Lattiero-Casearie (CRA-FLC), Italy*
[2] *Instituto de Lactología Industrial (INLAIN), Universidad Nacional del Litoral-CONICET, Argentina*

12.1. Introduction

Dairy starter culture research is moving from traditional microbiology to the use of functional genomics. Industrially interesting lactic acid bacteria (LAB) for dairy starter cultures may be selected by searching for required phenotypic traits at the genotypic level using high-throughput screening approaches. Nowadays, the available knowledge about the biology and genomics of microorganisms enables us to link phenotype directly with genotype through systems biology approaches. Searching among the natural diversity for strains with desirable industrial genotypes is a feasible solution, but food-grade genetic strategies should ensure novel solutions for starter culture improvement to produce highly desirable and innovative fermented dairy products, with functional properties, new sensory characteristics, and increased nutritional and health-promoting properties (Mills et al. 2010a).

12.2. Adjunct Cultures

The availability of "new" LAB strains to develop starter cultures with pro-technological functionalities is of great importance for companies that produce starter cultures as well as for the dairy industry, which is still making significant investments in searching for wild strains from natural sources to design new, industrially interesting starters. Nowadays, the availability of whole genome sequences for LAB is a powerful tool to search for new starter LAB for the production of tastier, healthier, and more convenient fermented dairy foods (Liu et al. 2008; Mills et al. 2010a). Starter adjuncts, which

* Corresponding author: Instituto de Lactología Industrial (INLAIN), Universidad Nacional del Litoral-CONICET, Santa Fe, Argentina. jorreinh@fiq.unl.edu.ar

Biotechnology of Lactic Acid Bacteria: Novel Applications, Second Edition. Edited by Fernanda Mozzi, Raúl R. Raya, and Graciela M. Vignolo.

are normally used for purposes other than lactic acid formation, can bring several benefits to the technological, sensory, safety, and health-promoting aspects of fermented dairy products (Giraffa 2012). Non Starter Lactic Acid Bacteria (NSLAB) are adventitious LAB that contaminate cheese, generally after milk pasteurization, and that originate from the raw milk, the ingredients used in cheese-making, or the environment. These bacteria are frequently facultatively heterofermentative *Lactobacillus* (FHL), mainly belonging to *Lact. casei*, *Lact. paracasei*, *Lact. plantarum*, and *Lact. curvatus*. The presence of enzymatic activities, such as proteolytic, aminopeptidolytic, lipolytic, and autolytic activities, are used to select FHL as adjunct cultures for cheese ripening (Nieto-Arribas et al. 2009).

12.2.1. Ripening Cultures

During the fermentation process, LAB influence the sensory properties of dairy products, including flavor development. Native milk microorganisms (as the NSLAB) and primary starters (SLAB), as well as milk or added enzymes, variously contribute to balance multiple flavor compounds belonging to different chemical classes, leading to the final aromatic pattern that characterizes each product. Many LAB species find wide industrial applications as complementary cultures in several food fermentations. Flavor compounds are formed by various processes, for example the conversions of lactose and citrate (glycolysis and pyruvate metabolism), fat (lipolysis), and protein (proteolysis, peptidolysis, and amino acids) catabolism (van Kranenburg et al. 2002; Hayes et al. 2007; Gonzales et al. 2010). SLAB and adventitious NSLAB are responsible for secondary proteolysis catalyzed by proteinases and a wide range of peptidases that act on the products of primary proteolysis and produce small peptides and free amino acids (FAA) (Upadhyay et al. 2004; Milesi et al. 2008, 2009). *Lact. helveticus*, traditionally used as a starter culture for hard cheese, is nowadays also used as an adjunct culture in many different cheese types. Its complex proteolytic system, consisting of cell-envelope associated proteases (CEPs) and intracellular peptidases that release FAA, is important for cheese ripening events (Jensen et al. 2009). The addition of NSLAB *Lact. plantarum* as adjunct culture in hard and soft cheese models neither contributed to acid production nor caused changes to the cheese composition, but caused changes to secondary proteolysis (i.e., modification of peptide profiles and increase of the total FAA), which were consistent with the acceleration of proteolysis (Milesi et al. 2008, 2009).

The catabolism of FAA plays a major role in flavor development during cheese ripening. FAA are transaminated or deaminated to α-ketoacids that are metabolized to several compounds, including fatty acids, alcohols, and aldehydes. Glutamate dehydrogenase (GDH) catalyzes the reversible deamination of glutamate, often present at elevated concentration in the curd during ripening, to α-ketoglutarate, the main amino group acceptor for the reaction of transamination (De Angelis et al. 2010). NSLAB that inherently possess high GDH activity are selected to favor an increase in the aroma intensity, but not all NSLAB expressing *in vitro* GDH activity are effective in cheese due either to the nature of GDH activity (NAD+ or NADP+ dependent) or to the capacity of fully expressing GDH activity under the environmental conditions during cheese ripening (Williams et al. 2006). De Angelis et al. (2010), studying the GDH activity of NSLAB *Lact. plantarum* DPPMA49, found that GDH enzyme activity and expression of *gdh* gene were affected by temperature, pH, and NaCl values used for cheese ripening.

NSLAB play a major role during cheese ripening, contributing positively to cheese flavor (Chamba and Irlinger 2004); however, some defects can be ascribed to their specific metabolic activities (Laleye et al. 1987). The contribution of NSLAB to ripening increases when prolonging the ripening period. The effects on flavor and texture can be positive, negative or neutral, depending on the predominant strains and their associated ripening properties. It is therefore necessary to carefully select NSLAB strains as adjuncts for the improvement and diversification of novel cheese flavor and, at the same time, to control other adventitious NSLAB to prevent undesirable effects (Nieto-Arribas et al. 2009). A careful selection of indigenous LAB isolates with high enzymatic potential for conversion of methionine (Met) to volatile sulfur compounds (VSCs) has been proposed to enhance cheese flavor development (Hanniffy et al. 2009). VSCs have a low detection threshold and strong odor activities, and their formation depends on both biosynthetic and catabolic pathways of Met leading to methanethiol (MTL) (Bustos et al. 2011). The assessment of the C-S lyase (cystathionine β- or γ-lyases), which catalyzes the simultaneous deamination and demethiolation of Met to MTL, could be used as a valuable approach for the selection of LAB strains with high VSC-producing abilities that are potentially interesting for industrial application.

Artisanal and Protected Designation of Origin (PDO) cheeses traditionally made with raw milk are increasingly manufactured from pasteurized milk with the addition of commercial starters, implying as a consequence a more standardized cheese quality but with impoverished sensory characteristics. Selection and technological characterization of wild LAB to be used as adjunct cultures to preserve the characteristics of traditional cheeses is constantly evolving. Bertolino et al. (2008) selected from artisanal Toma Piemontese cheese indigenous LAB showing proteolytic activity leading to the accumulation of varying levels of peptides and FAA. The study suggested that the same strains could be re-used for developing new starter cultures to improve the quality of industrial cheeses. To enhance the flavor and taste of industrial PDO Manchego cheese, mesophilic lactobacilli isolated from artisanal cheeses were successfully employed to restore the typical organoleptic characteristics of the products (Nieto-Arribas et al. 2009). Starter cultures from autochthonous strains of traditional cows' Mozzarella cheese have been set up to standardize the major traits distinguishing this cheese variety and to satisfy industrial requirements. These autochthonous multiple strain cultures improved texture and sensory properties, and were also useful to extend the shelf-life (De Angelis et al. 2008). Also enterococci, which are largely present in the microbiota of traditional cheeses mainly produced from raw ewes, goats or cows' milk, were considered for their role in cheese ripening due to their interesting proteolytic and lipolytic activities or for production of aroma metabolites like diacetyl, and have been proposed as NSLAB for different European cheeses (Giraffa 2003; Morandi et al. 2006; Nieto-Arribas et al. 2011). As the genus *Enterococcus* is a controversial group of LAB present in food products, because they are considered opportunistic pathogens for humans (Morandi et al. 2013), a specific safety evaluation for strains selected for food production is necessary to assess the risk associated with the intentional use of enterococci in the food chain (EFSA 2010).

The search for new strains to be used as adjunct cultures is a difficult and cumbersome task, involving high industrial costs. In this regard, the genomic era has revolutionized microbiology since genome sequence analysis enables us to predict and reconstruct metabolic pathways (i.e., flavor formation). This knowledge provides new tools to develop screening, cost-effective strategies based on genomic approaches for selection of LAB exhibiting desired properties that could be used for developing novel or improved dairy products (Teusink and Smid 2005; Liu et al. 2008; Mills et al. 2010a).

12.2.2. Protective Cultures

Within the adjunct or complementary cultures, a key role is played by the so-called "protective cultures," which are specialized cultures designed on the basis of a biological approach to preservation. Biological preservation, the use of a product's natural microbiota and its antibacterial products for protection against pathogens and spoilage, is a method of interest to improve microbial safety and to extend food shelf-life. In this context, LAB represent the microbial group most commonly associated with the protective cultures used in the dairy industry, as they are present in all fermented dairy products and have a long and recognized history of safe use. Protective effects are due to metabolic end products such as organic acids (and the concurrent reduction of pH), hydrogen peroxide, diacetyl, and reuterin, as well as more specific metabolites, such as fatty acids, bacteriocins, and related proteinaceous compounds, and antibiotics (Holzapfel et al. 1995; Caplice and Fitzgerald 1999; Topisirovic et al. 2011).

Based on the categories of antimicrobial activities, protective cultures from LAB can be generally split into two groups: primary starters and specialized cultures. Traditionally, primary starters are added to play a single role, ensuring a relevant contribution in terms of lactic acid production and concomitant pH decrease which, among other effects, prevents growth of potential pathogens and microbial spoilage. Schematically, primary starters are composed of *Lactococcus* spp. and *Leuconostoc* spp. among mesophilic LAB, and *Strep. thermophilus*, *Lact. delbrueckii*, and *Lact. helveticus* among thermophilic LAB (Leroy and De Vuyst 2004; Parente and Cogan 2004). The lower pH of cultured products, however, cannot be inhibitory to all unwanted microorganisms. Moreover, biological protection based on LAB fermentation cannot be applied to fresh, ready-to-eat dairy products, which can easily be contaminated by spoilage organisms. Therefore, the dairy industry has explored alternative approaches to complement biological preservation based on lactic acid fermentation, and the consequent pH decrease, introducing the use of specialized cultures carrying additional protection. Diacetyl (or 2,3-butanedione), which gives the distinctive buttery flavor, is a natural by-product of citrate fermentation carried out by species of the *Leuconostoc* genus and by *Lactococcus lactis* subsp. *lactis* var. *diacetylactis* (Chamba 2008). Diacetyl is characterized by a broad antimicrobial activity against spoilage or pathogenic species such as, among others, *Aeromonas* spp, *Escherichia coli*, *Staphylococcus aureus*, and *Listeria monocytogenes* (Jay 1982; Lanciotti et al. 2003). Reuterin (or β-hydroxypropionaldehyde) shows a broad spectrum of activity against both prokaryotes and eukaryotes. This molecule is an intermediate in the anaerobic metabolism of glycerol from *Lact. reuteri* and other LAB species (Sun et al. 2011). The antimicrobial activity of reuterin can open possibilities for its use as a food preservative within probiotic formulations or specialized cultures for dairy application. Supplementation of food with producer strains plus glycerol for the *in situ* production of reuterin could be also an alternative method to combine in a single culture preservation and probiotic traits. Finally, new possibilities to design protecting cultures for dairy preservation are emerging. For example, a study of Capozzi et al. (2012) describing the ability of *Lact. plantarum* to degrade biogenic amines in wine suggests the possibility of applying a similar approach to cheese.

Bacteriocins from LAB Within the extensive range of metabolites that are industrially used in the biopreservation of dairy products, bacteriocins are the most widely explored. Bacteriocins of LAB are exocellular, low-molecular-mass peptides or proteins, with an antibacterial mode of action usually restricted to related Gram-positive bacteria. Virtually all LAB species synthesize bacteriocins. Among these and in addition to nisin, pediocins, sakacins, enterocins, propionicins, and many other bacteriocins are found, exhibiting broad spectrum activity against *Clostridium botulinum*, *L. monocytogenes*, *S. aureus*, *Bacillus cereus*, and *Clostridium tyrobutyricum* among others (Galvez et al. 2011). Within protective cultures, bacteriocin-producing LAB are the most studied because of their food association. When combined with other antimicrobial substances or physical treatments (e.g., heat treatments, modified atmosphere packaging) bacteriocins can be applied as part of a hurdle technology, with also positive impact on the preservation of the product quality. Bacteriocins may be used by *in situ* synthesis (directly in the food) by producer strains or by directly adding the compounds obtained *ex situ*, isolated from the culture broth and purified (Leroy and De Vuyst 2004). The *in situ* synthesis offers advantages over purified bacteriocins allowing one to combine in a single commercial formulation polyfunctional cultures (e.g., displaying acidifying and protective abilities) (Gaggia et al. 2011). The eventual approval of use of this type of culture by food regulatory authorities is less problematic and in addition allows one to avoid bacteriocin production industrial costs. The *in situ* approach finally involves a greater acceptability to the consumer that would consider the addition of purified bacteriocin as an additive. On the other hand, the advantage of using pre-formed bacteriocins with known titer (such as nisin) would be linked to an easier modeling and control of the antimicrobial activity in the product.

Application of bacteriocins in dairy products In the scientific literature many examples of potential applications of LAB bacteriocins (or bacteriocin-producing strains) in dairy products are found. LAB bacteriocins have been suggested to inactivate bacterial pathogens in raw milk and yogurt, to control late blowing in cheese, and to accelerate cheese ripening (Galvez et al. 2011). However, industrial applications are still limited and are mainly concerned with nisin. Nisin is widely applied to control *C. botulinum* in processed cheeses and, secondarily, as an alternative to potassium nitrate to prevent blowing defects in cheese due to contamination by clostridia (Thomas and Delves-Broughton 2001). Other potential applications of bacteriocins involve the control of cheese flavor-altering contaminating bacteria (e.g., certain strains of mesophilic LAB or *L. lactis*), which are often responsible for cheese defects during cheese ripening (Leroy and De Vuyst 2004).

A recent and interesting application of bacteriocins in food preservation involves their inclusion in active packaging. Active packaging allows implementation of direct control on the complex phenomena occurring during food storage. Recent studies reported on the efficacy of bacteriocins, alone or in a hurdle approach, when directly applied to fresh dairy products by spreading onto the packaging materials that come in contact with the cheese, or by spraying, immersing, or dusting the products. For example, edible coatings containing nisin or a *Lact. curvatus* bacteriocin were reported to retard the growth of *L. monocytogenes* in ricotta, fresh, and Cheddar cheeses (Mauriello et al. 2004; Martins et al. 2010).

12.2.3. Probiotic Cultures

The incorporation of probiotic bacteria as adjuncts in different fermented milk products is an important topic with industrial and commercial consequences. Functional foods are foods like conventional ones consumed as part of the daily diet, displaying proved metabolic or physiological effects that are useful to maintain a good physical and mental health to help in the reduction of degenerative chronic disease risks, besides contributing to the food's basic nutritional functions. Apart from health-promoting properties, probiotic microorganisms in foods have to fulfil many other conditions. These include sufficient stability during production and storage, so that the probiotic content of the food during the whole shelf-life does not drop below the bacterial concentration required for a probiotic effect. Survival and bacterial counts of probiotic microorganisms in the food, and the maintenance of its probiotic activity, depend on the production process, the properties of the food matrix, and the bacterial physiological state. To avoid metabolically active probiotic cultures altering taste, flavour, texture, and shelf-life of the food product through post-acidification, lipolysis, and/or proteolysis, probiotic foods are normally stored at ≤ 8 °C. In recent years, there has been a remarkable increase in the number of dairy products carrying probiotic LAB, for example pasteurized and fermented milks, ice cream, and several cheese varieties (Tamime et al. 2005). The behavior of candidate probiotic strains of *Lact. plantarum* in fermented milk and cheese fermentation was studied (Georgieva et al. 2009; Milesi et al. 2009). *Lact. plantarum* had the capacity to survive in fermented milk over extended shelf-life (up to 28 days) at refrigerated temperatures and to maintain a good viability in the presence of food preservatives such as NaCl during cheese ripening (Georgieva et al. 2009). These technologically relevant properties make *Lact. plantarum* appropriate for diverse food applications and highlight the capability of this organism to be incorporated in dairy products. A *Lact. plantarum* cheese isolate with probiotic potential showed its usefulness as an adjunct starter when used in combination with one strain of *Lact. casei*. Both strains were able to grow to high counts in soft and semi-hard cheeses, causing a moderate increase of secondary proteolysis, without concomitant over-acidification, and with a positive influence on sensory profiles of cheeses (Milesi et al. 2009). Cheese represents a very attractive food for probiotic delivery. In the production of *petit-suisse* cheese the adjunct of oligofructose and/or inulin as prebiotics improved the viability of both *Lact. acidophilus* and

Bifidobacterium animalis subsp. *lactis* while the addition of eucalyptus honey reduced the viability level of these strains (Cardarelli et al. 2008). Interestingly, cheese produced with oligofructose and inulin demonstrated better consumer acceptability, indicating advantages of selected prebiotics not only in probiotic viability and stability, but also in improving sensory properties. This underlines the fact that if a probiotic strain shows positive impact on cheese quality, this could be an extra advantage besides health benefits. A recent investigation on the impact of the proteolytic activity of three probiotic strains of *Lact. acidophilus*, *Lact. paracasei*, and *B. lactis* used as adjunct cultures in semi-hard (Pategràs) cheese indicated that *Lact. acidophilus* played a major role in secondary proteolysis of probiotic cheeses; the probiotic strains produced acceptable functional cheese with similar acceptability scores with respect to the control (Bergamini et al. 2009).

Yogurts and fermented milks are considered the main vehicles to deliver probiotics and those containing fruits are popular among consumers. In a study of Kailasapathy et al. (2008) two strains of commercial probiotics (*Lact. acidophilus* LAFTI L10 and *B. animalis* ssp. *lactis* B49) showed no significant loss of viability after the addition of either 5 or 10 g/100 g fruit preparations in stirred yogurt, except for passion fruit and mixed berries. As a matter of fact, the recommended levels (10^6–10^7 cfu/g) of *Lact. acidophilus* and *B. animalis* ssp. *lactis* were contained in all the yogurts at the end of 35-day shelf life. The possible decrease of viability of probiotic bacteria in yogurts could be avoided by strain selection for properties such as acid and oxygen tolerance or by increasing the level of inoculum. The addition of protectants, such as whey proteins, may increase the buffer capacity and, therefore, enhance the viability of probiotics in yogurts. Positive effects of adding amylase maize starch, inulin, cysteine, and fructooligosaccharides on the viability of *Lact. acidophilus*, *Lact. casei*, *Lact. rhamnosus*, and *Bifidobacterium* spp. in yogurts were observed (Capela et al. 2006; Güler and Akin 2007). Ice cream is also considered a favorable probiotic carrier due to the lower storage temperature which, by minimizing the risk of temperature abuse during frozen storage, leads to higher viability of probiotics at the time of consumption (Cruz et al. 2009). Ice creams made with different levels of fat and sugar have shown different survival rates of the probiotic strain *Lact. johnsonii* La1 at 30 days of storage at -28 °C (Alamprese et al. 2002). Added inulin involved a significantly higher viability of *Lact. acidophilus* and *B. lactis* in ice cream; viability of these probiotics varied depending on the sugar levels of ice cream (Akin et al. 2007).

In conclusion, dairy products provide an effective delivery system for LAB probiotics because of their ability to grow in milk. In this regard, when probiotic organisms are delivered in dairy formulations, the influence of the substrate (milk, fermented milk, or cheese) on the strain growth and viability as well as their metabolic activity on the organoleptic properties of products must be evaluated. The market for probiotics is growing rapidly, as evidenced by the increasing interest of industry and consumers and the ever growing accumulation of scientific studies. The uncovered need for new strains with probiotic potential will expand the market opportunities for the food industry. The next generation of dairy products should contain emerging LAB species, in this context *Lact. helveticus* is of utmost interest as it is an important food-associated species traditionally used in the manufacture of Swiss-type and long-ripened Italian cheeses such as Emmental, Gruyere, Grana, and Provolone. Even though *Lact. helveticus* is not a gut microbe, it is intriguing that it is closely related to commensal GI lactobacilli. Consequently, it would be expected that it could have some probiotic potential and, in fact, some preliminary probiotic applications of this LAB species are already available in the literature (Slattery et al. 2010).

12.2.4. Exopolysaccharide-producing Starters

Some LAB produce polysaccharide polymers or exopolysaccharides (EPS) that consist of a repeating mono- or oligosaccharide subunit connected by varying glycosidic linkages, thereby generating homo- or heteropolymers, respectively. They can be capsular polysaccharides (CPS) that are tightly associated with the cell surface, or slime EPS, which are loosely attached to the outer cell structures or secreted into the environment. These extracellular polysaccharides are economically important because they can modify the physicochemical characteristics of food products by improving their sensory and structural properties (Badel et al. 2011). Apart from ecological function and well-demonstrated technological applications, it has been suggested that the EPS produced by some strains could exert functional effects.

LAB are GRAS (Generally Recognized as Safe) or QPS (Qualified Presumption of Safety) microorganisms that have a long history of safe use in food processing. Some strains with scientifically demonstrated probiotic properties are being included in the manufacture of "functional foods" (Salminen and Gueimonde 2004). Among the beneficial effects on human health, it has been suggested that EPSs produced by LAB act as protective agents against stomach ulcers (Rodríguez et al. 2009) and colitis (Sengül et al. 2005) and as prebiotics exerting a role as growth substrate for elements of the gut microbiota (Salazar et al. 2008, 2011; Fanning et al. 2012). Notably, EPSs – particularly in pathogens – are thought to be critical in host–microbe interactions, where they aid in adherence and colonization within the human host (Conover et al. 2010) and in immunomodulation (Xu et al. 2009). A serum lowering cholesterol ability, anticarcinogenic activity, and bile/acid protection have also been suggested (Rúas-Madiedo et al. 2002; Alp and Aslim 2010). Nowadays, in view of the increasing interest of modern consumers focusing on safe and healthy food without additives, new perspectives for development appear for these biopolymers. In particular, the GRAS status and, sometimes, the probiotic properties of certain strains of LAB give them more preference for consumable EPS production (De Vuyst and Degeest, 1999; Laws and Marshall 2001). One of their most described applications is their utilization as texturing agents naturally synthesized in the fermented food products, principally in the yogurt industry. The main drawbacks limiting their industrial expansion (particularly as food additives) are their low yields of production and the validation of their healthy allegations (Badel et al. 2011).

Yogurt is a well-known dairy product in which both bacterial species involved, *Strep. thermophilus* and *Lact. delbrueckii* subsp. *bulgaricus*, are capable of producing EPS in variable amounts. Stirred yogurts can be obtained by mild homogenization of the milk coagulum after fermentation but this mechanical treatment strongly affects the rheology of the product and facilitates syneresis (serum separation) by breakdown of the gel network. Diverse technological strategies to improve the quality of the product include an increase in milk solids such as fat, proteins or sugars (sucrose, fructose), and addition of stabilizers such as pectin, starch, alginate or gelatin, when permitted by national legislation. However, these approaches do not consider the consumer demand for products with low (or reduced) fat, sugar, and cost and with as few food additives as possible. A useful alternative to this challenge is to use EPS-producing functional starter cultures to enhance texture and mouthfeel, and to avoid syneresis during fermentation or upon storage of the yogurt (De Vuyst et al. 2001). The presence of these biopolymers may improve water retention and texture and confer thickness without altering the organoleptic characteristics of the final product. Thus, there is no need to add stabilizers, many of which are prohibited in a wide range of countries (Badel et al. 2011). Several studies have reported on the impact of EPS-producing starters on the physical properties of yogurt. Folkenberg et al. (2006) demonstrated that fermentation with EPS-producing strains resulted in yogurts with increased ropiness and mouth thickness and a tendency to be creamier than control yogurts, which in contrast had higher gel firmness and lower spontaneous syneresis. Other works have been carried out by comparing yogurts with EPS-producing strains of *Strep. thermophilus* and their genetic variants that only differed from the mother strain in their inability to produce EPS. For example, Hassan et al. (2003) showed that stirring affects a yogurt made with EPS-producing starter contrarily than yogurt made with its isogenic strain, avoiding the immediate syneresis. But the presence of EPS in the matrix of a yogurt and its effect on the texture properties does not always follow a direct correlation. It is well known that the physical properties of these molecules will be influenced by their mass and also by their subunit structure. Laws and Marshall (2001), by means of Nuclear Magnetic Resonance (NMR) spectroscopy, examined and compared EPS subunit structures in milk fermented with different EPS-producing LAB (lactococci, streptococci, and lactobacilli). They showed that the final texture of the product was influenced not only by the nature of the EPS but also by the amount produced, the composition and acidity of the milk, as well as the fermentation length. The complex interactions among bacterial cells (surrounded or not by EPS, encapsulated or not), the protein network (spatial structure, interactions, pH, ionic and gel strength), and the EPS (localization, aggregation, molecular mass, and structure) play an important role in texture build up (De Vuyst et al. 2003).

In cheese-making, despite the enormous market potential, the diffusion of low-fat cheeses over the past decades has been lower than expected, probably due to poor consumer perception based on inadequate texture and taste attributes. Fat not only has nutritional significance but also contributes to the sensory and functional features of the cheese. Most low-fat cheeses are considered dry, excessively firm, rubbery, difficult to chew, and with atypical flavor (e.g., off-flavors and bitterness). Removal of fat from the casein matrix leads to the formation of a much tighter para-casein network, which becomes excessively firm during syneresis. One of the main objectives for improving the texture of low-fat cheese is to increase moisture content. Among other technological alternatives, the use of EPS-producing starter cultures represents a useful strategy to improve textural, melting, and sensory characteristics of reduced-fat cheeses (Awad et al. 2005; Zisu and Shah 2005) by increasing the moisture levels, interfering with the association of casein micelles during coagulation, and affecting gel strength and the degree of syneresis. In Cheddar cheese, both capsular and ropy EPS-producing strains appeared to increase moisture retention and consequently cheese yield (Dabour et al. 2006). Both strains showed that they could modify and improve textural characteristics, particularly in freshly pressed reduced-fat Cheddar cheese. These desirable effects were less pronounced in aged cheeses. Likewise, no differences in hardness, springiness, and chewiness were found between young reduced-fat Cheddar cheese made with a ropy *Lactococcus lactis* subsp. *cremoris* when compared with full-fat controls (Awad et al. 2005; Hassan et al. 2007). No differences were detected by panelists between the textures of both types of products made with EPS-producing strains that were less rubbery or firm, curdy, and crumbly than all other reduced-fat cheeses. Similar results were described for Mozzarella cheeses where the use of EPS-producing starters improved texture and reduced maturation time (Zisu and Shah 2007). In this case, the control cheeses showed the lowest moisture content. Pre-acidification of the cheese milk did not influence moisture retention in the final product unless made with EPS-producing cultures, but improved texture during the early stages of storage. More recently, a combination of various technology approaches, such as supplementation with micro-particulated whey protein concentrate (MWPC), EPS-producing *Strep. thermophilus* (primary starter), and adjunct cultures for making low-fat Caciotta-type cheese, has been reported (Di Cagno et al. 2014). Comparison of full- and low-fat cheeses was done using an integrated approach, which considered compositional, microbiological, and biochemical, textures and sensory determinations. Although the use of MWPC markedly contributed to moisture retention, the presence of EPS seemed to positively affect texture of the low-fat Caciotta-type cheese. While the ability to enhance cheese yield (through increased moisture retention) and functionality are strong incentives to use EPS-producing cultures in cheese-making, concerns over EPS accumulation in cheese whey continue to limit commercial interest in EPS cheese starters. However, it is important to recognize that while EPS accumulation in cheese whey may be expected from ropy starters, capsular EPS-starters are less likely to affect whey viscosity because the polymer is covalently linked to the cell wall (Broadbent et al. 2003). For this reason, BAL that produce capsular EPS, where the extracellular material remains attached to the cell, are of particular interest as they have minimal adverse effects on whey processing.

Currently it is clear that diversification of functional foods, which have been scientifically validated as having beneficial properties, will increase in the near future and the EPS synthesized by LAB could have a place in the market as an ingredient of this type of food. To this purpose, EPS can be synthesized *in situ* by their producing strains or can be used in isolated form as a food grade additive. Their chances of reaching the market place will be improved by the discovery and utilization of new EPS-producing strains isolated from diverse BAL species (Mozzi et al. 2006). The discovery and characterization of new EPS producers isolated from food (e.g., processed meat products) and beverages (e.g., wine and cider) will increase the variety of EPS and the use of their producing strains for the elaboration of novel functional food.

12.3. Phage-Resistant Starters

Bacteriophage attack has been recognized as a major problem in dairy fermentative processes for almost 80 years, leading to total or partial inhibition of LAB present in the starter cultures. It is now accepted that the presence of phages is ubiquitous in dairies, thus several strategies have been designed to control rather than eliminate them (Briggiler et al. 2012). Culture rotation programs, direct vat-inoculation of starters, careful handling and disposal of whey (Coffey and Ross 2002), use of phage-inhibitory media, optimized sanitation, and use of starter cultures improved in their phage resistance, are among the approaches applied to minimize phage spreading in dairy plants (Moineau et al. 2002; Sturino and Klaenhammer 2004). Nevertheless, the aptitude of phages to survive thermal treatments (including pasteurization) and their resistance to sanitation measures have forced the industry to rely on bacteria with naturally occurring phage resistance systems for the formulation of starter cultures (Barrangou and Horvath 2012).

One strategy to obtain starter cultures with improved phage-resistance is the construction of genetically engineered strains. The designed genetic tools, intensively studied for the last three decades, are based on the characterization and exploitation of the LAB native phage defense mechanisms, as well as some phage genetic elements (Moineau 1999; Sturino and Klaenhammer 2004). These type of antiphage approaches include origin-derived phage-encoded resistance (O'Sullivan et al. 1993), antisense RNA technology (Walker and Klaenhammer 2000), and phage-triggered suicide systems (Djordjevic et al. 1997). Nevertheless, the lack of progress in the development of legislation regarding Genetically Modified Organisms has prohibited the use of these modified strains at the industrial level.

Phage infections are more problematic when probiotic LAB strains are the target, in particular when the manufacture of probiotic products includes a propagation of the strains as a starter (Saarela et al. 2000; Capra et al. 2006). As these strains usually grow slowly, they become highly vulnerable to phages (Capra et al. 2006). For probiotic phages, control strategies are restricted because these phages are resistant to thermal and biocide treatments (Capra et al. 2004); strain rotation is inappropriate; and also, many phages originate from lysogenic probiotic strains (Ventura et al. 2006; Durmaz et al. 2008). Furthermore, the development of phage-resistant probiotic starters is very limited, as the preservation of probiotic features is still a non-explored but crucial issue to be considered in practice.

L. lactis and *Strep. thermophilus* are the most extensively used LAB species in dairy fermentative processes, thus several phages infecting them have been isolated and characterized in detail (Suárez et al. 2002; Emond and Moineau 2007, Quiberoni et al. 2010). The extended co-survival of LAB and phages within the same environment has prompted the acquisition of a variety of native bacteriophage defense systems. These mechanisms include inhibition of phage adsorption, blocking of DNA injection, restriction/modification systems, abortive infection (Moineau and Lévesque 2005), and the novel clustered interspaced short palindromic repeats (CRISPR) systems.

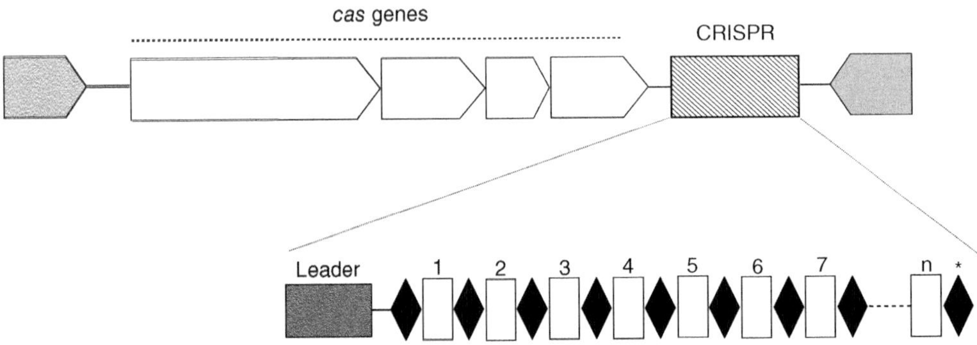

Figure 12.1. Simplified representation of CRISPR-*cas* system locus. CRISPR repeats are represented as black diamonds and spacers as numbered white boxes. The asterisk indicates the terminal repeat.

Phage-resistant strains can be selected from a sensitive population. They are known as bacteriophage-insensitive mutants (BIMs), which are spontaneous and natural derivatives that survive an exposure to virulent phages (Moineau and Lévesque 2005). In this sense, several researchers have described the isolation of spontaneous phage-resistant variants from sensitive strains of *Lactococcus* (Limsowtin and Terzaghi 1976; Weimer et al. 1993), *Strep. thermophilus* (Viscardi et al. 2003; Binetti et al. 2007), *Lact. helveticus* (Neviani et al. 1992; Carminati et al. 1993; Reinheimer et al. 1993; Quiberoni et al. 1999), and *Lact. delbrueckii* (Guglielmotti et al. 2006).

The mechanism involved in BIMs' generation was often attributed to mutations in the phage receptors (Viscardi et al. 2003), although recent studies (Barrangou et al. 2007; Deveau et al. 2008; Horvath et al. 2008; Sorek et al. 2008; Mills et al. 2010b) have demonstrated that CRISPRs play a role in the development of BIMs. CRISPR systems supply acquired, yet heritable, sequence-specific "adaptive" immunity against phage and other invasive genetic elements, such as plasmids. CRISPR loci normally consist of short and highly conserved DNA "repeat" sequences separated by variable sequences called "spacers" (Figure 12.1). The spacer sequences typically derive from foreign genetic fragments, such as viruses and plasmids, which are incorporated into the CRISPR locus during the immunization process (Horvath et al. 2008; Barrangou and Horvath 2012). CRISPRs, together with CRISPR-associated genes (*cas*), constitute the CRISPR-*cas* immune system. The immunization process is based on the incorporation of short DNA sequences from virulent phages or plasmids into the CRISPR locus. Afterward, CRISPR transcripts are processed into small interfering RNAs (crRNAs) that guide a multifunctional protein complex to recognize and cleave the complementary invading nucleic acid and targets this for degradation (Barrangou and Horvath 2012; Richter et al. 2012).

Although the initial discovery of a CRISPR structure was made fortuitously in *Escherichia coli* in 1987, the acronym was coined in 2002, after similar structures were observed in genomes of various bacteria and archaea. CRISPRs sequences have been extensively studied on BIMs isolated from *Strep. thermophilus* strains of technological importance, in which the acquisition of new spacers was well documented (Barrangou et al. 2007; Deveau et al. 2008, Horvath et al. 2008; Garneau et al. 2010; Mills et al. 2010b). These novel spacers are incorporated at the leader end of the locus, in response to challenges with virulent phages, which generated a phage-resistant phenotype. The polarized integration of novel repeat-spacer units at the leader end provides a chronological time line of spacer acquisition events (Barrangou and Horvath 2012).

The generation of BIMs is advantageous for the dairy industry since it offers a direct food-grade mechanism for producing strains with the capacity to evade phage attacks. Based on the CRISPR system, many potentially unique phage-resistant derivatives can be generated from a single strain (Mills et al. 2010b). CRISPR variants should be selected based on the acquisition of novel spacers with high homology with phage sequences (Deveau et al. 2008; Barrangou and Horvath 2012). Some authors cleverly suggested the creation of a bank of BIMs differing only in the CRISPR region, which could be used as part of a starter culture rotation strategy (Mills et al. 2010b). Due to the rapid evolution of phages (by single nucleotide mutations or deletions) in response to CRISPR (Deveau et al. 2008), various CRISPR variants (obtained from independent rounds of iterative phage challenges) can be selectively combined in starter rotation schemes for increased robustness and best industrial lifetime (Mills et al. 2010b; Barrangou and Horvath 2012).

The benefits of using CRISPR-based variants seem to be infinite and extremely promissory. In particular, the fermentative industry could be greatly benefited by the inclusion of phage-resistant variants to the pool of starter cultures. With this in mind, the DuPont group (DuPont Nutrition Biosciences ApS) was a pioneer in supporting numerous studies to increase the knowledge of CRISPR-*cas* systems. In fact, the research carried out has led to obtaining a European Patent (EP 2 325 332 B1, European Patent Specification, 31.10.2012 Bulletin 2012/44). Essentially, the claim of the patent describes a method for modulating/improving the resistance of a bacterial cell against a target nucleic acid or transcription product from a bacteriophage by the insertion of CRISPR systems. The improved CRISPR-mediated strains could be of high importance at the industrial level and be included in starter cultures. Examples of applications of this invention are countless, and include phage resistance, for diagnostics (to predict/determine the likely sensitivity of particular microbes to bacteriophage), and resistance to plasmid transfer, mobile genetic elements, antibiotic resistance genes, and genes encoding virulence factors, among others. At this time, there are available in the market some commercial starter cultures for pizza cheese manufacture (CHOOZIT™ SWIFT, DuPont™), composed of three to six *Strep. thermophilus* strains with improved phage-resistance, besides allowing reliable acidification during the cheese process. In addition, there is a company currently developing a yogurt starter using CRISPR technology (http://www.danisco.com/about-dupont/innovation/food-ingredients-innovation/innovation-in-taste-texture/).

12.4. New Sources of Starter Strains

Over the last years, researchers have tended to find new sources of isolation of LAB with industrially interesting characteristics to be used in fermented products. As is known, natural milk cultures from cow milk, used to manufacture traditional cheeses in Italy and Argentina, involve mainly *Strep. thermophilus* strains. In this sense, cheeses made with natural milk cultures could be an important source of wild strains with adequate technological characteristics. These wild strains could be used as starter cultures to manufacture fermented dairy products (Reinheimer et al. 1997; De Angelis et al. 2008). Besides cows' milk, milk from ewes or goats is used to make several types of traditional cheeses in mainly Southern European countries such as Spain (e.g., Roncal, Idiazábal, Manchego), Portugal (e.g., Serra da Estrela), France (e.g., Roquefort, Camembert), and Italy (e.g., Pecorino Romano, Robiola di Roccaverano) and constitute a valuable source of strains with technological potentiality (Arizcun et al. 1998; Medina and Nuñez 2004; Ortigosa et al. 2006; Bonetta et al. 2008).

The microbiota of raw milk and artisanal cheeses made with raw milk from ewes or goats involves mainly *L. lactis* (Hatzikamari et al. 1999; Durlu-Oskaya et al. 2001; Bonetta et al. 2008; Casalta et al. 2009), lactobacilli (principally *Lact. casei* and *Lact. plantarum*) (Macedo et al. 1995; Durlu-Oskaya et al. 2001; Psoni et al. 2003; Ortigosa et al. 2006; Casalta et al. 2009, Li et al. 2013), *Strep.*

thermophilus (Treu et al. 2014), and *Leuconostoc* (Macedo et al. 1995; Arizcun et al. 1998; Psoni et al. 2003). In addition, *Enterococcus* (*Ent. faecium*, *Ent. faecalis*, *Ent. durans*, *Ent. hirae*) strains were found at high counts in soft and semi-hard cheeses (Durlu-Oskaya et al. 2001; Psoni et al. 2003; Nikolaou et al. 2002). On the other hand, in these kind of cheeses there is a non-lactic acid microbiota formed by enterobacteria, filamentous fungi (*Mucor*) and yeasts (*Geotrichum*, *Kluyveromyces*) (Santos et al. 1996; Bonetta et al. 2008; Laurenčík et al. 2008).

As is known, the microbiota of cheeses as well as their microbial counts are affected by several factors, such as manufacturing processes, ripening stage, climatic conditions, and seasonality (Macedo et al. 1995; Psoni et al. 2003). As an example, Psoni et al. (2003) and Nikolaou et al. (2002) reported that enterococci and lactococci were abundant in goat's and ewe's milk in winter, while lactobacilli (mainly *Lact. plantarum*) were the most frequently genus found in spring and summer. Regarding the ripening stage, Psoni et al. (2003) found that *L. lactis* (*L. lactis* subsp. *lactis* and *L. lactis* subsp. *cremoris*) was found at high counts in milk and cheeses in early ripening stages. However, their counts decreased during the cheese ripening, possibly due to high salt concentrations in cheeses. Conversely, lactobacilli (*Lact. casei* and *Lact. plantarum*) were the most abundant group during the last stages of cheese ripening (Tornadijo et al. 1995; Hatzikamari et al. 1999; Psoni et al. 2003; Caridi et al. 2003).

Wild strains (*Lact. plantarum*, *Lact. paracasei*, *L. lactis* subsp. *lactis*) isolated from artisanal cheeses made with raw ewe's milk evidenced adequate acidifying and enzymatic activities to be used as starter/adjunct cultures in cheese-making. Thus, they may have a positive effect on cheese ripening and could contribute to taste and aroma development. On the other hand, wild strains may show inhibitory activity against pathogenic bacteria and therefore could be useful to control the growth of undesirable microorganisms during cheese production. Several authors reported that these strains contributed to the development of flavor and improved the microbiological quality of cheeses made with pasteurized milk (Xanthopoulos et al. 2000; Durlu-Oskaya et al. 2001; Caridi et al. 2003; Psoni et al. 2003; Psoni et al. 2006). However, the use of starter cultures and pasteurized milk instead of raw milk in cheese-making could result in cheeses with uniform characteristics, although some flavor properties could be decreased (Xanthopoulos et al. 2000).

Studies on the use of natural milk culture (developed following a standardized manufacture protocol) from ewe's milk to make semi-hard cheeses are in hand in Argentina. Preliminary results demonstrate that natural milk cultures would be a source of interesting *Strep. thermophilus* strains. Cheeses from diverse manufactures made with these natural milk cultures showed adequate microbiological quality and sensorial properties (Meinardi, Personal Communication).

12.5. Conclusions

Besides new strains to be used as primary starter cultures, an increasing attention is being given to the non-starter LAB (NSLAB) microbiota, which composes the so-called starter adjuncts (also defined as "subsidiary" or "complementary" cultures). Starter adjuncts can be used as ripening, protective, probiotic, and/or EPS-producing cultures. The aim of their use is to add new and useful characteristics to fermented products and, for this reason, the current research on LAB is also addressed to obtain and characterize new strains of NSLAB. Phage resistance in LAB and NSLAB is always a critical subject because of the activity loss of starter strains. So, phage resistance due to "new" resistance mechanisms should be investigated from a scientific and applied point of view. Microbiologists are always interested in obtaining new sources of wild LAB and NSLAB strains and, in this sense, natural cultures from ewe's or goat's milk offer interesting microbial diversity.

References

Akin, M.B., Akin, M.S., and Kirmaci, Z. (2007) Effects of inulin and sugar levels on the viability of yogurt and probiotic bacteria and the physical and sensory characteristics in probiotic ice-cream. *Food Chem* **104**, 93–99.

Alamprese, C., Foschino, R., Rossi, M., et al. (2002) Survival of *Lactobacillus johnsonii* La 1 and influence of its addition in retail-manufactured ice cream produced with different sugar and fat concentrations. *Int Dairy J* **12**, 201–208.

Alp, G. and Aslim, B. (2010) Relationship between the resistance to bile salts and low pH with exopolysaccharide (EPS) production of *Bifidobacterium* spp. isolated from infants feces and breast milk. *Anaerobe* **16**, 101–105.

Arizcun, C., Barcina, Y., and Torre, P. (1998) Identification of lactic acid bacteria isolated from Roncal and Idiazábal cheeses. *Lait* **77**, 729–736.

Awad, S., Hassan, A.N., and Muthukumarappan, K. (2005) Application of exopolysaccharide-producing cultures in reduced-fat Cheddar cheese: texture and melting properties *J Dairy Sci* **88**, 4204–4213.

Badel, S., Bernardi, T., and Michaud, P. (2011) New perspectives for lactobacilli exopolysaccharides. *Biotechnol Adv* **29**, 54–66.

Barrangou, R. and Horvath, P. (2012) CRISPR: New horizons in phage resistance and strain identification. *Ann Rev Food Sci Technol* **3**, 143–162.

Barrangou, R., Fremaux, C., Deveau, H., et al. (2007) CRISPR provides acquired resistance against viruses in prokaryotes. *Science* **315**, 1709–1712.

Bergamini, C.V., Hynes, E.R., Palma, S.B., et al. (2009) Proteolytic activity of three probiotic strains in semi-hard cheese as single and mixed cultures: *Lactobacillus acidophilus*, *Lactobacillus paracasei* and *Bifidobacterium lactis*. *Int Dairy J* **19**, 467–475.

Bertolino, M., Zeppa, G., Gerbi, V., and McSweeney, P.L.H. (2008) Study of proteolysis in miniature Toma Piemontese cheese made using wild bacteria. *Ital J Food Sci* **20**, 57–73.

Binetti, A.G., Suárez, V., Tailliez, P., and Reinheimer, J.A. (2007) Characterization of spontaneous phage-resistant variants of *Streptococcus thermophilus* by randomly amplified polymorphic DNA analysis and identification of phage-resistance mechanisms. *Int Dairy J* **17**, 1115–1122.

Bonetta, S., Bonetta, S., Carraro, E., et al. (2008) Microbiological characterization of Robiola di Roccaverano cheese using PCR-DGGE. *Food Microbiol* **25**, 786–792.

Briggiler Marcó, M., Moineau, S., and Quiberoni, A. (2012) Bacteriophages and dairy fermentations. Review. *Bacteriophage* **2**(3), 149–158.

Broadbent, J.R., McMahon, D.J., Welker, D.L., et al. (2003) Biochemistry, genetics, and applications of exopolysaccharide production in *Streptococcus thermophilus*: a review. *J Dairy Sci* **86**, 407–423.

Bustos, I., Martínez-Bartolomé, M.A., Achemchem, F., et al. (2011) Volatile sulphur compounds-forming abilities of lactic acid bacteria: C-S lyase activities. *Int J Food Microbiol* **148**, 121–127.

Capela, P., Hay, T.K.C., and Shah, N.P. (2006) Effect of cryoprotectants, prebiotics and microencapsulation on survival of probiotic organisms in yoghurt and freeze-dried yoghurt. *Food Res Int* **39**, 203–211.

Caplice, E. and Fitzgerald, G.F. (1999) Food fermentations: role of microorganisms in food production and preservation. *Int J Food Microbiol* **50**, 131–49.

Capozzi, V., Russo, P., Ladero, V., et al. (2012) Biogenic amines degradation by *Lactobacillus plantarum*: toward a potential application in wine. *Frontiers Microbiol* **3**, 1–6.

Capra, M.L., Quiberoni, A., and Reinheimer, J. (2004) Thermal and chemical resistance of *Lactobacillus casei* and *Lactobacillus paracasei* bacteriophages. *Lett Appl Microbiol* **38**, 499–504.

Capra, M.L., Quiberoni, A., and Reinheimer, J.A. (2006) Phages of *Lactobacillus casei/paracasei*: response to environmental factors and interaction with collection and commercial strains. *J Appl Microbiol* **100**, 334–342.

Cardarelli, H.R., Buriti, F.C.A., Castro, I.A., and Saad, S.M.I. (2008) Inulin and oligofructose improve sensory quality and increase the probiotic viable count in potentially synbiotic petit-suisse cheese. *LWT-Food Sci Technol* **41**, 1037–1046.

Caridi, A., Micari, P., Foti, F., et al. (2003) Ripening and seasonal changes in microbiological and chemical parameters of the artisanal cheese Carprino d'Aspromonte produced from raw or thermized goat's milk. *Food Microbiol* **20**, 201–209.

Carminati, D., Zennaro, R., Neviani, E., and Giraffa, G. (1993) Selezione e caratteristiche di mutanti fago-resistenti di *Lactobacillus helveticus*. *Scienza Tec Latt Casea* **44**, 33–48.

Casalta, E., Sorba, J.M., Aigle, M., and Ogier, J.C. (2009) Diversity and dynamics of the microbial community during the manufacture of Calenzana, an artisanal Corsican cheese. *Int J Food Microbiol* **133**, 243–251.

Chamba, J.F. and Irlinger, F. (2004) Secondary and adjunct cultures. In: Cheese: Chemistry, Physics and Microbiology ((eds) P.F. Fox, P.L.H. McSweeney, T.M. Cogan, and T.P. Guinee), 3rd edn, vol. **1**, Elsevier Academic Press, London, pp. 191–206.

Chamba, J.F. (2008) Application des bactéries lactiques lors des fabrication fromagères. In: Bactéries *lactiques – De la génétique aux ferments* ((eds) G. Corrieu and F.M. Luquet), Lavoisier, Paris, pp. 787–821.

Coffey, A. and Ross, R.P. (2002) Bacteriophage-resistance systems in dairy starter strains: Molecular analysis to application. *Antonie van Leeuwenhoek* **82**, 303–321.

Conover, M.S., Sloan, G.P., Love, C.F., et al. (2010) The Bps polysaccharide of *Bordetella pertussis* promotes colonization and biofilm formation in the nose by functioning as an adhesin. *Mol Microbiol* **77**, 1439–1455.

Cruz, A.G., Antunes, A.E.C., Sousa, A.L.O.P., et al. (2009) Ice-cream as a probiotic food carrier. *Food Res Int* **42**, 1233–1239.

Dabour, N., Kheadr, E., Benhamou, N., et al. (2006) Improvement of texture and structure of reduced-fat Cheddar cheese by exopolysaccharide-producing lactococci. *J Dairy Sci* **89**, 95–110.

De Angelis, M., Calasso, M., Di Cagno, R., et al. (2010) NADP-glutamate dehydrogenase activity in non starter lactic acid bacteria: effects of temperature, pH and NaCl on enzyme activity and expression. *J Appl Microbiol* **109**, 1763–1774.

De Angelis, M., de Candia, S., Calasso, M., et al. (2008) Selection and use of autochthonous multiple strain cultures for the manufacture of high-moisture traditional mozzarella cheese. *Int J Food Microbiol* **125**, 123–132.

De Vuyst, L. and Degeest, B. (1999) Heteropolysaccharides from lactic acid bacteria. *FEMS Microbiol Rev* **23**(2), 153–117.

De Vuyst, L., de Vin, F., Vaningelgem, F., and Degeest, B. (2001) Recent developments in the biosynthesis and applications of heteropolysaccharides from lactic acid bacteria. *Int Dairy J* **11**, 687–707.

De Vuyst, L., Zamfir, M., Mozzi, F., et al. (2003) Exopolysaccharide-producing strains as functional starter cultures in the production of fermented milks. *Int Dairy J* **13**, 707–717.

Deveau, H., Barrangou, R., Garneau, J.E., et al. (2008) Phage response to CRISPR-encoded resistance in *Streptococcus thermophilus*. *J Bacteriol* **190**, 1390–1400.

Di Cagno, R., De Pasquale, I., De Angelis, M., et al. (2014) Use of microparticulated whey protein concentrate, exopolysaccharide-producing *Streptococcus thermophilus*, and adjunct cultures for making low-fat Italian Caciotta-type cheese. *J Dairy Sci* **97**(1), 72–84.

Djordjevic, G.M., O'Sullivan, D.J., Walker, S.A., et al. (1997) A triggered-suicide system designed as a defense against bacteriophages. *J Bacteriol* **179**, 6741–6748.

Durlu-Oskaya, F., Xanthopoulos, V., Tunail, N., and Litopoulou-Tzanetaki, E. (2001) Technologically important properties of lactic acid bacteria isolates from Beyaz cheese made from raw ewes' milk. *J Appl Microbiol* **91**, 861–870.

Durmaz, E., Miller, M.J., Azcarate-Peril, M.A., et al. (2008) Genome sequence and characteristics of Lrm1, a prophage from industrial *Lactobacillus rhamnosus* strain M1. *Appl Environ Microbiol* **74**, 4601–4609.

EFSA, European Food Safety Authority. (2010) Scientific Opinion on the maintenance of the list of QPS biological agents intentionally added to food and feed (2010 update) The *EFSA J* **8**, 1944, 56 pp.

Emond, E. and Moineau, S. (2007) Bacteriophages and food fermentations. In: Bacteriophage: *Genetics and Molecular Biology* ((eds) S. McGrath and D. van Sinderen), Horizon Scientific Press/Caister Academic Press, Norfolk, UK, pp. 93–124.

Fanning, S., Hall, L., and van Sinderen, D. (2012) *Bifidobacterium breve* UCC2003 surface exopolysaccharide production is a beneficial trait mediating commensal-host interaction through immune modulation and pathogen protection. *PNAS* **109**, 2108–2113.

Folkenberg, D.M., Dejmek, P., Skriver, A., et al. (2006) Sensory and rheological screening of exopolysaccharide producing strains of bacterial yoghurt cultures. *Int Dairy J* **16**, 111–118.

Gaggia, F., Di Gioia, D., Baffoni, L., and Biavati, B. (2011) The role of protective and probiotic cultures in food and feed and their impact in food safety. *Trends Food Sci Technol* **22**, 58–66.

Galvez, A., Abriouel, H., Lucas, R., and Grande Burgos, M.J. (2011) Bacteriocins for bioprotection of foods. In: *Natural Antimicrobials in Food Safety and Quality* ((eds) M. Rai and M. Chikindas), CABI Int., Cambridge, MA, pp. 39–61.

Garneau, J.E., Dupuis, M.E., Villion, M., et al. (2010) The CRISPR/Cas bacterial immune system cleaves bacteriophage and plasmid DNA. *Nature* **468**, 67–71.

Georgieva, R., Iliev, I., Haertlé, T., et al. (2009) Technological properties of candidate probiotic *Lactobacillus plantarum* strains. *Int Dairy J* **19**, 696–702.

Giraffa, G. (2003) Functionality of enterococci in dairy products. *Int J Food Microbiol* **88**, 215–222.

Giraffa, G. (2012) Selection and design of lactic acid bacteria probiotic cultures. *Eng Life Sci* **12**, 391–398.

González, L., Sacristán, N., Arenas, R., et al. (2010) Enzymatic activity of lactic acid bacteria (with antimicrobial properties) isolated from a traditional Spanish cheese. *Food Microbiol* **27**, 592–597.

Guglielmotti, D.M., Reinheimer, J.A., Binetti, A.G., et al. (2006) Characterization of spontaneous phage-resistant derivatives of *Lactobacillus delbrueckii* commercial strains. *Int J Food Microbiol* **111**, 126–133.

Güler-Akin, M.B. and Akin, M.S. (2007) Effects of cysteine and different incubation temperatures on the microflora, chemical composition and sensory characteristics of bio-yogurt made from goat's milk. *Food Chem* **100**, 788–793.

Hanniffy, S.B., Peláez, C., Martínez-Bartolomé, M.A., et al. (2009) Key enzymes involved in methionine catabolism by cheese lactic acid bacteria. *Int J Food Microbiol* **135**, 223–230.

Hassan, A., Ipsen, R., Janzen, T., and Qvist, K.B. (2003) Microstructure and rheology of yogurt made with cultures differing only in their ability to produce exopolysaccharides. *J Dairy Sci* **86**, 1632–1638.

Hassan, A.N., Awad, S., and Mistry, V.V. (2007) Reduced fat process cheese made from young reduced fat Cheddar cheese manufactured with exopolysaccharide-producing cultures. *J Dairy Sci* **90**, 3604–3612.

Hatzikamari, M., Litopoulou-Tzanetaki, E., and Tzanetakis, N. (1999) Microbiological characteristics of Anevato: a traditional Greek cheese. *J Appl Microbiol* **87**, 595–601.

Hayes, M., Ross, R.P., Fitzgerald, G.F., and Stanton, C. (2007) Putting microbes to work: dairy fermentation, cell factories and bioactive peptides. Part I: overview. *Biotechnol J* **2**, 426–434.

Holzapfel, W.H., Geisen, R., and Schillinger, U. (1995) Biological preservation of foods with reference to protective cultures, bacteriocins and food-grade enzymes. *Int J Food Microbiol* **24**, 343–362.

Horvath, P., Romero, D.A., Coûté-Monvoisin, A.-C., Richards et al. (2008) Diversity, activity, and evolution of CRISPR loci in *Streptococcus thermophilus*. *J Bacteriol* **190**, 1401–1412.

Jay, J.M. (1982) Antimicrobial properties of diacetyl. *Appl Environ Microbiol* **44**, 525–532.

Jensen, M.P., Vogensen, F.K., and Ardo, Y. (2009) Variation in caseinolytic properties of six cheese related *Lactobacillus helveticus* strains. *Int Dairy J* **19**, 661–668.

Kailasapathy, K., Harmstorf, I., and Phillips, M. (2008) Survival of *Lactobacillus acidophilus* and *Bifidobacterium animalis* ssp. *lactis* in stirred fruit yogurts. *LWT-Food Sci Technol* **41**, 1317–1322.

Laleye, L.C., Simard, R.E., Lee, B.H., et al. (1987) Involvement of heterofermentative lactobacilli in development of open texture in cheeses. *J Food Prot* **50**, 1009–1012.

Lanciotti, R., Patrignani, F., Bagnolini, F., et al. (2003) Evaluation of diacetyl antimicrobial activity against *Escherichia coli*, *Listeria monocytogenes* and *Staphylococcus aureus*. *Food Microbiol* **20**, 537–543.

Laurenzick, M., Sulo, P., Sláviková, E., et al. (2008) The diversity of eukaryotic microbiota in the traditional Slovak sheep cheese-Bryndza. *Int J Food Microbiol* **127**, 176–179.

Laws, A. and Marshall, V. (2001) The relevance of exopolysaccharides to the rheological properties in milk fermented with ropy strains of lactic acid bacteria. *Int Dairy J* **11**, 709–721.

Leroy, F. and De Vuyst, L. (2004) Lactic acid bacteria as functional starter cultures for the food fermentation industry. *Trends Food Sci Technol* **15**, 67–78.

Li, X.R., Gong, F.M., Zheng, H.J., et al. (2013) Draft genome sequence of *Lactobacillus plantarum* strain AY01, isolated from the raw material of fermented goat milk cheese. *Gen Announ* **1**(5), 1.

Limsowtin, G.K.Y. and Terzaghi, B.E. (1976) Phage resistant mutants: their selection and use in cheese factories. *New Zealand J Dairy Sci Technol* **11**, 251–256.

Liu, M., Nauta, A., Francke, C., and Siezen, R.J. (2008) Comparative genomics of enzymes in flavor-forming pathways from amino acids in lactic acid bacteria. *Appl Environ Microbiol* **74**, 4590–4600.

Macedo, A.C., Malcata, F.X., and Hogg T.A. (1995) Microbiological profile in Serra ewes' cheese during ripening. *J Appl Bacteriol* **79**, 1–11.

Martins, J.T., Cerqueira, M.A., Souza, B.V.S., et al. (2010) Shelf life extension of ricotta cheese using coatings of galactomannans from nonconventional sources incorporating nisin against *Listeria monocytogenes*. *J Agric Food Chem* **58**, 1884–1891.

Mauriello, G., Ercolini, D., La Storia, A., et al. (2004) Development of polythene films for food packaging activated with an antilisterial bacteriocin from *Lactobacillus curvatus* 32Y. *J Appl Microbiol* **97**, 314–322.

Medina, M., and Nuñez, M. (2004) Cheeses made from ewes' and goats' milk. In: *Cheese: Chemistry, Physics and Microbiology* ((ed.) F.P. Fox). Volume 2: Major Cheese Groups, 3rd edn, Elsevier, Academic Press, London, pp. 279–299.

Milesi, M.M., McSweeney, P.L.H. and Hynes, E.R. (2008) Viability and contribution to proteolysis of an adjunct culture of *Lactobacillus plantarum* in two model cheese systems: Cheddar cheese-type and soft-cheese type. *J Appl Microbiol* **105**, 884–892.

Milesi, M.M., Vinderola, G., Sabbag, N., et al. (2009) Influence on cheese proteolysis and sensory characteristics of non-starter lactobacilli strains with probiotic potential. *Food Res Int* **42**, 1186–1196.

Mills, S., O'Sullivan, O., Hill, C., et al. (2010a) The changing face of dairy starter culture research: from genomics to economics. *Int J Dairy Technol* **63**, 149–170.

Mills, S., Griffin, C., Coffey, A., et al. (2010b) CRISPR analysis of bacteriophage-insensitive mutants (BIMs) of industrial *Streptococcus thermophilus* - implications for starter design. *J Appl Microbiol* **108**, 945–955.

Moineau, S. (1999) Applications of phage resistance in lactic acid bacteria. In: *Proceedings of the Sixth Symposium on Lactic Acid Bacteria: Genetics, Metabolism and Applications* ((eds) W.N. Konings, O.P. Kuipers, and J.H.J. Huis in't Veld), Kluwer Academic Publishers, Veldhoven, The Netherlands, pp. 377–382.

Moineau, S. and Lévesque, C. (2005) Control of bacteriophages in industrial fermentations. In: *Bacteriophages: Biology and Applications* ((eds) E. Kutter and A. Sulakvelidze), CRC Press, Boca Raton, FL, pp. 285–296.

Moineau, S., Tremblay, D. and Labrie, S. (2002) Phages of lactic acid bacteria: from genomics to industrial applications. *ASM News* **68**, 388–393.

Morandi, S., Brasca, M., Andrighetto, C., et al. (2006) Technological characterisation of enterococci isolated from north-west Italian dairy products. *Int Dairy J* **16**, 867–875.

Morandi, S., Silvetti, T. and Brasca, M. (2013) Biotechnological and safety characterization of *Enterococcus lactis*, a recently described species of dairy origin. *Antonie van Leeuwenhoek* **103**, 239–249.

Mozzi, F., Vaningelgem, F., Hébert, E.M., Van der Meulen, R., Foulquié Moreno, M.R., Font de Valdez, G., and De Vuyst, L. 2006. Diversity of heteropolysaccharide-producing lactic acid bacterium strains and their biopolymers. *Applied and Environmental Microbiology* **72**(6), 4431–4435.

Neviani, E., Carminati, D., and Giraffa, G. (1992) Selection of some bacteriophage and lysozyme-resistant variants of *Lactobacillus helveticus* CNRZ 892. *J Dairy Sci* **75**, 905–913.

Nieto-Arribas, P., Poveda, J.M., Seseña, S., et al. (2009) Technological characterization of *Lactobacillus* isolates from traditional Manchego cheese for potential use as adjunct starter cultures. *Food Control* **20**, 1092–1098.

Nieto-Arribas, P., Seseña, S., Poveda, J.M., et al. (2011) *Enterococcus* populations in artisanal Manchego cheese: biodiversity, technological and safety aspects. *Food Microbiol* **28**, 891–899.

Nikolaou, E., Tzanetakis, N., Litopoulou-Tzanetaki, E., and Robinson, R.K. (2002) Changes in the microbiological and chemical characteristics of an artisanal, low-fat cheese made from raw ovine milk during ripening. *Int J Dairy Technol* **55**, 12–17.

O'Sullivan, D.J., Hill, C., and Klaenhammer, T.R. (1993) Effect of increasing the copy number of bacteriophage origins of replication in *trans*, on incoming-phage proliferation. *Appl Environ Microbiol* **59**, 2449–2456.

Ortigosa, M., Arizcun, C., Irigoyen, A., et al. (2006) Effect of *Lactobacillus* adjunct cultures on the microbiological and physicochemical characteristics of Roncal-type ewes'-milk cheese. *Food Microbiol* **23**, 591–598.

Parente, E. and Cogan, T.M. (2004) Starter cultures: general aspects. In: *Cheese: Chemistry, Physics and Microbiology* ((eds) P.F. Fox, P.L.H. McSweeney, T.M. Cogan, and T.P. Guinee), 3rd edn, vol. **1**, Elsevier Academic Press, London, pp. 123–148.

Psoni, L., Tzanetakis, N., and Litopoulou-Tzanetaki, E. (2003) Microbiological characteristics of Batzos, a traditional Greek cheese from raw goat's milk. *Food Microbiol* **20**, 575–582.

Psoni, L., Tzanetakis, N., and Litopoulou-Tzanetaki, E. (2006) Characteristics of Batzos cheese made from raw, pasteurized and/or pasteurized standardized goat milk and a native culture. *Food Control* **17**, 533–539.

Quiberoni, A., Moineau, S., Rousseau, G.M., et al. (2010) *Streptococcus thermophilus* bacteriophages. Review. *Int Dairy J* **20**, 657–664.

Quiberoni, A., Reinheimer, J.A., and Suárez, V.B. (1999) Performance of *Lactobacillus helveticus* spontaneous phage-resistant mutants in hard cheese production. *Int Dairy J* **8**, 941–949.

Reinheimer, J.A., Binetti, A., Quiberoni, A., et al. (1997) Natural milk cultures for the production of Argentinian cheeses. *J Food Prot* **60**, 59–63.

Reinheimer, J.A., Morelli, R., Callegari, M.L., and Bottazzi, V. (1993) Phage resistance in *Lactobacillus helveticus* CNRZ 328. *MAN* **11**, 235–240.

Richter, C., Chang, J.T., and Fineran, P.C. (2012) Function and regulation of clustered regularly interspaced short palindromic repeats (CRISPR)/CRISPR associated (Cas) systems. *Viruses* **4**, 2291–2311.

Rodríguez, C, Medici, M., Rodríguez, A., et al. (2009) Prevention of chronic gastritis by fermented milks made with exopolysaccharide-producing *Streptococcus thermophilus* strains. *J Dairy Sci* **92**, 2423–2434.

Rúas-Madiedo, P., Hugenholtz, J., and Zoon, P. (2002) An overview of the functionality of exopolysaccharides produced by lactic acid bacteria. *Int Dairy J* **12**, 163–171.

Saarela, M., Mogensen, G., Fondén, R., et al. (2000) Probiotic bacteria: safety, functional and technological properties. *J Biotechnol* **84**, 197–215.

Salazar, N., Binetti, A., Gueimonde, G., et al. (2011) Safety and intestinal microbiota modulation by the exopolysaccharide-producing strains *Bifidobacterium animalis* IPLA R1 and *Bifidobacterium longum* IPLA E44 orally administered to Wistar rats. *Int J Food Microbiol* **144**, 342–351.

Salazar, N., Gueimonde, M., Hernández-Barranco, A., et al. (2008) Exopolysaccharides produced by intestinal *Bifidobacterium* strains act as fermentable substrates for human intestinal bacteria. *Appl Environ Microbiol* **74**(15), 4737–4745.

Salminen, S. and Gueimonde, M. (2004) Human studies on probiotics: what is scientifically proven. *J Food Sci* **69**(5), 1377–140.

Santos, J.A., López-Díaz, T.M., García-Fernández, M.C., et al. (1996) Characterization and extracellular activity of psychrotrophic bacteria isolated from Villalón cheese (fresh variety of Spanish sheep's milk cheese). *Int J Food Microbiol* **33**, 301–306.

Sengül, N., Aslím, B., Uçar, G., et al. (2005) Effects of exopolysaccharide-producing probiotic strains on experimental colitis in rats. *Dis Colon Rectum* **49**, 250–258.

Slattery, L., O'Callaghan, J., Fitzgerald, G.F., et al. (2010) *Lactobacillus helveticus* - a thermophilic dairy starter related to gut bacteria. *J Dairy Sci* **93**, 4435–4454.

Sorek, R., Kunin, V., and Hugenholtz, P. (2008) CRISPR - a widespread system that provides acquired resistance against phages in bacteria and archaea. *Nature Rev Microbiol* **6**, 181–186.

Sturino, J. and Klaenhammer, T. (2004) Bacteriophage defense systems and strategies for lactic acid bacteria. *Adv Appl Microbiol* **56**, 331–378.

Suárez, V.B., Quiberoni, A., Binetti, A.G., and Reinheimer, J.A. (2002) Thermophilic lactic acid bacteria phages isolated from Argentinean dairy industries. *J Food Prot* **65**, 1597–1604.

Sun, Y., Li, Y., Song, H. and Zhu, Y. (2011) Microbial fermentation for food preservation. In: *Natural Antimicrobials in Food Safety and Quality* ((eds) M. Rai and M. Chikindas), CABI Int., Cambridge, MA, pp. 77–94.

Tamime, A.Y., Saarela, M., Korslund Söndergaard, A., et al. (2005) Production and maintenance of viability of probiotic microorganisms in dairy products. In: *Probiotic Dairy Products* ((ed.) A.Y. Tamime), Blackwell Publishing Ltd., Oxford, UK, pp. 39–72.

Teusink, B. and Smid, E.J. (2005) Modelling strategies for the industrial exploitation of lactic acid bacteria. *Food Microbiol* **4**, 46–56.

Thomas, L.V. and Delves-Broughton, J. (2001) New advances in the application of the food preservative nisin. *Res Adv Food Sci* **2**, 11–22.

Topisirovic, L., Kojic, M., Strahinic, I., et al. (2011) A survey of antimicrobial activity in lactic acid bacteria of different origin. In: *Natural Antimicrobials in Food Safety and Quality* ((eds) M. Rai and M. Chikindas), CABI Int., Cambridge, MA, pp. 27–38.

Tornadijo, M.E., Fresno, J.M., Bernardo, A., et al. (1995) Microbiological changes throughout the manufacturing and ripening of a Spanish goat's raw milk cheese (Armada variety). *Lait* **75**, 551–570.

Treu, L., Vendramin,V., Bovo, B., et al. (2014) Whole-genome sequences of *Streptococcus thermophilus* strains TH1435 and TH1436, isolated from raw goat milk. *Genome Announ* **2**, 1–2.

Upadhyay, V.K., McSweeney, P.L.H., Magboul, A.A.A, and Fox, P.F. (2004) Proteolysis in cheese during ripening. In: *Cheese: Chemistry, Physics and Microbiology*. ((eds) P.F. Fox, P.L.H. McSweeney, T.M. Cogan, and T.P. Guinee), 3rd edn, vol. **1**, Elsevier Academic Press, London, pp. 391–433.

van Kranenburg, R., Kleerebezem, M., van Hylckama Vlieg, J., et al. (2002) Flavor formation from amino acids by lactic acid bacteria: predictions from genome sequence analysis. *Int Dairy J* **12**, 111–121.

Ventura, M., Canchaya, C., Bernini, V., et al. (2006) Comparative genomics and transcriptional analysis of prophages identified in the genomes of *Lactobacillus gasseri*, *Lactobacillus salivarius*, and *Lactobacillus casei*. *Appl Environ Microbiol* **72**, 3130–3146.

Viscardi, M., Capparelli, R., Di Matteo, R., et al. (2003) Selection of bacteriophages-resistant mutants of *Streptococcus thermophilus*. *J Microbiol Meth* **55**, 109–119.

Walker, S.A. and Klaenhammer, T.R. (2000) An explosive antisense RNA strategy for inhibition of a lactococcal bacteriophage. *Appl Environ Microbiol* **66**, 310–319.

Weimer, B.C., Blake, M., Hiller, A.J., and Davidson, B.E. (1993) Studies on the isolation of phage-resistant derivatives of *Lactococcus lactis* subsp. *cremoris* FG2 with phage sk1. *Aust J Dairy Technol* **48**, 59–61.

Williams, A.G., Withers, S.E., Brechany, E.Y., and Banks, J.M. (2006) Glutamate dehydrogenase activity in lactobacilli and the use of glutamate dehydrogenase-producing adjunct *Lactobacillus* spp. cultures in the manufacture of Cheddar cheese. *J Appl Microbiol* **101**, 1062–1075.

Xanthopoulos, V., Polychroniadou, A., Litopoulou-Tzanetaki, E., and Tzanetakis, N. (2000) Characteristics of Anevato cheese made from raw or heat-treated goat milk inoculated with a lactic starter. *LWT - Food Sci Technol* **33**, 483–488.

Xu, C.L., Wang, Y.Z., Jin, M.L., and Yang, X.Q. (2009) Preparation, characterization and immunomodulatory activity of selenium-enriched exopolysaccharide produced by bacterium *Enterobacter cloacae* Z0206. *Biores Technol* **100**, 2095–2097.

Zisu, B. and Shah, N.P. (2005) Textural and functional changes in low-fat Mozzarella cheeses in relation to proteolysis and microstructure as influenced by the use of fat replacers, pre-acidification and EPS starter. *Int Dairy J* **15**, 957–972.

Zisu, B. and Shah, N.P. (2007) Texture characteristics and pizza bake properties of low-fat Mozzarella cheese as influenced by pre-acidification with citric acid and use of encapsulated and ropy exopolysaccharide producing cultures. *Int Dairy J* **17**, 985–997.

Chapter 13

Lactobacillus sakei in Meat Fermentation

Marie-Christine Champomier-Vergès[1,2,*] and Monique Zagorec[3,4,*]

[1] *INRA, UMR 1319 Micalis, France*
[2] *AgroParisTech, UMR Micalis, France*
[3] *INRA, UMR 1014 Secalim, France*
[4] *LUNAM Université, Oniris, France*

13.1. Introduction

Lactobacillus sakei has a long history of use as starter for sausage manufacturing. It is regularly used in Western Europe in association with coagulase negative *Staphylococci* such as *Staphylococcus xylosus* or *Staphylococcus carnosus*. For many years, its metabolic capacities have been analyzed regarding its main role during fermentation technology, that is sugar fermentation and consecutive lactic acid production. Indeed, lactic acid synthesis and the associated pH drop both contribute to texture and microbial safety of sausages. Beside a need for high and fast acid production, efficient starters have also to face the stressing conditions encountered during the different processing steps of sausage production: presence of curing salts (NaCl, nitrite, and spices) that may contain bacteriostatic or bactericidal molecules, and temperatures of ripening. In addition, starters have to compete with the complex microflora naturally present in the meat matrix (for a review on the species, see Champomier-Vergès et al. 2002). During the last decade, development of "omics" technologies have allowed us to gain insight into food bacteria and particularly to understand the behavior of *Lact. sakei* during meat fermentation. Genome analyses reveal putative functions that may explain the adaptation of this species to the sausage environment and showed the wide intraspecies diversity. Post-genomics approaches, like proteomics and transcriptomics, and functional genomics or genetics were used to determine the functions important for *Lact. sakei* performance during meat fermentation or under conditions that may occur during meat processing. In parallel, genetic and molecular tool developments allowed us to investigate *Lact. sakei* physiology under sausage production conditions. Some genes specifically expressed during fermentation were highlighted, and molecular methods were used to selectively monitor *Lact. sakei* in sausage or to investigate the diversity of strains that may colonize meat.

This has provided knowledge about the functions that are important to consider for selecting efficient starter strains. However, as for the selection of protective cultures to be used to ensure

* Corresponding authors: Marie-Christine Champomier-Vergès. INRA, UMR 1319 Micalis, Domaine de Vilvert, F-78350 Jouy-en-Josas, France. AgroParisTech, UMR Micalis, F-78350 Jouy-en-Josas, France. marie-christine.champomier-verges@jouy.inra.fr. Tel: +33 1 34 65 22 92
Monique Zagorec. INRA, UMR 1014 Secalim, Rue de la Géraudière, CS82225, Nantes F-44307, France. LUNAM Université, Oniris, Nantes F-44307, France. monique.zagorec@oniris-nantes.fr. Tel: +33 2 51 78 55 42

Biotechnology of Lactic Acid Bacteria: Novel Applications, Second Edition. Edited by Fernanda Mozzi, Raúl R. Raya, and Graciela M. Vignolo.

meat microbial safety, tests performed under laboratory conditions are often informative but must be confirmed *in carnis*.

13.2. Genomics and Diversity of the Species *Lactobacillus sakei*

The first whole genome sequence of a *Lact. sakei* strain was that reported in 2005 for the 1,884-kb chromosome of *Lact. sakei* 23K, a plasmid cured strain derived from a sausage isolate (Chaillou et al. 2005). Recently, the draft genome sequence of a second strain (*Lact. sakei* LS25, a commercial starter for fermented sausage) has also been published (McLeod et al. 2013). The sequence reported a 2015-kb chromosome and a 20.5-kb plasmid.

A large genotypic and phenotypic diversity has been described within the species. McLeod et al. (2008) analyzed 10 strains and showed that their genome size ranged from 1.8 to 2.1 Mb, and that they belong to two different genotypic groups. However, these groups did not correlate with a clustering based on carbohydrate-fermenting profiles or on chromosome sizes. Chaillou et al. (2009), in a survey of 73 strains from various origins, reported that strains have indeed genomes of different sizes. This variability can reach 25% difference between strains, as the genome size of the strains varied from 1.8 to 2.3 Mb. The 73 isolates analyzed can be clustered into 10 different genomic groups based on the presence or absence of a subset of 60 gene markers belonging to the flexible gene pool of the species. Based on the genome sequence of *Lact. sakei* 23K, genomic islands could be mapped on the chromosome (Figure 13.1).

The 26 genomic islands are indicated by blue arrows. The average size per island is ~10.5 kb and represents ~11% of the 23K genome (~215 kbp).

This diversity was supposed to sustain the high fitness of the species in meat ecosystems (Figure 13.2).

Moreover, it was observed that strains, whether they were issued from fermented products or raw materials, were equally distributed within the different genomic groups. A genome comparison has been performed by comparative genome hybridization on a collection of 18 strains (Nyquist et al. 2011). *Lact. sakei* 23K appeared to harbor the smallest genome. The *Lact. sakei* 23K genes were used as a basis to construct a microarray which was implemented with additional genes originating from different strains. More recently, a Multi Locus Sequence Typing (MLST) approach, based on eight gene sequences, was used on a larger *Lact. sakei* collection (about 200 strains). This study indicated that the natural population of the species has three different lineages (Chaillou et al. 2013). Strains from different environments (various products, fermented or not, and from various meat and geographic origins) were found in the different lineages, indicating the absence of a clear link between strain ecological origin and lineage. It is noteworthy that on non-fermented meat an effect of storage conditions and possibly a seasonal effect are suspected to influence the lactic acid bacteria strains that develop on these products, therefore shaping the population fingerprint of meat products (Lucquin et al. 2012).

Following this genomic knowledge of the species, PCR-based methods have been developed for identifying and quantifying *Lact. sakei* strains in their meat natural environment. In a first study, real-time quantitative PCR (RT-q-PCR) using the sequence of rDNA genes allowed the specific quantification of *Lact. sakei* population in fermented sausages (Martín et al. 2006). More recently, gene markers belonging to the flexible gene pool of the species were used to quantify individual *Lact. sakei* strains, and the *katA* gene which is ubiquitous in *Lact. sakei* strains (Ammor et al. 2005; Najjari et al. 2008) was successfully used to quantify the species (Chaillou et al. 2014).

An IVET system (*in vivo* expression technology) was developed in order to identify the genes specifically expressed during sausage fermentation (Hüfner et al. 2007). In a meat fermentation model, 15 genes appeared to have their expression induced, in particular genes involved in stress response.

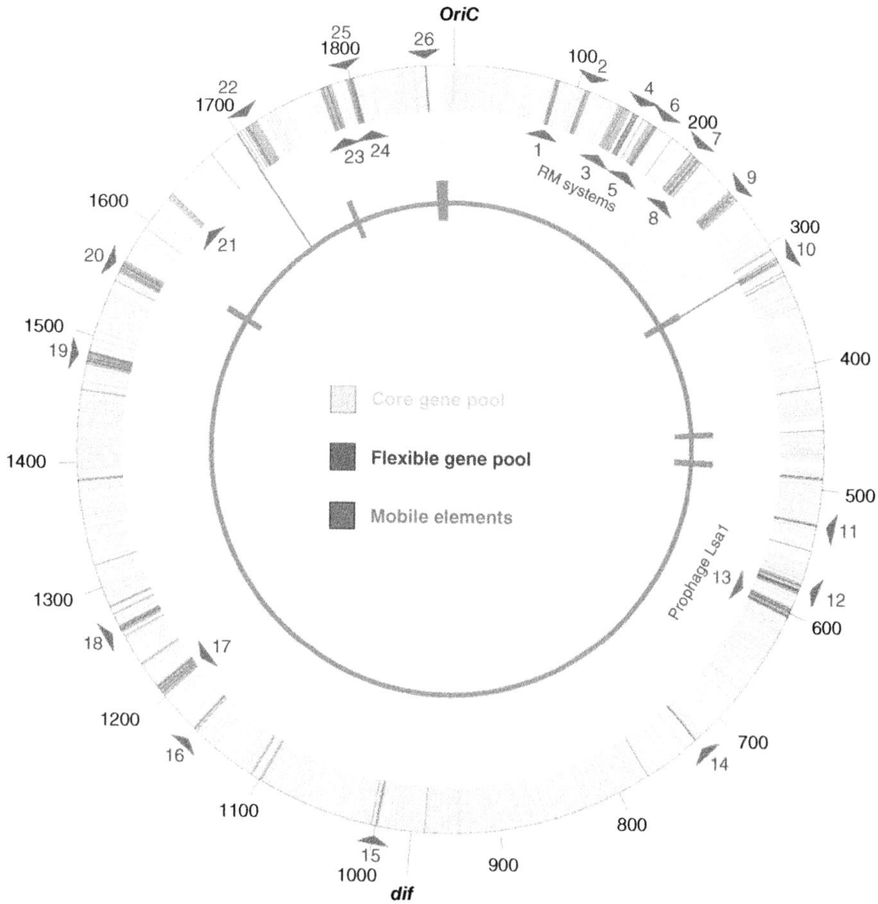

Figure 13.1. Accessory genome or flexible gene pool of *Lact. sakei.*

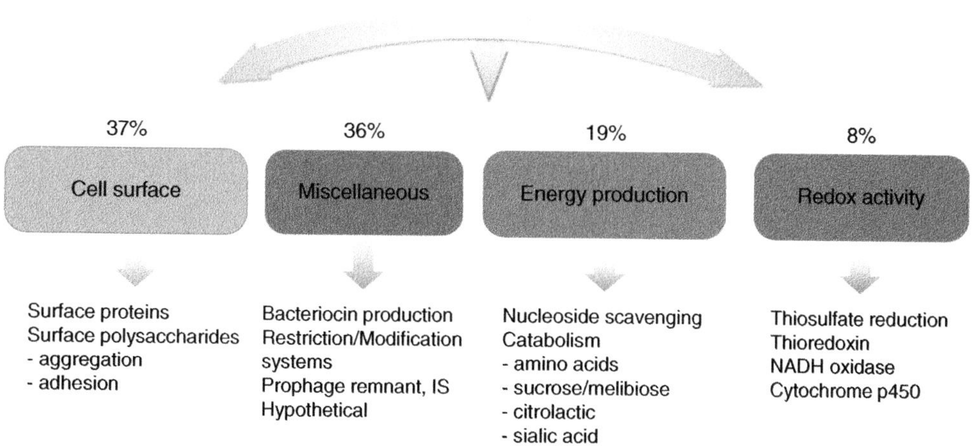

Figure 13.2. Main functional categories associated with the flexible gene pool of *Lact. sakei.*

13.3. Post-genomic Vision of Meat Fitness Traits of *Lactobacillus sakei*

In 2005, Eijsink and Axelson made a comment on the first whole genome *Lact. sakei* sequence that was aptly entitled "bacterial lesson in sausage making." In this, they underlined the main meat fitness traits revealed by this sequence:

- the ability to resist some stressing conditions encountered in meat production processes or during meat product storage (low temperature or oxidative stress, presence of curing salts);
- the ability to efficiently and competitively use nutrients and metabolites present in the meat matrix to ensure its growth;
- the ability to compete with other bacteria sharing the meat ecosystem, by producing antagonistic molecules (organic acids, bacteriocins, hydrogene peroxide).

13.3.1. Energy Sources

As mentioned above, the genome sequence of *Lact. sakei* 23K suggested an ability of the species to efficiently use nutrients from meat as energy sources. This was in accordance with previous hypotheses suggesting an ability to competitively use some sugars present on meat, such as ribose (Lauret et al. 1996; Stentz et al. 1997; Stentz and Zagorec 1999). Additionally, the importance of arginine utilization, already suspected to be important for *Lact. sakei* fitness in meat products (Zúñiga et al. 1998; Champomier-Vergès et al. 1999), was confirmed following genomic and post-genomic approaches.

13.3.1.1. Sugars Using a proteomic approach, McLeod et al. (2010) compared 10 different *Lact. sakei* strains grown either on glucose or ribose. The strains had similar patterns but one of them, LS25 a commercial starter, clearly stands apart from the others and reveals a more pronounced downregulation of glycolytic pathway when grown on ribose.

The same group (McLeod et al. 2011), using whole genome microarrays (Nyquist et al. 2011), also performed a transcriptomic comparison of three strains grown on ribose. The three strains responded differently during growth on ribose. It was shown that nucleoside catabolism was induced during growth on ribose, which had not been seen in the proteomic approach. An induction of glycerol metabolism in the presence of ribose was observed, both by the proteomic and microarray studies.

The genome sequence has revealed the presence of a metabolic pathway susceptible to lead to nucleoside utilization as an energy source for the growth of strain 23K (Chaillou et al. 2005). The functionality of this pathway was demonstrated in strain CTC494 by Rimaux et al. (2011a). These authors gave evidence that adenosine and inosine were transported into the cells, then their ribose-5P moiety was used in the pentose phosphate pathway, leading thus to energy production, whereas adenine and hypoxanthine were excreted. They also showed that this could occur under fermentation process conditions in a range of pH from 5 to 6.5.

One of the other features highlighted by the genome sequence was the potential capacity to use *N*-acetyl neuraminic acid as a carbon source. This sugar is abundant in eukaryotic cells and thus present in large amounts in meat. However, pathways for *N*-acetyl neuraminic acid utilization have been described mainly in pathogenic species such as *Escherichia coli*. Pathogenic bacteria can use this compound to produce extracellular polymers that participate in their protection against the immune system from the host after infection. It was shown that in *Lact. sakei N*-acetyl neuraminic acid is used as a carbon source to sustain growth (Anba-Mondoloni et al. 2013). Therefore, this feature could represent a competitive advantage for the use of *Lact. sakei* to fight against pathogenic species potentially present in meat and sausages.

13.3.1.2. Arginine Arginine degradation by arginine deiminase pathway (ADI) has long been considered a main taxonomical criterion for differentiating *Lact. sakei* from the closely related species *Lactobacillus curvatus*, also used as a starter for meat fermentation. The pathway was shown to provide energy (ATP) for survival during meat storage and fermentation (see references in Champomier-Vergès et al. 2002). Following genome analysis, recent studies led to new information on the structure and regulation of this pathway in different *Lact. sakei* strains. Actually two catabolic pathways can co-exist in *Lact. sakei*, which lead to ATP production from arginine consumption: the ADI pathway encoded by *arcABCTDR* operon (Zúñiga et al. 1998, 2002; Chaillou et al. 2005) and the agmatine deiminase pathway, both encompassing a carbamate kinase (Rimaux et al. 2012). In this second pathway, the agmatine deiminase is able to produce carbamyl phosphate from agmatine. Carbamyl-P is then transformed by carbamate kinase into NH_4 and CO_2 yielding ATP. This last step of agmatine catabolism is similar to the last step of ADI pathway *per se*.

In the ADI pathway, citrulline is produced under acidic conditions; thus, under fermentation conditions pH regulates the pathway (Rimaux et al. 2011b). Recently, it was also shown that PTP, an additional protein which gene is located just downstream from the *arcABCTDR* operon in *Lact. sakei*, is a citrulline/ornithine antiporter involved in citrulline-into-ornithine conversion through the ADI pathway (Rimaux et al. 2013). These authors also showed that this protein was differentially expressed depending on growth phase and external pH leading to a higher activity at acidic pH than at neutral pH (Rimaux et al. 2013).

13.3.2. Stress Response

As already mentioned, resistance to oxidative stress had also been pointed out as a potential fitness trait for *Lact. sakei* during meat processing. Guilbaud et al. (2012) have compared the response to various oxidative compounds of 12 *Lact. sakei* strains belonging to the different genetic groups defined by Chaillou et al. (2009). The strains exhibited various sensitivity phenotypes toward the components tested and their sensitivity also differed depending on the environmental conditions. For example, *Lact. sakei* 23K was more resistant to H_2O_2 in the presence of heme than could be another strain (*Lact. sakei* 64). Additionally, *Lact. sakei* 23 had a different behavior than its plasmid cured derivative *Lact. sakei* 23K, suggesting that some of the characteristics involved in resistance to oxidative stress may also have plasmidic localization. By a principal component analysis Guilbaud et al. (2012) could establish two main phenotypic groups, partially correlating with previously described genomic clusters. These authors also compared strain response to oxidative stress when grown alone or in mixtures with other strains. Interestingly, it appeared that a given strain may have a different behavior depending on the other strains present in the mixture. It seems that some strains may act as helpers in protecting the others from oxidative stress whereas others could be disadvantaged when grown in mixtures. These findings reveal that these various aptitudes in reaction to oxidative stress, alone or in the presence of others, should be taken into account for the constitution of starter strain cocktails for fermentation.

Other stressing treatments employed to improve the microbial stability of some meat products merit consideration. This is the case for high hydrostatic pressure (HHP) treatment in use in several countries for the preservation of various sausage or dry-cured ham. By a proteomic investigation, it was notably shown that *Lact. sakei* can react differently from other bacterial species after an HHP treatment (Jofré et al. 2007). In addition, flow cytometry was successfully used to test the ability of *Lact. sakei* cells to survive different stresses like acidic or high temperature stresses (Bonomo et al. 2013) illustrating one of the numerous tools available for starter culture selection.

13.4. Conclusions

Lact. sakei is one of the most emblematic starters used for sausage fermentation. The wide genomic diversity of the species that has been underlined by different approaches is being more and more studied in terms of physiology for explaining the various capacities of the different strains during meat processing. Moreover, besides its use in fermentation, the species is now regularly proposed as a protective agent for fresh meat products (Jones et al. 2010).

Finally, from the most recent results obtained on *Lact. sakei* one should keep in mind that:

- Its phenotypic intraspecies diversity, known for a long time, results from an important flexible gene pool. Strains harboring the largest chromosomes may thus encompass several hundreds of additional genes more than the strains having the smallest chromosomes.
- This diversity resides in both the redundancy of genes and on the regulation of their expression.
- The behavior of strains is therefore linked to their genomic content but, although the mechanisms are not yet known, it also depends on the nature of the other strains (and certainly species) sharing the same ecosystem.
- Important information about the metabolic functions essential for meat fermentation by *Lact. sakei* was recently gained. We have notably a better understanding of the various metabolic pathways this species needs and expresses under fermentation conditions.

All this should help further screening of new starter cultures. However, an additional feature not to be forgotten is to investigate these important functions within the complex microbial community that is present, together with *Lact. sakei*, in the meat ecosystem.

References

Ammor, S., Dufour, E., Zagorec, M., et al. (2005) Characterization and selection of *Lactobacillus sakei* strains isolated from traditional dry-sausage. *Food Microbiol* **22**, 529–538.

Anba-Mondoloni, J., Chaillou, S., Zagorec, M., and Champomier-Vergès, M.C. (2013) Catabolism of N-acetylneuraminic acid, a fitness function of the food-borne lactic acid bacterium *Lactobacillus sakei*, involves two newly characterized proteins. *Appl Environ Microbiol* **79**, 2012–2018.

Bonomo, M.G., Milella, L., Martelli, G., and Salzano, G. (2013) Stress response assessment of *Lactobacillus sakei* strains selected as potential autochthonous starter cultures by flow cytometry and nucleic acid double-staining analyses. *J Appl Microbiol* **115**, 786–795.

Chaillou, S., Champomier-Vergès, M.C., Cornet, M., et al. (2005) The complete genome sequence of the meat-borne lactic acid bacterium *Lactobacillus sakei* 23K. *Nat Biotechnol* **23**, 1527–1533.

Chaillou, S., Daty, M., Baraige, F., et al. (2009) Intraspecies genomic diversity and natural population structure of the meat-borne lactic acid bacterium *Lactobacillus sakei*. *Appl Environ Microbiol* **75**, 970–980.

Chaillou, S., Lucquin, I., Najjari, A., et al. (2013) Population genetics of *Lactobacillus sakei* reveals three lineages with distinct evolutionary histories. *PLoS One* **8**, e73253.

Chaillou, S., Christieans, S., Rivollier, M., et al. (2014) Quantification and efficiency of *Lactobacillus sakei* strain mixtures used as protective cultures in ground beef. *Meat Sci* **97**, 332–338.

Champomier-Vergès, M.C., Zuñiga, M., Morel-Deville, F., et al. (1999) Relationships between arginine degradation, pH and survival in *Lactobacillus sakei*. *FEMS Microbiol Lett* **180**, 297–304.

Champomier-Vergès, M.C., Chaillou, S., Cornet, M., and Zagorec, M. (2002) *Lactobacillus sakei*: recent developments and future prospects. *Res Microbiol* **153**, 115–123.

Eijsink, V.G. and Axelsson, L. (2005) Bacterial lessons in sausage making. *Nat Biotechnol* **23**, 1494–1495.

Guilbaud, M., Zagorec, M., Chaillou, S., and Champomier-Vergès, M.C. (2012) Intraspecies diversity of *Lactobacillus sakei* response to oxidative stress and variability of strain performance in mixed strains challenges. *Food Microbiol* **29**, 197–204.

Hüfner, E., Markieton, T., Chaillou, S., et al. (2007) Identification of *Lactobacillus sakei* genes induced during meat fermentation and their role in survival and growth. *Appl Environ Microbiol* **73**, 2522–2531.

Jofré, A., Champomier-Vergès, M.C., Anglade, P., et al. (2007) Protein synthesis in lactic acid and pathogenic bacteria during recovery from a high pressure treatment. *Res Microbiol* **158**, 512–520.

Jones, R.J., Wiklund, E., Zagorec, M., and Tagg, J.R. (2010) Evaluation of stored lamb bio-preserved using a three-strain cocktail of *Lactobacillus sakei*. *Meat Sci* **86**, 955–959.

Lauret, R., Morel-Deville, F., Berthier, F., et al. (1996) Carbohydrate utilization in *Lactobacillus sake*. *Appl Environ Microbiol* **62**, 1922–1927.

Lucquin, I., Zagorec, M., Champomier-Vergès, M., and M.C., Chaillou, S. (2012) Fingerprint of lactic acid bacteria population in beef carpaccio is influenced by storage process and seasonal changes. *Food Microbiol* **29**, 187–196.

Martín, B., Jofré, A., Garriga, M., et al. (2006) Rapid quantitative detection of *Lactobacillus sakei* in meat and fermented sausages by real-time PCR. *Appl Environ Microbiol* **72**, 6040–6048.

McLeod, A., Nyquist, O.L., Snipen, L., et al. (2008) Diversity of *Lactobacillus sakei* strains investigated by phenotypic and genotypic methods. *Syst Appl Microbiol* **31**, 393–403.

McLeod, A., Zagorec, M., Champomier-Vergès, M.C., et al. (2010) Primary metabolism in *Lactobacillus sakei* food isolates by proteomic analysis. *BMC Microbiol* **10**, 120.

McLeod, A., Snipen, L., Naterstad, K., and Axelsson, L. (2011) Global transcriptome response in *Lactobacillus sakei* during growth on ribose. *BMC Microbiol* **11**, 145.

McLeod, A., Brede, D.A., and Rud, I., Axelsson, L. (2013) Genome sequence of *Lactobacillus sakei* subsp. *sakei* LS25, a commercial starter culture strain for fermented sausage. *Genome Announc*, **1**, e00475-13.

Najjari, A., Ouzari, I., Boudabous, A., and Zagorec, M. (2008) Method for reliable isolation of a collection of *Lactobacillus sakei* strains originating from Tunisian seafood and meat products. *Int J Food Microbiol* **121**, 342–351.

Nyquist, O.L., McLeod, A., Brede, D.A., et al. (2011) Comparative genomics of *Lactobacillus sakei* with emphasis on strains from meat. *Mol Genet Genom* **285**, 297–311.

Rimaux, T., Vrancken, G., Vuylsteke, B., et al. (2011a) The pentose moiety of adenosine and inosine is an important energy source for the fermented-meat starter culture *Lactobacillus sakei* CTC 494. *Appl Environ Microbiol* **77**, 6539–6550.

Rimaux, T., Vrancken, G., Pothakos, V., et al. (2011b) The kinetics of the arginine deiminase pathway in the meat starter culture *Lactobacillus sakei* CTC 494 are pH-dependent. *Food Microbiol* **28**, 597–604.

Rimaux, T., Rivière, A., Illeghems, K., et al. (2012) Expression of the arginine deiminase pathway genes in *Lactobacillus sakei* is strain dependent and is affected by the environmental pH. *Appl Environ Microbiol* **78**, 4874–4883.

Rimaux, T., Rivière, A., Hebert, E.M., et al. (2013) A putative transport protein is involved in citrulline excretion and re-uptake during arginine deiminase pathway activity by *Lactobacillus sakei*. *Res Microbiol* **164**, 216–225.

Stentz, R., Lauret, R., Ehrlich, S.D., et al. (1997) Molecular cloning and analysis of the *ptsHI* operon in *Lactobacillus sakei*. *Appl Environ Microbiol* **63**, 2111–2116.

Stentz, R. and Zagorec, M. (1999) Ribose utilization in *Lactobacillus sakei*: analysis of the regulation of the *rbs* operon and putative involvement of a new transporter. *J Mol Microbiol Biotechnol* **1**, 165–173.

Zúñiga, M., Champomier-Vergès, M.C., Zagorec, M., and Pérez-Martinez, G. (1998) Structural and functional analysis of the gene cluster encoding the enzymes of the arginine deiminase pathway of *Lactobacillus sake*. *J Bacteriol* **180**, 4154–4159.

Zúñiga, M., Miralles, M.C., and Pérez-Martínez, G. (2002) The product of *arcR*, the sixth gene of the *arc* operon of *Lactobacillus sakei*, is essential for expression of the arginine deiminase pathway. *Appl Environ Microbiol* **68**, 6051–6058.

Chapter 14

Vegetable and Fruit Fermentation by Lactic Acid Bacteria

Raffaella Di Cagno*, Pasquale Filannino, and Marco Gobbetti

Department of Soil, Plant and Food Sciences, University of Bari Aldo Moro, Italy

14.1. Introduction

Fermented foods play an important role in human diet due to their health benefits. Fermentation is one of the oldest and most low-input technologies used in food bio-preservation. The health benefits of fermented foods are expressed either directly through the interactions of ingested live microorganisms with the host (probiotic effect) or indirectly as the result of the ingestion of microbial metabolites that are synthesized during fermentation (biogenic effect). Scientific evidence encourages the consumption of vegetables and fruits to prevent well-known chronic pathologies (He et al. 2007). In fact, the consumer trend towards fresh-like, high nutritional value, health-promoting, and rich flavor ready-to-eat or to-drink foods and beverages is increasing. To meet these requirements, lactic acid fermentation is considered a technological option to preserve and/or improve the safety, nutritional, sensory, and shelf-life features of vegetables and fruits (Di Cagno et al. 2013). The autochthonous microbiota may cause the spontaneous fermentation of several vegetables. Notwithstanding the reliable value of spontaneous fermentation to stabilize and preserve raw vegetables and fruits, a number of factors favor using selected starters. Selection of starter cultures within the autochthonous microbiota of vegetables and fruits should be recommended since indigenous cultures may ensure better performance compared to allochthonous/commercial strains (Di Cagno et al. 2013). Based on the above considerations, the exploitation of vegetables and fruits through lactic acid fermentation has generated increasing interest over the last decade.

14.2. Lactic Acid Bacteria Microbiota of Raw Vegetables and Fruits

Microorganisms are found in almost all ecological niches. Plants may provide a variety of nutrients and are, therefore, attractive hosts for microorganisms. Microorganisms colonize plant surfaces (epiphytes) or plant tissues (endophytes) (Vorholt 2012). Overall, phylloplane niches are characterized by extended variability, which is determined by several factors (Paramithiotis et al. 2012). The lack of stomata in some fruits (e.g., tomatoes) is an important factor to be considered. Non-injured fruit skins

* Corresponding author: Department of Soil, Plant and Food Sciences, University of Bari Aldo Moro, Via G. Amendola 165/a, 70126 Bari, Italy. raffaella.dicagno@uniba.it

Biotechnology of Lactic Acid Bacteria: Novel Applications, Second Edition. Edited by Fernanda Mozzi, Raúl R. Raya, and Graciela M. Vignolo.

are impermeable and thus microorganisms may acquire metabolizable compounds only through cracks, bruises, or sand scars. The phyllosphere, which encompasses the surface of the stem and leaves, is also considered a hostile environment for microorganisms, due to the limited availability of nutrients, strong exposure to sun irradiation, and variability of the water availability. Despite these conditions, the phyllosphere hosts a microbial population of remarkable complexity. Type, availability, and concentration of substrate, buffering capacity, competing microorganisms, and natural plant antagonisms provide a unique environment harboring a dominant and constant microbiota. Overall, the microbial population of vegetables and fruits is estimated to fluctuate between 5 to 7 log CFU g^{-1} (Spurr 1994). In particular, lactic acid bacteria represent a small part (2–4 log CFU g^{-1}) (Buckenhüskes 1997) of the autochthonous microbiota, and their cell density is mainly influenced by the species of vegetables, temperature, and harvesting conditions. Hetero- and homo-fermentative species belonging to *Leuconostoc*, *Lactobacillus*, *Weissella*, *Enterococcus*, and *Pediococcus* genera were variously identified, depending on the species of vegetables. Nevertheless, *Weissella cibaria/confusa* and, especially, *Lact. plantarum* were the species most frequently found. *Lact. plantarum* is considered to be a ubiquitous and metabolically versatile bacterium, which is largely found in fruits and vegetables (Di Cagno et al. 2013).

Leuconostoc spp., including *Leuc. mesenteroides* subsp. *mesenteroides*, mainly dominate the early spontaneous fermentation of carrots and leeks, and of raw vegetables such as white cabbages, carrots, green tomatoes, peppers, cucumbers, eggplants, lettuce, sauerkrauts, kimchi, cherries, and grape must (Di Cagno et al. 2013; Paramithiotis et al. 2013; Wouters et al. 2013). Pediococci and lactobacilli, mainly *Lact. plantarum*, were identified in many raw vegetables and, especially, in fermented vegetable juices. *Lact. plantarum*, *W. cibaria/confusa*, *Lact. brevis*, *P. pentosaceous*, *Lactobacillus* sp., and *E. faecium/faecalis* were found in raw red tomatoes (Di Cagno et al. 2008), *Leucoonostoc citreum*, *Lact. casei*, and *Lact. curvatus* in green tomatoes (Paramithiotis et al. 2013) and *Lact. curvatus*, *Leuc. mesenteroides*, *Lact. plantarum*, and *W. confusa* in red and yellow peppers (Di Cagno et al. 2009). Several species of LAB (e.g., *Lact. fermentum*, *Lact. brevis*, and *E. faecalis*) were identified in melon pod (Offonry and Achi 1998). Pineapple fruits harbored autochthonous lactic acid bacteria belonging to the species *Lact. plantarum* and *Lact. rossiae* (Di Cagno et al. 2010). Several cultivars of sweet cherry fruits harbored a low population of lactic acid bacteria at a cell density lower than that usually found in other fruits and vegetables (ca. 2.5 log CFU g^{-1}). *Lact. plantarum*, *Pediococcus acidilactici*, *P. pentosaceus*, and *Leuc. mesenteroides* subsp. *mesenteroides* were the only species identified in the eighth cultivars of sweet cherry. A similar diversity was found for other fruits such as blackberries, prunes, kiwifruits, and papaya, where *Lact. plantarum*, *Lact. pentosus*, and *W. cibaria* dominated. When spontaneously fermented, the same microbiota was harbored by cucumber, olives, pumpkins, peppers, carrots, and persimmons and eggplants (Di Cagno et al. 2013).

Recently, fructophilic lactic acid bacteria (FLAB), consisting of a special group of lactic acid bacteria, were described as the bacteria that prefer fructose to glucose as a growth substrate (Neveling et al. 2012). They are found in fructose-rich niches such as fruits, and fermented foods made from fruits and flowers. Only two *Lactobacillus* spp. were classified as fructophilic, that is *Lact. florum* and *Lact. kunkeei*. This latter was originally isolated from grape juice. All the other FLAB were grouped into the genus *Fructobacillus*, which contains *Fructobacillus durionis*, *F. ficulneus*, *F. fructosus*, *F. pseudoficulneus*, and *F. tropaeoli*. All these species are phylogenetically related to the genera *Leuconostoc*, *Oenococcus* and *Weissella*. *F. pseudoficulneus* were isolated from figs and banana (Endo et al. 2009). FLAB species are separated into two groups: obligately and facultatively fructophilic. Obligately FLAB grow well on fructose, and external electron acceptors such as fructose, pyruvate, and oxygen are needed for glucose metabolism.

Molecular typing of LAB isolated from vegetables and fruits should be useful to group them into several clusters, which reflect their distribution in the raw matrix and, presumptively, their way of entry. For instance, strains of *Lact. plantarum* were mainly isolated from the outer and inner ring of pineapple fruit, indicating these layers as their optimal niches (Di Cagno et al. 2010). Besides, intra-specific differentiation is the important preliminary step to select starter cultures. Technological, probiotic, sensory, and antimicrobial attributes are mainly strain-specific.

14.3. Fermentation of Vegetable Products

14.3.1. Spontaneous Fermentation

In raw vegetables and fruits, lactic acid fermentation may occur spontaneously, when anaerobic conditions, water activity, moisture, salt concentration, and temperature are favorable for the growth of the autochthonous lactic acid bacteria. In particular, LAB have to overgrow the higher number of non-lactic bacteria such as Gram (–) anaerobic *Enterobacteriaceae*, Gram (–) aerobic bacteria (e.g., *Pseudomonas* species), yeasts, and filamentous fungi. The addition of salt or submerging fresh vegetables in brine makes the environment more suitable for the growth of LAB. Brine also protects vegetables from light and oxygen. Physical air exclusion and the lack of oxygen, which is consumed by plant cells at the start of the fermentation, generate anaerobic conditions, which are also favorable for LAB (Hutkins 2006). The synthesis of carbon dioxide by hetero-fermentative LAB also contributes to their anaerobic growth. Furthermore, acidification of the vegetable matrix by LAB causes the death of the less acid-tolerant spoilage microbiota. Spontaneous fermentation may be optimized through back slopping, which is commonly used for sauerkraut production (e.g., inoculation of the raw material with a small quantity of a previously performed successful fermentation) (Di Cagno et al. 2013). Hence, back slopping results in the dominance of the best adapted strains and represents a way of using selected starter cultures to shorten the fermentation and reduce the risk of failure. When occurring spontaneously, the fermentation of vegetables and fruits is frequently characterized by a succession of hetero- and homo-fermentative LAB, together with or without yeasts, which are responsible for multi-step fermentation processes (Plengvidhya et al. 2004).

The spontaneous lactic acid fermentation of raw vegetables may require 4–6 days before the pH value undergoes a significant decrease. Consequently, in some cases spontaneous fermentation may be responsible for undesirable variations of the sensory and rheological properties of fresh vegetables and fruits, and/or it may occur too slowly to inhibit spoilage and pathogen microorganisms. Nevertheless, the production of fermented vegetables and fruits still relies on spontaneous fermentation. This appears to be in contrast with other fermented foods (e.g., dairy, meat and baked goods), where the use of starter cultures is a common practice. The main difficulties regarding the introduction of starter cultures for fermentation of vegetables and fruits are the typical succession of LAB species and the inability to decrease the autochthonous microbiota without causing drastic textural changes. Nowadays, the use of starter cultures is recommended to guarantee hygiene and safety, and to get products with consistent sensory and nutritional properties.

14.3.2. The Autochthonous Starters

The use of starter cultures selected within the autochthonous microbiota of vegetables and fruits is increasing. Compared to allochthonous/commercial LAB, autochthonous strains show better performances during fermentation and, especially, ensure higher viability during storage. Autochthonous cultures may also synthesize bacteriocins and positively contribute to the aroma. Recently, a number of vegetables and fruits were successfully subjected to fermentation by selected autochthonous LAB (Di Cagno et al. 2008–2010, 2013). Adaptation to the intrinsic features of the raw matrices guaranteed high survival (ca. 8.0–9.0 log CFU g^{-1}) during fermentation and storage (ca. 40 days).

Since they are very frequently isolated from spontaneous Almagro eggplant fermentations, *Lact. brevis*, *Lact. plantarum*, and *Lact. fermentum* were used as potential autochthonous starters.

The combination of *Lact. plantarum* and *Lact. brevis* ensured the most preferred end-products with regard to sensory properties. *Leuc. mesenteroides*, *Lact. sakei*, and *Lact. plantarum* dominated at different stages of leek fermentation, which made them good candidates for starter cultures. The mixed starter culture of *Lact. plantarum* IMDO 788 and *Leuc. mesenteroides* IMDO 1347 was the most promising (Wouters et al. 2013). It produced fermented leeks of high microbiological quality, the extensive consumption of carbohydrates, and various end-products were synthesized. When fermented with selected autochthonous strains of *Lact. plantarum*, *Leuc. mesenteroides*, and *P. pentosaceus*, carrots, French beans or marrows were characterized by a rapid decrease of pH, marked consumption of fermentable carbohydrates, and inhibition of *Enterobacteriaceae* and yeasts (Di Cagno et al. 2013). Almost the same was found when autochthonous strains of *Lact. plantarum*, *Lact. curvatus*, and *W. confusa* were used as mixed starters to ferment red and yellow peppers (Di Cagno et al. 2009). After 30 days of storage at room temperature, started vegetables positively differentiated from the unstarted ones also in higher firmness and color indexes. A protocol for minimally processing pineapple was set up, aiming at increasing the shelf-life and maintaining agreeable sensory and nutritional features. After 30 days of storage at 4 °C, pineapple fruits started with selected autochthonous strains of *Lact. plantarum* and *Lact. rossiae* had a number of LAB up to 1,000,000 times higher than the other processed pineapples, and the lowest number of yeasts (Di Cagno et al. 2010).

Fermentation of sweet cherry (*Prunus avium* L.) puree by selected autochthonous LAB starters showed that, despite the hostile environment (e.g., low pH, presence of phenolic compounds), the selected autochthonous strains of *P. pentosaceous* and *Lact. plantarum* grew well, showing peculiar metabolic traits. Consumption of carbohydrates was limited as well as the lactic acid fermentation, while consumption of organic acids (e.g., malic acid) and free amino acids was evident, especially, throughout storage. Both LAB remained viable during 60 days of storage at cell numbers, which exceeded those of potential probiotic beverages.

Overall, the use of autochthonous LAB starters to ferment vegetables and fruits ensures the preservation of various attributes: natural colors, firmness, antioxidant activities, and other health-promoting compounds. This effect may be the consequence of the modified profile of organic acids, the synthesis of lactic and acetic acids, and the metabolism of free amino acids. These modifications might have direct (pH) or indirect (redox potential) repercussions on the activity of endogenous browning enzymes, and oxidation and sensory properties (color, flavor, and aroma) of vegetables and fruits. Figure 14.1 shows an example of a biotechnology protocol to ferment raw vegetables and fruits.

Not all the strains that compose the LAB microbiota of vegetables and fruits guarantee the same performances during processing and storage. Therefore, selection is indispensable, which is based on peculiar metabolic traits (Di Cagno et al. 2013; Filannino et al. 2014). Although heat treatments could in part change the chemical composition of the raw matrix, sterilized/pasteurized vegetable or fruit juice media are considered to be the most suitable model systems to carry out the selection. The main criteria for selecting starters are usually divided into three main categories: (i) pro-technology; (ii) sensory; (iii) functional; and (iv) probiotic (Table 14.1). Rate of growth, rate and total production of acids which, in turn, affects the changes of pH, and environmental adaptation/tolerance are the main technological criteria. The predominance of LAB species is influenced by the chemical and physical environment, where they have to compete. *Lact. plantarum*, which predominates the later stage of vegetable fermentation due to its high acid tolerance and metabolic versatility, seems to be the better choice when homo-lactic fermentation is desired (Filannino et al. 2014). Moreover, the maintenance of high viability in the stationary phase of growth, under challenging and stressing environmental conditions, is an indispensable pre-requisite, which ensures prolonged shelf-life of fermented products, especially those containing functional starters. The synthesis of hydrogen peroxide by starters is another criterion to be considered for selection.

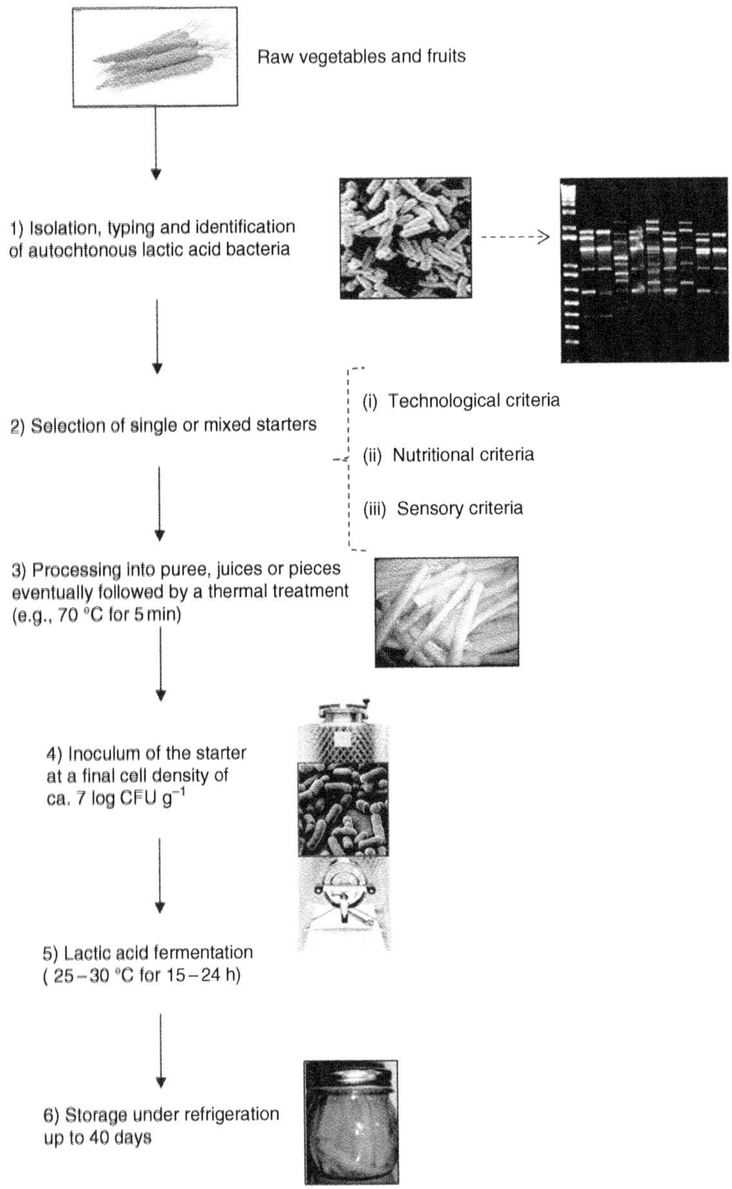

Figure 14.1. Example of biotechnology protocol to ferment raw vegetables and fruits.

Hydrogen peroxide, a strong oxidizing agent, may react with organic material and cause, aside from degradation of antioxidant components, an undesired loss of the product color. The synthesis of exopolysaccharides is another important metabolic trait, either for the viscosity of juices and purees or for nutritional properties (Di Cagno et al. 2008). In addition, the capacity of LAB to synthesize protopectinases, which may enhance the viscosity of the fruit matrices, may be an important characteristic for starters.

Table 14.1. Main criteria and corresponding metabolic traits to select starters for vegetable and fruit fermentation.

Criteria	Metabolic traits
Pro-technology	Growth rate
	Acidification rate
	Salt tolerance
	Growth at low values of pH
	Growth at low temperature
	Completeness of fermentation
	Synergy with other starters
	Metabolic adaptation during fermentation and storage
	Malolactic fermentation
	Metabolism of phenolics
	Synthesis of antimicrobial compounds
	Synthesis of hydrogen peroxide
	Pectinolytic activity
	Cell sedimentation
	Bacteriophage resistance
Sensory	Hetero-fermentative metabolism
	Synthesis of aroma compounds or their precursors
	Off-flavors production
Functional	Synthesis of bioactive compounds
	Synthesis of exo-polysaccharides
	Increase in bioavailability of functional compounds
	Increase of the antioxidant activity
	Synthesis of biogenic amine
	Release of bioactive peptides
	Decrease of toxic or antinutritive factors
Probiotic	Acid and bile salt resistance
	Inhibition of pathogenic bacteria
	Stimulation of immune-mediators
	Adherence to human intestinal cells
	Absence of any pathogenic characteristics
	Capability to hydrolyze nutritional constituents (e.g., galacto- and fructo-oligosaccharides)

14.4. Main Fermented Vegetable Products

Many fruits and vegetables are processed through lactic acid fermentation for the production of a wide range of traditional fermented products in various worldwide regions, mainly in the Mediterranean area. The manufacturing protocols for some traditional fermented vegetables are described below. Due to the wide literature describing the lactic acid fermentation of table olives, they are not considered in this chapter.

14.4.1. Sauerkrauts

Sauerkrauts, widely consumed in many European countries and the USA, is one of the most studied examples of fermented vegetables. Usually, they are manufactured by spontaneous lactic fermentation of white cabbage (*Brassica olereracea* L. var. *capitata*) (Di Cagno and Coda 2014). After removal of the core and outer leaves, fresh cabbages are shredded and mixed with salt, to obtain a final concentration of approximately 2% NaCl (wt/wt). The osmotic effect of NaCl leads to liquid release from the tissues. Cabbages are quickly surrounded by brine and covered with plastic sheeting draped over the tank to ensure air exclusion. Hetero-fermentative LAB contribute to realize anaerobic conditions in 24–48 h, through the exhaustion of oxygen and CO_2 production. The combination of salt, temperature (15–20 °C), and anaerobic conditions allows spontaneous lactic acid fermentation, which is characterized by the natural succession of LAB species. Variations of these conditions, such as the presence of air, too low or too high amount of NaCl, and fluctuating temperatures, affect the competitiveness of the autochthonous microorganisms, and may lead to spoilage of sauerkrauts. Typically, *Leuc. mesenteroides* and *Weissella* spp. dominate the early stages of fermentation, since they are numerous and have fast growth. The increase in the concentration of lactic acid promotes the growth of acid-tolerant species such as *Lact. brevis* and, in some cases, *Lact. curvatus*, *Lact. sakei*, *E. faecalis*, *L. lactis* subsp. *lactis*, and *P. pentosaceus*. *Lact. plantarum* prevails during late fermentation, leading to a further acidification with a final pH of ca. 3.5 (Di Cagno and Coda 2014). Strictly hetero-fermentative LAB metabolize carbohydrates to lactic acid, ethanol, CO_2, acetic acid, and mannitol. During the homo-fermentative phase, which is dominated by *Lact. plantarum*, carbohydrates are mainly metabolized to lactic acid. These microbial metabolites, together with other chemical compounds (e.g., acetaldehyde, diacetyl, and methanethiol), affect the sensory properties of sauerkrauts. Sauerkraut fermentation lasts for several weeks, depending on fermentation conditions, sensory attributes, and the autochthonous microbiota of cabbages and related sugar content. Secondary fermentations carried out by yeasts are undesirable. Yeasts, found in the late phases of sauerkraut fermentation, are generally related to spoilage and may produce off-flavors and cause texture defects. The total consumption of fermentable sugars is an indispensable prerequisite to prevent secondary fermentations. Sauerkrauts are commercialized as fresh products, whether or not with the addition of preserving agents, or as pasteurized products to better inhibit spoilage through secondary fermentation (Holzapfel et al. 2003). Although sauerkraut production mainly occurs through spontaneous fermentation, interest in starter cultures is increasing. Starter cultures were selected to carry out successful fermentations with a low addition of NaCl, and to increase the amount of bioactive compounds, such as allyl isothiocyanate, which result from glucosinolate breakdown (Johanningsmeier et al. 2007; Martinez-Villaluenga et al. 2009).

14.4.2. Kimchi

Kimchi is a traditional Korean fermented vegetable product that plays an essential role as a basic side dish for Korean meals. The word "kimchi" originated from "chimchae," the Chinese character, meaning salted vegetables. It was called "dimchae," "dimchi," and, then, kimchi in sequence by word-of-mouth transfer (Park et al. 2014). Kimchi fermentation in Korea represented an efficient method of preserving the fresh and crispy texture of vegetables during winter, when fresh vegetables are not available. Actually, kimchi is a globally popular food, and its market size is gradually increasing. Thanks to the physiological effects of its ingredients and to microbial metabolites, Kimchi is widely recognized as a health-promoting food. Kimchi is mainly manufactured with *baechu* cabbage (Chinese cabbage, *Brassica pekinensis*) and radish, but other seasoning ingredients such as garlic, green onion, ginger, red pepper, mustard, parsley, and carrot may be incorporated in the fermentation mixture (Park et al. 2014). There are many types of kimchi, more than 160, depending on the main ingredient, region, and preparation methods. *Baechu* kimchi is currently the most popular and commonly used in Korea and is generally made by lactic acid fermentation of *baechu* cabbage, radish, green onion, red pepper powder, garlic, ginger, and fermented sea foods (*jeotgal*) (Park et al. 2014). After soaking with water, vegetables are cut and placed in the salt solution (5–7%) for 12 h or for 3–7 h (15% of salt), until the salt content in the tissues reaches the value of ca. 2.0–4.0% of the total weight. Further, vegetables are rinsed several times with fresh water and drained. The mixed ingredients are then placed in fermentation vessels, covered with macerated cabbage leaves, and pressed by stones to ensure anaerobic conditions. Generally, the most suitable ripening requires low fermentation temperatures (2–6 °C), but sometimes higher temperatures (30 °C) are used to accelerate the process. Kimchi fermentation is carried out by various autochthonous microorganisms. The microbiota is largely affected by the fermentation conditions (e.g., anaerobiosis, ingredients, salt concentrations, temperature). Usually, *Leuc. mesenteroides* and *P. pentosaceus* grow in the first stage of fermentation. *Lact. sakei*, *Lact. plantarum*, and *W. koreensis* prevail during the later stages, as the kimchi fermentation environment changes to more anaerobic and acidic conditions. The temperature of fermentation affects the selection of the dominant species within the *Lactobacillus* genus. *Lact. plantarum* and *Lact. brevis* dominate when the fermentation is carried out at 20–30 °C, while *Lact. maltaromicus* and *Lact. bavaricus* prevail at 5–7 °C. *Lact. plantarum* allows a further decrease of the values of pH during the final stages. The optimal value of pH during kimchi fermentation is ca. 4.2–4.5 (Di Cagno and Coda 2014). Strongly acidified kimchi is not appreciated by consumers. Autochthonous strains of *Lact. sakei* were selected as starters for the manufacture of kimchi due to their mild acidifying properties (Di Cagno et al. 2013). The main microbial metabolites (lactic and acetic acids, CO_2, ethanol, and mannitol) with other minor chemical compounds (diacetyl, acetoin, acetaldehyde, secondary alcohols, esters, and lactones) affect the sensory properties of kimchi. For instance, the synthesis of mannitol is associated with a refreshing taste. The time for fermentation of kimchi may vary from a few days to several months, depending on the temperature and concentration of NaCl. After fermentation, kimchi is stored at low temperatures, although preserving treatment (e.g., pasteurization) may be used (Lee et al. 2011).

14.4.3. Pickled Cucumbers

Pickled cucumbers are the most popular pickled foods. Besides cucumbers, there are niche markets for many other pickled vegetables (e.g., artichokes, capers, garlic, peppers, okra, cauliflower, green tomatoes, and eggplants) mainly manufactured at local factories or household levels. Manufacture of pickled cucumbers requires that fully ripe cucumbers are washed and drained, and, eventually, sliced. Further, regular shaped cucumbers are dipped into brine (5–7% of NaCl). Industrial fermentations are carried out by spontaneous LAB. The fermentation lasts 2–3 weeks, depending on the temperature (20–27 °C), which allows it to reach a final pH of ca. 3.1–3.5. Because of the high concentration of NaCl, the growth of *Leuc. mesenteroides* and other less salt tolerant species is inhibited. The main species responsible for fermentation are homo-fermentative LAB, such as *P. pentosaceus* and *Lact. plantarum*. Strictly hetero-fermentative species are undesirable. The synthesis of CO_2 leads to texture defects and bloater damage, with serious economic losses. Because CO_2 results also from cucumber respiration or malolactic fermentation by *Lact. plantarum*, fermentation vessels may be purged with nitrogen gas (Di Cagno et al. 2013; Di Cagno and Coda 2014). Salt, consumption of fermentable carbohydrates, and low values of pH prevent undesirable secondary fermentations by yeasts and other LAB (e.g., *Lact. buchneri*), which often cause cucumber spoilage. Usually, yeasts are present in the upper part of the fermentation tanks, and are responsible for off-flavors and texture defects (Franco et al. 2012). Blanching of cucumbers is also used to decrease the concentration of NaCl (4% or less), partially inhibit the autochthonous microbiota, and improve the texture, through the inactivation of endogenous pectin esterase and glycosidases. Partial or complete replacement of NaCl by $CaCl_2$ was suggested as an efficient tool to prevent softening. $CaCl_2$ may be also added on the surface to gain a crisp texture during storage (Di Cagno et al. 2013).

14.5. Physiology and Biochemistry of LAB during Vegetable and Fruit Fermentation

More than other food ecosystems, raw fruits and some vegetables possess intrinsic chemical and physical parameters that make them particularly hostile environments for microorganisms. Mathematical models describing the growth and survival of *Lact. plantarum* in fruit and vegetable juices were developed. The pH and the concentration of organic acids and sugars were identified as the main affecting factors (Nualkaekul and Charalampopoulos 2011; Filannino et al. 2014). Because of the high number of components playing a role, it is difficult to develop mathematical models that are applicable to all plant matrices and bacterial strains. The diversity of plant environments and of bacterial enzyme activities makes the microbial adaptation to vegetable and fruit ecosystems markedly heterogeneous (Filannino et al. 2014). Adaptation depends on the microbial capacity to adopt specific metabolic pathways (e.g., use of non-conventional carbon sources, pathways allowing the homeostasis of intracellular pH), and to share the metabolic energy between biosynthesis (e.g., use of alternative substrates) and maintenance (e.g., global stress responses).

14.5.1. Metabolic Adaptation by LAB during Plant Fermentation

Recently, the adaptive growth and survival strategies of several *Lact. plantarum* strains during plant fermentation and storage were described, using a panel of various metabolomic approaches (Filannino et al. 2014). Different metabolic traits emerged depending on the strain and plant substrate, and on fermentation and storage processes. These traits resulted in alternative energy routes like the catabolism of free amino acids, malolactic fermentation, and synthesis of volatile compounds and volatile free fatty acids (Figure 14.2) (Filannino et al. 2014). The catabolism of free amino acids is a well-known mechanism for microbial adaptation. The concentration of branched chain amino acids (Val, Ile, and Leu) decreased during fermentation of vegetable and fruit juices by *Lact. plantarum*. They were converted into their respective 2-ketoacids, and in turn decarboxylated to corresponding aldehydes and reduced to branched alcohols (Figure 14.2). Consumption of branched chain amino acids into their corresponding 2-ketoacids leads to a gain of ATP allowing the regeneration of Glu from α-ketoglutarate. The reduction of aldehydes may contribute to the regeneration of the intracellular pool of NAD^+, which increases the efficiency of the energy producing pathways. During fermentation of vegetable and fruit juices by *Lact. plantarum* the activation of the acetate kinase route strictly relies on the availability of external acceptors of electrons like aldehydes, which are reduced into the corresponding alcohols. Additional metabolic routes include the conversion of Glu to γ-aminobutyric acid, which enhances the acid resistance, and the decarboxylation of His into histamine, which provides energy through the generation of proton motive force (Figure 14.2). Malic acid is one of the most abundant organic acids, which occurs in vegetables and fruits. At low pH, the decreased uptake of glucose and the preference for malic acid as the energy source was described for *Lact. plantarum*. Decarboxylation of malic acid provides energy advantages, due to the increased intracellular pH and the synthesis of reducing power (Figure 14.2). Pyruvate holds a key position in the carbohydrate metabolic pathways of LAB. The synthesis from pyruvate of neutral diacetyl, which is found at high levels in plant juices fermented by *Lact. plantarum*, is induced at the transcriptional level by acidic conditions, and, presumably, contributes to the regulation of the intracellular pH by decreasing the level of pyruvate.

14.6. Food Phenolic Compounds: Antimicrobial Activity and Microbial Responses

Phenolic compounds are secondary plant metabolites widely spread throughout the plant kingdom. These food components have attracted interest because of their health benefits, and their impact on multiple sensory attributes of foods.

14.6.1. Effect of Phenolics on the Growth and Viability of LAB

Most of the known plant phenols exert an antagonistic effect on the growth of a wide range of LAB. The inhibition was described to be concentration-dependent. Many studies were focused on *Lact.*

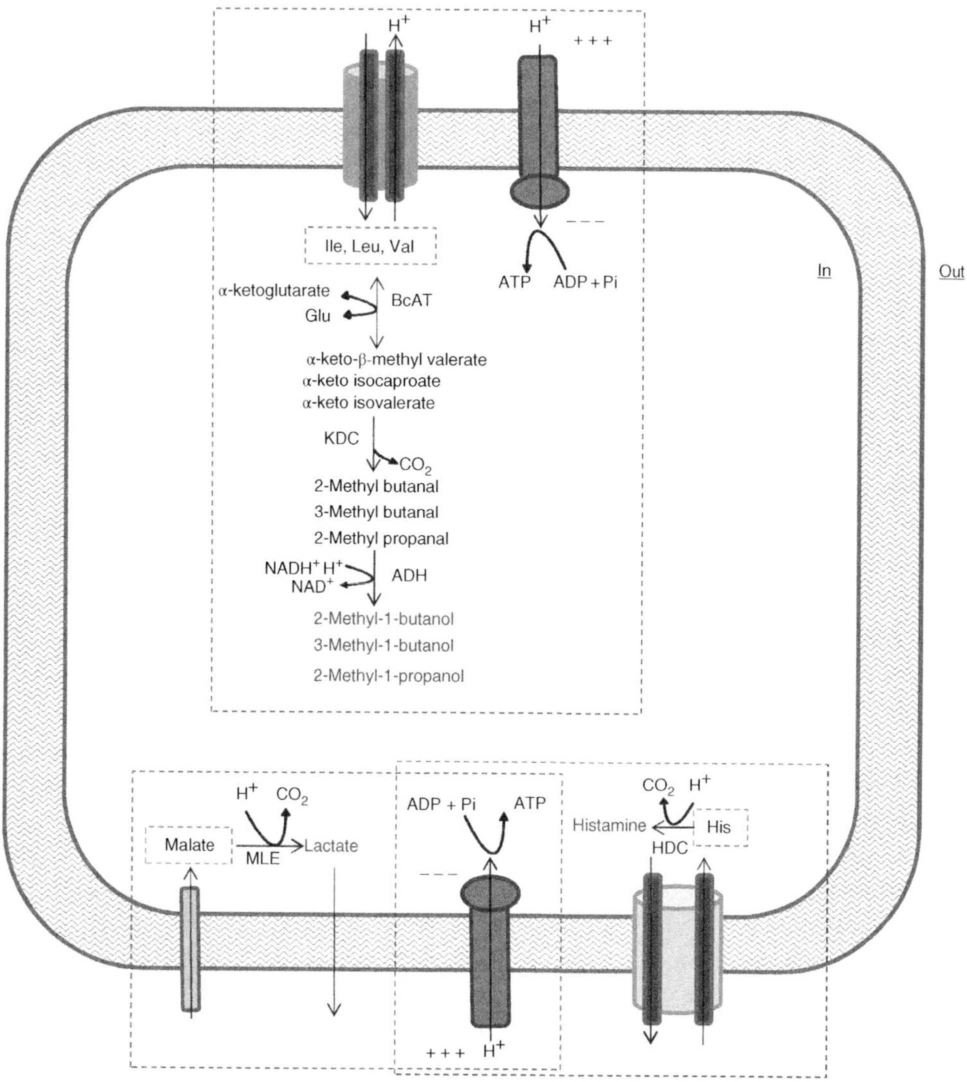

Figure 14.2. Example of some metabolic pathways adopted by *Lactobacillus plantarum* during fermentation of vegetable and fruit juices. Ile, isoleucine; Leu, leucine; Val, valine; His, histidine; Glu, glutamic acid; BcAT, branched-chain aminotransferase; KDC, α-keto acid decarboxylase; ADH, alcohol dehydrogenase; MLE, malolactic enzyme; HDC, histidine decarboxylase (Adapted from Filannino *et al.* 2014)

plantarum and wine species such as *Oenococcus oeni*, *Lact. hilgardii*, and *Lact. brevis*. The antimicrobial activity of wine phenolic compounds (catechin, gallic acid, gallate epicatechin, salicylic acid, methyl gallate, caffeic acid, ferulic acid, tryptophol, *p*-coumaric acid) was shown against *Lact. plantarum*. In addition, *O. oeni* is affected by these compounds depending on the phenolic type and concentration. Whereas many compounds have an inhibiting effect, gallic acid, free anthocyanins, catechin, and quercitin seem to activate the growth of and the rate of malic acid degradation by *O. oeni* (Rodrìguez et al. 2009).

Different mechanisms are involved in the bactericidal effect of phenolic compounds, alterations of the cell wall and the cytoplasmic membrane being the most important. The loss of cell wall integrity is due to the disruption of the peptidoglycan affecting the fatty acid composition of the cytoplasmic membrane of *Lact. plantarum*. Studies on grape must showed that phenolic acids cause an increase in the cell membrane permeability (Rodrìguez et al. 2009). The inhibitory activities of gallotannins are attributable to their affinity for iron and likely additionally related to the inactivation of membrane-bound proteins (Engels et al. 2011). The toxicity of hydroxybenzoic and hydroxycinnamic acids under acidic conditions was consistent with the dissipation of the ΔpH (Barthelmebs et al. 2000).

14.6.2. Metabolism of Phenolics by LAB

The capacity to tolerate and metabolize phenolic compounds is species- or strain-dependent. Compared to other genera (e.g., *Enterobacteria*, *Clostridium* spp. and *Bacteroides* spp.), LAB seem to be more resistant to phenolic compounds. Strains showing the capacity to metabolize or tolerate phenolics are usually isolated from fermented foods with a high content of these compounds (Sanchez-Maldonado et al. 2011). LAB use several mechanisms to overcome the inhibition by phenolics, which mainly includes the dissociation of polyphenol-substrate complexes, polyphenol inactivation by high-affinity binders, membrane modification/repair, metal ion sequestration, and phenolic modification/degradation. The conversion of phenolic compounds is based on: (i) glycosyl hydrolases, such as β-glucosidase and α-rhamnosidase, which convert flavonoid glycosides to the corresponding aglycones; (ii) esterases degrading methyl gallate, tannins, or phenolic acid esters; (iii) decarboxilases and reductases; and (iv) benzyl alcohol dehydrogenase, which catalyzes the reversible oxidation of some aromatic alcohols to aldehydes with the concomitant reduction of NAD^+ (Rodrìguez et al. 2009). These enzyme activities were largely studied in *Lact. plantarum* strains. Only a few reports described this metabolism for other LAB species (Table 14.2). Because the reduced or decarboxilated products of phenolic acids have lower antimicrobial activity compared to their precursors, the metabolism of phenolic acids was identified as an effective mechanism of detoxification (Sanchez-Maldonado et al. 2011). Besides, it could be assumed that the reduction of phenolic acids is involved in the reoxidation of the co-factor NADH, which provides a metabolic advantage through the NAD^+ regeneration (Silva et al. 2011).

14.7. Health-promoting Properties of Fermented Vegetables and Fruits

Vegetable- and fruit-based beverages manufactured by controlled fermentation of LAB are also new products, answering to the consumer's demand for minimally processed foods. A high proportion of protective substances that are contained in the raw material are preserved in these products. During the fermentation, LAB may also produce additional health-promoting components and allow a better preservation. LAB strains that were isolated from plant spontaneous fermentations were recently shown as promising probiotic candidates (Vitali et al. 2012). The micro-architecture of the vegetable surface (e.g., roughness) and the presence of prebiotic compounds (e.g., oligosaccharides) likely contribute to cell protection, and thus improve survival of bacteria (do Espirito Santo et al. 2011).

14.8. Alternative Sources of Novel Probiotics Candidates

Although the term probiotic cannot be used synonymously for beneficial members of the human gut microbiota, bacteria from this ecosystem are almost the only source from which probiotics are isolated. Nevertheless, the probiotic potential of bacteria from other ecosystems would not *a priori* be excluded. Studies on fermented foods, especially of plant origin, as alternative sources

Table 14.2. Metabolism of phenolic compounds by LAB.

Substrate	Product	Enzymatic activity	Lactic acid bacteria species
Benzyl alcohol	Benzaldehyde	Benzyl alcohol dehydrogenase	*Lact. plantarum*
Caffeic acid	Vinyl catechol	Decarboxylase	*Lact. plantarum; Lact. hammesii; Lact. fermentum; Lact. spicheri*
	Ethyl catechol	Reductase	*Lact. plantarum*
	Dihydrocaffeic acid	Reductase	*Lact. plantarum; Lact. fermentum; Lact. reuteri*
m-Coumaric acid	3-(3-hydroxyphenyl) propionic acid	Reductase	*Lact. plantarum*
p-Coumaric acid	Vinyl phenol	Decarboxylase	*Lact. plantarum; Lact. hammesii; Lact. fermentum; Lact. spicheri; Lact. reuteri; Lact. brevis; Pediococcus pentosaceus; Oenococcus oeni*
	Ethyl phenol	Reductase	*Lact. plantarum; Lact. collinoides*
	Phloretic acid	Reductase	*Lact. plantarum; Lact. fermentum; Lact. reuteri*
Ferulic acid	Vinyl guaiacol	Decarboxylase	*Lact. plantarum; Lact. brevis; P. pentosaceus; O. oeni*
	Ethyl guaiacol	Reductase	*Lact. plantarum*
	Dihydroferulic acid	Reductase	*Lact. plantarum; Lact. fermentum*
Vinyl guaiacol	Vanillin	Decarboxylase	*O. oeni; Lact. hilgardii; Lact. brevis; Lact. plantarum; Pediococcus damnosus*
Cinnamyl alcohol	Cinnamyl aldehyde	Benzyl alcohol dehydrogenase	*Lact. plantarum*
Coniferyl alcohol	Coniferyl aldehyde	Benzyl alcohol dehydrogenase	*Lact. plantarum*
Gallic acid	Pyrogallol	Decarboxylase	*Lact. plantarum; Lact. hilgardii*
Methyl gallate	Gallic acid	Tannase	*Lact. plantarum*
	Pyrogallol	Decarboxylase	*Lact. plantarum*
Oleuropein	Hydroxytyrosol; elenoic acid	β-Glucosidase; esterase	*Lact. plantarum*
Phenethylalcohol	Phenethylaldehyde	Benzyl alcohol dehydrogenase	*Lact. plantarum*
Protocatechuic acid	Catechol	Decarboxylase	*Lact. plantarum; Lact. hammesii*
Quinic acid	Catechol	Several enzymes	*Lact. plantarum*
	Dihydroshikimic acid	Several enzymes	*Lact. pastorianus; Lact. paracollinoides*
Shikimic acid	Catechol	Several enzymes	*Lact. plantarum*
	Dihydroshikimic acid	Several enzymes	*Lact. pastorianus; Lact. paracollinoides*
Tannic acid	Gallic acid	Tannase	*Lact. plantarum*
	Pyrogallol	Decarboxylase	*Lact. plantarum*
Chlorogenic acid	Caffeic acid; quinic acid	Esterase	*Lact. plantarum; Lact. gasseri; Lact. hammesii; Lact. reuteri; L. fermentum; L. paracollinoides*
Methyl ferulate	Ferulic acid	Esterase	*Lact. acidophilus*
Naringenin-7-O-glucoside	Naringenin	Glucosidase	*Lact. plantarum; Lact. casei*
Vanillin	Vanillin alcohol	Reductase	*O. oeni*
Geranyl glycoside	Geraniol	β-Glycosidase	*O. oeni*
Benzyl glycoside	Benzyl alcohol	β-Glycosidase	*O. oeni*
Daidzin	Daidzein	β-Glucosidase	*Lact. rhamnosus*
Genistin	Genistein	β-Glucosidase	*Lact. rhamnosus*
Hesperidin	Hesperetin	α-Rhamnosidase	*Lact. plantarum*
Rutin	Quercetin	α-Rhamnosidase	*Lact. plantarum*

(Adapted from Rodrìguez et al. 2009)

of novel probiotic candidates are accumulating. These ecosystems harbor functional strains that should possess resistance to gastric acidity and bile toxicity, and the capacity of adhesion to the gut epithelium, to hydrolyze nutritional constituents, which cannot be metabolized by the host (e.g., fructo-oligosaccharides, FOS), and to synthesize antimicrobial substances. More than other food ecosystems, raw fruits and some vegetables possess intrinsic chemical and physical parameters that, for some traits, mimic those of the human gastrointestinal tract, such as the extremely acid environment, buffering capacity, high concentration of indigestible nutrients (fiber, inulin, and FOS) and antinutritional factors (e.g., tannins and phenols). In most cases, the autochthonous microbiota of fruits and vegetables has to colonize and adhere to surfaces exerting antagonistic activity towards spoilage and pathogenic microorganisms. A large number of autochthonous LAB (*Lactobacillus*, *Weissella*, *Pediococcus*, *Enterococcus*, and *Leuconostoc* genera), isolated from carrots, French beans, cauliflower, celery, tomatoes, pineapples, kimchi, Japanese pickles, Chinese sauerkraut brine, and several ethnic fermented vegetables of the Himalayas were characterized for their functional features (Tamang et al. 2009; Vitali et al. 2012; Di Cagno et al. 2013). Adaptability to gastrointestinal environment, capacity for maintaining high cell densities, adherence to human or mouse intestinal cells, stimulation of immune-mediators, and antimicrobial activity, were shown. In particular, strains of *Lact. plantarum* showed high survival under simulated gastric and intestinal conditions, stimulated a large number of immune-mediators by peripheral blood mononuclear cells, including cytokines with pro- and anti-inflammatory activities, strongly adhered to Caco-2 cells, used FOS as the only carbon source, and inhibited pathogens from human sources. Similar results were found for autochthonous strains isolated from kimchi and belonging to the species *Lact. plantarum* and *Lact. sakei* (Lee et al. 2011). For these strains, a capacity to lower cholesterol was also suggested. Functional properties were also attributed to LAB strains, which were isolated from Nozawana-zuke, a traditional Japanese pickle made with *Brassica campestris* L. var. *rapa* (Kawahara and Otani 2006).

14.9. Vehicles for Delivering Probiotics

Fruit and vegetable juices are proposed as an ideal medium for functional health ingredients because they inherently contain beneficial nutrients, have taste profiles that are pleasant to humans of all ages, and are perceived as healthy and refreshing.

Beetroot and carrot juices enriched with brewer's yeast autolysate were subjected to lactic acid fermentation with probiotic *Lact. acidophilus* (Rakin et al. 2007). The addition of the autolysate increased the number of LAB during fermentation, decreased the time of fermentation, and enriched the vegetable juices with amino acids, vitamins, minerals, and antioxidants. The probiotic action of grape, cabbage, apple, orange, tomato, beta, and blackcurrant juices was also successfully reported. The juice made of a blend of carrots, celery, and apples was also a good matrix to grow *Lact. acidophilus* (Gobbetti et al. 2010). The effect of potential probiotics and autochthonous LAB on health-promoting and sensory properties of tomato juice was studied. Comparing unstarted and started tomato juices using allochthonous strains, properties such as viscosity, color, ascorbic acid, glutathione, total antioxidant activity, glutamic acid, and total free amino acids were positively affected by selected starters (Di Cagno et al. 2008). The protective effect towards several antioxidants and vitamins (e.g., white grape seed extract, green tea extract, vitamins B_2, B_3, B_6, C, and E) was shown in a model fruit juice system, using *Lact. rhamnosus* HN001, *Bifidobacterium lactis* strain HN001, and *Lact. paracasei* LPC 37 (Ferdousi et al. 2013).

The suitability of several vegetables and fruits as raw materials for manufacturing functional fermented juices has opened new perspectives.

14.10. Conclusions

Lactic acid fermentation undoubtedly represents one of the easiest and most suitable ways for increasing the daily consumption of fresh-like vegetables and fruits. Selection of autochthonous starters targeted for different vegetables and fruits has to be encouraged to gain new insights and to allow controlled fermentation processes. LAB tailored to the various intrinsic and extrinsic environmental conditions should exploit the potential of vegetables and fruits to enhance hygiene, sensory, nutritional, and shelf-life properties.

References

Barthelmebs L., Divies, C. and Cavin, J.F. (2000) Knockout of the *p*-coumarate decarboxylase gene from *Lactobacillus plantarum* reveals the existence of two other inducible enzymatic activities involved in phenolic acid metabolism. *Appl Environ Microbiol* **66**, 3368–3375.

Buckenhüskes, H.J. (1997) Fermented Vegetables. In: *Food Microbiology: Fundamentals and Frontiers* (ed. Doyle, P.D., Beuchat, L.R., and Montville T.J.) 2nd edn. pp. 595–609. Washington DC: ASM Press,

Di Cagno, R. and Coda, R. (2014) *Fermented Foods: Fermented Vegetable Products*. In: *Encyclopedia of Food Microbiology* ((eds) C.A. Batt and M.L. Tortorello, M.L.) vol 1, Elsevier Academic Press, London, pp. 875–883.

Di Cagno, R., Cardinali, G., Minervini, G., et al. (2010) Taxonomic structure of the yeasts and lactic acid bacteria microbiota of pineapple (*Ananas comosus* L. Merr.) and use of autochthonous starters for minimally processing. *Food Microbiol* **27**, 381–389.

Di Cagno, R., Coda, R., De Angelis, M. and Gobbetti, M. (2013) Exploitation of vegetables and fruits through lactic acid fermentation. *Food Microbiol* **33**, 1–10.

Di Cagno, R., Surico, R.F., Minervini, G., et al. (2009) Use of autochthonous starters to ferment red and yellow peppers (*Capsicum annum* L.) to be stored at room temperature. *Int J Food Microbiol* **130**, 108–116.

Di Cagno, R., Surico, R.F., Paradiso, A., et al. (2008) Effect of autochthonous lactic acid bacteria starters on health-promoting and sensory properties of tomato juices. *Int J Food Microbiol* **128**, 473–483.

do Espirito Santo, A.P., Perego, P., Converti, A. and Oliveira, M.N. (2011) Influence of food matrices on probiotic viability. a review focusing on the fruity bases. *Trends Food Sci Technol* **22**, 377–385.

Endo, A. Futagawa-Endo, Y., and Dicks, L.M.T. (2009) Isolation and characterization of fructophilic lactic acid bacteria from fructose-rich niches. *Syst Appl Microbiol* **32**, 593–600.

Engels, C., Schieber, A., and Gänzle, M.G. (2011) Inhibitory spectra and modes of antimicrobial action of gallotannins from mango kernels (*Mangifera indica* L.). *Appl Environ Microbiol* **77**, 2215–2223.

Ferdousi R., Rouhi, M., Mohammadi, R., et al. (2013) Evaluation of probiotic survivability in yogurt exposed to cold chain interruption. *Iran J Pharm Res* **12**, 139–144.

Filannino, P., Cardinali, G., Rizzello., C.G., et al. (2014) Metabolic responses of *Lactobacillus plantarum* strains during fermentation and storage of vegetable and fruit juices. *Appl Environ Microbiol* **80**, 2206–2215.

Franco, W., Pérez-Diaz, I.M., Johanningsmeier, S.D., and McFeeters, R.F. (2012) Characteristics of spoilage-associated secondary cucumber fermentation. *Appl Environ Microbiol* **78**, 1273–1284.

Gobbetti, M., Di Cagno, R. and De Angelis, M. (2010) Functional microorganisms for functional food quality. *Crit Rev Food Sci Nutr* **50**, 716–727.

He, F.J., Nowson, C.A., Lucas, M., and MacGregor, G.A. (2007) Increased consumption of fruit and vegetables is related to a reduced risk of coronary heart disease: meta-analysis of cohort studies. *J Human Hypert* **21**, 717–728.

Holzapfel, W., Schillinger, U. and Buckenhüskes, H. (2003) Sauerkraut. In: *Handbook of fermented functional foods* ((ed.) E.R. Farnworth), CRC Press, Boca Raton, pp. 395–412.

Hutkins, R.W. (2006) Fermented vegetables. In: *Microbiology and Technology of Fermented Foods* ((ed.) R.W. Hutkins), Blackwell, Oxford, pp. 223–259.

Johanningsmeier, S., McFeeters, R.F., Fleming, H.P., and Thompson, R.L. (2007) Effect of *Leuconostoc mesenteroides* starter culture on fermentation of cabbage with reduced salt concentrations. *J Food Sci* **72**, 166–172.

Kawahara, T. and Otani, H. (2006) Stimulatory effect of lactic acid bacteria from commercially available Nozawana-zuke pickle on cytokine expression by mouse spleen cells. *Biosci Biotechnol Biochem* **70**, 411–417.

Lee, H., Yoon, H., Ji, Y., et al. (2011) Functional properties of *Lactobacillus* strains from kimchi. *Int J Food Microbiol* **145**, 155–161.

Martinez-Villaluenga, C., Peñas, E., Frias, J., et al. (2009) Influence of fermentation conditions on glucosinolates, ascorbigen, and ascorbic acid content in white cabbage (*Brassica oleracea* var. *capitata* cv. Taler) cultivated in different seasons. *J Food Sci* **74**, C62–C67.

Neveling, D.P., Endo, A., and Dicks, L.M.T. (2012) Fructophilic *Lactobacillus kunkeei* and *Lactobacillus brevis* isolated from fresh flowers, bees and bee-hives. *Curr Microbiol* **65**, 507–515.

Nualkaekul, S. and Charalampopoulos, D. (2011) Survival of *Lactobacillus plantarum* in model solutions and fruit juices. *Int J Food Microbiol* **146**, 111–117.

Offonry, S.U. and Achi, O.K. (1998) Microbial populations associated with the retting of melon pods (*Colocynthis citrullus* L.) during seed recovery. *Plant Foods Human Nutr* **52**, 37–47.

Paramithiotis, S., Doulgeraki, A.I., Tsilikidis, I., et al. (2012) Fate of *Listeria monocytogenes* and *Salmonella typhimurium* during spontaneous cauliflower fermentation. *Food Control* **27**, 178–183.

Paramithiotis, S., Kouretas, K. and Drosinos, E.H. (2013) Effect of ripening stage on the development of the microbial community during spontaneous fermentation of green tomatoes. *J Sci Food Agric* **94**, 1600–1606.

Park, K.Y., Jeong, J.K., Lee, Y.E., and Daily III, J.W. (2014) Health benefits of Kimchi (Korean fermented vegetables) as a probiotic food. *J Med Food* **17**, 6–20.

Plengvidhya, V., Breidt, F. and Fleming, H.P. (2004) Use of RAPD-PCR as a method to follow the progress of starter cultures in sauerkraut fermentation. *Int J Food Microbiol* **93**, 287–296.

Rakin, M., Vukasinovic, M., Siler-Marinkovic, S. and Maksimovic, M. (2007) Contribution of lactic acid fermentation to improved nutritive quality vegetable juices enriched with brewer's yeast autolysate. *Food Chem* **100**, 599–602.

Rodrìguez, H., Curiel, J.A., Landete, J.M, et al. (2009) Food phenolics and lactic acid bacteria. *Int J Food Microbiol* **132**, 79–90.

Sanchez-Maldonado, A.F., Schieber, A. and Gänzle, M.G. (2011) Structure–function relationships of the antibacterial activity of phenolic acids and their metabolism by lactic acid bacteria. *J Appl Microbiol* **111**, 1176–1184.

Silva, I., Campos, F.M., Hogg, T. and Couto, J.A. (2011) Factors influencing the production of volatile phenols by wine lactic acid bacteria. *Int J Food Microbiol* **145**, 471–475.

Spurr, H.W. (1994) The microbial ecology of fruit and vegetable surfaces, its relationship to postharvest biocontrol. In: *Biological Control of Postharvest Diseases: Theory and Practice* ((eds) C. Wilson and M. Wisniewski), CRC Press, Boca Raton, pp. 11–23.

Tamang, J.P., Tamang, B., Schillinger, U., et al. (2009). Functional properties of lactic acid bacteria isolated from ethnic fermented vegetables of the Himalayas. *Int J Food Microbiol* **135**, 28–33.

Vitali, B., Minervini, G., Rizzello, C.G., et al. (2012) Novel probiotic candidates for humans isolated from raw fruits and vegetables. *Food Microbiol* **31**, 116–125.

Vorholt, J.A. (2012) Microbial life in the phyllosphere. *Nature* **10**, 828–840.

Wouters, D., Bernaert, N., Anno, N., et al. (2013) Application' and validation of autochthonous lactic acid bacteria starter cultures for controlled leek fermentations and their influence on the antioxidant properties of leek. *Int J Food Microbiol* **165**, 121–133.

Chapter 15

Lactic Acid Bacteria and Malolactic Fermentation in Wine

Aline Lonvaud-Funel*

University of Bordeaux, ISVV, France

15.1. Introduction

The transformation of grape juice into wine involves a combination of numerous biological and chemical reactions performed by a variety of microorganisms in interaction with their environment. An extremely complex microbiota, combining yeasts and bacteria, colonizes grape berries and is simplified over either sudden or gradual stages. After a phase of spontaneous selection due to the action of yeasts, the lactic acid bacteria (LAB) multiply and play their part in winemaking; they are responsible for the malolactic fermentation (MLF). The role of LAB has long been restricted to the deacidification resulting from the degradation of the malic acid, but the bacteria are also feared because they can cause damage. There are indeed "good" and "bad" bacteria and under certain conditions "good" bacteria can also spoil wine.

From all the species of lactobacilli and cocci identified in wine, *Oenococcus oeni* (known as *Leuconostoc oenos* until 1995, Dicks et al. 1995) dominates the system. Initially physiological and metabolic studies revealed the originality of *O.oeni* while recent genomic studies allow us to partly interpret it. Some unusual features of *O. oeni*, including the tolerance to acidity, ethanol, and other natural inhibitors of wine and some metabolic pathways, are also common to other species of the oenological niche.

Since the 1950s, studies on LAB have provided significant answers to requests from wine producer who have recognized the interest of MLF. Simple technological operations have been accurately described and implemented providing basic principles which are still very important to follow. Today, research in this area has become more sophisticated with regard to industrial demands concerning malolactic starter cultures and more strictly controlled winemaking.

15.2. The Lactic Acid Bacteria of Wine

15.2.1. Origin

The surface of the grape berry hosts many species of filamentous fungi, yeasts, and bacteria. At maturity, the microbiota is complex, so that after crushing, the grape must is naturally contaminated with a diverse microbiota. But this microbial system simplifies drastically when the grape must

* Corresponding author: University of Bordeaux, ISVV, 220 Chemin de Leysotte, 33882 Villenave d'Ornon, France. aline.lonvaud@u-bordeaux.fr

Biotechnology of Lactic Acid Bacteria: Novel Applications, Second Edition. Edited by Fernanda Mozzi, Raúl R. Raya, and Graciela M. Vignolo.

enters the tank due to sudden environmental changes. Only the microorganisms that can adapt to the new conditions remain, which are yeasts and bacteria belonging to various genera and species. The yeast responsible for most of the alcoholic fermentation (AF) belongs to the species *Saccharomyces cerevisiae*. The initial bacterial population of grapes consisting of Gram (+) and Gram (−), aerobic and microaerophilic bacteria is rapidly limited to lactic acid and acetic acid bacteria. But the latter cannot multiply as the environmental redox potential is too low. Overall, the LAB are well adapted to the conditions of winemaking.

15.2.2. Species

So far, all LAB isolated from wine belong to the four main genera *Lactobacillus*, *Pediococcus*, *Leuconostoc*, and *Oenococcus*. Except for the pediococci, they are all strict or facultative heterofermentative bacteria (Table 15.1). Molecular methods applied to their identification resulted in the extension of the list of lactobacilli species (i.e., *Lactobacillus diolivorans*, *Lact. nageli*), although some are very closely related. In addition, interspecies gene transfers make strains of different species share a closer physiology and metabolism than strains of the same species. Although LAB species are not equally present in wine, *O. oeni* dominates worldwide during MLF. While lactobacilli and pediococci are more abundant in less acidic wines, *Leuconostoc mesenteroides* is less common (Lonvaud-Funel 2006).

15.2.3. Identification

In spite of the heaviness of the methods and the ambiguity of results, the identification of LAB was decisive in the first studies of MLF in the 1950s. Phenotype methods following Bergey's manual for classification were used. From the early 1990s molecular methods have been gradually applied in oenology, leading to faster, more accurate and reliable results. However, the dominance of heterofermentative species has been confirmed. More specifically, there is a growing list of species, especially for lactobacilli, and the creation of the new genus *Oenococcus* (Dicks et al. 1995). At the beginning, DNA/DNA hybridization was the first molecular tool. It was very simple to find probes for *O. oeni* since their genomic DNA was specific enough. Colony hybridization allowed the identification and enumeration of species in a mixture (Lonvaud-Funel et al. 1991). The implementation of Polymerase Chain Reaction (PCR) alone or together with sequencing or restriction polymorphism led to spectacular advances. Today, the fastest and most reliable way of identification is 16S rDNA sequencing. Several protocols and methods, including multiplex PCR targeting several species, have been described (Rodas et al. 2003; Claisse et al. 2007; Bokulich and Mills 2012; Petri et al. 2013). Thus, more comprehensive inventories are achievable from samples of must or wine, or even from surface samples (Bokulich et al. 2013). Based on their variability, the *rpo*B gene and the V3 region of the 16S rDNA are targets well adapted for the identification of species by PCR-DGGE (Denaturing Gradient Gel Electrophoresis) without the need for previous bacterial isolation (Endo and Okada 2005; Renouf et al. 2006). In addition, quantitative PCR makes direct species enumeration possible (Cho et al. 2011).

Table 15.1. List of lactic acid bacteria species in the wine environment.

Morphology	Carbohydrate fermentation	Species
Bacilli	Heterofermentative facultative	*Lactobacillus plantarum, Lact. casei*
	Heterofermentative strict	*Lactobacillus hilgardii, Lact. brevis, Lact. fructivorans, Lact. diolivorans, Lact. buchneri, Lact. fermentum, Lact. jensenii, Lact. kunkei, Lact. nagelii, Lact. vermiforme*
Cocci	Homofermentative	*Pediococcus parvulus, P. damnosus, P. pentosaceus*
	Heterofermentative	*Leuconostoc mesenteroides*
		Oenococcus oeni

15.2.4. Typing at Strain Level

Besides ecological studies, the identification of *O. oeni* at strain level is important for the industry during the selection and the formulation of malolactic starters, and to verify the survival of the starter during the MLF. To date, typing methods applied to *O. oeni* species include RAPD-PCR (Random Amplified Polymorphic DNA) and AFLP (Amplified Fragment Length Polymorphism) techniques, (Zavaleta et al. 1997; Cappello et al. 2008); these methods are rapid and suitable but sometimes poorly reproducible. The restriction of the whole genome by using rare-cutting enzymes, like *Not*I or *Apa*I, generates strain specific patterns after the separation of digested fragments by pulsed gel electrophoresis (REA-PFGE). This method enabled us to distinguish *O. oeni* strains for the first time and is still frequently used (Daniel et al. 1993; Zapparoli et al. 2000; Larisika et al. 2008; Lopez et al. 2008). More recently, multi-locus sequence typing (MLST) has been successfully applied. By means of this technique strains from two *O. oeni* collections were identified using five (de Las Rivas et al. 2004) and eight allelic sequences (Bilhere et al. 2009). MLST was shown to be more discriminating and reliable than REA-PFGE, hitherto regarded as the reference method. In this approach, concatenated DNA sequences of seven or eight genes provide the signature of a particular strain.

Another different method is based on the presence of repeated sequences in the genome. Five loci that differ by the number of tandem repeats (variable number tandem repeat: VNTR) were selected and used to genotype a collection of 236 *O. oeni* strains. The VNTR typing proves to be more discriminatory than REA-PFGE and MLST (Claisse and Lonvaud-Funel 2012). It is a fast and reliable method since each strain has its unique VNTR type. The multiplexing and fluorescent labeling of the primers has improved the method, which in turn has become faster, easier, more accurate, and less expensive (Claisse and Lonvaud-Funel 2014). In addition, phylogenetic studies of collections are allowed.

15.2.5. Detection of Specific Strains

In any species, some strains may induce wine spoilage. They can easily be detected by PCR using primers based on the sequence of genes responsible for the undesirable metabolism. This is the case of acrolein-, glucan-, and biogenic amine-producing strains (Claisse and Lonvaud-Funel 2001; Gindreau et al. 2001; Nannelli et al. 2008).

15.3. The *Oenococcus Oeni* Species

As mentioned above *O. oeni* was identified as the dominant species responsible for MLF in wine. The analysis of its 16S rDNA sequence allowed its assignment as the new genus *Oenococcus* (Dicks et al. 1995) with only one species, *O. oeni*. Later, isolates from composted distilled shochu residues in Japan were included in this genus and assigned as a new species named *O. kitaharae*, distant enough from *O. oeni* (Endo and Okada 2006). Comparison of the complete genomes of both species suggests that the genome of *O. oeni* and of *O.kitaharae* evolved with different strategies: the first tolerating the harsh conditions of wine and the second one facing niche microbial competition (Borneman et al. 2012b).

Mills and colleagues published the first complete genome sequence of the *O. oeni* strain PSU-I (Mills et al. 2005). Comparison of the genomes of 14 LAB species showed that the *O. oeni* genome evolved significantly faster than the other LAB genomes (Makarova and Koonin 2007). The hypothesis for this high level of mutability is the lack of the MutS/MutL repair system, which is also the case for *O. kitaharae* (Marcobal et al. 2008). Using several genomic regions chosen after *in silico* subtractive hybridization between the genomes of two *O. oeni* strains of opposite oenological potential, a high diversity of strains was revealed (Delaherche et al. 2006). However, the species could be considered genetically homogenous as regards the sequence of the ribosomal regions. The genetic mapping of nine strains (Ze-Ze et al. 2008) and the MLST analysis of hundred-strain collections allowed definition of sub-groups A and B (Bilhere et al. 2009, Bridier et al. 2010). The allelic diversity showed that they had evolved separately without any exchange between them. Strains belonging to group A have more IS30 elements, recently acquired in the species, than group B (El Gharniti et al. 2012). It is thought that horizontal gene transfer caused the transposition and increased the copy numbers of IS30 in group A, helping the strain to cope with the wine environment. Within the species, the high allelic polymorphism is related to this adaptation. In the collection examined by Bridier et al. (2010) some phylogenetic groups appeared to be related to the geographical origin of the strains. Other strains were better clustered according to the type of products from which they were isolated. For example, cider strains were found exclusively in group B with specific sequence types. Taken together the results suggest that group A-strains could be better adapted to the oenological niche. By DNA subtractive hybridization between a high and a low oenological-performing strain in wine, six regions of genomic plasticity were isolated. None of them proved to be specifically present in high performing strains. However, there was a significant relationship between the adaptation and the expression, in stress conditions, of genes located in the three regions (Bon et al. 2009).

The comparison of 14 *O. oeni* genomes defined a 2846 ORF pangenome with 1165 ORF representing the core genome (Borneman et al. 2012a). The genomes differ in the three clusters of exopolysaccharide ORFs that may imply distinct cell wall composition polysaccharides and production of possibly different exopolysaccharides in wine. They also vary with respect to sugar utilization, especially pentoses. The genetic bases for the distinct capacity to ferment arabinose and xylulose are established. However, there is no information on the origin of the difference for xylose metabolism. Interestingly, in the past the fermentation of pentoses was a criterion for the distinction of four types of strains within the *O. oeni* (*Leuconostoc oenos*) species according to their capacity to ferment both, one of the two, or none of these pentoses, this having a possible oenological implication (Peynaud and Domercq 1968).

O. oeni plasmids, which are often related to stress tolerance, have been the subject of only a few studies. They could have an impact on the variability of strains in their adaptation to wine; however, small and medium plasmids (4–40kb) gave no indication of their possible role (Fremaux et al. 1993; Brito and Paveia 1999; Mesas et al. 2001). Another plasmid recently studied is not correlated with better growth under stress conditions. However, it carries genes that are frequently present in starters. Also, during the fermentation process indigenous strains that carry this plasmid are more frequently selected, indicating that it may contribute to the fitness of strains in the harsh wine conditions (Favier et al. 2012).

15.4. Evolution of Lactic Acid Bacteria during Winemaking

The grape must is contaminated by yeasts and bacteria where LAB are only a part of the system. The initial microbial concentration is very variable in number, about 10^2 to 10^4 CFU/ml, and depends on the climatic conditions during harvest, the pH of the must, and the sulfiting of the grapes. Several species and numerous strains within species are represented, making it difficult to inventory the microbial diversity of the niche. Today, this biodiversity is better estimated with the use of culture-independent methods. Thus, it is well established that lactobacilli are rather dominant before the AF begins (Renouf et al. 2006), then they gradually disappear depending on the wine. In the majority of the cases, the *O. oeni* is a minor species in the grape must and eventually dominates the bacterial system at the end of the AF (Figure 15.1); this selection is also observed within the species. For *O. oeni*, genetic markers, which result from the subtractive hybridization of two opposite strains of enological interest, are in greater numbers in the more stress-tolerant strains isolated at the end of vinification (Renouf et al. 2008).

During the AF, the LAB population changes drastically, depending on numerous factors (Lonvaud-Funel 2006). At the end of AF, the LAB cell count is about 10^2 CFU/ml, and then it reaches approximately 10^6 CFU/ml when the MLF is clearly visible. A few days or even several months can separate the end of the AF and the start of MLF. During this time, the wine must be monitored because yeast

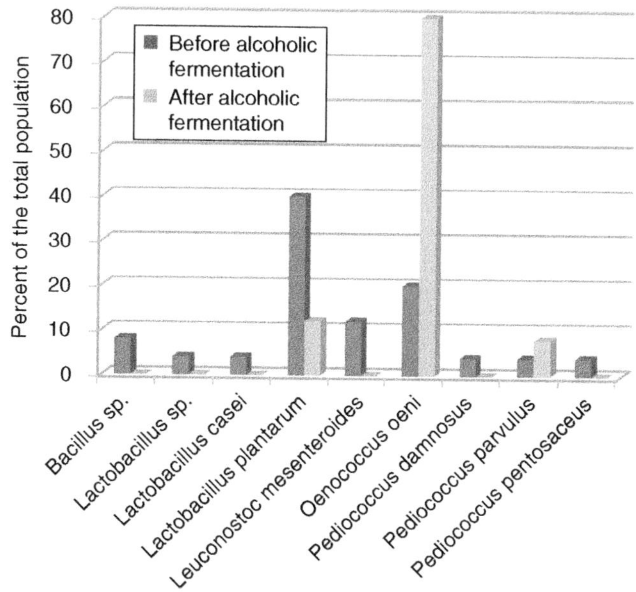

Figure 15.1. Example of the LAB species found before and after alcoholic fermentation as identified by rpoB/PCR-DGGE.

spoilage may also occur. During MLF, the population continues to increase or stays around 10^6 CFU/ml and then, when malic acid is no longer available, the microbial cell count is reduced by sulfite and racking. Red wines are often aged for several months in vats or barrels before being packaged. Throughout aging, low populations of LAB remain viable. However, it is not uncommon to observe the survival, and sometimes the multiplication, of strains that are particularly tolerant to the environmental conditions, which may cause wine alteration.

15.4.1. Interactions between Wine Microorganisms

As a result of yeast/LAB interactions, bacteria decrease during AF and multiply afterwards (Lonvaud-Funel 2006). At the beginning, yeasts prevail in the system and ferment hexoses to ethanol, releasing a wide range of metabolites that make up the finished wine. As a consequence of yeast growth, the medium is depleted of growth factors and nutrients that, together with the presence of toxic yeast metabolites such as ethanol and certain fatty acids, compromise bacterial growth during the AF process (Lonvaud-Funel et al. 1988). Only the most resistant cells survive. At the end of AF, the situation is reversed: yeasts die, autolyse and release vitamins, proteins, mannoproteins, peptides, amino acids, and other growth factors into the environment providing thus all the essential or beneficial amino acids and oligopeptides for bacterial growth (Feuillat 1977). Interestingly, peptides present in wine do not have a direct nutritional role, their transport into the cell being necessary, followed by the excretion of amino-acids, to provide the energy required by the cell to maintain its membrane pH gradient (Ritt et al. 2008). To sum up, those strains which survive the yeast-induced toxicity are well adapted to grow in wine, after an initial period of growth inhibition.

However, in the same grape must, interactions also depend on the strains. LAB growth can be much easier in the wines which are made when certain yeast strains are dominant (Fornachon 1968). Yeasts differences can be attributed to many factors such as the production of varying levels of SO_2 and fatty acids as well as specific proteinaceous molecules and mannoproteins. The last two compounds should act as inhibitors (Comitini et al. 2005) or activators, respectively (Diez et al. 2010).

In addition yeast cell walls, obtained by separating the soluble yeast extract from the insoluble cell wall after autolysis, detoxify the medium by absorbing the toxic fatty acids. They are added to wine for the promotion of MLF (Lafon-Lafourcade et al. 1983).

LAB also interact among themselves. Some *Pediococcus* and *Lactobacillus* strains were shown to be inhibitors of other lactobacilli and more often of *Leuc. mesenteroides* and *O. oeni*; also *Lact. plantarum* seems to be capable of inhibiting *O. oeni*. Bacteriocin production as well as the presence of sequences of the bacteriocin *pln* locus were established in inhibitory strains (Navarro et al. 2000; Knoll et al. 2008). In contrast, mutualism was demonstrated between a strain of *Pediococcus pentosaceus* and one of *O. oeni*. The high proteolytic activity of the *O. oeni* released essential amino acids for the *P. pentosaceus* strain (Fernandez and de Nadra 2006). All these phenomena are very complex involving a huge variety of strains that represent each of the species simultaneously present in the wine.

Finally, the last of the biological order interactions that might be active should be those of phages. *O. oeni* phages were studied when seeking reasons for MLF failures. Actually, some strains host a prophage that can induce the lytic cycle and destroy the population (Santos et al. 1998). But the situation is completely different in the laboratory and in a cellar. The population of phages in wine during MLF increases along with the bacteria, but remains low. The variety of bacterial strains and phages and the diversity of the resistance to phage attack mean that, in the end, they do not disrupt the process (Poblet-Icart et al. 1998). In winemaking, the phage attack would be problematic if the wine only contained one sensitive strain of *O. oeni*. In contrast, in the malolactic starter industry, phages can cause problems in production. But the factors for phage release are not known and it must be pointed out that starters have been manufactured with lysogenic strains for a long time.

15.4.2. Environmental Factors

Technologically speaking, the main factors that affect the growth of LAB in wine are the pH, the alcoholic strength of the wine, the temperature, and the concentration of SO_2. These factors have been known for a long time and have laid the foundations of oenological rules that every winemaker must observe. For wines displaying a pH range between 3.0 and 3.8, the higher the pH, the easier the growth. Regarding alcohol, from about 8%, the higher the alcoholic strength, the more difficult the growth, being almost impossible above 15%. However, strains belonging to different species such as *Lactobacillus fructivorans*, *O.oeni*, *Lact. plantarum*, and *Lact. hilgardii* have in fact been isolated from fortified wines at about 20% alcohol. Alcohol and pH act simultaneously. The ethanol tolerance is increased if the pH is higher, as is the case of wines from very ripe grape harvests which contain a lot of sugar and are low in acidity. Concerning temperature, although LAB are grown at 27–30 °C in the laboratory, the optimum temperature of wine is around 20 °C. Winemakers ensure that the temperature is maintained at least at 18 °C before fermentation is started. Cooling a few degrees slows fermentation down without stopping it.

Sulfur dioxide is used for its antioxidant and antiseptic properties, although today some winemakers no longer use it. When added to the grape must, it temporarily reduces some of the LAB population leaving the environment to the yeast. During the AF SO_2 is totally combined, mainly to acetaldehyde and oxo acids produced by yeast, becoming inactive. Conversely, some yeast strains produce significant amounts of SO_2 that may remain and contribute to the difficulties of growing bacteria. The combined SO_2 has a slight but real effect on bacteria, this is the reason why winemakers should limit the sulfiting of the must. After MLF the wine is sulfited so that free SO_2 remains active to eliminate all bacteria. However, their sensitivity is variable; generally *Lact. hilgardii* and *P. damnosus* are more resistant as this property is strain-dependent. Alcohol, pH, temperature, and SO_2 act in synergy or in opposition. The limiting effect of one of the factors is diminished if the others are favorable (Lafon-Lafourcade et al. 1983).

It is not uncommon that bacteria do not develop well even when all the mentioned parameters are optimal since the composition of wine is very complex. From a nutritional point of view, essential elements such as carbon, nitrogen, vitamins, and metal sources are appropriately represented in the wine. Pentoses and hexoses present after AF, in concentrations of about several hundred mg/l, are sufficient to produce the necessary biomass to carry out the MLF. In addition, the bacteria glycosidases and peptidases hydrolyze numerous glycosylated and peptide molecules into sugars and amino acids, which they then use. By contrast, grape molecules, such as some phenolic compounds, may be toxic but they may have positive effects as has also been demonstrated (Theobald et al. 2008; Campos et al. 2009).

A small number of genes have been studied but certainly not enough to explain the adaptation of bacteria, particularly of *O. oeni*, to wine. For example, some of these genes are those coding for OmrA: homologous to the ABC transporter; FtsH: a protein of the AAA family which protects against wine toxicity; and CtsR: a possible regulator of the expression of chaperone genes (Bourdineaud et al. 2003, 2004; Grandvalet et al. 2005). Many studies have focused on the genes involved in the stabilization and optimization of the membrane composition for the fluidity adjustment in stress conditions (Silveira et al. 2004; Grandvalet et al. 2008; Cafaro et al. 2014). The malolactic fermentability of wine is obviously the result of many parameters. During evolution, LAB have acquired physiological and metabolic functions to adjust to this complex niche (Makarova and Koonin 2007; Bon et al. 2009; Borneman et al. 2012b).

15.5. Lactic Acid Bacteria Metabolism and its Impact on Wine Quality (Table 15.2)

15.5.1. Sugars

Depending on the species to which they belong, wine LAB ferment hexoses and pentoses by following homo or heterofermentative pathways. However, the majority of them are obligatory or facultative heterofermentative. The diversity of sugars metabolized varies according to species. Heterofermentative bacteria always release acetic acid besides lactic acid from sugars; the amount produced being dependent on the type and concentration of the metabolized sugar. *O. oeni* ferments glucose by the pentose phosphate pathway; during growth less than 2 g/l glucose is required to increase its population from 10^2–10^3 CFU/ml to 10^8 CFU/ml. This growth is strongly stimulated in presence of a glucose : fructose mixture. When reduced to mannitol, fructose is an additional source for the re-oxidation of NAD(P)H produced in the pentose pathway, leading to an increased production of ATP and acetic acid (Richter et al. 2003). In winemaking conditions, after AF, LAB usually ferment only small amounts of sugar so that acetic acid production is low. If the AF is not correctly controlled it may slow down or stop, then heterofermentative LAB start their multiplication fermenting sugars in large amounts with the concomitant production of acetic acid. This fact results in an increase in the wine volatile acidity, an alteration known as "piqûre lactique," so controlling AF is essential to avoid this problem feared by winemakers.

15.5.2. Carboxylic Acids

Before MLF occurs, diverse organic acids from various origins are already present in the wine. Tartaric, malic, and citric acids come originally from grapes while gluconic acid, less abundant, originates from grapes affected by acetic bacteria. Other acids such as succinic and pyruvic acids are the result of yeasts' metabolism, although the concentration of pyruvic acid is very low.

L-malic acid is decarboxylated to *L*-lactic acid by LAB, a reaction known as MLF. Wine pH increases by several tenths of units depending on the initial malic acid concentration, which may range from 1 to 8 g/l. Although the MLF reaction has been known for a long time (Seifert 1901) only since the 1970s has the malolactic enzyme been purified and characterized, first from *Lact. plantarum*

Table 15.2. Main metabolisms known in wine

Substrates	Pathways/first enzyme	Products	Conditions	Impact on the quality
Carbohydrates				
Glucose	Heterofermentation	Acetic acid	Usual conditions of winemaking	Low
Fructose			Stuck alcoholic fermentation	High volatile acidity (piqure lactique)
Pentoses				
Glucose	Glucosyl transferase	Glucane	Some strains (Mainly *P.damnosus*)	Ropiness
Carboxylic acids				
Malic	Malolactic enzyme	L-Lactic acid	Almost all LAB strains	Increase of softness
Citric	Citrate lyase	Diacetyle Acetic acid	All *O.oeni* strains, during and after MLF	Role in the complexity of the flavor, no input on volatile acidity
Tartaric	Tartrate dehydratase	Acetic, lactic, succinic acids	Some strains of *L.brevis, L.plantarum*	Decrease of total acidity, increase of volatile acidity. Alteration "tourne"
Hydroxycinnamic	Decarboxylase	Vinyl/Ethyl phenols	Lactobacilli, Pediococci	Off flavors
Amino acids				
Arginine	Arginine deiminase	Citrulline, ornithine	Heterofermentative lactobacilli, some *O.oeni* strains	None
Methionine	Transaminase	Sulfur compounds	Lactobacilli and *O.oeni*	Flavor
Histidine	Histidine decarboxylase	Histamine	Strains of any species that carry the required gene clusters	Undesirable biogenic amines
Tyrosine	Tyrosine decarboxylase	Tyramine		
Ornithine	Ornithine decarboxylase	Putrescine		
Agmatine	Agmatine decarboxylase	Putrescine		
Others				
Glycosidic compounds	Osidases	Terpenes, esters etc....	All strains	Flavor
Esters and precursors	Esterases			
Glycerol	Glycerol dehydratase	Acrolein	Some strains carrying the gtf gene	Bitterness

and then from other species of different origin (Schutz and Radler 1973; Lonvaud-Funel and Strasser de Saad 1982). Unlike malic enzymes, which produce pyruvate from malic acid, the malolactic enzyme leads directly to *L*-lactate. Interestingly, the *mle*A gene sequence which encodes for this protein is very close to that coding for the malic enzyme from the yeast *S. cerevisiae* and *Schizosaccharomyces pombe* that degrades *L*-malate to pyruvate in wine (Lonvaud-Funel 2006). The malolactic operon of the *O. oeni* genome includes the *mle*A gene and that of a carrier *mle*P (Labarre et al. 1996). However, the usefulness of the carrier is probably not necessary because at the low wine pH malic acid is slightly ionized. The malolactic enzyme can only function in the intracellular environment due to its constraints for optimal pH and the need for co-enzymes. Since viable and intact bacteria are required for optimal malolatic enzyme activity, attempts to produce enzyme preparations for industrial application have all failed. Long regarded as a way for bacteria to improve their fitness by deacidification of the environment, the decarboxylation of malic acid is also a source of energy through the transmembrane exchanges of substrate and product, which generate a potential difference and a pH gradient (Poolman et al. 1991).

In wine, citric acid is present in much lower concentrations (≤0.3 g/l) than malic acid. Heterofermentative cocci and some lactobacilli degrade it during MLF but more slowly than malic acid. The metabolism begins with the activity of citrate lyase producing acetic acid and oxaloacetic acid, which in turn is decarboxylated to pyruvate. In addition to the carbohydrate fermentation, this pyruvate is directed towards the production of acetoinic compounds, acetic acid, and fatty acids for lipid synthesis. Among acetoinic compounds, diacetyl is the most important flavor compound due to its "buttery" taste; the sensitivity threshold being higher for red (about 7 mg/l) than for white wines (about 5 mg/l). When present in excess diacetyl is not always favorably accepted by consumers. However, this is rarely the case as its synthesis and its reduction to acetoin and butanediol, which have little or no odor, are simultaneous. Moreover, the amount of free diacetyl depends on the concentration of SO_2, which combines with carbonyl compounds reducing its concentration (Nielsen and Richelieu 1999).

The citrate lyase is a complex of several proteins whose six genes are organized in a cluster, preceded by a gene of organic acid transport, a putative gene for oxalate decarboxylase, and a regulator (Mills et al. 2005). The addition of alcohol in an acidic medium alters the transcription level of some of these genes in *O. oeni* suggesting that citrate metabolism plays a role in the stress response (Olguin et al. 2009). Citric acid is also a source of energy for *O. oeni*. In addition to the ATP that can be generated from acetyl phosphate via pyruvate, the uniport of the citrate uptake generates a pH gradient and a proton motive force as described in *Leuc. mesenteroides* (Marty-Teysset et al. 1996). It is important that citric and malic acids are no longer available to sustain bacterial growth at the end of winemaking.

The presence of tartaric acid in wine represents a serious deterioration (named "tourne") affecting the fixed acidity of wine. Tartaric acid is generally not utilized during MLF as few bacteria are capable of metabolizing it. The only study on this subject reported that four *Lact. plantarum* and one *Lact. brevis* out of 78 LAB strains were able to degrade tartaric acid (Radler and Yannissis 1972). The pathway, which is species dependent, starts with the activity of tartrate dehydratase and results in the production of acetic acid and lactic acid for *Lact. plantarum*, and acetic acid and succinic acid for *Lact. brevis*.

Interactions between LAB and phenolic compounds are very complicated. Among them, the decarboxylation of hydroxycinnamic acids (caffeic, p-coumaric, and ferulic) is of particular interest in oenology (Silva et al. 2011). These acids are esterified to tartaric acid and freed by the cinnamoyl esterase by yeasts and less often by bacteria. The decarboxylation releases vinyl phenols, which are reduced to ethyl phenols producing off-flavors. The second reaction can be achieved by bacteria to maintain the redox balance; however, it is mainly carried out by the spoilage yeast *Brettanomyces bruxellensis*, often present in wine.

15.5.3. Amino Acids

Depending on the species and on the strains, the amino acids required by wine LAB vary widely. While their concentration can be limiting during the active phase of the AF (arginine can disappear completely), at the end of the fermentation, and due to yeast autolysis, a wide variety and large amounts of amino acids are found. Amino acids, which are used for protein synthesis by the growing biomass, may also be metabolized during MLF, and even afterwards during wine aging. So far, studies have focused on arginine, histidine, ornithine, and tyrosine, whose conversion products to biogenic amines are considered undesirable. Moreover, methionine has been studied for the production of sulfur compounds, which impact the wine aroma.

Arginine Arginine is generally the most abundant amino acid in wine. It is metabolized by all heterofermentative lactobacilli by means of arginine deiminase (ADI), particularly by *Lact. hilgardii* and *Lact. brevis*. In *O. oeni*, arginine is not degraded by all strains as only some of them carry the Arc cluster of six genes comprising a regulator, the three genes encoding the three enzymes of the pathway (ADI, ornithine transcarbamylase, carbamate kinase), and two genes highly homologous to those encoding the antiporter arginine/ornithine in other Gram-(+) bacteria (Tonon et al. 2001; Divol et al. 2003). For lactobacilli and *O. oeni* the assimilation of arginine acts as an energy source that is important, both after AF since sugars are in low concentrations, and after MLF when malic and citric acids (also energy sources) are depleted. During wine aging, populations able to metabolize arginine may be favored (Tonon and Lonvaud-Funel 2000). In this pathway, ornithine, produced in equimolar amounts from arginine, leads to citrulline that, if released in wine, may be a precursor of ethyl carbamate. However, the ethyl carbamate synthesized is negligible as compared to the quantities generated from urea released by yeasts.

Utilization of ornithine, agmatine, tyrosine, histidine Biogenic amines are produced from amino acids. Although there are no official regulations on the allowed biogenic amine concentrations in wine, efforts to limit them are made by winemakers. Their level in wine varies depending on the wineries, the vintage, fermentation, and aging. Biogenic amine formation depends on the initial precursor concentration, which may be increased by contact with lees. In some cases, a decrease in their concentration occurs due to the metabolism of lactobacilli or pediococci (Garcia-Ruiz et al. 2011). *Lact. hilgardii*, *Lact. brevis*, and *O. oeni* strains can be isolated from biogenic amine-containing wines, which often harbor histamine, putrescine, tyramine, cadaverine, and phenylethylamine. Amine biogenesis is a common property among lactobacilli and is strain dependent.

In oenology, histamine was the first amine studied. A strain of *O. oeni* isolated from a histamine-rich wine, retained its ability to produce histamine in the poorest nutritional culture conditions (Lonvaud-Funel and Joyeux 1994). In some wine-producing areas the histamine-producing bacteria are very frequent (Lucas et al. 2008).The histidine decarboxylase (HDC) was characterized, purified, and the *hdc*A gene sequenced (Coton et al. 1998b). A PCR test showed a strict relationship between the presence of HDC+ bacteria and histamine in wine (Coton et al. 1998a). However the *hdc* cluster, comprising four genes including the gene coding for an histidine/histamine exchanger, was shown to be located on an unstable plasmid (Lucas et al. 2005); in *O.oeni* this character seems to be easily lost after strain isolation and culture in laboratory conditions. This fact should explain why, to date, no malolactic starter or candidate has ever shown this activity.

The tyrosine decarboxylase (TDC) has been characterized and purified and its gene sequenced in wine LAB, in particular *Lact. brevis* and *Lact. hilgardii* (Moreno-Arribas et al. 2000; Lucas and Lonvaud-Funel 2002). As for the HDC activity, a *tdc* operon comprises four genes including the *tdc*A coding for the enzyme and *tdc*B coding for a transporter that catalyzes the tyrosine/tyramine exchange. As a result of the TDC reaction, the decarboxylation and the transport provide a metabolic energy source (Wolken et al. 2006).

Putrescine is the most abundant amine in wine and is produced through two pathways: ornithine decarboxylation and agmatine deamination. The ODC system in *O. oeni* and *Lact. brevis* functions with an ornithine/putrescine transmembrane exchanger and should be involved in the metabolic energy generation and intracellular pH regulation (Bokulich et al. 2013). The agmatine deamination is very similar to the ADI pathway, which includes the agmatine/putrescine exchanger, agmatine deiminase, putrescine transcarbamylase, and carbamate kinase (Lucas et al. 2007). These bacteria, evolving in an acidic medium, have in their genome a block of several gene-clusters that allows them to recover metabolic energy and regulate their intracellular pH. Genomic studies on the biogenic amine-producing bacteria have allowed design of PCR primers for a rapid and quantitative determination of undesirable populations in wine (Nannelli et al. 2008).

15.5.4. Other Metabolisms with Sensorial Impact

Methionine metabolism Sulfur compounds, responsible for aroma in wine and typically related to the grape variety, are released by yeast during the AF. In addition, the metabolism of the sulfur-containing amino acid methionine has an impact on wine aroma. *Lact. brevis*, *Lact. plantarum*, and *O. oeni* strains, using a pathway similar to dairy LAB, catabolize methionine producing "light" volatile sulfur molecules such as methanethiol and dimethyldisulfide, and "heavy" volatile molecules such as 3-(methylsulphanyl) propan-1-ol and 3-(methylsulphanyl) propionic acid (Pripis-Nicolau et al. 2004; Weimer et al. 1999; Vallet et al. 2008). In wine, *O. oeni* strains produce more heavy compounds, mainly 3-(methylsulphanyl) propionic acid, than lactobacilli. In water 3-(methylsulphanyl) propionic acid descriptors are chocolate and roasted; but these notes are not found in wine where they are replaced by red fruit and earthy odors probably because of interactions with other wine components (Pripis-Nicolau et al. 2004).

Hydrolysis of glycosidic compounds Wine contains numerous glycosylated components of its color, structure, and aroma. Microorganisms such as yeast and bacteria can hydrolyze them and use the sugary moiety for microbial growth and release the aglycone. *O. oeni* glycosidases have been studied using a variety of *p*-nitrophenyl glycosides as substrates; several glycosidases are active at wine pH and their activity is strain-dependent (Grimaldi et al. 2005). Native wine glycosides extracted from Muscat wine are precursors of aromatic terpenes (Ugliano et al. 2003). Various glycosidases can metabolize oak wood glycosides and are responsible for the aromatic input when MLF is conducted in wood barrels instead of stainless steel tanks; the release of whiskylactone, eugenols, syringaldehyde, vanillin, acetophenone, and acetosyringone in variable amounts has been reported (De Revel et al. 2005; Gagne et al. 2011). However, aglycones may combine with polysaccharides, slightly changing the free aroma level during MLF (Boido et al. 2002).

Synthesis and hydrolysis of esters Sensorial and chemical analyses of more than 60 volatile compounds showed that during MLF of red wines the main character affected was the fruity note, which is related to fatty acid ethyl esters (Antalick et al. 2012). LAB are able to modify these compounds as they may degrade or synthesize them from fatty acids (Costello et al. 2013), with the concomitant reduction or intensification of the fruity note (Sumby et al. 2013). Actually, the phenomena underlying the variability of the impact of MLF on the fruity note of red wines are more complex than a simple strain effect. These variations reflect a bacteria–yeast–wine triptych in which each component is important. It is illusory to presume that a given aromatic impact can be ensured by the use of a particular bacterial strain (Antalick et al. 2012).

The "bitterness" disease The excessive bitterness of wine, described at the very beginning of wine microbiology, results from the glycerol metabolism, an abundant component of wine (7 to 10 g/l). In winemaking, degradation of glycerol is detrimental due its bad impact on taste. Several LAB species from all genera encountered in wine can degrade glycerol by two pathways; however, this character is restricted to some rare strains or environmental conditions. One pathway, catalyzed by the glycerol dehydrogenase, joins the glycolysis-producing lactic acid, thus increasing the total acidity of wine. It occurs in aerobiosis or microaerophilic conditions when fructose that influences the NAD/NADH balance is available, which can be the case in wine (Pasteris and Strasser de Saad 2009). The other pathway, catalyzed by the glycerol dehydratase (GDH), leads to 1,3 propanediol after reduction of the intermediate 3 hydroxypropionaldehyde (3HPA). While the former compound has no sensorial impact, a small amount of 3HPA (in a range of mg/l) can escape the reduction and be released into the wine inducing bitterness, as this intermediate or/and its derivative acrolein produce bitter complexes when combined with polyphenols. Today the bitterness problem is mainly encountered in wines and ciders that are further distilled into brandies. The detection of positive *gdh* strains is now possible by PCR reactions (Claisse and Lonvaud-Funel 2001). However, in addition to the presence of particular strains harboring this gene, specific environmental conditions are required for the alteration (Bauer et al. 2010).

The "ropiness" disease From small amounts of residual sugars, some LAB strains can produce a special β-glucan that increases the viscosity making the wine unpalatable, even at low concentrations of a few hundred mg/l. The alteration can occur after winemaking, during aging, and most of the time it happens in bottled wine after a long storage time. Even the very low population of specific strains that contaminate the wine at bottling can multiply slowly to reach concentrations of 10^5–10^6 CFU/ml, which is enough to produce ropiness. The strains responsible for this disease mainly belong to the *Pediococcus damnosus* or *P. parvulus* species, which are very well adapted to wine and are difficult to eliminate, even by sulfiting. The spoilage was early correlated with the presence of glucosyl transferase gene carried on a plasmid by *P. damnosus* and more recently by *O. oeni* (Walling et al. 2005; Dols-Lafargue et al. 2008). The detection of these strains by PCR test must be mandatory before bottling if previous problems have been encountered in the winery (Gindreau et al. 2001).

15.6. Controlling the Malolactic Fermentation

MLF is normally spontaneous when LAB selected during AF multiply. It normally follows AF after a delay of days or weeks. Among the parameters influencing bacterial multiplication, temperature is the easiest to control. Winemakers also use several other practices such as chemical deacidification to promote growth, or the addition of bacteria using pressed wine or lees from a tank where MLF has finished. Another alternative is the addition of malolactic starters, which at high concentrations of active bacteria quickly degrades malic acid, even if there are non-growing but surviving cells still active. The objective is to reduce the total time of winemaking by shortening, or even avoiding, the microbial growth phase after the AF. Starters can be indicated where there is a risk of biogenic amines; the added biomass dominates the system leaving little opportunity for native flora containing undesirable bacteria to grow. However, starters do not avoid the risk during storage because they do not eliminate indigenous strains.

In the 1980s, starter cultures were used from frozen but now they are mostly commercialized as lyophilized preparations. The first malolactic starters were successfully applied after a reactivation protocol was established, since cells reach a physiological state allowing them to cope with the stress of inoculation into wine (Lafon-Lafourcade et al. 1983). A proteomic approach for studying changes induced by reactivation showed the role of stress proteins and the importance, even at low concentrations, of sugars and amino acids in the reactivation medium (Cecconi et al. 2009). After the first generation of starters, industrial producers succeeded in the production of cultures for direct inoculation (Nielsen et al. 1996). However, survival, which greatly depends on the strains, still remains somewhat unpredictable. Research currently focuses on the identification of specific genetic markers best related to wine adaptation, although it is also necessary to optimize the various industrial steps. For a given strain, survival depends on growth conditions, the physiological state at the harvest, the protective agents in the suspension medium, the freezing conditions, and finally on the revival before use (Zhao and Zhang 2009). In addition, the time of inoculation is another important concern; the most logical time is after AF if the MLF process does not trigger naturally. In these conditions bacteria use only low concentrations of sugars for growth and no risk of acetic acid production occurs. Some winemakers prefer to add bacteria either at the beginning or at the end of the AF fermentation so bacteria can adapt gradually to the medium. However, if bacteria ferment larger quantities of sugars, in case AF slows down or stops, the risk of volatile acidity increases. This protocol is used by winemakers who have a perfect mastery on AF. Unavoidably, secondary product formation and therefore wine sensorial quality is different because substrates and conditions for their metabolism are not the same in a fermenting must as they are in a finished wine (Abrahamse and Bartowsky 2012; Knoll et al. 2012). Moreover, reacting in competition to the presence of bacteria, yeasts respond by modifying their physiology and metabolism as evidenced by the changes in gene expression (Rossouw et al. 2012). Selected strains may also be used as immobilized bacterial reactors (Genisheva et al. 2013).

For a long time it was not possible to control the predominance of starter cultures during MLF in the presence of the indigenous microbiota, which is now possible with new strain typing methods (REA-PFGE, VNTR). Moreover, a culture-independent method is now proposed, based on a quantitative PCR which targets a specific region of the strain genome (Solieri and Giudici 2010).

15.7. Conclusions

AF is the most efficient process for the selection of LAB from grape must. Selected strains, mainly belonging to the *O. oeni* species, are suited to grow in wine after yeasts have fermented almost all sugars and enriched the medium in amino acids, various peptides, and other growth factors. Although the acidity, ethanol, other toxic yeast metabolites, and some polyphenols are inhibitors for *O. oeni*, this species becomes predominant at the end of AF, after selection among the other LAB species less tolerant to inhibition. Phylogeny studies show that the *O.oeni* species comprises two different genomic groups that have evolved independently in the oenological niche. In winemaking terms, the selection at strain level and not only at species level is important. Indeed among *O. oeni*, only some strains are able to survive and to actively multiply in wine. Similarly, some spoilage lactobacilli and pediococci strains present at the beginning of AF are able to survive at all stages of winemaking, even better than *O. oeni*. Now it is possible to detect spoilage microorganisms by specific quantitative PCR tests. While the presence of spoilage strains is a risk factor, their detrimental activity is not necessarily expressed since several factors influencing wine alteration still remain unknown. Molecular study of metabolic pathways such as decarboxylation and hydrolysis of wine components have brought new interpretations and explanations about the physiology and growth of LAB in an environment often described as nutritionally poor. Genomic and transcriptomic approaches have begun to reveal some of the mechanisms of adaptation to oenological environmental conditions.

For millennia, the repeated use of the microbiota for the production of wine has led to the domestication of yeast strains and bacteria and their co-evolution in the oenological niche. However, it is to be feared that the increasingly systematic use of selected strains may lead to a loss of diversity and therefore a reduction of the gene pool available for evolution. At least for bacteria, the difficulty of strains that have undergone "industrialization" to settle permanently in the system is reassuring on this point. So, it is now imperative that the oenological microbial issues considering bacterial integration in a complex system are examined.

References

Abrahamse, C.E. and Bartowsky, E.J. (2012) Timing of malolactic fermentation inoculation in Shiraz grape must and wine: influence on chemical composition. *World J Microbiol Biotechnol* **28**, 255–265.

Antalick, G., Perello, M.C., and de Revel, G. (2012) Characterization of fruity aroma modifications in red wines during malolactic fermentation. *J Agric Food Chem* **60**, 12371–12383.

Bauer, R., du Toit, M., and Kossman, J. (2010) Influence of environmental parameters on production of the acrolein precursor 3-hydroxypropionaldehyde by *Lactobacillus reuteri* DSMZ 20016 and its accumulation by wine lactobacilli. *Int J Food Microbiol* **137**, 28–31.

Bergey's Manual of Systematic Babcteriology. 1986. John G. Holt Editor in Chief, Williamns and Wilkins, Baltimore, USA.

Bilhere, E., Lucas, P.M., Claisse, O., and Lonvaud-Funel, A. (2009) Multilocus sequence typing of *Oenococcus oeni*: detection of two subpopulations shaped by intergenic recombination. *Appl Environ Microbiol* **75**, 1291–1300.

Boido, E., Lloret, A., Medina, K., et al. (2002) Effect of beta-glycosidase activity of *Oenococcus oeni* on the glycosylated flavor precursors of Tannat wine during malolactic fermentation. *J Agric Food Chem* **50**, 2344–2349.

Bokulich, N.A. and Mills, D.A. (2012) Differentiation of mixed lactic acid bacteria communities in beverage fermentations using targeted terminal restriction fragment length polymorphism. *Food Microbiol* **31**, 126–132.

Bokulich, N.A., Ohta, M., Richardson, P.M., and Mills, D.A. (2013) Monitoring Seasonal Changes in Winery-Resident Microbiota. *PloS One* **8**, e66437.

Bon, E., Delaherche, A., Bilhere, E., et al. (2009) *Oenococcus oeni* genome plasticity is associated with fitness. *Appl Environ Microbiol* **75**, 2079–2090.

Borneman, A.R., McCarthy, J.M., Chambers, P.J., and Bartowsky, E.J. (2012a) Comparative analysis of the *Oenococcus oeni* pan genome reveals genetic diversity in industrially-relevant pathways. *BMC Genomics* **13**, 373.

Borneman, A.R., McCarthy, J.M., Chambers, P.J., and Bartowsky, E.J. (2012b) Functional divergence in the genus *Oenococcus* as predicted by genome sequencing of the newly-described species, *Oenococcus kitaharae*. *PloS One* **7**, e29626.

Bourdineaud, J.P., Nehme, B., Tesse, S., and Lonvaud-Funel, A. (2003) The ftsH gene of the wine bacterium *Oenococcus oeni* is involved in protection against environmental stress. *Appl Environ Microbiol* **69**, 2512–2520.

Bourdineaud, J.P., Nehme, B., Tesse, S., and Lonvaud-Funel, A. (2004) A bacterial gene homologous to ABC transporters protect *Oenococcus oeni* from ethanol and other stress factors in wine. *Int J Food Microbiol* **92**, 1–14.

Bridier, J., Claisse, O., Coton, M., et al. (2010) Evidence of distinct populations and specific subpopulations within the species *Oenococcus oeni*. *Appl Environ Microbiol* **76**, 7754–7764.

Brito, L. and Paveia, H. (1999) Presence and analysis of large plasmids in *Oenococcus oeni*. *Plasmid* **41**, 260–267.

Cafaro, C., Bonomo, M.G., and Salzano, G. (2014) Adaptive changes in geranylgeranyl pyrophosphate synthase gene expression level under ethanol stress conditions in *Oenococcus oeni*. *J Appl Microbiol* **116**, 71–80.

Campos, F.M., Figueiredo, A.R., Hogg, T.A., and Couto, J.A. (2009) Effect of phenolic acids on glucose and organic acid metabolism by lactic acid bacteria from wine. *Food Microbiol* **26**, 409–414.

Cappello, M.S., Stefani, D., Grieco, F., et al. (2008) Genotyping by amplified fragment length polymorphism and malate metabolism performances of indigenous *Oenococcus oeni* strains isolated from Primitivo wine. *Int J Food Microbiol* **127**, 241–245.

Cecconi, D., Milli, A., Rinalducci, S., et al. (2009) Proteomic analysis of *Oenococcus oeni* freeze-dried culture to assess the importance of cell acclimation to conduct malolactic fermentation in wine. *Electrophoresis* **30**, 2988–2995.

Cho, G.S., Krauss, S., Huch, M., et al. (2011) Development of a quantitative PCR for detection of Lactobacillus plantarum starters during wine malolactic fermentation. *J Microbiol Biotechnol* **21**, 1280–1286.

Claisse, O. and Lonvaud-Funel, A. (2001) Primers and a specific DNA probe for detecting lactic acid bacteria producing 3-hydroxypropionaldehyde from glycerol in spoiled ciders. *J Food Prot* **64**, 833–837.

Claisse, O. and Lonvaud-Funel, A. (2012) Development of a multilocus variable number of tandem repeat typing method for *Oenococcus oeni*. *Food Microbiol* **30**, 340–347.

Claisse, O. and Lonvaud-Funel, A. (2014) Multiplex variable number of tandem repeats for *Oenococcus oeni* and applications. *Food Microbiol* **38**, 80–86.

Claisse, O., Renouf, V., and Lonvaud-Funel, A. (2007) Differentiation of wine lactic acid bacteria species based on RFLP analysis of a partial sequence of rpoB gene. *J Microbiol Met* **69**, 387–390.

Comitini, F., Ferretti, R., Clementi, F., et al. (2005) Interactions between Saccharomyces cerevisiae and malolactic bacteria: preliminary characterization of a yeast proteinaceous compound(s) active against *Oenococcus oeni*. *J Appl Microbiol* **99**, 105–111.

Costello, P.J., Siebert, T.E., Solomon, M.R., and Bartowsky, E.J. (2013) Synthesis of fruity ethyl esters by acyl coenzyme A: alcohol acyltransferase and reverse esterase activities in *Oenococcus oeni* and *Lactobacillus plantarum*. *J Appl Microbiol* **114**, 797–806.

Coton, E., Rolan, G., Bertrand, A., and Lonvaud-Funel, A. (1998a) Histamine-producing lactic acid bacteria in wines. Early detection, frequency and distribution. *Amer J Enol Vit* **49**, 199–204.

Coton, E., Rollan, G.C., and Lonvaud-Funel, A. (1998b) Histidine decarboxylase of *Leuconostoc oenos* 9204: purification, kinetic properties, cloning and nucleotide sequence of the hdc gene. *J Appl Microbiol* **84**, 143–151.

Daniel, P., de Waele, E., and Hallet, J.-N. (1993) Optimization of transverse alternating field electrophoresis for strain identification of *Leuconostoc oenos*. *Appl Microbiol Biotechnol* **38**, 638–641.

de Las Rivas, B., Marcobal, A., and Munoz, R. (2004) Allelic diversity and population structure in *Oenococcus oeni* as determined from sequence analysis of housekeeping genes. *Appl Environ Microbiol* **70**, 7210–7219.

de Revel, G., Bloem, A., Augustin, M., et al. (2005) Interaction of *Oenococcus oeni* and oak wood compounds. *Food Microbiol* **22**, 569–575.

Delaherche, A., Bon, E., Dupe, A., et al. (2006) Intraspecific diversity of *Oenococcus oeni* strains determined by sequence analysis of target genes. *Appl Microbiol Biotechnol* **73**, 394–403.

Dicks, L.M., Dellaglio, F., and Collins, M.D. (1995) Proposal to reclassify *Leuconostoc oenos* as *Oenococcus oeni* [corrig.] gen. nov., comb. nov. *Int J System Bacteriol* **45**, 395–397.

Diez, L., Guadalupe, Z., Ayestaran, B., and Ruiz-Larrea, F. (2010) Effect of yeast mannoproteins and grape polysaccharides on the growth of wine lactic acid and acetic acid bacteria. *J Agric Food Chem* **58**, 7731–7739.

Divol, B., Tonon, T., Morichon, S., et al. (2003) Molecular characterization of *Oenococcus oeni* genes encoding proteins involved in arginine transport. *J Appl Microbiol* **94**, 738–746.

Dols-Lafargue, M., Lee, H.Y., Le Marrec, C., et al. (2008) Characterization of gtf, a glucosyltransferase gene in the genomes of *Pediococcus parvulus* and *Oenococcus oeni* two bacterial species commonly found in wine. *Appl Environ Microbiol* **74**, 4079–4090.

El Gharniti, F., Dols-Lafargue, M., Bon, E., Claisse et al. (2012) IS30 elements are mediators of genetic diversity in *Oenococcus oeni*. *Int J Food Microbiol* **158**, 14–22.

Endo, A. and Okada, S. (2005) Monitoring the lactic acid bacterial diversity during shochu fermentation by PCR-denaturing gradient gel electrophoresis. *J Biosci Bioeng* **99**, 216–221.

Endo, A. and Okada, S. (2006) *Oenococcus kitaharae* sp. nov., a non-acidophilic and non-malolactic-fermenting *oenococcus* isolated from a composting distilled shochu residue. *Int J System Evol Microbiol* **56**, 2345–2348.

Favier, M., Bilhere, E., Lonvaud-Funel, A., et al. (2012) Identification of pOENI-1 and related plasmids in *Oenococcus oeni* strains performing the malolactic fermentation in wine. *PloS One* **7**, e49082.

Fernandez, P.A. and de Nadra, M.C. (2006) Growth response and modifications of organic nitrogen compounds in pure and mixed cultures of lactic acid bacteria from wine. *Curr Microbiol* **52**, 86–91.

Feuillat, M. (1977) [Influence of nitrogen components of wine on the growth of lactic acid bacteria]. *Comptes rendus hebdomadaires des séances de l'Académie des sciences Serie D: Sciences naturelles* **285**, 1013–1016.

Fornachon, J.C. (1968) Influence of different yeasts on the growth of lactic acid bacteria in wine. *J Sci Food Agric* **19**, 374–378.

Fremaux, C., Aigle, M., and Lonvaud-Funel, A. (1993) Sequence analysis of *Leuconostoc oenos* DNA: organization of pLo13, a cryptic plasmid. *Plasmid* **30**, 212–223.

Gagne, S., Lucas, P.M., Perello, M.C., et al. (2011) Variety and variability of glycosidase activities in an *Oenococcus oeni* strain collection tested with synthetic and natural substrates. *J Appl Microbiol* **110**, 218–228.

Garcia-Ruiz, A., Gonzalez-Rompinelli, E.M., Bartolome, B., and Moreno-Arribas, M.V. (2011) Potential of wine-associated lactic acid bacteria to degrade biogenic amines. *Int J Food Microbiol* **148**, 115–120.

Genisheva, Z., Mussatto, S.I., Oliveira, J.M., and Teixeira, J.A. (2013) Malolactic fermentation of wines with immobilised lactic acid bacteria - influence of concentration, type of support material and storage conditions. *Food Chem* **138**, 1510–1514.

Gindreau, E., Walling, E., and Lonvaud-Funel, A. (2001) Direct polymerase chain reaction detection of ropy *Pediococcus damnosus* strains in wine. *J Appl Microbiol* **90**, 535–542.

Grandvalet, C., Assad-Garcia, J.S., Chu-Ky, S., et al. (2008) Changes in membrane lipid composition in ethanol- and acid-adapted *Oenococcus oeni* cells: characterization of the cfa gene by heterologous complementation. *Microbiology* **154**, 2611–2619.

Grandvalet, C., Coucheney, F., Beltramo, C., and Guzzo, J. (2005) CtsR is the master regulator of stress response gene expression in *Oenococcus oeni*. *J Bacteriol* **187**(16), 5614–5623.

Grimaldi, A., Bartowsky, E., and Jiranek, V. (2005) A survey of glycosidase activities of commercial wine strains of *Oenococcus oeni*. *Int J Food Microbiol* **105**, 233–244.

Knoll, C., Divol, B., and du Toit, M. (2008) Genetic screening of lactic acid bacteria of oenological origin for bacteriocin-encoding genes. *Food Microbiol* **25**, 983–991.

Knoll, C., Fritsch, S., Schnell, S., et al. (2012) Impact of different malolactic fermentation inoculation scenarios on Riesling wine aroma. *World J Microbiol Biotechnol* **28**, 1143–1153.

Labarre, C., Guzzo, J., Cavin, J.F., and Divies, C. (1996) Cloning and characterization of the genes encoding the malolactic enzyme and the malate permease of *Leuconostoc oenos*. *Appl Environ Microbiol* **62**, 1274–1282.

Lafon-Lafourcade, S., Lonvaud-Funel, A., and Carre, E. (1983) Lactic acid bacteria of wines: stimulation of growth and malolactic fermentation. *Antonie van Leeuwenhoek* **49**, 349–352.

Larisika, M., Claus, H., and Konig, H. (2008) Pulsed-field gel electrophoresis for the discrimination of *Oenococcus oeni* isolates from different wine-growing regions in Germany. *Int J Food Microbiol* **123**, 171–176.

Lonvaud-Funel, A. (2006) Lactic acid bacteria development in wine. In: *Handbook of Enology* ((eds) P. Ribereau-Gayon, D. Dubourdieu, B. Donèche, and A. Lonvaud), John Wiley & Sons Ltd, Chichester, England, pp.161–182.

Lonvaud-Funel, A. and Joyeux, A. (1994) Histamine production by wine lactic acid bacteria: isolation of a histamine-producing strain of *Leuconostoc oenos*. *J Appl Bacteriol* **77**, 401–407.

Lonvaud-Funel, A., Joyeux, A., and Desens, C. (1988) The inhibition of malolactic fermentation of wines by products of yeast metabolism. *J Sci Food Agric* **44**, 183–191.

Lonvaud-Funel, A., Joyeux, A., and Ledoux, O. (1991) Speciation of lactic acid bacteria in fermenting grape must and wine by colony hybridization with non-isotopic DNA probes. *J Appl Bacteriol* **71**, 501–508.

Lonvaud-Funel, A. and Strasser de Saad, A.M. (1982) Purification and properties of a malolactic enzyme from a strain of *Leuconostoc mesenteroides* isolated from grapes. *Appl Environ Microbiol* **43**, 357–361.

Lopez, I., Torres, C., and Ruiz-Larrea, F. (2008) Genetic typification by pulsed-field gel electrophoresis (PFGE) and randomly amplified polymorphic DNA (RAPD) of wild *Lactobacillus plantarum* and *Oenococcus oeni* wine strains. *Eur Food Res Technol* **227**, 547–555.

Lucas, P. and Lonvaud-Funel, A. (2002) Purification and partial gene sequence of the tyrosine decarboxylase of *Lactobacillus brevis* IOEB 9809. *FEMS Microbiol Lett* **211**, 85–89.

Lucas, P.M., Blancato, V.S., Claisse, O., et al. (2007) Agmatine deiminase pathway genes in *Lactobacillus brevis* are linked to the tyrosine decarboxylation operon in a putative acid resistance locus. *Microbiology* **153**, 2221–2230.

Lucas, P.M., Claisse, O., and Lonvaud-Funel, A. (2008) High frequency of histamine-producing bacteria in the enological environment and instability of the histidine decarboxylase production phenotype. *Appl Environ Microbiol* **74**, 811–817.

Lucas, P.M., Wolken, W.A., Claisse, O., et al. (2005) Histamine-producing pathway encoded on an unstable plasmid in *Lactobacillus hilgardii* 0006. *Appl Environ Microbiol* **71**, 1417–1424.

Makarova, K.S. and Koonin, E.V. (2007) Evolutionary genomics of lactic acid bacteria. *J Bacteriol* **189**, 1199–1208.

Marcobal, A.M., Sela, D.A., Wolf, Y.I., et al. (2008) Role of hypermutability in the evolution of the genus *Oenococcus*. *J Bacteriol* **190**, 564–570.

Marty-Teysset, C., Posthuma, C., Lolkema, J.S., et al. (1996) Proton motive force generation by citrolactic fermentation in *Leuconostoc mesenteroides*. *J Bacteriol* **178**, 2178–2185.

Mesas, J.M., Rodriguez, M.C., and Alegre, M.T. (2001) Nucleotide sequence analysis of pRS2 and pRS3, two small cryptic plasmids from *Oenococcus oeni*. *Plasmid* **46**, 149–151.

Mills DA, Rawsthorne H, Parker C, Tamir D, Makarova K (2005) Genomic analysis of *Oenococcus oeni* PSU-1 and its relevance to winemaking. *FEMS Microbiol Rev* **29**, 465–475

Moreno-Arribas, V., Torlois, S., Joyeux, A., et al. (2000) Isolation, properties and behaviour of tyramine-producing lactic acid bacteria from wine. *J Applied Microbiol* **88**, 584–593.

Nannelli, F., Claisse, O., Gindreau, E., et al. (2008) Determination of lactic acid bacteria producing biogenic amines in wine by quantitative PCR methods. *Lett Appl Microbiol* **47**, 594–599.

Navarro, L., Zarazaga, M., Saenz, J., et al. (2000) Bacteriocin production by lactic acid bacteria isolated from Rioja red wines. *J Appl Microbiol* **88**, 44–51.

Nielsen, J.C., Prahl, C., and Lonvaud-Funel, A. (1996) Malolactic fermentation in wine by direct inoculation with freeze-dried *Leuconostoc oenos* cultures. *Am J Enol Vitic* **47**, 42–48.

Nielsen, J.C. and Richelieu, M. (1999) Control of flavor development in wine during and after malolactic fermentation by *Oenococcus oeni*. *Appl Environ Microbiol* **65**, 740–745.

Olguin, N., Bordons, A., and Reguant, C. (2009) Influence of ethanol and pH on the gene expression of the citrate pathway in *Oenococcus oeni*. *Food Microbiol* **26**, 197–203.

Pasteris, S.E. and Strasser de Saad, A.M. (2009) Sugar-glycerol cofermentation by *Lactobacillus hilgardii* isolated from wine. *J Agric Food Chem* **57**, 3853–3858.

Petri, A., Pfannebecker, J., Frohlich, J., and Konig, H. (2013) Fast identification of wine related lactic acid bacteria by multiplex PCR. *Food Microbiol* **33**, 48–54.

Peynaud, E. and Domercq, S. (1968) Etude de 400 souches de coques hétérolactiques isolés de vin. *Ann Inst Pasteur* **19**, 159–169.

Poblet-Icart, M., Bordons, A., and Lonvaud-Funel, A. (1998) Lysogeny of *Oenococcus oeni* (syn. *Leuconostoc oenos*) and study of their induced bacteriophages. *Curr Microbiol* **36**, 365–369.

Poolman, B., Molenaar, D., Smid, E.J., et al. (1991) Malolactic fermentation: electrogenic malate uptake and malate/lactate antiport generate metabolic energy. *J Bacteriol* **173**, 6030–6037.

Pripis-Nicolau, L., de Revel, G., Bertrand, A., and Lonvaud-Funel, A. (2004) Methionine catabolism and production of volatile sulphur compounds by *Oenococcus oeni*. *J Appl Microbiol* **96**, 1176–1184.
Radler, F. and Yannissis, C. (1972) Decomposition of tartrate by lactobacilli. *Archiv fur Mikrobiologie* **82**, 219–239.
Renouf, V., Claisse, O., Miot-Sertier, C., and Lonvaud-Funel, A. (2006) Lactic acid bacteria evolution during winemaking: use of rpoB gene as a target for PCR-DGGE analysis. *Food Microbiol* **23**, 136–145.
Renouf, V., Delaherche, A., Claisse, O., and Lonvaud-Funel, A. (2008) Correlation between indigenous *Oenococcus oeni* strain resistance and the presence of genetic markers. *J Ind Microbiol Biotechnol* **35**, 27–33.
Richter, H., Hamann, I., and Unden, G. (2003) Use of the mannitol pathway in fructose fermentation of *Oenococcus oeni* due to limiting redox regeneration capacity of the ethanol pathway. *Arch Microbiol* **179**, 227–233.
Ritt, J.F., Guilloux-Benatier, M., Guzzo, J., et al. (2008) Oligopeptide assimilation and transport by *Oenococcus oeni*. *J Appl Microbiol* **104**, 573–580.
Rodas, A.M., Ferrer, S., and Pardo, I. (2003) 16S-ARDRA, a tool for identification of lactic acid bacteria isolated from grape must and wine. *Syst Appl Microbiol* **26**, 412–422.
Rossouw, D., Du Toit, M., and Bauer, F.F. (2012) The impact of co-inoculation with *Oenococcus oeni* on the trancriptome of *Saccharomyces cerevisiae* and on the flavour-active metabolite profiles during fermentation in synthetic must. *Food Microbiol* **29**, 121–131.
Santos, R., Sao-Jose, C., Vieira, G., et al. (1998) Genome diversity in temperate bacteriophages of *Oenococcus oeni*. *Arch Virol* **143**, 523–536.
Schutz, M. and Radler, F. (1973) The "malic enzyme" from *Lactobacillus plantarum* and Leuconostoc mesenteroides. *Archiv fur Mikrobiologie* **91**, 183–202.
Seifert, W. (1901) Uber die saureabnahme im wein und den dabei stattfinden gahrungsprozess. *Z Landwirstch Versuchsu Duet Oest* **4**, 980–992.
Silva, I., Campos, F.M., Hogg, T., and Couto, J.A. (2011) Wine phenolic compounds influence the production of volatile phenols by wine-related lactic acid bacteria. *J Appl Microbiol* **111**, 360–370.
Silveira, M.G., Baumgartner, M., Rombouts, F.M., and Abee, T. (2004) Effect of adaptation to ethanol on cytoplasmic and membrane protein profiles of *Oenococcus oeni*. *Appl Environ Microbiol* **70**, 2748–2755.
Solieri, L. and Giudici, P. (2010) Development of a sequence-characterized amplified region marker-targeted quantitative PCR assay for strain-specific detection of *Oenococcus oeni* during wine malolactic fermentation. *Appl Environ Microbiol* **76**, 7765–7774.
Sumby, K.M., Jiranek, V., and Grbin, P.R. (2013) Ester synthesis and hydrolysis in an aqueous environment, and strain specific changes during malolactic fermentation in wine with *Oenococcus oeni*. *Food Chem* **141**, 1673–1680.
Theobald, S., Pfeiffer, P., Zuber, U., and Konig, H. (2008) Influence of epigallocatechin gallate and phenolic compounds from green tea on the growth of *Oenococcus oeni*. *J Appl Microbiol* **104**, 566–572.
Tonon, T., Bourdineaud, J.P., and Lonvaud-Funel, A. (2001) The arcABC gene cluster encoding the arginine deiminase pathway of *Oenococcus oeni*, and arginine induction of a CRP-like gene. *Res Microbiol* **152**, 653–661.
Tonon, T. and Lonvaud-Funel, A. (2000) Metabolism of arginine and its positive effect on growth and revival of *Oenococcus oeni*. *J Appl Microbiol* **89**, 526–531.
Ugliano, M., Genovese, A., and Moio, L. (2003) Hydrolysis of wine aroma precursors during malolactic fermentation with four commercial starter cultures of *Oenococcus oeni*. *J Agric Food Chem* **51**, 5073–5078.
Vallet, A., Lucas, P., Lonvaud-Funel, A., and de Revel, G. (2008) Pathways that produce volatile sulphur compounds from methionine in *Oenococcus oeni*. *J Appl Microbiol* **104**, 1833–1840.
Walling, E., Gindreau, E., and Lonvaud-Funel, A. (2005) A putative glucan synthase gene dps detected in exopolysaccharide-producing *Pediococcus damnosus* and *Oenococcus oeni* strains isolated from wine and cider. *Int J Food Microbiol* **98**, 53–62.
Weimer, B., Seefeldt, K., and Dias, B. (1999) Sulfur metabolism in bacteria associated with cheese. *Antonie van Leeuwenhoek* **76**, 247–261.
Wolken, W.A., Lucas, P.M., Lonvaud-Funel, A., and Lolkema, J.S. (2006) The mechanism of the tyrosine transporter TyrP supports a proton motive tyrosine decarboxylation pathway in *Lactobacillus brevis*. *J Bacteriol* **188**, 2198–2206.
Zapparoli, G., Reguant, C., Bordons, A., et al. (2000) Genomic DNA fingerprinting of *Oenococcus oeni* strains by pulsed-field gel electrophoresis and randomly amplified polymorphic DNA-PCR. *Curr Microbiol* **40**, 351–355.
Zavaleta, A.I., Martinez-Murcia, A.J., and Rodriguez-Valera, F. (1997) Intraspecific genetic diversity of *Oenococcus oeni* as derived from DNA fingerprinting and sequence analyses. *Appl Environ Microbiol* **63**, 1261–1267.
Ze-Ze, L., Chelo, I.M., and Tenreiro, R. (2008) Genome organization in *Oenococcus oeni* strains studied by comparison of physical and genetic maps. *Int Microbiol: Offic J Span Soc Microbiol* **11**, 237–244.
Zhao, G. and Zhang, G. (2009) Influence of freeze-drying conditions on survival of *Oenococcus oeni* for malolactic fermentation. *Int J Food Microbiol* **135**, 64–67.

Chapter 16

The Functional Role of Lactic Acid Bacteria in Cocoa Bean Fermentation

Luc De Vuyst* and Stefan Weckx

Research Group of Industrial Microbiology and Food Biotechnology (IMDO), Department of Bioengineering Sciences, Faculty of Sciences and Bioengineering Sciences, Vrije Universiteit Brussel, Belgium

16.1. Introduction

The secrets of the cocoa tree, *Theobroma cacao* L. (literally "food of the gods"), were first exploited around 4000 years ago in the tropical rainforest of Latin-America. The fruits (cocoa pods) of this tree contain seeds (cocoa beans) that can be processed into cocoa mass, cocoa butter, and cocoa powder. Cocoa mass and cocoa butter are responsible for the unique flavor and melt-in-the-mouth properties of chocolate.

The story of how chocolate grew from a local and ritual Mesoamerican beverage gift of the gods, over a nutritious and medicinal product for wealthy European people, into a delicious sweet enjoyed by people around the world encompasses many different cultures and continents, wherein Latin-America, Africa and Asia (cocoa plantations), and Europe (cocoa trade, cocoa processing, and chocolate production) played a major role (Coe and Coe 2007). Today, approximately two-thirds of the world's cocoa production, which is estimated at over 4.8 million tons, comes from West-Africa (68%), followed by Latin-America (15%), and South-East Asia (17%) (www.worldcocoafoundation.org). At present, the seven largest cocoa-producing countries (in decreasing order of annual production size) are Ivory Coast (36%), Ghana (22%), Indonesia, Nigeria, Brazil, Cameroon, and Ecuador, which represent all together 95% of the cocoa world production (www.icco.org). Virtually all major cocoa production over the past five decades has involved bulk (basic or ordinary) cocoa. There are, however, many other smaller producers, particularly of fine (flavor) cocoas, which constitute about 5% of the cocoa world production. The best known is Nacional cocoa (Ecuador), which is the source of the fine-flavor cocoa called Arriba. Less known is Java Fine or Edel cocoa, a light-breaking cocoa from Java (Indonesia), which is mainly used for its color. Recent trends of cocoa use for chocolate production encompass, besides fair-trade chocolates, single-origin, high-cocoa solid content and other premium chocolates and chocolates obtained through controlled cocoa bean fermentation (www.tomsekstra.dk; www.terracacao.com).

About 90% of the world's cocoa is grown by smallholders, often in less well-developed tropical regions of the world. Chocolate production is mainly concentrated in Europe and the USA. Whereas Belgium is world famous for its extremely fine, excellent tasting, and high-quality chocolate, French

*Corresponding author: Research Group of Industrial Microbiology and Food Biotechnology (IMDO), Department of Bioengineering Sciences, Faculty of Sciences and Bioengineering Sciences, Vrije Universiteit Brussel, Pleinlaan 2, B-1050 Brussels, Belgium. ldvuyst@vub.ac.be

Biotechnology of Lactic Acid Bacteria: Novel Applications, Second Edition. Edited by Fernanda Mozzi, Raúl R. Raya, and Graciela M. Vignolo.

dark chocolate became renowned in French gastronomy and Swiss milk chocolate is appreciated worldwide for its seductive silky smoothness and purity.

16.2. Cocoa Crop Cultivation and Harvest

The cocoa tree is a small, evergreen tree, cultivated worldwide as a shade crop in rainforests in a 20° belt north and south of the equator (Wood and Lass 1985; Fowler et al. 1998; Bartley 2005; Afoakwa 2010). A large number of cocoa tree varieties exists, which are grouped according to morphological (fruits and seeds) and geographical (genetic origin) characteristics. This large cocoa variety is due to genetic factors (vast array of plant populations and genotypes, interbreeding, etc.) and human interventions (agricultural practices, re-crossing, etc.).

In principle, all cocoa varieties originate from the Criollo (which means "native" in Spanish) and Forastero varieties (which means "foreigner" in Spanish). These varieties are considered indigenous to the Upper Amazon basin (Amazonian region, Venezuela). Criollo varieties are most commonly grown in South/Central-America, the Sanchez variety for instance being grown in the Dominican Republic. Among the Forastero varieties, the melon-shaped Amelonado (Lower Amazon) and Amazon have been the most widely grown (Brazil, West-Africa, and Indonesia). Trinitario, a third cocoa group often referred to, originates from Trinidad and has been named so to distinguish it from the native cocoa of Venezuela. It is actually a hybrid between Criollo and Forastero (most often Lower Amazon) varieties, which is particularly suitable for cultivation (Motamayor et al. 2008). Today, Criollo, Forastero, and Trinitario varieties represent approximately 1, 94, and 5% of the world production of cocoa. Criollo and Trinitario are considered fine cocoas that are often sold at a higher price than Forastero and that are used for premium plain dark chocolates and couvertures. However, Forastero is much less prone to various diseases and overcomes several other weaknesses (yield, bearing, etc.) compared to Criollo, giving a more stable crop from year to year, and is thus preferred by farmers throughout the world for cultivation, yielding bulk cocoa (for milk and dark chocolates and other confectionary products). The exact origin of the Nacional type of cocoa is not known, but it is generally considered native to Ecuador. Molecular studies indicate that Nacional cocoa genotypes, first considered a Forastero variety, are genetically different from the Forastero, Criollo, or Trinitario cocoa groups (Crouzillat et al. 1996). In general, many cocoa farmers grow several varieties of cocoa and, due to cross-pollination and re-crossing, single trees with all the characteristics of a specific cocoa type are rare now and, hence, only hybrids are cultivated worldwide.

A vast body of work has been performed to improve the cocoa crop, including germplasm collection, plant breeding, crop yield, rapid propagation, disease resistance, full-sun cultivation, pollination through insects, development of improved seeds, pods with more and bigger beans, beans with elevated fat levels, and so on. Unfortunately, probably not the best varieties of *T. cacao* have been planted worldwide, as quality may have taken second place to quantity in the choice of the tree. Each year, more than 30% of the cocoa production is lost to plant diseases and pests by devastating fungal pathogens, viruses, and insects, and even greater losses will occur in the near future, due to further spread of diseases among the monocultures and decreasing yields from aged trees, if the necessary precautions such as good agricultural practices and replanting of trees are not taken in time (www.worldcocoafoundation.com; www.icco.org).

Cocoa trees carry buds, flowers, and fruits at the same time. In general, thousands of flowers adorn the trunk and major branches of the trees. However, only a few flowers will be fertilized through difficult cross-pollination and no more than 1–10% will develop into cocoa pods. The overall development of the pods takes 5–6 months from pollination of the flowers to full ripeness of the fruits. The oval, leathery fruits vary between varieties in size, shape, surface texture, and external color. They contain a pulpy endocarp (the cocoa pulp) wherein the seeds (the cocoa beans) are

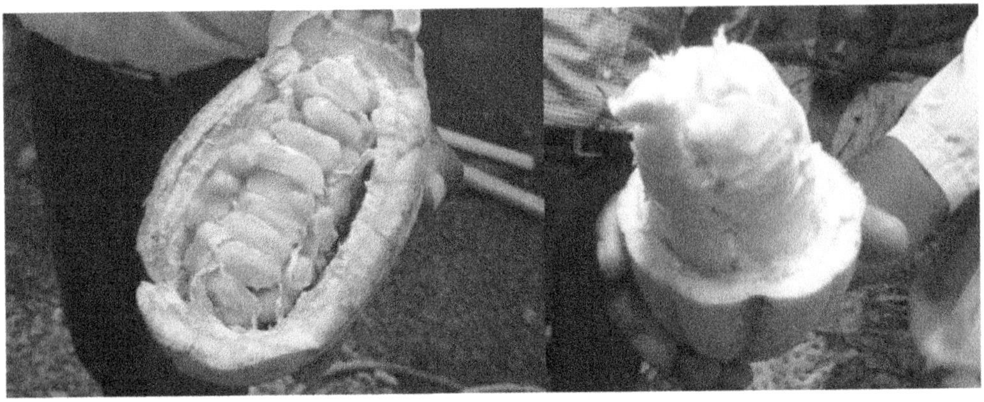

Figure 16.1. Opened cocoa pod with beans embedded in a mucilaginous pulp.

embedded (Figure 16.1). One flower or pod yields about 20–60 seeds, depending on the cocoa variety, which are attached to a central core or placenta. Cocoa pods are indehiscent and do not abscise; hence, the fruits have to be cut down at harvest. At harvest time, pods change color from green or dark red-purple to yellow, orange, or red, depending on the variety. As the pulp contains a germination inhibitor, germination of the seeds inside the pods does not occur, but once the pods are opened the pulp decomposes rapidly and germination of the seeds begins under appropriate humidity and temperature conditions.

Cocoa trees produce a few hundred to more than 3000 kg of dry beans per ha, depending on variety and agricultural and farming practices (www.cocoa-atlas.com). In West-Africa, healthy ripe pods are harvested twice a year, representing a main-crop and mid-crop season, with intervals of approximately three weeks within the peak seasons. It is common practice in many cocoa-producing regions to harvest the pods over 3–4 days or longer, according to the yield and size of the farm, before the collected pods are transported to a place suitable for subsequent handling, either at the farms in or at the edges of the plantations or at a central processing plant.

16.3. The Cocoa Pulp or Fermentation Substrate

Ripe cocoa pods measure 10–35 cm long and weigh (wet) from 200 g to more than 1 kg. They contain a considerable amount (approximately 40% of the pod fresh mass) of sweet, white, mucilaginous pulp surrounding the beans (Figure 16.1). This pulp is a rich medium for microbial growth and hence forms the raw material for the actual cocoa bean fermentation. Sufficient and ripe pulp (mature pods) guarantees a correct fermentation process. The degradation of the pulp during fermentation liberates pulp juices, referred to as "sweatings," which drain away. Therefore, the correct harvest time of mature pods is of utmost importance. In over-ripe pods, the amount of pulp is too small and the pulp appears to be dry; under-ripe pods possess a dense pulp.

Cocoa pulp is composed of (m/m) 82–87% water, 10–15% mono- and disaccharides, 1–5% pectin, 1–3% citric acid, 0.1–0.4% other non-volatile organic acids (e.g., malic acid, oxalic acid, and succinic acid), 0.5–0.7% proteins, and 8–10% minerals and oligoelements (Lehrian and Patterson 1983; Schwan et al. 1995; Barel 1998). Of the mono- and dissacharides present, about 60% is sucrose and 39% is a mixture of fructose and glucose, which are the pulp invertase hydrolysis products of sucrose. The concentration of sucrose, fructose, and glucose is a function of cultivar and fruit age; unripe pods contain a higher proportion of sucrose and ripe pods contain mainly fructose and glucose (Lehrian and Patterson 1983; Thompson et al. 2007, 2013). The amino acids, glutamic acid and aspartic acid, are present at high levels in the pulp compared with other amino acids. Vitamin C is the most important vitamin in cocoa pulp. Potassium and sulfate comprise the major cation and anion, respectively. The high contents of pectin and other polysaccharides (cellulose, hemicellulose, lignin) make the pulp viscous, sticky and cohesive. The pH of the pulp is relatively low (pH 3.0–4.0), mainly due to its citric acid content.

16.4. Fresh, Unfermented Cocoa Beans

Fresh, unfermented cocoa beans are oval (2–3 cm long, 1.0–1.7 cm wide, 0.7–1.2 cm thick) and weigh on average 2.5 g. The moisture content of the fresh beans is approximately 65% (m/m). Their pH is 6.3–7.0. The seeds basically consist of two parts: an outer part comprising the skin or testa (seed coat or shell; 10–14% of the dry mass of the beans), surrounding the beans, and an inner part comprising the embryo (germ) and two cotyledons (86–90%) contained within the testa and uniformed at a small embryonic axis (Figure 16.2a). The cotyledons are referred to as the nibs in the cured cocoa beans. The testa of the seeds is impermeable to large molecules (citric acid, carbohydrates, polyphenols, alkaloids), whereas small volatile molecules, such as ethanol and acetic acid, can easily penetrate it. Hence, this biological barrier controls the kinetics of fermentation and concomitant diffusion processes.

The cotyledons are basically made up of different types of parenchyma storage cells, namely white lipid/protein/starch-containing cells (reserve material cells) and (purple) polyphenol-containing cells (pigment cells), with parenchyma plasma forming a grid between the two types of cells (Figure 16.2b). The storage cells of Criollo beans are larger than those of Forastero beans, possibly contributing to their fine flavor quality (Elwers et al. 2010). Whereas the reserve material cells containing starch granules, protein vacuoles, and fat globules constitute the nutrients for the embryo and lipids further form a natural barrier around the water-soluble components (enzymes, substrates and inhibitors); the larger, vacuolated, polyphenol-storage cells contain unique cocoa bean compounds, encompassing alkaloids (the methylxanthines theobromine and caffeine) and polyphenols (including anthocyanins), both possibly involved in stress resistance responses (particularly light) of the plant.

The fat (cocoa butter) of the cocoa beans is important for chocolate production, as it will give chocolate its characteristic texture and mouthfeel, depending on the hardness and melting and solidification (crystallization) behavior of the fat (Asep et al. 2008; Beckett 2009; Afoakwa 2010). Fat makes up slightly more than 50% of the mass of dry unfermented cocoa bean cotyledons; Forastero beans possess a higher fat content than Criollo beans. The cocoa butter present in the cocoa beans is

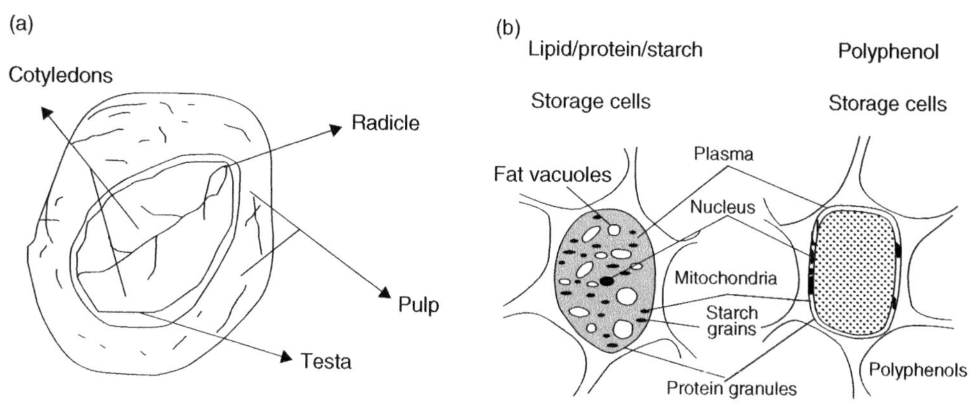

Figure 16.2. (a) Cocoa bean. (b) Cocoa bean storage cells.

a relatively simple fat, whereby triacylglycerols (up to 80% symmetrical and mono-unsaturated 1,3-distearoyl-2-oleoylglycerols) represent 97% of its total composition.

Cocoa bean quantities of alkaloids vary considerably in ripe seeds of different genetic origin, namely about 0.7–3.0% (m/m) of theobromine and 0.1–0.7% (m/m) of caffeine. Depending on the amount of anthocyanins, the polyphenol-storage or pigment cells, which constitute 10–20% (dry mass) of the cotyledons of the cocoa beans, are white to deep purple. Three groups of cocoa bean polyphenols (14–20%, m/m, in total) can be distinguished (Wollgast and Anklam 2000; Ortega et al. 2008): monomeric (epi)catechins or flavanols (ca. 37%, m/m), procyanidins or (epi)catechin oligomers with a degree of polymerization up to decamers and higher (ca. 58%, m/m) and anthocyanins or glycosylated derivatives of anthocyanidins (ca. 4%, m/m). The main catechin is (−)-epicatechin with up to 35% of the polyphenol content. In smaller amounts, (+)-catechin as well as traces of (+)-gallocatechin and (−)-epigallocatechin are found. Beans of the Forastero variety owe their characteristic violet color to their content of anthocyanins. White beans of the Criollo-type contain approximately two-thirds of the amount of polyphenols found in Forastero beans and do not possess anthocyanins. Both alkaloids and polyphenols are linked to cocoa bitterness.

Besides alkaloids and polyphenols, unfermented cocoa beans contain (sugar) alcohols, aldehydes, ketones, organic acids, esters, terpenes, and other compounds, some of which contribute to the cocoa flavor. It is important to notice that plant genetics (cultivar) and environmental conditions of crop cultivation (climate, soil quality and richness, water management, weeding, plant protection and phytosanitary management, shade canopy management, tree density and sanitary pruning, plant breeding, pollination, regular and complete harvest, etc.) are actually the first determinants that contribute to the final flavor, composition, and quality of the cocoa beans (Sukha et al. 2008; Kadow et al. 2013).

16.5. Cocoa Bean Fermentation

16.5.1. Rationale

Raw cocoa beans have an astringent, unpleasant, and bitter flavor and unpalatable taste and must be processed after harvest before they can be converted into good-tasting and full-flavored chocolates. Post-harvest processing of cocoa beans in the country of origin encompasses curing, grading, and storage, which determine the quality of the so-called commercial cocoa beans (Lehrian and Patterson 1983; Wood and Lass 1985; Fowler et al. 1998; Afoakwa 2010). Curing involves fermentation of the beans, followed by drying and roasting. Fermentation facilitates removal of the mucilaginous pulp surrounding the beans, as the pulp otherwise will inhibit the drying of the beans. Drying is necessary to get microbiologically stable moisture contents, which is in turn required for storage and shipping of the beans prior to chocolate manufacture. Also, fermentation lies at the basis of cocoa flavor and color development. However, it is the pulp surrounding the beans and not the beans themselves that undergoes microbial fermentation. Yet, during this fermentation process, exchange of compounds through diffusion takes place between the bean cotyledons and the environment. Actually, cocoa pulp fermentation enables the expression of the flavor potential (genetically and enzymatically) of the beans and is hence the second determinant of the final cocoa flavor, composition, and quality. Indeed, cocoa pulp fermentation (outside the beans) and endogenous enzymatic conversions (inside the beans) play a crucial role in the production of flavor precursors and pigment development.

In principle, the initiation of cocoa bean fermentation corresponds with an incipient germination phase, which is necessary for hydration of bean components through water uptake by the protein vacuoles within the storage cells and mobilization of the radical enzymes to prepare the growth of the seedling. Although the further germination phase is undesirable in cocoa beans destined for chocolate production, endogenous enzyme activities are necessary for cocoa flavor and color development. As there is no evidence to suggest that enzymes from the microorganisms involved in cocoa pulp fermentation penetrate the beans and create flavor compounds, proper curing of the beans mainly depends upon ethanol, acetic acid, and heat penetration into the beans, which are all provided during fermentation. This is necessary for inhibiting seed germination by killing the embryo (on the second day of fermentation) and activating hydrolytic enzymes (upon fermentation) inside the beans for flavor and color development. Also, this keeps the fat content of the bean cotyledons at high levels and avoids infection of the beans by leaving the testa closed. Accumulating acetic acid and ethanol in the beans cause swelling of the cotyledon tissue and breakdown of the biological barriers (membranes) between the storage and pigment cells within the bean cotyledons, so that various enzymes and substrates are free to mix (Biehl et al. 1982a,b). Ethanol, (volatile) acids, and water diffusing into the bean cotyledons act as solvents to transport compounds to sites of enzyme activity and vice versa and to enable the subsequent biochemical reactions to produce the flavor and color precursors. The penetrating acetic acid causes the cocoa bean pH to drop from an initial value of pH 6.3–7.0 to pH 4.0–5.5, a process determining the degree of the enzymatic activities. Also, the enzyme reaction rates are favored by an increasing temperature due to microbial activities during fermentation. When oxygen begins to enter the beans, at about 96 h into fermentation, a series of enzymatic, oxidative reactions are initiated that result in a browning of the cotyledons. After fermentation, flavor and color formation are further developed during the drying, roasting, and final processing steps of well-fermented cocoa beans, which represent the third determinant of the flavor, composition, and quality of the cocoa beans. Thus, there is no cocoa and chocolate flavor without fermentation, as cocoa bean roasting and chocolate processing depend on the composition of the fermented dry cocoa beans (Rohan 1964; Ziegleder and Biehl 1988; Owusu et al. 2011). Yet, the inherited characteristics of the beans set a limit on what can be achieved by fermentation and other processing. In other words, it is impossible to improve genetically inferior material by superior processing technologies. Alternatively, it is quite easy to ruin good-quality cocoa by inadequate curing of the beans.

16.5.2. Farming Practices

Various cocoa bean fermentation methods have been developed worldwide. The actual methods of fermentation vary in different cocoa-producing countries and regions and even from one cocoa farmer to another within a region or country. The different fermentation methods are heap (e.g., Ghana and Ivory Coast, Figure 16.3), box (e.g., Brazil, Indonesia, and Malaysia), basket (e.g., Nigeria and Ghana), tray (e.g., Ghana), sack (e.g., Ecuador), and platform fermentations (e.g., Ecuador). Approximately one-half of the cocoa is fermented in some type of box, for many years and worldwide, and the remaining half is fermented by using heaps or other traditional methods that are slowly disappearing. For a detailed description of these methods, see Lehrian and Patterson (1983), Wood and Lass (1985), Thompson et al. (2007, 2013), and Afoakwa (2010).

Before the start of the fermentation process, the cocoa pods are broken open with machetes, the husk and placenta of the pods are discarded, and the wet beans are scooped out manually. According to good practice, infected and black beans are removed and bean agglomerates are separated. Also, mechanical pod breaking and bean extraction exists. When healthy, undamaged pods are opened, the cocoa pulp is sterile, although the interior of ripe cocoa pods may contain a few hundred microorganisms (in particular yeasts) per gram (Maravalhas 1966; Jespersen et al. 2005; Crafack et al. 2013). When pulp and beans are removed from the opened pods, the cocoa pulp–bean mass is immediately contaminated with a variety of microorganisms from the environment. Environmental inoculation includes contamination from cocoa pod surfaces (probably the main inoculation source), unwashed baskets used for transient seed transport (also an important source), banana and plantain leaves, and sacks used to construct and cover the heaps, boxes, and baskets to retain the heat (to compensate for temperature variation during day and night) and to prevent the surface beans from drying and mildewing, machetes and knives used for opening of the pods, workers' hands, dried mucilage left

(a) (b) (c) (d) (e) (f)

Figure 16.3. (a–f) The cocoa bean heap fermentation is the most simple and most commonly used method of fermentation on small farms. It requires the simplest equipment at practically no cost, so that it can be run by a family. In a heap fermentation process the cocoa pulp–bean mass is piled on banana or plantain leaves, which are spread out in a circle on the ground, sometimes around a central hole in the ground or raised above soil level to allow easy pulp drainage. When the heap is complete, it is covered with more leaves and these are often held in place by small logs. The cover protects the fermenting cocoa pulp–bean mass against surface mold growth and keeps the heat inside. In general, sweatings are allowed to flow away and penetrate into the ground. The size of the heaps varies widely; heaps from about 25 to 2000 kg are common. This cocoa bean fermentation method is used throughout West-Africa and almost exclusively in Ghana, where farmers are fermenting their beans in heaps of 200 to 500 kg for six days. The heap fermentation method has been used to produce some of the world's most qualitative standard cocoa available, especially in Ghana.

on the walls of receptacles from previous fermentations, soil, dust, air, and insects (Ostovar and Keeney 1973; Jespersen et al. 2005; Camu et al. 2007, 2008a; Nielsen et al. 2007a; Lefeber et al. 2011b; Papalexandratou et al. 2011a,b,c, 2013). Hence, cocoa bean fermentation is a spontaneous on-farm fermentation process, to which many but not all of the microorganisms contaminating the cocoa pulp–bean mass contribute (Schwan and Wheals 2004; De Vuyst et al. 2010a). Cocoa bean fermentation takes place batchwise during 2 to 10 days, depending on the variety and local farm practices (fermentation method, methods of drainage and aeration, etc.), with 5 to 6 days being the most usual, whether or not accompanied by turning of the cocoa pulp–bean mass daily or every 72 h. Criollo beans ferment fast (2–3 days) and in the past have been reported to have a highly regarded aromatic (caramel, nut, or honey), slightly bitter, but usually smooth, cocoa flavor. Forastero beans ferment slowly (5–7 days) and their flavor produced upon proper processing is stronger (bitter and strong basic cocoa flavor). Trinitario beans provide a less intense but sufficient fine taste, which is intermediary between Criollo and Forastero beans. The Arriba-type beans (Nacional) produce cocoa with a fine flavor (floral and fruity notes), very aromatic and with a slight bitterness.

Besides the type of the cocoa beans, critical process parameters regarding agricultural and farm practices for successful cocoa bean fermentation processes are the healthy status of the pods (depending on appropriate agricultural practices such as well-maintained plantations, grafted cocoa trees, use of fertilizers and sanitary management); ripeness of the pods (harvest time); post-harvest pod storage/time (as a means of pulp pre-conditioning to influence pulp volume and acid production during fermentation); separation of healthy and infected pods before opening; handling and transportation of the cocoa pulp–bean mass; removal of the placenta and pod particles after pod opening; quantity of beans and pulp; bean packing or spreading; fermentation and drying method; batch size; control of the temperature of the fermenting cocoa pulp–bean mass; duration of fermentation and drying; turning of the fermenting cocoa pulp–bean mass; cleanness of the utensils and apparatus used for opening the pods, storage and transportation of the cocoa pulp–bean mass, and fermentation and drying; seasonal variations and weather conditions; and so on. All these practices affect the flavor, composition, and quality of the fermented dry cocoa beans and, hence, the chocolates produced thereof (Biehl et al. 1989; Ardhana and Fleet 2003; Camu et al. 2007, 2008a,b; Adeyeye et al. 2010; Caligiani et al. 2010; Garcia-Armisen et al. 2010; Papalexandratou et al. 2011a,b,c, 2013; Emmanuel et al. 2012; Afoakwa et al. 2013; Saltini et al. 2013). For instance, storing the harvested pods for some days before opening has been considered beneficial for fermentation, as it results in a faster fermentation because the pulp sucrose is already converted into fructose and glucose by pulp invertase activity (Tomlins et al. 1993; Schwan 1998; Schwan and Wheals 2004). However, this can also be achieved by harvesting the pods at the right maturation degree. Hence, the pod ripeness at harvest time is a crucial factor and appears superior to post-harvest storage regarding fermented cocoa bean quality (Camu et al. 2008a,b). Also, fresh pods and beans can vary considerably in the ratio of pulp to beans and in the amount of carbohydrates per bean, depending on cultivar, growing conditions, and post-harvest storage of pods. For instance, pulp and beans can contain more water during the wet season, which influences fermentation by affecting aeration of the fermenting cocoa pulp–bean mass and thus cocoa bean acidity. In such circumstances, up to 20% of the pulp of the total fresh cocoa pulp–bean mass can be removed, either through a natural "sweatings" drain, with mechanical depulpers, or through addition of pectinolytic enzymes or controlled fermentation, which enables reduction of the fermentation time from 7 to 4 days and hence lowers the acidity and enhances the flavor of the final cocoa (Schwan et al. 1995; Schwan and Wheals 2004; Leal et al. 2008; Crafack et al. 2013). As an example, in Malaysia, harvested pods are stored for up to 15 days before breaking and removal of the beans or the beans are pressed or pre-dried to reduce the pulp volume before fermentation (Lehrian and Patterson 1983). Unfortunately, Malaysian cocoa produced on large scale is of poor quality, because the beans are either not intentionally fermented at all or the fermentation and drying processes are not carefully carried out. Typically, fermented cocoa beans produced in Brazil and Malaysia present incomplete box fermentation with low pH (below the recommended value of pH 5.0–5.5), mainly due to low-care practices concerning pod choice and bean pre-processing. To ensure a uniform fermentation, to enhance growth of beneficial microorganisms, and to discourage surface mold growth, fermentations need to be carried out with healthy cocoa beans from ripe pods and heaps and boxes need to be turned or mixed every two days, which is done practically by forming another heap or by moving the beans from one box or box compartment to another, respectively (Baker et al. 1994; Leiter and Harding 2004). This turning aerates the cocoa pulp–bean mass and lets the temperature drop immediately after mixing, which later on rises again as more fermentation proceeds. However, turning of small heaps may favor growth of acetic acid bacteria (AAB) too much, and hence cause a too high acidity of the fermented dry cocoa beans (Camu et al. 2008a). Tray-fermented cocoa beans do not require turning, as air is allowed to circulate between the trays and sweatings drain away through the bottom of all trays. They form a promising alternative for heaps and boxes at farm scale (www.tomsekstra.dk). As turning of the fermenting cocoa pulp–bean mass seems to be a determining factor for good fermentation practices, fermentations in appropriate bioreactors have been proposed (Schwan and Wheals 2004; Pereira et al. 2013). Although mechanical turning is not a realistic alternative and forced aeration requires careful control, drum fermentors and stainless steel vessels have been constructed. Also, research has been performed with plastic vessels (Lefeber et al. 2011a; Ho et al. 2014). However, as fermentation of solid material is difficult to control, in particular with respect to homogeneity (gradient formation) and physical (e.g., temperature, pH, dissolved oxygen) and chemical factors (e.g., nutrient availability), new designs of fermentors are needed to have some of these process parameters under control, at least between certain limits (for instance through mixing and cooling), and to speed up and control the cocoa pulp–bean mass fermentation process accurately.

16.6. Succession of Microorganisms during Cocoa Bean Fermentation

16.6.1. The Spontaneous Three-phase Cocoa Bean Fermentation Process

Successful cocoa bean fermentation requires a succession of different microbial groups, or better, of particular microbial activities, reflecting the microbiological quality of the initial cocoa pulp–bean mass (microbial load and species diversity), environmental factors (temperature, pH, and oxygen tension), and the metabolism of substrates and metabolites available in the cocoa pulp–bean mass (Ardhana and Fleet 2003; Schwan and Wheals 2004; Camu et al. 2007; De Vuyst et al. 2010a). Indeed, microbial metabolism is held responsible for the pH increase and high temperatures that develop in the fermenting cocoa pulp–bean mass, the diffusion of important compounds into and out of the beans upon fermentation, and a pH decrease and enzymatic conversions of constituents inside the beans during fermentation and drying. Under optimal conditions, these microbial and enzymatic activities should not exceed four days (Lefeber et al. 2012; Papalexandratou et al. 2013).

Although overlap often occurs, three main fermentation phases can be considered consecutively (Figure 16.4): phase 1 or the anaerobic growth of yeasts, phase 2 or the microaerobic growth of lactic acid bacteria (LAB) and phase 3 or the aerobic growth of AAB (Schwan et al. 1995; Barel 1998; Ardhana and Fleet 2003; Schwan and Wheals 2004; Camu et al. 2007; Thompson et al. 2007, 2013; De Vuyst et al. 2010a). In general, during fermentation, the total count of microorganisms present in the cocoa pulp–bean mass increases during the first 24–36 h and then stabilizes or gradually reduces. The initial microbial communities are variable in numbers, type, and distribution, depending on cocoa cultivar, pod quality (pod healthiness, ripeness and storage), and bean quality/pre-processing (see above), and hence the microbial community dynamics vary between fermentation practices (Camu et al. 2007, 2008a; Nielsen et al. 2007a; Papalexandratou et al. 2011a,b,c, 2013; Moreira et al. 2013).

The microbiology of cocoa bean fermentations carried out in Africa (Ghana, Ivory Coast, Nigeria), Latin-America (Belize, Brazil, Dominican Republic, Ecuador, Mexico, Trinidad), South-East Asia

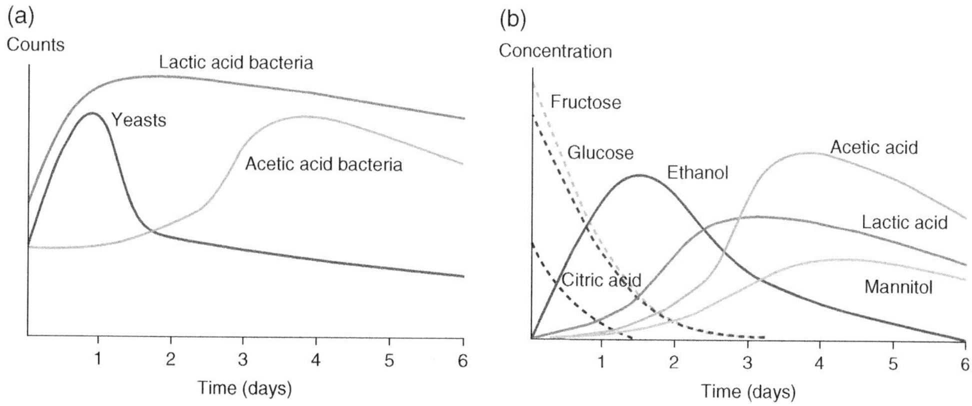

Figure 16.4. (a) Community dynamics and (b) metabolite kinetics during cocoa pulp–bean mass fermentation.

(Malaysia, Indonesia), and Oceania (Australia) has been studied in detail (Ostovar and Keeney 1973; Carr et al. 1979; Passos et al. 1984a,b; Ravelomanana et al. 1985, 1986; Schwan et al. 1995; Ardhana and Fleet 2003; Jespersen et al. 2005; Camu et al. 2007, 2008a,b; Nielsen et al. 2005, 2007a; Lagunes-Gálvez et al. 2007, Thompson et al. 2007, 2013; Kostinek et al. 2008; Daniel et al. 2009; Garcia-Armisen et al. 2010; Papalexandratou et al. 2011a,b,c, 2013; Papalexandratou and De Vuyst, 2011; Meersman et al. 2013; Ho et al. 2014). Up-to-date polyphasic studies have dealt with both culture-dependent (encompassing classical microbiological and state-of-the-art molecular identification techniques) and culture-independent methods (in particular denaturing gradient gel electrophoresis of polymerase chain reaction (PCR)-derived 16S rRNA (bacteria)/26S rRNA (yeast) gene amplicons (Nielsen et al. 2005, 2007a; Camu et al. 2007, 2008a; Papalexandratou et al. 2011a,b,c, 2013; Papalexandatrou and De Vuyst, 2011; Moreira et al. 2013), besides a 16S rDNA clone library (Garcia-Armisen et al. 2010) and a metagenomic sequencing study of (a) fermenting cocoa pulp–bean mass sample(s) (Illeghems et al. 2012)) as well as metabolite target analyses (Nielsen et al. 2007a; Camu et al. 2007, 2008a,b; Papalexandratou et al. 2011a,b,c, 2013; Moreira et al. 2013; Ho et al. 2014), including chocolate production (Camu et al. 2008a,b; Papalexandratou et al. 2011b,c, 2013; Ho et al. 2014). The DNA-based molecular methods applied have revealed that not only a number of different species of yeasts, LAB, and AAB is important for cocoa bean fermentation, but that within the different species particular strains may be involved and of importance as well, both often in a certain succession (Jespersen et al. 2005; Camu et al. 2007; Nielsen et al. 2007a).

Interestingly, it turned out that the fermenting cocoa pulp–bean mass is a treasure chest of new yeast, LAB, and AAB species, as exemplified by the isolation and new species descriptions of the yeasts *Candida awuaii* (Nielsen et al. 2010), *C. halmiae* (Nielsen et al. 2010), and *Geotrichum ghanense* (Nielsen et al. 2010), the LAB *Lactobacillus ghanensis* (Nielsen et al. 2007b), *Weissella ghanensis* (De Bruyne et al. 2008), *Lact. cacaonum* (De Bruyne et al. 2009), *Lact. fabifermentans* (De Bruyne et al. 2009), *W. fabaria* (De Bruyne et al. 2010), and *W. fabalis* (Snauwaert et al. 2013), the AAB *Acetobacter ghanensis* (Cleenwerck et al. 2007) and *A. fabarum* (Cleenwerck et al. 2008), and other not yet identified species (Papalexandratou et al. 2011c, 2013; Pereira et al. 2013).

16.6.2. Yeast Fermentation

During the first 24–48 h of cocoa bean fermentation (phase 1), yeasts convert saccharides (sucrose, glucose and/or fructose) into alcohol (ethanol) under conditions of high carbohydrate concentrations (characteristic for the fresh cocoa pulp–bean mass), limited oxygen availability (due to tight packing of the cocoa pulp–bean mass) and a pH below 4.0 (due to the high content of citric acid in the cocoa pulp-bean mass). They grow from an initial 2–7 to a maximum 6–9 log colony forming units (CFU) per gram of fermenting cocoa pulp–bean mass. The remaining pulp sucrose can be converted by yeast invertase into glucose and fructose, releasing 18.8 kJ of heat per mole of sucrose that is hydrolyzed. However, as mentioned above, sucrose concentrations of the pulp depend on pod ripeness/pulp pre-conditioning and hence pulp invertase activity. Further, glucose is preferentially converted into ethanol by anaerobic yeast fermentation, leaving fructose for a great part unfermented and producing glycerol as a side-product. Pulp ethanol concentrations as a result of yeast fermentation of 0.5–6.0% (m/m) have been reported in the literature. The production of ethanol is accompanied by a moderate temperature increase, as the conversion of saccharides into ethanol (and carbon dioxide, which maintains an anaerobic environment) is an exothermic process, producing 93.3 kJ of heat per mole of glucose or fructose, elevating the temperature of the fermenting cocoa pulp–bean mass from an ambient temperature of 25–30 °C to 35–40 °C within 48 h. Ethanol disappears upon fermentation due to its oxidation into acetic acid by AAB, oxidative consumption by aerobically growing yeasts upon cocoa bean fermentation, diffusion into the bean cotyledons, sweating, and evaporation. Also, yeasts produce organic acids (e.g., acetic acid and succinic acid), which may have a buffering capacity in the fermenting cocoa pulp–bean mass, as well as higher alcohols, aldehydes, ketones, and fatty acid esters, which may contribute to cocoa flavor or, more likely, to precursors of cocoa flavor, provided they diffuse from the pulp into the beans or are released inside the beans enzymatically.

Importantly, yeast activity causes reduction of viscosity and drainage of the fermenting cocoa pulp–bean mass, due to secretion of pectinolytic enzymes that break down the cell walls in the pulp (Gauthier et al. 1977; Sanchez et al. 1984, 1985; Schwan et al. 1995, 1997; Ardhana and Fleet 2003; Crafack et al. 2013). Optimal pectinolysis is required for the thickness of the shells of the resulting fermented dry cocoa beans (Crafack et al. 2013). Due to collapse of the parenchyma cells in the pulp, the spaces formed between the beans allow air to enter, which promotes the growth of bacteria, in particular LAB and AAB. To speed up the fermentation process and to improve the quality of the final products, pectinases may be added to the pulp or starter culture strains overproducing pectinolytic enzymes may be used to enhance sweating production (Schwan and Wheals 2004; Leal et al. 2008; Crafack et al. 2013).

Several studies have revealed the involvement of an abundant yeast diversity, encompassing species of *Candida* (including former *Torulopsis*), *Cryptococcus*, *Debaryomyces*, *Hanseniaspora* (and their anamorphs *Kloeckera*), *Kluyveromyces* (including *Lachancea thermotolerans*, formerly *Debaryomyces thermotolerans*), *Kodamaea*, *Meyerozyma* (formerly *Pichia*), *Millerozyma* (formerly *Pichia*), *Pichia* (including former *Issatchenkia* and *Hansenula*), *Rhodotorula*, *Saccharomyces* (including *Kazachstania exigua*), *Saccharomycopsis*, *Schizosaccharomyces*, *Torulaspora*, *Yamadazyma* (including former *Pichia*), *Yarrowia*, *Vanderwaltozyma* (including former *Kluyveromyces*), *Wickerhamomyces* (including former *Pichia*), and *Zygosaccharomyces* (Schwan et al. 1995; Fowler et al. 1998; Ardhana and Fleet 2003; Schwan and Wheals 2004; Jespersen et al. 2005; Lagunes-Gálvez et al. 2007; Nielsen et al. 2005, 2007a; Thompson et al. 2007, 2013; Daniel et al. 2009; Papalexandratou et al. 2011c, 2013; Papalexandratou and De Vuyst, 2011; Illeghems et al. 2012; Meersman et al. 2013; Ho et al. 2014). The most frequently mentioned yeast species are *Saccharomyces cerevisiae* (synonym *S. chevalieri*), *Hanseniaspora guilliermondii* (anamorph *Kloeckera apis*), *H. opuntiae*, *H. uvarum*, *Pichia kudriavzevii* (formerly *Issatchenkia orientalis*, anamorph *C. krusei*), *P. membranifaciens*, *P. manshurica*, *P. kluyveri*, *P. fermentans*, *P. anomala* and *Kluyveromyces marxianus*. Both culture-dependent and culture-independent investigations have shown the dominant role of *H. guilliermondii/opuntiae/uvarum*, *P. kudriavzevii*, *P. manshurica/membranifaciens*, and/or *S. cerevisiae* in several cocoa bean fermentation processes (Jespersen et al. 2005; Nielsen et al. 2005, 2007a; Daniel et al. 2009; Papalexandratou and De Vuyst 2011; Papalexandratou et al. 2011c, 2013; Meersman et al. 2013; Moreira et al. 2013). Moreover, a metagenomic study has unraveled the importance of *H. uvarum*, *H. opuntiae*, and *S. cerevisiae* (in decreasing order; Illeghems et al. 2012).

Besides a rather wide yeast species diversity, several strains of the same species may occur within one cocoa bean fermentation process. For instance, in the study of Daniel et al. (2009), strains of the three main species (*P. kudriavzevii*, *S. cerevisiae*, and *H. opuntiae*) and *P. kluyveri* var. *kluyveri* are represented by multiple strains, as minor sequence variation in rRNA gene cluster sequences and the actin (*ACT1*) gene among different isolates are found. Also, in the case of the dominant yeast species *P. kudriavzevii*, *S. cerevisiae*, *P. membranifaciens*, *P. kluyveri*, and *H. guilliermondii* in the study of Jespersen et al. (2005), strain variation has been examined by chromosome length polymorphism, indicating a considerable diversity of individual strains being present in cocoa bean fermentation processes, on the pod surfaces and at different production sites.

Some researchers have reported on several species that were not found in cocoa bean fermentation processes before (Daniel et al. 2009; Papalexandratou and De Vuyst 2011; Papalexandratou et al. 2011c; Meersman et al. 2013; Moreira et al. 2013). However, species solely isolated during single fermentations, at the start of the fermentations, and from the associated materials, might be considered as members of the surrounding ecosystems that occasionally become established during cocoa bean fermentation, if their physiological properties favor growth under the fermentation conditions in place. This holds true for bacteria as well.

In general, potentially due to its rapid growth, pectinolytic activity, and ethanol tolerance, the fermentative yeast *S. cerevisiae* is the most often isolated (culture-dependent) and detected species (PCR-DGGE) and also the most abundant species during cocoa bean fermentation, usually after an initial fermentation phase dominated frequently by *H. guilliermondii/opuntiae/thailandica/uvarum* (Nielsen et al. 2007a; Daniel et al. 2009; Papalexandratou and De Vuyst 2011; Papalexandratou et al. 2011c, 2013; Crafack et al. 2013; Meersman et al. 2013; Moreira et al. 2013; Pereira et al. 2012, 2013). For instance, Ardhana and Fleet (2003) have shown that *H. guilliermondii* grows rapidly at low pH during the first 24–36 h of fermentation, followed by more ethanol- and heat-tolerant *S. cerevisiae* and *C. tropicalis* at 48–120 h during wooden box fermentation of cocoa beans in East Java. Also, during the initial phase of cocoa bean heap fermentations carried out in Ghana, it has been shown that *H. guilliermondii/opuntiae/thailandica* plays an important role (Jespersen et al. 2005; Nielsen et al. 2005, 2007a; Daniel et al. 2009; Meersman et al. 2013), as is the case for cocoa bean box fermentations in the Dominican Republic (Lagunes-Gálvez et al. 2007). Upon prolonged fermentation, the more ethanol-tolerant yeast species, namely *S. cerevisiae*, *P. kudriavzevii*, and *P. membranifaciens*, become predominant in some cocoa bean fermentations (Ardhana and Fleet 2003; Jespersen et al. 2005; Nielsen et al. 2005, 2007a; Daniel et al. 2009; Papalexandratou et al. 2011c, 2013; Papalexandratou and De Vuyst 2011; Meersman et al. 2013; Ho et al. 2014). This yeast distribution is reflected in the fermentation capacity of yeasts and LAB and their co-existence in the beginning of the cocoa bean fermentation process (Daniel et al. 2009). Whereas LAB are responsible for a fast fructose and citrate consumption at the start of the cocoa bean fermentation process, *H. opuntiae* is not able to assimilate citrate and grows – together with other yeasts – on sucrose, glucose, and fructose as energy sources. The tolerance of low pH values of *H. opuntiae* is in accordance with environmental conditions prevailing during initial cocoa bean fermentation (low pH at the start of the fermentation process due to a high citrate concentration), *S. cerevisiae* being the most sensitive yeast species towards the lowest pH values and hence proliferating after *H. opuntiae* (Daniel et al. 2009; Papalexandratou and De Vuyst 2011). Alternatively, *P. kudriavzevii* is acid tolerant and may ferment citric acid and malic acid. Finally, *Pichia* and *Hanseniaspora* species may compete for the same substrates. Higher temperatures and ethanol, lactic acid, and acetic acid concentrations throughout a cocoa bean fermentation process, due to increased microbial activities, influence the survival of various cocoa-specific yeast species, as their tolerance towards these external factors are variable (Ardhana and Fleet 2003; Jespersen et al. 2005; Nielsen et al. 2005, 2007a;

Daniel et al. 2009; Papalexandratou and De Vuyst 2011; Papalexandratou et al. 2011c, 2013; Pereira et al. 2012; Crafack et al. 2013). As the variation of yeast species is much larger than that of bacterial species, the presence of the former may have profound effects on the fermentation efficiency and fermented cocoa bean quality, either directly or indirectly (Meersman et al. 2013).

However, the limited number of yeasts capable of growing at 45 °C and higher explains the disappearance of the yeast communities once ethanol oxidation by AAB, causing a substantial temperature increase during cocoa bean fermentation, has started, together with exhaustion of the appropriate energy sources, increasing pH, and greater aeration (Schwan and Wheals 2004; Camu et al. 2007; Daniel et al. 2009; De Vuyst et al. 2010a).

16.6.3. LAB Fermentation

Continuing yeast activity upon cocoa bean fermentation causes further ethanol production and more pulp draining away, allowing more air ingress, while both the temperature and pH of the fermenting cocoa pulp–bean mass increase up to 40 °C and pH 4.0, respectively. These conditions are favorable for the growth of bacteria. During phase 2, around 24–72 h into fermentation, the microaerobic and aciduric LAB present from the start of the fermentation and possibly interacting with yeasts (i.e., production of carbon dioxide by yeasts stimulating LAB growth, supply of simple saccharides (yeast invertase activity) and growth factors such as vitamins (upon yeast cell lysis)) grow rapidly, from an initial 3–5 to a maximum 7–9 log CFU per gram of fermenting cocoa pulp–bean mass. This coincides with a rapid decline in the dominance of the yeast communities.

LAB ferment a wide range of saccharides (mainly glucose and fructose), either homo- or hetero-fermentatively, into lactic acid, acetic acid, ethanol, and/or carbon dioxide, and convert some organic acids into lactic acid (for instance, citric acid and malic acid) or acetic acid (for instance citric acid, which leads to extra ATP generation). Hence, lactic acid and acetic acid are the main fermentation end-products of LAB metabolism during the cocoa bean fermentation process. Although most LAB species prefer glucose as energy source, some LAB species are fructophilic (preferential use of fructose as energy source; use of glucose in the presence of alternative external electron acceptors) or use fructose as an alternative external electron acceptor (heterofermentative LAB species) and convert it into mannitol (Camu et al. 2007, 2008a). The relative proportion of all these fermentation end-products will change the composition of the fermenting cocoa pulp–bean mass and consequently may influence the microbial successions. The lactic acid concentration of the fermenting cocoa pulp–bean mass peaks around 0.5% (m/m).

Early and classical microbiological studies have revealed a dominance of various species of *Lactobacillus* compared with species of *Enterococcus*, *Lactococcus*, *Leuconostoc*, *Pediococcus*, and/or *Weissella* (Ostovar and Keeney 1973; Carr et al. 1979; Passos et al. 1984a,b; Schwan et al. 1995; Thompson et al. 2007, 2013; Camu et al. 2007, 2008a; Lagunez-Gálvez et al. 2007; Nielsen et al. 2007a; Kostinek et al. 2008; Papalexandratou et al. 2011a,b,c, 2013; Lefeber et al. 2011b; Meersman et al. 2013; Ho et al. 2014; Table 16.1). In general, it appears that heterofermentative LAB occur in higher numbers than homofermentative LAB and that *Lactobacillus* spp. prevail during cocoa bean fermentation processes. Based on classical microbiological analysis, the main LAB species described are *Lact. fermentum* and *Lact. plantarum* (Carr et al. 1979; Camu et al. 2007, 2008a; Lagunez-Galvez *et al.* 2007; Nielsen et al. 2007a; Kostinek et al. 2008; Papalexandratou et al. 2011b,c, 2013; Lefeber et al. 2011b; Meersman et al. 2013; Moreira et al. 2013; Ho et al. 2014). For instance, Ardhana and Fleet (2003) have shown that *Lact. cellobiosus* (later synonym of *Lact. fermentum*) and *Lact. plantarum* are the dominant LAB species involved in Indonesian cocoa bean box fermentations. Whereas *Lact. plantarum* was especially significant during the first 24 h of fermentation, *Lact. cellobiosus* dominated 36–48 h and *Lact. hilgardii* died off by 48 h into fermentation. Cocoa bean heap and tray fermentations carried out in Nigeria were characterized by an

Table 16.1. Diversity of lactic acid bacteria associated with cocoa bean fermentations in different countries (original species names are used). *Ent.*, *Enterococcus*; *L.*, *Lactococcus*; *Lact.*, *Lactobacillus*; *Leuc.*, *Leuconostoc*; *O.*, *Oenococcus*; *Ped.*, *Pediococcus*; *W.*, *Weissella*.

Country – fermentation method	Microorganisms found	Reference
Trinidad – box	*L. lactis*, *Lact. fermentum*, *Lact. plantarum*, *Leuc. mesenteroides*	Ostovar and Keeney (1973)
Ghana – box	*Lact. collinoides*, *Lact. fermentum*, *Lact. mali*, *Lact. plantarum*	Carr et al. (1979)
Ghana – heap and tray	Culture-dependent analysis: *L. lactis* (heap only), *Lact. brevis*, *Lact. fermentum*, *Lact. ghanensis*[1] (heap only), *Lact. hilgardii* (heap only), *Lact. plantarum*, *Lact. rossii* (heap only), *Leuc. pseudoficulneum*, *Leuc. pseudomesenteroides*, *Ped. acidilactici*	Nielsen et al. (2007a)
	Culture-independent analysis (PCR-DGGE): *Lact. fermentum*, *Lact. plantarum*, *Leuc. pseudoficulneum*, *Leuc. pseudomesenteroides*, *Ped. acidilactici*	
Ghana – heap	Culture-dependent analysis: *Ent. casseliflavus*, *Enterococcus* spp., *L. lactis* subsp. *lactis*, *Lact. brevis*, *Lact. cacaonum*[2], *Lact. fabifermentans*[2], *Lact. fermentum*, *Lact. plantarum*, *W. fabaria*[3], *W. ghanensis*[4], *W. paramesenteroides*, *Weissella* spp.	Camu et al. (2007, 2008a)
	Culture-independent analysis (PCR-DGGE): *Lact. fermentum*, *Lact. plantarum*, *Leuc. pseudomesenteroides*, *W. ghanensis*[4]	
	Culture-independent analysis (16S rRNA gene clone library): *Lact. fermentum*	Garcia-Armisen et al. (2010)
Ghana – tray	Culture-independent analysis (PCR-DGGE): *Fruct. pseudoficulneus*, *L. lactis* subsp. *lactis*; *Lact. fermentum*	Crafack et al. (2013)
Malaysia – box	*Lact. collinoides*, *Lact. plantarum*	Carr et al. (1979)
Malaysia – box	Culture-dependent analysis: *Fruct. tropaeoli*, *Lact. fermentum*, *Lact. pentosus*, *Lact. plantarum*, *Leuc. pseudomesenteroides*, *Ped. acidilactici*, *W. paramesenteroides*	Papalexandratou et al. (2013)
	Culture-independent analysis (PCR-DGGE): *Lact. fermentum*, *Lact. plantarum*, *Leuc. pseudomesenteroides*, *W. paramesenteroides*	
Malaysia – box	*Lact. fermentum*, *Lact. plantarum*, *Leuc. mesenteroides*, *Leuc. pseudomesenteroides*	Meersman et al. (2013)
Brazil – box	*L. lactis*, *Lact. acidophilus*, *Lact. brevis*, *Lact. casei*, *Lact. delbrueckii*, *Lact. lactis*, *Lact. plantarum*, *Leuc. mesenteroides*, *Ped. acidilactici*, *Ped. dextrinicus*	Passos et al. (1984a,b)
Brazil – box	*Lact. durianis*, *Lact. fermentum*, *Lact. mali*, *Lact. nagelii*, *Lact. plantarum*, *Leuc. pseudomesenteroides*, *Ped. acidilactici*	Papalexandratou et al. (2011a)
	Culture-independent analysis (16S rRNA gene clone library): *Lact. brevis*, *Lact. cacaonum*[2], *Lact. durianis*, *Lact. fermentum*, *Lact. mali*, *Lact. nagelii*, *Lact. plantarum*	Garcia-Armisen et al. (2010)
Brazil – box	Culture-dependent analysis: *Fruct. pseudoficulneus*, *Lact. fabifermentans*[2], *Lact. fermentum*, *Lact. plantarum*, *Fruct. pseudoficulneus*, *Ped. acidilactici*, *Weissella fabalis*[5]	Papalexandratou et al. (2011b)
	Culture-independent analysis (PCR-DGGE): *Lact. fermentum*, *Lact. plantarum*, *Lact. reuteri*, *Leuc. pseudomesenteroides*	
	Culture-independent analysis (metagenome): *L. lactis*, *Lact. brevis*, *Lact. casei*, *Lact. fermentum*, *Lact. plantarum*, *Lact. reuteri*, *Lact. rhamnosus*, *Leuc. mesenteroides*, *O. oeni*	Illeghems et al. (2012)
	Culture-independent analysis (16S rRNA gene clone library): *Lact. fermentum*, *Lact. vaginalis*	Garcia-Armisen et al. (2010)

Table 16.1. *(Continued)*

Country – fermentation method	Microorganisms found	Reference
Brazil – plastic containers/ stainless steel tank	Culture-dependent analysis: *Lact. plantarum*, *Lact. fermentum* Culture-independent analysis (PCR-DGGE): *Lact. plantarum*, *Lact. fermentum*	Pereira et al. (2013)
Brazil – box	Culture-dependent analysis: *Lact. plantarum*, *Lact. fermentum* Culture-independent analysis (PCR-DGGE): *Lact. plantarum*, *Lact. fermentum*	Pereira et al. (2012)
Brazil – stainless steel tank	Culture-dependent analysis: *Lact. plantarum*, *Lact. fermentum*, *Weissella* sp. Culture-independent analysis (PCR-DGGE): *Lact. plantarum*, *Lact. fermentum*	Pereira et al. (2013)
Brazil – box	*L. lactis*, *Lact. casei*, *Lact. fermentum*, *Lact. rhamnosus*	Moreira et al. (2013)
Belize – box	*Lact. brevis*, *Lact. buchneri*, *Lact. casei*, *Lact. casei* subsp. *pseudoplantarum*, *Lact. cellobiosus*, *Lact. delbrueckii*, *Lact. fermentum*, *Lact. fructivorans*, *Lact. gasseri*, *Lact. kandleri*, *Lact. plantarum*, *Leuc. mesenteroides*, *Leuc. oenos*, *Leuc. paramesenteroides*	Thompson et al. (2007)
Indonesia – box	*Lact. cellobiosus*, *Lact. hilgardii*, *Lact. plantarum*	Ardhana and Fleet (2003)
Dominican Republic – box	*Lact. brevis*, *Lact. pentosus*, *Lact. paracasei* subsp. *paracasei*, *Lact.* (*para*)*plantarum*	Lagunes-Gálvez et al. (2007)
Nigeria – heap and tray	*Enterococcus* spp. (tray only), *Lact. brevis*, *Lact. fermentum*, *Lact. plantarum*, *Ped. acidilactici*	Kostinek et al. (2008)
Ivory Coast – box	Culture-independent analysis (PCR-DGGE): *Lact. fermentum*	Papalexandratou et al. (2011a)
Ivory Coast – heap	Culture-independent analysis (PCR-DGGE): *Lact. fermentum*, *Leuc. pseudomesenteroides*	Papalexandratou et al. (2011a)
Ivory Coast – plastic vessels	*Fruct. pseudoficulneus*, *Lact. fermentum*, *Lact. plantarum*	Lefeber et al. (2011b)
Ecuador – box	Culture-dependent analysis: *Fruct. tropaeoli*, *L. lactis* subsp. *lactis*, *Lact. fermentum*, *Lact. satsumensis*, *Leuc. pseudomesenteroides* Culture-independent analysis (PCR-DGGE): *Fruct. tropeaoli*, *Lact. fermentum*, *Leuc. pseudomesenteroides*, *W. beninensis*, *W. fabaria*[3]	Papalexandratou et al. (2011c)
Ecuador – platform	Culture-dependent analysis: *Fruct. tropaeoli*, *L. lactis* subsp. *lactis*, *Lact. cacaonum*[2], *Lact. fabifermentans*, *Lact. fermentum*, *Lact. nagelii*, *Leuc. pseudomesenteroides*, *W. fabaria*[5] Culture-independent analysis (PCR-DGGE): *Fruct. tropeaoli*, *Lact. cacaonum*, *Lact. fermentum*, *Lact. nagelii*, *Lact. plantarum*, *Leuc. pseudomesenteroides*, *W. fabaria*[3]	Papalexandratou et al. (2011c)
Australia – vessels	*L. lactis*, *Lact. fermentum*, *Lact. plantarum*	Ho et al. (2014)

[1] Nielsen et al. (2007b),
[2] De Bruyne et al. (2009),
[3] De Bruyne et al. (2010),
[4] De Bruyne et al. (2008),
[5] Snauwaert et al. (2013)

increasing *Lact. plantarum* community in the first half of the fermentation process and an increasing *Lact. fermentum*/*Lact. brevis* community (tray) and decreasing *Pediococcus acidilactici* (heap and tray) and *Lact. fermentum*/*Lact. brevis* community (heap) in the second half of the fermentation process (Kostinek et al. 2008). Recently, it has been shown by both culture-dependent and culture-independent methods that *Lact. plantarum* and *Leuconostoc pseudomesenteroides* dominate mainly at the onset of cocoa bean fermentation processes and *Lact. fermentum* towards the end (Camu et al. 2007, 2008a; Nielsen et al. 2007a; Garcia-Armisen *et al.* 2010; Papalexandratou et al. 2011a,b,c, 2013; Lefeber et al. 2011b; Pereira et al. 2013). Also, fructophilic LAB species whether or not citrate-converting (e.g., *Fructobacillus pseudoficulneus*, *Fruct. tropeaoli*, *Lact. cacaonum*, *Lact. fabifermentans*, and *Lact. plantarum*) generally constitute a small part of the LAB communities, most often at the start of the cocoa bean fermentation process when fructose is in excess (Camu et al. 2007a; Nielsen et al. 2007a; Lefeber et al. 2011b; Papalexandratou et al. 2011b,c; Snauwaert et al. 2012). All these investigations indicate a rather restricted LAB species diversity, in particular *Lact. plantarum* and *Lact. fermentum* dominating the cocoa bean fermentation process worldwide. Moreover, *Lact. fermentum* seems to play a significant role. Other LAB species have been found, but they have often been isolated as single strains or identified as rare bands in DGGE LAB community profiles, which rather reflects opportunistic contamination from the environment.

Although a few yeasts assimilate citric acid (Jespersen et al. 2005; Daniel et al. 2009), the LAB species are responsible for the breakdown of citric acid during the cocoa bean fermentation process (Camu et al. 2007, 2008a; Ho et al. 2014). Citric acid is converted via oxaloacetic acid into acetic acid and pyruvic acid, the latter in turn being converted into end-products of pyruvate metabolism, namely lactic acid, ethanol, formic acid, acetic acid, and/or flavor compounds such as acetaldehyde, diacetyl, acetoin, and/or 2,3-butanediol (Camu et al. 2007, 2008a; Lefeber et al. 2010, 2011a,b; Papalexandratou et al. 2011a,b,c, 2013; Adler et al. 2013). Consumption of citrate by LAB species prevailing at the early stages of the cocoa bean fermentation process, under low pH conditions, explains the importance of these bacteria during this fermentation stage, thereby avoiding competition with the depectinizing (citrate-negative) yeasts that degrade saccharides to ethanol anaerobically meanwhile.

Dominance of specific strains of *Lact. fermentum* throughout cocoa bean fermentation processes, usually in succession of *Lact. plantarum*, can be ascribed to their competitive metabolism (efficient heterofermentative use of glucose and fructose as energy source and external electron acceptor, respectively), their acid and ethanol tolerance (lower than for *Lact. plantarum*) and their oxygen tolerance (higher than for *Lact. plantarum*) (Camu et al. 2007, 2008a; Lefeber et al. 2010, 2011a; Adler et al. 2013). Indeed, monoculture fermentations in a cocoa pulp simulation medium (CPSM) for LAB indicate that cocoa-specific *Lact. fermentum* strains are well adapted to a fermenting cocoa pulp–bean mass environment, allowing heterolactate fermentation, citric acid conversion, and mannitol production, thereby producing less lactic acid than acetic acid (Lefeber et al. 2010, 2011a). Moreover, the metafluxome analysis of bacterial consortia composed of different strains of *Lact. fermentum* and *Lact. plantarum*, by means of isotope labeling experiments in CPSM for LAB, has revealed that the former species contributes to the overall carbon flux in the microbial communities dominantly through the phosphoketolase pathway, while the latter species exhibits a high glycolysis flux (Adler et al. 2013). However, significant differences among *Lact. fermentum* strains occur; some use citrate and fructose almost exclusively to maintain their redox balance (as well as to increase their intracellular pH and ATP generation) and channel only a low amount of fructose into central metabolism (heterolactate fermentation resulting in the production of lactic acid and acetic acid) as glucose is their energy source (favorable for cocoa bean fermentation processes), while others do not use fructose as an alternative external electron acceptor to produce mannitol but increase their citrate flux towards lactic acid to produce enough metabolic energy when glucose consumption is inefficient (Adler et al. 2013). Hence, a single strain of *Lact. fermentum* well-adapted to a fermenting

Table 16.2. Functional role of lactic acid bacteria associated with cocoa bean fermentation.

Functional property	Effect on cocoa bean fermentation
Consumption of citric acid	Causes a rapid start-up of the fermentation process under low-pH conditions, avoiding competition with sucrose- (glucose-) consuming, citrate-negative, depectinizing yeasts Causes a pH increase (replacement of citric acid by lactic acid and acetic acid), in turn controlling bacterial growth Beneficial for the production of pyruvate metabolites Stimulates amino acid conversions
Production of lactic acid	Plays a pH regulatory role Controls microbial contamination from the environment Beneficial for the growth of AAB (and yeasts)
Production of mannitol	Favors production of acetic acid (volatile) instead of lactic acid (non-volatile) out of fructose Beneficial for the growth of AAB
Production of flavor precursors (acetate, pyruvate catabolites, and amino acid conversion products)	Provides the necessary precursors for further cocoa flavor and color development during secondary processing
Creation of appropriate conditions for proteolysis, flavor precursor formation, and color development within the beans	Controls flavor (precursor) and color development in the beans

cocoa pulp-bean mass ecosystem may be able to dominate the microbial communities herein and hence drive the cocoa bean fermentation process.

Assimilation of citric acid causes the pH of the fermenting cocoa pulp–bean mass to increase from pH 3.5–4.0 to 4.2–5.0, allowing the growth of other bacteria (Camu et al. 2007, 2008a). As is the case for the yeasts, LAB numbers decline during later stages of the cocoa bean fermentation process, mainly due to exhaustion of energy sources, too high concentrations of ethanol, and a too high temperature.

To conclude, LAB do play an important role during the cocoa bean fermentation process, although differently to that thought in the past (Table 16.2).

16.6.4. AAB Fermentation

During phase 3, from 24–112 h into cocoa bean fermentation, aerobic AAB, which occur from the very early stages in the cocoa bean fermentation process and hence survive the initial steps of the fermentation (anaerobic conditions, lower pH than at the end), persist until the conditions for AAB growth are optimal. When more of the pulp is metabolized and drained away, aeration of the fermenting cocoa pulp–bean mass increases, and when the temperature rises above 37 °C, thermotolerant AAB grow from an initial 3–5 to a maximum 5–8 log CFU per gram of fermenting cocoa pulp–bean mass. Further, turning of large heaps and moving of the beans during cocoa bean box fermentation processes favor AAB growth (cell counts, acetic acid concentration) and increase the maximum fermentation temperature after a slight drop due to heat dissipation (Carr et al. 1979; Nielsen et al. 2007a; Camu et al. 2008a; Papalexandratou et al. 2011a,b,c 2013; Crafack et al. 2013). Concomitantly, turning of a small heap causes too much production of acetic acid, resulting in fermented dry cocoa beans with an elevated level of acetic acid, which in turn ends up in a sour-tasting chocolate (Camu et al. 2008a,b).

The main activity of AAB is the oxidation of ethanol, initially formed by the yeasts, into acetic acid by periplasmic pyrroloquinoline-quinone-dependent dehydrogenase activities coupled to the membrane-anchored respiratory chain that provide these bacteria with the necessary energy.

This exothermic reaction, producing 496 kJ of heat per mole of ethanol that is oxidized, is mainly responsible for the rise in temperature of the fermenting cocoa pulp–bean mass, which can reach 45–50 °C and even higher. When acetic acid production starts, a slight decrease in pH of the fermenting cocoa pulp–bean mass often takes place. Further, the decline of the concentration of ethanol coincides with a decline of the concentration of lactic acid, indicating simultaneous oxidation by AAB of ethanol into acetic acid and of lactic acid into acetic acid and/or acetoin (Adler et al. 2014; Moens et al. 2014). However, whereas strains of *A. pasteurianus* and *A. ghanensis* perform a fast oxidation of ethanol into acetic acid and of lactic acid into acetoin (*A. pasteurianus*) or a slow oxidation of lactic acid into acetoin plus acetic acid (*A. ghanensis*), strains of *A. fabarum* and *A. senegalensis* perform a slow oxidation of ethanol into acetic acid coupled with a fast oxidation of lactic acid into acetoin (*A. senegalensis*) or a slow oxidation of lactic acid into acetoin plus acetic acid (*A. fabarum*) (Moens et al. 2014). Acetoin can be considered an overflow metabolite of pyruvate out of lactic acid instead of the oxidation of pyruvate through the tricarboxylic acid (TCA) cycle, at least when ethanol is oxidized into acetic acid, not only due to repression of the TCA cycle by ethanol but also to speed up biomass formation, as lactic acid provides the cell with biomass precursor molecules (Adler *et al.* 2014). The acetic acid concentration of the fermenting cocoa pulp–bean mass peaks around 2.0% (m/m) and then it declines, probably due to its evaporation and further metabolism (Nielsen et al. 2007a; Camu et al. 2007, 2008a,b). Indeed, some AAB species are responsible for the over-oxidation of acetic acid into carbon dioxide and water, namely in the absence of ethanol and lactic acid (Moens et al. 2014), in turn liberating 1754 kJ of heat per mole of acetic acid that is over-oxidized. Oxidation of lactic acid and acetic acid also results in a pH increase.

As for the LAB species, the AAB species diversity associated with cocoa bean fermentation is restricted. Based on classical microbiological analysis, the AAB species involved are *A. aceti*, "*A. roseum*" and "*A. suboxydans*" in Trinidadian cocoa bean box fermentations (Ostovar and Keeney 1973); "*A. ascendens*", *A. rancens* (later synonym of *A. pasteurianus*), *A. xylinum* (now *Gluconacetobacter xylinus* subsp. *xylinus*) and *Gluconobacter oxydans* in Ghanaian cocoa bean heap fermentations (Carr et al. 1979); *A. lovaniensis*, *A. rancens*, *A. xylinum*, and *G. oxydans* in Malaysian cocoa bean box fermentations (Carr et al. 1979); *A. aceti* subsp. *liquefaciens* (now *Ga. liquefaciens*), *A. pasteurianus*, *A. peroxydans*, "*G. oxydans* subsp. *suboxydans*" (now *G. oxydans*) in Brazilian cocoa bean box fermentations (Passos and Passos 1985); and *Acetobacter* spp. and *G. oxydans* in Belizean cocoa bean box fermentations (Thompson et al. 2007, 2013). Applying molecular analysis, *A. pasteurianus* (first half of the cocoa bean fermentation), *A. syzygii* (first half) and *A. tropicalis* (later stages of the fermentation process), and occasionally *A. malorum* and *G. oxydans*, are the dominating AAB species during Ghanaian cocoa bean heap and tray fermentation processes (Nielsen et al. 2007a). *Acetobacter lovaniensis* is the sole species identified out of five isolates during a cocoa bean box fermentation process carried out in the Dominican Republic (Lagunes-Gálvez et al. 2007). Finally, *A. fabarum*, *A. ghanensis*, *A. senegalensis*, and *A. pasteurianus* are found during several cocoa bean fermentation processes, whereby most species are present at the beginning of the fermentation, while *A. pasteurianus* dominates upon fermentation (Nielsen et al. 2007a; Camu et al. 2007, 2008a; Papalexandratou et al. 2011b,c, 2013; Lefeber et al. 2011b; Meersman et al. 2013). Besides *A. pasteurianus*, *A. syzygii* and *A. tropicalis* occur frequently (Nielsen et al. 2007a; Crafack et al. 2013; Pereira et al. 2012, 2013). Next to *A. pasteurianus*, *G. frateurii* dominates North Queensland cocoa bean vessel fermentations (Ho et al. 2014). Also, other *Gluconobacter* (e.g. *G. oxydans*) and *Gluconacetobacter* species may be found (Schwan and Wheals 2004; Nielsen et al. 2007a; Garcia-Armisen et al. 2010; Papalexandratou et al. 2011b,c, 2013; Illeghems et al. 2012; Moreira et al. 2013; Pereira et al. 2013). In general, members of the genus *Acetobacter* are found more frequently than those of *Gluconobacter*.

The predominance of specific strains of *Acetobacter* (*pasteurianus*) over *Gluconobacter* can be ascribed to their preference for the oxidation of ethanol instead of glucose fermentation, their growth on lactic acid and mannitol, and their heat and ethanol tolerance (Ardhana and Fleet 2003; Camu et al. 2007, 2008a; Nielsen et al. 2007a; Papalexandratou et al. 2011b,c, 2013; Illeghems et al. 2013). Moreover, the oxidation of glucose, lactic acid, and mannitol by AAB is mediated by the presence of ethanol in the fermenting cocoa pulp–bean mass (Adler et al. 2014; Moens et al. 2014). Isolation of *Gluconobacter* may indicate poor fermentation of the cocoa pulp–bean mass, as glucose fermentation will dominate ethanol oxidation, resulting in organic acids other than acetic acid (e.g., gluconic acid) and other fermentation end-products by other bacteria that may proliferate on the available saccharides. While glucose should be exhausted at the time AAB prevail, the gluconic acid produced out of glucose by either *Gluconobacter* or *Acetobacter* may cause another yeast growth late into fermentation as well as deliver cocoa beans with off-flavors (Papalexandratou *et al.* 2011b; Ho et al. 2014; Moens et al. 2014). Also, ethanol oxidation by AAB too early into fermentation will result in too acid cocoa beans (Schwan 1998; Camu et al. 2008b; Papalexandratou et al. 2011a; Garcia-Armisen et al. 2010). Oxidation of mannitol (produced by LAB) into fructose is possible but undesirable, as this conversion also generates fermentable carbohydrates at the end of the cocoa bean fermentation process that subsequently can be used by undesirable spoilage bacteria and fungi that may proliferate when the cocoa bean fermentation process actually comes to an end (Papalexandratou et al. 2011c; Moens et al. 2014).

Finally, the exhaustion of ethanol and an increasing temperature of the fermenting cocoa pulp–bean mass cause the death of the AAB, which ultimately halts the whole cocoa bean fermentation process.

16.7. Biochemical Changes in the Cocoa Beans during Fermentation and Drying

Physico-chemical processes within the cocoa beans encompass diffusion, internal destruction (due to acetic acid ingress), and enzymatic processes (Figure 16.5), which are influenced by several environmental factors, such as timing of initial entry of molecules and heat, pH (determined by duration, speed and intensity of acidification), temperature (speed of fermentation), and moisture (speed of fermentation and drying) (Schwan et al. 1995; Thompson et al. 2007, 2013). Biochemical and physical changes within the bean cotyledons can be divided into two broad phases. The first

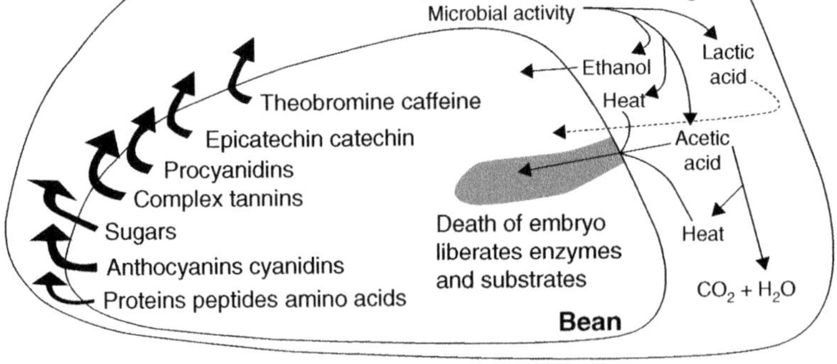

Figure 16.5. Biochemical changes and diffusion processes in pulp and beans during the cocoa bean fermentation process (after Lopez, 1986).

phase is the anaerobic hydrolytic phase that occurs during the first 3–4 days of the cocoa bean fermentation process and coincides with the yeast and LAB growth. It involves absorption of water and diffusion of substrates into the bean tissue during fermentation after seed death occurs. This allows hydrolysis of bean cotyledon proteins (by proteases), sucrose (invertase), and anthocyanins (glycosidases) (Biehl et al. 1993; Voight et al. 1994a,b; Voight and Biehl 1995; Amin et al. 1998; Hansen et al. 1998, 2000; Puziah et al. 1998a,b; Lercetau et al. 1999; de Brito et al. 2000; Laloi et al. 2002; Schwan and Wheals 2004; Afoakwa et al. 2008; Jinap et al. 2008; Kratzer et al. 2009). In particular, it has been shown that the concentration of free amino acids within the beans increases during fermentation (high-quality fermented cocoa beans contain approximately 8–14 mg of free amino acids per gram dry matter; Rohsius et al. 2006), which is attributed to proteolytic activities that degrade bean storage proteins; this involves liberation of hydrophilic peptides and hydrophobic amino acids from vicilin (globulin) through selective degradation and specific hydrolysis by means of an aspartic endoprotease and a serine carboxy-(exo)peptidase (Biehl et al. 1985; Kirchoff et al. 1989; Voight et al. 1994a,b; Amin et al. 1997; Hansen et al. 1998, 2000; Hashim et al. 1998; Buyukpamukcu et al. 2001; Rohsius et al. 2006). Whereas the bean invertase hydrolyses sucrose into glucose and fructose, glycosidases release saccharides (arabinose and galactose) from anthocyanins (cyanidin-3-α-*L*-arabinoside and cyanidin-3-β-*D*-galactoside, respectively) and glycoside-bound terpenes. Conversion of anthocyanins into anthocyanidins results in bleaching of the purple color of the beans. Liberated terpenes and terpene alcohols contribute floral notes to fermented cocoa beans, such as linalool (oxide) in fermented dry cocoa beans from Ecuador. The second phase is the aerobic oxidative phase, which coincides with the growth of AAB. Penetration of oxygen through the testa causes oxidative changes (oxidation of polyphenols to quinones by bean polyphenol oxidase) and complex formation of these very reactive compounds with other polyphenols and protein material (amino acids and proteins) at the surface of the bean cotyledons from the fourth day onwards. All enzymatic reactions are of short duration, as optimal conditions of pH (decreasing in the beans) and temperature (increasing in the beans) are transient upon fermentation. Some enzymes are strongly inactivated (aminopeptidase, invertase, and polyphenol oxidase) or partly inactivated (carboxypeptidase) upon cocoa bean fermentation, except for aspartic endoprotease and glycosidases that remain active during the whole cocoa bean fermentation process. For instance, if the pH of the bean cotyledons becomes too acid too soon ($pH < 4.5$), there will be both a final reduction in flavor precursors and an over-acidic final product. Indeed, a weak chocolate flavor is reported when cocoa bean proteolysis occurs at pH values below 5.0; the strongest chocolate flavor is produced in the pH range 5.0–5.5 (Biehl et al. 1985). Also, a certain amount of moisture (increasing during fermentation and decreasing during drying) is necessary to allow cocoa bean enzymes and their substrates to react to form the desirable products. Further, moisture helps the diffusion of ethanol, acetic acid, and to a lesser extent lactic acid inwards and polyphenols, alkaloids, and other compounds (e.g., saccharides and flavor compounds) outwards the beans, contributing not only to a reduction of astringency and bitterness but also to a loss of desirable compounds such as polyphenols (Camu et al. 2008b; Ho et al. 2014). Browning reactions (enzymatic and non-enzymatic) at the surface of the bean cotyledons, which are manifested towards the end of the cocoa bean fermentation process (aerobic phase) and are to some extent responsible for inhibiting enzyme activities, cause not only a brown color but also tanning of bean proteins (forming insoluble, high-molecular-mass compounds) and hence removal of astringency and bitterness. Also, the ability of tannin formation results in the reduction of off-flavors associated with the roasting of bean peptide and protein material. Finally, the oxidative phase contributes to the formation of ancillary flavors (e.g., fruity, floral, and spicy notes), in particular in fine cocoas.

Consequently, there is a need for a balance between the length of the cocoa bean fermentation process, environmental factors, and microbial activities that influence enzymatic activities within the bean cotyledons. Moreover, the heterogeneity of (large) heaps and slight, local differences in microbial counts and carbohydrates available for fermentation may determine flavor and color precursor formation within the beans. Alternatively, appropriate reducing sugars and free amino acids need to be available for Maillard reactions during roasting of fermented dry cocoa beans. All this may result in slight flavor differences, characterizing the resulting chocolate products (Camu et al. 2008a,b; Papalexandratou et al. 2011b,c, 2013; Ho et al. 2014). Controlled fermentation and addition of exogenous enzymes may contribute to steered flavor precursor formation (Yusep et al. 2002; Camu et al. 2008b; De Vuyst et al. 2010b; Crafack et al. 2013; Lefeber et al. 2012).

Apparently, there is no change in starch and fat content during cocoa bean fermentation and drying (Lehrian and Patterson 1983), although it is likely that fat content and composition of the beans are influenced not only by the cultivar but also by agricultural and farm practices (Lima et al. 2011). However, as the non-fat portion of the beans is the source of polyphenols, the polyphenol content and concomitant antioxidant properties of chocolate products is determined by the polyphenol levels in the cocoa solids of these products. Therefore, the flavor-enhancing properties of cocoa bean fermentation must be balanced against the polyphenol health effects, when polyphenol-enriched chocolate products are to be produced. The potential availability of cocoa beans with various degrees of fermentation provides the chocolate manufacturer with a first opportunity to produce chocolate products that may meet the diverse needs of the modern consumer (Thompson et al. 2007, 2013). For instance, if a chocolate with high polyphenol contents and soft flavor is desired, Forastero beans may be fermented for a short time. Alternatively, optimal processing of high-quality raw materials enables retention of the valuable polyphenols (www.acticoa.com).

16.8. Optimal Fermentation Course and End of Fermentation

The cocoa bean fermentation process lies at the basis of the entire chocolate-making process. An optimal fermentation course requires targeted microbial community dynamics, that is a correct succession of microbial and enzymatic activities, which are at the basis for the development of flavor precursors within the beans and their characteristic coloration. Therefore, an optimal internal pH of the beans and a normal temperature increase during fermentation has to be established. The minimum pH that gives acceptable cocoa mass is 5.2; however, the actual pH may be slightly lower (down to 4.8), which is related with acidity development during the cocoa bean fermentation process. The temperature increase of the fermenting cocoa pulp–bean mass reflects the actual functioning of the microorganisms. The rate of temperature increase is a quantitative measure for the type of fermentation (alcoholic, lactic and/or acetic), as it is correlated with the energy released and hence may be correlated with the eventual cocoa quality. Barel (1998) considers a rise in temperature to around 45 °C during the first 48 h of fermentation as satisfactory. This temperature increase is, however, mainly ascribed to the action of AAB. Therefore, according to Barel (1998), LAB are less desirable or not wanted at all during cocoa bean fermentation processes. If LAB are a high percentage of the total microbial communities during cocoa bean fermentation, high concentrations of lactic acid will be produced, as the breakdown of carbohydrates then contributes mainly to total acidity (more lactic acid formation by LAB and less ethanol production by yeasts) of the beans. Since lactic acid is not volatile, it will remain in the chocolate after manufacturing, producing excessive sourness that masks the chocolate flavour (Holm et al. 1993; Jinap 1994; Jinap and Zeslinda 1995). However, as described above, the functional role of LAB in developing well-fermented cocoa beans has been underestimated and a non-competitive cooperation between yeasts, LAB, and AAB exists in the fermenting cocoa pulp–bean mass; also, the diffusion of non-volatile lactic acid into the cocoa beans will be minimal (Camu et al. 2007). Finally, a well-performed cocoa bean fermentation process, as the result of good agricultural and farm practices and thus based on high-quality raw material, is dominated by a restricted yeast and bacterial species diversity, *in casu H. opuntiae/uvarum, S. cerevisiae, Lact. fermentum, and A. pasteurianus* (Illeghems et al. 2012; Lefeber et al. 2012; Papalexandratou et al. 2013).

As cocoa beans vary in their degree of fermentation, because of the unpredictable fermentation step due to its spontaneous nature, a fermentation index has been introduced as a quantitative measure for the degree of cocoa bean fermentation (Shamsuddin and Dimmick 1986; Romero et al. 2013). Cocoa beans can be under-fermented (too short fermentation or drying without fermentation, resulting in violet- or grey-colored beans, respectively, commonly designated as slaty beans), over-fermented (too long fermentation, resulting in black beans with a putrid or dull smell), or optimally fermented (exact fermentation time, resulting in brown-colored beans), with the concomitant sensory attributes of cocoa beans and chocolates made thereof. Traditionally, the cut test is performed (del Boca 1962). Cutting lengthwise 300 fermented cocoa beans taken randomly from a sample gives an idea of the quality of the beans, mainly based on intensity of color (grey, purple, brown, or black) and formation of a cracked interior (due to breakdown of the biological barriers), besides the occurrence of moldy beans (infected by molds), infested beans (insect-penetrated and -attacked beans), germinated beans or flat beans (no cotyledons or developed incompletely).

A key decision to obtain fermented cocoa beans of high quality is when to remove the beans from their fermentation environment to start the drying process. As mentioned above, when all of the ethanol is oxidized into acetic acid and then further over-oxidized into carbon dioxide and water, the fermentation stops, as energy sources for microbial growth are no longer available and temperature increases too high to allow survival of the desirable microorganisms. Extending the cocoa bean fermentation process can result in the production of off-flavors and putrefaction by undesirable microorganisms. The following factors may collectively indicate the status of the cocoa bean fermentation process: (i) a decrease in temperature of the fermenting cocoa pulp–bean mass when the microbial fermentation halts; (ii) a distinct smell of alcohol and acetic acid during the early and late stages of cocoa bean fermentation, respectively; (iii) a visual appearance of molds at the surface of the fermenting cocoa pulp–bean mass at late stages of the fermentation process; (iv) the external color of the cocoa beans varies from white-grey over purple to brown; (v) the internal color of the cocoa beans according to the bean cut test; and (vi) the internal pH of the cocoa beans. However, other factors may influence this decision, in particular when certain flavor and/or health-promoting compounds (e.g., polyphenols) of the fermented cocoa beans have to be valorized in the end-products. Therefore, there is a need for objective and reliable analytical methods to monitor the cocoa bean fermentation process (Aculey et al. 2010).

16.9. Further Processing of Fermented Cocoa Beans

16.9.1. Drying of Fermented Cocoa Beans

After 4–6 days of fermentation the drying process of the fermented cocoa beans can start. By then, the fermentation must be stopped completely, the moisture content of the cocoa beans must be brought from an initial 40–60% to 6–8% to avoid growth of molds within the beans during storage, and the major part of the acetic acid formed during fermentation must be eliminated (Thompson et al. 2007, 2013). The drying process relies on air movement to remove water. Both sun drying (5–10 days) and mechanical drying (maximum 60 °C for at least 48 h) are applied. Uniform drying is ensured by avoiding rain soaking and mixing the fermented cocoa beans regularly. Water removal results in a pronounced reduction in the number of viable microorganisms (Schwan and Wheals 2004). Furthermore, biochemical processes important for flavor and color development of the fermented cocoa beans, initiated during fermentation, continue during drying. A slow migration of moisture throughout the beans towards the shells will transport flavor precursors that were formed during fermentation. Hence, some oxidation will occur and some excess acids, in particular acetic acid, may volatilize through the shells, both of which phenomena are beneficial to obtain low-acid and less bitter fermented dry cocoa beans. Also, non-volatile lactic acid is partly transported by the water from the bean to the shell, reducing the perceived acid taste of fermented dry cocoa beans (Wood and Lass 1985; Nganhou et al. 2003; Thompson et al. 2007, 2013; Afoakwa 2010). Finally, rising temperatures and insufficient moisture become inhibiting factors of enzymatic activities. Also, the efficiency of the drying process will determine the shelf-life of the cured product. Dry fermented cocoa beans weigh on average one gram.

16.9.2. Roasting of Fermented Dry Cocoa Beans

Roasting (at 100–150 °C for 45–70 min) is the most important technological operation in the processing of fermented dry cocoa beans, as it produces the characteristic "cocoa flavor" out of flavor precursors, that is free (hydrophobic) amino acids, (hydrophilic) peptides, and reducing sugars (more fructose than glucose in fermented dry cocoa beans) formed during fermentation and drying, through non-enzymatic browning (Maillard) reactions, Strecker degradation, and/or thermal degradation (Ziegleder and Biehl 1988; de Brito et al. 2000; Granvogl et al. 2006; Beckett 2009; Afoakwa 2010). It brings about formation of the characteristic dark brown color, mild aroma, and texture of roasted beans. Concomitantly, the moisture content of the beans goes down to 2% and acetic acid is further liberated. Roasting conditions, mainly temperature and time, affect the properties of roasted beans, such as concentration of volatile flavor compounds, total acidity, polyphenols, and fat content. Pyrazines may be produced during fermentation, due to growth of bacilli (Hashim et al. 1998), but they are mainly formed during roasting and contribute earthy, roasty notes to the roasted cocoa beans; tetramethylpyrazines (milk coffee, mocha, roasted, green), together with 2,5-dimethylpyrazines (green, ether, rum), have been suggested as appropriate indicators of the roasting of cocoa beans (Jinap et al. 1998, 2008; Yusep et al. 2002). Other typical roasted cocoa flavor compounds include aldehydes (e.g., 2-methylpropanal, 2-methylbutanal, 3-methylbutanal, phenyl acetaldehyde, benzaldehyde, nonanal, hexanal), furans (e.g., 2-methyl-3-(methyldithio)furan), alcohols (e.g., 1,3- and 2,3-butanediol), higher alcohols (e.g., phenylethyl alcohol, isoamyl alcohol, 2-methyl-1-butanol, 3-methyl-1-butanol, 2-heptanol), ketones (e.g., 2,3-butanedione (diacetyl), 3-hydroxy-2-butanone (acetoin), 2-heptanone, 2-nonanone), organic acids (e.g., acetic acid, 2-methylbutanoic acid, 3-methylbutanoic acid, propanoic acid) and esters (e.g., isoamyl acetate, phenylethyl acetate, 2-methyl-1-butyl acetate), many of which contribute to pleasant fruity and floral aromas of cocoa (Ziegleder and Biehl 1988; Ziegleder 1991; Jinap et al. 1995; Puziah et al. 1998a,b; Frauendorfer and Schieberle 2006, 2008; Beckett 2009; Afoakwa 2010; Ho et al. 2014). Whereas the odor-active volatiles acetic acid (sour, vinegar-like note), 2-phenylethanol (flowery), 2- and 3-methylbutanoic acid (buttery, rancid, sweaty), 2- and 3-methylbutanal (malty, chocolate), and ethyl 2-methylbutanoate (fruity) are already formed during fermentation, others such as 2- and 3-methylbutanal, phenylacetaldehyde (sweet, honey-like) and 4-hydroxy-2,5-dimethyl-3(*2H*)-furanone (caramel-like, sweet) as the most important ones are intensified or formed during roasting and probably further steps of cocoa processing, depending on the chocolate recipe (Frauendorfer and Schieberle 2006, 2008; Afoakwa et al. 2008; Beckett 2009; Afoakwa 2010; Ho et al. 2014). Direct relationships are found between the initial composition of cocoa beans (depending on genotype and growth conditions of the crop), post-harvest processing (pod storage/pulp-preconditioning, cocoa pulp–bean mass preparation, fermentation, and drying), and subsequent processing (roasting, grinding, alkalization, and conching) and their effects on flavor formation and final flavor character in chocolate (Camu et al. 2008a,b; Afoakwa et al. 2008; Owusu et al. 2011; Lefeber et al. 2012; Papalexandratou et al. 2011b,c, 2013; Ho et al. 2014).

16.10. Use of Starter Cultures for Cocoa Bean Fermentation

16.10.1. Rationale

Technological challenges at the level of the natural cocoa bean fermentation process may have an impact on the quality of the cocoa beans and hence on the quality of the cocoa and chocolate made thereof. One may expect a possible impact not only on color and flavor development of the beans during fermentation, but also on carbohydrate and fat contents, on polyphenol and alkaloid contents, and so on. As cocoa bean fermentations are done in the field, in an artisan manner on a small scale or under often non-optimal conditions on a large scale, the results are variable in quality. In addition, there may be problems of acidity (due to a high content of organic acids because of poor fermentation or drying) or lack of cocoa flavor (due to poor or incomplete fermentation) and presence of off-flavors (due to over-fermented beans and spoilage, poor agricultural and fermentation practices, or poor mechanical drying), all of which lead to low crop value for the farmer. Therefore, it would be desirable to change cocoa bean fermentation from a wholly natural and unpredictable process to a controlled process, initiated with an appropriate starter culture, in which fermentation occurs more quickly and is steered. However, one of the reasons that chocolate quality has not been a priority for farmers is that it is possible with a relatively simple technology and minimal efforts to obtain appropriately fermented dry cocoa beans and there is no financial incentive to produce more consistent, predictable, and high-quality fermented cocoa beans. Over the last decades, multinational companies promoting the production of high-quality fermented dry cocoa beans have put most effort into two areas, namely encouraging farmers to maximize their production by socially, ethically, and environmentally sustainable farming practices and to improve their handling of the fermented dry cocoa beans. The fermentation process itself has been largely neglected.

The dairy, meat, and alcoholic beverage industries have largely replaced traditional natural fermentations with defined inocula of high-quality raw materials, strict control of the concomitant fermentation processes, better treatment of the final products, and diversification of the market in the second half of the twentieth century (Wood 1998, Hutkins, 2006; Leroy and De Vuyst 2004). Cocoa bean fermentations have still a long way to go before they reach that stage of development, but preliminary experiments in the use of defined starter cultures in such a complex fermentation process are promising, as long as good agricultural and farming practices and high-quality raw material are applied.

16.10.2. Experimental Use of Cocoa Bean Starter Cultures

Schwan (1998) used a defined microbial cocktail inoculum, consisting of one yeast species, *S. chevalieri*, two LAB species, *Lact. delbrueckii* subsp. *lactis* (formerly *Lact. lactis*) and *Lact. plantarum*, and two AAB species, *A. aceti* and *G. oxydans*, to perform cocoa bean fermentations in 200-kg wooden boxes with aseptically prepared cocoa beans, inoculated differently in time. The fermentation processes mimicked exactly the conditions in 800-kg boxes on Brazilian farms. With the zero-time inoculum the fermentation was almost identical to the spontaneous one. The fermentation with a phased-addition inoculum was similar, but slower and less pronounced, which led to a slightly poorer end-product. Although preliminary, these data show that the many common species of microorganisms found in spontaneous cocoa bean fermentations can be replaced by a judicious selection and concentration of members of each microbial group that is important for the process. There is of course need for improvement in the choice of species and the means of inoculation and fermentation when this technology is to be used commercially. For instance, independent studies on the pectinolytic enzyme, endopolygalacturonase, which is produced by *K. marxianus*, a yeast species isolated from spontaneous cocoa bean fermentations, suggest that this yeast species would be a better source of pectinase (Schwan et al. 1995, 1997). Therefore, Schwan (1998) proposed that a combination of

K. marxianus, as pectinase producer, and the naturally vigorous yeast, *S. cerevisiae*, as ethanol producer, might be a better yeast choice for future defined inocula. Moreover, inoculation of fresh cocoa pulp–bean mass with a *K. marxianus* hybrid yeast strain with increased pectinolytic activity has shown that it affects the microbial community structure during fermentation, increases by one third the volume of sweatings (and hence naturally increasing aeration), stimulates faster and improved cocoa bean protein degradation, results in reduced cocoa bean cotyledon acidity, and hence positively influences the fermented cocoa bean quality as well as the sensory quality of the chocolates produced thereof (Leal et al. 2008). The enhanced pectinolytic activity of *K. marxianus* has been confirmed during cocoa bean tray fermentation processes inoculated with a culture of this yeast species (Crafack et al. 2013).

There have been previous attempts to use yeast and bacterial starter cultures for cocoa bean fermentation processes, in particular yeasts for an enhanced production of pulp sweatings to produce pulp-derived products (wines, spirits, vinegars, jams), without influencing the quality of the fermented dry cocoa beans (Sanchez et al. 1985; Samah et al. 1992; Buamah et al. 1997; Dzogbefia et al. 1999). However, with respect to these earlier cocoa bean fermentation studies on starter cultures, no attempt was made to exclude the influence of the natural microbiota. Moreover, pectinolytic (*S. chevalieri*) and non-fermentative (*C. zeylanoides*) natural yeast isolates or yeasts from culture collections (*K. fragilis*) were tried and no bacteria were added or studied. Further, there was no evidence that the fermentation process could be accelerated or improved with these approaches regarding chocolate production. In the study of Samah et al. (1993), it has been shown that the AAB species *Ga. xylinus* subsp. *xylinus* (formerly *A. xylinum*), used as the sole inoculum, produces cocoa beans with higher pH and higher levels of acetic acid; however, a lower chocolate flavor is obtained, as compared with the control beans. Consequently, by setting the initial conditions, defined inocula could be considered as one of the ways that should lead to chocolate of more reliable and better quality.

One of the ultimate purposes of controlled cocoa bean fermentations will be not only to speed up the fermentation process but also to influence flavor and health properties of chocolate products through steered cocoa bean fermentation processes. Faster fermentation processes will be necessary to allow controlled, large-scale fermentations to increase the production capacity of fermented dry cocoa beans to respond to the growing demand for cocoa, although the cocoa crop is in danger worldwide. Flavor-steered fermentations will be necessary to produce marketable, tailor-made cocoa. Moreover, "steered" fermentations to maximize the (indigenous) level of health-promoting constituents will allow development of functional chocolate products. In this context, De Vuyst et al. (2010b) have shown that the choice of the starter culture does influence the flavour of chocolate made of the concomitant fermented dry cocoa beans, as the fermentation course (speed and community dynamics), substrates degraded, and metabolites produced are influenced. These researchers successfully used starter culture mixtures composed of a strain of *Lact. plantarum* (homolactic, citrate-non-fermenting, mannitol-non-producing, acid-tolerant), *Lact. fermentum* (heterolactic, citrate-fermenting, mannitol-producing, less acid-tolerant) and/or *A. pasteurianus* (ethanol- and lactic acid-oxidizing, acetic acid-over-oxidizing, acid-tolerant, heat-tolerant) to initiate cocoa bean heap fermentation processes in a natural (at the farm) and artificial environment (at a factory site) in Ghana. These strains were selected based on their kinetic analysis in cocoa pulp simulation media for LAB and AAB, respectively, during laboratory fermentations on 5-l scale (Lefeber et al. 2011b; Moens et al. 2014) and their competitiveness in the natural cocoa pulp–bean mass matrix during cocoa bean fermentation processes, using 20 kg of cocoa pulp–bean mass in plastic vessels, the latter being developed as screening device, successfully mimicking natural cocoa bean heap and box fermentation processes (Lefeber et al. 2011a). Moreover, these researchers have shown that a mixed starter culture, consisting of strains of *S. cerevisiae* (ethanol-tolerant, citric acid-non-converting, acid-tolerant, heat-tolerant), *Lact. fermentum*, and *A. pasteurianus* produces fermented dry cocoa beans of a superior quality compared to fermentation processes inoculated with a *Lact. fermentum*/*A. pasteurianus* starter culture, when applying 150-kg cocoa bean heap fermentation processes in the Ivory Coast and 500-kg cocoa bean box fermentation processes in Malaysia (Lefeber et al. 2012). With these uniformly fermented dry cocoa beans (uniform bulk cocoa displaying no variability in fermentation degree and flavor profiles), standard bulk milk and dark chocolates as regards the flavor profile could be produced, independently of the fermentation method and cocoa-producing region (and thus cocoa variety) but determined by good agricultural and fermentation practices and high-quality raw material. It should hence be possible to steer the flavor profile of fermented dry cocoa beans through fermentation. As an example, the use of highly aromatic yeasts, such as a *P. kluyveri* strain isolated from a spontaneous cocoa bean fermentation, together with strains of *Lact. fermentum* and *A. pasteurianus*, during a cocoa bean tray fermentation process, resulted in the production of chocolates (unconched) with a higher intensity of fruitiness and cocoa flavor, although the yeast culture did not dominate at any stage of the fermentation process (Crafack et al. 2013), indicating the importance of the ingress of flavor molecules into the cocoa beans during fermentation and/or drying. Finally, adding cocoa-adapted non-*Saccharomyces* yeast species to a yeast (*S. cerevisiae*)–LAB (*Lact. fermentum*)–AAB (*A. tropicalis*) starter culture to achieve a better control over cocoa bean fermentation processes (Pereira et al. 2012) has been suggested.

16.11. Concluding Remarks

It will be a challenge to introduce the application of starter cultures in cocoa bean fermentation practices, either to speed up and better control the fermentation process or to target the fermentation process towards predefined end-products. However, to scale up the initiation of cocoa bean fermentation processes with a defined inoculum to the farm level, whether applied to small farm holdings or large estates, will be an even greater challenge, especially as it relates to starter culture production and maintenance, inoculation and fermentation method, and last but not least farmers' training and extra costs.

References

Aculey, P.C., Snitkjaer, P., Owusu, M., et al. (2010) Ghanaian cocoa bean fermentation characterized by spectroscopic and chromatographic methods and chemometrics. *J Food Sci* **75**, S300–S307.

Adeyeye, E.I., Akinyeye, R.O., Ogunlade, I., et al. (2010) Effect of farm and industrial processing on the amino acid profile of cocoa beans. *Food Chem* **118**, 357–363.

Adler, P., Bolten, C.J., Dohnt, K., et al. (2013) Core fluxome and metafluxome of lactic acid bacteria under simulated cocoa pulp fermentation conditions. *Appl Environ Microbiol* **79**, 5670–5681.

Adler, P., Frey, L.J., Berger, A., et al. (2014) The key to acetate: metabolic fluxes of acetic acid acteria under cocoa pulp fermentation simulating conditions. *Appl Environ Microbiol* **80**, 4702–4716.

Afoakwa, E.O. (2010) *Chocolate Science and Technology*. Wiley-Blackwell, Ames, USA.

Afoakwa, E.O., Kongor, J.E., Takrama, J.F., and Budu, A.S. (2013) Changes in acidification, sugars and mineral composition of cocoa pulp during fermentation of pulp pre-conditioned cocoa (*Theobroma cacao*) beans. *Int J Food Res* **20**, 1215–1222.

Afoakwa, E.O., Paterson, A., Fowler, M., and Ryan, A. (2008) Flavour formation and character in cocoa and chocolate: a critical review. *Crit Rev Food Sc Nutr* **48**, 840–857.

Amin, I., Jinap, S., and Jamilah, B. (1997) Vicilin-class globulins and their degradation during cocoa fermentation. *Food Chem* **59**, 1–5.

Amin, I., Jinap, S., and Jamilah, B. (1998) Proteolytic activity (aspartic endoproteinase and carboxypeptidase) of cocoa beans during fermentation. *J Sci Food Agric* **76**, 123–128.

Ardhana, M. and Fleet, G. (2003) The microbial ecology of cocoa bean fermentations in Indonesia. *Int J Food Microbiol* **86**, 87–99.

Asep, E.K., Jinap, S., Tan, T.J., et al. (2008) The effects of particle size, fermentation and roasting of cocoa nibs on supercritical fluid extraction of cocoa butter. *J Food Eng* **85**, 450–458.

Baker, D.M., Tomlins, K.I., and Gray, C. (1994) Survey of Ghanaian cocoa farmer fermentation practices and their influence on cocoa flavour. *Food Chem* **51**, 425–431.

Barel, M. (1998) Première transformation du cacao. In *Cacao et Chocolat* ((ed.) J. Pontillion), Sciences et Techniques agroalimentaires, Lavoisier, Paris, France.

Bartley, B.G.D. (2005) *The Genetic Diversity of Cacao and its Utilization*, CABI Publishing, Cambridge, USA.

Beckett, S.T. (2009) *Industrial Chocolate Manufacture and Use*. 4th edn, John Wiley & Sons, Ltd., Chichester, United Kingdom.

Biehl, B., Brunner, E., Passern, D., et al. (1985) Acidification, proteolysis and flavor potential in fermenting cocoa beans. *J Sci Food Agric* **36**, 583–598.

Biehl, B., Heinrichs, J., Ziegeler-Berghausen, H., et al. (1993) The proteases of ungerminated cocoa seeds and their role in the fermentation process. *Angew Bot* **67**, 59–65.

Biehl, B., Meyer, B., Crone, G., et al. (1989) Chemical and physical changes in the pulp during ripening and post-harvest storage of cocoa pods. *J Sci Food Agric* **48**, 189–208.

Biehl, B., Passern, D., and Sagemann, W. (1982a) Effect of acetic acid on subcellular structures of cocoa bean cotyledons. *J Sci Food Agric* **33**, 1101–1109.

Biehl, B., Wewetzer, C., and Passern, D. (1982b) Vacuolar (storage) proteins of cocoa seeds and their degradation during germination and fermentation. *J Sci Food Agric* **33**, 1291–1304.

Buamah, R., Dzogbefia, V.P., and Oldham, J.H. (1997) Pure yeasts culture fermentation of cocoa (*Theobroma cacao* L.): Effect on yield of sweatings and cocoa bean quality. *World J Microbiol Biotechnol* **13**, 457–462.

Buyukpamukcu, E., Goodall, D.M., Hansen, C.E., et al. (2001) Characterization of peptides formed during fermentation of cocoa bean. *J Agric Food Chem* **49**, 5822–5827.

Caligiani, A., Acquotti, D., Cirlini, M., and Palla, G. (2010) ^{1}H NMR study of fermented cocoa (*Theobroma cacao* L.) beans. *J Agric Food Chem* **58**, 12105–12111.

Camu, N., De Winter, T., Addo, S. K., et al. (2008b) Fermentation of cocoa beans: influence of microbial activities and polyphenol concentrations on the flavour of chocolate. *J Sci Food Agric* **88**, 2288–2297.

Camu, N., De Winter, T., Verbrugghe, K., et al. (2007) Dynamics and biodiversity of lactic acid bacteria and acetic acid bacteria populations involved in spontaneous heap fermentations of cocoa beans in Ghana. *Appl Environ Microbiol* **73**, 1809–1824.

Camu, N., González, Á., De Winter, T., et al. (2008a) Influence of turning and environmental contamination on the dynamics of lactic acid bacteria and acetic acid bacteria populations involved in spontaneous cocoa bean heap fermentation in Ghana. *Appl Environ Microbiol* **74**, 86–98.

Carr, J.G., Davis, P.A., and Dougan, J. (1979) *Cocoa Fermentation in Ghana and Malaysia I*. Natural Resources Institute, Chatham, United Kingdom.

Carr, J.G. (1982) Cocoa. In: *Fermented Foods: Economic Microbiology* ((ed.) A.H. Rose), vol. 7, Academic Press, London, United Kingdom, pp. 275–292.

Cleenwerck, I., Camu, N., Engelbeen, K., et al. (2007) *Acetobacter ghanensis* sp. nov., a novel acetic acid bacterium isolated from traditional heap fermentations of Ghanaian cocoa beans. *Int J Syst Evol Microbiol* **57**, 1647–1652.

Cleenwerck, I., González, A., Camu, N., et al. (2008) *Acetobacter fabarum* sp. nov., an acetic acid bacterium from a Ghanaian cocoa bean heap fermentation. *Int J Syst Evol Microbiol* **58**, 2180–2185.

Coe, S. and Coe, M. (2007) *The True History of Chocolate*. Thames and Hudson, London, United Kingdom.

Crafack, M., Mikkelsen, M.B., Saerens, S., et al. (2013) Influencing cocoa flavour using *Pichia kluyveri* and *Kluyveromyces marxianus* in a defined mixed starter culture for cocoa fermentation. *Int J Food Microbiol* **167**, 103–116.

Crouzillat, D., Lerceteau, E., Petiard, V., et al. (1996) *Theobroma cacao* L: a genetic linkage map and quantitative trait loci analysis. *Theor Appl Genet* **93**, 1–2.

de Brito, E.S., Pezoa García, N.H., Gallão, M.I., et al. (2000) Structural and chemical changes in cocoa (*Theobroma cacao* L) during fermentation, drying and roasting. *J Sci Food Agric* **81**, 281–288.

Daniel, H.-M., Vrancken, G., Takrama, J.F., et al. (2009) Yeast diversity of Ghanaian cocoa bean heap fermentations. *FEMS Yeast Res* **9**, 774–783.

De Bruyne, K., Camu, N., De Vuyst, L., and Vandamme, P. (2009) *Lactobacillus fabifermentans* sp. nov. and *Lactobacillus cacaonum* sp. nov., isolated from Ghanaian cocoa fermentations. *Int J Syst Evol Microbiol* **59**, 7–12.

De Bruyne, K., Camu, N., De Vuyst, L., and Vandamme, P. (2010) *Weissella fabaria* sp. nov., from a Ghanaian cocoa fermentation. *Int J Syst Evol Microbiol* **60**, 1999–2005.

De Bruyne, K., Camu, N., Lefebvre, K., et al. (2008) *Weissella ghanensis* sp. nov., isolated from a Ghanaian cocoa fermentation. *Int J Syst Evol Microbiol* **58**, 2721–2725.

De Vuyst, L., Lefeber, T., Papalexandratou, Z. and Camu, N. (2010a) The functional role of lactic acid bacteria in cocoa bean fermentation. In: *Biotechnology of Lactic Acid Bacteria: Novel Applications* ((eds) F. Mozzi, R.R. Raya, and G.M. Vignolo). Wiley-Blackwell, Ames, USA, pp. 301–326.

De Vuyst, L., Takrama, J.F., Ampomah, Y.A., et al. (2010b) Influence of a lactic acid bacterium/acetic acid bacterium starter culture on cocoa bean heap fermentation dynamics and chocolate flavour. *Proceedings of the 16th International Cocoa Research Conference, Nusa Dua, Bali, Indonesia, November 16–20*, pp. 1325–1332.

del Boca, C. (1962) Cocoa beans: quality requirements and methods of assessment. *Rev Int Chocolaterie* **17**, 218–221.

Elwers, S., Zambrano, A., Rohsius, C., and Lieberei, R. (2010) Histologial features of phenolic compounds in fine and bulk cocoa seed (*Theobroma cacao* L.). *J Appl Bot Food Qual* **83**, 182–188.

Dzogbefia, V.P., Buamah, R., and Oldham, J.H. (1999) The controlled fermentation of cocoa (*Theobroma cacao* L.) using yeasts: enzymatic process and associated physico-chemical changes in cocoa sweatings. *Food Biotechnol* **13**, 1–12.

Emmanuel, O.A., Jennifer, Q., Agnes, S.B., et al. (2012) Influence of pulp-preconditioning and fermentation on fermentative quality and appearance of Ghanaian cocoa (*Theobroma cacao*) beans. *Int Food Res J* **19**, 127–133.

Fowler, M.S., Leheup, P. and Cordier, J.L. (1998) Cocoa, coffee and tea. In: *Microbiology of Fermented Foods* ((ed.) B.J.B. Wood), vol. **1**, Blackie Academic and Professional, London, United Kingdom.

Frauendorfer, F. and Schieberle, P. (2006) Identification of the key aroma compounds in cocoa powder based on molecular sensory correlations. *J Agric Food Chem* **54**, 5521–5529.

Frauendrofer, F. and Schieberle, P. (2008) Changes in key aroma compounds of Criollo cocoa beans during roasting. *J Agric Food Chem* **56**, 10244–10251.

Garcia-Armisen, T., Papalexandratou, Z., Hendrickx, H., et al. (2010) Diversity of the total bacterial community associated with Ghanaian and Brazilian cocoa bean fermentation samples as revealed by a 16S rRNA gene clone library. *Appl Microbiol Biotechnol* **87**, 2281–2292.

Gauthier, B., Guiraud, J., Vincent, J.C., et al. (1977) Comments on yeast flora from traditional fermentation of cocoa in the Ivory Coast. *Rev Ferment Ind Aliment* **32**, 160–163.

Gotsch, N. (1997) Cocoa biotechnology: status, constraints and future prospects. *Biotechnol Adv* **15**, 333–352.

Granvogl, M., Bugan, S. and Schieberle, P. (2006) Formation of amines and aldehydes from parent amino acids during thermal processing of cocoa and model systems: new insights into pathways of the Strecker reaction. *J Agric Food Chem* **54**, 1730–1739.

Hansen, C.E., del Olmo, M. and Burri, C. (1998) Enzyme activities in cocoa beans during fermentation. *J Sci Food Agric* **77**, 273–281.

Hansen, C.E., Mañez, A., Burri, C. and Bousbaine, A. (2000) Comparison of enzyme activities involved in flavor precursor formation in unfermented beans of different cocoa genotypes. *J Sci Food Agric* **80**, 1193–1198.

Hashim, P., Selamat, J., Muhammad, S.K.S. and Ali, A. (1998) Effect of mass and turning time on free amino acid, peptide-N, sugar and pyrazine concentration during cocoa fermentation. *J Sci Food Agric* **78**, 543–550.

Ho, V.T.T., Zhao, J. and Fleet, G. (2014) Yeasts are essential for cocoa bean fermentation. *Int J Food Microbiol* **174**, 72–84.

Holm, C.S., Aston, J.W. and Douglas, K. (1993) The effects of the organic acids in cocoa on the flavour of chocolate. *J Sci Food Agric* **61**, 65–71.

Hutkins, R.W. (2006) *Microbiology and Technology of Fermented Foods*. Blackwell, Ames, USA.

Illeghems, K., De Vuyst, L., Papalexandratou, Z., and Weckx, S. (2012) Phylogenetic analysis of a spontaneous cocoa bean fermentation metagenome reveals new insights into its bacterial and fungal community diversity. *PLoS ONE* **7**, e38040.

Illeghems, K., De Vuyst, L., and Weckx, S. (2013) Complete genome sequence and comparative analysis of *Acetobacter pasteurianus* 386B, a strain well-adapted to the cocoa bean fermentation ecosystem. *BMC Genomics* **14**, 526.

Jespersen, L., Nielsen, D.S., Hønholt, S., and Jakobsen, M. (2005) Occurrence and diversity of yeasts involved in fermentation of West African cocoa beans. *FEMS Yeast Res* **5**, 441–453.

Jinap, S. (1994) Organic acids in cocoa beans – a review. *ASEAN Food J* **9**, 3–12.

Jinap, S., Dimick, P.S., and Hollender, R. (1995) Flavour evaluation of chocolate formulated from cocoa beans from different countries. *Food Control* **6**, 105–110.

Jinap, S., Ikrawan, Y., Bakar, J., et al. (2008) Aroma precursors and methylpyrazines in underfermented cocoa beans induced by endogenous carboxypeptidase. *J Food Sci* **73**, 141–147.

Jinap, S., Rosli, W., Russly, A., and Nordin, L. (1998) Effect of roasting time and temperature on volatile component profiles during nib roasting of cocoa beans. *J Sci Food Agric* **77**, 441–448.

Jinap, S. and Zeslinda, A. (1995) Influence of organic acids on flavour perception of Malaysian and Ghanaian cocoa beans. *J Food Sci Technol* **32**, 153–155.

Jones, K.L. and Jones, S.E. (1984) Fermentations involved in the production of cocoa, coffee and tea. *Prog Ind Microbiol* **19**, 411–433.

Kadow, D., Bohlmann, J., Phillips, W., and Lieberei, R. (2013) Identification of main fine or flavor components in two genotypes of the cocoa tree (*Theobroma cacao* L.). *J Appl Bot Food Qual* **86**, 90–98.

Kirchoff, P.M., Biehl, B., and Crone, G. (1989) Peculiarity of the accumulation of free amino acids during cocoa fermentation. *Food Chem* **31**, 295–311.

Knight, I. (1999) *Chocolate Cocoa Health and Nutrition*. Blackwell Science Ltd., Oxford, United Kingdom.

Kostinek, M., Ban-Koffi, L., Ottah-Atikpo, M., et al. (2008) Diversity of predominant lactic acid bacteria associated with cocoa fermentation in Nigeria. *Curr Microbiol* **56**, 306–314.

Kratzer, U., Frank, R., Kalbacher, H., et al. (2009) Subunit structure of the vicilin-like globular storage protein of cocoa seeds and the origin of cocoa- and chocolate-specific aroma precursors. *Food Chem* **113**, 903–913.

Lagunes-Gálvez, S., Loiseau, G., Paredes, J.L., et al. (2007) Study on the microflora and biochemistry of cocoa fermentation in the Dominican Republic. *Int J Food Microbiol* **114**, 124–130.

Laloi, M., McCarthy, J., Morandi, O., et al. (2002) Molecular and biochemical characterization of two aspartic proteinases TcAP1 and TcAP2 from *Theobroma cacao* seeds. *Planta* **215**, 754–762.

Leal Jr., G.A., Gomes, L.H., Efraim, P., et al. (2008) Fermentation of cocoa (*Theobroma cacao* L.) seeds with a hybrid *Kluyveromyces marxianus* strain improved product quality attributes. *FEMS Yeast Res* **8**, 788–798.

Lefeber, T., Janssens, M., Camu, N., and De Vuyst, L. (2010) Kinetic analysis of strains of lactic acid bacteria and acetic acid bacteria in cocoa pulp simulation media toward development of a starter culture for cocoa bean fermentation. *Appl Environ Microbiol* **76**, 7708–7716.

Lefeber, T., Gobert, W., Vrancken, G., et al. (2011a) Dynamics and species diversity of communities of lactic acid bacteria and acetic acid bacteria during spontaneous cocoa bean fermentation in vessels. *Food Microbiol* **28**, 457–464.

Lefeber, T., Janssens, M., Moens, F., et al. (2011b) Interesting starter culture strains for controlled cocoa bean fermentation revealed by simulated cocoa pulp fermentations of cocoa-specific lactic acid bacteria. *Appl Environ Microbiol* **77**, 6694–6698.

Lefeber, T., Papalexandratou, Z., Gobert, W., et al. (2012) On-farm implementation of a starter culture for improved cocoa bean fermentation and its influence on the flavour of chocolates produced thereof. *Food Microbiol* **30**, 379–392.

Lehrian, D.W. and Patterson, G.R. (1983) Cocoa fermentation. In: *Biotechnology, a Comprehensive Treatise* ((ed.) G. Reed), vol. 5, Verlag Chemie, Basel, Switzerland, pp. 529–575.

Leiter, J. and Harding, S. (2004) Trinidad, Brazil, and Ghana: three melting moments in the history of cocoa. *J Rural Stud* **20**, 113–130.

Lercetau, E., Rogers, J., Pétiard, V. and Crouzillat, D. (1999) Evolution of cocoa bean proteins during fermentation: a study by two-dimensional electrophoresis. *J Sci Food Agric* **79**, 619–625.

Leroy, F. and De Vuyst, L. (2004) Lactic acid bacteria as functional starter cultures for the food industry. *Trends Food Sci Technol* **15**, 67–78.

Lima, L.J.R., Almeida, M.H., Nout, M.J.R., and Zwietering, M.H. (2011) *Theobroma cacao* L., "The food of the gods": quality determinants of commercial cocoa beans, with particular reference to the impact of the fermentation. *Crit Rev Food Sc Nutr* **52**, 731–761.

Lopez, A. S. (1986) *Proceedings of the Cacao Biotechnology Symposium* ((ed.) P.S. Dimick), The Pennsylvania State University, Pennsylvania, USA, pp. 19–53.

Lopez, A.S. and Dimick, P.S. (1995) Cocoa fermentation. In: *Enzymes, Biomass, Food and Feed*, ((eds) G. Reed and T.W. Nagodawithana), 2nd edn, VCH, Weinheim, Germany, pp. 561–577.

Maravalhas, N. (1966) Microbial deterioration of cocoa beans during fermentation and storage in Bahia. *Revue Int Chocolaterie* **21**, 375–378.

Meersman, E., Steensels, J., Mathawan, M., et al. (2013) Detailed analysis of the microbial population in Malaysian spontaneous cocoa pulp fermentations reveals a core and variable microbiota. *PLoS ONE* **8**, e81559.

Moens, F., Lefeber, T., and De Vuyst, L. (2014) Oxidation of metabolites highlights the microbial interactions and role of *Acetobacter pasteurianus* during cocoa bean fermentation. *Appl Environ Microbiol* **80**, 1848–1857.

Moreira, I.M., Miguel, M.G., Duarte, W.F., et al. (2013) Microbial succession and the dynamics of metabolites and sugars during the fermentation of three different cocoa (*Theobroma cacao* L.) hybrids. *Food Res Int* **54**, 9–17.

Motamayor, J.C., Lachenaud, P., da Silva e Mota, J.W., et al. (2008) Geographic and genetic population differentiation of the Amazonian chocolate tree (*Theobroma cacao* L.). *PLoS ONE* **3**, e3311.

Nganhou J, Njomo, D., Benet, J.C., et al. (2003) Perfecting a method of micro-analysis of water and acetic acid in a cocoa bean in the course of drying: applying to determine transportation coefficients. *Heat Mass Transfer* **39**, 797–803.

Nielsen, D.S., Jakobsen, M., and Jespersen L. (2010) *Candida halmiae* sp. nov., *Geotrichum ghanense* sp. nov. and *Candida awuaii* sp. nov., isolated from Ghanaian cocoa fermentations. *Int J Syst Evol Microbiol* **60**, 1460–1465.

Nielsen, D.S., Teniola, O.D., Ban-Koffi, L., et al. (2007a) The microbiology of Ghanaian cocoa fermentations analysed using culture-dependent and culture-independent methods. *Int J Food Microbiol* **114**, 168–186.

Nielsen, D.S., Schillinger, U., Franz, C.A.M.P., et al. (2007b) *Lactobacillus ghanensis* sp. nov., a motile lactic acid bacterium isolated from Ghanaian cocoa fermentations. *Int J Syst Evol Microbiol* **57**, 1468–1472.

Nielsen, D.S., Hønholt, S., Tano-Debrah, K., and Jespersen, L. (2005) Yeast populations associated with Ghanaian cocoa fermentations analysed using denaturing gradient gel electrophoresis (DGGE). *Yeast* **22**, 271–284.

Nielsen, D.S., Crafack, M., Jespersen, L., and Jakobsen, M. (2012) The microbiology of cocoa fermentation. In: *Chocolate in Health and Nutrition* ((eds) R.R. Watson, V.R. Preedy, and S. Zibadi.), Humana Press, New York, USA, pp. 39–60.

Ortega, N., Romero, M.-P., Macià, A., et al. (2008) Obtention and characterization of phenolic extracts from different cocoa sources. *J Agric Food Chem* **56**, 9621–9627.

Ostovar, K. and Keeney, P.G. (1973) Isolation and characterization of microorganisms involved in the fermentation of Trinidad's cocoa beans. *J Food Sci* **38**, 611–617.

Owusu, M., Petersen, M.A., and Heindal, H. (2011) Effect of fermentation method, roasting and conching conditions on the aroma volatiles of dark chocolate. *J Food Process Pres* **36**, 446–456.

Papalexandratou, Z., Camu, N., Falony, G., and De Vuyst, L. (2011a) Comparison of the bacterial species diversity of spontaneous cocoa bean fermentations carried out at selected farms in Ivory Coast and Brazil. *Food Microbiol* **28**, 964–973.

Papalexandratou, Z., Vrancken, G., De Bruyne, K., et al. (2011b) Spontaneous organic cocoa bean box fermentations in Brazil are characterized by a restricted species diversity of lactic acid bacteria and acetic acid bacteria. *Food Microbiol* **28**, 1326–1338.

Papalexandratou, Z. and De Vuyst, L. (2011) Assessment of the yeast species composition of cocoa bean fermentations in different cocoa-producing regions using denaturing gradient gel electrophoresis. *FEMS Yeast Res* **11**, 564–574.

Papalexandratou, Z., Falony, G., Romanens, E., et al. (2011c) Species diversity, community dynamics, and metabolite kinetics of the microbiota associated with traditional Ecuadorian spontaneous cocoa bean fermentations. *Appl Environ Microbiol* **77**, 7698–7714.

Papalexandratou, Z., Lefeber, T., Bahrim, B., et al. (2013) *Hanseniaspora opuntiae*, *Saccharomyces cerevisiae*, *Lactobacillus fermentum*, and *Acetobacter pasteurianus* predominate during well-performed Malaysian cocoa bean box fermentations, underlining the importance of these microbial species for a successful cocoa bean fermentation process. *Food Microbiol* **35**, 73–85.

Passos, F.M.L. and Passos, F.J.D. (1985) Descrição e classificação de bactérias acéticas isoladas da fermentação do cacau, com base em una analise numérica. *Rev Microbiol* **16**, 290–298.

Passos, F.M., Lopez, A.S., and Silva, D.O. (1984a) Aeration and its influence on the microbial sequence in cocoa fermentations in Bahia, with emphasis on lactic acid bacteria. *J Food Sci* **49**, 1470–1474.

Passos, F.M.L., Silva, D.O., Lopez, A., et al. (1984b) Characterization and distribution of lactic acid bacteria from traditional cocoa bean fermentations in Bahia. *J Food Sci* **49**, 205–208.

Pereira, G.V., Pedrozo, M.G., Ramos, C.L., and Schwan R.F. (2012) Microbiological and physicochemical characterization of small-scale cocoa fermentations and screening of yeast and bacterial strains to develop a defined starter culture. *Appl Environ Microbiol* **78**, 5395–5405.

Pereira, G.V., Magalhães, K.T., de Almeida, E.G., et al. (2013) Spontaneous cocoa bean fermentation carried out in a novel design stainless steel tank: influence on the dynamics of microbial populations and physical-chemical properties. *Int J Food Microbiol* **161**, 121–133.

Pontillon, J. (1998) Cacao et chocolat. Production, Utilisation, Caractéristiques. *Collection Sciences & Techniques Agroalimentaires*, Lavoisier Tec & Doc, Paris, France.

Puziah, H., Jinap, S., Sharifah, K.S.M., and Asbi, A. (1998a) Changes in free amino acids, peptide-N, sugar and pyrazine concentration during cocoa fermentation. *J Sci Food Agric* **78**, 535–542.

Puziah, H., Jinap, S., Sharifah, K.S.M., and Asbi, A. (1998b) Effect of mass and turning time on free amino acid, peptide-N, sugar and pyrazine concentration during cocoa fermentation. *J Sci Food Agric* **78**, 543–550.

Ravelomanana, R., Guiraud, J.-P., and Galzy, P. (1986) Isolation of a pectin-utilising yeast from cocoa beans. *Syst Appl Microbiol* **8**, 230– 233.

Ravelomanana, R., Guiraud, J.-P., Vincent, J.C., and Galzy, P. (1985) The yeast flora of cocoa bean fermentation in the Ivory Coast. *MIRCEN J* **1**, 319– 326.

Roelofsen, P.A. (1958) Fermentation, drying and storage of cocoa beans. *Adv Food Res* **8**, 225–296.

Rohan, T.A. (1964) The precursors of chocolate aroma: a comparative study of fermented and unfermented beans. *J Food Sci* **29**, 456–459.

Rohsius, C., Matissek, R., and Lieberei, R. (2006) Free amino acid amounts in raw cocoas from different origins. *Eur Food Res Technol* **222**, 432–438.

Romero, T., Salgado, M.A., Garica, P., et al. (2013) Relationship between fermentation index and other biochemical changes evaluated during the fermentation of Mexican cocoa (*Theobroma cacao*) beans. *J Sci Food Agric* **93**, 2596–2604.

Saltini, R., Akkerman, R., and Frosch, S. (2013) Optimizing chocolate production through traceability: a review of the influence of farming practices on cocoa bean quality. *Food Control* **29**, 167–187.

Samah, O.A., Ptih, M.F., and Selamat, J. (1992) Biochemical changes during fermentation of cocoa beans inoculated with *Saccharomyces cerevisiae* (wild strain). *J Food Sci Technol* **29**, 341–343.

Samah, O.A., Puteh, M.F., Selemat, J., and Alimon, H. (1993) Fermentation products in cocoa beans inoculated with *Acetobacter xylinum*. *ASEAN Food J* **8**, 22–25.

Sanchez, J., Guiraud, J.-P., and Gazy, P. (1984) A study of the polygalacturonase activity of several yeast strains isolated from cocoa. *Appl Microbiol Biotechnol* **20**, 262–267.

Sanchez, J., Daguenet, G., Guiraud, J.-P., et al. (1985) A study of the yeast flora and the effect of pure culture seeding during the fermentation of cocoa beans. *Lebensmitt Wiss Technol* **18**, 69–76.

Schwan, R.F. (1998) Cocoa fermentations conducted with a defined microbial cocktail inoculum. *Appl Environ Microbiol* **64**, 1477–1483.

Schwan, R.F., Rose, A.H., and Board, R.G. (1995) Microbial fermentation of cocoa beans, with emphasis on enzymatic degradation of the pulp. *J Appl Bacteriol Symp Suppl* **79**, 96S–107S.

Schwan, R.F. and Wheals, A.E. (2004) The microbiology of cocoa fermentation and its role in chocolate quality. *Crit Rev Food Sci Nutr* **44**, 205–221.

Schwan, R.F., Cooper, R.M., and Wheals, A.E. (1997) Endopolygalacturonase secretion by *Kluyveromyces marxianus* and other cocoa pulp-degrading yeasts. *Enzyme Microb Technol* **21**, 234–244.

Schwan, R.F., Vanetti, M.C.D., Silva, D.O., et al. (1986) Characterization and distribution of aerobic, spore-forming bacteria from cacao fermentations in Bahia. *J Food Sci* **51**, 1583–1584.

Shamsuddin, S.B. and Dimick, P.S. (1986) Qualitative and quantitative measurements of cocoa bean fermentation. In: *Proceedings of Cocoa Biotechnology* ((ed.) P.S. Dimick), Department of Food Science, The Pennsylvania State University, Pennsylvania, USA, pp. 55–78.

Snauwaert, I., Papalexandratou, Z., De Vuyst, L., and Vandamme, P. (2013) Characterization of strains of *Weissella fabalis* sp. nov. and *Fructobacillus tropaeoli* from spontaneous cocoa bean fermentations. *Int J Syst Evol Microbiol* **63**, 1709–1716.

Sukha, D.A., Butler, D.R., Umaharan, P. and Boult, E. (2008) The use of an optimised organoleptic assessment protocol to describe and quantify different flavour attributes of cocoa liquors made from Ghana and Trinitario beans. *Eur Food Res Technol* **226**, 405–413.

Thompson, S.S., Miller, K.B. and Lopez, A.S. (2007) Cocoa and coffee. In: *Food Microbiology: Fundamentals and Frontiers* ((eds) M.P Doyle, and L.R. Beuchat) 2nd edn, ASM Press, Washington, USA, pp. 837–850.

Thompson, S. S., Miller, K. B., Lopez, A. and Camu, N. (2013) Cocoa and coffee. In: *Food Microbiology: Fundamentals and Frontiers* ((eds) M.P Doyle and R.L. Beuchat), 4th edn, ASM Press, Washington, USA, pp. 881–889.

Tomlins, K.I., Baker, D.M., and Daplyn, P. (1993) Effect of fermentation and drying practices on the chemical and physical profiles of Ghana cocoa. *Food Chem* **46**, 257–263.

Voight, J. and Biehl, B. (1995) Precursors of the cocoa-specic aroma components are derived from the vicilin-class globulin of the cocoa seeds by proteolytic processing. *Botanica Acta* **167**, 283–289.

Voight, J., Heinrichs, H., Voight, G., and Biehl, B. (1994a) Cocoa-specific aroma precursors are generated by proteolytic digestion of the vicilin-like globulin of cocoa seeds. *Food Chem* **50**, 177–184.

Voight, J., Voight, G., Heinrichs, H., et al. (1994b) *In vitro* studies on the proteolytic formation of the characteristic aroma precursors of fermented cocoa seeds – the significance of endoprotease specificity. *Food Chem* **51**, 7–14.

Wollgast, J. and Anklam, E. (2000) Review on polyphenols in *Theobroma cacao*: changes in composition during the manufacture of chocolate and methodology for identification and quantification. *Food Res Int* **33**, 423–447.

Wood, B.J.B. (1998) *Microbiology of Fermented Foods*. Blackie Academic and Professional, London, United Kingdom.

Wood, G.A.R. and Lass, R.A. (1985) *Cocoa*. 4th edn. Longman Group Limited, London, United Kingdom.

Yusep, I., Jinap, S., Jamilah, B., and Nazamid, S. (2002) Influence of carboxypeptidases on free amino acid, peptide and methylpyrazine contents of under-fermented cocoa beans. *J Sci Food Agric* **82**, 1584–1592.

Ziegleder, G. (1991) Composition of flavour extracts of raw and roasted beans. *Z Lebensm Unters Forsch* **192**, 521–525.

Ziegleder, G. and Biehl, B. (1988) Analysis of cocoa flavour components and flavour precursors. In: *Methods in Plant Analysis* ((eds) H. Linskens and J.F. Jackson), vol. 8, Springer-Verlag, Berlin, Germany, pp. 321–393.

Chapter 17

B-Group Vitamins Production by Probiotic Lactic Acid Bacteria

Jean Guy LeBlanc, Jonathan Emiliano Laiño, Marianela Juárez del Valle, Graciela Savoy de Giori, Fernando Sesma, and María Pía Taranto*

Centro de Referencia para Lactobacilos (CERELA)-CONICET, Argentina

17.1. Introduction

Lactic acid bacteria (LAB) are a heterogeneous group of microorganisms that have a long history of use as starter cultures for the elaboration of fermented foods because these industrially important bacteria can improve the safety, shelf-life, nutritional value, flavor, and overall quality of fermented products. In addition, LAB have been shown to exert a large range of beneficial properties, which is why they are frequently used as probiotic microorganisms in a variety of novel products. In 2002, the Food Agriculture Organization (FAO) defined probiotics as "live microorganisms which, when administered in adequate amounts, confer a beneficial health effect on the host." The probiotic and beneficial aspects of LAB have been intensely reviewed elsewhere and will not be the subject of this chapter.

In addition to their intrinsic beneficial properties, certain strains of LAB have the capability of producing/releasing and/or increasing specific beneficial compounds in foods. These functional ingredients are sometimes referred to as nutraceuticals, a term that was first coined in 1989 to describe "a food (or part of a food) that provides medical or health benefits, including the prevention and/or treatment of a disease." These ingredients can be macronutrients (such as unsaturated fatty acids present in some oils), micronutrients (such as vitamins), or non-nutritive compounds (such as hydrolytic enzymes and flavonoids) and can be naturally present in foods (such as omega-3 fatty acids in fish or vitamin C in citrus fruits, etc.) or added (i.e., milks fortified with calcium and vitamin D, cereals fortified with folic acid, and others) (Hugenholtz and Smid 2002).

Since LAB are involved in the preparation of a wide range of fermented foods and because of their GRAS (Generally Recognized as Safe) status, the selection of strains delivering nutraceuticals is now the main objective of several research groups. Among these studies, vitamin production by LAB has gained the attention of the scientific community. It has been shown that certain foods fermented with LAB contain high levels of B-group vitamins as a result of microbial biosynthesis. For this reason, LAB (considered food-grade microorganisms) are the ideal candidates to deliver specific compounds such as vitamins into foods. The enrichment of vitamin produced by LAB in fermented foods is shown in Table 17.1.

* Corresponding author: Centro de Referencia para Lactobacilos (CERELA)-CONICET, Chacabuco 145, 4000 San Miguel de Tucumán, Tucumán, Argentina. ptaranto@cerela.org.ar

Biotechnology of Lactic Acid Bacteria: Novel Applications, Second Edition. Edited by Fernanda Mozzi, Raúl R. Raya, and Graciela M. Vignolo.

Table 17.1. Vitamin enrichment by lactic acid bacteria in fermented foods.

Strain	Vitamin	Increase (%)	Product	Reference
Lact. plantarum CRL 725	Riboflavin	226	Soymilk	Unpublished
Lact. plantarum CRL 725 (roseoflavin derivative)	Riboflavin	600	Soymilk	Unpublished
Lact. acidophilus ATCC 314	Riboflavin	657	Soymilk + vitamins	Ewe et al. 2010
Lact. acidophilus FTDC 8833	Riboflavin	243	Soymilk + vitamins	Ewe et al. 2010
Lact. acidophilus FTDC 8633	Riboflavin	113	Soymilk + vitamins	Ewe et al. 2010
Lactobacillus ssp + *Leuconostoc mesenteroides*	Riboflavin	345	Idli	Ghosh and Chattopadhyay 2012
Lact. plantarum UNIFG1 (roseoflavin resistant?)	Riboflavin	200	Pasta	Capozzi et al. 2011
Lact. plantarum UNIFG1 (roseoflavin derivative)	Riboflavin	300	Bread	Capozzi et al. 2011
Lact. fermentum PBCC11.5 (roseoflavin derivative)	Riboflavin	200	Soymilk	Russo et al. 2014
Lact. bulgaricus CRL 863 + *Strep. thermophilus* CRL 415 and 803	Folate	190	Yogurt	Laiño et al. 2013
Lact. bulgaricus CRL 863 + *Strep. thermophilus* CRL 415 and 803 + *Lact. amylovorous* CRL 887	Folate	290	Yogurt	Laiño et al. 2014
L. Lactis	Folate	600	Cucumber juice	Gangadharan and Nampoothiri 2011
L. Lactis	Folate	144	Watermelon juice	Gangadharan and Nampoothiri 2011
Lact. reuteri CRL 1098	B_{12}	100	Soymilk	Molina et al. 2013

Recently it has also been reported that LAB and other beneficial microorganisms have the potential to produce vitamins *in situ* in the gastrointestinal (GI) tract (LeBlanc et al. 2013).

In this chapter, an update of our previous reviews on the production of B-group vitamins by LAB from a biochemical and genetic point of view will be discussed (LeBlanc et al. 2010a, 2011). This information could be useful for the design of novel functional foods aimed at the prevention of specific vitamin deficiencies or to increase the production of vitamins in the GI tract of consumers.

17.2. B-Group Vitamins

Vitamins are micronutrients that are essential for the metabolism of all living organisms. They are found as precursors of intracellular co-enzymes that are necessary to regulate vital biochemical reactions in the cell. In view of the inability of mammalian cells to synthesize these vital biomolecules, an exogenous supply of most vitamins appears inevitable to prevent nutritional deficiencies. Although most vitamins are present in a variety of foods, vitamin deficiencies still exist in many countries including industrialized nations.

Thirteen vitamins are currently recognized as being essential to human health and are classified according to their solubility as either being fat- (vitamins A, D, E, and K) or water-soluble (vitamin C and B-group vitamins). The B-group (or B-complex) vitamins include thiamin, riboflavin, niacin, pyridoxin, pantothenic acid, biotin, folate, and cobalamin. Each B-group vitamin is chemically different and acts in synergy to maintain the body's homeostasis by playing major roles in metabolic processes such as energy production and red blood cell synthesis. B-group vitamins, normally present in many foods, can easily be removed or destroyed during cooking and food processing, so their deficiencies are present in many societies. For this reason, many countries have adopted laws requiring the fortification of certain foods with specific vitamins and minerals. For example, in Argentina the food industry is obligated to fortify all wheat flour for human consumption with iron, folic acid, thiamin, riboflavin, and niacin for the prevention of anemia and neural tube defects during embryo development. However, official reports have shown that because of a lack of regulatory controls, the fortification concentration is not always met and, in consequence, serum vitamin levels in the general population have only slightly been improved and subclinical deficiencies still persist (ENNyS 2007).

A detailed description of microbial synthesis of the three most relevant B-group vitamins (riboflavin, folate, and cobalamin) by LAB has been discussed in detail previously (LeBlanc et al. 2010a) and will not be the focus of this chapter. Here, the most recent applications of beneficial microorganisms in order to increase the concentration of vitamins through microbial biosynthesis will be discussed as an economically attractive alternative to mandatory fortification of foods.

17.2.1. Riboflavin (Vitamin B_2)

Riboflavin (vitamin B_2) plays an essential role in cellular metabolism, being the precursor of the co-enzymes flavin mononucleotide (FMN) and flavin adenine dinucleotide (FAD) that both act as hydrogen carriers in biological redox reactions involving enzymes such as NADH dehydrogenase. Once riboflavin is absorbed in the human body, the synthesis of these flavin co-enzymes is controlled by thyroid hormones that regulate the activities of the flavin biosynthetic enzymes (Rucker et al. 2001).

The recommended riboflavin requirements for humans vary with respect to sex, age, and physiological state (such is the case during pregnancy and lactation). Normal adults need to consume between 0.9 and 1.6 mg of this vitamin on a daily basis since the human body does not have deposits of riboflavin and an excess of vitamin intake is eliminated in urine (Institute of Medicine 1998). Although present in a wide variety of foods, riboflavin deficiency (ariboflavinosis) still occurs in both developing and industrialized countries (O'Brien et al. 2001; Blanck et al. 2002). Even though severe cases of ariboflavinosis are not common in most societies, subclinical manifestations are frequent and these are only detectable by measuring the blood vitamin concentrations.

Dairy products are not considered a good source of this essential vitamin since milk contains approximately 1.2 mg of riboflavin per liter, meaning that an average adult would need to consume about one liter per day to meet the daily requirement. This level of milk consumption is not normal amongst residents of industrialized countries such as the USA, where the daily per capita consumption of fresh milk is approximately 200 ml (Putman and Allshouse 2003). Thus, increasing the levels of riboflavin in milk is an interesting proposal to prevent deficiencies in populations where dairy consumption is low.

Riboflavin concentrations can sometimes vary in certain dairy products due to processing technologies and the action of microorganisms utilized during fermentation. The latter case can be beneficial since it has been shown that fermented dairy products such as buttermilk and yogurt have significantly higher riboflavin concentrations (1.7 and 2.0 mg/l, respectively) compared to unfermented milk (1.2 mg/l). This increase of vitamin B_2 would be due to the action of riboflavin-producing starter cultures during the fermentation process (LeBlanc et al. 2011). The proper selection of starter cultures can lead to increasing the concentration of specific water-soluble vitamins. It has been shown that certain yogurt starter cultures decreased riboflavin concentrations whereas others could increase the levels of this essential vitamin up to 60% of the initial concentration present in unfermented milk (Kneifel et al. 1992). The use of probiotic riboflavin-producing strains would be an interesting alternative to obtain products enriched with this vitamin in addition to conferring potential beneficial effects to the host.

Riboflavin biosynthesis has been described both in Gram (+) and Gram (−) bacteria, with more detail in *Bacillus* (*B.*) *subtilis* (Perkins and Pero 2002) and *Escherichia coli* (Bacher et al. 1996).

Microbial biosynthesis of riboflavin from the precursor guanosine triphosphate (GTP) and *D*-ribulose 5-phosphate occurs through seven enzymatic steps, which have been reviewed previously (Bacher et al. 2000).

Recently, the riboflavin producing capabilities of 179 strains of LAB, isolated from a variety of fermented dairy products, were evaluated (Juarez del Valle et al. 2014). Only 42 strains were able to grow in a commercial riboflavin-free medium after which the concentration of this vitamin was determined by HPLC. Five of these strains (*Lactobacillus fermentum* (two strains), *Lact. plantarum*, *Streptococcus thermophilus*, and *Lact. paracasei* subsp. *Paracasei*) were pre-selected for their capacity to produce elevated concentrations of riboflavin in this commercial B_2-free medium. These were then inoculated in soymilk to evaluate their ability to grow in this food matrix to increase its low riboflavin concentrations (309 ± 19 ng B_2/ml). Only *Lact. plantarum* CRL725 was able to significantly enhance the riboflavin levels in soymilk, more than doubling (700 ± 20 ng B_2/ml) the initial concentration.

In another study (Ewe et al. 2010), ten strains of *Lactobacillus* were evaluated for their viability and growth characteristics in soymilk supplemented with B-vitamin. These authors showed that the addition of B-vitamins promoted the growth of all the lactobacilli and also stimulated the synthesis of riboflavin and folic acid by lactobacilli in soymilk. *Lact. acidophilus* ATCC314 or *Lact. acidophilus* FTDC 8833, used as culture starters, increased riboflavin concentrations (657% and 243%, respectively) in soymilk, compared with its initial concentration prior to 24 h fermentation, whereas other strains, such as *Lact. acidophilus* FTDC 8833 and *Lact. gasseri* FTDC 8131, consumed this vitamin. These results confirm that riboflavin production/consumption is a strain-dependent trait.

Idli is a very popular fermented breakfast food consumed in India that is made mainly of rice and black gram. It was shown that riboflavin, thiamine, and folate concentrations were increased during the fermentation of *idli* batter due to the presence of yeast, *Lactobacillus* ssp., and *Leuconostoc mesenteroides* (Ghosh and Chattopadhyay 2012). In terms of riboflavin, the concentrations reached 7.6 ± 0.8 ng B_2/ml compared to the initial concentration (2.2 ± 1.1 mg B_2/ml). Further studies are required to identify the riboflavin-producing strains in this product.

In addition to the search for native riboflavin-producing strains, two strategies have been applied to overexpress riboflavin by microorganisms: (i) the construction of recombinant strains that overexpress riboflavin biosynthetic genes, and (ii) the use of roseoflavin-resistant strains. In the first strategy, a *Corynebacterium ammoniagenes* strain harboring a plasmid containing all of its riboflavin biosynthetic genes was engineered and accumulated 17-fold as much riboflavin as the host strain (Koizumi et al. 2000). In another study, a riboflavin consuming strain of *Lactococcus (L.) lactis* subsp. *cremoris* was converted to a riboflavin "factory" by overexpressing four of its biosynthesis genes (Burgess et al. 2004). Although interesting to understand the mechanisms involved in riboflavin production, these genetically engineered strains cannot yet be included in food products destined for human consumption.

On the other hand, it has been shown that mutants isolated on the basis of their resistance to the toxic riboflavin analog roseoflavin also exhibit a riboflavin-overproduction phenotype (Kukanova et al. 1982). Moreover, roseoflavin has been used to obtain constitutive riboflavin overproducing strains of *B. subtilis* (Perkins and Pero 2002), *L. lactis* (Burgess et al. 2004), *Lact. plantarum*, *Leuc. mesenteroides*, and *Propionibacterium freudenreichii* (Burgess et al. 2006). Some of these have been shown to provide beneficial effects in vitamin depleted animals and could be inserted as novel starter cultures (LeBlanc et al. 2005a,b, 2006)

LAB were obtained from durum wheat flour samples and screened for roseoflavin-resistant variants to isolate natural riboflavin-overproducing strains (Capozzi et al. 2011). Two riboflavin-overproducing strains of *Lact. plantarum* were used for the preparation of bread (by means of sourdough fermentation) and pasta (using a prefermentation step) to enhance their vitamin B_2 content. The applied approach resulted in a considerable increase of vitamin B_2 content (about 2- and

3-fold increases in pasta and bread, respectively), thus representing a convenient and efficient food-grade biotechnological application for the production of vitamin B_2-enriched bread and pasta. This methodology may be extended to a wide range of cereal-based foods, feed, and beverages.

In a recent study, one strain of *Lact. fermentum* isolated from sourdough was shown to be able to produce riboflavin (Russo et al. 2014). Spontaneous roseoflavin-resistant mutants were obtained by exposing the wild-type strain (named *Lact. fermentum* PBCC11) to increasing concentrations of roseoflavin. Fifteen spontaneous roseoflavin-resistant mutants were isolated, and the level of vitamin B_2 was quantified by HPLC. Seven mutant strains produced concentrations of vitamin B_2 higher than 1000 ng/ml, whereas three mutants were unable to overproduce riboflavin even though they were able to withstand the selective pressure of roseoflavin. Alignment of the *rib* leader region of PBCC11 and its derivatives showed only point mutations at two neighboring locations of the RFN element (also known as the FMN riboswitch). No mutations were detected in the derivative strains that did not have an overproducing phenotype. The best riboflavin overproducing strain, named *Lact. fermentum* PBCC11.5, and its parental strain were used to fortify bread. Bread produced using the co-inoculum yeast and *Lact. fermentum* PBCC11.5 led to an approximately 2-fold increase of final vitamin B_2 content compared to the use of the wild-type strain (Russo et al. 2014).

It was shown that *Lact. plantarum* CRL 725 was able to increase 2-fold the initial concentration of riboflavin in soymilk (Juarez del Valle et al. 2014). Roseoflavin-resistant variants of this strain were also obtained and evaluated in soymilk. One of the variants obtained was able to increase 6 times (1860 ± 20 ng B_2/ml) the basal riboflavin levels of soymilk. Roseoflavin-resistant strains capable of synthesizing riboflavin in soymilk constitute an interesting and economically feasible biotechnology strategy that could be adapted by the food industry to develop novel vitamin-bioenriched functional foods with enhanced consumer appeal.

Besides introducing an efficient riboflavin-producing strain, the design of an appropriate fermentation medium and optimal growth conditions are of crucial importance to improve efficiency and productivity. In a recent study, the effects of agitation speed on the performance of riboflavin production by recombinant *B. subtilis* RF1 was investigated in fed-batch fermentation (Man et al. 2014). During the initial phase of the fermentation process, a lower agitation speed (aeration) was beneficial for cell growth and riboflavin biosynthesis, whereas during the latter phase, higher agitation speed (900 rpm) was the most beneficial. Thus, a two-stage agitation speed control strategy was proposed based on kinetic analysis, in which the agitation speed was controlled at 600 rpm in the first 26 h and then switched to 900 rpm to maintain high cell growth and high riboflavin production during the entire fermentation process. The maximum concentration of riboflavin reached 9400 ng/ml in 48 h by applying this strategy, which was 20.5 and 21.4% higher than the best results controlled by constant agitation speeds (Man et al. 2014).

A good knowledge of the metabolic flux is also essential to improve riboflavin production. In one such study, the *gapB* (that codes for NADPH-dependent glyceraldehyde-3-phosphate dehydrogenase), and *fbp* (codes for fructose-1,6-bisphosphatase) were overexpressed in parental strain *B. subtilis* RH33. Significant enhancement of riboflavin concentrations (4890 ng/ml) was obtained in shake flask cultures when *gapB* and *fbp* were co-overexpressed (Wang et al. 2014). These results imply that deregulation of gluconeogenesis is an effective strategy for production of metabolites directly stemming from the pentose phosphate pathway as well as other NADPH demanding compounds with glucose as carbon source in *B. subtilis* (Wang et al. 2014).

Moreover, the human GI tract is colonized by a vast array of microorganisms known as the gut microbiota that apart from its impact on different human functions plays a pivotal role in food digestion and energy recovery, while it can also act as an important supplier of vitamins. In humans, it has been shown that members of the gut microbiota are able to synthesize vitamin K as well as most of the water-soluble B vitamins, such as biotin, cobalamin, folates, nicotinic acid, panthotenic acid, pyridoxine, riboflavin, and thiamine (Hill 1997). In contrast to dietary vitamins, which are absorbed in the proximal tract of the small intestine, the main uptake of microbial produced vitamins occurs in the colon (Ichihashi et al. 1992; Said and Mohammed 2006). So far, fragmentary information is available on the *de novo* synthesis of riboflavin by enteric bacteria. In the case of bifidobacteria, the enzymes needed for the biosynthesis of this vitamin seem to be partially or completely absent from the majority of currently available bifidobacterial genomes (Ventura et al. 2007). However, one cannot exclude the possibility that multiple, co-existing microbial species are capable of *de novo* synthesis (LeBlanc et al. 2013). With the expanding availability of genome sequences it is not only possible to identify potential vitamin-producing strains, but also to understand the intertwined mechanisms for their biosynthesis, all of which should be exploited to increase the vitamin-producing capacities in the GI tract of humans.

17.2.2. Folates (Vitamin B_9)

Human life cannot exist without folate since this B-group vitamin is involved in essential functions of cell metabolism such as DNA replication, repair, and methylation, and synthesis of nucleotides, other vitamins, and some amino acids. Folates possess antioxidant properties that protect the genome by inhibiting free radical attack of DNA, in addition to their role in DNA repair and replication mechanisms (Duthie et al. 2002). Folate deficiency has been implicated in a wide variety of disorders such as Alzheimer (Luchsinger et al. 2007), coronary heart diseases (Danesh and Lewington 1998), osteoporosis (Baines et al. 2007), increased risk of breast (Tjonneland et al. 2006) and colorectal cancers (Van Guelpen *et al.* 2006), poor cognitive performance (Durga et al. 2007a), hearing loss (Durga et al. 2007b), and neural tube defects (Czeizel and Dudas 1992).

The recommended daily intake (RDI) of folate in an adult is 200–400 μg (FAO/WHO 2002). For pregnant women, 400–600 μg is recommended. Although folate is present in a normal human diet, folate deficiencies still occur frequently, even in well-developed countries (Konings et al. 2001; O'Brien et al. 2001). It has been stated that folate intake levels are inadequate among various population groups including women of childbearing age (Morris and Tangney 2007). Studies such as these have obligated governing bodies of many countries to advocate mandatory supplementation of foods of mass consumption with folic acid. In Canada and the USA, for example, fortification of flours is mandatory since 1998 in order to reduce the incidences of neural tube malformations in newborns. Many other countries afterwards developed similar programs. As stated earlier, in Argentina the fortification of flours with specific vitamins, including folic acid, has been mandatory since 2002. Even with mandatory folic acid fortification, follow-up reports have shown that one-third of pregnant and lactating women may not be meeting their folate requirements from diet alone (Sherwood et al. 2006). On the other hand, many countries have not adopted a national folic acid fortification program because of the potential adverse effects of folic acid; in particular, that high intakes can delay the diagnosis of vitamin B_{12} deficiency (Bailey et al. 2003). The concern is that folic acid is added at levels where individuals with low folate intakes can meet folate RDI in order to prevent neural-tube defects (NTD) and/or lowering plasma homocysteine; however, others, with normal or higher folate intakes could inadvertently become exposed to excessive folic acid intake, which has the potential to mask the early hematological manifestations of vitamin B_{12} (Sweeney et al. 2007). Other safety considerations of excess folic acid consumption highlighted by the FDA (Food and Drug Administration 1996) include potential unknown risks for pregnant women, and persons on anti-epileptic and anti-folate medication. The FDA also noted the uncertainties regarding the effects of chronic elevated exposure in children, whose requirements for folate are lower than those of adults. Further concerns include the potential to promote cancer (Charles et al. 2004; Kim 2004) and the hypothesis that exposure of the fetus to excess folic acid may favor the selection of the methylentetrahydrofolate polymorphism, associated with a range of debilitating illnesses (Lucock and Yates 2005). Natural folates, such as those produced by plants and microorganisms, do not cause "masking" of pernicious anemia that occurs at high concentrations of folic acid and should thus be considered a viable alternative to folic acid fortification programs (Scott 1999). Due to the occurrence of problems associated with current folic acid fortification programs, researchers have been seeking novel methods to increase concentrations of naturally occurring folates in foods.

Milk is not a rich source of dietary folate compared to other foods; however, as is the case for riboflavin, folate concentrations can be significantly increased in many dairy products due to microbial fermentation. Among dairy products, fermented milks are considered a good potential matrix for folate fortification because folate-binding proteins present in milk improve folate stability and enhance the bioavailability of both 5-methyltetrahydrofolate (the most predominant natural form of the vitamin) and folic acid (Jones and Nixon 2002; Aryana 2003; Verwei et al. 2003). However, due to the potential risks of fortification with folic acid, the elaboration of fermented milks containing elevated levels of natural folates would be a better suited alternative.

The amount of folic acid found in cow's milk ranges from 20 to 60 μg/l, whereas its concentration in yogurts (normally between 50 and 70 μg/l) may be increased depending on the starter cultures used and on the storage conditions. This level depends on the strain of *Strep. thermophilus* and *Lact. delbrueckii* subsp. *bulgaricus* used, because the latter organism has been shown to use and to degrade folates during its growth (reviewed in LeBlanc et al. 2011). It is therefore of most importance to select the optimal combination of these strains leading to an organoleptically acceptable yogurt with a concomitantly increased folic acid concentration.

In a study performed by our group, *Lact. delbrueckii* subsp. *bulgaricus* and *Strep. thermophilus* strains were isolated from artisanal Argentinean yogurts and were grown in a folate-free culture medium and milk after which intracellular and extracellular folate production were evaluated (Laiño et al. 2012). From the initial 92 isolated LAB strains tested, 4 strains of *Lact. delbrueckii* subsp. *bulgaricus* and 32 *Strep. thermophilus* were able to grow in the absence of folate. From these, *Lact. delbrueckii* subsp. *bulgaricus* CRL863 and *Strep. thermophilus* CRL415 and CRL803 produced the highest extracellular folate levels (between 22 and 135 μg/l) in the folate-free medium. When used to inoculate milk, these strains were able to increase the initial folate concentrations by almost 190%. This is the first report where native strains of *Lact. delbrueckii* subsp. *bulgaricus* were shown to produce natural folate. In a follow-up study, the previously identified folate producing *Lact. delbrueckii* subsp. *bulgaricus* (three strains) and *Strep. thermophilus* (two strains) were combined and used to elaborate 15 different yogurts, each of which contained different folate concentrations (Laiño et al. 2013). The yogurt elaborated with *Lact. bulgaricus* CRL871 and two strains of *Strep. thermophilus* (CRL803 and CRL415) had a significantly higher folate concentration (180 μg/l) compared to milk (40 ± 10 μg/l) and conventional yogurts (80 ± 20 μg/l). These results show that the rational selection of strains was useful in obtaining a yogurt naturally bio-enriched in folate with a 4-fold increase in vitamin content compared to unfermented milk and a 2-fold increase compared to conventional yogurts. After these promising results, this same group demonstrated that the addition of a non-conventional starter culture (*Lact. amylovorus* CRL887) to the selected *Lact. bulgaricus* CRL871 and *Strep. thermophilus* strains was efficient in producing a new yogurt with even higher folate concentrations (260 μg/l) making this new bio-enriched product a very interesting alternative to fortification with the chemical folic acid (Laiño et al. 2014).

A recent study has shown that the effective manipulation of medium additives and cultivation conditions enhanced the folate level (164 ng/ml) in skim milk using *L. lactis* susp *cremoris*, an isolate from raw cow's milk (Gangadharan and Nampoothiri 2011). In this work, the importance of anoxygenic conditions on growth and stability of folate was demonstrated.

Besides fermented dairy products, microorganisms are able to increase folate content in a wide variety of other foods. For example, fermentation of rye dough to produce bread is often accompanied with an increase in folate content (Kariluoto et al. 2006). In these studies, the increase of folate content during fermentation was mainly due to folate synthesis by yeasts; LAB did not produce folates but rather consumed them. Variations in starter cultures led to great differences in folate content of sourdoughs but their effects have been scantly investigated. Proper strain selection, that is exchanging folate-consuming LAB with folate-producing ones, could significantly increase folate content in these breads.

Gangadharan and Nampoothiri (2011) also showed that the use of *L. lactis* proved to be an excellent choice to increase the folate content in cucumber and watermelon juice.

A more complete study was performed on oat bran where 20 bacteria were isolated and tested for their folate-producing capability (Kariluoto et al. 2014). The best producers were characterized as *B. subtilis* ON4, *Chryseobacterium* sp. NR7, *Janthinobacterium* sp. RB4, *Pantoea agglomerans* ON2, and *Pseudomonas* sp ON8. Marked differences in the distribution of folate among the bacterial strains were revealed by the HPLC analysis where the main forms produced were tetrahydrofolate, 5,10-methenyltetrahydrofolate, 5-methyltetrahydrofolate, and 5-formyltetrahydrofolate. Increase in the folate content during bacterial growth was accompanied by proportional increase in the 5-methyltetrahydrofolate (5-MTHF) content and decrease of 5-formyltetrahydrofolate. These forms of folate are those normally present in foods and do not cause unwanted secondary effects that are possible with the ingestion of the chemical form of the vitamin (folic acid). In a recent study from the same group, it was shown that the addition of glucose increased the 5-MTHF production by *Strep. thermophilus* ABM5097, *Lact. reuteri*, and *Propionibacterium* sp. ABM5378 (Kariluoto et al. 2014).

In a recent article, the extracellular production of folate and other vitamins was evaluated in cultures of LAB isolated from nukazuke, a traditional Japanese pickle (Masuda et al. 2012). Among the 180 LAB isolates tested, two strains of *Lact. sakei* (CN-3 and CN-28) and one strain of *Lact. plantarum* (CN-49) produced high levels of extracellular folates (higher than 100 μg/l) in a folate-free culture medium. However, the folate producing-strains did not show tolerance to high concentrations of salts and alcohol, greatly limiting their use in the elaboration of nukazuke, though they could be used in other fermented foods.

Additionally, the oral administration of folate-producing probiotic strains may confer a more efficient protection against inflammation and cancer, both by exerting the beneficial effects of the probiotics and by delivering folate to colonic-rectal cells (Pompei et al. 2007). In humans, folate produced by the microbiota in the small intestine is assimilated by the host (Camilo et al. 1996). Although it is believed that they supply only a minor percentage of all the absorbed folate in humans (Bates 1993), the contribution of the microbiota to the folate requirements of the high cell turnover intestinal epithelium is still unknown. A mechanism for luminal folate absorption by cells in the human colon has been reported (Dudeja et al. 1997), suggesting that folate produced *in situ* by the colonic microbiota may be utilized by cells in the colonic epithelium. It has recently been shown that bacterially synthesized folate can also be absorbed across the large intestine and incorporated into the liver and kidneys of piglets (Asrar and O'Connor 2005). These authors predicted that approximately 18% of the dietary folate requirement for the piglet could be met by folate absorption across the large intestine. Also, increased intestinal *Bifidobacterium* populations, induced by consumption of human breast milk, have been correlated with an enhanced folate status in rats (Krause et al. 1996). It is therefore possible that bacteria in the intestinal tract may be able to contribute to the folate requirement of colonic epithelial cells. However, further research is required to determine if these bacteria produce folate in the intestinal environment; the form in which this folate occurs; the availability of this folate for transport and utilization by colonocytes from the lumen; and the contribution of the intestinal microbiota to the total folate requirement of colonic epithelial cells.

17.3. Probiotics *In Situ*

D'Aimmo et al. (2012) screened 19 strains of bifidobacteria isolated mostly from human feces for their ability to produce folates. Ten strains grew in a synthetic folate-free culture medium showing folate autotrophy and suggesting folate auxotrophy of the remaining nine strains. These results suggest that bifidobacteria could contribute to the folate intake, either by producing this vitamin directly in foods or in the intestine.

Folate-producing *Lact. helveticus* CD6 isolated from fermented milk was found to produce 5-methyl tetrahydrofolate (5-MeTHF), although it must be stated that the levels of production were not disclosed in the publication (Ahire et al. 2013). This strain was then evaluated for probiotic properties following WHO and FAO guidelines. It showed 90.61% survival at highly acidic conditions (pH 2.0), 90.66% viability in presence of synthetic gastric juice, and 68% survivability at 0.5% bile concentration for 24 h. These properties make this strain an ideal candidate to produce folates directly in the harsh conditions of the GIT.

The application of lactobacilli as folate-producing probiotics has not yet been fully explored because *Lact. plantarum* WCFS1, a human GIT isolate, was shown to possess all the folate biosynthesis genes and has never been included in animal trials to provide evidence of their ability to produce folate *in vivo* (Rossi et al. 2011). Unlike lactobacilli, several folate-producing *Bifidobacterium* strains have been selected, but the release of high amounts of vitamin does not seem to be widespread within the genus. Animal trials confirmed that the administration of folate-producing bifidobacteria positively affected the plasmatic folate level, indicating that the vitamin is produced *in vivo* by the probiotic strains, and absorbed (Rossi et al. 2011). In a human trial, the administration of the same strains resulted in a significant increase of folate concentration in feces (Strozzi and Mogna 2008). Even though the effect on plasmatic levels has not been investigated, folate-producing bifidobacteria may provide a complementary endogenous source of the vitamin and may contribute to prevent folate deficiency.

According to the Integrated Microbial Genomes System (http://www.microbesonline.org/), more than 3707 microbial genomes (including more than 50 LAB genomes) have been sequenced to date, with several other projects ongoing and more in the process of being launched. With information from these sequencing projects, such as those published by the Joint Genome Institute Microbial Sequencing program (http://genome.jgi-psf.org/mic_home.html), it is now possible to analyze specific bacterial genomes *in silico* in order to determine if they contain the complete metabolic pathways necessary for *de novo* vitamin biosynthesis. A number of useful tools can be found on the Internet that are of great help in analyzing microbial genomes. One of the most useful sites is the Genome Analysis and System Modelling Group of the Life Sciences Division of Oak Ridge National Laboratory (http://genome.ornl.gov/microbial/). This site contains tools that allow organizing of genes into function of metabolic pathways, functional categories, taxonomic distribution, and so on. In the metabolism of co-factors and vitamins sub-category, it is possible to see which enzymes are present in the genome compared to the reference metabolic pathways, thus allowing the user to determine if a specific microorganism has the ability to biosynthesize B-vitamins (such as riboflavin, folates, and vitamin B_{12}) without performing laboratory assays. Similar tools are available at the KEGG (Kyoto Encyclopedia of Genes and Genomes) databases (http://www.genome.jp/kegg/).

The genes for folate biosynthesis have been identified in *L. lactis* (Sybesma et al. 2003), *Lact. plantarum* (Kleerebezem et al. 2003), and in *Lact. delbrueckii* susbp. *bulgaricus* (van de Guchte et al. 2006). The biosynthetic pathway in LAB includes seven consecutive steps to convert the precursor guanosine triphosphate into tetrahydrofolate. However, in some LAB which cannot synthesize folate, some of the genes involved in folate biosynthesis are lacking in the genome; this is the case for *Lact. gasseri* (Wegkamp et al. 2004), *Lact. salivarius* (Claesson et al. 2006), *Lact. acidophilus*, and *Lact. johnsonii* (van de Guchte et al. 2006).

It has been shown that metabolic engineering can be used to increase folate levels in *L. lactis* (Sybesma et al. 2003; Wegkamp et al. 2007) and *Lact. gasseri* (Wegkamp et al. 2004). The genetically engineered folate-overproducing strains of *L. lactis* were shown to be beneficial to folate-depleted animals (LeBlanc et al. 2010b). To our knowledge, this is the first *in vivo* study to demonstrate that folates produced by genetically engineered LAB represent a bioavailable source of this essential vitamin, since these were able to revert a partial megaloblastic anemia caused by folate deficiency and to increase the vitamin levels in blood. In a complementary study, it was shown that these genetically engineered folate-producing LAB were just as safe as the native strains and thus merit further studies to include them into the food chain (LeBlanc et al. 2010c).

17.3.1. Vitamin B_{12} (Cobalamin)

The term vitamin B_{12} (B_{12}) is generally used to describe a type of cobalt corrinoid, particularly of the cobalamin (cbl) group. In strict terms, vitamin B_{12} is the form of the vitamin that is obtained during industrial production and that does not exist naturally (Rucker et al. 2001). Cyanide stabilizes the molecule during the extraction procedure from microbial cultures forming cyanocobalamin. In its natural form, the vitamin is present principally as desoxyadenosilcobalamin (co-enzyme B_{12}), methylcobalamin, pseudocobalamin, among other forms. Structurally the cobalamin molecule can be divided into three parts: the central corrinic ring with the four ligands of a cobalt ion; a superior (or beta) ligand that is attached to adenosyl o methyl group; and the lower ligand (or alfa), normally dimethylbenzimidale (DMB); but in several anaerobic bacteria, adenine and other ligands can also be present, forming pseudocobalamin (or pseudo-B_{12}) and other active cofactors (Martens et al. 2002).

Animals, plants, and fungi are incapable of producing cobalamin; it is the only vitamin that is exclusively produced by microorganisms, particularly by anaerobes (Roth et al. 1996; Martens et al. 2002; Smith et al. 2007). Furthermore, biochemical and genomic data indicate that only a few bacteria and archaea possess the ability to produce this vitamin (Roth et al. 1996; Rodionov et al. 2003). Adult ruminant animals and strict vegetarians can obtain the vitamin in specialized bacteria present in the rumen. Humans do not have such microbiota in their small intestine and must absorb the co-enzyme from natural sources such as animal meats (especially liver and kidney), fish, eggs, and pharmaceutical products (Herbert 1996).

Vitamin B_{12} deficiency can cause different pathological manifestations that affect the hematopoietic, neurological, and cardiovascular system, among others. One of the extreme forms of B_{12} deficiency, known as pernicious anemia, is not normally associated with diet but rather with problems in the gastric system caused by a lack of production of a gastric glycoprotein called intrinsic factor, which facilitates the absorption of the vitamin in the small intestine (Beck 2001). Due to the complexity of B_{12} biosynthesis and to the limitations of the scope of this chapter, interested readers are invited to read excellent reviews that have been published on this subject (Roessner and Scott 2006; Escalante-Semerena 2007; Mera and Escalante-Semerena 2012).

With the constant publications of new microbial genomic sequences, it can be inferred that several microorganisms do not produce B_{12} but possess (using *in silico* evaluations) the capacity to transport and utilize B_{12} (Rodionov et al. 2003), although it is not always clear in which reactions this vitamin is involved.

As indicated above, only bacteria and archaea are able to synthesize B_{12}, although relatively few can synthesize it *de novo*. One of the first model organisms used for the study of B_{12} biosynthesis was *Propionibacterium freudenreichii* that is used in the industrial production of this vitamin. However, the first experiments with this bacterium were not fruitful due to the instability of the biosynthesis intermediates (Battersby 1994). In order to circumvent this problem, an important step was the use of the aerobic B_{12}-producing bacteria *Pseudomonas denitrificans* that allowed the isolation of numerous intermediates and the characterization of the majority of the genes involved in the biosynthesis of this vitamin (Battersby 1994; Thibaut et al. 1998). These studies concluded that the biosynthesis of cobalamin could be performed in either (i) aerobic or oxygen dependent or (ii) anaerobic or oxygen independent conditions.

The anaerobic route (Figure 17.1), observed in strains of *P. freudenreicchii*, *Salmonella enterica*, and *Bacillus megaterium*, was described (Warren et al. 2002; Warren 2006; Escalante-Semerena 2007). The initial characterization problems were due principally to the fact that the central cobalt ion was inserted into the corrinoid ring in an early step that generated unstable intermediates that were

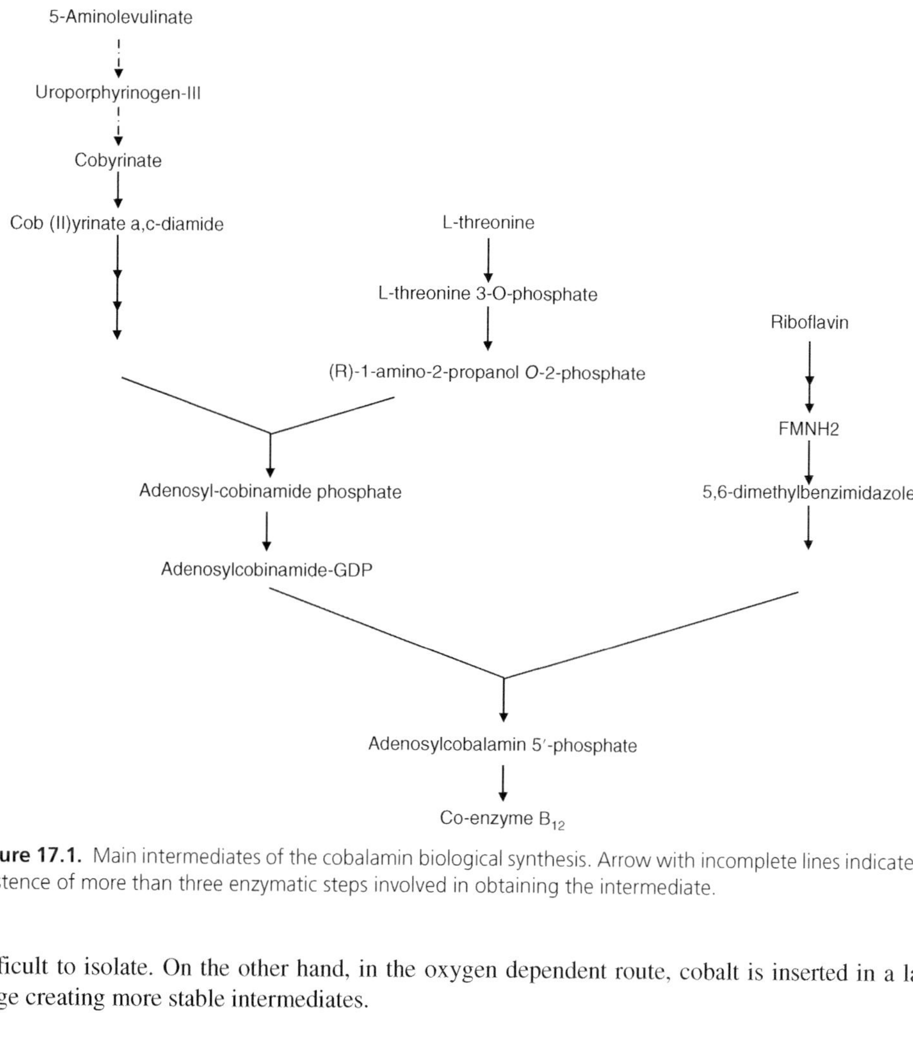

Figure 17.1. Main intermediates of the cobalamin biological synthesis. Arrow with incomplete lines indicate the existence of more than three enzymatic steps involved in obtaining the intermediate.

difficult to isolate. On the other hand, in the oxygen dependent route, cobalt is inserted in a later stage creating more stable intermediates.

17.3.2. Cobalamin Biosynthesis by Lactobacillus reuteri

It is well known that certain strains of *Lact. reuteri* are capable of metabolizing glycerol in co-fermentation with glucose. The first enzyme in this metabolic route is glycerol dehydratase that requires AdoCbl as a cofactor (Daniel et al. 1998). In this reaction, glycerol is converted to 3-HPA, which is subsequently reduced by NADH to 1,3-propanediol (1,3-PDL) mediated by the 1,3-PDL:NAD oxidoreductase. It was shown that *Lact. reuteri* CRL1098 was able to metabolize glycerol in a B_{12}-free medium, this being the first indirect evidence of the presence of intracellular cobalamin in *Lact. reuteri* (Taranto et al. 2003). The chromatographic analysis of intracellular extracts of *Lact. reuteri* CRL1098 revealed that this strain was able to produce a compound with an absorption spectrum closely identical to standard cobalamin but with a different elution time. The presence of cobalamin or of one of its intermediates in the intracellular extract of this strain was confirmed using different bioassays (Taranto et al. 2003).

Genetic evidence of cobalamin biosynthesis by *Lact. reuteri* CRL1098 was then obtained through the use of different molecular biology techniques. The genetic organization (*cob* and *cbi* genes) are very similar to those of *S. enterica* and *Listeria innocua* (Santos et al. 2007). It is important to note that the denomination of the genes for the biosynthesis of Cbl has been confused since some researchers have used different prefixes, *cbi* and *cob*, to describe the genes in *Ps. denitrificans* and *S. enterica*, respectively. Since some of the genes are present in both biosynthetic routes, these can have different denominations depending on which route is being described.

The genetic biosynthesis cluster of *Lact. reuteri* CRL1098 contains at least 30 genes that are involved in the *de novo* synthesis of the vitamin. One distinctive characteristic of the *cob* cluster of *Lact. reuteri* is the presence of *hem* genes in the middle of the cluster. In the respiratory organisms *Listeria* and *Salmonella* with similar *cob* clusters, the *hem* genes are described on other regions of their genome. The presence of the *hem* genes in the *cob* cluster is a characteristic that was only observed in certain genomes of *Clostridium* (Rodionov et al. 2003).

Additionally, transcriptional analysis performed in *Lact. reuteri* CRL1098 showed that the highest levels of expression of the *cob* cluster was observed in the late exponential growth phase in the absence of cobalamin, although the existence of exogenous cobalamin only partially repressed the expression of the operon. The transcription of a vast set of genes involved in cobalamin synthesis in sourdough prepared with strain *Lact. reuteri* ATCC 55730 was also described (Hüfner et al. 2008).

In addition to the strain CRL 1098, other *Lact. reuteri* strains were also shown to be capable of producing some corrinoids such as *Lact. reuteri* DCM 20016 (Santos et al. 2008), JCM1112 (Morita et al. 2008), and CRL1324 and 1327, strains isolated from vagina (Vannini et al. unpublished data).

Bioassays have been used to confirm that *Lact. reuteri* CRL 1098 produces Cbl in the order of 50 µg/l, which is a modest amount compared to the values obtained by the strains used in industrial production that surpass 200 mg/l (Martens et al. 2002). However, these values could be considered adequate when the daily requirement of vitamin B_{12} is taken into account (2–3 µg for an adult) because they could be sufficient for the development of B_{12} bio-enriched foods with. A possible drawback in the technological development is that the major form of the corrinoid produced in anaerobiosis by *Lact. reuteri* is pseudocobalamin (Santos et al. 2007) which, although present as a co-enzyme in many anaerobic and facultative anaerobic bacteria, is not active in animals and humans (Brandt et al. 1979; Rucker et al. 2001). Our studies indicate that in microaerophilic conditions *Lact. reuteri* CRL1098 not only produces pseudo-B_{12}, but also a small amount of other corrinoids including co-enzyme B_{12} (Vera 2007).

On the other hand, the supplementation with a B_{12}-free diet with *Lact. reuteri* CRL1098 was effective in preventing the pathologies caused by the nutritional deficiency of vitamin B_{12} both in pregnant and weaned young mice (Molina et al. 2009). Although these results are encouraging, our investigations are directed to maximize the synthesis of the vitamin, particularly in its active form in mammals. Therefore it is important to characterize some key enzymes in the biosynthesis of this vitamin by *Lact. reuteri* CRL 1098, in order to pave the way for future studies of metabolic engineering to increase the vitamin production and the amount of the active co-enzyme (see Figure 17.1 for the main metabolic pathway of this vitamin by LAB). The *pdu*O gene of this strain was overexpressed, and the coding protein PduO was crystallized and characterized (St Maurice et al. 2007). This enzyme is involved in one of the last steps of B_{12} biosynthesis: the corrinoid adenosylation, to give rise to the active co-enzyme adenosyl-cobalamin (Escalante-Semerena, 2007). Following these studies, Vannini et al. (2011) reported the cloning and heterologous expression of the uroporphyrinogen-III-methyltransferase-synthase gene (*cob*A/*hem*D), which codifies for a key enzyme in the early steps of cobalamin biosynthesis in *Lact. reuteri* CRL 1098.

With respect to B_{12} biotechnological application, the ability of a soymilk beverage containing *Lact. reuteri* CRL1098 to prevent the pathologies caused by a B_{12}-deficient diet was evaluated using an experimental murine model. Pregnant females were divided into four groups. Besides the B_{12}-deficient and the control group, the animals in the remaining two groups received non-fermented soymilk and soymilk fermented with *Lact. reuteri* CRL 1098 from the end of gestation to weaning. At the end of the trials, females and their corresponding offspring were sacrificed to determine hematological, immunological, and histological parameters. The results showed that the administration of fermented soymilk prevented the development of all symptoms observed as a consequence of nutritional B_{12} deficiency both in female mice and in their respective offspring. The design of soybean-based functional food bio-fortified with vitamin using a *Lactobacillus* strain able to produce compounds with B_{12} activity constitutes an interesting alternative therapy to prevent vitamin deficiency (Molina et al. 2013).

After the detection of the production of B_{12} by *Lact. reuteri* CRL1098 and the study of its cobalamin biosynthesis cluster, the genomic sequence of two strains of *Lact. reuteri*, with different characteristics, were released by the Joint Genome Institute: *Lact. reuteri* F275 (type stain DSM 20016) isolated from human feces that is unable to colonize mouse and *Lact. reuteri* 100-23 isolated from mouse intestines. Curiously, comparative genomic data revealed that the strain isolated from human feces (DSM 20016) presented the cobalamin biosynthesis cluster that was also associated with the anaerobic catabolism of glycerol (or 1,2-propanediol), whereas the mouse strain (100-23) did not contain the *cob* nor the glycerol metabolism genes.

Based on the horizontal transfer hypothesis of the *cob–pdu* cluster, it would be expected that other strains of LAB would have also received this genomic island by one of the many mechanisms of genetic transfer. Interestingly, the analysis of the sequenced LAB genomes shows that only *Lact. reuteri* (DSM20016/JCM1112 and ATCC55730) contain the *pdu–cob* genes. Martin et al. (2005) characterized a reuterin-producing strain of *Lact. coryniformis* isolated from goat milk. Using a microbiological method, these authors showed that the strain was able to produce a cobalamin-type compound. Preliminary genetic and biochemical data (Vannini et al. unpublished data), indicate that the *cob–pdu* cluster would be spread in other lactobacilli such as *Lact. coryniformis*, *Lact. murinus*, and *Lact. curvatus*. The availability of several vitamin B_{12}-producing strains and species of LAB would be important, not only for future basic studies on cobalamin production, but also in evolutionary studies related to the acquisition of the *cob–pdu* genomic island, and its potential application in the development of B_{12}-containing products.

The formation of water soluble vitamins (vitamin B_{12}, vitamin B_6, riboflavin, thiamine, nicotinic acid, and nicotinamide) during Tempe solid substrate fermentation was investigated. The addition of vitamin B_{12}-producing *Citrobacter freundii* and *Klebsiella pneumonia* resulted in an increase of physiologically active vitamin B_{12} (Keuth and Bisping 1993). The influence of *Rhizopus* on vitamin B_{12} formation by *Cit. freundii* was also investigated. The vitamin content of Tempe, obtained by fermentation with the mold and the bacterium, was three times as high as a control fermentation using *Cit. freundii* only.

In a recent study, the functional properties of *Lact. fermentum* CFR 2195, isolated from healthy infant feces, were evaluated (Basavanna and Prapulla 2013). This strain was able to adhere to HT-29 and Caco-2 cell-lines, showing high proteolytic activity and, more interestingly, vitamin B_{12} (29.45 ng/g dry biomass) production under submerged fermentation. In addition, vitamin production was evaluated in LAB isolated from nukazuke, a traditional Japanese pickle (Masudo et al. 2012). Among the 180 LAB isolates, one strain of *Lact. coryniformis* and one of *Lact. plantarum* were able to produce about 2 μg/l of vitamin B_{12}. Although these levels are not considered high, this study demonstrates that vitamin B_{12}-producing property should be evaluated in different ecological niches to find highly B_{12}-producing strains that could be used for developing new probiotic products aimed at preventing B_{12} deficiencies.

17.4. Conclusions

In this chapter, the newest studies describing B-group vitamin-producing LAB have been reviewed, showing that increasing the B-group vitamin concentrations in yogurts and fermented milks is possible through the judicious selection of the microbial species and cultivation conditions.

It is expected that the food industry will take the next step to use this information to select vitamin-producing strains to be used as starter cultures for vitamin bio-enriched fermented products destined to vitamin-deficient populations. Such products would provide economic benefits to food manufacturers since "naturally" increased vitamin concentrations would be an important added value without increasing production costs. Consumers would obviously benefit from such products increasing their vitamin intake while consuming them as part of their normal diet.

The *in situ* production of vitamins by the indigenous microbiota or probiotic strains has become an interesting new strategy to prevent vitamin deficiencies.

Acknowledgments

The authors would like to thank the Consejo Nacional de Investigaciones Científicas y Técnicas (CONICET) and the Agencia Nacional de Promoción Científica y Tecnológica (ANPCyT) for their financial support.

References

Ahire, J.J., Mokashe, N.U., Patil, H.J., and Chaudhari, B.L. (2013) Antioxidative potential of folate producing probiotic *Lactobacillus helveticus* CD6. *J Food Sci Technol* **50**, 26–34.

Aryana, K.J. (2003) Folic acid fortified fat-free plain set yoghurt. *Int J Dairy Technol* **56**, 219–222.

Asrar, F.M. and O'Connor, D.L. (2005) Bacterially synthesized folate and supplemental folic acid are absorbed across the large intestine of piglets. *J Nutr Biochem* **16**, 587–593.

Bacher, A., Eberhardt, S., and Richter, G. (1996) Biosynthesis of riboflavin. In: *Escherichia coli and Salmonella: Cellular and Molecular Biology* ((eds) F.C. Neidhardt, R. Curtiss III, J.L.L. Ingraham, et al.) ASM Press, Washington, D.C., pp. 657–664.

Bacher, A., Eberhardt, S., Fischer, M., et al. (2000) Biosynthesis of vitamin B_2 (riboflavin). *Ann Rev Nutr* **20**, 153–167.

Bailey, L.B., Rampersaud, G.C., and Kauwell, G.P. (2003) Folic acid supplements and fortification affect the risk for neural tube defects, vascular disease and cancer: evolving science. *J Nutr* **133**, 1961S–1968S.

Baines, M., Kredan, M.B., Usher, J., et al. (2007) The association of homocysteine and its determinants MTHFR genotype, folate, vitamin B_{12} and vitamin B_6 with bone mineral density in postmenopausal British women. *Bone* **40**, 730–736.

Basavanna, G. and Prapulla, S.G. (2013) Evaluation of functional aspects of *Lactobacillus fermentum* CFR 2195 isolated from breast fed healthy infants' fecal matter. *J Food Sci Technol* **50**, 360–366.

Bates, C.J. (1993) Folic Acid. In: *Encyclopaedia of Food Science Food Technology and Nutrition* ((eds) R. Macrae R.K. Robinson, and M.J. Sadler, M.J) Academic Press, London, pp.1936–1944.

Battersby, A.R. (1994) How nature builds the pigments of life: the conquest of vitamin B_{12}. *Science* **264**, 1551–1557.

Beck, W.S. (2001) Cobalamin (Vitamin B_{12}). In: *Handbook of Vitamins* ((eds) R. Rucker, J. Sutie, D. McCormick, and L.J. Machlin), 3rd Edn, Marcell Dekker, New York, pp. 463–512.

Blanck, H.M., Bowman, B.A., et al. (2002) Angular stomatitis and riboflavin status among adolescent Bhutanese refugees living in southeastern Nepal. *Am J Clin Nutr* **76**, 430–435.

Brandt, L.J., Goldberg, L., Bernstein, L.H., and Greenberg, G. (1979) The effect of bacterially produced vitamin B_{12} analogues (cobamides) on the in vitro absorption of cyanocabalamin. *Am J Clin Nutr* **32**, 1832–1836.

Burgess, C., O'Connell-Motherway, M., Sybesma, W., et al. (2004) Riboflavin production in *Lactococcus lactis*: potential for in situ production of vitamin-enriched foods. *Appl Environ Microbiol* **70**, 5769–5777.

Burgess, C.M., Smid, E.J., Rutten, G., and van Sinderen D. (2006) A general method for selection of riboflavin-overproducing food grade micro-organisms. *Microb Cell Fact* **5**, doi:10.1186/1475-2859-5-24

Camilo, E., Zimmerman, J., Mason, J.B., et al. (1996) Folate synthesized by bacteria in the human upper small intestine is assimilated by the host. *Gastroenterology* **110**, 991–998.

Capozzi, V., Menga, V., Digesu, et al. (2011) Biotechnological production of vitamin B2-enriched bread and pasta. *J Agric Food Chem* **59**, 8013–8020.

Claesson, M.J., Li, Y., Leahy, S., et al. (2006) Multireplicon genome architecture of *Lactobacillus salivarius*. *PNAS USA* **103**, 6718–6723.

Czeizel, A.E. and Dudas, I. (1992) Prevention of the first occurrence of neural-tube defects by periconceptional vitamin supplementation. *New Eng J Med* **327**, 1832–1835.

Charles, D., Ness, A.R., Campbell, D., et al. (2004) Taking folate in pregnancy and risk of maternal breast cancer. *Brit Med J* **329**, 1375–1376.

D'Aimmo, M.R., Mattarelli, P., Biavati, B., et al. (2012) The potential of bifidobacteria as a source of natural folate. *J Appl Microbiol* **112**, 975–984.

Danesh, J. and Lewington, S. (1998) Plasma homocysteine and coronary heart disease: systematic review of published epidemiological studies. *J Cardiov Risk* **5**, 229–232.

Daniel, R., Bobik, T.A., and Gottschalk, G. (1998) Biochemistry of coenzyme B_{12}-dependent glycerol and diol dehydratases and organization of the encoding genes. *FEMS Microbiol Rev* **22**, 553–566.

Dudeja, P.K., Torania, S.A., and Said, H.M. (1997) Evidence for the existence of a carrier–mediated folate uptake mechanism in human colonic luminal membranes. *Am J Physiol* **272**, 1408–1415.

Durga, J., van Boxtel, M.P., Schouten, E.G., et al. (2007a) Effect of 3-year folic acid supplementation on cognitive function in older adults in the FACIT trial: a randomised, double blind, controlled trial. *Lancet* **369**, 208–216.

Durga, J., Verhoef, P., Anteunis, L.J., Schouten, E., and Kok, F.J. (2007b) Effects of folic acid supplementation on hearing in older adults: a randomized, controlled trial. *Ann Int Med* **146**, 1–9.

Duthie, S.J., Narayanan, S., Brand, G.M., et al. (2002) Impact of folate deficiency on DNA stability. *J Nutr* **132**, 2444S–2449S.

ENNyS (2007) Encuesta Nacional de Nutrición y Salud. Dirección Nacional de Salud Materno Infantil - Ministerio de Salud – Argentina, http://msal.gov.ar/htm/Site/ennys/site/default.asp.

Escalante-Semerena, J.C. (2007) Conversion of cobinamide into adenosylcobamide in bacteria and archaea. *J Bacteriol* **189**, 4555–4560.

Ewe, J.A., Wan-Abdullah, W.N., and Liong, M.T. (2010) Viability and growth characteristics of *Lactobacillus* in soymilk supplemented with B-vitamins. *Int J Food Sci Nutr* **61**, 87–107.

FAO/WHO (2002) Human vitamin and mineral requirements. In: *Human Vitamin and Mineral Requirements*. Bangkok, Thailand.

Food and Drug Administration (1996) Food labeling: health claims and food label statements; folate and neural tube defects. *Federal Register* **61**, 8752–8807.

Gangadharan, D. and Nampoothiri, K.M. (2011) Folate production using *Lactococcus lactis* ssp *cremoris* with implications for fortification of skim milk and fruit juices. *LWT - Food Sci Technol* **44**, 1859–1864.

Ghosh D. and Chattopadhyay P. (2012) Preparation of idli batter, its properties and nutritional improvement during fermentation. *J Food Sci Technol* **48**, 610–615.

Herbert, V. (1996) Vitamin B_{12}. In: *Present Knowledge in Nutrition* ((eds) E.E. Ziegler and L.J. Filer, L.J), ILSI Press, Washington D.C., pp. 191–205.

Hill, M.J. (1997) Intestinal flora and endogenous vitamin synthesis. *Eur J Cancer Prev* **6**, S43–5.

Hüfner, E., Britton, R.A., Roos, S., et al. (2008) Global transcriptional response of *Lactobacillus reuteri* to the sourdough environment. *Syst Appl Microbiol* **31**, 323–338.

Hugenholtz, J. and Smid, E.J. (2002) Nutraceutical production with food-grade microorganisms. *Curr Opin Biotechnol* **13**, 497–507.

Ichihashi, T., Takagishi, Y., Uchida, K., and Yamada, H. (1992) Colonic absorption of menaquinone-4 and menaquinone-9 in rats. *J Nutr* **122**, 506–512.

Institute of Medicine (1998) *Dietary Reference Intakes for Thiamin, Riboflavin, Niacin, Vitamin B_6, Folate, Vitamin B_{12}, Panthothenic Acid, Biotin, and Choline*. National Academic Press, Washington, D.C.

Juarez del Valle, M., Laiño, J., Savoy de Giori, G., and LeBlanc, J.G. (2014) Use of lactic acid bacteria as a biotechnological strategy to increase riboflavin levels in soymilk. *Food Res Int* **62**, 1015–1019.

Jones, M.L. and Nixon, P.F. (2002) Tetrahydrofolates are greatly stabilized by binding to bovine milk folate-binding protein. *J Nutr* **132**, 2690–2694.

Kariluoto, S., Aittamaa, M., Korhola, M., et al. (2006) Effects of yeasts and bacteria on the levels of folates in rye sourdoughs. *Int J Food Microbiol* **106**, 137–143.

Kariluoto, S., Edelmann, M., Nystrom, L., et al. (2014) In situ enrichment of folate by microorganisms in beta-glucan rich oat and barley matrices. *Int J Food Microbiol* **176**, 38–48.

Keuth, S. and Bisping, B. (1993) Formation of vitamins by pure cultures of tempe moulds and bacteria during the tempe solid substrate fermentation. *J Appl Bacteriol* **75**, 427–434.

Kim, Y.I. (2004) Will mandatory folic acid fortification prevent or promote cancer? *Am J Clin Nutr* **80**, 1123–1128.

Kleerebezem, M., Boekhorst, J., van Kranenburg, R., et al. (2003) Complete genome sequence of *Lactobacillus plantarum* WCFS1. *PNAS USA* **100**, 1990–1995.

Kneifel, W., Kaufmann, M., Fleischer, A., and Ulberth, F. (1992) Screening of commercially available mesophilic dairy starter cultures: biochemical, sensory, and microbiological properties. *J Dairy Sci* **75**, 3158–3166.

Koizumi, S., Yonetani, Y., Maruyama, A., and Teshiba, S. (2000) Production of riboflavin by metabolically engineered *Corynebacterium ammoniagenes*. *Appl Microbiol Biotechnol* **53**, 674–679.

Konings, E.J., Roomans, H.H., Dorant, E., Goldbohm, R.A., et al. (2001) Folate intake of the Dutch population according to newly established liquid chromatography data for foods. *Am J Clin Nutr* **73**, 765–776.

Krause, L.J., Forsberg, C.W., and O'Connor, D.L. (1996) Feeding human milk to rats increases *Bifidobacterium* in the cecum and colon which correlates with enhanced folate status. *J Nutr* **126**, 1505–1511.

Kukanova, A., Zhdanov, V.G., and Stepanov, A.I. (1982) *Bacillus subtilis* mutants resistant to roseoflavin. *Genetika* **18**, 319–321.

Laiño, J.E., Juarez del Valle, M., Savoy de Giori, G. and LeBlanc J. G. (2014) Applicability of a Lactobacillus amylovorus strain as co-culture for natural folate bio-enrichment of fermented milk. *Int J Food Microb* **191**, 10–16 (2014)

Laiño, J.E., Juarez del Valle, M., Savoy de Giori, G., and LeBlanc, J.G. (2013) Development of a high folate concentration yogurt naturally bio-enriched using selected lactic acid bacteria. *LWT - Food Sci Technol* **54**, 1–5.

Laiño, J.E., LeBlanc, J.G., and Savoy de Giori, G. (2012) Production of natural folates by lactic acid bacteria starter cultures isolated from artisanal Argentinean yogurts. *Can J Microbiol* **58**, 581–588.

LeBlanc, J.G., Burgess, C., Sesma, F., et al. (2005a) *Lactococcus lactis* is capable of improving the riboflavin status in deficient rats. *Brit J Nutr* **94**, 262–267.

LeBlanc, J.G., Burgess, C., Sesma, F., et al. (2005b) Ingestion of milk fermented by genetically modified *Lactococcus lactis* improves the riboflavin status of deficient rats. *J Dairy Sci* **88**, 3435–3442.

LeBlanc, J.G., Laiño, J.E., Juarez del Valle, et al. (2011) B-group vitamin production by lactic acid bacteria - Current knowledge and potential applications. *J Appl Microbiol* **111**, 1297–309.

LeBlanc, J.G., Milani, C., Savoy de Giori, G., et al. (2013) Bacteria as vitamin suppliers to their host: a gut microbiota perspective. *Cur Opin Biotechnol* **24**, 160–168.

LeBlanc, J.G., Taranto, M.P., Molina, V., and Sesma, F. (2010a) B-group vitamins production by probiotic lactic acid bacteria. In: *Biotechnology of Lactic Acid Bacteria: Novel Applications* ((eds) F. Mozzi, R. Raya, and G. Vignolo), Wiley-Blackwell, Ames, IA, USA, pp. 211–232.

LeBlanc, J.G., Sybesma, W., Starrenburg, M., et al. (2010b) Supplementation with engineered *Lactococcus lactis* improves the folate status in deficient rats. *Nutrition* **26**, 835–41.

LeBlanc, J.G., Van Sinderen, D., Hugenholtz, J., et al. (2010c) Risk assessment of genetically modified lactic acid bacteria using the concept of substantial equivalence. *Cur Microbiol* **61**, 590–595.

LeBlanc, J.G., Rutten, G., Bruinenberg, P., et al. (2006) A novel dairy product fermented with *Propionibacterium freudenreichii* improves the riboflavin status of deficient rats. *Nutrition* **22**, 645–651.

Lucock, M. and Yates, Z. (2005) Folic acid - vitamin and panacea or genetic time bomb? *Nature Rev Gen* **6**, 235–240.

Luchsinger, J.A., Tang, M.X., Miller, J., et al. (2007) Relation of higher folate intake to lower risk of Alzheimer disease in the elderly. *Arch Neurol* **64**, 86–92.

Man, Z.W., Rao, Z.M., Cheng, Y.P., et al. (2014) Enhanced riboflavin production by recombinant *Bacillus subtilis* RF1 through the optimization of agitation speed. *World J Microbiol Biotechnol* **30**, 661–667.

Martens, J.H., Barg, H., Warren, M.J., and Jahn, D. (2002) Microbial production of vitamin B_{12}. *Appl Microbiol Biotechnol* **58**, 275–285.

Martin, R., Olivares, M., Marin, M.L., et al. (2005) Characterization of a reuterin-producing *Lactobacillus coryniformis* strain isolated from a goat's milk cheese. *Int J Food Microbiol* **104**, 267–277.

Masuda, M., Ide, M., Utsumi, H., et al. (2012) Production potency of folate, vitamin B(12), and thiamine by lactic acid bacteria isolated from Japanese pickles. *Bioscience, Biotechnology, and Biochemistry* **76**, 2061–2067.

Mera, P.E. and Escalante-Semerena, J.C. (2012) Multiple roles of ATP:cob(I)alamin adenosyltransferases in the conversion of B_{12} to coenzyme B_{12}. *Appl Microbiol Biotechnol* **88**, 41–48.

Molina, V., Medici, M., Taranto, M.P. and Font de Valdez, G. (2009) *Lactobacillus reuteri* CRL 1098 prevents side effects produced by a nutritional vitamin B12 deficiency. *J Appl Microbiol* **106**, 467–473.

Molina, V., Medici, M., Font de Valdez, G., and Taranto, M.P. (2013) Soybean-based functional food with vitamin B_{12}-producing lactic acid bacteria. *J Funct Food* **4**, 831–836.

Morita, H., Toh, H., Fukuda, S., et al. (2008) Comparative genome analysis of *Lactobacillus reuteri* and *Lactobacillus fermentum* reveal a genomic island for reuterin and cobalamin production. *DNA Res* **15**, 151–161.

Morris, M.C. and Tangney, C.C. (2007) Is dietary intake of folate too low? *Lancet* **369**, 166–167.

O'Brien, M.M., Kiely, M., Harrington, K.E., et al. (2001) The North/South Ireland Food Consumption Survey: vitamin intakes in 18–64-year-old adults. *Public Health Nutrition* **4**, 1069–1079.

Perkins, J.B. and Pero, J. (2002) Vitamin biosynthesis. In: *Bacillus subtilis and its Closest Relatives from Genes to Cells*, ((eds) A. Sonenshein, J. Hoch, and R. Losick), ASM Press, Washington, D.C., pp. 271–286.

Pompei, A., Cordisco, L., Amaretti, A., et al. (2007) Folate production by bifidobacteria as a potential probiotic property. *Appl Environ Microbiol* **73**, 179–185.

Putman, J. and Allshouse, J. (2003) Trends in U.S. per capita consumption of dairy products, 1909 to 2001: United States Department of Agriculture Economic Research Service.

Rodionov, D.A., Vitreschak, A.G., Mironov, A.A., and Gelfand, M.S. (2003) Comparative genomics of the vitamin B_{12} metabolism and regulation in prokaryotes. *J Biol Chem* **278**, 41148–41159.

Roessner, C.A. and Scott, A.I. (2006) Fine-tuning our knowledge of the anaerobic route to cobalamin (vitamin B_{12}). *J Bacteriol* **188**, 7331–7334.

Rossi, M., Amaretti, A., and Raimondi, S. (2011) Folate production by probiotic bacteria. *Nutrients* **3**, 118–34.

Roth, J.R., Lawrence, J.G., and Bobik, T.A. (1996) Cobalamin (coenzyme B_{12}): synthesis and biological significance. *Ann Rev Microbiol* **50**, 137–181.

Rucker, R.B., Suttie, J.W., McCormick, D.B., and Machlin, L.T. (2001) *Handbook of Vitamins*. Marcel Dekker Inc., New York.

Russo, P., Capozzi, V., Arena, M.P., et al. (2014) Riboflavin-overproducing strains of *Lactobacillus fermentum* for riboflavin-enriched bread. *Appl Microbiol Biotechnol* **98**, 3691–3700.

Said, H.M. and Mohammed, Z.M. (2006) Intestinal absorption of water soluble vitamins: an update. *Curr Opin Gastroenterol* **22**, 140–146.

Santos, F., Vera, J.L., Lamosa, P., et al. (2007) Pseudovitamin B($_{12}$) is the corrinoid produced by *Lactobacillus reuteri* CRL1098 under anaerobic conditions. *FEBS Lett* **581**, 4865–4870.

Santos, F., Vera, J.L., van der Heijden, R., et al J. (2008) The complete coenzyme B_{12} biosynthesis gene cluster of *Lactobacillus reuteri* CRL1098. *Microbiology* **154**, 81–93.

Scott, J.M. (1999) Folate and vitamin B_{12}. *P Nutr Soc* **58**, 441–448.

Sherwood, K.L., Houghton, L.A., Tarasuk, V., and O'Connor, D.L. (2006) One-third of pregnant and lactating women may not be meeting their folate requirements from diet alone based on mandated levels of folic acid fortification. *J Nutr* **136**, 2820–2826.

Smith, A.G., Croft, M.T., Moulin, M., and Webb, M.E. (2007) Plants need their vitamins too. *Cur Opin Plant Biol* **10**, 266–275.

St Maurice, M., Mera, P.E., Taranto, M.P., et al. (2007) Structural characterization of the active site of the PduO-type ATP:Co(I)rrinoidadenosyl transferase from *Lactobacillus reuteri*. *J Biol Chem* **282**, 2596–2605.

Strozzi, G.P. and Mogna, L. (2008) Quantification of folic acid in human feces after administration of Bifidobacterium probiotic strains. *J Clin Gastroenterol* **42**, S179–S184.

Sweeney, M.R., McPartlin, J., and Scott, J. (2007) Folic acid fortification and public health: report on threshold doses above which unmetabolised folic acid appear in serum. *BMC Public Health* **7**, 41–47.

Sybesma, W., Starrenburg, M., Kleerebezem, M., et al. (2003) Increased production of folate by metabolic engineering of *Lactococcus lactis*. *Appl Environ Microbiol* **69**, 3069–3076.

Taranto, M.P., Vera, J.L., Hugenholtz, J., et al. (2003) *Lactobacillus reuteri* CRL1098 produces cobalamin. *J Bacteriol* **185**, 5643–5647.

Thibaut, D., Blanche, F., Cameron, B., et al. (1998) Vitamin B_{12} biosynthesis in *Pseudomona denitrificans*. In: *Vitamin B_{12} and B12-Proteins* ((eds) B. Kräutler, D. Arigoni, D., and B.T. Golding), Wiley/VCH, Weinheim, pp. 63–67.

Tjonneland, A., Christensen, J., Olsen, A., et al. (2006) Folate intake, alcohol and risk of breast cancer among post-menopausal women in Denmark. *Eur J Clin Nutr* **60**, 280–286.

van de Guchte, M., Penaud, S., Grimaldi, C., et al. (2006) The complete genome sequence of *Lactobacillus bulgaricus* reveals extensive and ongoing reductive evolution. *PNAS USA* **103**, 9274–9279.

Van Guelpen, B., Hultdin, J., Johansson, I., et al. (2006) Low folate levels may protect against colorectal cancer. *Gut* **55**, 1461–1466.

Vannini, V., Rodríguez, A., Vera, J.L., et al. (2011) Cloning and heterologous expression of *Lactobacillus reuteri* uroporphyrinogen III synthase/methyltransferase gene (cobA/hemD): preliminary characterization. *Biotechnol Lett* **33**, 1625–1632.

Ventura, M., Canchaya, C., Fitzgerald, G.F., et al. (2007) Genomics as a means to understand bacterial phylogeny and ecological adaptation: the case of bifidobacteria. *Antonie Van Leeuwenhoek* **91**, 351–72.

Vera, J.L. (2007) Biosíntesis de cobalamina en *Lactobacillus reuteri* CRL 1098: Universidad Nacional de Tucumán (Argentina).

Verwei, M., Arkbage, K., Havenaar, R., et al. (2003) Folic acid and 5-methyltetrahydrofolate in fortified milk are bioaccessible as determined in a dynamic in vitro gastrointestinal model. *J Nutr* **133**, 2377–2383.

Wang, G., Bai, L., Wang, Z., et al. (2014) Enhancement of riboflavin production by deregulating gluconeogenesis in *Bacillus subtilis*. *World J Microb Biotechnol* **30**, 1893–1900.
Warren, M.J. (2006) Finding the final pieces of the vitamin B_{12} biosynthetic jigsaw. *PNAS USA* **103**, 4799–4800.
Warren, M.J., Raux, E., Schubert, H.L. and Escalante-Semerena, J.C. (2002) The biosynthesis of adenosylcobalamin (vitamin B_{12}). *Nat Prod Rep* **19**, 390–412.
Wegkamp, A., Starrenburg, M., de Vos, W.M., et al. (2004) Transformation of folate-consuming *Lactobacillus gasseri* into a folate producer. *Appl Environ Microbiol* **70**, 3146–3148.
Wegkamp, A., van Oorschot, W., de Vos, W.M. and Smid, E.J. (2007) Characterization of the role of para-aminobenzoic acid biosynthesis in folate production by *Lactococcus lactis*. *Appl Environ Microbiol* **73**, 2673–2681.

Chapter 18

Nutraceutics and High Value Metabolites Produced by Lactic Acid Bacteria

Elvira M. Hebert, Graciela Savoy de Giori, and Fernanda Mozzi*

Centro de Referencia para Lactobacilos (CERELA)-CONICET, Argentina

18.1. Introduction

Lactic acid bacteria (LAB) are a heterogeneous group of phylogenetically closely related microorganisms that produce lactic acid as the major or sole end-product from carbohydrate fermentation. The acidic environment provided by LAB has been empirically used to preserve raw food materials since ancient times. This feature, which allows prolonging the shelf-life of fermented foods and feeds, was successfully applied in the formulation of starter cultures for the elaboration of a wide variety of fermented foods and beverages of diverse origin including milk, vegetables, fruits, meat, fish, and cereal (Leroy and De Vuyst 2004; Ortiz et al. 2012). In addition to the rapid acidification traits of LAB, these microorganisms are capable of synthesizing other industrially interesting metabolites such as bacteriocins, exopolysaccharides (EPS), aroma compounds, sweeteners, vitamins, bioactive peptides, and so on, which in turn positively affect the organoleptic, texturizing, nutritional, and health-promoting properties of the final product (Hugenholtz 2008). Furthermore, LAB species belonging mainly to the genus *Lactobacillus*, known as probiotics, may exert beneficial health effects on the consumer (Bron and Kleerebezem 2011).

The role of glycolysis in LAB has been the target of numerous studies to control the flux needed not only for optimizing food fermentation processes but also for their use as microbial cell factories for the production of high value metabolites (Platteeuw et al. 1995; Hugenholtz 2008; Monedero et al. 2010; Boguta et al. 2014). The use of LAB as cell factories is mainly due to the following features: (i) their food-grade status and beneficial effects in the gastrointestinal tract; (ii) the availability of genetic tools for strain improvement and their small size genomes (1.8-3.4 Mbp); (iii) a relatively simple and fast metabolism; (iv) the possibility of using cheap substrates for growth; and finally (v) the capability of directly applying their metabolites in food products (Hugenholtz 2008; Gaspar et al. 2013). More recently, Mazzoli et al. (2014) proposed that LAB are among the most promising candidates for use in future biorefineries in converting plant-derived biomass (lignocellulosic feedstock) into high value-added products such as biofuels (ethanol and butanol), biodegradable plastic polymers, EPS, antimicrobial agents, health-promoting substances, and nutraceuticals. Although currently the use of probiotic LAB is preferable to supplementation with purified molecules as nutraceutical food additives, the challenge remains with respect to vitamins and amino acid derivatives.

* Corresponding author: Centro de Referencia para Lactobacilos (CERELA)-CONICET, Chacabuco 145, 4000 San Miguel de Tucumán, Tucumán, Argentina. fmozzi@cerela.org.ar

Biotechnology of Lactic Acid Bacteria: Novel Applications, Second Edition. Edited by Fernanda Mozzi, Raúl R. Raya, and Graciela M. Vignolo.

In past years, LAB applications have moved beyond the traditional food fermentation processes to use in delivery of molecules (Martin et al. 2014) and as microbial cell factories for producing value-added products (Boguta et al. 2014). In this chapter the capability of LAB to naturally produce nutraceutics and high value metabolites is discussed; the use of recombinant LAB for production of commodity chemicals is presented briefly.

18.2. Nutraceutics

In 1989 De Felici coined the term "Nutraceuticals" from "Nutrition" and "Pharmaceutical," and defined it as "a food or part of food (components) that provides medical or health benefits, including the prevention and/or treatment of a disease." Although the term nutraceutical is commonly used in marketing, it has no regulatory definition. For example, for the United States law this term has no meaning, while in Canada "nutraceutical" can either be marketed as a food or as a drug; the terms "nutraceutical" and "functional food" have no legal distinction and refer to "a product isolated or purified from foods that is generally sold in medicinal forms not usually associated with food and that possess a physiological benefit or provide protection against chronic disease." Kalra (2003) proposed redefining nutraceutics as a dietary supplement that should also aid in the prevention and/or treatment of disease and/or disorder.

Here, the production of nutraceutics (polyols, short-chain fatty acids, conjugated-linoleic acids, bioactive peptides, gamma-amino butyric acid, and vitamins) and other interesting metabolites such as exopolysaccharides by LAB is presented (Hugenholtz and Smid 2002; Hugenholtz 2008; Ortiz et al. 2012; Mazzoli et al. 2014).

18.2.1. Low-calorie Sugars

Polyols or low-calorie sugars are non-cyclic sugar alcohols in which the carbonyl group (aldehyde or ketone) of the precursor sugar is reduced to the corresponding primary or secondary alcohol (Cummings and Stephen 2007). These polyols occur naturally in fruits and vegetables and are synthesized by a variety of algae, plants, yeasts, filamentous fungi, and bacteria (Iwamoto and Shiraiwa 2005; Solomon et al. 2007). Although their physiological role in microorganisms is still not very well established, polyols may act as reservoirs of carbon and reducing power (Wisselink et al. 2002; Akinterinwa et al. 2008). In addition, it has been suggested that polyols are synthesized against adverse environmental conditions such as oxidative and osmotic stresses as compatible solutes and aid in preventing oxidative damage by scavenging of free reactive oxygen radicals (Chatuverdi et al. 1997; Wisselink et al. 2002; Hounsome et al. 2008).

Among the existing polyols, namely mannitol, sorbitol, erythritol, galactitol, xylitol, ribitol, lactitol and maltitol (Figure 18.1), only mannitol and erythritol are naturally produced by LAB (Vrancken et al. 2010).

Polyols have multiple industrial applications but they are mainly used as natural sweeteners in the food industry due to their low-caloric, low-glycemic, and anti-cariogenic properties being applicable in dietetic and diabetic food products (Livesey 2003). As sugar alcohols are partially or not absorbed in the small intestine, they may reach the colon where they can be degraded by certain bacteria (Schiweck et al. 1994) leading to the production of several short-chain fatty acids (SCFA), which have been claimed to confer health benefits to the host (van Munster and Nagengast 1993; Liong and Shah 2005). Moreover, due to their interesting physicochemical characteristics, polyols act as texturizing agents, softeners, color stabilizers, and humectants and do not take part in the Maillard reaction (Monedero et al. 2010). Mannitol, sorbitol, and erythritol have all received GRAS (Generally Recognized As Safe) status from the US Food and Drug Administration (Salminen and Hallikainen 2002). Mannitol

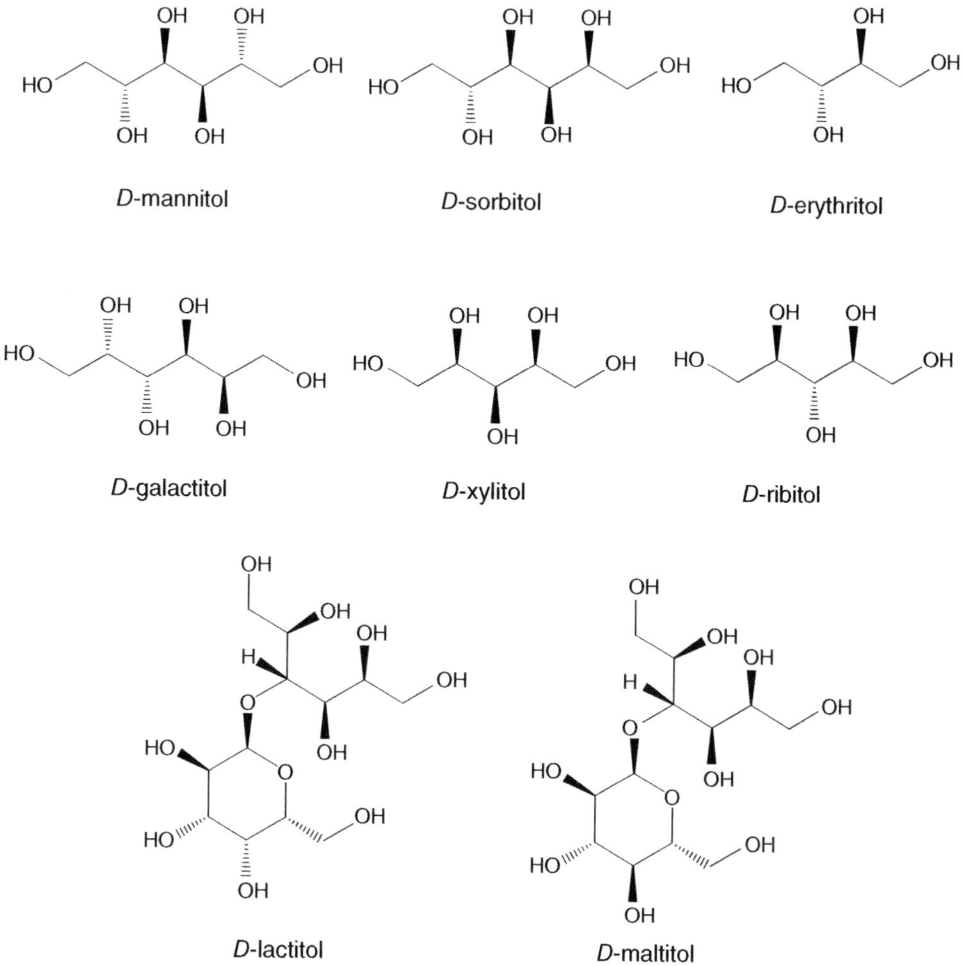

Figure 18.1. Chemical structures of *D*-mannitol, *D*-sorbitol, *D*-erythritol, *D*-galactitol, *D*-xylitol, *D*-ribitol, *D*-lactitol, and *D*-maltitol.

and erythritol were incorporated into the international world market of polyols in the 1990s and represent a lower production volume compared to sorbitol (von Weymarn et al. 2002).

Besides their application in the food industry, polyols are employed in the pharmaceutical and chemical industries and in medicine as a potent osmotic diuretic (Monedero et al. 2010; Saha and Racine 2010; Ortiz et al. 2013). In this respect, mannitol is applied to patients with kidney failure, brain edema, and intraocular pressure (von Weymarn et al. 2002).

Current industrial production of certain polyols (i.e., mannitol and sorbitol) is performed by catalytic reduction of a mixture of sugars with hydrogen gas and nickel catalyst using high temperatures and pressures (Kusserow et al. 2003). In addition to the high cost of this process due to the need for pure substrates and extreme conditions, it requires costly chromatographic purification steps. Over the last 15 years, biotechnological production of polyols has been sought as an alternative to the current chemical production. In this sense, heterofermentative LAB seem ideal microorganisms since they display a fermentative metabolism associated with an important redox modulation and a limited biosynthetic capacity (Monedero et al. 2010). While some heterofermentative LAB may naturally produce mannitol and erythritol under certain culture conditions, sorbitol and xylitol are only synthesized through metabolic engineering processes. Strains belonging to the genera *Lactobacillus*, *Oenococcus*, *Leuconostoc*, and *Weisella* have been reported to form **mannitol**. Interestingly, controlled culture conditions using selected LAB strains have yielded similar amounts of mannitol to those of current industrial chemical processes (von Weymarn et al. 2002).

Recently, Ortiz et al. (2012) demonstrated that *Lact. reuteri* CRL 1101 efficiently produces mannitol from sugarcane molasses as a carbon source; a maximum mannitol concentration of 178 mM was found using 7.5% molasses after incubating the cells at 37 °C for 24 h. Increasing the molasses concentration up to 10% (w/v) and the fermentation period up to 48 h did not significantly improve mannitol production. Under aerated culture conditions, high mannitol values (145 mM) were attained in shorter fermentation periods (8 h) as compared to static ones (6 mM). The highest activity value of mannitol 2-dehydrogenase (MDH), the enzyme responsible for reducing fructose into mannitol in heterofermentative LAB, was obtained during the log growth phase and no correlation between MDH activities and mannitol production was observed in the fermentations performed. At controlled pH = 5.0 mannitol production (228 mM = 41.5 g/l), volumetric productivity (Q_{Mtl}: 1.73 g/l h) and yield (Y_{Mtl}: 105%) were markedly improved after 24 h incubation using a minimized molasses-based culture medium (unpublished results).

The ability of other heterofermentative LAB strains to produce mannitol has been reported (von Weymarn et al. 2002; Fontes et al. 2009; Carvalheiro et al. 2011); the highly osmotolerant strain *Lact. intermedius* NRRL-3693 being the best mannitol producer so far with a yield higher than 200 g/l when grown on a high fructose concentration (300 g/l) medium (Saha and Nakamura 2003).

Sorbitol production by *Lact. casei* and *Lact. plantarum* strains has been achieved by reverting the sorbitol catabolic pathway due to the reversibility of the enzyme sorbitol 6-P dehydrogenase, which is responsible for the interconversion of sorbitol 6-P into fructose 6-P. While metabolic engineering strategies using homologous genes for sorbitol production by LAB have been applied, **xylitol** synthesis has been achieved by heterologous expression of the *D*-xylose reductase gene from the yeast *Pichia stipitis* CBS 5773 and of xylose transporter genes from the strain *Lact. brevis* ATCC 8287 (Nyyssölä et al. 2005). **Erythritol** is industrially produced by fermentation using osmophilic yeasts (Embuscado and Patil 2001) as erythrose is a very expensive substrate to be used for direct catalytic hydrogenation (Monedero et al. 2010). Erythritol production by some LAB such as *O. oeni*, *Leuc. mesenteroides*, and *Lact. sanfranciscencis* has been reported (Veiga-da-Cunha et al. 1993; Stolz et al. 1995; Richter et al. 2001).

18.2.2. Short-Chain Fatty Acids

In LAB short-chain fatty acids (SCFA) are not produced anabolically but as final catabolites of the energy metabolism. The most studied SCFA are **butyrate**, **propionate**, and **acetate,** which can be produced by LAB through pyruvate fermentation (Pessione 2012). Recently, Bosch et al. (2014) reported on the ability of three *Lact. plantarum* strains able to reduce blood cholesterol levels and to produce large quantities of propionate and butyrate. Several beneficial health properties have been attributed to these compounds. Besides their acidifying desirable characteristics, acetate may be metabolized by muscles and other tissues when entering the peripheral circulation while propionate is taken up by the liver. Both acids can lower glycemia improving insulin sensitivity (Pessione 2012). In contrast, butyrate may exert both beneficial (it can control oxidative and metabolic stresses at molecular level by enhancing repair responses and inhibiting the synthesis of pro-inflammatory (IL12, TNF, gamma IF) cytokines and detrimental effects (affects body weight and obesity since it represents an additional energy source) (Turnbaugh et al. 2006). When butyrate is absorbed in the large intestine, it is converted into fat within the liver and positively regulates host genes promoting lipogenesis as well as fat storage (DiBaise et al. 2008).

18.2.3. Conjugated Linoleic Acid (CLA)

Conjugated linoleic acids (CLA) are a type of biologically beneficial functional lipids that are positional and geometric isomers of the cis9, cis12 octa-deca-dienoicacid (linolenic acid, LA). The most frequently found isomers of LA have the unsaturation site at positions 9, 11 (namely, Δtrans 9, cis11-octa decadienoic acid) and 10, 12 (Δcis10-, trans12-linoleic acid). The human metabolism can be positively affected by these compounds as some CLA isomers may reduce cancer cell viability, inflammation, metabolic disorders, and cardiovascular diseases, as well as reducing body fat (Ewaschuk et al. 2006; Gorissen et al. 2013). However, these effects are dependent on the CLA isomers and doses (Pessione 2012; Gorissen et al. 2013) as some of them may lower the HDL/LDL ratio and modify eicosanoid production with detrimental health effects. The isomers for which evidence of prejudicial effects exist are the trans10 and cis12 isomers (Pessione 2012). Thus, the CLA dosage has to be maintained at the minimal active concentration to avoid undesired consequences (Ewaschuk et al. 2006).

Besides ruminal bacteria, other food grade microorganisms are able to synthesize CLA (Gorissen et al. 2013). Hitherto, the best CLA-producing LAB belong to the genera *Lactobacillus* and *Streptococcus*. Among food-grade bacteria, bifidobacteria and propionibacteria are also able to convert LA and α-linolenic acid (LNA) to CLA and conjugated linolenic acid (CLNA), respectively. Linoleate isomerase activity, responsible for this conversion, is strain-dependent and probably related to the capacity of the producing strain to tolerate the toxic effects of LA and LNA (Gorissen et al. 2011, 2013; Yang et al. 2014). Kishino et al. (2011) and more recently Yang et al. (2014) found a multicomponent linoleate isomerase system, composed of three enzymes existing in bacterial membrane and soluble fractions, involved in the CLA synthesis by LAB. As natural concentrations of CLA and CLNA in ruminal food products are relatively low to exert beneficial health effects, LAB with linoleate isomerase activity could be used as a starter or adjunct culture to formulate functional fermented dairy and meat products with increased CLA and CLNA levels or be included as probiotic cultures in fermented products (Gorissen et al. 2011).

18.2.4. Bioactive Peptides

Food proteins are source of several bioactive peptides, which are hidden within the native protein in a latent state, requiring enzymatic proteolysis for their release. The quality of a dietary protein depends not only on the amino acid composition, the absence of co-existing antinutrients that can limit protein digestibility, absorptive ability, and bioavailability, but also on the released peptides. **Bioactive peptides** are food-derived peptides that exert, beyond their nutritional value, a physiological, hormone-like effect in the body (Hebert et al. 2010; Saavedra et al. 2013). Bioactive peptides can be released from the parental protein by proteolysis *in vivo* during gastrointestinal digestion by the action of host (digestive enzymes) or microbial enzymes (derived from the human microbiota), *in vitro* or a combination of both. The *in vitro* production of bioactive peptides involves the enzymatic hydrolysis of the dietary protein by endogenous enzymes present in the food matrix as well as proteolysis occurring during food processing or ripening by action of starter cultures or by enzymes isolated from proteolytic microorganisms such as certain LAB (Hayes et al. 2007a; Hebert et al. 2010). Nowadays, microbial fermentation by LAB species such as *Lact. helveticus*, *Lact. delbrueckii* subsp. *lactis*, and *Lactococcus lactis* is one of the main processes to generate bioactive peptides in the dairy industry (Hayes et al. 2007a,b; Hebert et al. 2010; Espeche Turbay et al. 2012). Although bioactive peptides derive from a wide range of vegetable and animal proteins such as fish, egg, meat, soybean, rice, sunflower, cereals, and so on, milk proteins are currently the main source of a range of biologically active peptides (Scheirlinck et al. 2007; Kussmann and Van Bladeren 2011; Bouzerzour et al. 2012; Escudero et al. 2012; Senevirathne and Kim 2012). As milk contains only small amounts of amino acids and short peptides, LAB depend on a complex proteolytic system to obtain essential amino acids from caseins during growth in milk. This specialized proteolytic system consists of: (i) a cell envelope-associated proteinase (CEP); (ii) transport systems to allow uptake of the resulting peptides; and (iii) several intracellular peptidases, which degrade peptides into amino acids. The CEP is the key enzyme of this system as it is involved in the first step of casein degradation. In addition to its vital role for bacterial growth in milk, CEP also contributes to the development of flavor and texture of fermented dairy products, also promoting human health through the release of bioactive peptides during milk fermentation (Savijoki et al. 2006; Hebert et al. 2008). Comparative genome analysis of 22 LAB strains revealed that dairy lactobacilli are deficient in amino acid biosynthetic pathways, albeit to different extents (Liu et al. 2010); this fact being related to specific-niche adaptation. To compensate for the disability of these lactobacilli to synthesize most amino acids, their genomes contain several genes related to the proteolytic system, allowing them to take amino acids from proteins present in their environment (Hebert et al. 2008; Cai et al. 2009). Information about the distribution of proteolytic system genes could be used to improve the functional properties of dairy and other fermented food products based on the proteolytic potential of LAB strains.

Antimicrobial, antihypertensive, citomodulatory, antithrombotic, immunomodulatory, opioid, and antioxidant activities and enhancement of mineral absorption and/or bioavailability are some of the biological activities attributed to milk-derived peptides. Among the bioactive compounds derived from dietary proteins those with angiotensin-converting enzyme (ACE)-inhibitory activity have been studied extensively due to their potential use in the prevention of cardiovascular diseases (Phelan and Kerins 2011). Antihypertensive peptides are thought to be competitive substrates for ACE, an enzyme located in different tissues (e.g., plasma, lung, kidney, heart, skeletal muscle, pancreas, brain), which may influence several regulatory systems (Ondetti and Cushman 1982). ACE-inhibitory peptides from dairy origin usually contain up to 10 amino acids, strain selection being one of the main factors that affects the release of bioactive peptides in dairy fermentations (Hayes et al. 2007a; Saavedra et al. 2013). Recently, different antihypertensive peptides were characterized after α- and β-casein hydrolysis by *Lact. delbrueckii* subsp. *lactis* CRL 581 (Villegas et al. 2014). Two potent ACE inhibitory tri-peptides, VPP and IPP, derived from bovine casein from fermented milk with *Lact. helveticus* strains were isolated (Takano 2002; Hayes et al. 2007a,b). Different antihypertensive milk products containing peptides IPP and VPP such as Calpis® (Japanese sour fermented milk with a starter composed of *Lact. helveticus* and *Saccharomyces cerevisiae*), Ameal S™, and Evolus™ (fermented milks with *Lact. helveticus*) are currently available on the market (Fitzgerald and Murray 2006). These peptides were also identified in a β-casein hydrolysate generated by *Lact. delbrueckii* subsp. *lactis* CRL 581 (Hebert et al. 2008; Villegas et al. 2014).

Matar et al. (2001) demonstrated that immune-enhancing and antimutagenic peptides were released during milk fermentation by the proteolytic strain *Lact. helveticus* R389. Moreover, bioactive peptides released from milk proteins by this strain were able to stimulate the immune system and to inhibit the growth of an immunodependent fibrosarcoma in a mouse model (LeBlanc et al. 2002). Tellez et al. (2010) showed that a peptide-containing fraction with three peptides derived from β-casein and one from α-lactalbumin hydrolysis by using the strain *Lact. helveticus* LH-2 upregulated cytokines and nitric oxide production by macrophages as well as stimulating phagocytic activity of these cells, while a peptide derived from bovine β-casein (193–209) modulated functional properties of bone marrow-derived macrophages from germfree and human biota-associated mice; this peptide could enhance the antimicrobial activity of macrophages without proinflammatory effects (Sandré et al. 2001). In addition, a β-casein hydrolysate generated by *Lact. delbrueckii* subsp. *lactis* CRL 581 exerted a beneficial effect on acute intestinal inflammation by increased interleukin 10 and decreased IFN-γ production in the gut (Espeche Turbay et al. 2012).

Most antimicrobial peptides reported to date have been released from the parent milk protein following heat and/or alkali treatment or enzymatic hydrolysis (Hayes et al. 2007a,b). An antimicrobial peptide derived from β-casein (184–210) in a human sodium casein hydrolysate generated by a proteinase of *Lact. helveticus* PR4 has been reported (Minervini et al. 2003). This peptide showed a broad spectrum of inhibition against potentially pathogenic bacteria such as *Staphylococcus aureus*, *Enterobacter faecium*, *Yersinia enterocolitica*, and *Salmonella*. In addition, an antimicrobial peptide with the sequence RPKHPIKHQ derived from α_{s1}-casein (1–7) was isolated from yogurt (Hayes et al. 2007a,b).

Caseinphosphopeptides (CPP) derived from milk protein are abundant in cheeses and are known to increase calcium absorption *in vitro* by preventing calcium precipitation into an insoluble form (Hayes et al. 2007a,b).

To determine the biological activity of peptides, the enzymes used to hydrolyze the parental protein, the food-processing conditions, and the amino acid composition of the resulting peptides should be considered. Furthermore, it is necessary to follow up the bioactive peptides not only in the food matrix but also upon ingestion and absorption. Nowadays, the application of "omic"-based analytical strategies facilitates the characterization of bioactive peptides in food. In fact, *in silico* and empirical (*in vitro*/*in vivo*) approaches aimed at discovering and identifying bioactive peptides in food matrix are currently available (Figure 18.2) (Kussmann and Van Bladeren 2011). Bioinformatics analyses offer the possibility to simulate the hydrolysis of a potential precursor food protein allowing the prediction of possible bioactive peptides. These approaches provide a cost-effective strategy decreasing different steps of the traditional workflow. Quantitative structure activity relationship model (QSAR) approaches have been used in studies of ACE-inhibitory, antioxidant, and antimicrobial peptide activities (Gu et al. 2011; Toropova et al. 2012; Li and Li 2013; Sagardia et al. 2013; Shu et al. 2013). For these studies, peptide databases such as BIOPEP, Pepbank, BioPD, and SwePep are available.

On the other hand, the empirical approach involves a series of steps such as: (i) selection of a target food protein; (ii) release of the bioactive sequences by *in vitro* enzymatic hydrolysis; (iii) screening for a targeted bioactivity; (iv) fractionation and separation of the peptide mixture as well as determination of their biological activity; (v) peptide identification by mass spectrometry; (vi) design of synthetic structural analogs or peptide mimetic to corroborate bioactivity *in vitro* and *in vivo* (Figure 18.2). As stated, in addition to the *in vitro* enzymatic hydrolysis, the release of bioactive peptides from the parental food protein may occur by proteolysis during microbial fermentation and/or food processing as well as during gastrointestinal digestion (Rutherfurd-Markwick 2012). It is worth mentioning that to exert their biological role; bioactive peptides have to reach the target organs, their resistance to gastrointestinal digestion, absorption across the enterocytes, and bioavailability in target tissues being fundamental.

18.2.5. Gamma-aminobutyric Acid (GABA)

γ-Aminobutyric acid (GABA) is a non-protein amino acid that acts as the major inhibitory neurotransmitter in the sympathetic nervous system (Wong et al. 2003). GABA also shows hypotensive, tranquilizer, and diuretic effects (Jakobs et al. 1993; Kimura et al. 2002; Inoue et al. 2003) and can prevent diabetes (Adeghate and Ponery 2002), improve the concentration of plasma growth hormone and the rate of protein synthesis in brain (Tujioka et al. 2011). In addition, it has been claimed that GABA promotes the healing process of cutaneous wounds (Han et al. 2007) and enhances the survival of dermal fibroblasts exposed to oxidative stress (Ito et al. 2007).

GABA synthesis requires the activity of glutamate decarboxylase (GAD, EC 4.1.1.15), an intracellular pyridoxal 5′-phosphate-dependent enzyme, which catalyzes the irreversible α-decarboxylation of *L*-glutamate to produce GABA (Figure 18.3). GAD is largely distributed in higher plants, animals, and many microorganisms including bacteria (Di Cagno et al. 2010), fungi, and yeasts (Li and Cao 2010). In addition to the health-promoting properties in humans, bacterial GABA production

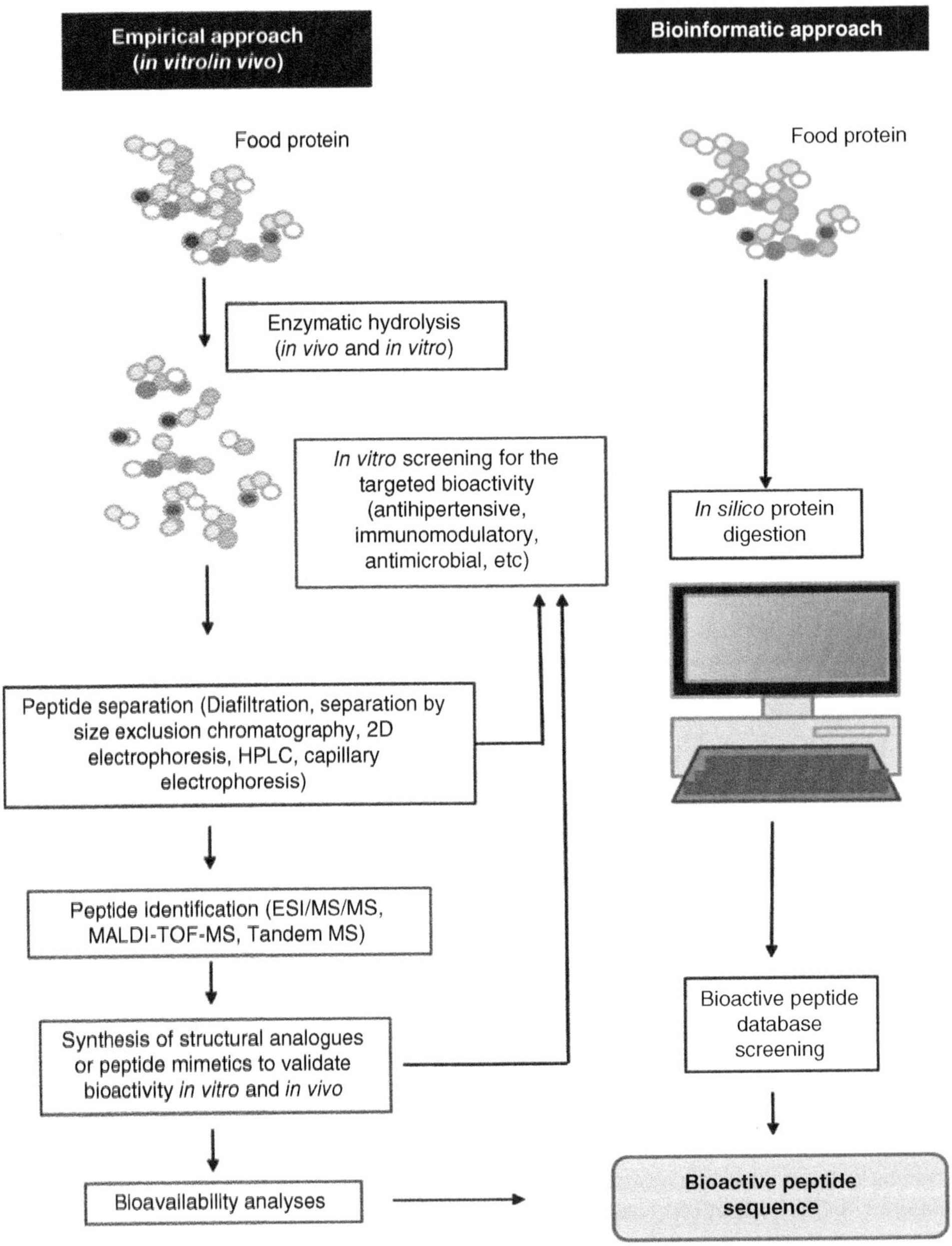

Figure 18.2. Discovering of bioactive peptides.

confers resistance to microbial cells under acidic environmental conditions. After uptake of *L*-glutamate by a specific transporter, GAD performs a proton-consuming decarboxylation reaction while the specific antiporter (GadC) promotes the $glutamate_{in}/GABA_{out}$ electrogenic antiporter (De Biase and Pennacchietti 2012). The net result is an increase in the intracellular pH related to the removal of hydrogen ions and an increase in the extracellular pH due to the exchange of extracellular glutamate for the more alkaline GABA.

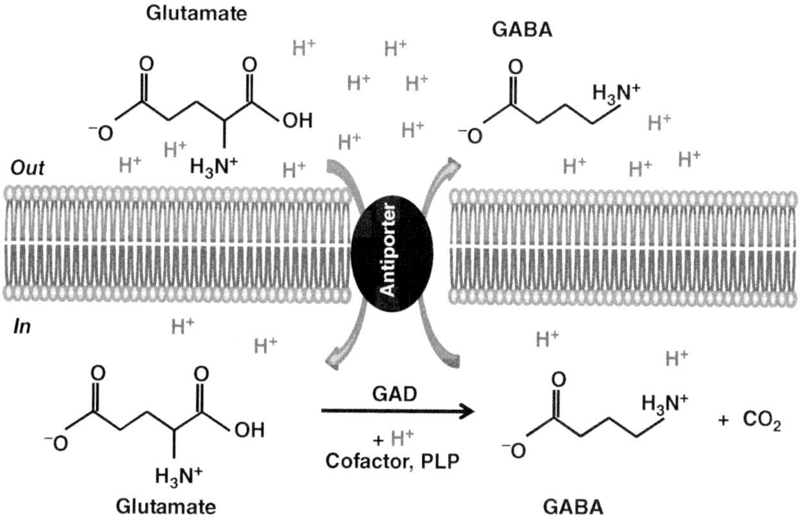

Figure 18.3. Schematic representation of GABA production. Glutamate is taken up by the electrogenic *L*-glutamate/GABA antiporter, the decarboxylation of *L*-glutamate via glutamate decarboxylase (GAD) consumes an intracellular H^+ and contributes to the generation of proton motive force by GABA. PLP: pyridoxal-5′-phosphate.

In LAB, the ability to produce and release GABA has been associated with the protection against acid stress, this allowing microbial survival through the passage of the gastrointestinal tract towards the intestine where LAB may exert a beneficial effect (Siragusa et al. 2007; De Biase and Pennacchietti 2012). In addition, GABA synthesis is associated with ATP production by coupling electrogenic antiport with glutamate decarboxylation (De Biase and Pennacchietti 2012).

Considering the potential of GABA as a bioactive component in foods and pharmaceuticals, the development of GABA-containing functional foods has attracted great attention over the last few years. Biosynthetic methods of GABA production are more promising than chemical ones since they have a simple reaction procedure, high catalytic efficiency, mild reaction conditions, and environmental compatibility (Dhakal et al. 2012). Several GABA-containing foods such as tea (Cherng et al. 2014), fermented grape (Di Cagno et al. 2010), Thai fermented pork sausage (Ratanaburee et al. 2013), sourdough (Rizzello et al. 2008), cheeses (Siragusa et al. 2007), Kimchi (Jeong et al. 2013), black raspberry juice (Kim et al. 2009), and rice bran (Kook et al. 2010) have been reported.

LAB possess the ability to produce and release GABA in the growth medium (Li and Cao 2010; Dhakal et al. 2012), this feature being strain-dependent. The main LAB capable of synthesizing GABA belong to the species *Lact. brevis*, *L. lactis*, *Lact. paracasei*, *Lact. buchneri*, *Lact. helveticus*, and *Lact. plantarum* (Siragusa et al. 2007; Li and Cao 2010; Dhakal et al. 2012). Among them, *Lact. brevis* produced the highest amount (345.83 mM) of GABA reported so far (Li et al. 2010a). However, many fermentation conditions including pH, incubation time, and temperature, medium components as well as different fermentation strategies can be optimized to achieve maximum GABA production by LAB strains.

18.2.6. Vitamins

Vitamins are essential micronutrients for the metabolism of living organisms. Due to insufficient food intake or unbalanced diets, human vitamin deficiencies are still very common.

Biotechnological production of vitamins using LAB as an alternative to chemical vitamin synthesis has been under study for more than a decade (Taranto et al. 2003; De Angelis et al. 2014; see also

Chapter 17 of this book). Furthermore, food fermentation with LAB offers unique opportunities to enhance the nutritional value of food products and the development of novel functional foods with improved vitamin content. As result of secondary metabolism, certain LAB strains synthesize group B vitamins such as B_{12} or **cobalamin**, B_{11} or **folate**, and B_2 or **riboflavin**. In this respect, Taranto et al. (2003) reported for the first time the synthesis of cobalamin, a complex corrin compound, by a LAB strain. The authors found that the sourdough strain *Lact. reuteri* CRL1098 was able to ferment the carbohydrate and glycerol present in vitamin B_{12}-free medium, reducing glycerol through a well-known cobalamin-dependent reaction with the formation of 1,3-propanediol as a final product; this was the first evidence of the ability of this strain to synthesize cobalamin. From a biotechnological viewpoint, this microorganism might be a good candidate to increase the cobalamin content in fermented food, although the main limitation is that cobalamin is not excreted to the environment, remaining in the cytoplasm of the cell producer.

Concerning folates, several industrially important LAB belonging to the species *L. lactis* and *Strep. thermophilus* are capable of synthesizing folate, as has been observed in fermented milks with high folate levels. More recently, Laiño et al. (2014) reported on the ability of a *Lact. amylovorus* strain that was successfully used as co-culture for natural folate bio-enrichment of fermented milk. With regard to riboflavin, its biosynthesis has been identified in several LAB species and some applications of riboflavin-producing LAB in dairy and cereal-based products have also been reported (LeBlanc et al. 2013). In general, vitamin production by LAB varies considerably being a species-specific or strain-dependent trait. This feature is generally related to the partial or complete interruption of the genetic information for vitamin biosynthesis (this topic is deeper covered in Chapter 17 by Taranto et al. in this book).

18.3. Exopolysaccharides

Certain LAB produce and secrete EPS, which are extracellular polymers of carbohydrates. According to their position relative to the cell, EPS can be classified as (i) capsular polysaccharides (CPS) if they are tightly associated with the cell surface or as (ii) slime EPS if they are loosely attached to the outer cell structures or secreted into the environment. Depending on their chemical composition, EPS are divided into homopolysaccharides (HoPS), which contain only one type of monosaccharide, and heteropolysaccharides (HePS) that are formed of repeating units composed of two or more different monosaccharides. Four groups of HoPS are produced by LAB: (i) α-*D*-glucans (dextrans, mutans, alternans, and reuterans); (ii) β-*D*-glucans; (iii) β-*D*-fructans; and (iv) polygalactans. In contrast, a wide diversity of HePS from LAB concerning their structure, molecular mass (MM), monomer composition, and glycosidic linkages has been reported; all these features conferring different physicochemical properties to the polymers (Vaningelgem et al. 2004; Mozzi et al. 2006). Mozzi et al. (2006) isolated and characterized the EPS produced by 31 thermophilic and mesophilic LAB strains and observed that the ability to synthesize CPS or to display ropiness by LAB was variable and strain-dependent. The structural characteristics of the HePS analyzed revealed a large diversity; a broad MM ranging from 8 to 5000 kDa was found, while the monosaccharides were composed of galactose, glucose, galactosamine, glucosamine, and rhamnose, albeit at different ratios. The isolated HePS were arbitrarily classified into nine groups, the number of different monomers varying from one to four and each group harboring strains of different LAB species.

The *in situ* synthesis of EPS by LAB during milk fermentation improves the viscosity and texture of the fermented product, especially in stirred yogurts (Ruas-Madiedo and de los Reyes Gavilán 2005) since these biopolymers act as thickeners and emulsifiers. Moreover, these EPS can be used as fat replacers in low-caloric (i.e., low-fat Mozzarella cheese) products (Hassan 2008). Although the effect of these compounds was firstly applied to dairy fermented foods, more recently EPS-producing

LAB and their polymers have been used in the manufacture of sourdough and cereal-based beverages (Zannini et al. 2013; Galle and Arendt 2014; Gänzle 2014). In this respect, Galle and Arendt (2014) have reported that EPS formed during sourdough fermentation by glycansucrase activity from sucrose influence the viscoelastic properties of the dough and beneficially affect the texture and shelf-life of bread. Consequently, EPS show potential to replace hydrocolloids commonly used as bread improvers and meet the consumer demands for a reduced use of food additives.

It has been claimed that EPS could help the producing bacteria to survive adverse environmental conditions; due to its water binding capacity EPS represents a protection factor against dessication and osmotic stresses (Kumar et al. 2007; Mozzi et al. 2009). Furthermore, these polymers could promote some benefit for human health (Ruas-Madiedo et al. 2008; Rodríguez et al. 2009; Nikolic et al. 2012; Ríos-Covian et al. 2013). In this respect, it has been suggested that EPS from LAB have cholesterol-lowering (Nakajima et al. 1992), antitumoral (Kitazawa et al. 1991), immunomodulatory (Patten et al. 2014; Shao et al. 2014), antiulcer (Nagaoka et al. 1994), and antigastritis (Rodríguez et al. 2009, 2010) activities, which might be related to their chemical structures (Ruas-Madiedo et al. 2002).

18.4. Commodity Chemicals

In the near future, food and energy will no longer be supplied by decreasing natural resources, thus alternative ways to produce high-value compounds by microbial bioconversion of cheap substrates or renewable raw materials have been sought (Boguta et al. 2014). Recently, the production of commodity chemicals such as lactic acid, 2,3-butanediol, 1,3-propanediol, bioethanol, and butanol by recombinant LAB has been reported (Pessione 2012; Gaspar et al. 2013; Boguta et al. 2014).

Traditionally **lactic acid** has been used in the food industry as a flavor additive and food preservative; however, its current use in the market is for the elaboration of biodegradable plastics (as a precursor of poly-lactic acid) and pharmaceuticals (Gaspar et al. 2013). Because of ability of several LAB species (*Lact. casei* mainly) to convert over 95% of sugar into lactic acid, these microorganisms have been employed industrially for this purpose. By applying metabolic engineering strategies and using renewable resources such as agro-industrial wastes as a substrate, other species such as *Lact. helveticus* and *Lact. plantarum* are able to efficiently form lactic acid either by achieving the deletion of the *L-ldh* (lactate dehydrogenase) gene or by introducing the *xyl*AB operon of *Lact. pentosus* in the *Lact. plantarum* genome. Production of pure *D*-lactic acid with $Y_{x/s}$= 0.8 g/g was achieved by these means (Okano et al. 2009).

Another important compound is **2,3-butanediol** (2,3-BDO), which is commonly used as antifreeze in the plastic, chemical, pharmaceutical, and cosmetic industries (Celinska and Grajek 2009). Furthermore, the buttery flavors acetoin and diacetyl can be formed by dehydrogenation of 2,3-BDO, which are important aroma compounds in some (dairy) food products. *Klebsiella* strains have been used for biotechnological production of 2,3-BDO, but due to the pathogenic status of this microorganism, the synthesis of 2,3-BDO by LAB has been sought as many species possess the biosynthetic pathway. Engineered strains of *Lact. plantarum* and *L. lactis* efficiently produced 2,3-BDO from inexpensive substrates such as whey permeate (Gaspar et al. 2011).

1,3-Propanediol (1,3-PDO) is a chemical compound with multiple applications in the production of polymers, solvents, resins, detergents, and cosmetics. Currently, this diol is industrially produced by chemical synthesis; however, its microbial production from glycerol has been intensively studied (Zeng and Sabra 2011). Again, the food grade status of LAB has gained attention for the synthesis of 1,3-PDO over *Klebsiella* and *Clostridium* strains, considered opportunistic pathogens and thus not applicable for food and biomedical products. Among the lactobacilli that naturally produce 1,3-PDO, the *Lact. reuteri* species has been reported to be the best so far (Vaidyanathan et al. 2011).

Biotechnological production of **ethanol** is maybe one of the most important sources of renewable energy as it is being used as an alternative biofuel to gasoline. Over the last few years, LAB have become interesting hosts for the production of ethanol from renewable biomass such as lignocellulose as these microorganisms can metabolize different sugars. So far, the most promising results were reported for a recombinant *ldh*-negative strain of *Lact. plantarum* expressing pyruvate decarboxylase from *Sarcina ventriculi* (Liu et al. 2006).

Also, **butanol** is a desirable solvent that can be an attractive fuel alternative. Interestingly, while butanol-producing strains cannot tolerate amounts higher than 2%, certain LAB members are considered potential hosts to apply genetic engineering strategies for butanol production as they are able to grow in the presence of 3% butanol (Li et al. 2010a,b). Interestingly, it was demonstrated that a *Lact. brevis* strain was able to produce butanol using some endogenous enzymes (Berezina et al. 2010).

18.5. Conclusions

Here, a portfolio of high-value metabolites produced by wild-type LAB displaying applications in the food, pharmaceutical, and chemical industries was presented. New approaches in the production of commodity chemicals, to replace natural resources, by recombinant LAB strains were briefly addressed. The versatility of this bacterial group demonstrates their potential to be used as cell factories beyond their classical use as food starter cultures, preservatives, or as probiotics. As shown, LAB are systematically screened for novel specific traits or applications thanks to the availability of novel molecular techniques and consumers' demands for healthier, tailor-made foods. It is clear that application of LAB as the workhorses of the food industry has expanded to their use as microbial factories to increase yields and product repertoire; which were earlier limited by the natural capacity of the existing bacterial biosynthetic pathways. Emerging fields include the design and creation of synthetic microbes to create novel metabolic pathways for new products. Will LAB be included in this next challenging approach?

References

Adeghate, E. and Ponery, A.S. (2002) GABA in the endocrine pancreas: cellular localization and function in normal and diabetic rats. *Tissue Cell* **34**, 1–6.

Akinterinwa, O., Khankal, R., and Cirino, P.C. (2008) Metabolic engineering for bioproduction of sugar alcohols. *Curr Opin Biotechnol* **19**, 461–467.

Berezina, O.V., Zakharova, N.V., Brandt, A., et al. (2010) Reconstructing the clostridial n-butanol metabolic pathway in *Lactobacillus brevis*. *Appl Microbiol Biotechnol* **87**, 635–646.

Boguta, A.M., Bringel, F., Martinussen, J., and Jensen, P.R. (2014) Screening of lactic acid bacteria for their potential as microbial cell factories for bioconversion of lignocellulosic feedstocks. *Microb Cell Fact* **13**, 97.

Bosch, M., Fuentes, M.C., Audivert, S., et al. (2014) *Lactobacillus plantarum* CECT 7527, 7528 and 7529: probiotic candidates to reduce cholesterol levels. *J Sci Food Agric* **94**, 803–809.

Bouzerzour, K., Morgan, F., Cuinet, I., Bonhomme et al. (2012) *In vivo* digestion of infant formula in piglets: protein digestion kinetics and release of bioactive peptides. *Br J Nutr* **108**, 1–10.

Bron, P.A. and Kleerebezem, M. (2011) Engineering lactic acid bacteria for increased industrial functionality. *Bioeng Bugs* **2**, 80–87.

Cai, H., Thompson, R., Budinich, M. F., et al. (2009) Genome sequence and comparative genome analysis of *Lactobacillus casei*: insights into their niche-associated evolution. *Genome Biol Evol* **1**, 239–257.

Carvalheiro, F., Moniz, P., Duarte, L.C., et al. (2011) Mannitol production by lactic acid bacteria grown in supplemented carob syrup. *J Ind Microbiol Biotechnol* **38**, 221–227.

Celińska, E. and Grajek, W. (2009) Biotechnological production of 2,3-butanediol-current state and prospects. *Biotechnol Adv* **27**, 715–725.

Chatuverdi, V., Bartiss, A., and Wong, B. (1997) Expression of bacterial *mtl*D in *Saccharomyces cerevisiae* results in mannitol synthesis and protects a glycerol-defective mutant from high-salt and oxidative stress. *J Bacteriol* **179**, 157–162.

Cherng, S.H., Huang, C.Y., Kuo, W.W., et al. (2014). GABA tea prevents cardiac fibrosis by attenuating TNF-alpha and Fas/FasL-mediated apoptosis in streptozotocin-induced diabetic rats. *Food Chem Toxicol* **65**, 90–96.

Cummings, J.H. and Stephen, A.M. (2007) Carbohydrate terminology and classification. *Eur J Clin Nutr* **61**, 5–18.

De Angelis, M., Bottacini, F., Fosso, B., et al. (2014) *Lactobacillus rossiae*, a vitamin B_{12} producer, represents a metabolically versatile species within the Genus *Lactobacillus*. *PLoS One* **9**, (9):e107232.

De Biase, D. and Pennacchietti, E. (2012) Glutamate decarboxylase-dependent acid resistance in orally acquired bacteria: function, distribution and biomedical implications of the gadBC operon. *Mol Microbiol* **86**, 770–786.

Dhakal, R., Bajpai, V.K., and Baek, K.H. (2012) Production of gaba (gamma - aminobutyric acid) by microorganisms: a review. *Braz J Microbiol* **43**, 1230–1241.

Di Cagno, R., Mazzacane, F., Rizzello, C.G., et al. (2010) Synthesis of gamma-aminobutyric acid (GABA) by *Lactobacillus plantarum* DSM19463: functional grape must beverage and dermatological applications. *Appl Microbiol Biotechnol* **86**, 731–741.

DiBaise, J.K., Zhang, H., Crowell, M.D., et al. (2008) Gut microbiota and its possible relationship with obesity. *Mayo Clin Proc* **83**, 460–469.

Embuscado, M.E. and Patil S.K. (2001) Erythritol. In: *Food Science and Technology*, ((ed.) Dekker M.) Vol. 17, Alternative Sweeteners, 3rd edn, New York, USA, pp. 235–254.

Escudero, E., Aristoy, M.C., and Nishimura, H., et al. (2012) Antihypertensive effect and antioxidant activity of peptide fractions extracted from Spanish dry-cured ham. *Meat Sci* **91**, 306–311.

Espeche Turbay, M.B., de Moreno de LeBlanc, A., and Perdigon, G., et al. (2012) β-Casein hydrolysate generated by the cell envelope-associated proteinase of *Lactobacillus delbrueckii* ssp. *lactis* CRL 581 protects against trinitrobenzene sulfonic acid-induced colitis in mice. *J Dairy Sci* **95**, 1108–1118.

Ewaschuk, J.B., Walker, J.W., Diaz, H., and Madsen, K.L. (2006) Bioproduction of conjugated linoleic acid by probiotic bacteria occurs in vitro and in vivo in mice. *J Nutr* **136**, 1483–1487.

Fitzgerald, G.F. and Murray, J.A. (2006) Bioactive peptides and lactic fermentations. *Int J Dairy Technol* **59**, 118–125.

Fontes, C.P.M.L., Honorato, T.L., Rabelo, M.C., and Rodrigues, S. (2009) Kinetic study of mannitol production using cashew apple juice as substrate. *Bioprocess Biosyst Eng* **32**, 493–499.

Galle, S. and Arendt, E.K. (2014) Exopolysaccharides from sourdough lactic acid bacteria. *Crit Rev Food Sci Nutr* **54**, 891–901.

Gänzle, M.G. (2014) Enzymatic and bacterial conversions during sourdough fermentation. *Food Microbiol* **37**, 2–10.

Gaspar, P., Carvalho, A.L., Vinga, S., et al. (2013) From physiology to systems metabolic engineering for the production of biochemicals by lactic acid bacteria. *Biotechnol Adv* **31**, 764–788.

Gaspar, P., Neves, A.R., Gasson, M.J., et al. (2011) High yields of 2,3-butanediol and mannitol in *Lactococcus lactis* through engineering of NAD^+ cofactor recycling. *Appl Environ Microbiol* **77**, 6826–6835.

Gorissen, L., Leroy, F., De Vuyst, L., et al. (2013) Bacterial production of conjugated linoleic and linolenic acid in foods: a technological challenge. *Crit Rev Food Sci Nutr* 2013 Oct 17 [Epub ahead of print] PMID: 24915316.

Gorissen, L., Weckx, S., Vlaeminck, B., et al. (2011) Linoleate isomerase activity occurs in lactic acid bacteria strains and is affected by pH and temperature. *J Appl Microbiol* **111**, 593–606.

Gu, C., Majumder, K., and Wu, J. (2011). QSAR-aided in silico approach in evaluation of food proteins as precursors of ACE inhibitory peptides. *Food Res Int* **44**, 2465–2474.

Han, D., Kim, H.Y., Lee, H.J., et al. (2007) Wound healing activity of gamma-aminobutyric Acid (GABA) in rats. *J Microbiol Biotechnol* **17**, 1661–1669.

Hassan, A.N. (2008) ADSA Foundation Scholar Award: possibilities and challenges of exopolysaccharide-producing lactic cultures in dairy foods. *J Dairy Sci* **91**, 1282–1298.

Hayes, M., Ross, R.P., Fitzgerald, G.F., and Stanton, C. (2007a) Putting microbes to work: dairy fermentation, cell factories and bioactive peptides. Part I: overview. *Biotechnol J* **2**, 426–434.

Hayes, M., Stanton, C., Fitzgerald, G.F., and Ross, R.P. (2007b) Putting microbes to work: dairy fermentation, cell factories and bioactive peptides. Part II: bioactive peptide functions. *Biotechnol J* **2** (4), 435–449.

Hebert, E. M., Mamone, G., Picariello, et al. (2008) Characterization of the pattern of as1- and b-casein breakdown and release of a bioactive peptide by a cell envelope proteinase from *Lactobacillus delbrueckii* subsp. *lactis* CRL 581. *Appl Environ Microbiol* **74**, 3682–3689.

Hebert, E.M., Saavedra, L., and Ferranti, P. (2010) Bioactive peptides derived from casein and whey proteins. In: *Biotechnology of Lactic Acid Bacteria: Novel Applications* ((eds) F. Mozzi, R. Raya, and G. Vignolo), Wiley-Blackwell, Ames, Iowa, USA, pp. 233–249.

Hounsome, N., Hounsome, B., Tomos, D., and Edwards-Jones, G. (2008) Plant metabolites and nutritional quality of vegetables. *J Food Sci* **73**, R48–R65.

Hugenholtz, J. (2008) The lactic acid bacterium as a cell factory for food ingredient production. *Int Dairy J* **18**, 466–475.
Hugenholtz, J. and Smid, E.J. (2002) Nutraceutical production with food- grade microorganisms. *Curr Opin Biotechnol* **13**, 497–507.
Inoue, K., Shirai, T., Ochiai, H., et al. (2003) Blood-pressure-lowering effect of a novel fermented milk containing gamma-aminobutyric acid (GABA) in mild hypertensives. *Eur J Clin Nutr* **57**, 490–495.
Ito, K., Tanaka, K., Nishibe, Y., et al. (2007) GABA-synthesizing enzyme, GAD67, from dermal fibroblasts: evidence for a new skin function. *Biochim Bioph Acta* **1770**, 291–296.
Iwamoto, K. and Shiraiwa, Y. (2005) Salt-regulated mannitol metabolism in algae. *Mar Biotechnol* **7**, 407–415.
Jakobs, C., Jaeken, J., and Gibson, K. M. (1993). Inherited disorders of GABA metabolism. *J Inherit Metab Dis* **16**, 704–715.
Jeong, S.H., Lee, S.H., Jung, J.Y., et al. (2013). Microbial succession and metabolite changes during long-term storage of Kimchi. *J Food Sci* **78**, 763–769.
Kalra, E.K. (2003) Nutraceutical-definition and introduction. *AAPS Pharm Sci* **5**, 1–2.
Kim, J.Y., Lee, M.Y., Ji, G.E., et al. (2009) Production of gamma-aminobutyric acid in black raspberry juice during fermentation by *Lactobacillus brevis* GABA100. *Int J Food Microbiol* **130**, 12–16.
Kimura, M., Hayakawa, K., and Sansawa, H. (2002) Involvement of gamma-aminobutyric acid (GABA) B receptors in the hypotensive effect of systemically administered GABA in spontaneously hypertensive rats. *Jpn J Pharmacol* **89**, 388–394.
Kishino, S., Park, S.B., Takeuchi, M., et al. (2011) Novel multi-component enzyme machinery in lactic acid bacteria catalyzing C=C double bond migration useful for conjugated fatty acid synthesis. *Biochem Biophys Res Commun* **416**, 188–193.
Kitazawa, H., Toba, T., Itoh, T., et al. (1991) Antitumoral activity of slime-forming encapsulated *Lactococcus lactis* subsp. *cremoris* isolated from Scandinavian ropy sour milk, "viili." *Anim Sci Technol* **62**, 277–283.
Kook, M.C., Seo, M.J., Cheigh, C.I., et al. (2010) Enhanced production of gamma-aminobutyric acid using rice bran extracts by *Lactobacillus sakei* B2-16. *J Microbiol Biotechnol* **20**, 763–766.
Kumar, A.S., Mody, K., and Jha, B. (2007) Bacterial exopolysaccharides-a perception. *J Basic Microbiol* **47**, 103–117.
Kusserow, C., Michels, S., and Schmidt-Erfurth, U. (2003). Chorioretinal anastomosis as unfavourable prognostic factor during photodynamic therapy. *Ophthalmologie* **100**, 197–202.
Kussmann, M. and Van Bladeren, P.J. (2011) The extended nutrigenomics - understanding the interplay between the genomes of food, gut microbes, and human host. *Front Genet* **2**, 21.
Laiño. J.E., Juarez Del Valle, M., Savoy de Giori. G., and LeBlanc J.G. (2014) Applicability of a *Lactobacillus amylovorus* strain as co-culture for natural folate bio-enrichment of fermented milk. *Int J Food Microbiol* **191**, 10–16.
LeBlanc, J.G., Matar, C., Valdez, J.C., et al. (2002) Immunomodulating effects of peptidic fractions issued from milk fermented with *Lactobacillus helveticus*. *J Dairy Sci* **85**, 2733–2742.
LeBlanc, J.G., Milani, C., de Giori, G.S., et al. (2013) Bacteria as vitamin suppliers to their host: a gut microbiota perspective. *Curr Opin Biotechnol* **24**, 160–168.
Leroy, F. and De Vuyst, L. (2004). Lactic acid bacteria as functional starter cultures for the food fermentation industry. *Trends Food Sci Technol* **15**, 67–78.
Li, H. and Cao, Y. (2010) Lactic acid bacterial cell factories for gamma-aminobutyric acid. *Amino Acids* **39**, 1107–1116.
Li, Y. and Li, B. (2013) Characterization of structure–antioxidant activity relationship of peptides in free radical systems using QSAR models: Key sequence positions and their amino acid properties. *J Theor Biol* **318**, 29–43.
Li, H., Qiu, T., Gao, D., and Cao, Y. (2010a) Medium optimization for production of gamma-aminobutyric acid by *Lactobacillus brevis* NCL912. *Amino Acids* **38**, 1439–1445.
Li, J., Zhao, J.B., Zhao, M., et al. (2010b) Screening and characterization of butanol-tolerant microorganisms. *Lett Appl Microbiol* **50**, 373–379.
Liong, M.T. and Shah, N.P. (2005) Production of organic acids from fermentation of mannitol, fructooligosaccharide and inulin by a cholesterol removing *Lactobacillus acidophilus* strain. *J Appl Microbiol* **99**, 783–793.
Liu, M., Bayjanov, J.R., Renckens, B., et al. (2010) The proteolytic system of lactic acid bacteria revisited: a genomic comparison. *BMC Genomics* **11**, 36.
Liu, S., Nichols, N.N., Dien, B.S., and Cotta, M.A. (2006) Metabolic engineering of a *Lactobacillus plantarum* double *ldh* knockout strain for enhanced ethanol production. *J Ind Microbiol Biotechnol* **33**, 1–7.
Livesey, G. (2003) Health potential of polyols as sugar replacers, with emphasis on low glycaemic properties. *Nutr Res Rev* **16**, 163–191.

Martín, R., Chain, F., Miquel, S., et al. (2014) Effects in the use of a genetically engineered strain of *Lactococcus lactis* delivering *in situ* IL-10 as a therapy to treat low-grade colon inflammation. *Hum Vaccin Immunother* **10**, 1611–1621.

Matar, C., Valdez, J.C., Medina, M., et al. (2001) Immunomodulating effects of milks fermented by *Lactobacillus helveticus* and its non-proteolytic variant. *J Dairy Res* **68**, 601–609.

Mazzoli, R., Bosco, F., Mizrahi, I., et al. (2014) Towards lactic acid bacteria-based biorefineries. *Biotechnol Adv* **32**, 1216–1236.

Minervini, F., Algaron, F., Rizzello, C.G., et al. (2003) Angiotensin I-converting-enzyme-inhibitory and antibacterial peptides from *Lactobacillus helveticus* PR4 proteinase-hydrolyzed caseins of milk from six species. *Appl Environ Microbiol* **69**, 5297–5305.

Monedero, V., Pérez-Martínez, G., and Yebra, M.J. (2010) Perspectives of engineering lactic acid bacteria for biotechnological polyol production. *Appl Microbiol Biotechnol* **86**, 1003–1015.

Mozzi, F., Gerbino, E., Font de Valdez, G., and Torino, M.I. (2009) Functionality of exopolysaccharides produced by lactic acid bacteria in an *in vitro* gastric system. *J Appl Microbiol* **107**, 56–64.

Mozzi, F., Vaningelgem, F., Hébert, E.M., et al. (2006) Diversity of heteropolysaccharide-producing lactic acid bacterium strains and their biopolymers. *Appl Environ Microbiol* **72**, 4431–4435.

Nagaoka, M., Hashimoto, S., Watanabe, T., et al. (1994) Anti-ulcer effects of lactic acid bacteria and their cell wall polysaccharides. *Biol Pharm Bull* **17**, 1012–1017.

Nakajima, H., Suzuki, H., Kaizu, H., and Hirota, T. (1992) Cholesterol lowering activity of ropy fermented milk. *J Food Sci* **57**, 1327–1329.

Nikolic, M., López, P., Strahinic, I., et al. (2012) Characterisation of the exopolysaccharide (EPS)-producing *Lactobacillus paraplantarum* BGCG11 and its non-EPS producing derivative strains as potential probiotics. *Int J Food Microbiol* **158**, 155–162.

Nyyssölä, A., Pihlajaniemi, A., Palva, A., et al. (2005) Production of xylitol from D-xylose by recombinant *Lactococcus lactis*. *J Biotechnol* **118**, 55–66.

Okano, K., Yoshida, S., Yamada, R., Tanaka, T., Ogino, C., Fukuda, H. and Kondo, A. (2009) Improved production of homo-D-lactic acid via xylose fermentation by introduction of xylose assimilation genes and redirection of the phosphoketolase pathway to the pentose phosphate pathway in L-Lactate dehydrogenase gene-deficient Lactobacillus plantarum. *Appl Environ Microbiol* **75**, 7858–7861

Ondetti, M.A. and Cushman, D.W. (1982) Enzymes of the renin-angiotensin system and their inhibitors. *Annu Rev Biochem* **51**, 283–308.

Ortiz, M.E., Bleckwedel, J., Raya, R.R., and Mozzi, F. (2013) Biotechnological and *in situ* food production of polyols by lactic acid bacteria. Mini-Review. *Appl Microbiol Biotechnol* **97**, 4713–4726.

Ortiz, M.E., Fornaguera, M.J., Raya, R.R., and Mozzi, F. (2012) *Lactobacillus reuteri* CRL 1101 highly produces mannitol from sugarcane molasses as carbon source. *Appl Microbiol Biotechnol* **95**, 991–999.

Patten, D.A., Leivers, S., Chadha, M.J., et al. (2014) The structure and immunomodulatory activity on intestinal epithelial cells of the EPSs isolated from *Lactobacillus helveticus* sp. *Rosyjski* and *Lactobacillus acidophilus* sp. 5e2. *Carbohyd Res* **384**, 119–127.

Pessione, E. (2012) Lactic acid bacteria contribution to gut microbiota complexity: lights and shadows. *Front Cell Infect Microbiol* **2**, 86.

Phelan, M. and Kerins, D. (2011) The potential role of milk-derived peptides in cardiovascular disease. *Food Funct* **2**, 153–167.

Platteeuw, C., Hugenholtz, J., Starrenburg, M., et al. (1995) Metabolic engineering of *Lactococcus lactis*: influence of the overproduction of alpha-acetolactate synthase in strains deficient in lactate dehydrogenase as a function of culture conditions. *Appl Environ Microbiol* **61**, 3967–3971.

Ratanaburee, A., Kantachote, D., Charernjiratrakul, W., and Sukhoom, A. (2013) Enhancement of gamma-aminobutyric acid (GABA) in Nham (Thai fermented pork sausage) using starter cultures of *Lactobacillus namurensis* NH2 and *Pediococcus pentosaceus* HN8. *Int J Food Microbiol* **167**, 170–176.

Richter, H., Vlad, D., and Unden, G. (2001) Significance of pantothenate for glucose fermentation by *Oenococcus oeni* and for suppression of the erythritol and acetate production. *Arch Microbiol* **175**, 26–31.

Rios-Covian, D., Arboleya, S., Hernandez-Barranco, A.M., et al. (2013) Interactions between *Bifidobacterium* and *Bacteroides* species in cofermentations are affected by carbon sources, including exopolysaccharides produced by bifidobacteria. *Appl Environ Microbiol* **79**, 7518–7524.

Rizzello, C.G., Cassone, A., Di Cagno, R., and Gobbetti, M. (2008) Synthesis of angiotensin I-converting enzyme (ACE)-inhibitory peptides and gamma-aminobutyric acid (GABA) during sourdough fermentation by selected lactic acid bacteria. *J Agric Food Chem* **56**, 6936–6943.

Rodríguez, C., Médici M., Rodríguez, A.V., et al. (2009) Prevention of chronic gastritis by fermented milks made with exopolysaccharide-producing *Streptococcus thermophilus* strains. *J Dairy Sci* **92**, 2423–2434.

Rodríguez, C., Medici, M., Mozzi, F., and Font de Valdez, G. (2010) Therapeutic effect of *Streptococcus thermophilus* CRL 1190-fermented milk on chronic gastritis. *World J Gastroenterol* **16**, 1622–1630.

Ruas-Madiedo, P. and de los Reyes-Gavilán, C.G. (2005) Invited review: methods for the screening, isolation, and characterization of exopolysaccharides produced by lactic acid bacteria. *J Dairy Sci* **88**, 843–56.

Ruas-Madiedo, P., Abraham, A., Mozzi, F. and de los Reyes-Gavilán C.G. (2008) Functionality of exopolysaccharides produced by lactic acid bacteria. In: *Molecular Aspects of Lactic Acid Bacteria for Traditional and New Applications* ((eds) B. Mayo, P. López, and G. Pérez-Martínez), Research Signpost, Kerala India, pp. 137–166.

Ruas-Madiedo, P., Hugenholtz, J., and Zoon, P. (2002) An overview of the functionality of exopolysaccharides produced by lactic acid bacteria. *Int Dairy J* **12**, 163–171.

Rutherfurd-Markwick, K.J. (2012). Food proteins as a source of bioactive peptides with diverse functions. *Br J Nutr* **108**, S149–157.

Saavedra, L., Hebert, E. M., Minahk, C., and Ferranti, P. (2013) An overview of "omic" analytical methods applied in bioactive peptide studies. *Food Res Int* **54**, 925–934.

Saha, B.C. and Nakamura, L.K. (2003) Production of mannitol and lactic acid by fermentation with *Lactobacillus intermedius* NRRL B-3693. *Biotechnol Bioeng* **82**, 864–871.

Saha, B.C. and Racine, F.M. (2010) Effects of pH and corn steep liquor variability on mannitol production by *Lactobacillus intermedius* NRRL B-3693. *Appl Microbiol Biotechnol* **87**, 553–560.

Sagardia, I., Roa-Ureta, R.H., and Bald, C. (2013). A new QSAR model, for angiotensin I-converting enzyme inhibitory oligopeptides. *Food Chem* **136**, 1370–1376.

Salminen, S. and Hallikainen, A. (2002) Sweeteners. In: *Food Additives* ((ed.) A.L. Branen, P.M. Davidson, S. Salminen, and J.H. Thorngate), Marcel Dekker, New York, pp. 451–456.

Sandré, C., Gleizes, A., Forestier, F., et al. (2001) A peptide derived from bovine beta-casein modulates functional properties of bone marrow-derived macrophages from germfree and human flora-associated mice. *J Nutr* **131**, 2936–2942.

Savijoki, K., Ingmer, H., and Varmanen, P. (2006) Proteolytic systems of lactic acid bacteria. *Appl Microbiol Biotechnol* **71**, 394–406.

Scheirlinck, I., Van der Meulen, R., Van Schoor, A., et al. (2007) *Lactobacillus namurensis* sp. nov., isolated from a traditional Belgian sourdough. *Int J Syst Evol Microbiol* **57**, 223–227.

Schiweck, H., Bär, A., Vogel, R., et al. (1994) Sugar alcohols. In: *Ullmann's Encyclopedia of Industrial Chemistry* ((eds) B. Elvers, S. Hawkins, and W. Russey), Wiley-VCH, New Jersey, USA, pp. 413–437.

Senevirathne, M. and Kim, S.K. (2012) Development of bioactive peptides from fish proteins and their health promoting ability. *Adv Food Nutr Res* **65**, 235–248.

Shao, L., Wu, Z., Zhang, H., et al. (2014) Partial characterization and immunostimulatory activity of exopolysaccharides from *Lactobacillus rhamnosus* KF5. *Carbohyd Polym* **107**, 51–56.

Shu, M., Yu, R., Zhang, Y., et al. (2013). Predicting the activity of antimicrobial peptides with amino acid topological information. *Med Chem* **9**, 32–44.

Siragusa, S., De Angelis, M., Di Cagno, R., et al. (2007) Synthesis of gamma-aminobutyric acid by lactic acid bacteria isolated from a variety of Italian cheeses. *Appl Environ Microbiol* **73**, 7283–7290.

Solomon, P.S., Waters, O.D., and Oliver, R.P. (2007) Decoding the mannitol enigma in filamentous fungi. *Trends Microbiol* **15**, 257–262.

Stolz, P., Bücker, G., Hammes, W.P., and Vogel, R.F. (1995) Utilization of electron acceptors by lactobacilli isolated from sourdough. *Z Lebensm Unters Forsch* **201**, 91–96.

Takano, T. (2002) Anti-hypertensive activity of fermented dairy products containing biogenic peptides. *Antonie Van Leeuwenhoek* **82**, 333–340.

Taranto, M.P., Vera, J.L., Hugenholtz, J., et al. (2003) *Lactobacillus reuteri* CRL1098 produces cobalamin. *J Bacteriol* **185**, 5643–5647.

Tellez, A., Corredig, M., Brovko, L.Y., and Griffiths, M.W. (2010) Characterization of immune-active peptides obtained from milk fermented by *Lactobacillus helveticus*. *J Dairy Res* **77**, 129–136.

Toropova, A.P., Toropov, A.A., Rasulev, B.F., et al. (2012). QSAR models for ACE-inhibitor activity of tri-peptides based on representation of the molecular structure by graph of atomic orbitals and SMILES. *Struct Chem* **23**, 1873–1878.

Tujioka, K., Ohsumi, M., Sakamoto, K., et al. (2011) Effect of dietary gamma-aminobutyric acid on the brain protein synthesis rate in hypophysectomized aged rats. *J Nutr Sci Vitaminol* **57**, 285–291.

Turnbaugh, P.J., Ley, R.E., Mahowald, M.A., et al. (2006) An obesity-associated gut microbiome with increased capacity for energy harvest. *Nature* **444**, 1027–1031.

Vaidyanathan, H., Kandasamy, V., Gopal Ramakrishnan, K., et al. (2011) Glycerol conversion to 1,3-propanediol is enhanced by the expression of a heterlogous alcohol dehydrogenase gene in *Lactobacillus reuteri*. *AMB express* **1**, 37.

van Munster, I.P. and Nagengast, F.M. (1993) The role of carbohydrate fermentation in colon cancer prevention. *Scand J Gastroenterol* Suppl **200**, 80–86.

Vaningelgem, F., Zamfir, M., Mozzi, F., et al. (2004) Biodiversity of exopolysaccharides produced by *Streptococcus thermophilus* strains is reflected in their production and their molecular and functional characteristics. *Appl Environ Microbiol* **70**, 900–912.

Veiga-da-Cunha, M., Santos, H., and van Schaftingen, E (1993) Pathway and regulation of erythritol formation in *Leuconostoc oenos*. *J Bacteriol* **175**, 3941–3948.

Villegas, J.M., Picariello, G., Mamone, G., et al. (2014) Milk-derived angiotensin-I-converting enzyme inhibitory peptides generated by *Lactobacillus delbrueckii* subsp. *lactis* CRL 581. *Peptidomics* **1**, 22–29.

von Weymarn, N. (2002) Process development for mannitol production by lactic acid bacteria. PhD Thesis. Helsinki University of Technology, Department of Chemical Technology. Helsinki, Finland.

von Weymarn, N., Hujanen, M., and Leisola, M. (2002) Production of D- mannitol by heterofermentative lactic acid bacteria. *Process Biochem* **37**, 1207–1213.

Vrancken, G., Rimaux, T., De Vuyst, L., and Mozzi, F. (2010) Low-calorie sugars produced by lactic acid bacteria. In: *Biotechnology of Lactic Acid Bacteria: Novel Applications* ((eds) F. Mozzi, R.R. Raya, and G.M. Vignolo), Wiley-Blackwell Publishing, Ames, IA, USA, pp. 193–209.

Wisselink, H.W., Weusthius, R.A., Eggink, G., et al. (2002) Mannitol production by lactic acid bacteria: a review. *Int Dairy J* **12**, 151–161.

Wong, C.G., Bottiglieri, T., and Snead, O.C. 3rd. (2003) GABA, gamma-hydroxybutyric acid, and neurological disease. *Ann Neurol* **54** Suppl 6, S3–12.

Yang, B., Chen, H., Gu, Z., et al. (2014) Synthesis of conjugated linoleic acid by the linoleate isomerase complex in food-derived lactobacilli. *J Appl Microbiol* **117**, 430–439.

Zannini, E., Mauch, A., Galle, S., et al. (2013) Barley malt wort fermentation by exopolysaccharide-forming *Weissella cibaria* MG1 for the production of a novel beverage. *J Appl Microbiol* **115**, 1379–1387.

Zeng, A.P. and Sabra, W. (2011) Microbial production of diols as platform chemicals: recent progresses. *Curr Opin Biotechnol* **22**, 749–757.

Chapter 19

Production of Flavor Compounds by Lactic Acid Bacteria in Fermented Foods

Anne Thierry[1,*], Tomislav Pogačić[2], Magalie Weber[1], and Sylvie Lortal[1]

[1] *INRA, AGROCAMPUS OUEST UMR1253 Science et Technologie du Lait et de l'OEuf, France*
[2] *Department of Dairy Science, Faculty of Agriculture University of Zagreb, Croatia*

19.1. Introduction

Fermented foods constitute a large part of human diet (Buckenhuskes 1993; Campbell-Platt 1994; Hugenholtz 2013) and have been consumed for millennia. The earliest evidence of cheese making, for example, dates from the sixth millennium BC in Northern Europe, as demonstrated by the presence of dairy fat in specialized pottery sieves used to separate curd from whey (Salque et al. 2013). At origin, fermentation occurred spontaneously from naturally contaminating microorganisms. Humans empirically controlled the processes of raw material fermentation for two main reasons: (i) to preserve the shelf-life of perishable products such as milk, meat, fish, cereals, and vegetables; and (ii) to increase and diversify the sensory properties of foodstuffs. Depending on locally available raw materials, every community has its specific fermented foods that have evolved over time and can be considered heritage foods (Buckenhuskes 1993; Tamang and Delwen 2010; van Hijum et al. 2013). Flavor is still today recognized as one of the most important food properties for the consumer. Flavor development is a complex process involving the biochemical conversion of varied food components, which should be fully understood to be able to steer flavor formation in foods.

Lactic acid bacteria (LAB) are present in almost all fermented foods, alone or more often in combination with other groups of bacteria, yeasts, and fungi. The main function of LAB in fermented foods is to rapidly acidify the food matrix, by fermenting carbohydrates to lactic acid as the main end-product, thus increasing the shelf-life and the microbial safety of products. Besides, most LAB possess a range of flavor-related activities.

This chapter will focus on the role of LAB in the production of flavor compounds in fermented foods. The main flavor-generating pathways in which LAB are involved are presented, as well as the biotic and abiotic factors impacting the effective and relative contribution of LAB to the formation of flavor. The new opportunities offered by the development of (meta)-omics approaches will also be

*Corresponding author: INRA, AGROCAMPUS OUEST UMR1253 Science et Technologie du Lait et de l'OEuf, 65 rue de Saint Brieuc, F-35042 Rennes, France. anne.thierry@rennes.inra.fr

Biotechnology of Lactic Acid Bacteria: Novel Applications, Second Edition. Edited by Fernanda Mozzi, Raúl R. Raya, and Graciela M. Vignolo.

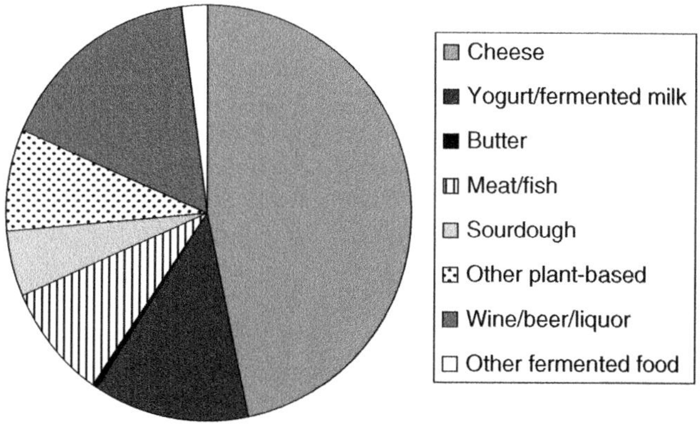

Figure 19.1. Numbers of papers published in relation to the formation of flavor compounds by lactic acid bacteria in different fermented foods. Interrogation of the ISI Web of Science database in June 2014, over the period 1994-2014.

summarized. This chapter will not review the use of LAB in modern biotechnology (development of genetically modified LAB and use in foods or *ex situ* to produce aroma compounds).

A bibliometric research of papers published over the last two decades in relation to the formation of flavor (compounds) by LAB in fermented foods shows that two-thirds of the studies concern dairy products (Figure 19.1). Therefore, many of the examples chosen as illustrations in this chapter refer to dairy products. Moreover, the involvement of LAB in winemaking, reviewed in Chapter 15, will not be considered.

19.2. Flavor and Aroma Compounds

Flavor is defined as the blend of taste and smell sensations evoked by a substance in the mouth and is evaluated using sensory analysis. It results from the perception of (i) a few water-soluble, taste compounds related to the five basic tastes (sweet, salty, bitter, sour, and umami), and (ii) a diversity of aroma compounds, which are odorous volatile compounds, responsible for the variety of flavors encountered in fermented foods.

19.2.1. Volatile Compounds: Diversity, Analytical Methods

Fermented foods contain high numbers of volatile compounds, exceeding 550 in cocoa (Mozzi et al. 2013), 200 in fermented sausages (Stahnke et al. 2002), 300 in bread (Pozo-Bayon et al. 2006), 200 in Cheddar cheese, and 300 in Comté cheeses (Maarse et al. 1994). Hundreds of volatile organic compounds can be produced by microorganisms (Lemfack et al. 2014). Their molecular weight in general lies between 100 and 300 g/mol. Volatile compounds are analyzed using gas chromatography coupled to mass spectrometry (GC-MS) techniques, after being extracted from the food matrix. The method of extraction impacts very significantly on the number and the nature of the volatiles extracted (Bosset et al. 1995). Head space techniques have become the most popular and widely used methods to extract food volatiles because they are (relatively) simple to use and are thought to be more connected to the results of sensory analysis, by extracting mainly the most volatile compounds (Bosset et al. 1995; Marsili 2011). Recent progress in GC-MS analysis include resolution improvement by two-dimensional gas chromatography and the speeding up of data analysis by using metabolomics-driven approaches (De Bok et al. 2011; Le Boucher et al. 2013; Pogačic et al. 2015).

19.2.2. Contribution of Volatile Aroma Compounds to Flavor

Only some volatiles are odorous and thus susceptible to contribute to flavor. Identification of the most important aroma compounds requires a combination of analytical and sensory methods. GC coupled to olfactometry (GC-O) is widely used to identify flavor-active compounds in food products, thus leading to the development of database of "key" aroma compounds (Molimard and Spinnler 1996; Curioni and Bosset 2002). Aroma compounds belong to many chemical classes such as acids, esters, aldehydes, ketones, sulfur-containing compounds, among others (Forss 1972; Engels 1997; Fox and Wallace 1997). Table 19.1 lists some of them produced by LAB in fermented foods and the odor notes associated. The aroma threshold values vary by several orders of magnitude (for example 1 and 10^5 ng/g in water for ethyl butanoate and ethanol, respectively). Threshold values in foods depend upon the threshold of the aroma in air, its solubility in oil and water, and food composition. The composition of food products influences the release of aroma compounds in the mouth and thus their perception. Important compositional parameters are the content of fat, which acts as a trap for most aroma compounds, and the pH, which influences the volatility of organic acids. As a result, the same mixture of aroma compounds, including those produced *in situ* by LAB, will be perceived differently according to the food matrix in which they are present.

The same aroma compound can contribute either positively or negatively to the flavor, depending on the food product. For example, short-chain free fatty acids are important aroma compounds in all cheeses (Table 19.1), where they contribute to pungent flavor notes. However, present in excessive levels, they are responsible for rancid flavor defects (Collins et al. 2003). Similarly, 3-methylbutanal, a flavor-active aldehyde coming from the catabolism of leucine, is considered desirable in some cheese such as Proosdij (Smit et al. 2005), but has been associated with flavor defects in Cheddar cheese (Morgan 1976). Furthermore, the most commonly occurring volatile organic compounds linked to the microbial spoilage of raw meat (Casaburi et al. 2015) are very similar to those positively involved in cheese flavor (Curioni and Bosset 2002).

19.2.3. Origin of Aroma Compounds

A great majority of the aroma compounds produced in fermented foods result from the activity of microorganisms. They derive from the conversion of food constituents via a series of biochemical reactions mainly catalyzed by microbial enzymes, although some of the enzymes either present in the raw material or added during food processing also have a role in the transformation. A few non-enzymatic reactions also occur, but they involve some compounds of microbial origin.

19.3. LAB of Fermented Foods and their Role in Flavor Formation

LAB are the most important bacteria used in food fermentations and many LAB species are involved. Their presence results from either a natural, spontaneous contamination derived from the raw materials themselves or the environment, the inoculation by foods from a previous batch (a practice referred to as back slopping), a deliberate addition of cultures (starters), or a combination of these different sources (Table 19.2). Most fermented foods consumed in developing countries are manufactured from spontaneous fermentation, without addition of any starters (Tamang and Delwen 2010). However, food fermentations have evolved towards a systematic use of starters to control fermentation, and nowadays starter cultures are widely used in the manufacture of many fermented products, such as dairy products, sourdough, and fermented dry sausages (Leroy et al. 2006).

Table 19.1. Origin and flavor notes of some important aroma compounds in fermented dairy products.

Volatile compound (trivial name widely used)	Precursor*	Odor description†	Dairy products in which they are flavor compounds‡	Produced by LAB as pure culture	
				LAB species	Culture medium (references)
Ethanoic acid (acetic acid)	C heterof	vinegar, sour, sharp, fresh cottage cheese	Yogurt	*Lact. plantarum*, *Lact. paracasei*, *Lact. casei*	fermented milk (Mugampoza 2013; Zareba et al. 2014), cheese-based medium (Budinich et al. 2011; Peralta et al. 2014)
Butanoic acid (butyric acid)	F	cheesy, rotten	butter, Cheddar, cream	*L. lactis* subsp. *lactis*, *Lact. plantarum*, *Lact. paracasei*, *Lact. casei*	fermented milk (Mugampoza 2013; Alemayehu et al. 2014; Zareba et al. 2014), cheese-based medium (Peralta et al. 2014)
Hexanoic acid (caproic acid)	F	popcorn, goaty	goat cheese	*Lact. paracasei*, *Lact. casei*, *L. lactis* subsp. *lactis*, *Lact. plantarum*	fermented milk (Mugampoza 2013; Zareba et al. 2014), cheese-based medium (Budinich et al. 2011; Peralta et al. 2014), cheese (Milesi et al. 2010)
Heptanoic acid	F	fatty, rancid		*L. lactis* subsp. *lactis*	fermented milk (Alemayehu et al. 2014)
Octanoic acid (caprylic acid)	F	sweaty, rancid	goat cheese	*Lact. casei*, *Lact. plantarum*, *L. lactis subsp. lactis*	fermented milk (Mugampoza 2013; Zareba et al. 2014), cheese (Milesi et al. 2010)
Decanoic acid	F	soapy		*Lact. casei*, *Lact. Plantarum*	cheese (Milesi et al. 2010)
2-methylbutanoic acid	AA	sweaty	Gruyère, Camembert	*L. lactis* subsp. *lactis*	fermented milk (Alemayehu et al. 2014)
3-methylbutanoic acid	AA	rancid, sweaty	Camembert	*Lact. plantarum*, *Lact. paracasei*, *Lact. casei*	fermented milk (Mugampoza 2013), cheese-based medium (Peralta et al. 2014), cheese (Milesi et al. 2010)
Ethanol	C heterof	Dry		*Lact. paracasei*, *L. lactis* subsp. *lactis*, *Lact. Casei*	Cheese-based medium (Budinich et al. 2011), fermented milk (Zareba et al. 2014)
1-propanol		sweet		*Lact. paracasei*, *L. lactis* subsp. *lactis* biovar *diacetylactis*, *Lact. plantarum*	Cheese-based medium (Budinich et al. 2011), fermented milk (Mugampoza 2013; Sharaf et al. 2013; Alemayehu et al. 2014)
1-butanol	C heterof	medicinal		*Lact. paracasei*, *L. lactis* subsp. *lactis*, *Lact. casei*, *Lact. plantarum*	Cheese-based medium (Budinich et al. 2011), fermented milk (Alemayehu et al. 2014), cheese (Milesi et al. 2010)
1-pentanol		sweet		*Lact. plantarum*, *Lact. casei*	cheese (Milesi et al. 2010)
1-hexanol		green		*Lact. paracasei*, *L. lactis* subsp. *lactis*, *Lact. casei*, *Lact. plantarum*	Cheese-based medium (Budinich et al. 2011), fermented milk (Alemayehu et al. 2014), cheese (Milesi et al. 2010)
1-heptanol				*Lact. casei*, *Lact. plantarum*	cheese (Milesi et al. 2010)

Table 19.1. *(Continued)*

Volatile compound (trivial name widely used)	Precursor[*]	Odor description[†]	Dairy products in which they are flavor compounds[‡]	Produced by LAB as pure culture: LAB species	Produced by LAB as pure culture: Culture medium (references)
2-propanol (isopropyl alcohol)	AA	fruity		*L. lactis* subsp. *lactis*, *Lact. casei*, *Lact. plantarum*, *Lact. paracasei*	fermented milk (Alemayehu et al. 2014), cheese-based medium (Budinich et al. 2011), cheese (Milesi et al. 2010)
2-butanol		wine		*L. lactis* subsp. *lactis*, *Lact. plantarum*, *Lact. casei*	fermented milk (Alemayehu et al. 2014), cheese (Milesi et al. 2010)
2-pentanol		fresh		*Lact. casei*, *Lact. plantarum*	cheese (Milesi et al. 2010)
2-heptanol		fruity, herbaceous	Grana Padano	*Lact. casei*, *Lact. plantarum*	cheese (Milesi et al. 2010)
2-methylpropanol	AA	wine-like		*L. lactis* subsp. *lactis*, *Lact. plantarum*	fermented milk (Alemayehu et al. 2014) (Mugampoza 2013)
2-methylbutanol	AA	alcoholic, green, fuel oil		*L. lactis* subsp. *lactis*, *Lact. plantarum*	fermented milk (Mugampoza 2013; Alemayehu et al. 2014)
3-methylbutanol	AA	alcoholic, green, fuel oil		*L. lactis* subsp. *lactis*, *Lact. plantarum*, *Lact. helveticus*, *Lact. paracasei*, *Lact. casei*	fermented milk (Mugampoza 2013; Alemayehu et al. 2014), cheese-based medium (Petersen et al. 2010; Peralta et al. 2014), cheese (Milesi et al. 2010)
2-ethyl-1-hexanol				*Lact. paracasei*	cheese-based medium (Budinich et al. 2011)
Phenol	AA			*Lact. paracasei*	cheese-based medium (Peralta et al. 2014)
Phenylethanol	AA	rose		*Lact. paracasei*, *Lact. casei*, *Lact. plantarum*	cheese-based medium (Peralta et al. 2014), cheese (Milesi et al. 2010)
2,3-butanediol	C heterof	fruity		*L. lactis* subsp. *lactis*, *Lact. paracasei*	fermented milk (Alemayehu et al. 2014), cheese-based medium (Peralta et al. 2014)
Acetaldehyde	AA, C heterof	green, fresh	Yogurt	*Lact. paracasei*, *Lact. acidophilus*, *Lact. crispatus*, *L. lactis* subsp. *lactis* biovar *diacetylactis*, *Lact. plantarum*, *Lact. delbrueckii* subsp. *bulgaricus*	cheese-based medium (Budinich et al. 2011); fermented milk (Celik 2007; Laniewska-Trokenheim et al. 2010; Sharaf et al. 2013)
Propanal				*Lact. paracasei*	cheese-based medium (Budinich et al. 2011)
Butanal				*L. lactis* subsp. *lactis* biovar *diacetylactis*, *Lact. Plantarum*	fermented milk (Mugampoza 2013; Sharaf et al. 2013; Alemayehu et al. 2014)
Hexanal	L	green, fruity		*Lact. paracasei*	cheese-based medium (Budinich et al. 2011)
Heptanal	L	soapy		*L. lactis* subsp. *lactis*	fermented milk (Alemayehu et al. 2014)
Octanal	L	fatty, green		*L. lactis* subsp. *lactis* biovar *diacetylactis*, *Lact. plantarum*	fermented milk (Mugampoza 2013; Sharaf et al. 2013)
Nonanal	L	soapy		*L. lactis* subsp. *lactis*	fermented milk (Alemayehu et al. 2014)

Decanal	L	green, nutty	*Lact. helveticus*	cheese-based medium (Petersen et al. 2010)
2-hydroxy benzaldehyde	AA		*Lact. paracasei*	cheese-based medium (Peralta et al. 2014)
2-methylpropanal	AA	banana, malty	*Lact. plantarum*, *Lact. paracasei*	fermented milk (Mugampoza 2013), cheese-based medium (Peralta et al. 2014)
2- methylbutanal	AA	malty,	*L. lactis* subsp. *lactis*, *Lact. plantarum*, *L. lactis* subsp. *lactis* biovar *diacetylactis*, *Lact. helveticus*	fermented milk (Mugampoza 2013; Sharaf et al. 2013; Alemayehu et al. 2014), cheese-based medium (Petersen et al. 2010)
3-methylbutanal	AA	malty, chocolate, caramel	*L. lactis* subsp. *lactis*, *Lact. plantarum*, *Lact. helveticus*, *Lact. paracasei*, *L. lactis* subsp. *lactis* biovar *diacetylactis*	fermented milk (Mugampoza 2013; Alemayehu et al. 2014), cheese-based medium (Petersen et al. 2010; Budinich et al. 2011), fermented milk (Sharaf et al. 2013)
Benzaldehyde	AA	almonds, sugar, burnt	*Lact. paracasei*, *L. lactis* subsp. *lactis*, *Lact. helveticus*, *Lact. casei*, *Lact. plantarum*	fermented milk (Alemayehu et al. 2014), cheese-based medium (Petersen et al. 2010; Budinich et al. 2011), cheese (Milesi et al. 2010)
Phenylacetaldehyde	AA	honey, rosy, flower	*Lact. paracasei*	cheese-based medium (Peralta et al. 2014)
Methional	AA	green, tallowy, animals	*L. lactis* subsp. *lactis*, *Lact. paracasei*	fermented milks (Mugampoza 2013; Sharaf et al. 2013), cheese-based medium (Peralta et al. 2014)
2-hexenal	L	green, balsam	*L. lactis* subsp. *lactis* biovar *diacetylactis*, *Lact. plantarum*	fermented milk (Mugampoza 2013; Sharaf et al. 2013)
Methyl acetate	C heterof		*Lact. casei*, *Lact. plantarum*	cheese (Milesi et al. 2010)
Ethyl acetate	C heterof	fruity, pineapple	*L. lactis* subsp. *lactis*, *Lact. paracasei*, *Lact. casei*, *Lact. plantarum*	fermented milk (Alemayehu et al. 2014), cheese-based medium (Milesi et al. 2010; Peralta et al. 2014)
Butyl acetate	C heterof, F	green, fruity	*L. lactis* subsp. *lactis*	fermented milk (Alemayehu et al. 2014)
Methyl propanoate			*Lact. plantarum*	fermented milks (Mugampoza 2013)
Ethyl butanoate (ethyl butyrate)	C heterof, F	sweet, fruity, apple	*L. lactis* subsp. *lactis* biovar *diacetylactis*, *Lact. plantarum*, *L. lactis* subsp. *lactis*, *Lact. paracasei*	fermented milks (Mugampoza 2013; Sharaf et al. 2013), cheese-based medium (Peralta et al. 2014)
Ethyl hexanoate	C heterof, F	fruity, melone	*L. lactis* subsp. *lactis*, *Lact. plantarum*	fermented milk (Mugampoza 2013; Alemayehu et al. 2014)
Ethyl octanoate	C heterof, F	green, fruity	*L. lactis* subsp. *lactis*, *Lact. casei*, *Lact. plantarum*	fermented milk (Alemayehu et al. 2014), cheese-based medium (Milesi et al. 2010)

Table 19.1. *(Continued)*

Volatile compound (trivial name widely used)	Precursor*	Odor description†	Dairy products in which they are flavor compounds‡	Produced by LAB as pure culture	
				LAB species	Culture medium (references)
Ethyl 2-methylbutanoate	C heterof, AA			*L. lactis* subsp. *lactis*	fermented milk (Alemayehu et al. 2014)
Ethyl 3-methylbutanoate	C heterof, AA	fresh cheese		*L. lactis* subsp. *lactis*	fermented milk (Alemayehu et al. 2014)
3-methylbutyl acetate (isoamyl acetate)	C heterof, AA	banana, fruity		*Lact. paracasei*, *Lact. plantarum*	cheese-based medium (Peralta et al. 2014) fermented milk (Mugampoza 2013)
2-propanone (acetone)		wood pulp, hay		*Lact. paracasei*, *L. lactis* subsp. *lactis*, *Lact. plantarum*, *Lact. casei*	cheese-based medium (Budinich et al. 2011), fermented milk (Alemayehu et al. 2014), cheese (Milesi et al. 2010)
2-butanone		soap, acetone		*Lact. paracasei*, *L. lactis* subsp. *lactis*, *Lact. helveticus*, *Lact. casei*, *Lact. plantarum*	cheese-based medium (Budinich et al. 2011), fermented milk (Alemayehu et al. 2014), cheese-based medium (Petersen et al. 2010), (Milesi et al. 2010) fermented milk (Zareba et al. 2014)
2-pentanone		fruity, acetone		*L. lactis* subsp. *lactis*, *Lact. plantarum*, *Lact. casei*, *Lact. Rhamnosus*	fermented milk (Mugampoza 2013; Alemayehu et al. 2014), (Zareba et al. 2014), cheese-based medium (Sgarbi et al. 2013)
2-hexanone		fruity		*L. lactis* subsp. *lactis*, *Lact. plantarum*, *Lact. casei*	fermented milk (Mugampoza 2013; Alemayehu et al. 2014), cheese (Milesi et al. 2010)
2-heptanone	F	soapy, varnish		*L. lactis* subsp. *lactis*, *Lact. plantarum*, *Lact. rhamnosus*, *Lact. helveticus*, *Lact. casei*	fermented milk (Mugampoza 2013; Alemayehu et al. 2014; Zareba et al. 2014), cheese-based medium (Sgarbi et al. 2013), cheese-based medium (Petersen et al. 2010)
2-octanone		fruity		*Lact. helveticus*	cheese-based medium (Petersen et al. 2010)
2-nonanone	F	fruity, floral, hot milk		*L. lactis* subsp. *lactis*, *Lact. plantarum*, *Lact. helveticus*, *Lact. casei*	fermented milk (Mugampoza 2013; Alemayehu et al. 2014; Zareba et al. 2014), cheese-based medium (Petersen et al. 2010)
2,3-butanedione (diacetyl)	C, AA	creamy, buttery, sour milk	butter, cheese, fermented milks	*Lact. delbrueckii*, *Lact. acidophilus*, *Lact. casei*, *Strep. thermophilus*, *L. lactis* subsp. *lactis* var *diacetylactis*	(Imhof and Bosset 1994)

3-hydroxy-2-butanone (acetoin)	C, AA	sour milk	butter, cheese, fermented milks	*Lact. rhamnosus, L. lactis* subsp. *lactis* biovar *diacetylactis, Lact. acidophilus, Lact. crispatus, L. lactis* subsp. *lactis, Lact. plantarum, Lact. helveticus, Lact. paracasei, Lact. plantarum*	fermented milk (Laniewska-Trokenheim et al. 2010; Alemayehu et al. 2014) (Mugampoza 2013), cheese-based medium (Petersen et al. 2010) (Sgarbi et al. 2013; Peralta et al. 2014)
2,3-pentanedione	C, AA		butter, cheese, fermented milks	*L. lactis* subsp. *lactis* biovar *diacetylactis, Lact. plantarum, Lact. casei, Lact. rhamnosus, L. lactis* subsp. *lactis, Lact. paracasei, Lact. plantarum*	fermented milk (Sgarbi et al. 2013; Mugampoza 2013; Sharaf et al. 2013; Alemayehu et al. 2014), cheese-based medium (Peralta et al. 2014), cheese (Milesi et al. 2010)
Acetophenone				*Lact. paracasei*	cheese-based medium (Peralta et al. 2014)
Cyclohexanone				*Lact. paracasei*	cheese-based medium (Budinich et al. 2011)
Dimethyl sulphone	AA	sulfurous, burnt		*L. lactis* subsp. *lactis*	fermented milk (Alemayehu et al. 2014)
Carbon disulphide	AA			*L. lactis* subsp. *lactis*	fermented milk (Alemayehu et al. 2014)
Dimethyl sulfide (DMS)	AA	sulfurous		*L. lactis* subsp. *lactis*	fermented milk (Alemayehu et al. 2014)
Dimethyl disulfide (DMDS)	AA	sulfurous		*L. lactis* subsp. *lactis*	fermented milk (Alemayehu et al. 2014)
Dimethyl trisulfide (DMTS)	AA	sulfurous		*L. lactis* subsp. *lactis*	fermented milk (Alemayehu et al. 2014)
Methanethiol	AA	cabbage, sulfur		*L. lactis* subsp. *lactis*	fermented milk (Alemayehu et al. 2014)
4-hydroxy-2,5-dimethyl-3(2H)-furanone, 2-ethyl-4-hydroxy-5-methyl-3(2H)-furanone	AA and C heterof	caramel-like, sweet	cheese	*Lact. helveticus* and *Lact. delbrueckii*	(Kowalewska et al. 1985; Preininger and Grosch 1995; Preininger et al. 1996)

[*] Precursor: carbohydrates, homofermented (C homof) or heterofermented (C heterof), amino acids (AA), or lipids degradation (L)
[†] (Curioni and Bosset 2002) and http://www.odour.org.uk/odour/index.html
[‡] (Curioni and Bosset 2002)

Table 19.2. Main LAB species in some examples of fermented foods and their role in the formation of flavor.

Raw material	Fermented food products	Main LAB species	Other micro-organisms			Main role of LAB*	Origin of LAB†	References
			bacteria‡	yeast	fungi			
Fish	fish sauce	*T. halophilus*	yes	no	no	flav	backsl	(Udomsil et al. 2010)
Meat	fermented sausage	*Lact. sakei*, *Lact. curvatus*	yes	variable	variable	acidif	spont/start	(Berdagué et al. 1993; Leroy et al. 2006; Ravyts et al. 2012)
Meat	Bologna type sausages	*Lactobacillus*, *Streptococcus*, and *Enterococcus spp*	nd	yes	no	flav	spont	(Schrampf and Leitner 11 A.D.)
Meat	pastrma	*Lact. plantarum*, *Lact. sakei*, *Lact. pentosus*, *Lact. curvatus*, *Lact. fermentum*	nd	yes	no	flav	spont	(Kaban 2013)
Meat	Slavonski kulen (fermented sausage)	*Leuc. mesenteroides*, *Lact. acidophilus*	nd	no	no	acidif/flav	spont	(Babic et al. 2011)
Meat	Sucuk	*Lact. plantarum*, *Lact. sakei*, *Lact. pentosus*, *Lact. curvatus*, *Lact. fermentum*	nd	yes	no	flav	spont	(Kaban 2013)
Milk	Cabrales cheese	*Lact. plantarum*, *Lact. paraplantarum*, *Leuc. mesenteroides*	nd	yes	no	flav	spont	(Belen Florez et al. 2006)
Milk	Camembert cheese	*L. lactis*	yes	yes	yes	acidic	start	(Broadbent et al. 2011a) (Banks and Williams 2004)
Milk	Emmental cheese	*Lact. helveticus* and/or *Lact delbrueckii* subsp. *lactis*, *Strep. thermophilus*, *Lact. pacacasei* and other *Lactobacillus* spp	yes	no	no	acidif/flav	start/spont	(Grappin et al. 1999; Fröhlich-Wyder and Bachmann 2004)
Milk	Parmigianno Reggiano	*Lact. rhamnosus*, *Lact. paracasei*, *Lact. casei*	nd	no	no	flav	spont	(Sgarbi et al. 2013; Gatti et al. 2014)
Milk	suusac (fermented camel milk)	*Strep. thermophilus*, *L. lactis* subsp. *lactis*, *Strep. infantarius* subsp. *infantarius*	nd	yes	no	acidif/flav	spont	(Jans et al. 2012)
Peaches	Yan-Taozih	*Leuc. mesenteroides*	nd	no	no	flav	spont	(Chen et al. 2013)
Cucumber	khalpi	*Lact. plantarum*, *P. pentosaceus*	nd	no	no	acidif/flav	spont	(Tamang and Delven 2010)

Cereals	noodle	*Lact. casei, Lact. cellobiosus, Lact. fermentum*	nd	no	no	acidic	spont	(Rhee et al. 2011)
Cocoa bean	cocoa bean	*Lact. plantarum, Lact. fermentum, Leuc. pseudomesenteroides, Ent. casseliflavus*	nd	yes	no	acidic	spont	(Camu et al. 2007)
Cucumber	burongpipino	*Leuc. mesenteroides, Lact. brevis, Lact. plantarum, P. cerevisisae*	nd	no	no	acidif/flav	spont	(Banaay et al. 2013)
Leek	fermented leek	*Lact. plantarum, Lact. sakei, Leuc. mesenteroides*	nd	yes	no	acidif/flav	start	(Wouters et al. 2013)
Maize	Magou	*Lact. delbrueckii*	nd	no	no	acidif/flav	spont	(Beuchat 2008)
Cabbage	sauerkraut	*Leuc. mesenteroides, Lact. brevis, Lact. plantarum, P. pentosaceus*	nd	yes	no	acidif/flav	spont	(Tamang and Delwen 2010)
Mustard	Burong-mustala	*Lact. brevis, P. cerevisiae*	nd	no	no	acidic	spont	(Rhee et al. 2011)
Olive	fermented table olive	*Lact. casei, Lact. plantarum, Lact. brevis, Ent. faecium*	nd	yes	no	acidif/flav	spont	(Randazzo et al. 2004; Hurtado et al. 2012)
Rice or millet	Busa	*Lactobacillus* spp.	nd	yes	no	acidic	spont	(Beuchat 2008)
Mungbean	noodle	*Leuc. mesenteroides*	nd	no	no	acidic	spont	(Rhee et al. 2011)
Soja	stinky tofu	*Lactobacillus, Leuconostoc*	nd	no	no	acidif/flav	spont	(Chao et al. 2008)
Rice and black gram	Idli	*Leuc. mesenteroides*	nd	yes	no	acidif/flav	backsl	(Tamang and Delwen 2010)
Cabbage, Radish, other Vegetables	Kimchi	*Leuc. mesenteroides, Lact. brevis, Lact. plantarum, P. cerevisiae*	yes	yes	no	acidic	spont	(Chang et al. 2008; Rhee et al. 2011)
Rice, shrimp	Balao-Balao	*Leuc. mesenteroides, Lact. plantarum, P. cerevisisae*	nd	no	no	acidif/flav	spont	(Banaay et al. 2013)

[*] main role of LAB: acidifying (acid), flavoring (flav), spoilage (spoil)
[†] origin of LAB: added as starter (start), back slopping (backsl), spontaneous (spont)
[‡] bacteria other than LAB present in high population density: *Propionibacterium freudenreichii* in Emmental cheese; coagulase negative staphylococci and Kocuriae in fermented sausage;
nd: non determined

Lactobacillus appears as the most ubiquitous genus, with some species present in nearly all fermented foods, such as *Lact. plantarum* (De Vuyst et al. 2009; Todorov and Franco 2010; Guyot 2012; Hurtado et al. 2012). Many other LAB species are present in foods, such as *Lactococcus lactis*, *Streptococcus thermophilus* and *Strep. gallolyticus* subsp. *macedonicus*, *Leuconostoc mesenteroides* and *Leuc. lactis*, *Carnobacterium maltaromaticum*, *Pediococcus* sp., and *Enterococcus* sp. in dairy products, *Oenococcus oeni* in wine and beer, *Leuc. mesenteroides* in sauerkraut (Buckenhuskes 1993; Afzal et al. 2010), *Weissella* in cereals (Leroy and De Vuyst 2004; Guyot 2012), *Tetragenococcus halophilus* in fish sauce (Udomsil et al. 2010). Food microbiology has benefited from the advances in molecular techniques and sequencing approaches that are now commonly used to describe food microbial communities, leading to new findings (van Hijum et al. 2013; Ercolini et al. 2013). Microbial community profiling of fermented foods based on PCR-amplification and sequencing of rDNA gene regions emerged a decade ago, followed by metagenome sequencing, that is total DNA sequencing (van Hijum et al. 2013). These approaches have been recently applied to fermented foods, such as *kimchi* (Jung et al. 2011) and cocoa beans (Illeghems et al. 2012). New microorganisms were identified even in products such as cheese, previously thoroughly investigated using culture-dependent approaches (Quigley et al. 2012; Montel et al. 2014).

The primary function of LAB is to ensure the initial acidification of the product via the fermentation of carbohydrates into lactic acid. The latter contributes to the basic, pleasant acidic taste of many fermented foods. The presence of LAB does not necessarily imply that they have a determinant contribution to the formation of other aroma compounds. Indeed the actual contribution of LAB to flavor formation depends: (i) on their metabolic capacities to convert food components into aroma compounds; (ii) on their aroma-producing activities *in situ*; and, finally, (iii) on their relative contribution compared to that of other microorganisms, or enzymes, simultaneously active in the food. These points will be detailed in the following sections.

19.3.1. Biochemical Processes of Flavor Compound Formation in Food and Potential of LAB

LAB possess enzymes capable of hydrolyzing many food constituents, including (complex) carbohydrates, proteins, peptides, and lipids, into aroma precursors, and to further convert them into a variety of aroma compounds. Identifying pathways and especially metabolic bottlenecks in the formation of (important) aroma compounds is necessary to identify (key) genes and enzymes. It opens the way to a systematic mining of genomes to identify the potential of species and strains within species.

The breakdown of complex carbohydrates and proteins occurs outside the cell and is catalyzed by secreted or cell-bound enzymes. Most reactions are intracellular, and for some of them, for example esterification, the localization is unknown. It is interesting to note here that although flavor compounds include some end-products of primary metabolism (carbohydrate fermentation), most of them are secondary metabolites, for which the physiological role for bacteria is often unknown and the pathways of formation can be only partially known (Liu et al. 2008, 2014).

The main known metabolic routes in LAB leading to the formation of flavor compounds are detailed in this section and summarized in Figure 19.2.

19.3.2. Flavor Compounds Produced from Carbohydrate Fermentation by LAB

The primary substrates of LAB in fermented foods are the sugars of raw materials, either present (e.g., lactose in milk; glucose, fructose and sucrose in plant material (Daeschel et al. 1987) or

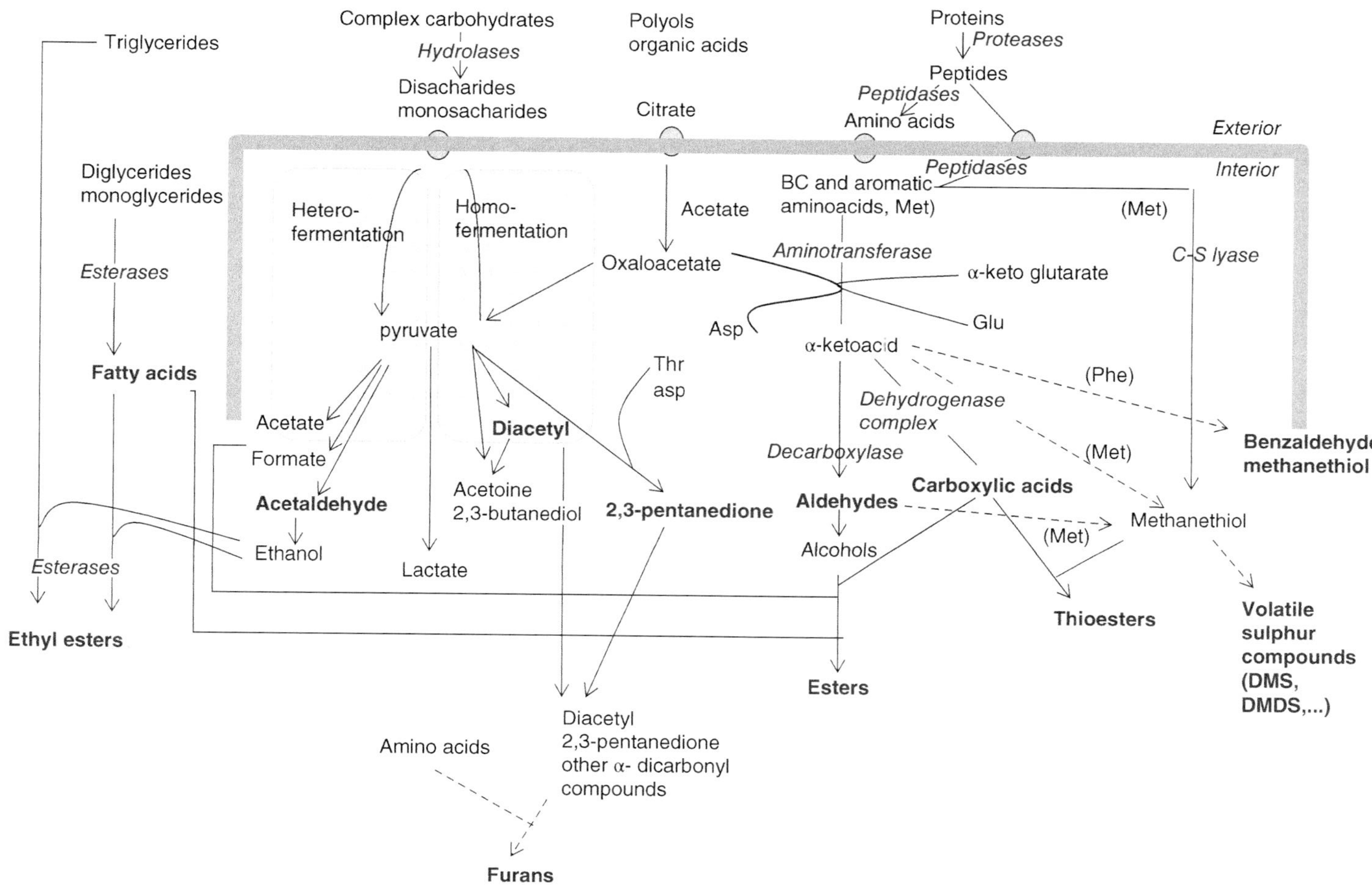

Figure 19.2. Schematic overview of conversion pathways leading to the formation of the main flavor compounds by lactic acid bacteria in fermented foods. Enzymes: extracellular enzymes or intracellular enzymes released from lysed cells; broken arrows: chemical (non enzymatic) reactions; red, bold: flavor compounds; blue, italics: main enzymes involved.

added (e.g., glucose in fermented sausages). Sugars can also be released from the hydrolysis of complex carbohydrates by LAB. Starch can be hydrolyzed by amylolytic LAB strains isolated from starchy fermented foods such as *Lact. plantarum*, *Lact. fermentum*, or other species (Guyot 2010). Other substrates can also be used, such as organic acids (citric and malic acids), polyols, and amino acids, which can be either co-metabolized with sugars or used as sole carbon source (Daeschel et al. 1987; Liu 2003). In some cases, the carbon sources used by LAB to grow are still debated, such as those used by non-starter LAB (NSLAB) that grow during cheese ripening after total lactose consumption by starter LAB (Williams et al. 2000; Broadbent et al. 2011a; Sgarbi et al. 2014). The substrates that LAB can use as carbon and energy source for growth and survival determine the composition of LAB population and its dynamic during food processing.

Carbon substrates undergo homo- or heterofermentation, depending on the LAB species, the substrate, and the environmental conditions. Homofermentative pathways generate lactic acid as an almost unique end-product, whereas heterofermentative pathways result in the formation of several aroma compounds or aroma precursors (Table 19.1, Figure 19.2): acetaldehyde is an important aroma volatile in yogurt and in cheese; acetic acid contributes to cheese flavor; ethanol has importance as a precursor of ethyl esters present in many fermented foods (Friedrich and Acree 1998; Ott et al. 1999; Curioni and Bosset 2002).

Citrate is present in milk, fruit, and vegetables. It can be co-metabolized with sugars by citrate-utilizing LAB. Citrate utilization results in an excess of pyruvate, which is thus converted to diacetyl (2,3-butanedione), acetoin (2-hydroxy-3-butanone), and 2,3-butanediol to equilibrate the redox balance of cellular metabolism (Collins 1972; Bartowsky and Henschke 2004). Some LAB can also synthesize 2,3-pentanedione from pyruvate and threonine (Ott et al. 2000). Diacetyl and 2,3-pentanedione are associated with a buttery aroma, which positively contributes to the flavor of a range of fermented dairy products such as butter (Mallia et al. 2008), yogurt (Routray and Mishra 2011), and cheese (Curioni and Bosset 2002). Diacetyl also contributes to wine style, while it is responsible for flavor defects in beer. Diacetyl is widely produced by LAB, including species of the *Lactococcus*, *Streptococcus*, *Leuconostoc*, *Lactobacillus*, *Pediococcus*, and *Oenococcus* genera.

α-Dicarbonyl compounds (diacetyl and 2,3-pentanedione, but also glyoxal and methylglyoxal) are also important as aroma precursors. A variety of highly odorous compounds can be formed from non-enzymatic Maillard and Strecker reactions between α-dicarbonyl compounds and amino acids in mild conditions, similar to those prevailing in wine and cheese (Griffith and Hammond 1989; Slaughter 1999; Pripis-Nicolau et al. 2000). For example, some furans associated with a sweet, caramel-like aroma have been identified in different cheeses (Preininger and Grosch 1994; Kubickova and Grosch 1998) and in cultures of *Lact. helveticus* and *Lact. delbrueckii* (Kowalewska et al. 1985; Preininger and Grosch 1995).

19.3.3. Flavor Compounds from Amino Acid Conversion by LAB

LAB are able to produce many important aroma compounds from amino acid (AA) catabolism (Table 19.1). LAB possess a powerful proteolytic system, in relation to their multiple auxotrophy for AA, which has been reviewed in detail (Fox and Wallace 1997; Christensen et al. 1999; Smit et al. 2005; Steele et al. 2013). LAB hydrolyze proteins into peptides and amino acids, which impact the taste of foods. Large hydrophobic peptides are associated with (undesirable) bitter taste, while peptides and AA contribute to the basic taste of cheese. The hydrolysis of large hydrophobic peptides can be accelerated by the release of intracellular peptidases from lysed LAB cells in cheese paste, in which they remain active during the ripening period, thus decreasing cheese bitterness (McSweeney and Sousa 2000).

LAB also possess AA-converting enzymes that generate a variety of aroma compounds from branched-chain AA (Leu, Ile, Val), aromatic AA (Phe and Tyr), methionine, and threonine (Yvon and Rijnen 2001; Martinez-Cuesta et al. 2013). AA conversion is initiated by transamination and/or elimination for Met and Thr, catalyzed by aminotransferases and lyases, respectively (Figure 19.2).

The bottleneck in the series of reactions of proteolysis, peptidolysis, and AA conversion to aroma compounds occurring during cheese ripening is transamination, as elegantly demonstrated by Yvon et al. (Yvon et al. 1998; Rijnen et al. 2000). Transamination requires an α-keto acid as a co-substrate, with α-ketoglutarate as the preferred keto donor for LAB aminotransferases (Figure 19.2). Hence, the conversion of amino acid to potent aroma compounds in cheese was greatly increased either by adding exogenous α-ketoglutarate to cheese curd, or by using a genetically modified *L. lactis* strain capable of producing α-ketoglutarate from glutamate, a reaction catalyzed by glutamate dehydrogenase (Gdh) (Yvon et al. 1998; Rijnen et al. 2000). Gdh activity thus appears to be a major criterion for the selection of aroma compounds producing LAB strains (Tanous et al. 2002). The role of CitP, the citrate uptake system in LAB, in the uptake of external α-ketoglutarate was recently demonstrated (Pudlik and Lolkema 2013). The use of oxaloacetate as keto donor in the transamination reaction was also recently shown in *L. lactis* (Pudlik and Lolkema 2012).

The α-keto acids produced from transamination are not odorant, but can further undergo different enzymatic reactions: either a reduction generating flavorless α-hydroxy acids, or a decarboxylation forming aldehydes, which can further reduce to alcohols, or an oxidative decarboxylation forming acylcoA and then carboxylic acids (Figure 19.2). AA-converting enzymes have been described in many comprehensive reviews (Christensen et al. 1999; Yvon and Rijnen 2001; Smit et al. 2005). Their activity is highly strain-and species-dependent (see below). Some α-keto acids, in particular those corresponding to aromatic AA and Met, can also be chemically oxidized in certain conditions, leading to the formation of aroma compounds. LAB can contribute to these chemical reactions by providing catalyzing cations such as manganese ions, which are released from lysed cells (Nierop Groot and Debont 1999; Smit et al. 2005).

In conclusion, the conversion of AA by LAB generates a variety of aroma compounds, including branched-chain acids (cheesy) and aldehydes (malty), aromatic alcohols and aldehydes (floral), and volatile sulfur compounds (garlic, cabbage, boiled potato) (Table 19.1). The expression of LAB enzymes *in situ* is influenced by a number of factors detailed in Section 19.5.

19.3.4. Flavor Compounds from Lipids in LAB

Food lipids are also a source of aroma compounds. The hydrolysis of lipids releases free fatty acids (FA), including short- and medium-chain FA, which are important aroma compounds in most ripened cheeses, associated with pungent, rancid aroma notes (Collins et al. 2003). A small level of fat hydrolysis (i.e., 0.5% in cheese) releases free FA in concentrations high enough to impact flavor. Free FA are also precursors of other aroma compounds, including esters, methyl ketones, secondary alcohols, and lactones (Molimard and Spinnler 1996; Collins et al. 2003). Lipid hydrolysis is catalyzed by lipolytic esterases (also referred to as lipases), defined as carboxylesterases active on long-chain acylglycerols insoluble in water (Verger 1997).

Most LAB possess only intracellular esterases, except *Lact. fermentum*, in which a surface-associated esterase has been isolated (Holland et al. 2005; Yvon 2006). As a result, most LAB esterases cannot hydrolyze food lipids until their release from lysed cells. Moreover, LAB esterases are only weakly lipolytic. Their activity has mainly been studied on soluble synthetic esters, and rarely on glycerides. Cell extracts of *L. lactis* and *Strep. thermophilus* show little activity on triglycerides of FA of longer than C4 and on diglycerides of FA longer than C6. However, they are active on monoglycerides of FA of medium- or long-chain FA (Holland et al. 2005). As a result, the lipolytic activity of LAB could vary according to the nature of the lipids present and their degree of pre-hydrolysis, for example in milk by lipoprotein lipase.

Esterases of LAB can also catalyze the direct synthesis of esters from glycerides and alcohols via a transferase reaction (alcoholysis) in aqueous systems, that is the transfer of fatty acyl groups from glycerides to alcohols. This topic requires further research (Holland et al. 2005), as well as the pathways of formation of other lipid-derived aroma compounds in LAB.

19.3.5. Synthesis of Esters

A range of acids and alcohols are generated from the different pathways detailed above, thereby providing potential substrates for esterification reactions. Many esters, most generally associated with fruity flavors, are found in fermented foods. LAB esterases can catalyze the formation of esters from acids and alcohols (Liu et al. 2004; Holland et al. 2005). In cheese, acids are present in relatively high concentrations (10 to 10^4 µg/g), while ethanol, the main alcohol detected (McGugan et al. 1975), is present in low amounts. Ethanol has been shown to be in concentrations limiting the synthesis of ethyl esters in cheeses such as Emmental (Richoux et al. 2008) and Cheddar cheese (Urbach 1995; Liu et al. 2004).

19.3.6. Interspecies and Intraspecies Variations of Aroma Compound Production

Flavor-related activities of LAB depend on the species, with some specific activities found only in a small number of species. The branched-chain α-keto acid decarboxylase activity involved in the formation of malty branched-chain aldehydes from branched-chain AA, for example, has been found only in *L. lactis* and not in the *Lactobacillus* and *Leuconostoc* strains tested (Fernandez De Palencia et al. 2006). Most LAB, however, display a great strain-to-strain variability, on the genomic level and/or at the phenotypic level. Table 19.3 gives some examples of intra- and interspecies variability of flavor-related LAB properties, such as proteolytic activities, AA-converting enzymatic activities, ester synthesis, and autolysis.

As a result, strategies have been developed to explore this diversity on both genomic and phenotypic levels. New insights into the potential of LAB to produce aroma compounds are expected from the comparison of complete genome sequences of a number of members of the same bacterial group, genus or species (Garrigues et al. 2013). The number of sequenced bacterial genomes increases exponentially along with the reduction in cost and time of sequencing methods, with more than 140 draft or finished LAB genome sequences publicly available (Steele et al. 2013). However, the activity *in situ* cannot always be predicted from the genomics or transcriptomics approaches. This is the case, for example, for esterases, for which the activity on food lipids such as triglycerides should be biochemically assessed. As an example, only one of the 12 predicted esterases identified in the genome of *Propionibacterium freudenreichii* was able to hydrolyze milk triglycerides (Dherbécourt et al. 2008; Abeijon Mukdsi et al. 2014).

The ultimate aim is to select combinations of species and strains with complementary activities to produce tailored flavor development in fermented foods (Garrigues et al. 2013). Nice demonstrations of the potential of the relevant combination of strains harboring complementary amino acid catabolic routes have been done for *L. lactis* in a cheese context. The formation of branched-chain aroma compounds from branched-chain AA, in particular 3-methylbutanal, was enhanced by associating *L. lactis* strains exhibiting high proteolytic, aminotransferase, or α-keto acid decarboxylase activity, resulting in higher aroma compound formation than isolated strains (Ayad et al. 2001; Kieronczyk et al. 2003; Pelaez and Requena 2005; Smit et al. 2005; Amarita et al. 2006).

Table 19.3. Examples of strain and species-dependent traits related to the formation of aroma compounds in fermented foods.

LAB species	Gene or activity targeted, conditions	Main results	References
Lact. helveticus	Proteolysis	Genes for cell envelope-associated proteinases vary widely	(Genay et al. 2009; Broadbent et al. 2011b)
Lact. helveticus	Autolysis	Same equipment in peptidoglycan hydrolases; strain-dependence of lysis probably due to differences in cell wall composition	(Jebava et al. 2011; Vinogradov et al. 2013)
Lactococcus, lactobacillus, Streptococcus, and *Leuconostoc*	Leucine catabolism	Leu aminotransferase activity: detected in the 4 LAB genera tested (*Lactococcus, Lactobacillus, Streptococcus, Leuconostoc*); highly strain-dependent BC keto acid decarboxylase activity: only detected in 1 genus (*Lactococcus*) out of the 5 LAB genera tested; highly strain-dependent.	(Smit et al. 2004)
Lactococcus, Lactobacillus, and *Leuconostoc* (156 cheese isolates)	AA -converting enzymes: -aromatic (AraT) and branched-chain (BcaT) aminotransferase activities, -glutamate dehydrogenase (Gdh) -branched-chain alphaketo acid (Kivd)	BcaT and AraT activity in all isolates, with activity higher in *Lactococcus* compared to *Lactobacillus* and *Leuconostoc;* high intraspecies variability, for ex. AraT activity of *L. lactis* varied by a factor > 13 Gdh activity: 50% of positive isolates among *Leuconostoc, 14-46* in *Lactobacillus* spp., and 24% in *L. lactis*; high intra-species variability, for ex. Gdh activity of *Lact plantarum* and *Lact. brevis* varied by a factor 10-20; Kivd activity: only found in *L. lactis*, in 4 out of 45 strains, with high strain-to-strain variability (factor 30) within the positive isolates	(Fernandez De Palencia et al. 2006)
Lactococcus, Lactobacillus, and *Leuconostoc* (14 strains)	Met-converting enzymatic activities	Large natural biodiversity in enzyme capabilities and high inter- and intra-species variability: high C-S lyase and transaminase activity in *Lactococcus* compared to *Lactobacillus* and *Leuconostoc* higher glutamate dehydrogenase activities among *Leuconostoc* and *Lactobacillus* spp.; a-keto acid decarboxylase activities highly strain-dependent, mainly detected in *Lactobacillus*	(Hanniffy et al. 2009)
L. Lactis (102 isolates)	Partial sequence of 3 flavor-related genes	strains very closely related to each other at the protein sequence level for the genes *bcaT* (branched chain amino acid aminotransferase), *pepN* (aminopeptidase N), and *pepX* (X-prolyl dipeptidyl peptidase)	(Rademaker et al. 2007)

Table 19.3. *(Continued)*

LAB species	Gene or activity targeted, conditions	Main results	References
L. Lactis (84 strains)	AA converting activity (cell extracts from cells grown in 2 media)	Activity of branched chain aminotransferase (BcaT), alpha-hydroxyisocaproic acid dehydrogenase (HicDH), and esterase: individual strains showed very diverse regulatory responses, with the activities in the 2 media in general poorly correlated	(Bachmann et al. 2009)
Lact. casei, Lact. plantarum and *L. lactis*	Met conversion:	methional production from 4-methylthio-2-ketobutanoate: methional produced by 5 strains (4 *L. lactis* and 1 *Lact. casei*) out of 20, with only 1 *L. lactis* strain producing high amounts	(Amarita et al. 2001)
Lact. plantarum and *other Lactobacillus Streptococcus, Ent faecium* (20 strains)	in vitro ester synthesis by cell extracts: from triglycerides plus ethanol (alcoholysis) and fatty acid plus ethanol (esterification)	alcoholysis: higher activity in *Ent. faecium* compared to *Lact*, with great strain-to-strain variations (factors of 7 and 30 within strains of *Ent. faecium* and *Lact. plantarum*, respectively) esterification: most strains active; great intraspecies variability (factors of 60 and 20 within strains of *Ent. faecium* and *Lact. plantarum*, respectively)	(Liu et al. 1998, 2003; Abeijon Mukdsi et al. 2009)
Lact sakei, Lact curvatus	AA catabolism	Profile of volatiles; only slight differences between the strains of the two species studied	(Freiding et al. 2011)
Lact. sakei	response to various oxidative compounds	extremely variable from strain-to-strain	(Guilbaud et al. 2012)
Pediococcus and *Lactobacillus*	genes involved in starch metabolism	most isolates (75%) possess *amy* gene (alpha-amylase gene) and 54% *dexC* (neopullulanase gene)	(Turpin et al. 2011)

19.4. Biotic and Abiotic Factors Modulating the Contribution of LAB to Flavor Formation

19.4.1. General Scheme of Flavor Formation in Fermented Foods In Situ

The development of flavor in fermented foods results from the activity of microorganisms, in interaction with each other and with the abiotic environment. This ecosystem is characterized by the combination of raw material(s) and processing conditions. Raw materials provide the substrates to microorganisms, but are also a source of endogenous enzymes and adventitious microorganisms. Processing conditions comprise a succession of operations that orientate the fermentation, including the deliberate addition of microbial cultures (starter or adjunct LAB and/or other microorganisms) and exogenous enzymes. Figure 19.3 illustrates the interactions occurring between microorganisms, product, and processing conditions relevant for flavor formation in fermented foods.

The flavor-forming enzymes of LAB are expressed by growing, resting, and even non- culturable cells (Pelaez and Requena 2005). Active enzymes are also released from lysed cells (see above). Both abiotic and biotic factors modulate the expression of the flavor-related activities of LAB in food ecosystems, by affecting LAB growth, metabolic activity, survival, or lysis in foods, and by providing the conditions and substrates required for aroma-forming reactions. A population of ~10^6/g viable bacterial cells is generally considered the minimum level to impact food flavor. Finally, the actual contribution of LAB to flavor will also depend on the presence and activity of other, potentially more active, flavor-producing agents in the food ecosystem.

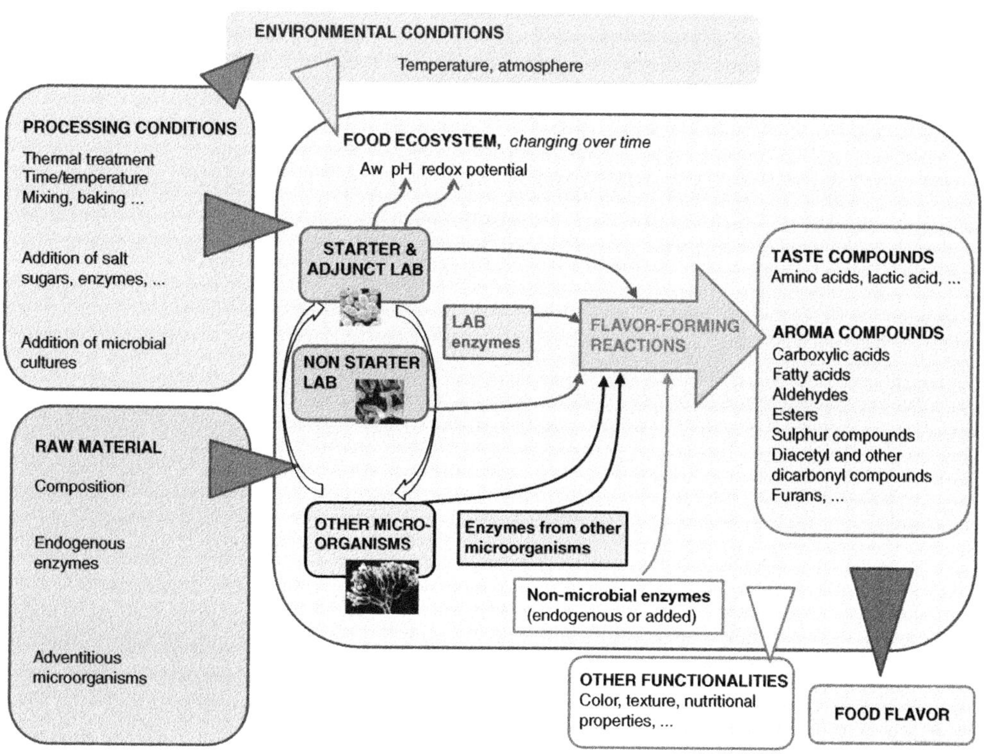

Figure 19.3. General scheme illustrating the dynamic of flavor development in fermented foods and the interactions within this ecosystem that influence flavor formation.

19.4.2. Factors Modulating the Expression of the Flavor-related Activities of LAB

Many abiotic factors of fermented foods influence the formation of flavor compounds by LAB. The composition of food (possible limitations in nutriments, pH, water activity), the physical conditions (temperature, atmosphere), and their changes over the manufacturing process, constitute a "funnel" which selects, among the microorganisms present (including LAB), the ones capable of growing and remaining active. The processing conditions can include steps that impact microbial metabolism and survival, such as mixing (increase of O_2 concentration), thermal treatments, salting and/or drying (decrease of water activity).

The exhaustion of the main carbon source is an important factor modulating LAB metabolism. For example, carbon starvation induces a non-culturable state in *L. lactis* associated with new metabolic capabilities supporting survival, including the catabolism of branched-chain AA (Ganesan et al. 2006, 2007). The activity of non-proliferating microorganisms including LAB generally greatly impacts the formation of aroma compounds (Pelaez and Requena 2005). The presence of viable but non-culturable (VBNC) LAB has been demonstrated in foods, for example in cheese (Falentin et al. 2012), and plants (Todorov and Franco 2010). These findings suggest that the flavor-related activity should be considered for all LAB populations, including the non-proliferating or non-cultivable population of LAB encountered during food aging.

Alternatively, LAB can die and lyse. Autolysis is provoked by an enzymatic hydrolysis of the cell wall peptidoglycan by endogenous enzymes called autolysins or peptidoglycan hydrolases (Lortal and Chapot-Chartier 2005). Lysed cells release their intracellular enzymes, thus contributing differently to flavor formation. The lysis of LAB is important during cheese ripening because it enhances the hydrolysis of some hydrophobic bitter peptides. In other foods such as salami, LAB lysis is also thought to contribute to the flavor development (Zambonelli et al. 2002). How lysis impact the formation of aroma compounds has been investigated in (model) cheese by using bacteriocin-producing *L. lactis* strains or isogenic *L. lactis* strains differing by their autolytic properties. The results observed varied according to the experimental conditions and the aroma compound considered. It was observed that lysis enhanced the first step of AA conversion, that is transamination, and the formation of some AA-derived aroma compounds. In contrast, some reactions requiring intact LAB cells, such as oxidative or reduction reactions, were reduced at a high lysis level. The impact of induced-lysis on cheese flavor was generally positive (Garde et al. 1997, 2005; Bourdat-Deschamps et al. 2004). Associating autolytic and non-autolytic strains appears a suitable solution.

Salt is also an important factor modulating the aroma-related activity of LAB. *T. halophilus* was isolated in fish sauce containing 25% NaCl, and its presence associated to higher concentrations of some free AA and the corresponding aldehydes, in a strain-dependant manner (Udomsil et al. 2010, 2011).

Biotic factors also modulate the expression of the flavor-related activities of LAB. Microbial interactions, which can be beneficial (cooperation), neutral or adverse (inhibition), determine the succession of microbial communities in fermented foods (Pelaez and Requena 2005; Irlinger and Mounier 2009). Interactions occur within the strains and species of the LAB starter, between starter and non-starter LAB (Martley and Crow 1993; Grappin et al. 1999; Crow et al. 2001), and between LAB and other microbial groups (Narvhus and Gadaga 2003). Microbial interactions can also impact cell death. As an example, the lysis of strains of *Lact. helveticus* in Swiss cheeses induces a concomitant lysis of *Strep. thermophilus* (Deutsch et al. 2002)

The intensity and richness of fermented food flavor is generally associated with high microbial diversity, at both the species and strain levels. This has been widely demonstrated for cheeses by comparing the sensory profiles of cheeses manufactured from the same milk either used raw, pasteurized or microfiltered (Grappin et al. 1999; Montel et al. 2014). The role of NSLAB on flavor formation has been widely investigated. NSLAB grow during the ripening of most cheeses, often reaching cell densities of 10^8 CFU/g, including those manufactured from pasteurized milk (Broadbent et al. 2011a). NSLAB are considered desirable for the development of mature Cheddar cheese flavor development, but can also negatively influence cheese flavor, depending on the species, the strains, and the type of cheese (Crow et al. 2001; Banks and Williams 2004; Broadbent et al. 2011a). Therefore, adjunct LAB cultures have been widely proposed to control the development of NSLAB in cheese and thus limit the variations in cheese quality induced by the adventitious NSLAB (Ross et al. 2000; Crow et al. 2001; Broadbent et al. 2011a)

Microbial interactions can directly impact the flavor-related activities of LAB, by providing substrates for aroma-generating reactions. For example, the production of ethanol by heterofermentative LAB or yeasts can increase ethyl ester production (Das et al. 2005). Another example is the impact of wild LAB strains harboring Gdh activity, which provide α-ketoglutarate and thus enhance transamination reactions and aroma compound formation (Tanous et al. 2002). The knowledge acquired opens the way for strategies of combinations of strains that complete metabolic pathways to enhance flavor (Pelaez and Requena 2005). In most cases, differences are observed between the profiles of volatile aroma compounds in pure and mixed cultures but the molecular mechanisms are not elucidated (Narvhus and Gadaga 2003).

19.4.3. Factors Determining the Real Contribution of LAB to Food Flavor

The level of flavor-related activities of the different microorganisms and non-microbial enzymes present in food ecosystems determines the actual contribution of LAB to flavor. Lipolysis in cheese is a good example. Although they possess esterases, LAB have no or very little contribution to lipolysis in cheese and probably in most fermented foods. In the cheeses ripened exclusively by LAB, such as Cheddar cheese, the degree of lipolysis of ripened cheese is close to that of milk (FA/fat 0.4%). It can be estimated that ~80% of the final amount of free FA are present before the ripening (Hickey et al. 2007) and likely result from the activity of milk lipase during the first steps of milk processing. In cheeses ripened using both LAB and propionibacteria, such as Emmental cheese, lipolysis mainly occurs during the ripening and results from propionibacteria activity (Dherbécourt et al. 2010; Abeijon Mukdsi et al. 2014), whereas in blue cheeses and other fungi-ripened cheeses, the high degree of lipolysis results from fungi activity. In fermented sausage, the activity of bacterial protease against meat protein is weak compared to that of meat endogenous protease (Leroy et al. 2006), whereas the aroma-forming activities of *Lact. sakei* are low compared to that of coagulase negative staphylococci (Berdagué et al. 1993; Ravyts et al. 2012). The latter results are in agreement with the fact that *Lact. sakei* can be used in meat or seafood preservation without undesirable impact on its flavor.

In summary the contribution of LAB to flavor formation is decisive in some fermented foods, for example fermented milks and cheeses fermented only with LAB. However, their direct contribution to flavor formation can be low when LAB are part of a complex microbial community including in particular yeasts or fungi with high enzymatic activities, or when food includes flavoring constituents, such as in some baked foods or *kimchi*. In all cases, LAB are essential as primary agents of the acidification of the food matrix. They are also associated with other changes such as a decrease of redox potential, which likely influences flavor-related reactions (Urbach 1995). Table 19.2 gives some examples of the flavor-related role of LAB in different fermented foods.

19.5. Conclusions and Research Perspectives

Fermented foods are recipients of a renewal of interest from consumers and food companies (Hugenholtz 2013) and offer an opportunity to support sustainable food systems in the context of increased pressure on food resources (van Hijum et al. 2013). Steering flavor quality and/or diversification requires more research to further identify the important aroma compounds in the final products, and to determine the main biological agents involved in their formation and the interactions within these ecosystems.

The final contribution of LAB to the formation of fermented food flavor depends on their intrinsic potential to produce aroma compounds and of the way this potential will (or not) be "revealed" during the fermentation process. The potential of LAB species/strains can be investigated using a series of complementary approaches, ranging from the *in silico* mining of the genomic information, the assessment of expression at transcriptomics and/or proteomics levels, to the (high-throughput) screening of the volatiles produced using metabolomics-based approaches. For all *in vitro* studies, the choice of the culture conditions determines the results observed, and thus it is absolutely essential to investigate the main factors that determine and orientate the activity of LAB *in situ* in foods, and to take them into consideration to design relevant screening approaches. The results of screening experiments should be systematically subjected to benchmarking against "real" foods. Good correlations between the results observed *in vitro* and in real cheeses were obtained both in cheese-relevant culture conditions (Thierry et al. 2004; Abeijon Mukdsi et al. 2014), and in a nice model of "microcheese" recently proposed (Bachmann et al. 2009). Today, it still remains a challenge to link the knowledge gained on isolated LAB strains and the final result in the product in terms of aroma compounds, in particular because of numerous interactions within the microbial community. In addition to their intrinsic aroma-forming activities, the ability of LAB strains to compete with spontaneous non-starter bacteria should also be targeted (Leroy et al. 2006).

Besides, many LAB (and non LAB) strains have been isolated from high quality (traditional) fermented foods and from wild environmental sources. It is fundamental to preserve the patrimonial resource they constitute in perennial collections, such as Biological Resource Centers (BRCs), in which strains can be deposited to guarantee their long-term conservation. In fermented foods such as cheeses for which (simplified) starter cultures have been used for decades, along with the constant improvement of the sanitary conditions in milk production and processing, a decrease of the intensity and richness of flavor has often be observed. In this context, BRCs offer a reservoir of potentially interesting strains for flavor diversification.

The development of efficient flavoring (LAB) cultures will benefit from actions in several directions. Microbial diversity can now be described with an increasing specificity and new insights on the active non-viable population, through the use of both targeted and non-targeted molecular approaches. Metagenomics and metatranscriptomics are increasingly applied in food microbiology, whereas application of metabolomics approaches to food products enlarge our view of *in situ* microbial metabolism and of the variety of metabolites. Such approaches will also benefit from improvements of functional annotation of the key enzymes in the formation of aroma compounds (Liu et al. 2008), and advances in the integration of the data resulting from these different approaches, which is still a challenge.

References

Abeijon Mukdsi, M.C., Falentin, H., Maillard, M.B., et al. (2014) The secreted esterase of *Propionibacterium freudenreichii* has a major role in cheese lipolysis. *Appl Environ Microbiol* **80**, 751–756.

Abeijon Mukdsi, M.C., Medina, R.B., Alvarez, M.D., and Gonzalez, S.N. (2009) Ester synthesis by lactic acid bacteria isolated from goat's and ewe's milk and cheeses. *Food Chem* **117**, 241–247.

Afzal, M.I., Jacquet, T., Delaunay, S., et al. (2010) *Carnobacterium maltaromaticum*: identification, isolation tools, ecology and technological aspects in dairy products. *Food Microbiol* **27**, 573–579.

Alemayehu, D., Hannon, J.A., McAuliffe, O., and Ross, R.P. (2014) Characterization of plant-derived lactococci on the basis of their volatile compounds profile when grown in milk. *Int J Food Microbiol* **172**, 57–61.

Amarita, F., de la Plaza, M., Fernandez De Palencia, P., et al. (2006) Cooperation between wild lactococcal strains for cheese aroma formation. *Food Chem* **94**, 240–246.

Amarita, F., Fernandez-Espla, D., Requena, T., and Pelaez, C. (2001) Conversion of methionine to methional by *Lactococcus lactis*. *FEMS Microbiol Lett* **204**, 189–195.

Ayad, E.H.E., Verheul, A., Engels, W.J.M., et al. (2001) Enhanced flavour formation by combination of selected lactococci from industrial and artisanal origin with focus on completion of a metabolic pathway. *J Appl Microbiol* **90**, 59–67.

Babic, I., Markov, K., Kovacevic, D., et al. (2011) Identification and characterization of potential autochthonous starter cultures from a Croatian "brand" product "Slavonski kulen". *Meat Sci* **88**, 517–524.

Bachmann, H., Starrenburg, M.J.C., Dijkstra, A., et alV. (2009) Regulatory phenotyping reveals important diversity within the species *Lactococcus lactis*. *Appl Environ Microbiol* **75**, 5687–5694.

Banaay, C.G.B., Baolong, M.P., and Elgado, F.B. (2013) Lactic acid bacteria in Philippine traditional fermented foods. In: *Lactic Acid Bacteria - R and D for Food, Health and Livestock Purposes* ((ed.) M. Kongo), InTech, pp. 571–588.

Banks, J.M. and Williams, A.G. (2004) The role of the nonstarter lactic acid bacteria in Cheddar cheese ripening. *Int J Dairy Technol* **57**, 145–152.

Bartowsky, E.J. and Henschke, P.A. (2004) The 'buttery' attribute of wine-diacetyl-desirability, spoilage and beyond. *Int J Food Microbiol* **96**, 235–252.

Belen Florez, A., Lopez-Diaz, T.M., Alvarez-Martin, P. and Mayo, B. (2006) Microbial characterisation of the traditional Spanish blue-veined Cabrales cheese: identification of dominant lactic acid bacteria. *Eur Food Res Technol* **223**, 503–508.

Berdagué, J.L., Monteil, P., Montel, M.C. and Talon, R. (1993) Effects of starter cultures on the formation of flavor compounds in dry sausage. *Meat Sci* **35**, 275–287.

Beuchat, L.R. (2008) Indigenous fermented foods. In: *Biotechnology Set* ((eds) H.-J. Rehm, and G. Reed), Wiley-VCH Verlag GmbH, Weinheim, Germany, pp. 506–512.

Bosset, J.O., Gauch, R., Mariaca, R., and Klein, B. (1995) Comparison of various sample treatments for the analysis of volatile compounds by GC-MS: application to the Swiss Emmental cheese. *Mitt Gebiete Lebensm Hyg* **86**, 672–698.

Bourdat-Deschamps, M., Le Bars, D., Yvon, M., and Chapot-Chartier, M.P. (2004) Autolysis of *Lactococcus lactis* AM2 stimulates the formation of certain aroma compounds from amino acids in a cheese model. *Int Dairy J* **14**, 791–800.

Broadbent, J.R., Budinich, M.F., and Steele, J.L. (2011a) Non-starter lactic acid bacteria. In: *Encyclopedia of Dairy Science* ((eds) J.W. Fuquay, P.F. Fox, and P. McSweeney), Elsevier, London, pp. 639–644.

Broadbent, J.R., Cai, H., Larsen, R.L., et al. (2011b) Genetic diversity in proteolytic enzymes and amino acid metabolism among *Lactobacillus helveticus* strains. *J Dairy Sci* **94**, 4313–4328.

Buckenhuskes, H.J. (1993) Selection criteria for lactic acid bacteria to be used as starter cultures for various food commodities. *FEMS Microbiol Rev* **12**, 253–272.

Budinich, M.F., Perez-Diaz, I., Cai, H., et al. (2011) Growth of *Lactobacillus paracasei* ATCC 334 in a cheese model system: a biochemical approach. *J Dairy Sci* **94**, 5263–5277.

Campbell-Platt, G. (1994) Fermented foods - a world perspective. *Food Res Int* **27**, 253–257.

Camu, N., De Winter, T., Verbrugghe, K., et al. (2007) Dynamics and biodiversity of populations of lactic acid bacteria and acetic acid bacteria involved in spontaneous heap fermentation of cocoa beans in ghana. *Appl Environ Microbiol* **73**, 1809–1824.

Casaburi, A., Piombino, P., Nychas, G.J., et al. (2015) Bacterial populations and the volatilome associated to meat spoilage. *Food Microbiol* **45**, 83–102.

Celik, E.S. (2007) Determination of aroma compounds and exopolysaccharides formation by lactic acid bacteria isolated from traditional yoghurts. pp. 1–99. The Graduate School of Engineering and Sciences of Izmir Institute of Technology.

Chang, H.W., Kim, K.H., Nam, Y.D., et al. (2008) Analysis of yeast and archaeal population dynamics in kimchi using denaturing gradient gel electrophoresis. *Int J Food Microbiol* **126**, 159–166.

Chao, S.H., Tomii, Y., Watanabe, K., and Tsai, Y.C. (2008) Diversity of lactic acid bacteria in fermented brines used to make stinky tofu. *Int J Food Microbiol* **123**, 134–141.

Chen, Y.S., Wu, H.C., Pan, S.F., et al. (2013) Isolation and characterization of lactic acid bacteria from yan-taozih (pickled peaches) in Taiwan. *Ann Microbiol* **63**, 607–614.

Christensen, J.E., Dudley, E.G., Pederson, J.A., and Steele, J.L. (1999) Peptidases and amino acid catabolism in lactic acid bacteria. *Antonie Van Leeuwenhoek* **76**, 217–246.

Collins, E.B. (1972) Biosynthesis of flavor compounds by microorganisms. *J Dairy Sci* **55**, 1022–1028.

Collins, Y.F., McSweeney, P.L.H., and Wilkinson, M.G. (2003) Lipolysis and free fatty acid catabolism in cheese: a review of current knowledge. *Int Dairy J* **13**, 841–866.

Crow, V., Curry, B., and Hayes, M. (2001) The ecology of non-starter lactic acid bacteria (NSLAB) and their use as adjuncts in New Zealand Cheddar. *Int Dairy J* **11**, 275–283.

Curioni, P.M.G. and Bosset, J.O. (2002) Key odorants in various cheese types as determined by gas chromatography-olfactometry. *Int Dairy J* **12**, 959–984.

Daeschel, M.A., Andersson, R.E., and Fleming, H.P. (1987) Microbial ecology of fermenting plant materials. *FEMS Microbiol Lett* **46**, 357–367.

Das, S., Holland, R., Crow, V.L., et al. (2005) Effect of yeast and bacterial adjuncts on the CLA content and flavour of a washed-curd, dry-salted cheese. *Int Dairy J* **15**, 807–815.

De Bok, F.A.M., Janssen, P.W.M., Bayjanov, J.R., et al. (2011) Volatile compound fingerprinting of mixed-culture fermentations. *Appl Environ Microbiol* **77**, 6233–6239.

De Vuyst, L., Vrancken, G., Ravyts, F., et al. (2009) Biodiversity, ecological determinants, and metabolic exploitation of sourdough microbiota. *Food Microbiol* **26**, 666–675.

Deutsch, S.M., Ferain, T., Delcour, J., and Lortal, S. (2002) Lysis of lysogenic strains of *Lactobacillus helveticus* in Swiss cheeses and first evidence of concomitant *Streptococcus thermophilus* lysis. *Int Dairy J* **12**, 591–600.

Dherbécourt, J., Falentin, H., Jardin, J., et al. (2010) Identification of a secreted lipolytic esterase in *Propionibacterium freudenreichii*, a ripening process bacterium Involved in Emmental cheese lipolysis. *Appl Environ Microbiol* **76**, 1181–1188.

Dherbécourt, J., Maillard, M.B., Catheline, D., and Thierry, A. (2008) Production of branched-chain aroma compounds by *Propionibacterium freudenreichii*: links with the biosynthesis of membrane fatty acids. *J Appl Microbiol* **105**, 977–985.

Engels, W.J.M. (1997) Volatile and non-volatile compounds in ripened cheese their formation ant their contribution to flavour. Wageningen Agricultural University, Wageningen, Netherlands, pp. 1–127.

Ercolini, D., Pontonio, E., De Filippis, F., et al. (2013) Microbial ecology dynamics during rye and wheat sourdough preparation. *Appl Environ Microbiol* **79**, 7827–7836.

Falentin, H., Henaff, N., Le Bivic, P., et al. (2012) Reverse transcription quantitative PCR revealed persistency of thermophilic lactic acid bacteria metabolic activity until the end of the ripening of Emmental cheese. *Food Microbiol* **29**, 132–140.

Fernandez De Palencia, P., de la Plaza, A., Amarita, F., et al. (2006) Diversity of amino acid converting enzymes in wild lactic acid bacteria. *Enz Microbial Technol* **38**, 88–93.

Forss, D.A. (1972) *Odor and Flavor Compounds from Lipids*. Pergamon Press, Oxford, pp. 177–258.

Fox, P.F. and Wallace, J.M. (1997) Formation of flavor compounds in cheese. *Adv Appl Microbiol* **45**, 17–85.

Freiding, S., Gutsche, K.A., Ehrmann, M.A., and Vogel, R.F. (2011) Genetic screening of *Lactobacillus sakei* and *Lactobacillus curvatus* strains for their peptidolytic system and amino acid metabolism, and comparison of their volatilomes in a model system. *Syst Appl Microbiol* **34**, 311–320.

Friedrich, J.E. and Acree, T.E. (1998) Gas chromatography olfactometry (GC/O) of dairy products. *Int Dairy J* **8**, 235–241.

Fröhlich-Wyder, M.T. and Bachmann, H.P. (2004) Cheeses with propionic acid fermentation. In: *Cheese. Chemistry, Physics and Microbiology* ((eds) P.F. Fox, P.L.H. McSweeney, T.M. Cogan, and T.P. Guinee), Elsevier, London, pp. 141–156.

Ganesan, B., Dobrowolski, P., and Weimer, B.C. (2006) Identification of the leucine-to-2-methylbutyric acid catabolic pathway of *Lactococcus lactis*. *Appl Environ Microbiol* **72**, 4264–4273.

Ganesan, B., Stuart, M.R., and Weimer, B.C. (2007) Carbohydrate starvation causes a metabolically active but nonculturable state in *Lactococcus lactis*. *Appl Environ Microbiol* **73**, 2498–2512.

Garde, S., Avila, M., Medina, M., and Nunez, M. (2005) Influence of a bacteriocin-producing lactic culture on the volatile compounds, odour and aroma of Hispanico cheese. *Int Dairy J* **15**, 1034–1043.

Garde, S., Gaya, P., Medina, M., and Nunez, M. (1997) Acceleration of flavour formation in cheese by a bacteriocin-producing adjunct lactic culture. *Biotechnol Lett* **19**, 1011–1014.

Garrigues, C., Johansen, E., and Crittenden, R. (2013) Pangenomics - an avenue to improved industrial starter cultures and probiotics. *Curr Opin Biotechnol* **24**, 187–191.

Gatti, M., Bottari, B., Lazzi, C., et al. (2014) Invited review: Microbial evolution in raw-milk, long-ripened cheeses produced using undefined natural whey starters. *J Dairy Sci* **97**, 573–591.

Genay, M., Sadat, L., Gagnaire, V., and Lortal, S. (2009) prtH2, not prtH, is the ubiquitous cell wall proteinase gene in *Lactobacillus helveticus*. *Appl Environ Microbiol* **75**, 3238–3249.

Grappin, R., Beuvier, E., Bouton, Y., and Pochet, S. (1999) Advances in the biochemistry and microbiology of Swiss-type cheese. *Lait* **79**, 3–22.

Griffith, R. and Hammond, E.G. (1989) Generation of Swiss cheese flavor components by the reaction of amino acids with carbonyl compounds. *J Dairy Sci* **72**, 604–613.

Guilbaud, M., Zagorec, M., Chaillou, S., and Champomier-Verges, M.C. (2012) Intraspecies diversity of *Lactobacillus sakei* response to oxidative stress and variability of strain performance in mixed strains challenges. *Food Microbiol* **29**, 197–204.

Guyot, J.P. (2010) Fermented cereal products. In: *Fermented Foods and Beverages of the World* ((eds) J.P. Tamang, and K. Kailasapathy), CRC Press, New York, pp. 247–261.

Guyot, J.P. (2012) Cereal-based fermented foods in developing countries: ancient foods for modern research. *Int J Food Sci Technol* **47**, 1109–1114.

Hanniffy, S.B., Pelaez, C., Martinez-Bartolome, M.A., et al. (2009) Key enzymes involved in methionine catabolism by cheese lactic acid bacteria. *Int J Food Microbiol* **135**, 223–230.

Hickey, D.K., Kilcawley, K.N., Beresford, T.P., et al. (2007) Starter strain related effects on the biochemical and sensory properties of Cheddar cheese. *J Dairy Res* **74**, 9–17.

Holland, R., Liu, S.Q., Crow, V.L., et al. (2005) Esterases of lactic acid bacteria and cheese flavour: Milk fat hydrolysis, alcoholysis and esterification. *Int Dairy J* **15**, 711–718.

Hugenholtz, J. (2013) Traditional biotechnology for new foods and beverages. *Curr Opin Biotechnol* **24**, 155–159.

Hurtado, A., Reguant, C., Bordons, A., and Rozes, N. (2012) Lactic acid bacteria from fermented table olives. *Food Microbiol* **31**, 1–8.

Illeghems, K., De Vuyst, L., Papalexandratou, Z., and Weckx, S. (2012) Phylogenetic analysis of a spontaneous cocoa bean fermentation metagenome reveals new insights into its bacterial and fungal community diversity. *PLoS ONE* **7**, 38040.

Imhof, R. and Bosset, J.O. (1994) Relationships between microorganisms and formation of aroma compounds in fermented dairy products - review. *Z Lebensm Unters Forsch* **198**, 267–276.

Irlinger, F. and Mounier, J. (2009) Microbial interactions in cheese: implications for cheese quality and safety. *Curr Opin Biotechnol* **20**, 142–148.

Jans, C., Bugnard, J., Njage, P.M.K., et al. (2012) Lactic acid bacteria diversity of African raw and fermented camel milk products reveals a highly competitive, potentially health-threatening predominant microflora. *LWT - Food Sci Technol* **47**, 371–379.

Jebava, I., Plockova, M., Lortal, S., and Valence, F. (2011) The nine peptidoglycan hydrolases genes in *Lactobacillus helveticus* are ubiquitous and early transcribed. *Int J Food Microbiol* **148**, 1–7.

Jung, J.Y., Lee, S.H., Kim, J.M., et al. (2011) Metagenomic analysis of kimchi, a traditional korean fermented food. *Appl Environ Microbiol* **77**, 2264–2274.

Kaban, G. (2013) Sucuk and pastirma: Microbiological changes and formation of volatile compounds. *Meat Sci* **95**, 912–918.

Kieronczyk, A., Skeie, S., Langsrud, T., and Yvon, M. (2003) Cooperation between *Lactococcus lactis* and nonstarter lactobacilli in the formation of cheese aroma from amino acids. *Appl Environ Microbiol* **69**, 734–739.

Kowalewska, J., Zelazowska, H., Babuchowski, A., et al. (1985) Isolation of aroma-bearing material from *Lactobacillus helveticus* culture and cheese. *J Dairy Sci* **68**, 2165–2171.

Kubickova, J. and Grosch, W. (1998) Quantification of potent odorants in Camembert cheese and calculation of their odour activity values. *Int Dairy J* **8**, 17–23.

Laniewska-Trokenheim, L., Olszewska, M., Miks-Krajnik, M., and Zadernowska, A. (2010) Patterns of survival and volatile metabolites of selected *Lactobacillus* strains during long-term incubation in milk. *J Microbiol* **48**, 445–451.

Le Boucher, C., Courant, F., Jeanson, S., et al. (2013) First mass spectrometry metabolic fingerprinting of bacterial metabolism in a model cheese. *Food Chem* **141**, 1032–1040.

Lemfack, M.C., Nickel, J., Dunkel, M., et al. (2014) mVOC: a database of microbial volatiles. *Nucleic Acids Res* **42**, D744–748.

Leroy, F. and De Vuyst, L. (2004) Lactic acid bacteria as functional starter cultures for the food fermentation industry. *Trends Food Sci Technol* **15**, 67–78.

Leroy, F., Verluyten, J., and De Vuyst, L. (2006) Functional meat starter cultures for improved sausage fermentation. *Int J Food Microbiol* **106**, 270–285.

Liu, M., Bienfait, B., Sacher, O., et al. (2014) Combining chemoinformatics with bioinformatics: in silico prediction of bacterial flavor-forming pathways by a chemical systems biology approach “Reverse pathway engineering”. *PLoS ONE* **9**, 4769.

Liu, M., Nauta, A., Francke, C., and Siezen, R.J. (2008) Comparative genomics of enzymes in flavor-forming pathways from amino acids in lactic acid bacteria. *Appl Environ Microbiol* **74**, 4590–4600.

Liu, S.Q. (2003) Practical implications of lactate and pyruvate metabolism by lactic acid bacteria in food and beverage fermentations. *Int J Food Microbiol* **83**, 115–131.

Liu, S.Q., Holland, R., and Crow, V.L. (1998) Ethyl butanoate formation by dairy lactic acid bacteria. *Int Dairy J* **8**, 651–657.

Liu, S.Q., Holland, R., and Crow, V.L. (2003) Ester synthesis in an aqueous environment by *Streptococcus thermophilus* and other dairy lactic acid bacteria. *Appl Microbiol Biotechnol* **63**, 81–88.

Liu, S.Q., Holland, R., and Crow, V.L. (2004) Esters and their biosynthesis in fermented dairy products: a review. *Int Dairy J* **14**, 923–945.

Lortal, S. and Chapot-Chartier, M.P. (2005) Role, mechanisms and control of lactic acid bacteria lysis in cheese. *Int Dairy J* **15**, 857–871.

Maarse, H., Visscher, C.A., Willemsens, L.C., et al. (1994) *Volatile Compounds in Food: Qualitative and Quantitative Data*. TNO Nutrition and Food Research, Zeist, The Netherlands.

Mallia, S., Escher, F., and Schlichtherle-Cerny, H. (2008) Aroma-active compounds of butter: A review. *Eur Food Res Technol* **226**, 315–325.

Marsili, R. (2011) Flavors and off-flavors in dairy foods. In: Encyclopedia of Dairy Science ((eds) J.W. Fuquay, P.F. Fox, and P. McSweeney), Elsevier, London, pp. 533–551.

Martinez-Cuesta, M.D., Pelaez, C., and Requena, T. (2013) Methionine metabolism: major pathways and enzymes involved and strategies for control and diversification of volatile sulfur compounds in cheese. *Crit Rev Food Sci Nutr* **53**, 366–385.

Martley, F.G. and Crow, V.L. (1993) Interactions between non-starter microorganisms during cheese manufacture and ripening. *Int Dairy J* **3**, 461–483.

McGugan, W.A., Blais, J.A., Boulet, M., et al. (1975) Ethanol, ethyl esters, and volatile fatty acids in fruity Cheddar cheese. *Can Inst Food Sci Technol J* **8**, 196–198.

McSweeney, P.L.H. and Sousa, M.J. (2000) Biochemical pathways for the production of flavour compounds in cheeses during ripening: a review. *Lait* **80**, 293–324.

Milesi, M.M., Wolf, I.V., Bergamini, C.V., and Hynes, E.R. (2010) Two strains of nonstarter lactobacilli increased the production of flavor compounds in soft cheeses. *J Dairy Sci* **93**, 5020–5031.

Molimard, P. and Spinnler, H.E. (1996) Review: compounds involved in the flavor of surface mold-ripened cheeses: origins and properties. *J Dairy Sci* **79**, 169–184.

Montel, M.C., Buchin, S., Mallet, A., et al. (2014) Traditional cheeses: Rich and diverse microbiota with associated benefits. *Int J Food Microbiol* **177**, 136–154.

Morgan, M.E. (1976) The chemistry of some microbially induced flavor defects in milk and dairy foods. *Biotechnol Bioeng* **18**, 953–965.

Mozzi, F., Ortiz, M.E., Bleckwedel, J., et al. (2013) Metabolomics as a tool for the comprehensive understanding of fermented and functional foods with lactic acid bacteria. *Food Res Int* **54**, 1152–1161.

Mugampoza, D. (2013) Characterization of the non-starter bacterial flora of Stilton cheese. PhD thesis, University of Notttingham.

Narvhus, J.A. and Gadaga, T.H. (2003) The role of interaction between yeasts and lactic acid bacteria in African fermented milks: a review. *Int J Food Microbiol* **86**, 51–60.

Nierop Groot, M.N. and Debont, J.A. M. (1999) Involvement of manganese in conversion of phelylalanine to benzaldehyde by lactic acid bacteria. *Appl Environ Microbiol* **65**, 5590–5593.

Ott, A., Germond, J.E., Baumgartner, M., and Chaintreau, A. (1999) Aroma comparisons of traditional and mild yogurts: headspace gas chromatography quantification of volatiles and origin of alpha-diketones. *J Agric Food Chem* **47**, 2379–2385.

Ott, A., Germond, J.E., and Chaintreau, A. (2000) Vicinal diketone formation in yogurt: 13C precursors and effect of branched-chain amino acids. *J Agric Food Chem* **48**, 724–731.

Pelaez, C. and Requena, T. (2005) Exploiting the potential of bacteria in the cheese ecosystem. *Int Dairy J* **15**, 831–844.

Peralta, G.H., Wolf, I.V., Bergamini, C.V., Perotti, M.C., and Hynes, E.R. (2014) Evaluation of volatile compounds produced by *Lactobacillus paracasei* I90 in a hard-cooked cheese model using solid-phase microextraction. *Dairy Sci Technol* **94**, 73–81.

Petersen, M.A., Kristensen, H.T., Bakman, M., et al. (2010) Aroma formation in a cheese model system by different *Lactobacillus helveticus* strains. Zurich University of Applied Sciences.

Pozo-Bayon, M.A., Guichard, E., and Cayot, N. (2006) Flavor control in baked cereal products. *Food Rev Int* **22**, 335–379.

Pogačic, T., Maillard, M.B., Leclerc, A., Hervé, C., Chuat, V., Yee, A., Valence, F., and Thierry, A. (2015) A methodological approach to screen diverse cheese-related bacteria for their ability to produce aroma compounds. *Food Microbiol* **46**, 145–153.

Preininger, M. and Grosch, W. (1994) Evaluation of key odorants of the neutral volatiles of Emmentaler cheese by the calculation of odour activity values. *Lebensm Wiss Technol* **27**, 237–244.

Preininger, M. and Grosch, W. (1995) Determination of 4-hydroxy-2,5-dimethyl-3(2H)-furanone (HDMF) in cultures of bacteria. *Z Lebensm Unters For* **201**, 97–98.

Preininger, M., Warmke, R., and Grosch, W. (1996) Identification of the character impact flavour compounds of Swiss cheese by sensory studies of models. *Z Lebensm Unters Forsch* **202**, 30–34.

Pripis-Nicolau, L., de Revel, G., Bertrand, A., and Maujean, A. (2000) Formation of flavor components by the reaction of amino acid and carbonyl compounds in mild conditions. *J Agric Food Chem* **48**, 3761–3766.

Pudlik, A.M. and Lolkema, J.S. (2012) Rerouting citrate metabolism in *Lactococcus lactis* to citrate-driven transamination. *Appl Environ Microbiol* **78**, 6665–6673.

Pudlik, A.M. and Lolkema, J.S. (2013) Uptake of alpha-ketoglutarate by citrate transporter CitP drives transamination in *Lactococcus lactis*. *Appl Environ Microbiol* **79**, 1095–1101.

Quigley, L., O'Sullivan, O., Beresford, T.P., et al. (2012) High-throughput sequencing for detection of subpopulations of bacteria not previously associated with artisanal cheeses. *Appl Environ Microbiol* **78**, 5717–5723.

Rademaker, J.L., Herbet, H., Starrenburg, M.J., et al. (2007) Diversity analysis of dairy and nondairy *Lactococcus lactis* isolates, using a novel multilocus sequence analysis scheme and (GTG)5-PCR fingerprinting. *Appl Environ Microbiol* **73**, 7128–7137.

Randazzo, C.L., Restuccia, C., Romano, A.D., and Caggia, C. (2004) *Lactobacillus casei,* dominant species in naturally fermented Sicilian green olives. *Int J Food Microbiol* **90**, 9–14.

Ravyts, F., De Vuyst, L., and Leroy, F. (2012) Bacterial diversity and functionalities in food fermentations. *Eng Life Sci* **12**, 356–367.

Rhee, S.J., Lee, J.E., and Lee, C.H. (2011) Importance of lactic acid bacteria in Asian fermented foods. *Microb Cell Fact* **10**, 13.

Richoux, R., Maillard, M.B., Kerjean, J.R., et al. (2008) Enhancement of ethyl ester and flavour formation in Swiss cheese by ethanol addition. *Int Dairy J* **18**, 1140–1145.

Rijnen, L., Courtin, P., Gripon, J.C., and Yvon, M. (2000) Expression of a heterologous glutamate dehydrogenase gene in *Lactococcus lactis* highly improves the conversion of amino acids to aroma compounds. *Appl Environ Microbiol* **66**, 1354–1359.

Ross, R.P., Stanton, C., Hill, C., et al. (2000) Novel culture for cheese improvement. *Trends Food Sci Technol* **11**, 96–104.

Routray, W. and Mishra, H.N. (2011) Scientific and Technical Aspects of Yogurt Aroma and Taste: A Review. *Comprehensive Rev Food Sci Food Safe* **10**, 208–220.

Salque, M., Bogucki, P.I., Pyzel, J., et al. (2013) Earliest evidence for cheese making in the sixth millennium bc in northern Europe. *Nature* **493**, 522–525.

Schrampf, E. and Leitner, E. (11 A.D.) Monitoring of bacteria causing off-flavours in Bologna type sausage by SPME-GC-MS. pp. 147. Laboratory for Aroma Analysis and Enology (LAAE), University of Zaragoza, Spain.

Sgarbi, E., Bottari, B., Gatti, M., and Neviani, E. (2014) Investigation of the ability of dairy nonstarter lactic acid bacteria to grow using cell lysates of other lactic acid bacteria as the exclusive source of nutrients. *Int J Dairy Technol* http://dx.doi.org10.1111/1471-0307.12132.

Sgarbi, E., Lazzi, C., Tabanelli, G., et al. (2013) Nonstarter lactic acid bacteria volatilomes produced using cheese components. *J Dairy Sci* **96**, 4223–4234.

Sharaf, O.M., Shahein, M.S., Shaaban, H.A., et al. (2013) Production of cheese flavor and antifungal substances by *Lactobacillus plantarum* and *Lactococcus lactis* subsp *diacetylactis* and their application in bakery products. *World Appl Sci J* **24**, 900–907.

Slaughter, J.C. (1999) The naturally occurring furanones: formation and function from pheromone to food. *Biol Rev* **74**, 259–276.

Smit, B.A., Engels, W.J.M., and Wouters, J.T.M. (2004) Diversity of L-leucine catabolism in various microorganisms involved in dairy fermentations, and identification of the rate-controlling step in the formation of the potent flavour component 3-methylbutanal. *Appl Microbiol Biotechnol* **64**, 396–402.

Smit, G., Smit, B.A., and Engels, W.J. (2005) Flavour formation by lactic acid bacteria and biochemical flavour profiling of cheese products. *FEMS Microbiol Rev* **29**, 591–610.

Stahnke, L.H., Holck, A., Jensen, A., et al. (2002) Maturity acceleration by *Staphyloccocus carnosus* in fermented sausage -relatonship between maturity and flavor compounds. *J Food Sci* **67**, 1914–1921.

Steele, J., Broadbent, J., and Kok, J. (2013) Perspectives on the contribution of lactic acid bacteria to cheese flavor development. *Curr Opin Biotechnol* **24**, 135–141.

Tamang, J.P. and Delwen, S. (2010) Dietary cultures and antiquity of fermented foods and beverages. In: *Fermented Foods and Beverages of the World* ((eds) J.P. Tamang and K. Kailasapathy), CRC Press, New York, pp. 42–84.

Tanous, C., Kieronczyk, A., Helinck, S., et al. (2002) Glutamate dehydrogenase activity: a major criterion for the selection of flavour-producing lactic acid bacteria strains. *Antonie Van Leeuwenhoek* **82**, 271–278.

Thierry, A., Richoux, R., Kerjean, J.R., and Lortal, S. (2004) A simple screening method for isovaleric acid production by *Propionibacterium freudenreichii* in Swiss cheese. *Int Dairy J* **14**, 697–700.

Todorov, S.D. and Franco, B.D.G.D. (2010) *Lactobacillus plantarum*: characterization of the species and application in food production. *Food Rev Int* **26**, 205–229.

Turpin, W., Humblot, C., and Guyot, J.P. (2011) Genetic screening of functional properties of lactic acid bacteria in a fermented pearl millet slurry and in the metagenome of fermented starchy foods. *Appl Environ Microbiol* **77**, 8722–8734.

Udomsil, N., Rodtong, S., Choi, Y.J., et al. (2011) Use of *Tetragenococcus halophilus* as a starter culture for flavor improvement in fish sauce fermentation. *J Agric Food Chem* **59**, 8401–8408.

Udomsil, N., Rodtong, S., Tanasupawat, S., and Yongsawatdigul, J. (2010) Proteinase-producing halophilic lactic acid bacteria isolated from fish sauce fermentation and their ability to produce volatile compounds. *Int J Food Microbiol* **141**, 186–194.

Urbach, G. (1995) Contribution of lactic acid bacteria to flavour compound formation in dairy products. *Int Dairy J* **5**, 877–903.

van Hijum, S.A., Vaughan, E.E., and Vogel, R.F. (2013) Application of state-of-art sequencing technologies to indigenous food fermentations. *Curr Opin Biotechnol* **24**, 178–186.

Verger, R. (1997) 'Interfacial activation' of lipases: facts and artifacts. *Tibtech* **15**, 32–38.

Vinogradov, E., Valence, F., Maes, E., et al. (2013) Structural studies of the cell wall polysaccharides from three strains of *Lactobacillus helveticus* with different autolytic properties: DPC4571, BROI, and LH1320. *Carbohydr Res* **379**, 7–12.

Williams, A.G., Withers, S.E., and Banks, J.M. (2000) Energy sources of non-starter lactic acid bacteria isolated from Cheddar cheese. *Int Dairy J* **10**, 17–23.

Wouters, D., Grosu-Tudor, S., Zamfir, M., and De Vuyst, L. (2013) Applicability of *Lactobacillus plantarum* IMDO 788 as a starter culture to control vegetable fermentations. *J Sci Food Agr* **93**, 3352–3361.

Yvon, M. (2006) Key enzymes for flavour formation by lactic acid bacteria. *Aust J Dairy Technol* **61**, 88–96.

Yvon, M., Berthelot, S., and Gripon, J.C. (1998) Adding alpha-ketoglutarate to semi-hard cheese curd highly enhances the conversion of amino acids to aroma compounds. *Int Dairy J* **8**, 889–898.

Yvon, M. and Rijnen, L. (2001) Cheese flavour formation by amino acid catabolism. *Int Dairy J* **11**, 185–201.

Zambonelli, C., Chiavari, C., Benevelli, M., and Coloretti, F. (2002) Effects of lactic acid bacteria autolysis on sensorial characteristics of fermented foods. *Food Technol Biotechnol* **40**, 347–351.

Zareba, D., Ziarno, M., Scibisz, I., and Gawron, J. (2014) The importance of volatile compound profile in the assessment of fermentation conducted by *Lactobacillus casei* DN-114 001. *Int Dairy J* **35**, 11–14.

Chapter 20

Lactic Acid Bacteria Biofilms: From their Formation to their Health and Biotechnological Potential

Jean-Christophe Piard* and Romain Briandet

INRA and AgroParisTech, UMR1319 Micalis, France

20.1. Lactic Acid Bacteria Biofilms are Ubiquitous in a Wide Variety of Environments from Nature to Domesticated Settings

Biofilms are formed by sessile bacterial communities associated to a surface and of a self-produced matrix made of extracellular polymeric substances (EPS) such as proteins, DNA, and polysaccharides (Costerton et al. 1999; Flemming 2011). Although prevailing in nature (Costerton 2007), the biofilm bacterial lifestyle has been extensively studied for just two decades, mostly through pioneering work in the study of *Pseudomonas aeruginosa* biofilms involved in cystic fibrosis (Costerton et al. 1995, 1999). Biofilms may form on a large variety of biotic or abiotic supports. For microbial cells, biofilms are a shelter that allows improved protection against both hostile environments and environment fluctuations such as pH or oxygen changes, presence of biocides or antibiotics, and host response in the case of biotic biofilms. As an example, it has been assessed that sessile bacteria present in biofilms can stand 1000 times higher biocide concentrations than their planktonic counterparts (Bridier et al. 2011). The high resistance of biofilm bacteria to otherwise lethal conditions is due to different factors including the biofilm matrix hampering the diffusion/reaction of biocides, some preferentially expressed genes in biofilm bacteria involved in stress response or in antibiotic efflux, and the appearance of biofilm-derived variants with enhanced ability to survive adverse conditions (Boles et al. 2004; Mah 2012; Van Acker et al. 2014).

Lactic acid bacteria (LAB) form a group of Gram (+), non-spore-forming cocci, coccobacilli, or rods present in food or feed products, with a nucleotide composition of less than 50 mol% GC and with lactic acid as the major carbohydrate catabolism end-product (Stiles and Holzapfel 1997). As such, LAB include the genera *Abiotrophia*, *Aerococcus*, *Dolosicoccus*, *Eremococcus*, *Facklamia*, *Globicatella*, *Ignavigranum*, *Alkalibacterium*, *Allofustis*, *Alloiococcus*, *Atopobacter*, *Atopococcus*, *Atopostipes*, *Carnobacterium*, *Desemzia*, *Dolosigranulum*, *Granulicatella*, *Isobaculum*, *Lacticigenium*, *Marinilactibacillus*, *Pisciglobus*, *Trichococcus*, *Bavariicoccus*, *Catellicoccus*, *Enterococcus*, *Melissococcus*, *Pilibacter*, *Tetragenococcus*, *Vagococcus*, *Lactobacillus*, *Pediococcus*, *Leuconostoc*, *Fructobacillus*, *Oenococcus*, *Weissella*, *Lactococcus*, *Lactovum*, and *Streptococcus* (Felis et al., see Chapter 2, Vandamme et al. 1996). These bacteria have been isolated from the numerous ecological

* Corresponding author: INRA and AgroParisTech, UMR1319 Micalis, Domaine de Vilvert, 78350-Jouy-en-Josas, France. jean-christophe.piard@jouy.inra.fr

Biotechnology of Lactic Acid Bacteria: Novel Applications, Second Edition. Edited by Fernanda Mozzi, Raúl R. Raya, and Graciela M. Vignolo.

niches present in the global food web including soil (Magnusson et al. 2002; Chen et al. 2005), plants (Kelly et al. 1998; Yang et al. 2010; Pang et al. 2012), silages (Tohno et al. 2012; Chen et al. 2013; Nguyen et al. 2013), milk (Salama et al. 1995; Fernandez et al. 2013), fermented food products (Prajapati and Nair 2003; Liu et al. 2011), and both man and animal gastrointestinal tract (GIT) (Arumugam et al. 2011; Mann et al. 2014). While much less studied than pathogen biofilms, LAB biofilms have been logically reported in different biotopes in which LAB occur. This covers the food chain (plants, milk, meat products, and food processing plants), the gastrointestinal and vaginal animal and human mucosae, as well as industrial and domesticated settings (Table 20.1). Therefore, LAB biofilms appear to colonize a continuum of habitats that likely cross-inoculate each other as schemed in Figure 20.1.

Most LAB present in food agricultural products are considered safe and benefit of both the GRAS (*G*enerally *R*ecognized *A*s *S*afe) status from the FDA in the USA and of the QPS (*Q*ualified *P*resumption of *S*afety) assessment from EFSA in Europe. Based on this and given that LAB play crucial functions in the transformation of many agricultural products into processed food (see Chapters 12–16, 19), the autochthonous LAB biota present in native products is often reinforced by addition of bacterial starters consisting of cocktails of selected LAB in order to accelerate and better control fermentation processes. This practice leads to the presence of an important LAB biota in food products (ca. 10^9 CFU/g) that is prone to form biofilms on food matrices, on food manufacture settings, and ultimately on the mucosae of man and animals upon ingestion of live LAB through the consumption of fermented products. LAB biofilms are also regarded as beneficial and one of their major described functions is their barrier effect against the development of spoilage and pathogen organisms. Interesting examples in food processing plants come from some traditional cheese manufacture technologies (Ragusano in Italy and PDO (Protected Designation of Origin) Salers in France) making use of wooden vats for milk fermentation (Licitra et al. 2007; Didienne et al. 2012). These cheeses are made with raw milk whose complex microbiota rapidly forms biofilms on the vat wood. The resident biofilm behaves as a reservoir of strains whose release in milk allows milk fermentation and cheese ripening without the addition of exogenous starters. LAB biofilms are also often considered as beneficial *in vivo* in man and animals. They have only been little explored in man because of sampling restrictions and difficulties in preserving biofilms upon biopsy sampling and sample preparation (Bollinger et al. 2007). In the human vagina, the occurrence of health beneficial lactobacilli biofilms has been demonstrated (Domingue et al. 1991; Strus et al. 2005; Terraf et al. 2012; Malik et al. 2013) (see below). In human GIT, LAB biofilms consisting of lactobacilli may cover food residues present in the gut and participate in their digestion (Macfarlane and Macfarlane 2006). While the microbiota composition of the food residue associated biofilms is similar to the non-adhering microbiota, the two communities appear metabolically distinct in both their efficiency to catabolize different carbohydrates and the type of short-chain fatty acid that they produce the most. In animals, LAB communities resembling biofilms have long been observed in rodents, chicken, pigs, and more recently in insects such as honeybees (Savage et al. 1968; Fuller and Brooker 1974; Fuller et al. 1978; Vasquez et al. 2012). The bee crop (a central organ between the esophagus and ventriculus) used for collection and transport of nectar to hive is inhabited by a complex LAB biofilm consisting of various *Lactobacillus* strains (Vasquez et al. 2012). This biofilm is attached to the wall of the crop and its partial release inoculates honey, bee pollen, and bee bread with high levels of LAB. Members of this LAB microbiota have been shown to produce antimicrobials active against transient microorganisms present on flowers frequently visited by honeybees. This antimicrobial arsenal produced by the LAB microbiota has also been shown to inhibit two major bee pathogens *in vivo* (*Melissococcus plutonius* and *Paenibacillus larvae*) highlighting the pivotal role played by LAB biofilms in honeybee health (Forsgren et al. 2010; Vasquez et al. 2012). Interestingly, this crucial LAB biota is sensitive to antibiotics used in apiculture to control some bacterial diseases.

On the other hand, certain LAB may be detrimental in food products such as some meat products (Bjorkroth and Korkeala 1997; Johansson et al. 2011; Jaaskelainen et al. 2013), pickles (Perez-Diaz

Table 20.1. Some examples of LAB biofilms identified along the food chain and in domesticated environments.

Ecological niche	Strains involved	Special features	Beneficial/detrimental effects** and functions	References
Food and food processing plants				
Olives	*Lact. pentosus**	Mixed bacteria-yeast biofilms	B; extension of shelf life and improved sensory quality	(Arroyo-Lopez et al. 2012; Dominguez-Manzano et al. 2012)
Sausages	*Lact. sakei*	43% of *Lact. sakei* strains are biofilm-formers	B; extension of shelf life and improved sensory quality	(Landeta et al. 2013)
Cheddar cheese	*Lact. curvatus*	Biofilms resist to vat surface cleaning	D; conversion of $L(+)$- to $D(-)$ lactic acid yielding calcium lactate crystallization	(Somers et al. 2001; Agarwal et al. 2006)
Cheese wooden vat	*Strep. thermophilus*, *Lactobacillus* sp., *Lactococcus* sp., *Leuconostoc* sp., *Enterococcus* sp.	LAB biofilms from the vat form a reservoir of starter LAB	B; no commercial starters needed; protection against the development of food-borne pathogens	(Licitra et al. 2007; Lortal et al. 2009; Didienne et al. 2012)
Cheese wooden shelves	*Leuconostoc* sp., *Lactobacillus* sp.	Multispecies biofilms with high composition stability over time	B; biofilms on wooden shelves maintain the undesirable flora at low levels	(Mariani et al. 2007)
Ice cream plant	*Streptococcus* sp., *Leuconostoc* sp., *Pediococcus* sp.	Biofilm present on equipment resist cleaning	D; source of bacterial contamination of the product and low transfer in heat exchangers	(Gunduz and Tuncel 2006)
Food processing plants	*L. lactis* *Lact. plantarum* *Ent. durans*	Biofilms isolated from floor drains	B; LAB biofilms inhibit *L. monocytogenes*	(Zhao et al. 2004; Zhao et al. 2013)
Sugar refineries	*Leuconostoc* sp.	Biofilm formation ability appeared unrelated to glucansucrase production	D; *Leuconostoc* biofilms produce dextran that clog filters and inhibit sugar crystallization	(Leathers and Bischoff 2011)

Table 20.1. *(Continued)*

Ecological niche	Strains involved	Special features	Beneficial/detrimental effects** and functions	References
Gastrointestinal and genital tracts				
Food residues from human gut	*Lact. plantarum*, *Lact. acidophilus*, *Lact. paracasei*	Same flora composition between biofilms attached to food residues and non-adherent populations	Distinct patterns of saccharide utilization and of fermentation product formation between biofilm and non-adhering bacteria	(Macfarlane and Macfarlane 2006)
Mice forestomach	*Lact. reuteri*	Biofilms are formed with rodent specific strains only	Bacterial genes for adhesion, specialized protein transport, cell aggregation, environmental sensing, and cell lysis are involved in biofilm formation	(Walter et al. 2011; Frese et al. 2013)
Chicken crop	*Lact. brevis*	Poultry probiotic strain	Biofilms occur in specific niche (crop)	(Lebeer et al. 2011)
Human vagina	*Lactobacillus* sp.	Biofilm formation is mediated by bacterial surface proteins	B; inhibitory activity of lactobacilli against *Candida albicans* involved in vaginitis	(Strus et al. 2005; Terraf et al. 2012; Malik et al. 2013)
Honeybee crop	*Lactobacillus* sp.	The dominant LAB biofilm member is *Lact. kunkeei*	B; LAB biofilms inhibit many bacteria encountered on flowers and two major honeybee pathogens	(Vasquez et al. 2012)
Other				
Built human environments	*Lactococcus* sp. *Lactobacillus* sp.	Biofilms in dust, airborne particles, domestic settings	Biofilm composition in spaces highly occupied by humans reflects human microbiota	(Kembel et al. 2014; Meadow et al. 2014)
Fuel ethanol plant	*Lactococcus* sp. *Lactobacillus* sp. *Leuconostoc* sp. *Pediococcus* sp.	Dominance of lactobacilli in biofilms	D; create a drain on carbon sources, compete for growth factors with yeast, and produce yeast-inhibiting organic acids	(Skinner-Nemec et al. 2007)

*For convenience, the unconventional initials *Lact.* will be used throughout the chapter to designate the *Lactobacillus* genus;

**if relevant; B: beneficial; D: detrimental

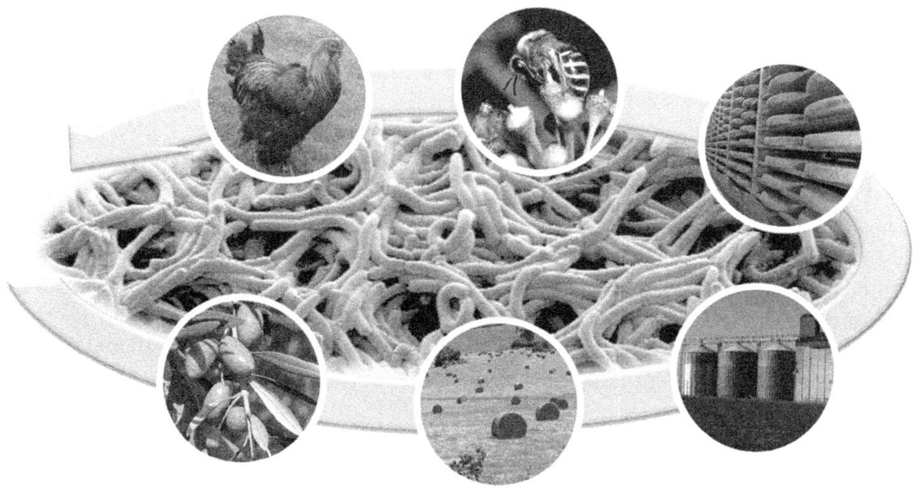

Figure 20.1. The ubiquity of LAB biofilms in nature including plants, silage, cheese manufacturing environment, and mucosae of man and animals. Images embedded in the arrow are from Jean Weber ©INRA, Alain Fraval ©INRA, Jean-Marie Bossennec ©INRA, Christian Slagmulder ©INRA, Serge Carre ©INRA, and Sophie Normant ©INRA and were obtained through the INRA media library (http://mediatheque.inra.fr/). The central image showing a lactobacilli biofilm is from Thierry Meylheuc and Alexis Canette (MIMA2 platform, http://www6.jouy.inra.fr/mima2_eng/).

et al. 2007), sake, and beer (Suzuki et al. 2008) because of their production of unwanted flavors and slime (Ray and Bhunia 2008). In some cases, such as wine, the problem is tricky as the LAB biota is beneficial up to a certain population level and is detrimental above it, thus requiring a subtle control of the LAB population (Lonvaud-Funel 1999; Bartowsky 2009). Aside from their possible spoiling of some food products, LAB biofilms may also contribute to equipment biofouling, and subsequent corrosion and reduction in heat transfer (Faille and Carpentier 2009) (Table 20.1). In Cheddar cheese, some non-starter LAB (NSLAB) such as *Lact. curvatus* biofilms produce a racemase converting *L* (+) -lactic acid to the less soluble *D* (-)-lactic acid yielding calcium lactate crystals (Somers et al. 2001; Agarwal et al. 2006). These are harmless but the consumers mistake the white haze for mold growth and reject such cheeses. In sugar refineries, *Leuconostoc* sp. are a big threat as they produce glucansucrases converting sucrose into polymeric glucans. Apart from causing significant losses of sucrose, insoluble glucans contribute to clogging of filters, pipelines, and tanks, while soluble glucans modify the viscosity of sucrose syrups and retard the rate of sucrose crystallization (Naessens et al. 2005). Also, glucans, which are important components of the biofilm matrix of leuconostocs themselves but also of other bacteria, may boost biofilm formation on equipment (Flemming and Wingender 2010; Leathers and Bischoff 2011). Lastly, LAB formation of mixed species biofilms of both pathogens and LAB can raise food safety problems because of their higher resistance than single species biofilms to disinfection (van der Veen and Abee 2011).

Altogether, these studies show that, albeit much less characterized yet than pathogen microbiota, LAB biofilms occur in the food chain in which they may have dual beneficial and detrimental properties. Similarly to pathogen biofilms, LAB biofilms show increased resistance to environmental stresses than their planktonic counterparts. This applies to resistance to different inhibitors and preservatives such as organic acids (namely, acetic acid, citric acid, lactic acid, and malic acid) and ethanol used in the food industry (Kubota et al. 2008, 2009). This also applies to resistance to sanitizers used for equipment cleaning in the food industry (Somers et al. 2001). Given that LAB may face numerous hostile environmental conditions in nature, in food processing, and in the gut of man and animals, such high resistance to stresses is either a chance in the case of beneficial biofilms or a challenge in the case of detrimental biofilms. Therefore, the need to control their formation in different environments requires unraveling the factors governing biofilm formation in LAB.

20.2. Biofilm Life Cycle and Bacterial Factors Involved in LAB Biofilm Lifestyle

Microbial biofilms have been compared to human cities in which individuals choose a city, select the neighborhood that best suits their needs, set up their homes amongst others, and occasionally leave when life conditions degrade (Watnick and Kolter 2000). This anthropomorphic view of biofilms refers to planktonic bacterial cells that settle in a niche, adhere to a surface, produce an extracellular matrix yielding a growing 3D-mature biofilm that can be regarded as a bacterial building (Flemming 2011), and occasionally leave the biofilm to settle a new community in another more favorable niche (Figure 20.2). Multispecies biofilms, which are the prevailing form of biofilms in nature, are governed by societal features: a biofilm contains heterogeneous microenvironments with different oxygen concentrations, pH-values, redox potential, and so on (Stewart and Franklin 2008). This leads to biofilm heterogeneity and to a functionally structured system in which bacterial groups distribute themselves according to their preferred particular microenvironment and to symbiotic relationships between them (Watnick and Kolter 2000; Flemming 2011). This seemingly societal life of bacteria within biofilms is influenced by exogeneous environmental factors such as the physicochemical properties of the (a)biotic surface, bacterial nutrition supply, and prevailing hydrodynamic conditions. To organize within biofilms, bacteria utilize intercellular communication through a process termed "quorum sensing" (QS), allowing them to sense the local environmental conditions and to adapt through the regulation of specific genes. QS utilizes a chemical language consisting of small self-produced molecules whose accumulation enable bacteria to sense when the minimal number of cells or "quorum" is reached for a concerted response to be initiated. This mode of communication through small diffusible molecules is particularly well suited to bacteria in a diffusion-limited environment such as a biofilm, and has indeed been reported to impact and shape biofilm formation in bacteria (Parsek and Greenberg 2005; Di Cagno et al. 2011). In LAB, two types of signaling molecules, that is small post-translationally modified peptides (here referred to as AIP for auto-inducing peptides) and LuxS derived autoinducer-2 (AI-2), have been shown to impact biofilm production. An AIP system designated Lam for *Lactobacillus agr*-like modules modulates adherence and biofilm formation in *Lact. plantarum* WCFS1 (Sturme et al. 2005). The involved AIP is a cyclic pentapeptide showing a ring structure similar to those of AIP of the staphylococcal *agr* system (Ji et al. 1995). It is not yet known which genes are involved in the impaired biofilm formation ability of a *lam* deletion

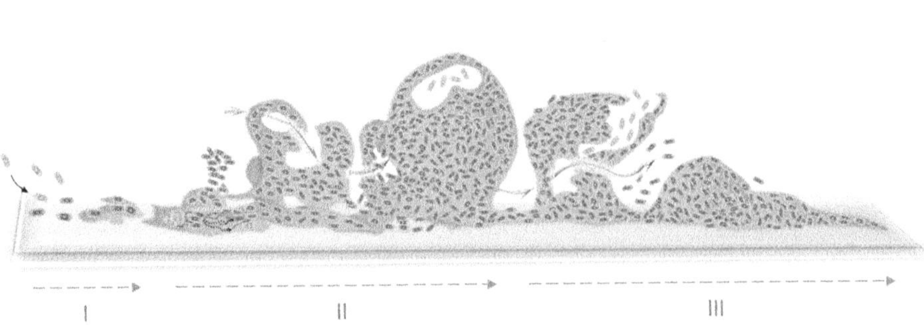

Figure 20.2. Biofilm lifecycle schemed as three main steps: (I) bacterial adhesion, (II) biofilm maturation through matrix biosynthesis and cell proliferation, and (III) biofilm decline and cell dispersal. Arrows within the biofilm matrix represent water channels.

mutant, although several genes linked to surface polysaccharides biosynthesis have been shown to be targets of Lam and might be involved. A similar *fsr* QS system involved in biofilm formation is present in *E. faecalis* (Nakayama et al. 2002; Hancock and Perego 2004). The other AI-2 QS system has been described in *Lact. rhamnosus* GG and is involved in both adherence and biofilm formation as shown in a *luxS* deletion mutant impaired in these traits (Lebeer et al. 2007, 2008a). However, this deletion results in pleiotropic effects and the AI-2 targets involved in biofilm formation are not known yet. Aside from these regulators that appear pivotal in the biofilm lifestyle of LAB, other bacterial factors involved in the formation of LAB biofilm are described below.

Pili are bacterial appendages involved in biofilm formation in both Gram (−) and Gram (+) organisms (Klausen et al. 2003; Mandlik et al. 2008). Their recent discovery in LAB (Nallapareddy et al. 2006; Kankainen et al. 2009; Oxaran et al. 2012) has revealed their major contribution to the first steps of LAB biofilm formation (Table 20.2). Typically, pili are heteropolymers of three pilins, one of which is the major pilin subunit and forms the pilus shaft, while the other ancillary pilin subunits are basically located at the base and at the tip of the pilus, respectively (Hendrickx et al. 2011). Pilin subunits are covalently linked by dedicated transpeptidases termed sortases C. LAB pili have been shown to drive both adhesion (Kankainen et al. 2009; von Ossowski et al. 2010; Lebeer et al. 2012; Le et al. 2013; Tripathi et al. 2013) and aggregation (Oxaran et al. 2012) of bacterial cells, two features considered prerequisite steps in biofilm formation. In the model *L. lactis* IL1403 strain, pili are not produced under standard culture conditions, although this bacterium harbors a pilus operon (Figure 20.3). Overexpression of this operon in *L. lactis* IL1403 yielded piliated cells showing an autoaggregation phenotype and the development of thicker and more aerial biofilms than those of non-piliated counterparts (Oxaran et al. 2012). Partial molecular dissection revealed that the autoaggregation phenotype was mediated by the surface display of oligomers of the shaft pilin while the peculiar biofilm structure was driven by the presence of oligomers of the shaft pilin. Microscopic observations of piliated lactococci revealed physical associations between pili that might be involved in the observed bacterial aggregates. Such homophilic interactions between pili have been recently described by Atomic Force Microscopy (AFM) analyses in another piliated LAB, *Lact. rhamnosus* GG (Tripathi et al. 2013). In this strain as well as in *Ent. faecalis* (Lebeer et al. 2012; Sillanpaa et al. 2013), the molecular determinants associated with the biofilm phenotype are borne by the tip pilin. Adhesion of piliated LAB has been depicted on both abiotic and biotic surfaces. Pili were shown to be associated with the adhesion of a natural *L. lactis* isolate to gastric mucin (Le et al. 2013) and with that of *Lact. rhamnosus* GG to collagen (a protein of the extracellular matrix) (von Ossowski et al. 2010; Lebeer et al. 2012; Tripathi et al. 2013). This is in agreement with the presence of functional binding domains, such as the type A domain of von Willebrand factor and collagen-binding domains in some pilins (Tripathi et al. 2013). Collectively, these observations indicate that pili are important factors involved in the colonization of biotic and abiotic surfaces. They should be taken into account in strategies aiming to either eradicate LAB biofilms or favor these communities. In anti-LAB biofilm strategies, compounds that prevent pilus polymerization through sortase C inhibition (Maresso and Schneewind 2008) or that decrease the adhesive properties of pili (Pinkston et al. 2014) could be used. Conversely, in both probiotics and protecting biofilm strategies in which LAB should colonize biotic and abiotic surfaces, respectively, strains endowed with pilus biogenesis machinery should be preferably selected (Kankainen et al. 2009).

Other cell surface factors are involved in adhesion and subsequent biofilm formation in LAB. In *Ent. faecalis*, the housekeeping sortase (sortase A, SrtA) is involved in the initial adhesion of the bacterium and in subsequent biofilm formation (Guiton et al. 2009). While moderate under static growth conditions, the biofilm formation defect of the *srtA* mutant was drastic under hydrodynamic conditions. Housekeeping sortases A anchor the majority of bacterial proteins termed sortase-dependent proteins (SDPs) as well as nascent pili to the bacterial cell wall. SDPs harbor a cell wall sorting signal at their *C*-terminus that comprises a pentapeptide signature (LPxTG) followed by a

Table 20.2. Bacterial factors involved in different steps of biofilm formation in LAB.

Bacterial factors	LAB strain(s)	Relevant features	References
Initial steps of biofilm formation (adhesion, aggregation)			
Pili	*L. lactis*, *Lact. rhamnosus* GG, *Ent. faecalis* OG1RF	Involved in both adhesion and aggregation; the pilins involved are the cap and/or the shaft pilins	(Lebeer et al. 2012; Oxaran et al. 2012; Sillanpaa et al. 2013; Pinkston et al. 2014)
Sortase A (SrtA)	*Ent. faecalis* OG1RF	SrtA involved in initial bacterial adhesion and biofilm formation	(Guiton et al. 2009)
	Lact. plantarum CMPG5300	SrtA involved in autoaggregation of *Lact. plantarum* and adhesion on vaginal epithelial cells	(Malik et al. 2013)
Sortase-dependent protein (SDP)	*Lact. rhamnosus* GG	Protein MabA involved in adhesion to murine GIT and biofilm formation	(Velez et al. 2010)
Mannose-specific adhesion (Msa)	*Lact. plantarum*	Drives co-aggregation with yeasts and mixed-species biofilm formation	(Pretzer et al. 2005; Furukawa et al. 2012)
D- alanylation of teichoic acids	*Ent. faecalis* 12030	Involved in adhesion to epithelial cells and in biofilm formation	(Fabretti et al. 2006)
	Lact. reuteri 100-23	Not involved in initial adhesion; associated with *in vivo* biofilm formation; involved in resistance against both low pH and defensins	(Walter et al. 2007)
Capsular polysaccharides	*Lact. plantarum* WCFS1	Adversely affect adhesion and biofilm formation	(Sturme et al. 2005)
Glycosystransferase	*Lact. rhamnosus* GG	Idem; the produced polysaccharides would shield adhesins such as pili	(Lebeer et al. 2009)
Autolysins	*Ent. faecalis* OG1RF, *L. lactis* MG1363	Involved in initial adhesion	(Mercier et al. 2002; Mohamed et al. 2004; Guiton et al. 2009)
Biofilm maturation			
glucosystransferase and inulosucrase	*Lact. reuteri* TMW1.106	Glucans and fructans associated with cell aggregation, biofilm formation, and murine GIT colonization	(Walter et al. 2008)
Enterococcal polysaccharide antigen (Epa)	*Ent. faecalis* OG1RF	Associated with primary adhesion and biofilm formation	(Mohamed et al. 2004; Teng et al. 2009)
Glycosystransferases BgsA and BgsB	*Ent. faecalis* 12030	Not involved in initial adhesion; associated with biofilm formation	(Theilacker et al. 2009; Theilacker et al. 2011)
Autolysins	*Ent. faecalis* OG1RF	Released eDNA involved in biofilm maturation	(Mohamed et al. 2004; Guiton et al. 2009)
	L. lactis MG1363	Involved in the 3D-structure of biofilms	(Habimana et al. 2009)
Biosurfactants	*L. lactis* 53, *Lact. fermentum* ATCC9338, *Lact. paracasei*, *Strep. thermophilus* A	Anti-adhesive and inhibition properties against some pathogens	(Rodrigues et al. 2004; Rodrigues et al. 2006a; b; Gudina et al. 2010; Tahmourespour et al. 2011)

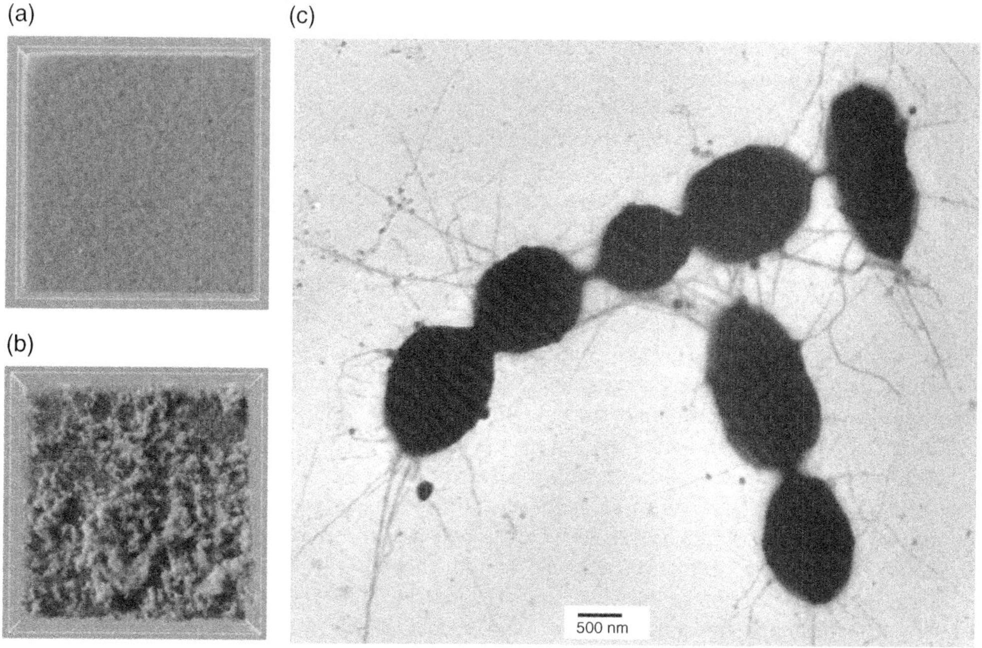

Figure 20.3. The pili of the lactic acid bacterium *Lactococcus lactis* affect the architecture of biofilms. (a) Non-piliated *L. lactis* biofilm and (b) piliated *L. lactis* biofilm obtained from confocal image series. (c) *L. lactis* pili observed using negative staining and transmission electron microscopy. Images (a) and (b) are from Julien Deschamps and image (c) from Alexis Cannette (http://www6.jouy.inra.fr/mima2_eng/).

hydrophobic domain and a cationic tail. In *Ent. faecalis* pangenome, about 40 SDPs are predicted some of which harbor putative functional adhesive domains (http://www.cmbi.ru.nl/locatep-db/cgi-bin/locatepdb.py). In a *Lact. plantarum* human vaginal isolate, SrtA has also been shown to be involved in autoaggregation, adhesion to vaginal epithelial cells, and biofilm formation (Malik et al. 2013). In *Lact. rhamnosus* GG, a protein termed MabA (for *modulator of adhesion and biofilm*) harboring an LPxTG signature was involved in adhesion to murine GIT and in biofilm formation (Velez et al. 2010). A derivative in which the corresponding gene was deleted showed a 2-fold reduction in adhesion to CaCo-2 cells and in biofilm formation. This is compared to the *Lact. rhamnosus* pilin mutant that almost completely lost both adhesive properties and biofilm formation abilities (Lebeer et al. 2012). This suggests that MabA could act as a modulator of both adhesion and biofilm formation and play a role in the later step of the adhesion processes, that is the pili would initiate long-distance contacts and MabA would make short-distance contacts. MabA is ca. 9-fold less expressed in the *Lact. rhamnosus* LC705 strain than in the GG strain and this, in addition to the absence of pili production in the former strain, explains the low adhesive properties of *Lact. rhamnosus* LC705 (Savijoki et al. 2011). Interestingly, when biofilms of *Lact. rhamnosus* GG and LC705 were treated with a broad-spectrum proteinase, a complete disintegration of the attached cells was observed, suggesting that proteins such as MabA and pili (in the case of the GG strain) are key players in biofilm formation in *Lact. rhamnosus* (Savijoki et al. 2011). Mucus-binding proteins (MBP) are also sortase A substrates that are important factors involved in host colonization of LAB (van Pijkeren et al. 2006; Lebeer et al. 2008b; Meyrand et al. 2013; Jensen et al. 2014). Although little is known yet about their involvement in the *in vivo* biofilm development in man and animal mucosae, a recent metatranscriptomic approach comparing transcript levels in murine forestomach biofilms and in the hindgut lumen microbiota showed that MBP corresponding transcripts were overproduced in forestomach biofilms (Schwab et al. 2014). Altogether, these data suggest a pivotal role of SDPs, most of which have unknown functions, in mechanisms leading to biofilm formation in the different niches in which LAB may live.

LAB may coexist with yeasts in a number of traditional fermented foods such as vinegar, sake, olives, and cheese. This may lead to the development of mixed-species biofilms (Furukawa et al. 2010) whose formation relies on a prerequisite co-aggregation step (Furukawa et al. 2011). A molecular basis of such a LAB–yeast co-aggregation event has been studied in *Lact. plantarum.* Both biochemical and molecular biology approaches revealed that the interaction is based on mannose specific adhesins that target mannan, a cell wall mannose polymer in yeasts (Pretzer et al. 2005; Hirayama et al. 2012) (Table 20.2). One of these adhesins has been shown to be an LPxTG substrate of *Lact. plantarum* sortase A. As *Lact. plantarum* is a versatile LAB that is also encountered in the human GIT, mannose specific adhesins could also be relevant in the first step of LAB biofilm formation within the GIT as suggested by previous studies in various Gram (−) bacteria, showing that these adhesins mediate colonization of the gut (Adlerberth et al. 1996).

D-Alanylation of teichoic acids (TA) and lipoteichoic acids (LTA) partially shapes the bacterial global charge as *D*-alanine esterification brings positive charge to the anionic TA polymers. As such, the *dlt* operon involved in *D*-alanylation has been shown to impact adhesion and/or biofilm formation in LAB by modulating the net negative charge of bacterial surface (Table 20.2). In *E. faecalis* 12030, both adherence to epithelial cells and biofilm formation of a *dltA* deletion mutant were reduced compared to those of the wild-type strain (Fabretti et al. 2006). In *Lact. reuteri 100-23*, similar adhesion to the murine forestomach epithelium was detected in a *dltA* mutant and in the wild-type strain, although the mutant failed to form biofilms *in vivo* on the forestomach epithelium (Walter et al. 2007). In a metatranscriptomic approach of murine forestomach biofilms, *dlt* transcripts were 368-fold more abundant in forestomach biofilms than in the hindgut lumen microbiota (Schwab et al. 2014). In contrast, in *Lact. rhamnosus* GG neither adhesion to human intestinal epithelial cells nor biofilm formation was affected in a *dlt* mutant (Perea Velez et al. 2007). Noteworthy, the latter is piliated while the former is not and probably, as observed above with MabA, pili might be the major bacterial component involved in both adhesion and biofilm formation traits.

Once bacteria have adhered onto a surface, the following step leading to definite attachment and to biofilm formation is mediated by the production of extracellular polymers (polysaccharides, proteins, and DNA) that are the bricks of the building bacterial house (Stoodley et al. 2002). Polysaccharides are major players in biofilm development as they are involved in many processes including adhesion, aggregation, biofilm cohesion, retention of water, and protection against biocides and host defense upon infection (Flemming 2011). Polysaccharides are produced by many LAB (De Vuyst and Degeest 1999). Three types, that is capsular, cell wall, and excreted polysaccharides have been distinguished (Delcour et al. 1999) and seem to exhibit different effects in biofilm formation (Table 20.2). Although most research on LAB polysaccharides has focused on their texturing properties in food, recent evidence showed that they may, as expected, impact biofilm formation in LAB. Capsular polysaccharides may adversely affect biofilm formation, probably through a reduced bacterial adherence. This has been observed in *Lact. plantarum* WCFS1 in which increased expression of the genes for capsular polysaccharides is associated with decreased adherence to a glass surface and reduced biofilm formation (Sturme et al. 2005; Remus et al. 2012). It is assumed that these polysaccharides could shield bacterial adhesion factors involved in the adhesion primary step required for biofilm formation, as was previously observed in *Staphylococcus aureus* (Pohlmann-Dietze et al. 2000). The same conclusion was made in *Lact. rhamnosus* GG in which adhesion to mucus and CaCo-2 cells, and biofilm formation were increased in a deletion mutant of a gene encoding a priming glycosystransferase (Lebeer et al. 2009). This raised the hypothesis that some polysaccharides could shield pili that have been shown to be essential in the adhesive and biofilm properties of this strain. In contrast, glucans and to a lesser extent fructans produced in a *Lact. reuteri* strain by a glucosystransferase (GftA) and an inulosucrase (Inu), respectively, mediate aggregation and subsequent biofilm formation (Walter et al. 2008). A *gftA* mutant strain showed delayed colonization of the murine gut although final colonization of the epithelium was similar to that of the wild-type strain. In their metatranscriptomic approach of murine forestomach biofilms, Schwab et al. (Schwab et al. 2014) found that transcripts for lactobacilli-like fructan- and glucansucrase were ca. 50-fold more abundant in forestomach biofilms than in the hindgut lumen microbiota. Altogether, this highlights the role of exopolysaccharides in biofilm formation and in the ecological performance of some lactobacilli, although it is not clear whether they intervene at the step of surface colonization or as a component of the biofilm matrix. In *Ent. faecalis* OG1RF, an 18–gene cluster termed *epa* for *e*nterococcal *p*olysaccharide *a*ntigen encodes enzymes and transporters involved in polysaccharide biosynthesis and is involved in primary bacterial adhesion and biofilm formation (Mohamed et al. 2004; Teng et al. 2009). Two other glycosystransferases, BgsA and B (for *b*iofilm-associated *g*lycolipid *s*ynthesis), driving synthesis of di- and mono-glucosysdiacylglycerol, respectively, are involved in biofilm formation (Theilacker et al. 2009; Theilacker et al. 2011). Both *bgsA* and *bgsB* deletion mutants showed impaired biofilm formation although adhesion level on polystyrene plates was unaffected. This suggests that, in contrast to other glycosystransferases, the corresponding BgsA and BgsB enzymes are not involved in the initial surface attachment of bacteria and are instead associated with the accumulation of bacteria in the growing biofilm. Interestingly, LTA were overproduced in those mutants thus conferring a higher negative charge to the cell envelope that could impair bacterial clustering within the biofilm through electrostatic repulsion. Also, some of these LTA were released into the culture supernatant and could act as biosurfactants promoting release of bacterial cells from the biofilm.

Extracellular DNA (eDNA) is an important component of the biofilm matrix and mostly results from bacterial autolysis driven by autolysins (Allesen-Holm et al. 2006; Habimana et al. 2009) some of which have been identified in LAB (Layec et al. 2008). In *Ent. faecalis* OG1RF, autolysin mutant strains showed impaired biofilm formation (Mohamed et al. 2004; Guiton et al. 2009) through two mechanisms. Primary bacterial adhesion appeared as autolysin-dependent and DNA-independent. This was shown through the study of autolysin mutants whose adhesion was reduced compared to wild-type strain. This could not be attributed to the low level of released DNA as the wild-type strain treated with DNAse I was not affected in primary adhesion. It was proposed that the longer chaining of bacteria in autolysin mutants led to reduced binding sites and a lower cell deposition rate (Guiton et al. 2009). In contrast, biofilm maturation is DNA dependent. Under both static and hydrodynamic conditions, DNA appeared as a major biofilm-cementing component as both DNAse-treated biofilms and biofilms of the autolysin mutant failed to form 3D structures observed with native biofilms of wild-type strains (Guiton et al. 2009). In another LAB strain, *L. lactis* MG1363, an autolysin mutant strain, also adhered poorly to a glass surface (Mercier et al. 2002) while it produced, under static conditions, biofilms showing a porous filamentous 3D structure that was attributed to both the chaining of the mutant and the low DNA content present in these biofilms (Habimana et al. 2009).

Biosurfactants are amphiphilic compounds, mostly lipopeptides and glycopeptides, produced by microorganisms with surface and emulsifying activities (Van Hamme et al. 2006). As such, they are involved in the dynamics of biofilms at different levels, for example microbial adhesion and desorption, swarming of bacteria, creation of channels within biofilms allowing nutrient supply in the lower biofilm layers. Biosurfactants are also key factors in the bacterial detachment from biofilms allowing bacteria to return to planktonic life and to settle a new community in another microenvironment (Boles et al. 2005). Biosurfactants can be produced in LAB and their partial characterization showed both anti-adhesion and inhibitory activities towards pathogens such as *Escherichia coli*, *S. epidermidis*, *S. aureus*, and yeasts such as *Candida* sp. (Rodrigues et al. 2004, 2006a, b; Gudina et al. 2010). Such biosurfactants from LAB are likely to modulate biofilm formation. *Lact. fermentum* ATCC9338 produced a biosurfactant that negatively impacted biofilm formation in *Strep. mutans* through the reduction in the expression of glucosystransferase encoding genes (Tahmourespour et al. 2011). These enzymes are involved in the production of extracellular glucans that are essential in tooth surface colonization and biofilm formation leading to dental caries. This suggests that LAB biosurfactants might potentialize LAB biofilms as surface protective biofilms against pathogens. However, molecular characterization of these compounds is lacking and should be studied in order to dissect their production, mode of action, and activity in LAB biofilms as well as in unwanted biofilms.

20.3. Health and Biotechnological Potential of LAB Biofilms and Underlying Mechanisms

Although many LAB are considered probiotics, little work has focused on their activities in relation to their biofilm *in vivo* lifestyle in humans and animals. The human vagina is inhabited by LAB biofilms, mostly formed by *Lact. iners* and *Lact. crispatus*, representing 90% of the microbiota (Shipitsyna et al. 2013). This lactobacilli biota is believed to protect the urogenital tract from pathogen invasion through the production of antimicrobials, host defense stimulation, and the barrier effect (Malik et al. 2013). Indeed, there is high prevalence of vaginal dysbiosis worldwide leading to bacterial or yeast vaginosis associated with a shift of the lactobacilli population down to 7% of the total microbiota (Liu et al. 2013; Shipitsyna et al. 2013). This depletion of LAB biota on vaginal mucosa along with the concomitant increase in the population of vaginal pathogens suggested that probiotic administration through oral or vaginal route could offer an alternative sustainable treatment or an additive treatment to the use of antimicrobials (Reid 2001). Several works down this line reported on the selection of LAB with relevant *in vitro* phenotypes for inhibition of vaginal pathogens, including autoaggregation, adhesion, biofilm formation ability, and hydrogen peroxide and bacteriocin production (Domingue et al. 1991; Strus et al. 2005; Terraf et al. 2012; Malik et al. 2013). However, *in vivo* studies in man and animal yielded varied results (Zhou et al. 2009; Menard 2011; Vicariotto et al. 2012; Vujic et al. 2013; Daniele et al. 2014; Homayouni et al. 2014). *Gardnerella vaginalis*, the major pathogen involved in bacterial vaginosis, is a highly competitive biofilm former, able to both adhere on the lactobacilli-colonized vaginal epithelium and displace part of the lactobacilli microbiota (Machado et al. 2013). Therefore, vaginosis dysbiosis results from the competition among several biofilm formers (Saunders et al. 2007; McMillan et al. 2011). As *G. vaginalis* biofilms showed enhanced resistance to the inhibitory metabolites produced by lactobacilli compared to their planktonic counterparts (Patterson et al. 2007), attempts at eradication through bactericidal action may prove to be unsuccessful, while strategies focusing on the inhibition of *G. vaginalis* adhesion and biofilm displacement might be more promising (Spurbeck and Arvidson 2011).

LAB have been proposed for their promising immunomodulating properties in the treatment of a number of GIT diseases including enteric infections and inflammatory bowel diseases (IBD) (Lebeer et al. 2008b; Macho Fernandez et al. 2011). Lactobacilli strains may be either immunosuppressive or immunostimulatory and both have potential in therapeutic issues. Immunosuppressive strains reduced inflammation in an induced murine model of IBD (Pena et al. 2005) while stimulation of the mucosal immune system may protect against enteric infections (Jones and Versalovic 2009). In this field of immunomodulation, recent studies revealed that biofilm lactobacilli were more efficient than their planktonic counterparts in the reduction of the pro-inflammatory cytokine TNFα produced by LPS-activated human monocytic cells (Rieu et al. 2014). Attempts to characterize the effectors produced by these lactobacilli biofilms showed that the chaperonin GroEL was involved in the observed anti-inflammatory effect. These observations suggest that future research into the *in vivo* functions of LAB should consider LAB as putative biofilm formers fitted with functional properties that may be different to those of planktonic cells. In future attempts to explore the potential of LAB biofilms in probiotic issues, adequate human LAB strains should be selected as there is increasing evidence for host-specificity of LAB, indigeneous strains outperforming exogeneous strains when competing *in vivo* (Denou et al. 2008; Walter et al. 2011; Buhnik-Rosenblau et al. 2012; Tannock et al. 2012).

LAB biofilms also have potential in food/feed safety issues. With the final objective of developing protective biofilms against the development of *Listeria monocytogenes* in food processing facilities, Zhao et al. (Zhao et al. 2004) successfully isolated *Listeria*-inhibiting LAB strains from floor drains of poultry processing plants with a recent history of no detectable *L. monocytogenes*. Some of these LAB strains were highly effective in competition-exclusion assays in lowering the population of *L. monocytogenes* biofilms below detection levels. A first field experiment consisting in repeated applications of these strains to *Listeria*-contaminated floor drains showed a drastic decrease (ca. 4 $\log_{10}$ CFU/100 cm^2) in the pathogen population (Zhao et al. 2006). In a second field experiment, a combination of two biofilm formers (*L. lactis* C-1-92 and *Ent. durans* 152) showing high inhibition against *L. monocytogenes* was repeatedly applied (four times during the first week of experiment) to *Listeria*-contaminated floor drains. This led to the elimination of the *Listeria* biofilm population below detection levels and prevented recontamination of floor drains by *Listeria* for 13 weeks following the application (Zhao et al. 2013).

As mentioned before, traditional cheese-making technologies used for Ragusano in Italy and PDO salers in France employ wooden vats for raw milk fermentation. The biofilms that rapidly developed on wood showed a predominance of thermophilic LAB and, notably, neither pathogen nor spoilage organisms were present, highlighting the possible barrier effect of LAB biofilms (Licitra et al. 2007; Lortal et al. 2009; Didienne et al. 2012). The same type of protecting biofilms were described on the wooden shelves used in the ripening of the French Reblochon cheese (Mariani et al. 2007). Deposited populations of *Listeria* decreased during cheese ripening on shelves harboring natural biofilms while the same experiment performed on shelves harboring dead biofilms showed a sharp increase in *Listeria* population upon cheese ripening. In another study, Woo and Ahn (2013) showed the potential of several probiotic LAB strains to outcompete, exclude, or displace biofilms of *Listeria* and *Salmonella* Typhymurium with variable reduction levels of pathogen biofilms according to the probiotic strain.

Several mechanisms are involved in the protective effect of LAB biofilms. One is linked to the competition for nutrients and is termed the *Jameson effect*. It has been reported in the inhibition of *Listeria* by natural biofilm microbiota from cheese ripening wooden shelves (Guillier et al. 2008). The *L. monocytogenes* growth rate in co-culture with the biofilm was not reduced compared to *Listeria* monocultures, but the pathogen growth stopped as soon as the biofilm bacteria reached the stationary phase leading to a reduction in maximum population density of the pathogen. This was explained by nutrient consumption and exhaustion by the biofilm bacteria. Bacteriocin production (see Chapter 5) is another mechanism involved in protection provided by LAB biofilms. Among the highly protecting strains isolated above from floor drains, one *L. lactis* isolate appeared as a nisin producer strain (Piard and Desmazeaud 1992; Zhao et al. 2004). Such nisin-producing LAB biofilms are indeed able to reduce *Listeria* biofilms below detection limit in simultaneous adhesion of *Listeria* and *L. lactis* as well as in deferred adhesion of the pathogen over a one-day *L. lactis* biofilm (Leriche et al. 1999). Another LAB bacteriocin, sakacin1, produced by *Lact. sakei* 1, is active against *L. monocytogenes* and biofilms of the producer strain have been shown to decrease the *Listeria* biofilm population on stainless steel coupons (Winkelströter et al. 2011). A third mechanism that may intervene in the protection provided by LAB biofilms relates to spatial competition. This has been shown in the eradication of biofilms of *L. monocytogenes* LO-28 and EGDe strains by *L. lactis* MG1363 (Habimana et al. 2011). The dynamics of dual-*Listeria/L. lactis* continuous-flow biofilms monitored using confocal laser scanning microscopy revealed that *Listeria* cells were confined to the biofilm base in contact with the support and were covered by *L. lactis* cells. Modeling the development of this dual-species biofilm using individual-based modeling (IBM), it was proposed that the spatial competition occurring during the initial phase of biofilm formation was governed by the growth parameters of the two strains. *L. lactis* showing shorter latency and generation time than *L. monocytogenes* outcompetes for occupying the surface layer of the biofilm. The biofilm further develops with an advantage in favor of *L. lactis* that has primary access to nutrient supply, while *Listeria* cells, embedded in the bottom layers of the biofilm, are deprived of nutrients and are maintained in a survival lifestyle.

The high cell density in LAB biofilms, the innocuousness of LAB, and the natural stress resistance of LAB biofilms have boosted attempts to exploit these bacterial communities in biotechnological issues. In lactic acid production, packed bed biofilm reactors (PBBR) of *Lact. rhamnosus* RS93 produced more acid than reactors harboring chemically immobilized bacteria (Dagher et al. 2010). In another system consisting of a *Lact. delbrueckii* NCIM 2365 PBBR coupled to a tank bioreactor, the biofilm-released cells from the PBBR inoculated the bioreactor and sugar fermentation occurred in the dual reactor by both planktonic and biofilm bacteria (Rangaswamy and Ramakrishna 2008). This allowed higher lactic acid productivity than in a single bioreactor, while it showed a long-term sustainability and robustness, allowing continuous fermentation for 1000 h without the need for novel inoculum. This suggests that LAB biofilms are alternative and competitive cell immobilization systems in biotechnological processes.

The ability of *Lact. plantarum* strains to agglutinate yeasts and to form mixed-species biofilms has been assessed for its potential in ethanol production. Mixed *S. cerevisiae–Lact. plantarum* HM23 biofilms formed on cellulose beads allowed good ethanol production levels and high robustness during 10 successive batch cultures of 24 h each. Comparison between mixed-species biofilms and *S. cerevisiae* single biofilms showed that the former showed an increased resistance to washing and an enhanced protection against bacterial contaminants, probably because of lactate production by *Lact. plantarum* that maintained the pH low (Abe et al. 2013).

Lactic acid bacteria have been proposed as cell factories in a number of applications targeting either *in vitro* production and subsequent purification of biomolecules, or *in vivo* delivery of biomolecules in hosts by the LAB itself (Hugenholtz et al. 2002; LeBlanc et al. 2005; Morello et al. 2008). The bacteriocin nisin, whose industrial use as additive has been approved in a number of countries worldwide (Cleveland et al. 2001), is produced by some *L. lactis* strains. Attempts to boost nisin production using *L. lactis* biofilm reactors allowed both a ca. 2-fold increase in nisin production compared to the level obtained in a suspended-cell reactor and a drastic reduction in the lag phase of nisin production (Pongtharangkul and Demirci 2006).

20.4. Conclusions

In contrast to biofilms of pathogens, whose disruption and eradication is a tremendous challenge for medical and food microbiologists (Bridier et al. 2014), biofilms of LAB have fantastic potential in a number of issues. One foreseen promising application deals with protective biofilms able to shield biotic and abiotic surfaces from the occurrence of pathogen biofilms. Examples taken from the current state of the art in the field showed that LAB are able to form multispecies biofilms in a number of niches and efficiently keep away pathogens through competition, antagonism, and inhibition mechanisms. This suggests that LAB biofilms are potent candidates in the development of environmentally friendlier methodologies as potential alternatives to the alarming use of biocides and antibiotics in human and animal health as well as in industry worldwide.

A considerable number of bacterial factors involved in biofilm formation or prevention have already been identified or characterized in LAB. This is of utmost interest in order to develop reliable methodologies to exploit LAB biofilms in different applications. It is noteworthy though that most of these characterized factors are involved in the initial steps of biofilm formation, and knowledge is still lacking in bacterial effectors that contribute to biofilm maturation through the building up (and occasionally the deconstruction) of the matrix. The matrix is formed of EPS that comprise mainly exopolysaccharides, proteins, eDNA, and their degradation products. It is still termed "the dark matter of biofilms" as it shows a complex and variable chemical structure and high heterogeneity in space and time. Exopolysaccharides that are crucial EPS components have been mostly studied in LAB for their texturing properties in food, but their contribution to the biofilm matrix edification in relation to their chemical properties has been little explored. Also EPS-degrading enzymes are important (i) in the breaking down of polymers and in the release of lower molecular-mass products that can be utilized as carbon and energy sources by the biofilm population and (ii) in the detachment and dispersion of cells from the biofilm, which allows new biofilms to be formed elsewhere. Further work down those lines should provide a deeper insight into the biology of LAB biofilms and of the various exogeneous and endogenous factors involved in their dynamics. This is required for a better control of the cohesiveness and dispersion of LAB biofilms in the different environments in which these bacteria are or could be used.

Acknowledgments

We thank Thierry Meylheuc for his precious help in figure preparation and we are grateful to our colleagues from the Micalis B2HM group for helpful discussions.

References

Abe, A., Furukawa, S., Watanabe, S., and Morinaga, Y. (2013) Yeasts and lactic acid bacteria mixed-specie biofilm formation is a promising cell immobilization technology for ethanol fermentation. *Appl Biochem Biotechnol* **171**, 72–79.

Adlerberth, I., Ahrne, S., Johansson, M.L., et al. (1996) A mannose-specific adherence mechanism in *Lactobacillus plantarum* conferring binding to the human colonic cell line HT-29. *Appl Environ Microbiol* **62**, 2244–2251.

Agarwal, S., Sharma, K., Swanson, B.G., et al. (2006) Nonstarter lactic acid bacteria biofilms and calcium lactate crystals in Cheddar cheese. *J Dairy Sci* **89**, 1452–1466.

Allesen-Holm, M., Barken, K.B., Yang, L., et al. (2006) A characterization of DNA release in *Pseudomonas aeruginosa* cultures and biofilms. *Mol Microbiol* **59**, 1114–1128.

Arroyo-Lopez, F.N., Bautista-Gallego, J., Dominguez-Manzano, J., et al. (2012) Formation of lactic acid bacteria-yeasts communities on the olive surface during Spanish-style Manzanilla fermentations. *Food Microbiol* **32**, 295–301.

Arumugam, M., Raes, J., Pelletier, E., et al. (2011) Enterotypes of the human gut microbiome. *Nature* **473**, 174–180.

Bartowsky, E.J. (2009) Bacterial spoilage of wine and approaches to minimize it. *Lett Appl Microbiol* **48**, 149–156.

Bjorkroth, K.J. and Korkeala, H.J. (1997) Use of rRNA gene restriction patterns to evaluate lactic acid bacterium contamination of vacuum-packaged sliced cooked whole-meat product in a meat processing plant. *Appl Environ Microbiol* **63**, 448–453.

Boles, B.R., Thoendel, M., and Singh, P.K. (2004) Self-generated diversity produces "insurance effects" in biofilm communities. *Proc Natl Acad Sci USA* **101**, 16630–16635.

Boles, B.R., Thoendel, M., and Singh, P.K. (2005) Rhamnolipids mediate detachment of *Pseudomonas aeruginosa* from biofilms. *Mol Microbiol* **57**, 1210–1223.

Bollinger, R.R., Barbas, A.S., Bush, E.L., et al. (2007) Biofilms in the normal human large bowel: fact rather than fiction. *Gut* **56**, 1481–1482.

Bridier, A., Briandet, R., Thomas, V., and Dubois-Brissonnet, F. (2011) Resistance of bacterial biofilms to disinfectants: a review. *Biofouling* **27**, 1017–1032.

Bridier, A., Sanchez-Vizuete, P., Guilbaud, M., et al. (2015) Biofilm-associated persistence of food-borne pathogens. *Food Microbiol* **45** 159–300.

Buhnik-Rosenblau, K., Matsko-Efimov, V., Jung, M., et al. (2012) Indication for Co-evolution of *Lactobacillus johnsonii* with its hosts. *BMC Microbiol* **12**, 149.

Chen, Y.S., Liou, M.S., Ji, S.H., et al. (2013) Isolation and characterization of lactic acid bacteria from yan-tsai-shin (fermented broccoli stems), a traditional fermented food in Taiwan. *J Appl Microbiol* **115**, 125–132.

Chen, Y.S., Yanagida, F., and Shinohara, T. (2005) Isolation and identification of lactic acid bacteria from soil using an enrichment procedure. *Lett Appl Microbiol* **40**, 195–200.

Cleveland, J., Montville, T.J., Nes, I.F., and Chikindas, M.L. (2001) Bacteriocins: safe, natural antimicrobials for food preservation. *Int J Food Microbiol* **71**, 1–20.

Costerton, J.W. (2007) *The Biofilm Primer*. Springer-Verlag, Berlin and Heidelberg.

Costerton, J.W., Lewandowski, Z., Caldwell, D.E., et al. (1995) Microbial biofilms. *Annu Rev Microbiol* **49**, 711–745.

Costerton, J.W., Stewart, P.S., and Greenberg, E.P. (1999) Bacterial biofilms: a common cause of persistent infections. *Science* **284**, 1318–1322.

Dagher, S.F., Ragout, A.L., Sineriz, F., and Bruno-Barcena, J.M. (2010) Cell immobilization for production of lactic acid biofilms do it naturally. *Adv Appl Microbiol* **71**, 113–148.

Daniele, M., Pascual, L., and Barberis, L. (2014) Curative effect of the probiotic strain *Lactobacillus fermentum* L23 in a murine model of vaginal infection by *Gardnerella vaginalis*. *Lett Appl Microbiol. Jul;59(1):93–8. doi: 10.1111/lam.12249. Epub 2014 Apr 9*

De Vuyst, L. and Degeest, B. (1999) Heteropolysaccharides from lactic acid bacteria. *FEMS Microbiol Rev* **23**, 153–177.

Delcour, J., Ferain, T., Deghorain, M., et al. (1999) The biosynthesis and functionality of the cell-wall of lactic acid bacteria. *Antonie Van Leeuwenhoek* **76**, 159–184.

Denou, E., Pridmore, R.D., Berger, B., et al. (2008) Identification of genes associated with the long-gut-persistence phenotype of the probiotic *Lactobacillus johnsonii* strain NCC533 using a combination of genomics and transcriptome analysis. *J Bacteriol* **190**, 3161–3168.

Di Cagno, R., De Angelis, M., Calasso, M., and Gobbetti, M. (2011) Proteomics of the bacterial cross-talk by quorum sensing. *J Proteomics* **74**, 19–34.

Didienne, R., Defargues, C., Callon, C., et al. (2012) Characteristics of microbial biofilm on wooden vats ('gerles') in PDO Salers cheese. *Int J Food Microbiol* **156**, 91–101.

Domingue, P.A., Sadhu, K., Costerton, J.W., et al. (1991) The human vagina: normal flora considered as an in situ tissue-associated, adherent biofilm. *Genitourin Med* **67**, 226–231.

Dominguez-Manzano, J., Leon-Romero, A., Olmo-Ruiz, C., et al. (2012) Biofilm formation on abiotic and biotic surfaces during Spanish style green table olive fermentation. *Int J Food Microbiol* **157**, 230–238.

Fabretti, F., Theilacker, C., Baldassarri, L., et al. (2006) Alanine esters of enterococcal lipoteichoic acid play a role in biofilm formation and resistance to antimicrobial peptides. *Infect Immun* **74**, 4164–4171.

Faille, C. and Carpentier, B. (2009) *Biofilms in the Food and Beverage Industries*. Woodhead Publishing Limited, Cambridge, UK.

Fernandez, L., Langa, S., Martin, V., et al. (2013) The human milk microbiota: origin and potential roles in health and disease. *Pharmacol Res* **69**, 1–10.

Flemming, H.C. (2011) The perfect slime. *Colloids Surf B Biointerfaces* **86**, 251–259.

Flemming, H.C. and Wingender, J. (2010) The biofilm matrix. *Nat Rev Microbiol* **8**, 623–633.

Forsgren, E., Olofsson, T., Vásquez, A., and Fries, I. (2010) Novel lactic acid bacteria inhibiting *Paenibacillus* larvae in honey bee larvae. *Apidologie* **41**, 99–108.

Frese, S.A., Mackenzie, D.A., Peterson, D.A., et al. (2013) Molecular characterization of host-specific biofilm formation in a vertebrate gut symbiont. *PLoS Genet* **9**, e1004057.

Fuller, R., Barrow, P.A., and Brooker, B.E. (1978) Bacteria associated with the gastric epithelium of neonatal pigs. *Appl Environ Microbiol* **35**, 582–591.

Fuller, R. and Brooker, B.E. (1974) Lactobacilli which attach to the crop epithelium of the fowl. *Am J Clin Nutr* **27**, 1305–1312.

Furukawa, S., Nojima, N., Nozaka, S., al. (2012) Mutants of *Lactobacillus plantarum* ML11-11 deficient in co-aggregation with yeast exhibited reduced activities of mixed-species biofilm formation. *Biosci Biotechnol Biochem* **76**, 326–330.

Furukawa, S., Nojima, N., Yoshida, K., et al. (2011) The importance of inter-species cell-cell co-aggregation between *Lactobacillus plantarum* ML11-11 and *Saccharomyces cerevisiae* BY4741 in mixed-species biofilm formation. *Biosci Biotechnol Biochem* **75**, 1430–1434.

Furukawa, S., Yoshida, K., Ogihara, H., et al. (2010) Mixed-species biofilm formation by direct cell-cell contact between brewing yeasts and lactic acid bacteria. *Biosci Biotechnol Biochem* **74**, 2316–2319.

Gudiña E.J., Teixeira, J.A. and Rodrigues, L.R. (2010) Isolation and functional characterization of a biosurfactant produced by *Lactobacillus paracasei*. *Colloids Surf B Biointerfaces* **76**, 298–304.

Guillier, L., Stahl, V., Hezard, B., et al. (2008) Modelling the competitive growth between *Listeria monocytogenes* and biofilm microflora of smear cheese wooden shelves. *Int J Food Microbiol* **128**, 51–57.

Guiton, P.S., Hung, C.S., Kline, K.A., et al. (2009) Contribution of autolysin and Sortase A during *Enterococcus faecalis* DNA-dependent biofilm development. *Infect Immun* **77**, 3626–3638.

Gunduz, G.T. and Tuncel, G. (2006) Biofilm formation in an ice cream plant. *Antonie Van Leeuwenhoek* **89**, 329–336.

Habimana, O., Guillier, L., Kulakauskas, S., and Briandet, R. (2011) Spatial competition with *Lactococcus lactis* in mixed-species continuous-flow biofilms inhibits *Listeria monocytogenes* growth. *Biofouling* **27**, 1065–1072.

Habimana, O., Meyrand, M., Meylheuc, T., et al. (2009) Genetic features of resident biofilms determine attachment of *Listeria monocytogenes*. *Appl Environ Microbiol* **75**, 7814–7821.

Hancock, L.E. and Perego, M. (2004) The *Enterococcus faecalis* fsr two-component system controls biofilm development through production of gelatinase. *J Bacteriol* **186**, 5629–5639.

Hendrickx, A.P.A., Budzik, J.M., Oh, S.-Y., and Schneewind, O. (2011) Architects at the bacterial surface - sortases and the assembly of pili with isopeptide bonds. *Nat Rev Microbiol* **9**, 166–176.

Hirayama, S., Furukawa, S., Ogihara, H., and Morinaga, Y. (2012) Yeast mannan structure necessary for co-aggregation with *Lactobacillus plantarum* ML11-11. *Biochem Biophys Res Commun* **419**, 652–655.

Homayouni, A., Bastani, P., Ziyadi, S., et al. (2014) Effects of probiotics on the recurrence of bacterial vaginosis: a review. *J Low Genit Tract Dis* **18**, 79–86.

Hugenholtz, J., Sybesma, W., Groot, M.N., et al. (2002) Metabolic engineering of lactic acid bacteria for the production of nutraceuticals. *Antonie Van Leeuwenhoek* **82**, 217–235.

Jaaskelainen, E., Johansson, P., Kostiainen, O., et al. (2013) Significance of heme-based respiration in meat spoilage caused by *Leuconostoc gasicomitatum*. *Appl Environ Microbiol* **79**, 1078–1085.

Jensen, H., Roos, S., Jonsson, H., et al. (2014) Role of *Lactobacillus reuteri* cell and mucus-binding protein A (CmbA) in adhesion to intestinal epithelial cells and mucus in vitro. *Microbiology* **160**, 671–681.

Ji, G., Beavis, R.C., and Novick, R.P. (1995) Cell density control of staphylococcal virulence mediated by an octapeptide pheromone. *Proc Natl Acad Sci USA* **92**, 12055–12059.

Johansson, P., Paulin, L., Sade, E., et al. (2011) Genome sequence of a food spoilage lactic acid bacterium, *Leuconostoc gasicomitatum* LMG 18811T, in association with specific spoilage reactions. *Appl Environ Microbiol* **77**, 4344–4351.

Jones, S.E. and Versalovic, J. (2009) Probiotic *Lactobacillus reuteri* biofilms produce antimicrobial and anti-inflammatory factors. *BMC Microbiol* **9**, 35.

Kankainen, M., Paulin, L., Tynkkynen, S., et al. (2009) Comparative genomic analysis of *Lactobacillus rhamnosus* GG reveals pili containing a human- mucus binding protein. *Proc Natl Acad Sci USA* **106**, 17193–17198.

Kelly, W.J., Davey, G.P., and Ward, L.J. (1998) Characterization of lactococci isolated from minimally processed fresh fruit and vegetables. *Int J Food Microbiol* **45**, 85–92.

Kembel, S.W., Meadow, J.F., O'Connor, T.K., et al. (2014) Architectural design drives the biogeography of indoor bacterial communities. *PLoS One* **9**, e87093.

Klausen, M., Heydorn, A., Ragas, P., et al. (2003) Biofilm formation by *Pseudomonas aeruginosa* wild type, flagella and type IV pili mutants. *Mol Microbiol* **48**, 1511–1524.

Kubota, H., Senda, S., Nomura, N., et al. (2008) Biofilm formation by lactic acid bacteria and resistance to environmental stress. *J Biosci Bioeng* **106**, 381–386.

Kubota, H., Senda, S., Tokuda, H., et al. (2009) Stress resistance of biofilm and planktonic *Lactobacillus plantarum* subsp. *plantarum* JCM 1149. *Food Microbiol* **26**, 592–597.

Landeta, G., Curiel, J.A., Carrascosa, A.V., et al. (2013) Technological and safety properties of lactic acid bacteria isolated from Spanish dry-cured sausages. *Meat Sci* **95**, 272–280.

Layec, S., Decaris, B., and Leblond-Bourget, N. (2008) Diversity of Firmicutes peptidoglycan hydrolases and specificities of those involved in daughter cell separation. *Res Microbiol* **159**, 507–515.

Le, D.T., Tran, T.L., Duviau, M.P., et al. (2013) Unraveling the role of surface mucus-binding protein and pili in muco-adhesion of *Lactococcus lactis*. *PLoS One* **8**, e79850.

Leathers, T.D. and Bischoff, K.M. (2011) Biofilm formation by strains of *Leuconostoc citreum* and *L. mesenteroides*. *Biotechnol Lett* **33**, 517–523.

Lebeer, S., Claes, I., Tytgat, H.L., et al. (2012) Functional analysis of *Lactobacillus rhamnosus* GG pili in relation to adhesion and immunomodulatory interactions with intestinal epithelial cells. *Appl Environ Microbiol* **78**, 185–193.

Lebeer, S., Claes, I.J., Verhoeven, T.L., et al. (2008a) Impact of luxS and suppressor mutations on the gastrointestinal transit of *Lactobacillus rhamnosus* GG. *Appl Environ Microbiol* **74**, 4711–4718.

Lebeer, S., De Keersmaecker, S.C., Verhoeven, T.L., et al. (2007) Functional analysis of luxS in the probiotic strain *Lactobacillus rhamnosus* GG reveals a central metabolic role important for growth and biofilm formation. *J Bacteriol* **189**, 860–871.

Lebeer, S., Vanderleyden, J., and De Keersmaecker, S.C. (2008b) Genes and molecules of lactobacilli supporting probiotic action. *Microbiol Mol Biol Rev* **72**, 728–764.

Lebeer, S., Verhoeven, T.L., Claes, I.J., et al. (2011) FISH analysis of *Lactobacillus* biofilms in the gastrointestinal tract of different hosts. *Lett Appl Microbiol* **52**, 220–226.

Lebeer, S., Verhoeven, T.L., Francius, G., et al. (2009) Identification of a gene cluster for the biosynthesis of a long, galactose-rich exopolysaccharide in *Lactobacillus rhamnosus* GG and functional analysis of the priming glycosyltransferase. *Appl Environ Microbiol* **75**, 3554–3563.

LeBlanc, J.G., Burgess, C., Sesma, F., et al. (2005) *Lactococcus lactis* is capable of improving the riboflavin status in deficient rats. *Br J Nutr* **94**, 262–267.

Leriche, V., Chassaing, D., and Carpentier, B. (1999) Behaviour of *L. monocytogenes* in an artificially made biofilm of a nisin-producing strain *of Lactococcus lactis*. *Int J Food Microbiol* **51**, 169–182.

Licitra, G., Ogier, J.C., Parayre, S., et al. (2007) Variability of bacterial biofilms of the "tina" wood vats used in the ragusano cheese-making process. *Appl Environ Microbiol* **73**, 6980–6987.

Liu, M.B., Xu, S.R., He, Y., et al. (2013) Diverse vaginal microbiomes in reproductive-age women with vulvovaginal candidiasis. *PLoS One* **8**, e79812.

Liu, S.-n., Han, Y. and Zhou, Z.J. (2011) Lactic acid bacteria in traditional fermented Chinese foods. *Food Res Int* **44**, 643–651.

Lonvaud-Funel, A. (1999) Lactic acid bacteria in the quality improvement and depreciation of wine. *Antonie Van Leeuwenhoek* **76**, 317–331.

Lortal, S., Di Blasi, A., Madec, M.N., et al. (2009) Tina wooden vat biofilm: a safe and highly efficient lactic acid bacteria delivering system in PDO Ragusano cheese making. *Int J Food Microbiol* **132**, 1–8.

Macfarlane, S. and Macfarlane, G.T. (2006) Composition and metabolic activities of bacterial biofilms colonizing food residues in the human gut. *Appl Environ Microbiol* **72**, 6204–6211.

Machado, A., Jefferson, K.K., and Cerca, N. (2013) Interactions between *Lactobacillus crispatus* and bacterial vaginosis (BV)-associated bacterial species in initial attachment and biofilm formation. *Int J Mol Sci* **14**, 12004–12012.

Macho Fernandez, E., Pot, B., and Grangette, C. (2011) Beneficial effect of probiotics in IBD: are peptidogycan and NOD2 the molecular key effectors? *Gut Microbes* **2**, 280–286.

Magnusson, J., Jonsson, H., Schnurer, J., and Roos, S. (2002) *Weissella soli* sp. nov., a lactic acid bacterium isolated from soil. *Int J Syst Evol Microbiol* **52**, 831–834.

Mah, T.F. (2012) Biofilm-specific antibiotic resistance. *Future Microbiol* **7**, 1061–1072.

Malik, S., Petrova, M.I., Claes, I.J., et al. (2013) The high auto-aggregative and adhesive phenotype of the vaginal *Lactobacillus plantarum* strain CMPG5300 is sortase-dependent. *Appl Environ Microbiol* **79**, 4576–4585.

Mandlik, A., Swierczynski, A., Das, A., and Ton-That, H. (2008) Pili in Gram-positive bacteria: assembly, involvement in colonization and biofilm development. *Trends Microbiol* **16**, 33–40.

Mann, E., Schmitz-Esser, S., Zebeli, Q., et al. (2014) Mucosa-associated bacterial microbiome of the gastrointestinal tract of weaned pigs and dynamics linked to dietary calcium-phosphorus. *PLoS One* **9**, e86950.

Maresso, A.W. and Schneewind, O. (2008) Sortase as a target of anti-infective therapy. *Pharmacol Rev* **60**, 128–141.

Mariani, C., Briandet, R., Chamba, J.F., et al. (2007) Biofilm ecology of wooden shelves used in ripening the French raw milk smear cheese Reblochon de Savoie. *J Dairy Sci* **90**, 1653–1661.

McMillan, A., Dell, M., Zellar, M.P., et al. (2011) Disruption of urogenital biofilms by lactobacilli. *Colloids Surf B Biointerfaces* **86**, 58–64.

Meadow, J.F., Altrichter, A.E., Kembel, S.W., et al. (2014) Bacterial communities on classroom surfaces vary with human contact. *Microbiome* **2**, 7.

Menard, J.P. (2011) Antibacterial treatment of bacterial vaginosis: current and emerging therapies. *Int J Womens Health* **3**, 295–305.

Mercier, C., Durrieu, C., Briandet, R., et al. (2002) Positive role of peptidoglycan breaks in lactococcal biofilm formation. *Mol Microbiol* **46**, 235–243.

Meyrand, M., Guillot, A., Goin, M., et al. (2013) Surface proteome analysis of a natural isolate of *Lactococcus lactis* reveals the presence of pili able to bind human intestinal epithelial cells. *Mol Cell Proteomics* **12**, 3935–3947.

Mohamed, J.A., Huang, W., Nallapareddy, S.R., et al. (2004) Influence of origin of isolates, especially endocarditis isolates, and various genes on biofilm formation by *Enterococcus faecalis*. *Infect Immun* **72**, 3658–3663.

Morello, E., Bermudez-Humaran, L.G., Llull, D., et al. (2008) *Lactococcus lactis*, an efficient cell factory for recombinant protein production and secretion. *J Mol Microbiol Biotechnol* **14**, 48–58.

Naessens, M., Cerdobbel, A., Soetaert, W., and Vandamme, E.J. (2005) Leuconostoc dextransucrase and dextran: production, properties and applications. *J Chem Technol Biotechnol* **80**, 845–860.

Nakayama, J., Kariyama, R., and Kumon, H. (2002) Description of a 23.9-kilobase chromosomal deletion containing a region encoding *fsr* genes which mainly determines the gelatinase-negative phenotype of clinical isolates of *Enterococcus faecalis* in urine. *Appl Environ Microbiol* **68**, 3152–3155.

Nallapareddy, S.R., Singh, K.V., Sillanpaa, J., et al. (2006) Endocarditis and biofilm-associated pili of *Enterococcus faecalis*. *J Clin Invest* **116**, 2799–2807.

Nguyen, D.T., Van Hoorde, K., Cnockaert, M., et al. (2013) A description of the lactic acid bacteria microbiota associated with the production of traditional fermented vegetables in Vietnam. *Int J Food Microbiol* **163**, 19–27.

Oxaran, V., Ledue-Clier, F., Dieye, Y., et al. (2012) Pilus biogenesis in *Lactococcus lactis*: molecular characterization and role in aggregation and biofilm formation. *PLoS One* **7**, e50989.

Pang, H., Tan, Z., Qin, G., et al. (2012) Phenotypic and phylogenetic analysis of lactic acid bacteria isolated from forage crops and grasses in the Tibetan Plateau. *J Microbiol* **50**, 63–71.

Parsek, M.R. and Greenberg, E.P. (2005) Sociomicrobiology: the connections between quorum sensing and biofilms. *Trends Microbiol* **13**, 27–33.

Patterson, J.L., Girerd, P.H., Karjane, N.W., and Jefferson, K.K. (2007) Effect of biofilm phenotype on resistance of *Gardnerella vaginalis* to hydrogen peroxide and lactic acid. *Am J Obstet Gynecol* **197**, 170 e171–177.

Pena, J.A., Rogers, A.B., Ge, Z., et al. (2005) Probiotic Lactobacillus spp. diminish *Helicobacter hepaticus*-induced inflammatory bowel disease in interleukin-10-deficient mice. *Infect Immun* **73**, 912–920.

Perea Velez, M., Verhoeven, T.L., Draing, C., et al. (2007) Functional analysis of D-alanylation of lipoteichoic acid in the probiotic strain *Lactobacillus rhamnosus* GG. *Appl Environ Microbiol* **73**, 3595–3604.

Perez-Diaz, I.M., Kelling, R.E., Hale, S., et al. (2007) Lactobacilli and tartrazine as causative agents of red-color spoilage in cucumber pickle products. *J Food Sci* **72**, M240–245.

Piard, J.C. and Desmazeaud, M. (1992) Inhibiting factors produced by lactic acid bacteria. 2. Bacteriocins and other antibacterial substances. *Lait* **72**, 113–142.

Pinkston, K.L., Singh, K.V., Gao, P., et al. (2014) Targeting pili in enterococcal pathogenesis. *Infect Immun* **82**, 1540–1547.

Pohlmann-Dietze, P., Ulrich, M., Kiser, K.B., et al. (2000) Adherence of *Staphylococcus aureus* to endothelial cells: influence of capsular polysaccharide, global regulator agr, and bacterial growth phase. *Infect Immun* **68**, 4865–4871.

Pongtharangkul, T. and Demirci, A. (2006) Effects of pH profiles on nisin production in biofilm reactor. *Appl Microbiol Biotechnol* **71**, 804–811.

Prajapati, J.B. and Nair, B.M. (2003) *The History of Fermented Foods*. CRC Press, Boca Raton, FA.

Pretzer, G., Snel, J., Molenaar, D., et al. (2005) Biodiversity-based identification and functional characterization of the mannose-specific adhesin of *Lactobacillus plantarum*. *J Bacteriol* **187**, 6128–6136.

Rangaswamy, V. and Ramakrishna, S.V. (2008) Lactic acid production by *Lactobacillus delbrueckii* in a dual reactor system using packed bed biofilm reactor. *Lett Appl Microbiol* **46**, 661–666.

Ray, B. and Bhunia, A. (2008) Spoilage of specific food groups. In: *Fundamental Food Microbiology* ((ed.) C. Press), Taylor & Francis Group, Boca Raton, FL, pp. 209–223.

Reid, G. (2001) Probiotic agents to protect the urogenital tract against infection. *Am J Clin Nutr* **73**, 437S–443S.

Remus, D.M., van Kranenburg, R., van, S., II, et al. (2012) Impact of 4 *Lactobacillus plantarum* capsular polysaccharide clusters on surface glycan composition and host cell signaling. *Microb Cell Fact* **11**, 149.

Rieu, A., Aoudia, N., Jego, G., et al. (2014) The biofilm mode of life boosts *Lactobacillus* anti-inflammatory properties of *Lactobacillus*. *Cell Microbiol* **16**, 1836–1853.

Rodrigues, L., van der Mei, H., Teixeira, J.A., and Oliveira, R. (2004) Biosurfactant from *Lactococcus lactis* 53 inhibits microbial adhesion on silicone rubber. *Appl Microbiol Biotechnol* **66**, 306–311.

Rodrigues, L.R., Teixeira, J.A., van der Mei, H.C., and Oliveira, R. (2006a) Isolation and partial characterization of a biosurfactant produced by *Streptococcus thermophilus* A. *Colloids Surf B Biointerfaces* **53**, 105–112.

Rodrigues, L.R., Teixeira, J.A., van der Mei, H.C., and Oliveira, R. (2006b) Physicochemical and functional characterization of a biosurfactant produced by *Lactococcus lactis* 53. *Colloids Surf B Biointerfaces* **49**, 79–86.

Salama, M.S., Musafija-Jeknic, T., Sandine, W.E., and Giovannoni, S.J. (1995) An ecological study of lactic acid bacteria: isolation of new strains of *Lactococcus* including *Lactococcus lactis* subspecies *cremoris*. *J Dairy Sci* **78**, 1004–1017.

Saunders, S., Bocking, A., Challis, J., and Reid, G. (2007) Effect of *Lactobacillus* challenge on *Gardnerella vaginalis* biofilms. *Colloids Surf B Biointerfaces* **55**, 138–142.

Savage, D.C., Dubos, R., and Schaedler, R.W. (1968) The gastrointestinal epithelium and its autochthonous bacterial flora. *J Exp Med* **127**, 67–76.

Savijoki, K., Lietzen, N., Kankainen, M., et al. (2011) Comparative proteome cataloging of *Lactobacillus rhamnosus* strains GG and Lc705. *J Proteome Res* **10**, 3460–3473.

Schwab, C., Tveit, A.T., Schleper, C., and Urich, T. (2014) Gene expression of lactobacilli in murine forestomach biofilms. *Microb Biotechnol* **7**, 347–359.

Shipitsyna, E., Roos, A., Datcu, R., et al. (2013) Composition of the vaginal microbiota in women of reproductive age-sensitive and specific molecular diagnosis of bacterial vaginosis is possible? *PLoS One* **8**, e60670.

Sillanpaa, J., Chang, C., Singh, K.V., et al. (2013) Contribution of individual Ebp Pilus subunits of *Enterococcus faecalis* OG1RF to pilus biogenesis, biofilm formation and urinary tract infection. *PLoS One* **8**, e68813.

Skinner-Nemec, K.A., Nichols, N.N., and Leathers, T.D. (2007) Biofilm formation by bacterial contaminants of fuel ethanol production. *Biotechnol Lett* **29**, 379–383.
Somers, E.B., Johnson, M.E., and Wong, A.C. (2001) Biofilm formation and contamination of cheese by nonstarter lactic acid bacteria in the dairy environment. *J Dairy Sci* **84**, 1926–1936.
Spurbeck, R.R. and Arvidson, C.G. (2011) Lactobacilli at the front line of defense against vaginally acquired infections. *Future Microbiol* **6**, 567–582.
Stewart, P.S. and Franklin, M.J. (2008) Physiological heterogeneity in biofilms. *Nat Rev Microbiol* **6**, 199–210.
Stiles, M.E. and Holzapfel, W.H. (1997) Lactic acid bacteria of foods and their current taxonomy. *Int J Food Microbiol* **36**, 1–29.
Stoodley, P., Sauer, K., Davies, D.G., and Costerton, J.W. (2002) Biofilms as complex differentiated communities. *Annu Rev Microbiol* **56**, 187–209.
Strus, M., Kucharska, A., Kukla, G., et al. (2005) The in vitro activity of vaginal *Lactobacillus* with probiotic properties against *Candida*. *Infect Dis Obstet Gynecol* **13**, 69–75.
Sturme, M.H., Nakayama, J., Molenaar, D., et al. (2005) An agr-like two-component regulatory system in *Lactobacillus plantarum* is involved in production of a novel cyclic peptide and regulation of adherence. *J Bacteriol* **187**, 5224–5235.
Suzuki, K., Asano, S., Iijima, K., and Kitamoto, K. (2008) Sake and beer spoilage lactic acid bacteria — a review. *J Inst Brewing* **114**, 209–223.
Tahmourespour, A., Salehi, R., Kermanshahi, R.K. and Eslami, G. (2011) The anti-biofouling effect of Lactobacillus fermentum-derived biosurfactant against *Streptococcus mutans*. *Biofouling* **27**, 385–392.
Tannock, G.W., Wilson, C.M., Loach, D., et al. (2012) Resource partitioning in relation to cohabitation of *Lactobacillus* species in the mouse forestomach. *ISME J* **6**, 927–938.
Teng, F., Singh, K.V., Bourgogne, et al. (2009) Further characterization of the epa gene cluster and Epa polysaccharides of *Enterococcus faecalis*. *Infect Immun* **77**, 3759–3767.
Terraf, M.C., Juarez Tomas, M.S., Nader-Macias, M.E., and Silva, C. (2012) Screening of biofilm formation by beneficial vaginal lactobacilli and influence of culture media components. *J Appl Microbiol* **113**, 1517–1529.
Theilacker, C., Sanchez-Carballo, P., Toma, I., et al. (2009) Glycolipids are involved in biofilm accumulation and prolonged bacteraemia in *Enterococcus faecalis*. *Mol Microbiol* **71**, 1055–1069.
Theilacker, C., Sava, I., Sanchez-Carballo, P., et al. (2011) Deletion of the glycosyltransferase bgsB of *Enterococcus faecalis* leads to a complete loss of glycolipids from the cell membrane and to impaired biofilm formation. *BMC Microbiol* **11**, 67.
Tohno, M., Kobayashi, H., Nomura, M., et al. (2012) Genotypic and phenotypic characterization of lactic acid bacteria isolated from Italian ryegrass silage. *Anim Sci J* **83**, 111–120.
Tripathi, P., Beaussart, A., Alsteens, D., et al. (2013) Adhesion and nanomechanics of pili from the probiotic *Lactobacillus rhamnosus* GG. *ACS Nano* **7**, 3685–3697.
Van Acker, H., Van Dijck, P., and Coenye, T. (2014) Molecular mechanisms of antimicrobial tolerance and resistance in bacterial and fungal biofilms. *Trends Microbiol* **22**, 326–333.
van der Veen, S. and Abee, T. (2011) Mixed species biofilms of *Listeria monocytogenes* and *Lactobacillus plantarum* show enhanced resistance to benzalkonium chloride and peracetic acid. *Int J Food Microbiol* **144**, 421–431.
Van Hamme, J.D., Singh, A., and Ward, O.P. (2006) Physiological aspects. Part 1 in a series of papers devoted to surfactants in microbiology and biotechnology. *Biotechnol Adv* **24**, 604–620.
van Pijkeren, J.P., Canchaya, C., Ryan, K.A., et al. (2006) Comparative and functional analysis of sortase-dependent proteins in the predicted secretome of *Lactobacillus salivarius* UCC118. *Appl Environ Microbiol* **72**, 4143–4153.
Vandamme, P., Pot, B., Gillis, M., et al. (1996) Polyphasic taxonomy, a consensus approach to bacterial systematics. *Microbiol Rev* **60**, 407–438.
Vasquez, A., Forsgren, E., Fries, I., et al. (2012) Symbionts as major modulators of insect health: lactic acid bacteria and honeybees. *PLoS One* **7**, e33188.
Velez, M.P., Petrova, M.I., Lebeer, S., et al. (2010) Characterization of MabA, a modulator of *Lactobacillus rhamnosus* GG adhesion and biofilm formation. *FEMS Immunol Med Microbiol* **59**, 386–398.
Vicariotto, F., Del Piano, M., Mogna, L., and Mogna, G. (2012) Effectiveness of the association of 2 probiotic strains formulated in a slow release vaginal product, in women affected by vulvovaginal candidiasis: a pilot study. *J Clin Gastroenterol* **46** Suppl, S73–80.
von Ossowski, I., Reunanen, J., Satokari, R., et al. (2010) Mucosal adhesion properties of the probiotic *Lactobacillus rhamnosus* GG SpaCBA and SpaFED pilin subunits. *Appl Environ Microbiol* **76**, 2049–2057.
Vujic, G., Jajac Knez, A., Despot Stefanovic, V., and Kuzmic Vrbanovic, V. (2013) Efficacy of orally applied probiotic capsules for bacterial vaginosis and other vaginal infections: a double-blind, randomized, placebo-controlled study. *Eur J Obstet Gynecol Reprod Biol* **168**, 75–79.

Walter, J., Britton, R.A., and Roos, S. (2011) Host-microbial symbiosis in the vertebrate gastrointestinal tract and the *Lactobacillus reuteri* paradigm. *Proc Natl Acad Sci USA* **108** Suppl 1, 4645–4652.

Walter, J., Loach, D.M., Alqumber, M., et al. (2007) D-alanyl ester depletion of teichoic acids in *Lactobacillus reuteri* 100-23 results in impaired colonization of the mouse gastrointestinal tract. *Environ Microbiol* **9**, 1750–1760.

Walter, J., Schwab, C., Loach, D.M., et al. (2008) Glucosyltransferase A (GtfA) and inulosucrase (Inu) of *Lactobacillus reuteri* TMW1.106 contribute to cell aggregation, in vitro biofilm formation, and colonization of the mouse gastrointestinal tract. *Microbiology* **154**, 72–80.

Watnick, P. and Kolter, R. (2000) Biofilm, city of microbes. *J Bacteriol* **182**, 2675–2679.

Winkelströter, L.K., Gomes, B.C., Thomaz, M.R.S., et al. (2011) Lactobacillus sakei 1 and its bacteriocin influence adhesion of *Listeria monocytogenes* on stainless steel surface. *Food Control* **22**, 1404–1407.

Woo, J. and Ahn, J. (2013) Probiotic-mediated competition, exclusion and displacement in biofilm formation by foodborne pathogens. *Lett Appl Microbiol* **56**, 307–313.

Yang, J., Cao, Y., Cai, Y., and Terada, F. (2010) Natural populations of lactic acid bacteria isolated from vegetable residues and silage fermentation. *J Dairy Sci* **93**, 3136–3145.

Zhao, T., Doyle, M.P., and Zhao, P. (2004) Control of *Listeria monocytogenes* in a biofilm by competitive-exclusion microorganisms. *Appl Environ Microbiol* **70**, 3996–4003.

Zhao, T., Podtburg, T.C., Zhao, P., et al. (2013) Reduction by competitive bacteria of *Listeria monocytogenes* in biofilms and Listeria bacteria in floor drains in a ready-to-eat poultry processing plant. *J Food Prot* **76**, 601–607.

Zhao, T., Podtburg, T.C., Zhao, P., et al. (2006) Control of Listeria spp. by competitive-exclusion bacteria in floor drains of a poultry processing plant. *Appl Environ Microbiol* **72**, 3314–3320.

Zhou, X., Westman, R., Hickey, R., et al. (2009) Vaginal microbiota of women with frequent vulvovaginal candidiasis. *Infect Immun* **77**, 4130–4135.

Index

Biotechnology of Lactic Acid Bacteria: Novel Applications, Second Edition. Edited by Fernanda Mozzi, Raúl R. Raya, and Graciela M. Vignolo.

Printed and bound by CPI Group (UK) Ltd, Croydon, CR0 4YY

17/09/2026

14964304-0002

PROCEEDINGS OF A
SYMPOSIUM IN APPLIED MATHEMATICS
OF THE AMERICAN MATHEMATICAL SOCIETY

Held in New York City
April 30–May 2, 1963

COSPONSORED BY
THE SOCIETY FOR INDUSTRIAL AND APPLIED MATHEMATICS

Richard Bellman
EDITOR

PROCEEDINGS OF
SYMPOSIA IN APPLIED MATHEMATICS

VOLUME XVI

STOCHASTIC PROCESSES IN MATHEMATICAL PHYSICS AND ENGINEERING

AMERICAN MATHEMATICAL SOCIETY
190 HOPE STREET, PROVIDENCE, RHODE ISLAND

1964

Prepared by the American Mathematical Society under Grant No. AF-AFOSR-66-63 with the United States Air Force Office of Scientific Research and Contract No. DA-31-124-ARO(D)-81 with the U.S. Army Research Office (Durham).

International Standard Serial Number 0160-7634
International Standard Book Number 0-8218-1316-1
Library of Congress Catalog Number 64-18128

Printed in the United States of America

CONTENTS

INTRODUCTION

In the study of the classification and recognition of mathematical processes, dichotomies can be introduced in several different fashions. Two of the most useful have been "deterministic vs. stochastic" and "descriptive vs. variational." In recent years, we have seen greater and greater emphasis laid upon stochastic and decision processes, and even a blending of the two in the form of modern control theory.

The gradual shift of interest from the classical deterministic descriptive process has been largely a consequence of the continuing challenges to the mathematician to explain phenomena, produce numbers, and guide research in such fields as physics, engineering, economics, biology, medicine and operations research. Combinations of complexity and uncertainty have forced a more frequent use of probabilistic concepts.

In this volume, the emphasis is upon stochastic processes, divided a bit unevenly between descriptive and decision aspects. In dealing with random effects, we can use deterministic or stochastic equations, just as in dealing with deterministic effects we can employ stochastic or deterministic equations or in treating descriptive processes, we can use either descriptive or variational equations. The choice of a particular analytic formulation is a tribute to the existing analytic techniques and the computational resources available, as well as a reflection of the underlying physical process and the prevalent scientific philosophy.

The papers by Adomian, Bharucha-Reid, Hoffman, Keller, Richardson, and Twersky deal with various and overlapping aspects of the general theory of stochastic equations, principally ordinary and partial differential equations. Much of this work is motivated by particular equations arising in wave propagation and scattering theory, as the papers by Hoffman, Keller, and Twersky indicate. The objective of obtaining specific analytic results leads to the development of perturbation techniques and the consideration of questions of closure, as, for example, in Richardson's paper.

The paper by Bellman deals with stochastic iteration, a topic which arises from the consideration of stochastic differential equations in very much the same way as classical iteration theory arises from deterministic differential equations. A particularly important type of stochastic iteration arises in the theory of stochastic approximation. This approach is used by Gray to study a design problem arising from circuits with random components. In their papers, Blackwell and Derman use dynamic programming, the first to obtain estimates in probability theory by identification with a stochastic decision process, and the second to treat an interesting class of sequential decision processes of Markovian type.

Classical estimation and identification problems arising from stochastic processes are treated by Parzen and Root using new and powerful approaches.

Finally, the papers by Montroll, Rota, and Wong consider deterministic equations associated respectively with random walk, turbulence, and diffusion theory.

Looking over the papers in this volume, we see many new, fascinating, and difficult classes of problems, ideal areas for the young analyst to test his talents, develop his techniques, and plant his flag.

RICHARD BELLMAN
Santa Monica, California
1964

STOCHASTIC GREEN'S FUNCTIONS

BY

G. ADOMIAN

1. **Introduction.** The purpose of this paper is a brief introduction to an approach [**1**] to the theory of random equations which appears fruitful. Such equations arise in many connections. Physical constants for equations representing dynamical systems are usually the mean value of a set of experimental observations so in reality are not constants at all but random variables whose values are subject to a probability distribution. Often it may be that the random equation actually is more likely to bring theory and experiment into agreement but the problem has been simplified to avoid the complication of randomness. In many-body problems it may be desirable to consider a stochastic Hamiltonian operator to represent the effects of many other bodies on the body of interest trading the complexity of $3n$ coordinates for 3 coordinates but with randomness in the operator. In wave propagation problems with random media and in optimization problems with complex systems involving random parameters, and in many other connections, such random equations arise. In the work to follow, we intend to be more concerned with vigor than rigor although a rigorous reformulation more in line with the work of the Prague probabilists [**2**; **3**] and methods of probabilistic functional analysis will hopefully follow.

The remainder of this paper is divided into three sections. In §2 we present a basic summary of necessary ideas and definitions sufficient for the level of presentation in the remainder of this paper. The paper of Bharucha-Reid in this volume will present on a rigorous level the definitions and theorems. §§3 and 4 respectively are devoted to a treatment of two types of physical problems. In §3, a linear transformation or operator theory is developed which allows the operators as well as the operand to be stochastic. Statistical measures and distributions are found for the transformed function under general conditions. The effects of a propagation medium, or of a physical observation or measuring process, or of some control system with statistically varying parameters can be represented in terms of such stochastic operators where instead of a Green's function for the kernel of the operator inverting the differential operator, one is interested in a stochastic Green's function which maps a desired statistical measure from the space of the original random function to the space of the transformed function. §4 is devoted to a consideration of the far more difficult problem of the stochastic differential operator. We are interested in relating this problem to the operator formulation to give statistical measures for the dependent variable.

Some early applications appear in [1], and more extensive treatments are to be published elsewhere. Other applications are readily suggested by the papers of Keller [4] and Hoffman [5]. We will begin by reviewing fundamentals without worrying about rigorous aspects.

2. A summary of the basic ideas.

2.1. THE NATURE OF STOCHASTIC PROCESSES. Certain phenomena in nature are characterized by unpredictable changes in time; they exhibit variations from one observation to the next which no amount of effort or control can remove. They will be called random or stochastic processes and are represented by random functions. If these processes were completely unpredictable in every sense, no theory at all could be formulated. However, they often show regularities or stabilized properties as the number of observations increases. These regularities are called statistical properties and our theories will apply to these. Since the theory of probability is the mathematical theory for statistical properties, our methods will be based on such concepts. Obviously these methods will be appropriate to physical systems whose individual initial states and transition laws cannot be specified in such a way as to yield unique descriptions of final states, but whose behavior in the large (or statistical behavior) is predictable. To summarize, a physical phenomenon is interpreted as a stochastic process if the future cannot be determined from the present except in a statistical way using probability theory. In other words, whenever a process develops in time in a manner controlled by probabilistic rather than deterministic laws, we have a stochastic process. Examples are provided by the path of a particle in Brownian motion, growth of a bacterial population, fluctuating numbers of electrons in a cosmic ray shower, etc. Some interesting examples have been discussed by Kac [6].

2.2. JOINT DISTRIBUTION FUNCTIONS. In order to form some intuitive notions before attacking the general problem, let us refer briefly to an excellent treatment by Lawson and Uhlenbeck [7], of the mathematical description of noise output of receivers. When no signal is present, the output is not always zero but fluctuates randomly around some average value. These fluctuations are called noise and are of great concern because they often prevent detection of weak signals. Observation of the output $y(t)$ of a single receiver over a period of time does not make it possible to predict the output for any later time or to predict the output as a function of time for another receiver identical with the first. If we imagine now a great number of macroscopically identical receivers (an "ensemble" of receivers) all turned on simultaneously, we can observe the noise outputs $y_1(t), y_2(t), \cdots$. All these functions are different. At a definite time t, it can be observed for what fraction of the total number of cases y occurs in a given interval between y and $y + \Delta y$. This fraction now depends on y and t and is proportional to Δy when Δy is small. It is written $F_1(y, t)\, dy$. Next can be considered all the pairs of values of y occurring at two given times t_1 and t_2. The fraction of the total number of pairs in which y occurs in the range $(y_1, y_1 + \Delta y_1)$ at t_1 and in the range $(y_2, y_2 + \Delta y_2)$ at t_2 is written $F_2(y_1, t_1; y_2, t_2)\, dy_1\, dy_2$. We can continue in this manner determining all the triples of values of y at three given times to arrive at F_3, etc. The joint probability (density) F_n is called the nth probability distribution for the random process.

The general theory of stochastic processes is of great interest in many other areas than noise theory alone. Applications exist in many branches of physics, astronomy, engineering, biology, economics, and in mathematics itself. Therefore, let us rephrase what we have said in somewhat more general terms [**8**].

To define a stochastic process $u(P)$ where P is a variable point in a given set, we specify the joint probability distributions $F_n(u_1, u_2, \cdots, u_n; P_1, P_2, \cdots, P_n)$, where $u_i = u(P_i)$. These distributions F_n with increasing n describe the stochastic process in increasing detail. Some relations immediately follow.

1. $F_n \geqq 0$ (since the F_n are probability densities).
2. F_n is a symmetric function of its arguments (u_k, P_k) since F_n is a joint probability distribution.
3.

$$F_m(u_1, \cdots, u_m; P_1, \cdots, P_m) = \int_{u_{m+1}} \cdots \int_{u_n} F_n(u_1, \cdots, u_n; P_1, \cdots, P_n)\, du_{m+1} \cdots du_n$$

and

$$\int_{u_1} \cdots \int_{u_n} F_n\, du_1 \cdots du_n = 1.$$

(Therefore each function F_n must imply all the previous F_m with $m < n$.) In the noise theory example the expression for F_m in (3) would be

$$\int \cdots \int F_n(y_1, \cdots, y_n; t_1, \cdots, t_n)\, dy_{m+1} \cdots dy_n$$

and it is clear that any one of the F_m with $m < n$ can be determined from F_n.

Wang and Uhlenbeck [**9**] have classified stochastic processes by the order of n required for a complete description. In general, joint probability distributions of unbounded order are required to describe general processes. It is of interest to consider some special, or simple, stochastic processes in which only first or second probability distribution functions are sufficient to determine the process. Among these are:

a. *Purely random or independent process.* In this case the successive values of u are statistically independent. Hence the first probability distribution function is sufficient to determine all probability distribution functions, i.e., for $(P_i \neq P_j)$

$$F_n(u_1, \cdots, u_n; P_1, \cdots, P_n) = \prod_{j=1}^{n} F_1(u_j; P_j).$$

b. *Simple Markov process.* This is determined by the second probability distribution function. The simple Markov process is defined by the equation

$$p_n(u_n, P_n | u_1, \cdots, u_{n-1}; P_1, \cdots, P_{n-1}) = p_2(u_n, P_n | u_{n-1}, P_{n-1})$$

(assuming $P_1 < P_2 < \cdots$) where $p(a|b)$ means the conditional probability of a where b is given **[10]**. (We may describe this intuitively as a short memory situation in which only the preceding value matters.)

Processes are also classified according to the type of their distribution functions, i.e., the probability law involved. For example we speak of a Gaussian process, a Poisson process, etc. The probability density function for the Gaussian process, which is of particular significance because of the Central Limit Theorem **[11]** is

$$F_n\,du_1 \cdots du_n = (2\pi)^{-n/2}B^{-1/2}\exp\left\{-(2B)^{-1}\sum_{j,k}^{n} B_{jk}u_ju_k\right\} du_1 \cdots du_n$$

where B is the determinant of B_{jk}. Thus in our noise example we have

$$F_n = F(y_1, y_2, \cdots, y_n)$$

obtained from the above by substituting y's for the u's, assuming $\langle y_i \rangle = 0$. The matrix is symmetric and positive definite. B_{jk} is the cofactor of the matrix element b_{jk} defined by the quadratic average or "moment"

$$b_{jk} = \langle y_jy_k \rangle = \int_{-\infty}^{\infty} \cdots \int_{-\infty}^{\infty} y_jy_kF(y_1, \cdots, y_n)\,dy_1 \cdots dy_n.$$

The one-dimensional form is more familiar and often occurs in applications

$$F(y) = [1/\sigma(2\pi)^{1/2}]\exp\{-(y - \bar{y})^2/2\sigma^2\}$$

where the average value is $\langle y \rangle$ or $\bar{y}$ and the variance $\sigma^2 = \langle (y - \bar{y})^2 \rangle$.

2.3. STATIONARITY AND ERGODICITY. We classify linear and nonlinear processes with customary definitions. Further, a stochastic process is called a statistically nonhomogeneous or nonstationary process if its distribution functions depend on the points $P_1, P_2, \cdots, P_n$ as we have so far written. If the distribution functions depend only on differences $P_i - P_j$ with $(i, j = 1, 2, \cdots, n)$ for $i \neq j$, the process is said to be statistically homogeneous or stationary.

Consider again our noise example of an ensemble of receivers. One could object that an observation on the ensemble could not be made in this example. (In other areas, e.g., statistical mechanics or thermodynamics, this is the only type of observation which can be made; we cannot follow the motion of a single system.) When all transients have damped out so that the receiver is in a "stationary state" all the statistics are obtainable from a sufficiently long record of the output of a single receiver.[1] If this record is cut into pieces of duration T where T is long compared to any periodicities in the process, each piece can be considered part of an ensemble of observations from which the distributions can be determined. We note immediately that the distributions are simpler now. The first example $F_1(y, t)$ is now $F_1(y)$, i.e., the probability of finding y between y and $y + \Delta y$ no longer depends upon t. The second distribution F_2 depends only on the time difference $t_2 - t_1$ so we write $F_2(y_1, y_2, t)$. Continuing, $F_3(y_1, y_2, y_3, t_1, t_2)\,dy_1\,dy_2\,dy_3$ is

[1] Implicitly assumes the ergodic property to be discussed later.

the joint probability of finding a triple of values of y in the ranges dy_1, dy_2, dy_3 where dy_1 and dy_2 are the time interval t_1 apart and dy_2 and dy_3 are the time interval t_2 apart, etc. This is an example of stationarity. For a stochastic process $x(t)$, it clearly means an invariance of the distribution functions to a translation of the time origin. (This is sometimes called stationarity in the strict sense to distinguish it from stationarity in the wide sense, to be defined later. Processes which are stationary in the strict sense are also stationary in the wide sense.)

A property implicitly assumed in the discussion so far of the stationary random process involved in the noise example is the ergodic property. Thus we stated the sequence of samples drawn from a single function (the pieces of length T of the record of the output of a single receiver) can be taken as an ensemble. We implicitly used the ergodic hypothesis **[12]** in saying this latter ensemble possesses the same statistical characteristics as the original ensemble. An equivalent statement of the ergodic property is that all statistical averages of any random variable are equal to the time averages of every member of the ensemble. Thus for a stationary random process $x(t)$, the ergodic property holds if for any random variable $V(x)$ such as $x^2(t)$, $x(t_1)x(t_2)$, etc., the expectation value $\langle V\{x(t)\}\rangle$ is equal to

$$\lim_{T\to\infty} \frac{1}{2T}\int_{-T}^{T} V\{x(t+\tau)\}\, d\tau$$

except for a set of measure zero.

If the process is ergodic, statistical averages across the ensemble are equal to the time averages of the individual sample functions. Again in an intuitive way, the essence of the ergodicity condition is that each sample function must eventually take on all the modes of behavior of every other sample function in order for the process to be ergodic, i.e., for time averages to equal statistical or ensemble averages. Ergodicity implies stationarity, i.e., the invariance of the statistics (across the ensemble) to time translation, but the converse need not be true. (See *Ergodic theorem* of Birkhoff **[12]**.) Finally if a process is stationary, a statistical average will be time-independent. If it is ergodic as well, a time average of a member of the ensemble will be the same as for any other member of the ensemble. Thus the two ways of averaging give the same result. Whether ergodicity holds or not, time averaging an already statistically averaged process gives the same result as the operations in inverse order.

2.4. Random variables and random processes. Some more general mathematical concepts are often valuable in the treatment of random functions. Thus we may understand the randomness of a stochastic function by considering $u(P)$ to be a function of a random point in the sample space, or event, ω, as well as the deterministic point P. We write $u = u(P; \omega)$ assuming ω ranges over a sample space (or probability space) or a set of elementary events, Ω. For each ω a probability is defined. To be more precise, we would say u is a p-measurable function of ω, and the p-measure, or probability measure, is defined on the Borel field generated by subsets of Ω. This notation conveniently gives an ordinary function for a fixed value of ω, say ω_1, which represents an experiment or event. Thus $u(P; \omega_1) = u(P)$ is a realization or realized value or sample function of u. For a fixed value of P, on the other hand, we have $u(P_1; \omega) = u(\omega)$ as a random variable on the probability space Ω.

An example of such a random variable is the outcome of a particular random experiment E which may be repeated under similar conditions a large number of times. Let the result of this experiment on each trial be given by a single real quantity ω. (In tossing a die, ω represents the numbers 1 to 6 which can come up.) Thus we introduce a variable point ω in a probability, or measure, or sample space S representing the possible outcomes. (In the case of the die, S is a set of 6 discrete points.) When ω has a particular outcome we say ω has the value associated with a sample point in S. We associate a measure or probability with each sample point and we speak of the probability that ω has a particular value.

A random (or stochastic) process is an ensemble of functions in time together with a probability measure by which we can determine probabilities of observable properties. Thus at each sample point we now have a time function $u(t)$, or $u(P)$ in general. The ensemble of such functions over the space is the random process and could be symbolized by $\{u(P)\}$ although we frequently omit the brackets when the meaning is clear. We can also look upon a stochastic process as a family of random variables involving t as a parameter, i.e., for each value of t we get an observed value of the random variable. Each realization of the process is the result of an observation on the process. If we fix the parameter t we get an ordinary random variable. The family of such variables arising from considering t as a parameter can be regarded as the process.

The member functions of the random process $u(t)$ may be defined for discrete instants of time only as well as for a continuum in t in which case we speak of a random series. Depending then on whether the member functions are a continuum or a discrete set, we then speak of a continuous or discrete random series. An important physical example of a continuous random series is the classical problem of the random walk where the steps or displacements at any one move may take a value $y^{(j)}(t_k)$ from a continuum $(a < y < b)$ but these moves are made only at discrete instants $t_1, t_2, \cdots, t_k, \cdots$. An example of a discrete random series is provided by a sequence of tossed coins. The t_k represent the times at which the coin is tossed; the two possible values $y^{(1)}(t_k)$ and $y^{(2)}(t_k)$ correspond to "heads" and "tails" respectively. Finally we have discrete random processes where t represents a continuum but y can only take discrete values as in counters for nuclear particles. The general case is a continuous random process and a Dirac δ formalism can be used for special cases. Finally we may remark that all representations of an ensemble need not be entirely random but may exhibit periodicities. In fact, sets of functions exhibiting no random behavior are included. These are ensembles of a single unit where all members are identical. Thus there is a definite functional dependence upon time and the future behavior of any representative is completely known by a measurement. This is a deterministic process.

2.5. AVERAGE VALUES AND COVARIANCE FUNCTIONS. The stochastic process $u(P)$ is taken now as $u(t)$ and the probability distributions are

$$F_n(u_1, u_2, \cdots, u_n;\ t_1, t_2, \cdots, t_n)$$

where $u_i = u(t_i)$. From the first distribution $F_1(u, t)$, the average value

$$\langle u \rangle = \int u F_1(u, t)\, du$$

can be determined. It clearly depends in general on the time t, unless stationarity holds. This is a statistical or ensemble average and can be regarded as the average at a time t of the ensemble of functions represented by u. On the other hand, the time average

$$[u]_{\text{av}} = \lim \frac{1}{2T} \int_{-T}^{T} dt\, u(t) \qquad (T \to \infty)$$

is independent of time but, in general, different for each member of the ensemble of u. For an ergodic process the two ways of averaging give the same result. Then $\langle u \rangle$ is independent of time and $[u]_{\text{av}}$ is the same for each member of the ensemble.

We can now define the moments

$$m_n = E\{u^n\} = \langle u^n \rangle = \int u^n F_1(u)\, du$$

and the fluctuation, or variance, or dispersion

$$\sigma^2 = \langle (u - m_1)^2 \rangle = \int (u - m_1)^2 F_1(u)\, du.$$

An often useful concept called the characteristic function is defined by

$$\phi_1(\lambda) = \langle e^{i\lambda u} \rangle = \int e^{i\lambda u} F_1(u)\, du$$

and if all the moments are finite, we can write[2]

$$\phi_1 = \sum_{n=0}^{\infty} \frac{(i\lambda)^n}{n!} \langle u^n \rangle = \sum_{n=0}^{\infty} \frac{(i\lambda)^n}{n!} m_n$$

where

$$F_1(u) = \frac{1}{2\pi} \int_{-\infty}^{\infty} d\lambda\, e^{-i\lambda u}\, \phi_1(\lambda)$$

so that the probability distribution is determined uniquely[3] by the characteristic function.

---

[2] Assuming convergence.

[3] In most cases. See Loève, *Probability theory* **[15]**.

From the second distribution $F_2(u_1, u_2; t_1, t_2)$ an important average is $\langle u_1u_2\rangle$ or $E\{u_1u_2\}$ defined by

$$\langle u_1u_2\rangle = \int\int u_1u_2F_2(u_1, u_2; t_1, t_2)\, du_1\, du_2$$

in general, a function of t_1 and t_2. Letting $t_2 = t_1 + \tau$ and time-averaging over t_1 we have

$$R(\tau) = \lim_{T\to\infty} \frac{1}{2T}\int_{-T}^{T} \langle u_1(t_1)u_2(t_1+\tau)\rangle\, dt_1 = [\langle u_1u_2\rangle]_{\text{av}}.$$

If we take

$$[u(t_1)u(t_1+\tau)]_{\text{av}} = \lim_{T\to\infty} \frac{1}{2T}\int_{-T}^{T} u(t_1)u(t_1+\tau)\, dt_1$$

and now average over the ensemble, clearly, we get $R(\tau)$ again. However, if the process is ergodic either $\langle u_1u_2\rangle$ or $[u(t_1)u(t_1+\tau)]_{\text{av}}$ gives the same result. $R(\tau)$ is a measure of the correlation between successive values of u and is therefore called the correlation function.

The conditional probability distribution $P_2(u_2, t_2|u_1, t_1)$ gives the probability of finding u between u_2 and $u_2 + du_2$ at time t_2 given that $u = u_1$ at t_1. Clearly,

$$F_2(u_1, t_1; u_2, t_2) = F_1(u_1, t_1)P_2(u_2, t_2|u_1, t_1);$$

also

$$\int P_2(u_2, t_2|u_1, t_1)\, du_2 = 1,$$

$$\int F_1(u_1, t_1)P_2(u_2, t_2|u_1, t_1)\, du_1 = F_1(u_2, t_2),$$

since

$$\int F_2\, du_2 = F_1(u_1, t_1),$$

$$\int F_2\, du_1 = F_1(u_2, t_2).$$

The notion of the characteristic function is easily extended to multivariate characteristic functions. Thus the bivariate characteristic function of the random variables u_1 and u_2 is

$$\begin{aligned}\phi_2(\lambda_1, \lambda_2) &= \langle e^{i(\lambda_1u_1+\lambda_2u_2)}\rangle \\ &= \int\int e^{i(\lambda_1u_1+\lambda_2u_2)}\, F_2(u_1, u_2)\, du_1\, du_2 \\ &= \sum_{j=0}^{\infty}\sum_{k=0}^{\infty} \frac{(i\lambda_1)^j}{j!}\frac{(i\lambda_2)^k}{k!}\langle u_1^ju_2^k\rangle.\end{aligned}$$

The quantity $\langle u_1^j u_2^k \rangle$ is a joint moment analogous to m_n in the expression for $\phi_1(\lambda)$. F_2 is found from ϕ_2 by a two-dimensional Fourier transform. Finally the n-dimensional general case gives

$$\phi_n(\lambda) = \int_{-\infty}^{\infty} \cdots \int_{-\infty}^{\infty} \exp\left[i \sum_{j=1}^{n} \lambda_j u_j\right] F_n(u_1; t_1, u_2; t_2, \cdots, u_n; t_n)\, du_1 \cdots du_n$$

and the multivariate density function

$$F_n(u_1; t_1, u_2; t_2, \cdots, u_n; t_n) = \frac{1}{(2\pi)^n} \int_{-\infty}^{\infty} \cdots \int_{-\infty}^{\infty} \exp\left[-i \sum \lambda_j u_j\right] \phi_n\, du_1 \cdots du_n.$$

If the functions ϕ_n can be determined for the output the problem is solved by taking the inverse Fourier transform.

There are several methods for doing this in very special cases. The Kac–Siegert approach **[16]** exists for cases in which the input to a deterministic linear system is a squared Gaussian process and the Fortet–Siegert **[17–20]** approach assumes the input is either a Markov process or one component of a multidimensional Markov process. Samuels **[21]** shows ϕ_n can also be derived from knowledge of the correlation function as we suspect by looking at ϕ_1 and ϕ_2 and the generalized expression. We will return to this point in a moment.

Return now to our expression $\langle u_1 u_2 \rangle$ also written as $E\{u_1 u_2\}$ and examine it more closely. $u(t)$ is a random process. u_1 and u_2 are a pair of random variables referring to possible values which can be assumed at the given instants t_1 and t_2, respectively, by the sample functions of the process $u(t)$. However, the joint probability distribution of u_1 and u_2 may change as t_1 and t_2 change, in general. Therefore, the average denoted by $\langle u_1 u_2 \rangle$ may well depend upon both time instants.[4] In general then we write the correlation function as $R_u(t_1, t_2)$ and call it the autocorrelation because it involves only the u process. Similarly we could consider random variables $u(t)$ and $v(t)$ and define $\langle uv \rangle$ as a cross-correlation. Also used is the covariance function defined by $\langle (u - \bar{u})(v - \bar{v}) \rangle$ for random variables u and v. Thus in the joint moments $\langle u_1^j u_2^k \rangle$ already defined, the case $j = k = 1$ or second order joint moment is the correlation function for u_1 and u_2 (autocorrelation if u_1 and u_2 are derived from u and cross-correlation if they are different random variables). If $\bar{u}_1 = \bar{u}_2 = 0$ the second order joint moment is also the covariance function. In generalizing to nth order correlations, knowledge of all the correlations implies knowledge of all the joint-moments since some of the time variables in the correlation functions can be made equal to obtain all the joint moments.

The various joint moments can be obtained from their joint characteristic function by successive differentiations **[22]**; or a series expansion of the joint characteristic function in terms of the various joint moments is easily obtained as in the cases of ϕ_1 and ϕ_2.

If a given random process u is stationary, the joint probability distribution for u_1 and u_2 depends only on the time difference $\tau = t_2 - t_1$ and not on the

[4] If the process is complex, $R_u = \langle u_1 u_2^* \rangle$.

particular values of t_1 and t_2. Then $R_u(t_1, t_2)$ can be written $R_u(\tau)$ for any t. Sometimes a process is not stationary in the strict sense defined earlier, i.e., the probability distributions are not invariant under time translation, yet the autocorrelation can be written as $R_u(\tau)$ and the various means are constant functions of time. It is then said to be stationary in the wide sense. Obviously, processes stationary in the strict sense are also stationary in the wide sense.[5] This implies an obvious idealization. No physical process can start at $-\infty$ or continue till ∞ without changes in the underlying mechanism. One always deals in actuality with finite samples, and what is of importance is the correlation time relative to the sample length.

We now consider a random field **[23]** $v(x, t)$ where x represents the three space coordinates and t is the time. Following previous definitions we define the nth order correlation R_n by the natural generalization

$$\int \cdots \int v_1 v_2 \cdots v_n F_n(v_1, v_2, \cdots, v_n; t_1, \cdots, t_n, x_1, \cdots, x_n)\, dv_1 \cdots dv_n$$

$$= \langle v_1 v_2 \cdots v_n \rangle = E\{v(x_1, t_1; \omega) v(x_2, t_2; \omega) \cdots v(x_n, t_n; \omega)\}$$

$$= R_n(x_1, x_2, \cdots, x_n; t_1, t_2, \cdots, t_n),$$

the integral being over the domain of the random variables $v_1, v_2, \cdots, v_n$ and the x_j and t_j being ordered. (E is sometimes called the expectation operator.) We note knowledge of all the R_n implies knowledge of all the joint moments.

We can now define the n-dimensional characteristic function

$$\phi_n(\lambda; t, x) = \int e^{i\lambda \cdot v} F_n(v; t, x)\, dv$$

where λ, v, x are n-dimensional vectors and the integration is over the domain of v.

Now since F_n and ϕ_n are n-dimensional Fourier transform pairs, F_n can be determined from ϕ_n (see Levy's inversion theorem) **[24]**. Knowledge of all the F_n for the output process gives us complete knowledge, i.e., solves the problem. Now ϕ_n can be written as a series involving all the correlation functions as is clear in principle from what has been said so far. Writing out ϕ_n in this form as given by Samuels and Eringen,

$$\phi_n(\lambda) = 1 + \sum_{k=1}^{\infty} \frac{(i)^k}{k!} \left\{ \sum_{m_j} R_k(x_1^{(m_1)}, x_2^{(m_2)}, \cdots; t_1^{(m_1)}, \cdots, t_n^{(m_n)})(\lambda_1^{m_1} \cdots \lambda_n^{m_n}) \right\},$$

where the summation over m_j means the sum of such expressions for all values of the individual m's subject to $\sum_{j=1}^{n} m_j = k$. Now, e.g., ϕ_1 becomes

$$\phi_1(\lambda) = 1 + \sum_{k=1}^{\infty} \frac{(i\lambda)^k}{k!} R_k(x^{(k)}; t^{(k)}),$$

[5] The converse is not true except for Gaussian processes where knowledge of the covariance and the mean functions completely determines the process.

the R_k clearly representing all the moments so we get the same expression as given earlier. Similarly,

$$\phi_2(\lambda_1, \lambda_2) = 1 + \sum_{k=1}^{\infty} \frac{(i)^k}{k!} \left\{ \sum_{m_j} R_k(x_1^{(m_1)}, x_2^{(m_2)}; t_1^{(m_1)}, t_2^{(m_2)}) \lambda_1^{m_1} \lambda_2^{m_2} \right\}$$

where $m_1 + m_2 = k$.

The value of this expression as Samuels and Eringen point out is the fact that in linear deterministic systems, the correlation functions of the output are simply related to those of the input if Green's functions can be found. Thus knowledge of the correlation functions of the input allows writing those for the output from which ϕ_n and finally F_n can be determined.

However, we can go even further. For linear stochastic systems we can write output autocorrelations in terms of input autocorrelations by using appropriate stochastic Green's functions. In principle at least, then we can similarly write all the correlations of the output (i.e., any order correlation) in terms of those of the input. The stochastic kernels are by no means the same; however, they are determinable in the same manner. Thus we have outlined a method in principle for the determination of all the distribution functions of the output knowing those for the input and, therefore, in this sense a solution of the most general statistical problem for a linear system.

2.6. Spectral density. We wish now to define a function called the power spectral density or simply spectral density (alternatively, spectrum or power spectrum) which gives the distribution in frequency of the power of a signal or noise and is closely related to the autocorrelation.

Suppose $u(t)$ is any function and assume that $u(t)$ is zero outside the time interval $[-T, T]$. The function can now be developed in a Fourier integral

$$u(t) = \int_{-\infty}^{\infty} df\, A(f)\, e^{2\pi i f t},$$

i.e., $u(t)$ and $A(f)$ are a Fourier transform pair. $A(f) = A^*(-f)$ if $u(t)$ is real. By the Parseval theorem

$$\int_{-\infty}^{\infty} u^2(t)\, dt = \int_{-T}^{T} u^2(t)\, dt = \int_{-\infty}^{\infty} |A(f)|^2\, df$$

assuming the last integral is not unbounded (Plancherel's theorem). We note that $|A(f)|^2$ is an even function, and letting $T \to \infty$,

$$[u^2]_{\text{av}} = \lim_{T\to\infty} \frac{1}{2T} \int_{-T}^{T} u^2(t)\, dt = \int_0^{\infty} G(f)\, df$$

where[6]

$$G(f) = \lim_{T\to\infty} \frac{|A(f)|^2}{T}$$

[6] This definition requires care in use. Integration before limiting processes avoids difficulties because the means converge. See Davenport and Root [22] for a discussion of this point.

is called the spectral density or power spectrum of the function $u(t)$ whenever the limit exists. Now instead of $[u(t)u(t)]_{av}$ take $[u(t)u(t+\tau)]_{av}$

$$[u(t)u(t+\tau)]_{av} = \lim_{T\to\infty} \frac{1}{2T}\int_{-T}^{T} u(t)u(t+\tau)\,dt.$$

Developing $u(t)$ and $u(t+\tau)$ in Fourier integrals as before

$$[u(t)u(t+\tau)]_{av} = \int_{-\infty}^{\infty} e^{-2\pi i f\tau}\, G(f)\,df.$$

Assume now $u(t)$ is actually a random process so the above functions u are members of the ensemble of the process u, i.e., sample functions or representatives. Then each of these representatives has a corresponding $G(f)$ which we can now average over the ensemble. The resulting $\langle G(f)\rangle$ or $\Phi(f)$ is the spectral density or power spectrum of the random process. We saw earlier that $\langle [u(t)u(t+\tau)]_{av}\rangle$ is the correlation function $R(\tau)$ for the process, so we have

$$R(\tau) = \int_{-\infty}^{\infty} e^{-2\pi i f\tau}\, \Phi(f)\,df.$$

We defined $G(f)$ as a time spectrum (the spectrum corresponding to a time function), and then averaged over the ensemble of such functions to get $\Phi(f)$. Thus $\langle G(f)\rangle = \Phi(f)$. We note $G(f)$ is the Fourier transform of the correlation function $\mathscr{R}(\tau)$ where $\mathscr{R}(\tau) = [u(t)u(t+\tau)]_{av}$ and $\Phi(f)$ is the Fourier transform of $R(\tau)$ where $R(\tau) = \langle \mathscr{R}(\tau)\rangle$.

The time autocorrelation $\mathscr{R}(\tau)$ (i.e., the autocorrelation of a sample function of the random process) is equal to the statistical autocorrelation $R(\tau)$ if the process is ergodic (with probability one).

Finally for nonstationary processes, the (statistical) autocorrelation must be written $R(t,\tau)$ and its Fourier transform $\Phi(t,f)$ might be considered a power spectrum for a limited interval. If the limit exists as $T\to\infty$ of the quantity $(2T)^{-1}\int_{-T}^{T} \Phi(t,f)\,dt$, and this average is equal to nearly all the measurements of $\Phi(t,f)$ for all values of t, it can be inferred the process is stationary.

3. Stochastic operations on random processes: operator formulation.

3.1. CONCEPT OF A STOCHASTIC OPERATOR AND STOCHASTIC GREEN'S FUNCTIONS. In many physical problems of interest, we are concerned with the properties of a function obtained by transformation of a given function by a known operator. In others, the inverse problem arises, where the quantity of interest is the dependent variable in a differential equation. Often the real physical problem involves randomness which has been ignored to make equations tractable. Unavoidable random effects are present or a lack of definite knowledge exists. These problems can be considered by generalizing the above operators to be linear *stochastic* operators which reduce to ordinary operators as randomness vanishes and parameters become deterministic. Analogous to the usual representation of operators by matrices, stochastic operators will be represented by stochastic matrices or matrices whose elements have a distribution. Thinking of a given deterministic operator as a realization of a stochastic operator, or of a stochastic operator as an ensemble of such deterministic operators, one immediately thinks of stochastic mappings between generalized spaces in mathematics; of quantum theory and many-body problems, statistical mechanics, or wave propagation in random media in the area of physics; or of randomly time-varying systems, communication theory, or control system theory in the area of engineering.

In applications, stochastic operators arise in the processing of an input quantity (which may itself be a random process) by a linear system, observation, or measurement, in which certain parameters are random or stochastic variables. Problems involving random inputs to fixed linear systems (i.e., where all the a_ν are constants and only x is stochastic) are well understood. A great deal of attention has been devoted also to deterministic time variations in linear systems (linear differential equations with time-varying coefficients). When random variations of system parameters (or of the a_ν in the differential equation example) occur, the analysis is more subtle. The output now for either a deterministic or a random input is a stochastic process. A complete description of the output would involve the determination of all multivariate probability distributions of the output from similar knowledge of the input. Such a complete description is usually neither possible nor necessary.

It is instructive to consider a subclass of stochastic differential operators given by

$$L = \sum_{\nu=0}^{n} a_\nu(t) \frac{d^\nu}{dt^\nu}$$

and differential equations of the form $Ly = x$, where the nonhomogeneous term $x(t)$, called the input or forcing function, and the operator L, by virtue of the coefficients $a_\nu(t)$, are, in general, stochastic quantities. The converse problem involving the conceptual inversion of the matrix for the random differential operator L is especially interesting. For appropriate operators, we think of y as the result of operating on the input by a stochastic operator symbolically denoted by H or by L^{-1} and otherwise undefined except to say it would include the action of any accompanying boundary conditions. We wish to keep the differential equation formulation in mind in the analysis, although in this section we are primarily interested in general operators H acting on a process x with as little restriction as possible.

Suppose that $x(t)$ is a known input process or the operand. We wish to determine the result of an operation on x by the stochastic operator H. Thus, x might be a communication signal into the proverbial black box which then represents the operator H. The box may be a circuit, filter, servo, observation, experiment, or measurement, or a transmission medium which has random properties. All real media are inhomogeneous, i.e., the mean physical properties vary from one point to another. When, in addition, there is a variance from the mean at each point, the medium is a random inhomogeneous medium. When the properties or parameters of the medium are constants and $x(t)$ is nonstochastic, the output $y = L^{-1}x$ is often found without difficulty (depending on the nature of the operator). When the x is stochastic, and certainly when the properties of the medium vary randomly with time, we can only hope to obtain various statistical measures or estimates of y in terms of similar measures of x. If the inverse H of the stochastic matrix L exists, we have the general problem again given by $y = Hx$ (where we emphasize the product is not multiplicative but denotes the action of the stochastic operator H on the process x) and the differential equation could be solved as well as the general problem Hx. However, the inversion of a stochastic matrix is clearly to be avoided. Since we are asking only for an appropriate statistical measure of y rather than y itself, we may be able to solve the differential equation at the same time as the general problem without the inversion. A well-known and convenient statistical measure is the power spectral density or power spectrum, or by the Wiener–Khinchin theorem, the autocorrelation function [7]. Thus, if $\Phi_x(f)$ is the power spectrum of the input and $\Phi_y(f)$ is the power spectrum of the output, we would like to obtain Φ_y in terms of Φ_x. Such an expression can be obtained as an integral equation whose kernel K_H involves the random parameters of the operator H. We will then clarify the interpretation of the stochastic kernel K_H and show its derivation from the L as well, providing certain restrictions on L are met. The constant-parameter system is handled easily in terms of Green's functions (weighting functions in control system theory) or frequency response functions and transfer functions, respectively, the Fourier and Laplace transforms of the system Green's function. We will (in the stochastic case) obtain a stochastic Green's function that depends on the statistical measure chosen for a solution and which reduces to the ordinary expressions when the parameters are constants or nonstochastic.

3.2. Examples of stochastic operators. In many cases of physical interest (2) can be determined directly from the nature of the process. Thus we have an input $x(t)$ and a "stochastically filtered" or processed output $F_{\alpha,\beta,\cdots}\{x(t)\}$ where the stochastic operator $F_{\alpha,\beta,\cdots}$ depends on the distributions of its parameters $\alpha, \beta, \cdots$. For example, (a) a function $x(t)$ is translated by amounts α_n at intervals of time τ so $F\{x(t)\} = x(t - \alpha_n)$ for $n\tau < t < (n + 1)\tau$, (b) $x(t)$ is multiplied by β_n over intervals of length τ. Thus $F\{x(t)\} = \beta_n x(t)$ for $n\tau < t < (n + 1)\tau$. These are linear processes and the random variables in question are the α_n and the β_n. The distribution functions of these parameters determine the statistical properties of the process. An example (c), of greater interest arising in an application to be considered is the operation $S\{x(t)\} = x(t_n)$ for $t_n < t < t_{n+1}$ for all n where the intervals $l_n = t_{n+1} - t_n$ are random variables with a distribution $P(l)\, dl$. Thus S might be called a "random sampling operator" (where the last sample value is held until another is taken). The S operator gives the correct results for the much-used case of regular sampling as randomness vanishes. A closely related example (d) is $P\{x(t)\}$ which samples randomly as before but inverts alternate samples. Thus $P\{x(t)\} = (-1)^n x(t_n)$ for $t_n < t < t_{n+1}$ where the intervals $l_n = t_{n+1} - t_n$ are specified by $P(l)\, dl$.

The Green's functions for these latter processes ((c) and (d)) can be written by inspection

$$s_i(t, \tau) = \sum_k \Delta(t_k, l_k; t)\delta(\tau - t_k),$$

$$p_i(t, \tau) = \sum_k (-1)^k \Delta(t_k, l_k; t)\delta(\tau - t_k),$$

where

$$\Delta(t_k, l_k; t) = \begin{cases} 1 & \text{if } t_k < t < t_k + l_k, \\ 0 & \text{otherwise.} \end{cases}$$

The spectral densities for the results of operations on x by S or P are considered in the referenced dissertation in connection with a generalized random-walk problem, first studied jointly with DuBois. We state the results here as examples.

$$\Phi_{Sx}(f) = \mathrm{Re}\left\{\frac{1 - z(f)}{2\pi^2 f^2 \bar{l}} \int_{-\infty}^{\infty} ds \Phi_x(s - f)[1 - z(s - f)] \sum_0^{\infty} z^n(s)\right\},$$

$$\Phi_{Px}(f) = \mathrm{Re}\left\{\frac{1 - z(f)}{2\pi^2 f^2 \bar{l}} \int_{-\infty}^{\infty} ds\, \Phi_x(s - f)[1 + z(s - f)] \sum_0^{\infty} (-1)^n z^n(s)\right\}$$

where $z(f)$ is the characteristic function of the distribution for the sampling intervals $P(l)$, i.e.,

$$z(f) = \int_0^{\infty} dl\, P(l)\, e^{-2\pi i f l},$$

$$\bar{l} = \int_0^{\infty} lP(l)\, dl = -\frac{1}{2\pi i} z'(0),$$

relations which are clearly of the same form as (3) with rather complicated kernels dependent on the probability laws in this case just $P(l)$ for the parameters which are random. Evaluation of the kernel in terms of distributions of the random parameters is obtained directly from the nature of the process on the input, i.e., from Hx.

3.3. Development of a stochastic Green's function. Now consider a linear operation $H_{\alpha,\beta,\cdots}$, or H for brevity, on an input process x for which the spectral density Φ_x is known. The operation represents a medium or processing system with parameters $\alpha(t)$, $\beta(t)$, $\cdots$, one or more of which are random. Thus H is determined only by the probability distributions of its parameters, i.e., in the sense of an ensemble of possible operations H_i which have a distribution determined by the parameters. Each of the H_i corresponds to an ordinary nonstochastic matrix. Each member H_i of the ensemble H has associated with it a Green's function $h_i(t, \tau)$ which represents the response at time t of the system H_i to a unit impulse $\delta(t - \tau)$ applied at time τ. More specifically, h_i should be written $h_i(\alpha, \beta, \cdots, t; \tau)$ where α, β, $\cdots$ are the parameters of H subject to probability laws. The response of the H_i system to the input $x(t)$, thinking of x as a defined continuous member of the x process for the moment, can now be given in terms of the Green's function by

$$H_i x(t) = \int_{-\infty}^{\infty} h_i(\alpha, \beta, \cdots, t; \tau) x(\tau)\, d\tau,$$

or for brevity, where parameter dependence is not used,

$$H_i x(t) = \int_{-\infty}^{\infty} h_i(t; \tau) x(\tau)\, d\tau, \tag{1}$$

where for causal (physically realizable) systems, $h_i(t, \tau)$ is zero for $t < \tau$. The upper limit can also be written t without change. The implication of the lower limit is that all past values are significant. We can write the spectral density of $H_i x(t)$ by first developing this function in a Fourier integral. Thus assuming stationarity and supposing $H_i x$ to be zero outside some interval $[-T, T]$ then taking the limit at $T \rightarrow \infty$

$$H_i x(t) = \int_{-\infty}^{\infty} df A_i(f)\, e^{2\pi i f t}$$

$$\Phi_{H_i}(f) = \lim_{T \rightarrow \infty} \frac{1}{T} |A_i(f)|^2.$$

Our stationarity assumption can be modified somewhat in that the $A_i(f)$ can depend on T, thus $A_i^T(f)$, as long as the limit exists. The above expression holds for all $H_i x$ which are members of the ensemble Hx. Each $H_i x$ is developed in a Fourier integral as we have shown. The corresponding spectra Φ_{H_i} are averaged over the ensemble. The result which could have been written immediately is $\Phi_{Hx}(f)$, the spectral density of the output. Thus,

$$\Phi_{Hx}(f) = \lim_{T \rightarrow \infty} \frac{1}{T} \left\langle \left| \int_{-T}^{T} dt\, H_i\{x(t)\}\, e^{-2\pi i f t} \right|^2 \right\rangle,$$

where the average is over the ensembles of H and of x.

$$\Phi_{Hx}(f) = \lim_{T \rightarrow \infty} \frac{1}{T} \int_{-T}^{T} \int_{-T}^{T} dt_1\, dt_2 \langle H_i\{x(t_1)\} H_i^*\{x(t_2)\} \rangle\, e^{-2\pi i f(t_1 - t_2)}.$$

The quantity in brackets is evidently an autocorrelation function. Let us define the autocorrelation of the Hx as

$$R_{Hx}(\tau) = \langle Hx(t) H^* x(t + \tau) \rangle,$$

where the random process may be complex and we have limited ourselves to stationary input $x(t)$, though it is possible the operator H may be nonstationary. On using (1), this equation becomes

$$R_{Hx}(\tau) = \iint_{-\infty}^{\infty} d\tau_1\, d\tau_2 \langle h(t; \tau_1) h^*(t + \tau; \tau_2) x(\tau_1) x(\tau_2) \rangle.$$

We note the h's correspond to different times t_1 and t_2 (or t and $t + \tau$). Thus we have $h(\alpha_1, \beta_1, \cdots, t_1; \tau_1)$ and $h^*(\alpha_2, \beta_2, \cdots, t_2; \tau_2)$ so all parameters are different.

In many cases of physical interest, the $x(t)$ and the $a_\nu(t)$ are independent. In other words, the input process and the "filtering" process are independent. Transmission of a signal through a randomly varying medium (*a stochastic filter*) is such a case. The result of the action of the medium on x is the stochastically filtered signal Hx. The term in brackets in the expression for $R_{Hx}(\tau)$ can now be factored. Indicating the autocorrelation for x alone by R_x

$$R_x(\tau_2 - \tau_1) = \langle x(\tau_1)x^*(\tau_2)\rangle,$$

with averaging obviously over the ensemble of x, we now have

$$R_{Hx}(\tau) = \int d\tau_1 \int d\tau_2\, R_x(\tau_2 - \tau_1)\langle h(t; \tau_1)h^*(t + \tau; \tau_2)\rangle.$$

The corresponding spectral density is

$$\begin{aligned}\Phi_{Hx}(f) &= \int d\tau\, e^{2\pi i f\tau} R_{Hx}(\tau)\\ &= \int d\tau_1 \int d\tau_2 \int d\tau\, e^{2\pi i f\tau} R_x(\tau_2 - \tau_1)\langle h(t; \tau_1)h^*(t + \tau; \tau_2)\rangle\\ &= \int d\tau_1 \int d\tau_2 \int d\tau \int ds\, e^{2\pi i\tau f}\, e^{-2\pi i(\tau_2 - \tau_1)s}\langle h(t; \tau_1)h^*(t + \tau; \tau_2)\rangle\Phi_x(s),\end{aligned}$$

where $\Phi_x(s)$ is the (power) spectral density of $x(t)$. The last equation is now rewritten in the convenient form

$$\Phi_{Hx}(f) = \int_{-\infty}^{\infty} ds\, K_H(s, f)\Phi_x(s), \tag{2}$$

where the integral kernel or "stochastic kernel"

$$K_H(s, f) = \int_{-\infty}^{\infty} d\tau_1 \int_{-\infty}^{\infty} d\tau_2 \int_{-\infty}^{\infty} d\tau\, e^{2\pi i\tau f}\, e^{-2\pi i(\tau_2 - \tau_1)s}\langle h(t; \tau_1)h^*(t + \tau; \tau_2)\rangle \tag{3}$$

is a "stochastic Green's function." The form of the kernel or stochastic Green's function depends on the statistical measure chosen for the solution. In solving $Ly = x(t)$, we can express y in terms of x by the use of a Green's function $G(t, \tau)$. However, we are now solving not for y but Φ_y in terms of Φ_x and a "Green's function" appropriate to such measures. This is our kernel K_H or what we call a stochastic Green's function. Since it depends upon the statistical measure, we should properly identify the kernel K_H as a stochastic Green's function for power spectral density measures or a spectral density Green's function.

Whereas H and the above results are quite general, the L in the differential equation was very special, for convenience, though still an important form. No difficulty in principle would be expected in generalizing to systems of differential equations with many dependent variables (multiple inputs and outputs) or to random fields $V(x, t)$.

To the spectral density derived for the "output" we should add any contributions due to a mean value since the derivation tacitly assumed a zero-mean random process x. When the time average of the process is nonzero, there are singular peaks of the well-known Dirac δ-function type in the spectral density. Thus, to the $\Phi_x(f)$, which represents only the spectrum of what we could call the a.c. part in electrical language, we should add a term $2\bar{x}^2\delta(f)$. Equivalently, to $\Phi_y(f)$ we add a term $2\bar{x}^2K_H(0, f)$.

3.4. General statistical problem. Now, in principle at least, we can express all the correlations of the output by similar expressions involving correlations of the input for general linear stochastic operations on arbitrary random inputs. But the n-dimensional multivariate characteristic function ϕ_n of the "output" or transformed process can be given in terms of all the correlations of the output, and by the above reasoning can now be given in terms of all the correlations of the input. These, in turn, are clearly determinable from knowledge of all the joint distribution functions of the input process or operand. Finally the n-dimensional joint distribution function for the output is obtained by Fourier inversion of ϕ_n. Thus we have suggested a method, at least in principle, for solving the most general statistical problem for a linear system.

3.5. Linear properties. Equation (2) shows that the power spectrum (or autocorrelation function by use of the Wiener–Khinchin theorem) of the output can be given as a transform of the corresponding measure of the input and a kernel which, as we shall see, involves the probability laws for the parameters. Thus the spectral density of the output is related to the spectral density of the input by a linear operation. The solution of any problem of this type involves finding the appropriate kernel. We write (2) symbolically as

$$\Phi_{Hx} = K_H \cdot \Phi_x$$

understanding K_H to be the appropriate integral operator. We note some properties which would reasonably be expected.

If $x = ax_1 + bx_2 + \cdots$, where the x_i are mutually independent, then $Hx = aHx_1 + bHx_2 + \cdots$ and $\Phi_{Hx} = K_H \cdot \Phi_x = a^2K_H \cdot \Phi_{x1} + b^2K_H \cdot \Phi_{x2} + \cdots$. Also if $H = H_1 + H_2$, then $\Phi_{Hx} = (K_{H_1} + K_{H_2}) \cdot \Phi_x$.

3.6. Interpretation of the kernel in special cases. Consider first a special case of (2) corresponding to a constant-parameter medium. For such a case (differential equations with constant coefficients and random forcing function), the Green's function depends only on the time interval between application of the impulse and observation of the output. The averaging is obviously unnecessary now, so that (3) can be rewritten

$$K_H(s, f) = \int_{-\infty}^{\infty} d\tau_1 \int_{-\infty}^{\infty} d\tau_2 \int_{-\infty}^{\infty} d\tau\, e^{2\pi i\tau f}\, e^{-2\pi i(\tau_2-\tau_1)s} h(t - \tau_1)h(t + \tau - \tau_2),$$

which with appropriate changes of variables becomes

$$K_H(s, f) = \int_{-\infty}^{\infty} d\tau \, e^{-2\pi i(f-s)\tau} \int_0^{\infty} d\tau_2 \, e^{2\pi i \tau_2 s} h(\tau_2) \int_0^{\infty} d\tau_1 \, e^{-2\pi i \tau_1 s} h(\tau_1)$$

$$= | Y(s)|^2 \delta(f - s),$$

where

$$Y(s) \doteq \int_0^{\infty} d\tau \, e^{-2\pi i \tau s} h(\tau),$$

i.e., the Fourier transform of the Green's function. Then (2) becomes

$$\Phi_{Hx}(f) = | Y(f)|^2 \Phi_x(f),$$

a well-known result.

Occasionally, the notion of frequency response function is extended to linear time-varying systems by defining:

$$Y(s, t) \doteq \int_0^{\infty} h(t; t - \tau) \, e^{-2\pi i s \tau} \, d\tau.$$

The general kernel $K_H(s, f)$ becomes

$$K_H = \int_{-\infty}^{\infty} d\tau \, e^{2\pi i \tau f} \int_{-\infty}^{\infty} d\tau_2 \, e^{-2\pi i \tau_2 s} \int_{-\infty}^{\infty} d\tau_1 \, e^{2\pi i \tau_1 s} \langle h(t; \tau_1) h^*(t + \tau; \tau_2) \rangle$$

$$= \int_{-\infty}^{\infty} d\tau \, e^{2\pi i \tau f} \left\langle \int_{-\infty}^{\infty} d\tau_2 \, e^{-2\pi i \tau_2 s} h^*(t + \tau; \tau_2) \int_{-\infty}^{\infty} d\tau_1 \, e^{2\pi i \tau_1 s} h(t; \tau_1) \right\rangle$$

$$= \int_{-\infty}^{\infty} d\tau \, e^{2\pi i \tau f} \langle Y^*(s, t + \tau) \, e^{-2\pi i s \tau} \, Y(s, t) \rangle$$

$$= \int_{-\infty}^{\infty} d\tau \, e^{2\pi i \tau (f-s)} \langle Y^*(s, t + \tau) \, Y(s, t) \rangle.$$

This is compatible with our definition for the constant-parameter case since the above expression then reduces to

$$\int d\tau \, e^{2\pi i \tau (f-s)} \langle | Y(s)|^2 \rangle = \delta(f - s) | Y(s)|^2.$$

So far, our stochastic kernel has been quite general. The matrix H has elements with any distribution. However, some statistical properties of matrices whose elements have normal, or Gaussian, distributions are well known. We can make further identifications by specializing to the case of such distributions (possibly first determining the eigenvalues of the operator then the Green's functions, and consequently the kernel). To see how the distributions enter, it is noted that the Green's functions involve the random parameters $\alpha, \beta, \cdots$, which have distributions $p(\alpha), p(\beta), \cdots$ that must be used in obtaining the average or expectation value involved in the kernel. Thus, for ergodic operators

$$\langle h(\alpha, \beta, \cdots, t; \tau_1)h^*(\alpha, \beta, \cdots, t + \tau; \tau_2)\rangle$$
$$= \lim_{T\to\infty} \frac{1}{2T}\int_{-T}^{T} du \int_0^\infty d\alpha \int_0^\infty d\beta \int \cdots p(\alpha)p(\beta) \cdots$$
$$\times\ h(\alpha, \beta, \cdots, t + u; \tau_1)h^*(\alpha, \beta, \cdots, t + \tau + u; \tau_2)$$

and we can specialize to certain distributions or even to the case where the distribution is given by a δ function and the parameter is not random.

3.7. STATISTICAL (MEASURE) OPERATIONS. A class of operations on random processes including, e.g., the ensemble average of a random process, the autocorrelation, mean squared, or higher means, can be called statistical or measure operations. A statistical operator or measure in this sense indicates a (nonstochastic) deterministic operation on a random process. It is distinguished from ordinary deterministic operations on ordinary functions only in that the operand is now a random process rather than a function, so that the averages are with respect to a random variable. Clearly, any ordinary deterministic operator, e.g., a Laplace (or other) transform should commute with such statistical operators and one could consider a possibility of appropriate (statistical) transforms of random processes since this would just mean the ordinary transform of the result of the statistical operation on the random process, or a transform of a mean.

Stochastic operators present much more difficulty. Let us consider a random function or process $y(t)$. It can be regarded as a generalized vector in an infinite dimensional continuum or a (infinite) continuous (column) matrix, since it consists of an entire ensemble of sample functions or representatives. The elements of the matrix vary according to the distribution to which $y(t)$ is subject. A (stochastic) operator represented by a (stochastic) matrix (with elements subject to an appropriate distribution) now acts on the matrix of the input random process. Clearly, we expect only to find some statistical property of the result. Let us indicate the statistical measure by Γ. Thus, emphasizing the random processes by brackets $\{\ \}$,

$$\{y(t)\} = H\{x(t)\},$$
$$\Gamma(y(t)\} = \Gamma H\{x(t)\}.$$

If we choose spectral density as our statistical measure Γ, we have $\Gamma\{y(t)\} = \Phi_y(f)$ and $\Gamma\{x(t)\} = \Phi_x(f)$. Noting that Γ and H do not commute (unless H is deterministic), we have derived Φ_y in terms of Φ_x in the integral relation of equation (2), or

$$\Phi_y = \Gamma H\{x(t)\} = \int K_H(s, f)\Phi_x(s)\, ds.$$

When the stochastic operation H and the random process x are statistically independent, the statistical operator Γ must act separately on each, i.e., providing ΓH is defined (and it should be if H is representable by a matrix) and if neither

ΓH nor Γx is zero (eliminating a zero-mean random process if Hx does not also have a zero mean) then

$$\Gamma Hx = \Gamma H \cdot \Gamma x.$$

This means the appropriate statistical operation over each ensemble, i.e., $\Gamma_H H \cdot \Gamma_x x$. Thus Γx is found by forming $\langle x(t)x^*(t+\tau)\rangle$ then making the transform $\int e^{2\pi i f\tau}\langle\ \rangle\, d\tau$. Similarly ΓH is given by

$$\Gamma H = \int e^{2\pi i f\tau}\langle H(t)H^*(t+\tau)\rangle\, d\tau.$$

Thus we can compute an appropriate autocorrelation and then a transform to get ΓH, or a spectral density of H, if we are given the stochastic matrix of H. Thus we can write

$$\Phi_y = \Phi_H \cdot \Phi_x = \int K_H(s, f)\Phi_x(s)\, ds.$$

If we can find Green's functions, the kernel K_H can be found and Φ_y determined. We can always do this in principle for deterministic operators, as we shall show, and at least for stochastic operators having the ergodicity property, as mentioned earlier. If H is given as a stochastic matrix, K_H and therefore Φ_y correspond to the transform of the autocorrelation of the matrices.

Finally, if H is a deterministic operator $\Gamma Hx = H\Gamma x$ or $\Phi_y = H\Phi_x$. Thus if H is the Laplace transform operator, i.e.,

$$H(\ ) = \int_0^\infty e^{-st}(\ )\, dt,$$

then

$$\Gamma Hx = \Gamma \int_0^\infty e^{-st}x(t)\, dt = \int_0^\infty e^{-st}\Gamma x(t)\, dt = H\Gamma x.$$

This clearly is not $\Gamma X(s)$ where $X(s)$ is the transform. Γ acts only on the random function $x(t)$. Further, even for stationary $x(t)$, Hx or $X(s)$ would not be stationary and $\Gamma X(s)$ would be undefined. But (supposing the statistical measure Γ to be the spectral density) we have

$$H\Gamma x = \int_0^\infty e^{-st}\Gamma x(t)\, dt = \int_0^\infty e^{-st}\Phi_x(s)\, dt,$$

noting the variable of Φ_x should be the same as the transform variable s. Thus

$$H\Gamma x = \Phi_x(s)\int_0^\infty e^{-st}\, dt = \frac{\Phi_x(s)}{s}$$

and we see that a statistical measure Γ (in this case the spectral density) of the (Laplace) transform of a random process $x(t)$ is given by a pole of magnitude Γx (the spectral density of $x(t)$ in the transform plane).

3.8. Stochastic Green's functions for correlation measures of input and output. The stochastic Green's functions or stochastic kernels for other statistical measures of input and output can be found in the same way as carried out for spectral density measures. Some of these kernels, e.g., those for correlation functions, are particularly valuable, both because we frequently want to determine the correlation matrix for a random process and because of our suggestion for determination of the nth joint distribution function, if we can first find all the output correlations in terms of the input correlations.

The kernel for the autocorrelation measure of input and output for the stationary case can be seen from the derivation of K_H in §1. Thus, the relation of the autocorrelations R_{Hx} of the output and R_x of the input for an operation H is

$$R_{Hx}(\beta) = \int_{-\infty}^{\infty} d\sigma G_H(\beta, \sigma) R_x(\sigma),$$

where the kernel G_H is given by

$$G_H(\beta, \sigma) = \int_{-\infty}^{\infty} d\tau \langle h(t, \tau) h^*(t + \beta, \tau + \sigma)\rangle.$$

Transformation of R_{Hx} and R_x to spectral densities again gives the relation involving the spectral density kernel $K_H(s, f)$. Because G_H is simpler than K_H, it is usually desirable to work with correlations although this investigation has emphasized spectra.

3.9. NONSTATIONARY PROCESSES.[7] If we had not restricted ourselves to stationary processes in writing $R_{Hx}(\tau)$ (early in §2), we would instead write $R_{Hx}(t_1, t_2)$ or simply $R(t_1, t_2)$ given by

$$\begin{aligned} R(t_1, t_2) &= \int\int \langle h_i(t_1, \sigma) x(\sigma) h_i(t_2, \tau) x(\tau)\rangle \, d\sigma \, d\tau \\ &= \int\int R_x(\sigma, \tau) \langle h_i(t_1, \sigma) h_i(t_2, \tau)\rangle \, d\sigma \, d\tau, \end{aligned}$$

where $R_x(\sigma, \tau)$ is the correlation of the input. Thus,

$$R(t_1, t_2) = \int\int R_x(\sigma, \tau) H(t_1, t_2, \sigma, \tau) \, d\sigma \, d\tau,$$

where $H(t_1, t_2, \sigma, \tau)$ can be considered a kernel for the autocorrelation where the processes need not be stationary. When the processes are stationary, $R_x(\sigma, \tau) = R_x(\sigma - \tau)$, and H becomes $H(t_1 - \sigma, t_2 - \tau)$, a function of two variates as in $K_H(s, f)$.

3.10. MAPPING OF SPACES. If the measure chosen for $x(t)$ and $y(t)$ is the spectrum Φ, we have

$$\Phi_y(f) = \int K(s, f) \Phi_x(s) \, ds.$$

If the measure chosen is the autocorrelation we have

$$R(\beta) = \int G(\beta, \sigma) R_x(\sigma) \, d\sigma.$$

---

[7] This treatment was suggested by Professor Balakrishnan, of the University of California at Los Angeles.

A mathematically more elegant formulation is to define a general statistical measure μ.[8] Thus the measure of y is $\mu(y)$ and the measure of x is $\mu(x)$. Now $\mu(y)$ may be $\Phi(f)$ or $R(\tau)$, for example. It means a chosen measure defined over the space of the random y. Similarly, $\mu(x)$ is defined over the space of $x(t)$ and may, for example, be the spectrum $\Phi_x(s)$. Now we can define

$$\Phi_x(s)\, ds = d\mu(x)$$

and write (1) as

$$\mu(y) = \int K(y, x)\, d\mu(x).$$

This form is now appropriate to any chosen measure operation. The kernel $K(y, x)$ maps from the space of x to the space of y and depends on the chosen μ.

A stochastic process $X(t)$,[9] can be described in terms of n-dimensional random variables $X(t_1), X(t_2), \cdots, X(t_n)$ by means of their families of distribution functions $F_n(t_1; X_1, t_2; X_2, \cdots, t_n; X_n)$. However, such a law will be meaningless on the quantum level since it implies the observation of a physical phenomenon at an instant determined by a point on a time scale, or generalizing immediately to random fields for precise measurements of position, momentum, or other variables. Such precise or instantaneous observations are inconsistent with basic postulates so a reinterpretation becomes necessary. However, it is not essential to consider random processes as families of random variables depending upon t (or $\mathbf{t}$). Instead, we think of X as an element in a function space. Each realization of the process $X(t)$ becomes one point in the function space. We define a general random element X with values in a space χ in the following way. Let $x(u)$ be a mapping from a measurable space of elementary events U into the space χ. The random element X is the result of the mapping. The elements of χ are the values which X can take. Fortet **[25]** gives some good examples. If X is a random variable, χ is the space R of real numbers. If X is a k-dimensional random variable, then χ is a k-dimensional Euclidean space. If X is a Gaussian random function of t on an interval (t_1, t_2) (with continuous covariance) then χ may be either the Hilbert space of those functions on (t_1, t_2) whose square is integrable or the space of all functions on (t_1, t_2).

Now we can extend the definition of stochastic operators to abstract spaces. Thus, the stochastic operator L acting on the random function X can be defined as the transformation which maps the space χ into a space κ representing values which LX can take. Any operator satisfies the definition so far; now we include linear transformations which themselves require statistical treatment in terms of realizations such as we gave $X(t)$. A single realization of a stochastic operator yields an ordinary operator or nonrandom matrix. It is possible that a formulation of quantum theory using stochastic operators may couple the advantages of the

---

[8] Called Γ in the section on statistical measure operations.

[9] We can generalize to a random field by writing $X(t, r)$ or simply letting t be a vector, i.e., X can be a function of several variables.

Feynman formulation with calculable expressions and a more natural appearance of probabilities. This work is intended to lay the foundation for later use in this connection. The extension will be of the nature of a general transformation theory involving topological spaces and functional analysis.

3.11. SOME GENERAL MATHEMATICAL PROPERTIES OF KERNELS AND OPERATORS. We note in passing some general mathematical properties of the kernel and its relations to stochastic operations and statistical measure operations. If $y = Hx$, where H is a stochastic operation on the stochastic process or random function x and we take the statistical measure μ of both sides of this equation, we have

$$\mu(y) = \mu(Hx) = \int K(y, x)\, d\mu(x) \doteq K\mu(x)$$

(it being assumed that μ does not assign finite measure to a point). In order that the integral operator K exists, μ must have certain properties with respect to H. From the above equation we can write the operator equation

$$\mu H - K\mu = 0.$$

Let us assume that the quantity $H\mu$ is operationally defined in some sense. We have already seen in the section on measure operations that μ does not commute with stochastic operations, i.e.,

$$\mu H \neq H\mu.$$

Since we have assumed that $H\mu$ is defined, we may define the operator ζ by

$$\mu H = H\mu + \zeta\mu.$$

Substituting this into the equation $\mu H - K\mu = 0$, we have

$$(H + \zeta - K)\mu = 0.$$

Thus, under the assumption that $H\mu$ exists, we have

$$K = H + \zeta + \sigma,$$

where σ is any operator such that

$$\sigma\mu = 0.$$

Conversely, if K exists and $H\mu$ is defined, then the equation $\mu H - K\mu = 0$ is satisfied if and only if there exists an operator ζ such that

$$\mu H = H\mu + \zeta\mu.$$

Combining the above considerations, we obtain the following result:

THEOREM. *If $H\mu$ is defined, there exists an operator K which satisfies the operator equation*

$$\mu H - K\mu = 0$$

if and only if there exists an operator ζ such that

$$\mu H = H\mu + \zeta\mu.$$

If these conditions are satisfied, then K is given by

$$K = H + \zeta + \sigma,$$

where σ is any operator such that $\sigma\mu = 0$.

Now assume that σ is zero so $H + \zeta$ represents K. The procedure is clear in principle. For a given stochastic operation and chosen measure operation, a kernel K exists if $\mu H - K\mu$ is a set of measure zero over the space of x on which H acts. The commutator of μ with H determines ζ. Then $K = H + \zeta$.

If we consider μ and H as operators each with a unique set of eigenvectors and suppose these sets are nonparallel, i.e., μ and H do not commute, then ζ is orthogonal to H, i.e., the eigenvectors of ζ are orthogonal to the eigenvectors of H. We are defining an operator ζ such that the eigenvectors of $H + \zeta$ commute with K. For H and μ such that ζ is sufficiently simple, it is conceivable that one could determine K from $H + \zeta$.

Now consider the inverse problem $Ly = x$ where we are particularly, but not exclusively, interested in the (unbounded) differential operator. Again we suppose that $\mu(x)$ is a well-defined measure given for x and we ask for $\mu(y)$. Write $\mu(Ly) = \mu(x)$. L and y are not statistically independent, so μ does not act separately on L and y as in the case Hx. Again the measure operation μ does not commute with the stochastic operation L. Suppose the commutation relation is

$$\mu L = L\mu + \chi\mu,$$

where again χ may be, and in general is, quite complicated. Now supposing $\mu(y)$ exists,

$$L\mu(y) + \chi\mu(y) = \mu(x)$$

or

$$LK\mu(x) + \chi K\mu(x) = \mu(x),$$

and we can write the operator rèlation

$$(L + \chi)K = 1.$$

Thus, given the measure operator μ and the stochastic operator L, the operator K is defined and exists when χ exists or when $\mu(y)$ and $\mu(x)$ exist. However this is not a method of calculating K since we do not know how to calculate the inverse of a stochastic matrix. We may also note that even if $\mu(x)$ exists, $\mu(y)$ may not exist. L is an unbounded operator (for the differential operators) which may be nonmeasure-preserving. The space of y must be suitably restricted. If a measure on x space is mapped into the same measure on y space, then it is clear that not all measures will work for an arbitrary L.

This shows $(L + \chi)(H + \zeta) = 1$ or $H + \zeta$ is inverse to $L + \chi$, but we started with the supposition that H was inverse to L by writing $y = Hx$ as the solution for $Ly = x$. This implies that $\chi H + L\zeta + \chi\zeta = 0$ or that χ and ζ are related for a particular L and its inverse H.

We will now generalize our remarks in the following manner. We have seen that we can write

$$\Phi_y(f) = \int ds\, K_H(s,f)\Phi_x(s)$$
$$\doteq W(s,f)\Phi_x(s)$$

defining $W(s,f)$ to mean the integral operator $\int ds K_H(s,f)$. The above expression is equivalent to

$$\Gamma Hx = W(s,f)\Gamma x,$$

where Γ represents the statistical measure operation of forming the spectral density. To keep track of the arguments f or s we use subscripts on the Γ operator. Thus, $\Phi_y(f) = \Gamma_f Hx$ and $\Phi_x(s) = \Gamma_s x$. Now

$$\Gamma_f Hx = W(s,f)\Gamma_s x = \Gamma H \cdot \Gamma_f x$$
$$= \Gamma H\delta(s-f)\Gamma_s x$$

so that

$$W(s,f) = \Gamma H\delta(s-f).$$

Also, knowing ΓH is a noncommuting pair, we write

$$\Gamma_f H = H\Gamma_f + \eta\Gamma_f,$$
$$\eta\Gamma_f = \Gamma_f H - H\Gamma_f$$
$$= \Gamma H\delta(s-f)\Gamma_s - H\delta(s-f)\Gamma_s$$
$$= [W(s,f) - H\delta(s-f)]\Gamma_s.$$

Thus, $\eta\Gamma_f$ is an essentially singular operation. η is related to our kernel operation $W(s,f)$ by an additive Dirac matrix. We have shown that η exists and have shown its specific structure. If Γ were to represent an ordinary averaging operation and H a stochastic operation, then ΓHx is a defined operation. $H\Gamma x$ is undefined, but the Dirac function behavior of η is just such as to eliminate the point in space from which the indeterminism results.

We can now state the following as a theorem:

THEOREM. *There exists a kernel $K(s,f)$ or the integral operator*

$$W(s,f) = \int_{-\infty}^{\infty} ds K_H(s,f)$$

if and only if there exists an η satisfying the commutation relation $\Gamma H - H\Gamma = \eta\Gamma$, such that

$$W(s,f) = \Gamma H\delta(s-f) + \chi,$$

where χ is an arbitrary operator such that $\chi\Gamma x = 0$, i.e., χ annihilates Γ.

We have shown $K(s,f)$ exists under the assumption $\Gamma Hx = \Gamma H \cdot \Gamma x$. Now one could consider all Γ, for example, that result in in $\Gamma Hx = \Gamma H \cdot \Gamma x$ plus some other term and again find a necessary and sufficient condition for existence of a kernel.

3.12. PHYSICAL MEASUREMENTS AND ESTIMATES OF STATISTICS. With a priori knowledge of probability distributions of given random processes we have calculated various statistical measures (moments, characteristic functions, spectral densities, etc.) for these distributions.

In physical measurements of random processes with unknown distributions (where there is no possibility of taking advantage of the Central Limit Theorem), a question of great importance we may ask is whether from a set of measurements we can now determine some of the statistical properties (mean, variance, etc., . . .).

Suppose we have available N measurements of a basic experiment defining a process or N observations or measurements of the value of a sample function of a given random process at N different instants of time. This set of N values is a sample point in an N-dimensional space characterizing the entire experiment. Clearly, this single point will not uniquely determine the statistical property we seek. In this situation the statistician speaks of making an estimate of the particular statistical property by finding a function or statistic of the set of results or values which gives a reasonably close estimate of the statistical property for the process. This is called sampling theory and is discussed by Davenport [**22**], Cramer [**11**, Part III], and others. It is clear from the theory and our own results on the random sampling example, that we can now give more generalized sampling theorems for situations in which it may matter that the N instants of time at which measurements are made may not be regularly spaced but are subject to a distribution.[10] Such a case might occur, for example, if the observations were to be made whenever an incoming particle triggered the observing apparatus.

The application of the theory to the statistical description of mechanical systems of macroscopic dimensions is a clear possibility. We can assume a stochastic process corresponding to the measurement of an n-tuple of macroscopic observables. Then a theory of repeated observations or measurements can be made using stochastic operators, where some parameter of the observation is unknown and can only be given a distribution.

4. The inverse problem: stochastic differential equations.

4.1. DIFFERENTIAL OPERATORS. We have discussed in §3 various statistical measures for the action of a linear stochastic operator on a random process, or equivalently, a linear (stochastic) transformation of a random process. A particular measure of interest, the spectral density measure, involves a stochastic kernel or spectral density Green's function K_H for a stochastic operator H. This formulation is useful in some problems involving the statistical optimization of complex systems and in the synthesis of models for some physical phenomena. In other

[10] Since the term random sampling is already used by statisticians to indicate that the sample is chosen from a general population in a random manner, it might be better to call this stochastic sampling.

physical problems, however, the convenient formulation involves a "stochastic differential equation" rather than a given operator. Generally, linear operators, linear transformation, linear filters, linear networks, and linear control systems are equivalent mathematical systems describable by systems of linear integro-differential equations. The forcing functions are the inputs. The outputs or dependent variables appear as the operands. For simplicity, we consider a single input, single output equation in the form

$$Ly = x$$

where

$$L = \sum_{\nu=0}^{n} a_\nu(t) \frac{d^\nu}{dt^\nu}$$

and both $a_\nu(t)$ and $x(t)$ are stochastic. Samuels [**26**], Sundstrom [**27**], and others have pointed out that such problems arise in the study of linear servomechanisms in which several parameters are undergoing noise modulations [**28**; **29**] and in certain propagation problems in stochastically varying media [**30–33**] or media with random inhomogeneities.

We now consider the equation $Ly = x$, where L is the previously defined stochastic operator and $x(t)$ is a random function. Defining H as the operator inverse to L, we investigate the possibility of determining the kernel K_H for H by examining L and avoiding the problem of inversion, i.e., not determining H. We assume reasonable restrictions on L.

1. The inverse must exist. Thus L must be nonsingular, i.e., the determinant of the matrix corresponding to L must never become zero as the elements vary according to some distribution.

2. K_H is stationary or, at least, the coefficients in L can be translated, i.e., have stationary properties.

Now, it is possible that differential equations can be solved in this sense, i.e., in terms of a statistical measure, where the coefficients are stochastic as well as where the coefficients can be considered to be chosen from random processes. The previous solution for a statistical measure of y, given H, is the solution to the differential equation if the kernel can be determined from L, i.e., in terms of given statistics of the random coefficients. Our stationarity condition means simply that the probability distributions of the random parameters are such that the stochastic coefficients (time functions) in the differential equation describe stationary processes during a time interval of interest, i.e., the statistical characteristics during the interval are constants. In order to have a full description of the processes, the time interval must be at least so long that the correlation between values at the limit points can be neglected. Actually, widesense stationarity is sufficient for the coefficients.

4.2. EVALUATION OF STOCHASTIC GREEN'S FUNCTION FOR A DIFFERENTIAL OPERATOR. A series of papers by Sundstrom [**34**] leads to the following approach for expressing the kernel in terms of the coefficients of the differential equation. Treating x not as a random process but as a defined continuous member of the x

process in the interval $[-T/2, T/2]$, and similarly treating y and a_ν, we define Fourier transforms of the "input" x, the "output" y, and the coefficients a_ν.

$$x(t) = \int_{-\infty}^{\infty} X(f)\, e^{2\pi i f t}\, df,$$

$$y(t) = \int_{-\infty}^{\infty} Y(f)\, e^{2\pi i f t}\, df, \tag{4}$$

$$a_\nu(t) = \int_{-\infty}^{\infty} A_\nu(f)\, e^{2\pi i f t}\, df.$$

Now if $y(t)$ is differentiated ν times, we have

$$y^{(\nu)}(t) = \int_{-\infty}^{\infty} (2\pi i f)^\nu\, Y(f)\, e^{2\pi i f t}\, df.$$

Substituting in the differential equation

$$\sum_{\nu=0}^{n} a_\nu(t) y^{(\nu)}(t) = x(t).$$

Multiplying by $e^{-2\pi i s t}$ and integrating from $-T/2$ to $T/2$, we obtain:

$$\int_{-\infty}^{\infty} df\, \mathscr{A}(f, s)\, Y(f) = X(s) \tag{5}$$

as a formal relation between input and output transforms where:

$$\mathscr{A}(f, s) = \sum_{\nu} (2\pi i f)^\nu A_\nu(s - f). \tag{6}$$

However, it is clear that we need an expression for $Y(f)$ in terms of $X(s)$, since ensemble averages to be taken later must be separable and, while the x can be statistically independent of the a_ν, the output y cannot, except in a trivial case, and except for one special case of some interest considered by Samuels **[26]**. He analyzes random linear systems containing one or more nonindependent parameters under the restriction that the parameter processes and the solution or output process have very widely separated spectra. He considers an equation of our form $Ly = x$ with the $a_\nu(t)$ and the $x(t)$ as prescribed random functions, supposing as in our treatment that $x(t)$ is statistically independent of any $a_\nu(t)$ while the $a_\nu(t)$ may or may not be correlated with each other. Following his treatment we assume the mathematical expectation[11] $\langle a_\nu(t)\rangle$ of each coefficient a_ν exists and write

$$a_\nu(t) = \langle a_\nu(t)\rangle + \alpha_\nu(t)$$

so that α_ν represents the randomly fluctuating part of $a_\nu(t)$. The differential equation can now be written

$$Hy = \sum_0^n a_\nu(t) \frac{d^\nu y}{dt^\nu} = x(t) - \sum_0^n \alpha_\nu \frac{d^\nu y}{dt^\nu}.$$

[11] Mathematical expectation or statistical average or mean or ensemble average of a continuous random variable $g(x)$ is $E\{g(x)\} = \int_{-\infty}^{\infty} g(x)p(x)\, dx = \langle g(x)\rangle$.

Now the left side can again be viewed as an operator acting on y, but this is now a nonrandom operator H, and a Green's function can immediately be written, say $h(t, \tau)$, for the left side alone. Now the solution is

$$y(t) = \int_{t_0}^{t} h(t, \tau)x(\tau)\, d\tau - \int_{t_0}^{t} h(t, \tau) \sum_{\nu=0}^{n} \frac{d^\nu y(\tau)}{d\tau^\nu} \alpha_\nu(\tau)\, d\tau + \sum_{\nu=1}^{n} c_\nu \varphi(t),$$

where the φ_ν are a fundamental set of independent solutions of the homogeneous nonrandom equation $Hy = 0$ and the c_ν are arbitrary constants. The upper limit can be ∞ for a realizable system. The lower limit is a matter of memory. Using t_0 instead of $-\infty$ implies simply that all earlier values are not significant. Considering infinite limits

$$y(t) = F(t) - \int_{-\infty}^{\infty} K(t, \tau)y(\tau)\, d\tau,$$

where

$$K(t, \tau) = h(t, \tau) \sum_{\nu=0}^{n} \alpha_\nu(\tau) \frac{d^\nu}{d\tau^\nu}$$

$$= \sum_{\nu=0}^{n} (-1)^\nu \frac{\partial^\nu}{\partial\tau^\nu} [\alpha_\nu(\tau)h(t, \tau)],$$

$$F(t) = \int_{-\infty}^{\infty} h(t, \tau)x(\tau)\, d\tau + \sum_{\nu=1}^{n} c_\nu \varphi_\nu(t).$$

Now multiply $y(t_1)$ and $y^*(t_2)$ and average.

$$\langle y(t_1)y^*(t_2)\rangle = \langle F(t_1)F^*(t_2)\rangle - \int_{-\infty}^{\infty} \langle K(t, \tau)F^*(t_2)y(\tau)\rangle\, d\tau$$

$$- \int_{-\infty}^{\infty} \langle K^*(t_2, \tau)F(t_1)y^*(\tau)\rangle\, d\tau$$

$$+ \iint_{-\infty}^{\infty} \langle K(t_1, \tau_1)K^*(t_2, \tau_2)y(\tau_1)y^*(\tau_2)\rangle\, d\tau_1\, d\tau_2.$$

Assume the α_ν are stationary and ergodic and the $\langle a_\nu\rangle$ are constants so $h(t, \tau)$ is $h(t - \tau)$. Then it is true that $y(t)$ and $\alpha_\nu(t)y(t)$, etc., are stationary and ergodic. Also, Samuels states if the $\langle a_\nu(t)\rangle$ approach constants as $t \to \infty$ then the products in question are "almost ergodic" for large t. We then assume both K and y are ergodic. Now we replace ensemble averages by time averages,[12] e.g.,

$$E\{G_1(\xi)G_2(\xi)\} = \lim_{x\to\infty} \frac{1}{2x} \int_{-x}^{x} G_1(\xi)G_2(\xi)\, d\xi = \bar{G}_1\bar{G}_2$$

[12] Time averages are indicated by a bar over a single letter without exponents. For more complicated expressions we will use the symbol $[\ ]_{\text{av}}$.

if the rates at which the functions change are reasonably far apart, e.g., if G_1 is a rapidly changing quantity and G_2 is slowly varying, in which case the plot of the autocorrelation of G_2 is wide and that of G_1 is very narrow, so the spectral densities are well separated narrow peaks. (G_2 close to zero frequency and G_1 farther out.) This is also shown by integrating by parts assuming one rate, say $G_2'(\xi) \simeq 0$. Now we can make the separations

$$\langle K(t,\tau)F^*(t_2)y(\tau)\rangle = \langle K(t,\tau)\rangle\langle F^*(t_2)y(\tau)\rangle,$$
$$\langle K^*(t_2,\tau)F(t_1)y^*(\tau)\rangle = \langle K^*(t_2,\tau)\rangle\langle F(t_1)y^*(\tau)\rangle,$$
$$\langle K(t_1,\tau_1)K^*(t_2,\tau_2)y(\tau_1)y^*(\tau_2)\rangle = \langle K(t_1,\tau_1)K^*(t_2,\tau_2)\rangle\langle y(\tau_1)y^*(\tau_2)\rangle.$$

Substituting and noting $\langle K\rangle = 0$, since $\langle \alpha_\nu\rangle = 0$,

$$\langle y(t_1)y^*(t_2)\rangle = R_y(t_1, t_2),$$
$$\langle F(t_1)F^*(t_2)\rangle = R_F(t_1, t_2),$$

we now have

$$R_y(t_1, t_2) = R_F(t_1, t_2) + \iint_{-\infty}^{\infty} \langle K(t_1,\tau_1)K^*(t_2,\tau_2)\rangle R_y(\tau_1,\tau_2)\, d\tau_1\, d\tau_2.$$

The above kernel is given by

$$\langle K(t_1,\tau_1)K^*(t_2,\tau_2)\rangle = \sum_{j,k=0}^{n} (-1)^{j+k} \frac{\partial^{j+k}}{\partial\tau_1^k \partial\tau_2^j}\{\rho_{kj}(\tau_1,\tau_2)H(t_1,\tau_1)H(t_2,\tau_2)\},$$

which is a convenient form when the various cross correlations ρ_{kj} are given. If the cross-spectral densities are specified and $\langle a_\nu(t)\rangle$ are constants, a more convenient form is obtained by taking a double Fourier transform of $R_y(t_1, t_2)$ thus defining

$$\iint_{-\infty}^{\infty} e^{i(\zeta_1 t_1 + \zeta_2 t_2)} R_y(t_1, t_2)\, dt_1\, dt_2 = \Phi_y(\zeta_1, \zeta_2)$$

with inverse

$$R_y(t_1, t_2) = \frac{1}{4\pi^2} \iint_{-\infty}^{\infty} e^{-i(\zeta_1 t_2 + \zeta_2 t_2)} \Phi_y(\zeta_1, \zeta_2)\, d\zeta_1\, d\zeta_2.$$

We can now write

$$\Phi_y(\zeta_1, \zeta_2) = \Phi_F(\zeta_1, \zeta_2) + \sum_{k,j=0}^{n} \frac{(-1)^{k+j}}{4\pi^2} \iint_{-\infty}^{\infty} \Phi_H(\zeta_1)$$
$$\times\, \Phi_H(\zeta_2)\beta_1^k\beta_2^j \Phi_{\rho_{kj}}(\zeta_1 - \beta_1;\ \zeta_2 - \beta_2)\Phi_y(\beta_1, \beta_2)\, d\beta_1\, d\beta_2,$$

where $\Phi_{\rho_{kj}}$ is the cross-spectral density of the coefficients. If the α_ν are wide-sense stationary,

$$\Phi_{\rho_{kj}}(\zeta_1 - \beta_1; \zeta_2 - \beta_2) = 2\pi\Phi_{\rho_{kj}}(\zeta_2 - \beta_2)\delta(\zeta_1 - \beta_1 + \zeta_2 - \beta_2).$$

Now we have a second-order linear integral equation which can be solved in closed form. Samuels develops a mean square stability theory and considers some special cases, e.g., an RLC circuit with randomly varying capacitance. He applies Kirchoff's laws to get a differential equation, then considers mean-square stability, i.e., regions where

$$\lim_{t\to\infty} \langle y^2(t)\rangle < M$$

where M is a finite constant. Mean-square capacity deviation is plotted against damping of the circuit to show regions of stable operation.

These results, though interesting, are not sufficient for our purpose since they apply to a special case where the ensemble averages can be separated on the basis of rate. Furthermore, we would really like a solution for the output in terms of the input (see §1) and the parameter variation rather than in terms of an integral involving the output again.

At this point, we will first note that some interesting approximation methods are possible for special cases where $a_\nu(t)$ are given by

$$a_\nu(t) = \langle a_\nu(t)\rangle + \alpha_\nu(t) = a_\nu + \alpha_\nu(t)$$

where we suppose the a_ν are constant and the $\alpha_\nu(t)$ are zero-mean random processes such that $\langle \alpha_\nu(t)\rangle = 0$. Thus α_ν represents the randomly fluctuating part of $a_\nu(t)$. Now the differential equation $Ly = x$ becomes

$$Hy = \sum_{\nu=0}^{n} a_\nu \frac{d^\nu y}{dt^\nu} = x(t) - \sum_{\nu=0}^{n} \alpha_\nu \frac{d^\nu y}{dt^\nu}$$

where the left side is again an operator acting on y but a nonrandom operator for which a Green's function $h(t, \tau)$, actually $h(t - \tau)$ if a_ν is a constant as stated, can be written. The solution is

$$y(t) = \int_{-\infty}^{\infty} h(t, \tau)x(\tau)\,d\tau - \int_{-\infty}^{\infty} h(t, \tau)\sum_{\nu=0}^{n} \frac{d^\nu y(\tau)}{d\tau^\nu}\,\alpha_\nu(\tau)\,d\tau$$

if we neglect the solution of the homogeneous nonrandom equation $Hy = 0$. Let us call the first integral $F(t)$ and rewrite the second as $\int K(t, \tau)y(\tau)\,d\tau$. A Born approximation, or method of successive approximations, is promising now. Thus suppose $y(t)$ can be expanded in a series which is convergent with probability 1. Let

$$y(t) = \sum_{i=0}^{\infty} (-1)^i y_i(t).$$

Then

$$y(t) = F(t) - \int K(t, \tau)\sum_{i=0}^{\infty} (-1)^i y_i(\tau)\,d\tau.$$

If we write $y_0 = F(t) = \int h(t, \tau)x(\tau)\,d\tau$, we see each y_i is given in terms of the preceding one; thus,

$$y_i = \int K(t, \tau)y_{i-1}(\tau)\,d\tau \qquad (i = 1, 2, \cdots).$$

Convergence properties remain to be considered but one notes immediately the separation of the ensemble averages involving the y and the α

$$\langle K(t, \tau)F^*(t_2)y(\tau)\rangle = \langle K(t, \tau)F^*(t_2)[y_0(\tau) - y_1(\tau) + y_2(\tau) - \cdots]\rangle.$$

The first term is

$$\langle K(t, \tau)F^*(t_2)y_0(\tau)\rangle = \langle K(t, \tau)\rangle\langle F^*(t_2)F(\tau)\rangle = 0.$$

The second term

$$\langle K(t, \tau)F^*(t_2)y_1(\tau)\rangle = \int\langle K(t, \tau)K(\tau, \sigma)\rangle F^*(t_2)F(\sigma)\, d\sigma$$

etc. Where $L = \langle L\rangle + \Delta L$ and $\Delta L << \langle L\rangle$ so there is only a small amount of randomness, the problem simplifies considerably. A trivial case shows that $\langle y\rangle$ can be written immediately in terms of x and a kernel depending only on averages of L. A more interesting result is obtained by considering a second order differential equation [35].

$$\frac{d^2y}{dt^2} + 2n\frac{dy}{dt} + [1 + \varepsilon\xi(t)]y = 0 \qquad (n << 1)$$

where $\xi(t)$ is a random function and ε is a small parameter. We assume the correlation function $R(\Delta t) = \langle\xi(t + \Delta t)\xi(t)\rangle$ is known and $\langle\xi(t)\rangle = 0$. By the substitution of variables

$$z = e^{nt}y, \qquad \tau = t\sqrt{(1 + n^2)}$$

this equation is reduced to

$$\frac{d^2z}{d\tau^2} + z = -\varepsilon'\xi(\tau)z \qquad \varepsilon' = \frac{\varepsilon}{1 + n^2}$$

which is now in a form for successive approximations. Thus we see we can handle first order linear equations whose integration can be carried out [36] and second order equations where assumptions can be made of small fluctuations in the parameters. We are interested though in further consideration of the problem of finding the statistical characteristics of solutions directly and in avoiding the restriction to perturbation methods.

Now returning to equation (5), we define a new function $B(\sigma, s)$ by the following orthogonality requirement

$$\int_{-\infty}^{\infty} ds\, \mathscr{A}(f, s)B(\sigma, s) = \delta(f - \sigma). \tag{7}$$

From (5) and (7) we immediately obtain

$$Y(f) = \int_{-\infty}^{\infty} ds\, B(f, s)X(s). \tag{8}$$

The power spectrum (of the representative member of the ensemble we have chosen) is then given by

$$G_v(f) = \lim_{T\to\infty} \frac{2}{T}\iint_{-\infty}^{\infty} d\sigma\, ds\, B(\sigma, f)B^*(s, f)X(\sigma)X^*(s).$$

The spectrum of the process is $\langle G_v(f)\rangle$ or

$$\Phi_v(f) = \lim_{T\to\infty} \frac{2}{T}\iint_{-\infty}^{\infty} \langle B(\sigma, f)B^*(s, f)X(\sigma)X^*(s)\rangle\, d\sigma\, ds.$$

But B depends only on the coefficients a_v not on the input x so that the parameter processes and the input process are statistically independent. Thus,

$$\Phi_v(f) = \lim_{T\to\infty} \frac{2}{T}\iint_{-\infty}^{\infty} \langle B(\sigma, f)B^*(s, f)\rangle\langle X(\sigma)X^*(s)\rangle\, d\sigma\, ds,$$

$$\Phi_v(f) = \lim_{T\to\infty} \frac{2}{T}\iint_{-\infty}^{\infty} \langle B(\sigma, f)B^*(s, f)\rangle\left\langle\int_{-T/2}^{T/2} x(t)\, e^{-2\pi i\sigma t} dt \int_{-T/2}^{T/2} x^*(t)\, e^{2\pi i s t} dt\right\rangle d\sigma\, ds,$$

$$\Phi_v(f) = \lim_{T\to\infty} \frac{2}{T}\iint_{-\infty}^{\infty}\iint_{-T/2}^{T/2} d\sigma\, ds\, du\, dv\, \langle B(\sigma, f)B^*(s, f)\rangle\langle x(u)x^*(v)\rangle\, e^{2\pi i(sv - \sigma u)}.$$

If the input is stationary ($v = u + \tau$) this becomes

$$\lim_{T\to\infty} \frac{2}{T}\int\cdots\int d\sigma\, ds\, du\, d\tau\langle B(\sigma, f)B^*(s, f)\rangle\langle x(u)x^*(u + \tau)\rangle\, e^{2\pi i(s-\sigma)u}\, e^{2\pi i s\tau}.$$

But, $\langle x(u)x^*(u + \tau)\rangle = R_x(\tau)$ and $\int R_x(\tau)\, e^{2\pi i s\tau}\, d\tau = \Phi_x$ in the limit $T \to \infty$. Thus we can write our earlier result from §2

$$\Phi_v(f) = \int ds K_H(s, f)\Phi_x(s)$$

with K_H now defined by

$$\begin{aligned} K_H(s, f) &= \int_{-\infty}^{\infty} d\sigma \int_{-\infty}^{\infty} du\, \langle B(\sigma, f)B^*(s, f)\rangle\, e^{2\pi i(s-\sigma)u} \\ &= \int d\sigma\, \langle B(\sigma, f)B^*(s, f)\rangle \int du\, e^{2\pi i(s-\sigma)u} \\ &= \int d\sigma\, \delta(s - \sigma)\langle B(\sigma, f)B^*(s, f)\rangle \end{aligned}$$

or

(9) $$K_H(s, f) = \langle |B(s, f)|^2\rangle,$$

so that the stochastic kernel has been expressed in terms of the coefficients of the differential equation. While this result establishes the existence of the connection we were seeking in a surprisingly simple relationship, it can only be useful in solving problems or specific equations if we can evaluate it, using the definitions (6) and (7). Now the difficulties become clear, for if $\mathscr{A}$ and B are represented by matrices, these matrices must be orthogonal and the two are obviously not independent so we cannot make separations in any ensemble averaging. Thus, we have not shown a prescription for the solution of differential equations unless a method can be shown for finding the ensemble mean given by (9) in terms of the given statistics of the $a_\nu(t)$.

We return now to our condition defining B in terms of $\mathscr{A}$, i.e.,

$$\int_{-\infty}^{\infty} ds\, \mathscr{A}(f, s) B(\sigma, s) = \delta(f - \sigma).$$

If we write $B(s, \sigma)$ instead of $B(\sigma, s)$ (this was an arbitrary choice) and conceive the function $\mathscr{A}(f, s)$ to be a matrix by imagining the f as representing a continuous row index and s a continuous column index, the left-hand side looks like a matrix product while the right-hand side is similar to the identity matrix except that the Dirac δ in place of a Kronecker δ implies ∞ on the diagonal rather than a series of 1's. By analogy we are justified in applying the term "inverse" to B. More specifically, we call B an inverse functional of $\mathscr{A}$. Now, it is obvious that $\langle \mathscr{A}\mathscr{A}^* \rangle$ can be given explicitly in terms of the correlations R_{a_ν} for the random coefficients $a_\nu(t)$ in $\mathscr{L}$. To see this we use the definition of A_ν and write

$$\langle A_\nu(s - f) A_\nu^*(s - f) \rangle = \iint_{-T/2}^{T/2} \langle a_\nu(u) a_\nu(v) \rangle \, du\, dv.$$

Let $v = u + \tau$, then the above quantity becomes

$$\int_{-T/2}^{T/2} du \int_{-T/2-u}^{T/2-u} d\tau\, R_{a_\nu}(\tau).$$

Finally,

$$\langle \mathscr{A}(f, s) \mathscr{A}^*(f, s) \rangle = \sum_\mu \sum_\nu (-1)^\mu (2\pi i f)^{\mu+\nu} \int_{-T/2}^{T/2} du \int_{-T/2-u}^{T/2-u} d\tau\, R_{a_\nu}(\tau). \tag{10}$$

If we could write B in terms of $\mathscr{A}$, form $\langle BB^* \rangle$ and possibly expand in a series of functions or powers of $\langle \mathscr{A}\mathscr{A}^* \rangle$, we might have a computational method of interest. Of course, we are still suggesting the inversion of a matrix, but a much simpler one than the problem of inversion of the matrix corresponding to $\mathscr{L}$. Our matrices are continuous and infinite here but can be made discrete by originally using series instead of transforms. Consider the case of discrete, ordinary or deterministic matrices and the matrix product $AB = I$, where I is the identity matrix. Now B is called the inverse or reciprocal matrix and is indicated by A^{-1}. To see how A^{-1} is written in terms of A let A be represented by the array of elements:

$$A = \begin{Vmatrix} a_{11} & a_{12} \cdots a_{1n} \\ a_{21} & a_{22} \cdots \\ \cdot & \\ \cdot & \\ \cdot & \\ a_{n1} & a_{n2} \cdots a_{nn} \end{Vmatrix}.$$

We know, of course, that the determinant of A, or $|A|$, can be expanded by cofactors as

$$|A| = \sum_j a_{ij}A_{ij} = \sum_i a_{ij}A_{ij},$$

if we define A_{ij} as the cofactor of a_{ij}. We know further that $|A|$ vanishes if any two rows or columns of A are the same. Thus,

$$\sum_j a_{kj}A_{ij} = \sum_j a_{ji}A_{jk} = 0 \qquad i \neq k.$$

Combining these statements

$$\sum_j a_{ij}A_{kj} = \sum_j a_{ji}A_{jk} = |A|\,\delta_{ki}, \tag{11}$$

which says simply that expansion by a row or column i in terms of the cofactors of a row or column k vanishes if $i \neq k$ and is equal to $|A|$ when $i = k$.

Now equation (11) looks like a matrix product. Set $\hat{a}_{ij} = A_{ji}$ and (11) becomes

$$\sum_j a_{ij}\hat{a}_{jk} = \sum_j \hat{a}_{kj}a_{ji} = |A|\,\delta_{ki},$$

which can be written in matrix form as

$$A\hat{A} = \hat{A}A = |A|I$$

if

$$\hat{A} = \|\hat{a}_{ij}\| = \begin{Vmatrix} A_{11} & A_{21} \cdots & A_{n1} \\ A_{12} \cdots & & \\ \cdot & & \\ \cdot & & \\ \cdot & & \\ A_{1n} \cdots & & A_{nn} \end{Vmatrix}.$$

Now following Hadley [37] we define

$$A^{-1} = (1/|A|)\hat{A}. \tag{12}$$

Then

$$AA^{-1} = A^{-1}A = A\hat{A}/|A| = I.$$

If A is nonsingular, it has an inverse defined by (12) and further, the inverse is unique.

To sum up we say if $AB = I$ then $B = \hat{A}/|A|$ where we write $\hat{A}$ to mean the so-called adjoint matrix of A where this overworked term is not to be confused with the Hermitian adjoint $A^{\dagger}$ since $\hat{A}$ is found from the transpose of the cofactors of the elements of A not the transpose of the complex conjugates of the elements of A.

Thus, in our own problem with the defined (but discretized) $\mathscr{A}$ and B we write

$$\langle BB^*\rangle = \left\langle\frac{\hat{\mathscr{A}}\hat{\mathscr{A}}^*}{|\mathscr{A}||\mathscr{A}|^*}\right\rangle = \left\langle\frac{\hat{\mathscr{A}}\hat{\mathscr{A}}^*}{|\mathscr{A}\mathscr{A}^{\dagger}|}\right\rangle,$$

which means precisely the average of the following array

$$\frac{\begin{Vmatrix} A_{11} & A_{21} & \cdots & A_{n1} \\ A_{12} & \cdots & & \\ \cdot & & & \\ \cdot & & & \\ \cdot & & & \\ A_{1n} & \cdots & & A_{nn} \end{Vmatrix} \begin{Vmatrix} A^*_{11} & A^*_{21} & \cdots & A^*_{n1} \\ \cdot & & & \\ \cdot & & & \\ \cdot & & & \\ A^*_{1n} & \cdots & & A^*_{nn} \end{Vmatrix}}{\begin{vmatrix} a_{11} & a_{12} & \cdots & a_{1n} \\ a_{21} & a_{22} & \cdots & \\ \cdot & & & \\ \cdot & & & \\ \cdot & & & \\ a_{n1} & \cdots & & a_{nn} \end{vmatrix} \begin{vmatrix} a^*_{11} & a^*_{12} & \cdots & a^*_{1n} \\ \cdot & & & \\ \cdot & & & \\ \cdot & & & \\ a^*_{n1} & \cdots & & a^*_{nn} \end{vmatrix}},$$

where a_{ij} are the elements of $\mathscr{A}$ and A_{ij} are the cofactors of a_{ij}.

The difficulty in finding $|\mathscr{A}|$ and the above expression is avoided in quantum theory by using approximate representations in which $\mathscr{A}$ is diagonal. The general problem here appears to be enormously difficult even with arbitrary limitations of the number of rows and columns for successive approximations. Since we do know $\langle\mathscr{A}\mathscr{A}^*\rangle$ explicitly it is logical to ask if we can write our expression in some series of powers of $\langle\mathscr{A}\mathscr{A}^*\rangle$ or some function of $\langle\mathscr{A}\mathscr{A}^*\rangle$ which could then be written as a function of R_{a_ν}. It is not clear how the desired expansion is to be made so the problem is not completed. In the referenced dissertation a differential equation with a random Gaussian coefficient is studied at length and a solution is obtained. Perhaps then if the expansion is used for the same problem, we will be able to see how it can be done or if $\langle\mathscr{A}\mathscr{A}^*\rangle$ is already the first term. We know the series terminates in that problem and we can see that the kernel is a function of the correlation of the random coefficient alone which is easily calculable. It seems reasonable to expect that our desired calculation can be made for that same (Gaussian) case.

Whether this general problem can be solved is still a question. However much else can be done and many possibilities exist for approximation and asymptotic methods. Further, we note that for equations where ordinary Green's functions can be found once explicit time dependences are obtained by making Karhunen–Loève expansions of the random coefficients, we can then obtain stochastic Green's functions. Conditions under which this can be done have been discussed in a recent paper [1] and in a dissertation submitted in April 1961. However, much work remains to be done in investigation of the connections between this and the matrix method. Also, it is often true that knowledge of a Green's function for a simple nonrandom problem can be used to generate stochastic Green's functions or approximate stochastic Green's functions for randomized versions of the same problem, i.e., the class of known Green's functions can be thought of as generating stochastic Green's functions for more difficult problems involving random media or properties. Examples have been discussed elsewhere [1] and more extensive treatments in terms of the general matrix problem will appear in a forthcoming book. A rigorous reformulation will be desirable connecting with much recent work in obtaining probabilistic analogues of well-known results in the theory of deterministic operator equations in studying properties of mappings in generalized spaces, existence, uniqueness, convergence, measurability properties, and applications to problems in physics.

References

1. G. Adomian, *Linear stochastic operators*, Rev. Modern Phys. **35** (1963), 185–207.
2. A. Spaček, *Zufallige Gleichungen*, Czechoslovak Math. J. **5** (**80**) (1955), 462–466.
3. O. Hanš, *Random operator equations*, Proc. 4th Berkeley Sympos. on Math. Statist. and Prob., Vol. II, pp. 185–202, Univ. of California Press, Berkeley, Calif., 1961.
4. J. B. Keller, *Propagation through random media*, Proc. Sympos. Appl. Math., Vol. 16, pp. 145–170, Amer. Math. Soc., Providence, R. I., 1964.
5. W. C. Hoffman, *The electromagnetic field in a randomly inhomogeneous medium*, Proc. Sympos. Appl. Math., Vol. 16, pp. 117–144, Amer. Math. Soc., Providence, R. I., 1964.
6. M. Kac, *Lectures in probability and related topics in physical sciences*, Interscience, New York, 1959.
7. J. L. Lawson and G. E. Uhlenbeck, *Threshold signals*, Massachusetts Institute of Technology Radiation Laboratory Series, Vol. 24, McGraw-Hill, New York, 1950.
8. A. Kolmogorov, *Grundbegriffe der Wahrscheinlichkeitsrechnung*, Springer, Berlin, 1933; *Foundations of the theory of probability*, Chelsea, New York, 1948.
9. M. C. Wang and G. E. Uhlenbeck, *On the theory of the brownian motion*. II, Rev. Modern Phys. **17** (1945), 323–342.
10. A. Markoff, *Extension of the law of large numbers to dependent events*, Bull. Soc. Phys. Math. Kazan (USSR) (2) **15** (1906), 135–156.
11. H. Cramer, *Mathematical methods of statistics*, Princeton Univ. Press, Princeton, N. J., 1946.
12. G. D. Birkhoff, *Proof of the ergodic theorem*, Proc. Nat. Acad. Sci. U. S. A. **17** (1931), 650–655, 656–660.
13. A. Ya. Khinchin, *Mathematical foundations of statistical mechanics*, Chapters 2, 3, Dover, New York, 1953.
14. J. L. Doob, *Stochastic processes*, Chapters 10, 11, Wiley, New York, 1953.
15. M. Loève, *Probability theory*, Chapter 9, Van Nostrand, Princeton, N. J., 1955.
16. M. Kac and A. J. F. Siegert, *On the theory of noise in radio receivers with square law detectors*, J. Appl. Phys. **18** (1947), 383–397.
17. A. Blanc-Lapierre and R. Fortet, *Théorie des fonctions aléatoires*, Masson et Cie, Paris, 1953.

18. A. J. F. Siegert and D. A. Darling, *On the distribution of certain functionals of Markov processes*, Rand report P429, 1954.

19. A. J. F. Siegert, *Passage of stationary processes through linear devices*, IRE Trans. **PGIT-3** (1954).

20. D. A. Darling and A. J. F. Siegert, *A systematic approach to a class of problems in the theory of noise and other random phenomena*, IRE Trans. on Information Theory, **3** (1957), no. 1, 32–37; 38–43. March, 1957.

21. J. C. Samuels and A. C. Eringen, *On stochastic linear systems*, Tech. Rep. No. 11, Division of Engineering Sciences, Purdue University, Lafayette, Ind., 1957.

22. W. B. Davenport and W. L. Root, *An introduction to the theory of random signals and noise*, McGraw-Hill, New York, 1958.

23. J. E. Moyal, *Stochastic processes and statistical physics*, J. Roy. Statist. Soc. Ser. B **11** (1949), 150–210.

24. P. Lévy, *Processus stochastiques et mouvement brownien*, Gauthier-Villars, Paris, 1948.

25. R. Fortet, *Recent advances in probability theory*, Some aspects of analysis and probability, pp. 169–240, Wiley, New York, 1958.

26. J. C. Samuels, IRE Trans. Circuit Theory, CT–6 (Special Supplement), 1959.

27. M. Sundstrom, *Some statistical problems in the theory of servomechanisms*, Ark. Mat. **2** (1951), 52.

28. A. Rosenbloom, *Analysis of randomly time-varying linear systems*, Ph.D. thesis, University of California, Los Angeles, Calif., 1954 (unpublished).

29. J. F. Buchan and R. S. Raven, *Gain modulation of servomechanisms*, IRE (Wescon) Convention Record, Part 4, August 1957 (unpublished).

30. P. G. Bergman, *Propagation of radiation in a medium with random inhomogeneities*, Phys. Rev. (2) **70** (1946), 486.

31. D. Mintzer, *Wave propagation in randomly inhomogeneous medium*. III, J. Acoust. Soc. Amer. **26** (1954), 186–190.

32. L. A. Chernov, *Wave propagation in a random medium*, McGraw-Hill, New York, 1960.

33. F. Villars and V. F. Weisskopf, *The scattering of electromagnetic waves by turbulent atmospheric fluctuations*, Phys. Rev. (2) **94** (1954), 232–240.

34. M. Sundstrom, Tech. Notes TN 45–51, Royal Institute of Technology, Stockholm, Sweden (unpublished).

35. B. Chelpanov, *Vibration of a second-order system with a randomly varying parameter*, Appl. Math. Mech. **26** (1962), no. 4, 1145.

36. V. I. Tihonov, *The effect of fluctuation action in the simplest parametric systems*, Automat. Remote Control **19** (1958), 705.

37. G. Hadley, *Linear algebra*, Addison-Wesley, Reading, Mass., 1961.

HUGHES AIRCRAFT COMPANY
CULVER CITY, CALIFORNIA

ON THE THEORY OF RANDOM EQUATIONS

BY

A. T. BHARUCHA-REID[1]

1. **Introduction.** The purpose of this paper is to present a brief survey of the theory of random equations. Random equations arise in virtually every branch of physics and engineering, as well as in other areas of science and technology; and many specific types of random equations have been studied by physicists and engineers. In order to understand how random equations arise, and why they are important, we have only to look at the experimental procedures used to investigate various physical phenomena and the mathematical equations, or models, used to describe these phenomena.

The mathematical equation, or equations, used to describe a given physical phenomenon usually contains one or more coefficients, or parameters, which have some definite physical interpretation. For example, in the theory of diffusion we have the diffusion coefficient, in the theory of wave propagation the propagation coefficient, and in the theory of elasticity the modulus of elasticity. In all of the cases mentioned, the magnitudes of the coefficients are experimentally determined. When solving the appropriate mathematical equation, and in subsequent calculations, it is usually the *mean value* of a set of experimental observations that is used as the value of the coefficient or parameter. In some instances this may provide an adequate description; but in many instances the variance may be sufficiently large to warrant consideration. Hence, when we talk about physical constants we are not, in many instances, talking about constants at all, but random variables whose values are determined by some probability distribution. Therefore a more realistic approach would be to introduce the coefficient or parameter as a random variable, obtain the "random solution" of the mathematical equation, and then study its statistical properties.

Random equations arise in many other ways; for example, in the study of wave propagation in stochastic media, in the study of random vibration problems, in magnetohydrodynamics when the magnetic field is subject to random fluctuations, in electrical circuit theory when, for example, the capacitor is subject to random variation, in the study of differential equations with random initial and/or boundary conditions, and in the study of integral equations with random kernels, or defined over random domains. We could go on, but the above examples should suffice.

[1] Supported by National Science Foundation Grant G–24334 and U. S. Army Research Office (Durham) Grant No. DA–ARO(D)–31–124–G124.

Although random equations arise in many different branches of physics and engineering, and have been studied by many workers, it is only recently that attempts have been made to develop a theory of random equations. The study of random equations, employing the methods of functional analysis, was initiated by the Prague school of probabilists under the direction of the late Antonín Špaček. Their research was motivated by the importance of random equations in applied mathematics, and as a concrete application of results in probabilistic functional analysis.[2]

The remainder of this paper is divided into six sections. In §2 we present some theory of random equations. The main purposes of this section are to give some definitions and theorems from probabilistic functional analysis that are used in the theory of random equations, and to give some basic results concerning random equations and their solutions. §§3–6 are devoted to different classes of random equations, and in these sections we outline some of the results that have been obtained for random algebraic equations, random difference equations, random differential equations, and random integral equations. In §7, the last section, we say a few words about stochastic processes associated with random equations.

We make no attempt in this paper to discuss the numerous applications of random equations. A reasonably complete and unified treatment of random equations and their applications, together with a complete bibliography, will be given in our monograph **[19]**.

2. Some theory of random equations.

2.1. GENERALIZED RANDOM VARIABLES AND RANDOM OPERATORS. In this subsection we give certain basic definitions and theorems which will be used throughout this section and the remainder of the paper. Let $(\Omega, \mathscr{A}, \mu)$ denote a probability measure space, i.e., Ω is a nonempty abstract set, $\mathscr{A}$ is a σ-algebra of subsets of Ω, and μ is a complete probability measure on $\mathscr{A}$; and let $(\mathscr{X}, \mathscr{B})$ be a measurable space, where $\mathscr{X}$ is a complex, separable Banach space, and $\mathscr{B}$ is the σ-algebra of all Borel subsets of $\mathscr{X}$.

DEFINITION 2.1.[3] *A mapping $x(\omega)$ of Ω into $\mathscr{X}$ is called a generalized random variable (with values in the Banach space $\mathscr{X}$) if $\{\omega: x(\omega) \in B\} \in \mathscr{A}$ for all $B \in \mathscr{B}$.*

DEFINITION 2.2. *A mapping $T(\omega)$ of the Cartesian product space $\Omega \times \mathscr{X}$ into a Banach space $\mathscr{Y}$ is called a random transformation if $T(\omega)[x]$ is for every $x \in \mathscr{X}$ a generalized random variable with values in $\mathscr{Y}$.*

DEFINITION 2.3. *A random transformation $T(\omega)$ is said to be*

(i) *linear, if $T(\omega)[\alpha x_1 + \beta x_2] = \alpha T(\omega)[x_1] + \beta T(\omega)[x_2]$ for every $\omega \in \Omega$, x_1, $x_2 \in \mathscr{X}$ and $\alpha, \beta \in R$;*

---

[2] Probabilistic functional analysis is concerned with the application, and extension, of the methods of functional analysis in the study of the various concepts, processes, and structures that arise in probability theory and its applications.

[3] In applications where a parameter set Θ, say $\Theta = \{t: t \geqq 0\}$, is introduced, we will say that the mapping $x(\omega, t)$ on $\Omega \times \Theta$ into $\mathscr{X}$ is a generalized random variable if $\{\omega: x(\omega, t) \in B\} \in \mathscr{A}$ for all $B \in \mathscr{B}$ and for every fixed $t \in \Theta$.

(ii) *bounded, if there exists a mapping* $c(\omega)$ *on* Ω *into* R *(i.e.,* $c(\omega)$ *is a real-valued random variable) such that for all* $\omega \in \Omega$ *and* $x \in \mathscr{X}$, $\|T(\omega)x\| \leqq c(\omega)\|x\|$.

DEFINITION 2.4. *A random transformation* $T(\omega)$ *on* $\Omega \times \mathscr{X}$ *into* $\mathscr{X}$ *is called a random operator; and a bounded linear random operator on* $\Omega \times \mathscr{X}$ *to* $\mathscr{X}$ *is called a random endomorphism of* $\mathscr{X}$.

Several examples might clarify the above definitions.

EXAMPLE 1. The $n \times n$ matrix $A = (a_{ij})$, where the matrix elements $a_{ij} = a_{ij}(\omega)$, $\omega \in \Omega$. $A(\omega)$ is then a random operator on $\Omega \times R_n$ to R_n.

EXAMPLE 2. The ordinary differential operator $Lx = \sum_{i=0}^{n} a_i(t)d^i x/dt^i$, $x \in R$, say, where the coefficients $a_i(t) = a_i(\omega, t)$, $\omega \in \Omega$. $L(\omega)$ is then a random operator on $\Omega \times R$ to R.

EXAMPLE 3. The partial differential operator $Hu = D\partial^2 u/\partial x^2$, $u \in C[-\infty, \infty]$, where the coefficient $D = D(\omega)$, $\omega \in \Omega$. $H(\omega)$ is then a random operator on $\Omega \times C[-\infty, \infty]$ to $C[-\infty, \infty]$.

EXAMPLE 4. The integral operator

$$Tx = \int_a^b K(t, u)x(u)\, du, \qquad x \in C[a, b],$$

where the kernel $K(t, u) = K(\omega, t, u)$, $\omega \in \Omega$. $T(\omega)$ is then a random operator on $\Omega \times C[a, b]$ to $[Ca, b]$.

We will denote by $E(\mathscr{X})$ the Banach algebra of endomorphisms of a Banach space $\mathscr{X}$, and by $\mathscr{C}$ the σ-algebra of all Borel subsets of the space $E(\mathscr{X})$, provided the norm topology is assumed.

DEFINITION 2.5. *The mapping* $S(\omega)$ *is said to be the inverse of the random operator* $T(\omega)$ *if* $\mu\{\omega: T(\omega)[S(\omega)x] = x \text{ for every } x \in \mathscr{X}\} = 1$.

DEFINITION 2.6. *Let* $\mathscr{X}^*$ *denote the adjoint space of* $\mathscr{X}$. *The mapping* $T^*(\omega)$ *of the Cartesian product space* $\Omega \times \mathscr{X}^*$ *into* $\mathscr{X}^*$ *is called the random adjoint of* $T(\omega)$ *if, for every* $\omega \in \Omega$, *the equality*

$$y^* = T^*(\omega)[x^*]$$

is equivalent to the equality

$$x^*(T(\omega)[y]) = y^*(y)$$

for every $y \in \mathscr{X}$.

We now state two basic theorems due to O. Hanš [**39**; **42**].

THEOREM 2.1. *If* $x_1(\omega), x_2(\omega), \cdots$ *is a sequence of generalized random variables with values in* $\mathscr{X}$ *converging almost surely to a mapping* $x(\omega)$ *of* Ω *into* $\mathscr{X}$, *then* $x(\omega)$ *is a generalized random variable with values in* $\mathscr{X}$.

THEOREM 2.2. *Let* $x(\omega)$ *be a generalized random variable with values in* $\mathscr{X}$, *and let* $T(\omega)$ *be a continuous random operator on* $\Omega \times \mathscr{X}$ *to* $\mathscr{X}$. *Then the mapping* $y(\omega)$ *of* Ω *into* $\mathscr{X}$ *defined for every* $\omega \in \Omega$ *by* $y(\omega) = T(\omega)[x(\omega)]$ *is a generalized random variable with values in* $\mathscr{X}$.

2.2. RANDOM EQUATIONS. Let us first consider the *deterministic operator equation*

$$T[\xi] = y, \tag{2.1}$$

where ξ, y are elements of a Banach space $\mathscr{X}$, and T is an operator on $\mathscr{X}$. Let S denote the set

$$S = \{x: T[x] = y\}.$$

We will call S the *solution set* of equation (2.1), and say that an element x of S is a *solution* of equation (2.1). If $S = \varnothing$ (the empty set), we say that equation (2.1) does not possess a solution; and if $S \neq \varnothing$ we say that equation (2.1) is solvable. If S has only one element we say that equation (2.1) has a unique solution.

We now consider some stochastic analogues of equation (2.1). Firstly, we might consider $y = y(\omega)$ to be a generalized random variable with values in $\mathscr{X}$. Hence equation (2.1) becomes

$$T[x] = y(\omega). \tag{2.2}$$

In this case, if T^{-1} exists, $x(\omega) = T^{-1}[y(\omega)]$ will also be a generalized random variable with values in $\mathscr{X}$. Secondly, we might take $T = T(\omega)$ to be a random operator on $\Omega \times \mathscr{X}$ to $\mathscr{X}$, and y to be a deterministic element. In this case

$$T(\omega)[x] = y \tag{2.3}$$

will be called a *random operator equation*; and $x(\omega) = T^{-1}(\omega)[y]$ will be, for the set of ω's for which $T^{-1}(\omega)$ exists, a generalized random variable with values in $\mathscr{X}$. Thirdly, we have the most general case of a random operator equation in which $T(\omega)$ is a random operator on $\Omega \times \mathscr{X}$ to $\mathscr{X}$ and $y = y(\omega)$ is a generalized random variable with values in $\mathscr{X}$. Hence equation (2.1) becomes

$$T(\omega)[x] = y(\omega); \tag{2.4}$$

and the solution $x(\omega) = T^{-1}(\omega)[y(\omega)]$ will be a generalized random variable with values in $\mathscr{X}$.

With respect to solutions of equation (2.4), we introduce the following definitions.

DEFINITION 2.7. *Every mapping $x(\omega)$ of Ω into $\mathscr{X}$ satisfying the equality $T(\omega)[x(\omega)] = y(\omega)$ for every $\omega \in \Omega_0$, where $\mu(\Omega_0) = 1$, is said to be a wide sense solution of equation* (2.4).

If, in addition, the wide sense solution is measurable, we have

DEFINITION 2.8. *Every generalized random variable $x(\omega)$ with values in $\mathscr{X}$ satisfying the condition*

$$\mu\{\omega: T(\omega)[x(\omega)] = y(\omega)\} = 1 \tag{2.5}$$

is said to be a random solution of the random operator equation (2.4).

The following is an example of a wide sense solution which is not a random solution (cf. **[42]**). Let $\mathscr{X} = R$ (the space of all real numbers), E a nonmeasurable subset of Ω, and let $T(\omega)$ be a random operator on $\Omega \times R$ into R, defined for every $\omega \in \Omega$ and $x \in R$ by $T(\omega)[x] = x^2 - 1$. Then the mapping $\xi(\omega)$ of Ω into R defined by

$$\xi(\omega) = \begin{cases} 1, & \text{for } \omega \in E, \\ -1, & \text{for } \omega \in \Omega - E \end{cases}$$

is a wide sense solution, but not a random solution of the homogeneous random operator equation $T(\cdot)[x(\cdot)] = 0$.

2.3. EXISTENCE, UNIQUENESS, AND MEASURABILITY OF THE RANDOM SOLUTION OF RANDOM OPERATOR EQUATIONS. In order to study the existence, uniqueness, and measurability of the random solutions of random operator equations we will use probabilistic analogues of some well-known results in the theory of deterministic operator equations. It is well known that the principle of contraction mappings and fixed point theorems play a very important role in connection with the existence and uniqueness theory of linear (and nonlinear) operator equations [**68**]. Let $\mathscr{X}$ be an arbitrary metric space with distance function $d(x_1, x_2)$. A mapping T of $\mathscr{X}$ into itself is said to be a *contraction mapping* if there exists a number $c < 1$ such that $d(Tx_1, Tx_2) \leqq cd(x_1, x_2)$ for any two elements $x_1, x_2 \in \mathscr{X}$. The *Principle of Contraction Mappings* can be stated as follows: *Every contraction mapping defined on a complete metric space $\mathscr{X}$ has one and only one fixed point; that is, if T is a contraction on $\mathscr{X}$, the equation $T[x] = x$ has one and only one solution.*

Early in the development of the theory of random equations workers in probabilistic functional analysis realized the importance of obtaining stochastic analogues of classical fixed point theorems; and in this connection O. Hanš [**38**; **40**] and A. Špaček [**84**] initiated a theory of random contraction operators, and obtained a stochastic generalization of the Banach–Cacciopoli fixed point theorem. It would also be of interest to obtain a stochastic analogue of the Schauder fixed point theorem [**80**], which can be stated as follows:[4] *Let T be a completely continuous operator which transforms a bounded, closed, convex set E of a Banach space $\mathscr{X}$ into a subset of E. Then there exists at least one point $x \in E$ such that $T[x] = x$.*

DEFINITION 2.9. *A random operator $T(\omega)$ on $\Omega \times \mathscr{X}$ to $\mathscr{X}$ is said to be a random contraction operator if there exists a real-valued random variable $c(\omega)$ such that $c(\omega) < 1$ for all $\omega \in \Omega$, and such that*

$$\|T(\omega)[x_1] - T(\omega)[x_2]\| \leqq c(\omega)\|x_1 - x_2\| \tag{2.6}$$

for every $\omega \in \Omega$ and for every pair of generalized random variables $x_1, x_2 \in \mathscr{X}$. If $c(\omega) = c$ (a constant) for all $\omega \in \Omega$, we say that $T(\omega)$ is a uniform random contraction operator.

The stochastic analogue of the Banach–Cacciopoli theorem can be stated as follows.

[4] For applications of Schauder's theorem in the theory of integral equations we refer to [**69**].

THEOREM 2.3. *Let $T(\omega)$ be a random contraction operator on $\Omega \times \mathscr{X}$ to $\mathscr{X}$. Then there exists a generalized random variable $\xi(\omega)$ with values in $\mathscr{X}$ such that*

$$\mu\{\omega: T(\omega)[\xi(\omega)] = \xi(\omega)\} = 1. \tag{2.7}$$

The generalized random variable $\xi(\omega)$ is unique in the sense that if $\psi(\omega)$ is another generalized random variable satisfying (2.7), then $\mu\{\omega: \xi(\omega) = \psi(\omega)\} = 1$. Further, the random fixed point $\xi(\omega)$ can be obtained by successive approximations starting from an arbitrary generalized random variable $x_0(\omega) \in \mathscr{X}$.

Theorem 2.3 has been widely used in the theory of random equations; for if conditions can be found such that a given random operator is a contraction, then Theorem 2.3 can be employed to give the existence, uniqueness, and measurability of the solution, and the theorem also asserts that this solution can be obtained by the method of successive approximations.

Finally, we state a random fixed point theorem, also due to O. Hanš [**40**], which involves a sequence of random operators.

THEOREM 2.4. *Let $T_0(\omega), T_1(\omega), T_2(\omega), \cdots$ be a sequence of continuous random operators on $\Omega \times \mathscr{X}$ to $\mathscr{X}$ satisfying the following conditions:*

(i) *there exists a mapping $c(\omega)$ on Ω to R with $0 \leqq c(\omega) < 1$ for every $\omega \in \Omega$ and a generalized random variable $\xi(\omega)$ such that for every $x \in \mathscr{X}$*

$$\mu\{\omega: d(T_0(\omega)[x], \xi(\omega)) \leqq c(\omega)\, d(x, \xi(\omega))\} = 1;$$

(ii) *the sequence $T_1(\omega), T_2(\omega), \cdots$ converges uniformly (with respect to $\mathscr{X}$) almost surely to $T_0(\omega)$.*

Let $y_0(\omega)$ be an arbitrary generalized random variable with values in $\mathscr{X}$, and define for every $n = 0, 1, \cdots$ the mapping $y_{n+1}(\omega)$ of Ω into $\mathscr{X}$ for every $\omega \in \Omega$ as follows:

$$y_{n+1}(\omega) = T_n(\omega)[y_n(\omega)]. \tag{2.8}$$

Then $y_1(\omega), y_2(\omega), \cdots$ is a sequence of generalized random variables which converges almost surely to the generalized random variable $\xi(\omega)$.

Theorem 2.4 asserts that if an arbitrary sequence of continuous random operators converges uniformly almost surely to a random operator which is a contraction at its fixed point, then the sequence of generalized random variables defined by (2.8) converges almost surely to a generalized random variable which is the fixed point of the limit random contraction operator. As we shall point out later, Theorem 2.4 and analogous theorems are required for the study of approximate solutions of random operator equations.

For other random fixed point theorems we refer to O. Hanš [**40**], and O. Hanš and A. Špaček [**43**].

2.4. RESOLVENT AND SPECTRUM OF A RANDOM OPERATOR. Let U be a bounded closed subset in finite dimensional Euclidean space, and let $x(u)$ be a Lebesgue measurable function defined on U. A large number of operator equations can be written in the form

$$(T - \lambda I)x(u) = y(u), \tag{2.9}$$

where x and y are elements of a concrete Banach space $\mathscr{X}$, T is a linear, not necessarily bounded, operator with domain and range in $\mathscr{X}$, I the identity operator, and λ a scalar.

In this section we consider the stochastic analogue of equation (2.9), namely the random operator equation

$$(T(\omega) - \lambda I)x(\omega, u) = y(\omega, u). \tag{2.10}$$

The main problem we wish to consider is that of the existence and measurability of the random resolvent operator associated with the random linear operator $T(\omega)$. Let

$$T_\lambda(\omega) = T(\omega) = \lambda I, \tag{2.11}$$

and let $(\Lambda, \mathscr{F})$ be a Borel space.[5]

DEFINITION 2.10. *Consider the Borel space $(\Lambda, \mathscr{F})$ where Λ is the complex plane, and $\mathscr{F}$ is the σ-algebra of all Borel sets of complex numbers. The set of those pairs $(\omega, \lambda) \in \Omega \times \Lambda$ for which $T_\lambda(\omega)$ has a linear bounded inverse, say $R(\lambda; T(\omega))$, with domain dense in $\mathscr{X}$, is called the resolvent set $\rho(T(\omega))$ of $T(\omega)$. The operator*

$$R(\lambda; T(\omega)) = T_\lambda^{-1}(\omega) = (T(\omega) - \lambda I)^{-1} \tag{2.12}$$

will be called the resolvent of $T(\omega)$. The set of pairs (ω, λ) for which no inverse of $T_\lambda(\omega)$ exists is called the point spectrum $P\sigma(T(\omega))$ of $T(\omega)$. The set of pairs (ω, λ) for which $T_\lambda(\omega)$ has an unbounded inverse with domain in $\mathscr{X}$ is called the continuous spectrum $C\sigma(T(\omega))$ of $T(\omega)$; and the set of pairs (ω, λ) for which $T_\lambda(\omega)$ has an inverse whose domain is not dense in $\mathscr{X}$ is called the residual spectrum $R\sigma(T(\omega))$ of $T(\omega)$.

We now state two theorems due to C. Ryll-Nardzewski [75].

THEOREM 2.5. *Let Z be a measurable[6] subset of $\Omega \times \Lambda$, and let Z_λ and Z_ω denote the sets $\{\lambda: (\omega, \lambda) \in Z\}$ and $\{\omega: (\omega, \lambda) \in Z\}$, respectively. If the sets Z_ω are countable for almost all $\omega \in \Omega$ (in the sense of the measure μ, where μ is σ-finite), then $\mu(Z_\lambda) = 0$ for all $\lambda \in \Lambda$ with the exception of a countable number of λ's; that is, the cardinality of the set $\{\lambda: \mu(Z_\lambda) > 0\} \leqq \aleph_0$.*

Theorem 2.5, which is of independent interest, has been used to prove the following theorem, which gives sufficient conditions that the point spectrum of a random operator be countable.

---

[5] A *Borel space* is an abstract set Λ for which there exists a 1-1 point transformation φ on a Borel set E (lying in a complex metric space), and the class $\mathscr{F}$ of subsets of Λ is mapped by φ on the class of all Borel subsets of E.

[6] Here measurability is with respect to the product σ-algebra $\mathscr{A} \times \mathscr{F}$; that is, the smallest σ-algebra containing all rectangles $A \times F$, where $A \in \mathscr{A}$ and $F \in \mathscr{F}$.

THEOREM 2.6. *Let $T(\omega)$ be a random operator on $\Omega \times \mathscr{X}$ to $\mathscr{X}$. If* (1) *the operator-function $\{T(\omega)\}$ is weakly measurable (i.e., the function $x^*(T(\omega)[x])$, as a function of ω is measurable for all $x^* \in \mathscr{X}^*$ and $x \in \mathscr{X}$), and* (2) *for almost every $\omega \in \Omega$, the set of complex numbers for which a bounded inverse of $T_\lambda(\omega)$ does not exist is countable, then the set of λ's for which a bounded inverse $T_\lambda^{-1}(\omega)$ on any set of ω's of positive measure does not exist is also countable.*

The next theorem, due to O. Hanš, describes the resolvent set $\rho(T(\omega))$ of a random endomorphism $T(\omega)$.

THEOREM 2.7. *Let $T(\omega)$ be a random endomorphism of $\mathscr{X}$, $E(\mathscr{X})$ the Banach algebra of endomorphisms of $\mathscr{X}$, $\mathscr{C}$ the σ-algebra of Borel subsets of $E(\mathscr{X})$, and Λ the complex plane. If $T(\omega)$ is such that $\{\omega: T(\omega) \in C\} \in \mathscr{A}$ for every $C \in \mathscr{C}$, then for every $\lambda \in \Lambda$, $\{\omega: (\omega, \lambda) \in \rho(T(\omega))\} \in \mathscr{A}$.*

The spectral properties of some concrete random operators, in particular, random matrices have been studied by several workers. We refer to the papers of F. Dyson **[26]**, R. Engleman **[28]**, U. Grenander **[34]**, M. L. Mehta and M. Gaudin **[63]**, and E. Wigner **[92]**.

The existence and measurability of the random resolvent operator $R(\lambda; T(\omega))$ is given by the following theorem (cf. **[14**; **42]**).

THEOREM 2.8. *Let $T(\omega)$ be a random endomorphism on $\Omega \times \mathscr{X}$ to $\mathscr{X}$; and let the random operator $T_\lambda(\omega)$ be invertible for each ω separately in the set*

$$\Omega_0(\lambda) = \{\omega: \|T(\omega)\| < |\lambda|, \lambda \neq 0\}.$$

Then the resolvent operator $R(\lambda; T(\omega))$ exists for every $\omega \in \Omega_0(\lambda)$, and $R(\lambda; T(\omega))$ is measurable with respect to the reduced σ-algebra $\Omega_0(\lambda) \cap \mathscr{A}$.

The above theorem can now be used to obtain the existence, uniqueness and measurability of the solution of equation (2.10). We have

THEOREM 2.9. *For every $\omega \in \Omega_0(\lambda)$ the solution $x(\omega, u)$ of equation* (2.10) *is, for every generalized random variable $y(\omega, u)$ with values in $\mathscr{X}$, given by*

$$x(\omega, u) = R(\lambda; T(\omega))[y(\omega, u)]; \tag{2.13}$$

and $x(\omega, u)$ is, for every fixed $u \in U$, measurable with respect to the σ-algebra $\Omega_0(\lambda) \cap \mathscr{A}$.

2.5. EXPECTED SOLUTIONS OF RANDOM EQUATIONS. In applications of the theory of random equations it is often of interest to determine the moments (in particular, the mean and variance) of the solution of a given random equation. With regard to the expectation (or mean), variance and higher moments of a generalized random variable, we introduce the following definitions.

DEFINITION 2.11. *The expectation $\bar{x}$ of a generalized random variable $x(\omega)$ is given by the Bochner integral*

$$\bar{x} = \mathscr{E}\{x(\omega)\} = \int_\Omega x(\omega)\, d\mu(\omega), \tag{2.14}$$

provided the integral exists. The variance of $x(\omega)$ is given, if it exists, by the integral

$$\mathscr{D}^2\{x(\omega)\} = \int_\Omega \|x(\omega) - \bar{x}\|^2\, d\mu(\omega). \tag{2.15}$$

Similarly, higher moments of order k of $x(\omega)$ are given, if they exist, by the integral

$$\mathscr{D}^k\{x(\omega)\} = \int_\Omega \|x(\omega) - \bar{x}\|^k\, d\mu(\omega). \tag{2.16}$$

We also give the definition of the conditional expectation of a generalized random variable.

DEFINITION 2.12. *Let $x(\omega)$ be a generalized random variable, the expectation (2.14) of which exists. Let $\mathscr{A}_0$ and $\mathscr{A}'_0$ be σ-algebras of subsets of Ω, where $\mathscr{A}'_0$ is either in $\mathscr{A}_0$ or differs from $\mathscr{A}_0$ by a set of measure zero. Then the conditional expectation of $x(\omega)$ relative to the σ-algebra $\mathscr{A}_0$, written $\mathscr{E}\{x(\omega)|\mathscr{A}_0\}$, is defined as a $\mathscr{A}'_0$-measurable, Bochner integrable function which satisfies the equation*

$$\int_A \mathscr{E}\{x(\omega)|\mathscr{A}_0\}\, d\mu(\omega) = \int_A x(\omega)\, d\mu(\omega), \tag{2.17}$$

for every $A \in \mathscr{A}_0$.

In §2.2 we introduced three types of random equations and indicated their formal solutions. For the expected solution of equation (2.2), we have, from (2.14)

$$\bar{x} = T^{-1}\int_\Omega y(\omega)\, d\mu(\omega) = T^{-1}[\bar{y}]. \tag{2.18}$$

For the expected solution of equation (2.3) we have

$$\bar{x} = \int_\Omega T^{-1}(\omega)[y]\, d\mu(\omega) = \overline{T^{-1}[y]}. \tag{2.19}$$

Similarly, we have for the expected solution of equation (2.4)

$$\bar{x} = \int_\Omega T^{-1}(\omega)[y(\omega)]\, d\mu(\omega) = \overline{T^{-1}[y]}. \tag{2.20}$$

While (2.19) and (2.20) are formally the same, in concrete situations they would most likely be different. The integral in (2.19) can only be defined over that subset of Ω for which $T^{-1}(\omega)$ exists and for which $T^{-1}(\omega)[y]$ defines a generalized random variable with values in $\mathscr{X}$. Similarly, the integral in (2.20) can only be defined for that subset of Ω for which $T^{-1}(\omega)$ exists and for which $T^{-1}(\omega)[y(\omega)]$ is a generalized random variable.

In §2.4 we considered the random operator equation $(T(\omega) - \lambda I)x(\omega, u) = y(\omega, u)$. In this case the expected solution $\bar{x}(u)$ is given by

$$\bar{x}(u) = \int_{\Omega_0(\lambda)} R(\lambda; T(\omega))[y(\omega, u)]\, d\mu(\omega). \tag{2.21}$$

So much for formalities. Many workers in a number of studies, too numerous to mention here, have investigated the mean and variance, and in some cases higher moments, of the solutions of some concrete random equations of the first three types considered above. In most of these studies the probability measure μ was assumed to be Gaussian; however, other assumptions have also been introduced. In the general case the problem of determining the moments of random solution seems to be rather difficult, and much more work needs to be done in this area.

In connection with random operator equations an important question arises, namely: What is the relationship between the expected solution, if it exists, of a given random operator equation and the solution of the associated deterministic operator equation? To make clear what we mean by the deterministic operator equation we give the definition of the expectation of a random operator. If $T(\omega)$ is a random operator, we have from Definitions 2.2 and 2.11 that $\mathscr{E}\{T(\omega)\} = \bar{T}$ is given by

$$\bar{T}[x] = \int_{\Omega} T(\omega)[x]\, d\mu(x), \tag{2.22}$$

where the Bochner integral is assumed to exist for all $x \in \mathscr{X}$. The above question can now be stated as follows: If $T(\omega)[x(\omega)] = y(\omega)$ is a random equation, and if ξ is the solution of the deterministic (mean) operator equation $\bar{T}[x] = \bar{y}$, under what conditions will

$$\mathscr{E}\{x(\omega)\} = \bar{x} = \xi ? \tag{2.23}$$

It is clear that in general the above relation will not obtain. However, we can give examples for which (2.23) does hold. The most trivial one is to consider a random equation with the probability measure μ Dirac measure; that is, $\mu(\{\omega_0\}) = 1$ and $\mu(\Omega - \{\omega_0\}) = 0$. If (2.23) did obtain in all cases we would conclude that the random solution simply takes into consideration the random fluctuations about the deterministic solution ξ, which, in view of (2.23), is identical with the expected solution of the random equation.

To date very little progress has been made toward the resolution of the general problem posed above. However, we would like to refer to the theory of Reynolds operators and suggest that this theory may have some connection with our problem. The notion of a Reynolds operator had its origin in connection with the problem of solving the Navier–Stokes equation. In 1895 O. Reynolds introduced a simplified method for studying the solutions of the Navier–Stokes equations. His method involved taking expected values of the (presumed) solutions of the Navier–Stokes equations, and determining the partial differential equations (the Reynolds equations) satisfied by these expected values.

In our terminology, his method involved regarding the expected solution $\bar{x}$ of the Reynolds equations as the result of a transformation of the random solution $x(\omega)$ of the Navier–Stokes equations (cf. [**73**]). The operator effecting the passage from the Navier–Stokes equations to the Reynolds equations is called a *Reynolds operator*. For a discussion of the characterization and properties of Reynolds operators we refer to [**74**]. We feel that the theory of Reynolds operators is of importance in the general theory of random equations and that their role in connection with the general problem posed above should be investigated.

For other studies concerning the solutions of deterministic operator equations and their relation to the solutions of random operator equations we refer to [**14**; **24**; **42**].

2.6. APPROXIMATE SOLUTIONS OF RANDOM OPERATOR EQUATIONS. In recent years numerical analysts have been utilizing the methods of functional analysis to obtain approximate solutions of linear, and nonlinear, functional and operator equations. For discussions of numerical analysis and functional analysis we refer to [**4**; **5**; **53**]. The iteration procedure developed in various areas of applied mathematics are, in the main, applications of the Banach–Cacciopoli fixed point theorem on successive approximations, or similar fixed point theorems. Because of the importance of iteration procedures for obtaining solutions of many deterministic operator equations, it is to be expected that analogous iteration procedures be considered for random operator equations in applied mathematics.

In [**16**] we considered the stochastic analogue of the Newton–Kantorovitch iteration procedure for the operator equation $y = T[x] = 0$ in a Banach space $\mathscr{X}$. The Newton–Kantorovitch sequence of approximate solutions $\{x_n\}$ is defined by

$$x_{n+1} = x_n - (T'[x_0])^{-1}T[x_n], \qquad n \geqq 0, \tag{2.24}$$

where T' is the Fréchet differential of T, and $x_0 \in \mathscr{X}$ is the initial approach to the solution. In the case of a random operator equation $T(\omega)[x(\omega)] = 0$ in a separable Banach space of generalized random variables, we consider a sequence of generalized random variables defined by

$$x_{n+1}(\omega) = x_n(\omega) - (T'(\omega)[x_0(\omega)])^{-1}T[x_n], \qquad n \geqq 0, \tag{2.25}$$

where the Fréchet differential of $T(\omega)$ is a random endomorphism of $\mathscr{X}$, and $x_0(\omega)$ is an arbitrary generalized random variable with values in $\mathscr{X}$. We can rewrite (2.25) as

$$x_{n+1}(\omega) = S_0(\omega)[x_n(\omega)], \tag{2.26}$$

where

$$S_0(\omega)[x(\omega)] = x(\omega) - (T'(\omega)[x_0(\omega)])^{-1}T(\omega)[x(\omega)].$$

Under appropriate assumptions, and using random fixed point theorems due to Hanš [**40**], we showed that the sequence $\{x_n(\omega)\}$ converges almost surely to a generalized random variable $\xi(\omega)$ which is a solution of $T(\omega)[x(\omega)] = 0$. Problems of uniqueness and rate of convergence were also considered.

For other studies involving the use of stochastic methods in numerical analysis, we refer to the papers of W. Uhlmann [**87**; **88**] who developed a stochastic approach to the Newton–Cotes scheme, and utilized this method in studying the Poisson differential equation.

3. **Random algebraic equations.** Many applied problems lead to the study of algebraic equations whose coefficients are random variables. Algebraic equations with random coefficients will arise, for example, if the coefficients are subject to random error. This might occur when the coefficients result from experimental data, or when numerical procedures require that the coefficients be rounded off to a specified number of decimal places.

Consider the algebraic equation

$$F_n(x) = \xi_0 + \xi_1 x + \cdots + \xi_n x^n = 0, \tag{3.1}$$

where the coefficients $\xi_i = \xi_i(\omega)$, $i = 0, 1, \cdots, n$, and $\omega \in \Omega$, are real- (or complex-) valued random variables. We shall call $F_n(x) = F_n(\omega, x) = 0$ a *random algebraic equation of degree n*.

Within the past 25 years several papers appeared which were devoted to the study of equation (3.1). In this section we shall give a very brief survey of the studies concerned with random algebraic equations.

In [**58**; **59**] J. E. Littlewood and A. C. Offord studied equation (3.1) under the assumption that the coefficients $\xi_i(\omega)$ are independent and identically distributed real-valued random variables. They considered the following three cases:

(i) The coefficients are normally distributed.

(ii) The coefficients are uniformly distributed in $(-1, 1)$.

(iii) The coefficients assume the values $+1$ and -1 with equal probabilities.

Let $N_n = N_n(\xi_0, \xi_1, \cdots, \xi_n)$ denote the number of real roots of equation (3.1). The results of Littlewood and Offord are as follows.

THEOREM 3.1.[7] *In each of the three cases listed above, and for $n \geqq n_0$, the probability that equation* (3.1) *has more than* $25(\log n)^2$ *real roots is at most* $(12 \log n)/n$.

COROLLARY. *For $n \geqq n_0$, $\mathscr{E}\{N_n\} \leqq 25(\log n)^2 + 12 \log n$.*

THEOREM 3.2. *In each of the three cases listed above, the probability that equation* (3.1) *has less than* $\alpha\{\log n/(\log \log n)^2\}$ *real roots is less than $A/\log n$, where α and A are absolute constants.*

The results of Littlewood and Offord established lower and upper bounds for N_n, as well as an upper bound for $\mathscr{E}\{N_n\}$, in each of the three cases listed.

M. Kac [**47**; **48**; **49**] improved the results of Littlewood and Offord by showing that in the case of a random algebraic equation of degree $n - 1$ with normally distributed coefficients

$$\mathscr{E}\{N_n\} = \frac{4}{\pi} \int_0^1 \frac{(1 - \Phi_n^2)^{1/2}}{1 - x^2}\, dx, \tag{3.2}$$

[7] In particular, the theorem is true for $n_0 = 2{,}000$.

where

$$\Phi_n = nx^{n-1}\left(\frac{1-x^2}{1-x^{2n}}\right). \tag{3.3}$$

He also obtained the asymptotic result

$$\mathscr{E}\{N_n\} \sim \frac{2}{\pi}\log n, \tag{3.4}$$

and the estimate

$$\mathscr{E}\{N_n\} \leqq \frac{2}{\pi}\log n + \frac{14}{\pi}, \quad \text{for } n \geqq 2 \tag{3.5}$$

and showed that (3.4) also holds when the coefficients are uniformly distributed in $(-1, 1)$.

P. Erdös and A. C. Offord [29] have studied the equations

$$F_n(\omega, x) = 1 + \xi_1(\omega)x + \xi_2(\omega)x^2 + \cdots + \xi_n(\omega)x^n = 0 \tag{3.6}$$

in which each coefficient $\xi_i(\omega)$, $i = 1, 2, \cdots, n$ is $+1$ or -1 with equal probability. In this case they proved the following:

THEOREM 3.3. *The number of real roots of most of the equations* (3.6) *is* $(2/\pi)\log n + o\{(\log n)^{2/3}\log(\log n)\}$; *and the exceptional set does not exceed a proportion* $o\{(\log\log n)^{-1/3}\}$ *of the total number of equations.*

More recently, G. Samal [76] has studied equation (3.1) under the following conditions: the coefficients $\xi_i(\omega)$ are independent and identically distributed real-valued random variables with expectations zero, and whose variances and third absolute moments are finite and nonzero. His results can be summarized as follows:

THEOREM 3.4. *The number of real roots of most of equations* (3.1) *satisfies the inequality*

$$\varepsilon_n \log n \leqq N_n \leqq \alpha(\log n)^2, \tag{3.7}$$

where α *is a positive absolute constant, and* $\{\varepsilon_n\}$ *is a sequence of numbers tending to zero but such that* $\varepsilon_n \log n$ *tends to infinity. The measure of the exceptional set tends to zero as n tends to infinity.*

In an important paper [37], J. Hammersley gives a complete formal solution to the problem of determining the distribution of the roots of equation (3.1) in the case where the coefficients $\xi_i(\omega)$ are complex-valued random variables with a given joint distribution. The main tool in his investigation is the *condensed distribution of the roots.* This can be described as the measure which assigns to a Borel subset S of the complex plane $1/n$ times the expected number of roots of equation (3.1) falling in S. Explicit formulae for the condensed distribution of roots are given for several cases; in particular, when the coefficients $\xi_i(\omega)$ are normally distributed real-valued random variables, a generalization of the result of Kac (cf. (3.2)) is obtained. Hammersley's formula contains more information than the formula of Kac in that it not only gives $\mathscr{E}\{N_n\}$, but it also shows how the roots are distributed along the real axis.

The most recent study concerning random algebraic equations is that of D. I. Šparo and M. G. Šur [**85**]. They considered the equation

$$F_n(\omega, z) = \xi_0 + \xi_1(\omega)z + \cdots + \xi_n(\omega)z^n = 0, \tag{3.8}$$

where z is a complex variable, and the coefficients $\xi_i(\omega)$ are independent and identically distributed complex-valued random variables. Their results can be summarized as follows:

THEOREM 3.5. *Let α and β be arbitrary numbers such that $0 \leqq \alpha < \beta < 2\pi$; and let $N_n(\alpha, \beta)$ and $R_n(\delta)$, $\delta > 0$, denote the number of roots of equation* (3.8) *in the sector $\alpha < \arg z < \beta$, and the ring $1 - \delta \leqq |z| \leqq 1 + \delta$, respectively. Then, as $n \to \infty$, $N_n(\alpha, \beta)/n$ tends to $(\beta - \alpha)/2\pi$ in probability, and $R_n(\delta)/n$ tends to* 1 *in probability.*

In closing this section we remark that random algebraic equations are also encountered in the study of random determinants, and in the study of random Fredholm integral equations with separable kernels.

4. **Random difference equations.** A linear difference equation of order n with constant coefficients has the form

$$a_0x(t + n) + a_1x(t + n - 1) + \cdots + a_nx(t) = y(t), \tag{4.1}$$

or

$$T[x(t)] = y(t), \tag{4.2}$$

where the linear operator T is defined by

$$T = a_0\tau^n + a_1\tau^{n-1} + \cdots + a_n\tau^0. \tag{4.3}$$

In (4.3) τ denotes the translation operator, that is, $\tau^n[x(t)] = x(t + n)$, $n = 0, 1, 2, \cdots$.

Linear random difference equations, which are analogous of equation (4.1), are of three types; these types being special cases of equations (2.2), (2.3), and (2.4). In the first we have

$$T[x(t)] = y(\omega, t), \tag{4.4}$$

where T is defined by (4.3). In the second case, which arises when the coefficients are randomly variables (i.e., $a_i = a_i(\omega)$), we have

$$T(\omega)[x(t)] = y(t). \tag{4.5}$$

Finally, in the third case, we have

$$T(\omega)[x(t)] = y(\omega, t). \tag{4.6}$$

Equations of the form (4.4) have been studied rather extensively in mathematical economics and econometrics **[54; 94]**, and in connection with stationary time series [33; 93]. A stochastic process $\{x(\omega, t),\ t = 0,\ \pm 1,\ \pm 2, \cdots\}$ satisfying equation (4.4) is called an *autoregressive process.* To obtain the solution of equation (4.4) we can proceed as follows. Let us assume that the $y(\omega, t)$ have mean zero, variance σ^2, and are uncorrelated. If the process $\{y(\omega, t)\}$ is weakly stationary it admits the representation

$$y(\omega, t) = \int_{-\pi}^{\pi} e^{i\lambda t}\, dz_y(\omega, \lambda), \tag{4.7}$$

where $z_y(\omega, \lambda)$ is an orthogonal set function defined on the Borel sets of the interval $(-\pi, \pi)$. Since

$$\mathscr{E}\{y(\omega, t_1)\overline{y(\omega, t_2)}\} = \sigma^2 r(t_1, t_2) = \frac{\sigma^2}{2\pi}\int_{-\pi}^{\pi} e^{i\lambda(t_1 - t_2)}\, d\lambda \tag{4.8}$$

the process $\{y(\omega, t)\}$ has an absolutely continuous spectrum with (constant) spectral density $f(\lambda) = \sigma^2/2\pi$. Hence $\{y(\omega, t)\}$ is a so-called *white noise process.*

If $x(\omega, t)$ is a stationary solution of equation (4.4), we can write

$$x(\omega, t) = \int_{-\pi}^{\pi} e^{i\lambda t}\, dz_x(\omega, \lambda), \tag{4.9}$$

and

$$\mathscr{E}\{x(\omega, t_1)\overline{x(\omega, t_2)}\} = \int_{-\pi}^{\pi} e^{i\lambda(t_1 - t_2)}\, dF_x(\omega, \lambda), \tag{4.10}$$

where $F_x(\omega, \lambda)$ is the spectral decomposition of the process $\{x(\omega, t)\}$. From (4.4), (4.7), and (4.9) we have

$$\begin{aligned} T[x(\omega, t)] &= \int_{-\pi}^{\pi} \sum_{k=0}^{n} a_k\, e^{i\lambda(n-k)}\, e^{i\lambda t}\, dz_x(\omega, \lambda) \\ &= \int_{-\pi}^{\pi} e^{i\lambda t}\, dz_y(\omega, \lambda). \end{aligned} \tag{4.11}$$

The unique stationary solution of equation (4.4) can then be shown to be (cf. [33])

$$x(\omega, t) = \int_{-\pi}^{\pi} e^{i\lambda t}[\varphi(e^{i\lambda})]^{-1}\, dz_y(\omega, \lambda), \tag{4.12}$$

where

$$\varphi(r) = \sum_{k=0}^{n} a_k r^{(n-k)}. \tag{4.13}$$

Random difference equations arise in mathematical physics and engineering in connection with differential equations whose coefficients are random but are piecewise constant in the independent variable. The random difference equations which arise are of the form

$$x_k = \Phi_k x_{k-1}, \qquad k = 1, 2, \cdots, \tag{4.14}$$

where x_k is an n-vector, and Φ_k is an $n \times n$ random matrix. Equations of the form (4.14) have been studied in connection with randomly sampled linear systems (cf. **[50]**); and in quantum mechanics in connection with energy levels in random lattices. Equation (4.14) will be considered in §5, when we discuss random differential equations.

5. **Random differential equations.** The literature on random equations contains more studies on random differential equations than any other type of random equation. This is not at all surprising, for mathematical models formulated as nonrandom differential equations have long played a prominent role in mathematical physics and engineering. And, as workers attempted to make their models more realistic and take into account the random nature of many of the systems and media they study, it is only natural that models formulated as random differential equations would play an increasingly active role. For a discussion of random differential equations and their role in physics, engineering, and technology we refer to the paper of I. Babuška **[6]**.

The remainder of this section is divided into two parts, which are devoted to random ordinary and partial differential equations, respectively. Random differential equations have attracted the attention of probabilists for quite some time, the work of S. Bernstein **[8]** being the first in this area. Since the work of Bernstein, other workers have continued in the same tradition; we refer, in particular, to the studies of I. I. Gihman, and G. Maruyama. These studies will not be discussed here, since we restrict our attention, in the main, to the random equations of mathematical physics and engineering. We refer to **[19]** for a discussion of their work. Throughout this section we restrict our attention to linear differential equations.

5.1. Random ordinary differential equations. We will restrict our attention to only two types of random ordinary differential equations which have been studied. Brief reference will be made to some studies of particular random equations in mathematical physics. For a general discussion of random ordinary differential equations and their applications in engineering, we refer to the books of V. L. Lebedev **[57]** and V. S. Pugachev **[71]**.

The first type of equation we consider is the random analogue of a linear differential equation of order n. This equation is of the form

$$T(\omega)x(t) = a_n(\omega, t)\frac{d^n x}{dt^n} + a_{n-1}(\omega, t)\frac{d^{n-1}x}{dt^{n-1}} + \cdots + a_0(\omega, t)x = y(\omega, t), \tag{5.1}$$

where $a_i(\omega, t)$, $i = 0, 1, \cdots, n$, $y(\omega, t)$ are given random functions. Hence equation (5.1) is a random operator equation of the form $T(\omega)[x] = y(\omega)$. We assume that $y(\omega, t)$ is independent of any of the coefficients $a_i(\omega, t)$; however, the coefficients may be correlated with each other. Equation (5.1) has been studied by a number of workers; we refer, in particular, to the papers of J. C. Samuels [77; **78**], and J. C. Samuels and A. C. Eringen [79] (cf. also, U. Grenander **[34]**).

The method used by Samuels and Eringen to solve equation (5.1) can be described as follows: We write the coefficients $a_i(\omega, t)$ in the form

$$a_i(\omega, t) = \alpha_i(t) + \varepsilon\beta_i(\omega, t), \qquad i = 0, 1, \cdots, n, \tag{5.2}$$

where $\alpha_i(t) = \mathscr{E}\{a_i(\omega, t)\}$, and the $\beta_i(\omega, t)$ are random variables with $\mathscr{E}\{\beta_i(\omega, t)\} = 0$. With the coefficients as defined by (5.2), a solution of equation (5.1) is sought in the form

$$x(\omega, t) = \sum_{i=0}^{\infty} \varepsilon^i x_i(\omega, t). \tag{5.3}$$

If we now substitute (5.3) in equation (5.1) and equate the coefficients of ε^i, we obtain a system of operator equations

$$\begin{cases} T_\alpha x_0 = y, \\ T_\alpha x_i = -T_\beta x_{i-1}, \qquad i = 1, 2, \cdots, \end{cases} \tag{5.4}$$

where the differential operators T_α and T_β denote the operator T with the coefficients $a_i(\omega, t)$ replaced by $\alpha_i(t)$ and $\beta_i(\omega, t)$, respectively. Hence T_α is a nonrandom operator and T_β is a random operator. The formal solution of (5.4) is

$$\begin{cases} x_0(\omega, t) = \sum_{i=1}^{n} c_i \xi_i(t) + \int_0^t G(t, \tau) y(\omega, \tau)\, d\tau, \\ x_i(\omega, t) = -\int_0^t G(t, \tau) T_\beta(\omega) x_{i-1}(\omega, \tau)\, d\tau, \qquad i = 1, 2, \cdots, \end{cases} \tag{5.5}$$

where $\{\xi_i(t)\}$ is a set of fundamental solutions of the operator equation $T_\alpha \xi = 0$, the c_i are constants to be determined by the initial conditions, and $G(t, \tau)$ is the one-sided Green's function associated with the operator T_α. While (5.5) constitutes a solution to equation (5.1) of the form (5.4), it is highly desirable that other methods be developed for solving equations of the form (5.1). For an approach to the study of random ordinary differential equations utilizing stochastic Green's functions, we refer to the studies of G. Adomian [**1**; **2**; **3**]. For some concrete applications of equation (5.1) we refer to [**79**] and related papers.

A problem of great theoretical and practical importance is that of the stability of linear systems described by equations of the form (5.1). This problem is, however, too involved to go into here; hence we will not attempt to summarize the results that have been obtained concerning the stability of random linear systems. For definitions of the various types of stability that have been introduced and studied, and for some concrete applications of stability criteria, we refer to [**9**; **10**; **67**; **72**; **77**; **78**; **79**; **81**].

The next type of equation we consider is the system of random differential equations with piecewise constant coefficients:

$$\frac{dx(\omega, t)}{dt} = A_k x(\omega, t), \quad t \in [t_{k-1}, t_k), k = 1, 2, \cdots, \tag{5.6}$$

where $x(\omega, t)$ is a random n-vector, and A_k is a constant $n \times n$ matrix. Our discussion here is based on the thesis of B. H. Bharucha [**10**]. Given the initial vector $x(\omega, t_0) = x_0(\omega)$ the solution of equation (5.6) in the interval $[t_0, t_1)$ is

$$x(\omega, t) = e^{A_1(t-t_0)}x_0(\omega). \tag{5.7}$$

Proceeding in this way, we obtain

$$x(\omega, t) = e^{A_k(t-t_{k-1})}\, e^{A_{k-1}(t_{k-1}-t_{k-2})} \cdots e^{A_1(t_1-t_0)}x_0(\omega), \tag{5.8}$$

for $t \in [t_{k-1}, t_k)$, $k = 1, 2, \cdots$. The solution $x(\omega, t)$ given by (5.8) satisfies the initial condition $x(\omega, t_0) = x_0(\omega)$, and satisfies equation (5.6) for all values of t except $t = t_1, t_2, \cdots$. It is clear that the behavior of the solution is completely determined by the random sequence $\{A_k(t_k - t_{k-1})\}$.

Let

$$\Phi_k = e^{A_k(t_k-t_{k-1})}; \tag{5.9}$$

then the solution (5.8) can be written as

$$x(\omega, t) = e^{A_k(t-t_{k-1})}\Phi_{k-1}\Phi_{k-2} \cdots \Phi_1 x_0(\omega), \tag{5.10}$$

for $t \in [t_{k-1}, t_k)$, $k = 1, 2, \cdots$. If we restrict our attention to the solution at times $t = t_k$ only, (5.10) becomes

$$x(\omega, t_k) = x_k(\omega) = \Phi_k\Phi_{k-1} \cdots \Phi_1 x_0(\omega). \tag{5.11}$$

Hence the $x_k(\omega)$ satisfy the random difference equation (first mentioned in §4)

$$x_k(\omega) = \Phi_k x_{k-1}(\omega). \tag{5.12}$$

From (5.9) it is clear that the behavior of $x_k(\omega)$ is determined by the sequence of random matrices Φ_k. We remark that from the continuity, with respect to $x_0(\omega)$, of solutions of equation (5.6) it follows that stability of the (trivial) solution of equation (5.6) can be determined by studying the properties of the sequence $\{x_k(\omega)\}$, provided the Φ_k are bounded. In [**10**] this approach to stability is utilized, and stability is studied in the following cases: (i) the Φ_k are independent, identically distributed random matrices, and (ii) the Φ_k form a Markov chain.

In closing this subsection we refer to some other studies on random ordinary differential equations which are of interest in mathematical physics and engineering. S. K. Srinivasan [**86**] has studied equations of the form

$$\frac{dx(\omega, t)}{dt} + Ax(\omega, t) = y(\omega, t), \tag{5.13}$$

where $x(\omega, t)$ and $y(\omega, t)$ are random n-vectors, and A is an $n \times n$ matrix. He investigates the solution of equation (5.13) under the assumption that the components y_i of $y(\omega, t)$ are independent random processes. The theory developed is applied to the Langevin equation, and to some first, second, and third order equations arising in various physical and engineering applications.

D. A. Edwards and J. E. Moyal [**27**], and J. E. Moyal [**66**], have studied several random equations in mathematical physics, in particular, the Langevin equation for simple Brownian motion and the equation for the Brownian oscillator. The random differential equation for the amplitude of a random linear oscillator has been considered by R. H. Kraichnan [**55**]. K. Urbanik [**90**] in his studies on generalized stochastic processes as distributions in the sense of Mikusiński has studied the Langevin equation

$$\frac{dv(\omega, t)}{dt} + \alpha v(\omega, t) = \Phi(\omega, t), \tag{5.14}$$

where $v(\omega, t)$ is the velocity of a Brownian particle, α is a constant, and $\Phi(\omega, t)$ is the generalized derivative of a Gaussian process.

Differential equations with random coefficients and their applications have been studied by D. Maravall Casesnoves [**61**; **62**]; in particular, he has studied the random equation associated with a "stochastic" pendulum.

5.2. RANDOM PARTIAL DIFFERENTIAL EQUATIONS. The study of random solutions of partial differential equations was initiated by J. Kampé de Fériet ([**51**] and earlier papers). In particular, he has studied random solutions of the heat equation for an infinite rod when the initial temperature at the point x on the rod is a random function $f(\omega, x)$. Let $u(\omega, t, x)$ denote the random solution of the equation

$$\begin{cases} \dfrac{\partial u}{\partial t} = \dfrac{\partial^2 u}{\partial x^2}, \\ u(\omega, t, x) = f(\omega, x). \end{cases} \qquad -\infty < x < \infty, \tag{5.15}$$

Kampé de Fériet showed that the solution of (5.15) is

$$u(\omega, t, x) = \int_{-\infty}^{\infty} K(x - \xi, t) f(\omega, \xi)\, d\xi, \tag{5.16}$$

where $K(x, t) = (4\pi t) -\frac{1}{2}\, e^{-x^2/4t}$.

The problem investigated by Kampé de Fériet is a special case of what can be called a stochastic boundary value problem. The notion of a stochastic boundary problem is due to Kampé de Fériet; and it was introduced in order to provide a framework within which to study boundary value problems for which the boundary conditions cannot be expressed by a single well-determined or known function $\gamma(u)$, say. In these cases one introduces a set Γ of realizable boundary conditions which is the range of a random variable. Hence an event consists of the choice of an element $\omega_0 \in \Omega$, which in turn specifies the boundary condition $\gamma(\omega_0, u) \in \Gamma$ that is to be imposed.

We would now like to discuss, in some detail, an interesting paper by I. Babuška [7] which studies the Dirichlet and Poisson problems for Laplace's equation with random boundary conditions and random right-hand side, respectively. Here we restrict our attention to the random Dirichlet problem.

Let E_n, $n = 1, 2, \cdots$, be Euclidean n-space, K the open unit parallelepiped in E_n, and $\bar{K}$ its closure in E_n. As before, let $(\Omega, \mathscr{A}, \mu)$ denote a probability space, where the elements ω are real-valued functions defined on $\bar{K}$. Functions defined on Ω we denote by f, g, etc., and for $x \in \bar{K}$, $f_x(\omega) = \omega(x)$ will be the function defined on $\Omega \times \bar{K}$. We assume that for every $x \in \bar{K}$ the function $f_x(\omega)$ is μ-measurable, and that

$$\int_\Omega f_x^2(\omega)\, d\mu \leqq M^2 < \infty \tag{5.17}$$

(M independent of x); and also that

$$\lim_{x \to y} \int_\Omega (f_x(\omega) - f_y(\omega))^2\, d\mu = 0. \tag{5.18}$$

Now, let L^{**} denote the set of all functions $f_x(\omega)$ defined on Ω for every $x \in \bar{K}$, together with the function $z_0(\omega) = 1$. L^* denotes the linear module over L^{**} with the reals as coefficient domain. We now define a scalar product and norm in L^* by

$$(g_1, g_2) = \int_\Omega g_1(\omega) g_2(\omega)\, d\mu, \qquad g_1, g_2 \in L^*,$$

$$\|g\| = (g_1, g_2)^{1/2}.$$

Finally, we denote by L the completion, with respect to the norm just defined, of the linear space L^*.

To formulate the stochastic boundary value problem we proceed as follows. Let γ be a continuous function defined on $\bar{K}$, and let U denote a region, with $\bar{U} \subset K$. Denote by $W(\gamma, x)$ the function defined on $\bar{K}$ as follows:

(i) $W(\gamma, x) = \gamma(x)$ for $x \notin U$,

(ii) for $x \in U$, $W(\gamma, x)$ is the generalized (Wiener) solution[8] of the Dirichlet problem for Laplace's equation on U, with the boundary condition defined by $\gamma(x)$ on the boundary of U. The function $W(\gamma, x)$ will be called the solution of the Dirichlet's problem for the boundary function $\gamma(x)$.

We remark that $W(\gamma, x)$ is dependent only on the values which γ assumes on $\tilde{U}$, the boundary of U.

We now introduce the following definition.

DEFINITION 5.1. *A random function $w(\omega, x)$ defined on $\Omega \times \bar{K}$ will be called a random Wiener solution of the Dirichlet problem on U, with boundary condition defined by $f_x(\omega)$, if $w(\omega, x) \in L$ for every $x \in K$ and, for every $z \in L$,*

$$(z, w) = W((z, f_x(\omega)), x). \tag{5.19}$$

Babuška's first result can be stated as follows.

[8] Cf. [21].

THEOREM 5.1. *There is precisely one random solution (in the sense of Definition* 5.1) *of the Dirichlet problem.*

We now define the *covariance function* $\rho(x, y)$ as follows: $\rho(x, y) = (f_x(\omega), f_y(\omega))$. From (5.18) it follows that $\rho(x, y)$ is a continuous function if x and y on $\bar{R} \times \bar{R}$. Since, by Theorem 5.1, there is only one random solution of the Dirichlet problem, we will denote this solution by $w(\omega, x)$. For every x, $w(\omega, x) \in L$; hence we can introduce the function

$$R(x, y) = (w(\omega, x), w(\omega, y)). \tag{5.20}$$

$R(x, y)$ will be called the *covariance function of the random solution of the Dirichlet problem on U.*

The next theorem gives an expression for $R(x, y)$ in terms of the solution W.

THEOREM 5.2. *Let* $h_x(x, y) = W(\rho(x, y), x)$, *and* $h_y(x, y) = W(\rho(x, y), y)$. *Then* $R(x, y) = W(h_x(x, y), y) = W(h_y(x, y), x)$.

Finally, we consider the expectation of the random solution $w(\omega, x)$. From (2.14) we have

$$\bar{w}(x) = \mathscr{E}\{w(\omega, x)\} = \int_\Omega w(\omega, x)\, d\mu. \tag{5.21}$$

The function $\bar{f}(x) = \int_\Omega f_x(\omega)\, d\mu$ will be called the *expected value of the random boundary condition.* The following theorem is true.

THEOREM 5.3. *Let* $w(\omega, x)$ *be the random solution (in the sense of Definition* 5.1) *of the Dirichlet problem. Then*

$$\bar{w}(x) = W(\bar{f}(x), x). \tag{5.22}$$

This result, which is germane to our discussion in §2.5, states that (deterministic) calculations starting with expected values determine the expected value of random solutions.

We next mention some applications of random equations in quantum mechanics. R. H. Kraichnan [**55**] has studied random solutions of the Schrödinger equation for a particle in a random potential. Similarly, H. L. Frisch and S. P. Lloyd [**31**], and J. A. Morrison [**64**], in connection with their studies on electron levels in one-dimensional random lattices, have considered the Schrödinger equation with random potential function

$$V(\omega, x) = V_0 \sum_{i=-\infty}^{\infty} \delta(x - x_i(\omega)), \qquad -\infty < x < \infty, \tag{5.23}$$

where $V_0 < 0$ and $\{x_i(\omega)\}$ is a sequence of real-valued random variables with a Poisson distribution. Let $\psi(x)$ denote the solution of the Schrödinger equation. In [**31**] it was shown that the random variables

$$z_i = \frac{\psi'(x_i - 0)}{\psi(x_i)}, \qquad -\infty < i < \infty$$

form a stationary Markov process.

In another interesting paper, H. Primas [70] has considered quantum mechanical systems with random Hamiltonians of the form $H(\omega, t) = H_0(t) + \lambda V(\omega, t)$, where H_0 and V are Hermitean operators, with H_0 being a nonrandom operator, and λ is a real number. In particular, he considers solutions of the random equation

$$\frac{\partial \rho(\omega, t)}{\partial t} = -i[H(\omega, t), \rho(\omega, t)], \tag{5.24}$$

where $\rho(\omega, t)$ is the ordinary density matrix. The theory developed is used to outline a new foundation for F. Bloch's relaxation theory.

In order to study random solutions of partial differential equations using semigroup theory, we have initiated the study of semigroups of random operators, and formulated the stochastic analogue of the abstract Cauchy problem[9] [18]. This theory, when developed, should enable us to solve partial differential equations when the differential operator (infinitesimal generator of the semigroup) is a random operator.

In closing, we mention two other interesting studies. W. E. Boyce [20] in his studies on random vibration problems has considered what might be called a random Sturm–Liouville problem. In this case the coefficients multiplying the dependent variable and its derivatives are random variables. M. Ullrich [89] has developed a theory of random Mikusinski operators, and has used this theory to study random partial differential equations with constant coefficients.

6. Random integral equations.

Studies on random integral equations have, in the main, proceeded along two different, but we do not think independent, lines. We have the work of K. Ito ([45; 46]; cf. also [23]) devoted to the study of random integral equations associated with Markov processes; and we have the studies of O. Hanš [42] and the author [13; 14; 15; 17] which consider solutions of some classical integral equations with random kernels, or defined over random domains. Random integral equations also arise in connection with certain problems in control theory. The random integral equations of control theory will be treated in [19].

6.1. THE ITO RANDOM INTEGRAL EQUATION. The random integral equation referred to as the *Ito equation* is of the form

$$x(\omega, t) = x(\omega, t_0) + \int_{t_0}^{t} m(\tau, x(\omega, \tau))\, d\tau + \int_{t_0}^{t} \sigma(\tau, x(\omega, \tau))\, dw(\omega, \tau), \tag{6.1}$$

where $x(\omega, t) \in R$, $m(t, x)$, and $\sigma(t, x)$ are real functions continuous in x, uniformly in t, and $w(\omega, t)$ is a Wiener process. Equation (6.1) arises in the following way. Let $w(\omega, t)$ be a process corresponding to the differential operator $d^2/d\xi^2$, and let $x(\omega, t)$ be a process corresponding to the operator $\frac{1}{2}\sigma(\xi)\, d^2/d\xi^2 + m(\xi)\, d/d\xi$. Ito showed that the two processes can be so chosen that the relationship between them

[9] We refer to E. Hille and R. S. Phillips [44] for a discussion of the abstract Cauchy problem.

is given by equation (6.1). Equation (6.1) can also be written in the form of the random differential equation

$$dx(\omega, t) = m(t, x)\, dt + \sigma(t, x)\, dw(\omega, t). \tag{6.2}$$

The Ito approach enables us to replace the study of the partial differential equations for the transition probabilities of a Markov process by the study of the random ordinary differential equation of its realizations. Equation (6.2) can be reduced to equation (6.1), and the solution of equation (6.1) can be obtained using the method of successive approximations.

In this paper we will not give a detailed discussion of the Ito equation. The interested reader will find a complete treatment of the Ito equation and its applications in [**19**]. It is, however, of interest to refer to the following recent theoretical studies and applications of the Ito equation: (1) K. Z. Dambis ([**22**]; cf. also [**25**]) has studied random integral equations in Hilbert space; (2) I. V. Girsanov [**32**] has discussed the uniqueness of solutions of the Ito equation; (3) A. V. Skorohod [**82**] has utilized the Ito equation in the study of diffusion processes; (4) M. I. Freĭdlin [**30**] has employed the Ito equation in the study of elliptic differential equations; and (5) T. Nagai [**67**] has investigated the stability of a random linear system of first order using an Ito equation.

6.2. RANDOM SOLUTIONS OF SOME CLASSICAL INTEGRAL EQUATIONS. The studies of Hanš and the author on random solutions of some classical integral equations were motivated by: (1) the widespread occurrence of integral equations in mathematical physics and other branches of applied mathematics, (2) the existence of a well-developed theory, and (3) the relationship between integral equations and differential equations, systems of linear algebraic equations, boundary value problems, etc. In our studies we have considered the random solutions to be elements of an Orlicz space (generalized Lebesgue space) since much of the recent work in the theory of integral equation (cf. [**56**; **95**]) has been carried out in these spaces.[10]

Let U be a bounded or unbounded interval in k-dimensional Euclidean space R_k (hence we can have $U \equiv R_k$), and let m be Lebesgue measure on U. Let $(\Omega, \mathscr{A}, \mu)$ be a probability space, and let $(L_\Phi, \mathscr{B})$ be a measurable space, where L_Φ is the Orlicz space $L_\Phi(U, m)$ and $\mathscr{B}$ is the σ-algebra of all Borel subsets of L_Φ.

The random version of the nonhomogeneous Fredholm integral equation of second kind, i.e.,

$$\int_U K(t, u)x(u)\, dm(u) - \lambda x(t) = y(t), \tag{6.3}$$

when random boundary conditions are imposed, can be written as

$$\int_{D_\omega} K(t, u)x(\omega, u)\, dm(u) - \lambda x(\omega, t) = y(\omega, t). \tag{6.4}$$

[10] For a detailed discussion of Orlicz spaces and their properties we refer to [**56**; **95**]; we refer to [**11**] for a discussion of generalized random variables with values in an Orlicz space.

In equation (6.4) the unknown function $x(\omega, t)$ and the known function $y(\omega, t)$ are generalized random variables with values in the separable[11] Orlicz space L_Φ. We shall assume that $y(\omega, t) = y(t)$ for all $\omega \in \Omega$, and that the kernel, which is deterministic (i.e., it does not depend on ω) is an $m \times m$-measurable function in $U \times U$. The parameter λ is an arbitrary complex number.

When considering random boundary value problems for Fredholm integral equations, the choice of an element $\omega \in \Omega$ specifies an interval $D_\omega \subseteq U$ which is the fundamental domain over which the integral is defined. Hence D_ω is a *random set*, and m is now random Lebesgue measure with $0 < m(D_\omega) \leqq m(U)$. We assume that the integral in (6.4) exists in the sense of Lebesgue for almost all $\omega \in \Omega$.

It is also possible to consider equation (6.3) when the kernel is a function of ω. In our case we have

$$\int_U K(\omega, t, u)x(\omega, u)\, dm(u) - \lambda x(\omega, t) = y(t), \tag{6.5}$$

where

$$K(\omega, t, u) = \begin{cases} K(t, u), & \text{for } u \in D, \\ 0, & \text{for } u \in U - D. \end{cases} \tag{6.6}$$

In operator form equation (6.4) becomes

$$(T(\omega) - \lambda I)x = y, \tag{6.7}$$

where I is the identity operator, and

$$T(\omega)x = \int_{D_\omega} K(t, u)x(u)\, dm(u) \tag{6.8}$$

is an operator on the Orlicz space $L_\Phi(D_\omega, m(D_\omega))$.

We now discuss some properties of the operator $T(\omega)$ defined by (6.8). In particular, we want to know if $T(\omega)$ is a random transformation; that is, $T(\omega)$, as a function of ω, is measurable.

Let us rewrite (6.8) in the form

$$\int_U K(t, u)h(\omega, u)\, dm(u), \tag{6.9}$$

where

$$h(\omega, u) = \begin{cases} x(\omega, u), & \text{for } u \in D, \\ 0, & \text{for } u \in U - D_\omega. \end{cases} \tag{6.10}$$

The function $h(\omega, u)$ can also be written as

$$h(\omega, u) = \mathscr{X}_{D_\omega}(u)x(\omega, u), \tag{6.11}$$

---

[11] The Orlicz space under consideration will always be separable, since we shall assume that the convex function $\Phi(r)$ defining the Orlicz space satisfies the Δ_2-condition, and the measure m is Lebesgue measure.

where $\mathscr{X}_{D_\omega}(u)$ is the indicator, or characteristic function, of the random set D_ω. Now, since $K(t, u)$ is a continuous function of $t, u \in D_\omega$, and is by hypothesis an $m(D_\omega) \times m(D_\omega)$-measurable function on $D_\omega \times D_\omega$, and if we assume that $x(\omega, u)$ is measurable (in the sense of Theorem 2.9), we see that the following proposition is true: *A sufficient condition for $h(\omega, u)$, and consequently $T(\omega)$, to be measurable is that $\mathscr{X}_{D_\omega}(u)$ be measurable,*[12] *and also the random solution $x(\omega, u)$.*

The results given in **[14]** concerning the Fredholm operator can be summarized as follows: *The Fredholm operator, defined by* (6.8), *on the Orlicz space $L_\Phi(D_\omega, m(D_\omega))$ will be a random endomorphism (i.e., a measurable linear bounded operator) of L_Φ if* (i) *the kernel $K(t, u)$ has the property that*

$$r(\omega, t) = \int_{D_\omega} |K(t, u)x(\omega, u)|\, dm(u)$$

is a generalized random variable with values in L_Φ, and (ii) *the solution $x(\omega, t)$ and the indicator of the random set D_ω are measurable with respect to the same σ-algebra of Borel subsets of Ω.*

In **[42]** Hanš, by introducing a random kernel, establishes the measurability of the Fredholm operator without requiring that the solution of the integral equation be measurable.

Finally, application of Theorems 2.2 and 2.9, and the classical theory, shows that the solution of (6.4) is

$$x(\omega, t) = R(\lambda; T(\omega))[y(t)] \in L_\Phi(D_\omega, m(D_\omega)), \tag{6.12}$$

and that the solution is unique. The theory of random contraction operators can be used to prove the following

THEOREM 6.1. *Consider the random Fredholm integral equation* (6.4) *in the Orlicz space $L_\Phi(D_\omega, m(D_\omega))$, where the kernel $K(t, u)$ is measurable on $D_\omega \times D_\omega$, with $|K(t, u)| < M$, and for all $\omega \in \Omega$ the function $y(t) \in L_\Phi$ is known. If $|\lambda| > m(D_\omega)M$, where $m(D_\omega)$ is the Lebesgue measure of D_ω, then for every fixed $u \in D_\omega$ there exists a mapping $\xi_\lambda(\omega)$ from Ω into L_Φ which is a generalized random variable, and the solution of equation* (6.4).

We remark that the solution of equation (6.4) can be found by the method of successive approximations, the nth approximation being given by

$$x_n(\omega, t) = \frac{1}{\lambda}\left\{\int_{D_\omega} K(t, u)x_{n-1}(\omega, u)\, dm(u) - y(t)\right\}, \tag{6.13}$$

where $x_0(\omega, u)$ is an arbitrary generalized random variable with values in L_Φ.

Similar results can be obtained for the random Volterra integral equations, and for systems of random integral equations. For a discussion of these results we refer to **[14; 42]**. In **[15]** we have considered the random Fredholm integral

---

[12] $\mathscr{X}_{D_\omega}(u)$ will be measurable if D_ω is measurable.

equations with separable (degenerate, or finite-rank) kernels; that is, we consider equation (6.4) with[13]

$$K(t, u) = \sum_{i=1}^{n} \alpha_i(t)\beta_i(u). \tag{6.14}$$

In this case we obtain the following analogue of a classical result in the theory of Fredholm integral equations: *The solution of a random Fredholm integral equation with separable kernel of rank n is equivalent to the solution of a system of n random linear algebraic equations in n unknowns.* The theory of random contraction operators was used to establish the existence and uniqueness of the system of algebraic equations, and as a consequence the solution of the integral equation.

In [17] random solutions of some nonlinear integral equations are considered. In particular, the random Hammerstein equation

$$\int_{D_\omega} K(t, u)q[u, x(\omega, u)]\, dm(u) - \lambda x(t) = y(t) \tag{6.15}$$

is studied. While the analysis is a little more involved than in the case of random linear integral equations, fixed point methods are applicable, and were used to establish the existence and uniqueness of solutions.

7. Stochastic processes associated with random equations. In most applied situations we are primarily interested in obtaining the solution of a given random equation and studying its statistical properties. However, from the point of view of the theory of stochastic processes, an interesting problem is to study the stochastic processes generated by solutions of random equations. For example, in the case of equation (2.2), we can ask the following question: If the input $y(\omega, t)$ is a given stochastic process and T is a nonrandom operator, what type of stochastic process $\{x(\omega, t), t \geqq 0\}$ is generated by the solution $x(\omega, t) = T^{-1}[y(\omega, t)]$? This is clearly the simplest case; and many results have been obtained in this case.

We can ask similar questions for the solutions of equations (2.3) and (2.4). However, the analysis of the stochastic processes in these cases can be more involved. In the case of random equations of the form $T(\omega)[x] = y$, some results are available. For example, B. H. Bharucha **[10]** has considered the stochastic process associated with the random difference equation (4.14) when the matrices Φ_k are independent, identically distributed random operators. In this case the process $\{x_k\}$ is an infinite Markov chain. Studies on ordinary differential equations with random coefficients have, in the main, been concerned with the stability (in some sense) of the random solution when the coefficients are, say white noise processes, and not with the stochastic processes generated by the solutions. In the case of random partial differential equations very little is known. We have the

[13] It is also possible to consider equation (6.4) defined on the domain U with a random separable kernel $K(\omega, t, u) = \sum_{i=1}^{n} \alpha_i(t)\beta_i(\omega, u)$, where $\beta_i(\omega, u) = \beta_i(u)$ for $u \in D_\omega$, and zero otherwise.

result of J. Kampé de Fériet [**51**] who showed that when the random initial temperature $f(\omega, x)$ is a stationary (wide sense) process, then the solution $u(\omega, t, x)$ of the heat equation, as a function of x is a stationary (wide sense) process, and as a function of t is exponentially convex.

For random integral equations, we have already pointed out that the solution of the Ito equation is a Markov process; however, nothing is known about the processes associated with the random solutions of the classical integral equations considered in §6.2. In control theory we encounter random integral equations of the form

$$x(\omega, t) = x_0 + \int_{t_0}^{t} K(x, u, \xi)\, d\xi, \tag{7.1}$$

where $x(\omega, t)$ is the state variable and $u(\omega, t)$ is random disturbance. The initial state x_0 may also be a random variable. It is known that if $u(\omega, t)$ is a white noise process, then the solution of equation (7.1) is a Markov process.

In the case of random solutions of the form $x(\omega, t) = R(\lambda; T(\omega))[y(\omega, t)]$ (cf. (2.13)), we feel that a problem of interest is the study of the stochastic processes $\{x(\omega, t)\}$ generated by the process $\{y(\omega, t)\}$ and the random operator $T(\omega)$; in particular, the relationship between the process $\{x(\omega, t)\}$ and the spectral properties of the random operator $T(\omega)$.

Bibliography

1. G. Adomian, *Linear stochastic operators*, Rev. Modern Phys. **35** (1963), 185–207.

2. ———, *Stochastic Green's functions*, Proc. Sympos. Appl. Math. Vol. 16, pp. 1–39, Amer. Math. Soc., Providence, R. I., 1964.

3. ———, *Linear stochastic operators*, Academic Press, New York (to appear).

4. M. Altman, *Lectures on approximation methods in functional analysis*, California Institute of Technology, Pasadena, Calif., 1959 (unpublished).

5. H. A. Antosiewicz and W. C. Rheinboldt, *Numerical analysis and functional analysis*, Survey of Numerical Analysis, pp. 485–517, McGraw-Hill, New York, 1962.

6. I. Babuška, *Random differential equations*, Apl. Mat. **4** (1959), 227–232. (Czech)

7. ———, *On randomized solutions of Laplace's equations*, Časopis Pěst. Mat. **86** (1961), 269–279.

8. S. Bernstein, *Equations différentielles stochastique*, Actualités Sci. Ind., No. 738, Hermann, Paris, 1938, 5–31.

9. J. E. Bertram and P. E. Sarachik, *Stability of circuits with random time-varying parameters*, IRE Trans. PGIT-5 (1959), Special suppl. 260–270.

10. B. H. Bharucha, *On the stability of randomly varying systems*, Ph.D. thesis, Univ. of California, Berkeley, Calif., 1961.

11. A. T. Bharucha-Reid, *On random elements in Orlicz spaces*, Bull. Acad. Polon. Sci. Cl. III **4** (1956), 655–657.

12. ———, *On random operator equations in Banach spaces*, Bull. Acad. Polon. Sci. Ser. Sci. Math. Astronom. Phys. **7** (1959), 561–564.

13. ———, *On random solutions of Fredholm integral equations*, Bull. Amer. Math. Soc. **66** (1960), 104–109.

14. ———, *On random solutions of integral equations in Banach space*, Trans. 2nd Prague Conf. Information Theory, pp. 27–48, Publ. House Czechoslovak Acad. Sci., Prague, 1960.

15. ———, *Sur les équations intégrales aléatoires de Fredholm à noyaux séparables*, C. R. Acad. Sci. Paris **250** (1960), 454–456, 657–658.

16. ———, *Approximate solutions of random operator equations*, Notices Amer. Math. Soc. **7** (1960), 361.

17. ———, *Random solutions of nonlinear integral equations* (to appear).

18. ———, *Semigroups of random operators* (to appear).

19. ———, *Random equations*, Academic Press, New York (to appear).

20. W. E. Boyce, *Random vibration of elastic strings and bars*, Proc. 4th U. S. Nat. Congr. Appl. Mech. (Univ. of California, Berkeley, Calif., 1962), Vol. 1, pp. 77–85, Amer. Soc. Mech. Engrs., New York, 1962.

21. R. Courant, *Methods of mathematical physics*, Vol. II, Interscience, New York, 1962.

22. K. Z. Dambis, *Stochastic integral equations in Hilbert space*, Diploma thesis, Moscow State University, Moscow, 1959. (Russian)

23. J. L. Doob, *Stochastic processes*, Wiley, New York, 1953.

24. M. Driml and O. Hanš, *On experience theory problems*, Trans. 2nd Prague Conf. Information Theory, pp. 93–111, Publ. House Czechoslovak Acad. Sci., Prague, 1960.

25. E. B. Dynkin, *Markov processes and analysis problems connected with them*, Uspehi Mat. Nauk **15** (1960), no. 2 (92), 3–24. (Russian)

26. F. J. Dyson, *A Brownian-motion model for the eigenvalues of a random matrix*, J. Mathematical Phys. **3** (1962), 1191–1198.

27. D. A. Edwards and J. E. Moyal, *Stochastic differential equations*, Proc. Cambridge Philos. Soc. **51** (1955), 663–677.

28. R. Engleman, *The eigenvalues of a randomly distributed matrix*, Nuovo Cimento (10) **10** (1958), 615–621.

29. P. Erdös and A. C. Offord, *On the number of real roots of a random algebraic equation*, Proc. London Math. Soc. (3) **6** (1956), 139–160.

30. M. I. Freĭdlin, *Mixed boundary value problem for elliptic differential equations of second order with small parameter*, Dokl. Akad. Nauk SSSR **143** (1962), 1300–1303. (Russian)

31. H. L. Frisch and S. P. Lloyd, *Electron levels in a one-dimensional random lattice*, Phys. Rev. (2) **120** (1960), 1175–1189.

32. I. V. Girsanov, *Ein Beispiel der uneindeutlichkeit der Lösung der stochastischen Integralgleichung von K. Ito*, Teor. Verojatnost. i Primenen. **7** (1962), 336–342. (Russian)

33. U. Grenander and M. Rosenblatt, *Statistical analysis of stationary time series*, Wiley, New York, 1957.

34. U. Grenander, *Some non linear problems in probability theory*, Probability and statistics, pp. 108–129, Wiley, New York, 1959.

35. ———, *Stochastic groups*, Ark. Mat. **4** (1960), 163–183.

36. ———, *Stochastic groups and related structures*, Proc. 4th Berkeley Sympos. Math. Statist. and Prob. (1960), Vol. II, pp. 171–184, Univ. of California Press, Berkeley, Calif., 1961.

37. J. Hammersley, *The zeros of a random polynomial*, Proc. 3rd Berkeley Sympos. Math. Statist. and Prob. (1955), Vol. II, pp. 89–111, Univ. of California Press, Berkeley, Calif., 1956.

38. O. Hanš, *Reduzierende zufällige Transformationen*, Czechoslovak Math. J. **7** (1957), 154–158.

39. ———, *Generalized random variables*, Trans. 1st Prague Conference on Information Theory, Statistical Decision Functions, and Random Processes (1956), pp. 61–103, Czech. Acad. Sci., Prague, 1957.

40. ———, *Random fixed point theorems*, ibid., pp. 105–125.

41. ———, *Inverse and adjoint transforms of linear bounded random transforms*, ibid., pp. 127–133.

42. ———, *Random operator equations*, Proc. 4th Berkeley Sympos. Math. Statist. and Prob. (1960), Vol. II, pp. 185–202, Univ. of California Press, Berkeley, Calif., 1961.

43. O. Hanš, and A. Špaček, *Random fixed point approximations by differentiable trajectories*, Trans. 2nd Prague Conference on Information Theory, pp. 203–213, Publ. House Czechoslovak Acad. Sci., Prague, 1960.

44. E. Hille and R. S. Phillips, *Functional analysis and semi-groups*, Amer. Math. Soc. Colloq. Publ. Vol. 31, Amer. Math. Soc., Providence, R. I., 1957.

45. K. Ito, *On a stochastic integral equation*, Proc. Japan Acad. **22** (1946), 32–35.

46. ———, *On stochastic differential equations*, Mem. Amer. Math. Soc. No. 4 (1951), 51 pp.

47. M. Kac, *On the average number of real roots of a random algebraic equation*, Bull. Amer. Math. Soc. **49** (1943), 314–320, 938.

48. ———, *On the average number of real roots of a random algebraic equation*. II, Proc. London Math. Soc. (2) **50** (1948), 390–408.

49. ———, *Probability and related topics in physical sciences*, Interscience, New York, 1959.

50. R. E. Kalman, *Control of randomly varying linear dynamical systems*, Proc. Sympos. Appl. Math. Vol. 13, pp. 287–298, Amer. Math. Soc., Providence, R. I., 1962.

51. J. Kampé de Fériet, *Random solutions of partial differential equations*, Proc. 3rd Berkeley Sympos. Math. Statist. and Prob. (1955), Vol. III, pp. 199–208, Univ. of California Press, Berkeley, Calif., 1956.

52. ———, *Statistical mechanics of continuous media*, Proc. Sympos. Appl. Math. Vol. 13, pp. 165–198, Amer. Math. Soc., Providence, R. I., 1962.

53. L. V. Kantorovič, *Functional analysis and applied mathematics*, Uspehi Mat. Nauk **3** (1948), no. 6 (28), 89–185. (Russian)

54. T. Koopmans, *Statistical inference in dynamic economic models*, Cowles Commission for Research in Economics, Monograph No. 10, Wiley, New York, 1950.

55. R. H. Kraichnan, *Dynamics of nonlinear stochastic systems*, J. Mathematical Phys. **2** (1961), 124–148.

56. M. A. Krasnosel'skiĭ and Ya. B. Rutickiĭ, *Convex functions and Orlicz spaces*, Gosudarstv. Izdat. Fiz.-Mat. Lit, Moscow, 1958. (Russian)

57. V. L. Lebedev, *Random processes in electrical and mechanical systems*, Gosudarstv. Izdat. Fiz.-Mat. Lit., Moscow, 1958. (Russian)

58. J. E. Littlewood and A. C. Offord, *On the number of real roots of a random algebraic equation*, J. London Math. Soc. **13** (1938). 288–295.

59. ———, *On the number of real roots of a random algebraic equation*. II, Proc. Cambridge Philos. Soc. **35** (1939), 133–148.

60. J. L. Lumley, *An approach to the Eulerian-Lagrange problem*, J. Mathematical Phys. **3** (1962), 309–312.

61. D. Maravell Casenoves, *Algunos nuevos procesos estocásticos y sus aplicaciones*, Rev. Acad. Ci. Madrid **53** (1959), 435–489, 659, 726.

62. ———, *Ingeniería de las oscilaciones*, Dossat, Madrid, 1959.

63. M. L. Mehta and M. Gaudin, *On the density of eigenvalues of a random matrix*, Nuclear Phys. **18** (1960), 420–427.

64. J. A. Morrison, *On the number of electron levels in a one-dimensional random lattice*, J. Math. Phys. **3** (1962), 1023–1027.

65. E. Mourier, *Eléments aléatoires dans un espace de Banach*, Ann. Inst. H. Poincaré **13** (1953), 161–244.

66. J. E. Moyal, *Stochastic processes and statistical physics*, J. Roy. Statist. Soc. Ser. B **11** (1949), 150–210.

67. T. Nagai, *Stability problem of random linear system of the first order*, Mem. Fac. Sci. Kyushu Univ. Ser. A **16** (1962), 47–59.

68. V. V. Nemytskiĭ, *The method of fixed points in analysis*, Uspehi Mat. Nauk **1** (1936), 141–174. (Russian)

69. W. Pogorzelski, *Integral equations and their applications*, Vols. I and II, Pánstwowe Wydawnictwo Naukowe, Warsaw, 1953, 1958. (Polish)

70. H. Primas, *Über quantenmechanische Systeme mit einem stochastischen Hamiltonoperator*, Helv. Phys. Acta **34** (1961), 36–57.

71. V. S. Pugachev, *Theory of random functions and its application to problems of automatic control*, Gosudarstv. Izdat. Tekh.-Teor. Lit., Moscow, 1957. (Russian)

72. A. Rosenbloom, *Analysis of random time-varying linear systems*, Ph.D. thesis, Univ. of California, Los Angeles, Calif., 1954.

73. G.-C. Rota, *On the passage from the Navier–Stokes equations to the Reynolds equations*, Proc. Internat. Conf. on Partial Differential Equations and Continuum Mechanics, pp. 374–376, Univ. of Wisconsin Press, Madison, Wis., 1961.

74. ———, *Linear operators satisfying algebraic identities*, Proc. Sympos. Appl. Math. Vol. 16, pp. 70–83, Amer. Math. Soc., Providence, R. I., 1964.

75. C. Ryll-Nardzewski, *An analogue of Fubini's theorem and its application to random linear equations*, Bull. Acad. Polon. Sci. Ser. Sci. Math. Astronom. Phys. **8** (1960), 511–513.

76. G. Samal, *On the number of real roots of a random algebraic equation*, Proc. Cambridge Philos. Soc. **58** (1962), 433–442.

77. J. C. Samuels, *On the mean square stability of random linear equations*, IRE Trans. **PGIT-5** (1959), Special suppl., 248–259.

78. ———, *On the stability of random systems and the stabilization of deterministic systems with random noise*, J. Acoust. Soc. Amer. **32** (1960), 594–601.

79. J. C. Samuels and A. C. Eringen, *On stochastic linear systems*, J. Math. and Phys. **38** (1959), 83–103.

80. J. Schauder, *Der Fixpunktsatz in Funktionalräumen*, Studia Math. **2** (1930), 171–180.

81. O. Šefl, *On stability of a randomized linear system*, Sci. Sinica **7** (1958), 1027–1034.

82. A. V. Skorohod, *Stochastic equations for diffusion processes in a bounded region*, Teor. Verojatnost. i Primenen **6** (1961), 287–298. (Russian)

83. ———, *On the existence and uniqueness of solutions of stochastic differential equations*, Sibirsk Mat. Ž. **2** (1961), 129–137.

84. A. Špaček, *Zufällige Gleichungen*, Czechoslovak Math. J. **5** (**80**) (1955), 462–466.

85. D. I. Šparo and M. G. Šur, *On the distribution of roots of random polynomials*, Vestnik Moskov. Univ. Ser. I Mat. Meh. **1962**, 40–43. (Russian)

86. S. K. Srinivasan, *On a class of stochastic differential equations*, Z. Angew. Math. Mech. **43** (1963), 259–265.

87. W. Uhlmann, *Eine Wahrscheinlichkeitstheorestische Begründung der Integrationsformeln von Newton-Cotes*, Z. Angew. Math. Phys. **10** (1959), 189–207.

88. ———, *Über harmonische und isotrope stochastische Prozesse mit Fehterschätzung für ein Differenzenverfahren*, Z. Angew. Math. Mech. **41** (1961), 428–447.

89. M. Ullrich, *Random Mikusiński operators*, Trans. 2nd Prague Conf. Information Theory, pp. 639–659, Publ. House Czechoslovak Acad. Sci., Prague, 1960.

90. K. Urbanik, *Generalized stochastic processes*, Studia Math. **16** (1958), 268–334.

91. G. A. van Lear, Jr. and G. E. Uhlenbeck, *The Brownian motion of strings and elastic rods*, Phys. Rev. (2) **38** (1931), 1583–1598.

92. E. P. Wigner, *On the distribution of the roots of certain symmetric matrices*, Ann. of Math. (2) **67** (1958) 325–327.

93. H. Wold, *A study in the analysis of stationary time series*, Almquist and Wiksells, Stockholm, 1950.

94. H. Wold and L. Juréen, *Demand analysis*, Wiley, New York, 1952.

95. A. C. Zaanen, *Linear analysis*, Interscience, New York, 1953.

WAYNE STATE UNIVERSITY,
DETROIT, MICHIGAN

REYNOLDS OPERATORS

BY

GIAN-CARLO ROTA[1]

DEDICATED TO PROFESSOR JOSEPH KAMPÉ DE FÉRIET ON THE OCCASION OF HIS RETIREMENT

1. **Introduction.** Let A be an algebra over a field of characteristic not equal to 2. A *Reynolds operator* (or Reynolds endomorphism) is a linear map of A into A which satisfies the following identity, for any f and g in A:

$$R(fg) = RfRg + R((f - Rf)(g - Rg)).$$

We call this identity the *Reynolds identity*. It occurs for the first time in a famous paper of Osborne Reynolds[2] on turbulence theory.

A related identity, introduced almost contemporarily by Kolmogorov and Kampé de Fériet and studied by Garrett Birkhoff, is the *averaging identity*

$$A(fAg) = (Af)(Ag).$$

A linear operator A satisfying the averaging identity is usually called an *averaging operator*. In this paper we discuss the relationship between averaging operators and Reynolds operators, and we give a representation of a Reynolds operator when A is an algebra of bounded measurable functions, provided the operator satisfies a certain mild continuity restriction.

There is an extensive literature on averaging operators, motivated largely by the fact that they characterize and generalize the operation of conditional expectation in probability theory. It is easy to see that

PROPOSITION 1. *If the algebra A has an identity* 1 *and A is an averaging operator such that* $A1 = 1$, *then A is a Reynolds operator.*

The proof is immediate: setting $f = 1$ in the averaging identity, we get $A^2 = A$. Expanding the Reynolds identity, we get the equivalent identity

$$R(fRg + gRf) = RfRg + R(RfRg), \tag{1}$$

from which it is clear that if A is an averaging operator such that $A^2 = A$, then $A = R$ is a Reynolds operator.

The converse of Proposition 1 is not true, and we shall see that there is an interesting variety of Reynolds operators other than conditional expectation. In fact, our first question is the following:

---

[1] This work was begun under contract NSF-GP-149, continued under Contract with the Office of Naval Research, and concluded while the author was a Fellow of the Sloan Foundation.

[2] Authors' names refer to the bibliography at the end.

QUESTION 1. Under what conditions is a Reynolds operator an averaging operator?

The answer (Theorem 1) is found, unexpectedly, to depend on whether or not the range of R is a closed subspace, in a suitable topology. Of course there are Reynolds operators without a closed range.

We shall be concerned with Reynolds operators on the algebra $L_\infty(S, \Sigma, m)$ of bounded measurable functions on a measure space (S, Σ, m). This choice is motivated in part by the idea of applying the results to probability and ergodic theory, in part by the fact that this algebra has a particularly simple structure. We shall not assume the measure space to be finite; it suffices to assume enough so that the Radon-Nikodym theorem is applicable; to avoid verbal detours, we shall simply assume that the measure space is σ-finite.

The choice of a topology in $L_\infty(S, \Sigma, m)$ is an old and thorny question, which we shall resolve in the easiest way. Norm-convergence in $L_\infty(S, \Sigma, m)$ bears notoriously little relation to the measure-theoretic structure. On the other hand, there is a topology which is the natural one for L_∞, namely, the topology of L_1-convergence, and *this is the topology we shall use.* Thus, all Reynolds operators are assumed continuous in this topology, as well as, of course, everywhere defined in $L_\infty(S, \Sigma, m)$. With this proviso, an answer to Question 1 is the following:

THEOREM 1. *A Reynolds operator in $L_\infty(S, \Sigma, m)$ is an averaging operator if and only if its range is closed in the L_1-topology.*

Of course the continuity restriction on R can be relaxed, but only at the cost of introducing pathological cases, even for operators acting on the sequence space l_∞, and even in the case of "conditional expectation."

It seems unexpected at first that the answer to a question which appears to be purely algebraic should depend upon a topological condition, but we hope that the proof below shows the "reason" for this phenomenon.

The second question we shall attack is the following:

QUESTION 2. Find an analytic representation for Reynolds operators.

To answer this question, we drop the assumption of continuity in the L_1-topology, but in exchange make a more stringent assumption on R, namely the following:

$$\int_S |Rf|^2 \, dm \leqq \int_S |f|^2 \, dm \tag{2}$$

for any f in L_∞. Such an assumption could be justified on "physical grounds" (Reynolds operators should not "increase the energy"); mathematically, it enables us to use the machinery of spectral theory in Hilbert space. At any rate, assumption (2) is more innocuous than the assumption of positivity of R ($Rf \geqq 0$ if $f \geqq 0$) that has often been made and that we shall *not* make. Strangely enough, we found that (2) implies that R is positive, but this fact is gotten only after the representation theorem. Under assumption (2), our final result (Theorem 2 below) gives a complete classification of Reynolds operators in terms of conditional expectation and one-parameter semigroups of measure-preserving transformations.

2. **Motivation and history.** Reynolds operators first arose in connection with a problem in partial differential equations which has never been fully settled. Let $N(f) = g$ be the Navier-Stokes equations, with an external acting force g (we do not write them down explicitly because their specific form is irrelevant for the present discussion). The problem is to find all operators R such that if f is a solution, then Rf is a solution of $N(Rf) = g'$, possibly with a different external acting force g'. The idea is that the function Rf should depend on "fewer degrees of freedom"; it should be in some sense an average of f. A simple computation shows that this transformation is possible by a linear operator R only if R satisfies the Reynolds identity. One thus expects the Reynolds identity to express an averaging, or smoothing, property of the operator.

Interesting though this connection with partial differential equations is, we shall not study it in the present work; our concern is largely classificatory. We hope to show in the process that the Reynolds identity is also interesting for reasons other than those that originated it.

In recent years there has been considerable interest in linear operators satisfying algebraic identities. The most spectacular developments were those obtained by Andersen-Atkinson-Baxter-Spitzer and several others, with abstract Wiener-Hopf methods. The operators considered by these authors satisfy an identity very close to the Reynolds identity; this fact alone provides a strong motivation for the study of Reynolds operators. In fact, we hope to show in a separate work (and it is easy to verify formally) that Reynolds operators are "infinitesimal analogs" of those considered by Atkinson-Baxter, etc.

To clarify the "meaning" of the Reynolds identity, we consider some examples. The simplest is conditional expectation, which in an infinite measure space must be defined with some care. Let Σ' be totally σ-finite σ-subfield of Σ (see Halmos, *Measure theory*), and let $A: f \to f'$ be the operator mapping a real function f into the Radon-Nikodym derivative of f relative to Σ'. Then A is a Reynolds operator and an averaging operator. The Radon-Nikodym derivative is the function f', unique in virtue of the Radon-Nikodym theorem, which is measurable relative to Σ' and which has the property that

$$\int_E f'\,dm = \int_E f\,dm$$

for every E of finite measure in Σ'. Of course f is assumed integrable over every set of finite measure.

Conditional expectation is the typical example of an averaging operator, and in fact, every Reynolds operator satisfying $R^2 = R$ is a generalization of conditional expectation (see the author's Padova Rendiconti paper).

Our second example is somewhat more sophisticated. Let T^t be a one-parameter semigroup of measure-preserving transformations of (S, Σ, m), and define $V^t f(S) = f(T^t S)$, as usual. Since $\|V^t\| = 1$ in L_∞, the operator

$$R = \int_0^\infty e^{-t} V^t \, dt \tag{3}$$

is well defined. To dispose of integrability questions, we may assume that V^t is L_1-continuous, and let the integral be a Bochner integral. We claim that R is a Reynolds operator. To establish this claim, we shall make use of two well-known results:

(a) V^t is a semigroup of operators "arising from" a semigroup of measure-forming transformations in the manner described above if and only if for each t, the operator V^t is a *homomorphism* of L_∞, that is, if and only if $V^t(fg) = (V^t f)(V^t g)$ for all essentially bounded f and g (see Halmos, *Ergodic theory*, for a proof of this result due to von Neumann). (b) If V^t is *any* one-parameter semigroup of bounded operators and R is defined as in (3), then $dV^t R/dt = V^t(R - I)$ (see Hille-Phillips or Dunford-Schwartz, Chapter VIII). Using these two results, we find that

$$\frac{d}{dt} V^t(RfRg) = (V^t(Rf - f))(V^t Rg) + (V^t Rf)(V^t(Rg - g)).$$

Now, integrating by parts identity (3) and using this, we find

$$\int_0^\infty e^{-t} V^t(RfRg)\, dt = [e^{-t} V^t(RfRg)]_0^\infty + \int_0^\infty e^{-t} V^t(Rg(Rf - f) + Rf(Rg - g))\, dt.$$

Suitably translated, this reads

$$R(RfRg) = -RfRg + R(Rg(Rf - f) + Rf(Rg - g)),$$

which is trivially equivalent to the Reynolds identity.

Operators of the form (3) are a generalization of conditional expectation. Roughly speaking, one "mixes" before averaging. In fact, we have pointed out elsewhere that many of the properties of conditional expectation can be extended to operators of the form (3). For example, the martingale convergence theorems (for *closed* martingales) extend to operators of the form (3), giving the Abel version of the classical ergodic theorems, and in fact giving a unified approach to martingales and ergodic theorems, with some "mixed" cases to boot.

Our second result (Theorem 2) shows that under assumption (2), all Reynolds operators "split" into a product of two commuting operators, one of which is a conditional expectation and the other is of the form (3). It would seem that this result reduces the study of Reynolds operators to ergodic theory. Actually, the converse is just as true. Several questions of ergodic theory, for example, questions of mixing and existence of invariant measure, can be conveniently formulated in terms of the Reynolds operators, where they become easier to attack. In fact, identity (3) shows that Reynolds operators are the *potentials*, in the language of probabilistic potential theory, of semigroups of measure-preserving transformations.

Finally, the Reynolds identity bears a close connection with the identity $D(fg) = fDg + gDf$ characterizing derivations. If R is a Reynolds operator and R^{-1} exists, perhaps as an unbounded operator, then $R^{-1} - I$ is a derivation. Unfortunately, R^{-1} seldom exists, as the example of conditional expectation shows, and Theorem 1 can be compared to the theorem of Singer and Wermer on the nonexistence of bounded derivations, at least for the algebra L_∞ (a version of the Singer-Wermer Theorem for L_∞ is a corollary of Theorem 1 and their entire theorem could be proved by these methods).

Algebraically, the Reynolds identity combines a feature of homomorphisms ($V(fg) = VfVg$), since the range of R is evidently a subalgebra; and a feature of derivations, namely, the "local operator" property, since if $Rf = 0$ then $Rg = 0$ for any g in the ideal generated by f in the range of R that is, any g of the form fRh, as follows easily from the Reynolds identity. These unusual properties make the Reynolds identity in many respects richer and more inclusive than either the "homomorphism" or the "derivation" identities.

A word about the history of the subject. Since Osborne Reynolds, it was Kampé de Fériet who first recognized the importance of studying the averaging and Reynolds identities, and began their study in a series of papers stretching over a period of thirty years. Curiously enough, the averaging identity was being studied at about the same time as Kolmogorov's *Foundations of probability* became known, but the connection with conditional expectation was brought out only many years later in Mrs. Moy's thesis. The Reynolds identity, on the other hand, first made its appearance because of the difficulty of producing averaging operators that commuted with differentiation in Euclidean spaces. The first algebraic study of the identity is Mme. Dubreil's. The first study of averaging operators by the methods of functional analysis is due to Garrett Birkhoff.

3. **Analytic preliminaries.** The two lemmas we shall use from analysis are: the "discontinuous" version of the Stone-Weierstrass Theorem, and the operator characterization of conditional expectation, suitably extended to infinite measure spaces. In both cases the extensions to infinite measure proceed along much the same lines as is done in the author's note in the Padova Rendiconti, and we shall only sketch the arguments here.

LEMMA 1. *Let* (S, Σ, m) *be a* σ*-finite measure space, and let* A *be a subalgebra of* $L_\infty(S, \Sigma, m)$. *Then the closure of* A *in the* L_1*-topology of* L_∞ *is a subalgebra of the form* $L_\infty(E, \Sigma', m)$, *for some measurable subset* $E \subseteq S$.

If in addition $m(S) < \infty$ *and* A *contains the identity function, then the closure of* A *in* $L_2(S, \Sigma, m)$ *is of the form* $L_2(S, \Sigma', m)$, *for some* σ*-subfield* Σ'.

SKETCH OF PROOF. Let E be a measurable set of finite measure. Consider the subalgebra $\chi_E A$, where χ_E is the characteristic function (or indicator) of the set E. Then an argument like that of Lemma 1 on page 55 of the mentioned paper shows that the L_1-closure of $\chi_E A$ is of the form $L_\infty(E, \Sigma_E, m)$, where Σ_E is a σ-subfield of the restriction of Σ to E. If F is a measurable set and $F \supseteq E$, then it follows easily that Σ_E is the restriction of Σ_F to E. Thus, we can define Σ' as the inductive limit of the Σ_E as E runs through an increasing sequence of sets of finite measure whose union is S. It is then easy to see that the L_1-closure of A is $L_\infty(E, \Sigma', m)$ for some subset $E \in \Sigma$.

Note that the σ-subfield Σ' need not be totally σ-finite, but will evidently be totally σ-finite if A has a set of generators which are square-integrable.

The following lemma is essentially due to Blackwell:

LEMMA 2. *Let A be a selfadjoint projection in $L_2(S, \Sigma, m)$ whose range is $L_2(S, \Sigma', m)$ with a totally σ-finite σ-subfield Σ'. Then A is the conditional expectation relative to Σ'.*

Again the proof results from routine extensions of the arguments given in the above-quoted paper. The reduction to the finite-measure case is accomplished as follows. Let $E, F \in \Sigma'$ be sets of finite measure, and let $g \in L_2(S, \Sigma, m)$. Then, by the selfadjointness of A,

$$\int_S \chi_E \chi_F A g \, dm = \int_S \chi_E \chi_F g \, dm = \int_S \chi_E A(\chi_F g) \, dm.$$

Since this holds for all E, and since Ag and $A(\chi_F g)$ both belong to $L_2(S, \Sigma', m)$, it follows that $\chi_F Ag = A(\chi_F g)$. This reduces the proof to the case where S is of finite measure, and in fact proves that the operator A satisfies the averaging identity.

A word about notation. If E is a subset of S and Σ' a σ-subfield of Σ, we write $L_p(E, \Sigma', m)$ to denote the restriction of Σ' to E.

All Banach spaces are taken over the complex field, unless otherwise specified. If R is a Reynolds operator defined on L_∞ over the *real* field, then define $R_1 f = Ru + iRv$ for $f = u + iv$. It is easily verified that R_1 is again a Reynolds operator over the complex field. Similarly for averaging operators.

All our conventions relative to the use of functional-analytic details follow Dunford-Schwartz or Halmos.

4. **Reduction to averaging operators.**

We begin by deriving some algebraic identities which are consequences of the Reynolds identity.

PROPOSITION 1. *Let R be a Reynolds operator in any commutative algebra over a field of characteristic zero, and let n be an integer $\geqq 1$. Then, for any element g the following identity holds:*

$$nR(g(Rg)^{n-1}) = (n-1)R((Rg)^n) + (Rg)^n. \tag{*}$$

PROOF. For $n = 1$ the identity is trivial; for $n = 2$ it follows from the Reynolds identity by setting $f = g$. For $n > 2$ we proceed by induction: assuming the identity

$$(n-1)R(g(Rg)^{n-2}) = (n-2)R((Rg)^{n-1}) + (Rg)^{n-1},$$

we establish (*). In the Reynolds identity, set $f = -(n-2)(Rg)^{n-1} + (n-1)g(Rg)^{n-2}$. It follows from the induction hypothesis that $Rf = (Rg)^{n-1}$. Therefore, substituting this choice of f in the Reynolds identity, we obtain

$$R(-(n-2)(Rg)^n + (n-1)g(Rg)^{n-1} + g(Rg)^{n-1}) = (Rg)^n + R((Rg)^n).$$

Simplifying, identity (*) follows at once.

PROPOSITION 2. *Let R be as in Proposition* 1, *let $p(x)$ be any polynomial in the variable x without constant term. Then the following identity holds for any Reynolds operator, for g in A:*

$$R(p(Rg)) = p(Rg) + R(Rgp'(Rg)) - R(gp'(Rg)),$$

where p' is the derivative of p.

PROOF. If $p(x) = x^n$ the statement reduces to that of Proposition 1. The general case follows at once by linearity.

PROPOSITION 3. *Let R be a continuous Reynolds operator in $L_\infty(S, \Sigma, m)$, and let $q(x)$ be a continuously differentiable function of a real variable defined for x in a compact interval such that $q(0) = 0$. For f in L_∞, let the range of f be contained in the domain of q. Then the function $q(f)$ is in L_∞, and*

$$R(q(Rf)) = q(Rf) + R(Rfq'(Rf)) - R(fq'(Rf)).$$

PROOF. Let D be the domain of q, a compact interval. By a variant of the Stone-Weierstrass Theorem, we can find a sequence p_n of polynomials without constant term such that $p_n \to q$ and $p_n' \to q'$ uniformly on D. In view of the assumed continuity of R, the result follows at once from the preceding proposition.

REMARK. The preceding result is not the most general of its kind, one can easily extend it to the case of several elements, to functions q which are not continuously differentiable, etc.

COROLLARY 1. *Let f be in the range of R and let q be as above. Then $q(f)$ is in the range of R.*

COROLLARY 2. *Let $f(s) \geqq \varepsilon > 0$ for s in $E \subseteq S$, and let f be in the range of R. Then there is a function g in the range of R such that $g(s) \geqq \delta > 0$ for almost all s in E, and $g(s) \geqq 0$ elsewhere.*

PROOF. Let the range of f be contained in a compact interval J containing 0. Choose a continuously differentiable function q such that $q(0) = 0$ and $q(x) > 0$ for $x \neq 0$ in J. Such a function obviously exists. Then $g = q(f)$ is the desired function by Corollary 1.

COROLLARY 3. *If for every $F \subseteq S$ with $m(F) > 0$ there is a function f in the range of R and an $\varepsilon > 0$ such that if $f(s) \geqq \varepsilon$ on F, then there is a g such that $Rg(s) \geqq \varepsilon > 0$ for almost all s in S.*

PROOF. Here we use for the first time the fact that the algebra on which R acts is L_∞. It is convenient to use the compactification of L_∞, so that we can talk about the value of a function at a point. Thus, we may and will assume that R acts on $C(E)$, the space of all continuous functions on an extremally disconnected compact Hausdorff space E. For every s in E there is an open set G_s containing s and nonnegative g_s such that $Rg_s(t) \geqq \varepsilon$ for t in G_s. The family G_s as s ranges through E forms a covering of E, hence it has a finite subcovering $G_{s_2}, \cdots, G_{s_n}$. Then the function $g = g_{s_1} + \cdots + g_{s_n}$ will do the trick.

COROLLARY 4. *Let R be as in the preceding corollary on the space $L_\infty(E, \Sigma, m)$. Then $R\chi_E = \chi_E$, where χ_E is the indicator of E.*

PROOF. Choose a function $f = Rg$ as in the preceding Corollary 3, and choose q so that $q(0) = q'(0) = 0$ and $q(x) = 1, q'(x) = 0$ for $x \neq 0$ in the range of f. Then $q(f) = \chi_E$, and from Proposition 3 it follows that

$$R\chi_E = \chi_E + R((f-g)q'(f)) = \chi_E, \qquad \text{q.e.d.}$$

THEOREM 1. *Let R be a continuous Reynolds operator in $L_\infty(S, \Sigma, m)$. Then R is an averaging operator if and only if it has closed range.*

PROOF. If R is an averaging operator, as well as a Reynolds operator, then $R^2 = R$, so R is a projection and hence its range is closed.

Conversely, suppose R is a Reynolds operator and its range is closed. Since the range is also closed under products, it follows from Lemma 1 that the range of R is of the form $L_\infty(E, \Sigma', m)$, where Σ' is a σ-subfield of Σ. It follows that R maps $L_\infty(E, \Sigma, m)$ into itself, and by Corollary 4 that $R\chi_E = \chi_E$. Hence $R(\chi_E f) = \chi_E Rf$, and we can and will assume that $S = E$ and $R1 = 1$. The proof results from the following steps:

(1) Suppose that Σ' is the trivial σ-subfield, containing only the null set and the whole space. Then Rf is a constant function, say α, and hence $R^2f = \alpha R1 = \alpha = Rf$. Hence R is a projection.

(2) Suppose now that Σ' is not the trivial σ-subfield, and let A be a set of positive measure belonging to Σ'. Let χ_A be the indicator of A. By assumption, $Rf = \chi_A$ for some f. We shall prove that $R\chi_A = \chi_A$. To this end, let $g(x)$ be a continuously differentiable function of the real variable x, for $-1 \leqq x \leqq 2$, with the following properties: (a) $g(x) = 0$ for $-1 \leqq x \leqq \frac{1}{2}$; (b) $g(x) = 1$ for $\frac{3}{4} \leqq x \leqq 2$. Then $g(\chi_A) = \chi_A$, since χ_A takes only the values 0 and 1. Hence, by Proposition 3 we have

$$R(\chi_A) = \chi_A + R((Rf - f)g'(\chi_A))$$

where f is a function such that $Rf = \chi_A$. But $g'(\chi_A)$ vanishes identically, because $g'(0) = g'(1) = 0$. This concludes the proof of this step.

(3) If $R\chi_A = \chi_A$ for the indicator of every set A, it follows by linearity that $Rf = f$ for every step function f in the range of R, and by continuity that $Rg = g$ for every function g in the range of R. But this amounts to saying that $R^2f = Rf$

for every f. Hence R is idempotent. (Since in addition we are assuming that $R1 = 1$, we obtain by setting $g = 1$ in the Reynolds identity that $R^2 = R$.) Now set $g = Rf$ in the Reynolds identity, and get $R(fRf) = (Rf)^2$. Polarizing, the averaging identity follows.

COROLLARY. *Let the algebra $L_\infty(S, \Sigma, \mu)$ be finite-dimensional. Then every Reynolds operator is an averaging operator.*

Similar results can be obtained for operators acting on the sequence space l_∞.

REMARK. The reasoning that led to Theorem 1 can be generalized without much difficulty (but with lengthy technical details) to $C(S)$, the space of all continuous functions on a compact Hausdorff space S. One then obtains a generalization of the Singer-Wermer Theorem on derivations.

5. **Representation.** In the following the continuity of R in L_∞ need not be assumed; we do, however, assume that R is everywhere defined in L_∞, and maps L_∞ into L_∞.

THEOREM 2. *Let R be a Reynolds operator in $L_\infty(S, \Sigma, m)$, $m(S) < \infty$, satisfying the condition $R1 = 1$ and*

$$(*) \qquad \int_S |Rf|^2\, dm \leqq \int_S |f|^2\, dm, \qquad f \in L_\infty(S, \Sigma, m).$$

Then there exist:

(1) *A unique σ-subfield Σ_1 of Σ;*

(2) *A unique positive conditional expectation operator A onto $L_\infty(S, \Sigma_1, m)$;*

(3) *A unique semigroup V^t of measure-preserving transformations defined for $t \geqq 0$ on $L_\infty(S, \Sigma_1, m)$, such that:*

(a) *The algebra $L_\infty(S, \Sigma_1, m)$ is invariant under R;*

(b) *The operators A and R commute; and the range of A is the closure of the range of R;*

(c) *The action of R for f in $L_\infty(S, \Sigma, m)$ is given by the formula*

$$Rf = \int_0^\infty e^{-t} V^t A f\, dt.$$

PROOF. (1) From assumption (*) it follows that the domain of R can be uniquely extended to the space $L_2(S, \Sigma, m)$. We denote again by R the resulting operator, since this can cause no confusion. In view of (*), R is an operator of norm one. We now apply a theorem of Nagy (cf. Riesz and Nagy), and infer that $R^*1 = 1$, where R^* is the adjoint operator of R, again acting on $L_2(S, \Sigma, m)$.

(2) It follows from step (1) that the operator $2R - I$ is isometric: that is, for all f in $L_2(S, \Sigma, m)$,

$$\int_S |2Rf - f|^2\, dm = \int_S |f|^2\, dm.$$

This can be established as follows. By the Reynolds identity applied to essentially bounded f and g we have

$$\int_S R(fRg + gRf)\, dm = \int_S [RfRg + R(RfRg)]\, dm.$$

Passing to adjoints, we have

$$\int_S (R^*1)(fRg + gRf)\, dm = \int_S [RfRg + (R^*1)(RfRg)]\, dm.$$

Using the conclusion of (1), this simplifies to

$$\int_S (fRg + gRf)\, dm = \int_S 2RfRg\, dm,$$

and passing to adjoints again (f and g may be assumed real without loss of generality)

$$\int_S g(R^*f + Rf)\, dm = \int_S 2gR^*Rf\, dm.$$

Since this is true for all functions g, we infer that

$$R^*f + Rf = 2R^*Rf,$$

for all f in L_∞, and hence, by continuity, also for all f in L_2. Now let $U = 1 - 2R$; then we obtain from the above identity that $U^*U = 1$. Hence $\int (Uf)^2\, dm = \int fU^*Uf\, dm = \int f^2\, dm$, which shows that U is isometric, at least on real functions. The extension to complex functions is immediate.

(3) Since the operator U is isometric in L_2, the ergodic theorem applies to it (cf. Dunford-Schwartz, Riesz-Nagy) and gives an orthogonal projection operator $I - A$ as a strong limit of the Cesàro averages $(1 + U + U^2 + \cdots + U^n)/(n+1)$. The projection $I - A$ has the following properties. (a) The range of $I - A$ consists of the fixpoints of $U = 1 - 2R$, that is, the set of all f such that $Rf = 0$; (b) the range of the orthogonal complement $A = I - (I - A)$ consists of the closed subspace spanned by functions of the form $f + Uf$, that is, the closure of the range R; in particular, since $R1 = 1$, we have $A1 = 1$, (c) A commutes with R. But the range of R, restricted to L_∞, is a subalgebra of L_∞. Hence we can apply Lemma 1 of §3, and conclude that the range of R is a subspace of the form $L_2(S, \Sigma_1, m)$, where Σ_1 is a σ-subfield of Σ. By Lemma 2 of §3, it now follows that A is the conditional expectation operator relative to Σ_1.

(4) It follows from the preceding step that R, restricted to the invariant subspace $L_2(S, \Sigma_1, m)$ is a one-to-one operator. More is actually true, namely, the range of R, restricted to $L_2(S, \Sigma_1, m)$ is dense in $L_2(S, \Sigma_1, m)$. For if it were not, then there would be a function f in $L_2(S, \Sigma_1, m)$ orthogonal to it: for such a function, $R^*f = 0$, that is, $U^*f = f$. But U^* is of norm one, hence, again by the Nagy theorem quoted above, $U^{**}f = Uf = f$. This in turn implies that $Rf = 0$, which is impossible if f lies in $L_2(S, \Sigma_1, m)$.

(5) Restrict R to the invariant subspace $L_2(S, \Sigma_1, m)$, from now on. We first verify that the unbounded closed operator $D = 1 - R^{-1}$ is antisymmetric, that is, D^* is an extension of $-D$. From the identity, already established in step (2), namely,

$$\int_S (Rf - f)Rg\, dm = -\int_S Rf(Rg - g)\, dm$$

we obtain, writing $p = Rf$ and $q = Rg$,

$$\int_S (1 - R^{-1})pq\, dm = -\int_S p(1 - R^{-1})q\, dm.$$

This holds for all p and q in the domain of the operator $D = I - R^{-1}$, and proves that $1 - R^{-1}$ is antisymmetric and the domain of $1 - R^{-1*}$ includes the domain of $I - R^{-1}$.

(6) The spectrum of D lies in the half plane $y \leq 0$ in the complex plane ($z=x+iy$). This follows from a variant of the spectral mapping theorem: the spectrum of U is inside the unit disk, hence the spectrum of R lies inside the circle of radius $\frac{1}{2}$ centered at $x = \frac{1}{2}$, and $D = I - R^{-1}$.

(7) From (5) and (6), together with the Hille-Yosida theorem, we conclude that D is a semigroup generator. Let V^t be the one-parameter semigroup generated by D. Since $R = (1 - D)^{-1}$, it follows from a well-known formula of semigroup theory (cf. Hille-Phillips or Dunford-Schwartz) that

$$Rf = \int_0^\infty e^{-t} V^t f\, dt, \qquad f \in L_2(S, \Sigma_1, m).$$

(8) We now prove that $V^t fg = V^t f V^t g$ for bounded f and g. Since the range of R is dense, it will suffice to show this for $f = Rp$ and $g = Rq$.

First, recall that if V^t is *any* one-parameter semigroup of bounded operators and R is defined as in (6), then $dV^tR/dt = V^t(R - I)$ (cf. Hille-Phillips), as follows easily from the relationship between R and V^t given in (6). We adapt to the present setup an argument due to Feller. Consider the function

$$h(s, t) = V^s((V^tRf)(V^tRg)),$$

for fixed f and g and variable s and t. An easy computation, using the Reynolds identity, the above remark, and the fact that R commutes with V^t, shows that $\partial h(s, t)/\partial s = \partial h(s, t)/\partial t$, hence $h(s, t) = q(s + t)$ for some function q. The conclusion amounts to saying that $h(t, 0) = h(0, t)$.

(9) The representation in (6) was obtained for functions of class L_2; however, it is easily seen that it can be extended to all functions of class $L_p(S, \Sigma, m)$, $1 \leqq p \leqq \infty$. This comes from the fact that V^t, being a semigroup of measure-preserving transformations, is norm-preserving in all L_p.

(10) The proof of the theorem is now almost complete, except for the passage from the σ-subfield Σ_1 to the σ-field Σ. The projection A is a conditional expectation; therefore it can be extended to any L_p so that it remains bounded. Clearly

$Rf = RAf = ARf$ for any f for which the expression is defined. Therefore on the right of the displayed formula in (7) we can replace f by Af, q.e.d.

Remark (*Added in proof*). The restriction $m(S) < \infty$ is not essential, and can be removed with a little extra work.

Bibliography

1. J. Arbault, *Nouvelles propriétés des transformations de Reynolds*, C.R. Acad. Sci. Paris, **239** (1954), 858–860.
2. ———, *Sur les transformations de Reynolds quasi régulières*, C.R. Acad. Sci. Paris, **239** (1954), 949–951.
3. ———, *Transformations de Reynolds sur les ensembles finis*, Conferenze, Centro Internazionale di Matematica Estivo, Varenna, September, 1957.
4. R. R. Bahadur, *Measurable subspaces and subalgebras*, Proc. Amer. Math. Soc. **6** (1955), 565–570.
5. Martin Billik and Gian-Carlo Rota, *On Reynolds operators in finite-dimensional algebras*, J. Math. Mech. **9** (1960), 927–932.
6. G. Birkhoff, *Moyennes des fonctions bornées*, Colloque d'Algèbre et de Théorie des Nombres, pp. 143–153, Centre National de la Recherche Scientifique, Paris, 1949.
7. Barron Brainerd, *On the structure of averaging operators*, J. Math. Anal. Appl. **5** (1962), 347–377.
8. ———, *Sur la structure des opérateurs moyennes*, C.R. Acad. Sci. Paris **252** (1960), 2058–2060.
9. J. Dieudonné, *Sur le théorème de Lebesgue-Nikodym*. III, Ann. Univ. Grenoble Sect. Sci. Math. Phys. (N.S.) **23** (1948), 25–53.
10. M. L. Dubreil-Jacotin, *Propriétés algébriques des transformations de Reynolds*, C.R. Acad. Sci. Paris **236** (1953), 1950–1951.
11. ———, *Propriétés algébriques des transformations de Reynolds*, C.R. Acad. Sci. Paris **236** (1953), 1136–1138.
12. ———, *Propriétés générales des transformations de Reynolds*, C.R. Acad. Sci. Paris **239** (1954), 856–858.
13. ———, *Étude algébrique des transformations de Reynolds*, Colloque d'Algèbre Supérieure, Brussels, December, 1956.
14. ———, *Sur le passage des équations de Navier-Stokes aux équations de Reynolds*, C.R. Acad. Sci. Paris **244** (1957), 2887–2890.
15. ———, *Sur les axiomes des moyennes*, Corsô sulla teoria della turbolenza Vol. 1, pp. 107–114, Centro Internazionale di Matematica Estivo, Libreria Editrice Universitaria Levrotto e Bella, Turin, 1957.
16. N. Dunford and J. Schwartz, *Linear operators*, Interscience, New York, 1958.
17. P. R. Halmos, *Lectures on ergodic theory*, Publications of the Mathematical Society of Japan, no. 3, The Mathematical Society of Japan, 1956.
18. ———, *Measure theory*, rev. ed., Van Nostrand, New York, 1952.
19. ———, *On a theorem of Dieudonné*, Proc. Nat. Acad. Sci. U.S.A. **35** (1949), 38–42.
20. Rudi Hirschfeld, *Sur les semi-groupes de transformations de Reynolds*, C. R. Acad. Sci. Paris **245** (1957), 1493–1495.
21. G. C. Hufford, *One-parameter semigroups of maps*, Duke Math. J. **24** (1957), 443–453.
22. J. Kampé de Fériet, *L'état actuel du problème de la turbulence*, La science Aeriennce **3** (1934), 9–34; II, ibid. **4** (1935), 12–52.
23. ———, *Les fonctions aléatoires stationnaires et la théorie statistique de la turbulence homogène*, Ann. Soc. Sci. Bruxelles Sér. I **59** (1939), 145–194.
24. ———, *La notion de moyenne dans les équations du mouvement turbulent d'un fluide*, 6ième Congrès International de Mécanique Appliquée, Paris, 1946.
25. ———, *Sur un problème d'algèbre abstraite posé par la définition de la moyenne dans la théorie de la turbulence*, Ann. Soc. Sci. Bruxelles Sér. I **63** (1949), 165–180.

26. ———, *Turbulencia*, Vol. I, Instituto Nacional Esteban Terradas de Tecnica Aeronautica, Madrid (to appear).

27. ———, *Averaging processes and Reynolds equations in atmospheric turbulence*, J. Meteorol. **8** (1951), 358–361.

28. ———, *Construction des transformations de Reynolds régulières*, C.R. Acad. Sci. Paris **239** (1954), 934–936.

29. ———, *Transformations de Reynolds opérant dans un ensemble de fonctions mesurables non negatives*, C.R. Acad. Sci. Paris **239** (1954), 787–789.

30. ———, *La notion de moyenne dans la théorie de la turbulence*, Rend. Sem. Mat. Fis. Milano **27** (1955–1956).

31. J. L. Kelley, *Averaging operators on $C_\infty(X)$*, Illinois J. Math. **2** (1958), 214–223.

32. A. Kolomogoroff, *Grundbegriffe der Wahrscheinlichkeitsrechnung*, Springer, Berlin, 1933.

33. V. J. Mizel and M. M. Rao, *Nonsymmetric projections in Hilbert space*, Dept. of Math. Tech. Rep. No. 47, Carnegie Institute of Technology, Pittsburgh, Pa., 1960.

34. Italico Molinaro, *Détermination d'une R-transformation de Reynolds*, C.R. Acad. Sci. Paris **244** (1957), 2890–2893.

35. ———, *Sur les endomorphismes de Reynolds de fonctions définies sur un ensemble fini*, Publ. Sci. Univ. Alger. Sér. A **4** (1957), 87–101.

36. Shu-Teh Chen Moy, *Characterizations of conditional expectations as a transformation on function spaces*, Pacific J. Math. **4** (1954), 47–63.

37. M. Nakamura, *A remark on the integral decomposition of a measure*, Mem. Osaka Univ. Lib. Arts Ed. Ser. B **3** (1954), 25–28.

38. C. W. Oseen, *Das Turbulenzproblem*, 3ième Congrès International de Mécanique Appliquée, Stockholm, tome 1, pp. 3–18, 1930.

39. O. Reynolds, *On the dynamical theory of incompressible viscous fluids and the determination of the criterion*, Philos. Trans. Roy. Soc. Ser. A **186** (1895), 123–164.

40. Gian-Carlo Rota, *On the representation of averaging operators*, Rend. Sem. Mat. Univ. Padova, **30** (1960), 52–64.

41. ———, *Remarks on stochastic solutions of differential equations*, Proc. Sympos. Partial Differential Equations, U.S. Army Research Center, Madison, Wis., 1960.

42. ———, *Spectral theory of smoothing operators*, Proc. Nat. Acad. Sci. U.S.A. **46** (1960), 863–868.

43. ———, *Une généralisation de l'espérance mathématique conditionnelle qui se présente dans la théorie statistique de la turbulence*, C.R. Acad. Sci. Paris **251** (1960), 624–626.

44. ———, *Une théorie unifiée des martingales et des moyennes ergodiques*, C.R. Acad. Sci. Paris **252** (1961), 2064–2066.

45. I. E. Segal, *Equivalence of measure spaces*, Amer. J. Math. **73** (1951), 275–313.

46. ———, *Abstract probability space and a theorem of Kolmogoroff*, Amer. J. Math. **76** (1954), 721–732.

47. Zbyněk Sidak, *On relations between strict sense and wide sense conditional expectations*, Teor. Verojatnost. i Primenen. **2** (1957), 283–288.

48. I. M. Singer and J. Wermer, *Derivations on commutative normed algebras*, Math. Ann. **129** (1955), 260–264.

49. J. Sopka, *On the characterization of Reynolds'-operators on the algebra of all continuous functions on a compact Hausdorff space*, Thesis, Harvard Univ., Cambridge, Mass., 1950.

50. F. B. Wright, *Generalized means*, Trans. Amer. Math. Soc. **98** (1961), 187–203.

51. E. Sparre Andersen, *On the fluctuations of random variables*, Math. Scand. **1** (1953), 263–285; II, ibid. **2** (1954), 195–223.

52. F. Spitzer, *A combinatorial lemma and its applications to probability theory*, Trans. Amer. Math. Soc. **82** (1956), 323–337.

53. G. Baxter, *An operator identity*, Pacific J. Math. **8** (1958), 649–663.

54. ———, *An analytic problem whose solution follows from a simple algebraic identity*, Pacific J. Math. **10** (1960), 731–762.

55. F. V. Atkinson, *Some aspects of Baxter's functional equation*, J. Math. Anal. Appl. 7 (1963), 1–30.

56. E. Hille and R. Phillips, *Functional analysis and semigroups*, Amer. Math. Soc. Colloq. Publ. Vol. 31, Amer. Math. Soc., Providence, R.I., 1958.

57. F. Riesz and B. Nagy, *Leçons d'analyse fonctionnelle*, 2nd ed., Akademiai Kiado, Budapest, 1953.

Massachusetts Institute of Technology,
Cambridge, Massachusetts

Indiana University,
Bloomington, Indiana

ON PROPAGATION IN RANDOM MEDIA OF DISCRETE SCATTERERS

BY

VICTOR TWERSKY

1. **Introduction.** General introductions and bibliographies for the present topic exist in the literature [**1**; **2**; **3**; **4**; **5**], and the physical significance of the subject is discussed there in detail. This section is therefore restricted to stating the primary problem at hand and to summarizing several results. Some introductory analytical material on scattering problems, and on random distributions of discrete scatterers is included in subsequent sections.

Consider the solution Ψ for the scattering of a wave by a given configuration of N scatterers, say $\Psi(\mathbf{1}, \mathbf{2}, \cdots, \mathbf{N})$, where $\mathbf{1}$ stands for all significant properties of scatterer "one" (including its location $\mathbf{r}_1$), etc. We introduce an ensemble of configurations in terms of an appropriate probability distribution function $W(\mathbf{1}, \mathbf{2}, \cdots, \mathbf{N})$, and define the average of Ψ, as

$$(1) \qquad \langle\Psi\rangle = \int \cdots \int \Psi(\mathbf{1}, \mathbf{2}, \cdots, \mathbf{N}) W(\mathbf{1}, \mathbf{2}, \cdots, \mathbf{N})\, d\mathbf{1}\, d\mathbf{2} \cdots d\mathbf{N};$$

we call $\langle\Psi\rangle$ the "coherent field." Similarly, we average the "intensity" (or "energy density") $|\Psi|^2$ over the ensemble, and write the "average total intensity" as

$$(2) \qquad \langle|\Psi|^2\rangle = |\langle\Psi\rangle|^2 + V,$$

where the "coherent intensity" $|\langle\Psi\rangle|^2$ is obtained from (1), and where the "incoherent intensity" V is the average absolute squared deviation of Ψ from its mean value $\langle\Psi\rangle$. We also consider the average of the corresponding normalized "flux" (e.g., the Poynting vector in electromagnetics) $\mathbf{J} = \mathrm{Re}\,(\bar{\Psi}\nabla\Psi/ik)$, where the bar indicates complex conjugate; we write

$$(3) \qquad \langle\mathbf{J}\rangle = \mathrm{Re}\,\langle\bar{\Psi}\nabla\Psi/ik\rangle = \mathrm{Re}\,[\langle\bar{\Psi}\rangle\nabla\langle\Psi\rangle/ik] + \mathbf{I} = \mathbf{C} + \mathbf{I},$$

where $\mathbf{C}$ and $\mathbf{I}$ are the "coherent" and incoherent" components of the flux. (The present definition of $\mathbf{J}$ suffices for observation in "free k-space"; more generally, other physical parameters than the propagation parameter k are also required.)

In general, we proceed by stating conditions on $\Psi(\mathbf{1}, \mathbf{2}, \cdots, \mathbf{N})$ to make the scattering problem for one configuration determinate, then introduce an ensemble of configurations in terms of $W(\mathbf{1}, \mathbf{2}, \cdots, \mathbf{N})$, and then seek the average values defined in (1), (2), and (3). [The average of Ψ^2, and the moments higher than the second are also of interest [**6**]; we will use a formalism for (2) which will enable us to obtain $\langle \Psi^2 \rangle$ by inspection. However, we do not consider the higher moments in the present paper. We also do not consider here the central statistical problem of obtaining explicit forms of W for the various "random distributions" of physical interest [7]; we assume that W is known, and subsequently use the simplest form of W for illustrative purposes.] We have essentially two kinds of representations [**8**] for Ψ: "compact representations" in terms of the solutions of a set of simultaneous integral equations (say Ψ_c), and "expanded" series representations (say Ψ_e). If we work with Ψ_e, then we obtain $\langle \Psi \rangle = \langle \Psi_e \rangle$, etc., by integration; the integrated forms of the terms of the series Ψ_e may be sufficiently simpler than the original and lead to closed or relatively compact forms for $\langle \Psi_e \rangle$. (See for example [**9**].) More usually [**2**; **3**; **4**], we average the compact representation Ψ_c, proceed heuristically to construct a set of "average equations" that are simpler than the originals, and then solve these to obtain say $\langle \Psi_c \rangle \approx \langle \Psi \rangle$. Most of the explicit results in the literature are of the form $\langle \Psi_c \rangle$; some of these have been shown adequate in specific ranges of parameters by comparison with experiment. (See for example [**10**; **11**].)

A more general introduction than the above, one that applies both to a random distribution of scatterers and to a randomly perturbed continuum, is given by Keller's [**5**] parallel discussion of "honest" and "dishonest" methods; Keller [**5**] uses both methods to derive results for scintillations of star-light arising from inhomogeneities in the atmosphere. The conditions for which $\langle \Psi_c \rangle$ approaches $\langle \Psi_e \rangle$ for a class of one-dimensional scattering problems have been determined by Bazer [**12**]. However, for three-dimensional distributions of discrete scatterers, the relationship of the heuristic c-forms of (1, 2, 3) considered by Foldy [**2**], Lax [**3**], Twersky [**4**], and others [**1**], to their corresponding e-forms is largely an open problem area. The present paper deals primarily with this topic.

In §2, we sketch the essentials of the formalism and notation for the scattering of a wave ϕ by a single object, and list several representations for the scattered wave u (surface integral, complex integral, and infinite series). Then, given a set of such "single-scattered" waves u_s, we construct analogous representations of the "multiple-scattered" set U_s for a configuration of scatterers. For brevity, we use operational forms to relate the U's to the presumably known u's: we have a compact "self-consistent" representation in the form of a set of inhomogeneous algebraic equations, and (by iteration) an expanded "successive scattering series" in terms of products of u's ("orders of scattering"). The corresponding representations of $\Psi = \phi + \sum U_s$, are labeled Ψ_c and Ψ_e respectively.

In §3, we introduce some notation and definitions relating to the probabilistic formalism for an ensemble of configurations. This is quite brief, since detailed discussions are given elsewhere, e.g., in [**1**; **2**; **3**].

In §4.1, we consider the averages of the expanded representations $\langle \Psi_e \rangle$ and $\langle |\Psi_e|^2 \rangle$. In §4.2, we give compact heuristic representations $\langle \Psi_c \rangle$ in the integral equation forms obtained by Foldy [**2**], Lax [**3**], and Twersky [**4**], and compare them with $\langle \Psi_e \rangle$. The essential simplification (as discussed previously in [**9**]) is that $\langle \Psi_c \rangle$ neglects terms corresponding to common path differences in a given order of scattering, i.e., each scatterer is "used only once" in a given order of scattering of $\langle\Psi_c\rangle$, so that, e.g., the back and forth interactions between a pair are neglected. An alternative heuristic compact form for $\langle\Psi\rangle$ is constructed to restore those of the "two-body" processes ignored by $\langle\Psi_c\rangle$. In §4.3, we start with a heuristic approximation for $\langle|\Psi|^2\rangle$ determined by $\langle\Psi_c\rangle$ and the "energy theorem" for $\langle\mathbf{J}\rangle$. Then we give heuristic integral equation forms $\langle\bar{\Psi}_c(\mathbf{r})\Psi_c(\mathbf{r}')\rangle$ derived by Foldy **[2]** and Lax **[3]**, and construct an alternative heuristic form that takes into account additional two-body processes. Comparison with the average of the product of the exact series indicates that these heuristic forms $\langle\bar{\Psi}_c\Psi_c\rangle$ are poorer approximations of $\langle\bar{\Psi}_e\Psi_e\rangle$ than the $\langle\Psi_c\rangle$ are of $\langle\Psi_e\rangle$. In order to obtain a form of $\langle\bar{\Psi}_c\Psi_c\rangle$ consistent with the approximation $\langle\Psi_c\rangle$, we introduce into $\langle\bar{\Psi}_e\Psi_e\rangle$ the same simplification that converts $\langle\Psi_e\rangle$ to $\langle\Psi_c\rangle$, e.g., we neglect any "chain" of u's or $\bar{u}$'s or $\bar{u}u$'s involving back and forth interactions between pairs, etc. This simplification in the series $\langle\bar{\Psi}_e\Psi_e\rangle$ leads directly to a relatively simple integral equation whose inhomogeneous term is $\langle\bar{\Psi}_c\rangle\langle\Psi_c\rangle$, and whose kernel involves a "new single scatterer" $u\varepsilon$; the function $u\varepsilon$ satisfies the integral equation analogous to $\langle\Psi_c\rangle$ whose inhomogeneous term is u, i.e., $u\varepsilon$ is the corresponding "Green's function" (in the sense the term is used in physics) for an angle dependent source u. (For the case of "monopole scatterers," Foldy **[2]** obtained two series representations for the kernel function in his integral equation for $\langle\bar{\Psi}_c\Psi_c\rangle$; the initial term of one of these series corresponds to the new integral equation specialized to monopoles.)

In §5, we consider an illustration that exhibits the relationships between the heuristic and exact forms explicitly. We restrict consideration to a very simple class of scattering problems: (1) the radiation field of the individual scatterers are significant only in the "forward half-space," particularly near the forward scattering direction; (2) the scatterers are located within a slab region of space $0 \leqq z \leqq d$ on which a plane wave is normally incident; (3) the scatterers are identical (or, equivalently, the averages over isolated scatterer parameters are identical), and their position statistics are those of a "rare gas." For the e-forms, we work in general with a finite number of scatterers (N) in a finite available volume ($\mathscr{V}$); for the corresponding c-forms, N and $\mathscr{V}$ are infinite but their ratio is bounded.

The illustrative problem is constructed around scatterers which do not need to be taken into account more than once in each order of scattering. Thus, for this case, the heuristic form $\langle\Psi_c\rangle$ used explicitly by Foldy **[2]**, Lax **[3]**, Twersky **[4]**, and used implicitly by practically all papers in the earlier physics literature **[1]**, is the correct limit of $\langle\Psi_e\rangle$ for $N \to \infty$, $\mathscr{V} \to \infty$. However, the situation is quite different for $\langle|\Psi|^2\rangle$: the heuristic form constructed with the aid of the energy theorem **[4]** gives the correct value $\langle|\Psi_e|^2\rangle$ in the appropriate limit; the analog of Foldy's **[2]** heuristic integral equation $\langle|\Psi_c|^2\rangle$, for which we obtain the kernel function in closed form, does not; the "$\langle\Psi_c\rangle$-consistent" heuristic integral equation for $\langle|\Psi_c|^2\rangle$ is identically the limit of $\langle|\Psi_e|^2\rangle$, and this is also the initial term of the analog of one of Foldy's two series **[2]**.

The result for $\langle\Psi_c\rangle$ is of the form $e^{\alpha}\phi$, and the corresponding result $\langle\Psi_e\rangle$ for finite N is simply $(1 + x/N)^N\phi$. We also obtain a simple form involving polynomials as a practical approximation for $\langle|\Psi'_e|^2\rangle$. Corresponding results for $\langle\Psi'^2\rangle$ are given in the form of integral equations.

2. Scattering formalism.

2.1. ONE SCATTERER. The time-independent scattering problem for an incident wave

$$\phi(\mathbf{i}) = e^{i\mathbf{k}\cdot\mathbf{r}}, \qquad \mathbf{k} = k\mathbf{i} = |k|\mathbf{i}, \qquad \mathbf{r} = r\mathbf{o}, \tag{4}$$

is specified outside the scatterer's surface S by a solution of

$$(\nabla^2 + k^2)\psi(\mathbf{r}) = 0, \qquad \nabla^2 = \partial_x^2 + \partial_y^2 + \partial_z^2, \qquad k = |k|, \tag{5}$$

subject to prescribed conditions at S and subject to conditions at large distances from S (i.e., $r \to \infty$) which insure

$$\psi \to \phi, \qquad \psi - \phi \equiv u \sim f(\mathbf{o}, \mathbf{i})\frac{e^{ikr}}{r}; \qquad r \to \infty. \tag{6}$$

Here u is the "scattered wave," f (which is independent of r) is the "scattering amplitude," and the origin of coordinates is the center of the smallest sphere enclosing the scatterer (henceforth "scatterer center"). The weakest "radiation condition" on u and its normal derivative $\partial_n u$ leading to (6) appears to be that considered by Wilcox [13], i.e., that $\int|\partial_n u - iku|^2\, dS$ vanishes for a closed surface at infinity.

Various "boundary conditions" and "transition conditions" at S are of interest to physics, particularly those that satisfy the "reciprocity relation"

$$\int_S [\psi(\mathbf{i})\partial_n\psi(\mathbf{j}) - \psi(\mathbf{j})\partial_n\psi(\mathbf{i})]\, dS = 0; \qquad \psi(\mathbf{i}) = \psi(\mathbf{r}; \mathbf{i}) = \phi(\mathbf{r}; \mathbf{i}) + u(\mathbf{r}; \mathbf{i}). \tag{7}$$

Equation (7) holds for example if either ψ, or $\partial_n\psi$, or a linear combination $\psi + C\partial_n\psi$ vanishes on S; it also holds if $\psi = A\psi'$ and $\partial_n\psi = B\partial_n\psi'$ on S, and $(\nabla^2 + k'^2)\psi = 0$ inside, such that the discontinuity coefficients A and B are constants while the internal propagation coefficient k' may be a function of $\mathbf{r}$, etc. Special conditions insure that the scatterer is "lossless" in the sense that

$$\int_S [\bar{\psi}(\mathbf{i})\partial_n\psi(\mathbf{j}) - \psi(\mathbf{j})\partial_n\bar{\psi}(\mathbf{i})]\, dS = 0, \tag{8}$$

e.g., any of the conditions mentioned after (7) for real values of the coefficients C, A, B, and k'. If (7) and (8) hold on the scatterer's surface, then from Green's theorem and (5), they hold for any surface outside the scatterer.

Various analytical representations exist for u. We obtain the surface integral form by applying Green's theorem to $u(\mathbf{r})$ and

$$\frac{e^{ik|\mathbf{r}-\mathbf{r}'|}}{|\mathbf{r}-\mathbf{r}'|} \equiv E(k|\mathbf{r} - \mathbf{r}'|)$$

in the region outside the scatterer:

$$
\begin{aligned}
u(\mathbf{r}) &= -\frac{1}{4\pi}\int [E(k|\mathbf{r}-\mathbf{r}'|)\partial_n u(\mathbf{r}') - u\partial_n E]\, dS(\mathbf{r}') \\
&\equiv \{E(k|\mathbf{r}-\mathbf{r}'|), u(\mathbf{r}'; \mathbf{i})\} = \{E(k|\mathbf{r}-\mathbf{r}'|), \psi(\mathbf{r}'; \mathbf{i})\},
\end{aligned}
\tag{9}
$$

where $S(\mathbf{r}')$ is any surface isolating $\mathbf{r}$ from the scatterer. For the conditions leading to (7), integral equations specifying ψ may be obtained from (9). Comparison of the asymptotic form of (9) for $|\mathbf{r}-\mathbf{r}'| \to \infty$ with (6) leads to the surface integral representation for the scattering amplitude:

$$
f(\mathbf{o}, \mathbf{i}) = \{e^{-ik\mathbf{o}\cdot\mathbf{r}'}, u(\mathbf{r}'; \mathbf{i})\} = \{e^{-ik\mathbf{o}\cdot\mathbf{r}'}, \psi(\mathbf{r}'; \mathbf{i})\}.
\tag{10}
$$

Definition (10), (6), Green's theorem, plus the reciprocity relation (7) lead to

$$
f(\mathbf{i}, \mathbf{j}) = f(-\mathbf{j}, -\mathbf{i}).
\tag{11}
$$

Similarly for lossless scatterers, it follows from (8) that

$$
\begin{aligned}
-\frac{2\pi i}{k}[f(\mathbf{j}, \mathbf{i}) - \bar{f}(\mathbf{i}, \mathbf{j})] &= \int_\pi \bar{f}(\mathbf{o}, \mathbf{j}) f(\mathbf{o}, \mathbf{i})\, d\Omega_o; \\
\frac{4\pi}{k}\operatorname{Im} f(\mathbf{i}, \mathbf{i}) &= \int_\pi |f(\mathbf{o}, \mathbf{i})|^2\, d\Omega \equiv \sigma_s(\mathbf{i}),
\end{aligned}
\tag{12}
$$

where σ_s is the "total scattering cross-section," and where $\int_\pi d\Omega_o$ means integrate over all directions $\mathbf{o}$.

The surface integral form for a scatterer of "type" s whose "center" is located at $\mathbf{r}_s$ scattering to a point $\mathbf{r}_a$ under excitation $e^{i\mathbf{k}\cdot(\mathbf{r}-\mathbf{r}_s)}$ may be written

$$
u_s^a(\mathbf{i}) = u_s(\mathbf{r}_a - \mathbf{r}_s; \mathbf{i}) = \{E(k|\mathbf{r}_a - \mathbf{r}_s - \mathbf{r}'|), u(\mathbf{r}'; \mathbf{i})\},
\tag{13}
$$

where $\mathbf{r}'$ is measured from the "scatterer's center" $\mathbf{r}_s$ to a point on an inclosing surface. Using essentially Weyl's representation of $E(kr)$ as a complex integral of plane waves, we rewrite (13) in terms of f of (10) as [8]

$$
u_s^a(\mathbf{i}) = \frac{ik}{2\pi}\int e^{ik\mathbf{p}\cdot\mathbf{r}_{as}} f_s(\mathbf{p}, \mathbf{i})\, d\Omega_p, \qquad \mathbf{r}_{as} = \mathbf{r}_a - \mathbf{r}_s = r_{as}\mathbf{o}_{as}.
\tag{14}
$$

Here, in general, $\int d\Omega_p$ represents a double integral over complex paths, the complex solid angle Ω_p being specified by two angles associated with the unit vector $\mathbf{p}$ (each path of the form used by Sommerfeld for the Hankel function of the first kind). The plane wave form is valid at least outside the circumscribing sphere; see [8] for weaker restriction. The asymptotic form of u as in (6) follows from a saddle point approximation of (14). More generally, (14) leads to the Sommerfeld convergent series in inverse powers of r and angular derivatives of f [8]:

$$u_s^a(\mathbf{i}) = u_s(r_{as}\mathbf{o}_{as}; \mathbf{i}) = E_{as}[r_{as}; D(\mathbf{o}_{as})]f_s(\mathbf{o}_{as}, \mathbf{i}),$$

$$(15)\quad E(r; D) = \frac{e^{ikr}}{r}\left\{1 + \frac{i}{2kr} D + \left(\frac{i}{2kr}\right)^2 \frac{D(D-2)}{2!} + \cdots + \left(\frac{i}{2kr}\right)^n \frac{D(D - 1\cdot 2)\cdots(D - [n-1]\cdot n)}{n!} + \cdots\right\},$$

$$D(\mathbf{o}) = (-1/\sin^2\theta)[\partial_\omega^2 + \sin\theta\,\partial_\theta(\sin\theta\,\partial_\theta)],$$

where $D(\mathbf{o})$ is Beltrami's operator on the polar angle (θ) and azimuthal angle (ω) associated with the unit vector $\mathbf{o}$. The series converges outside the circumscribing sphere [13]. We may also represent u in terms of a symbolic operation on $\phi(\mathbf{i})$ [3], as a Bessel–Legendre series ("multipole expansion") [8], etc. In particular if f is independent of directions ("monopole scatterers") then E of (15) reduces to its leading term.

For an isolated scatterer at $\mathbf{r}_s$ scattering to $\mathbf{r}_a$, under excitation by $e^{ik\mathbf{i}\cdot\mathbf{r}}$, the scattered field is

$$(16)\qquad u_s^a(\mathbf{i})\phi_s(\mathbf{i}), \qquad \phi_s(\mathbf{i}) = e^{ik\mathbf{i}\cdot\mathbf{r}_s},$$

where any of the above representations may be used for u_s^a. In general, in order to maintain a relatively simple (if not particularly explicit) notation, we indicate (16) symbolically by

$$(17)\qquad u_s^a\phi_s = E_{as}f_s\phi_s.$$

If necessary, auxiliary notation will be introduced.

2.2. Many scatterers. For a configuration of N, not necessarily identical, scatterers located at $\mathbf{r}_1, \mathbf{r}_2, \cdots, \mathbf{r}_N$, we represent the solution outside the scatterers' surfaces by

$$(18)\qquad \Psi(\mathbf{r}) = \phi + \mathscr{U} = \phi + \sum_{s=1}^{N} U_s(\mathbf{r} - \mathbf{r}_s), \qquad (\nabla^2 + k^2)\Psi = 0, \qquad k = |k|,$$

where Ψ satisfies (7) for all surfaces S_s (although the boundary conditions may be different on different surfaces). Using the forms of the previous section operationally, we have [8]

$$(19)\qquad U_s^a = U_s(\mathbf{r}_a - \mathbf{r}_s) = E_{as}F_s,$$

where U_s is the "multiple scattered wave" of scatterer s, and F_s is its "multiple scattered amplitude." Equation (19) serves to indicate that U_s and F_s are related to each other as indicated in (15), (14), (10), (6), etc. If the other scatterers recede to infinite distances from s, then U_s and F_s reduce to their known analogs for scatterer s in isolation, i.e., to their "single scattered" values:

$$(20)\qquad U_s^a \to u_s^a\phi_s, \qquad F_s \to f_s\phi_s; \qquad r_{st} = |\mathbf{r}_s - \mathbf{r}_t| \to \infty \quad \text{for fixed } |\mathbf{r}_s - \mathbf{r}_a|.$$

We need not rework any isolated scatterer problems to determine the solution for the configuration. The problem is linear, and since we know the response of scatterer s to $\phi_s(\mathbf{i})$, we can determine its response U_s^a to the present "exciting field"

$$\Phi_s = \phi_s + \sum_t{}' U_t^s, \qquad \sum_t{}' = \sum_{t\neq s} \tag{21}$$

by applying the superposition principle with $U_t^s = E_{st}F_t$ in the plane wave form (14); symbolically we write $U_s^a = u_s^a\Phi_s$, and represent (18) compactly by the operational forms (at a point $\mathbf{r}_c$):

$$\Psi_c = \phi_c + \sum u_s^c\Phi_s, \qquad \Phi_s = \phi_s + \sum_t{}' u_t^s\Phi_t. \tag{22}$$

Equivalently, we have

$$\begin{aligned} \Psi_c &= \phi_c + \sum E_{cs}F_s = \phi_s + \sum E_{cs}f_s\Phi_s, \\ F_s &= f_s\Phi_s = f_s\phi_s + \sum_t{}' E_{st}f_sF_t. \end{aligned} \tag{23}$$

As an explicit illustration, in terms of the plane wave operator E implicit in (14), we have

$$\begin{aligned} \Psi_c &= e^{ik\mathbf{i}\cdot\mathbf{r}_c} + \frac{ik}{2\pi}\sum_s \int e^{ik\mathbf{p}\cdot(\mathbf{r}_c-\mathbf{r}_s)}F_s(\mathbf{p}, \mathbf{i})\, d\Omega_p, \qquad \mathbf{p} = \mathbf{p}(s); \\ F_s(\mathbf{o}, \mathbf{i}) &= f_s(\mathbf{o}, \mathbf{i})\, e^{ik\mathbf{i}\cdot\mathbf{r}_s} + \frac{ik}{2\pi}\sum_t{}' \int f_s(\mathbf{o}, \mathbf{p})\, e^{ik\mathbf{p}\cdot(\mathbf{r}_s-\mathbf{r}_t)}F_t(\mathbf{p}, \mathbf{i})\, d\Omega_p, \end{aligned} \tag{24}$$

where (24) (which holds at least outside mutually exclusive circumscribing spheres) represents the set F_s as the solutions of a system of N integral equations involving the known functions f_s as inhomogeneous terms and in the kernels. Similarly, we may represent E in (23) by the series in (15). See [8] for discussions, applications, and other representations.

When no questions of convergence arise, we iterate the "compact" forms (22) and (23) in terms of the single scattered value and write the "expanded" forms Ψ_e (at a point $\mathbf{r}_e$) as a series of "orders of scattering":

$$\begin{aligned} \Psi_e &= \phi_e + \sum u_s^e\phi_s + \sum u_s^e\sum{}' u_t^s\phi_t + \sum u_s^e\sum{}' u_t^s\sum{}' u_m^t\phi_m + \cdots \\ &= \phi_e + \sum E_{es}f_s\phi_s + \sum E_{es}\sum{}' E_{st}f_sf_t\phi_t + \cdots, \end{aligned} \tag{25}$$

which may be converted to more explicit forms by inspection of (24), etc.

The corresponding "intensity" for two "observation points" $\mathbf{r}_c$ and $\mathbf{r}_a$, in the form we consider subsequently, is

$$\bar{\Psi}_c\Psi_a = \bar{\phi}_c\phi_a + \bar{\phi}_c\mathscr{U}_a + \phi_a\bar{\mathscr{U}}_c + \sum\sum \bar{U}_s^cU_t^a. \tag{26}$$

The "flux"

$$\mathbf{J}_{ca} = (\bar{\Psi}_c\nabla_a\Psi_a - \Psi_a\nabla_c\bar{\Psi}_c)/ik, \qquad \mathbf{J}_{cc} = \mathbf{J}_c = \mathbf{J}, \tag{27}$$

may be obtained from (26). For lossless scatterers, it follows from (8) applied to $\bar{\Psi}_c(\mathbf{i})$ and $\Psi_c(\mathbf{i})$, that

$$\int_S \mathbf{J} \cdot d\mathbf{S} = 0, \tag{28}$$

where S incloses any number of scatterers.

In §3 we introduce the probabilistic formalism, and in §4 we apply it to obtain the averages of (22) and (25), and of the corresponding product forms. The operational representations with their lack of detailed structure are adequate for comparison of the essential features of the c and e forms of present interest. However, in general, more structure is required in order to apply such forms to specific problems. To facilitate restoring the arguments of the functions, we keep redundant factors of ϕ, etc., in subsequent manipulations; the appropriate arguments can be determined by starting with ϕ and working from right to left in the "chains" of products of u's that arise. The redundant factors of ϕ, etc., also enable us to obtain the forms for $\langle\Psi^2\rangle$ by inspection of $\langle\bar{\Psi}\Psi\rangle$.

3. **Probabilistic formalism.** The general distribution function $W(\mathbf{1}, \mathbf{2}, \cdots, \mathbf{N})$ used in (1) satisfies $\int \cdots \int W\, d\mathbf{1} \cdots d\mathbf{N} = 1$, when integrated over the full domains of all variables; $W\, d\mathbf{1} \cdots d\mathbf{N}$ is the probability of finding the scatterers in a configuration in the "volume" between $(\mathbf{1}, \mathbf{2}, \cdots, \mathbf{N})$ and $(\mathbf{1}, \mathbf{2}, \cdots, \mathbf{N}) + d\mathbf{1}\, d\mathbf{2} \cdots d\mathbf{N}$. We may factor W into such forms as

$$W(\mathbf{1}, \mathbf{2}, \cdots, \mathbf{N}) = w_s(\mathbf{s})P_s(\mathbf{s};\, \text{avb}\ \mathbf{s}) = w_{st}(\mathbf{s}, \mathbf{t})P_{st}(\mathbf{s}, \mathbf{t};\, \text{avb}\ \mathbf{s}, \mathbf{t}), \tag{29}$$

etc., where "avb" stands for "all variables besides." The one-particle distribution function $w_s(\mathbf{s}) = \int W\, d(\text{avb}\ \mathbf{s}) \equiv \langle 1\rangle_s$, may be written $w_s(\mathbf{r}_s, \boldsymbol{\xi}_s)$, where $\mathbf{r}_s$ is the position variable, and $\boldsymbol{\xi}_s$ represents various parameters of scatterer s (size, shape, dielectric properties, etc.). Thus $w_s(\mathbf{r}_s) = \int w_s(\mathbf{r}_s, \boldsymbol{\xi}_s)\, d\boldsymbol{\xi}_s$ (i.e., $w_s(\mathbf{s})$ integrated over all values of $\boldsymbol{\xi}_s$) times $d\mathbf{r}_s$ is the probability of finding scatterer s (the center of its circumscribing sphere) in the volume element $d\mathbf{r}_s$ around the point $\mathbf{r}_s$, etc. Similarly $w_{st}(\mathbf{s}, \mathbf{t}) = \int W\, d(\text{avb}\ \mathbf{s}, \mathbf{t}) \equiv \langle 1\rangle_{st}$ is the two-particle probability distribution function, etc. The P's are the corresponding conditional probability functions, $\int P_s\, d(\text{avb}\ \mathbf{s}) = \int P_{st}\, d(\text{avb}\ \mathbf{s}, \mathbf{t}) = 1$, etc. The two-particle function (the "pair distribution function") may be written

$$w_{st}(\mathbf{s}, \mathbf{t}) = w_s(\mathbf{s})p_{st}(\mathbf{s};\, \mathbf{t}) = w_t(\mathbf{t})p_{ts}(\mathbf{t};\, \mathbf{s}); \tag{30}$$

the conditional probability p_{st} for identical particles whose distribution depends only on $r_{st} = |\mathbf{r}_s - \mathbf{r}_t|$ is the "radial distribution function" discussed in [7].

If W factors into a product of N one-particle functions $w_s(\mathbf{s})$, then we may say we are dealing with a "gas." If the w's are identical functions of the parameters, then it is essentially a "single-component" gas. If the w's are independent of $\mathbf{r}$, then the gas is "homogeneous."

If we are dealing with an "inhomogeneous gas" in the sense that the location and characteristics of each scatterer is independent of the locations and characteristics of the others, and is the same for all scatterers, then

$$W(\mathbf{1}, \mathbf{2}, \cdots, \mathbf{N}) = w(\mathbf{1})w(\mathbf{2}) \cdots w(\mathbf{N}), \qquad w(\mathbf{s}) = w(\mathbf{r}_s, \boldsymbol{\xi}_s). \tag{31}$$

The corresponding "particle density" function for scatterers of type $\boldsymbol{\xi}$ is

$$\rho(\mathbf{s}) = \rho(\mathbf{r}, \boldsymbol{\xi}) = Nw(\mathbf{r}, \boldsymbol{\xi}), \tag{32}$$

such that $\rho(\mathbf{r}) = \int \rho(\mathbf{r}, \boldsymbol{\xi})\, d\boldsymbol{\xi}$ is the number of scatterers per unit volume at $\mathbf{r}$, and $\int \rho(\mathbf{r})\, d\mathbf{r} = N$ is the total number of scatterers; the volume integration is over the available volume $\mathscr{V}$. If ρ is independent of $\mathbf{r}$, then we use $\rho = N/\mathscr{V}$; we may have $N \to \infty$ and $\mathscr{V} \to \infty$ provided that $N/\mathscr{V}$ remains bounded. Gases of scatterers are considered for example in [**2**; **3**; **4**]. A simple model for some phenomena observed in scattering by liquids may be based on a W which factors into products of identical two-particle functions. This was introduced for single-scattering of x-rays by liquids by Zernike and Prins [**14**], and is also considered in [**3**; **9**], and by Mazur, Mandel, Jansen [**15**], and by others.

4. **Scattering by random distributions.** In general, the ensemble average of the total scattered field at a point $\mathbf{r}_c$ as in (18) may be written

$$\langle \Psi_c \rangle = \phi_c + \langle \mathscr{U}_c \rangle = \phi_c + \sum \langle U_s^c \rangle, \tag{33}$$

where as in (1), $\langle\ \rangle$ means multiply by W and integrate over all variables. Knowing the coherent field $\langle \Psi_c \rangle$, we write the "incoherent intensity" for $\mathbf{r}_c$ and $\mathbf{r}_a$ in the form

$$V_{ca} \equiv \langle \bar{\Psi}_c \Psi_a \rangle - \langle \bar{\Psi}_c \rangle \langle \Psi_a \rangle = \sum\sum [\langle \bar{U}_s^c U_t^a \rangle - \langle \bar{U}_s^c \rangle \langle U_t^a \rangle], \tag{34}$$

from which we ultimately obtain the total average intensity $\langle |\Psi|^2 \rangle = |\langle \Psi \rangle|^2 + V$ and the corresponding flux $\langle \mathbf{J} \rangle = \mathrm{Re}[\bar{\Psi} \nabla \Psi / ik]$. For lossless scatterers, we average (28) to obtain

$$\int \langle \mathbf{J} \rangle \cdot d\mathbf{S} = 0 \tag{35}$$

over a surface inclosing the distribution.

4.1. Expanded representation. Multiplying the series (25) by W as in (29) and integrating over all variables, we obtain

$$\begin{aligned} \langle \Psi_e \rangle &= \phi_e + \langle \mathscr{U}_e \rangle = \phi_e + \sum_s \langle u_s^e \phi_s \rangle + \sum_s \sum_t{}' \langle u_s^e u_t^s \phi_t \rangle \\ &\quad + \sum_s \sum_t{}' \sum_m{}' \langle u_s^e u_t^s u_m^t \phi_m \rangle + \cdots \\ &= \phi_e + \sum_s \int u_s^e \phi_s w_s\, d\mathbf{s} + \sum_s \sum_t{}' \int\!\!\int u_s^e u_t^s [\phi_t + u_s^t \phi_s + \cdots] w_{st}\, d\mathbf{s}\, d\mathbf{t} \\ &\quad + \sum_s \sum_t{}' \sum_m{}'' \int\!\!\int\!\!\int u_s^e u_t^s u_m^t [\phi_m + \cdots] w_{stm}\, d\mathbf{s}\, d\mathbf{t}\, d\mathbf{m} + \cdots, \end{aligned} \tag{36}$$

where $\sum_m''$ means $m \neq t$, $m \neq s$. The original triple sum $\sum_s \sum_t' \sum_m'$ of the third order of scattering contributes essentially two kinds of terms when averaged, i.e., the fact that $m \neq t$ and $t \neq s$ does not rule out $s = m$. Thus there are $N(N-1)$ terms for $m = s$ of the form $\langle u_s^e u_t^s u_s^t \phi_s \rangle$ which are functions only of two variables, and these wave forms differ essentially from the $N(N-1)$ terms $\langle u_s^e u_t^s \phi_t \rangle$ of the second order, and also from the remaining $N(N-1)(N-2)$ terms $\langle u_s^e u_t^s u_m^t \phi_m \rangle$ of the third order. Similarly for the higher orders. The "chains" in which the subscripts of the u's are all different form a sequence, each term of which contains $N!/(N-\nu)!$ terms; this sequence of terms in which each scatterer appears only once terminates with $\nu = N$, the Nth order of scattering. The first three terms of this sequence are shown in (36), which also gives the starting term of the sequence in which one scatterer appears twice. The fourth order of scattering provides a term for each of these sequences and starts a sequence in which two scatterers are used twice. There are an infinite number of such sequences.

Similarly the quadratic functions follow from

$$
\begin{aligned}
\langle \bar{\Psi}_e \Psi_a \rangle = \bar{\phi}_e \phi_a + \bar{\phi}_e \langle \mathscr{U}_a \rangle + \phi_a \langle \bar{\mathscr{U}}_e \rangle + \sum \int \overline{u_s^e \phi_s} u_s^a \phi_s w_s \, d\mathbf{s} \\
+ \sum\sum{}' \int\int \{ \overline{u_s^e \phi_s} [u_t^a \phi_t + u_s^a u_t^s \phi_t + u_t^a u_s^t \phi_s + u_t^a u_s^t u_t^s \phi_t + u_s^a u_t^s u_s^t \phi_s + \cdots] \\
+ \overline{u_s^e u_t^s \phi_t} [u_s^a \phi_s + u_t^a \phi_t + u_s^a u_t^s \phi_t + u_t^a u_s^t \phi_s + u_t^a u_s^t u_t^s \phi_t + \cdots] \\
+ \overline{u_s^e u_t^s u_s^t \phi_s} [u_s^a \phi_s + \cdots] + \cdots \} w_{st} \, d\mathbf{s} \, d\mathbf{t} + \cdots,
\end{aligned} \tag{37}
$$

which may be rewritten in terms of $u = Ef$ and the previous representations for E, and specialized to particular distributions. In principle, the functions in (36) and (37) are known so that the evaluation of $\langle \Psi_e \rangle$ and $\langle \bar{\Psi}_e \Psi_a \rangle$ corresponds to carrying out the integrations explicitly and constructing compact forms for the results. However, since even the case $N = 2$ involves an infinite series of orders of scattering (an infinite series inside the double integral containing w_{st}), this program is practical only for special cases. Because of this, most of the work in this subject has been based on heuristic manipulations of the original compact forms which generated the series of orders of scattering.

4.2. Compact representations for the coherent field. Using (29) in (33), we integrate implicitly avb s and write

$$\langle \Psi_c \rangle = \phi_c + \sum \int \langle U_s^c \rangle_s w_s(\mathbf{s}) \, d\mathbf{s}, \tag{38}$$

where $\langle\ \rangle_s$, which means an average with $\mathbf{s}$ fixed, is a function only of $\mathbf{s}$. Using (22) in (38), we have

$$\langle \Psi_c \rangle = \phi_c + \sum \int u_s^c \langle \Phi_s \rangle_s w_s(\mathbf{s}) \, d\mathbf{s} = \phi_c + \int \rho(\mathbf{s}) u_s^c \langle \Phi_s \rangle_s \, d\mathbf{s} \tag{39}$$

with

$$\langle \Phi_s \rangle_s = \phi_s + \sum{}' \langle u_t^s \Phi_t \rangle_s = \phi_s + \sum{}' \int u_t^s \langle \Phi_t \rangle_{st} p_{st}(\mathbf{s}; \mathbf{t}) \, d\mathbf{t}, \tag{40}$$

where $\langle\ \rangle_{st}$ is the average with both **s** and **t** fixed. Similarly the expression for $\langle\Phi_t\rangle_{st}$ introduces $\langle\Phi_m\rangle_{stm}$, etc., to generate a hierarchy of equations which truncates when N sets of variables are held fixed. Formal reduction leads to (36), or to more compact series, e.g., with the terms in the first brackets of (36) compacted operationally to $[1 - u_s^t u_t^s]^{-1}[\phi_t + u_s^t \phi_s]$, etc.

Foldy's form. For a gas as in (31) with $N \to \infty$, Foldy's approximation [2]

$$\langle\Phi_s\rangle_s \approx \langle\Psi_s\rangle \tag{41}$$

reduces (39) to the form of an integral equation

$$\langle\Psi_c\rangle = \phi_c + \int \rho(\mathbf{s}) u_s^c \langle\Psi_s\rangle\, ds. \tag{42}$$

Equations of this form for various scattering problems are considered in practically all the cited references and in the earlier literature [1]. For monopoles, $u_s^c = fE_{as}$ with f independent of directions and E_{as} equal to the free-space Green's function $E(k|\mathbf{r}_a - \mathbf{r}_s|) = [e^{ik|\mathbf{r}_a - \mathbf{r}_s|}]/|\mathbf{r}_a - \mathbf{r}_s|$, Foldy applied $\nabla^2 + k^2$ to both sides of the resulting (42) to obtain

$$[\nabla^2 + K^2(\mathbf{r})]\langle\Psi_c(\mathbf{r})\rangle = 0, \quad K^2(\mathbf{r}) = k^2 + 4\pi g(\mathbf{r}), \quad g(\mathbf{r}) = \int \rho(\mathbf{r}, \boldsymbol{\xi}) f(\boldsymbol{\xi})\, d\boldsymbol{\xi}. \tag{43}$$

Similarly, for more general scatterers, Lax [3] used (41) (with an unknown constant c) and obtained (43) in terms of the forward scattered values of cf with the amplitude f "evaluated in K-space." Alternative representations for K [4] for slab regions ($0 < z < d$) of homogeneous distributions of arbitrary scatterers have been obtained in terms of k-space amplitudes, and in terms of amplitudes corresponding to excitation in K-space and radiation into k-space (i.e., as suggested by the meaning of the terms in the kernel of (42)); in general, these forms involve the forward scattered value of f, and also the value corresponding to scattering in the direction of geometrical reflection from the slab's face. [The slab geometry for a homogeneous distribution is particularly simple: for this case $\langle\Psi_c\rangle$ consists essentially of only two plane waves (whose directions of propagation are images in the slab face), and consequently the original complicated operation $u_s^c\langle\Phi_s\rangle_s$ approximated by $u_s^c\langle\Psi_s\rangle$ entails at most manipulating a two by two matrix [4]. For monopoles (and dipoles), all forms reduce essentially to (43). The different forms for arbitrary scatterers correspond essentially to different interpretations of the discontinuity represented by a finite sized scatterer within the "synthetic medium" specified by K plus the boundary conditions on $\langle\Psi_c\rangle$ and $\partial_n\langle\Psi_c\rangle$ at interfaces of K- and k-spaces. No rigorous results for finite sized scatterers exist in the literature, and except for limiting cases the relations between various heuristic procedures have not been discussed. We do not consider this matter in the present paper; it is secondary to the relation of the c and e forms, since essentially the same details of explicit integration arise in both forms.]

Lax's form. Approximation (41), if appropriate for a "gas" is not appropriate for a "liquid" specified by w_{st}. For this case the corresponding approximation is

$$\langle \Phi_t \rangle_{st} \approx \langle \Phi_t \rangle_t. \tag{44}$$

Lax [3] used (44) in an analog of (40) to obtain essentially

$$\langle \Phi_s \rangle_s = \phi_s + \int u_t^s \langle \Phi_t \rangle_t \mu(\mathbf{s}; \mathbf{t})\, d\mathbf{t}, \qquad \mu(\mathbf{s}; \mathbf{t}) = \sum_t{}' p_{st}(\mathbf{s}; \mathbf{t}), \tag{45}$$

which together with (39) makes the problem determinate. (Essentially this approximation was also discussed by Twersky [**9**; **4**], who applied it in detail to planar distributions of cylinders [**9**] specified by the one-dimensional form of $\mu(\mathbf{s}; \mathbf{t})$ introduced by Zernike and Prins [**14**].) Using (39) in (45) gives

$$\langle \Phi_s \rangle_s = \langle \Psi_s \rangle - \int u_t^s \langle \Phi_t \rangle_t [\rho(\mathbf{t}) - \mu(\mathbf{s}; \mathbf{t})]\, d\mathbf{t}, \tag{46}$$

which, if $\mu(\mathbf{r}_s; \mathbf{r}_t) \to \rho$ for $|\mathbf{r}_s - \mathbf{r}_t| \to \infty$ (see [7]) is often more convenient to work with. For the case of a "gas" in the sense $\mu = \rho$, (46) reduces to (41).

Alternative form. The physical significance of the above heuristic approximations (41) and (45) was discussed in the report in [**9**] and their limitations were considered for a specific problem. Their limitations are indicated by comparing the iterated form of (42) with the average of the iterated series given in (36). Thus (42) gives

$$\begin{aligned} \langle \Psi_c \rangle = \phi_c &+ \sum \int u_s^c \phi_s w_s \, d\mathbf{s} + \sum\sum \int\int u_s^c u_t^s \phi_t w_s w_t \, d\mathbf{s}\, d\mathbf{t} \\ &+ \sum\sum\sum \int\int\int u_s^c u_t^s u_m^t \phi_m w_m w_s w_t \, d\mathbf{s}\, d\mathbf{t}\, d\mathbf{m} + \cdots \end{aligned} \tag{47}$$

In view of the discussion after (36), we see that (47) includes only the first of the infinite number of different sequences involved in "successive scattering," i.e., only chains in which all scatterers are different appear in (47); no common path differences can occur in higher orders, since (42) corresponds to successive scatterings of the average of the first order. In the best case, for special scatterers such that only the "main line sequence" is significant, the numerical coefficients in (47) are N^ν as compared to $N!/(N - \nu)!$, so that this difference disappears for $N \to \infty$. Essentially the same limitations are shown by the heuristic system (42) plus (45). In addition, whereas the exact series would involve w_{stm}, etc., the iterated integrals give only chains of one and two particle functions $w_s p_{st} p_{tm}$, etc., and this also introduces errors. Thus if we worked with (36), a more realistic value for w_{stm} obtained by the "superposition approximation" [7] would include p_{sm}, etc.

An alternative approximation for $\langle \Phi_s \rangle_s$ which restores "two body" processes lost in the above may be constructed by returning to the definition of Φ_t. Instead of using (44) we start with $\langle \Phi_t \rangle_{st} = \phi_t + u_s^t \langle \Phi_s \rangle_{st} + \sum_m'' \langle U_m^t \rangle_{st}$, where the double prime means $m \neq t$ and $m \neq s$. Thus the rigorous form of $\langle \Phi_t \rangle_{st}$ shows that the excitation of the fixed scatterer t consists essentially of two types of terms: $\langle U_s^t\rangle_{st}$ corresponds to a spherical wave traveling from a fixed scatterer s to excite t; and ϕ_t plus the sum is more or less the average excitation $\langle\Phi_t\rangle_t$. This suggests that we solve the simultaneous equations for $\langle\Phi_t\rangle_{st}$ and $\langle\Phi_s\rangle_{st}$ in order to eliminate $\langle\Phi_s\rangle_{st}$, and then approximate the resulting two-body solution:

$$(48)\quad \langle\Phi_t\rangle_{st} = \frac{[\phi_t + \sum''\langle U_m^t\rangle_{st}] + u_s^t[\phi_s + \sum''\langle U_n^s\rangle_{st}]}{1 - u_s^t u_t^s} \approx \frac{\langle\Phi_t\rangle_t + u_s^t\langle\Phi_s\rangle_s}{1 - u_s^t u_t^s},$$

where the terms of the expanded denominators operate from the left on the numerators. Using (48) in (40), we obtain

$$(49)\quad \begin{aligned}\langle\Phi_s\rangle_s &= \phi_s + \int u_t^s\left[\frac{\langle\Phi_t\rangle_t\mu(\mathbf{s};\mathbf{t})}{1 - u_s^t u_t^s}\right] d\mathbf{t} + \int u_t^s\left[\frac{u_s^t\mu(\mathbf{s};\mathbf{t})}{1 - u_s^t u_t^s}\right] d\mathbf{t}\langle\Phi_s\rangle_s \\ &= \frac{\phi_s + \int u_t^s\left[\dfrac{\langle\Phi_t\rangle_t\mu(\mathbf{s};\mathbf{t})}{1 - u_s^t u_t^s}\right] d\mathbf{t}}{1 - \int u_m^s\left[\dfrac{u_s^m\mu(\mathbf{s};\mathbf{m})}{1 - u_s^m u_m^s}\right] d\mathbf{m}},\end{aligned}$$

which together with (39) provides a determinate system.

4.3. Compact representations for intensity. For the quadratic function (34), we use (29) and integrate implicitly avb **s**, **t** to obtain

$$(50)\quad \begin{aligned}V_{ca} &= \int \sum\langle\bar{U}_s^c U_s^a\rangle_s w_s(\mathbf{s})\, d\mathbf{s} \\ &+ \int\int \sum[\sum'\langle\bar{U}_s^c U_t^a\rangle_{st} w_{st}(\mathbf{s},\mathbf{t}) - \sum\langle\bar{U}_s^c\rangle_s\langle U_t^a\rangle_t w_s(\mathbf{s})w_t(\mathbf{t})]\, d\mathbf{s}\, d\mathbf{t};\end{aligned}$$

here $\langle\bar{U}_s^c U_t^a\rangle_{st} = \bar{u}_s^c u_t^a\langle\bar{\Phi}_s\Phi_t\rangle_{st}$, etc. Using $w_{st}(\mathbf{s},\mathbf{t}) = w_s(\mathbf{s})p_{st}(\mathbf{s};\mathbf{t})$, $\sum w_s(\mathbf{s}) = \rho(\mathbf{s})$, and $\sum_t' p_{st}(\mathbf{s};\mathbf{t}) = \mu(\mathbf{s};\mathbf{t})$, we rewrite (50) as

$$(51)\quad \begin{aligned}V_{ca} &= \int \rho(\mathbf{s})\bar{u}_s^c u_s^a\langle|\Phi_s|^2\rangle_s\, d\mathbf{s} \\ &+ \int\int \rho(\mathbf{s})\bar{u}_s^c u_t^a[\langle\bar{\Phi}_s\Phi_t\rangle_{st}\mu(\mathbf{s};\mathbf{t}) - \langle\bar{\Phi}_s\rangle_s\langle\Phi_t\rangle_t\rho(\mathbf{t})]\, d\mathbf{s}\, d\mathbf{t}.\end{aligned}$$

Expressions for $\langle|\Phi_s|^2\rangle_s$ and $\langle\bar{\Phi}_s\Phi_t\rangle_{st}$ analogous to (40) may be constructed directly. Thus

$$(52)\quad \begin{aligned}\langle|\Phi_s|^2\rangle_s &= \bar{\phi}_s\langle\Phi_s\rangle_s + \phi_s\langle\bar{\Phi}_s\rangle_s - \bar{\phi}_s\phi_s + \sum'\int\langle|U_t^s|\rangle_{st} p(\mathbf{s};\mathbf{t})\, d\mathbf{t} \\ &+ \sum'\sum''\int\int\langle\bar{U}_t^s U_n^s\rangle_{stn} p(\mathbf{s};\mathbf{t},\mathbf{n})\, d\mathbf{t}\, d\mathbf{n},\end{aligned}$$

$$\langle \Phi_s \Phi_t \rangle_{st} = \bar{\phi}_s \langle \Phi_t \rangle_{st} + \phi_t \langle \bar{\Phi}_s \rangle_{st} - \bar{\phi}_s \phi_t + \langle \bar{U}_t^s U_s^t \rangle_{st}$$

$$+ \sum'' \int [\langle \bar{U}_t^s U_n^t \rangle_{stn} + \langle \bar{U}_n^s U_s^t \rangle_{stn} + \langle \bar{U}_n^s U_n^t \rangle_{stn}] p(\mathbf{s}, \mathbf{t}; \mathbf{n}) \, d\mathbf{n}$$

$$+ \sum'' \sum''' \int \int \langle \bar{U}_n^s U_m^t \rangle_{stnm} p(\mathbf{s}, \mathbf{t}; \mathbf{n}, \mathbf{m}) \, d\mathbf{n} \, d\mathbf{m}.$$

Energy approximation. Perhaps the simplest heuristic approximation for the form of $V = V_{cc}$ for lossless scatterers is based on using the energy theorem. Outside a slab region of space $(0 < z < d)$ containing a homogeneous distribution of lossless scatterers, (35) for plane wave excitation gives **[4]**

$$\text{(53)} \quad \text{Re}[\bar{\Psi} \nabla \Psi / ik] \cdot \mathbf{z}/z = \langle \mathbf{J} \rangle \cdot \mathbf{n} = (\mathbf{C} + \mathbf{I}) \cdot \mathbf{n} = |\langle \Psi \rangle|^2 \mathbf{i} \cdot \mathbf{n} + \mathbf{I} \cdot \mathbf{n} = \text{constant}.$$

For the slab geometry we have explicit approximations for the form of $\langle \Psi \rangle$ in terms of free space amplitudes f. Using the single scattering theorems for f, we then construct the simplest form of $\mathbf{I}$ which together with $|\langle \Psi \rangle|^2$ satisfies (53), and obtain the corresponding form of V by inspection. The result **[4; 9]** for a gas

$$\text{(54)} \qquad V \approx \sum \int |\langle U_s \rangle|^2 w_s(\mathbf{s}) \, d\mathbf{s} = \int \rho(\mathbf{s}) |\langle U_s \rangle|^2 \, d\mathbf{s}$$

is also equivalent to using simply $\langle \Phi_s \Phi_t \rangle_{st} \approx \langle \Psi_s \rangle \langle \Psi_t \rangle$ in (50).

Foldy's form. For a gas of monopoles, Foldy **[2]** used (41) plus the heuristic approximations

$$\text{(55)} \qquad \langle |\Phi_s|^2 \rangle_s \approx \langle |\Psi_s|^2 \rangle, \qquad \langle \bar{\Phi}_s \Phi_t \rangle_{st} \approx \langle \bar{\Psi}_s \Psi_t \rangle,$$

to reduce (51) to

$$\text{(56)} \qquad V_{ca} = \int \rho(\mathbf{s}) \bar{u}_s^c u_s^a \langle |\Psi_s|^2 \rangle \, d\mathbf{s} + \int \int \rho(\mathbf{s}) \rho(\mathbf{t}) \bar{u}_s^c u_t^a V_{st} \, d\mathbf{t}.$$

Iterating the integral equation for V_{ca}, starting with the single-integral inhomogeneous term, and then using the definition $V_{ca} = \langle \Psi_c \Psi_a \rangle - \langle \bar{\Psi}_c \rangle \langle \Psi_a \rangle$ he obtained an integral equation form for $\langle \bar{\Psi}_c \Psi_a \rangle$:

$$\langle \bar{\Psi}_c \Psi_a \rangle = \langle \bar{\Psi}_c \rangle \langle \Psi_a \rangle + \int L(c, a; s) \langle |\Psi_s|^2 \rangle \, d\mathbf{s};$$

$$\text{(57)} \qquad L(c, a; s) = \sum_{n=0}^{\infty} L_n(c, a; s);$$

$$L_0(c, a; s) = \bar{u}_s^c u_s^a \rho(\mathbf{s});$$

$$L_n(c, a; s) = \int \int \bar{u}_t^c u_m^a \rho(\mathbf{t}) \rho(\mathbf{m}) L_{n-1}(t, m; s) \, d\mathbf{t} \, d\mathbf{m}.$$

Foldy suppressed the $\boldsymbol{\xi}$ integrations by using g as in (43), and a corresponding average of $|f(\boldsymbol{\xi})|^2$. He also recast the kernel function L as a series in the K-space Green's function $E(K|\mathbf{r}_a - \mathbf{r}_s|)$. With ds in (57) replaced by $d\mathbf{r}_s$, he used

$$L = \sum M_n;$$

(58)
$$M_0(c, a; s) = G(\mathbf{r}_s)\bar{E}(K|\mathbf{r}_c - \mathbf{r}_s|)E(K|\mathbf{r}_a - \mathbf{r}_s|), \qquad G(\mathbf{r}) = \int \rho(\mathbf{r},\boldsymbol{\xi})|f(\boldsymbol{\xi})|^2\, d\boldsymbol{\xi};$$

$$M_n(c, a; s) = \frac{1}{4\pi}\iint\{\bar{g}(\mathbf{r}_t)[\nabla_m^2 + K^2(\mathbf{r}_m)] + g(\mathbf{r}_m)[\nabla_t^2 + \bar{K}^2(\mathbf{r}_t)]\}$$
$$\times M_{n-1}(t, m; s)\bar{E}(K|\mathbf{r}_c - \mathbf{r}_t|)E(K|\mathbf{r}_a - \mathbf{r}_m|)\, d\mathbf{r}_t\, d\mathbf{r}_m,$$

to obtain

$$(59)\quad \langle|\Psi_c|^2\rangle = |\langle\Psi_c\rangle|^2 + \int G(\mathbf{r}_s)\langle|\Psi_s|^2\rangle\left\{|E(K|\mathbf{r}_c - \mathbf{r}_s|)|^2 + \sum_1^\infty M_n(c, c; s)\right\} d\mathbf{r}_s.$$

The resulting integral equation obtained by keeping only the first term of the kernel function in (59) is quite plausible physically and was used by Foldy as a practical approximation for comparison with experiment **[10]**; however, the procedure leading to (59) cannot justify neglecting $\sum_1^\infty M_n$.

Lax's form. Lax [3] used (55) and (41) (except for a constant multiplier) to obtain essentially

(60)
$$V_{ca} = \int \rho(\mathbf{s})\bar{u}_s^c u_s^a\langle|\Psi_s|^2\rangle\, ds$$
$$+\iint \rho(\mathbf{s})\bar{u}_s^c u_t^a\{\langle\bar{\Psi}_s\Psi_t\rangle[\mu(\mathbf{s};\mathbf{t}) - \rho(\mathbf{t})] + V_{st}\rho(\mathbf{t})\}\, ds\, dt,$$

which is an integral equation form analogous to (56) but having a more complicated "inhomogeneous term."

Alternative form. Alternative equations may be obtained by using a different approximation than (55). Thus more or less as for (48), we may group terms in $\langle\Phi_s\Phi_t\rangle_{st}$ to isolate the different wave forms involved, and then introduce approximations. Equivalently, we start by solving the two-body problem corresponding to $\Phi_t = \phi_t + \sum'' U_m^t + U_s^t \equiv R_t(s) + u_s^t\Phi_s$ and $\Phi_s = R_s(t) + u_t^s\Phi_t$, and then average the product of the resulting forms:

(61)
$$\langle\Phi_s\Phi_t\rangle_{st} = \left\langle\left[\frac{\overline{R_s(t) + u_t^s R_t(s)}}{1 - u_t^s u_s^t}\right]\left[\frac{R_t(s) + u_s^t R_s(t)}{1 - u_s^t u_t^s}\right]\right\rangle_{st},$$
$$R_s(t) = \Phi_s - U_t^s = \phi_s + \sum'' U_m^s.$$

In order to obtain a form consistent with (42), we drop terms which would lead to a scatterer being "used twice" in either an "a-chain" or a "c-chain" of u's (i.e., which would lead to a chain of u's or $\bar{u}$'s having at least one subscript in common) and work with

$$(62)\qquad \langle\Phi_s\Phi_t\rangle_{st} \approx \langle\bar{R}_s(t)R_t(s)\rangle_{st} + \bar{u}_t^s\langle\bar{R}_t(s)R_t(s)\rangle_{st} + u_s^t\langle\bar{R}_s(t)R_s(t)\rangle_{st}.$$

A simple approximation of (62) more general than (55) is

$$\langle \Phi_s \Phi_t \rangle_{st} \approx \langle \bar{\Psi}_s \Psi_t \rangle + \bar{u}_t^s \langle \bar{\Psi}_t \Psi_t \rangle + u_s^t \langle \bar{\Psi}_s \Psi_s \rangle. \tag{63}$$

Substituting into (54) for $\mu(\mathbf{s};\mathbf{t}) = \rho(\mathbf{t})$, we obtain

$$\begin{aligned} V_{ca} &= \int \rho(s)\bar{u}_s^c u_s^a \langle |\Psi_s|^2 \rangle \, d\mathbf{s} \\ &\quad + \int\int \rho(s)\rho(t)\bar{u}_s^c u_t^a [u_s^t \langle |\Psi_s|^2 \rangle + u_t^s \langle |\Psi_t|^2 \rangle + V_{st}] \, d\mathbf{s}\, d\mathbf{t}, \end{aligned} \tag{64}$$

which we regard as an integral equation for V.

$\langle \Psi_c \rangle$*-consistent form.* We could determine the physical significance of the above heuristic forms by means of the same procedure we used in (47) for $\langle \Psi \rangle$. Thus the integral equations for $V_{ca} = \langle \bar{\Psi}_c \Psi_a \rangle - \langle \bar{\Psi}_c \rangle \langle \Psi_a \rangle$ could be iterated to eliminate V, and the resulting integral equations for $\langle \bar{\Psi}_c \Psi_a \rangle$ could then be iterated in terms of $\langle \bar{\Psi}_c \rangle \langle \Psi_a \rangle$, and then $\langle \Psi_c \rangle$ could be replaced by the series (47) to finally obtain a form to compare with the exact series (37). This procedure is too tedious (and unrewarding) to attempt for the general case. (Subsequently, we follow a more or less equivalent procedure for the equations that arise in a simple illustration such that we get L of (57) in closed form.) However, it is clear that the $\langle \bar{\Psi}_c \Psi_a \rangle$ are poorer approximations of $\langle \bar{\Psi}_e \Psi_a \rangle$ than the $\langle \Psi_c \rangle$ are of $\langle \Psi_e \rangle$.

It is also tedious to introduce more and more of the omitted features into such forms as (56) to obtain results consistent with (42) or (46). Instead, we go back to the exact result (37), drop the terms that are omitted by (42) (as we have determined by comparing (47) and (36)) as well as the new analogs, and obtain a compact form for the limit of the result for $N \to \infty$.

Thus in (37) we drop all terms containing a backscattered path $u_s^t u_t^s$, $\bar{u}_s^t u_t^s$, etc., replace all coefficients $N!/(N-\nu)!$ by N^ν, and then compact and set aside the expansions of the coherent term $\langle \bar{\Psi}_c \rangle \langle \Psi_a \rangle$ (known from (47)). For the gas, writing $Nw(\mathbf{s}) = \rho(\mathbf{s})$, etc., we obtain

$$\begin{aligned} \langle \bar{\Psi}_e \Psi_a \rangle - \langle \bar{\Psi}_e \rangle \langle \Psi_a \rangle &= \int \rho(\mathbf{s}) [\overline{u_s^e \phi_s} \cdot u_s^a \phi_s]_1 \, d\mathbf{s} \\ &\quad + \int\int \rho(\mathbf{s})\rho(\mathbf{t}) \{ [\overline{u_s^e \phi_s} \cdot (u_s^a u_t^s \phi_t + u_t^a u_s^t \phi_s) + \overline{(e \leftrightarrow a)}]_1 + [\overline{u_s^e u_t^s \phi_t} \cdot u_s^a u_t^s \phi_t]_2 \} \, d\mathbf{s}\, d\mathbf{t} \\ &\quad + \int\int\int \rho(\mathbf{s})\rho(\mathbf{t})\rho(\mathbf{m}) \{ [\overline{u_s^e \phi_s} \cdot (u_s^a u_t^s u_m^t \phi_m + u_t^a u_s^t u_m^s \phi_m + u_t^a u_m^t u_s^m \phi_s) + \overline{(e \leftrightarrow a)}]_1 \\ &\quad + [\overline{u_s^e u_t^s \phi_t} \cdot (u_s^a u_m^s \phi_m + u_m^a u_t^m \phi_t + u_t^a u_m^t \phi_m + u_m^a u_s^m \phi_s)]_1 \\ &\quad + [\overline{u_s^e u_t^s \phi_t} \cdot (u_m^a u_s^m u_t^s \phi_t + u_s^a u_m^s u_t^m \phi_t + u_s^a u_t^s u_m^t \phi_m) + \overline{(e \leftrightarrow a)}]_2 \\ &\quad + [\overline{u_s^e u_t^s u_m^t \phi_m} \cdot u_s^a u_t^s u_m^t \phi_m]_3 \} \, d\mathbf{s}\, d\mathbf{t}\, d\mathbf{m} + \cdots, \end{aligned} \tag{65}$$

where $\overline{(e \leftrightarrow a)}$ means include the complex conjugate of the analogous terms with e and a interchanged; the subscripts on the brackets will be explained shortly.

The above expansion with terms grouped by the number of integrations (or, equivalently, powers of ρ) is not particularly revealing. However, the structure of the equation becomes transparent if we regroup terms on the basis of the number of common subscripts in such products as $\overline{u_s^e u_t^s u_s^a u_m^s}$, i.e., essentially on the basis of the number of incoherent scattering processes involved. It is this second grouping that is indicated by the subscripts on the brackets in (65).

If we compare the set of one-brackets with the set of two-brackets, with the term in the three-bracket, and keep in mind the expansion of $\langle\Psi\rangle$ given in (47), we see that these three sets are respectively the leading terms of the first three iterations of an integral equation for $\langle\bar{\Psi}_c\Psi_a\rangle$ in terms of $\langle\bar{\Psi}_c\rangle\langle\Psi_a\rangle$. Explicitly, we have

$$\begin{aligned}\langle\bar{\Psi}_c\Psi_a\rangle &= \langle\bar{\Psi}_c\rangle\langle\Psi_a\rangle + \int \rho(\mathbf{s})\overline{u_s^c\varepsilon_{cs}\langle\Psi_s\rangle}u_s^a\varepsilon_{as}\langle\Psi_s\rangle\, ds \\ &\quad + \iint \rho(\mathbf{s})\rho(\mathbf{t})\overline{u_s^c\varepsilon_{cs}u_t^s\varepsilon_{st}\langle\Psi_t\rangle}u_s^a\varepsilon_{as}u_t^s\varepsilon_{st}\langle\Psi_t\rangle\, ds\, dt + \cdots \\ &= \langle\bar{\Psi}_c\rangle\langle\Psi_a\rangle + \int \rho(\mathbf{s})\overline{u_s^c\varepsilon_{cs}}u_s^a\varepsilon_{as}\langle|\Psi_s|^2\rangle\, ds;\end{aligned} \tag{66}$$

$$u_s^c\varepsilon_{cs} = u_s^c + \int \rho(\mathbf{t})u_t^c(u_s^t\varepsilon_{ts})\, dt, \qquad \langle\Psi_c\rangle = \phi_c + \int \rho(\mathbf{s})u_s^a\langle\Psi_s\rangle\, ds,$$

where $u_s^c\varepsilon_{cs}$ is essentially a new single-scatterer function, say v_s^c, such that

$$u_s^c\varepsilon_{cs} = v_s^c = u_s^c + \int \rho(\mathbf{t})u_t^c v_s^t\, dt \tag{67}$$

satisfies the integral equation analogous to $\langle\Psi_c\rangle$ whose inhomogeneous term is u, i.e., v_s^c is the "Green's function" for an angle dependent source u_s^c. Two special cases are of particular interest:

$$\begin{aligned}\langle|\Psi_c|^2\rangle &= |\langle\Psi_c\rangle|^2 + \int \rho(\mathbf{s})|u_s^c\varepsilon_{cs}|^2\langle|\Psi_s|^2\rangle\, ds, \\ \langle\Psi_c^2\rangle &= \langle\Psi_c\rangle^2 + \int \rho(\mathbf{s})(u_s^c\varepsilon_{cs})^2\langle\Psi_s^2\rangle\, ds;\end{aligned} \tag{68}$$

the first is the average total intensity, and the second plus the first give all second moments of Re Ψ and Im Ψ [6].

For monopole scatterers, we have $u_s^a\varepsilon_{as} = fE(k|\mathbf{r}_a - \mathbf{r}_s|)\varepsilon_{as} = fE(K|\mathbf{r}_a - \mathbf{r}_s|)$ with $E(K|\mathbf{r}_a - \mathbf{r}_s|)$ as the Green's function for the synthetic medium; for this case the leading term of Foldy's M-series of (58) gives the same result as (66) (but the rest of the terms of the M-series are not necessarily negligible). For more general distributions than the gas, we could use (39) plus (45) for $\langle\Psi_s\rangle$ (which, it should be understood, involve heuristic simplifications of the n-body probability functions of the exact series (36)), and similarly replace $\rho(\mathbf{t})$ in the kernel of $u_s^c\varepsilon_{cs}$ by $\mu(\mathbf{s};\mathbf{t})$ (which introduces additional heuristic simplifications), or else construct a more complete representation. The analog of (39) plus (49) may be constructed by isolating the two-body processes in the original series by essentially the formalism introduced for (61).

The simple form (66) indicates the source of difficulties in seeking such forms through the previous heuristic formalism, i.e., $L(c, a; s)$ of (57) should be a product $\bar{v}_s^c v_s^a$ each factor of which satisfies a relatively simple integral equation. Working backwards from the above we construct the integral equation for $P = \bar{v}v$ and iterate it to compare with (57). Thus

$$V_{ca} = \int P(c, a; s)\langle|\Psi_s|^2\rangle\, ds; \qquad P = \sum P_n, \qquad P_0 = L_0,$$

(69) $$P_n(c, a; s) = \int \rho(\nu)\bar{u}_\nu^c P_{n-1}(\nu, a; s)\, d\nu + \int \rho(\mu) u_\mu^a P_{n-1}(c, \mu; s)\, d\mu - \int\int \rho(\nu)\rho(\mu)\bar{u}_\nu^c u_\mu^a P_{n-1}(\nu, \mu; s)\, d\nu\, d\mu.$$

(Note that (64) is generated in the course of the first iteration of (69).) Equivalently (69) follows from the gas form of (50) by using

(70) $$\langle \bar{U}_s^c U_t^a\rangle_{st} - \langle \bar{U}_s^c\rangle_s \langle U_t^a\rangle_t = \bar{u}_s^c V_{sa} + u_t^a V_{ct} - \bar{u}_s^c u_t^a V_{st}.$$

5. **Illustration.** The previous sections introduced the general scattering and statistical formalism, and gave certain "compact" (c) and "expanded" (e) representations for several functions. The c-forms based on heuristic approximations are relatively easier to apply to specific problems but we have no analytical criteria for their validity. The e-forms involve no heuristic assumptions, but they are in general too cumbersome to work with. By comparing expansions of c-forms with corresponding e-forms, we can see what terms are omitted by the approximations, and this provides a general avenue that can be followed for obtaining criteria. This should be relatively easy to do for $\langle\Psi\rangle$ (see (47)), but the $\langle|\Psi|^2\rangle$ forms are considerably more complicated. Since $\langle\Psi_c\rangle$ of (42) has proved so useful for interpreting experiments, we now construct an illustration such that $\langle\Psi_c\rangle$ equals the limit of $\langle\Psi_e\rangle$ for $N \to \infty$; this allows us to concentrate on $\langle|\Psi|^2\rangle$ and determine the corresponding $\langle|\Psi_c|^2\rangle$. We first show that it is only the $\langle\Psi_c\rangle$-consistent form among the previous general forms $\langle|\Psi_c|^2\rangle$ that satisfies the energy theorem, and then show that it is also the limit of $\langle|\Psi_e|^2\rangle$ for $N \to \infty$. We consider the e-results in some detail, because even for special distributions such that the terms omitted from the correct c-forms are negligible for $N \to \infty$, we are interested in the differences between e and c results, and in the corrections required for finite N.

In order to exhibit explicitly the relationships between the heuristic c-forms and the exact e-forms we restrict consideration to a very simple class of scattering problems. We restrict the "statistics," we restrict the geometry, and we restrict the type of scatterers. Subject to these restrictions, we also introduce analytical simplifications in evaluating certain integrals that arise. Since *all* simplifications appear equally in the c and e forms, they do not affect the primary aim of determining the significance of the heuristic approximations.

Ensemble. The ensemble we consider is that of a "rare gas" of N identical scatterers, i.e.,

$$W(\mathbf{1}, \mathbf{2}, \cdots, \mathbf{N}) = w^N,$$

where w is independent of position. The scatterers u_s we deal with are thus either identical (e.g., identical ellipsoids which remain at constant orientation), or represent an identical average value over scatterer characteristics (average over orientation, size, etc.); for the second case we incorporate this independent averaging in the definitions of u and f. Thus for either case, we need consider only integrations over position variables explicitly. Treating the scatterers as points uniformly distributed in an available volume $\mathscr{V}$, we take the part of w which we consider explicitly to equal

$$w = 1/\mathscr{V}$$

The corresponding density function

$$\rho = N/\mathscr{V}$$

is the average number of scatterers in unit available volume. We are interested in finite values of N and $\mathscr{V}$, and also in values $N \to \infty$ and $\mathscr{V} \to \infty$ such that ρ remains bounded.

Geometry. We restrict the scatterers to a slab region of space of thickness d, normal to the direction of incidence $\mathbf{i} = \mathbf{n} = \mathbf{z}/z$ (i.e., $0 \leq z \leq d$). The dimensions along x and y will be regarded as infinite in order to preserve symmetry perpendicular to $\mathbf{n}$. Infinite limits on x and y also enable us to do various integrations exactly **[4]**; however, this is not significant in the present treatment since we evaluate all integrals by the method of stationary phase (which requires only that the limits be very large compared to k^{-1}, and this we assume). The available volume of the slab region (which is in general a function of the number of scatterers and their individual volumes) will be treated as finite as long as we are dealing with finite N.

Scatterers. We restrict consideration to "forward-type scatterers" such that

$$|f(\mathbf{i}, \mathbf{i})| \gg\!> |f(-\mathbf{i}, \mathbf{i})|, \qquad \mathbf{i} = \mathbf{n} = \mathbf{z}/z \tag{71}$$

and neglect all radiation into the half-space back toward the source $\phi = e^{ikz}$. Because of (71), we may replace the "forward scattering theorem" for lossless scatterers in (12) by

$$\frac{4\pi}{k} \operatorname{Im} f(\mathbf{i}, \mathbf{i}) \approx \int_{\pi/2} |f(\mathbf{o}, \mathbf{i})|^2 \, d\Omega_0 \approx \sigma_s \tag{72}$$

where $\int_{\pi/2}$ indicates the forward half-space of angles around $\mathbf{o} = \mathbf{i} = \mathbf{n}$. Where convenient, in order to illustrate certain relations explicitly, we may assume that $|f(\mathbf{o}, \mathbf{i})|^2$ is "tightly peaked" around $\mathbf{o} = \mathbf{i}$ and neglect slow variations of factors such as the scalar product $\mathbf{i} \cdot \mathbf{o}$. More generally, we assume that the significant directions $\mathbf{o}$ make small enough angles from the slab normal $\mathbf{n} = \mathbf{i}$ to insure that

$$|f(\mathbf{o}, \mathbf{o})| \gg\!> |f(\mathbf{o}', \mathbf{o})|, \qquad \mathbf{o}' = 2\mathbf{o}\cdot\mathbf{n}\mathbf{n} - \mathbf{o}, \tag{73}$$

where $\mathbf{o}'$ is the image of $\mathbf{o}$ in the slab's face. (These restrictions correspond for example to scatterers whose dimensions are very large compared to k^{-1}, particularly if their index of refraction, etc., differ little from those of the external region. However, we consistently treat the scatterers as points in carrying out essentially the same integrations in c and e forms. In order to avoid discussion of convergence, we assume the too strong condition $\rho\sigma\, d < 1$. Our procedure may suggest that we require $|f(\mathbf{i}, \mathbf{i})|^N \gg |f(-\mathbf{i}, \mathbf{i})|$, and this would be incompatible with convergence for $N \to \infty$. However, if we introduce more structure into the problem, e.g., a simple $\mu(\mathbf{s}; \mathbf{t})$ "hole," we also introduce additional factors which may be adjusted to insure convergence subject to (71). For present purposes, we ignore $f(-\mathbf{i}, \mathbf{i})$.)

In carrying out integrations, in general, we apply the method of stationary phase and use the asymptotic form of the scattered wave

$$u_s^a(\mathbf{i}) \sim f(\mathbf{o}_{as}, \mathbf{i}) \frac{e^{ikr_{as}}}{r_{as}}, \qquad \mathbf{r}_{as} = \mathbf{r}_a - \mathbf{r}_s = r_{as}\mathbf{o}_{as}, \tag{74}$$

i.e., we take E_{as} in the operational forms to be simply the free k-space Green's function and interpret the symbolic forms as ordinary algebraic ones. (For f independent of angles, the form is then that of a monopole as treated by Foldy; however, restriction (71) makes the present scatterers the simpler ones.) Using (74) in the integrals of the previous section, we evaluate them by means of

$$\iint G(x, y)\, e^{ikg(x,y)}\, dx\, dy \sim \frac{2\pi i}{k} \frac{[G\, e^{ikg}]}{[g_{xx}g_{yy} - g_{xy}^2]^{1/2}}, \tag{75}$$

where the brackets indicate that the functions g, G, and the second derivatives of g (i.e., g_{xx}, etc.) are evaluated at the stationary points for which $g_x = g_y = 0$. The integrals we deal with all have two stationary points, but the restriction (73) allows us to neglect the one corresponding to "geometrical reflection" and keep only the one meaning essentially "forward scattering."

Because of the above, we may drop all terms which involve back and forth scatterings between a pair, i.e., we need consider only the "main line sequence" mentioned after (36), and shown in (47) and (65): any terms containing $u_t^s u_s^t$ or $\bar{u}_t^s u_s^t$, etc., are proportional to a back scattered value $f(\mathbf{o}_{st}, \mathbf{o}_{ts}) = f(-\mathbf{o}_{ts}, \mathbf{o}_{ts}) \approx 0$.

5.1. THE COHERENT FIELD.

Compact representation. For the present problem, (42) reduces to

$$\begin{aligned}
&\langle \Psi_c \rangle = \phi_c(\mathbf{i}) + \rho \iiint u_s^c \langle \Psi_s \rangle\, dx_s\, dy_s\, dz_s, \qquad \rho = N/\mathscr{V};\\
&\phi_c(\mathbf{i}) = e^{i\mathbf{k}\cdot\mathbf{r}_c}, \qquad \mathbf{k} = k\mathbf{i} = k\mathbf{z}/z,\\
&u_s^c = f(\mathbf{o}_{cs}, \mathbf{i}) \frac{e^{ik|\mathbf{r}_c - \mathbf{r}_s|}}{|\mathbf{r}_c - \mathbf{r}_s|},\\
&\mathbf{r}_c - \mathbf{r}_s = [(x_c - x_s)^2 + (y_c - y_s)^2 + (z_c - z_s)^2]^{1/2}\mathbf{o}_{cs},
\end{aligned} \tag{76}$$

where we integrate over a slab region ($0 \leqq z \leqq d$, perpendicular to $\mathbf{i} = \mathbf{n} = \mathbf{z}/z$) whose faces may be treated essentially as infinite planes.

It is clear from symmetry that $\langle \Psi \rangle$ depends only on z:

$$\langle \Psi_s \rangle = \langle \Psi(\mathbf{r}_s) \rangle = \langle \Psi(z_s) \rangle : \tag{77}$$

The integrals over x_s and y_s in (76) for arbitrary scatterers, may be evaluated exactly for infinite limits by using the complete forms of u given in §2 (see [4]). For present purposes, we obtain the same result by using the method of stationary phase (75) and the asymptotic form of u, i.e.,

$$\iint \frac{e^{ik|\mathbf{r}_c - \mathbf{r}_s|}}{|\mathbf{r}_c - \mathbf{r}_s|} f(\mathbf{o}_{cs}, \mathbf{i})\, dx_s\, dy_s \sim \frac{2\pi i}{k} e^{\pm ik(z_c - z_s)} f(\pm \mathbf{i}, \mathbf{i}) \quad \text{for } z_c \lessgtr z_s, \tag{78}$$

for $\mathbf{i} = \mathbf{z}/z$. Thus the remaining integration over z_s splits into two ranges; the integral for $z_s < z_c$ has a factor $f(\mathbf{i}, \mathbf{i})$ corresponding to forward scattering, and the remaining range has a factor $f(-\mathbf{i}, \mathbf{i})$ corresponding to geometrical reflection. Because of (71) we drop the reflected term, and reduce (76) to

$$\langle \Psi(z_c) \rangle = e^{ikz_c} \left[1 + \frac{i2\pi f(\mathbf{i}, \mathbf{i})}{k} \rho \int_0^{z_c} \langle \Psi(z_s) \rangle\, e^{-ikz_s}\, dz_s \right]. \tag{79}$$

The solution of the integral equation (obtained by inspection or by iteration) equals

$$\langle \Psi(z_c) \rangle = e^{ikz_c + i\Delta z_c}, \qquad \Delta = \frac{2\pi f(\mathbf{i}, \mathbf{i})}{k} \rho, \quad \rho = \frac{N}{\mathscr{V}}, \tag{80}$$

i.e., the "internal coherent wave" is the plane wave

$$\langle \Psi_c \rangle = \phi_c\, e^{i\Delta z_c} \equiv e^{iKz_c}, \qquad K = k + \Delta. \tag{81}$$

Thus the procedure has synthesized a "medium" in which

$$(\nabla^2 + K^2)\langle \Psi_c \rangle = 0, \qquad K = k + \frac{2\pi \rho f(\mathbf{i}, \mathbf{i})}{k}. \tag{82}$$

The field transmitted through the slab is

$$\langle \Psi_c \rangle_{\text{trans}} = e^{ikz_c + i\Delta d}, \qquad z > d, \tag{83}$$

and the corresponding "coherent intensity" is

$$|\langle \Psi(d) \rangle|^2 = e^{-2\,\mathrm{Im}\,\Delta d}, \quad 2\,\mathrm{Im}\,\Delta d = \left[\frac{4\pi}{k} \mathrm{Im}\, f(\mathbf{i}, \mathbf{i}) \right] \rho d. \tag{84}$$

For lossless scatterers, in view of (72), we have

$$2\,\mathrm{Im}\,\Delta d = \sigma_s \rho d = \rho d \int_{\pi/2} |f(\mathbf{o}, \mathbf{i})|^2\, d\Omega, \tag{85}$$

so that the "coherent attenuation coefficient" $2\,\mathrm{Im}\,\Delta$ is simply the total scattering cross section of one scatterer times the number of scatterers per unit volume. Since the scatterers are lossless (i.e., there are no sinks for **J**), the "losses" shown in (84) must be made up by the "incoherent component" which we consider subsequently.

The above is a very sketchy discussion of matters treated elsewhere in detail [**4**]. In basing the discussion on free-space scattering amplitudes $f(\mathbf{i}, \mathbf{i})$, we are avoiding a central problem of the subject: clearly our treatment of the immediate neighborhood of the scatterers, and of the face planes $z = 0$ and d are cursory. However, these are secondary questions pertaining to parameters that play the same roles in both the c and e forms and will not be discussed in the following.

Expanded representation. For the present problem, (36) with $w = 1/\mathscr{V}$ reduces to

$$\begin{aligned}\langle \Psi_e \rangle = \phi_e &+ Nw \int u_s^e \phi_s \, d\mathbf{r}_s + N(N-1)w^2 \int\int u_s^e u_t^s \phi_t \, d\mathbf{r}_s \, d\mathbf{r}_t \\ &+ N(N-1)(N-2)w^3 \int\int\int u_s^e u_t^s u_m^t \phi_m \, d\mathbf{r}_s \, d\mathbf{r}_t \, d\mathbf{r}_m + \cdots,\end{aligned} \tag{86}$$

where, because of (71), we retained only the "main line sequence"; to be more explicit, we could write $u_s^e(\mathbf{o}_{st}) u_t^s(\mathbf{i}) \phi_t(\mathbf{i})$, etc. Using the forward scattered value of (78), i.e.,

$$w \int\int u_t^s \phi_t \, dx_t \, dy_t = \frac{2i\pi w}{k} f(\mathbf{i}, \mathbf{i}) \phi_s \equiv i\delta\phi_s = i\frac{\Delta}{N}\phi_s \tag{87}$$

which states simply that a uniform plane of spherical scatterers excited by a plane wave regenerates the plane wave, we reduce (86) to

$$\begin{aligned}\langle \Psi_e \rangle &= \phi_e \left\{ 1 + iN\delta \int_0^{z_e} dz_s + N(N-1)(i\delta)^2 \int_0^{z_e} dz_s \int_0^{z_s} dz_t + \cdots \right\} \\ &= \phi_e \left\{ 1 + Ni\delta z_e + N(N-1)\frac{(i\delta z_e)^2}{2} + \cdots + \frac{N!}{(N-\nu)!}\frac{(i\delta z_e)^\nu}{\nu!} + \cdots \right\} \\ &= \phi_e (1 + i\delta z_e)^N.\end{aligned} \tag{88}$$

The present result obtained without the heuristic approximation $\langle \Phi_s \rangle_s \approx \langle \Psi_s \rangle$ which lead to (81) is valid down to $N = 1$ (the only case for which the single scattering approximation is exact, and also measurable in microwave experiments on a moving scatterer).

For comparison with (81), we use $N\delta = \Delta$ and write

$$\langle \Psi \rangle = \phi (1 + i\delta z)^N = \phi \left(1 + i\frac{\Delta}{N} z \right)^N, \tag{89}$$

where $Nw = N/\mathscr{V}$ in Δ is bounded for $N \to \infty$. Since $(1 + i\Delta z/N)^N \to e^{i\Delta z}$ for $N \to \infty$, we see that the heuristic approximation $\langle \Phi_s \rangle \approx \langle \Psi \rangle$ is valid for large N for the present problem, i.e., the corresponding $\langle \Psi_c \rangle$ is the limit of $\langle \Psi_e \rangle$ for $N \to \infty$.

If we replace $N!/(N - \nu)!$ in (86) and (88) by N^ν, we get the series that follow from iterating (76) and (79) respectively. We also note for subsequent manipulations of more complicated series than (86) (in which, essentially, the present ϕ is replaced by an unknown function), that we could just as well have started with the left hand "head" instead of the right hand "tail" of a chain of integrals. Thus, for example, in the double integral of (86) we may integrate first over $\mathbf{r}_s$ and then over $\mathbf{r}_t$. Using (75), we integrate over the $x_s y_s$-plane and (because of (73)) keep only the forward stationary point to obtain

$$\begin{aligned} w\int\int u_s^a(\mathbf{o}_{st})u_t^s(\mathbf{i})\,dx_s\,dy_s &= w\int\int f(\mathbf{o}_{as}, \mathbf{o}_{st})f(\mathbf{o}_{st}, \mathbf{i})E_{as}E_{st}\,dx_s\,dy_s \\ &\sim E_{at}\,f(\mathbf{o}_{at}, \mathbf{i})\left[\frac{i2\pi w f(\mathbf{o}_{at}, \mathbf{o}_{at})}{k\cos\theta_{at}}\right] = u_t^a(\mathbf{i})i\delta_{at}, \end{aligned} \tag{90}$$

where δ_{at} is the propagation factor (analogous to $\delta_i = \delta$ of (87)) associated with the direction $\mathbf{o}_{at}$, and $\cos\theta_{at} = \mathbf{n}\cdot\mathbf{o}_{at}$; analogous to (87), the present result states that a uniform plane of spherical scatterers excited by a spherical source regenerates the spherical wave of the source in the forward half-space. Multiplying (90) by ϕ_t we integrate over x_t and y_t by using (87) (which also converts δ_{at} of (90) to $\delta_i = \delta$), and are left with $N(N-1)[i\delta]^2\int_0^{z_e} dz_t \int_{z_t}^{z_e} dz_s$ which of course gives the same result as the corresponding term of (88).

We also note that we may obtain the "transmission coefficient" $T_N = (1 + i\delta z)^N$ of (89) algebraically from elementary physical considerations, i.e., we have essentially

$$T_N = T_1^N, \qquad T_1 = 1 + i\delta z = 1 + S_1, \tag{91}$$

where S_1 represents the forward scattering amplitude for a "planar scatterer" whose back scattering is zero. From elementary "transmission line" considerations, we write $T_N = T_1^N$ directly. The binomial coefficients of the individual terms in the expansion of T_N, i.e., the coefficient in $[N!/(N-\nu)!\nu!]S^\nu$ merely indicates the number of possible order-preserving forward-scattering combinations for ν out of N scatterers. From this viewpoint, we would not expect S to depend on z (i.e., we have factored out the plane wave excitation, and we are dealing simply with in-phase forward scattering), and it actually does not. For planar scatterers the available volume for an observation point z within the distribution is $\mathscr{V} = \mathscr{A}z$, where "$\mathscr{A}$" is a redundant area factor. Thus $S = i\delta z$ reduces to $S = i2\pi f/k\mathscr{A}$; this same value follows if our planar scatterers have finite thickness "b," since z in both numerator and denominator would then be replaced by $z - Nb$.

5.2. Total intensity.

Energy approximation. For the present problem, (54) equals

$$V_a = \langle|\Psi_a|^2\rangle - |\langle\Psi_a\rangle|^2 \approx \rho\int |u_s^a|^2|\langle\Psi_s\rangle|^2\, d\mathbf{r}_s$$
(92)
$$= \rho\iiint \frac{|f(\mathbf{o}_{as}, \mathbf{i})|^2}{|\mathbf{r}_a - \mathbf{r}_s|^2} e^{-2\operatorname{Im}\Delta z_s} dx_s\, dy_s\, dz_s.$$

Using

$$\frac{\cos\theta_{as}\, dx_s\, dy_s}{|\mathbf{r}_a - \mathbf{r}_s|^2} = d\Omega_{as},$$

we integrate over z_s for constant θ_{as} and obtain

$$V_a \approx \rho\int \frac{|f(\mathbf{o}, \mathbf{i})|^2}{\cos\theta}\, d\Omega \int_0^d e^{-2\operatorname{Im}\Delta z}\, dz = \frac{(1 - e^{-2\operatorname{Im}\Delta d})}{2\operatorname{Im}\Delta}\, \rho\int \frac{|f(\mathbf{o}, \mathbf{i})|^2}{\cos\theta}\, d\Omega. \tag{93}$$

For practical measurement purposes, we would restrict the integration over Ω to a cone of, say, half-angle α corresponding to the receiver's main lobe. (More generally we would work with a spherical source and introduce angular factors corresponding to antenna factors for transmitter and receiver horns **[11]**.)

The corresponding incoherent flux is

$$\mathbf{I} \approx \operatorname{Re}\left[\frac{\rho}{ik}\int \bar{u}_s^a \nabla_a u_s^a |\langle\Psi_s\rangle|^2\, d\mathbf{r}_s\right] \dot{\approx} \rho\int \frac{|f(\mathbf{o}_{as}, \mathbf{i})|^2}{|\mathbf{r}_a - \mathbf{r}_s|^2} e^{-2\operatorname{Im}\Delta z_s} \mathbf{o}_{as}\, d\mathbf{r}_s$$
(94)
$$= (1 - e^{-2\operatorname{Im}\Delta d})(\rho/2\operatorname{Im}\Delta)\int_\alpha |f(\mathbf{o}, \mathbf{i})|^2 (\mathbf{o}/\cos\theta)\, d\Omega,$$

where we differentiated only the exponent of E_{as} and then proceeded as for (93); in the final form, α is the "receiver's cone angle" mentioned above. The coherent flux transmitted through the slab (from (83)) is simply

$$\mathbf{C} = \operatorname{Re}\left[\frac{\langle\bar{\Psi}\rangle\nabla\langle\Psi\rangle}{ik}\right] = |\langle\Psi\rangle|^2\mathbf{i} = e^{-2\operatorname{Im}\Delta d}\mathbf{i}, \qquad \mathbf{i} = \mathbf{n}. \tag{95}$$

The component of the total flux (in a cone α) normal to the slab for $z > d$ is thus

$$\langle\mathbf{J}\rangle\cdot\mathbf{n} = (\mathbf{C} + \mathbf{I})\cdot\mathbf{n} = e^{-2\operatorname{Im}\Delta d} + \frac{(1 - e^{-2\operatorname{Im}\Delta d})}{2\operatorname{Im}\Delta}\, \rho\int_\alpha |f(\mathbf{o}, \mathbf{i})|^2\, d\Omega, \tag{96}$$

where we used $\mathbf{n}\cdot\mathbf{o}_{as} = \cos\theta_{as}$. The incident flux, the only component of flux for $z < 0$, is

$$\mathbf{J} = \operatorname{Re}(\bar{\phi}\nabla\phi/ik) = \mathbf{i}. \tag{97}$$

For lossless scatterers, theorem (53) states that the component of average total flux normal to the slab is a constant in the external region; from (97), the constant must equal $\mathbf{i}\cdot\mathbf{n} = 1$. For $\alpha = \pi/2$ (a "point receiver"), we use (85) in (96) for lossless scatterers to reduce the incoherent term to $(1 - e^{-2\operatorname{Im}\Delta d})$. Thus (96) gives $\langle\mathbf{J}\rangle\cdot\mathbf{n} = 1$ in accord with (97).

The basis for the form (92) is thus the energy theorem for lossless scatterers. There are no total energy losses in such problems, and loss effects shown by the coherent component must be compensated by the corresponding incoherent term. We use simply the coherent intensity as excitation in (92) and (94) (instead of the total) and then let the scatterers radiate directly to free space without further multiple scattering; however, this must yield a result in accord with the energy principle. There can be no additional "coherent" losses in the problem; since the "incoherent scattering" cannot become "coherent," it may be treated as radiating directly to free space.

We note that V of (93) differs from $\mathbf{I}\cdot\mathbf{n}$ of (96) only by the factor $\cos\theta$, and this will also hold for more complicated forms we consider subsequently. To facilitate checking whether approximations for V are in accord with the energy theorem for lossless scatterers, we may use $\cos\theta \approx 1$ (which is appropriate if $|f(\mathbf{o},\mathbf{i})|^2$ is peaked sharply around $\mathbf{o}=\mathbf{n}=\mathbf{i}$), and use simply

$$(98)\quad \mathbf{I}\cdot\mathbf{n} \approx V, \qquad \mathbf{C}\cdot\mathbf{n} = |\langle\Psi\rangle|^2, \qquad \langle\mathbf{J}\rangle\cdot\mathbf{n} = 1 \approx |\langle\Psi\rangle|^2 + V = \langle|\Psi|^2\rangle.$$

Thus any proposed approximation for V, in the limit of tightly peaked lossless $|f|^2$ must reduce as follows:

$$(99)\quad V = \langle|\Psi|^2\rangle - |\langle\Psi\rangle|^2 = \int L\langle|\Psi|^2\rangle\, d\mathbf{r}_s \to V = 1 - |\langle\Psi\rangle|^2 = \int L\, d\mathbf{r}_s.$$

In view of (92) ff., the simplest limiting forms of L satisfying (99) are

$$(100)\qquad L(c,c;s) = \rho|u_s^c|^2\, e^{-2\,\mathrm{Im}\,\Delta z_s}, \qquad \rho|u_s^c|^2\, e^{-2\,\mathrm{Im}\,\Delta(z_c - z_s)}.$$

We use (99) and (100) as checks in the following.

Foldy's integral equation form. For the present problem, the kernel function L required for Foldy's form (57) can be evaluated explicitly by using the results of the integration (90), and performing the remaining elementary integrations over z. Thus

$$L_0(c,a;s) = \rho\bar{u}_s^c(\mathbf{i})u_s^a(\mathbf{i}),$$

$$L_1(c,a;s) = \rho^2\iint \bar{u}_t^c(\mathbf{o}_{ts})u_m^a(\mathbf{o}_{ms})[\rho\bar{u}_s^t(\mathbf{i})u_s^m(\mathbf{i})]\, d\mathbf{r}_m\, d\mathbf{r}_t$$

$$= \overline{\rho u_s^c i\Delta_{cs}\, u_s^a i\Delta_{as}}\int_{z_s}^{z_c} dz_t \int_{z_s}^{z_a} dz_m$$

$$= \overline{\rho u_s^c i\Delta_{cs}(z_c - z_s)u_s^a i\Delta_{as}(z_a - z_s)},$$

$$(101)\quad L_2(c,a;s) = \rho^2\iint \bar{u}_t^c(\mathbf{o}_{ts})u_m^a(\mathbf{o}_{ms})[\overline{\rho u_s^t(\mathbf{i})i\Delta_{ts}(z_t - z_s)u_s^m(\mathbf{i})i\Delta_{ms}(z_m - z_s)}]\, d\mathbf{r}_m\, d\mathbf{r}_t$$

$$= \overline{\rho u_s^c(\mathbf{i})(i\Delta_{cs})^2 u_s^a(\mathbf{i})(i\Delta_{as})^2}\int_{z_s}^{z_c}(z_t - z_s)\, dz_t \int_{z_s}^{z_a}(z_m - z_s)\, dz_m$$

$$= \rho\bar{u}_s^c(\mathbf{i})\,\frac{\overline{[i\Delta_{cs}(z_c - z_s)]^2}}{2}\cdot u_s^a(\mathbf{i})\,\frac{[i\Delta_{as}(z_a - z_s)]^2}{2},$$

$$L_n(c,a;s) = \rho\bar{u}_s^c(\mathbf{i})\,\frac{\overline{[i\Delta_{cs}(z_c - z_s)]^n}}{n!}\, u_s^a(\mathbf{i})\,\frac{[i\Delta_{as}(z_a - z_s)]^n}{n!},$$

where

$$\Delta_{as} = N\delta_{as} = \frac{\rho 2\pi f(\mathbf{o}_{as}, \mathbf{o}_{as})}{k\cos\theta_{as}}.$$

Since the zeroth order Bessel function $J_0(x)$ may be written

$$\sum_{n=0}^{\infty}\left[\frac{ix}{2}\right]^{2n}\frac{1}{(n!)^2},$$

and since

$$I_0(x) = J_0(ix) = \sum\left(\frac{x}{2}\right)^{2n}\frac{1}{(n!)^2},$$

we may rewrite (101) in the closed form

$$L(c, a; s) = \sum_{n=0}^{\infty} L_n = \rho\bar{u}_s^c(\mathbf{i})u_s^a(\mathbf{i})I_0(2\sqrt{\Delta_{cs}(z_c - z_s)\Delta_{as}(z_a - z_s)}). \tag{102}$$

In particular, for $c = a$, we use $L(c, c; s)$ of (102) in the corresponding form of (57) to obtain

$$\begin{aligned}\langle|\Psi_c|^2\rangle - |\langle\Psi_c\rangle|^2 &= \rho\int|u_s^c(\mathbf{i})|^2\langle|\Psi_s|^2\rangle I_0(2|\Delta_{cs}|(z_c - z_s))\,d\mathbf{r}_s \\ &= \rho\int[|f(\mathbf{o},\mathbf{i})|^2/\cos\theta]\,d\Omega\int\langle|\Psi_s|^2\rangle I_0(2|\Delta_{cs}|(z_c - z_s))\,dz_s.\end{aligned} \tag{103}$$

Thus we have "solved" the original integral equation form, to obtain an actual integral equation for $\langle|\Psi|^2\rangle$ in terms of $|\langle\Psi\rangle|^2$. However, we do not regard (103) as correct: the limiting form of the kernel function for lossless tightly peaked $|f|^2$ is not in accord with (100). More explicitly, in this limit we have $\langle|\Psi|^2\rangle \to 1$, $\cos\theta \to 1$, and the right and left hand sides give different values for

$$V = \langle|\Psi|^2\rangle - |\langle\Psi\rangle|^2 = 1 - |\langle\Psi\rangle|^2;$$

the right hand side of (103) gives $2\,\mathrm{Im}\,\Delta\int_0^{z_c} I_0(2|\Delta|(z_c - z_s))\,dz_s$ which does not equal $1 - e^{-2\,\mathrm{Im}\,\Delta z_s}$ obtained from the left side.

The $\langle\Psi_c\rangle$-consistent integral equation. In order to apply $\langle\bar{\Psi}_c\Psi_a\rangle$ of (66), we must first determine the new scattering function $u_s^c\varepsilon_{cs}$. We have

$$\begin{aligned}u_s^c(\mathbf{j})\varepsilon_{cs} &= u_s^c(\mathbf{j}) + \rho\int u_t^c(\mathbf{o}_{ts})u_s^t(\mathbf{j})\varepsilon_{ts}\,d\mathbf{r}_t = u_s^c(\mathbf{j})\left[1 + i\Delta_{cs}\int_{z_s}^{z_c}\varepsilon_{ts}\,dz_t\right], \\ \varepsilon_{cs} &= 1 + \Delta_{cs}\int_{z_s}^{z_c}\varepsilon_{ts}\,dz_t = e^{i\Delta_{cs}(z_c - z_s)},\end{aligned} \tag{104}$$

where we used (90) and then followed essentially the same procedure as for $\langle\Psi\rangle$ of (79) in solving the integral equation for ε. Using (104) in (66), we obtain

$$\begin{aligned}\langle|\Psi_c|^2\rangle &= |\langle\Psi_c\rangle|^2 + \rho\int|u_s^c e^{i\Delta(z_c - z_s)}|^2\langle|\Psi_s|^2\rangle\,d\mathbf{r}_s \\ &= e^{-2\,\mathrm{Im}\,\Delta z_s} + \rho\int\frac{|f(\mathbf{o}_{cs},\mathbf{i})|^2}{\cos\theta_{cs}}\,d\Omega_{cs}\int\langle|\Psi_s|^2\rangle\,e^{-2\,\mathrm{Im}\,\Delta(z_c - z_s)}\,dz_s,\end{aligned} \tag{105}$$

which clearly satisfies the energy theorem in the appropriate limit (see (99) and (100)). We also show subsequently that (105) is the required limit of the corresponding explicit approximation of $\langle|\Psi_e|^2\rangle$ for $N\to\infty$.

From (104), we see that we have also constructed the K-space Green's function for the present problem. In terms of $z_{as} = z_a - z_s = \cos\theta_{as} r_{as}$, we have

$$E(K_{as}r_{as}) = E(kr_{as})\varepsilon_{as} = \frac{e^{ikr_{as}+i\Delta_{as}r_{as}\cos\theta_{as}}}{r_{as}} \equiv \frac{e^{iK_{as}r_{as}}}{r_{as}},$$
$$K_{as} = k + \Delta_{as}\cos\theta_{as} = k + 2\pi\rho f(\mathbf{o}_{as}, \mathbf{o}_{as})/k, \tag{106}$$

where $K_{as} = K(\mathbf{o}_{as})$ differs from $K = K(\mathbf{i})$ of (82) only in the argument of f. Thus if $f(\mathbf{i}, \mathbf{i})$ is independent of direction (i.e., if the scatterer is spherically symmetric, or if u represents a symmetrical average over all orientations of ellipsoids, etc.) then $E(Kr)$ is also independent of direction.

Foldy's recast kernel function. Foldy obtained $\sum M_n$ of (58) by operating on $\sum L_n$ of (57) with the wave equation (43). The propagation parameter of (43) equals $K^2 = k^2 + 4\pi\rho f$ for identical (or averaged) monopoles, while the present K^2 (as in (82)) has an additional term in f^2; to facilitate comparison, we assume suitable restrictions for neglecting the difference and apply $E(Kr_{as})$ of (106) to (58).

Since $\sum L_n$ gives an inadequate result, so does the recast form $\sum M_n$. However, because the initial term M_0 gives the correct $\langle\Psi_c\rangle$-consistent form of $\langle|\Psi_c|^2\rangle$, it is of interest to consider (58) further. Using essentially the same procedure as for (101), we obtain

$$M_0(c, a; s) = \rho\bar{u}_s^c(\mathbf{i})u_s^a(\mathbf{i})\, e^{\overline{i\Delta_{cs}z_{cs}}+i\Delta_{as}z_{as}}, \qquad z_{cs} = z_c - z_s;$$

$$\begin{aligned} M_1(c, a; s) &= -\rho^2 f(\mathbf{o}_{as}, \mathbf{i})E(Kr_{as})\int \overline{f(\mathbf{o}_{ct}, \mathbf{o}_{ts})f(\mathbf{o}_{ts}, \mathbf{i})E(Kr_{ct})E(Kr_{ts})}\, d\mathbf{r}_t \\ &\quad - \rho^2\overline{f(\mathbf{o}_{cs}, \mathbf{i})E(Kr_{cs})}\int f(\mathbf{o}_{am}, \mathbf{o}_{ms})f(\mathbf{o}_{ms}, \mathbf{i})E(Kr_{am})E(Kr_{ms})\, d\mathbf{r}_t \\ &= -M_0(c, a; s)[\overline{i\Delta_{cs}z_{cs}} + i\Delta_{as}z_{as}]; \end{aligned} \tag{107}$$

$$M_2(c, a; s) = M_0(c, a; s)[\tfrac{1}{2}(\overline{i\Delta_{cs}z_{cs}})^2 + 2(\overline{i\Delta_{cs}z_{cs}})(i\Delta_{as}z_{as}) + \tfrac{1}{2}(i\Delta_{as}z_{as})^2].$$

From (107), we have symbolically

$$\begin{aligned} \sum M_n &= L_0\, e^{A+B}[1 - A - B + \tfrac{1}{2}A^2 + 2AB + \tfrac{1}{2}B^2 + O(A^3)] \\ &= L_0[1 + AB + O(A)^3], \end{aligned} \tag{107'}$$

so that M_1 provides a first order "correction" to M_0, and successive terms serve to cancel additional essential terms of the exponent. Comparison with the result of (101) for the same number of iterations, i.e.,

$$\sum L_n = L_0[1 + AB \mp \tfrac{1}{4}A^2B^2 + O(A^6)], \tag{101'}$$

shows that $\sum L_n$ is the more rapidly converging series in A, B. Both series converge to $\sum L_n = \sum M_n = L$ of (102), which is an incorrect form for the energy theorem.

Other representations. Similarly the heuristic form (64) may be rewritten in the form (57) with the kernel function L replaced by

$$N = L_0[1 + A + B + AB + \tfrac{1}{2}AB(A + B) + \tfrac{1}{4}A^2B^2 + O(A^5)], \tag{108}$$

where we followed essentially the same procedure as for (101).

Comparing the above expansions with the analogous two stages of the expansion of the kernel function of (105), i.e., with

$$\rho\bar{u}_s^e u_s^a \bar{\varepsilon}_{cs}\varepsilon_{as} = L_0 e^{A+B} = L_0[1 + A + B + AB + \tfrac{1}{2}A^2 + \tfrac{1}{2}B^2 + O(A^3)], \tag{109}$$

we see that (108) is the only expansion that has the proper leading terms to provide the first corrections to the single scattering approximation.

In the above we showed that the $\langle\Psi_c\rangle$-consistent form of $\langle|\Psi_c|^2\rangle$ gave results in accord with the energy theorem. In the following we show explicitly that this form is also the limit of $\langle|\Psi_e|^2\rangle$ for $N \to \infty$ (as is, of course, required by the development leading to (66) subject to the present restrictions).

Expanded representation. We use (37) with $w = 1/\mathscr{V}$ but because of the present restrictions, we drop all terms involving back-scattered values of f; thus we obtain a finite sequence. We group the terms on the basis of their numerical coefficient in N: analogous to (65), we have

$$\begin{aligned}\langle\bar{\Psi}_e\Psi_a\rangle = \bar{\phi}_e\phi_a &+ Nw\int(\bar{\phi}_e u_s^a\phi_s + \overline{u_s^e\phi_s}\phi_a + \overline{u_s^e\phi_s}u_s^a\phi_s)\,d\mathbf{r}_s \\ &+ N(N-1)w^2\int\int[\overline{u_s^e\phi_s}u_t^a\phi_t + \bar{\phi}_e u_s^a u_t^s\phi_t + \overline{u_s^e u_t^s\phi_t}\phi_a \\ &\qquad + \overline{u_s^e\phi_s}(u_s^a u_t^s\phi_t + u_t^a u_s^t\phi_s) + \overline{(u_s^e u_t^s\phi_t + u_t^e u_s^t\phi_s)}u_s^a\phi_s \\ &\qquad + \overline{u_s^e u_t^s\phi_t}u_s^a u_t^s\phi_t]\,d\mathbf{r}_s\,d\mathbf{r}_t + \cdots\end{aligned} \tag{110}$$

where (as usual, for brevity) the suppressed directions of incidence in the u's are determined by the terms on their right. Using (87) and (90) and performing the integrations over x, y, we evaluate all integrals of the form considered in previous sections. Thus in terms of

$$\begin{aligned}&\delta = \frac{2\pi f(\mathbf{i},\mathbf{i})}{k}w, \quad \delta_{as} = \frac{2\pi f(\mathbf{o}_{as},\mathbf{o}_{as})}{k\cos\theta_{as}}w, \quad z_{es} = z_e - z_s; \\ &W_{ea}(s) = \frac{w}{\bar{\phi}_e\phi_a}\int\overline{u_s^e\phi_s}u_s^a\phi_s\,d\mathbf{r}_s, \quad W_{ee}(s) = w\int\frac{|f(\mathbf{o}_{es},\mathbf{i})|^2}{\cos\theta_{es}}\int_0^{z_e}dz_s = Q(\mathbf{o}_{es},\mathbf{i})\int_0^{z_e}dz_s,\end{aligned} \tag{111}$$

we obtain initially

$$\frac{\langle \Psi_e^* \Psi_a \rangle}{\phi_e^* \phi_a} = 1 + N\{(\overline{i\delta z_e} + i\delta z_a) + W_{ea}(s)\}$$
$$+ N(N-1)\{\tfrac{1}{2}(\overline{i\delta z_e} + i\delta z_a)^2$$
$$+ W_{ea}(s)[\overline{i\delta_{es} z_{es}} + i\delta_{as} z_{as} + 2 \operatorname{Re} i\delta z_s] + W_{ea}(s) W_{ss}(t)\}$$
$$+ N(N-1)(N-2)\left\{\frac{1}{3!}(\overline{i\delta z_e} + i\delta z_a)^3\right. \tag{112}$$
$$+ W_{ea}(s)\tfrac{1}{2}[\overline{i\delta_{es} z_{es}} + i\delta_{as} z_{as} + 2 \operatorname{Re} i\delta z_s]^2$$
$$+ W_{ea}(s) W_{ss}(t)[\overline{i\delta_{es} z_{es}} + i\delta_{as} z_{as} + 2 \operatorname{Re} i\delta_{st} z_{st} + 2 \operatorname{Re} i\delta z_t]$$
$$\left. + W_{ea}(s) W_{ss}(t) W_{tt}(m)\right\} + \cdots.$$

Succeeding may be obtained by inspection, e.g., the next term is

$$\frac{N!}{(N-4)!}\left\{\frac{1}{4!}(\overline{i\delta z_e} + i\delta z_a)^4 + W_{ea}(s)\frac{1}{3!}[\overline{i\delta_{es} z_{es}} + i\delta_{as} z_{as} + 2 \operatorname{Re} i\delta z_s]^3\right.$$
$$+ W_{ea}(s) W_{ss}(t)\tfrac{1}{2}[\overline{i\delta_{es} z_{es}} + i\delta_{as} z_{as} + 2 \operatorname{Re} i\delta_{st} z_{st} + 2 \operatorname{Re} i\delta z_t]^2 \tag{112'}$$
$$+ W_{ea}(s) W_{ss}(t) W_{tt}(m)[\overline{i\delta_{es} z_{es}} + i\delta_{as} z_{as} + 2 \operatorname{Re} i\delta_{st} z_{st} + 2 \operatorname{Re} i\delta_{tm} z_{tm}$$
$$+ 2 \operatorname{Re} i\delta z_m]$$
$$\left. + W_{ea}(s) W_{ss}(t) W_{tt}(m) W_{mm}(n)\right\}.$$

As a check on the above, and for its own interest, we temporarily specialize (112) to tightly peaked $|f|^2$. For this case, all the δ's are identical and the chains of δz's in the square brackets in (112) collapse to $\overline{i\delta z_e} + i\delta z_a$; similarly, all Q's in $W_{ss}(t)$, etc., are identical, and we may integrate over all the remaining z's, except z_s which is involved in $W_{ea}(s)$. The relationship of the resultant form to the binomial expansion of the Nth power of 1 plus the first term in braces in (112) is that expected from elementary physical considerations. To make the relationship explicit, we then let $\mathbf{r}_a = \mathbf{r}_e$, integrate over the remaining z_s, and obtain simply

$$\langle |\Psi|^2 \rangle = 1 + N(-2 \operatorname{Im} \delta + Q)z + \frac{N(N-1)}{2}(-2 \operatorname{Im} \delta + Q)^2 z^2 + \cdots$$
$$= (1 - 2 \operatorname{Im} \delta z + Qz)^N. \tag{113}$$

We thus have an intensity transmission coefficient $\mathscr{T}_N = \mathscr{T}_1^N$ analogous to the wave coefficient T_N, and the result may be rederived by the elementary "transmission line" algebraic procedure mentioned after (91). For the present case of peaked $|f|^2$, we may use $\cos\theta \approx 1$ in Q of (111). Thus for lossless scatterers, we have $Q \approx \sigma w = 2 \operatorname{Im} \delta$, and (113) reduces to $\langle |\Psi|^2 \rangle = 1$ as required by the energy theorem.

More generally, we regroup the terms of (112) into the coherent component plus other components corresponding to products of the operator W:

$$\frac{\langle \bar{\Psi}_e \Psi_a \rangle}{\bar{\phi}_e \phi_a} = (1 + \overline{i\delta z_e} + i\delta z_a)^N + NW_{ea}(s)(1 + \overline{i\delta}_{es} z_{es} + i\delta_{as} z_{as} + 2 \operatorname{Re} i\delta z_s)^{N-1}$$

$$(114) \qquad + N(N-1)W_{ca}(s)W_{ss}(t)(1 + \overline{i\delta}_{es} z_{es} + i\delta_{as} z_{as} + 2 \operatorname{Re} i\delta_{st} z_{st} + 2 \operatorname{Re} i\delta z_t)^{N-2} + \cdots.$$

We first specialize this to a more practical form for finite N, and then proceed to the limit $N \rightarrow \infty$.

For a finite relatively small number of elements, we would compute the intensity for $\mathbf{r}_e = \mathbf{r}_a$ from

$$\begin{aligned} \langle |\Psi_e|^2 \rangle &= \frac{\langle \bar{\Psi}_e \Psi_e \rangle}{\bar{\phi}_e \phi_e} \\ &= (1 + 2 \operatorname{Re} i\delta z_e)^N + NW_{ee}(s)(1 + 2 \operatorname{Re} i\delta_{es} z_{es} + 2 \operatorname{Re} i\delta z_s)^{N-1} \\ &\quad + N(N-1)W_{ee}(s)W_{ss}(t)(1 + 2 \operatorname{Re} i\delta_{es} z_{es} + 2 \operatorname{Re} i\delta_{st} z_{st} + 2 \operatorname{Re} i\delta z_t)^{N-2} + \cdots \\ &\quad + \frac{N!}{(N-\nu)!}\left[\prod_{n=1}^{\nu} W_{e(n-1)e(n-1)}(e^{(n)})\right]\left[1 + 2 \operatorname{Re} \sum_{m=1}^{\nu} i\delta_{e(m-1)e(m)} z_{e(m-1)e(m)} + 2 \operatorname{Re} i\delta z_{e^{(\nu)}}\right]^{N-\nu} \end{aligned} \tag{115}$$

(from which we also obtain $\langle \Psi^2 \rangle / \phi_e^2$ by dropping Re, and using the first form of W of (111) without bars). For practical purposes, for peaked $|f|^2$ (but not going quite to the limiting form of (113)) we regard the δ's as identical and integrate over all z's except z_s to obtain the transmitted intensity ($z_e > d$)

$$\begin{aligned} \langle \mathbf{J} \cdot \mathbf{i} \rangle \approx \langle |\Psi|^2 \rangle &= (1 - 2 \operatorname{Im} \delta d)^N + NW_{dd}(s)\{(1 - 2 \operatorname{Im} \delta d)^{N-1} + (N-1)(1 - 2 \operatorname{Im} \delta d)^{N-2} Q z_s + \cdots\} \\ &= (1 - 2 \operatorname{Im} \delta d)^N + NQ_\alpha \int_0^d (1 - 2 \operatorname{Im} \delta d + Q z_s)^{N-1} \, dz_s \end{aligned} \tag{116}$$

where the subscript α on Q corresponds to the half-angle of the cone of observation. Integrating over z_s, we obtain

$$\begin{aligned} \langle \mathbf{J} \cdot \mathbf{i} \rangle &\approx \langle |\Psi|^2 \rangle \\ &= (1 - 2 \operatorname{Im} \delta d)^N + \frac{Q_\alpha}{Q}[(1 - 2 \operatorname{Im} \delta d + Qd)^N - (1 - 2 \operatorname{Im} \delta d)^N], \end{aligned} \tag{117}$$

which we leave in the present grouping to facilitate comparison with measurements, i.e., we can measure the coherent term and the incoherent term (proportional to

Q_α/Q) separately. For practical measurements involving a spherical source of finite cone angle α_t, both Q_α and Q would also depend on α_t; for lossless scatterers, the corresponding $Q(\alpha_t)$ would equal 2 Im δ only in an appropriate limit of α_t. The form (117) may also prove useful as a first approximation for all scatterers very large compared to wavelength, i.e., even for "nontenuous" scatterers provided that $|f(\mathbf{i}, \mathbf{i})| \gg |f(-\mathbf{i}, \mathbf{i})|$. For this case and plane wave excitation, $Q \neq w\sigma_s$ should still represent a large fraction of the total scattering cross section of one object.

Returning to (114) we introduce the symbols,

$$\begin{aligned} &\Delta = N\delta, \qquad \rho = Nw = N/\mathscr{V}, \qquad P_{ea}(s) = NW_{ea}(s), \\ &P_{ee}(s) = NW_{ee}(s) = q(\mathbf{o}_{es}, \mathbf{i}) \int_0^{z_e} dz_s, \qquad q_\alpha = NQ_\alpha, \end{aligned} \tag{118}$$

and proceed to the limit $N \to \infty$ by using $N!/(N-\nu)! \to N^\nu$, $(1 + x/N)^N \to e^x$, etc. Thus (114) reduces to

$$\begin{aligned} \frac{\langle \bar{\Psi}_e \Psi_a \rangle}{\bar{\phi}_e \phi_a} &= e^{i\bar{\Delta} z_e + i\Delta z_a} + P_{ea}(s)\, e^{i\bar{\Delta}_{es} z_{es} + i\Delta_{as} z_{as} + 2\,\mathrm{Re}\, i\Delta z_s} \\ &\quad + P_{ea}(s) P_{ss}(t)\, e^{i\bar{\Delta}_{es} z_{es} + i\Delta_{as} z_{as} + 2\,\mathrm{Re}\, i\Delta_{st} z_{st} + 2\,\mathrm{Re}\, i\Delta z_t} + \cdots, \end{aligned} \tag{119}$$

where succeeding terms correspond to successive iterations of the coherent term $e^{i\bar{\Delta} z_e + i\Delta z_a}$ in the operator $P_{ea}(s)\, e^{i\bar{\Delta}_{es} z_{es} + i\Delta_{as} z_{as}}$. Compacting the representation, we obtain the integral equation

$$\begin{aligned} &\langle \bar{\Psi}_c \Psi_a \rangle = \langle \bar{\Psi}_c \rangle \langle \Psi_a \rangle + \rho \int \bar{u}_s^e(\mathbf{i}) u_s^a(\mathbf{i})\, e^{i\bar{\Delta}_{es} z_{es} + i\Delta_{as} z_{as}} \langle |\Psi_s|^2 \rangle\, d\mathbf{r}_s, \\ &\langle \Psi_a \rangle = e^{i\mathbf{k}\cdot\mathbf{r}_a + i\Delta z_a}, \quad u_s^a(\mathbf{i})\, e^{i\Delta_{as} z_{as}} = f(\mathbf{o}_{as}, \mathbf{i}) \frac{e^{ikr_{as} + i\Delta_{as} z_{as}}}{r_{as}}, \end{aligned} \tag{120}$$

which is of course the special case of (66) obtained by using the present u and (104). For $\mathbf{r}_c = \mathbf{r}_a$, (120) reduces to the $\langle \Psi_c \rangle$-consistent form $\langle |\Psi_c|^2 \rangle$ of (105).

In terms of the total cross section for one scatterer $(4\pi/k)\,\mathrm{Im}\, f(\mathbf{i}, \mathbf{i}) = \sigma = \sigma_a + \sigma_s$ (absorption plus scattering), we rewrite (105) as

$$\langle |\Psi_a|^2 \rangle = e^{-\rho\sigma(\mathbf{i}) z_a} + \rho \int \frac{|f(\mathbf{o}_{as}, \mathbf{i})|^2}{r_{as}^2} \langle |\Psi_s|^2 \rangle\, e^{-\rho\sigma(\mathbf{o}_{as}) r_{as}}\, d\mathbf{r}_s \tag{121}$$

where $\sigma(\mathbf{o}) = (4\pi/k)\,\mathrm{Im}\, f(\mathbf{o}, \mathbf{o})$. For lossless scatterers, $\sigma(\mathbf{i})$ reduces to the total scattering cross section $\sigma_s(\mathbf{i}) = \int |f(\mathbf{o}, \mathbf{i})|^2\, d\Omega$.

The limit of the practical form (116) is also of interest. For this case, we have

$$\begin{aligned} \langle \mathbf{J}\cdot\mathbf{i} \rangle \approx \langle |\Psi|^2 \rangle &= e^{-2\,\mathrm{Im}\,\Delta d} + q_\alpha \int_0^{d} e^{-2\,\mathrm{Im}\,\Delta d + q\zeta}\, d\zeta \\ &= e^{-2\,\mathrm{Im}\,\Delta d} + q_\alpha \int_0^{d} e^{-2\,\mathrm{Im}\,\Delta (d-\zeta)} [e^{(-2\,\mathrm{Im}\,\Delta + q)\zeta}]\, d\zeta, \end{aligned} \tag{122}$$

where

$$q = NQ = \rho \int \frac{|f(\mathbf{o}, \mathbf{i})|^2}{\cos\theta}\, d\Omega \approx \sigma_s \rho.$$

The function in brackets is the present approximation for the total average internal intensity $\langle|\Psi(\zeta)|^2\rangle$, and the remainder of the integrand is the remains of the Green's function. Integrating over ζ, we obtain the analog of (117)

$$(123) \qquad \langle \mathbf{J}\cdot\mathbf{i}\rangle \approx e^{-\rho\sigma d} + \frac{q_\alpha}{q}(e^{-\rho\sigma d + qd} - e^{-\rho\sigma d}).$$

For the present lossless scatterers, $\rho\sigma = q = 2\,\mathrm{Im}\,\Delta$, and (122) reduces to the result given previously in (96). However, as mentioned after (117), the present forms may also provide general first approximations for very large, nontenuous scatterers.

The average of $\langle\Psi^2\rangle$. As mentioned in the introduction, averages of functions other than Ψ, $|\Psi|^2$, and $\mathbf{J}$ are also of interest, e.g., Ψ^2, $(\mathrm{Re}\,\Psi)^2$, etc. [6]. For completeness as far as quadratic functions are concerned, we note that $\langle\Psi_e\Psi_a\rangle$ may be obtained by inspection of (114), (120), etc., i.e., by simply removing the bar throughout all definitions and replacing $2\,\mathrm{Re}\, i\delta$, etc., by $2i\delta$, etc.; it was because of this extension (used in (68) and mentioned after (115)) that we kept the redundant factors of ϕ in the definitions of W and P in (111) and (118). The corresponding analog of (121) is the integral equation

$$(124) \qquad \langle\Psi_a^2\rangle = \langle\Psi_a\rangle^2 + \rho\int [u_s^a(\mathbf{i})\, e^{i\Delta_{as}(z_a - z_s)}]^2 \langle\Psi_s^2\rangle\, d\mathbf{r}_s,$$

where $\langle\Psi_a\rangle = \phi_a\, e^{i\Delta z_a}$. We may work with

$$(125) \qquad \langle\Psi_a^2\rangle = \langle\Psi_a\rangle^2 + \rho\int^* [f(\mathbf{o}_{as}, \mathbf{i})\phi_s E(K|\mathbf{r}_a - \mathbf{r}_s|)]^2 \langle\Psi_s^2\rangle\, d\mathbf{r}_s$$

where * indicates an artifice to avoid the singularity at $r_{as} = 0$, e.g., a minimum separation for the points. For forward peaked $|f|^2$ and small angles, we use (75) to obtain

$$(126) \qquad \begin{aligned} \langle\Psi_a^2\rangle &= e^{i2\mathbf{k}\cdot\mathbf{r}_a + i2\Delta z_a}\left\{1 + \frac{i2\pi\rho f^2(\mathbf{i}, \mathbf{i})}{2k}\int^* \frac{\langle\Psi^2(z_s)\rangle}{z_a - z_s}\, e^{-i2\mathbf{k}\cdot\mathbf{r}_s - i2\Delta z_s}\, dz_s\right\} \\ &= \langle\Psi_a\rangle^2\left\{1 + \frac{if(\mathbf{i}, \mathbf{i})\Delta}{2}\int^* \frac{\langle\Psi^2(z_s)\rangle}{\langle\Psi_s\rangle^2(z_a - z_s)}\, dz_s\right\}. \end{aligned}$$

The transmitted form of (125) for $z_a \gg d = (z_s)_{\max}$ equals

$$(127) \qquad \langle\Psi^2\rangle = \phi^2\, e^{i2\Delta d}\left\{1 + \frac{if\Delta}{2z}\int_0^d F(\zeta)\, d(\zeta)\right\},$$

where $F(\zeta) = \langle\Psi^{\prime 2}(\zeta)\rangle/\langle\Psi^{\prime}(\zeta)\rangle^2$ satisfies the integral equation

$$F(\zeta) = 1 + \frac{if(\mathbf{i}, \mathbf{i})\Delta}{2} \int^{*} \frac{F(\zeta)}{z - \zeta}\, d\zeta. \tag{128}$$

The leading term of (127) is simply

$$\langle\Psi^{\prime 2}\rangle \sim \phi^2\, e^{i2\Delta d}\left(1 + \frac{if\Delta d}{2z}\right). \tag{129}$$

References

1. V. Twersky, *Multiple scattering of waves*, J. Res. Nat. Bur. Standards Sect. D **64** (1960), 715–730; J. E. Burke and V. Twersky, *On scattering of waves by many bodies*, Report EDL–L23, Sylvania Electronic Defense Labs., Mt. View, Calif., 1963; J. Res. Nat. Bur. Standards Sect. D (1964) (to appear).

2. L. L. Foldy, *The multiple scattering of waves*, Phys. Rev. (2) **67** (1945), 107–119.

3. M. Lax, *Multiple scattering of waves*. I, Rev. Modern Phys. **23** (1951), 287–310; II. *The effective field in dense systems*, Phys. Rev. (2) **88** (1952), 621–629.

4. V. Twersky, *On scattering of waves by random distributions*. I. *Free-space scattering formalism*, J. Mathematical Phys. **3** (1962), 700–715; II. *Two-space scatterer formalism*, ibid. **3** (1962), 724–734.

5. J. B. Keller, *Wave propagation in random media*, Proc. Sympos. Appl. Math. Vol. 13, pp. 227–246, Amer. Math. Soc., Providence, R. I., 1962; *Stochastic equations and wave propagation in random media*, Proc. Sympos. Appl. Math. Vol. 16, pp. 145–170, Amer. Math. Soc., Providence, R. I., 1964.

6. V. Twersky, *Signals, scatterers, and statistics*, Report EDL–E72, Sylvania Electronic Defense Labs., Mt. View, Calif., 1962; IEEE Trans. **AP–11** (1963), 668–680.

7. G. H. A. Cole, *Classical theory of the equilibrium liquid pair distribution*, Advances in Phys. **7** (1962), 224–251.

8. V. Twersky, *Multiple scattering by arbitrary configurations in three dimensions*, J. Mathematical Phys. **3** (1962), 83–91.

9. ———, *Multiple scattering of waves by planar random distributions of parallel cylinders and bosses*, Research Report EM–58, Courant Inst. Math. Sci., New York Univ., New York, 1953; *Scattering by quasi-periodic and quasi-random distributions*, IRE Trans. **AP–7** (1959), S307–319.

10. E. I. Carstensten and L. L. Foldy, *Propagation of sound through a liquid containing bubbles*, J. Acoust. Soc. Amer. **19** (1947), 481.

11. C. I. Beard and V. Twersky, *Forward coherent and incoherent scattering from volume distributions of spheres*, Report EDL–E46, Sylvania Electronic Defense Labs., Mt. View, Calif., 1960; *Off-forward scattering from volume distributions of spheres*, Report EDL–E47, ibid.

12. J. Bazer, *Multiple scattering in one dimension*, Research Report EM–122, Courant Inst. Math. Sci., New York Univ., New York, 1959.

13. C. H. Wilcox, *A generalization of theorems of Rellich and Atkinson*, Proc. Amer. Math. Soc. **7** (1956), 271.

14. F. Zernike and J. A. Prins, *Die Beugung von Röntegen Strahlen in Flussigkeiten als Effekt der Mölekulanordnung*, Z. Physik **41** (1927), 184.

15. P. Mazur and M. Mandel, *On the theory of molecular polarization in gases*, Physica **22** (1956), 289; L. Jansen and P. Mazur, Physica **21** (1955), 193.

Technion, Israel Institute of Technology,
Haifa, Israel
Sylvania Electronic Defense Laboratories,
Mountain View, California

WAVE PROPAGATION IN A GENERAL RANDOM CONTINUOUS MEDIUM

BY

W. C. HOFFMAN

1. **Introduction.** The problem of wave motion in a randomly inhomogeneous medium is of considerable importance in mathematical physics, where it arises naturally in the following contexts: propagation of electromagnetic or acoustic waves through a turbulent medium, the statistical mechanics of a continuous medium, and the molecular scattering of light. The present paper is devoted primarily to the mathematics associated with the first of these three topics, i.e., the theoretical problem of wave motion in a medium which, while isotropic, is characterized by having a random point function for refractive index. This problem will henceforth be referred to as the "random continuum" problem, and for a continuous medium is equivalent to "turbulent scatter."

An admirably complete and lucid treatment of the statistical mechanics of a continuous medium has been given by Kampé de Fériet in another of these symposium volumes [**1**]. The third type of phenomenon, that involving scattering from random configurations of elementary scatterers, is discussed elsewhere in this symposium volume [**2**].

Theoretical approaches to the random continuum problem have generally been devised in connection with the problem of turbulent scatter, and may be broadly subsumed under the following heads: perturbation theory (including in particular the Born approximation), the integral equation formulation, transformation to a spatial analogue of the Riccati equation,[1] and the diffusion equation approach (which is meant to include ray theory, transport equations, and the Fokker–Planck equation). It is clear from a casual consideration of these headings that most investigators have approached the problem more from the point of view of differential equations and electromagnetic waves than from the stochastic aspects, and in point of fact probabilistic considerations have often languished in comparison to those of classical analysis. As examples one may cite the fact that the correlation function $\exp\{-\alpha|x|\}$ was used for some time in engineering formulations of a turbulent scatter theory in spite of the fact that it was known from a theorem of Doob [3] that such a correlation function could only correspond to a Brownian motion process, i.e., one whose first derivative is everywhere discontinuous, an unfortunate feature for a hydrodynamic flow. Again, as Kampé de Fériet has pointed out [**4**], the application of time or spatial averaging to integral

[1] This is referred to in the Russian literature as Rytov's method.

representations of field quantities, as has been commonly done in most engineering analyses, requires justification in the form of an as yet nonexistent ergodic theorem, and what should actually be calculated therefore are mathematical expectations rather than time or volume averages, the statistical inference problem being left for later on.

We shall therefore pay special attention in the present discussion to the probabilistic aspects of the problem, in particular to the meaning of derivative or integral of a random function and to the concept of random integral of a partial differential equation, to the various forms of spectral representation for the random functions involved, and the appropriate aspects of functional analysis.

Kampé de Fériet **[1]** has recognized three aspects of random integrals of partial differential equations: (i) random integrals of particular initial value or boundary value problems (e.g., the heat equation, Laplace equation, wave equation, etc.); (ii) the matter of defining a probability measure on the function space of regular integrals; and (iii) partial differential equations with random functions for coefficients. It is this third aspect that occupies us here, since our concern lies mainly with the reduced wave equation

$$\Delta U + k^2(\mathbf{x})U(\mathbf{x}) = 0 \tag{1.1}$$

which follows from the wave equation $\Delta u - v^{-2}(\mathbf{x})u_{tt} = 0$ under the assumption of harmonic time dependence: $e^{i\omega t}$. The essential feature of (1.1) is that $k(\mathbf{x})$ is a random function of position, and so also therefore is $U(\mathbf{x})$. Thus the field quantity is defined as a functional of the random refractive index process.

Nearly all studies to date of wave propagation in a random continuum have assumed harmonic time dependence, i.e., the reduced wave equation (1.1). Silver **[5]** has pointed out that the applicability of this assumption (or that of quasi-monochromaticity) requires verification. A medium that is swirling just as rapidly as the wave propagates through it will clearly give rise to frequency shifts that invalidate the assumption of monochromaticity. Thus hot or very energetic media are excluded. In general, use of the reduced wave equation appears proper whenever the time variation in the refractive properties of the medium is of a much lower order than the propagation time of the wave.

2. **Mathematical analysis for random functions.** The analysis of random functions can be approached in several different ways **[6]**, but for our purposes that based on mean square convergence suffices. Thus *regular random functions* are those for which

$$E\{|X(\omega, \xi)|^2\} < \infty, \tag{2.1}$$

and by mean square convergence of a sequence $\{X_n\}$ we mean

$$\lim_{n\to\infty} E\{|X_n(\omega) - X(\omega)|^2\} = 0 \iff \operatorname{l.i.m.} X_n = X$$

and mean square convergence of random functions means

$$\lim_{\xi\to\xi_0} E\{|X(\omega, \xi) - X(\omega, \xi_0)|^2\} = 0 \iff \operatorname*{l.i.m.}_{\xi\to\xi_0} X(\omega, \xi) = X(\omega, \xi_0). \tag{2.2}$$

The connection between mean square convergence and the covariance function $\Gamma(\xi_1, \xi_2) = E\{X(\omega, \xi_1)X^*(\omega, \xi_2)\}$ (where it has been assumed that $EX(\omega, \xi) = 0$) is immediately clear from (2.2):

$$
\begin{aligned}
&E\{|X(\omega, \xi + h) - X(\omega, \xi)|^2\} \\
&\qquad = E\{[X(\omega, \xi + h) - X(\omega, \xi)][X^*(\omega, \xi + h) - X^*(\omega, \xi)]\} \\
&\qquad = 2\Gamma(0) - \Gamma(h) - \Gamma(-h) = 2 \operatorname{Re}[\Gamma(0) - \Gamma(h)].
\end{aligned}
$$

It follows that continuity of the covariance function at $\xi_1 = \xi_2$ in the ordinary sense is enough to guarantee the continuity of $X(\omega, \xi)$ in the mean square sense, and it is a general rule that mean square properties of a random function are implied by the corresponding ordinary calculus properties of its covariance function Γ. Thus for mean square properties to hold, $X(\omega, \xi)$ must have finite variance.

If stronger properties than mean square ones are desired, e.g., ordinary continuity or differentiability almost everywhere (or, in probability measure, "almost certainly") then we are faced with a somewhat more difficult problem, but one which can be resolved on the basis of the following criterion for almost certain convergence **[6]**: If

$$
E\{|X(\omega, \xi) - X(\omega, \xi_0)|^{p+1}\} < C|\xi - \xi_0|^{q+1} \qquad (p, q > 0), \tag{2.3}
$$

then

$$
X(\omega, \xi) \to X(\omega, \xi_0) \text{ almost certainly as } \xi \to \xi_0.
$$

It follows from this criterion that $X(\omega, \xi)$ is almost certainly continuous if it is once mean square differentiable, and that a sufficient condition for $X(\omega, \xi)$ to be n times almost certainly differentiable is that it be $(n + 1)$-times m. s. differentiable.

The mean square partial derivative $\partial X(\omega, \xi_1, \cdots, \xi_n)/\partial \xi_j$ of $X(\omega, \boldsymbol{\xi}_n)$ will exist at $\boldsymbol{\xi}_n^0$ if

$$
\lim_{h \to 0} E\left\{\left|\frac{X(\omega, \xi_1, \cdots, \xi_j + h, \cdots, \xi_n) - X(\omega, \boldsymbol{\xi}_n)}{h} - \frac{\partial X}{\partial \xi_j}\right|^2\right\} = 0,
$$

and similarly for higher derivatives. A n. a. s. c. for the mean square existence of the partial derivative of order p is that the corresponding ordinary derivative of order $2p$ of the covariance function of $X(\omega, \boldsymbol{\xi}_n)$ exist at the origin **[6]**.

Alternatively, if $X(\omega, \boldsymbol{\xi}_n)$ has a spectral representation

$$
X(\omega, \boldsymbol{\xi}_n) = \int_{R_n} e^{i\boldsymbol{\sigma} \boldsymbol{\xi}_n} dZ(\omega, \boldsymbol{\sigma}_n)
$$

(which will be defined more fully below) then a n. a. s. c. for the existence of a mean square derivative of order p is the existence of the multiple integral

$$
\int_{R_n} [D_{\boldsymbol{\xi}_n}^{2p} e^{i\boldsymbol{\sigma}_n \cdot \boldsymbol{\xi}_n}]_{\boldsymbol{\xi}_n = 0} \, dZ(\omega, \boldsymbol{\sigma}_n),
$$

where

$$
D_{\boldsymbol{\xi}_n}^{2p} = \prod_1^n \frac{\partial^{2p_i}}{\partial \xi_i^{2p_i}}, \qquad \sum_1^n p_i = p.
$$

Thus for our purposes the reduced wave equation can be taken as equivalent to the condition

$$E\left\{\left|\sum_1^3 \frac{\partial^2 U}{\partial x_i^2} + k^2(\mathbf{x})U(\mathbf{x})\right|^2\right\} = 0,$$

or, if we wish to deal with realizations $U(\mathbf{x}_3)$ that are almost certainly twice continuously differentiable, then $U(\mathbf{x}_3)$ must be three times mean square differentiable, and

$$E\{[\Delta U(\mathbf{x}) - (-k^2(\mathbf{x}+\mathbf{h}))U(\mathbf{x}+\mathbf{h})][\Delta U^*(\mathbf{x}) + k^{*2}(\mathbf{x}+\mathbf{h})U^*(\mathbf{x}+\mathbf{h})]\} \leq C|\mathbf{h}|^2$$

uniformly in $\mathbf{x}$.

We note that the rules for differentiation in ordinary calculus carry over to mean square differentiation, including the rule for partial and total differentiation of functions of functions, provided certain general conditions on mean square majorants are satisfied [**6**].

Stochastic Riemann, Stieltjes, and Lebesgue integrals of a random function are defined as the limits in mean square of the corresponding approximating sums, when they exist. If $X(\omega, \xi)$ is a regular random function with continuous covariance function, then apart from possible equivalences, the mean square indefinite integral

$$Y(\omega, \xi) = \int_a^\xi X(\omega, \eta)\, d\eta$$

exists and is mean square differentiable with mean square derivative $X(\omega, \xi)$. Further, if $\mathscr{Y}(\omega, \xi)$ is a primitive of $X(\omega, \xi)$ in mean square, then $Y(\omega, \xi) = \mathscr{Y}(\omega, \xi) - \mathscr{Y}(\omega, a)$ [**7**]. We thus have a basis for the formulation of mean square Volterra and Fredholm integral equations.

3. **Spectral representations.** The extension of the usual probability concepts associated with random variables to stochastic processes is accomplished via the probability field $(\Omega, \mathfrak{F}, P)$, where Ω is the space of all real functions, $\mathfrak{F}$ is a Borel field of sets generated in the usual way from the elementary events S in the space, and P is a probability measure defined on $\mathfrak{F}$. Any function $X(\omega, \xi) \in \Omega$ is called, for fixed ω, a *realization*, or sample function of the process, and is often written simply $X(\xi)$. However, the argument ω will often be omitted from the process itself in the sequel if the meaning is clear from the context. For any finite set $\xi_1, \cdots, \xi_n$ of parameter values of the process $\{X(\omega, \xi), \xi \in R\}$, the corresponding multivariate distribution F_n of the random variables $X(\omega, \xi_1), \cdots, X(\omega, \xi_n)$ is called a *finite-dimensional* distribution of the process and plays a fundamental role in the definition of a stochastic process after Kolmogoroff. According to Kolmogoroff's fundamental theorem [**8**; **9**], if the F_n satisfy the *consistency relations*

$$F_m(X(\omega, \xi_1), \cdots, X(\omega, \xi_m)) = F_n(X(\omega, \xi_1), \cdots, X(\omega, \xi_m), \infty, \cdots, \infty)$$
$$(m < n, \quad -\infty < \xi_i < \infty),$$

then the family $\{F_n\}$ of distribution functions uniquely determines the probabilities $P(S)$ of all events $S \in \mathfrak{F}$, and conversely any system $\{F_n\}$ of distribution functions defines a probability field $(\Omega, \mathfrak{F}, P)$, provided the distribution functions satisfy the consistency relations.

Kolmogoroff's theorem makes possible a mathematically rigorous treatment of stochastic processes, but, due to the complexity of the calculations involved in the general multivariate distribution, is not of much use in practice. If the structure of the process is of first interest, e.g., its continuity or measurability or that it satisfies a certain functional equation, then the process is best analyzed either as an abstract function by imbedding it in a suitable Banach space or else in terms of a spectral representation. The class of second order, or *regular*, functions that we have encountered above is such as to admit spectral representations, and in view of the suitability of regular functions for random analysis, this class of functions seems particularly well suited to our purposes. We shall therefore adopt this approach in the sequel except where explicitly stated to the contrary.

A general form of the spectral theorem for regular processes has been given by Blanc-Lapierre and Fortet **[10]**. This asserts the existence of the following spectral representation for the (not necessarily stationary) stochastic process $X(\boldsymbol{\xi})$:

$$X(\boldsymbol{\xi}) = \int_{R_3} e^{i\boldsymbol{\xi}\cdot\boldsymbol{\eta}}\, dZ(\boldsymbol{\eta}),$$

where $Z(\boldsymbol{\eta})$ is a stochastic process in $\boldsymbol{\eta}$-space such that the covariance has the representation

$$\operatorname{cov}\,[X(\boldsymbol{\xi}), X(\boldsymbol{\xi}')] = \int_{R_3\times R_3} e^{i(\boldsymbol{\xi}\cdot\boldsymbol{\eta}-\boldsymbol{\xi}'\cdot\boldsymbol{\eta}')}\, d^2\Sigma(\boldsymbol{\eta}, \boldsymbol{\eta}'),$$

and

$$d^2\Sigma(\boldsymbol{\eta}, \boldsymbol{\eta}') = E\{dZ(\boldsymbol{\eta})\, dZ^*(\boldsymbol{\eta}')\}$$

is such as to be absolutely integrable over all of $(\boldsymbol{\eta}, \boldsymbol{\eta}')$-space. These conditions require that the covariance function be uniformly bounded and continuous over the product space $R_3 \times R_3$, which in turn implies that $X(\boldsymbol{\xi})$ is continuous in mean square in the whole space.

For stationary second order processes, the spectral representation is somewhat more straightforward, and we have that

$$Y(\boldsymbol{\xi}) = \int_{R_3} e^{i\boldsymbol{\xi}\cdot\boldsymbol{\eta}}\, d\Pi(\boldsymbol{\eta})$$

where $\Pi(\boldsymbol{\eta})$ is a stochastic process with orthogonal increments.

Further, apart possibly from a set of realizations of probability zero, the $\Pi(\boldsymbol{\eta})$ process is determined *uniquely* [9] in terms of the $Y(\boldsymbol{\xi})$ process by the equation

$$\tfrac{1}{2}[\Pi(\boldsymbol{\eta}_2+) + \Pi(\boldsymbol{\eta}_2-)] - \tfrac{1}{2}[\Pi(\boldsymbol{\eta}_1+) + \Pi(\boldsymbol{\eta}_1-)]$$

$$= (2\pi)^{-3}\, \underset{A\to\infty}{\text{l.i.m.}} \iiint_{-A}^{A} \prod_{j=1}^{3} \frac{e^{-i\eta_{2j}\xi_j} - e^{-i\eta_{1j}\xi_j}}{-i\xi_j}\, Y(\boldsymbol{\xi})\, d\boldsymbol{\xi}$$

where η_1 and η_2 range over all of $(\eta_1 \times \eta_2)$-space. This gives us a uniqueness theorem (up to a set of measure zero) for what are essentially Fourier transforms of second order stationary random functions.

Silverman **[11]** has drawn a distinction between what he calls a globally stationary process, meaning one which is stationary in the wide sense over the whole space, and locally stationary processes, which he defines as follows: A centered random process $X(\omega, \xi)$, defined for all real ξ, is said to be locally stationary (wide sense) if its covariance $\Gamma(\xi_1, \xi_2) = E\{X(\omega, \xi_1)X^*(\omega, \xi_2)\}$ can be put in the form

$$\Gamma(\xi_1, \xi_2) = \Gamma_1(\tfrac{1}{2}(\xi_1 + \xi_2))\Gamma_2(\xi_1 - \xi_2),$$

where Γ_1 is a non-negative function and Γ_2 is a stationary covariance function. Γ_1 describes the global structure of $X(\xi)$, i.e., its behavior over the whole real line, while Γ_2, which is invariant under translation, can be thought of as describing its local behavior. The Fourier transform of $\Gamma(\xi_1, \xi_2)$, i.e., the corresponding two-dimensional spectral density, is factorable into a non-negative function and the Fourier transform of a non-negative function. The harmonic analysis of locally stationary random processes appears more suited to physical problems than that for globally stationary processes, since the Fourier transform of the latter will not exist in general except as a stochastic integral. The Fourier transform of a locally stationary process can also be interpreted as an ensemble of integrals over distinct realizations. For globally stationary processes the variance $\Gamma(\xi, \xi)$ is a positive constant and the Fourier transform over the whole line of the sample function $X(\omega_0, \xi)$ will therefore fail to exist, with probability one. Locally stationary processes will probably prove to be of considerable significance in physical applications since their use removes the unphysical assumption that, with probability one, the sample functions (realizations) possess infinite energy. Locally stationary processes (though not under this name) have recently been studied in connection with spectral estimation for turbulence theory by Kampé de Fériet and Frenkiel **[12]**, and this type of process certainly deserves further investigation.

The following table of covariance functions and associated spectra is taken from Wheelon **[13]**:

TABLE I

Covariance Function	Spectral Density
$\exp\{-\rho/\rho_0\}$	$8\pi \dfrac{E\{X^2\}\rho_0^3}{[1 + k^2\rho_0^2]^2}$
$\exp\{-\rho^2/\rho_0^2\}$	$\pi^{3/2}E\{X^2\}\rho_0^3 \exp\{-1/4\, k^2\rho_0^2\}$
$(\rho/\rho_0)K_1(\rho/\rho_0)$	$6\pi^2 \dfrac{E\{X^2\}\rho_0^3}{[1 + k^2\rho_0^2]^{5/2}}$
$[1 + (\rho/\rho_0)^2]^{-2}$	$\pi^2E\{X^2\}\rho_0^3 \exp\{-k\rho_0\}$.

These are the covariance functions that have been assumed in theoretical analyses of turbulent scattering of electromagnetic waves. They have no basis in any physical theory of turbulence,[2] but do aid materially in evaluating the integrals that arise (each one is a well-known kernel). The question of the proper covariance function for an atmospheric refractive index subject to turbulent fluctuations is an important problem that still remains open.

Bugnolo **[14]** has given a heuristic discussion of the question of stationarity when the refractive index of the medium constitutes a stochastic process in space and time. He concludes that, in the application to turbulent scattering in the troposphere at least, the refractive index process must be nonstationary, and therefore proposes a redefinition of the expected value of the process that would be consistent with the result for a deterministic inhomogeneous medium.

4. **Ergodicity.** Ergodic properties refer to those features of a stochastic process which are equivalent under either of the operations: mathematical expectation or sample average. Thus ergodic theorems consitute the basis for statistical estimation and inference, for in any given practical situation we observe only realizations, and the ergodic property constitutes the bridge connecting theory and experiment. The principal area in which the development of ergodic theory has taken place is that of statistical mechanics. There the statement of ergodic theorems generally involves the shift operator and the trajectories in phase space. The situation is somewhat different with respect to the reduced wave equation, which represents a different type of flow.

An *invariant set* is a measurable set E such that $TE = E$; an *invariant function*, a measurable function such that $f(x) = f(T_\xi x)$ almost everywhere in Ω. The transformation T_ξ is said to be metrically transitive if the only invariant sets are the null set and the whole space Ω (in the case of functions, if the only functions invariant with probability one are constants). In statistical mechanics metric transitivity means that with probability one a trajectory covers almost all of Ω. For stationary stochastic processes metric transitivity is entirely equivalent to the ergodic property, and may be expressed by the Birkhoff–Khintchine theorem: For a measurable, stationary, metrically transitive, continuous parameter process $\{X(\omega, t)\}$, $-\infty < t < \infty$, with finite expectation,

$$EX = \lim_{T\to\infty} \frac{1}{2T} \int_{-T}^{T} X(t)\, dt$$

with probability 1. That a stationary process is not necessarily ergodic is shown by the discrete-parameter process

$$X_t = a\varepsilon + Y_t, \qquad -\infty < t < \infty$$

[2] And some could only correspond to a Brownian motion, not to a turbulent motion in a continuous medium.

where a is a nonrandom constant, the Y_t are independent and equidistributed and are further independent of the random variable ε. (In this example $n^{-1}\sum_1^n \rho_\nu = a^2/(1 + a^2) \neq 0$ as $n \to \infty$, which violates a n. a. s. c. for ergodicity.) This limiting condition on the correlation function is equivalent to the requirement that there be no line of frequency zero in the spectrum of the process, or alternatively that the spectrum of the process be continuous at $\omega = 0$.

Though ergodicity has received considerable investigation over the past few years [**7**; **15**], usable criteria for ergodic properties are still somewhat limited in scope. A particularly useful result, but one which applies only to Gaussian processes, is due to G. Maruyama [**16**]. It states that a Gaussian process is ergodic if and only if its spectral function is continuous. This theorem has been applied by Kampé de Fériet [**1**] to the important case of random solutions of $u_{tt} = c^2 u_{xx}$ in a homogeneous medium.

The ergodic properties of wave propagation through a random time-varying medium has been investigated by T. J. Skinner [**17**], who showed that for the solution of (1.1) the time-varying medium can be replaced by a time independent probability space provided that on the boundary of the time-varying medium the scalar field $u(\mathbf{x}, t) = A(\mathbf{x}, t) \exp\{i\phi(\mathbf{x}, t) + i\omega t\}$ has the properties:

(i) $A(\mathbf{x}, t)$ and $\phi(\mathbf{x}, t)$ are ergodic;

(ii) the absolute value of any frequency component of either $A(\mathbf{x}, t) \sin \phi(\mathbf{x}, t)$ or $A(\mathbf{x}, t) \cos \phi(\mathbf{x}, t)$ is less than $\omega/(2\pi)$, the mean frequency of the source; and

(iii) $u(\mathbf{x}, t)$ is quasi-monochromatic.

5. **The radiation problem for an unbounded random medium.** After this prefatory material on the probabilistic aspects of the problem we come now to the particular functional equations involved. Keller [**18**] has formulated the radiation problem corresponding to equation (1.1) as follows:

$$(5.1)\qquad \begin{cases} \Delta U + k_0^2 n^2(\mathbf{x}) U(\mathbf{x}) = -4\pi\delta(\mathbf{x}) \\ \lim\limits_{|\mathbf{x}|\to\infty} |\mathbf{x}| \left(\dfrac{\partial U}{\partial |\mathbf{x}|} + i k_0 n(\mathbf{x}) U \right) = 0 \end{cases}$$

where k_0 is the free-space wave number and $k^2(\mathbf{x}) = k_0^2 n^2(\mathbf{x})$, $n(\mathbf{x})$ being an as yet unspecified random function of position $\mathbf{x}$. The second of equations (5.1) is the Sommerfeld radiation condition which guarantees that the waves from the unit source $\delta(\mathbf{x})$ at the origin will be outgoing at infinity.

One can proceed in several different directions from the general problem (5.1). As mentioned in the introduction, the following approaches have been considered: (i) perturbation theory [**18**; **19**]; (ii) transformation to an equivalent spatial Riccati equation via the so-called Rytov transformation [**20**; **21**]; (iii) the integral equation formulation [**19**; **21**; **22**]; and (iv) what is in fact a special case of (i) or (iii), the Born approximation [**13**; **23**]. The simplification resulting from taking the reduced wave equation to be one-dimensional

$$(5.2)\qquad Y'' + k^2(x) Y(x) = 0,$$

i.e., the transmission line equation, has also received considerable attention [**24**; **25**]. One can also more or less bypass equations (5.1) in the form given and instead adopt what is essentially a transport equation approach, e.g., ray-tracing [**18**; **20**; **26**; **27**; **28**]; invariant imbedding [**29**]; and diffusion of the electromagnetic field by multiple scattering [**30**]. In the transport context attention is directed not so much at the field quantity itself but at some function associated with it, e.g., spectral-density, ray paths, transition probabilities, or the like [**31**; **32**; **33**]. In other connections, notably quantum field theory, still other approaches have been adopted, in particular integration in functional spaces (the so-called Wiener integrals) [**34**; **35**; **36**], and these techniques, properly generalized appear to have promise for problem (5.1). A start in a similar direction has been made by Kraichnan [**37**].

6. **Perturbation theory.** If the refractive index of the medium exhibits only small departures from its equilibrium value, then a perturbation expansion in terms of a small parameter ε may be adopted:

$$n(\mathbf{x}) = 1 + \sum_{1}^{\infty} \nu_m(\mathbf{x})\varepsilon^m, \qquad \varepsilon \ll 1. \tag{6.1}$$

If the field variable also has a perturbation expansion

$$U = \sum_{0}^{\infty} U_m(\mathbf{x})\varepsilon^m, \qquad \varepsilon \ll 1 \tag{6.2}$$

then (5.1) gives rise to the following ascending hierarchy of radiation problems (in powers of ε):

$$\begin{cases} \Delta U_0 + k_0^2 U_0 = -4\pi\delta(\mathbf{x}), \\ \lim\limits_{|\mathbf{x}|\to\infty} |\mathbf{x}|\left(\dfrac{\partial U_0}{\partial|\mathbf{x}|} + ik_0 U_0\right) = 0, \end{cases} \tag{6.3}$$

$$\begin{cases} \Delta U_1 + k_0^2 U_1 = -2k_0^2 \nu_1 U_0(\mathbf{x}), \\ \lim\limits_{|\mathbf{x}|\to\infty} |\mathbf{x}|\left(\dfrac{\partial U_1}{\partial|\mathbf{x}|} + ik_0 U_1 + 2ik_0\nu_1 U_0\right) = 0, \end{cases} \tag{6.4}$$

$$\vdots \qquad\qquad \vdots$$

$$\begin{cases} \Delta U_p + k_0^2 U_p = -k_0^2 \sum\limits_{q=1}^{p} U_{p-q} \sum\limits_{m=0}^{q} \nu_{q-m}\nu_m, \\ \lim\limits_{|\mathbf{x}|\to\infty} |\mathbf{x}|\left(\dfrac{\partial U_p}{\partial|\mathbf{x}|} + ik_0 U_p + ik_0 \sum\limits_{m=0}^{p-1} \nu_{p-m} U_m\right) = 0. \end{cases} \tag{6.5}$$

The solution of (6.3) is given by the well-known expression for an outgoing spherical wave

$$U_0(\mathbf{x}) = \frac{e^{-ik_0|\mathbf{x}|}}{|\mathbf{x}|}. \tag{6.6}$$

Then Green's theorem yields the solution of (6.4) immediately in the form

$$U_1(\mathbf{x}) = 2k_0^2 \int_{R_3} \frac{e^{-ik_0|\mathbf{x}-\mathbf{x}'|}}{|\mathbf{x}-\mathbf{x}'|} \nu_1(\mathbf{x}')U_0(\mathbf{x}')\, d\mathbf{x}' \tag{6.7}$$

and in general

$$U_p(\mathbf{x}) = k_0^2 \int_{R_3} \frac{e^{-ik_0|\mathbf{x}-\mathbf{x}'|}}{|\mathbf{x}-\mathbf{x}'|} \sum_{q=1}^{p} \left(\sum_{m=0}^{q} \nu_{q-m}(\mathbf{x}')\nu_m(\mathbf{x}') \right) U_{p-q}(\mathbf{x}')\, d\mathbf{x}'. \tag{6.8}$$

(In particular, for $p = 2$,

$$U_2(\mathbf{x}) = k_0^2 \int_{R_3} \frac{e^{-ik_0|\mathbf{x}-\mathbf{x}'|}}{|\mathbf{x}-\mathbf{x}'|} ([\nu_1^2 + 2\nu_2]U_0(\mathbf{x}') + 2\nu_1(\mathbf{x}')U_1(\mathbf{x}'))\, d\mathbf{x}'.) \tag{6.9}$$

Each of the above integral formulas is to be understood in the mean square sense, as is also the convergence of the perturbation series (6.2). We recognize in formula (6.7) the so-called Born approximation to the solution of (5.1). This formula has served as the basis for nearly all the past work on turbulent scattering [**13**]. A good summary of the mathematical rationale of this approach may be found in [**18**]. The formulas (6.7)–(6.9) have proved particularly useful in determining the moments of $U(\mathbf{x})$. Keller's principal result is that to terms $O(\varepsilon^2)$ the mean value of the field quantity U propagates as if it traveled through an "effective" medium with complex refractive index

$$n_{\text{eff.}} = \left\{ 1 + \varepsilon^2 \left(1 - 2ik_0 \int_0^\infty [e^{2ik_0 r} - 1]\Gamma(r)\, dr \right) \right\}^{1/2}.$$

A similar conclusion has also been obtained by Ament [**19**].

7. **Born approximation.** The first approximation in the iteration-perturbation solution of the scattering integral equation is called the Born approximation. In the electromagnetic case the Born approximation expresses the total field $\mathbf{E}(\mathbf{x})$ (incident plus scattered) in terms of the unperturbed incident field $\mathbf{E}_0(\mathbf{x})$ as follows [**13**]:

$$\mathbf{E}(\mathbf{x}) = \mathbf{E}_0(\mathbf{x}) - k_0^2 \int d^3\mathbf{x}' G_0(\mathbf{x}, \mathbf{x}')\Delta\varepsilon(\mathbf{x}'; t)\mathbf{E}_0(\mathbf{x}') \sin \mathrm{X} \tag{7.1}$$

where V is the randomly inhomogeneous volume comprised in the intersection of transmitter and receiver beams, X is the angle between $\mathbf{E}_0$ and the Poynting vector of the scattered wave, and $\Delta\varepsilon$ is the random residual for the "dielectric constant," or permittivity. Equation (7.1) or the equivalent mean square power equation

$$\mathbf{P}_{\text{rec.}} = A_r E\{|\mathbf{E}_{sc}|^2\} = P_t(\pi/\lambda_0)^2 \int\int_{V\times V} \frac{e^{-i(k_0-k_1)(|\mathbf{x}-\mathbf{x}'|-|\mathbf{x}-\mathbf{x}''|)}}{|\mathbf{x}-\mathbf{x}'|^2|\mathbf{x}-\mathbf{x}''|^2} \sqrt{G_r G_t G_r' G_t'} \cdot \Gamma_{\Delta\epsilon}(\mathbf{x}'-\mathbf{x}'')\, d\mathbf{x}'\, d\mathbf{x}'' \tag{7.2}$$

where A_r is the capture cross-section of the receiving antenna, $\mathbf{E}_{sc} \equiv \mathbf{E} - \mathbf{E}_0$, $\mathbf{P}_t$ is the transmitted power, with associated gain function $G_t(\theta, \varphi)$, V the common volume of the two beams, $\mathbf{k}_1$ is the propagation vector of the transmitted wave, and $G_r(\theta, \varphi)$ is the gain function of the receiving antenna. The polarization factors involving sin X have been dropped. In the far-zone approximation (7.2) becomes

$$\mathbf{P}_{\text{rec.}} = \mathbf{P}_t(\pi/\lambda_0)^2 \int_V \frac{G_r G_t}{|\mathbf{x}-\mathbf{x}'|^2|\mathbf{x}-\mathbf{x}''|^2} \int_V e^{i(\mathbf{k}_1-\mathbf{k}_2)\cdot\boldsymbol{\rho}}\, \Gamma_{\Delta\epsilon}(\boldsymbol{\rho})\, d\boldsymbol{\rho}\, d\mathbf{r} \tag{7.3}$$

where

$$\boldsymbol{\rho} = \mathbf{x}' - \mathbf{x}'', \qquad \mathbf{r} = \mathbf{x}' + \mathbf{x}''.$$

Equation (7.3) has provided the basis for nearly all theoretical analyses of "forward scatter" of radio waves in a turbulent atmosphere. We note that it is subject to the following hypotheses: (i) multiple scattering can be neglected, (ii) the scattering is isotropic, and (iii) the random variations of the permittivity are small and slowly varying in time. The first and second hypotheses appear somewhat unrealistic when applied to the terrestrial atmosphere and the third can apply only in a "cool" medium wherein particle energies are low (for example, stellar atmospheres are excluded, as are also "hot" plasmas in general). Attempts to remove the restriction to single-scattering phenomena have been made by Wheelon [**13**], who introduced the second approximation to the iteration-perturbation solution, and by Ament, who employed ray-tracing between successive scatterers [**26**], Bremmer, who based his analysis on diffusion of the joint probability density for position and direction of a unit of radiation energy [**30**], and Bugnolo, who considered diffusion of the spectral density function of an electromagnetic field under multiple scattering [**31**]. The latter approaches will receive further discussion below under the head of transport equations.

Generalization of (7.2) to the case where the refractive index is anisotropic (as, e.g., when the scale of turbulence is different in the horizontal and vertical directions) has been carried out by Staras [**38**] and Booker [**39**]. The covariance function is simply assumed to be of the form

$$\Gamma(\mathbf{x} - \mathbf{x}') = \Gamma\left(\frac{x - x'}{l_0}, \frac{y - y'}{l_0}, \frac{z - z'}{l_v}\right), \tag{7.4}$$

and its introduction into (7.2) makes the problem anisotropic. It presently appears that partial reflections from flattened atmospheric cells ("platelets" or "feuillets") constitute the most tenable explanation of tropospheric "forward scatter" phenomena, and the distinction between the "feuillet" approach and that embodied in (7.4) is worthy of note. In the former it is not scattering from turbulence but rather partial reflections from stable horizontal laminae having gradients of refractive index that are large relative to the turbulent medium in which they are imbedded that give rise to "forward scatter." In principle the anisotropic scattering theory could be extended to take account of "feuillet" phenomena, but such a generalization is lacking at present. The "feuillet" approach has received considerable attention, starting with the pioneering work of Friis, Crawford and Hogg [**40**]. The interested reader is referred to the report by Bremmer [**30**] for an excellent description of the investigations in this field.

Other work on scattering by an anisotropic randomly inhomogeneous medium has been carried out by Bowhill [**41**; **42**] and Yeh [**43**]. Bowhill considered the case of a medium with stationary Gaussian inhomogeneities having different scales along three space axes and the effect this would work upon the usual model of a succession of equivalent diffracting screens. He attempted to employ for this purpose the theory of emergent angular power spectrum but found that the diffractive phase shifts as the wave passed from one inhomogeneity to a neighboring, correlated one rendered such an approach invalid. However, spatial spectra and correlation functions for the emergent wave were determined for anisotropic "random shallow phase screens" for both vertical and oblique incidence.

Yeh's analysis [**43**], based upon Karavainikov's perturbation theory for a spherical wave in an inhomogeneous medium,[3] prescribes the phase and amplitude fluctuations of a spherical wave propagating through a medium containing anisotropic (elongated) inhomogeneities. The medium is assumed to have a refractive index whose correlation function is ellipsoidally symmetric and of Gaussian form in the arguments. Yeh thus obtains first order approximations to the covariance functions for phase fluctuations and logarithmic amplitude.

8. **Formulation as a random integral equation.** Let us consider instead of (5.1) the corresponding radiation problem for the inhomogeneous reduced wave equation

$$\begin{cases} \Delta V(\mathbf{x}) + k_0^2 n^2(\mathbf{x}) V(\mathbf{x}) = -f(\mathbf{x}), \\ \lim_{|\mathbf{x}|\to\infty} |\mathbf{x}| \left(\dfrac{\partial V}{\partial |\mathbf{x}|} + i k_0 V(\mathbf{x}) \right) = 0. \end{cases} \tag{8.1}$$

The forcing function $f(\mathbf{x})$, arising perhaps from a current distribution or acoustic pressure, is assumed to vanish everywhere outside some finite region of space, and the form of the radiation condition in (8.1) requires that $n(\mathbf{x}) \to 1$ as $|\mathbf{x}| \to \infty$. Green's theorem applied to the pair of functions $V(\mathbf{x})$ and the free-space Green's function $G_0(\mathbf{x}, \mathbf{x}') = \exp\{-ik_0|\mathbf{x} - \mathbf{x}'|\}/|\mathbf{x} - \mathbf{x}'|$ yields the integral equation

$$V(\mathbf{x}) - \Lambda_0 \int_{R_3} G_0(\mathbf{x}, \mathbf{x}') \nu(\mathbf{x}') V(\mathbf{x}') \, d\mathbf{x}' = F(\mathbf{x}) \tag{8.2}$$

[3] Very similar in approach to Rytov's method leading to equations (9.19).

where

$$F(\mathbf{x}) = \frac{1}{4\pi} \int G_0(\mathbf{x}, \mathbf{x}') f(\mathbf{x}')\, d\mathbf{x}', \tag{8.3}$$

$$\nu(\mathbf{x}) = n^2(\mathbf{x}) - 1, \tag{8.4}$$

$$\Lambda_0 = k_0^2/(4\pi).$$

It is clear from (8.2)–(8.4) that $V(\mathbf{x})$ is given by the usual integral formula over the region where $n(\mathbf{x}) = 1$. Let D denote the region where $\nu(\mathbf{x})$ is different from zero (the compact support of ν). In view of the radiation condition, D must be finite.[4] Consider then the integral equation for $\mathbf{x}, \mathbf{x}' \in D$:

$$V(\mathbf{x}) - \Lambda_0 \int_D G_0(\mathbf{x}, \mathbf{x}') \nu(\mathbf{x}') V(\mathbf{x}')\, d\mathbf{x}' = F(\mathbf{x}). \tag{8.5}$$

Equation (8.5) differs from a standard Fredholm integral in two respects: (i) in general the kernel $G_0(\mathbf{x}, \mathbf{x}')\nu(\mathbf{x}')$ is not in L_2, and (ii) $\nu(\mathbf{x})$ and $V(\mathbf{x})$ are random functions. Provided $\nu(\mathbf{x})$ is uniformly bounded with probability 1 in D (a not unrealistic hypothesis), the first difficulty can be removed in the usual way via introduction of the iterated kernel:[5]

$$K_2(\mathbf{x}, \mathbf{x}'') = \int_D G_0(\mathbf{x}, \mathbf{x}') \nu(\mathbf{x}') G_0(\mathbf{x}', \mathbf{x}'') \nu(\mathbf{x}'')\, d\mathbf{x}'. \tag{8.6}$$

Thus the following integral equation is nonsingular with probability 1 and is equivalent to (8.5):

$$V(x) - \Lambda_0^2 \int_D K_2(\mathbf{x}, \mathbf{x}') V(\mathbf{x}')\, d\mathbf{x}' = H(\mathbf{x}), \tag{8.7}$$

where

$$H(\mathbf{x}) = F(\mathbf{x}) + \Lambda_0 \int_D G_0(\mathbf{x}, \mathbf{x}') \nu(\mathbf{x}') F(\mathbf{x}')\, d\mathbf{x}'. \tag{8.8}$$

---

[4] However, so large that the radiation condition permits one to neglect the surface integral that would otherwise appear in (8.5). This assumes that $\nu(\mathbf{x})$ decreases smoothly near the boundary of D.

[5] Since $\Pr\{|\nu(\mathbf{x})| < m;\ \mathbf{x} \in D\} = 1$, we have with probability 1 that

$$\int_{D\times D} |G_0(\mathbf{x}, \mathbf{x}')\nu(\mathbf{x}')G_0(\mathbf{x}', \mathbf{x}'')\nu(\mathbf{x}'')V(\mathbf{x}'')|\, d\mathbf{x}'\, d\mathbf{x}'' \leqq m^2 \int_{D\times D} |G_0(\mathbf{x}, \mathbf{x}')G_0(\mathbf{x}', \mathbf{x}'')V(\mathbf{x}'')|\, d\mathbf{x}'\, d\mathbf{x}''.$$

Now $\Pr\{|V| > \eta\} \leqq \eta^{-2}E|V|^2$, and since V is regular in D by hypothesis, $\Pr\{|V| > \eta\}$ can be made arbitrarily small by taking η large enough. Therefore almost everywhere in D, with probability 1,

$$\int_{D\times D} |G_0(\mathbf{x}, \mathbf{x}')\nu(\mathbf{x}')G_0(\mathbf{x}', \mathbf{x}'')\nu(\mathbf{x}'')V(\mathbf{x}'')|\, d\mathbf{x}'\, d\mathbf{x}'' \leqq m^2 V_0 \int_{D\times D} |G_0(\mathbf{x}, \mathbf{x}')G_0(\mathbf{x}', \mathbf{x}'')|\, d\mathbf{x}'\, d\mathbf{x}''.$$

The integral on the right with respect to $d\mathbf{x}'$ is known to be in L_2. Therefore the original integrand is absolutely integrable with probability 1, and Fubini's theorem permits the inversion of integration leading to (8.6) almost everywhere in the probability space.

The probabilistic features of regularity (second order random functions) and boundedness with probability 1 enter strongly in the above argument. These hypotheses in essence enable us to refer the problem to the corresponding deterministic one. The kernel in (8.7) is not symmetric as it stands, but can easily be made so by multiplying (8.7) through by $\sqrt{\nu(\mathbf{x})}$ and introducing the new functions

$$W(\mathbf{x}) = \sqrt{\nu(\mathbf{x})}V(\mathbf{x}), \qquad L(\mathbf{x}) = \sqrt{\nu(\mathbf{x})}H(\mathbf{x}) \tag{8.9}$$

and new kernel

$$K(\mathbf{x}, \mathbf{x}') = \sqrt{\nu(\mathbf{x})}K_2(\mathbf{x}, \mathbf{x}')\sqrt{\nu(\mathbf{x}')}. \tag{8.10}$$

The Fredholm theory applies in any case,[6] and a regular solution can be determined by the usual techniques for Fredholm equations.

For the case that the refractive index constitutes a Gaussian random process, and so will be unbounded, Hoffman **[21]** has shown that (8.1) is solved by the mean square convergent Neumann series

$$V(\mathbf{x}) = F(\mathbf{x}) + \sum_{m=1}^{\infty} \Lambda_0^m \int_D d\mathbf{x}_1 G_0(\mathbf{x}, \mathbf{x}_1)\nu(\mathbf{x}_1) \int_D d\mathbf{x}_2 \cdots \int_D G_0(\mathbf{x}_{m-1}, \mathbf{x}_m)\nu(\mathbf{x}_m)F(\mathbf{x}_m)\, d\mathbf{x}_m$$

provided (i) the Neumann series solution for the corresponding deterministic equation

$$V(\mathbf{x}) = F(\mathbf{x}) + \Lambda_0 \int_D G_0(\mathbf{x}, \mathbf{x}')V(\mathbf{x}')\, d\mathbf{x}'$$

converges in the ordinary way; and (ii)

$$|\Lambda_0 \sqrt{\operatorname{var} \nu}\, B| < 1.$$

where

$$\left| \int_D G_0(\mathbf{x}, \mathbf{x}')F(\mathbf{x}')\, d\mathbf{x}' \right|^2 < B.$$

9. **Equivalent spatial Riccati equation.** The so-called Rytov transformation **[20]**

$$U(\mathbf{x}) = A_0 \exp\{-ik_0\psi(\mathbf{x})\} \tag{9.1}$$

takes (1.1) into the nonlinear differential equation

$$(\nabla\psi)^2 + \frac{i}{k_0}\nabla^2\psi = n^2(\mathbf{x}), \tag{9.2}$$

where

$$\psi(\mathbf{x}) = S(\mathbf{x}) + \frac{i}{k_0}\ln[A(\mathbf{x})/A_0], \tag{9.3}$$

$A(\mathbf{x})$ and $S(\mathbf{x})$ being the "amplitude" and "phase" associated with representation of $U(\mathbf{x})$ as

$$U(\mathbf{x}) = A(\mathbf{x}) \exp\{-ik_0 S(\mathbf{x})\}. \tag{9.4}$$

[6] Although the Hilbert–Schmidt theory may not, since the kernel is not necessarily definite.

The following approach is suggested by an analysis of Burgers **[44]** for the one-dimensional case. Let

$$\boldsymbol{\Theta} = \nabla\psi. \tag{9.5}$$

Equation (9.2) then becomes the three-dimensional analogue of the Riccati equation:

$$\nabla \cdot \boldsymbol{\Theta} - ik_0\Theta^2 = -ik_0 n^2(\mathbf{x}). \tag{9.6}$$

If we now introduce the asymptotic expansions

$$n^2(\mathbf{x}) = En^2(\mathbf{x}) + \sum_1^\infty \nu_m(\mathbf{x})\varepsilon^m,$$

and

$$\boldsymbol{\Theta}(\mathbf{x}) \sim \sum_{m=0}^\infty \boldsymbol{\Theta}_m(\mathbf{x})\varepsilon^m,$$

the following system of equations results

$$\begin{cases} \dfrac{i}{k_0}\nabla \cdot \boldsymbol{\Theta}_0 + \Theta_0^2 = En^2(\mathbf{x}), \\ \dfrac{i}{k_0}\nabla \cdot \boldsymbol{\Theta}_m + \sum\limits_{p=0}^{m} \boldsymbol{\Theta}_p \cdot \boldsymbol{\Theta}_{m-p} = \nu_m(\mathbf{x}) \qquad (m = 1, 2, \cdots). \end{cases} \tag{9.7}$$

Provided $\nabla\psi_0 = \boldsymbol{\Theta}_0$, which corresponds to the case of a refractive index whose deterministic component is everywhere constant, we may introduce the spectral representations

$$\nu_1(\mathbf{x}) = \int_{R_3} e^{i\boldsymbol{\alpha}\cdot\mathbf{x}}\, d\eta_1(\boldsymbol{\alpha}) \quad \text{and} \quad \boldsymbol{\Theta}_1(\mathbf{x}) = \int_{R_3} e^{i\boldsymbol{\alpha}\cdot\mathbf{x}}\boldsymbol{\gamma}_1(\boldsymbol{\alpha})\, d\xi_1(\boldsymbol{\alpha}),$$

where $\eta_1(\boldsymbol{\alpha})$ and $\xi_1(\boldsymbol{\alpha})$ are processes with orthogonal increments and $\boldsymbol{\gamma}_1(\boldsymbol{\alpha})$ is a sure vector function of $\boldsymbol{\alpha}$. The Fourier transform of (9.7) with $m = 1$ then yields the following relation **[21]** connecting the spectral density functions of the refractive index and gradient of generalized phase function $\boldsymbol{\Theta}$:

$$d\xi_1(\boldsymbol{\alpha}) = k_0 \frac{d\eta_1(\boldsymbol{\alpha})}{2k_0\boldsymbol{\Theta}_0 \cdot \boldsymbol{\gamma}_1(\boldsymbol{\alpha}) - \boldsymbol{\alpha} \cdot \boldsymbol{\gamma}_1(\boldsymbol{\alpha})}. \tag{9.8}$$

The covariance function of $\boldsymbol{\Theta}_1$ is then given by the following formula, provided the Fourier transform exists:

$$\Gamma_1(\mathbf{x} - \mathbf{x}') = k_0^2 \int_{R_3(\boldsymbol{\alpha})} e^{i(\mathbf{x}-\mathbf{x}')\cdot\boldsymbol{\alpha}} \frac{d\sigma_1(\boldsymbol{\alpha})}{[2k_0\boldsymbol{\Theta}_0 \cdot \boldsymbol{\gamma}_1(\boldsymbol{\alpha}) - \boldsymbol{\alpha} \cdot \boldsymbol{\gamma}_1(\boldsymbol{\alpha})]^2} \tag{9.9}$$

where

$$d\sigma_1(\boldsymbol{\alpha}) = E\{d\eta_1(\boldsymbol{\alpha})\, d\eta_1^*(\boldsymbol{\alpha}')\}.$$

For nonconstant $\nabla\psi_0$, a similar approach **[21]** leads to the following integral equation connecting the respective spectral density functions

$$(\boldsymbol{\alpha}\cdot\boldsymbol{\alpha})d\xi(\boldsymbol{\alpha}) = k_0^2\int_{R_3(\beta)}\{m(\boldsymbol{\alpha}-\boldsymbol{\beta}) + d\eta(\boldsymbol{\alpha}-\boldsymbol{\beta})\}\,d\xi(\boldsymbol{\beta}), \tag{9.10}$$

where $m(\boldsymbol{\alpha})$ is the Fourier transform of $En^2(\mathbf{x})$.

The above approach has the virtue of making full use of the probabilistic structure of the process but perhaps falls short in terms of analytical results. Following Rytov **[20]**, we introduce the deviation $\Psi = \psi - \psi_0$ of the solution of (9.2) from that of the deterministic equation (first of (9.7)), and obtain the differential equation

$$\nabla(\Psi + 2\psi_0)\cdot\nabla\Psi + \frac{i}{k_0}\,\Delta\Psi = \nu(\mathbf{x}) \equiv n^2(\mathbf{x}) - En^2(\mathbf{x}). \tag{9.11}$$

Provided

$$E|\nabla\Psi|^2 \ll 4, \qquad En^2 = 1, \qquad \psi_0 = x, \tag{9.12}$$

(9.11) reduces to

$$2\frac{\partial\Psi}{\partial x} + \frac{i}{k_0}\,\Delta\Psi = \nu(\mathbf{x}). \tag{9.13}$$

These conditions correspond physically to a weakly inhomogeneous medium. If now (9.13) has a solution of the form

$$\Psi(\mathbf{x}) = e^{+ik_0x}w(\mathbf{x}), \tag{9.14}$$

then w will satisfy the differential equation

$$\Delta_3 w + k_0^2 w = -ik_0\, e^{-ik_0x}\nu(\mathbf{x}) = -4\pi f(\mathbf{x}). \tag{9.15}$$

Equation (9.14) has the well-known solution

$$w(\mathbf{x}) = \int_{R_3}\frac{e^{-ik_0|\mathbf{x}-\mathbf{x}'|}}{|\mathbf{x}-\mathbf{x}'|}\,f(\mathbf{x}')\,d\mathbf{x}', \tag{9.16}$$

which, expressed in terms of Ψ, reads as follows

$$\Psi(\mathbf{x}) = \frac{i}{2\lambda_0}\int_{R_3}\frac{e^{-ik_0(R-[x-x'])}}{R}\,\nu(\mathbf{x}')\,d\mathbf{x}'. \tag{9.17}$$

The first of conditions (9.12) requires that

$$\int_{R_3\times R_3} E\{|\nu(\mathbf{x}')\nu(\mathbf{x}'')|\}\left(\frac{2k_0}{R}+\frac{1}{R^2}\right)\left(\frac{2k_0}{R'}+\frac{1}{R'^2}\right)d\mathbf{x}'\,d\mathbf{x}'' \ll 16\lambda_0^2 \tag{9.18}$$

in order that formula (9.17) be applicable. If (9.18) holds, then, to first order, the phase and amplitude fluctuations are described respectively by

(9.19)
$$S(\mathbf{x}) = x + \frac{1}{2\lambda_0}\int_{R_3} \frac{\sin\{k_0(R - [x - x'])\}}{R} \nu(\mathbf{x}')\, d\mathbf{x}',$$
$$A(\mathbf{x}) = A_0 \exp\left(\frac{\pi}{\lambda_0^2}\int_{R_3} \frac{\cos\{k_0(R - [x - x'])\}}{R} \nu(\mathbf{x}')\, d\mathbf{x}'\right).$$

Expressions (9.19) are taken to hold in the mean square sense.

A one-dimensional form of the spatial Riccati equation (9.6) results from the "ansatz"

$$\boldsymbol{\Theta} = \Theta_1(x)\mathbf{i} + \Theta_2(y)\mathbf{j} + \Theta_3(z)\mathbf{k},$$

and

$$-ik_0 n^2(x, y, z) = f_1(x) + f_2(y) + f_3(z).$$

In that event (9.6) reduces to a system of three independent equations of the form

$$\Theta_j' - ik_0\Theta_j^2 = f_j(x_j), \qquad (j = 1, 2, 3).$$

This equation is essentially the classical Riccati equation

(9.20)
$$y'(x) + y^2(x) = -p(x)$$

where

$$x = -ik_0 x_j, \qquad y = \Theta_j, \qquad p = \frac{1}{ik_0} f_j(x_j).$$

Equation (9.20) has been the object of much study. The following iterative solution

(9.21)
$$y(x) = -\underset{n\to\infty}{\text{l.i.m.}}\left(\int_0^x d\xi_{n-1}\left(\int^{\xi_{n-1}} d\xi_{n-2}\cdots\int^{\xi_3} d\xi_2\left(\int_0^{\xi_2} d\xi_1 P^2(\xi_1) + P(\xi_2)\right)^2 + \cdots + P(\xi_{n-1})\right)^2 + P(x)\right)$$

to the nonlinear integral equation for $z(x) = y^2(x)$, $y(0) = 0$:

(9.22)
$$z(x) = \left[\int_0^x z(\xi)\, d\xi + P(x)\right]^2, \qquad P(x) = \int_0^x p(\xi)\, d\xi$$

appears to offer some potential value in the probabilistic case.

If $p(x)$ is an analytic random function, so that

$$p(x) = \sum_0^\infty p_m x^m,$$

the p_m being random variables, then (9.20) can be solved by the following recurrence relation for the random coefficients η_m occurring in the power series for $y(x)$:

$$m\eta_m = -\sum_{l=0}^{m-1} \eta_l \eta_{m-1-l} - p_{m-1} \qquad (m = 1, 2, \cdots).$$

Knowledge of the sure coefficient η_0 is therefore enough to specify the wave function for the analytic case.

10. **The one-dimensional problem.** The second order linear ordinary differential equation (5.2) with random wave number $k(x)$ has also received considerable attention, both because it occurs naturally as the transmission line equation and also because it is the simplest case of (1.1).

Kay and Silverman [24] have studied the multiple scattering of a plane wave by a stack of N dielectric slabs, all of the same thickness d, but each of which has permittivity $1 \pm \Delta\varepsilon$, with probability 1/2 for each value. If a scalar plane wave e^{-ik_0z} is incident from $-\infty$ on such a stack of Bernoulli slabs, the resultant field satisfies the reduced one-dimensional wave equation

$$\left\{\frac{d^2}{dz^2} + k_0^2[1 + \Delta\varepsilon(z)]\right\} Y(z) = 0. \tag{10.1}$$

A Neumann series solution is then obtained for the corresponding Fredholm integral equation:

$$Y(z) = e^{-ik_0z} + \sum_{n=1}^{\infty} (\tfrac{1}{2}ik_0)^n \int_0^{Nd} \cdots \int_0^{Nd} \exp\left[-ik_0\{|z - z_1| + \cdots + |z_{n-1} - z_n|\} - ik_0z_n\right] \cdot \prod_{m=1}^{n} \Delta\varepsilon(z_m)\, dz_m. \tag{10.2}$$

Kay and Silverman next define transmission and reflection coefficients T and R in terms of integral expressions like (10.2) according to the relations

$$Y(z) = \begin{cases} T\, e^{-ik_0z}, & z \geq Nd, \\ e^{-ik_0z} + R\, e^{+ik_0z}, & z < 0. \end{cases}$$

The integrals in T and R are then estimated and used to show the existence of stochastic majorants for the Neumann series for $E|R|^2$ and $E|T|^2$. For those terms in the Neumann series whose order is $<N$ the convergence is considerably more rapid than in the corresponding deterministic case. This feature is attributed to an enhanced attenuation by multiple scattering arising out of the random character of the medium, and is adduced by the authors in support of the common use of the Born approximation, or single-scattering model, in "forward scatter" theories.

11. **Wave propagation as a transport process.**

11.1. GEOMETRIC OPTICS OF A RANDOM MEDIUM. Geometric optics constitutes a "ballistic" (or zero wavelength) approximation to wave propagation phenomena. The trajectories along which field disturbances propagate are thus the ray paths. The geometrical optics approximation involves an assumption that wave-fronts behave locally like plane waves. This implies that the only admissible solutions of a reduced wave equation are those of the form (9.4), whence we obtain the eikonal equation for the optical path length $S(\mathbf{x})$:

$$(\nabla S)^2 + o(\lambda_0) = n^2 \tag{11.1}$$

while for the amplitude function $A(\mathbf{x})$:

$$\nabla \ln A \cdot \nabla S + \tfrac{1}{2}\nabla^2 S = 0. \tag{11.2}$$

Now introduce the unit vector in the ray direction at $\mathbf{x}$:

$$\hat{\mathbf{s}} = \nabla S/n. \tag{11.3}$$

Equation (11.3) is equivalent to (11.1) to within terms $o(\lambda_0)$. Letting $\hat{\boldsymbol{\rho}}$ denote a unit vector at $\mathbf{x}$ directed along the radius of curvature ρ, we have, since

$$\frac{d\mathbf{s}}{ds} = (\mathbf{s} \cdot \nabla)\mathbf{s} = -\mathbf{s} \times (\nabla \times \mathbf{s}),$$

that

$$\frac{\hat{\boldsymbol{\rho}}}{\rho} = -\hat{\mathbf{s}} \times (\nabla \times \hat{\mathbf{s}}),$$

and the ray curvature at $\mathbf{x}$ is therefore given by

$$\frac{1}{\rho} = \hat{\boldsymbol{\rho}} \cdot \nabla \ln n. \tag{11.4}$$

Keller [18] has formulated the initial value problem for (11.1) in the well-posed form

$$\begin{aligned} (\hat{\mathbf{s}} \cdot \nabla)(n\hat{\mathbf{s}}) &= \nabla n, \\ S(\mathbf{0}) &= 0, \\ \hat{\mathbf{s}}(\mathbf{0}) &= \mathbf{1}. \end{aligned} \tag{11.5}$$

The first of these equations can also be derived variationally from Fermat's principle [20].

We now consider $n^2(x)$ to be a regular random function $n^2(\omega; x)$, whence it follows that $S(\omega; x)$ is a differentiable regular random function. Then mean square differentiation and integration can be invoked and the preceding equations (and their consequences to follow) have a meaning in the case of regular random functions.

The pioneering work in the application of geometric optics to random continuous media appears to be that of P. G. Bergmann [28]. Bergmann began with the eikonal equation (11.1) together with the equation

$$\nabla \cdot \left(\frac{\phi}{n} \nabla S\right) = 0, \tag{11.6}$$

where $\phi(\mathbf{x})$ is the field intensity (flux density) at $\mathbf{x}$, and assumed a refractive index of the form $n = 1 + \varepsilon\nu$, $\varepsilon \ll 1$. He thus obtained integral formulas for the first order solutions of (11.1) and (11.6) from which the mean square deviation of optical path length from the geometrical path length and the variance of signal level fluctuations can be determined.

Since 1946 the geometrical optics of a random medium has received considerable further investigation (see the extensive bibliographies in [30], [20], [45], and [13]). Keller [18] has given a thorough and penetrating treatment of the problem, and he as well as Chernov [20] and Kharanen [27] have analyzed ray propagation in a random medium under the hypothesis that the refractive index constitutes a Markov process. Keller's form of the analysis will be followed here.

Keller denotes our $\mathbf{s}$ by $\mathbf{x}'$ and is then able to write equations (11.5) as

$$\begin{aligned} (n\mathbf{x}')' &= \nabla n, \\ \mathbf{x}(0) &= 0, \\ \mathbf{x}'(0) &= \mathbf{u}, \end{aligned} \tag{11.6'}$$

where $\mathbf{x}(s, \varepsilon)$ is the position vector of the terminus of a ray that departed from the origin in the direction of the unit vector $\mathbf{u}$. Assuming that the perturbation expansion of n extends only to the first order in ε:

$$n(\mathbf{x}, \varepsilon) = 1 + \varepsilon\mu(\mathbf{x}), \tag{11.7}$$

with the usual form of the perturbation series for $\mathbf{x}(s, \varepsilon)$, one is led to the second order expansion

$$\mathbf{x}(s, \varepsilon) = \mathbf{x}_0(s) + \varepsilon\mathbf{x}_\varepsilon(s) + \tfrac{1}{2}\varepsilon^2\mathbf{x}_{\varepsilon\varepsilon}(s) + o(\varepsilon^2), \tag{11.8}$$

where

$$\mathbf{x}_0(s) = s\mathbf{u}, \tag{11.9}$$

$$\mathbf{x}_\varepsilon(s) = \int_0^s (s - t)\nabla_T\mu(t\mathbf{u})\, dt, \tag{11.10}$$

$$\begin{aligned} \mathbf{x}_{\varepsilon\varepsilon}(s) = 2\int_0^s (s - t)\Big[(\mathbf{x}_\varepsilon \cdot \nabla_T)\nabla_T\mu - \tfrac{1}{2}\nabla_T\mu^2 - \mathbf{u}\nabla_T\mu \cdot \int_0^t \nabla_T\mu(\tau\mathbf{u})\, d\tau \\ - \mathbf{u} \cdot \nabla\mu \int_0^t \nabla_T\mu(\tau\mathbf{u})\, d\tau\Big]\, dt \end{aligned} \tag{11.11}$$

and ∇_T denotes the gradient transverse to the unit vector $\mathbf{x}' = \mathbf{u}$; i.e.,

$$\nabla_T = \nabla - \mathbf{x}_0'(\mathbf{x}_0' \cdot \nabla).$$

It is then a straightforward matter to obtain moments of various orders under the hypothesis that the μ-process is stationary in the wide sense:

$$E\{\mu(\mathbf{x}_1)\mu(\mathbf{x}_2)\} = \gamma(|\mathbf{x}_1 - \mathbf{x}_2|) = E|\mu^2|\rho(|\mathbf{x}_1 - \mathbf{x}_2|). \tag{11.12}$$

Thus

$$E\mathbf{x}(s, \varepsilon) = s\mathbf{u} + \varepsilon^2 E\mu^2 \int_0^s (s - t)^2 t^{-1}\rho(t)\, dt + o(\varepsilon^2), \tag{11.13}$$

while for the variance of the transverse displacement δ of the endpoint of a ray of length s from the corresponding extension of the initial unit vector $\mathbf{u}$:

$$E\delta^2 \approx -\tfrac{4}{3}\varepsilon^2 E\mu^2 s^3 \int_0^\infty r^{-1}\rho(r)\, dr + o(\varepsilon^2), \tag{11.14}$$

and for the mean square value of $\mathbf{x}' - \mathbf{u}$:

$$E\{(\mathbf{x}' - \mathbf{u})^2\} = 2[1 - E\cos\alpha(s, \varepsilon)] \approx -4\varepsilon^2 E\mu^2 s \int_0^\infty r^{-1}\rho(r)\, dr + o(\varepsilon^2). \tag{11.15}$$

We now pass to a consideration of the Markovian ray tracing, i.e., $\mathbf{x}'(s, \varepsilon)$ is assumed to constitute a continuous parameter Markov process. This appears physically plausible, since Huyghen's principle is the optical analogue of the Chapman–Kolmogorov equation. However, it is clear that $\mathbf{x}(s, \varepsilon)$ in the form (11.8) could not in general constitute a wide-sense Markov process, for instance, since the necessary and sufficient condition for the process to be Markovian (wide sense) is that [9]

$$\Gamma(s, \sigma) = \Gamma(s, s')\Gamma(s', \sigma), \tag{11.16}$$

($\Gamma(s, \sigma)$ being the covariance function of $\mathbf{x}'(s, \varepsilon)$ and $\mathbf{x}'(\sigma, \varepsilon)$) is violated. The left-hand side of (11.16) is $O(\varepsilon^2)$ whereas the right-hand side is $O(\varepsilon^4)$, and in addition the integrand of the right-hand side involves the product $\kappa(t, t')\kappa(t'', \tau)$, $t' \neq t''$, where $\kappa(t, t')$ is the covariance function of $\nabla_T\mu(t)$ and $\nabla_T\mu(t')$. Even if the ε's do not enter and the $\nabla\mu$ process is wide sense Markovian, the equation

$$\int_0^s \int_0^{s'} \kappa(\sigma, \sigma')\, d\sigma\, d\sigma' = \int_0^s \int_0^t \int_0^t \int_0^{s'} \kappa(\sigma, \tau)\kappa(\tau', \sigma')\, d\sigma\, d\tau\, d\tau'\, d\sigma' \tag{11.17}$$

would have to hold if the $x(s, \varepsilon)$ process were to be Markovian.

Even so, it is plausible that, insofar as its next position is concerned, only the present location of the end of the ray path is required and not the entire previous trajectory, and as mentioned above, several authors have adopted the Markov process approach to the geometric optics of a randomly inhomogeneous medium. The state space as well as the parameter space are continuous, the state at s being the direction of the ray path there:

$$\alpha(s) = (\theta, \varphi).$$

If the Markovian hypothesis is tenable, then the transition probability $p(s, \alpha; s + \Delta s, \alpha + \Delta\alpha)$ would satisfy the Fokker–Planck equation [20]

$$
\begin{aligned}
\frac{\partial p}{\partial s} = \frac{1}{2}\frac{\partial^2}{\partial\theta^2}\left(\lim_{\Delta s\to 0}\frac{E(\Delta\theta)^2}{\Delta s}\sin\theta\, p\right) + \frac{\partial^2}{\partial\theta\partial\varphi}\left(\lim_{\Delta s\to 0}\frac{E\{(\Delta\theta)(\Delta\varphi)\}}{\Delta s}\sin\theta\, p\right) \\
+ \frac{1}{2}\frac{\partial^2}{\partial\varphi^2}\left(\lim_{\Delta s\to 0}\frac{E(\Delta\varphi)^2}{\Delta s}\sin\theta\, p\right) - \frac{\partial}{\partial\theta}\left(\lim_{\Delta s\to 0}\frac{E(\Delta\theta)}{\Delta s}\sin\theta\, p\right) \\
- \frac{\partial}{\partial\varphi}\left(\lim_{\Delta s\to 0}\frac{E(\Delta\varphi)}{\Delta s}\sin\theta\, p\right).
\end{aligned}
\tag{11.18}
$$

After some trigonometric reductions (11.18) can be reduced to the following form (which is formally the same as the Fokker–Planck equation for the rotational motion of a Brownian particle):

$$
\frac{\partial p}{\partial s} = D\left\{\frac{1}{\sin\theta}\frac{\partial}{\partial\theta}\left(\sin\theta\frac{\partial p}{\partial\theta}\right) + \frac{1}{\sin^2\theta}\frac{\partial^2 p}{\partial\varphi^2}\right\},
\tag{11.19}
$$

where

$$
D = \lim_{\Delta s\to 0}\frac{E\alpha^2}{4\Delta s} + o(\varepsilon^2).
$$

For the case that p is independent of φ, the solution of the initial value problem consisting of (11.19) and the initial absolute probability $p(0, \alpha) = \delta(\theta)$ is given by

$$
p(0, \alpha; s, \alpha) = \frac{1}{4\pi}\sum_{n=0}^{\infty}(2n + 1)P_n(\cos\theta)\, e^{-n(n+1)Ds}.
\tag{11.20}
$$

The expectation of $\cos\alpha$ obtained from (11.20) apparently differs from that which follows from (11.15). However, this results from different ranges of validity of the two results, that based on (11.15) holding well for small s while that following from (11.20) applies best for large values of s. The two results do, however, agree to terms of second order in the intermediate, overlapping range of s.

11.2. Diffusion and transport equations for multiple scattering. Another approach to wave propagation that involves a transport equation rather than the wave equation [31] is based on the connection between the electromagnetic energy flux and the energy density associated with the photon distribution function. According to the differential form of Poynting's theorem, we have that

$$
\operatorname{div}\mathbf{P} + \frac{\partial U}{\partial t} = 0,
\tag{11.21}
$$

where U is the total electromagnetic energy, related to the photon flux as follows:

$$
U = \hbar\int_0^\infty \omega N_\omega\, d\omega = \int d\omega\int_{\text{vel. space}} \omega f_\omega\, dV_v
\tag{11.22}
$$

and

N_ω = number of photons per unit volume in the frequency band $(\omega, \omega + d\omega)$,

f_ω = the probability density function for the distribution of photons in positional velocity space.

Now introduce the spectral density function

$$\Sigma(\mathbf{x}, \mathbf{v}, \omega) = \omega f_\omega(\mathbf{x}, \mathbf{v}), \tag{11.23}$$

and supplement (11.21) by a term taking into account the scattering of photons out of $(\omega, \mathbf{v})$. After some reductions one is led to the integro-differential equation

$$\mathbf{v} \cdot \nabla\Sigma = \int\int p(\omega, \omega' | \mathbf{v}, \mathbf{v}')\Sigma(\mathbf{x}, \omega', \mathbf{v}')\, d\omega'\, dV_{v'} - \int Q_T \Sigma(\mathbf{x}, \omega')\, d\omega' \tag{11.24}$$

where

$$p(\omega, \omega' | \mathbf{v}, \mathbf{v}')\, d\omega\, dV_v = \Pr\{\text{a photon at } (\boldsymbol{\omega}', \mathbf{v}') \text{ is scattered into the range } (\omega + d\omega, \mathbf{v} + d\mathbf{v})\}, \tag{11.25}$$

$$Q_T = \text{scattering cross section}$$

$$= \frac{1}{2\pi} \int_{-\infty}^{\infty} e^{-i(\omega - \omega')\tau}\, d\tau \int_{\mathcal{O}} \sigma(\mathbf{K}, 0)\, d\mathcal{O}, \quad \mathbf{K} = \mathbf{k} - \mathbf{k}', \tag{11.26}$$

$$\sigma(\mathbf{K}, \tau) = \frac{\pi^2}{\lambda^4} \sin^2 \gamma \int e^{-i\mathbf{K}\cdot\mathbf{r}}\Gamma(\mathbf{r}, \tau)\, d\mathbf{r}, \tag{11.27}$$

$\Gamma(\mathbf{r}, \tau)$ being the covariance function for dielectric noise (turbulent scattering). It follows from (11.25)–(11.27) that

$$p(\omega, \omega' | \mathbf{v}, \mathbf{v}') = \frac{1}{2\pi} \int_{-\infty}^{\infty} e^{-i(\omega - \omega')\tau}\, \sigma(\mathbf{K}, \tau)\, d\tau.$$

For a source radiating into a semi-infinite half-space, Bugnolo was able to reduce (11.24) to the partial differential equation

$$\frac{\partial\Sigma}{\partial\xi} \cong \sum_{n=1}^{N} (-1)^n \gamma_{2n}^2 \frac{\partial^{2n}\Sigma}{\partial\omega^{2n}} + \sum_{m=0}^{M} (-1)^m \frac{\alpha_{2m}^2}{4} \left(\frac{\partial^{2m+2}}{\partial\delta^2 \partial\omega^{2m}} + \frac{1}{\delta}\frac{\partial^{2m+1}}{\partial\delta\partial\omega^{2m}}\right)\Sigma, \tag{11.28}$$

which was then solved by transform methods to obtain the solution

$$\begin{aligned}\Sigma(\xi, \theta, \omega) = \frac{1}{2\pi} \int_{-\infty}^{\infty} e^{i\omega s} \int_0^{\infty} J_0(g\theta) \\ \times \exp\left\{\sum_{n=1}^{N} (-i)^{2n}\gamma_{2n}^2 s^2 + \frac{1}{4}\sum_{m=0}^{M} (-i)^{2m}\alpha_{2m}^2 s^{2m} g^2\right\} \xi\Psi(g, s) g\, dg\, ds\end{aligned} \tag{11.29}$$

where

$$\Psi(g, s) = \int_{-\infty}^{\infty} e^{-i\omega s} \int_{-\infty}^{\infty} s(\beta, \omega) J_0(g\beta)\beta\, d\beta\, d\omega. \tag{11.30}$$

(For definition of the parameters in (11.28), (11.29), and (11.30), see the original reference [**31**].)

Bugnolo has also studied the application of these results to tropospheric and ionospheric forward scatter of radio waves [**32**; **33**]. He finds that the spectral density is Gaussian, to the first order at least, with a variance proportional to penetration distance x

$$\text{(11.31)} \qquad \operatorname{var}_\omega(x, \omega_0) = x\left(\int (\mathbf{K}\cdot\mathbf{V})^2\sigma(\theta)\,d\Omega\right)^2,$$

where $\sigma(\theta)$ is the scattering cross section per unit solid angle per unit volume and $\mathbf{V}$ is the mean wind velocity. This realistic feature, viz., the increase in width of the spectral density function with increasing distance, is lacking in the single scatter theories. A mean free path for scattering is also defined:

$$\text{(11.32)} \qquad d_0 = \left(\int \sigma(\theta, \varphi)\,d\Omega\right)^{-1} \sim \lambda^2\varepsilon_0^2/[lE(\Delta\varepsilon)^2],$$

l being the average scale of turbulence, and, assuming that the scattering constitutes a stationary Markov process, the following expression is determined for the probability that a ray is scattered n times in a distance R:

$$\frac{Q_s^n}{\Gamma(n)}\int_0^R r^{n-1}\,e^{-Q_s r}\,dr,$$

Q_s being the total scattering cross-section per unit volume at wave number k. It turns out that for the odds to be even that a ray will be scattered twice in a distance R, R must be of the order of 1.7 mean free paths.

Another transport equation approach, based on multiply scattered rays, has been given by Ament [**26**], who finds that multiple scattering can take place for path lengths even smaller than the mean free path d_0 under certain circumstances (large cross-sections or large angular deviations). The distribution in angle of the scattered power at distance x is assumed to follow a compound Poisson process

$$P(x; y', z') = \sum_{m=0}^{\infty} (Qx)^m\,e^{-Qm}P_m(y', z')/m!.$$

Ament also gives several diffusion equations which apply to the phase coherence of multiply scattered rays, together with a brief discussion of the appropriate Wiener integrals. We shall discuss the functional integral approach further below.

Bremmer [**30**] has also considered from the standpoint of the diffusion model the probabilistic properties of radiation traversing a turbulent medium. He uses for this purpose the joint probability density $h_z(x, y; \xi, \eta)$ that an energy unit departing from the origin parallel to the z-axis will intersect a transverse plane at distance z in an element of area $dxdy$ in a direction with direction cosines ξ and η. This probability density (actually a transition probability with initial value at the origin) is termed the multiple-scattering diffusion function and is taken to satisfy a Chapman–Kolmogorov equation. The latter leads, once a differential cross-section profile $\sigma(\xi, \eta)$ has been assigned, to a diffusion equation in operator form

$$\left(\xi \frac{\partial}{\partial x} + \eta \frac{\partial}{\partial y} + \frac{\partial}{\partial z}\right) h_z(x, y; \xi, \eta) \tag{11.33}$$
$$= \frac{1}{d_0} \ln \left\{d_0 \int\!\!\int_{-\infty}^{\infty} d\xi'\, dy' \sigma(\xi', \eta') \exp\left[-\xi' \frac{\partial}{\partial \xi} - \eta' \frac{\partial}{\partial \eta}\right]\right\} h_z(x, y; \xi, \eta).$$

The corresponding probability density $h_{z,m}(x, y; \xi, \eta)$ for consecutive scattering at m equidistant planes, each with separation distance z/m, has the bivariate Laplace transform [30]:

$$p_1 q_1 p_2 q_2 \exp \left\{\frac{m}{z} \int_0^z d\zeta \ln \left(\frac{z}{m} \int\!\!\int_{-\infty}^{\infty} d\xi\, d\eta \sigma(\xi, \eta) \cdot \exp\left[-\xi(p_1\zeta + p_2) - \eta(q_1\zeta + q_2)\right]\right)\right\}. \tag{11.34}$$

Bremmer has determined $h_{z,m}$ for the cases of a Gaussian cross-section profile and a so-called Norton profile (the third in Table I). The interested reader is referred to Bremmer's paper [30] for further details.

11.3. INVARIANT IMBEDDING. Invariant imbedding, developed by Bellman and others as an application of dynamic programming, has enjoyed conspicuous success in such transport phenomena as neutron scattering and radiative transfer in stellar atmospheres [46; 47]. Appeal is made to the duality between wave mechanics and particle mechanics to carry over the transport process to wave propagation phenomena. In the deterministic single slab case the method leads to the well-known Riccati equations governing the reflection and transmission coefficients of an inhomogeneous transmission line. In the stochastic case expressions are obtained for reflection and transmission coefficients, R_N and T_N respectively, for a plane wave incident from the right upon a stack of N slabs, each of which has a random wave number which is constant (for any particular realization) in that slab. The approach is as follows: the problem for fixed N is first of all imbedded in the class of problems for $N = 1, 2, \cdots$. Next appeal is made to the so-called localization principle (i.e., partial reflection and transmission occur at each plane section of the stratified medium as the wave passes through) in order to obtain iterative relations between R_N and R_{N-1}, and T_N and R_{N-1} and T_{N-1}. A brief discussion is given in [25] of how to approach the actual probabilistic problem of determining the probability distribution functions associated with these formal expressions, but no real attack on this aspect of the problem appears to have been yet carried out.

12. **Relation to more general theories.** Much of what we have discussed is a good deal more closely allied to classical analysis and what is now classical probability theory than to the modern functional analysis approach. This reflects,

however, the fact that those portions of functional analysis that would be most useful in connection with (5.1) remain in large part still to be developed. Equations (5.1) constitute a non-self-adjoint problem [**48**], even for deterministic constant k^2, and the mathematical apparatus appropriate to spectral representations for non-self-adjoint equations and integration in the appropriate functional spaces is still largely lacking. Though the method of integration in functional spaces [**34**], has enjoyed marked success in problems of quantum mechanics, that is to say for Hilbert space, Wiener integrals would have to be generalized in two directions, (i) to the class of twice continuously differentiable functions and (ii) to more general stochastic processes than the Brownian motion process, to be of use in the present context.

Many of the recently developed techniques of modern partial differential equation theory would also seem to carry over to the random case, e.g., parametrix methods using either a deterministic or simple random function as fundamental solution. The possibilities in this direction are largely unexplored, although the basic constant-coefficient differential equation has been studied by Kampé de Fériet [**1**]. The general question of random equations is discussed elsewhere in this symposium [**49**]. Prior to this work such approaches have in general applied only to first order differential equations or to equations with constant coefficients, the element of randomness arising from the presence of a random forcing function or random boundary conditions. Much the same applies to past studies of random operators on Banach spaces [**50**; **51**; **52**], although the use of Orlicz spaces introduced in this connection may offer considerable promise.

References

1. J. Kampé de Fériet, *Statistical mechanics of continuous media*, Proc. Sympos. Appl. Math. Vol. 13, pp. 165–198, Amer. Math. Soc., Providence, R. I., 1962.

2. V. Twersky, *Propagation in media of discrete scatterers*, Proc. Sympos. Appl. Math. Vol. 16, pp. 84–116, Amer. Math. Soc., Providence, R. I., 1964.

3. J. L. Doob, *The Brownian movement and stochastic equations*, Ann. of Math. (2) **43** (1942), 351–369.

4. J. Kampé de Fériet, *Spectral tensor of homogeneous turbulence*, Symposium on turbulence, U.S. Naval Ord. Lab. Rep. No. 1136, Silver Springs, Md., July 1, 1949.

5. S. Silver, *Delineations and methodology in the subject of scattering by a statistically inhomogeneous medium*, Electromagnetic theory and antennas, E. C. Jordan, ed., Pergamon, London, 1963.

6. J. E. Moyal, *Stochastic processes and statistical physics*, J. Roy. Statist. Soc. **11** (1949), 150–210.

7. M. Loève, *Probability theory*, 2nd ed., Van Nostrand, Princeton, N. J., 1960.

8. A. Kolmogoroff, *Grundebgriffe der Wahrscheinlichkeitsrechnung*, Chelsea, New York, 1946.

9. J. L. Doob, *Stochastic processes*, Wiley, New York, 1953.

10. A. Blanc-Lapierre and R. Fortet, *Théorie des fonctions aléatoires*, Masson, Paris, 1953.

11. R. A. Silverman, *Locally stationary random processes*, Res. Rep. MME-2, New York University Institute of Mathematical Sciences Div. Electromag. Res., New York, 1957.

12. J. Kampé de Fériet and F. N. Frenkiel, *Correlation and spectra for non-stationary random functions*, Math. Comp. **16** (1962), 1–21.

13. A. D. Wheelon, *Radio-wave scattering by tropospheric irregularities*, J. Res. Nat. Bureau Standards Sect. D **63D** (1959), 205–233.

14. D. S. Bugnolo, *Stochastic processes and beyond-the-horizon propagation*, IRE Trans. **AP–9** (1961), 226–228.

15. K. Jacobs, *Neuere Methoden und Ergebnisse der Ergodentheorie*, Springer, Berlin, 1960.

16. G. Maruyama, *The harmonic analysis of stationary stochastic processes*, Mem. Fac. Sci. Kyūsyū Univ. Ser. A **4** (1949), 45–106.

17. T. J. Skinner, *Ergodic theorem in the solution of the scalar wave equation with statistical boundary conditions*, J. Opt. Soc. Amer. **51** (1961), 1246–1251.

18. J. B. Keller, *Wave propagation in random media*, Proc. Sympos. Appl. Math. Vol. 13, pp. 227–246, Amer. Math. Soc., Providence, R. I., 1962.

19. W. S. Ament, *The effect of atmospheric inhomogeneities on radio propagation velocity*, Naval Research Laboratory, Branch Memo. WPR TA–5/50, Washington, D.C., 1950.

20. L. A. Chernov, *Wave propagation in a random medium*, McGraw-Hill, New York, 1960.

21. W. C. Hoffman, *The electromagnetic field in a randomly inhomogeneous medium*, IRE Trans. **AP–7** (1959), Special Suppl., S301–S306.

22. C. Müller, *Zur mathematischen Theorie elektromagnetischer Schwingungen*, Abh. Deutsch. Akad. Wiss. Berlin Kl. Math.-Nat. Klasse 1945/46, No. 3 (1950), 5–56.

23. R. A. Silverman, *Scattering of plane waves by locally homogeneous dielectric noise*, Proc. Cambridge Philos. Soc. **54** (1958), 530–537.

24. I. Kay and R. A. Silverman, *Multiple scattering by a random stack of dielectric slabs*, Nuovo Cimento (10) **9** (1958), Suppl., 626–645.

25. R. Bellman and R. Kalaba, *Invariant imbedding and wave propagation in stochastic media*, Electromagnetic wave propagation, Desirant and Michaels, eds., pp. 243–252, Academic Press, New York, 1960.

26. W. S. Ament, *Modification of a ray-tracer for Monte Carlo prediction of multiple-scattered radio fields*, Statistical methods in radio wave propagation, W. C. Hoffman, ed., pp. 184–196, Pergamon, London, 1960.

27. V. Ya. Kharanen, *Sound propagation in a medium with random fluctuations of the refractive index*, National Science Foundation Translation NSF-tr-8, U. S. Dept. Commerce, Washington 25, D.C., June, 1953.

28. P. G. Bergmann, *Propagation of radiation in a medium with random inhomogeneities*, Phys. Rev. (2) **70** (1946), 486–492.

29. R. Bellman and R. Kalaba, *Wave branching processes and invariant imbedding*, Proc. Nat. Acad. Sci. U. S. A. **47** (1961), 1507–1509.

30. H. Bremmer, *Scattering by a perturbed continuum*, Electromagnetic Wave Theory and Antennas, E. C. Jordan, ed., Pergamon, London, 1963.

31. D. S. Bugnolo, *Transport equation for the spectral density of a multiple-scattered electromagnetic field*, J. Appl. Phys. **31** (1960), 1176–1182.

32. ———, *On the question of multiple scattering in the troposphere*, J. Geophys. Res. **65** (1960), 879–884.

33. ———, *Radio star scintillation and multiple scattering in the ionosphere*, IRE Trans. **AP–9** (1961), 89–96.

34. I. M. Gel'fand and A. M. Yaglom, *Integration in functional space and its applications in quantum physics*, J. Math. and Phys. **1** (1960), 48–69.

35. R. P. Feynman, *An operator calculus having applications in quantum electrodynamics*, Phys. Rev. (2) **84** (1951), 108–128.

36. T. I. Seidman, *A survey of functional integration with applications of the Wiener integral*, Rep. 55 G–1, Lincoln Lab., Massachusetts Institute of Technology, Bedford, Mass., 1961.

37. R. H. Kraichnan, *Dynamics of nonlinear stochastic systems*, J. Mathematical Phys. **2** (1961), 124–148.

38. H. Staras, *Forward scattering of radio waves by anisotropic turbulence*, Proc. IRE **43** (1955), 1374–1380.

39. H. G. Booker, *Radio scattering in the lower ionosphere*, J. Geophys. Res. **64** (1959), 2164–2177.

40. H. T. Friis, A. B. Crawford, and D. C. Hogg, *A reflection theory for propagation beyond the horizon*, Bell System Tech. J. **36** (1957), 627–644.

41. S. A. Bowhill, *The scattering of radio waves by an extended randomly refracting medium*, J. Atmos. Terrest. Phys. **20** (1961), 9–18.

42. ———, *Statistics of a radio wave diffracted by a random ionosphere*, J. Res. Nat. Bureau Standards Sect. D **65D** (1961), 275–292.

43. K. C. Yeh, *Propagation of spherical waves through an ionosphere containing anisotropic irregularities*, J. Res. Nat. Bureau Standards Sect. D **66D** (1962), 621–636.

44. G. K. Batchelor, *The theory of homogeneous turbulence*, Section 4.1, Cambridge Univ. Press, Cambridge, 1956.

45. V. I. Tatarski, *Wave propagation in a turbulent medium*, McGraw-Hill, New York, 1961.

46. R. Bellman, R. Kalaba, and G. M. Wing, *On the principle of invariant imbedding and neutron transport theory*. I. *One-dimensional case*, J. Math. Mech. **7** (1958), 149–162.

47. ———, *Invariant imbedding and neutron transport theory*. IV. *Generalized transport theory*, J. Math. Mech. **8** (1959), 575–584.

48. C. Dolph, *Recent developments in some non-self-adjoint problems of mathematical physics*, Bull. Amer. Math. Soc. **67** (1961), 1–69.

49. A. T. Bharucha-Reid, *On the theory of random equations*, Proc. Sympos. Appl. Math. Vol. 16, pp. 40–69, Amer. Math. Soc., Providence, R. I., 1964.

50. ———, *Über die konvergenz der folgen von verallgemeinerten zufälligan grössen in Orlicz'schen raümen*, Bull. Acad. Polon. Sci. Sér. Sci. Math. Astronom. Phys. **7** (1959), 425–427.

51. ———, *On random operator equations in Banach spaces*, Bull. Acad. Polon. Sci. Sér. Sci. Math. Astronom. Phys. **7** (1959), 561–564.

52. ———, *On random solutions of integral equations in Banach spaces*, Trans. 2nd Prague Conf. Information Theory, Statistical Decision Functions and Random Processes (1959), pp. 27–48, Publ. House Czechoslovak Acad. Sci., Prague, 1960.

BOEING SCIENTIFIC RESEARCH LABORATORIES,
SEATTLE, WASHINGTON

STOCHASTIC EQUATIONS AND WAVE PROPAGATION IN RANDOM MEDIA[1]

BY

JOSEPH B. KELLER

1. **Introduction.** A stochastic equation is a family of equations depending upon a parameter α which ranges over a space A in which a probability density $p(\alpha)$ is defined. The probability density $p(\alpha)$ determines the probability of a given value of α and therefore of the corresponding equation of the family. If for each value of α there is determined a unique solution $u(\alpha)$ of the equation then $u(\alpha)$ is a random variable (or function) and its probability density is also $p(\alpha)$. An example of a stochastic equation is a linear differential equation in which some of the coefficients and the inhomogeneous term are random, i.e., depend upon α. The objective of the theory of stochastic equations is the determination of the probability distribution of u or of various statistical properties of u such as its expectation, its variance and its higher moments.

Wave propagation in a continuous random medium is often governed by a stochastic linear differential equation. The coefficients in the equation characterize the propagation medium. They may be its refractive index $n(x)$, its density $\rho(x)$, its dielectric constant, etc. A random medium is a family of media, each labeled by one value of α, and each with its own index $n(x, \alpha)$, density $\rho(x, \alpha)$, etc. The probability $p(\alpha)$ gives the probability of each member of the family. The inhomogeneous term represents the source of the waves, which may be either random or sure (i.e., not random). A random medium may also consist of a random distribution of discrete scattering objects distributed throughout some fixed medium within which waves can propagate. This latter description is appropriate when matter is viewed on a molecular scale. We shall see that propagation in such a medium can also be analyzed in terms of stochastic equations.

The practical reason for studying stochastic equations is the belief that their solutions represent physical phenomena which could not be satisfactorily investigated in any other way. Let us consider, for example, the propagation of an electromagnetic wave through air. Because the wave travels so fast, it is adequate to assume that the air molecules do not move during the passage of the wave. If the locations of all the molecules were known, one could in principle determine how they scatter the wave. However, the full details of the scattered wave would be too complex to be useful. Furthermore it is out of the question to determine

[1] This research was supported by the Air Force Cambridge Research Laboratories, under Contract No. AF 19(604)3495. Reproduction in whole or in part is permitted for any purpose of the U.S. Government.

the locations of all the molecules. In view of these two difficulties, it is hoped that the observable features of the wave scattering will be determined by the gross features of the molecular distribution. Therefore the actual distribution of molecular positions is replaced by a random medium, i.e., a collection of molecular distributions with a probability associated with each. The mean particle density and some other statistics of the random medium are adjusted to equal the corresponding properties of the actual particle distribution. Then the mean scattered wave and other statistics of the random wave are presumed to equal the corresponding properties of the actual wave. Of course the success of this procedure, like that of all other theories of natural phenomena, can only be judged by comparison of its predictions with observations.

Once the concept of a stochastic equation is understood, it becomes clear that most, if not all, equations which arise in science and engineering may be viewed as stochastic. This is because the various coefficients and parameters entering such equations are subject to random variations due to noise, thermal fluctuations, etc. Consequently all problems involving such equations should be reexamined from the stochastic viewpoint. In most cases the random variations are small and this is fortunate for analytic reasons, since then the random part of the solution can be determined by perturbation or expansion methods. In the next section a general perturbation method will be developed for the determination of the mean solution of a linear stochastic equation. This is a generalization of a method which has been used before [1] on the reduced wave equation for a random medium. It can be used to study propagation in many kinds of continuous and discrete media, as we shall see.

When the random variations are not small, perturbation methods become inconvenient and other methods of analysis must be sought. One such method, considered in §5, is that of direct determination of equations satisfied by the various moments of the solution. One finds that the equation for a moment of any order involves moments of higher order. Therefore an infinite system of equations must be considered for the simultaneous determination of all moments. These may be called the hierarchy equations, in analogy with the similar equations in statistical mechanics. One method of obtaining finite sets of equations is by assuming that some moment can be expressed in a particular way in terms of moments of lower order. Such a closure procedure is the basis for most of the so-called "dishonest" [1] methods of analyzing stochastic equations. Boltzmann's molecular chaos assumption in the kinetic theory of gases, Kirkwood's superposition approximation in the kinetic theory of liquids, Millionstchikov's quasinormality approximation in the theory of isotropic turbulence, the effective field approximation in the theory of multiple scattering by discrete scatterers, used by L. Foldy [2], M. Lax [3] and V. Twersky [4], and the local independence hypothesis of R. C. Bourret [5] and W. S. Ament [6] are all closure assumptions of this type. We shall see that the latter two approximations follow from the perturbation analysis of §2, up to terms of a certain order. Thus to the appropriate order they can be deduced. Therefore postulating them for large random variations amounts to using a perturbation expansion for large values of the expansion parameter. Although apparently unjustified, this procedure often yields accurate results. Hierarchy equations and closure approximations in the theory of turbulence have been discussed and analyzed by R. H. Kraichnan [7].

The hierarchy equations can be combined into a single functional differential equation for the generating functional of the moments of the solution and the random coefficients. This procedure is illustrated in §5. It was introduced into statistical mechanics in 1947 by N. N. Bogolyubov [**8**] and into the theory of turbulence by E. Hopf [**9**] in 1952. Methods for solving Bogolyubov's equation have been given by B. Zumino [**10**] and by R. M. Lewis and J. B. Keller [**11**], and for Hopf's equation by R. M. Lewis and R. H. Kraichnan [**12**]. These methods all involve expansions with respect to parameters.

An apparently quite independent description of wave propagation in a random medium is provided by the theory of radiative transfer, which is also called transport theory. The concept of a wave does not occur in it, nor do the related notions of phase and interference. In §6 we shall show how this theory can be deduced from a stochastic reduced wave equation when appropriate conditions are satisfied.

2. **Perturbation theory of linear stochastic equations.** Let us consider a linear operator M which transforms elements of some linear space into itself. Suppose $M = M(\alpha)$ depends upon a parameter α which ranges over a set A (measure space) and that $p(\alpha)$ is a probability density in A. Then $M(\alpha)$ is a stochastic operator. If g is a given element of the linear space and u is an unknown element, we may consider the following stochastic equation for u,

$$M(\alpha)u = g. \tag{1}$$

If this equation has a unique solution $u(\alpha)$ for each α, the solution depends upon α. Since $p(\alpha)$ is the probability density of α, it determines the probability density of $u(\alpha)$, and therefore $u(\alpha)$ is a random solution of (1). We shall attempt to obtain an equation for $\langle u \rangle$, the expected value of $u(\alpha)$, where the expected value of any function $f(\alpha)$ is defined by

$$\langle f \rangle = \int_A f(\alpha)p(\alpha)\, d\alpha. \tag{2}$$

To make this problem tractable, and at the same time to cover a large variety of practical cases, we shall assume that $M(\alpha, \varepsilon)$ depends upon a small parameter ε and that for $\varepsilon = 0$, M reduces to a sure (nonstochastic) operator L. Upon expanding $M(\alpha, \varepsilon)$ in powers of ε we may rewrite (1) in the form

$$[L + \varepsilon L_1(\alpha) + \varepsilon^2 L_2(\alpha) + O(\varepsilon^3)]u(\alpha, \varepsilon) = g. \tag{3}$$

The stochastic operators $L_1(\alpha)$ and $L_2(\alpha)$ represent stochastic perturbations of the sure operator L. We shall denote by u_0 a particular solution of the sure equation obtained by setting $\varepsilon = 0$ in (3)

$$Lu_0 = g. \tag{4}$$

Now assuming that L^{-1} is defined, we can rewrite (3) in the form

$$u = u_0 - L^{-1}(\varepsilon L_1 + \varepsilon^2 L_2)u + O(\varepsilon^3). \tag{5}$$

To obtain the solution of (5) we can employ the method of iterations or successive substitutions which yields

$$u = u_0 - \varepsilon L^{-1}L_1u_0 + \varepsilon^2 L^{-1}(L_1L^{-1}L_1 - L_2)u_0 + O(\varepsilon^3). \tag{6}$$

We now take the expectation value of (6) and obtain

$$\langle u\rangle = u_0 - \varepsilon L^{-1}\langle L_1\rangle u_0 + \varepsilon^2 L^{-1}(\langle L_1L^{-1}L_1\rangle - \langle L_2\rangle)u_0 + O(\varepsilon^3). \tag{7}$$

To obtain an equation for $\langle u\rangle$ we shall eliminate u_0 from (7) by first observing from (7) that

$$\begin{aligned} u_0 &= \langle u\rangle + \varepsilon L^{-1}\langle L_1\rangle u_0 + O(\varepsilon^2) \\ &= \langle u\rangle + \varepsilon L^{-1}\langle L_1\rangle\langle u\rangle + O(\varepsilon^2). \end{aligned} \tag{8}$$

We now use (8) in (7) and find an equation for $\langle u\rangle$,

$$\begin{aligned} \langle u\rangle = u_0 &- \varepsilon L^{-1}\langle L_1\rangle\langle u\rangle \\ &+ \varepsilon^2 L^{-1}[\langle L_1L^{-1}L_1\rangle - \langle L_1\rangle L^{-1}\langle L_1\rangle - \langle L_2\rangle]\langle u\rangle + O(\varepsilon^3). \end{aligned} \tag{9}$$

Upon multiplying (9) by L we obtain another form of equation for $\langle u\rangle$ which is

$$\{L + \varepsilon\langle L_1\rangle + \varepsilon^2[\langle L_1\rangle L^{-1}\langle L_1\rangle - \langle L_1L^{-1}L_1\rangle + \langle L_2\rangle]\}\langle u\rangle = g + O(\varepsilon^3). \tag{10}$$

This equation simplifies, when $\langle L_1\rangle = 0$, to

$$\{L + \varepsilon^2[\langle L_2\rangle - \langle L_1L^{-1}L_1\rangle]\}\langle u\rangle = g + O(\varepsilon^3). \tag{11}$$

By omitting the $O(\varepsilon^3)$ term in (9), (10) or (11) we obtain an explicit equation for $\langle u\rangle$, which is what we were seeking.

Let us now consider the case in which $u = u(x)$ is an n component vector function of a vector variable x. Then the operators L, L_1 and L_2 are nth order matrices, each element of which is an operator which may be a differential or integral operator. The inverse operator L^{-1} is also an nth order matrix which we shall represent as an integral operator. The kernel $G(x, x')$ is called the Green's matrix and is defined by the equation

$$LG(x, x') = I\delta(x - x'). \tag{12}$$

In (12) $\delta(x - x')$ denotes the Dirac delta function and I is the nth order unit matrix. With the aid of G, L^{-1} can be written as the integral operator

$$L^{-1}f(x) = \int G(x, x')f(x')\, dx'. \tag{13}$$

Now (10) can be written in the following form

$$(14)\quad \begin{aligned} L(x)\langle u(x)\rangle &+ \varepsilon\langle L_1(x)\rangle\langle u(x)\rangle + \varepsilon^2\Big[\langle L_1(x)\rangle\int G(x,x')\langle L_1(x')\rangle\langle u(x')\rangle\,dx' \\ &- \Big\langle L_1(x)\int G(x,x')L_1(x')\langle u(x')\rangle\,dx'\Big\rangle + \langle L_2(x)\rangle\langle u(x)\rangle\Big] = g(x) + O(\varepsilon^3). \end{aligned}$$

In (14) the explicit dependence of the operators upon x has been indicated. When $\langle L_1(x)\rangle = 0$, (14) becomes

$$(15)\quad \begin{aligned} L(x)\langle u(x)\rangle + \varepsilon^2\Big[\langle L_2(x)\rangle\langle u(x)\rangle - \Big\langle L_1(x)\int G(x,x')L_1(x')\langle u(x')\rangle\,dx'\Big\rangle\Big] \\ = g(x) + O(\varepsilon^3). \end{aligned}$$

Equations (14) and (15) become equations for $\langle u(x)\rangle$ when the $O(\varepsilon^3)$ terms are omitted. The resulting equations are the main results of this section.

Before applying our results to special cases, let us consider an alternative "dishonest" derivation of (11). We shall set $L_2 = 0$ and assume that $\langle L_1\rangle = 0$. Then upon taking expectation values in (3) we obtain

$$(16)\quad L\langle u\rangle + \varepsilon\langle L_1 u\rangle + O(\varepsilon^3) = g.$$

To determine $\langle L_1 u\rangle$ we multiply (15) by L_1 and take expectations to obtain

$$(17)\quad \langle L_1 u\rangle = -\varepsilon\langle L_1 L^{-1} L_1 u\rangle + O(\varepsilon^3).$$

In deriving (17) we used the hypothesis that $\langle L_1\rangle = 0$. Now we assume that the expectation on the right side of (17) splits into a product as follows:

$$(18)\quad \langle L_1 L^{-1} L u\rangle = \langle L_1 L^{-1} L_1\rangle\langle u\rangle.$$

When (18) is used in (17) and the result substituted into (16), (16) becomes

$$(19)\quad L\langle u\rangle - \varepsilon\langle L_1 L^{-1} L_1\rangle\langle u\rangle = g + O(\varepsilon^3).$$

This is just our result (11) when $L_2 = 0$. This derivation is essentially that of R. C. Bourret **[5]** who called (18) the assumption of local independence. Practically the same derivation was given by W. S. Ament **[6]** for the special case of Maxwell's equations for an electromagnetic wave in a random medium. Our derivation of (11) avoids the assumption (18) by effectively proving that it is correct within $O(\varepsilon)$, and this is all that is needed in the derivation when $L_2 = 0$.

3. **Applications to wave propagation in continuous media.** Let us consider a scalar function $u(x, t)$ which satisfies the wave equation in a medium having a random propagation speed $c[1 + \varepsilon\mu(x, t, \alpha)]^{-1}$. Then the wave equation is

$$(1)\quad \Delta u - c^{-2}[1 + 2\varepsilon\mu + \varepsilon^2\mu^2]u_{tt} = 0.$$

Upon comparing this equation with (2.3) we see that $g = 0$ while L, L_1 and L_2 are given by

$$L = \Delta - c^{-2}\partial_t^2, \tag{2}$$

$$L_1 = -2c^{-2}\mu\partial_t^2, \tag{3}$$

$$L_2 = -c^{-2}\mu^2\partial_t^2. \tag{4}$$

If $\langle\mu(x, t)\rangle = 0$ then $\langle L_1\rangle = 0$ and (2.11) applies to the mean wave, which is also called the "coherent" wave. In it L^{-1} occurs. If x denotes a point in three-dimensional space, then L^{-1} is given by

$$L^{-1}f(x, t) = \frac{-1}{4\pi}\int r^{-1}f(y, t - c^{-1}r)\,dy, \qquad r = |x - y|. \tag{5}$$

This choice of L^{-1} is such that $L^{-1}f$ tends to zero as t tends to $-\infty$ provided that f vanishes sufficiently rapidly at $t = -\infty$.

The term in (2.11) involving L^{-1} is

$$\begin{aligned}\langle L_1L^{-1}L_1\rangle\langle u(x, t)\rangle &= \Big\langle -2c^{-2}\mu(x, t)\,\partial_t^2\frac{(-1)}{4\pi}\int r^{-1}[-2c^{-2}\mu(y, t - c^{-1}r) \\ &\qquad \times \partial_t^2\langle u(y, t - c^{-1}r)\rangle]\,dy\Big\rangle \\ &= \frac{-1}{\pi c^4}\Big\langle \mu(x, t)\int r^{-1}[\mu_{tt}(y, t - c^{-1}r)\langle u(y, t - c^{-1}r)\rangle_{tt} \\ &\qquad + 2\mu_t(y, t - c^{-1}r)\langle u(y, t - c^{-1}r)\rangle_{ttt} \\ &\qquad + \mu(y, t - c^{-1}r)\langle u(y, t - c^{-1}r)\rangle_{tttt}]\,dy\Big\rangle.\end{aligned} \tag{6}$$

If the factor $\mu(x, t)$ is taken under the integral sign in (6), the expectation value of each term of the integrand can be expressed in terms of the correlation function R defined by

$$\langle\mu(x, t)\mu(y, s)\rangle = R(|x - y|, |t - s|). \tag{7}$$

In (7) we have written R as a function of $|x - y|$ and $|t - s|$ which we assume to be the case. This is appropriate when the medium is statistically homogeneous and isotropic in space and stationary in time. When (7) is used in (6) and the result inserted into (2.11), the final equation is

$$\begin{aligned}&[\Delta - c^{-2}(1 + \varepsilon^2\langle\mu^2\rangle)\,\partial_t^2]\langle u(x, t)\rangle + \frac{\varepsilon^2}{\pi c^4}\int r^{-1}[R_{\tau\tau}(r, r/c)\langle u(y, t - c^{-1}r)\rangle_{tt} \\ &\quad - 2R_\tau(r, r/c)\langle u(y, t - c^{-1}r)\rangle_{ttt} + R(r, r/c)\langle u(y, t - c^{-1}r)\rangle_{tttt}]\,dy = 0.\end{aligned} \tag{8}$$

In (8) R_τ denotes the derivative of R with respect to its second argument and the $O(\varepsilon^3)$ term has been omitted. It is to be noted that from (7), $\langle\mu^2\rangle = R(0, 0)$. If the medium is not statistically isotropic then R depends upon the vector $x - y$ rather than just its length. In this case (8) still applies if the first argument of R

is understood to denote this vector. R. C. Bourret [**5**] derived the special form of (8) which applies when $\mu(x)$ is independent of t so that $R_\tau = R_{\tau\tau} = 0$, but he omitted the term $-c^{-2}\langle\mu^2\rangle\,\partial_t^2\langle u\rangle$.

Let us now apply (8) to a coherent wave $\langle u\rangle$ which is time harmonic with angular frequency ω

$$\langle u(x, t)\rangle = e^{-i\omega t}v(x). \tag{9}$$

If we introduce k_0 defined by $k_0 = \omega/c$, the resulting equation for $v(x)$ becomes

$$\begin{aligned}&[\Delta + k_0^2(1 + \varepsilon^2\langle\mu^2\rangle)]v(x)\\ &\quad + \frac{\varepsilon^2 k_0^2}{\pi c^2}\int r^{-1}\, e^{ik_0 r}[\omega^2 R(r, r/c) - i\omega R_\tau(r, r/c) - R_{\tau\tau}(r, r/c)]v(y)\, dy = 0.\end{aligned} \tag{10}$$

W. C. Meecham [**13**] and F. C. Karal and J. B. Keller [**14**] have derived the special form of (10) in which $R_\tau = R_{\tau\tau} = 0$ by applying the present method to the reduced wave equation for a time independent random medium. (Meecham did not include the term in $\varepsilon^2\mu^2$ in his original equation, so $\varepsilon^2\langle\mu^2\rangle$ does not occur in his result.)

An important consequence of (10) is the effective propagation constant k of the random medium. We may find it by considering a plane wave solution with propagation vector $\mathbf{k}$, with $k = |\mathbf{k}|$. Thus let

$$v(x) = e^{i\mathbf{k}\cdot\mathbf{x}}. \tag{11}$$

To evaluate the volume integral in (10) we may first integrate over the surface of a sphere of radius r centered at x. This integral can be evaluated by using a mean value theorem which applies to any solution v of the reduced wave equation

$$(\Delta + k^2)v = 0. \tag{12}$$

It yields

$$\int v(\mathbf{x} + \mathbf{r})\, dS = \frac{4\pi r}{k}\sin kr \cdot v(x). \tag{13}$$

Upon inserting (11) into (10) and using (13), we obtain

$$\begin{aligned}k^2 = k_0^2(1 + \varepsilon^2\langle\mu^2\rangle) + \frac{4\varepsilon^2 k_0^2}{kc^2}\int_0^\infty & e^{ik_0 r}\sin kr[\omega^2 R(r, r/c)\\ &- i\omega R_\tau(r, r/c) - R_{\tau\tau}(r, r/c)]\, dr.\end{aligned} \tag{14}$$

This is an equation for k. For the special case in which $R_\tau = R_{\tau\tau} = 0$, it was obtained previously [**13**; **14**]. Any solution v of (10) also satisfies (12) with k determined by (14) as we can see from the above derivation of (14).

We can solve (14) for k when ε is small by replacing k by k_0 on the right side. When $R_\tau = R_{\tau\tau} = 0$, so that R depends only upon its first argument, this procedure yields the following known result [**1**; **13**; **14**]

$$(k/k_0)^2 = 1 + \varepsilon^2\langle\mu^2\rangle - 2i\varepsilon^2 k_0\int_0^\infty (e^{2ik_0 r} - 1)R(r)\, dr. \tag{15}$$

The imaginary part of k is the attenuation coefficient for the coherent wave. We shall now prove from (15) that the attenuation coefficient is positive and also that $\operatorname{Re} k > (1 + \varepsilon^2\langle\mu^2\rangle) \operatorname{Re} k_0$, which shows that the phase velocity of the coherent wave is less when $\varepsilon \neq 0$ than when $\varepsilon = 0$. This is in accordance with expectations, since the scattering of the wave should increase the distance which it travels in going from one point to another, and therefore decrease its effective velocity. Our demonstration is essentially that of W. C. Meecham **[13]**.

To prove these things we introduce $\rho(s)$, the three-dimensional Fourier transform of $R(r)$, and utilize (13) to rewrite it as follows:

$$\rho(s) = \int e^{i\mathbf{s}\cdot\mathbf{r}} R(r)\, d\mathbf{r} = \int_0^\infty \frac{4\pi r}{s} \sin sr R(r)\, ds. \tag{16}$$

Since $\rho(s)$ is the Fourier transform of a correlation function, $\rho(s) > 0$. The inverse Fourier sine transform yields

$$rR(r) = \frac{1}{2\pi^2} \int_0^\infty s\rho(s) \sin sr\, ds. \tag{17}$$

Upon using (17) in the integral in (15), that integral becomes

$$\begin{aligned}\int_0^\infty (e^{2ik_0 r} - 1)R(r)\, dr &= \frac{1}{2\pi^2} \int_0^\infty \int_0^\infty (e^{2ik_0 r} - 1) r^{-1} s\rho(s) \sin sr\, ds\, dr \\ &= -\frac{1}{4\pi} \int_0^{2k_0} s\rho(s)\, ds \\ &\quad + \frac{i}{4\pi^2} \int_0^\infty s\rho(s) \log |(s + 2k_0)/(s - 2k_0)|\, ds.\end{aligned} \tag{18}$$

In (18) we have made use of known Fourier sine transforms. Now (15) and (18) yield

$$\begin{aligned}k = k_0 + k_0\varepsilon^2\langle\mu^2\rangle &+ \frac{\varepsilon^2 k_0^2}{4\pi^2} \int_0^\infty s\rho(s) \log |(s + 2k_0)/(s - 2k_0)|\, ds \\ &+ \frac{i\varepsilon^2 k_0^2}{4\pi} \int_0^{2k_0} s\rho(s)\, ds.\end{aligned} \tag{19}$$

Both integrals in (19) are positive, which proves the statements in the preceding paragraph. It is also of interest to note from (19) that only those Fourier components of the correlation with $s < 2k_0$ contribute to the attenuation. Thus the long wavelength components of the correlation attenuate the wave while the short wavelength ones merely scatter it and slow it down.

It is instructive to consider another derivation of (15) which is of quite general applicability and is based upon physical considerations. Let us consider a time harmonic wave $u_0(x) = e^{ik_0 x}$ incident along the x-axis upon a slab of scattering material lying between the planes $x = 0$ and $x = \delta x$. Let the wave scattered by unit volume of the scattering material be $U_s(x) = f(\theta) r^{-1} e^{ik_0 r}$ at large distances r

from the scattering material, where θ is the polar coordinate of x with the x-axis as polar axis. Then the total field $u(x) = u_0(x) + u_s(x)$ is given by

$$u(x) = e^{ik_0x} + \int_0^{\delta x}\int_0^{\infty} f(\theta)r^{-1}\, e^{ik_0r} 2\pi\rho \, d\rho \, dx'. \tag{20}$$

Here $r^2 = \rho^2 + (x-x')^2$. If $k_0\delta x << 1$ we may set $x' = 0$ in (20) and then the x' integration merely yields multiplication by δx. Now $\rho d\rho = rdr$ and $\theta = \cos^{-1}(x/r)$ so (20) becomes for $x > 0$,

$$\begin{aligned} u(x) &= e^{ik_0x} + 2\pi\delta x\int_x^{\infty} f[\cos^{-1}(x/r)]\, e^{ik_0r}\, dr \\ &\approx e^{ik_0x} - \frac{2\pi\delta x}{ik_0} f(0)\, e^{ik_0x}. \end{aligned} \tag{21}$$

The last form of (21) is obtained by integration by parts and then neglecting the contribution from infinity, which is valid if k_0 has a positive imaginary part, and neglecting the remaining integral. Actually the last form of (21) is a solution of the reduced wave equation, and it is exactly correct when the exact form of $U_s(x)$ is used rather than the far field form.

When δx is small we may rewrite (21) at $x = \delta x$ in the form

$$u(x) = e^{i[k_0 + 2\pi f(0)/k_0]\delta x}. \tag{22}$$

Thus the scattering material behaves as if it had an effective propagation constant k given by

$$k = k_0 + 2\pi f(0)/k_0. \tag{23}$$

Upon squaring both sides of this equation, and assuming that $f(0)$ is small, we obtain

$$k^2 = k_0^2 + 2\pi f(0). \tag{24}$$

Equation (23) was essentially obtained by Rayleigh **[15]** and (24) by F. Reiche **[16]**, L. Foldy **[2]**, and M. Lax **[3]** for discrete scatterers of various types. The quantity $f(0)$ is called the forward scattering amplitude per unit volume. From (23) we have $\operatorname{Im} k = (2\pi/k_0) \operatorname{Im} f(0)$. The right side of this relation is just the total energy scattered by unit volume of scatterer, according to the so-called cross section theorem. Thus the attenuation coefficient is equal to this energy. Direct calculations of the scattered energy have yielded the expression for $\operatorname{Im} k$ contained in (15). That calculation and relevant references are given by L. A. Chernov **[15]**.

Let us apply the preceding considerations to the time harmonic coherent wave $\langle u(x, t)\rangle = e^{-i\omega t}v(x)$. The coherent scattered wave, defined to be $\langle u\rangle - u_0$, can be found from (2.9). When $\langle L_1\rangle = 0$ it is given by

$$\langle u\rangle - u_0 = \varepsilon^2 L^{-1}[\langle L_1 L^{-1} L_1\rangle - \langle L_2\rangle]\langle u\rangle + O(\varepsilon^3). \tag{25}$$

The quantity $\langle L_1 L^{-1} L_1\rangle\langle u\rangle$ has been computed in (6) and specialized to a statistically homogeneous isotropic medium in (8), where it is minus the term involving the integral. It is further specialized to a time harmonic wave in (10) and is again the integral term multiplied by $-e^{-i\omega t}$. The term $\langle L_2\rangle\langle u\rangle$, also given in (10), is $k_0^2\langle\mu^2\rangle v(x)\, e^{-i\omega t}$. Thus the right side of (25) is L^{-1} applied to $-e^{-i\omega t}$ times the ε^2 term in (10). When $v(x) = e^{i\mathbf{k}\cdot\mathbf{x}}$ it is L^{-1} applied to $-e^{-i\omega t+i\mathbf{k}\cdot\mathbf{x}}$ times the ε^2 term in (14), and that term is just a constant which we shall denote by α. Thus (25) becomes

$$(26) \qquad \langle u\rangle - u_0 = -\alpha L^{-1}\, e^{-i\omega t+i\mathbf{k}\cdot\mathbf{x}} = \frac{\alpha e^{-i\omega t}}{4\pi}\int_V r^{-1}\, e^{i\mathbf{k}\cdot\mathbf{y}+ik_0 r}\, dy.$$

Here V is the scattering volume. If we set $k = k_0$ in the integral, and evaluate it at a point x far from V in the forward direction (i.e., in the direction of $\mathbf{k}$) we find that the exponent is constant. If $y = 0$ is a point in V and $r = |x|$ then (26) becomes

$$(27) \qquad \langle u(x, t)\rangle - u_0(x, t) = \frac{\alpha V}{4\pi r}\, e^{-i\omega t+ikr}.$$

Thus the forward scattering amplitude per unit volume is $f(0) = \alpha/4\pi$. When α, the coefficient of ε^2, is replaced by $4\pi f(0)$ in (14), (14) becomes identical with (24). Thus (24), which was obtained by a physical argument, also follows from our formal result (14). In the case in which $R_\tau = R_{\tau\tau} = 0$, this was shown by W. C. Meecham [**13**].

It should be observed that the solution of (14) given by (15) becomes infinite, and is therefore not valid when the correlation function $R(r)$ is a constant independent of r. This is the case in which the wavelength is small compared to the correlation length, defined as the distance beyond which the correlation function is practically zero. In this case if $R_\tau = R_{\tau\tau} = 0$ and $R(r) = \langle\mu^2\rangle$, (14) becomes

$$(28) \qquad k^2 = k_0^2(1 + \varepsilon^2\langle\mu^2\rangle) + 2\varepsilon^2\langle\mu^2\rangle k_0^4 k^{-1}\left[\frac{1}{k_0 + k} - \frac{1}{k_0 - k}\right].$$

The solution of (28) for k is

$$(29) \qquad k = k_0(1 + \varepsilon\sqrt{\langle\mu^2\rangle}) + O(\varepsilon^2).$$

We see from (29) that the correction to k_0 is of order ε when the correlation length is long compared to the wavelength, while from (15) it is of order ε^2 when the correlation length is not long compared to the wavelength. The result (29) is also given by F. C. Karal and J. B. Keller [**14**], who give similar results for other types of waves.

The method of §2, which we have employed in this section, has been applied to the equations of elastic wave propagation and to Maxwell's equations by F. C. Karal and J. B. Keller [**14**]. It has also been applied under less general conditions to the wave equation and to Maxwell's equations by R. C. Bourret [**5**], as well as to various other equations.

4. **Wave propagation in media of discrete scatterers.** Let us now consider wave propagation in a homogeneous medium containing a random distribution of discrete scatterers. If the material constituting the individual scatterers differs very little from the surrounding medium, the present case can be treated by specializing the preceding analysis. However, if the individual scatterers are quite different from the surrounding medium, that analysis does not suffice in its present form. Therefore we shall give a modification of it which is applicable then provided a suitably defined collision operator associated with each scatterer is small.

We begin by assuming that the equation for the field u is of the form

$$\left(L - \sum_i V_i\right) u = 0. \tag{1}$$

Here the operator V_i represents the effect of the ith scatterer. Let u_0 denote a particular solution of the equation which applies when no scatterers are present, namely

$$Lu_0 = 0. \tag{2}$$

Now we may rewrite (1) for a particular solution in the form

$$u = u_0 + L^{-1} \sum_i V_i u. \tag{3}$$

Let us define u^i by

$$u^i = u - L^{-1} V_i u. \tag{4}$$

Then u^i represents the field incident upon the ith scatterer. In terms of u^i we define the ith transition operator T_i' by

$$T_i' u^i = V_i u. \tag{5}$$

From (4) and (5) T_i' is given by

$$T_i' = V_i(1 - L^{-1}V_i)^{-1} = V_i(L - V_i)^{-1}L. \tag{6}$$

We shall henceforth assume that T_i' is known since it depends only upon scattering by one scatterer, and shall continue our analysis in terms of it.

By using (5) and (4) we see that

$$u^i = (1 + L^{-1}T_i')^{-1}u. \tag{7}$$

Now (7) and (5) permit us to write (3) in the form

$$u = u_0 + L^{-1} \sum_i T_i'(1 + L^{-1}T_i')^{-1}u. \tag{8}$$

Upon applying L to (8) it becomes

$$\left[L - \sum_i T_i'(1 + L^{-1}T_i')^{-1}\right] u = 0. \tag{9}$$

To make explicit our hypothesis that T_i' is small we shall write $T_i' = \varepsilon T_i$. Upon expanding the quantity $(1 + \varepsilon L^{-1}T_i)^{-1}$ in powers of ε, we may write (9) in the form

$$\left[L - \varepsilon \sum_i T_i' + \varepsilon^2 \sum_i T_i L^{-1} T_i + O(\varepsilon^3)\right] u = 0. \tag{10}$$

This is the form of the equation for u which we sought.

We now observe that (10) is of the form (2.3) with L_1 and L_2 given by

$$L_1 = -\sum_i T_i, \tag{11}$$

$$L_2 = \sum_i T_i L^{-1} T_i. \tag{12}$$

Thus the analysis and results of §2 apply to (10), and $\langle u \rangle$ satisfies (2.10). When the T_i all have the same mean and there are N of them, we have from (11) and (12)

$$\langle L_1 \rangle = -N\langle T_1 \rangle, \tag{13}$$

$$\langle L_2 \rangle = N\langle T_1 L^{-1} T_1 \rangle, \tag{14}$$

$$\langle L_1 \rangle L^{-1} \langle L_1 \rangle = N^2 \langle T_1 \rangle L^{-1} \langle T_1 \rangle. \tag{15}$$

If $\langle T_i L^{-1} T_j \rangle$ is independent of i and j when $i \neq j$ and independent of i when $i = j$ then we also have

$$\langle L_1 L^{-1} L_1 \rangle = \sum_{i,j} \langle T_i L^{-1} T_j \rangle = N\langle T_1 L^{-1} T_1 \rangle + N(N-1)\langle T_1 L^{-1} T_2 \rangle. \tag{16}$$

Now (2.10) becomes

$$\{L - \varepsilon N\langle T_1 \rangle - (\varepsilon N)^2[(1 - N^{-1})\langle T_1 L^{-1} T_2 \rangle - \langle T_1 \rangle L^{-1} \langle T_1 \rangle]\}\langle u \rangle = O(\varepsilon^3). \tag{17}$$

This is the equation for $\langle u \rangle$ which we shall utilize.

Let us now assume that the ith scatterer, represented by the transition operator T_i, is characterized by a position x_i and an internal coordinate or label f_i. Then we may write $T = T(x_i, f_i)$. Let us further assume that each f_i is independent of all other variables so the mean of T_i with respect to f_i is a function of x_i only which we denote by $T(x_i)$. In addition let each x_i be uniformly distributed over a region of volume v which contains the N scatterers. Then if we let N and v become infinite with $N/v = \rho$ remaining constant, and drop the $O(\varepsilon^3)$ term, (17) becomes

$$\begin{aligned}\Big\{L - \varepsilon\rho \int T_1(x)\, dx - \varepsilon^2\rho^2 \Big[\iint T(x_1)L^{-1}T(x_2)F(x_1, x_2)\, dx_1\, dx_2 \\ - \int T(x)\, dx L^{-1} \int T(x)\, dx\Big]\Big\}\langle u \rangle = 0.\end{aligned} \tag{18}$$

In (18) $F(x_1, x_2)$ denotes v^2 times the probability density of scatterer 1 being at x_1 and scatterer 2 at x_2. We shall assume that $F(x_1, x_2)$ depends only upon $r = x_1 - x_2$ and define the pair correlation function $g(r)$ by $g(r) = f(r) - 1$. When $g(\mathrm{r}) = g(r)$, g is called the radial distribution function.

As an example of the use of (18) let us consider the case in which u is a scalar and $L = \Delta + k_0^2$. Then the Green's function of L is $G(|x - x'|) = G(r) = -e^{ik_0 r}/4\pi r$. Let us assume that the scatterers are small isotropic scatterers so that the wave scattered by a scatterer at x_1 is $L^{-1}(x)T(x_1, f_1)u = -4\pi f_1 G(|x - x_1|)u(x_1)$, a spherical wave emanating from x_1. Upon applying $L(x)$ to this equation we find that T is given by $T(x_1, f_1) = -4\pi f_1 \delta(x - x_1)$. Let us denote the mean value of f_1 by f. Then (18) becomes

$$(19)\qquad (\Delta + k_0^2 + 4\pi\varepsilon\rho f)\langle u\rangle - (4\pi\varepsilon\rho f)^2\left[\int G(|x - x'|)F(x - x')\langle u(x')\rangle\, dx' - \int G(|x - x'|)\langle u(x')\rangle\, dx'\right] = 0.$$

We now assume that $\langle u(x)\rangle$ satisfies (3.12) or is a plane wave with propagation constant k, that $F(\mathbf{r}) = F(r)$ and then use (3.13) to simplify the integral in (19). Then (19) yields the following equation for k^2

$$(20)\qquad k^2 = k_0^2 + 4\pi\varepsilon\rho f + (4\pi\varepsilon\rho f)^2 k^{-1}\int_0^\infty e^{ik_0 r} g(r)\sin kr\, dr.$$

The solution of (20) valid to order ε^2 is

$$(21)\qquad k^2 = k_0^2 + 4\pi\varepsilon\rho f + (4\pi\varepsilon\rho f)^2 k_0^{-1}\int_0^\infty e^{ik_0 r} g(r)\sin k_0 r\, dr.$$

Let us compare this result with that obtained from a random continuous medium for which u satisfies

$$(22)\qquad [\Delta + k_0^2 + \varepsilon\mu(x)k_0^2]u = 0.$$

From (22) we see that $L = \Delta + k_0^2$, $L_1 = \mu k_0^2$, $\langle L_1\rangle = k_0^2\langle\mu\rangle$, and $L_2 = 0$. By using these values in (2.14) and assuming that $\langle\mu\rangle$ is a constant, we obtain

$$(23)\qquad (\Delta + k_0^2 + \varepsilon k_0^2\langle\mu\rangle)\langle u\rangle + \varepsilon^2 k_0^4\langle\mu\rangle^2\int G(x - x')\langle u(x')\rangle\, dx' - \varepsilon^2 k_0^4\int G(x - x')\langle\mu(x)\mu(x')\rangle\langle u(x')\rangle\, dx' = 0.$$

We now assume that the correlation function $\langle\mu(x)\mu(x')\rangle$ is a function only of $r = |x - x'|$ and in terms of it we define $g_c(r)$ by

$$(24)\qquad g_c(r) = \frac{\langle\mu(x)\mu(x')\rangle}{\langle\mu\rangle^2} - 1.$$

Then if $\langle u(x)\rangle$ is a plane wave or satisfies (3.12) with propagation constant k, we can simplify the integrals in (23) by using (3.13). In this way from (23) we obtain for k the equation

$$(25)\qquad k^2 = k_0^2 + \varepsilon k_0^2\langle\mu\rangle + (\varepsilon k_0^2\langle\mu\rangle)^2 k^{-1}\int_0^\infty e^{ik_0 r} g_c(r)\sin kr\, dr.$$

Upon comparing (25) with (20) we see that they become identical if $\langle\mu\rangle = 4\pi\rho f/k_0^2$ and if $g(r) = g_c(r)$.

We may interpret this result by noting that the discrete random medium can be described as a continuous medium with refractive index $n = k/k_0$, where k is given by (20) or (21). This description is appropriate when considering the mean wave $\langle u\rangle$ which satisfies the equation $(\Delta + k_0^2 n^2)\langle u\rangle = 0$. In this description the effective refractive index n is a constant which is given up to order ε^2 by using (21) for k. Alternatively we could use only the terms up to order ε in (21) and define a different equivalent continuous medium with refractive index $1 + \varepsilon\mu/2 = (k_0^2 + 4\pi\varepsilon\rho f)^{1/2}/k_0 = 1 + 2\pi\varepsilon\rho f k_0^{-2}$ so that $\mu = 4\pi\rho f k_0^{-2}$. Then if we let the density of scatterers be a random function of position, $\rho = \rho(x)$, it follows that $\mu = \mu(x)$, so the refractive index is also a random function. Then the mean wave in this medium has a propagation constant given by (25) and its effective refractive index is the constant k/k_0. As we have seen, this effective index is the same as the constant n obtained directly from the discrete case, provided $g(r) = g_c(r)$. Now $g_c(r) = \langle\rho(x)\rho(x')\rangle\langle\rho\rangle^{-2} - 1$. If $\rho(x)$ denotes the probable density at x, this expression is the same as $g(r)$.

To consider more general scatterers let us return to (18) and assume that $T(x)$ differs from the operator $T(0)$ by a translation, which we express by

$$T(x) = e^{-\mathbf{x}\cdot\nabla} T(0)\, e^{\mathbf{x}\cdot\nabla}. \tag{26}$$

Let us also assume that L is independent of x. Then the operator in (18) is translationally invariant so (18) has plane wave solutions. To find them we may assume that $\langle u\rangle$ is a plane wave and insert it into (18). When u is a scalar the result can be expressed in terms of "matrix elements" of the operator in (18). For any operator such as L they are defined by

$$L_{k'k} = \lim_{v\to\infty} \frac{1}{v}\int_v e^{-i\mathbf{k}'\cdot\mathbf{x}} L\, e^{i\mathbf{k}\cdot\mathbf{x}}\, dx. \tag{27}$$

However, for $T(0)$ we define them by

$$T_{k'k}(0) = \int e^{-i\mathbf{k}'\cdot\mathbf{x}} T(0)\, e^{i\mathbf{k}\cdot\mathbf{x}}\, dx. \tag{28}$$

Now if (18) has a plane wave solution $e^{i\mathbf{k}\cdot\mathbf{x}}$ then it follows from (18), (27) and (28) that k must satisfy the equation

$$\begin{aligned} L_{kk} - \varepsilon\rho T_{kk}(0) - \varepsilon^2\rho^2 \Big[\int T(x_1) L^{-1} T(x_2) F(x_1, x_2)\, dx_1\, dx_2 \\ - \int T(x)\, dx L^{-1} \int T(x)\, dx \Big]_{kk} = 0. \end{aligned} \tag{29}$$

If we use the facts that L^{-1} is a diagonal matrix and that $F(x_1, x_2) = F(x_1 - x_2)$ we may evaluate the last matrix element in (29) by using the Fourier integral, or equivalently the plane wave representation of the operators. The result after some calculation is that (29) becomes the following equation for k:

$$L_{kk} - \varepsilon\rho T_{kk}(0) + \varepsilon^2\rho^2 \frac{1}{(2\pi)^3} \int T_{ks}(0) L_{ss}^{-1} T_{sk}(0) \int e^{i(\mathbf{s}-\mathbf{k})\cdot\mathbf{r}} g(\mathbf{r})\, d\mathbf{r}\, d\mathbf{s} = 0. \tag{30}$$

The equation (30) for k may be rewritten in terms of the operator $f = -T(0)/4\pi$ and becomes

$$L_{kk} + 4\pi\varepsilon\rho f_{kk} + \frac{2}{\pi}(\varepsilon\rho)^2 \int f_{ks} L_{ss}^{-1} f_{sk} \int e^{i(\mathbf{s}-\mathbf{k})\cdot\mathbf{r}} g(\mathbf{r})\, d\mathbf{r}\, d\mathbf{s} = 0. \tag{31}$$

This result is identical with that of M. Lax **[3b]** which is obtained from his equations (3.19) and (4.10) provided his factor $c = (1 - J)^{-1}$ is expanded in the form $c \approx 1 + J$, which is appropriate for weak scatterers. His method of derivation, which differs from ours, involves an assumption which will be considered later. We note that the last integrand in (30) involves only the $\mathbf{s} - \mathbf{k}$ Fourier component of the pair correlation function $g(\mathbf{r})$. This is the component of $g(\mathbf{r})$ which yields first order Bragg reflection of a wave with the propagation vector $\mathbf{k}$ into one with the propagation vector $\mathbf{s}$. It also involves only those double scattering processes which lead from a wave with the propagation vector $\mathbf{k}$ into one with the same propagation vector, i.e., which combine to yield forward scattering without change of wavelength. When we specialize (30) to the case in which $L = \Delta + k_0^2$, $g(\mathbf{r}) = g(r)$ and $T(0) = -4\pi f\delta(x)$ we obtain the previous result (19), which provides a check on our calculations.

Let us now see how our original equation (17) simplifies when all V_i are small. To this end we shall indicate that V_i is small by replacing V_i by εV_i in (6), from which it then follows that

$$T_i = V_i + \varepsilon V_i L^{-1} V_i + O(\varepsilon^2). \tag{32}$$

When (32) is used in (17), (17) becomes

$$\begin{aligned}\{L - \varepsilon N\langle V_1\rangle - (\varepsilon N)^2[(1 - N^{-1})\langle V_1 L^{-1} V_2\rangle - \langle V_1\rangle L^{-1}\langle V_1\rangle \\ + N^{-1}\langle V_1 L^{-1} V_1\rangle]\}\langle u\rangle = O(\varepsilon^3).\end{aligned} \tag{33}$$

This equation also follows directly from (1) in which $L_1 = \sum_i V_i$ and $L_2 = 0$. When these expressions are used in (2.10) and $\langle V_i\rangle$ is independent of i while $\langle V_i L^{-1} V_j\rangle$ is independent of i and j for $i \neq j$ and is independent of i for $i = j$, (33) follows at once. This shows that our original analysis applies directly when the scatterers are "weak." If N is large the terms in N^{-1} can be omitted and (33) simplifies to

$$\{L - \varepsilon N\langle V_1\rangle - (\varepsilon N)^2(\langle V_1 L^{-1} V_2\rangle - \langle V_1\rangle L^{-1}\langle V_1\rangle)\}\langle u\rangle = O(\varepsilon^3). \tag{34}$$

If the N^{-1} term is omitted in (17), then (17) has the same form as (34) with T_i replacing V_i. Thus results deduced for weak scatterers from (34) can be converted into results valid for strong scatterers by replacing V_i by T_i in them.

A "dishonest" derivation of (17), which does not yield the ε^2 term, has been used in special cases by L. Foldy [2] and V. Twersky [4], and in a general form by M. Lax [3]. It is based upon a set of equations for the u^i which are derived by first subtracting $L^{-1}V_j u$ from each side of (3) to yield

$$u - L^{-1}V_j u = u_0 + L^{-1}\sum_{i\neq j} V_i u. \tag{35}$$

From (4) the left side of (35) is u^j while the right side can be expressed in terms of the u^i with the aid of (5). Then (35) becomes

$$u^j = u_0 + L^{-1}\sum_{i\neq j} T_i' u^i. \tag{36}$$

There is one equation of the form (36) for each j and these N equations determine the u^j self-consistently. It is easy to show that when the u^i satisfy the set (36), a solution of (3) is

$$u = u_0 + L^{-1}\sum_{i} T_i' u^i. \tag{37}$$

Upon taking the expectation value of (37) we obtain

$$\langle u\rangle = u_0 = L^{-1}\sum_{i}\int T_i'\langle u^i\rangle_i \, dp(V_i). \tag{38}$$

Here $\langle u^i\rangle_i$ denotes the conditional expectation of u^i with V_i fixed, i.e., the expectation with respect to the conditional probability distribution of all the V_j with $j \neq i$ and $dp(V_i)$ denotes the probability distribution of V_i. To determine $\langle u^i\rangle_i$ we take the expectation of (36) keeping V_j fixed. This yields

$$\langle u^j\rangle_j = u_0 + L^{-1}\sum_{i\neq j}\int T_i'\langle u^i\rangle_{ij} \, dp(V_i \mid V_j). \tag{39}$$

Here $\langle u^i\rangle_{ij}$ is the conditional expectation of u^i with V_i and V_j fixed and $dp(V_i \mid V_j)$ is the conditional distribution of V_i with V_j fixed. We see that (39) involves $\langle u^i\rangle_{ij}$ and to determine it we could take the expectation of (36) with V_i and V_j fixed but this would introduce new conditional expectations. Equations (38) and (39) are the first two in a sequence of equations for successive conditional expectations called the hierarchy equations.

In order to avoid solving the hierarchy equations Foldy and Twersky assume that the "effective" field $\langle u^i\rangle_i$ incident upon scatterer i is the same as the average field $\langle u\rangle$. Lax introduces a factor c and writes

$$\langle u^i\rangle_i = c\langle u\rangle. \tag{40}$$

Then (38) becomes

$$\langle u\rangle = u_0 + L^{-1}cN\langle T_1'\rangle\langle u\rangle. \tag{41}$$

Upon applying L to (41) it becomes

$$[L - cN\langle T_1'\rangle]\langle u\rangle = 0. \tag{42}$$

When $c = 1$ this is the same as (17), except for the ε^2 terms and this completes the derivation. Lax [**3b**] has also used (39), assuming that $\langle u^i \rangle_{ij} = \langle u^i \rangle_i$ together with a variational procedure to obtain an expression for c. When his value of c is used in (42) it leads to (31), as pointed out below (31). Lax [**3**] has concluded from (31) that in (42) T_1' should be the scattering operator appropriate to scattering in the "new" medium governed by (42). Twersky [**4**] instead takes T_1' to be the scattering operator which results when an incident wave appropriate to the new medium is scattered into the original medium. Both of these choices as well as the effective field hypothesis (40) are unproved. Our derivation of (17) confirms the result (42) of the effective field hypothesis with $\varepsilon = 1$ through terms of order ε, but it does not confirm the hypothesis itself. Our result (31) confirms Lax's calculation of c to order ε^2.

To examine the validity of the hypothesis (40) with $c = 1$, we proceed as follows. Let us use (5) in (4) and then take the expectation of the resulting equation with V_i fixed. We obtain

$$\langle u^i \rangle_i = \langle u \rangle_i - L^{-1} T_i' \langle u^i \rangle_i. \tag{43}$$

Solving (43) for $\langle u^i \rangle_i$ yields

$$\langle u^i \rangle_i = (1 + L^{-1} T_i')^{-1} \langle u \rangle_i. \tag{44}$$

Upon setting $T_i' = \varepsilon T_i$ and expanding, we obtain from (44)

$$\langle u^i \rangle_i = [L - \varepsilon L^{-1} T_i + \varepsilon^2 L^{-1} T_i L^{-1} T_i + O(\varepsilon^3)] \langle u \rangle_i. \tag{45}$$

Performing the same expansion in (8) and solving by iterations yields

$$u = u_0 + \varepsilon L^{-1} \sum_j T_j u_0 + \varepsilon^2 L^{-1} \sum_{j \neq k}\sum T_j L^{-1} T_k u_0 + O(\varepsilon^3). \tag{46}$$

The mean of (46) is

$$\langle u \rangle = u_0 + \varepsilon N L^{-1} \langle T_1 \rangle u_0 + \varepsilon^2 N(N-1) L^{-1} \langle T_1 L^{-1} T_2 \rangle u_0 + O(\varepsilon^3). \tag{47}$$

The mean of (46) with T_i fixed, and $i \neq 1, 2$ for notational convenience, is

$$\begin{aligned} \langle u \rangle_i = u_0 &+ \varepsilon[(N-1) L^{-1} \langle T_1 \rangle_i + L^{-1} T_i] u_0 + \varepsilon^2 (N-1)[(N-2) L^{-1} \langle T_1 L^{-1} T_2 \rangle_i \\ &- L^{-1} \langle T_1 L^{-1} T_1 \rangle_i + L^{-1} \langle T_1 \rangle_i L^{-1} T_i + L^{-1} T_i L^{-1} \langle T_1 \rangle_i] u_0 + O(\varepsilon^3). \end{aligned} \tag{48}$$

Now (45) and (48) yield

$$\begin{aligned} \langle u^i \rangle_i &= \langle u_i \rangle - \varepsilon^2 (N-1) L^{-1} T_i L^{-1} \langle T_1 \rangle_i u_0 + O(\varepsilon^3) \\ &= u_0 + \varepsilon[(N-1) L^{-1} \langle T_1 \rangle_i + L^{-1} T_i] u_0 + \varepsilon^2 (N-1)[(N-2) L^{-1} \langle T_1 L^{-1} T_2 \rangle_i \\ &\quad - L^{-1} \langle T_1 L^{-1} T_1 \rangle_i + L^{-1} \langle T_1 \rangle_i L^{-1} T_i] u_0 + O(\varepsilon^3). \end{aligned} \tag{49}$$

To test the effective field hypothesis (40) with $c = 1$ we must compare (49) with (47). To do so we form the difference

$$\begin{aligned} \langle u^i \rangle_i - \langle u \rangle = {} & \varepsilon N L^{-1} [(1 - N^{-1}) \langle T_1 \rangle_i - \langle T_1 \rangle + N^{-1} T_i] u_0 \\ & + \varepsilon^2 N(N-1) L^{-1} [(1 - 2N^{-1}) \langle T_1 L^{-1} T_2 \rangle_i - \langle T_1 L^{-1} T_2 \rangle \\ & - N^{-1} \langle T_1 L^{-1} T_1 \rangle_i + N^{-1} \langle T_1 \rangle_i L^{-1} T_i] u_0 + O(\varepsilon^3). \end{aligned} \tag{50}$$

Even if the terms in N^{-1} are dropped from (50) when N is large, the difference between $\langle u^i \rangle_i$ and $\langle u \rangle$ is still $O(\varepsilon)$. This order of difference leads to a result (42) which is correct through terms of order ε when $c = 1$, but fails to yield the ε^2 terms correctly. The intuitive argument for (40) is that keeping V_i fixed affects only $(1/N)$th of the scattered field, which is negligible for large N. This only pertains to the terms in (50) proportional to N^{-1} but it fails to account for the effect of keeping V_i fixed upon the distribution of the other scatterers, which affects the mean fields scattered by them. That effect is responsible for the terms $\varepsilon N(\langle T_1 \rangle_i - \langle T_1 \rangle)$, which would vanish if the scatterers were independently distributed. In that case the ε^2 term in (17) is proportional to N rather than to N^2, so that the effective field hypothesis then yields a result which is practically correct through terms of order ε^2.

One might also be led to believe the hypothesis in the case of independently identically distributed scatterers on the basis of its self-consistency, which we shall now demonstrate. By this we mean that the assumption $\langle u^i \rangle_{ij} = \langle u^i \rangle_i$ implies $\langle u^i \rangle_i = \langle u \rangle$ when N is large. To show this we first note that $\langle u^i \rangle_i = \langle u^1 \rangle_1$ by symmetry, and that $dp(V_i | V_j) = dp(V_i)$ by independence. Therefore the second hierarchy equation (39) can be written as

$$\begin{aligned} \langle u^1 \rangle_1 &= u_0 + L^{-1} \sum_{i \neq j} \int T_i' \langle u^i \rangle_i \, dp(V_i) \\ &= u_0 + (N-1) L^{-1} \int T_1' \langle u^1 \rangle_1 \, dp(V_1). \end{aligned} \tag{51}$$

From (38) we have

$$L^{-1} \int T_1' \langle u^1 \rangle_1 \, dp(V_1) + N^{-1}(\langle u \rangle - u_0). \tag{52}$$

Therefore (51) becomes

$$\langle u^1 \rangle_1 = \langle u \rangle - N^{-1}(\langle u \rangle - u_0). \tag{53}$$

For large N this is just (40) with $c = 1$, proving the asserted self-consistency.

Before concluding the present discussion, we shall consider an exact consequence of the second hierarchy equation (39), based on a consideration applicable to all hierarchy equations. In our previous analysis of (39) we noted that it contained both $\langle u^i \rangle_i$ and $\langle u^i \rangle_{ij}$ and we eliminated the latter by making a hypothesis. Instead we shall now eliminate the former, which can be done exactly since by definition

$$\langle u^i \rangle_i = \int \langle u^i \rangle_{ij} \, dp(V_j | V_i). \tag{54}$$

By using (54), and noting that $\int u_0 \, dp(V_i | V_j) = u_0$, we can write (39) as an equation for $\langle u^i \rangle_{ij}$

$$\int \langle u^j \rangle_{ji} \, dp(V_i | V_j) = \int u_0 \, dp(V_i | V_j) + L^{-1} \sum_{i \neq j} \int T_i' \langle u^i \rangle_{ij} \, dp(V_i | V_j). \tag{55}$$

If the V_i are symmetrically distributed it follows that $\langle u^i\rangle_{ij} = \langle u^1\rangle_{12}$ and (55) can be written as

$$\int [\langle u^2\rangle_{21} - u_0 - (N-1)L^{-1}T_1'\langle u^1\rangle_{12}]\, dp(V_1|V_2) = 0. \tag{56}$$

From (56) it follows that

$$\langle u^2\rangle_{21} - u_0 - (N-1)L^{-1}T_1'\langle u^1\rangle_{12} = w_{12}. \tag{57}$$

Here w_{12} is some function which satisfies the equation

$$\int w_{12}\, dp(V_1|V_2) = 0. \tag{58}$$

Since (57) involves both $\langle u^2\rangle_{21}$ and $\langle u^1\rangle_{12}$ we rewrite it with 1 and 2 interchanged

$$\langle u^1\rangle_{12} - u_0 - (N-1)L^{-1}T_2'\langle u^2\rangle_{21} = w_{21}. \tag{59}$$

We may now solve (57) and (59) for $\langle u^2\rangle_{21}$, obtaining

$$\langle u^2\rangle_{21} = [1 - (N-1)^2L^{-1}T_1'L^{-1}T_2']^{-1}[u_0 + w_{12} + (N-1)L^{-1}T_1'(u_0 + w_{21})]. \tag{60}$$

Now (54) yields $\langle u^1\rangle_1$ and then (38) yields $\langle u\rangle$ in terms of the random function w_{12}, which is known only to satisfy (58). This interesting expression for $\langle u\rangle$ has not yet been used.

5. Hierarchy equations and equations for generating functionals. When the stochastic quantities in an equation are not small, the perturbation method used in the three preceding sections is not useful and a different method of analysis must be employed. We shall now describe two such methods, namely the method of hierarchy equations, which we have already touched upon, and the method involving a functional equation for a certain generating functional. The perturbation method involves solving the equation first and then averaging the solution while the hierarchy equations result from averaging first and then attempting to solve. In order to solve, it has been necessary to make unverified assumptions about the solution, some of which have been described above. Therefore we shall not discuss further the solution of these equations, but just their derivation. For this purpose we shall employ a simple example.

Let us consider a scalar function $u(x)$ satisfying the stochastic equation

$$L_x u + n(x)u(x) = g(x). \tag{1}$$

Here L_x is a given operator, $g(x)$ is a given function and $n(x)$ is a random function. Therefore $u(x)$ is also random. Upon taking the expectation of (1) in order to get an equation for $\langle u(x)\rangle$ we find

$$L_x\langle u\rangle + \langle n(x)u(x)\rangle = g(x). \tag{2}$$

Since (2) involves $\langle n(x)u(x)\rangle$ we may try to get an equation for this quantity by multiplying (1) by $n(x)$ and averaging. But this leads to $\langle nL_x u\rangle$ rather than $L_x\langle nu\rangle$.

Therefore since $n(x_1)$ commutes with L_x we multiply (1) instead by $n(x_1)$ and average to obtain

$$L_x\langle n(x_1)u(x)\rangle + \langle n(x_1)n(x)u(x)\rangle = \langle n(x_1)\rangle g(x). \tag{3}$$

Once $\langle n(x_1)u(x)\rangle$ is found from (3), we may evaluate it at $x_1 = x$ and use it in (2) to get an equation for $\langle u(x)\rangle$. Since (3) involves a new moment we must obtain another equation for it, etc. If we also try to determine higher moments of u, we are led to the following infinite set of equations

$$\begin{aligned} &L_x\langle u(x)u(x_1)\cdots u(x_j)n(x_{j+1})\cdots n(x_{j+k})\rangle \\ &\quad + \langle n(x)u(x)u(x_1)\cdots u(x_j)n(x_{j+1})\cdots n(x_{j+k})\rangle \\ &\quad = g(x)\langle u(x_1)\cdots u(x_j)n(x_{j+k})\rangle, \qquad j, k = 0, 1, \cdots. \end{aligned} \tag{4}$$

These equations for the hierarchy of moments are called the hierarchy equations. Such equations have been discussed in the introduction and in the preceding sections.

Whenever an infinite set of moments occurs, it is helpful to seek a function which generates them. Since the moments in (4) are functions, they will be generated by a functional. We shall now show that they are generated by the functional $F[\xi, \eta]$, a functional of the two functions $\xi(x)$ and $\eta(x)$ defined by

$$F[\xi, \eta] = \left\langle \exp\left[\int \{\xi(x')u(x') + \eta(x')n(x')\}\, dx'\right]\right\rangle. \tag{5}$$

Functional differentiation of (5) yields

$$\begin{aligned} \frac{\delta^{j+k}F}{\delta\xi(x_1)\cdots\delta\xi(x_j)\delta\eta(x_{j+1})\cdots\delta\eta(x_{j+k})} &= \Big\langle u(x_1)\cdots u(x_j)n(x_{j+1})\cdots n(x_{j+k}) \\ &\quad \times \exp\left[\int\{\xi(x')u(x') + \eta(x')n(x')\}\, dx'\right]\Big\rangle. \end{aligned} \tag{6}$$

Evaluation of (6) at $\xi(x) \equiv 0$, $\eta(x) \equiv 0$ leads to the desired result

$$\begin{aligned} &\left[\frac{\delta^{j+k}F}{\delta\xi(x_1)\cdots\delta\xi(x_j)\delta\eta(x_{j+1})\cdots\delta\eta(x_{j+k})}\right]_{\xi=\eta=0} \\ &\quad = \langle u(x_1)\cdots u(x_j)n(x_{j+1})\cdots n(x_{j+k})\rangle. \end{aligned} \tag{7}$$

Thus the functional derivatives of $F[\xi, \eta]$ at $\xi = \eta = 0$ are the moments occurring in the hierarchy equations. In terms of the moments, $F[\xi, \eta]$ can be expressed as a functional Taylor series

$$\begin{aligned} F[\xi, \eta] = \sum_{j,k=0}^{\infty} \frac{1}{j!k!} \int &\langle u(x_1)\cdots u(x_j)n(x_{j+1})\cdots n(x_{j+k})\rangle \xi(x_1)\cdots\xi(x_j) \\ &\cdots \eta(x_{j+1})\cdots\eta(x_{j+k})\, dx_1\cdots dx_{j+k}. \end{aligned} \tag{8}$$

We could also define a generating functional of the form (5) with the integral multiplied by i, which might exist for a larger class of distributions. But then factors of i would occur in many equations so we have omitted the i since our analysis is entirely formal.

From the hierarchy equations (4) and the Taylor series (8) one can deduce an equation satisfied by the functional $F[\xi, \eta]$. However, it is simpler to derive this equation directly from (1) by first noting that

$$\begin{aligned} L_x \frac{\delta F}{\delta \xi(x)} + \frac{\delta^2 F}{\delta \xi(x) \delta \eta(x)} &= \left\langle \{L_x u(x) + u(x) n(x)\} \exp\left[\int \{\xi u + \eta n\}\, dx'\right]\right\rangle \\ &= \left\langle g(x) \exp\left[\int \{\xi u + \eta n\}\, dx'\right]\right\rangle. \end{aligned} \tag{9}$$

The last form of (9) follows from (1). Since $g(x)$ is not random it can be taken outside the expectation and (9) becomes

$$L_x \frac{\delta F[\xi, \eta]}{\delta \xi(x)} + \frac{\delta F[\xi, \eta]}{\delta \xi(x) \delta \eta(x)} = g(x) F[\xi, \eta]. \tag{10}$$

This is a functional differential equation for the generating functional F. By differentiating (10) with respect to $\xi(x_1), \cdots, \xi(x_j)$, $\eta(x_{j+1}) \cdots \eta(x_{j+k})$, then setting $\xi(x) \equiv \eta(x) \equiv 0$ and using (7), the hierarchy equations (4) result. Thus (10) implies those equations, and it also follows from them. We note that when $g(x) \equiv 0$, F does not occur in (10) and it may be considered as an equation for the functional $\delta F/\delta \xi(x)$.

An additional condition on F can be obtained by setting $\xi(x) \equiv 0$ in (5), which yields

$$F[0, \eta] = \left\langle \exp\left[\int \eta(x') n(x')\, dx'\right]\right\rangle. \tag{11}$$

Since the distribution of $n(x)$ is supposed to be known, the right side of (11) is known. Thus (11) is a "boundary condition" on F at the boundary $\xi(x) \equiv 0$. Other conditions on F can be obtained from information about u or conditions on u. For example, if $u(x) = 0$ for x on a surface S, then it follows from (5) that

$$\frac{\delta F[\xi, \eta]}{\delta \xi(x)} = 0 \quad \text{for } x \text{ on } S. \tag{12}$$

Similarly, vanishing of $u(x)$ at infinity implies the vanishing there of $\delta F/\delta \xi(x)$.

Some methods for solving equations of the type (10) with additional conditions on F of the form (11) or (12) are described in the papers referred to in the introduction. Other possible methods, such as those involving functional integration and functional Fourier transforms, have not been exploited much.

6. **Transport theory of incoherent intensity and flux.** So far we have mainly considered the expected or mean wave. Now we shall examine the variance of the

wave as well as other quadratic quantities involving it. To this end let us consider a time harmonic field $u(x)$ produced by a source distribution $j(x)$ in a medium of refractive index $n(x)$. We assume that u satisfies the equation

$$[\Delta + k^2n^2(x)]u = -j. \tag{1}$$

In this equation the constant k is the propagation constant of the medium. In terms of $u(x)$ two quantities of practical interest, the intensity $I(x)$ and the flux $\mathbf{S}(x)$ are defined by

$$I(x) = u(x)\bar{u}(x). \tag{2}$$

$$\mathbf{S}(x) = \frac{1}{2i}(\bar{u}\nabla u - u\nabla\bar{u}) = \text{Im } \bar{u}\nabla u. \tag{3}$$

Here the overbar denotes the complex conjugate. It follows at once from (1) that I and $\mathbf{S}$ are related by the energy equation

$$\nabla\cdot\mathbf{S} = -I\,\text{Im } k^2n^2 + \text{Im } u\bar{j}. \tag{4}$$

The first term on the right represents energy absorption by the medium and the second term represents work done by the source.

We shall now assume that u, n and j are random functions and denote the expectation value of any quantity q by $\langle q\rangle$. Then the coherent field u_c and the incoherent field u_i are defined by

$$u_c = \langle u\rangle, \tag{5}$$

$$u_i = u - \langle u\rangle = u - u_c. \tag{6}$$

The coherent intensity and flux I_c and $\mathbf{S}_c$ are defined by

$$I_c = u_c\bar{u}_c, \tag{7}$$

$$\mathbf{S}_c = \text{Im }(\bar{u}_c\nabla u_c). \tag{8}$$

The incoherent intensity and incoherent flux I_i and $\mathbf{S}_i$ are defined by

$$I_i = \langle I\rangle - I_c, \tag{9}$$

$$\mathbf{S}_i = \langle\mathbf{S}\rangle - S_c. \tag{10}$$

From these definitions it follows that $\langle u_i\rangle = 0$, $I_i = \langle\bar{u}_iu_i\rangle$ and $\mathbf{S}_i = \text{Im }\langle\bar{u}_i\nabla u_i\rangle$. The energy equation (4) yields

$$\nabla\cdot\mathbf{S}_c + \nabla\cdot\mathbf{S}_i = -\langle I\,\text{Im } k^2n^2\rangle + \text{Im }\langle u\bar{j}\rangle. \tag{11}$$

Upon using the above definitions in (1) we find the following equations for u_c and u_i

$$[\Delta + k^2\langle n^2\rangle]u_c = -\langle j\rangle - k^2\langle n^2u_i\rangle, \tag{12}$$

$$[\Delta + k^2n^2]u_i = k^2\langle n^2u_i\rangle + k^2[\langle n^2\rangle - n^2]u_c - j + \langle j\rangle. \tag{13}$$

Equation (12) shows that we may regard $\langle j \rangle$ as the source of u_c and $-k^2\langle n^2 u_i \rangle$ as a negative source or sink. Similarly the right side of (13) shows which terms are sources of u_i. It should be noted that $\langle I \rangle$ is a second moment of u and that $\langle \mathbf{S}(x) \rangle$ can be computed from $\langle \bar{u}(y)u(x) \rangle$, the two point second moment of u, by the relation

$$\mathbf{S}\langle (x) \rangle = \lim_{y \to x} \nabla_x \langle \bar{u}(y)u(x) \rangle. \tag{14}$$

We shall now specialize the foregoing considerations to waves in a uniform nonrandom medium for which $n(x) \equiv 1$. Then (1), (12) and (13) become

$$(\Delta + k^2)u = -j, \tag{15}$$

$$(\Delta + k^2)u_c = -\langle j \rangle, \tag{16}$$

$$(\Delta + k^2)u_i = -j + \langle j \rangle. \tag{17}$$

For k real, the solution of (15) which satisfies the radiation condition is

$$u(x) = \frac{1}{4\pi} \int \frac{e^{ik|x-x'|}}{|x - x'|} j(x')\, dx'. \tag{18}$$

If Im $k > 0$ this is the bounded solution. The bounded or radiating solutions of (16) and (17) are given by (18) with $\langle j \rangle$ and $j - \langle j \rangle$ in place of j, respectively.

From the result of the form (18) for u_i we shall now calculate the incoherent intensity and flux. Upon using the fact that $I_i = \langle \bar{u}_i u \rangle$ and writing the product of two integrals as a double integral we have

$$I_i(x) = \left\langle \frac{1}{(4\pi)^2} \int [j(x') - \langle j(x') \rangle][\bar{j}(x'') - \langle \bar{j}(x'') \rangle] \frac{e^{ik|x-x'| - i\bar{k}|x-x''|}}{|x - x'||x - x''|}\, dx'\, dx'' \right\rangle. \tag{19}$$

This may be written in terms of $N(x', x'')$ the two point correlation function of the source, which is defined by

$$N(x', x'') = \langle [j(x') - \langle j(x') \rangle][\bar{j}(x'') - \langle \bar{j}(x'') \rangle] \rangle. \tag{20}$$

Now (19) becomes

$$I_i(x) = \frac{1}{(4\pi)^2} \int \frac{N(x', x'')}{|x - x'||x - x''|} e^{ik|x-x'| - i\bar{k}|x-x''|}\, dx'\, dx''. \tag{21}$$

Similarly we find

$$\begin{aligned} \mathbf{S}_i(x) &= \frac{1}{(4\pi)^2} \operatorname{Im} \int \frac{N(x', x'')}{|x - x'||x - x''|} \\ &\quad \times e^{ik|x-x'| - i\bar{k}|x-x''|} \left(ik - \frac{1}{|x - x'|} \right) \nabla(|x - x'|)\, dx'\, dx''. \end{aligned} \tag{22}$$

These results (21) and (22) acquire an interesting form when the source is uncorrelated at distinct points, so that N is given by

$$N(x', x'') = (4\pi)^2 M(x') \delta(x' - x''). \tag{23}$$

Then (21) becomes

$$I_i(x) = \int \frac{M(x')}{|x - x'|^2} e^{-2|x-x'|\mathrm{Im}\, k}\, dx'. \tag{24}$$

Upon introducing polar coordinates r, ω with their origin at x, (24) may be written as

$$I_i(x) = \int_\Omega I_i(x, \omega)\, d\omega. \tag{25}$$

In (25) ω denotes a point on the unit sphere Ω and $I_i(x, \omega)$ is defined by

$$I_i(x, \omega) = \int_0^\infty M(x - r\omega)\, e^{-2r\,\mathrm{Im}\, k}\, dr. \tag{26}$$

Similarly (22) becomes

$$\mathbf{S}_i(x) = (\mathrm{Re}\, k) \int_\Omega \boldsymbol{\omega} I_i(x, \omega)\, d\omega. \tag{27}$$

The results (25) and (27) may be interpreted in terms of an incoherent field of intensity $I_i(x, \omega)$ and flux $\boldsymbol{\omega} I_i(x, \omega)$ Re k traveling through x in each direction ω. The intensities and fluxes of these fields add together without interference, so we say they combine incoherently. Each source point on the line $x - r\omega$ contributes to $I_i(x, \omega)$ an amount equal to M at the source point multiplied by the decay factor $e^{-2r\,\mathrm{Im}\, k}$, which results from absorption in the medium.

A different way of finding I_i and $\mathbf{S}_i$ is based upon the differential equations satisfied by the moments. Any moment of $u(x)$ satisfies a linear equation with the corresponding moment of $j(x)$ as inhomogeneous term. Thus the moment $\langle u(x_1) \cdots u(x_s) j(x_{s+1}) \cdots j(x_r)\rangle$ satisfies the equation

$$L_1 \cdots L_s \langle u(x_1) \cdots u(x_s) j(x_{s+1}) \cdots j(x_r)\rangle = (-1)^s \langle j(x_1) \cdots j(x_r)\rangle, \tag{28}$$

where $L_s = \Delta_s + k^2$ and Δ_s is the Laplacian with respect to x_s. The set of equations (28) for all the moments of u and j may be called the hierarchy equations. In the present case they are uncoupled in the sense that each equation contains only one moment involving u. For nonuniform media the equations are coupled, as we shall see. The bounded or radiating solution of (28) is the multiple integral

$$\langle u(x_1) \cdots u(x_s) j(x_{s+1}) \cdots j(x_r)\rangle = \frac{1}{(4\pi)^s} \int \frac{e^{ik(|x-x_1| + \cdots + |x-x_s|)}}{|x - x_1| \cdots |x - x_s|} \times \langle j(x_1) \cdots j(x_s)\rangle\, dx_1 \cdots dx_s. \tag{29}$$

This result can also be obtained by multiplying together the solutions (18) for u and the factors j and then averaging. Similar considerations apply to the solution of any linear equation in which the inhomogeneous term and the initial and/or boundary data are random while the linear operator (e.g., coefficients, etc.) is "sure."

Let us now try to obtain equations for the intensity and flux in a random medium with a nonrandom source. To this end we assume that the random inhomogeneity in the medium is small and write $n(x) = 1 + \varepsilon\mu(x)$. Then since $\langle j \rangle = j$, we obtain from (13) the following equation for u_i, in which terms through order ε are shown explicitly

$$(\Delta + k^2)u_i = 2\varepsilon k^2[\langle \mu u_i \rangle - \mu u_i + (\langle \mu \rangle - \mu)u_c] + O(\varepsilon^2). \tag{30}$$

From (30) we see that u_i is $O(\varepsilon)$ so (30) may be further simplified to

$$(\Delta + k^2)u_i = 2\varepsilon k^2(\langle \mu \rangle - \mu)u_c + O(\varepsilon^2). \tag{31}$$

The solution of (31) is

$$u_i(x) = 2\varepsilon k^2 \int G(x, x')[\langle \mu(x') \rangle - \mu(x')]u_c(x')\, dx' + O(\varepsilon^2). \tag{32}$$

We now multiply (32) by its complex conjugate and take the mean of the result, obtaining

$$\begin{aligned} I_i(x) = 4\varepsilon^2 k^2 \int & G(x, x')G(x, x'') \langle [\langle \mu(x') \rangle - \mu(x')][\langle \mu(x'') \rangle - \mu(x'')] \rangle \\ & \times u_c(x')u_c(x'')\, dx'\, dx'' + O(\varepsilon^3). \end{aligned} \tag{33}$$

The correlation function $R(x', x'')$ is defined to be the mean value in the integrand in (33), so (33) can be rewritten as

$$I_i(x) = 4\varepsilon^2 k^2 \int G(x, x')G(x, x'')u_c(x')u_c(x'')R(x', x'')\, dx'\, dx'' + O(\varepsilon^3). \tag{34}$$

Let us now suppose that the refractive index variations at distinct points are uncorrelated. Then we may write $R(x', x'')$ in the form

$$R(x', x'') = M(x')\delta(x' - x''). \tag{35}$$

Now (34) becomes

$$I_i(x) = \frac{\varepsilon^2 k^2}{4\pi^2} \int \frac{M(x')}{|x - x'|^2} I_c(x') \exp[-2|x - x'| \operatorname{Im} k]\, dx' + O(\varepsilon^3). \tag{36}$$

This is an explicit expression for I_i in terms of I_c. Since I_i is $O(\varepsilon^2)$ and $\langle I \rangle = I_c + I_i$, we may replace I_c by $\langle I \rangle$ on the right side of (36). If we then add $I_c(x)$ to each side and omit $O(\varepsilon^3)$, (36) becomes

$$\langle I(x) \rangle = I_c(x) + \frac{\varepsilon^2 k^2}{4\pi^2} \int \frac{M(x')}{|x - x'|^2} \langle I(x) \rangle\, e^{-2|x - x'| \operatorname{Im} k}\, dx'. \tag{37}$$

This is an integral equation for $\langle I(x) \rangle$ with $I_c(x)$ as the inhomogeneous term. The study of the incoherent radiation by means of integral equations such as this is called the transport theory of radiation.

The integral in (37), which is equal to $I_i(x)$, can be written in the form (25) with $I_i(x, \omega)$ defined by

$$I_i(x, \omega) = \frac{\varepsilon^2 k^2}{4\pi^2} \int_0^\infty M(x - r\omega)\langle I(x - r\omega) \rangle\, e^{-2r \operatorname{Im} k}\, dr. \tag{38}$$

The incoherent flux $\mathbf{S}_i(x)$ can be computed from (32) in the same way as $I_i(x)$ was. One finds that it can be written in the form (27) with $I_i(x, \omega)$ given by (38).

Similar considerations to those described here have been applied to the incoherent field in media of discrete scatterers. Other work has concerned the solution of boundary value problems by means of the integral equations of transport theory. These and related matters are discussed in works on transport theory.

References

1. J. B. Keller, *Wave propagation in random media*, Proc. Sympos. Appl. Math. Vol. 13, Amer. Math. Soc., Providence, R. I., 1960.

2. L. L. Foldy, *The multiple scattering of waves*, Phys. Rev. (2) **67** (1945), 107.

3a. M. Lax, *Multiple scattering of waves*, Rev. Modern Phys. **23** (1951), 287.

3b. ———, *Multiple scattering of waves*. II, Phys. Rev. (2) **85** (1952), 621.

4a. V. Twersky, *On multiple scattering of waves*, No. EDL–L16, Electronic Defense Labs., Mt. View, Calif., April, 1960.

4b. ———, *On scattering of waves by a slab region of randomly distributed objects*, No. EDL–E26, Electronic Defense Labs., Mt. View, Calif., January, 1958.

4c. ———, *On propagation in random media of discrete scatterers*, Proc. Sympos. Appl. Math. Vol. 17, Amer. Math. Soc., Providence, R. I., 1964.

5a. R. C. Bourret, *Bilocal theory of randomly perturbed fields*, Research Report No. 215, Hughes Research Labs., Malibu, Calif., September, 1961.

5b. ———, *The scattering of sound in a turbulent and thermally random medium by the method of bilocal perturbations*, Research Report No. 193, Hughes Research Labs., Malibu, Calif., February, 1962.

5c. ———, *Stochastically perturbed fields, with applications to wave propagation in random media*, Hughes Research Labs., Malibu, Calif.

6. W. S. Ament, *Wave propagation in suspensions*, Report 5307, U. S. Naval Research Lab., 1952.

7. R. H. Kraichnan, *The closure problem of turbulence theory*, Proc. Sympos. Appl. Math. Vol. 13, Amer. Math. Soc., Providence, R. I., 1960.

8. N. N. Bogolyubov, *Problems of a dynamical theory in statistical physics*, English transl., Providence College, Providence, R. I., 1959.

9. E. Hopf, *Statistical hydromechanics and functional calculus*, J. Rational Mech. Anal. **1** (1952), 87.

10. B. Zumino, *Formal solution of the equations of statistical equilibrium*, Phys. Fluids **2** (1959), 20.

11. R. M. Lewis and J. B. Keller, *Solution of the functional differentiation equation for the statistical equilibrium of a crystal*, Phys. Rev. (2) **121** (1961), 1022.

12. R. M. Lewis and R. H. Kraichnan, *A space-time functional formalism for turbulence*, Comm. Pure Appl. Math. **15** (1962), 397.

13. W. C. Meecham, *On radiation in a randomly inhomogeneous medium*, Space Technology Labs., Inc., Los Angeles 45, Calif., September, 1961.

14. F. C. Karal and J. B. Keller, *Elastic, electromagnetic and other waves in a random medium*, J. Mathematical Phys. (1964) (to appear).

15. Lord Rayleigh, *Transmission of light through an atmosphere containing small particles in suspension*, Philos. Mag. **47** (1899), 375.

16. F. Reiche, *Zur Theorie der Dispersion in Gasen und Dämpfen*, Ann. Physics **50** (1916), 1; 121.

17. L. A. Chernov, *Wave propagation in a random medium*, English transl., McGraw-Hill, New York, 1960.

New York University
New York, New York

STOCHASTIC TRANSFORMATIONS AND FUNCTIONAL EQUATIONS

BY

RICHARD BELLMAN

1. **Introduction.** A large amount of attention devoted to the study of the linear stochastic transformation

$$s_n = s_{n-1} + r_n, \qquad s_0 = 0, \tag{1.1}$$

or

$$s_n = \sum_{i=1}^{n} r_i,$$

where the r_i are independent random variables, has been amply rewarded. It is to be expected, in view of the richness of this special case, that the study of more general stochastic transformations of the form

$$x_{n+1} = g(x_n, r_n), \tag{1.2}$$

will reveal equally interesting phenomena. These transformations arise quite naturally in many branches of mathematical physics, in modern control theory, in mathematical psychology in connection with learning processes, in stochastic approximation, in dynamic programming, and in many other mathematical settings as well.

There are considerable difficulties in obtaining general results. In this paper, we wish to discuss three particular problems. The first leads us to functional equations generalizing those of Abel–Schroder; the second involves a use of quasilinearization, the third arises in connection with the produce of random matrices.

2. **The Abel–Schroder equation.** Let $g(x)$ be a function which is analytic in the neighborhood of the origin

$$g(x) = a_1 x + a_2 x^2 + \cdots, \qquad |x| \leq c_1. \tag{2.1}$$

We say that the transformation represented by $g(x)$ is *linearized* by $f(x)$ if

$$f(g(x)) = a_1 f(x). \tag{2.2}$$

Does there exist a function $f(x)$, analytic in some neighborhood of the origin, which accomplishes this objective? Of the many paths that can be followed in answer to this question, let us pursue one based upon iteration. Write $g^{(n)}$ for the nth iterate of g,

$$g^{(0)} = x, \qquad g^{(n+1)} = g(g^{(n)}). \tag{2.3}$$

The functions $g^{(n)}$ enjoy the semi-group property

$$g^{(m+n)} = g^{(m)}(g^{(n)}), \tag{2.4}$$

$m, n = 0, 1, 2, \cdots$ Consider the new sequence of functions defined by

$$f_n(x) = g^{(n)}(x)/a_1, \tag{2.5}$$

where a_1 is, as in (2.1), $g'(0)$. Assume that $0 < |a_1| < 1$. Observe that

$$f_{n+1}(x) = g^{(n+1)}/a_1^{n+1} = \frac{f_n(g(x))}{a_1}. \tag{2.6}$$

Hence, if the sequence $\{f_n(x)\}$ converges as $n \to \infty$ to a function $f(x)$, we have

$$f(g(x)) = a_1 f(x). \tag{2.7}$$

Thus, provided $f(x)$ exists, it is a linearizing function. The existence of $f(x)$ and the validity of the preceding formalism can be readily established, in the one-dimensional case, Koenigs [**1**], the multidimensional case, Bellman [**2**], where many other references may be found, and in more general cases arising in the theory of branching processes, Harris [**3**].

3. **Stochastic iteration.** Let us now consider a stochastic version of the foregoing. Let

$$g(x) = g(x, a) = a_1 x + a_2 x^2 + \cdots, \tag{3.1}$$

where the a_i are random variables with a joint distribution $dG(a_1, a_2, \cdots) \equiv dG(a)$.

Let $g_1(x)$ represent the function obtained with the random variables $[a_1^{(1)}, a_2^{(1)}, \cdots]$, $g_2(x)$ the function obtained with set $[a_1^{(2)}, a_2^{(2)}, \cdots]$, and so on. We wish to consider the function

$$h_n(x) = \underset{a^{(i)}}{\exp}\, g_n(g_{n-1}(\cdots g_1(x) \cdots)). \tag{3.2}$$

It is easy to see that

$$h_n(x) = \int h_{n-1}(g(x, a))dG(a), \tag{3.3}$$

$n \geqq 1$, with $h_0(x) = x$, as before.

A particularly interesting case is that where

$$\begin{aligned} g(x, a) &= t_1(x) = b_{11}x + b_{12}x^2 + \cdots, \text{ with probability } p_1 \\ &= t_2(x) = b_{21}x + b_{22}x^2 + \cdots, \text{ with probability } 1 - p_1. \end{aligned} \tag{3.4}$$

Then (3.3) becomes

$$h_n(x) = p_1 h_{n-1}(t_1(x)) + (1 - p_1)h_{n-1}(t_2(x)). \tag{3.5}$$

Proceeding formally, write

$$e_1 = \int a_1 \, dG(a), \tag{3.6}$$

and introduce the functions

$$f_n(x) = \frac{h_n(x)}{e_1^n}. \tag{3.7}$$

We see that

$$f_{n+1}(x) = \frac{1}{e_1} \int f_n(g(x, a))dG(a). \tag{3.8}$$

If $f_n(x)$ converges to $f(x)$ as $n \to \infty$, we have

$$f(x) = \frac{1}{e_1} \int f(g(x, a))dG(a), \tag{3.9}$$

an average linearization.

Following the argument in Koenigs [**1**], it is easy to show that if $t_1(x)$ and $t_2(x)$ are analytic in some neighborhood of the origin and if $|t_1'(0)|$, $|t_2'(0)| < 1$, then the sequence $\{f_n(x)\}$ converges uniformly to an analytic function $f(x)$ satisfying (3.9).

Setting

$$f(x) = x + c_2x^2 + \cdots + c_nx^2 + \cdots, \tag{3.10}$$

we can determine the coefficients recurrently from (3.9).

4. **Asymptotic behavior.** Returning to the deterministic case discussed in §2, suppose that we wish to study the asymptotic behavior of $g^{(n)}(x)$ as $n \to \infty$. Since (2.7) holds, we have

$$f(g^{(n)}(x)) = a_1^n f(x), \tag{4.1}$$

and thus

$$g^{(n)}(x) = f^{-1}(a_1^n f(x)). \tag{4.2}$$

This yields a complete asymptotic description as $n \to \infty$, since $f(x)$ is a known function.

It follows that we have a complete description of the asymptotic behavior of the solution of the difference equation

$$u_{n+1} = g(u_n), \qquad u_0 = x, \tag{4.3}$$

provided that $|x|$ is sufficiently small. The multidimensional analogues of these results are valid, but require considerably more effort; cf. [**2**], where other references may be found. Similar results may be obtained for differential equations [**4**]. These represent extensions of the classical work of Poincaré–Lyapunov concerning the stability of the null solution of nonlinear differential equations [**5**].

The results of this paper were obtained in the course of attempts to derive similar results for differential and difference equations with stochastic elements. Let us see what the results of §3 yield. From (3.7), we see that if

$$u_{n+1} = g(u_n, a^{(n)}), \qquad u_0 = x, \tag{4.4}$$

then

$$\exp_a (u_n) \sim e_1^n f(x), \tag{4.5}$$

where $f(x)$ is determined by (3.9), and e_1 is as in (3.6).

The multidimensional versions of (4.5) are valid, and one can obtain analogous results for all moments, a topic we shall discuss in more detail below in connection with the product of random matrices.

Observe that (4.5) is very much weaker than (4.2). It is an open question as to whether an analogue of (4.2) exists for stochastic iteration.

5. **Random matrices.** Consider the vector difference equation

$$x_{n+1} = R_n x_n, \qquad x_0 = c, \tag{5.1}$$

where x_n is an N-dimensional vector and R_n is an $N \times N$ matrix whose elements are stochastic variables. Clearly,

$$x_n = R_n R_{n-1} \cdots R_0 c. \tag{5.2}$$

Let us assume at this point that the R_i are independent.

We wish to determine various moments of the components of x_n. First moments are easy. We have

$$E(x_n) = E(R)E(x_{n-1}) = E(R)^n c. \tag{5.3}$$

One way to obtain higher moments is to introduce the Kronecker product of two matrices [**6**]. To see the way in which this arises, consider the 2×2 case. We have, componentwise,

$$\begin{aligned} u_{n+1} &= r_{11}u_n + r_{12}v_n, \\ v_{n+1} &= r_{21}u_n + r_{22}v_n. \end{aligned} \tag{5.4}$$

Hence,

$$\begin{aligned} u_{n+1}^2 &= r_{11}^2 u_n^2 + 2r_{11}r_{12}u_nv_n + r_{12}^2 v_n^2, \\ u_{n+1}v_{n+1} &= r_{11}r_{21}u_n^2 + \cdots \\ &\ \ \vdots \end{aligned} \tag{5.5}$$

In vector-matrix notation,

$$\begin{pmatrix} u_{n+1}^2 \\ u_{n+1}v_{n+1} \\ v_{n+1}u_{n+1} \\ v_{n+1}^2 \end{pmatrix} = \begin{pmatrix} r_{11}R & r_{12}R \\ & \\ r_{21}R & r_{22}R \end{pmatrix} \begin{pmatrix} u_n^2 \\ u_nv_n \\ v_nu_n \\ v_n^2 \end{pmatrix} \tag{5.6}$$

The 4×4 matrix $(r_{ij}R)$ is the Kronecker product of R with itself, written $R \otimes R$ or $R^{(2)}$. It plays an important role in the study of the stability of linear systems of difference or differential equations with random coefficients; see Zadeh [7].

Higher Kronecker powers will yield the higher moments; see Bellman [8].

6. **Correlation.** In place of the introduction of Kronecker products, we can employ the following procedure which applies in the more complex case of correlation between the R_i. If the R_i are independent, introduce the function

$$f_n(c) = \exp (x_n, Bx_n). \tag{6.1}$$

Then

$$f_n(c) = \int f_{n-1}(Rc)dG(R), \tag{6.2}$$

with $f_0(c) = (c, Bc)$. If the R_i are correlated in a Markovian fashion, where the distribution of R_i depends only on the matrix R_{i-1}, we introduce the function

$$f_n(c, S) = \exp (x_n, Bx_n), \tag{6.3}$$

where $R_{-1} = S$. Then

$$f_n(c, S) = \int f_{n-1}(Rc, R)dG(R, S), \tag{6.4}$$

with $f_0(c, S) = (c, Bc)$.

Turning to (6.2), we write $f_n(c) = (c, Q_n c)$, a quadratic form in c. We obtain the matrix relation

$$Q_n = \int RQ_{n-1}RdG(R). \tag{6.5}$$

The relations connecting the elements of Q_n with those of Q_{n-1} introduce the Kronecker product in a direct fashion.

7. **Asymptotic behavior.** There is no difficulty in determining the asymptotic behavior of the moments of x_n. However, the study of the distribution of a random variable such as $\log x_{1n}$, where x_{1n} is the first component of x_n, is a matter of considerably greater difficulty. If the R_i are independent, one would expect that $\log x_{1n}$ suitably normalized will behave like a Gaussian variable. Under various assumptions, this is demonstrated by Kesten–Furstenberg [9].

8. **Dynamic programming.** A characteristic functional equation in the theory of dynamic programming is

$$f_n(x) = \max_y f_{n-1}(T(x, y)), \qquad f_0(x) = x. \tag{8.1}$$

Suppose that

$$T(x, y) = a_1(y)x + a_2(y)x^2 + \cdots, \tag{8.2}$$

where $|a_1(y)| < 1$ for all y. Let

$$b = \max_y a_1(y). \tag{8.3}$$

Under various conditions which we will not enter into here, it can be shown that

$$\lim_{n\to\infty} \frac{f_n(x)}{b^n} = \phi(x) \tag{8.4}$$

uniformly for $|x| \leqq b_2$ and that

$$b\phi(x) = \max_y \phi(T(x, y)). \tag{8.5}$$

These results have applications in the study of the asymptotic behavior of variational processes.

If we consider stochastic variational processes, we obtain an extension of the foregoing results as well as those obtained above; see also [**10**; **11**].

9. **Quasilinearization.** Let us now briefly indicate a new technique that can be applied to the study of differential equations with random elements. Consider the scalar equation

$$\frac{du}{dt} = g(u, r(t)), \qquad u(0) = c, \tag{9.1}$$

where $r(t)$ is a random function of t and $g(u, r)$ is convex in u for any value of r. An important class of functions of this type is given by

$$\frac{du}{dt} = u^k + r(t), \qquad k > 1. \tag{9.2}$$

Applying the technique of quasilinearization [**12**; **13**], we consider the associated linear equation

$$\frac{dv}{dt} = g(w, r(t)) + (v - w)g_w(w, r(t)), \qquad v(0) = c, \tag{9.3}$$

where w is a random function as yet unspecified. Since

$$g(u, r) = \max_w [g(w, r) + (v - w)g_w(w, r)], \tag{9.4}$$

by virtue of convexity, we see that

$$\frac{du}{dt} = \max_w [g(w, r) + (u - w)g_w(w, r)], \qquad w(0) = c. \tag{9.5}$$

It now follows readily that

$$v(t) \leqq u(t) \tag{9.6}$$

for all choices of w, with equality if $w = u$. Hence,

$$\text{Prob}\,(v(t) \geqq z) \leqq \text{Prob}\,(u(t) \geqq z) \tag{9.7}$$

for all w, and thus

$$\text{Prob}\,(u(t) \geqq z) = \max_w \text{Prob}\,(v(t) \geqq z). \tag{9.8}$$

With suitable choices of w, we can obtain excellent approximations to the left-hand side. A case of particular interest is that where $k = z$.

REFERENCES

1. M. G. Koenigs, *Nouvelles recherches sur les équations fonctionnelles*, Ann. Sci. Ecole Norm. Sup. **2** (1885), 385–404.

2. R. Bellman, *The iteration of power series in two variables*, Duke Math. J. **19** (1952), 339–347.

3. T. E. Harris, *Branching processes*, Springer, Berlin, 1963.

4. R. Bellman and J. M. Richardson, *On the asymptotic behavior of solutions of nonlinear differential equations*, J. Math. Anal. Appl. **4** (1962), 470–474.

5. ———, *Stability theory of differential equations*, McGraw-Hill, New York, 1954.

6. ———, *Introduction to matrix analysis*, McGraw-Hill, New York, 1960.

7. L. Zadeh, *Time-varying networks*, Proc. IRE. **49** (1961), 1488–1503.

8. R. Bellman, *Limit theorems for noncommutative operations*. I, Duke Math. J. **21** (1954), 491–500.

9. J. Kesten and H. Furstenberg, *Products of random matrices*, Ann. Math. Statist. **31** (1960), 457–469.

10. R. Bellman, *A Markovian decision process*, J. Math. Mech. **6** (1957), 679–684.

11. ———, *Functional equations in the theory of dynamic programming*. XI. *Limit theorems*, Rend. Circ. Mat. Palermo **8** (1959), 1–3.

12. R. Kalaba, *On nonlinear differential equations, the maximum operation and monotone convergence*, J. Math. Mech. **8** (1959), 519–573.

13. R. Bellman, *Functional equations in the theory of dynamic programming*. V. *Positivity and quasilinearity*, Proc. Nat. Acad. Sci. U.S.A. **41** (1955), 743–746.

14. U. Grenander, *Stochastic groups*, Ark. Mat. **4** (1961), 163–183.

15. J. C. Samuels and A. C. Eringen, *On stochastic linear systems*, J. Math. Phys. **38** (1959), 83–103.

16. A. Dvoretsky, *On stochastic approximation*, Proc. 3rd Berkeley Sympos. Math. Statist. and Prob. (1954–1955), Vol. 1, pp. 39–56, Univ. of California Press, Berkeley, Calif., 1956.

17. R. R. Bush and F. Mosteller, *Stochastic models for learning*, Wiley, New York, 1955.

THE RAND CORPORATION,
SANTA MONICA, CALIFORNIA

THE APPLICATION OF STOCHASTIC APPROXIMATION TO THE OPTIMIZATION OF RANDOM CIRCUITS

BY

K. B. GRAY

1. **Problem.** The problem is that of designing a circuit, whose configuration is fixed in advance, to achieve a certain desired output. The circuit input may be stochastic and there may be internal stochastic disturbances. Moreover, many circuit component values cannot (or need not) be deterministically fixed. Instead, we may specify the statistical distribution of the circuit's component values in order that the circuit output will have desired statistical properties.

The problem arises in the real world because circuit components exhibit variations from their design or nominal values. Therefore, it is generally unrealistic to design circuits or systems solely on the basis of design values. Thus, one does not order, say, 10 ohm resistors. Rather, one specifies the distribution of the resistors. Frequently, many of the circuits are going to be used in larger circuits or systems. Hence, one is interested in the statistical aspects of the circuit performance [**1**].

2. **Example.** Let the circuit output $Y_t(\omega)$ be a random function that satisfies the differential equation

$$R(\omega)\frac{dY_t(\omega)}{dt} + \frac{1}{C(\omega)}Y_t(\omega) = 0 \tag{2.1}$$

with initial value $Y_0(\omega)$ being a random variable. ω denotes a point in the basic probability space. From here on we will not explicitly exhibit the functional form of the random variables involved. Rather, we use the shortened notation

$$R\frac{dY_t}{dt} + \frac{1}{C}Y_t = 0. \tag{2.2}$$

Call R and C the resistance and capacitance, respectively, of the circuit. The circuit output is, of course, the random function

$$Y_t = Y_0 e^{-t/RC}. \tag{2.3}$$

Suppose we desire to select the distribution of R and C so that in some statistical sense Y_t is close to $e^{-t/10}$ for $0 \leq t \leq 5$. That is, a significant difference between Y_t and $e^{-t/10}$ represents a loss.

Let $U(a, b)$ denote the rectangular density over the interval $[a, b]$. Suppose that the densities of R and C are $U(u_1, u_1 + 1/u_2)$ and $U(u_3, u_3 + 1/u_4)$, respectively,

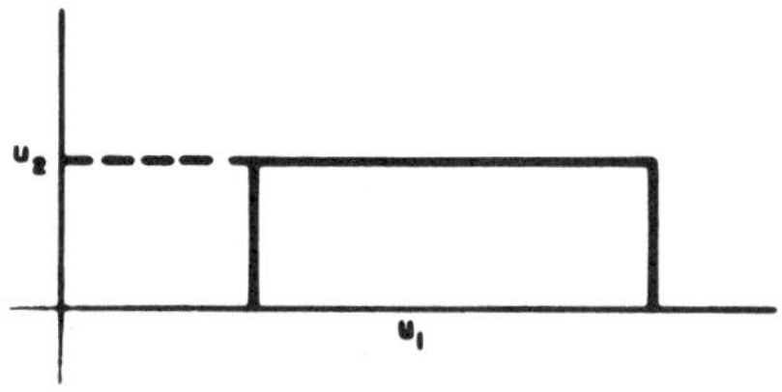

FIGURE 1. Density of R.

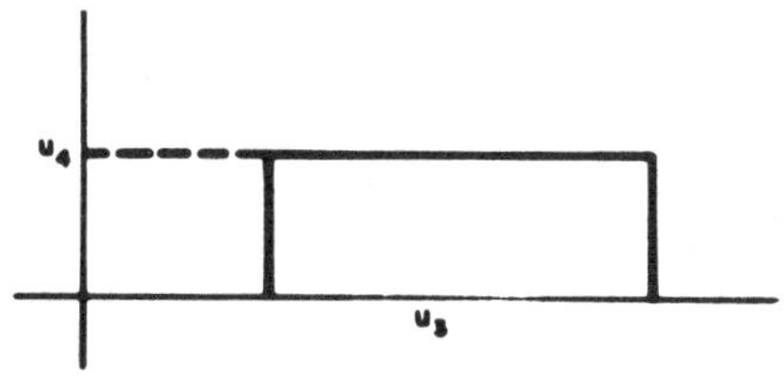

FIGURE 2. Density of C.

with $u_1, u_3 \geqq 0$ and $u_2, u_4 > 0$. Note that u_2 and u_4 measure the dispersion of R and C about their mean values.

Define our loss to be

$$L(Y, \mathbf{u}) = \sum_{t=0}^{5} |Y_t - e^{-t/10}| + d_2u_2 + d_4u_4 \tag{2.4}$$

where $d_2, d_4 \leqq 0$ indicate the cost relationship between output errors and the cost of decreasing the dispersion of R and C. We seek to minimize the average loss, or risk,

$$R(\mathbf{u}) = EL(Y, \mathbf{u}) \tag{2.5}$$

subject to a rectangular constraint on the vector $\mathbf{u}$. The constraint is of the form

$$0 < a_i \leqq u_i \leqq b_i < \infty \qquad \text{for } i = 1, \cdots, 4. \tag{2.6}$$

3. **Traditional approaches to statistical circuit design.** The simplest approach is to ignore the statistical aspects of the design problem and proceed under the assumption that component values do not vary from their nominal values. The problem is then deterministic. This method may be a fair way to make an initial guess at the optimum setting for the parameter $\mathbf{u}$, but it obviously gives no information about the output tolerances or deviations.

A somewhat more sophisticated design technique is so-called "worst-case" design [**2**]. This assumes tolerances for all component values. The tolerance for the output is computed by letting the components take their "worst" values. It is generally assumed that the "worst" values lie at the end points of the tolerances. "Worst-case" design frequently results in significant "over-design."

Certain relatively simple circuits, with distribution assumptions about the circuit components and input, allow one to approximate analytically the probability distribution of the output [**3**; **4**]. In order to solve complex problems by this method it is usually necessary to make significantly unrealistic assumptions.

If the parameter $\mathbf{u}$, defining the probability distribution of the component values, is fixed in advance, then one can use Monte Carlo simulation to estimate the probability that the output will be out of tolerance [**5**]. This technique becomes impractical as a design aid as the dimension of the parameter $\mathbf{u}$ increases.

It is clear that none of the techniques described really solve the problem. An approach is needed that will allow numerical implementation for a general class of circuits to find optimum parameter settings by the use of circuit simulation. The Monte Carlo method utilizes circuit simulation to estimate reliability. In order to use Monte Carlo to find an optimum setting for **u** it would be necessary to estimate the reliability at many points in the parameter space. The number of points required would increase exponentially with q, the dimension of **u**.

A further criticism of the above techniques is that they do not allow sufficiently general criteria of loss. It would seem desirable to consider criteria other than that of being out of tolerance. Also, one should consider the cost of selecting a particular setting for **u**.

4. **Mathematical model of problem.** Let the random variables $X_{\mathbf{u}}^1, X_{\mathbf{u}}^2, \cdots, X_{\mathbf{u}}^p$ represent the various component values of the circuit. The vector $\mathbf{u} = (u_1, u_2, \cdots, u_q)$ determines the distribution of these random variables. Inputs and internal disturbances can be represented as random variables $X_{\mathbf{u}}^1$ whose distributions are independent of **u**. Let the circuit output be

$$Y_{\mathbf{u}} = T[X_{\mathbf{u}}^1, X_{\mathbf{u}}^2, \cdots, X_{\mathbf{u}}^p] \tag{4.1}$$

where T is a scalar valued measurable function and $Y_{\mathbf{u}}$ is a random variable whose distribution depends on **u**.

We say that if the output is $Y_{\mathbf{u}}$ and **u** is the parameter setting, we accrue a loss

$$L(Y_{\mathbf{u}}, \mathbf{u}) = L_1(Y_{\mathbf{u}}) + L_2(\mathbf{u}). \tag{4.2}$$

For example, L_1 could represent the cost of the output being out of tolerance and L_2 the cost of selecting the parameter **u**. In some problems L_2 might reflect the fact that component cost usually increases as the allowable component tolerance decreases.

We consider the problem of minimizing the expected loss

$$R(\mathbf{u}) = EL(Y_{\mathbf{u}}, \mathbf{u}), \tag{4.3}$$

subject to a constraint

$$a_i \leqq u_i \leqq b_i; \qquad i = 1, 2, \cdots, q.$$

If the exact form of the risk function $R(\mathbf{u})$ were known, the problem would be straightforward. However, it is not. Suppose we can, by simulation or some other technique, sample the random variable $L(Y_{\mathbf{u}}, \mathbf{u})$ for prescribed settings of **u**. We propose applying a stochastic steepest ascent method, based on the sample observations of $L(Y_{\mathbf{u}}, \mathbf{u})$, that will converge in a probabilistic sense to the optimum setting for **u**. The Kiefer–Wolfowitz [**6; 7; 8**] stochastic approximation procedure is such a technique.

5. **Multidimensional Kiefer–Wolfowitz procedure.** The setting of the discussion is R^q, q-dimensional real coordinate space. Let $\mathbf{x} = (x_1, \cdots, x_q) \in R^q$. We define the norm of $\mathbf{x}$ by

$$\|\mathbf{x}\| = \left(\sum_{i=1}^{q} x_i^2\right)^{1/2}. \tag{5.1}$$

The inner product of $\mathbf{x}$ and $\mathbf{y}$, both elements of R^q, is

$$(\mathbf{x}, \mathbf{y}) = \sum_{i=1}^{q} x_i y_i. \tag{5.2}$$

Suppose there is a family $\{Z_\mathbf{u}\}$ of random variables $Z_\mathbf{u}$ defined for all $\mathbf{u} \in R^q$. We assume that if $\mathbf{u}$ is specified it is possible to sample any of the random variables $Z_\mathbf{u}$. Let

$$f(\mathbf{u}) = EZ_\mathbf{u} \tag{5.3}$$

be the expected value of $Z_\mathbf{u}$. If the scalar-valued function f defined on R^q has a unique maximum at $\boldsymbol{\theta}$ and satisfies various other technical assumptions, then the Kiefer–Wolfowitz (KW) procedure provides a stochastic sequence $\mathbf{X}_n$ that converges to $\boldsymbol{\theta}$ in the sense that $E\|\mathbf{X}_n - \boldsymbol{\theta}\|^2 \to 0$ and $P[\lim \mathbf{X}_n = \boldsymbol{\theta}] = 1$.

DEFINITION 5.1. Let the nonnegative, real sequences $\{b_n\}$ and $\{c_n\}$ defined for $n = 1, 2, 3, \cdots$ satisfy

$$\sum_{n=1}^{\infty} b_n = \infty, \tag{5.4}$$

$$\lim_{n\to\infty} c_n = 0, \tag{5.5}$$

and

$$\sum_{n=1}^{\infty} \frac{b_n^2}{c_n^2} < \infty. \tag{5.6}$$

Let $\mathbf{e}_1 = (1, 0, \cdots, 0)$, $\mathbf{e}_2 = (0, 1, 0, \cdots, 0), \cdots, \mathbf{e}_q = (0, \cdots, 0, 1)$ be an orthonormal basis for R^q. For $c > 0$, define the random vectors

$$\mathbf{Z}_{\mathbf{u},c}^{+} = (Z_{\mathbf{u}+c\mathbf{e}_1}, \cdots, Z_{\mathbf{u}+c\mathbf{e}_q}) \tag{5.7}$$

and

$$\mathbf{Z}_{\mathbf{u},c}^{-} = (Z_{\mathbf{u}-c\mathbf{e}_1}, \cdots, Z_{\mathbf{u}-c\mathbf{e}_q}). \tag{5.8}$$

The expectations of the above random vectors are

$$\mathbf{f}^{+}(\mathbf{u}, c) = (f(\mathbf{u} + c\mathbf{e}_1), \cdots, f(\mathbf{u} + c\mathbf{e}_q)) \tag{5.9}$$

and

$$\mathbf{f}^{-}(\mathbf{u}, c) = (f(\mathbf{u} - c\mathbf{e}_1), \cdots, f(\mathbf{u} - c\mathbf{e}_q)). \tag{5.10}$$

It is assumed that, for any n, all $2qn$ random observations made in n iterations are independent. The KW procedure begins by guessing $\mathbf{X}_1$. The succeeding elements of the stochastic sequence are defined by

$$\mathbf{X}_{n+1} = \mathbf{X}_n + \frac{b_n}{c_n}(\mathbf{Z}^{+}_{\mathbf{X}_n,c_n} - \mathbf{Z}^{-}_{\mathbf{X}_n,c_n}), \tag{5.11}$$

for $n = 1, 2, 3, \cdots$.

We shall state a theorem (Theorem 5.2) that gives conditions for convergence of the multidimensional KW procedure. Sacks [9] has considered the asymptotic distribution of $\mathbf{X}_n$ as $n \to \infty$. The q-dimensional procedure considered here and discussed by Sacks uses $2q$ sample observations of Z at each iteration. Blum [10] gives a procedure that uses $q + 1$ sample observations at each iteration. The Blum procedure does not converge as rapidly as the multidimensional KW procedure. Originally the scalar KW procedure was shown to converge in probability by Kiefer and Wolfowitz [6]. Blum [7] strengthened the result to almost sure convergence. Dvoretzky [8] derived general theorems that yield both almost sure and expected norm squared convergence as special cases.

THEOREM 5.1 (DVORETZKY). *Let $\{\alpha_n\}$ be a nonnegative real sequence, defined for $n = 1, 2, \cdots$, satisfying*

$$\limsup_{n\to\infty} \alpha_n \leqq \alpha; \tag{5.12}$$

suppose that $\gamma_n(\mathbf{r}_1, \cdots, \mathbf{r}_n)$ is a sequence of nonnegative scalar-valued functions, on $\mathbf{r}_1, \cdots, \mathbf{r}_n$, satisfying

$$\sum_{n=1}^{\infty} \gamma_n(\mathbf{r}_1, \cdots, \mathbf{r}_n) = \infty \tag{5.13}$$

uniformly for all sequences $\mathbf{r}_1, \mathbf{r}_2, \cdots$ such that

$$\sup_n \|\mathbf{r}_n\| < L, \tag{5.14}$$

where $L > 0$ is arbitrary. Let $\boldsymbol{\theta} \in R^q$ and suppose $\{\mathbf{T}_n\}$ are measurable vector-valued transformations satisfying

$$\|\mathbf{T}_n(\mathbf{r}_1, \cdots, \mathbf{r}_n) - \boldsymbol{\theta}\| \leqq \max\,[\alpha_n, \|\mathbf{r}_n - \boldsymbol{\theta}\| - \gamma_n] \tag{5.15}$$

for all sequences $\mathbf{r}_1, \mathbf{r}_2, \cdots, \mathbf{r}_n$. Define

$$\mathbf{X}_{n+1}(\omega) = \mathbf{T}_n[\mathbf{X}_1(\omega), \cdots, \mathbf{X}_n(\omega)] + \mathbf{Y}_n(\omega) \tag{5.16}$$

for $n = 1, 2, \cdots$, where ω is a point in the fundamental probability space Ω. Let $\mathbf{X}_1$ and $\mathbf{Y}_n$ be random variables satisfying

$$E\|\mathbf{X}_1\|^2 < \infty, \tag{5.17}$$

$$\sum_{n=1}^{\infty} E\|\mathbf{Y}_n\|^2 < \infty, \tag{5.18}$$

and

$$E\{\mathbf{Y}_n | \mathbf{x}_1, \cdots, \mathbf{x}_n\} = 0 \text{ almost surely.} \tag{5.19}$$

Then,

$$\limsup_{n\to\infty} E\|\mathbf{X}_n - \boldsymbol{\theta}\|^2 \leqq \alpha^2 \tag{5.20}$$

and

$$P\left[\limsup_{n\to\infty} \|\mathbf{X}_n - \boldsymbol{\theta}\| \leqq \alpha\right] = 1.$$

PROOF. The theorem is a statement of the relevant results in Dvoretzky's Theorem [**8**], its extension, and Generalization 3. The multidimensional extension of his results is straightforward.

Armed with Theorem 5.1 we are prepared to prove the convergence of the KW procedure. The method of proof is the same as Dvoretzky's proof for the scalar case. The conditions of Theorem 5.2 are somewhat different from the conditions used by Dvoretzky. His conditions do not allow a convenient multidimensional extension. The conditions of this paper, for the case $q = 1$, are at least as weak as Dvoretzky's.

THEOREM 5.2. *Let*

$$E|Z_{\mathbf{u}} - f(\mathbf{u})|^2 \leqq \sigma^2 < \infty \tag{5.21}$$

for all $\mathbf{u} \in R^q$. *Suppose that for some* $\varepsilon_0 > 0$ *there exist constants* A, B *such that* $0 \leqq \varepsilon \leqq \varepsilon_0 \Rightarrow$ *for all* $\mathbf{u} \in R^q$,

$$\|\mathbf{f}^+(\mathbf{u}, \varepsilon) - \mathbf{f}^-(\mathbf{u}, \varepsilon)\| < A\|\mathbf{u} - \boldsymbol{\theta}\| + B. \tag{5.22}$$

Also, suppose that for arbitrary constants ρ *and* L $(0 < \rho < L)$ *there exists* $\varepsilon_1 > 0$ *such that* $0 < \varepsilon \leqq \varepsilon_1$ *and* $\rho < \|\mathbf{u} - \boldsymbol{\theta}\| < L \Rightarrow$ *for* $1/\sqrt{2} < \kappa$,

$$(\mathbf{u} - \boldsymbol{\theta}, \mathbf{f}^+(\mathbf{u}, \varepsilon) - \mathbf{f}^-(\mathbf{u}, \varepsilon)) \leqq -\kappa\|\mathbf{u} - \boldsymbol{\theta}\| \cdot \|\mathbf{f}^+(\mathbf{u}, \varepsilon) - \mathbf{f}^-(\mathbf{u}, \varepsilon)\|, \tag{5.23}$$

and

$$\liminf_{0\neq\delta\to 0} \frac{\|\mathbf{f}^+(\mathbf{u}, \delta) - \mathbf{f}^-(\mathbf{u}, \delta)\|}{2\delta} \geqq \gamma > 0. \tag{5.24}$$

Then,

$$P\left[\lim_{n\to\infty} \mathbf{X}_n = \boldsymbol{\theta}\right] = 1 \tag{5.25}$$

and

$$\lim_{n\to\infty} E\|\mathbf{X}_n - \boldsymbol{\theta}\|^2 = 0. \tag{5.26}$$

PROOF. We apply Theorem 5.1. Let the transformation $\mathbf{T}_n$ be defined by

$$\mathbf{T}_n(\mathbf{r}_n) = \mathbf{r}_n + \frac{b_n}{c_n}(\mathbf{f}^+(\mathbf{r}_n, c_n) - \mathbf{f}^-(\mathbf{r}_n, c_n)). \tag{5.27}$$

Define the random variable $\mathbf{Y}_n$ by

$$\mathbf{Y}_n = \frac{b_n}{c_n}(\mathbf{Z}^+_{\mathbf{r}_n,c_n} - \mathbf{Z}^-_{\mathbf{r}_n,c_n}). \tag{5.28}$$

Then,

$$\mathbf{X}_{n+1} = \mathbf{T}_n[\mathbf{X}_1, \cdots, \mathbf{X}_n] + \mathbf{Y}_n. \tag{5.29}$$

It is obvious from the definitions of $\mathbf{T}_n$ and $\mathbf{Y}_n$ that (5.19) holds. Also,

$$\begin{aligned} E\|\mathbf{Y}_n\|^2 \leqq \frac{b_n^2}{c_n^2}\{E\|\mathbf{Z}^+_{\mathbf{r}_n,c_n} - \mathbf{f}^+(\mathbf{r}_n, c_n)\|^2 \\ + 2E(\mathbf{Z}^+_{\mathbf{r}_n,c_n} - \mathbf{f}^+(\mathbf{r}_n, c_n), \mathbf{Z}^-_{\mathbf{r}_n,c_n} - \mathbf{f}^-(\mathbf{r}_n, c_n)) \\ + E\|\mathbf{Z}^-_{\mathbf{r}_n,c_n} - \mathbf{f}^-(\mathbf{r}_n, c_n)\|^2\} \\ \leqq 2\frac{b_n^2}{c_n^2}q\sigma^2, \end{aligned} \tag{5.30}$$

since the expected value of the inner product term is zero and the variance is bounded by (5.21). Hence,

$$\sum_{n=1}^{\infty} E\|\mathbf{Y}_n\|^2 \leqq 2q\sigma^2 \sum_{n=1}^{\infty} \frac{b_n^2}{c_n^2} < \infty, \tag{5.31}$$

by (5.6).

It is clear that (5.17) holds for the KW procedure.

We have yet to verify (5.15). There is no loss of generality by considering the case that $\boldsymbol{\theta} = 0$. Let $\rho > 0$ be arbitrary. Either $\|\mathbf{r}_n\| \leqq \rho$ or $\|\mathbf{r}_n\| > \rho$. Consider the case that $\|\mathbf{r}_n\| \leqq \rho$. Now

$$\begin{aligned} \|\mathbf{T}_n(\mathbf{r}_n)\| &= \left\|\mathbf{r}_n + \frac{b_n}{c_n}(\mathbf{f}^+(\mathbf{r}_n, c_n) - \mathbf{f}^-(\mathbf{r}_n, c_n))\right\| \\ &\leqq \|\mathbf{r}_n\| + \frac{b_n}{c_n}\|\mathbf{f}^+(\mathbf{r}_n, c_n) - \mathbf{f}^-(\mathbf{r}_n, c_n)\|. \end{aligned} \tag{5.32}$$

(5.22) implies that, for n sufficiently large,

$$\|\mathbf{T}_n(\mathbf{r}_n)\| \leqq \rho + \frac{b_n}{c_n}(A\|\mathbf{r}_n\| + B). \tag{5.33}$$

Since $b_n/c_n \to 0$, for n sufficiently large and $\|\mathbf{r}_n\| \leqq \rho$ we have that

$$\|\mathbf{T}_n(\mathbf{r}_n)\| \leqq 2\rho. \tag{5.34}$$

Next, consider the case that $\|\mathbf{r}_n\| > \rho$. Let $L > \rho$ be the arbitrary constant mentioned in Theorem 5.1 and assume that $\rho < \|\mathbf{r}_n\| \leqq L$. Then,

$$\|\mathbf{T}_n(\mathbf{r}_n)\|^2 = \left\|\mathbf{r}_n + \frac{b_n}{c_n}(\mathbf{f}^+(\mathbf{r}_n, c_n) - \mathbf{f}^-(\mathbf{r}_n, c_n))\right\|^2$$

$$= \|\mathbf{r}_n\|^2 + 2\frac{b_n}{c_n}(\mathbf{r}_n, \mathbf{f}^+(\mathbf{r}_n, c_n) - \mathbf{f}^-(\mathbf{r}_n, c_n)) \tag{5.35}$$

$$+ \frac{b_n^2}{c_n^2}\|\mathbf{f}^+(\mathbf{r}_n, c_n) - \mathbf{f}^-(\mathbf{r}_n, c_n)\|^2.$$

By (5.23),

$$\|\mathbf{T}_n(\mathbf{r}_n)\|^2 \leqq \|\mathbf{r}_n\|^2 - 2\kappa\frac{b_n}{c_n}\|\mathbf{r}_n\| \cdot \|\mathbf{f}^+(\mathbf{r}_n, c_n) - \mathbf{f}^-(\mathbf{r}_n, c_n)\| \tag{5.36}$$

$$+ \frac{b_n^2}{c_n^2}\|\mathbf{f}^+(\mathbf{r}_n, c_n) - \mathbf{f}^-(\mathbf{r}_n, c_n)\|^2.$$

Hence,

$$\|\mathbf{T}_n(\mathbf{r}_n)\| \leqq \left|\, \|\mathbf{r}_n\| - \kappa\frac{b_n}{c_n}\|\mathbf{f}^+(\mathbf{r}_n, c_n) - \mathbf{f}^-(\mathbf{r}_n, c_n)\| \,\right| \tag{5.37}$$

$$+ \frac{b_n}{c_n}\sqrt{(1 - \kappa^2)}\|\mathbf{f}^+(\mathbf{r}_n, c_n) - \mathbf{f}^-(\mathbf{r}_n, c_n)\|.$$

We need

LEMMA 5.1. *For n sufficiently large and $\rho < \|\mathbf{r}_n\| < L$,*

$$\frac{b_n}{c_n}\|\mathbf{f}^+(\mathbf{r}_n, c_n) - \mathbf{f}^-(\mathbf{r}_n, c_n)\| < \|\mathbf{r}_n\|. \tag{5.38}$$

PROOF. By (5.22)

$$\frac{b_n}{c_n}\|\mathbf{f}^+(\mathbf{r}_n, c_n) - \mathbf{f}^-(\mathbf{r}_n, c_n)\| \leqq \frac{b_n}{c_n}(A\|\mathbf{r}_n\| + B), \tag{5.39}$$

for n sufficiently large. Equation (5.6) implies that $b_n/c_n \to 0$. Hence,

$$\frac{b_n}{c_n}(A\|\mathbf{r}_n\| + B) \leqq \frac{b_n}{c_n}(AL + B) < \rho \tag{5.40}$$

for n sufficiently large.

Applying the lemma, we see that we can remove the absolute value signs in (5.37) for n sufficiently large. Thus,

$$\|\mathbf{T}_n(\mathbf{r}_n)\| \leqq \|\mathbf{r}_n\| - (\kappa - \sqrt{(1 - \kappa^2)})\frac{b_n}{c_n}\|\mathbf{f}^+(\mathbf{r}_n, c_n) - \mathbf{f}^-(\mathbf{r}_n, c_n)\| \tag{5.41}$$

for n sufficiently large. Now, for $1/\sqrt{2} < \kappa$, $\lambda = \kappa - \sqrt{(1 - \kappa^2)} > 0$. Therefore, (5.24) implies that

$$\frac{1}{c_n} \|\mathbf{f}^{+}(\mathbf{r}_n, c_n) - \mathbf{f}^{-}(\mathbf{r}_n, c_n)\| \geqq \gamma > 0. \tag{5.42}$$

This together with (5.41) yields

$$\|\mathbf{T}_n(\mathbf{r}_n)\| \leqq \|\mathbf{r}_n\| - \lambda\gamma b_n \tag{5.43}$$

for n sufficiently large and $\rho < \|\mathbf{r}_n\| < L$.

Combining this with (5.34) we have for n sufficiently large and $\|\mathbf{r}_n\| < L$,

$$\|\mathbf{T}_n(\mathbf{r}_n)\| \leqq \max\,[2\rho, \|\mathbf{r}_n\| - \lambda\gamma b_n] \tag{5.44}$$

where $\rho > 0$ is an arbitrary constant. We have thus satisfied the hypotheses of Theorem 5.1. Hence,

$$E\|\mathbf{X}_n - \boldsymbol{\theta}\|^2 \leqq 4\rho^2 \tag{5.45}$$

and

$$P[\|\mathbf{X}_n - \boldsymbol{\theta}\| \leqq 2\rho] = 1,$$

for all $\rho > 0$. Therefore,

$$E\|\mathbf{X}_n - \boldsymbol{\theta}\|^2 \to 0, \quad \text{as } n \to \infty, \tag{5.46}$$

and

$$P[\mathbf{X}_n \to \boldsymbol{\theta}] = 1.$$

This completes the proof of Theorem 5.1.

6. **Kiefer–Wolfowitz procedure over a rectangular subset of R^q.** The multidimensional KW procedure assumes that $z_{\mathbf{u}}$ can be sampled for all $\mathbf{u} \in R^q$, and that the conditions of Theorem 5.2 hold for all $\mathbf{u} \in R^q$. In the application we are concerned with, it is possible to sample $z_{\mathbf{u}}$ only for $\mathbf{u}$ in a certain subset

$$C(\mathbf{a}, \mathbf{b}) = \{\mathbf{x}\colon a_i \leqq x_i \leqq b_i, \quad i = 1, \cdots, q\} \tag{6.1}$$

of R^q. Call the restriction of $\mathbf{u}$ to $C(\mathbf{a}, \mathbf{b})$ a rectangular constraint. We consider a modified multidimensional KW procedure that samples $Z_{\mathbf{u}}$ only for $\mathbf{u} \in C(\mathbf{a}, \mathbf{b})$.

If $f(\mathbf{u})$ has a unique maximum over $C(\mathbf{a}, \mathbf{b})$ at $\boldsymbol{\theta}'$ and satisfies various other technical assumptions, the procedure $\{\mathbf{X}_n'\}$, to be defined, converges almost surely and in mean norm squared.

DEFINITION 6.1. If $\mathbf{x} = (x_1, \cdots, x_q)$, define the vector-valued function $(\mathbf{x})^* = (x_1^*, \cdots, x_q^*)$ by

$$x_i^* = \begin{cases} a_i & \text{if } x_i < a_i, \\ x_i & \text{if } a_i \leqq x_i \leqq b_i, \\ b_i & \text{if } b_i < x_i \end{cases} \tag{6.2}$$

for $i = 1, \cdots, q$.

Using this notation the constrained KW procedure can be conveniently defined so that it stays inside of $C(\mathbf{a}, \mathbf{b})$ and all sampling is done inside of $C(\mathbf{a}, \mathbf{b})$.

DEFINITION 6.2. Let the sequences $\{b_n\}$ and $\{c_n\}$ be as in Definition 5.1. For $c > 0$, define the random variables

$$\mathbf{Z}_{\mathbf{u},c}^{+,*} = (Z_{(\mathbf{u}+c\mathbf{e}_1)^*}, \cdots, Z_{(\mathbf{u}+c\mathbf{e}_q)^*}),$$
$$\mathbf{Z}_{\mathbf{u},c}^{-,*} = (Z_{(\mathbf{u}-c\mathbf{e}_1)^*}, \cdots, Z_{(\mathbf{u}-c\mathbf{e}_q)^*}). \tag{6.3}$$

It is assumed that all $2qn$ random observations made in n iterations are independent for all n. The procedure begins by guessing $\mathbf{X}_1' \in C(\mathbf{a}, \mathbf{b})$. The succeeding elements of the sequence are given by the recursive relation

$$\mathbf{X}_{n+1}' = \left(\mathbf{X}_n' + \frac{b_n}{c_n}(\mathbf{Z}_{\mathbf{X}_n',c_n}^{+,*} - \mathbf{Z}_{\mathbf{X}_n',c_n}^{-,*})\right)^* \tag{6.4}$$

for $n = 1, 2, \cdots$. Before stating a theorem giving conditions for convergence of $\mathbf{X}_n'$, it is convenient to define the random variable $\mathbf{D}(\mathbf{u}, \varepsilon, \delta) = (D_1, \cdots, D_q)$, for $\varepsilon, \delta > 0$, by defining its components

(6.5)
$$D_i(\mathbf{u}, \varepsilon, \delta) = \begin{cases} \dfrac{1}{\delta}(a_i - u_i) & \text{if } Z_{(\mathbf{u}+\varepsilon\mathbf{e}_i)^*} - Z_{(\mathbf{u}-\varepsilon\mathbf{e}_i)^*} < \dfrac{1}{\delta}(a_i - u_i), \\ Z_{(\mathbf{u}+\varepsilon\mathbf{e}_i)^*} - Z_{(\mathbf{u}-\varepsilon\mathbf{e}_i)^*} & \text{if } \dfrac{1}{\delta}(a_i - u_i) \leqq Z_{(\mathbf{u}+\varepsilon\mathbf{e}_i)^*} - Z_{(\mathbf{u}-\varepsilon\mathbf{e}_i)} \leqq \dfrac{1}{\delta}(b_i - u_i), \\ \dfrac{1}{\delta}(b_i - u_i) & \text{if } \dfrac{1}{\delta}(b_i - u_i) < Z_{(\mathbf{u}+\varepsilon\mathbf{e}_i)^*} - Z_{(\mathbf{u}-\varepsilon\mathbf{e}_i)^*}. \end{cases}$$

Define the mean of $\mathbf{D}(\mathbf{u}, \varepsilon, \delta)$ to be

$$\mathbf{M}(\mathbf{u}, \varepsilon, \delta) = E\mathbf{D}(\mathbf{u}, \varepsilon, \delta). \tag{6.6}$$

Using this notation we can state the relation of (6.4) in the form

$$\mathbf{X}_{n+1}' = \mathbf{X}_n' + \frac{b_n}{c_n}\mathbf{D}\left(\mathbf{X}_n', c_n, \frac{b_n}{c_n}\right). \tag{6.7}$$

Also, let

$$\mathbf{f}^{+,*}(\mathbf{u}, \varepsilon) = (f[(\mathbf{u} + \varepsilon\mathbf{e}_1)^*], \cdots, f[(\mathbf{u} + \varepsilon\mathbf{e}_q)^*]),$$
$$\mathbf{f}^{-,*}(\mathbf{u}, \varepsilon) = (f[(\mathbf{u} - \varepsilon\mathbf{e}_1)^*], \cdots, f[(\mathbf{u} - \varepsilon\mathbf{e}_q)^*]). \tag{6.7a}$$

THEOREM 6.1. *Let*

$$E|Z_{\mathbf{u}} - f(\mathbf{u})|^2 \leqq \sigma^2 < \infty \tag{6.8}$$

for all $\mathbf{u} \in C(\mathbf{a}, \mathbf{b})$, *and suppose that for positive* ε *sufficiently small there exist constants* A *and* B *such that*

$$\|\mathbf{f}^{+,*}(\mathbf{u}, \varepsilon) - f^{-,*}(\mathbf{u}, \varepsilon)\| < A\|\mathbf{u} - \boldsymbol{\theta}'\| + B. \tag{6.9}$$

Assume that for arbitrary $\rho > 0$ *there exists* $\kappa > 1/\sqrt{2}$ *such that if positive* ε *and* δ *are sufficiently small,* $\mathbf{u} \in C(\mathbf{a}, \mathbf{b})$, *and* $\rho < \|\mathbf{u} - \boldsymbol{\theta}'\|$, *it follows that*

$$(\mathbf{u} - \boldsymbol{\theta}', \mathbf{M}(\mathbf{u}, \varepsilon, \delta)) \leqq -\kappa\|\mathbf{u} - \boldsymbol{\theta}'\| \cdot \|\mathbf{M}(\mathbf{u}, \varepsilon, \delta)\|. \tag{6.10}$$

Further, suppose that for arbitrary $\rho > 0$, *there exists* $\varepsilon_0 > 0$ *such that*

$$\inf_{\substack{0<\varepsilon,\delta<\varepsilon_0 \\ \rho<\|\mathbf{u}-\boldsymbol{\theta}'\|}} \frac{\|\mathbf{M}(\mathbf{u}, \varepsilon, \delta)\|}{2\varepsilon} > 0. \tag{6.11}$$

Then,

$$P[\|\mathbf{X}_n' - \boldsymbol{\theta}'\| \to 0] = 1 \tag{6.12}$$

and

$$E\|\mathbf{X}_n' - \boldsymbol{\theta}'\|^2 \to 0. \tag{6.13}$$

PROOF. It is sufficient to modify certain parts of the proof of Theorem 5.2. Let

$$\mathbf{T}_n'(\mathbf{r}_n) = \mathbf{r}_n + \frac{b_n}{c_n}\mathbf{M}\left(\mathbf{r}_n, c_n, \frac{b_n}{c_n}\right) \tag{6.14}$$

and

$$\mathbf{Y}_n' = \frac{b_n}{c_n}\left(\mathbf{D}\left(\mathbf{r}_n, c_n, \frac{b_n}{c_n}\right) - \mathbf{M}\left(\mathbf{r}_n, c_n, \frac{b_n}{c_n}\right)\right).$$

Then

$$\mathbf{X}_{n+1}' = \mathbf{T}_n'(\mathbf{r}_n) + \mathbf{Y}_n', \tag{6.15}$$

where $\mathbf{r}_n$ is a number denoting the value taken by $\mathbf{X}_n'$.

As in the proof of Theorem 5.2, we must verify that the conditions of Theorem 5.1 are satisfied.

Clearly, (5.19) holds.

The following lemma will be used several times in checking that the conditions of Theorem 5.1 hold.

LEMMA 6.1. *For positive* ε, δ *sufficiently small, there exist constants* A', B' *such that*

$$\|\mathbf{M}(\mathbf{u}, \varepsilon, \delta)\| \leqq A'\|\mathbf{u} - \boldsymbol{\theta}'\| + B', \tag{6.16}$$

for all $\mathbf{u} \in C(\mathbf{a}, \mathbf{b})$.

PROOF. Let M_i be a component of $\mathbf{M}(\mathbf{u}, \varepsilon, \delta)$. Since $\mathbf{M}$ is the expected value of $\mathbf{D}$,

$$
\begin{aligned}
|M_i| = \Bigg| &\frac{1}{\delta}(a_i - u_i)P\left[Z_{(\mathbf{u}+\varepsilon\mathbf{e}_i)*} - Z_{(\mathbf{u}-\varepsilon\mathbf{e}_i)*} < \frac{1}{\delta}(a_i - u_i)\right] \\
&+ \int_{\{(a_i-u_i)/\delta < Z_{(\mathbf{u}+\varepsilon\mathbf{e}_i)*} - Z_{(\mathbf{u}-\varepsilon\mathbf{e}_i)*} < (b_i-u_i)/\delta\}} (Z_{(\mathbf{u}+\varepsilon\mathbf{e}_i)*} - Z_{(\mathbf{u}-\varepsilon\mathbf{e}_i)*})dP \\
&+ \frac{1}{\delta}(b_i - u_i)P\left[\frac{1}{\delta}(b_i - u_i) < Z_{(\mathbf{u}+\varepsilon\mathbf{e}_i)*} - Z_{(\mathbf{u}-\varepsilon\mathbf{e}_i)*}\right]\Bigg| \\
\leqq |f[(\mathbf{u} &+ \varepsilon\mathbf{e}_i)^*] - f[(\mathbf{u} - \varepsilon\mathbf{e}_i)^*]| \\
&+ \int_{\{Z_{(\mathbf{u}+\varepsilon\mathbf{e}_i)*} - Z_{(\mathbf{u}-\varepsilon\mathbf{e}_i)*} < (a_i-u_i)/\delta\}} \left|\frac{1}{\delta}(a_i - u_i) - (Z_{(\mathbf{u}+\varepsilon\mathbf{e}_i)*} - Z_{(\mathbf{u}-\varepsilon\mathbf{e}_i)*})\right| dP \\
&+ \int_{\{(b_i-u_i)/\delta < Z_{(\mathbf{u}+\varepsilon\mathbf{e}_i)*} - Z_{(\mathbf{u}-\varepsilon\mathbf{e}_i)*}\}} \left|\frac{1}{\delta}(b_i - u_i) - (Z_{(\mathbf{u}+\varepsilon\mathbf{e}_i)*} - Z_{(\mathbf{u}-\varepsilon\mathbf{e}_i)*})\right| dP.
\end{aligned}
\tag{6.17}
$$

Consider the second term above.

$$
\int_{\{Z_{(\mathbf{u}+\varepsilon\mathbf{e}_i)*} - Z_{(\mathbf{u}-\varepsilon\mathbf{e}_i)*} < (a - u_i)/\delta\}} \left|\frac{1}{\delta}(a_i - u_i) - (Z_{(\mathbf{u}+\varepsilon\mathbf{e}_i)*} - Z_{(\mathbf{u}-\varepsilon\mathbf{e}_i)*})\right| dP
\tag{6.18}
$$

$$
\leqq \frac{1}{\delta}|a_i - u_i| P\left[Z_{(\mathbf{u}+\varepsilon\mathbf{e}_i)*} - Z_{(\mathbf{u}-\varepsilon\mathbf{e}_i)*} < \frac{1}{\delta}(a_i - u_i)\right] + E|Z_{(\mathbf{u}+\varepsilon\mathbf{e}_i)*} - Z_{(\mathbf{u}-\varepsilon\mathbf{e}_i)*}|.
$$

The first term above is zero if $a_i = u_i$. If not, then by Loève's Basic Inequality [**11**, p. 157] it is

$$
\leqq \frac{1}{\delta}|a_i - u_i| \frac{E|Z_{(\mathbf{u}+\varepsilon\mathbf{e}_i)*} - Z_{(\mathbf{u}-\varepsilon\mathbf{e}_i)*}|}{|a_i - u_i|/\delta}.
\tag{6.19}
$$

Thus, (6.18) is

$$
\leqq 2E|Z_{(\mathbf{u}+\varepsilon\mathbf{e}_i)*} - Z_{(\mathbf{u}-\varepsilon\mathbf{e}_i)*}|.
\tag{6.20}
$$

Similar reasoning applies to the third term in (6.17). Therefore,

$$
|M_i| \leqq |f[(\mathbf{u} + \varepsilon\mathbf{e}_i)^*] - f[(\mathbf{u} - \varepsilon\mathbf{e}_i)^*]| + 4E|Z_{(\mathbf{u}+\varepsilon\mathbf{e}_i)*} - Z_{(\mathbf{u}-\varepsilon\mathbf{e}_i)*}|.
\tag{6.21}
$$

Thus,

$$
|M_i| \leqq |f[(\mathbf{u} + \varepsilon\mathbf{e}_i)^*] - f[(\mathbf{u} - \varepsilon\mathbf{e}_i)^*]| + C
\tag{6.22}
$$

and

$$
|M_i|^2 \leqq |f[(\mathbf{u} + \varepsilon\mathbf{e}_i)^*] - f[(\mathbf{u} - \varepsilon\mathbf{e}_i)^*]|^2 + C
\tag{6.23}
$$

where all of the C's here are not necessarily equal. Equation (6.23) implies that

$$
\|\mathbf{M}(\mathbf{u}, \varepsilon, \delta)\| \leqq \|\mathbf{f}^{+,*}(\mathbf{u}, \varepsilon) - \mathbf{f}^{-,*}(\mathbf{u}, \varepsilon)\| + C.
\tag{6.24}
$$

Now, (6.9) implies that

$$\|\mathbf{M}(\mathbf{u}, \varepsilon, \delta)\| \leqq A'\|\mathbf{u} - \boldsymbol{\theta}'\| + B' \tag{6.25}$$

for some constants A', B'. This completes the proof of the lemma.

Next we verify condition (5.18) of Dvoretzky's Theorem. By the definition of $\mathbf{Y}'_n$,

$$E\|\mathbf{Y}'_n\|^2 = \frac{b_n^2}{c_n^2} E\left\|\mathbf{D}\left(\mathbf{r}_n, c_n, \frac{b_n}{c_n}\right) - \mathbf{M}\left(\mathbf{r}_n, c_n, \frac{b_n}{c_n}\right)\right\|^2. \tag{6.26}$$

Since $\sum b_n^2/c_n^2 < \infty$, it is sufficient to show that $E\|\mathbf{D} - \mathbf{M}\|^2$ is uniformly bounded for n sufficiently large. Now, let $r_{n,i}$ denote the ith component of $\mathbf{r}_n$. Then,

$$\begin{aligned} E|D_i - M_i|^2 \leqq & \left|\frac{c_n}{b_n}(a_i - r_{n,i}) - M_i\right|^2 P\left[Z_{(\mathbf{r}_n+c_n\mathbf{e}_i)^*} - Z_{(\mathbf{r}_n-c_n\mathbf{e}_i)^*} < \frac{1}{\delta}(a_i - r_{n,i})\right] \\ & + \int_{\{(a_i-r_{n,i})\leq b_n(Z_{(\mathbf{r}_n+c_n\mathbf{e}_i)^*}-Z_{(\mathbf{r}_n-c_n\mathbf{e}_i)^*})/c_n\leq(b_i-r_{n,i})\}} |Z_{(\mathbf{r}_n+c_n\mathbf{e}_i)^*} - Z_{(\mathbf{r}_n-c_n\mathbf{e}_i)^*} - M_i|^2 \\ & + \left|\frac{c_n}{b_n}(b_i - r_{n,i}) - M_i\right|^2 P\left[\frac{c_n}{b_n}(b_i - r_{n,i}) < Z_{(\mathbf{r}_n+c_n\mathbf{e}_i)^*} - Z_{(\mathbf{r}_n-c_n\mathbf{e}_i)^*}\right]. \end{aligned} \tag{6.27}$$

Condition (6.9) and Lemma 6.1 together with the fact that $\|\mathbf{u}\|$ is bounded ($\mathbf{u} \in C(\mathbf{a}, \mathbf{b})$) yield that the middle term above is bounded.

Consider the first term of (6.27). It is

$$\begin{aligned} \leqq & \frac{c_n^2}{b_n^2}|a_i - r_{n,i}|^2 P\left[Z_{(\mathbf{r}_n+c_n\mathbf{e}_i)^*} - Z_{(\mathbf{r}_n-c_n\mathbf{e}_i)^*} < \frac{c_n}{b_n}(a_i - r_{n,i})\right] \\ & + |M_i|^2 P\left[Z_{(\mathbf{r}_n+c_n\mathbf{e}_i)^*} - Z_{(\mathbf{r}_n-c_n\mathbf{e}_i)^*} < \frac{c_n}{b_n}(a_i - r_{n,i})\right]. \end{aligned} \tag{6.28}$$

The first term of (6.28) is bounded by argument used in the proof of Lemma 6.1. The second term is bounded by the lemma. Thus

$$E\|\mathbf{Y}'_n\|^2 \leqq \frac{b_n^2}{c_n^2}\text{const} \tag{6.29}$$

and, hence,

$$E(\sum\|\mathbf{Y}'_n\|^2) \leqq \text{const} \sum \frac{b_n^2}{c_n^2} < \infty. \tag{6.30}$$

This completes the verification of (5.18) of Theorem 5.1.

The remainder of the proof of the theorem is a straightforward modification of the proof of Theorem 5.2.

7. **Application to circuit design.** In §4 a model was indicated for a class of circuit design problems. We wished to minimize the risk

$$R(\mathbf{u}) = EL(Y_\mathbf{u}, \mathbf{u}) = EL_1(Y_\mathbf{u}) + L_2(\mathbf{u}) \tag{7.1}$$

subject to a rectangular constraint

$$\mathbf{u} \in C(\mathbf{a}, \mathbf{b}). \tag{7.2}$$

By considering the equivalent problem of maximizing

$$-R(\mathbf{u}) = E[-L(Y_{\mathbf{u}}, \mathbf{u})], \tag{7.3}$$

we can apply the techniques of §§5 and 6.

In order to implement these techniques, we must have available a means of sampling the random variable

$$Z_{\mathbf{u}} = -L(Y_{\mathbf{u}}, \mathbf{u}) \tag{7.4}$$

for any parameter setting $\mathbf{u} \in C(\mathbf{a}, \mathbf{b})$. This can be accomplished by computer simulation. Thus, for the example of §2 it would be sufficient to generate random variables R and C whose uniform distributions are determined by the four components of $\mathbf{u}$. Using the relations of (7.3) and (7.4) it would be possible to make random observations of $Z_{\mathbf{u}} = -L(Y_{\mathbf{u}}, \mathbf{u})$.

Consider the application of the technique to a practical problem. The first task is to define the loss function L_1, giving the contribution to the total loss resulting from an output $Y_{\mathbf{u}}$. L_2 reflects the cost of selecting the parameter setting $\mathbf{u}$.

Next, one could use the design values and possibly some "worst-case" analysis for a first guess $\mathbf{X}_1'$ at the optimum parameter setting $\boldsymbol{\theta}'$. Also, one faces the problem of defining a suitable constraint for the problem. This would likely be governed by such considerations as the availability and the cost of components with different distributions on their values.

Frequently, for a realistically defined loss function $L(Y_{\mathbf{u}}, \mathbf{u})$, the function $-R(\mathbf{u})$ to be maximized has more than one *local* maximum, although it may satisfy the conditions of Theorem 6.1 locally for various points. The problem, of course, is that generally $L(Y_{\mathbf{u}}, \mathbf{u})$ is defined explicitly and the form of the function $-R(\mathbf{u})$ is unknown. Thus, in order for stochastic approximation to be useful, the technique must be used experimentally. That is, one may suspect that within $C(\mathbf{a}, \mathbf{b})$ there is more than one local maximum. But, he may also suspect that the conditions of §6 are satisfied locally. Suppose the modified KW procedure is applied blindly starting at the best guess $\mathbf{X}_1'$ that one's intuition will yield. Possibly several independent KW sequences will converge to, say, two or three points. If, no matter where in $C(\mathbf{a}, \mathbf{b})$ the process is started, it invariably converges to these points, then one would suspect that these were the only local maxima in $C(\mathbf{a}, \mathbf{b})$. Monte Carlo estimation could then be used at the local maxima to find which was a maximum for $C(\mathbf{a}, \mathbf{b})$.

The point of the above discussion is that in practice it may be difficult, or impossible, to verify the assumptions of §6, but the technique may still be a useful tool for searching for a maximum of $-R(\mathbf{u})$.

References

1. J. R. Rosenblatt, *On prediction of system performance from information on component performance*, Proc. Western Joint Computer Conf., I.R.E., February, 1957.

2. W. D. Ashcroft and W. Hochwald, *Design by worst-case analysis: A systematic method to approach specified reliability requirements*, IRE Trans. **RQC–10** (1961), 15–21.

3. J. W. Tukey, *The propagation of errors, fluctuations and tolerances. Basic generalized formulas*, Princeton Univ. Tech. Rep. No. 10, AD 155 082, Princeton University, Princeton, N. J.

4. H. J. Gray, Jr., *An application of piecewise approximations to reliability and statistical design*, Proc. IRE **47** (1959), 1226–1231.

5. S. I. Firstman, *Reliability estimating by the use of random sampling simulation*, Rep. No. P–1521, The RAND Corp., Santa Monica, Calif., 1958.

6. J. Kiefer and J. Wolfowitz, *Stochastic estimation of the maximum of a regression function*, Ann. Math. Statist. **23** (1952), 462–466.

7. J. R. Blum, *Approximation methods which converge with probability one*, Ann. Math. Statist. **25** (1954), 382–386.

8. A. Dvoretzky, *On stochastic approximation*, Proc. 3rd Berkeley Sympos. Math. Statist. and Prob., Vol. I, pp. 39–55, Univ. of California Press, Berkeley, Calif., 1956.

9. J. Sacks, *Asymptotic distribution of stochastic approximation procedures*, Ann. Math. Statist. **29** (1958), 373–405.

10. J. R. Blum, *Multidimensional stochastic approximation procedures*, Ann. Math. Statist. **25** (1954), 737–744.

11. M. Loève, *Probability theory*, Van Nostrand, New York, 1955.

Hughes Research Laboratories,
Malibu, California

RANDOM WALKS ON LATTICES

BY

ELLIOTT W. MONTROLL

The theory of random walks on periodic space lattices has been dealt with sporadically since the turn of the century. In all university courses on the theory of probability or statistical mechanics the lecturer finds an excuse for discussing the one dimensional (1 D) lattice walk by a walker who with every step has a probability p of going to the right and $q = 1 - p$ of going to the left. In keeping with tradition this problem will also be treated here even though the emphasis will be on walks on lattices in higher dimensional spaces.

The century opened with the classic paper of Bachelier [1][1] which presented a random walk model of the stock market. It was there shown that if a large number of stocks is normalized to the same value, the dispersion of the prices of the stocks from their mean value grows as the square root of the time just as the dispersion from the mean position of a random walker varies with the time. This paper, which preceded Einstein's work on Brownian motion (1905), contained many of the ideas which evolved into the modern theory of stochastic processes. In his review of the theory of stochastic processes Chandrasekhar [2] briefly touches on 1 D walks in the presence of reflecting and absorbing barriers. Lattice walks are a special case of his discussion of the theory of random flights.

In more recent times problems of solid state physics have stimulated either explicitly or implicitly a number of investigations on lattice walks. The motions of defects in crystals [3] are direct examples of such walks while the dynamics of various excitations such as spin waves [4; 5] are mathematically equivalent to certain lattice walk problems.

The aim of this paper is to give a self-contained review of the theory of lattice walks. As a preview we list below several of the topics to be considered.

The existence of a qualitative difference between random walks on 3 D lattices and those on 1 D or 2 D lattices was first noticed by Pólya [6] who showed that a random walker on a 1 D or 2 D lattice is certain to return to his starting point while one on a 3 D lattice has a nonvanishing escape probability. The probability of escape was estimated by McCrea and Whipple [7] with an improved calculation having been made by Domb [8]. The author has derived exact values for walks on simple cubic, body centered and face centered cubic lattices [9; 10]. Here we will discuss the Pólya problem as well as the closely connected problem of the average number of steps required to reach a given lattice point for the first time.

[1] *The random walk or Brownian motion of the stock market* was rediscovered independently some years later by M. F. M. Osborne, Operations Res. 7 (1959), 145, 808. B. Mandelbrot has emphasized the importance of long tails in certain distribution functions.

Dvoretsky and Erdös [11] and Vineyard [12] have discussed the number of distinct lattice points visited by a lattice walker. We rederive their results in a more direct manner by employing a certain Tauberian theorem.

The importance of Green's function in the theory of random walks on lattices was recognized by R. Courant [13] and his collaborators [14]. Our work in this paper is based completely on the Green's function technique. As one might expect there is considerable connection between random walks on lattices and discrete potential theory [15]. The use of Green's functions is especially applicable to random walks on lattices with a small number of defective points [16]. The analysis of such situations is quite similar to that of the effect of defects on lattice vibrations [17; 18; 19; 20],[2] and will be presented in §IV.

A 3 D lattice is defined as follows. Consider a set of three unit vectors, $\boldsymbol{i}_1$, $\boldsymbol{i}_2$, and $\boldsymbol{i}_3$ which are not coplanar. The end points of the vectors

$$\boldsymbol{l} \equiv (l_1, l_2, l_3) \equiv l_1\boldsymbol{i}_1 + l_2\boldsymbol{i}_2 + l_3\boldsymbol{i}_3$$

which are generated by letting the l's range through the integers $0, \pm 1, \pm 2, \cdots$ form a space lattice of three dimensions. Periodic lattices of $1, 2, 4, 5, \cdots$ dimensions can be defined in a similar manner.

We begin our discussion with the basic calculation of

I. **The probability of going from a given lattice point to another in t steps.** We discuss this first in the context of walks on the one dimensional lattice of Figure 1 such that at each step a walker steps to the right with a probability p and to

$$\dot{-2} \quad \dot{-1} \quad \dot{0} \quad \dot{1} \quad \dot{2}$$

FIGURE 1

the left with a probability $q = 1 - p$. The coefficient p of $e^{i\phi}$ in the expression

$$(p\, e^{i\phi} + q\, e^{-i\phi})$$

represents the probability of the first step being to the right while the coefficient q of $e^{-i\phi}$ represents the probability of the first step being to the left. The coefficient p^2 of $e^{2i\phi}$ in

$$(p\, e^{i\phi} + q\, e^{-i\phi})^2$$

represents the probability of a walker being at $l = 2$ after two steps, the coefficient $2pq$ of $e^{0i\phi}$ the probability that he has returned to the origin, and the coefficient q^2 of $e^{-2i\phi}$ the probability that he ends at $l = -2$ after two steps. Generally

$$\begin{aligned} P_t(l) &= \text{probability that walker is at } l \text{ after } t \text{ steps} \\ &= \text{coefficient of } e^{-il\phi} \text{ in } (p\, e^{i\phi} + q\, e^{-i\phi})^t \qquad (1) \\ &= \frac{1}{2\pi}\int_{-\pi}^{\pi} (p\, e^{i\phi} + q\, e^{-i\phi})^t\, e^{-il\phi}\, d\phi \end{aligned}$$

[2] [18] contains many earlier Russian references.

since the identity

$$\frac{1}{2\pi}\int_{-\pi}^{\pi} e^{-i\phi m}\, d\phi = \delta_{m,0}$$

implies that the integral operator

$$\frac{1}{2\pi}\int_{-\pi}^{\pi} \exp(-i\phi l)\, d\phi\,\cdot$$

filters out the coefficient of $\exp il\phi$ in the Fourier series representation of a periodic function of period 2π. In the symmetrical case with $p = q = 1/2$

(2) $$P_t(l) = \frac{1}{2\pi}\int_{-\pi}^{\pi} (\cos\phi)^t e^{-i\phi l}\, d\phi = 2^{-t}\frac{t!}{l!(t-l)!}.$$

The appearance of the binomial coefficient $t!/l!(t-l)!$ could of course have been predicted from purely combinational arguments.

The above argument can be generalized immediately to the investigation of walks on a two dimensional square lattice in which

p_1 = prob. of step in positive x direction,
q_1 = prob. of step in negative x direction,
p_2 = prob. of step in positive y direction,
q_2 = prob. of step in negative y direction.

The probability of going from the origin to lattice point $\boldsymbol{l} \equiv (l_1, l_2)$ is

(3a) $$\begin{aligned} P_t(\boldsymbol{l}) &= P_t(l_1, l_2) \\ &= \frac{1}{(2\pi)^2}\iint_{-\pi}^{\pi} [p_1 e^{i\phi_1} + q_1 e^{-i\phi_1} + p_2 e^{i\phi_2} + q_2 e^{-i\phi_2}]\, e^{-i\boldsymbol{\phi}\cdot \boldsymbol{l}}\, d\phi_1\, d\phi_2, \end{aligned}$$

where

(3b) $$p_1 + q_1 + p_2 + q_2 = 1,$$

(3c) $$\phi_1 l_1 + \phi_2 l_2 = \boldsymbol{\phi}\cdot\boldsymbol{l}.$$

In the symmetrical case $p_1 = q_1 = p_2 = q_2 = 1/4$ and

(4) $$P_t(\boldsymbol{l}) = \frac{1}{(2\pi)^2}\iint_{-\pi}^{\pi} [\tfrac{1}{2}(\cos\phi_1 + \cos\phi_2)]^t\, e^{-i\boldsymbol{\phi}\cdot\boldsymbol{l}}\, d\phi_1\, d\phi_2.$$

Indeed, in general, on a hypercubic lattice of s dimensions with $2s$ possible steps of equal weight from each lattice point to a nearest neighbor lattice point [**6**; **10**]

(5) $$P_t(\boldsymbol{l}) = \frac{1}{(2\pi)^s}\int\cdots\int_{-\pi}^{\pi}\left[\frac{1}{s}(c_1 + c_2 + \cdots + c_s)\right]^t e^{-i\boldsymbol{\phi}\cdot\boldsymbol{l}}\, d^s\phi,$$

where

(6) $$c_j = \cos\phi_j \quad\text{and}\quad d^s\phi = d\phi_1\cdots d\phi_s.$$

If one analyzes more exotic walks than those which involve successive steps to nearest neighboring points on a simple cubic lattice he can easily get dizzy trying to generate walks in the elementary manner described above. A more systematic approach to the problem involves the difference equation [10]

$$P_{t+1}(\boldsymbol{l}) = \sum_{\boldsymbol{l}'} p(\boldsymbol{l} - \boldsymbol{l}')P_t(\boldsymbol{l}'), \tag{7a}$$

where $p(\boldsymbol{l})$ = probability of step from $\boldsymbol{l}'$ to $\boldsymbol{l}'' \equiv \boldsymbol{l}' + \boldsymbol{l}$ (independently of t), i.e., of a displacement $\boldsymbol{l}$ at any given step and

$$\sum p(\boldsymbol{l}) = 1. \tag{7b}$$

We consider first the case of a finite $N \times N \times \cdots \times N$ lattice with periodic boundary conditions. In the 1 D case this would correspond to a ring of regularly spaced points. In the 2 D case it corresponds to square grid of points wrapped into the form of a torus, etc. Then

$$\begin{aligned} P_t(l_1, l_2, \cdots, l_s) &= P_t(l_1 + N, l_2, \cdots, l_s) \\ &= P_t(l_1, l_2 + N, l_3, \cdots, l_s) = \text{etc.} \end{aligned} \tag{8}$$

If we assume that $P_t(\boldsymbol{l})$ can be expressed as a linear combination of the complete set of functions

$$\exp 2\pi i(k_1 l_1 + \cdots + k_s l_s)/N \tag{9a}$$

with

$$k_j = 0, 1, \cdots, N - 1 \tag{9b}$$

and that

$$\begin{aligned} \lambda(2\pi \boldsymbol{k}/N) &\equiv \lambda(2\pi k_1/N, \cdots, 2\pi k_s/N) \\ &= \sum_{\boldsymbol{l}} p(\boldsymbol{l}) \exp(2\pi i \boldsymbol{l} \cdot \boldsymbol{k}/N), \end{aligned} \tag{10}$$

then

$$\begin{aligned} \sum_{\boldsymbol{l}} P_{t+1}(\boldsymbol{l}) e^{2\pi i \boldsymbol{k}\cdot \boldsymbol{l}/N} &= \sum_{\boldsymbol{l},\boldsymbol{l}'} p(\boldsymbol{l} - \boldsymbol{l}')\, e^{2\pi i(\boldsymbol{l}-\boldsymbol{l}')\cdot \boldsymbol{k}/N} P_t(\boldsymbol{l}')\, e^{2\pi i \boldsymbol{l}'\cdot \boldsymbol{k}/N} \\ &= \lambda(2\pi \boldsymbol{k}/N) \sum_{\boldsymbol{l}} P_t(\boldsymbol{l})\, e^{2\pi i \boldsymbol{l}\cdot \boldsymbol{k}/N}. \end{aligned} \tag{11}$$

If we use the initial condition that

$$P_0(\boldsymbol{l}) = \delta_{\boldsymbol{l},0} \tag{12}$$

it is clear by inverting (11) that

$$P_t(\boldsymbol{l}) = N^{-s} \sum_{\boldsymbol{k}} [\lambda(2\pi \boldsymbol{k}/N)]^t \exp(-2\pi i \boldsymbol{k} \cdot \boldsymbol{l}/N). \tag{13}$$

In the limit of an infinite lattice we let $\phi_j = 2\pi k_j/N$ and $d\phi_j = 2\pi/N$. Then as $N \to \infty$

$$(14) \qquad P_t(\boldsymbol{l}) = \frac{1}{(2\pi)^s} \int_{-\pi}^{\pi} \cdots \int [\lambda(\boldsymbol{\phi})]^t \, e^{-i\boldsymbol{\phi}\cdot\boldsymbol{l}} \, d^s\phi.$$

With this formula we can easily reproduce (5). For example consider the simple cubic lattice with $s = 3$ and

$$p(\boldsymbol{l}) = \begin{cases} \frac{1}{6} & \text{for} \quad l = (0, 0, \pm 1), (0, \pm 1, 0) \quad \text{and} \quad (\pm 1, 0, 0), \\ 0 & \text{otherwise.} \end{cases}$$

Then

$$(15) \qquad \begin{aligned} \lambda(\phi_1, \phi_2, \phi_3) &= \tfrac{1}{6}(e^{i\phi_1} + e^{-i\phi_1} + e^{i\phi_2} + e^{-i\phi_2} + e^{i\phi_3} + e^{-i\phi_3}) \\ &= \tfrac{1}{3}(c_1 + c_2 + c_3). \end{aligned}$$

Expressions for $\lambda(\phi_1, \phi_2, \phi_3)$ for a 3 D face centered and body centered cubic lattices are obtained as follows.

The random walk between nearest neighbor points on a face centered cubic lattice is equivalent to one on a simple cubic lattice of lattice spacing 1 on which a walker can only take steps which correspond to the twelve displacements $(\pm 1, \pm 1, 0)$, $(\pm 1, 0, \pm 1)$, $(0, \pm 1, \pm 1)$.

Then

$$(16) \qquad \lambda(\boldsymbol{\phi}) \equiv \tfrac{1}{3}(c_1c_2 + c_2c_3 + c_3c_1).$$

The random walk between nearest neighbor points on a body centered cubic lattice is equivalent to one on a simple cubic lattice of lattice spacing 1 on which a walker can only take steps which correspond to the eight displacement vectors $(\pm 1, \pm 1, \pm 1)$ which result from all possible selections of plus and minus signs. Then if each possible step can occur with the same probability, 1/8,

$$(17) \qquad \lambda(\boldsymbol{\phi}) = c_1c_2c_3.$$

Various moments of the distribution function $P_t(\boldsymbol{l})$ can be obtained by inverting (14). The inverse function $\Lambda_t(\boldsymbol{\phi})$ or characteristic function of $P_t(\boldsymbol{l})$ is

$$(18) \qquad \Lambda_t(\boldsymbol{\phi}) \equiv [\lambda(\boldsymbol{\phi})]^t = \sum_{\boldsymbol{l}} P_t(\boldsymbol{l})\, e^{i\boldsymbol{\phi}\cdot\boldsymbol{l}}.$$

The mathematical expectation of l_j after t steps is just (we use the physicist's notation $\langle \boldsymbol{l} \rangle$ rather than the mathematician's $E(\boldsymbol{l})$ for the average value of $\boldsymbol{l}$)

$$(19) \qquad \begin{aligned} \langle l_j \rangle &= \sum_{\boldsymbol{l}} l_j P_t(\boldsymbol{l}) = -\lim_{\phi \to 0} it[\lambda(\boldsymbol{\phi})]^{t-1} \partial\lambda/\partial\phi_j \\ &= t \sum_{\boldsymbol{l}} l_j p(\boldsymbol{l}) \equiv t\mu_j \end{aligned}$$

since $\lambda(0) = 1$. As a special example consider the 1 D walk with

$$p(1) = p, \quad p(-1) = q = 1 - p,$$
$$p(l) = 0 \quad \text{if} \quad l \neq 1 \quad \text{or} \quad -1.$$

Then

$$\langle l \rangle = t(p - q). \tag{20}$$

Second moments can be found in a similar manner

$$\begin{aligned} \langle l_j l_k \rangle = \sum_{l} l_j l_k P_t(\mathbf{l}) &= -\lim_{\boldsymbol{\phi}\to 0} \partial^2 \Lambda / \partial\phi_j \partial\phi_k \\ &= -t \frac{\partial^2 \lambda}{\partial\phi_j \partial\phi_k}\bigg]_{\boldsymbol{\phi}=0} - t(t-1) \frac{\partial \lambda}{\partial\phi_j} \frac{\partial \lambda}{\partial\phi_k}\bigg]_{\boldsymbol{\phi}=0} \\ &= t\mu_{jk} + t(t-1)\mu_j \mu_k, \end{aligned} \tag{21}$$

where

$$\mu_{jk} = \sum_{l} l_j l_k p(\mathbf{l}). \tag{22}$$

Also

$$\langle l_j l_k \rangle - \langle l_j \rangle \langle l_k \rangle = t(\mu_{jk} - \mu_j \mu_k). \tag{23}$$

By an application of the central limit theorems for a sum of vectors, each with the same distribution function, one can show that the joint distribution of $l_1, l_2, \cdots, l_s$ after t steps is an s dimensional Gaussian distribution which is characterized by the mean values (19) and the correlations (21), (22).

II. **Random walk generating functions.** Most of the questions discussed in the remainder of this paper can be derived from random walk generating functions. The most important of these is the generating function of all walks which end at lattice point $\mathbf{l}$ independently of t:

$$U(z, \mathbf{l}) = \sum_{t=0}^{\infty} z^t P_t(\mathbf{l}). \tag{1}$$

On an infinite lattice one finds from §I, (1), §I, (5), and §I, (14):

(a) for 1 D walk

$$U(z, l) = \frac{1}{2\pi} \int_{-\pi}^{\pi} \frac{\exp(-i\phi l)\, d\phi}{1 - z(p\, e^{i\phi} + q\, e^{-i\phi})}; \tag{2}$$

(b) for s D symmetric walk on a simple cubic lattice (all steps being taken to nearest neighbor points)

$$U(z, \mathbf{l}) = \frac{1}{(2\pi)^s} \int \cdots \int_{-\pi}^{\pi} \frac{\exp(-i\boldsymbol{\phi} \cdot \mathbf{l})\, d^s\phi}{1 - \dfrac{z}{s}(c_1 + \cdots + c_s)}; \tag{3}$$

(c) for arbitrary s D lattice [**6**; **10**; **21**]

$$U(z, \boldsymbol{l}) = \frac{1}{(2\pi)^s} \int_{-\pi}^{\pi} \cdots \int \frac{\exp(-i\boldsymbol{\phi} \cdot \boldsymbol{l})\, d^s\phi}{1 - z\lambda(\boldsymbol{\phi})}. \tag{4}$$

In particular the generating function for all walks which start and end at the same point (say the origin) is

$$U(z, 0) = \frac{1}{(2\pi)^s} \int_{-\pi}^{\pi} \cdots \int \frac{d^s\phi}{1 - z\lambda(\boldsymbol{\phi})}. \tag{5}$$

A set of difference equations for $U(z, \boldsymbol{l})$ can be obtained from §I, (7), and §I, (12) by noting that

$$\begin{aligned} U(z, \boldsymbol{l}) &= \delta_{l,0} + \sum_{t=0}^{\infty} z^{t+1} P_{t+1}(\boldsymbol{l}) \\ &= \delta_{l,0} + z \sum_{\boldsymbol{l}'} p(\boldsymbol{l} - \boldsymbol{l}') \sum_{t=0}^{\infty} P_t(\boldsymbol{l}'), \end{aligned}$$

which implies

$$U(z, \boldsymbol{l}) - z \sum_{\boldsymbol{l}'} p(\boldsymbol{l} - \boldsymbol{l}') U(z, \boldsymbol{l}') = \delta_{l,0}. \tag{6}$$

This is a Green's function type of equation.

A simple expression for $\sum_l U(z, \boldsymbol{l})$ follows from (6) by summing over all $\boldsymbol{l}$ for (remembering §I, (7b))

$$\begin{aligned} \sum_{\boldsymbol{l}} U(z, \boldsymbol{l}) &= 1 + z \sum_{\boldsymbol{l},\boldsymbol{l}'} p(\boldsymbol{l} - \boldsymbol{l}') U(z, \boldsymbol{l}') \\ &= 1 + z \sum_{\boldsymbol{l}''} p(\boldsymbol{l}'') \sum_{\boldsymbol{l}} U(z, \boldsymbol{l} - \boldsymbol{l}'') \\ &= 1 + z \sum_{\boldsymbol{l}} U(z, \boldsymbol{l}) \end{aligned}$$

so that

$$\sum_{\boldsymbol{l}} U(z, \boldsymbol{l}) = 1/(1 - z), \tag{7}$$

a result which is also valid for a finite lattice with periodic boundary conditions.

The analogue of (4) for a finite lattice with periodic boundary conditions is (from §I, (13))

$$U_N(z, \boldsymbol{l}) = \frac{1}{N^s} \sum_{k_1=0}^{N-1} \cdots \sum_{k_s=0}^{N-1} \frac{\exp -2\pi i(l_1 k_1 + \cdots + l_s k_s)/N}{1 - z\lambda(2\pi \boldsymbol{k}/N)}. \tag{8}$$

From §I, (7a), and §I, (10) one finds

$$\lambda(0) = 1 \tag{9}$$

and no other $\lambda(\boldsymbol{k})$ are unity. Hence $U_N(z, \boldsymbol{l})$ can be expressed as a sum of a singular and a nonsingular part.

$$U_N(z, \boldsymbol{l}) = \frac{1}{N^s(1 - z)} + \phi(z, \boldsymbol{l}), \tag{10a}$$

where the nonsingular part $\phi(z, \boldsymbol{l})$ is the sum

$$\phi(z, \boldsymbol{l}) = \frac{1}{N^s} \sum_{\boldsymbol{k}}{}' \frac{\exp -2\pi i \boldsymbol{k} \cdot \boldsymbol{l}/N}{1 - z\lambda(2\pi \boldsymbol{k}/N)} \tag{10b}$$

with the term corresponding to $\boldsymbol{k} = 0$ omitted.

On the infinite lattice ($N \to \infty$) the singular part inversely proportional to N^s vanishes but $\phi(z, \boldsymbol{l})$, which then becomes identical to $U(z, \boldsymbol{l})$, develops new singularities of a weaker type. Since the detailed character of $U(z, \boldsymbol{l})$ as $z \to 1$ will be of considerable importance, we consider its properties here. Since the remainder of this section is essentially a table of integrals, the reader may prefer to go directly to §III, referring back to formulae here as indicated in the later sections.

The integral (2) in the symmetric case $p = q = 1/2$ can be found in various tables of definite integrals. Its value is

$$\begin{aligned} U(z, l) &= \{1 - (1 - z^2)^{1/2}\}^{|l|}/z^l(1 - z^2)^{1/2} \\ &= x^{|l|}/(1 - z^2)^{1/2} \end{aligned} \tag{11a}$$

with

$$x = [1 - (1 - z^2)^{1/2}]/z. \tag{11b}$$

As is shown in Appendix 1, the generating function for a ring of N points is

$$U_N(z, l) = \left\{\frac{x^l + x^{N-l}}{1 - x^N}\right\}\Big/(1 - z^2)^{1/2}. \tag{12}$$

Since as $z \to 1$

$$x \sim 1 - [2(1 - z)]^{1/2} + (1 - z) - \frac{3\sqrt{2}}{4}(1 - z)^{3/2} + (1 - z)^2 - \cdots, \tag{13}$$

we find

$$U_N(z, 0) \sim \{N(1 - z)[1 - \tfrac{1}{6}(1 - z)(N^2 - 1) + O(1 - z)^2]\}^{-1} \tag{14}$$

while for $l \geqq 0$

$$\begin{aligned} U_N(z, l)/U_N(z, 0) \sim 1 &- l(N - l)(1 - z) \\ &+ \tfrac{1}{6}l(N - l)\{(N^2 - 5) + l(N - l)\}(1 - z)^2 - \cdots \end{aligned} \tag{15}$$

and

$$U_N(z, l) = U_N(z, -l). \tag{16}$$

In the 2 D infinite square lattice case with

$$p(1, 0) = p(-1, 0) = p_1,$$
(17) $$p(0, 1) = p(0, -1) = p_2,$$
$$p(0, 0) = r \quad \text{and} \quad p(l_1, l_2) = 0 \text{ for other } (l_1, l_2)$$

and with

$$2p_1 + 2p_2 + r = 1$$

one finds

(18)
$$U(z; l_1, l_2) = \frac{1}{(2\pi)^2} \iint_{-\pi}^{\pi} \frac{\exp(-i\boldsymbol{l} \cdot \boldsymbol{\phi})\, d^3\phi}{(1 - zr) - 2zp_1 \cos\phi_1 - 2zp_2 \cos\phi_2}$$
$$= \frac{1}{(2\pi)^2} \int_0^{\infty} dx \iint_{-\pi}^{\pi} d^2\phi \exp\{-i\boldsymbol{l} \cdot \boldsymbol{\phi} - x(1 - zr) + 2xz(p_1 \cos\phi_1 + p_2 \cos\phi_2)\}$$

(19) $$= \int_0^{\infty} e^{-x(1-zr)} I_{l_1}(2zxp_1) \mathrm{I}_{l_2}(2zxp_2)\, dx,$$

where $I_l(x)$ is the lth Bessel function of purely imaginary argument. When $l_1 = l_2 = m \geqq 0$ a closed form expression exists for $U(z; m, m)$ (also note that $U(z; l_1, l_2) = U(z; -l_1, l_2) = U(z; l_1, -l_2) = U(z; -l_1, -l_2)$) **[22]**

(20) $$U(z; m, m) = \frac{1}{2\pi z(p_1 p_2)^{1/2}} Q_{m-1/2}\left(\frac{(1 - zr)^2 - 4z^2(p_1^2 + p_2^2)}{8z^2 p_1 p_2}\right),$$

where $Q_n(x)$ is the nth Legendre function of the second kind. Also

(21) $$Q_{-1/2}(\cosh \eta) = 2\, e^{-\eta/2} K(e^{-\eta}),$$

where $K(k)$ is the complete elliptic integral of the second kind. Note that

$$(1 - zr)^2 - 4z^2(p_1^2 + p_2^2) = 8z^2 p_1 p_2 + (1 - z)\{(1 + z) + 2zr\}.$$

Since as $\varepsilon \to 0$

$$Q_{m-1/2}(1 + \varepsilon) \sim -\tfrac{1}{2} \log(\tfrac{1}{2}\varepsilon) - \gamma - \psi(m + \tfrac{1}{2}) + O(\varepsilon),$$

where $\gamma = 0.57721$ is Euler's gamma and $\psi(x) = d \log \Gamma(x)/dx$, we find

(22)
$$U(z; m, m) = \frac{1}{2\pi z(p_1 p_2)^{1/2}} \left\{-\tfrac{1}{2} \log \frac{(1 - z)[(1 + z) + 2rz]}{16z^2 p_1 p_2} - \gamma - \psi(m + \tfrac{1}{2}) + O(1 - z)\right\}$$

which has a logarithmic divergence as $z \to 1$. Note also that since

$$\psi(n+x) - \psi(x) = \frac{1}{x} + \frac{1}{1+x} + \cdots + \frac{1}{x+n+1}, \tag{23}$$

$$-U(z;m,m) + U(z;0,0) = \left\{1 + \frac{1}{3} + \frac{1}{5} + \cdots + \frac{1}{2m-1}\right\} \Big/ \pi z(p_1 p_2)^{1/2} + O(1-z).$$

It can also be shown that for large l_1, l_2, l_3, and l_4 one has the limit result

$$U(1;l_1,l_2) - U(1;l_3,l_4) = \frac{1}{2\pi(p_1p_2)^{1/2}} \log\left(\frac{p_2 l_3^2 + p_1 l_4^2}{p_2 l_1^2 + p_1 l_2^2}\right). \tag{24}$$

In the 3 D case as $N \to \infty$ the integral $U(z;l)$ converges as $z \to 1$. The limit has been evaluated by Watson for $U(1;0)$ in the case of the three cubic lattices on which a walker can step only to nearest neighbor points with an equal a priori probability of stepping to each such point. He found [23]:

(a) simple cubic lattice

$$\begin{aligned} U(1,0) &= \frac{1}{(2\pi)^3} \iiint_{-\pi}^{\pi} \frac{d^3\phi}{1 - \frac{1}{3}(c_1 + c_2 + c_3)} \\ &= \frac{4}{3\pi^2}[18 + 12\sqrt{2} - 10\sqrt{3} - 7\sqrt{6}]K^2[(2-\sqrt{3})(\sqrt{3}-\sqrt{2})] \\ &= 1.5163860591; \end{aligned} \tag{25}$$

(b) face centered cubic lattice

$$\begin{aligned} U(1,0) &= \frac{1}{(2\pi)^3} \iiint_{-\pi}^{\pi} \frac{d^3\phi}{1 - \frac{1}{3}(c_1c_2 + c_2c_3 + c_3c_1)} \\ &= 9\{\Gamma(\tfrac{1}{3})\}^6/2^{14/3}\pi^4 = 1.3446610732; \end{aligned} \tag{26}$$

(c) body centered cubic lattice

$$\begin{aligned} U(1,0) &= \frac{1}{(2\pi)^3} \iiint_{-\pi}^{\pi} \frac{d^3\phi}{1 - c_1c_2c_3} \\ &= \frac{1}{4\pi^3}\{\Gamma(\tfrac{1}{4})\}^4 = 1.3932039297. \end{aligned} \tag{27}$$

The correction to $U(1,0)$ as z decreases from 1 is proportional to $(1-z)^{1/2}$. For example, in the simple cubic case [24],

$$U(z,0) \sim 1.51638 \cdots - \frac{3}{\pi}\sqrt{\frac{3}{2}}(1-z)^{1/2} + 0.131625\,(1-z) - \cdots. \tag{28}$$

In Table 1 we exhibit some values of $U(1, l)$ as defined by (3) for the 3 D case. More extensive tables of $U(z, l)$ for both the symmetric and several unsymmetric cases $0 \leq z \leq 1$ can be found in reference **[22]**. By methods indicated in references **[15]** and **[24]** one finds

$$(29) \qquad U(1, l) \sim \frac{3}{2\pi l}\left[1 + \frac{1}{8l^2}\left\{-3 + \frac{5(l_1^4 + l_2^4 + l_3^4)}{l^4}\right\} + \cdots\right]$$

as $l = (l_1^2 + l_2^2 + l_3^2)^{1/2} \to \infty$.

(l_1, l_2, l_3)	$U(1, l)$	(l_1, l_2, l_3)	$U(1, l)$
001	0.516387	023	0.132451
002	0.257336	111	0.261470
003	0.165271	112	0.191792
011	0.331149	113	0.144196
012	0.215590	122	0.156953
013	0.153139	123	0.126946
022	0.168331	222	0.135908

TABLE 1. VALUES OF $U(1, l)$ FOR A SQUARE LATTICE WHEN $l^2 = l_1^2 + l_2^2 + l_3^2 < 15$. THESE NUMBERS CORRESPOND TO THE SYMMETRICAL CASE WITH $U(1; l_1, l_2, l_3) = U(1; l_2, l_1, l_3) = \cdots$ ETC.

III. **Recurrence and first passage times.** §I was concerned with the probability $P_t(\boldsymbol{l})$ of a walker going from the origin to the point $\boldsymbol{l}$ in t steps without asking whether the walker had already been at $\boldsymbol{l}$ at an earlier time. We now consider the probability $f_t(\boldsymbol{l})$ of a walker arriving at $\boldsymbol{l}$ *for the first time* after t steps. The generating function of all those walks which reach $\boldsymbol{l}$ for the first time independently of t,

$$(1) \qquad F(z, \boldsymbol{l}) = \sum_{t=1}^{\infty} z^t f_t(\boldsymbol{l})$$

is related to $U(z, \boldsymbol{l})$ through

$$(2) \qquad F(z, \boldsymbol{l}) = \frac{1}{U(z, 0)}\{U(z, \boldsymbol{l}) - \delta_{0,l}\}.$$

This equation is derived by noting that

$$(3) \qquad P_0(\boldsymbol{l}) = \delta_{l,0},$$

$$(4) \qquad P_t(\boldsymbol{l}) = \sum_{j=1}^{t} F_j(\boldsymbol{l}) P_{t-j}(0) \quad \text{if} \quad t > 0$$

since in order for our walker to arrive at $\boldsymbol{l}$ in t steps he must have first arrived at $\boldsymbol{l}$ at some step j and then returned to $\boldsymbol{l}$ after $t - j$ steps. Then

$$(5) \qquad \begin{aligned} U(z, \boldsymbol{l}) &= \sum_{t=0}^{\infty} z^t P_t(\boldsymbol{l}) = \delta_{l,0} + \sum_{t=1}^{\infty}\sum_{j=1}^{\infty} z^j F_j(\boldsymbol{l}) z^{t-j} P_{t-j}(0) \\ &= \delta_{l,0} + F(z, \boldsymbol{l}) U(z, 0) \end{aligned}$$

which is equivalent to (2).

Note that $F(1, \boldsymbol{l})$ is just the probability that the walker ever reaches the point $\boldsymbol{l}$ since $\Sigma f_t(\boldsymbol{l})$ is the sum of probabilities of the independent ways the walker can reach l for the first time. If $U(1, 0) = \infty$ then $F(1, 0) = 1$ and the walker is certain to return to the origin. If $U(1, 0) = u < \infty$ then the probability $F(1, 0) = 1 - u^{-1}$ of return is less than 1 by u^{-1} which represents the probability of escape without return to origin. The discussion of the effect of dimensionality on escape was first given by Pólya **[6]** who showed that when steps are made only to nearest neighbor sites walkers are certain to return to the origin on 1 D and 2 D lattices while a certain escape probability exists on s D lattices with $s \geqq 3$. An exposition of this result and others concerning the theory of recurrent events has also been given by Feller **[25]**. We repeat calculations made by the author **[9; 10]** elsewhere for escape probabilities.

Before proceeding with the exact calculation of probabilities of return to the origin let us examine the qualitative behavior of §II, (5) when $z = 1$:

$$U(1, 0) = \frac{1}{(2\pi)^s} \int_{-\pi}^{\pi} \cdots \int \frac{d^s\phi}{1 - \lambda(\boldsymbol{\phi})}. \tag{6}$$

The convergence of this integral depends on the behavior of the denominator of the integrand. As $(\phi_1, \phi_2, \cdots, \phi_s) \to 0$

$$1 - \lambda(\boldsymbol{\phi}) = \sum_{l} p(\boldsymbol{l})[1 - e^{i\boldsymbol{\phi}\cdot\boldsymbol{l}}] \to 0.$$

In the symmetric case with $p(\boldsymbol{l}) = p(-\boldsymbol{l})$ and $p(l_x) = p(l_y) = p(l_z)$

$$\sum_{l} l_x^{2k+1} p(\boldsymbol{l}) = 0$$

and

$$\sum l_x^2 p(\boldsymbol{l}) = \tfrac{1}{3} \sum l^2 p(\boldsymbol{l}) = \tfrac{1}{3}\mu_2.$$

Hence as $(\phi_1, \cdots, \phi_s) \to 0$

$$1 - \lambda(\boldsymbol{\phi}) \sim \tfrac{1}{6}\phi^2 \mu_2. \tag{7}$$

The integral $U(1, 0)$ can be expressed as the sum of two components, the first being over an s dimensional sphere of a small radius, a, whose center is at the origin and the second the integral over our s D hypercube of volume $(2\pi)^s$ with the central small sphere excluded. In a region not including the origin the integrand is well behaved; no divergence can come from the second component of the integral. We calculate the contribution of the integral over a small spherical shell about the origin of our ϕ space in the neighborhood of $\boldsymbol{\phi} = 0$ but omit the contribution of the sphere of radius ε and then let $\varepsilon \to 0$. As long as the exterior radius of the shell is small, the integrand depends only on r so we can use polar coordinates in our integration and $d^s\phi$ is proportional to $r^{s-1}\,dr$. Our required integral is then proportional to

$$I(\varepsilon) = \int_{\varepsilon}^{a} \frac{r^{s-1}}{r^2}\,dr = \int_{\varepsilon}^{a} r^{s-3}\,dr = \begin{cases} \varepsilon^{-2} - a^{-2} & s = 1, \\ -\log \varepsilon/a & s = 2, \\ (a^{s-2} - \varepsilon^{s-2})/(s - 2) & s \geqq 3. \end{cases} \tag{8}$$

As $\varepsilon \to 0$, so that the neighborhood of the origin is included in the integration, $I(\varepsilon) \to \infty$ when $s = 1$ or 2 while $I(\varepsilon) < \infty$ when $s \geqq 3$. Hence $U_s(1, 0)$ *diverges when* $s = 1, 2$ *and converges when* $s \geqq 3$. This is equivalent to Pólya's result: A random walker who walks in the manner prescribed above is certain to return to his point of origin if he walks on a 1 D or 2 D lattice. A nonvanishing escape probability exists in lattices of more than three dimensions.

We now go beyond Pólya to the exact calculation of $U(1, 0)$ for the three 3 D cubic lattices; simple cubic (S.C.), body centered cubic (B.C.C.) and face centered cubic (F.C.C.). By good fortune G. N. Watson **[23]** has done all the work for us. He evaluated the three integrals §II, (25), (26), and (27) which are needed here. Since the probability of eventual return to the origin is

$$F(1, 0) = 1 - [U(1, 0)]^{-1} \tag{9}$$

we find that on the various cubic lattices is

$$\text{Probability of return to origin} = \begin{cases} 0.340537330 & \text{S.C.,} \\ 0.282229985 & \text{B.C.C.,} \\ 0.256318237 & \text{F.C.C.} \end{cases}$$

The number of nearest neighbors to a lattice point in a simple cubic lattice is 6, in a B.C.C. 8, and in a F.C.C. 12. It is not surprising that the return probability diminishes as the number of nearest neighbors increases since more ways for escape exists. The escape probability also increases with dimensionality for a fixed number of nearest neighbors to each lattice point. For example, the probability of return in a 4 D simple cubic lattice **[9]** (8 nearest neighbors to each lattice point) is 0.20 as compared with the 0.28 figure mentioned above. Earlier estimates in return probabilities on simple cubic lattices have been 0.35 (McCrea and Whipple **[7]**) and 0.34054 (Domb **[8]**).

Since on an s D lattice for $s \geqq 3$ there is a nonvanishing probability that a walker does not return to his point of origin, the average time required for the return is infinite. However, even in the 1 D and 2 D cases in which a walker eventually returns to his point of origin the mean recurrence time is still infinite. In the 1 D symmetrical walk (see §II, (2), with $p = q = 1/2$)

$$\begin{aligned} U(z, l) &= \frac{1}{2\pi} \int_{-\pi}^{\pi} \frac{\exp(-i\phi l)\, d\phi}{1 - z\cos\phi}, \\ &= x^{|l|}(1 - z^2)^{-1/2} \end{aligned} \tag{10a}$$

where

$$x = [1 - (1 - z^2)^{1/2}]/z. \tag{10b}$$

Hence from (2) the generating functions for all walks which return to the origin is

$$\begin{aligned} F(z, 0) &= \sum_{t=1}^{\infty} z^t f_t(0) \\ &= 1 - (1 - z^2)^{1/2}. \end{aligned} \tag{11}$$

The average number of steps required to return to the origin for the first time is

(12)
$$\langle t \rangle = \frac{\partial F}{\partial z}\bigg]_{z=1} = \sum_{t=1}^{\infty} t f_t(0)$$
$$= \lim_{z\to 1} z(1 - z^2)^{-1/2} = \infty.$$

One obtains the same result for a 2 D lattice. These results have already been obtained by Feller **[25]**.

The generating function $U_N(z, l)$ for walks from the origin to the point l on a ring of N particles is

(13)
$$U_N(z, l) = \frac{1}{N}\sum_{j=0}^{N-1} \frac{\exp 2\pi i l j/N}{1 - z \cos 2\pi j/N}$$

while on an s D simple cubic lattice of $N \times N \times \cdots \times N$ points wrapped into an $s + 1$ D torus (periodic boundary conditions) it is given by §II, (8). It is convenient to express $U_N(z, \mathbf{l})$ as §II, (10a)

(14)
$$U_N(z, \mathbf{l}) = \phi(z) + 1/N^s(1 - z),$$

where $\phi(z)$ is a finite sum which has no singularity in the neighborhood of $z = 1$ (see §II, (10a) and (10b)). As $z \to 1$, $\phi(z)$ being a sum of a finite number of finite terms remains bounded while $(1 - z)^{-1} \to \infty$ and therefore $U_N(z, \mathbf{l}) \to \infty$. Hence from (9) one finds $F(1, 0) = 1$ for all dimensions s and all N so that, as one would expect, every walker eventually returns to his starting point on a finite lattice (with periodic boundary conditions).

We now show that (with periodic boundary conditions) *on a finite lattice the average number of steps required by the walker to return to his point of origin is finite and exactly equal to the number of lattice points.* From (2) we see that the generating functions for returns to the origin for the first time is given on an s D lattice by

(15)
$$F(z, 0) = 1 - [U(z, 0)]^{-1}$$
$$= 1 - \frac{(1 - z)N^s}{1 + N^s(1 - z)\phi(z)}.$$

Hence the average number of steps required to return to the origin for the first time is

(16)
$$\langle t \rangle = \frac{\partial}{\partial z} F(z, 0)\bigg]_{z=1}$$
$$= \lim_{z\to 1} \frac{N^s + 2(1 - z)N^{2s}\phi(z) - (1 - z)^2 N^{2s}\phi'(z)}{[1 + N^s(1 - z)\phi(z)]^2}$$
$$= N^s.$$

This result is actually a special case of a general theorem on recurrence times in Markoff chains. Consider a system which can achieve any one of n states. If: (a) at each successive trial the probability of a transition from the ith state to the jth state is the constant p_{ji} (i.e., transitions are generated by a Markoff chain); (b) on every trial a transition to a *new* state occurs; and (c) the matrix P whose elements are p_{ji} has a single characteristic value of absolute value unity, then it can be shown that (a) this characteristic value is unity; (b) an equilibrium distribution which is independent of the initial distribution develops for the occupation probability of the various states; and (c) if P_j is the equilibrium occupation probability of state j, and one transition occurs per unit time, then the mean recurrence time for state i is

$$\langle t_i \rangle = 1/P_i. \tag{17}$$

It is also known that if $p_{ji} = p_{ij}$ (which is the case for our lattice walks) is added to the conditions listed above, the equilibrium distribution $\{P_i\}$ is the uniform distribution, $P_i = 1/n$ for all i. In this case $t_i = n =$ total number of states, which is equivalent to (16).

The second moment of the distribution $\{f_n\}$ depends on the detailed form of the probability of a displacement $\boldsymbol{l}$, $p(\boldsymbol{l})$, from a given lattice point in one step. We have

$$\langle n^2 \rangle = \langle n(n-1) \rangle + \langle n \rangle$$
$$= N^s + \frac{\partial^2}{\partial z^2} F_s(z, 0) \Big]_{z=1}$$

From §II, (10a) we find

$$\langle n^2 \rangle = N^s + \lim_{z \to 1} 2N^{2s}[\phi(z) - (1-z)\phi'(z)]. \tag{18}$$

It can be shown from a careful analysis of $\phi(z)$ which will not be given here that

$$\lim_{z \to 1} (1-z)\phi'(z) = 0.$$

Hence

$$\langle n^2 \rangle = N^s + 2N^{2s}\phi(1). \tag{19}$$

In the 1 D case we find from §II, (14)

$$\phi(1) = \lim_{z \to 1} \left[U_\Lambda(z, 0) - \frac{1}{N(1-z)} \right] = (N^2 - 1)/6N$$

so that

$$\langle n^2 \rangle = N + \tfrac{1}{3}N(N^2 - 1), \tag{20a}$$

from which we find the dispersion to be

$$\langle (n - \bar{n})^2 \rangle = \tfrac{1}{3}N(N-1)(N-2). \tag{20b}$$

In the 3 D case $\phi(1)$ has the limit values (25), (26), and (27) for the various cubic lattices as $N \to \infty$. Hence for example in the S.C. case

$$\lim \langle n^2 \rangle / N^6 = 2.03279 \tag{21a}$$

so that

$$\lim \langle (n - \bar{n})^2 \rangle / \langle n \rangle^2 = 1.03279. \tag{21b}$$

As was pointed out in the beginning of this section, the probability that a walker ever reaches $\boldsymbol{l}$ is just

$$F(1, \boldsymbol{l}) = \frac{U(1, \boldsymbol{l})}{U(1,0)}. \tag{22}$$

In the 1 D case one finds from §II, (11a) that

$$F(z, l) = [1 - (1 - z^2)^{1/2}]^{|l|}/z^{|l|} \tag{23}$$

which, as $z \to 1$ approaches unity. Hence our walker is certain to reach any lattice point $\boldsymbol{l}$ on a 1 D lattice. In the 2 D case we first note from §II, (22) which gives $U(z; m, m)$ as $z \to 1$, that the divergent contribution is proportional to $-\log(1 - z)$ independently of $\boldsymbol{l} \equiv (m, m)$. Hence $F(1; m, m) = 1$ and a walker is certain to reach a point (m, m). It is not difficult to generalize this result to all lattice points on a 2 D lattice provided that $p(\boldsymbol{l}') \to 0$ sufficiently rapidly as $|\boldsymbol{l}'| \to \infty$.

In the 3 D case $U(1, 0)$ is finite while for large l (for simple cubic lattice)

$$U(1, \boldsymbol{l}) \sim \frac{3}{2\pi l} + \frac{3}{16\pi l^3}\left\{-3 + \frac{5(l_1^4 + l_2^4 + l_3^4)}{l^4}\right\} + O(1/l^5).$$

Hence the probability of eventually reaching the point $\boldsymbol{l}$,

$$F(1, \boldsymbol{l}) \sim \frac{3}{2\pi l U(1,0)}\left\{1 + \frac{1}{8l^2}\left[-3 + \frac{5(l_1^4 + l_2^4 + l_3^4)}{l^4}\right] + \cdots\right\} \tag{24}$$

diminishes as $1/l$ for large l. $F(1, \boldsymbol{l})$ for small values of l is given in Table 2.

(l_1, l_2, l_3)	$F(1, l)$	(l_1, l_2, l_3)	$F(1, l)$
001	0.3405	023	0.0873
002	0.1697	111	0.1724
003	0.1090	112	0.1265
011	0.2184	113	0.0951
012	0.1422	122	0.1035
013	0.1010	123	0.0837
022	0.1110	223	0.0896

TABLE 2. PROBABILITY, $F(1, l)$, THAT A WALKER WHO STARTS FROM THE ORIGIN EVER REACHES A POINT $l = (l_1, l_2, l_3)$ ON A SIMPLE CUBIC LATTICE. IN THIS SYMMETRICAL WALK $F(1; l_1, l_2, l_3) = F(1; l_2, l_1, l_3)$, ETC.

As in the case of return to the origin, the mean number of steps required to reach the point $\boldsymbol{l}$ for the first time is infinite for infinite lattices of any number of dimensions. On the other hand, when the number of lattice points is finite this mean first passage time is proportional to the number of lattice points. As in the case of return to the origin we consider our lattice to have periodic boundary conditions.

In the 1 D case we find from §II, (15) that

$$F_N(z, l) = \frac{U_N(z, l)}{U_N(z, 0)} \sim 1 - l(N - l)(1 - z) \tag{25}$$
$$+ \tfrac{1}{6}l(N - l)\,\{(N^2 - 5) + l(N - l)\}(1 - z)^2 - \cdots$$

so that the mean number of steps to reach l for the first time is

$$\langle n \rangle = \sum n z^n f_n(l) = \frac{\partial}{\partial z} F_N(z, l)\Big]_{z=1} \tag{26a}$$
$$= l(N - l)$$

so that as $N \to \infty$

$$\langle n \rangle / N \sim l \tag{26b}$$

if $l > 0$. By symmetry $\lim \langle n \rangle / N = |l|$ for all l.

A corresponding expression for $\lim \langle n \rangle / N^s$ for the $s = 2$ and 3 cases can be obtained without using formulae as detailed as (25). We go back to §II, (10):

$$U_N(z, \mathbf{l}) = \frac{1}{N^s(1 - z)} + \phi(z, \mathbf{l}),$$

where the nonsingular term $\phi(z, l)$ is bounded as $z \to 1$. Note that

$$F_N(z, \mathbf{l}) = \frac{1 + N^s(1 - z)\phi(z, \mathbf{l})}{1 + N^s(1 - z)\phi(z, 0)}$$

so that

$$\lim_{N \to \infty} \langle n \rangle / N^s = \lim_{z \to 1, N \to \infty} N^{-s} \partial F_N / \partial z \tag{27}$$
$$= \lim_{N \to \infty} [\phi(1, 0) - \phi(1, \mathbf{l})] = \lim_{z \to 1} [U(z, 0) - U(z, \mathbf{l})].$$

We first verify our 1 D result from §II, (11),

$$U(z, 0) - U(z, l) \sim (1 - z^2)^{-1/2}\{|l|(1 - z^2)^{1/2} + O(1 - z)\} \to |l|$$

as $z \to 1$ which yields (26b).

In the 2 D case consider the point $\mathbf{l} = (l_1, l_2)$ and its "closest" diagonal (m, m) defined so that $|l_1^2 + l_2^2 - 2m^2| =$ minimum (for example, if $(l_1, l_2) = (17, 45)$, then $m = 33$). Also let

$$2\alpha m^2 = l_1^2 + l_2^2,$$

α being 1.007 in the case (17, 45), and generally of $O(1)$. Then from §II, (23) and (24) we see that for large $|l| = (l_1^2 + l_2^2)^{1/2}$

$$U(1; 0, 0) - U(1; l_1, l_2) = U(1; 0, 0) - U(1; m, m) + U(1; m, m) - U(1; l_1, l_2)$$
$$= \frac{4}{\pi}\left[1 + \frac{1}{3} + \frac{1}{5} + \cdots + \frac{1}{2m - 1}\right] + \frac{2}{\pi}\log\left(\frac{l_1^2 + l_2^2}{2m^2}\right)$$
$$\sim \frac{1}{2\pi}\log|l| + O(1).$$

Hence, in the 2 D case the average number of steps required to go from the origin to (l_1, l_2) is, on an $N \times N$ lattice with periodic boundary conditions,

$$\langle n\rangle/N^2 \sim \frac{1}{2\pi} \log |l|. \tag{28}$$

A remarkable property of the 3 D case is that as $|l| \to \infty$, $\langle n\rangle/N^3 \to$ constant. This follows from the fact that (see §II, (25) and (29))

$$U(1;0,0,0) - U(1;l_1,l_2,l_3) \sim 1.516\cdots - (3/2\pi l) + O(1/l^2)$$

which implies that

$$\langle n\rangle/N^3 \sim 1.516 - (3/2\pi l) \tag{29}$$

so that on the average, the number of steps required to reach for the first time any point sufficiently far from the origin is independent of the exact location of the point. Values of $\langle n\rangle/N^3$ for small l are given in Table 3. Further discussion on the material in the last half of this section can be found in **[26]**.

(l_1, l_2, l_3)	$\langle n\rangle/N^3$	(l_1, l_2, l_3)	$\langle n\rangle/N^3$
001	1.000	023	1.384
002	1.259	111	1.255
003	1.351	112	1.325
011	1.185	113	1.372
012	1.301	122	1.359
013	1.363	123	1.389
022	1.348	222	1.380

TABLE 3. RATIO OF THE AVERAGE NUMBER OF STEPS REQUIRED TO REACH THE LATTICE POINT $\boldsymbol{l}$ FOR THE FIRST TIME, $\langle n\rangle$, TO N^3, THE TOTAL NUMBER OF LATTICE POINTS ON A SIMPLE CUBIC LATTICE WITH PERIODIC BOUNDARY CONDITIONS. THE LIMIT OF THIS RATIO AS $N \to \infty$ IS TABULATED.

IV. **The number of lattice points visited after *n* steps.** As a lattice walker continues his walk the number of distinct points visited increases. This number can be determined as shown below from the properties of the generating function $U(z, 0)$. This problem was first considered by Dvoretzky and Erdös **[11]**, who showed that the number of points visited after n steps as $n \to \infty$ is

(1a) for 1 D lattice $(8n/\pi)^{1/2}$,

(1b) for 2 D lattice $\pi n/\log n$,

(1c) for s D lattice with $s \geqq 3$, cn,

where the constant c depends on the topology of the lattice. The value of the constant c has been evaluated by Vineyard **[12]** for the 3 D simple cubic, F.C.C. and B.C.C. lattices.

Let S_n be the average number of distinct lattice points visited after n steps and let $F_n(\boldsymbol{l})$ be the probability that the lattice point $\boldsymbol{l}$ has been visited at least once in the first n steps of the walker. Then

$$F_n(\boldsymbol{l}) = f_1(\boldsymbol{l}) + f_2(\boldsymbol{l}) + \cdots + f_n(\boldsymbol{l}), \tag{2}$$

where, as defined in §I, $f_j(\boldsymbol{l})$ is the probability that our walker occupied the point l for the first time on the jth step. Clearly

$$S_n = \sum_l F_n(\boldsymbol{l}) \tag{3}$$

so that

$$\Delta_n \equiv S_n - S_{n-1} = \sum_l [F_n(\boldsymbol{l}) - F_{n-1}(\boldsymbol{l})] = \sum_l f_n(\boldsymbol{l}), \tag{4}$$

where the summation extends over all lattice points $\{\boldsymbol{l}\}$.

We now relate the generating function of Δ_n to $U(z, 0)$:

$$\begin{aligned} \Delta(z) &= \sum_{n=1}^{\infty} z^n \Delta_n = \sum_l \sum_{n=1}^{\infty} z^n f_n(\boldsymbol{l}) \\ &= \sum_l F(z, \boldsymbol{l}) = \frac{1}{U(z, 0)} \left\{-1 + \sum_l U(z, \boldsymbol{l})\right\}, \end{aligned} \tag{5}$$

where we have used (4) and §III, (7). Since it has already been shown in §II, (7) that

$$\sum_l U(z, \boldsymbol{l}) = (1 - z)^{-1},$$

we find

$$\Delta(z) = z/\{(1 - z)U(z, 0)\}, \tag{6}$$

from which it follows that

$$\Delta_n = \text{coef. of } z^n \text{ in } \Delta(z) = \frac{1}{2\pi i} \int_c \frac{dz}{z^n(1 - z)U(z, 0)},$$

the contour being a counter-clockwise one surrounding the origin but not including $z = 1$.

The average number of distinct points visited after n steps by our walker is just (see (4))

$$\begin{aligned} &S_0 = 1, \quad S_1 = 2, \\ &S_n = 1 + \Delta_1 + \Delta_2 + \cdots + \Delta_n, \qquad n \geq 1. \end{aligned} \tag{7}$$

When n is large the asymptotic properties of S_n can be derived from the asymptotic properties of $\Delta(z)$ as $z \to 1$ by employing the Tauberian theorem (a special case of Theorem 108 in [27]: Let the sum $S(y) = \sum a_n \exp(-ny)$ be convergent for all $y > 0$ and let $a_n > 0$ for all n. If as $y \to 0$

$$S(y) \sim \phi(y^{-1}), \tag{8a}$$

where

$$\phi(x) \equiv x^\sigma L(x) \tag{8b}$$

is a positive increasing function of x (when x is greater than some fixed x_0) which tends to ∞ as $x \to \infty$, and if $\sigma \geqq 0$ and $L(cx) \sim L(x)$ as $x \to \infty$; then as $n \to \infty$

$$a_1 + a_2 + \cdots + a_n \sim \phi(n)/\Gamma(\sigma + 1). \tag{8c}$$

Inserting the asymptotic properties of $U(z, 0)$ as $z \to 1$ (the constants u and u_1 in the 3 D case depend on the lattice structure and in particular the values of u are given in §II, (25)–(27),

$$U(z, 0) = (1 - z^2)^{-1/2} \qquad 1\text{ D}, \tag{9a}$$

$$U(z, 0) \sim \begin{cases} -\dfrac{1}{\pi}\log(1 - z) & 2\text{ D}, \quad (9b) \\ u + u_1(1 - z)^{1/2} + \cdots & 3\text{ D}, \quad (9c) \end{cases}$$

into (6) and letting $z = \exp(-y)$ we find that as $y \to 0$

$$\Delta(z) \sim \begin{cases} (2/y)^{1/2} & 1\text{ D}, \\ (\pi/y)/\log(1/y) & 2\text{ D}, \\ 1/uy & 3\text{ D} \quad (\text{and generally } s\text{ D with } s \geqq 3). \end{cases}$$

The Tauberian theorem given above applies directly to our problem if we choose

$$\begin{aligned} &1\text{ D } \sigma = \tfrac{1}{2}, && L(x) = 2^{1/2} = \text{constant}, \\ &2\text{ D } \sigma = 1, && L(x) = \pi/\log x, \\ &3\text{ D } \sigma = 1, && L(x) = 1/u = \text{constant}. \end{aligned}$$

With this choice we obtain the results of Erdös and Dvoretsky for the number of distinct lattice points visited by our walker after n steps as $n \to \infty$

$$S_n \sim \begin{cases} (8n/\pi)^{1/2} & 1\text{ D}, \\ \pi n/\log n & 2\text{ D}, \\ n/u & 3\text{ D}. \end{cases} \tag{1}$$

The values of $1/u$ for various cubic lattices are

$$1/u = \begin{cases} 0.629\,462\,670 & \text{S.C.}, \\ 0.717\,770\,010 & \text{B.C.C.}, \\ 0.743\,681\,763 & \text{F.C.C.} \end{cases} \tag{10}$$

They yield values of S_n in the 3 D case in agreement with those of Vineyard.

Asymptotic expansions for S_n (of which (1) is the leading term) are given in Appendix 2 for the 1 D and 3 D cases. These expansions are valid for small as well as large n.

V. **Random walks on slightly defective lattices.** So far we have been concerned with random walks on perfectly periodic lattices. Sometimes one wishes to investigate the influence of a small number of defective lattice points on random walks. For example one might analyze the effect of a trapping point on the probability of a walker eventually going from his origin to a given lattice point. Defects have been studied in the theory of semiconductors, lattice vibrations, and various other aspects of solid state physics. The methods discussed below are based on those developed for these fields **[17–20]**.

Let us consider the effect of a few defect points on random walks. Using the notation of §I the probability that a random walker is at a point $\boldsymbol{l}$ after t steps, $P_t(\boldsymbol{l})$, is determined from a set of equations

$$P_{t+1}(\boldsymbol{l}) = \sum_{\boldsymbol{l}} p(\boldsymbol{l}, \boldsymbol{l}')P_t(\boldsymbol{l}'), \tag{1}$$

where $p(\boldsymbol{l}, \boldsymbol{l}')$ is the probability of a step from $\boldsymbol{l}'$ to $\boldsymbol{l}$ by a walker known to be at $\boldsymbol{l}'$. In a perfect lattice $p(\boldsymbol{l}, \boldsymbol{l}')$ was postulated to be a function only of $(\boldsymbol{l} - \boldsymbol{l}')$. In an imperfect lattice some points differ from others so this is no longer the case. However, we can express $p(\boldsymbol{l}, \boldsymbol{l}')$ as

$$p(\boldsymbol{l}, \boldsymbol{l}') = p(\boldsymbol{l} - \boldsymbol{l}') + q(\boldsymbol{l}, \boldsymbol{l}'), \tag{2}$$

the component $p(\boldsymbol{l} - \boldsymbol{l}')$ being that of the perfect lattice and $q(\boldsymbol{l}, \boldsymbol{l}')$ the perturbation. After any step a walker is certain to be somewhere. Hence

$$\sum_{\boldsymbol{l}} p(\boldsymbol{l}, \boldsymbol{l}') = 1 \text{ for all } \boldsymbol{l}'. \tag{3a}$$

This corresponds to the conservation of walkers. In the perfect lattice it was expressed as

$$\sum_{\boldsymbol{l}} p(\boldsymbol{l} - \boldsymbol{l}') = 1. \tag{3b}$$

Hence

$$\sum_{\boldsymbol{l}} q(\boldsymbol{l}, \boldsymbol{l}') = 0 \text{ for all } \boldsymbol{l}'. \tag{3c}$$

Let $G(z, \boldsymbol{l})$ be the generating function for $\{P_t(\boldsymbol{l})\}$:

$$G(z, \boldsymbol{l}) = \sum_{t=0}^{\infty} z^t P_t(\boldsymbol{l}). \tag{4}$$

On the supposition that at $t = 0$, before the first step, the walker is at the origin

$$P_0(\boldsymbol{l}) = \delta_{\boldsymbol{l},0}. \tag{5}$$

Multiplication of (1) by z^{t+1}, summation from $t = 0$ to ∞, and incorporation of (2) and (5) yields

$$G(z, \boldsymbol{l}) - \delta_{\boldsymbol{l},0} = z\sum_{\boldsymbol{l}'} p(\boldsymbol{l}, \boldsymbol{l}')G(z, \boldsymbol{l}')$$

or

$$G(z, \boldsymbol{l}) - z\sum_{\boldsymbol{l}'} p(\boldsymbol{l} - \boldsymbol{l}')G(z, \boldsymbol{l}') = \delta_{\boldsymbol{l},0} + z\sum_{\boldsymbol{l}'} q(\boldsymbol{l}, \boldsymbol{l}')G(z, \boldsymbol{l}'). \tag{6}$$

This set of linear equations can be solved in terms of the Green's function $U(z, \mathbf{l})$, which is the solution of

$$U(z, \mathbf{l}) - z\sum_{\mathbf{l}'} p(\mathbf{l} - \mathbf{l}')U(z, \mathbf{l}') = \delta_{\mathbf{l},0}. \tag{7}$$

The exact formula for $U(z, \mathbf{l})$, has been given for nearest neighbor walks on various lattices in §I. The inhomogeneous equation

$$G(z, \mathbf{l}) - z\sum_{\mathbf{l}'} p(\mathbf{l} - \mathbf{l}')G(z, \mathbf{l}') = F(\mathbf{l}) \tag{8}$$

has the solution

$$G(z, \mathbf{l}) = \sum_{\mathbf{l}'} U(z, \mathbf{l} - \mathbf{l}')F(\mathbf{l}'), \tag{9}$$

as can be verified by substituting this expression into the left side of (8) and employing (7) to obtain the right hand side of (8). Now if we let

$$F(\mathbf{l}) = \delta_{\mathbf{l},0} + z\sum_{\mathbf{l}'} q(\mathbf{l}, \mathbf{l}')G(z, \mathbf{l}')$$

in (8), then (9) becomes [**16**]

$$G(z, \mathbf{l}) = U(z, \mathbf{l}) + z\sum_{\mathbf{l}',\mathbf{l}''} U(z, \mathbf{l} - \mathbf{l}')q(\mathbf{l}', \mathbf{l}'')G(\mathbf{l}'', z). \tag{10}$$

Generally this equation can be solved by iteration. However, under the special conditions that $q(\mathbf{l}, \mathbf{l}')$ vanishes except at a few points, a closed form solution can be derived [**16**].

First consider the effect of a single defective point on a random walk which on each step the walker would normally go to one of his nearest neighbors (all nearest neighbor steps being given the same weight). Suppose one point, $\mathbf{l}_1$, differs from the rest so that there is a probability ε that a walker pauses at $\mathbf{l}_1$ each time he normally would have stepped away. If Z is the number of nearest neighbors to any point

$$\begin{aligned} &q(\mathbf{l}_1, \mathbf{l}_1) = \varepsilon, \\ &q(\mathbf{l}, \mathbf{l}_1) = -\frac{\varepsilon}{Z} \quad \text{if } l \text{ is an n.n. to } l_1, \\ &q(\mathbf{l}, \mathbf{l}') = 0 \quad \text{if } \mathbf{l}' \neq \mathbf{l}_1 \quad \text{or if } \mathbf{l}' \equiv \mathbf{l}_1 \quad \text{and } \mathbf{l} \text{ is not an n.n. to } \mathbf{l}_1. \end{aligned} \tag{11}$$

Then

$$G(z, \mathbf{l}) = U(z, \mathbf{l}) + z\varepsilon U(z, \mathbf{l} - \mathbf{l}_1) - \frac{z\varepsilon}{Z}\sum_{\{\mathbf{l}_k\}} U(z, \mathbf{l} - \mathbf{l}_k)G(z, \mathbf{l}_1), \tag{12}$$

where $\{\mathbf{l}_k\}$ is the set of points which are nearest neighbors to $\mathbf{l}_1$. One finds from (7) that with our choice of unperturbed transition probabilities

$$(z/Z)\sum_{\{\mathbf{l}_k\}} U(z, \mathbf{l} - \mathbf{l}_k) = U(z, \mathbf{l} - \mathbf{l}_1) - \delta_{\mathbf{l},\mathbf{l}_1} \tag{13}$$

so that

$$G(z, \mathbf{l}) = U(z, \mathbf{l}) + \varepsilon\{(z - 1)U(z, \mathbf{l} - \mathbf{l}_1) + \delta_{\mathbf{l},\mathbf{l}_1}\}G(z, \mathbf{l}_1). \tag{14}$$

An expression can be found for $G(z, \boldsymbol{l}_1)$ by setting $\boldsymbol{l} = \boldsymbol{l}_1$ in (14). Then

$$G(z, \boldsymbol{l}_1) = U(z, \boldsymbol{l}_1) + \varepsilon\{(z - 1)U(z, 0) + 1\}G(z, \boldsymbol{l}_1) \tag{15}$$

so that

$$G(z, \boldsymbol{l}_1) = U(z, \boldsymbol{l}_1)/[(1 - \varepsilon) + \varepsilon(1 - z)U(z, 0)] \tag{16}$$

and generally if $\boldsymbol{l} \neq \boldsymbol{l}_1$

$$G(z, \boldsymbol{l}) = U(z, \boldsymbol{l}) - \frac{\varepsilon(1 - z)U(z, \boldsymbol{l} - \boldsymbol{l}_1)U(z, \boldsymbol{l}_1)}{1 - \varepsilon + \varepsilon(1 - z)U(z, 0)}. \tag{17}$$

In particular if $\boldsymbol{l}_1$ is a trap, $\varepsilon = 1$, so that

$$G(z, \boldsymbol{l}) = U(z, \boldsymbol{l}) - U(z, \boldsymbol{l} - \boldsymbol{l}_1)U(z, \boldsymbol{l}_1)/U(z, 0). \tag{18}$$

Actually this formula could have been derived more directly if one notices that (see §III, (2))

$$U(z, \boldsymbol{l}_1)/U(z, 0) = F(z, \boldsymbol{l}_1)$$

is the generating function for all walks which go through $\boldsymbol{l}_1$ for the first time at various times. Then $U(z, \boldsymbol{l} - \boldsymbol{l}_1)F(z, \boldsymbol{l}_1)$ is the generating function for all walks which start at the origin, proceed through $\boldsymbol{l}_1$ and end at $\boldsymbol{l}$. If $\boldsymbol{l}_1$ is a trap the generating function $G(z, \boldsymbol{l})$, of all those walks which take the walker to $\boldsymbol{l}$ without being trapped is exactly the difference between the generating function for all walks and that for trapping walks.

As an application of (18) let us calculate the effect of a trap at $\boldsymbol{l}_1$ on the probability of a walker eventually returning to the origin. From §III, (2) the probability of eventual return to the origin is

$$F(1, 0) = 1 - \frac{1}{G(1, 0)}. \tag{19}$$

But from (18) and §II, (11)

$$G(z, 0) = (1 - x^{|2l_1|})/(1 - z^2)^{1/2}, \tag{20}$$

where x is $[1 - (1 - z^2)^{1/2}]/z$. Then as $z \to 1$, $x^{|2l_1|} \sim 1 - 2|l_1|\sqrt{(1 - z^2)}$ so that

$$G(1, 0) = |2l_1| \tag{21}$$

and the probability of eventual return to the origin is

$$F(1, 0) = 1 - 1/|2l_1|. \tag{22}$$

The largest probability of a walker being trapped before return to the origin is of course with a trap at $l_1 = \pm 1$ in which case $F(1, 0) = \frac{1}{2}$.

Although the analogue of (18) can also be derived without resorting to our formalism in the case of two or more defect points, the bookkeeping gets quite tedious. We apply the formalism to the 2-defect problem with one pausing or trapping point at $\boldsymbol{l}_1$ and the other at $\boldsymbol{l}_2$. Then (12) becomes

$$\begin{aligned} G(z, \boldsymbol{l}) = U(z, \boldsymbol{l}) &+ \varepsilon\{(z - 1)U(z, \boldsymbol{l} - \boldsymbol{l}_1) + \delta_{\boldsymbol{l},\boldsymbol{l}_1}\}G(z, \boldsymbol{l}_1) \\ &+ \varepsilon\{(z - 1)U(z, \boldsymbol{l} - \boldsymbol{l}_2) + \delta_{\boldsymbol{l},\boldsymbol{l}_2}\}G(z, \boldsymbol{l}_2). \end{aligned} \tag{23}$$

If one successively sets l equal to l_1 and l_2 he obtains a pair of simultaneous equations for $G(z, l_1)$ and $G(z, l_2)$. Upon solution in the trapping case $\varepsilon = 1$ this yields

$$G(z, l_1) = \frac{U(z, l_1)U(z, 0) - U(z, l_2)U(z, l_1 - l_2)}{(1 - z)[\{U(z, 0)\}^2 - |U(z, l_1 - l_2)|^2]} \tag{24}$$

and the same expression with l_1 and l_2 interchanged for $G(z, l_2)$. Then from (23)

$$G(z, 0) = U(z, 0) - \frac{\begin{array}{c}\{U(z, -l)[U(z, 0)U(z, l) - U(z, l_1 - l_2)U(z, l_2)] \\ + U(z, -l_2)[U(z, 0)U(z, l_2) - U(z, l_2 - l_1)U(z, l_1)]\}\end{array}}{[U^2(z, 0) - U^2(z, l_1 - l_2)]}. \tag{25}$$

From this formula we can find the effect of two traps on the probability of a walker eventually returning to his starting point. We let $l_2 < 0 < l_1$ on a 1 D lattice. Then from the expressions §II, (11) derived for $U(z, l)$ in the 1 D case we obtain

$$G(z, 0) = \frac{1}{\sqrt{(1 - z^2)}} \frac{(1 - x^{2l_1})(1 - x^{-2l_2})}{[1 - x^{2(l_1 - l_2)}]}$$

so that

$$G(1, 0) = \frac{2l_1(-l_2)}{l_1 - l_2}. \tag{26}$$

The probability of a walker returning to the origin before he is trapped is

$$F(1, 0) = 1 - \frac{1}{G(1, 0)} = 1 - \frac{(l_1 - l_2)}{2l_1(-l_2)}. \tag{27}$$

As the new trap at $-l_2$ recedes to $-\infty$ this reduces to the one trap formula (22).

(l_1, l_2, l_3)	$F(1, 0)$	(l_1, l_2, l_3)	$F(1, 0)$
001	0.254	023	0.335
002	0.321	111	0.320
003	0.333	112	0.330
011	0.305	113	0.334
012	0.327	122	0.333
013	0.334	123	0.336
022	0.332	222	0.335

TABLE 4. PROBABILITY OF A WALKER EVENTUALLY RETURNING TO THE ORIGIN, $F(1, 0)$, IN THE PRESENCE OF A TRAP AT POINT (l_1, l_2, l_3). THESE NUMBERS ARE TO BE COMPARED WITH THE VALUE 0.341 FOR A TRAPLESS WALK.

In the 3 D case the probability of eventual return to the origin when a trap is located at l is

$$F(1, 0) = 1 - \frac{U(1, 0)}{[U(1, 0)]^2 - [U(1, l)]^2}. \tag{28}$$

This quantity is tabulated in Table 4 (using $U(1, l)$ values of Table 1) for various l values with $(l_1^2 + l_2^2 + l_3^2) < 15$. When l is large one finds that $U(1, \mathbf{l})$ is much smaller than $U(1, 0)$ so that (28) becomes (with $u \equiv U(1, 0)$)

$$F(1, 0) = 1 - \frac{1}{u}\left[1 + \left(\frac{U(1, \mathbf{l})}{u}\right)^2 + \cdots\right]. \tag{29}$$

Then the asymptotic formula §III, (29) implies

$$F(1, 0) = 1 - \frac{1}{u}\left[1 + \left(\frac{3}{2\pi ul}\right)^2 + \cdots\right]. \tag{30}$$

It can be shown from (25) and (19) that when l_1, l_2 and $|l_1 - l_2|$ are all large in the 2-defect case, the probability of return to the origin is diminished in an additive way by the influence of the traps at $\mathbf{l}_1$ and $\mathbf{l}_2$. Then

$$F(1, 0) \sim 1 - \frac{1}{u}\left[1 + \left(\frac{3}{2\pi u}\right)^2 \Sigma l_j^{-2} + \cdots\right], \tag{31}$$

this formula being also valid for any number of widely separated distant traps. Of course interference terms appear when some of the l_j's are close to each other or when they might be near the origin. These terms can be obtained from (25) in the 2-trap case and from the generalization of (25) in the n-trap case.

Appendix 1. Generating Function $U_N(z, l)$ for a Ring of N Equally Spaced Points

This is a derivation of §II, (12) for the generating function $U_N(z, l)$ which satisfies the boundary condition $U_N(z, l) = U_N(z, l + N)$ appropriate for a ring of N points. From §II, (11a), the formula for $U(z, l)$ for an infinite lattice is

$$U_\infty(z, l) = x^{|l|}(1 - z^2)^{-1/2},$$

where x is defined by §II, (11b). This is the generating function for walks from 0 to l on an infinite lattice. The generating function for walks from 0 to l on a ring of N points is equivalent to the sum of the generating function for all those walks on an infinite line which represent walks from 0 to l, from 0 to $l + N$, from 0 to $l + 2N$, etc., as well as from 0 to $l - N$, $l - 2N$, etc. Hence

$$\begin{aligned} U_N(z, l) &= U_\infty(z, l) + U_\infty(z, l + N) + U_\infty(z, l + 2N) + \cdots \\ &\quad + U_\infty(z, l - N) + U_\infty(z, l - 2N) + \cdots \\ &= (1 - z^2)^{-1/2}[x^l + x^{l+N} + x^{l+2N} + \cdots + x^{N-l} + x^{2N-l} + \cdots] \\ &= \left(\frac{x^l + x^{N-l}}{1 - x^N}\right)\Big/(1 - z^2)^{1/2} \end{aligned}$$

which is exactly §II, (12).

Appendix 2

The Tauberian theorem used in §IV to determine the number of distinct points visited gives asymptotic results easily for long walks. In this appendix we develop a more systematic procedure that yields results applicable to short as well as long walks. Using the notation of §IV, S_n is the average number of distinct points visited in a walk of n steps. The generating function for $\Delta_n = S_n - S_{n-1}$ was (§IV, (6))

$$\Delta(z) = \sum_1^\infty \Delta_n z^n = z/(1-z)U(z,0). \tag{1}$$

If we let $S_0 = 1$ (the starting point being occupied before the walk starts), then the generating function for $s_n = S_n - 1$ is (see §IV, (7))

$$\begin{aligned} s &= \sum_1^\infty s_n z^n = \sum_{n=1}^\infty (\Delta_1 + \Delta_2 + \cdots + \Delta_n)z^n = (1-z)^{-1}\Delta(z) \\ &= z/\{(1-z)^2 U(z,0)\}. \end{aligned} \tag{2}$$

In the 1 D case

$$s = z(1-z^2)^{1/2}/(1-z)^2 \tag{3}$$

$$= 2^{1/2}\frac{[1-(1-z)]}{(1-z)^{3/2}}(1-\tfrac{1}{2}[1-z])^{1/2}$$

$$= 2^{1/2}\{(1-z)^{-3/2} - \tfrac{5}{4}(1-z)^{-1/2} + \tfrac{7}{32}(1-z)^{1/2} + \cdots\}. \tag{4}$$

Our required coefficient s_n of z^n in the power series expansion of this quantity can be obtained by adding together the coefficients of z^n in the expansion of each term in the bracket. One finds

$$S_n - 1 = s_n = \frac{(2n+1)!}{2^{2n-1/2}n!n!}\left\{1 - \frac{5}{4(2n+1)} + \frac{7}{32(4n^2-1)} + \cdots\right\}. \tag{5}$$

Exact values of S_n are obtained for $n \leq 10$ by considering the coefficient of z^n in (3) for $n = 1, 2, \cdots, 10$. These are given below.

n	S_n	n	S_n
1	2	6	19/4
2	3	7	81/16
3	7/2	8	43/8
4	4	9	723/128
5	35/8	10	379/64

When $n = 8$, $S_8 = 5.375$ while from (5), using the three terms explicitly exhibited one finds $S_8 \simeq 5.376$. Even for n as small as 2 we find that $S_2 \simeq 3.02$ from (5) as compared with the exact value 3. When n is large the various factorials can be approximated by employing Stirling's formula so that

$$(6)\qquad S_n \sim 1 + \left(\frac{8n}{\pi}\right)^{1/2}\left[1 - \frac{1}{4n} + \frac{5}{64n^2} - \cdots\right].$$

We proceed in essentially the same way to improve on the asymptotic 3 D formula given by §IV, (1). It has been shown in reference [24] that in the 3 D case as $z \to 1$

$$(7)\qquad U(z, 0) = u[1 + c_1(1 - z)^{1/2} + c_2(1 - z) + c_3(1 - z)^{3/2} + \cdots].$$

Then from (2) the 3 D analogue of (4) is

$$s = u^{-1}\{(1 - z)^{-2} - c_1(1 - z)^{-3/2} + (c_1^2 - c_2 - 1)(1 - z)^{-1} + (2c_1c_2 + c_1 - c_1^3 - c_3)(1 - z)^{-1/2} + \cdots\}.$$

By adding together the coefficient of z^n from each term we find

$$S_n = (n/u) - (c_1/u)\frac{(2n + 1)!}{2^{2n}n!n!} + [1 + (c_1^2 - c_2 - 1)u^{-1}] + u^{-1}(2c_1c_2 + c_1 - c_1^3 - c_3)\frac{(2n)!}{2^{2n}n!n!} + \cdots.$$

When n is sufficiently large we employ Stirling's formula for the various factorials and find

$$(8)\qquad S_n \sim (n/u)\left\{1 - \frac{2c_1}{(\pi n)^{1/2}} + \frac{1}{n}[u + c_1^2 - c_2 - 1] + \frac{\pi}{(n\pi)^{3/2}}[2c_1c_2 + \tfrac{1}{4}c_1 - c_1^3 - c_3] + \cdots\right\}.$$

In the case of a simple cubic lattice $1/u$ is given by §IV, (10) while from reference [24]

$$c_1 = -(3/\pi u)(3/2)^{1/2}.$$

Hence

$$(9)\qquad S_n \sim 0.629462n\left\{1 + \frac{4}{u}\left(\frac{3}{2\pi}\right)^{3/2} n^{-1/2} + \cdots\right\}.$$

References

1. L. Bachelier, *Théorie de la spéculation*, Ann. Sci. École Norm. Sup. (3) **17** (1900), 21.

2. S. Chandrasekhar, *Stochastic problems in physics and astronomy*, Rev. Modern Phys. **15** (1943), 1–89.

3. A. C. Damask and C. J. Dienes, *Point defects in metals* (to appear).

4. F. J. Dyson, *General theory of spin-wave interactions*, Phys. Rev. **102** (1956), 1217–1230; *Thermodynamic behavior of an ideal ferromagnet*, ibid., 1230–1244.

5. W. Opachowski, *On the calculation of the partition function of a ferromagnetic crystal*, Physica **25** (1959), 476–486.

6. G. Pólya, *Über Eine Aufgabe der Wahrscheinlichkeitrechnung Betreffena die Irrfahrt in Stratzennetz*, Math. Ann. **84** (1921), 149.

7. W. H. McCrea and F. J. W. Whipple, *Random paths in two and three dimensions*, Proc. Roy. Soc. Edinburgh **60** (1939), 281.

8. C. Domb, *On multiple returns in the random-walk problem*, Proc. Cambridge Philos, Soc. **50** (1954), 586–591.

9. E. W. Montroll, *Random walks in multidimensional spaces, especially on periodic lattices*, J. Soc. Indust. Appl. Math. **4** (1956), 241–260.

10. ———, *Topics in statistical mechanics of interacting particles*, La théorie des gaz neutres et ionisés (Grenoble, 1959), pp. 15–148, p. 17, Hermann, Paris, 1960.

11. A. Dvoretzky and P. Erdös, *Some problems on random walk in space*, Proc. 2nd Berkeley Sympos. Math. Statist. and Prob. (1950), pp. 353–367, Univ. of California Press, Berkeley, Calif., 1951.

12. G. H. Vineyard, *The number of distinct sites visited in a random walk on a lattice*, J. Mathematical Phys. **4** (1963), 1191–1193.

13. R. Courant, Vol. **3.** p. 83, Atti Congresso Internazionale Dei Matematica Bologna, Bologna, 1928.

14. R. Courant, K. Friedrichs, and H. Lewy, Math. Ann. **100** (1928), 32.

15. R. J. Duffin, *Discrete potential theory*, Duke Math. J. **20** (1953), 223–251.

16. E. W. Montroll, *Applied combinatorial mathematics*, E. F. Beckenbach, ed. (to appear).

17. E. W. Montroll and R. Potts, *Effect of defects on lattice vibrations*, Phys. Rev. **100** (1955), 525–543; **102** (1956), 72.

18. I. M. Lifshitz, Nuovo Cimento Supp. (X) **3** (1956), 716.

19. A. Maradudin, E. Montroll, and G. Weiss, *Lattice vibrations in the harmonic approximation*, Academic Press, New York, 1963.

20. M. Toda, ed., *Proceedings of symposium on the theory of lattice vibrations of imperfect crystals held at Kyoto University*, Supplement of the Progress of Theoretical Physics **23** (1962).

21. F. G. Foster and I. G. Good, *On a generalization of Pólya's random-walk theorem*, Quart, J. Math. Oxford Ser. (2) **4** (1953), 120–126.

22. A. Erdélyi, W. Magnus, F. Oberhettinger, and F. Tricomi, *Tables of integral transforms*, Vol. 1, p. 183, McGraw-Hill, New York, 1954.

23. G. N. Watson, Quart. J. Math. Oxford Ser. **10** (1939), 266.

24. A. Maradudin, E. Montroll, G. Weiss, R. Herman, and H. Milnes, *Green's functions for monatomic simple cubic lattices*, Acad. Roy. Belg. Cl. Sci. Mém. Coll. in-4° (2) **14** (1960), no. 7, 176 pp.

25. W. Feller, *An introduction to probability theory and its applications*, Wiley, New York, 1951.

26. E. Montroll and G. Weiss, J. Mathematical Phys. (to appear).

27. G. H. Hardy, *Divergent series*, Oxford, at the Clarendon Press, 1949.

UNIVERSITY OF MARYLAND,
COLLEGE PARK, MARYLAND

ON STATISTICAL SPECTRAL ANALYSIS[1]

BY

EMANUEL PARZEN

DEDICATED TO PROFESSOR CHARLES LOEWNER ON HIS 70TH BIRTHDAY

1. **Introduction.** Statistical spectral analysis may be defined as a theory which seeks to provide a method whereby from observed records, of finite length, of an empirical phenomenon one can interpret the time function representing the phenomenon as a superposition (finite or infinite linear combination) of harmonics cos ωt and sin ωt. In other words, statistical spectral analysis attempts to provide a modern solution to what classically has been called "the search for hidden periodicities" in a phenomenon. In modern terms, statistical spectral analysis is concerned with the estimation of the "spectrum" of observed time series.

Spectral analysis seems to be one of the most useful tools available for time series analysis. The theory of time series analysis is concerned with the analysis and the synthesis of stochastic models which may be used to describe (and perhaps also to control) the mechanisms generating a time series and relating various time series. Because the concepts of time series analysis possess meanings of their own which do not depend on particular applications, the theory of time series analysis provides mathematical procedures which can be used to solve problems arising in diverse fields. In my view, time series analysis is an interdisciplinary field since it provides a medium for the exchange of ideas between research workers concerned with quite different subject matters.

Much of the impetus for the recent development of statistical spectral analysis has come from noise theory. This might lead one to believe that the problem of statistical spectral analysis is best discussed in the language of communication engineering, using such notions as "black box" and "filter." This is clearly the right point of view as far as interpreting the spectrum is concerned. But I believe that in developing a theory of *estimation* of the spectrum it is necessary to bring to bear in addition other points of view, derived from approximation theory and from statistical theory. In particular, it seems to me that the problem involved in estimating the spectrum is to some extent a special case of the problem of how to best approximate a function given only its first N Fourier coefficients.

It is not my aim to review the extensive development of the theory of statistical spectral analysis in the last decade. Fortunately there are available a number of review papers which do this (see [**5**; **8**; **11–13**; **18–20**; **22**]). Rather, my aim is to

[1] Prepared under Contract Nonr 225(21) for the Office of Naval Research.

sketch (without proofs) the theory of statistical spectral analysis adopting what may be called a *covariance approach* rather than a *spectral approach*. There are four sections: definition of the spectrum, estimation of the spectrum, estimation of spectral density, estimation of spectral jumps. Applications of spectral analysis are not discussed. But it should be noted that spectral analysis can be applied to one of the central problems of noise theory, namely signal detection and estimation in the presence of stationary noise of unspecified structure. Similarly, spectral analysis seems to be applicable to one of the central problems of economic time series analysis, namely how to seasonally adjust time series and how to determine the effect of the seasonal adjustment procedures that are employed.

2. **Definition of the spectrum.** Early researchers envisaged a time series $\{X(t),\ t = 1, 2, \cdots\}$ as satisfying the model

$$X(t) = \sum_{j=1}^{q} C_j \cos(\omega_j t + \phi_j) + \eta(t), \qquad t = 1, 2, \cdots, \tag{2.1}$$

in which the following assumptions are made. The number of terms q, the amplitudes C_j, the angular frequencies ω_j, and the phases ϕ_j are constants, some of which are given while the others are to be estimated. The sequence $\{\eta(t),\ t = 1, 2, \cdots\}$ is a white noise process; that is, it consists of independent identically distributed random variables with means 0 and common variance σ_η^2. A technique, known as the *periodogram*, was introduced by Schuster **[21]** as a method for discovering the hidden frequencies ω_j (or, equivalently, the hidden periods $2\pi/\omega_j$).

Let $\{X(t),\ t = 1, 2, \cdots, T\}$ be a finite sample of a time series. In order to detect the presence in the time series $\{X(t),\ t = 1, 2, \cdots\}$ of a harmonic component,

$$C\cos(\omega t + \phi) = A\cos\omega t + B\sin\omega t, \tag{2.2}$$

of frequency ω one would form the (discrete) *Fourier coefficients*

$$\begin{aligned} A_T(\omega) &= \frac{2}{T}\sum_{t=1}^{T} X(t)\cos\omega t, \\ B_T(\omega) &= \frac{2}{T}\sum_{t=1}^{T} X(t)\sin\omega t \end{aligned} \tag{2.3}$$

and then form the *intensity*

$$S_T(\omega) = A_T^2(\omega) + B_T^2(\omega) = \left|\frac{2}{T}\sum_{t=1}^{T} X(t)\, e^{-i\omega t}\right|^2 \tag{2.4}$$

or equivalently the periodogram

$$I_T(\omega) = \frac{T}{2} S_T(\omega) = \frac{2}{T}\left|\sum_{t=1}^{T} X(t)\, e^{-i\omega t}\right|^2. \tag{2.5}$$

If there were no noise fluctuations present in the time series, then the intensity $S_T(\omega)$ would, for large values of T be approximately equal to $A^2 + B^2$, no matter what other harmonics were present. Therefore to test the hypothesis that there is no component of frequency ω present in the time series one would test whether the observed value of $S_T(\omega)$ is small enough as to be compatible with $A^2 + B^2 = 0$. Under the assumption that the noise process $\eta(t)$ is white, the distribution of $S_T(\omega)$ is *exponential* (or chi-square with two degrees of freedom) with mean

$$E[S_T(\omega)] = \frac{4\sigma^2}{T}. \tag{2.6}$$

Using this fact, significance tests for the presence of hidden periodicities in a time series were developed by Schuster **[21]**, Walker **[23]**, and Fisher **[4]**. By the decades of the 1940's, however, these methods were discredited for practical use since when applied to empirical time series they seemed to provide evidence for the existence of "spurious cycles" (for an example of this phenomenon on a constructed time series, see Brooks and Carruthers **[3**, Chapters 19–22]).

The development of a theory of spectral analysis of empirically observed time series which would provide consistent and interpretable results is made possible by introducing the notion of the *sample covariance function* of a time series $\{X(t), t = 1, 2, \cdots, T\}$:

$$R_T(v) = \begin{cases} \dfrac{1}{T}\displaystyle\sum_{t=1}^{T-v} X(t)X(t+v), & v = 0, 1, \cdots, T-1; \\ 0, & v \geqq T; \\ R_T(-v), & v < 0. \end{cases} \tag{2.7}$$

I hope to indicate in this paper the central role of the sample covariance function in the theory of statistical spectral analysis. There are those who reject this "covariance" approach in favor of what may be called the "spectral" approach to spectral analysis. In any event I believe it fair to claim that the "covariance" approach corresponds to the attitude adopted by Wiener **[25]** in his monumental paper on generalized harmonic analysis and represents the "radical recasting" which Wiener says (p. 118) is necessary in order "to make the Schuster theory assume a form suitable for extension and generalization."

It is worth noting that one can even adopt a "correlation" approach. A theory of statistical spectral analysis of somewhat wider applicability can be developed using the sample *correlation* function, defined by

$$\rho_T(v) = \frac{R_T(v)}{R_T(0)}. \tag{2.7'}$$

Rearranging the double sum involved in (2.5), one sees that the periodogram is essentially the Fourier transform of $R_T(v)$:

$$I_T(\omega) = 2 \sum_{|v|<T} e^{-i\omega v} R_T(v) \tag{2.8}$$

so that

$$R_T(v) = (4\pi)^{-1}\int_{-\pi}^{\pi} e^{iv\omega} I_T(\omega)\, d\omega. \tag{2.9}$$

In order to obtain Fourier transformation formulas which are somewhat more familiar than (2.8) and (2.9) we introduce the *sample spectral density function* $f_T(\omega)$, defined by

$$f_T(\omega) = \frac{1}{4\pi} I_T(\omega) = \frac{1}{2\pi T}\left|\sum_{t=1}^{T} e^{-i\omega t} X(t)\right|^2. \tag{2.10}$$

Then

$$\begin{aligned} f_T(\omega) &= \frac{1}{2\pi}\sum_{|v|<T} e^{-iv\omega} R_T(v) \\ &= \frac{1}{2\pi} R_T(v) + \frac{1}{\pi}\sum_{v=1}^{T-1} \cos v\omega R_T(v); \end{aligned} \tag{2.11}$$

$$R_T(v) = \int_{-\pi}^{\pi} e^{iv\omega} f_T(\omega)\, d\omega. \tag{2.12}$$

The tools are now at hand to give a general definition of the spectrum of a time series $\{X(t),\ t = 1, 2, \cdots\}$. One assumes that there exists a function $R(v)$, called the covariance function, such that

$$\lim_{T\to\infty} E[R_T(v)] = R(v), \qquad v = 0, \pm 1, \pm 2, \cdots. \tag{2.13}$$

From (2.12) and (2.13) it follows (by the continuity theorem of probability theory) that there exists a bounded nondecreasing function of a real variable ω, denoted $F(\omega)$ and called the spectral distribution function, such that

$$R(v) = \int_{-\pi}^{\pi} e^{iv\omega}\, dF(\omega), \qquad v = 0, \pm 1, \pm 2, \cdots, \tag{2.14}$$

and such that for every continuous function $g(\omega)$

$$\lim_{T\to\infty} \int_{-\pi}^{\pi} g(\omega) E[f_T(\omega)]\, d\omega = \int_{-\pi}^{\pi} g(\omega)\, dF(\omega). \tag{2.15}$$

If further the sample covariance function $R_T(v)$ is a consistent estimate of $R(v)$, in the sense that for each $v = 0, \pm 1, \cdots$

$$\lim_{T\to\infty} R_T(v) = R(v) \tag{2.16}$$

as a limit in mean square then for every continuous function $g(\omega)$

$$\lim_{T\to\infty} \int_{-\pi}^{\pi} g(\omega) f_T(\omega)\, d\omega = \int_{-\pi}^{\pi} g(\omega)\, dF(\omega) \tag{2.17}$$

as a limit in mean square.

One important case in which (2.13) holds is when $\{X(t), t = 0, 1, \cdots\}$ has zero means and its covariance kernel $E[X(s)X(t)]$ is stationary in the sense that

$$E[X(s)X(t)] = R(|s - t|) \tag{2.18}$$

for some even function $R(v)$; this property is called *covariance stationarity* in Parzen [**14**, p. 70]. Then

$$E[R_T(v)] = \left(1 - \frac{|v|}{T}\right) R(v) \rightarrow R(v). \tag{2.19}$$

Consequently the spectral representation, equation (2.14), holds. From the spectral representation for the covariance function of a stationary time series one obtains a spectral representation for the time series itself:

$$X(t) = \int_{-\pi}^{\pi} e^{it\omega}\, dZ(\omega), \qquad t = 0\ \pm 1, \pm 2, \cdots, \tag{2.20}$$

where $\{Z(\omega), -\pi \leqq \omega \leqq \pi\}$ is a process with orthogonal increments and

$$E[|dZ(\omega)|^2] = dF(\omega). \tag{2.21}$$

From (2.21) one obtains an intuitive interpretation of the spectral distribution function $F(\omega)$; its increment across a frequency band is a measure of the mean square average of the Fourier components of the time series in this frequency band.

The problem of statistical spectral analysis can now be defined as the problem of estimating the spectral distribution function of a time series which satisfies (2.13). It is to be emphasized that this problem is well defined not only for stationary time series but also for asymptotically stationary ones (see Parzen [**12**]).

Like a probability distribution function, $F(\omega)$ can be uniquely written as the sum,

$$F(\omega) = F_{d}(\omega) + F_{sc}(\omega) + F_{ac}(\omega) \tag{2.22}$$

of three distribution functions which are respectively discrete (or purely discontinuous), singular continuous, and absolutely continuous. $F_{ac}(\omega)$ is the integral of a non-negative function called the *spectral density function* of the time series,

$$F_{ac}(\omega) = \int_{-\pi}^{\omega} f(\lambda)\, d\lambda, \qquad -\pi \leqq \omega \leqq \pi. \tag{2.23}$$

$F_{d}(\omega)$ may be written as a sum over the discontinuity points $\{\omega_j\}$ of $F(\omega)$ of its jumps

$$J(\omega) = F(\omega + 0) - F(\omega - 0) \tag{2.24}$$

by the formula

$$F(\omega) = \sum_{\substack{\omega_j \text{ such that} \\ J(\omega_j) > 0 \text{ and } \omega_j \leqq \omega}} J(\omega_j). \tag{2.25}$$

Observed time series are usually assumed to have a *mixed spectrum*, in the sense that the singular continuous part of the spectral distribution function vanishes. The spectrum of the time series is said to be *continuous* if the spectral jump function $J(\omega)$ vanishes for all ω, and to be *discrete* if the spectral density function $f(\omega)$ vanishes for all ω.

An important example of a time series with a mixed spectrum is one satisfying the model (2.1), but with a more general assumption on the *residuals* or *fluctuations* $\eta(t)$. It is assumed that $\{\eta(t), t = 1, 2, \cdots\}$ is a *normal stationary* stochastic process with zero means,

$$E[\eta(t)] = 0, \tag{2.26}$$

and covariance kernel satisfying

$$E[\eta(s)\eta(t)] = R_\eta(s-t) = \int_{-\pi}^{\pi} \cos \omega(s-t)\, f(\omega)\, d\omega. \tag{2.27}$$

The observed time series $\{X(t), t = 0, 1, \cdots\}$ then satisfies (2.13) with

$$\begin{aligned} R(v) &= \frac{1}{2}\sum_{j=1}^{q} |A_j|^2 \cos \omega_j v + R_\eta(v) \\ &= \sum_{j=1}^{q} J(\omega_j) \cos \omega_j v + \int_{-\pi}^{\pi} \cos v\omega\, f(\omega)\, d\omega \end{aligned} \tag{2.28}$$

defining

$$J(\omega) = \begin{cases} \frac{1}{2}|A_j|^2 & \text{if } \omega = \omega_j, \\ 0 & \text{if } \omega \text{ is not equal to } \omega_1, \cdots, \omega_q. \end{cases} \tag{2.29}$$

The spectral distribution function is thus of *mixed type*; its absolutely continuous part is given by the spectral density function $f(\omega)$ of the fluctuation process $\{\eta_t\}$ and its discrete part has jumps at the frequencies $\omega_1, \cdots, \omega_q$, the respective jumps being of size $J(\omega_1), \cdots, J(\omega_q)$.

It should be noted that since we are using (for mathematical convenience) both negative and positive frequencies, in defining in (2.29) the jump function $J(\omega)$ we perhaps should define it as an even function as follows:

$$J(\omega) = \begin{cases} \frac{1}{4}|A_j|^2 & \text{if } \omega = \omega_j \text{ or } \omega = -\omega_j \text{ and } \omega_j \neq 0; \\ \frac{1}{2}|A_j|^2 & \text{if } \omega = \omega_j = 0; \\ 0 & \text{otherwise.} \end{cases} \tag{2.30}$$

However, we will not employ this definition in this paper.

3. **Estimation of the spectrum.** If one knew the covariance function $R(v)$, one could recover from it (by well-known inversion formulas) the spectral jump function $J(\omega)$ and the spectral density function $f(\omega)$. For ease of exposition,

we assume in this section that the spectrum is continuous. Then if one forms a linear combination of covariances,

$$\sum_{v=-\infty}^{\infty} a(v)R(v) = \int_{-\pi}^{\pi} \left\{ \sum_{v=-\infty}^{\infty} a(v) \cos v\omega \right\} f(\omega)\, d\omega, \tag{3.1}$$

with coefficients $a(v)$ one obtains a *spectral average*

$$J(A) = \int_{-\pi}^{\pi} A(\omega) f(\omega)\, d\omega \tag{3.2}$$

corresponding to a *spectral window*,

$$A(\omega) = \sum_{v=-\infty}^{\infty} a(v) \cos v\omega, \tag{3.3}$$

which is the Fourier transform of the coefficients. Note that conversely

$$a(v) = \frac{1}{\pi} \int_{-\pi}^{\pi} A(\omega) \cos v\omega\, d\omega. \tag{3.4}$$

Alas we do not know $R(v)$ but only its sample estimates $R_T(v)$. Therefore let us introduce

$$D_T(v) = R_T(v) - R(v), \qquad v = 0, \pm 1, \pm 2, \cdots,$$

the *deviations* of the sample covariance function about the true covariance function. We may then write

$$R_T(v) = R(v) + D_T(v). \tag{3.5}$$

For a stationary normal time series with zero means and continuous spectrum, the deviation time series $\{D_T(v), v = 0, \pm 1, \cdots, \pm T\}$ may be shown (for references, see Parzen [**13**, p. 982]) to be approximately normal for large values of T with means

$$E[D_T(v)] = E[R_T(v)] - R(v) \to 0 \tag{3.6}$$

as $T \to \infty$ and covariance kernel satisfying (for non-negative v_1 and v_2)

$$\begin{aligned} T \operatorname{Cov}[D_T(v_1), D_T(v_2)] &\to \sum_{u=-\infty}^{\infty} R_\eta(u)\{R_\eta(u + v_1 - v_2) + R_\eta(u + v_1 + v_2)\} \\ &= 4\pi \int_{-\pi}^{\pi} \cos \lambda v_1 \cos \lambda v_2 f^2(\lambda)\, d\lambda \end{aligned} \tag{3.7}$$

as $T \to \infty$, where $f(\omega)$ is the spectral density function of $\{\eta_t\}$.

From (3.7) we obtain heuristically a conclusion (which can be shown rigorously; see Parzen [**10**]) of some interest, namely while at each v the sample covariance $R_T(v)$ is a consistent estimate of $R(v)$, the entire function $\{R_T(v), v = 0, \pm 1, \cdots\}$

is not a satisfactory estimate of the entire covariance function $\{R(v), v = 0, \pm 1, \cdots\}$ since in particular

$$E\left[\sum_{v=-(T-1)}^{T-1} \{R_T(v) - R(v)\}^2\right] = \sum_{|v|<T} E[D_T^2(v)]$$

$$= \frac{4\pi}{T} \sum_{|v|<T} \int_{-\pi}^{\pi} \cos^2 \lambda v f^2(\lambda)\, d\lambda \tag{3.8}$$

$$\to \left\{2\pi \int_{-\pi}^{\pi} f^2(\lambda)\, d\lambda\right\} > 0 \quad \text{as} \quad T \to \infty.$$

By introducing suitable weights $\{k_T(v),\ v = 0,\ 1, \cdots, T\}$ one can obtain estimates

$$R_T^*(v) = k_T(v) R_T(v) \tag{3.9}$$

of the covariance functions which do have the property that

$$E\left[\sum_{v=0}^{T} |R_T^*(v) - R(v)|^2\right] \to 0 \quad \text{as} \quad T \to \infty. \tag{3.10}$$

In a sense (3.10) indicates the statistical role of weights in spectral analysis; they help damp out fluctuations in the sample covariance function. *We next indicate the role that weights play in estimating spectral averages from the point of view of approximation theory.*

A linear combination of sample covariances may be written

$$\sum_{|v|<T} a(v) R_T(v) = \sum_{|v|<T} a(v) R(v) + \sum_{|v|<T} a(v) D_T(v)$$

$$= \int_{-\pi}^{\pi} \left\{\sum_{|v|<T} a(v) \cos \omega v\right\} dF(\omega) + \sum_{|v|<T} a(v) D_T(v). \tag{3.11}$$

We hereafter assume that the deviation time series may be considered to be normal with zero means and covariance given by (3.7). Then the statistic $\sum_{|v|<T} a(v) R_T(v)$ is normal with *mean*

$$\int_{-\pi}^{\pi} \left\{\sum_{|v|<T} a(v) R(v)\right\} dF(\omega) \tag{3.12}$$

and *variance*

$$\sum_{|v_1|<T} \sum_{|v_2|<T} a(v_1) a(v_2) \operatorname{Cov}[D_T(v_1), D_T(v_2)] = \int_{-\pi}^{\pi} \left|\sum_{|v|<T} a(v) \cos v\lambda\right|^2 f^2(\lambda)\, d\lambda \tag{3.12}$$

assuming, as we do hereafter, that

$$a(-v) = a(v). \tag{3.13}$$

One thus sees that if one desires to consider

$$\sum_{|v|<T} a(v) R_T(v) \tag{3.14}$$

as an estimate of the spectral average

$$J(A) = \sum_{v=-\infty}^{\infty} a(v)R(v) = \int_{-\infty}^{\infty} A(\omega)\, dF(\omega) \tag{3.15}$$

that there are two kinds of error to be considered. One type of error, called *bias*, arises because the mean of (3.14) is not equal to (3.15) but rather is equal to the spectral average

$$J\left(\sum_{|v|<T} a(v) \cos \omega v\right) \tag{3.16}$$

whose spectral window is the truncated Fourier series of the desired spectral window $A(\omega)$. The other type of error, the *variance*, is given by (3.12).

To make these considerations more meaningful, let us consider the case of an important spectral average, namely (for fixed ω and $\Delta > 0$) the *spectral mass in the interval* $(\omega - \Delta, \omega + \Delta)$ represented by

$$\frac{1}{2}\left\{\int_{-\omega-\Delta}^{-\omega+\Delta} dF(\omega') + \int_{\omega-\Delta}^{\omega+\Delta} dF(\omega')\right\}, \tag{3.17}$$

which corresponds to the spectral window

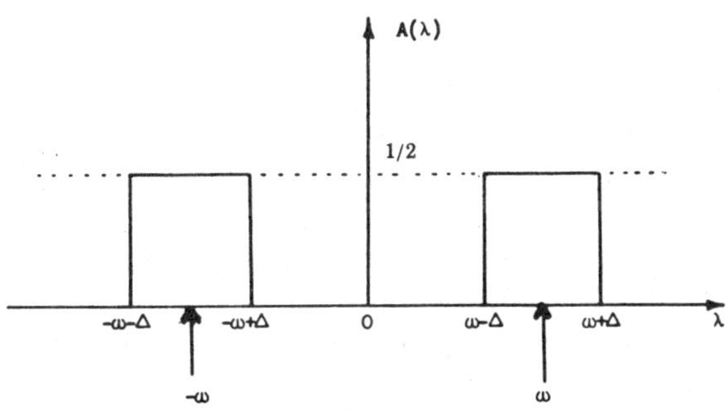

The Fourier coefficients of the even function $A(\lambda)$ are given by

$$\begin{aligned} a(v) &= \frac{1}{\pi}\int_{-\pi}^{\pi} A(\lambda) \cos \lambda v \, d\lambda = \frac{1}{\pi}\int_{\omega-\Delta}^{\omega+\Delta} \cos \lambda v \, d\lambda \\ &= \frac{\sin(\omega + \Delta)v - \sin(\omega - \Delta)v}{\pi v} \\ &= \cos \omega v \frac{\sin \Delta v}{\pi v}. \end{aligned} \tag{3.18}$$

It would seem that to estimate the spectral average $J(A)$ one would adopt the estimate

$$\sum_{|v|<T} \cos \omega v \frac{\sin \Delta v}{\pi v} R_T(v). \tag{3.19}$$

However, it is a well-known fact that because of the Gibbs phenomenon (see Hamming [**6**, p. 295], Lanczos [**9**, p. 225]) the truncated Fourier series

$$\sum_{|v|<T} \cos \omega v \frac{\sin \Delta v}{\pi v} \cos \lambda v \tag{3.20}$$

can be improved as a pointwise (rather than least squares) approximation to $A(\lambda)$ by introducing *weights* $k_T(v)$ and using the function

$$A^*_{T;k}(\lambda) = \sum_{|v|<T} \cos \omega v \frac{\sin \Delta v}{\pi v} \cos \lambda v \, k_T(v) \tag{3.21}$$

as an approximation to $A(\lambda)$.

The Fourier transform

$$K_T(z) = \frac{1}{2\pi} \sum_{|v|<T} \cos \omega v \, k_T(v) \tag{3.22}$$

is defined to be a *window*, since one may write

$$A^*_{T;k}(\omega) = \int_{-\pi}^{\pi} \tfrac{1}{2}\{K_T(\omega - \lambda) + K_T(\omega + \lambda)\}A(\lambda)\, d\lambda. \tag{3.23}$$

In view of (3.23) we regard $A^*_{T;k}(\omega)$ as the impression obtained of the function $A(\omega)$ when it is viewed through a window (or channel) of variable transmission properties given by the window $K_T(z)$. In other words, $K_T(z)$ is the window through which $A^*_{T;k}(\omega)$ views $A(\omega)$.

A variety of possible weights $k_T(v)$ may be considered. These are listed in Table 3A, together with their corresponding windows. They are graphed in Figure 3A for $N = 6$. Since we are not considering any precise criterion by which to judge these windows, it may be a matter of debate which is most desirable. It is my opinion that *different windows are best compared by comparing window generating functions* (this notion is defined at the end of this section).

Windows are used in spectral analysis in three important ways.

(i) When forming estimates of the spectral mass in a band of frequencies, instead of (3.19) one would use the windowed estimate

$$\sum_{|v|<T} \cos \omega v \frac{\sin \Delta v}{\pi v} k_T(v) R_T(v). \tag{3.24}$$

This use of windows helps guard against the leakage into the mean of our estimate of spectral mass from outside the band under study.

TABLE 3A. WEIGHTS $k_N(n)$ AND WINDOWS $K_N(z)$

$$k_N(0) = 1, \quad k_N(-n) = k_N(n), \quad k_N(n) = \int_0^{2\pi} \cos nz K_N(z)\, dz; \; K_N(z) = \frac{1}{2\pi} \sum_{j=-N}^{N} k_N(j) \cos jz$$

$k_N(n)$	$2\pi K_N(z)$
I. Truncated Fourier Series $1, n = 0, 1, \cdots, N.$	$2\pi D_N(z) = \dfrac{\sin (N + \frac{1}{2})}{\sin \frac{1}{2}z}$
II. Modified Truncated Fourier Series $1, n = 0, 1, \cdots, N - 1;$ $\frac{1}{2}, n = N.$	$2\pi D_N^*(z) = \dfrac{\sin Nz}{\tan \frac{1}{2}z}$
III. Fejer weights $1 - \dfrac{\lvert n\rvert}{N + 1}, n = 0, 1, \cdots, N.$	$2\pi F_N(z) = \dfrac{1}{N + 1}\left\{\dfrac{\sin (N + 1)\frac{1}{2}z}{\sin \frac{1}{2}z}\right\}^2$
IV. Tukey–Hamming weights $\dfrac{1}{2}\left(1 + \cos \dfrac{\pi n}{N}\right), n = 0, 1, \cdots, N - 1.$	$2\pi T_N(z) = 2\pi\left\{\dfrac{1}{2} D_N^*(z) + \dfrac{1}{4} D_N^*\left(z + \dfrac{\pi}{N}\right) + \dfrac{1}{4} D_N^*\left(z - \dfrac{\pi}{N}\right)\right\}$ $= \sin Nz \left\{\dfrac{1}{2} \cot \dfrac{1}{2} z - \dfrac{1}{4} \cot \dfrac{1}{2}\left(z + \dfrac{\pi}{N}\right) - \dfrac{1}{4} \cot \dfrac{1}{2}\left(z - \dfrac{\pi}{N}\right)\right\}$
V. Lanczos weights $\dfrac{\sin\left(\dfrac{\pi n}{N}\right)}{\dfrac{\pi n}{N}}, n = 0, 1, \cdots, N.$	$2\pi L_N(z) = N \displaystyle\int_{-\pi/N}^{\pi/N} D_N(z + u)\, du$
VI. Parzen weights $1 - 6\left(\dfrac{n}{N}\right)^2 + 6\left(\dfrac{n}{N}\right)^3, n = 0, 1, \cdots, \dfrac{N}{2}.$ $2\left(1 - \dfrac{n}{N}\right)^3, \quad n = \dfrac{N}{2} + 1, \cdots, N.$	$2\pi P_N(z) = \dfrac{3}{4N^3}\left\{\dfrac{\sin \frac{1}{4} Nz}{\frac{1}{2} \sin \frac{z}{2}}\right\}^4 \left\{1 - \dfrac{2}{3}\left(\sin \dfrac{z}{2}\right)^2\right\}$ for N even

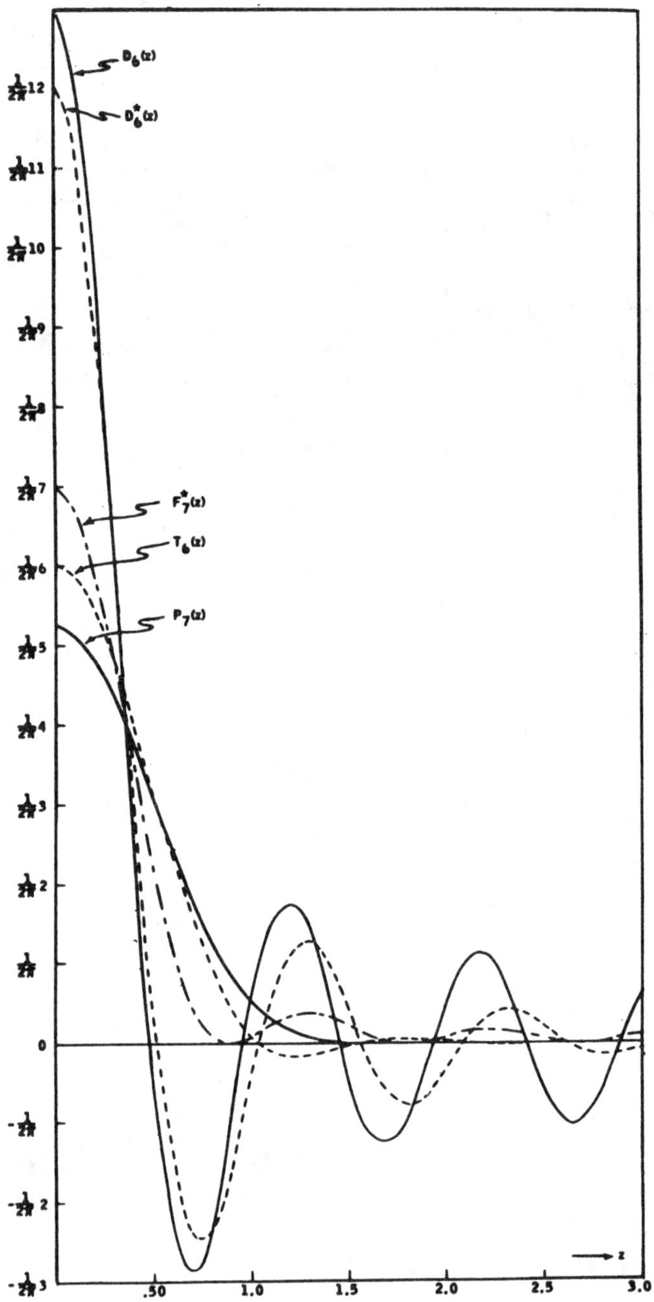

FIGURE 3A. Windows.

(ii) If the weights $k_T(v)$ in (3.24) are chosen so that for some $M < T$

$$k_T(v) = 0 \tag{3.25}$$

for $|v| > M$, then a computational saving is made in forming the estimate since the sample covariances $R_T(v)$ need not be computed for $|v| > M$. If (3.25) holds, we say that the estimate is truncated at M, and call M the *truncation point* of the estimate.

(iii) Weights are used to form estimates of the spectral density function (see §4).

A general way of generating weights which lead to truncated estimates is to consider a function $k(u)$ of a real variable with the following properties:

$$\begin{aligned} &\text{(i)} \quad k(0) = 1, \\ &\text{(ii)} \; k(-u) = k(u), \\ &\text{(iii)} \quad k(u) = 0 \qquad \text{for } |u| > 1. \end{aligned} \tag{3.26}$$

Next, choose a real number M (in theory, M will be a member of a sequence M_T of numbers which tend to ∞ as T tends to ∞). One may define weights by

$$k_T(v) = k\left(\frac{v}{M}\right). \tag{3.27}$$

Many of the weights listed in Table 3A may be generated in this way, as is shown in Table 3B. Therefore we call a function $k(u)$, satisfying (3.26), a *weight* or *lag window generator.*

Its Fourier transform

$$\begin{aligned} K(z) &= \frac{1}{2\pi}\int_{-\infty}^{\infty} e^{-iuz}\, k(u)\, du \\ &= \frac{1}{2\pi}\int_{-1}^{1} e^{-iuz} k(u)\, du, \qquad -\infty < z < \infty, \end{aligned} \tag{3.28}$$

is called a *window generator.* It may be shown that the Fourier transform

$$K_M(z) = \frac{1}{2\pi}\sum_{|v|<T} \cos vz\, k\left(\frac{v}{M}\right) \tag{3.29}$$

approximately satisfies, for large values of M,

$$K_M(z) = MK(Mz). \tag{3.30}$$

From (3.30) it follows that *the essential features of the graph of $K_M(z)$ can be obtained from a graph of $K(z)$.* An exposition of these facts is given in Parzen **[16]**.

TABLE 3B. WINDOW GENERATORS $K(z)$

Weight generating function $k(u)$	$k_T(v) = k\left(\frac{v}{M}\right)$	$K(z) = \frac{1}{2\pi}\int_{-\infty}^{\infty} e^{-iuz}k(u)\,du$	Approximation $2\pi K_M(z) = 2\pi MK(Mz)$
I. Dirichlet			
$k_D(u) = \begin{cases} 1, & \lvert u\rvert \leqq 1; \\ 0, & \lvert u\rvert > 0. \end{cases}$	$k_T(v) = \begin{cases} 1, & v = 0, \pm 1, \cdots, \pm M; \\ 0, & \text{otherwise.} \end{cases}$	$K_D(z) = \frac{1}{\pi}\frac{\sin z}{z}$	$\frac{\sin Mz}{\frac{z}{2}}$
II. Modified Dirichlet			
$k_D^*(u) = \begin{cases} 1, & \lvert u\rvert < 1; \\ \frac{1}{2}, & \lvert u\rvert = 1; \\ 0, & \lvert u\rvert > 1. \end{cases}$	$k_T(v) = \begin{cases} 1, & v = 0, \pm 1, \cdots, \pm(M-1); \\ \frac{1}{2}, & v = M; \\ 0, & \text{otherwise.} \end{cases}$	$\frac{1}{\pi}\frac{\sin z}{z}$	$\frac{\sin Mz}{\frac{z}{2}}$
III. Fejer			
$k_F(u) = \begin{cases} 1 - \lvert u\rvert, & \lvert u\rvert \leqq 1; \\ 0, & \lvert u\rvert > 0. \end{cases}$	$k_T(v) = \begin{cases} 1 - \frac{\lvert v\rvert}{M}, & n = 0, \cdots, \pm M; \\ 0, & \text{otherwise.} \end{cases}$	$K_F(z) = \frac{1}{2\pi}\left\{\frac{\sin \frac{z}{2}}{\frac{z}{2}}\right\}^2$	$\frac{1}{M}\left\{\frac{\sin \frac{Mz}{2}}{\frac{z}{2}}\right\}^2$
IV. Tukey–Hamming			
$k_T(u) = \frac{1}{2}(1 + \cos \pi u), \quad \lvert u\rvert \leqq 1$	$k_T(v) = \begin{cases} \frac{1}{2}\left(1 + \cos \frac{\pi v}{M}\right), & v = 0, \pm 1, \cdots, \pm M; \\ 0, & \text{otherwise.} \end{cases}$	$K_T(z) = \frac{1}{2\pi}\frac{\sin z}{z}\frac{\pi^2}{\pi^2 - z^2}$	$\frac{\sin Mz}{z}\frac{\pi^2}{\pi^2 - (Mz)^2}$

V. Lanczos $k_L(u) = \begin{cases} \dfrac{\sin \pi u}{\pi u}, & \lvert u \rvert \leq 1 \\ 0, & \lvert u \rvert > 1 \end{cases}$	$k_T(v) = \begin{cases} \dfrac{\sin \dfrac{\pi v}{M}}{\dfrac{\pi v}{M}}, & v = 0, \pm 1, \cdots, \pm M; \\ 0, & \text{otherwise.} \end{cases}$	$K_L(z) = \dfrac{1}{2\pi^2} \{Si(z + \pi) - Si(z - \pi)\},$ where $Si(x) = \begin{cases} \displaystyle\int_0^x \frac{\sin u}{u}\, du, \\ \displaystyle\int_0^1 \frac{\sin xu}{u}\, du, \end{cases}$ $Si(-x) = -Si(x),$	$\dfrac{M}{\pi} \{Si(Mz + \pi) - Si(Mz - \pi)\}$
VI. Parzen $k_P(u) = \begin{cases} 1 - 6u^2 + 6\lvert u \rvert^3, & \lvert u \rvert \leq \frac{1}{2} \\ 2(1 - \lvert u \rvert)^3, & \frac{1}{2} \leq \lvert u \rvert \leq 1 \end{cases}$	$k_T(v) = 1 - 6\left(\dfrac{v}{M}\right)^2 + 6\left\lvert\dfrac{v}{M}\right\rvert^3,$ $v = 0, \pm 1, \cdots, \pm \dfrac{M}{2};$ $= 2\left(1 - \left\lvert\dfrac{v}{M}\right\rvert\right)^3,$ $v = \pm\left(\dfrac{M}{2} + 1\right), \cdots, \pm M.$	$K_P(z) = \dfrac{3}{8\pi} \left\{\dfrac{\sin \dfrac{z}{4}}{\dfrac{z}{4}}\right\}^4$	$\dfrac{3}{4M^3} \left\{\dfrac{\sin \dfrac{Mz}{4}}{\dfrac{z}{4}}\right\}^4$

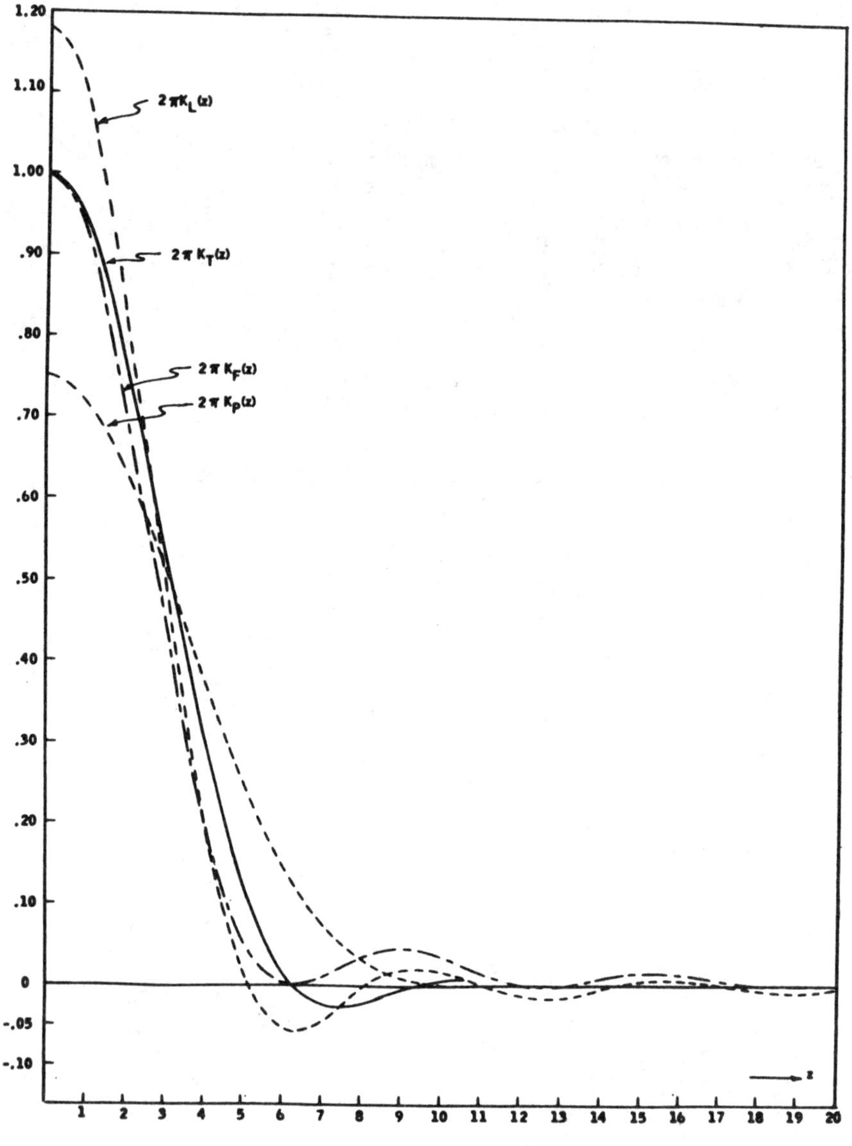

FIGURE 3B. Amplitude window generators.

4. Estimation of spectral density. The problem which up to recently seems to have received most attention in the literature of statistical spectral analysis has been that of estimating the spectral density function $f(\omega)$ under the assumption that the observed time series possesses a continuous spectrum. For sufficiently small values of Δ, the problem of estimating $f(\omega)$ at a frequency ω is equivalent to the problem of estimating the average spectral mass

$$\frac{1}{2\Delta}\int_{\omega-\Delta}^{\omega+\Delta} dF(\omega') = \frac{1}{2\Delta}\int_{\omega-\Delta}^{\omega+\Delta} f(\omega')\,d\omega' = f(\omega) \tag{4.1}$$

in a very narrow band of width 2Δ, centered at ω. To estimate the expression in (4.1), one might use the estimate (3.24) divided by 2Δ, and weights of the form of (3.27). We would then have the estimate

$$\frac{1}{2\pi}\sum_{|v|<T} \cos v\omega \,\frac{\sin \Delta v}{\Delta v}\, k\left(\frac{v}{M}\right) R_T(v). \tag{4.2}$$

Estimates of this form have not been explicitly examined in the statistical literature although I wonder if they are not worth consideration. What has been considered is the version of (4.2) obtained by setting $\Delta = 0$.

We are thus led to estimates of the spectral density which we denote

$$f_{T;k,M}(\omega) = \frac{1}{2\pi} R(0) + \frac{1}{\pi}\sum_{v=1}^{T-1} \cos v\omega R_T(v)\, k\left(\frac{v}{M}\right), \tag{4.3}$$

where M is a positive constant (called the truncation point of the estimate) and $k(u)$ is a kernel satisfying (3.26). If $k(u)$ is identically one, and $M = T$, then $f_{T;k,M}(\omega)$ is the sample spectral density function which is well known *not* to be a consistent estimate of $f(\omega)$.

The mean of the estimate (4.3), given by

$$\begin{aligned} E[f_{T;k,M}(\omega)] &= \int_{-\pi}^{\pi} d\lambda f(\lambda)\left\{\frac{1}{2\pi}\sum_{|v|<T} \cos v\omega\, k\left(\frac{v}{M}\right) \cos v\lambda\right\} \\ &= \int_{-\pi}^{\pi} d\lambda\, f(\lambda)\tfrac{1}{2}\{K_M(\omega-\lambda) + K_M(\omega+\lambda)\}, \end{aligned} \tag{4.4}$$

essentially views the true spectral density $f(\omega)$ through the window $K_M(\omega)$. In interpreting computed values of this estimate, this fact should always be borne in mind. Nevertheless for theoretical purposes it is useful to note the following asymptotic evaluation of the mean [which holds for a twice differentiable spectral density and twice differentiable kernel $k(u)$]:

$$\lim_{M\to\infty} M^2\{E[f_{T;k,M}(\omega)] - f(\omega)\} = -k_2 f''(\omega) \tag{4.5}$$

where

$$k_2 = -\frac{1}{2}k''(0) = \frac{1}{2}\int_{-\infty}^{\infty} \omega^2 K(\omega)\,d\omega, \tag{4.6}$$

assuming absolute convergence of the integral in (4.6).

The variance of $f_{T;k,M}(\omega)$, denoted by

$$D(\omega) = \text{Var}\,[f_{T;k,M}(\omega)]$$

and approximately given by

$$D(\omega) = \frac{\pi}{T}\int_{-\pi}^{\pi} d\lambda\, f^2(\lambda)\{K_M(\omega+\lambda) + K_M(\omega-\lambda)\}^2, \tag{4.7}$$

may be shown to satisfy, as $T \to \infty$,

$$D(\omega) \sim \begin{cases} \dfrac{M}{T}\displaystyle\int_{-\infty}^{\infty} k^2(u)\,du & \text{if } 0 < \omega < \pi, \\ 2\dfrac{M}{T}\displaystyle\int_{-\infty}^{\infty} k^2(u)\,du & \text{if } \omega = 0 \text{ or } \omega = \pi. \end{cases} \tag{4.8}$$

Heuristically (4.8) may be obtained as follows:

$$\begin{aligned} TD(\omega) &\doteq 2\pi f^2(\omega)\int_{-\pi}^{\pi} K_M^2(z)\,dz \\ &= 2\pi f^2(\omega)\int_{-\infty}^{\infty} M^2K^2(Mz)\,dz \\ &= Mf^2(\omega)2\pi\int_{-\infty}^{\infty} K^2(z)\,dz \\ &= Mf^2(\omega)\int_{-\infty}^{\infty} k^2(u)\,du. \end{aligned} \tag{4.9}$$

From the point of view of the practical use of the theory of spectral estimation the fundamental questions to be settled concerns the actual computing program to be used to estimate spectra. Briefly some of the issues involved in this question are as follows (see also Parzen [**12**, p. 233]):

(i) How to choose the number M of lag values at which the sample covariance function $R_T(v)$ is to be computed (one may want to use several values of M).

(ii) At what values of ω in the interval $0 \leqq \omega \leqq \pi$ should one compute the estimate $f_{T;k,M}(\omega)$? Even though the estimate is determined by its values at the frequencies $\omega_j = \pi j/M$, for $j = 0, 1, \cdots, M$, does it suffice to plot it only at these frequencies in order to fairly represent it as a function defined at all frequencies ω?

(iii) What lag window, or kernel, $k(u)$ should one employ? Two kernels frequently considered seem to be the kernel suggested by the author (and related to a kernel known in the theory of approximation as the Jackson-de la Vallée Poussin kernel; see Achieser [**1**, p. 119]):

$$k_P(u) = \begin{cases} 1 - 6u^2 + 6|u|^3, & |u| \leqq \frac{1}{2}, \\ 2(1-|u|)^3, & \frac{1}{2} \leqq |u| \leqq 1, \\ 0, & \text{otherwise}; \end{cases} \tag{4.10}$$

and a kernel suggested by Tukey, called "hanning" or (slightly modified) "Hamming" (see Blackman and Tukey [2, p. 14]):

$$k_H(u) = \begin{cases} \frac{1}{2}(1 + \cos \pi u), & |u| \leqq 1, \\ 0, & \text{otherwise.} \end{cases} \tag{4.11}$$

We refer to $k_P(u)$ as the Parzen kernel and to $k_H(u)$ as the Tukey–Hamming kernel.

One difference between these kernels is that, when used with a positive definite sample covariance function, (4.10) leads to non-negative estimates while (4.11) may yield negative estimates. Is this a consideration to be borne in mind when deciding between kernels?

Another criterion to be used in comparing two kernels is the variability of the estimates to which they lead. It may be shown that

$$\int_{-\infty}^{\infty} k_P^2(u)\, du = 0.54, \tag{4.12}$$

$$\int_{-\infty}^{\infty} k_H^2(u)\, du = 0.75. \tag{4.13}$$

Using (4.8) one sees that, if used with the same truncation point M the Parzen kernel leads to an estimate whose variance is $54/75 = 72\%$ that of the Hamming–Tukey kernel. However, this comparison is not quite fair, as one discovers by looking at the corresponding windows. At 0 the windows have values satisfying

$$2\pi M K_P(0) = \tfrac{3}{4}M, \tag{4.14}$$

$$2\pi M K_H(0) = M. \tag{4.15}$$

If one uses the Parzen kernel with a truncation point $\frac{4}{3}M$, where M is the truncation point used with the Hamming–Tukey kernel, it turns out (see Figure 4A) that the resulting estimates correspond to rather similar windows. The variance of the Parzen kernel with the increased truncation point is

$$\frac{4}{3}\frac{M}{T}(0.54) = \frac{M}{T}(0.72) \tag{4.16}$$

which is 96% of the variance $0.75\,(M/T)$ of the Hamming–Tukey kernel.

In general, to evaluate a kernel $k(u)$ from the point of view of its variability, one evaluates the ratio

$$\frac{\int_{-\infty}^{\infty} k^2(u)\, du}{2\pi K(0)} = \frac{\int_{-\infty}^{\infty} k^2(u)\, du}{\int_{-\infty}^{\infty} k(u)\, du} \tag{4.17}$$

which (when multiplied by M/T) is a measure of its asymptotic variance when the window $K(z)$ is normalized to have height $(2\pi)^{-1}$ at $z = 0$ (as well as unit area). The

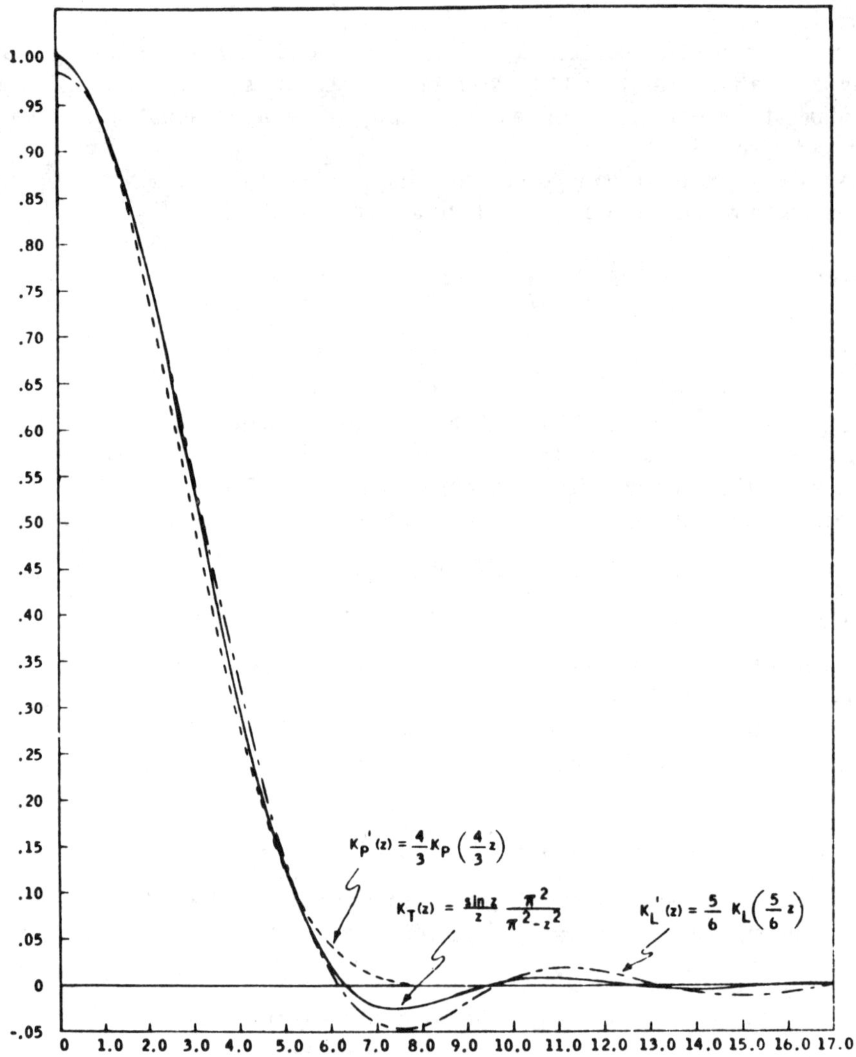

FIGURE 4A. Window generators normalized to have approximately same height at center frequency.

value of the ratio in (4.17) is 0.72 for the Parzen kernel, and 0.75 for the Tukey–Hamming kernel. For comparison we list its value for a kernel suggested by Lanczos (and known as Lanczos' σ factors; see Lanczos [**9**, p. 221] and Hamming [**6**, p. 298]):

$$k_L(u) = \begin{cases} \dfrac{\sin \pi u}{\pi u}, & |u| \leq 1, \\ 0, & \text{otherwise}; \end{cases} \tag{4.18}$$

$$\int_{-\infty}^{\infty} k_L^2(u)\, du = \frac{2}{\pi} \int_0^{2\pi} \frac{\sin x}{x}\, dx; \tag{4.19}$$

$$2\pi K_L(0) = \frac{2}{\pi} \int_0^{\pi} \frac{\sin x}{x}\, dx, \tag{4.20}$$

$$\frac{\displaystyle\int_{-\infty}^{\infty} k^2(u)\, du}{\displaystyle\int_{-\infty}^{\infty} k(u)\, du} = \frac{1.42}{1.85} = 0.77. \tag{4.21}$$

5. **Estimation of spectral jumps.** The problem of estimating the discrete as well as the continuous spectrum of a time series possessing a mixed spectrum has only recently been attacked. A basic contribution to the solution of this problem has recently been made by Priestley [**17–19**] who also summarizes previous results (Whittle [**24**], Hannan [**7**]). Further work on this problem is now being done at Stanford by my student George Hext who has advanced the idea of computing the weighted sample spectral density for several values (say, $M_1 < M_2 < M_3 < M_4$) of the truncation point M, and then to perform at each frequency ω a regression analysis on the resulting estimates $f_{T;k,M_i}(\omega)$ by means of which one can hope in theory to simultaneously obtain a measure of (i) the jump in the spectrum at ω, (ii) the spectral density at ω, and even (iii) the second derivative of the spectral density at ω. A brief outline of this approach is given below.

In any event, it is my belief that *computing programs for spectral analysis should at each frequency print out the estimate corresponding to several truncation points.* I hope that in the near future we will have quantitative, as well as qualitative, procedures for interpreting the spectra obtained from several choices of M. We may then finally have an answer to one of the central questions which arises in any computing program for statistical spectral analysis, namely how to choose the truncation point.

It appears to me that ultimately the answer will turn out to be of the following form. Given an observed time series $\{X(t), t = 1, 2, \cdots, T\}$ of length T, choose a maximum truncation point M which will be an integer between 20% and 40% of the sample size T. Then choose several (say four) integers

$$M_1 < M_2 < M_3 < M_4 = M$$

and write

$$\alpha_1 = \frac{M_1}{M}, \quad \alpha_2 = \frac{M_2}{M}, \quad \alpha_3 = \frac{M_3}{M}, \quad \alpha_4 = \frac{M_4}{M}. \tag{5.1}$$

Compute the weighted sample spectral density function $f_{T;k,M}(\omega)$ for the four values of M (that is, for $M = M_1, M_2, M_3, M_4$) and for $Q + 1$ equidistant values of ω;

$$\omega = 0, \frac{\pi}{Q}, 2\frac{\pi}{Q}, \cdots, \pi. \tag{5.2}$$

The integer Q can be chosen at will. Often one lets $Q = M$, the maximum truncation point. Usually, Q will be an integer, approximately equal to M, such that the frequencies which are multiples of π/Q are of physical interest. Thus if one had 240 observations of an economic time series observed at monthly intervals one might choose $M = 64$ and $Q = 60$. With $Q = 60$, and monthly data, the following multiples of π/Q correspond to interesting periods:

k	Period
10	one year
20	six months
30	four months
40	three months
60	two months.

For a given value of ω, we then have four estimates to be denoted $f_i(\omega)$:

$$f_i(\omega) = f_{T;k,M_i}(\omega), \qquad i = 1, 2, 3, 4. \tag{5.3}$$

We next obtain approximate expressions for the mean and variance of the weighted sample spectral density function for a time series satisfying the model (2.1) with normal stationary residuals.

For the mean we obtain, approximately for large values of T,

$$\begin{aligned} E[f_{T;k,M}(\omega)] &= \frac{1}{2\pi} \sum_{|v|<T} \cos v\omega \, k\left(\frac{v}{M}\right) E[R_T(\omega)] \\ &= \sum_{j=1}^{k} J(\omega_j) \left\{\frac{1}{2\pi} \sum_{|v|<T} \cos v\omega \, k\left(\frac{v}{M}\right) \cos v\omega_j\right\} \\ &\quad + \int_{-\pi}^{\pi} d\lambda f(\lambda) \left\{\frac{1}{2\pi} \sum_{|v|<T} \cos v\omega \, k\left(\frac{v}{M}\right) \cos v\lambda\right\} \\ &= \sum_{j=1}^{k} J(\omega_j) \tfrac{1}{2}\{K_M(\omega + \omega_j) + K_M(\omega - \omega_j)\} \\ &\quad + \int_{-\pi}^{\pi} d\lambda f(\lambda) \tfrac{1}{2}\{K_M(\omega + \lambda) + K_M(\omega - \lambda)\}. \end{aligned} \tag{5.4}$$

In words, the mean of the weighted sample spectral density functions is composed of two parts, one an average over the continuous spectrum, represented by the spectral density function $f(\omega)$, and the other an average over the discrete spectrum, represented by the spectral jump function $J(\omega)$. The terms on the right-hand side of (5.4) will be called respectively the *discrete mean* E_d and the *continuous mean* E_c.

George Hext has obtained an approximate expression for variance:

$$\begin{aligned} T \operatorname{Var}[f_{T;k,M}(\omega)] &= 4\pi \int_{-\pi}^{\pi} d\lambda f^2(\lambda) \tfrac{1}{4}\{K_M(\omega+\lambda) + K_M(\omega-\lambda)\}^2 \\ &\quad + 8\pi \sum_{j=1}^{q} J(\omega_j) f(\omega_j) \tfrac{1}{4}\{K_M(\omega+\lambda) + K_M(\omega-\lambda)\}^2. \end{aligned} \tag{5.5}$$

The continuous mean and the variance are evaluated as in §4. Assuming that for all truncation points M_i under consideration, the spectral density function $f(\omega)$ is varying slowly compared to the window $K_{M_i}(\omega)$, it follows that to a first approximation the continuous mean is independent of M,

$$E_c[f_{T;k,M}(\omega)] = f(\omega), \tag{5.6}$$

while to a second approximation

$$E_c[f_{T;k,M}(\omega)] = f(\omega) + \frac{1}{M^2} k_2 f''(\omega). \tag{5.7}$$

We next obtain an approximate expression for the discrete mean, and find that *its contribution, if nonzero, increases linearly with truncation point* M:

$$\begin{aligned} E_d(\omega) &= \sum_{j=1}^{k} J(\omega_j) \tfrac{1}{2}\{K_M(\omega+\omega_j) + K_M(\omega-\omega_j)\} \\ &= \begin{cases} M\tfrac{1}{2}J(\omega_j)K(0) & \text{if } \omega = \omega_j \text{ and } \omega_j \neq 0; \\ MJ(\omega_j)K(0) & \text{if } \omega = \omega_j \text{ and } \omega_j = 0 \text{ or } \pi; \\ 0 & \text{otherwise.} \end{cases} \end{aligned} \tag{5.8}$$

We now sum up these results from the point of view of their dependence on M for fixed T, ω, and $k(\cdot)$:

$$f_{T;k,M}(\omega) = E_M(\omega) + \varepsilon_M(\omega), \tag{5.9}$$

where $E_M(\omega)$ and $\varepsilon_M(\omega)$ are to begin with defined by

$$E_M(\omega) = E[f_{T;k,M}(\omega)], \qquad \varepsilon_M(\omega) = f_{T;k,M}(\omega) - E_M(\omega).$$

For each M, $\varepsilon_M(\omega)$ is a random variable. The covariance between $\varepsilon_{M_1}(\omega)$ and $\varepsilon_{M_2}(\omega)$ for two distinct values M_1 and M_2 of M is approximately given by (for $0 < \omega < \pi$)

$$
\begin{aligned}
T \operatorname{Cov}[\varepsilon_{M_1}(\omega), \varepsilon_{M_2}(\omega)] &= 2\pi f^2(\omega) M_1 M_2 \int_{-\infty}^{\infty} K(M_1 z) K(M_2 z)\, dz \\
&\quad + 2\pi J(\omega) f(\omega) M_1 M_2 K^2(0).
\end{aligned} \tag{5.10}
$$

The tools are now at hand to examine the properties of the estimates $f_i(\omega)$ defined by (5.3).

If we adopt the first approximation (5.6) to the mean of the weighted sample spectral density function, we may write

$$
\begin{aligned}
f_i(\omega) &= f(\omega) + M_i^{\frac{1}{2}} J(\omega) K(0) + \varepsilon_i(\omega) \\
&= f(\omega) + \alpha_i M^{\frac{1}{2}} J(\omega) K(0) + \varepsilon_i(\omega),
\end{aligned} \tag{5.11}
$$

where $\varepsilon_i(\omega)$ denotes the fluctuation of $f_i(\omega)$ about its mean. The random variables $\varepsilon_1(\omega), \cdots, \varepsilon_4(\omega)$ may be assumed to be jointly normal with covariance matrix (for $0 < \omega < \pi$)

$$
\{\operatorname{Cov}[\varepsilon_i(\omega), \varepsilon_j(\omega)]\} = \frac{1}{T} C, \tag{5.12}
$$

where the entries of the matrix

$$
C = \begin{bmatrix} C_{11} & C_{12} & C_{13} & C_{14} \\ C_{21} & C_{22} & C_{23} & C_{24} \\ C_{31} & C_{32} & C_{33} & C_{34} \\ C_{41} & C_{42} & C_{43} & C_{44} \end{bmatrix} \tag{5.13}
$$

may be shown to be approximately given by

$$
\begin{aligned}
C_{ij} &= M_i M_j \{J(\omega) K^2(0)\}\{2\pi f(\omega)\} \\
&\quad + \sqrt{(M_i M_j)}\, f^2(\omega) \int_{-\infty}^{\infty} k\left(u \sqrt{\frac{M_i}{M_j}}\right) k\left(u \sqrt{\frac{M_j}{M_i}}\right) du.
\end{aligned} \tag{5.14}
$$

To facilitate computation of the C matrix, define

$$
I(\beta) = \int_{-\infty}^{\infty} k(v) k(\beta v)\, dv. \tag{5.15}
$$

This function is given in Table 5A for the Parzen and Tukey kernels.

Estimates of $f(\omega)$ and $J(\omega)$ can now be obtained by regression analysis, using the theory of minimum variance unbiased estimation of a straight line in the presence of correlated residuals. Because the parameters being estimated also appear in the covariance matrix C, an iterative procedure is necessary.

The foregoing remarks are necessarily incomplete. Further work along these lines will be reported by George Hext in his thesis.

TABLE 5A. FUNCTION $I(\beta) = \int_{-\infty}^{\infty} k(u)k(\beta u)\,du$

Parzen's kernel

$$I_P(\beta) = \begin{cases} \frac{3}{4} - \beta^2\frac{3}{8} + \beta^3\frac{93}{560} & \text{for } 0 \leqq \beta \leqq \frac{1}{2}; \\ \frac{3}{4} - \beta^2\frac{3}{8} + \beta^3\frac{93}{560} - \frac{(2\beta-1)^7}{\beta^4}\frac{1}{560} & \text{for } \frac{1}{2} \leqq \beta \leqq 1. \end{cases}$$

Tukey's kernel

$$I_T(\beta) = \begin{cases} \frac{1}{2}\left\{1 + \frac{\sin \pi\beta}{\pi\beta}\frac{1}{1-\beta^2}\right\} & \text{for } 0 \leqq \beta < 1; \\ \frac{3}{4} & \text{for } \beta = 1. \end{cases}$$

REFERENCES

1. N. I. Achieser, *Theory of approximation*, Ungar, New York, 1956.

2. R. B. Blackman and J. W. Tukey, *The measurement of power spectra from the point of view of communications engineering*, Dover, New York, 1950.

3. C. E. P. Brooks and N. Carruthers, *Handbook of statistical methods in meteorology*, H.M. Stationery Office, London, 1953.

4. R. A. Fisher, *Tests of significance in harmonic analysis*, Proc. Roy. Soc. Ser. A **125** (1929), 54.

5. N. R. Goodman, *Spectral analysis of multiple stationary time series*, Proc. Sympos. Time Series Analysis (Brown Univ., 1962), pp. 260–266, Wiley, New York, 1963.

6. R. W. Hamming, *Numerical methods for scientists and engineers*, McGraw-Hill, New York, 1962.

7. E. J. Hannan, *Testing for a jump in the spectral function*, J. Roy. Statist. Soc. Ser. B **23** (1961), 394–404.

8. G. M. Jenkins, *General considerations in estimation of spectra*, Technometrics **3** (1961), 133–166.

9. C. Lanczos, *Applied analysis*, Prentice Hall, Englewood Cliffs, N. J., 1956.

10. E. Parzen, *On asymptotically efficient consistent estimates of the spectral density function of a stationary time series*, J. Roy. Statist. Soc. Ser. B **20** (1958), 303–322.

11. ———, *Mathematical considerations in estimation of spectra*, Technometrics **3** (1961), 167–190.

12. ———, *Spectral analysis of asymptotically stationary time series*, Bull. Inst. Internat. Statist. 33rd Session, Vol. 39, Book 2, pp. 87–104, Paris, 1962.

13. ———, *An approach to time series analysis*, Ann. Math. Statist. **32** (1961), 951–989.

14. ———, *Stochastic processes*, Holden Day, San Francisco, 1962.

15. ———, *On spectral analysis with missing observations and amplitude modulation*, Sankhyā, December 1963.

16. ———, *Notes on Fourier analysis and spectral windows*, Tech. Rep. No. 48, Statistics Dept., Stanford Univ., Stanford, Calif., 1963.

17. M. B. Priestley, *The analysis of stationary processes with mixed spectra.* I, J. Roy. Statist. Soc. Ser. B **24** (1962), 215–233.

18. ———, *The analysis of stationary processes with mixed spectra.* II, J. Roy. Statist. Soc. Ser. B **24** (1962), 511–529.

19. ———, *Basic considerations in the estimation of spectra*, Technometrics **4** (1962), 551–564.

20. M. Rosenblatt, *Statistical analysis of stochastic processes with stationary residuals*, Probability and statistics: The Harold Cramér volume, pp. 246–273, Wiley, New York, 1959.

21. A. Schuster, *On the investigation of hidden periodicities with application to a supposed 26-day period of meteorological phenomena*, Terrestrial Magnetism **3** (1897), 13.

22. J. W. Tukey, *An introduction to the measurement of spectra*, Probability and statistics: The Harold Cramér volume, pp. 300–330, Wiley, New York, 1959.

23. G. Walker, *On the criteria for the reality of relationships or periodicities*, Calcutta Ind. Met. Memo. **21** (1914), Part 9.

24. P. Whittle, *The simultaneous estimation of a time-series harmonic components and covariance structure*, Trabajos Estadíst. **3** (1952), 43–57.

25. N. Wiener, *Generalized harmonic analysis*, Acta Math. **55** (1930), 117–258.

STANFORD UNIVERSITY,
STANFORD, CALIFORNIA

STABILITY IN SIGNAL DETECTION PROBLEMS[1]

BY

WILLIAM L. ROOT

A device for detecting the presence of a signal in noise may be represented, obviously, as a two-valued function on an appropriate function space. Usually the operation performed by such a detector amounts first to calculating the value of a real-valued functional, which we shall call the test functional, and then comparing this value with a prescribed threshold. It is now common engineering practice to determine as well as possible the statistics of the noise, and from this knowledge and that of the structure of the signal to try to design the detector to be optimum or nearly optimum according to some statistical criterion. How well the detector operates (how accurate the decisions are) depends on the distributions of the test functional and on the threshold. In the special case that the signal is a known function and the probability measure of the noise process is known, the induced distributions of the test functional as determined on either hypothesis, signal present or signal absent, are known, at least in principle. The probabilities of the two kinds of decision error can be calculated. If, however, the actual probability measure of the noise process is different from what was assumed, so are the induced distributions of the test functional and so are the error probabilities. There is the possibility that a slight difference in the actual statistics of the noise from the assumed will result in a large change in the error probabilities. Such a situation will be termed *unstable*. Instability of this kind is obviously undesirable to the system designer, who almost certainly cannot know the noise statistics precisely, but who would like some guarantee of the performance of his system.

This concept of instability, or more precisely a notion of degree of stability, is appropriate in a more general class of problems than the one indicated. It applies to a study of any system in which a statistical inference is made from the value or values of a functional on a sample space, where the functional is carefully chosen from an assumed a priori knowledge of the statistics and is critically dependent on them.

In this paper are developed some preliminary results concerning stability in signal detection and signal measurement. The discussion is limited to hypotheses-test and parameter-estimation problems;[2] filtering and prediction are more or less

[1] This work was supported by the National Aeronautics and Space Administration under research grant NsG–2–59.

[2] Comprehensive treatments of the kinds of problems referred to are given, e.g., by Parzen [10], Hajek [6], Balakrishnan [2].

arbitrarily excluded. Both sure-signal-in-noise and noise-in-noise cases are discussed. All of what might be considered general theory is a direct application of a few well-known results about the weak convergence of probability measures on function spaces.

Models for detection problems. We shall say a signal-detection or signal-measurement problem is of the *sure-signal-in-noise* type if the observed waveform can be represented as the sum of a function known to within finitely many parameters and a sample function from a stochastic process. We write

$$y(t) = s(t; \alpha) + x(t) \tag{1}$$

where $y(t)$ is a real- (or complex-, or vector-) valued function of a real variable t, representing the observed waveform; where for each α $s(t; \alpha)$ is a function of t representing a possible signal, and where $x(t)$ is a sample function from a stochastic process representing the noise. The time variable t belongs to a linear set, which here will always be taken to be a finite interval, the observation interval. The parameter α may have finitely many components, $\alpha = (\alpha_1, \cdots, \alpha_k)$, each of which is either an index taking on finitely or countably many values, or is real- or complex-valued. For example, in the simplest detection problem, $\alpha = 0$ or 1, and $s(t; 0) \equiv 0$, $s(t; 1) = f(t)$, a known function. More generally, in radar and radio communication problems the α_i may determine amplitude, phase, etc., of different portions of the received signal. The process of signal detection or signal measurement consists of making required inferences about the α_i, whose true values are unknown.

Usually the noise process $x(t)$ is assumed to be Gaussian with mean zero and continuous autocorrelation, often stationary. This seems justifiable in many instances because of the nature of the physical origin of the noise.

We shall say a signal-detection or signal-measurement problem is of the *noise-in-noise* type if the observed waveform can be represented as the sum of sample functions from two stochastic processes, one carrying the desired information, the other being unwanted noise. Again we write

$$y(t) = s(t; \alpha) + x(t)$$

but $s(t; \alpha)$ is now a sample function from a process indexed by α. For example, in radiometry, α might be the power level of noise-like radiation from a certain solid-angle of the sky in a certain frequency band.

Again, usually the noise process $x(t)$ is assumed to be Gaussian, and often the signal process $s(t; \alpha)$ is also. However, $s(t; \alpha)$ may have a much simpler structure; it may be a family of functions of t parameterized in some simple way by α, but with α taken to be a random variable. Sometimes in such a case it is more natural to treat the problem as a modified sure-signal-in-noise problem than as noise-in-noise.

In part of what follows we shall impose intuitively reasonable restrictions on the class of processes which are admissible to characterize noise and signals. The most important of these is the requirement that the sample functions be continuous. If a process is stationary and Gaussian (and separable, with continuous autocorrelation function) then it is known from results of Dobrushin **[4]** and Belyaev **[3]**, respectively, that the alternative to the sample functions being continuous with probability one is that with probability one they have discontinuities of the second kind everywhere, and in fact are unbounded in every finite interval. As long, then, as one is discussing stationary Gaussian noise processes the assumption of continuity of sample functions is strongly justified; if a noise process is only nearly stationary and Gaussian it is perhaps still plausible. Kolmogorov's well-known sufficient condition for continuous sample functions is available as a verifiable hypothesis.

A second condition sometimes required concerning the signal and noise taken together is that any two measures on the sample space corresponding to different statistical hypotheses about the signal be equivalent, that is, the statistical inference problems be nonsingular. Arguments have been given that for a Gaussian-noise model of a realizable communication system this must be true.[3]

For stationary second-order processes continuity of sample functions requires of course, that, roughly speaking, the distribution of power at high frequencies falls off sufficiently fast. Non-singularity of the measures implies, roughly, that the distribution of noise power falls off more slowly at high frequencies than that of the signal. For example, Parzen's sufficient condition[4] that noise and noise-plus-signal processes be relatively nonsingular on any finite interval when noise and signal processes are real, stationary, independent, Gaussian, and with spectral densities $S_n(f)$ and $S_s(f)$, respectively, is that

$$\int_{-\infty}^{\infty} \frac{S_s(f)}{S_n(f)}\, df < \infty.$$

An analogous sufficient condition[5] for sure-signal in noise replaces $S_s(f)$ with the square of the modulus of the Fourier transform of the signal, which is supposed to be of integrable square and vanish outside a finite interval.

Most solutions to problems involving signals in Gaussian noise are based on a knowledge of the autocorrelation function or spectrum (if the noise is stationary). In fact, of course, as mentioned before, they are rarely if ever known precisely. The noise spectrum may be determined from a knowledge of the structure of the generating process, as, for example, when thermal noise is passed through a filter of known characteristics, or it may be estimated from measurements. In the first instance errors result because of imprecise characterization of the generating process, in the second because of errors inherent in statistical measurements.

Let us digress slightly to consider one point that applies particularly to the

---

[3] See Parzen **[9]** and Root **[13]**. For a general survey of the theory of equivalence of Gaussian measures, see also Yaglom **[15]**.

[4] See Parzen **[9, Example 4]**.

[5] See Kelly, Reed and Root **[8]**.

situation where the spectrum or autocorrelation is determined empirically. At best the autocorrelation can be determined only for a finite interval. Even if it be known exactly for this interval, the spectrum cannot be known in principle unless the autocorrelation has a unique extrapolation as a positive definite function. There are necessary and sufficient conditions that a finite section of a positive definite function have a unique positive definite extrapolation,[6] but none that I know are in an easily applicable form. However, one sufficient condition for a nonunique extrapolation is a condition which one might normally expect to be satisfied by electrical noise. *If a second-order stationary process with absolutely continuous spectrum is nondeterministic, then a restriction of its autocorrelation function to any finite interval has a nonunique positive definite extrapolation.*[7]

In fact, let $S(f)$ be the spectral density of such a process and $R(t)$ its autocorrelation function. Then $S(f)$ is absolutely integrable and satisfies the condition

$$\int_{-\infty}^{\infty} \frac{|\log S(f)|}{1+f^2}\, df < \infty. \tag{2}$$

Let

$$S_1(f) = \begin{cases} S(f), & S(f) \leqq 1 \\ S^{1/2}(f), & S(f) > 1. \end{cases}$$

Then $S_1(f)$ is of integrable square on the interval $(-\infty, \infty)$ and also satisfies the condition (2). By a well-known theorem of Paley and Wiener $(1/3)S_1(f)$ is the modulus of a function $G(f)$ belonging to L_2 (and here, also to L_1) with a Fourier transform $g(t)$ which vanishes for $t < -a$, a arbitrarily chosen greater than zero. Let

$$h(t) = g(t) + \overline{g(-t)}.$$

Then $h(t)$ vanishes for $-a < t < a$. Let $H(f)$ be the Fourier transform of $h(t)$. We have that $H(f)$ is real and

$$|H(f)| \leqq 2|G(f)| \leqq \tfrac{2}{3}S_1(f) < S(f).$$

Hence $S(f) + H(f)$ is non-negative and integrable and is itself the spectral density of a process with autocorrelation function $R(t) + h(t)$, which agrees with $R(t)$ on $(-a, a)$.

If $R(t)$ is real, the construction described will yield real nonunique extensions, since $g(t)$ may be chosen to be real.

To a certain extent the fact just noted is irrelevant to the consideration of a signal-in-Gaussian-noise problem, because knowledge of the autocorrelation function over just the observation interval completely determines the statistics of the problem; that is, there is in principle no reason to introduce the spectrum of the noise at all. However, since such a great amount of the technique of design in engineering is

[6] See Akutowicz [**1**].

[7] J. Chover has an independent proof of this result. He has pointed out to me further that it is in Krein's fundamental work.

based on the idea of frequency spectra, it is practically important whenever possible to state conditions in terms of frequency spectra.

Detection stability for sure signals in Gaussian noise. In this section we discuss two examples, the simple detection of a known signal in Gaussian noise, and the estimation of linear regression coefficients of a signal in Gaussian noise. In each example (and obviously in other similar ones) there is a natural notion of stability that can be treated quite simply and directly. We start without a formal definition of stability and introduce a definition in context. Both examples are phrased in terms of the Karhunen–Loève expansion as a matter of personal choice and convenience.

Let G be a set of real-valued Gaussian random processes, each with parameter set the interval $[a, b]$, mean-value zero, and continuous autocorrelation function. That is, G is a set of measure spaces $(X, \mathscr{B}, \mu_g)$, where X is a fixed set of real-valued functions $x(\cdot)$ on $[a, b]$, $\mathscr{B}$ is a Borel field of subsets of X and $\{\mu_g\}$ is a family of Gaussian probability measures on $\mathscr{B}$ such that the mean-value and autocorrelation functions are as prescribed. One of the measures, μ_0, is called the *nominal* one; expectations with respect to this measure are denoted by E_0, and the autocorrelation function of the corresponding process is

$$R_0(t, s) = E_0 x(t)x(s).$$

G is to be thought of as a set of possible noise processes for the signal detection problem. The actual noise process is unknown, but the one which has been assumed in the design of the detector is the nominal process; or in other words, the autocorrelation function of the noise has been assumed to be $R_0(t, s)$.

We take $\{\lambda_n, \phi_n\}$ to be the eigenvalues and corresponding eigenfunctions (real and orthonormal) of the integral operator R_0 given by

$$R_0 f(t) = \int_a^b R_0(t, s) f(s)\, ds, \qquad a \leqq t \leqq b.$$

Then for the nominal process,

$$x(t) = \sum_n x_n \phi_n(t),$$

where

$$x_n = \int_a^b x(t)\phi_n(t)\, dt, \qquad E_0 x_n \bar{x}_k = \lambda_n \delta_{nk},$$

and the convergence is in the sense of mean-square with respect to μ_0 for each t.

Consider now the first example. The received waveform is

$$y(t) = s_i(t) + x(t), \qquad a \leqq t \leqq b,\ i = 0, 1,$$

where $s_0(t) \equiv 0$ and $s_1(t) = a(t)$ is a known real-valued function of integrable square on $[a, b]$. The detector is to indicate signal present when $i = 1$, no signal when $i = 0$. The maximum-likelihood test[8] for this decision problem on the

[8] See Grenander [5].

assumption $x(t)$ is a sample from the nominal noise process can be written in terms of the test functional

(3)
$$f(y) = \sum_k \frac{y_k a_k}{\lambda_k},$$

where

$$a_k = \int_a^b s_1(t)\phi_k(t)\,dt,$$

$$y_k = \int_a^b y(t)\phi_k(t)\,dt.$$

The test is:

accept hypothesis "signal present" if $f(y) > \eta$

reject hypothesis "signal present" if $f(y) \leqq \eta$,

where η is a prescribed constant. We shall assume the hypothesis-test problem with the nominal noise sample is nonsingular, which is equivalent to the condition[8]

(4)
$$\sum \frac{a_k^2}{\lambda_k} < \infty.$$

This condition guarantees that the series for $f(y)$ converges with probability one with respect to μ_0.

Suppose now that the actual signal $s_1(t)$ is exactly as specified, but the noise, although Gaussian with mean zero, has autocorrelation

$$R_1(t, s) = R_0(t, s) + A(t, s),$$

so that $A(t, s)$ represents the difference between the true autocorrelation function and the nominal one. Because of the boundedness of R_0 and R_1 on $[a, b] \times [a, b]$, the double norm of A, denoted by $|A|$, and defined by

$$|A|^2 = \int_0^1 \int_0^1 |A(t, u)|^2\,dt\,du,$$

is necessarily finite. We denote the measure on $\mathscr{B}$ corresponding to R_1 by μ_1, and expectations with respect to μ_1 by $E_1(\cdot)$. We shall determine the effect on the error probabilities of the change from μ_0 to μ_1. The test functional, $f(y)$ is to remain unchanged, of course, since the object here is to find the effect on detection of the use of incorrect noise statistics in the design of the detector.

Then,

$$E_1 f(y) = E_0 f(y) = \sum_k \frac{a_k^2}{\lambda_k} = \beta,$$

where β is defined by this equation. Also,

$$E_1 x_k x_j = \lambda_k \delta_{kj} + (A\phi_k, \phi_j),$$

where A is the integral operator on $L_2[a, b]$ with kernel $A(t, u)$, and the inner product is that of $L_2[a, b]$. Consider a truncated test functional,

$$f_N(y) = \sum_{k=1}^{N} \frac{y_k a_k}{\lambda_k}.$$

The variance of f_N is

$$\sigma^2(f_N) = E_1 \sum_{k=1}^{N} \frac{y_k a_k}{\lambda_k} \sum_{j=1}^{N} \frac{y_j a_j}{\lambda_j} - \left(\sum_{k=1}^{N} \frac{a_k^2}{\lambda_k}\right)^2$$

$$= \sum_{k=1}^{N} \frac{a_k^2}{\lambda_k} + \sum_{\substack{k=1\\ j=1}}^{N} \frac{a_k a_j}{\lambda_k \lambda_j} (A\phi_k, \phi_j).$$

If now it is required that

(5) $$\sum_k \frac{a_k^2}{\lambda_k^2} < \infty$$

then

$$\Delta = \sum_{\substack{k=1\\ j=1}}^{\infty} \frac{a_k a_j}{\lambda_k \lambda_j} (A\phi_k, \phi_j)$$

exists, and it is easily justified that $\sigma^2(f) = \beta + \Delta$. Under this assumption the error probabilities are:

$$P\{\text{signal detected} | \text{no signal sent}\} = \frac{1}{\sqrt{(2\pi)}} \int_{\eta/\sqrt{(\beta+\Delta)}}^{\infty} e^{-u^2/2}\, du,$$

$$P\{\text{no signal detected} | \text{signal sent}\} = \frac{1}{\sqrt{(2\pi)}} \int_{-\infty}^{(\eta-\beta)/\sqrt{(\beta+\Delta)}} e^{-u^2/2}\, du.$$

The change in noise statistics influences the error probabilities only through the number Δ. As $\Delta \to 0$, the changes in error probabilities approach zero. One has,

$$|\Delta| \leq \sum^{\infty} \frac{a_k^2}{\lambda_k^2} \|A\| \leq \sum^{\infty} \frac{a_k^2}{\lambda_k^2} |A|$$

so that in the presence of the condition (5) the "stability parameter," Δ, is dominated by a constant times the root-mean-square error in the determination of the autocorrelation function.

If, on the other hand,

$$\sum_k \frac{a_k^2}{\lambda_k^2} = +\infty$$

then, given an arbitrary $\varepsilon > 0$, one can choose, by the Banach–Steinhaus theorem, a sequence of real numbers b_k such that $\sum b_k^2 \leq \varepsilon$ and

$$\sum_k \frac{a_k b_k}{\lambda_k} = +\infty.$$

Let $A(t, u) = \sum_{k,j} b_k b_j \phi_k(t)\phi_j(u)$. Then $\|A\| \leqq \varepsilon$, and

$$\Delta = \sum_{k,j} \frac{a_k}{\lambda_k}\frac{a_j}{\lambda_j}(A\phi_k, \phi_j) = +\infty.$$

Furthermore A, and hence R_1, are positive definite.

Thus, if the condition (5) is satisfied, the error probabilities approach their nominal values as $\|A\|$ approaches zero; if (5) is not satisfied, they do not necessarily. We shall say that in the first case the detector is stable, that in the second case it is not. Obviously the smaller the value of the sum $\sum(a_k^2/\lambda_k^2)$, the more stable the detector.

There is a trivial generalization of the above to allow for imperfect knowledge of the signal. If the true signal function $a(t)$ differs from the nominal one by $\delta(t)$, where

$$\|\delta\|^2 = \int_0^1 |\delta(t)|^2\, dt < \infty$$

then the mean of the test functional is changed by an amount less than or equal to $\|\delta\|(\sum(a_k^2/\lambda_k^2))^{1/2}$, where the a_k are the Fourier coefficients of the nominal signal. The variance of f is as above. Again condition (5) is necessary and sufficient for stability.

It is of interest to re-formulate the stability condition for this simple but basic example. First, of course, the condition

$$\sum \frac{a_k^2}{\lambda_k^2} < \infty$$

is equivalent to the condition that the function $s_1(t)$ belong to the domain of the (unbounded) inverse of R_0, $s_1 \in \mathscr{D}(R_0^{-1})$, or in other words that

$$\int_a^b R_0(t, u)g(u)\, du = s_1(t), \qquad a \leqq t \leqq b$$

have a solution $g \in L_2[a, b]$. If it does then the test functional can be written

$$f(y) = \int_a^b y(t)g(t)\, dt,$$

and the quantity Δ which determines the degree of stability is given by

$$\Delta = (Ag, g).$$

If $R_0(t, u) = \rho(t - u)$ is a stationary autocorrelation function which is absolutely integrable on E^1, then it necessarily has a bounded spectral density $S(f)$, and it also belongs to $L_2(-\infty, \infty)$ (henceforth written L_2). In this case the stability condition may be written as follows (for convenience $[a, b]$ is taken to be $[-T, T]$):

Let $a(t)$, $t \in [-T, T]$, belong to $L_2[-T, T]$. Then $a \in \mathscr{D}(R_0^{-1})$ if and only if $a(t)$ has an extension $\tilde{a}(t)$ to the real number line such that

(1) $\tilde{a}(t) \in L_2$;

(2) *the Fourier transform $\tilde{A}(f)$ of $\tilde{a}(t)$ satisfies the conditions that $\tilde{A}(f)/S(f) \in L_2$ and is an integral function of exponential type with index $\leqq 2\pi T$.*

In fact, suppose there is a function $\phi(t) \in L_2[-T, T]$ which satisfies

$$\int_{-T}^{T} \rho(t - s)\phi(s)\, ds = a(t), \qquad -T \leqq t \leqq T. \tag{6}$$

Consider $a(t)$ and $\phi(t)$ to be defined for all real numbers t, by putting them equal to zero for $|t| > T$. Then (6) can be rewritten,

$$c_T(t)\int_{-\infty}^{\infty} \rho(t - s)\phi(s)\, ds = a(t), \tag{7}$$

where $c_T(t)$ is the indicator function of the interval $[-T, T]$. We denote the Fourier transform of c_T by $D_T(f)$, and the Fourier transforms of $\phi(t)$, $a(t)$ by $\Phi(f)$, $A(f)$ respectively. Then

$$\int_{-\infty}^{\infty} D_T(f - g)S(g)\Phi(g)\, dg = A(f), \tag{8}$$

where $\Phi(g)$ is necessarily an element of L_2 and is of exponential type with index $\leqq 2\pi T$. Let $\Psi(g) = S(g)\Phi(g)$. Then $\Psi(g) \in L_1 \cap L_2$, and the fact that it satisfies the equation

$$\int_{-\infty}^{\infty} D_T(f - g)\Psi(g)\, dg = A(f) \tag{9}$$

implies that its inverse Fourier transform $\psi(t)$ satisfies $c_T(t)\psi(t) = a(t)$. Thus $\psi(t)$ is an extension of $a(t)$ which meets the conditions for $\tilde{a}(t)$.

Conversely, suppose $\psi(t)$ is an extension of $a(t)$ with the required properties. Then (9) is satisfied by its transform $\Psi(f)$. Define $\Phi(g) = \Psi(g)/S(g)$. Then (8) is satisfied, and since $A(f) \in L_2$, (7) is also satisfied. Since $\Phi(g)$ is of exponential type with index $\leqq T$, (7) is equivalent to (6).

In particular it is sufficient that $a(t)$ satisfy the conditions on $\tilde{a}(t)$; this happens, for example, if $S(f)$ is the reciprocal of a polynomial and $A(f)/S(f) \in L_2$. The original condition applies, of course, to the $S(f)$ determined by any integrable positive definite extension of $\rho(t)$, $-T \leqq t \leqq T$.

For a second example we consider the estimation of the unknown real parameters β_k which appear in

$$y(t) = \sum^{K} \beta_k s_k(t) + x(t), \qquad a \leqq t \leqq b. \tag{10}$$

Again $y(t)$ represents the observed waveform; the $s_k(t)$, $k = 1, \cdots, K$ represent known signals, and $x(t)$ is Gaussian noise. Let $a_{ki} = (s_k, \phi_i)$, assume $\sum_i (a_{ki}^2/\lambda_i) < \infty$ for $k = 1, \cdots, K$, and define the matrix

$$B = [B_{jk}], \qquad B_{jk} = \sum_i \frac{a_{ji}a_{ki}}{\lambda_i}, \qquad j, k = 1, \cdots, K$$

and the column vector

$$Z = [Z_j], \qquad Z_j = \sum_i \frac{a_{ji} y_i}{\lambda_i}, \qquad j = 1, \cdots, K.$$

The series for B_{jk} converges, and the series for Z_j converges with probability one with respect to the nominal measure μ_0. We assume B is nonsingular. The maximum-likelihood estimator $\hat{\beta}$ for the column vector $\beta = [\beta_k]$, $k = 1, \cdots, K$, with respect to μ_0, is

$$\hat{\beta} = B^{-1} Z.$$

The estimator $\hat{\beta}$ is Gaussian, unbiased and has covariance matrix with respect to μ_0,

$$E\hat{\beta}\hat{\beta}' = B^{-1}.$$

If now, as in the first example, the autocorrelation function of $x(t)$ is changed to $R_1(t, s) = R_0(t, s) + A(t, s)$ while the mean remains zero, the estimator $\hat{\beta}$ with respect to the new Gaussian measure μ_1 is unbiased but has covariance matrix

$$E_1(\hat{\beta}\hat{\beta}') = B^{-1} + B^{-1} \Delta B^{-1}$$

where the matrix Δ is given by

$$\Delta = [\Delta_j k], \qquad \Delta_j k = \sum_{i,n} \frac{a_{ji} a_{kn}}{\lambda_i \lambda_n} (A\phi_i, \phi_n).$$

If stability is to mean that the distribution of $\hat{\beta}$ with respect to μ_1 measure necessarily approaches the distribution of $\hat{\beta}$ with respect to μ_0 measure as $\|A\| \to 0$, then the condition for it is

$$\sum_i \frac{|a_{ji}|^2}{\lambda_i^2} < \infty.$$

The discussion of the stability criterion for the first example re-applies with trivial modification.

A general stability criterion. Since in the examples of the previous section the functionals are linear and the family of measures is restricted to include only Gaussian ones, the functionals have Gaussian distributions which are determined by a routine calculation of moments. In general the estimators or test functionals are not linear, as, e.g., in the problem of estimating the range of a radar target, and their distributions, even with Gaussian noise and perhaps Gaussian signals, are either unknown or known only in a form ill-adapted to further calculation. In any case, the restriction to Gaussian measures were better removed, even though the noise is nearly Gaussian, if the stability criteria are to be believable in practice. One would like, therefore, to treat detection stability in greater generality, with the expectation, of course, of less specific results.

The following informal definition is suggested as being both obvious and natural. Suppose the decision or inference made by a signal detecting or measuring device (henceforth, simply *detector*) depends on the value taken on by a functional f on a sample space of functions. The detector is *stable* if the induced probability distribution of f converges in the usual sense to its nominal distribution as the actual probability measure on the sample space converges in some reasonable sense to the nominal measure. In terms of this definition the study of detector stability becomes a direct application of results from the theory of weak convergence of probability measures on function spaces. There is available a choice of function spaces and topologies for which a well-developed theory of weak convergence of measures exists.[9] For reasons previously indicated, it seems appropriate to take the space of continuous real-valued functions, say on [0, 1], in the uniform topology. Also, $C[0, 1]$ has the attraction that it provides a relatively simple theory. We shall use certain results of Prokhorov,[10] which are stated below for convenience (the base space is specialized to be $C[0, 1]$).

We take $\mathscr{B}$ to be the Borel field of subsets of $C[0, 1]$ generated by the closed sets, and $\mathscr{M}$ to be the set of regular probability measures μ on $\mathscr{B}$. A sequence of measures $\{\mu_n\}$ from $\mathscr{M}$ converges weakly to μ, $\mu_n \Rightarrow \mu$, if for any continuous bounded functional g on $C[0, 1]$, $\int g d\mu_n \rightarrow \int g d\mu$. If ξ is a μ-almost-everywhere continuous mapping of $C[0, 1]$ into a metric space $\mathscr{R}$, it induces a regular probability measure μ^ξ in $\mathscr{R}$. For the convergence $\mu_n \Rightarrow \mu$ it is necessary and sufficient that for any real-valued μ-almost-everywhere continuous function ξ, $\mu_n^\xi \Rightarrow \mu^\xi$. The Lévy distance $L(\mu_1, \mu_2)$ between two measures μ_1 and μ_2 in $\mathscr{M}$ is defined as follows: let F be a closed subset of $C[0, 1]$ and let F^ε be the open ε-neighborhood of F,

$$F^\varepsilon = \left\{x: \inf_{x_0 \in F} |x - x_0| < \varepsilon\right\}$$

where $|x|$ is the $C[0, 1]$ norm of x. Let ε_{12} and ε_{21} be the lower bounds, respectively, of those ε for which for all closed sets $F \subset C[0, 1]$,

$$\mu_2(F) < \mu_1(F^\varepsilon) + \varepsilon \quad \text{and} \quad \mu_1(F) < \mu_2(F^\varepsilon) + \varepsilon.$$

Then $L(\mu_1, \mu_2)$ is defined to be $\max(\varepsilon_{12}, \varepsilon_{21})$. Convergence of probability measures on $\mathscr{B}$ with respect to the Lévy distance is equivalent to weak convergence. Finally, a sufficient condition that a sequence $\{\mu_n\}$, $\mu_n \in \mathscr{M}$, converge weakly to a measure $\mu \in \mathscr{M}$ is that:

(1) the finite-dimensional distributions induced by μ_n converge to limit distributions, and

(2) for all n, all $t_1, t_2 \in [0, 1]$,

$$E|x(t_2) - x(t_1)|^a \leqq k|t_2 - t_1|^\alpha$$

for some fixed $a > 0$, $k > 0$, $\alpha > 1$ (Kolmogorov's condition for continuity).

Thus, if we accept the Lévy distance as a suitable indication of the closeness of two measures, a sufficient condition that a detector be stable in the sense described is that the functional f be continuous on $C[0, 1]$. This formulation does not allow much hope for actually calculating useful bounds on errors, but it does allow

[9] See, e.g., Prokhorov [12], Skorokhod [14], and Kallianpur [7].

[10] See Prokhorov [12].

comparison between what might be called the degrees of stability of two detectors. In particular, Prokhorov gives the following estimate: if for arbitrary $h > 0$ a compact subset K of $C[0, 1]$ is such that both the μ_0- and μ_1-measure of its complement is less than $h/2$, and if $\delta < h/2$ is sufficiently small that $|f(x_1) - f(x_2)| < h/2$ for all $x_1, x_2 \in K$ satisfying $|x_1 - x_2| < \delta$, then $L(\mu_0, \mu_1) < \delta$ implies that the Lévy distance between μ_0^f and μ_1^f is less than h. Thus, a rough comparison of the stability of two test functionals can be obtained from a comparison of their moduli of continuity on compact sets.

Application of the stability criterion. First, let us look again at the simple detection example already discussed. We suppose the nominal probability measure μ_0 of the noise process to be Gaussian, but no longer restrict all the family of possible measures to be Gaussian. According to the criterion proposed, the detector is stable if $f(y) = \sum(a_k y_k/\lambda_k)$ is continuous on $C[0, 1]$. The condition used previously amounted to $f(y)$ being continuous on $L_2[0, 1]$, which of course implies its continuity on $C[0, 1]$. If one for the moment thinks of the case where all the possible noise measures are Gaussian, the weakening of the condition for stability simply reflects the fact that a tighter notion of closeness of autocorrelation function is now in effect. Probably neither notion is exactly appropriate if the actual nature of estimational errors of autocorrelation is taken into account. The topology used for the space of sample functions is arbitrary; so the stability criteria developed, although they may be precise mathematically, cannot be regarded as being sharp from a practical point of view. This precautionary comment applies to all our examples.

It is likely in practice that the actual implementation of a test functional can be represented directly as an integral,

$$f(y) = \int_0^1 y(t)\, dg(t)$$

where the weighting function $g(t)$ is of bounded variation. In this case the detector is necessarily stable, and the smaller the total variation of $g(t)$, the more stable the detector. Essentially the same remarks as these apply to the linear parameter estimation problem.

The estimation of time delay of a known signal, as in the measurement of range with a radar, is a common example in which the maximum-likelihood estimator is not a linear functional. We now describe a class of problems which is a slight generalization of it and make some preliminary remarks about stability. Let

$$y(t) = bs(t, \alpha) + x(t), \qquad 0 \leqq t \leqq 1,$$

where $y(t)$ is the observed waveform, $s(t, \alpha)$ is for each α a known real-valued continuous function (signal), $x(t)$ is a noise process with mean zero, and b, α are unknown real parameters with α required to be in some closed finite interval. It is desired to estimate α. Assume that with the nominal measure, μ_0, $x(t)$ is

Gaussian with autocorrelation function $R_0(t, s)$. Signal and noise are expanded in terms of the eigenfunctions $\phi_n(t)$ as before, and we write

$$s_n(\alpha) = \int_0^1 s(t, \alpha)\phi_n(t)\, dt.$$

Since it is required that the estimation problem be nonsingular for the nominal noise process, the sum

$$\sum \frac{s_n^2(\alpha)}{\lambda_n}$$

must converge for all α, and without loss of generality the signal may be normalized so that this sum is equal to one for all α. Then a maximum-likelihood estimator for α is that value $\hat{\alpha}$ which maximizes

$$z(\alpha) = \sum \frac{s_n(\alpha)y_n}{\lambda_n} \tag{11}$$

if such exists. Let b^0 and α^0 denote the true values of b and α. Then, with the measure induced from μ_0, $z(\alpha)$ is a Gaussian process with some closed finite interval as parameter set, and with mean-value and covariance functions

$$Ez(\alpha) = b^0 \sum \frac{s_n(\alpha)s_n(\alpha^0)}{\lambda_n} = b^0 q(\alpha, \alpha^0) \tag{12}$$

and

$$E[z(\alpha) - Ez(\alpha)][z(\alpha') - Ez(\alpha')] = \sum \frac{s_n(\alpha)s_n(\alpha')}{\lambda_n} = q(\alpha, \alpha') \tag{13}$$

respectively, where $q(\alpha, \alpha')$[11] is defined by the last equation. Since it is convenient to treat $\hat{\alpha}$ as a functional on the space of functions z, a natural first step in analyzing stability is to establish conditions so that a sequence of probability measure μ_n converging weakly to the nominal measure μ_0 on the space of sample functions induces a weakly convergent sequence of measures μ_n^z on the space of functions z. The Kolmogorov continuity condition is satisfied for the nominal induced measure on z if

$$|q(\alpha', \alpha^0) - 2q(\alpha, \alpha') + q(\alpha, \alpha^0) + (b^0)^2[q^2(\alpha', \alpha^0) - 2q(\alpha, \alpha^0)q(\alpha', \alpha^0) + q^2(\alpha, \alpha^0)]| \leq k|\alpha' - \alpha|^p$$

for some $k > 0$, $p > 1$. Then it is sufficient, by a theorem of Prokhorov quoted in the preceding section, for $\mu_n^z \Rightarrow \mu_0^z$ that the Kolmogorov condition be met by all the μ_n with $a = 2$ (second moment form of the condition) and the same k and p as above, and that, say,

$$\sum \frac{s_n^2(\alpha)}{\lambda_n^2} < C = \text{constant for all } \alpha.$$

[11] See Kelly, Reed and Root [8].

Under these conditions $z(\alpha)$ may be taken to be a process with continuous sample functions, and the measures μ_0^z and μ_n^z are regular. In a sense the stage is set for analyzing the behavior of $\hat{\alpha}(z)$ as $\mu_n \Rightarrow \mu_0$, but this is not attempted here. One simple observation, however, is the following. The function $q(\alpha, \alpha^0)$ has a local maximum at $\alpha = \alpha^0$. If for given $\delta > 0$, $M > 0$, $q(\alpha, \alpha^0) + M < q(\alpha^0, \alpha^0) - M$ whenever $|\alpha - \alpha^0| > \delta$, then

$$\text{Prob}\{|\hat{\alpha} - \alpha^0| < \delta\} \geqq \text{Prob}\{|z(\alpha)| \leqq M\}, \tag{14}$$

an estimate for the error of $\hat{\alpha}$ appropriate to the case of high signal-to-noise ratio (large b^0). Now if μ_n^z converges weakly to μ_0^z, $\mu_n^z\{|z(\alpha)| \leqq M\}$ converges to $\mu_0^z\{|z(\alpha)| \leqq M\}$ for any M for which $\mu_0^z\{\max_\alpha |z(\alpha)| = M\} = 0$, making the right side of the error estimate (14) continuous except for at most a countable set of values of M.

As a final example consider the detection of a noise-like signal in noise. The observed waveform $y(t)$ is a sample from one of two random processes, the first representing, say, noise alone, the second, signal plus noise. The nominal probability measure for the first process is μ_0, for the second, ν_0. We assume both μ_0 and ν_0 are Gaussian and have mean zero, and that they are strongly equivalent. We want to find a condition for the stability of the maximum-likelihood test between the two hypotheses. A formula for the likelihood ratio given by Pitcher[12] is taken as the starting point.

Let R_0 be the integral operator on $L_2[0, 1]$ previously defined, associated with μ_0, and Q_0 the corresponding integral operator associated with ν_0. Since the processes are equivalent, $Q_0^{1/2}R_0^{-1/2}$ has a bounded extension X. Again, since the processes are equivalent, $I - X^*X$ is Hilbert–Schmidt; and we let λ_n and $\psi_n(t)$ be the eigenvalues and associated (orthonormal) eigenfunctions of X^*X. $R_0^{-1/2}$ is densely-defined, so there is a sequence $\{\psi_{nj}\}_j$, for each n, such that $\psi_{nj} \in \mathscr{D}(R_0^{-1/2})$ and $\psi_{nj} \to \psi_n$ (strongly, in $L_2[0, 1]$) as $j \to \infty$. Define

$$\phi_{nj}(x) = \int_0^1 y(t)[R_0^{-1/2}\psi_{nj}](t)\,dt. \tag{15}$$

Then, for each n, the ϕ_{nj} converge in mean square with respect both to μ_0 and ν_0 to limit $\theta_n(x)$. The likelihood ratio (derivative of μ_0 with respect to ν_0) is

$$\frac{d\mu_0}{d\nu_0}(y) = \exp\left[\frac{1}{2}\sum\theta_n^2(y)\left(\frac{1}{\lambda_n} - 1\right) + \log\lambda_n\right]$$

or, if $\sum|1/\lambda_n - 1| < \infty$, the case of strong equivalence,

$$\log\frac{d\mu_0}{d\nu_0}(y) = \sum\log\lambda_n + \frac{1}{2}\sum\theta_n^2(y)\left(\frac{1}{\lambda_n} - 1\right). \tag{16}$$

[12] See Pitcher [11], also Root [13].

It is therefore sufficient to consider the functional

$$f(y) = \sum \theta_n^2(y)\left(\frac{1}{\lambda_n} - 1\right). \tag{17}$$

We show now that if the sum

$$\sum_n \left(\int_0^1 |[R_0^{-1/2}\psi_{nj}](t)|\,dt\right)^2 \left|\frac{1}{\lambda_n} - 1\right| \tag{18}$$

converges uniformly for $j = 1, 2, \cdots$ to a limit $B_j < B$, then $f(y)$ as given by (17) is a stable test functional.

Since the ϕ_{nj} converge in mean square to θ_n with respect to μ_0 and ν_0, one can find a subsequence of the j's (which will not be relabeled) and a subset C_0' of $C[0, 1]$ of μ_0- and ν_0-measure one such that $\phi_{nj}(y)$ converges to $\theta_n(y)$, for every $y \in C_0'$ and for all n. Since $\sum|1/\lambda_n - 1|$ converges, and since the θ_k are independent Gaussian random variables with respect to μ_0 with $E_0\theta_k = 0$, $E_0\theta_k^2 = 1$, the sum $\sum\theta_n^2(1/\lambda_n - 1)$ converges with probability one with respect to μ_0, and hence with respect to ν_0. Hence there is a set $C_0 \subset C_0'$, $\mu_0 C_0 = \nu_0 C_0 = 1$, such that $\sum\theta_n^2(y)(1/\lambda_n - 1)$ converges for all $y \in C_0$.

The sums

$$f_j(y) = \sum_n \phi_{nj}^2(y)\left(\frac{1}{\lambda_n} - 1\right), \qquad j = 1, 2, \cdots, \tag{19}$$

converge uniformly in j for all $y \in C[0, 1]$ by the condition (18) since

$$|\phi_{nj}(y)| \leqq |y| \int_0^1 |[R^{-1/2}\psi_{nj}](t)|\,dt.$$

Furthermore, for $y, \eta \in C[0, 1]$,

$$|\phi_{nj}^2(y + \eta) - \phi_{nj}^2(y)| \leqq 2|\eta||y|\left(\int_0^1 |[R^{-1/2}\psi_{jn}](t)|\,dt\right)^2 + |\eta|^2\left(\int_0^1 |[R^{-1/2}\psi_{jn}](t)|\,dt\right)^2$$

so that

$$\left|\sum_n \phi_{nj}^2(y)\left(\frac{1}{\lambda_n} - 1\right) - \sum_n \phi_{nj}^2(y + \eta)\left(\frac{1}{\lambda_n} - 1\right)\right| \leqq 2|\eta|\,|y|B + |\eta|^2 B. \tag{20}$$

The functionals $f_j(y)$ converge a.e. with respect to μ_0 to $f(y)$. In fact, for $y \in C_0$ and arbitrary $\varepsilon > 0$, one can choose N_0 sufficiently large that

$$\left|\sum_{N_0}^{\infty} \theta_n^2(y)\left(\frac{1}{\lambda_n} - 1\right)\right| < \frac{\varepsilon}{3},$$

$$\left|\sum_{N_0}^{\infty} \phi_{nj}^2(y)\left(\frac{1}{\lambda_n} - 1\right)\right| < \frac{\varepsilon}{3}, \qquad j = 1, 2, \cdots,$$

and then choose j_0 sufficiently large that

$$|\theta_n^2(y) - \phi_{nj}^2(y)| < \frac{\varepsilon}{3N_0\left(\dfrac{1}{\lambda_n} - 1\right)}, \; n = 1, \cdots, N_0, \; j > j_0.$$

Then $|f(y) - f_j(y)| < \varepsilon, j > j_0$.

Now suppose there is a sequence of probability measures $\{\mu_n\}$ on $C[0, 1]$ which converge weakly to μ_0 and all of which are equivalent to μ_0. Then, given $h > 0$, one can choose a compact set $K_h \subset C[0, 1]$ such that $\mu_n(C[0, 1] - K_h) < h/2$ for all n. Let $|y| < M$ for $y \in K_h$. Then choose $\varepsilon > 0$ small enough so that $2\varepsilon MB + \varepsilon^2 B < h/2$. It follows that if the Lévy distance between μ_n and μ_0 is less than min $(\varepsilon, h/2)$, then the Lévy distance between the corresponding induced distributions of $f_j(y)$, $\mu_n^{f_j}$ and $\mu_0^{f_j}$ respectively, is less than h for all $j = 1, 2, \cdots$. Since the μ_n are all equivalent to μ_0, the functionals f_j converge a.e. $[\mu_n]$ to f, and hence the distributions $\mu_n^{f_j}$ converge in Lévy metric to μ_n^f. Finally then, if $L(\mu_n, \mu_0)$ is less than min $(\varepsilon, h/2)$, $L(\mu_n^f, \mu_0^f) \leqq h$. The same argument shows that if $\{\nu_n\}$ is a sequence of probability measures on $C[0, 1]$ which converge weakly to ν_0, and are equivalent to ν_0, then $\nu_n^f \Rightarrow \nu_0^f$.

The condition (18) is unfortunately impractical, largely because the formula for the likelihood ratio has already considerable implicit complexity. In special cases where the derivative of the two nominal Gaussian measures can be written more explicitly (as, e.g., when the θ_n's are given directly) it can obviously be simplified.

For example, in the relatively trivial case that R_0 and Q_0 have the same eigenfunctions with eigenvalues $\{\mu_n\}$, $\{\eta_n\}$, respectively, the condition (18) reduces to the condition that

$$\sum_n \left| \frac{\mu_n - \eta_n}{\mu_n \eta_n} \right| < \infty.$$

Concluding remark. The phenomenon that a statistical inference procedure delicately adjusted to certain underlying statistics may be very sensitive to changes in those statistics is well recognized. It suggests the question, of interest to statisticians, of when to forego parametric methods altogether. We have not, of course, touched on this question here, but have concerned ourselves only with the damage incurred by parametric inferences as the underlying statistics are changed. In many detection-theory problems it is desirable to introduce some degree of parametrization.

The actual appearance of instability as defined here seems nearly impossible. The examples indicate that the infinite-dimensionality of the observations must be fully exploited in ways which can hardly be implemented for the detector to be completely unstable.

Bibliography

1. E. J. Akutowicz, *On extrapolating a positive definite function from a finite interval*, Math. Scand. 7 (1959), 157–169.

2. A. V. Balakrishnan, *Estimation and detection theory for multiple stochastic processes*, J. Math. Anal. Appl. **1** (1960), 386–410.

3. Yu. K. Belyaev, *Continuity and Hölder's conditions for sample functions of stationary Gaussian processes*, Proc. 4th Berkeley Sympos. Math. Statist. and Prob., Vol. II, pp. 23–25, Univ. of California Press, Berkeley, Calif., 1960.

4. R. D. Dobrushin, *Properties of sample functions of a stationary Gaussian process*, Theor. Probability Appl. **5** (1960), 120–122.

5. U. Grenander, *Stochastic processes and statistical inference*, Ark. Mat. **1** (1960), 195–277.

6. J. Hajek, *On linear statistical problems in stochastic processes*, Czechoslovak Math. J. **12** (**87**) (1962), 404–444.

7. G. Kallianpur, *The topology of weak convergence of probability measures*, J. Math. Mech. **10** (1961), 947–969.

8. E. J. Kelly, I. S. Reed and W. L. Root, *The detection of radar echoes in noise*. I, II, J. Soc. Indust. Appl. Math. **8** (1960), 309–341, 481–507.

9. E. Parzen, *Probability density functionals and reproducing kernel Hilbert spaces*, Proc. Sympos. Time Series Analysis (Brown Univ., 1962), pp. 155–169, Wiley, New York, 1963.

10. ———, *An approach to time series analysis*, Ann. Math. Statist. **32** (1961), 951–989.

11. T. Pitcher, Unpublished memorandum, 1960.

12. Yu. V. Prokhorov, *Convergence of random processes and limit theorems in probability theory*, Theor. Probability Appl. **1** (1956), 157–214.

13. W. L. Root, *Singular Gaussian measures in detection theory*, Proc. Sympos. Time Series Analysis (Brown Univ., 1962), pp. 292–316, Wiley, New York, 1963.

14. A. V. Skorokhod, *On the differentiability of measures which correspond to stochastic processes*. I, Theor. Probability Appl. **2** (1957), 407–432.

15. A. M. Yaglom, *On the equivalence or perpendicularity of two gaussian probability measures in function space*, Proc. Sympos. Time Series Analysis (Brown Univ., 1962), pp. 327–349, Wiley, New York, 1963.

University of Michigan,
Ann Arbor, Michigan

THE CONSTRUCTION OF A CLASS OF STATIONARY MARKOFF PROCESSES

BY

EUGENE WONG

Introduction. We take a forward, or Fokker–Planck, equation of diffusion theory

$$\frac{\partial^2}{\partial x^2}[B(x)p] - \frac{\partial}{\partial x}[A(x)p] = \frac{\partial p}{\partial t}, \qquad p = p(x|x_0, t), \qquad 0 < t < \infty, \tag{1}$$

on an interval I with end points x_1 and x_2, and consider the solution $p(x|x_0, t)$ corresponding to an initial value $\delta(x - x_0)$, i.e., the principal solution. The function $B(x)$ can be interpreted as a variance [**15**], and is therefore non-negative on (x_1, x_2). In general, we take, as boundary conditions, those corresponding to reflecting barriers [**9**],

$$\frac{\partial}{\partial x}[B(x)p] - A(x)p = 0, \qquad x = x_1, x_2. \tag{2}$$

The unique solution $p(x|x_0, t)$ is the density function for the transitional probability of a stationary Markoff process $X(t)$ in one dimension,

$$\Pr\{X(t + \tau) \in E | X(\tau) = x_0\} = \int_E p(x|x_0, t)\, dx, \qquad x_0 \in I, \qquad E \subseteq I. \tag{3}$$

Next, we consider the class of first-order probability density functions $W(x)$ which satisfy the Pearson equation [**4**]

$$\frac{dW(x)}{dx} = \frac{ax + b}{cx^2 + dx + e} W(x). \tag{4}$$

In §1 we shall show that by suitably identifying the functions $A(x)$ and $B(x)$ in (1) with the polynomials $ax + b$ and $cx^2 + dx + e$ in (4), a class of stationary Markoff processes is constructed for which

$$\lim_{t \to \infty} p(x|x_0, t) = \int_{x_1}^{x_2} W(x_0) p(x|x_0, t)\, dx_0 = W(x), \tag{5}$$

where $W(x)$ satisfies (4). The identification scheme is such that the Pearson equation (4) uniquely specifies the Fokker–Planck equation (1). In a sense, the Pearson equation serves a rather natural role in bridging the gap between the first-order statistical description characterized by $W(x)$ and the transitional properties represented by $p(x|x_0, t)$.

In §2 we shall exhibit some specific processes of this class. These examples include several well-known ones, as well as some that appear not to have been investigated previously.

In §3 we shall examine some properties of processes of this class and give some physical interpretation to processes in §2. Finally, the distribution of some functionals involving processes of this class are discussed in §4.

1. **Construction.** We note that a straightforward application of separations of variables to (1) yields an equation of the Sturm–Liouville type

$$\frac{d}{dx}\left[B(x)\rho(x)\frac{d\varphi(x)}{dx}\right] + \lambda\rho(x)\varphi(x) = 0. \tag{6}$$

The boundary conditions (2) imply that

$$B(x)\rho(x)\frac{d\varphi(x)}{dx} = 0, \qquad x = x_1, x_2. \tag{7}$$

The function $\rho(x)$ is taken to be non-negative on (x_1, x_2) and satisfies

$$\frac{d}{dx}[B(x)\rho(x)] - A(x)\rho(x) = 0. \tag{8}$$

A comparison of (8) and (4) shows that they become identical if we make the following identifications:

$$\rho(x) = W(x), \tag{9}$$

$$B(x) = \beta(cx^2 + dx + e) \tag{10}$$

and

$$A(x) = \frac{dB(x)}{dx} + \beta(ax + b), \tag{11}$$

subject to the condition that $B(x)$ be non-negative on (x_1, x_2). It is clear that through (10) and (11), $A(x)$ and $B(x)$ are uniquely specified by the Pearson equation (4) up to a common positive multiplicative constant β, which represents a scaling factor in the variable t.

With the substitution of $W(x)$ for $\rho(x)$, (6) and (7) become

$$\frac{d}{dx}\left[B(x)W(x)\frac{d\varphi(x)}{dx}\right] + \lambda W(x)\varphi(x) = 0, \tag{12}$$

and

$$B(x)W(x)\frac{d\varphi(x)}{dx} = 0, \qquad x = x_1, x_2. \tag{13}$$

From the classical Sturm–Liouville theory it is known that the spectrum of (12) is discrete if x_1 and x_2 are finite. If one or both of the boundaries are infinite, then a continuous range of eigenvalues may be present. In terms of the solutions of (12) the principal solution $p(x|x_0, t)$ of (1) can be written as

$$p(x|x_0, t) = W(x)\left\{\sum_n e^{-\lambda_n t}\varphi_n(x_0)\varphi_n(x) + \int e^{-\lambda t}\varphi(\lambda, x_0)\varphi(\lambda, x)\, d\lambda\right\}, \tag{14}$$

where the summation is taken over all discrete eigenvalues and the integral is taken over the continuous range of eigenvalues. The eigenfunctions $\varphi(x)$ in (14) are assumed to be normalized so that corresponding to discrete eigenvalues we have

$$\int_{x_1}^{x_2} W(x)\varphi_m(x)\varphi_n(x) = \delta_{mn}, \tag{15}$$

and corresponding to continuous eigenvalues

$$\int_{x_1}^{x_2} W(x)\varphi(\lambda, x)\varphi(\lambda', x)\, dx = \delta(\lambda - \lambda'). \tag{16}$$

It can be verified directly that corresponding to boundary conditions (13), the Sturm–Liouville equation (12) has at least one discrete eigenvalue, namely, $\lambda = 0$ with corresponding eigenfunction $\varphi(x) = 1$. In addition, if we assume that the spectrum is discrete, then

$$\begin{aligned} \lambda_n &= -\int_{x_1}^{x_2} \varphi_n(x)\frac{d}{dx}\left[B(x)W(x)\frac{d\varphi_n(x)}{dx}\right] dx \\ &= \int_{x_1}^{x_2} B(x)W(x)\left[\frac{d\varphi_n(x)}{dx}\right]^2 dx, \end{aligned} \tag{17}$$

where we have made use of the boundary conditions (13). Since $B(x)W(x)[d\varphi(x)/dx]^2$ is non-negative on (x_1, x_2), it follows from (17) that

$$\lambda_n \geqq 0. \tag{18}$$

Furthermore, the case where a continuous range of eigenvalues is present can be arrived at by taking cases with finite boundaries, thus with discrete spectra, and considering the limiting situation when one or both of the boundaries becomes infinite. It is clear that, in general, we have

$$\lambda \geqq 0. \tag{19}$$

Further, even in the limiting cases $\lambda = 0$ must remain a discrete eigenvalue, since the corresponding eigenfunction $\varphi = 1$ is always square-integrable with respect to the density function $W(x)$. It follows, from the fact that $\lambda = 0$, $\varphi = 1$ is always a discrete solution, that

$$\int_{x_1}^{x_2} p(x|x_0, t)W(x_0)\, dx_0 = W(x). \tag{20}$$

Further, from the fact that $\lambda = 0$ is also the minimum eigenvalue, it follows that

$$\lim_{t\to\infty} p(x|x_0, t) = W(x). \tag{21}$$

The construction procedure is seen to consist of two parts. First, starting with a first-order probability density function $W(x)$ and a corresponding Pearson equation, we identify the functions $A(x)$ and $B(x)$ in (1) with the polynomials in the Pearson equation according to (10) and (11). Secondly, the density function $p(x|x_0, t)$ of the transitional probability is obtained as the principal solution to (1). (For a brief discussion along similar lines, see Kolmogorov [**13**].)

2. **Some specific processes.** If we consider the roots of the equation $B(x) = 0$, five distinct situations are possible, namely: (a) no root, in which case $B(x)$ is a constant, (b) a single real root, (c) two unequal real roots, (d) two equal real roots, and (e) a pair of complex-conjugate roots. In this section we shall consider six specific processes which, while by no means exhausting all the possibilities, include at least one example of each of the five situations corresponding to the roots of $B(x) = 0$. In every case except one, the choice of boundaries will be a natural one in that each boundary is either at a real root of "$B(x) = 0$" or is infinite. Without loss of generality we shall set $\beta = 1$, thus normalizing the time scale.

A. Consider the first-order density function

$$W(x) = e^{-x}, \qquad 0 \leq x < \infty, \tag{22}$$

with a corresponding Pearson equation

$$\frac{dW(x)}{dx} = -\frac{1}{1} W(x).$$

The resulting Fokker–Planck equation is

$$\frac{\partial^2 p}{\partial x^2} + \frac{\partial p}{\partial x} = \frac{\partial p}{\partial t}$$

with boundary condition

$$\frac{d}{dx} p(x|x_0, t) = 0, \qquad x = 0.$$

The principal solution $p(x|x_0, t)$ is given by

$$p(x|x_0, t) = e^{-x}\left\{1 + e^{-t/4}\int_0^\infty e^{-\mu^2 t}\psi(\mu, x_0)\psi(\mu, x)\, d\mu\right\}, \tag{23}$$

where

$$\psi(\mu, x) = \sqrt{\frac{2}{\pi}}\, \frac{2\mu}{\sqrt{1 + 4\mu^2}}\left[\cos \mu x - \frac{1}{2\mu}\sin \mu x\right] e^{x/2}.$$

The function $p(x|x_0, t)$ can also be written as

$$\begin{aligned} p(x|x_0, t) = {} & \frac{1}{2\sqrt{\pi t}} \exp\left[-\tfrac{1}{2}(x - x_0)\right] \exp\left[-\tfrac{1}{4}t\right]\left\{\exp\left[-\frac{(x - x_0)^2}{4t}\right]\right. \\ & \left. + \exp\left[-\frac{(x + x_0)^2}{4t}\right]\right\} + \frac{1}{\sqrt{\pi}} e^{-x} \int_{(x+x_0-t)/2\sqrt{t}}^{\infty} e^{-z^2}\, dz. \end{aligned} \tag{24}$$

B. Let the first-order density function be given by

$$W(x) = \frac{1}{\sqrt{2\pi}} \exp\,[-\tfrac{1}{2}x^2], \qquad -\infty < x < \infty, \tag{25}$$

with an associated Pearson equation

$$\frac{dW(x)}{dx} = -\frac{x}{1}\,W(x).$$

The resulting Fokker–Planck equation can be written as

$$\frac{\partial^2 p}{\partial x^2} + \frac{\partial}{\partial x}(xp) = \frac{\partial p}{\partial t}.$$

The density function $p(x|x_0, t)$ is given by

$$p(x|x_0, t) = \frac{1}{\sqrt{2\pi}}\, e^{-x^2/2} \sum_{n=0}^{\infty} \frac{e^{-nt}}{n!}\, H_n(x_0)H_n(x), \tag{26}$$

where $H_n(x)$ are the Hermite polynomials **[14]**

$$H_n(x) = (-1)^n\, e^{-x^2/2}\, \frac{d^n}{dx^n}\,(e^{-x^2/2}).$$

By the use of Mehler's formula we can write $p(x|x_0, t)$ as

$$p(x|x_0, t) = \frac{1}{\sqrt{2\pi(1 - e^{-2t})}} \exp\left[-\frac{1}{2(1 - e^{-2t})}(x - x_0\, e^{-t})^2\right]. \tag{27}$$

C. Consider the first-order density function

$$W(x) = \frac{x^\alpha}{\Gamma(\alpha + 1)}\, e^{-x}, \qquad \alpha > -1,\ 0 \leq x < \infty \tag{28}$$

with corresponding Pearson equation

$$\frac{dW(x)}{dx} = \left(\frac{\alpha - x}{x}\right) W(x).$$

The associated Fokker–Planck equation is

$$\frac{\partial^2}{\partial x^2}(xp) - \frac{\partial}{\partial x}[(\alpha + 1 - x)p] = \frac{\partial p}{\partial t}.$$

The principal solution $p(x|x_0, t)$ is given by

$$p(x|x_0, t) = x^\alpha\, e^{-x} \sum_{n=0}^{\infty} \frac{n!}{\Gamma(n + \alpha + 1)}\, e^{-n\beta t}\, L_n^\alpha(x_0)L_n^\alpha(x), \tag{29}$$

where

$$L_n^\alpha(x) = \frac{1}{n!}\, x^{-\alpha}\, e^{x}\, \frac{d^n}{dx^n}\,(x^\alpha\, e^{-x}),$$

are the Laguerre polynomials **[14]**. The function $p(x|x_0, t)$ can also be written as

$$p(x|x_0, t) = \frac{1}{1 - e^{-t}} \left(\frac{x}{x_0 \exp[-t]}\right)^{\alpha/2} \cdot \exp\left[-\frac{1}{(1 - e^{-t})}(x + x_0 e^{-t})\right] I_\alpha\left(\frac{2 e^{-t/2}\sqrt{x_0 x}}{1 - \exp[-t]}\right), \tag{30}$$

where $I_\alpha(z)$ is the modified Bessel function.

D. Corresponding to the situation when "$B(x) = 0$" has two unequal real roots, we take $W(x)$ to be

$$W(x) = \frac{\Gamma(\alpha + \gamma + 2)}{\Gamma(\alpha + 1)\Gamma(\gamma + 1)} \frac{(1 + x)^\alpha(1 - x)^\gamma}{2^{\alpha+\gamma+1}}, \quad \alpha, \gamma \geqq -1, \ -1 \leqq x \leqq +1, \tag{31}$$

with associated Pearson equation

$$\frac{dW(x)}{dx} = \frac{(\alpha - \gamma) - (\alpha + \gamma)x}{(1 - x^2)} W(x).$$

The resulting Fokker–Planck equation becomes

$$\frac{\partial^2}{\partial x^2}[(1 - x^2)p] + (\alpha + \gamma + 2)\frac{\partial}{\partial x}[xp] - (\alpha - \gamma)\frac{\partial p}{\partial x} = \frac{\partial p}{\partial t}.$$

The solution $p(x|x_0, t)$ is given by

$$p(x|x_0, t) = \frac{(1 + x)^\alpha(1 - x)^\gamma}{2^{\alpha+\gamma+1}} \sum_{n=0}^{\infty} e^{-n(n+\alpha+\gamma+1)t} A_n P_n^{\alpha,\gamma}(x_0) P_n^{\alpha,\gamma}(x), \tag{32}$$

where $P_n^{\alpha,\gamma}(x)$ are the Jacobi polynomials[1] **[14]**

$$P_n^{\alpha,\gamma}(x) = \frac{(-1)^n}{2^n}(1 + x)^\alpha(1 - x)^\gamma \frac{d^n}{dx^n}[(1 + x)^{\alpha+n}(1 - x)^{\gamma+n}],$$

and

$$A_n = \frac{(2n + \alpha + \gamma + 1)\Gamma(n + \alpha + \gamma + 1)}{\Gamma(n + \alpha + 1)\Gamma(n + \gamma + 1)n!}.$$

E. Corresponding to a pair of complex conjugate-roots of "$B(x) = 0$," we take

$$W(x) = \frac{\Gamma(\alpha + \frac{1}{2})}{\Gamma(\frac{1}{2})\Gamma(\alpha)}(1 + x^2)^{-(\alpha+1/2)}, \qquad \alpha > 0, \ -\infty < x < \infty, \tag{33}$$

and associate with it a Pearson equation

$$\frac{dW(x)}{dx} = -\frac{(2\alpha + 1)x}{(1 + x^2)} W(x).$$

[1] The normalization for $P_n^{\alpha,\gamma}$ here is not the conventional one.

The Fokker–Planck equation for this case becomes

$$\frac{d^2}{dx^2}[(1+x^2)p] + (2\alpha - 1)\frac{d}{dx}(xp) = \frac{\partial p}{\partial t}.$$

The Sturm–Liouville equation (12) in this case has $N+1$ discrete eigenvalues ($\alpha - 1 \leqq N < \alpha$), and a continuous range of eigenvalues. More precisely, we have

$$\lambda_n = n(2\alpha - n), \qquad n = 0, 1, 2, \cdots, N,$$

and

$$\lambda = \alpha^2 + \mu^2, \qquad \mu \geqq 0.$$

The solution $p(x|x_0, t)$ can be written as

$$\begin{aligned} p(x|x_0, t) = (1+x^2)^{-(\alpha+1/2)} \Bigg\{ &\frac{1}{\pi}\sum_{n=0}^{N} \frac{(\alpha - n)}{n!\,\Gamma(2\alpha + 1 - n)} e^{-n(2\alpha - n)t}\theta_n(x_0)\theta_n(x) \\ &+ \frac{1}{2\pi}\int_0^\infty e^{-(\alpha^2+\mu^2)t}[\psi(\mu, x_0)\psi(-\mu, x) + \psi(-\mu, x_0)\psi(\mu, x)]\,d\mu \Bigg\}, \end{aligned} \tag{34}$$

where

$$\theta_n(x) = 2^{\alpha-n}\Gamma(\alpha - n + \tfrac{1}{2})(-1)^n(1+x^2)^{\alpha+1/2}\frac{d^n}{dx^n}[(1+x^2)^{n-\alpha-1/2}], \tag{35}$$

are polynomials of degree n, and $\psi(\mu, x)$ is given by

$$\psi(\mu, x) = (x + \sqrt{1+x^2})^{i\mu}(1+x^2)^{1/2}{}_2F_1\left(-\alpha, \alpha+1; 1+i\mu; \frac{1}{2} + \frac{1}{2}\frac{x}{\sqrt{1+x^2}}\right),$$

${}_2F_1$ being the Gauss hypergeometric series [7].

For $\alpha = K$, a positive integer, $p(x|x_0, t)$ can be written somewhat more explicitly as

$$\begin{aligned} p(x|x_0, t) = \frac{1}{(1+x^2)^{K+1/2}} \Bigg\{ &[(1+x_0^2)(1+x^2)]^{K/2}\frac{1}{2\sqrt{\pi t}} e^{-K^2 t} e^{-u^2} \\ &+ \frac{1}{\pi}\sum_{n=0}^{K-1}\frac{(K-n)}{n!\,\Gamma(2K+1-n)} e^{-n(2K-n)t}\theta_n(x_0)\theta_n(x)f_n(x_0, x, t) \Bigg\}, \end{aligned} \tag{36}$$

where

$$u = u(x_0, x, t) = \frac{\operatorname{arc\,sinh} x - \operatorname{arc\,sinh} x_0}{2\sqrt{t}},$$

$$f_n(x_0, x, t) = \frac{1}{\sqrt{\pi}}\int_{u-(K-n)\sqrt{t}}^{u+(K-n)\sqrt{t}} e^{-z^2}\,dz,$$

and θ_n are polynomials defined by (35).

F. Corresponding to the situation where "$B(x) = 0$" has a double real root, we consider the first-order density function

$$W(x) = \frac{1}{\Gamma(2\alpha)} x^{-(2\alpha+1)} e^{-1/x}, \qquad \alpha > 0, 0 \leqq x < \infty, \tag{37}$$

with corresponding Pearson equation

$$\frac{dW(x)}{dx} = \frac{1 - (2\alpha + 1)x}{x^2} W(x).$$

The resulting Fokker–Planck equation is given by

$$\frac{\partial^2}{\partial x^2} [x^2 p] - \frac{\partial}{\partial x} \{[1 - (2\alpha - 1)x]p\} = \frac{\partial p}{\partial t}.$$

The solution of the Sturm–Liouville equation for this case consists of $N + 1$ discrete eigenvalues ($\alpha - 1 \leqq N < \alpha$), and a continuous range of eigenvalues:

$$\lambda_n = n(2\alpha - n), \qquad n = 0, 1, \cdots, N,$$
$$\lambda = \alpha^2 + \mu^2, \qquad \mu \geqq 0.$$

The transitional probability density function $p(x|x_0, t)$ is found to be

$$\begin{aligned} p(x|x_0, t) = x^{-(2\alpha+1)} e^{-1/x} \bigg\{ & \sum_{n=0}^{N} \frac{2(\alpha - n)}{\Gamma(2\alpha + 1 - n)} \frac{1}{n!} e^{-n(2\alpha-n)t} \theta_n(x_0)\theta_n(x) \\ & + \frac{1}{2\pi} \int_0^\infty e^{-(\alpha^2+\mu^2)t} A(\mu)\psi(\mu, x_0)\psi(\mu, x)\, d\mu \bigg\}. \end{aligned} \tag{38}$$

Here, $\theta_n(x)$ are orthogonal polynomials of degree n,

$$\theta_n(x) = (-1)^n x^{2\alpha+1} e^{1/x} \frac{d^n}{dx^n} (x^{2n-2\alpha-1} e^{-1/x}),$$

and

$$\psi(\mu, x) = {}_2F_0(-\alpha - i\mu, -\alpha + i\mu, -x),$$

${}_2F_0$ being the generalized hypergeometric series [7]. Alternative representations of $\psi(\mu, x)$ in terms of confluent hypergeometric functions follow from properties of ${}_2F_0$, e.g.,

$${}_2F_0(-\alpha - i\mu, -\alpha + i\mu, -x) = x^{(\alpha+i\mu)} \Psi\left(-\frac{\alpha}{2} - i\mu, 1 - 2i\mu, \frac{1}{x}\right),$$

where Ψ is the hypergeometric Ψ-function [7]. The quantity $A(\mu)$ is a normalization factor given by

$$A(\mu) = \frac{\Gamma(-\alpha + i\mu)\Gamma(-\alpha - i\mu)}{\Gamma(i\mu)\Gamma(-i\mu)}.$$

3. **Physical interpretation and properties.** The processes outlined in (A) through (D) of §2 are familiar processes with some well-known interpretations. The process of (2 A) represents Brownian motion in a constant force-field (e.g., a gravitational field) with a reflecting barrier [**3**; **11**]. The process of (2 B) is the Ornstein–Uhlenbeck process.

The process of (2 C) has been studied in connection with population growth [**8**]. For $\alpha = N/2 - 1$, $N = 1, 2, \cdots$, it can also be considered as the sum of the squares of N independent and identical Ornstein–Uhlenbeck processes, or alternatively as the square of the radial component of a Brownian motion in N dimensions with radial restoring force proportional to the distance from the origin [**12**].

The process of (2 D), whose transitional density function involves Jacobi polynomials, has been studied by Bochner, and Karlin and McGregor [**2**; **12**]. For special values of the parameters α and γ, $\alpha = \gamma = (N - 3)/2$, $N = 2, 3, \cdots$, a geometric interpretation of the Jacobi diffusion process in terms of Brownian motion on a unit sphere in N dimensions has been given by Karlin and McGregor. It has also been used in connection with biological applications (see Karlin and McGregor).

The process of (2 E) and (2 F) appear to be new. It is of some interest to note that the density function $W(x)$ in (2 E) and (2 F), for $\alpha = \frac{1}{2}$ and $\alpha = \frac{1}{4}$ respectively, become $1/\pi(1 + x^2)$ and $(1/\sqrt{\pi})x^{-3/2} e^{-1/x}$. Both of these are known to be density functions for stable distributions [**10**].

The process $X(t)$ of (2 E) can be interpreted in terms of a Langevin type stochastic differential equation. Specifically, if we make the transformation $Y(t) = \sinh X(t)$, then $Y(t)$ is the solution of a differential equation [**6**]

$$dY(t) + 2\alpha \tanh Y(t)\, dt = dU(t),$$

where $U(t)$ is a Brownian motion process. The term $2\alpha \tanh Y(t)$ is interesting from the point of view of application, since it approximates a saturating linear element (e.g., a saturating linear amplifier).

For $\alpha = K$ an integer, the form of $p(x|x_0, t)$ given by (36) suggests that a simple interpretation of the process (2 E) in terms of Brownian motion should be possible. However, no such interpretation has yet been found.

The process $X(t)$ of (2 F) can also be interpreted in terms of a differential equation

$$dY + [2\alpha - \exp(-Y)]\, dt = dU(t),$$

where $Y(t) = \ln X(t)$ and $U(t)$ is a Brownian motion process.

A feature of processes of this class is that the transitional probability density function can always be represented in terms of relatively simple classical orthogonal functions. This is due to the fact that $B(x)$ and $A(x)$ are chosen to be polynomials of second and first degrees, thus simplifying the resulting Sturm–Liouville equation. In particular, we note the presence of classical orthogonal polynomials. The transitional probability density function $p(x|x_0, t)$ given by (26), (29) and (32) have the following common form of representation:

$$p(x|x_0, t) = W(x)\sum_{n=0}^{\infty} e^{-\lambda_n t}\theta_n(x_0)\theta_n(x), \tag{39}$$

where $\theta_n(x)$ are normalized orthogonal polynomials. Representations of second-order probability density functions as a single sum of orthogonal polynomials were studied by Barrett and Lampard [1], where they discussed the properties and application of such density functions. It has been shown [16] that if $p(x|x_0, t)$ satisfies a Fokker–Planck equation of the form given by (1), and if

$$\lim_{t\to\infty} p(x|x_0, t) = W(x),$$

then $p(x|x_0, t)$ has the representation given by (39) if and only if

(1) $B(x) = ax^2 + bx + c$, $A(x) = dx + e$,
(2) $B(x_1)W(x_1) = B(x_2)W(x_2) = 0$,
(3) $\int_{x_1}^{x_2} W(x)x^n\,dx < \infty$, $x = 0, 1, 2, \cdots$.

In effect, (2 B), (2 C) and (2 D) exhaust all such possibilities.

A property, satisfied by most of the processes discussed in §2, is that the normalized covariance function is an exponential function of t, i.e.,

$$E\left(\frac{X(t+\tau)-m}{\sigma}\right)\left(\frac{X(\tau)-m}{\sigma}\right) = e^{-\beta t}, \qquad \beta > 0,\ t \geqq 0, \tag{40}$$

where m and σ^2 are the mean and variance respectively. This property has long been known for Ornstein–Uhlenbeck process, and is closely connected with the question of representing $p(x|x_0, t)$ in terms of orthogonal polynomials. It can be shown that if the transitional probability density function $p(x|x_0, t)$ of a Markoff process $X(t)$ satisfies (1) and if $\lim_{t\to\infty} p(x|x_0, t) = W(x)$, then (40) is satisfied if and only if

(1) $B(x) = ax^2 + bx + c$, $A(x) = dx + e$,
(2) $B(x_1)W(x_1) = B(x_2)W(x_2) = 0$,
(3) $\int_{x_1}^{x_2} W(x)x^n\,dx < \infty$, $n = 0, 1, 2$.

In addition to (2 B), (2 C) and (2 D), the process of (2 E) for $\alpha > 1$, and the process of (2 F) for $\alpha > 1$, also satisfy (40). In each of these two cases the representation of $p(x|x_0, t)$ is in part in terms of a sum of orthogonal polynomials.

4. **The distribution of functionals of Markoff processes.** We consider a functional of the form

$$Y(t) = \int_0^t f[X(\tau)]\,d\tau, \tag{41}$$

where $X(\tau)$ is a stationary Markoff process of the class being considered in this paper. Darling and Siegert [5] have defined a function

$$r(x|x_0, t, \eta) \equiv E\{e^{-\eta Y(t)}|X(0) = x_0, X(t) = x\}\cdot p(x|x_0, t),$$

which in our case is the principal solution of

$$\frac{\partial^2}{\partial x^2}[B(x)r] - \frac{\partial}{\partial x}[A(x)r] - \eta f(x)r = \frac{\partial r}{\partial t}. \tag{42}$$

The corresponding Sturm–Liouville equation becomes

$$\frac{d}{dx}\left[B(x)W(x)\frac{d\varphi(x)}{dx}\right] + W(x)[\lambda - \eta f(x)]\varphi(x) = 0. \tag{43}$$

A comparison of (43) and (12) shows that the only difference is the presence of a term $\eta f(x)$ in (43). If the addition of the term $\eta f(x)$ does not significantly complicate the Sturm–Liouville equation, then it is to be expected that the distribution of the corresponding functional is not too difficult to find.

With a standard transformation, (43) can be rewritten in the form

$$\frac{d^2\psi}{dz^2} + [\lambda - V(z) - \eta f]\psi = 0, \tag{44}$$

with

$$z = \int \frac{1}{\sqrt{B(x)}}\,dx,$$

$$q = [B(x)W^2(x)]^{1/4},$$

$$\psi = q\varphi,$$

$$V(z) = q^{-1}(z)\frac{d^2}{dz^2}q(z)$$

and

$$f = f(x(z)).$$

If the kernel f of the functional is such that

$$V(z) + \eta f(x(z)) = a^2 V(az + b) + c, \tag{45}$$

where a, b, c are constants (i.e., independent of z), then (44) can be rewritten as

$$\frac{d^2\psi}{d\zeta^2} + \left[\left(\frac{\lambda - c}{a^2}\right) - V(\zeta)\right]\psi = 0, \tag{46}$$

with

$$\zeta = az + b.$$

In that case the solutions of (43) follow immediately from the solution of (12).

As an example, consider the functional

$$Y(t) = \int_0^t X(\tau)\,d\tau, \tag{47}$$

where $X(\tau)$ is the process of (2 C) with first-order probability density function

$$W(x) = \frac{1}{\Gamma(\alpha + 1)} x^\alpha e^{-x}.$$

For this case (43) becomes

$$\frac{d}{dx}\left[x^{\alpha+1} e^{-x} \frac{d\varphi}{dx}\right] + x^{\alpha} e^{-x}[\lambda - \eta x]\varphi(x) = 0. \tag{48}$$

It can be verified that $f(x) = x$ for this case satisfies (45) with

$$\begin{aligned} a &= (1 + 4\eta)^{1/4}, \\ b &= c = 0. \end{aligned} \tag{49}$$

The solutions of (48) are

$$\lambda_n = a^2(n + \alpha + 1) - (\alpha + 1) \tag{50}$$

and

$$\varphi_n(x) = e^{(1-a^2)x/2} L_n^{\alpha}(a^2 x), \tag{51}$$

with

$$L_n^{\alpha}(x) = \frac{1}{n!} x^{-\alpha} e^{x} \frac{d^n}{dx^n}(x^{n+\alpha} e^{-x}).$$

The function $r(x|x_0, t; \eta)$ can be written as

$$r(x|x_0, t; \eta) = a^2(a^2x)^{\alpha} e^{-x} \sum_{n=0}^{\infty} \frac{n!}{\Gamma(n + \alpha + 1)} \varphi_n(x_0)\varphi_n(x). \tag{52}$$

From (52) we find

$$\begin{aligned} F(\eta, t) = E(e^{-\eta Y(t)}) &= \int_0^{\infty}\int_0^{\infty} W(x_0) r(x|x_0, t, \eta)\, dx_0\, dx \\ &= \left[\frac{4a^2 e^{-(a^2-1)t/2}}{(a^2+1)^2 - (a^2-1)e^{-a^2 t}}\right]^{\alpha+1}. \end{aligned} \tag{53}$$

Finally, the probability density function for $Y(t)$ can be found from (53) using the inversion integral of Laplace transform. Specifically, let $W(y, t)$ be the density function. Then

$$W(y, t) = \frac{1}{2\pi i}\int_{c-i\infty}^{c+i\infty} e^{\eta y} F(\eta, t)\, d\eta, \qquad c > -\tfrac{1}{4}. \tag{54}$$

REFERENCES

1. J. F. Barrett and D. G. Lampard, *An expansion for some second-order probability distributions and its application to noise problems*, IRE Trans. **IT–1** (1955), 10–15.

2. S. Bochner, *Sturm-Liouville and heat equations whose eigenfunctions and ultraspherical polynomials or associated Bessel functions*, Proc. Conf. Differential Equations, pp. 23–48, University of Maryland Book Store, College Park, Md., 1955.

3. S. Chandrasekhar, *Stochastic problems in physics and astronomy*, Rev. Modern Phys. **15** (1943), 1–89.

4. H. Cramer, *Mathematical methods of statistics*, Princeton Univ. Press, Princeton, N. J., 1946.

5. D. A. Darling and A. J. F. Siegert, *A systematic approach to a class of problems in the theory of noise and other random phenomena.* I, IRE Trans. **IT–3** (1957), 32–37.

6. J. L. Doob, *The Brownian movement and stochastic equations*, Ann. of Math. (2) **43** (1942), 351–369.

7. A. Erdelyi, ed., *Higher transcendental functions*, Vol. I, McGraw-Hill, New York, 1953.

8. W. Feller, *Diffusion processes in genetics*, Proc. 2nd Berkeley Sympos. Math. Statist. and Prob., pp. 227–246, Univ. of California Press, Berkeley, Calif., 1951.

9. ———, *Diffusion processes in one dimension*, Trans. Amer. Math. Soc. **77** (1954), 1–31.

10. B. V. Gnedenko and A. N. Kolmogorov, *Limit distributions for sums of independent random variables*, Addison-Wesley, Cambridge, 1954.

11. M. Kac, *Random walk and the theory of Brownian motion*, Amer. Math. Monthly **54** (1947), 369–391.

12. S. Karlin and J. McGregor, *Classical diffusion processes and total positivity*, J. Math. Anal. Appl. **1** (1960), 163–183.

13. A. N. Kolmogorov, *Über die analytische Methoden in der Wahrscheinlichkeitsrechrung*, Math. Ann. **104** (1931), 415–458.

14. G. Szegö, *Orthogonal polynomials*, Amer. Math. Soc. Colloq. Publ. Vol. 23, Amer. Math. Soc., Providence, R. I., 1939.

15. M. C. Wang and G. E. Uhlenbeck, *On the theory of the Brownian motion*. II, Rev. Modern Phys. **17** (1945), 323–342.

16. E. Wong and J. B. Thomas, *On polynomial expansions of second-order distributions*, J. Soc. Indust. Appl. Math. **10** (1962), 507–516.

UNIVERSITY OF CALIFORNIA,
BERKELEY, CALIFORNIA

PROBABILITY BOUNDS VIA DYNAMIC PROGRAMMING

BY

DAVID BLACKWELL

1. **A general theorem.** Let M be any set of probability measures m on the Borel sets of n-space. A sequence $X_1, X_2, \cdots$ of n-dimensional random variables will be called an M-sequence if (a) the distribution of X_1 is an element of M and (b) the conditional distribution of X_{k+1} given $X_1, \cdots, X_k$ is an element of M, with probability 1, for $k = 1, 2, \cdots$. Thus if M consists of the single element m, the M-sequences are independent with common distribution m; if M consists of all distributions with mean 0, the M-sequences are those for which $\{S_k = X_1 + \cdots + X_k\}$ is a martingale with mean 0.

Fix a subset R of n-space, and a point $r \in R$. For any M-sequence $\{X_k\}$, define $S_k = r + X_1 + \cdots + X_k$, $k = 0, 1, \cdots$. Define N as the smallest k for which $S_k \notin R$. The value $+\infty$ is not excluded. As $\{X_k\}$ varies over all M-sequences, the least upper bound of EN is, for fixed M, R, a function of r. We denote this function by v and define $v = 0$ for $r \notin R$. Our problem is to find interesting upper bounds for v.

This problem can be considered as a dynamic programming or gambling problem, in which we start in state r, and choose any displacement distribution m from M. An x_1 is selected from m and we move to $r + x_1$. We continue as long as the state remains in R, receiving one unit of income per play. Play stops as soon as a state not in R is reached. $v(s)$ is then our maximum expected income, starting from s. Indeed the main purpose of this paper is to show how a dynamic programming idea can be applied. The idea is described in the following theorem.

THEOREM. *For any non-negative f on n-space and any probability measure m on n-space, let $T_m f$ be the function defined by*

$$T_m f(r) = \begin{cases} 1 + \int f(r + x)\, dm(x) & \text{for } r \in R, \\ 0 & \text{for } r \notin R, \end{cases}$$

and define $Tf = \sup_{m \in M} T_m f$. If $Tf \leqq f$, then $v \leqq f$.

A very general form of this theorem appears in Dubins and Savage [**4**]. See also Blackwell [**7**].

We sketch the proof. For $f_0 \equiv 0$, we show by induction on j that, for every

D-sequence $\{X_k\}$, $E \min(N, j) \leqq T^j f_0$. This is true for $j = 1$. If true for j_0, we have, starting from $r \in R$

$$E \min(N, j_0 + 1 | X_1) = 1 + E \min(N', j_0), \tag{1}$$

where N' is the N for the M-sequence $X_2, X_3, \cdots$ conditioned by X_1 and with starting point $S_1 = r + X_1$. The inductive hypothesis implies

$$E \min(N', j_0) \leqq T^{j_0} f_0(r + x_1).$$

Taking expectation in (1) yields $E \min(N, j_0 + 1) \leqq 1 + E T^{j_0} f_0(r + X_1) \leqq T^{j_0+1} f_0(r)$.

The operator T is monotone in the sense that $f_1 \leqq f_2$ for all r implies $Tf_1 \leqq Tf_2$ for all r. Thus $f \geqq 0 \equiv f_0$ implies $T^j f \geqq T^j f_0$ for all j. If in addition $Tf \leqq f$, then $T^j f \leqq f$ for all j, so that $f \geqq E \min(N, j)$ for all j. Letting $j \to \infty$ yields $f \geqq v$.

We remark that, as will be familiar to workers in dynamic programming, $T^j f(r)$ is the maximal income, starting from r when play is truncated after j moves even if the state r' is still inside R, and you receive a final payment $f(r')$.

2. **Applications.** (A) $n = 1$, $R = \{|r| \leqq H\}$, M consists of all distributions concentrated on $|x| \leqq 1$, with mean 0 and second moment $s > 0$. Take any quadratic function $q = A - Cr^2$ which is $\geqq 0$ for $|r| \leqq H + 1$, and take $f = \max(q, 0)$. We have, for $|r| \leqq H$, $Tf(r) = 1 + A - C(r^2 + s)$, so that $Tf \leqq f$ if $C \geqq 1/s$. For $C_0 = 1/s$, the smallest A which makes $q \geqq 0$ for $|r| \leqq H + 1$ is $C_0(H + 1)^2$. We conclude that

$$v(r) \leqq \frac{(H + 1)^2 - r^2}{s}. \tag{2}$$

This bound is in fact sharp for integral r, H, and is achieved for the M-sequence with X_k independent, $\Pr\{X_k = 1\} = \Pr\{X_k = -1\} = s/2$, $\Pr\{X_k = 0\} = 1 - s$. For independent identically distributed M-sequences, the bound (2) is known as a consequence of Wald's identity (see Doob [**2**] or Wolfowitz [**6**]). The arguments given to obtain this consequence actually use only the present hypotheses so could also be used to obtain (2).

(B) $n = 1$, $R = \{r \leqq 0\}$, M consists of all distributions concentrated on $|x| \leqq 1$ with mean $m > 0$. Take any linear function $q = A - Bt$ with $q \geqq 0$ for $t \leqq 1$, and take $f = \max(q, 0)$. For $r \leqq 0$,

$$Tf(r) = 1 + A - Br - Bm = f(r) + 1 - Bm,$$

so that $Tf \leqq f$ if $B \geqq 1/m$. For $B = 1/m$, the smallest A which makes $q \geqq 0$ for $t \leqq 1$ is $A = 1/m$. We conclude

$$v(r) \leqq \frac{1 - r}{m}. \tag{3}$$

Again the bound is exact for integral r, and is attained, for example, for $\{X_k\}$ independent with $\Pr\{X_k = \pm 1\} = (1 \pm m)/2$. The bound (3) can be obtained directly from the fact that $\{S_k - mk\}$, stopped when $S_k > 0$, is a martingale bounded above (by 1).

(C) $n = 1$, $R = \{|r| \leqq H\}$, M consists of all distributions concentrated on $|x| \leqq 1$ with non-negative mean and second moment at least $s_0 > 0$. For a quadratic function $q = A - Bt - Ct^2$ with $q \geqq 0$ for $|t| \leqq H + 1$, and $f = \max(q, 0)$, the condition $Tf \leqq f$ becomes $1 - Bm - 2Ctm - Cs \leqq 0$ for $m \geqq 0$, $s \geqq s_0$, $|t| \leqq H$. For $m = 0$, $s = s_0$ we obtain $C \geqq 1/s_0$. For $t = -H$, $s = s_0$, $C = 1/s_0$, $M = 1$, we obtain $B \geqq 2H/s_0$. The values $C_0 = 1/s_0$, $B_0 = 2H/s_0$ satisfy the conditions, and the smallest A which makes $q \geqq 0$ for $|t| \leqq H + 1$ is $B_0(H + 1) + C_0(H + 1)^2$. We conclude

$$v(r) \leqq \frac{(H + 1)^2 - r^2 + 2H(H + 1 - r)}{s_0}. \tag{4}$$

For $r = 0$, H integral, the bound $(H + 1)(3H + 1)/s_0$ is exact, and is achieved by the random walk with displacements $-1, 0, 1$ with probabilities $s_0/2$, $1 - s_0$, $s_0/2$ except when position $-H$ is reached, when the displacements are 0, 1 with probabilities $1 - s_0$, s_0. The expected income is clearly equivalent to that in the random walk with boundaries $-2H$, $2H$ with starting point H and displacements $-1, 0, 1$ with probabilities $s_0/2$, $1 - s_0$, $s_0/2$. As noted in (A), the exact expected income is $((2H + 1)^2 - H^2)/s_0 = (H + 1)(3H + 1)/s_0$.

(D) For any positive integer n, let M be the set of distributions concentrated within the unit sphere $|x| \leqq 1$, with mean 0 and covariance matrix $\sigma^2 I$, σ^2 fixed and positive. Let $R = \{|r| \leqq \rho\}$. Take $q(t) = A((\rho + 1)^2 - \sum_1^n t_i^2)$, $f = \max(q, 0)$. $Tf \leqq f$ reduces to $A \geqq 1/n\sigma^2$. We conclude

$$v(r) \leqq \frac{(\rho + 1)^2 - |r|^2}{n\sigma^2}. \tag{5}$$

I do not know how sharp (5) is.

(E) Even in problems where v is infinite, the method can be used to obtain bounds on the distribution of N, by introducing a discount factor $\beta < 1$. If the state is still inside R at time k, we receive an income increment β^k, so that the total expected reward is bounded above by $1/(1 - \beta)$. The operator T becomes

$$Tf(r) = \begin{cases} 1 + \beta \sup_{m \in M} \int f(r + x)\, dm(x) & \text{for } r \in R, \\ 0 & \text{for } r \notin R, \end{cases}$$

and the theorem and proof are as before, with

$$v = \sup \sum_{k=0}^{\infty} \beta^k \Pr N \geqq k,$$

where the sup is over all M-sequences $\{X_k\}$. We illustrate with $n = 1$, M consisting of the single distribution on ± 1 with equal probabilities, and $R = \{r \geqq 0\}$.

It is known that N is finite with probability 1, but EN is infinite. For any $f \geqq 0$ on $t \geqq -1$, $Tf \leqq f$ becomes

$$2 + \beta[f(t+1) + f(t-1)] \leqq 2f(t) \quad \text{for } t \geqq 0.$$

For

$$f(t) = A(1 - e^{-C(t+1)}), \quad \text{where } A > 0,\ C > 0,$$

the required inequality becomes

$$2 + 2\beta A(1 - (\cosh C)\, e^{-C(t+1)}) \leqq 2A(1 - e^{-C(t+1)}).$$

We get equality for $A = 1/(1-\beta)$, $\cosh C = 1/\beta$, and conclude:

$$v(r) \leqq \frac{1 - e^{-C(r+1)}}{1-\beta}, \quad \text{where } C > 0,\ \cosh C = 1/\beta. \tag{6}$$

For integral r, the bound is exact. For β near 1, $C \cong (1-\beta)^{1/2}$, so that, for integral r, as noted in [5],

$$v(r) \cong \frac{r+1}{(1-\beta)^{1/2}}.$$

The same approach can be used to bound the probability that $\{S_n\}$ ever escapes from R [1]. The operator T becomes

$$TF(r) = \begin{cases} \sup_{m \in M} EF(r+x) & \text{for } r \in R, \\ 1 & \text{for } r \notin R. \end{cases}$$

For another application of the same approach, see Dubins [3].

References

1. D. Blackwell, *On optimal systems*, Ann. Math. Statist. **25** (1954), 394–397.
2. J. L. Doob, *Stochastic processes*, pp. 350–352, Wiley, New York, 1953.
3. L. E. Dubins, *Rises and upcrossings of non-negative martingales*, Illinois J. Math. **6** (1962), 226–241.
4. L. E. Dubins and L. J. Savage, *How to gamble if you must* (to appear).
5. W. Feller, *An introduction to probability theory and its applications*, 2nd ed., p. 87, Wiley, New York, 1957.
6. J. Wolfowitz, *The efficiency of sequential estimates and Wald's equation for sequential processes*, Ann. Math. Statist. **18** (1947), 215–230.
7. D. Blackwell, *On the functional equation of dynamic programming*, J. Math. Anal. Appl. **2** (1961), 273–276.

University of California,
Berkeley, California

MARKOVIAN SEQUENTIAL DECISION PROCESSES[1]

BY

CYRUS DERMAN

1. **Introduction.** We are concerned with a dynamic stochastic system which is observed at times $t = 0, 1, \cdots$, and classified into one of a finite number of states $0, 1, \cdots, L$. After each observation the system is "controlled" by making one of a finite number of decisions $1, \cdots, K$. Let $\{Y_t\}$, $t = 0, 1, \cdots$, denote the sequence of observations and $\{\Delta_t\}$, $t = 0, 1, \cdots$, the sequence of decisions. We assume that

$$P(Y_{t+1} = j \mid Y_0, \Delta_0, \cdots, Y_t = i, \Delta_t = k) = q_{ij}(k),$$

for $t = 0, 1, \cdots$; $i, j = 0, \cdots, L$; $k = 1, \cdots, K$, where the $q_{ij}(k)$'s are given non-negative numbers such that

$$\sum_{j=0}^{L} q_{ij}(k) = 1, \qquad i = 0, \cdots, L; \ k = 1, \cdots, K.$$

We shall say a rule R is a collection of functions

$$D_k(Y_0, \Delta_0, \cdots, \Delta_{t-1}, Y_t), \qquad t = 0, 1, \cdots; \ k = 1, \cdots, K$$

of the observations and decisions such that

$$D_k(\cdot) \geqq 0, \qquad k = 1, \cdots, K$$

and

$$\sum_{k=1}^{K} D_k(\cdot) = 1.$$

A rule R is used to control the system by setting

$$P(\Delta_k = k \mid Y_0, \Delta_0, \cdots, \Delta_{t-1}, Y_t) = D_k(Y_0, \Delta_0, \cdots, Y_t), \qquad k = 1, \cdots, K$$

for every possible set of observations and decisions $Y_0, \Delta_0, \cdots, Y_t$, $t = 0, 1, \cdots$. Once the initial probabilities $P(Y_0 = i)$, $i = 0, \cdots, L$ are given and a rule R is designated we may speak of the sequence $\{Y_t\}$, $t = 0, 1, \cdots$ as being a stochastic process with state space $0, 1, \cdots, L$.

Processes of this sort were first discussed by Bellman [**1**; **2**]. The monograph by Howard [**17**] and papers by d'Epenoux [**13**], Manne [**19**], and Dantzig and Wolfe [**6**], providing computational methods, have stimulated recent activity in this area [**4**; **8**; **9**; **10**; **15**; **16**; **18**; **20**].

---

[1] Work sponsored in part by the Office of Naval Research under contract NR–042–099.

The problems of interest center around determining optimal rules. In a number of problems the criterion to be optimized is expressible in terms of $\Phi_T(i)$, as $T \to \infty$, where $\Phi_T(i)$ is the vector of components

$$\phi_{Tjk}(i) = \frac{1}{T+1} \sum_{t=0}^{T} P(Y_t = j,\ \Delta_t = k \mid Y_0 = i), \qquad j = 0, \cdots, L;\ k = 1, \cdots, K,$$

for a given i.

We give several examples.

EXAMPLE 1. Suppose $Y_0 = i$ with probability 1. Let

$$Z_t = w_{jk}, \quad \text{if } Y_t = j,\ \Delta_t = k,$$

for $j = 0, \cdots, L$; $k = 1, \cdots, K$; $t = 0, 1, \cdots$. We can think of Z_t as being the cost or conditional expected cost ascribed to the system at time t given it is in state j and decision k is made. The expected average cost up to time T is

$$Q_T(i) = E \frac{1}{T+1} \sum_{t=0}^{T} Z_t = \sum_{j=0}^{L} \sum_{k=1}^{K} w_{jk} \phi_{Tjk}(i).$$

A problem of interest is to find that rule R which minimizes $\limsup_{T\to\infty} Q_T(i)$.

EXAMPLE 2. Let $Q_T(i)$ be as in Example 1. A second problem is to find the rule R which maximizes $\limsup_{T\to\infty} Q_T(i)$. This problem may arise when the $q_{ij}(k)$'s represent the parameters associated with a given system design or statistical procedure; e.g. a continuous sampling plan, and the rule R denotes the behavior of "Nature." In these cases it is of interest to evaluate a procedure in terms of the maximum possible expected cost it may incur.

EXAMPLE 3. Suppose again $Y_0 = i$ with probability 1. Let

$$\tau = \min \{t \mid Y_t = i,\ t \geqq 1\}.$$

Then $\sum_{t=1}^{\tau} Z_t$, where Z_t is defined as in Example 1, is the cost associated with $\{Y_t\}$, $1 \leqq t \leqq \tau$. Consider the problem of choosing R, of all possible R such that $E\tau < \infty$, which minimizes

$$S(i) = E \sum_{t=1}^{\tau} Z_t.$$

Eaton and Zadeh **[12]**, except for slight differences, have referred to such problems as "pursuit" problems; the idea being that at time τ (a random variable) a certain target state is reached for the first time. The aim is to reach the target state in the most efficient manner.

In order to express $S(i)$ as a function of $\Phi_T(i)$, as $T \to \infty$ we first notice that

$$S(i) = \sum_{k=1}^{K} \sum_{j=0}^{L} w_{jk} \eta_{jk}$$

where η_{jk}, $j = 0, \cdots, L$; $k = 1, \cdots, K$, is the expected number of t, $1 \leqq t \leqq \tau$, such that $Y_t = j$, $\Delta_t = k$. For the purposes of this problem it is clear that we may restrict ourselves to rules which repeat themselves whenever the system returns to state i; i.e., the process $\{Y_t\}$, $t = 0, 1, \cdots$, under such a rule is a recurrent event process (see Feller [14]) with the recurrent event identified as "entrance of the system into state i." When this is the case, standard recurrent event type arguments lead to

$$\lim_{T\to\infty} \phi_{Tjk}(i) = \frac{\eta_{jk}}{E\tau}, \qquad j = 0, \cdots, L;\ k = 1, \cdots, K.$$

However, by definition of τ,

$$\sum_{k=1}^{K} \eta_{ik} = 1,$$

from which it follows that

$$E\tau = \frac{1}{\lim\limits_{T\to\infty} \sum\limits_{k=1}^{K} \phi_{Tik}(i)}.$$

On substitution we get

$$S(i) = \frac{\sum\limits_{j=0}^{L} \sum\limits_{k=1}^{K} w_{jk} \lim\limits_{T\to\infty} \phi_{Tjk}(i)}{\lim\limits_{T\to\infty} \sum\limits^{K} \phi_{Tik}(i)}.$$

EXAMPLE 4. Suppose $Y_0 = i$ with probability 1 and τ is defined as in Example 3. Further, suppose there are states $j_1, \cdots, j_r$ different from i such that for every rule R

$$P(Y_{t+1} = i \mid Y_t = j_v) = 1, \qquad v = 1, \cdots, r;\ t = 1, 2, \cdots.$$

That is, whenever the system enters state j_v, $v = 1, \cdots, r$, its next state is i. Let

$$\alpha_{j_v} = P(Y_t \neq j_v,\ 1 \leqq t \leqq \tau \mid Y_0 = i), \qquad v = 1, \cdots, r.$$

α_{j_v} is the probability of avoiding state j_v within the first recurrence cycle of i. Consider the problem of finding that R, of all considered in Example 3, such that $E\tau$ is maximized subject to the constraints

$$\alpha_{j_v} \geqq c_v, \qquad v = 1, \cdots, r,$$

where c_v, $v = 1, \cdots, r$ are given constants. This problem may arise in replacement type situations where τ is the length of time until a certain component, say, is replaced and states $j_1, \cdots, j_r$ stand for events of possible catastrophic proportions.

$E\tau$ is as in Example 3. Since $Y_t = j_v$ for at most one t, $1 \leqq t \leqq \tau$, it follows that

$$1 - \alpha_{j_v} = \sum_{k=1}^{K} \eta_{j_v k}, \qquad v = 1, \cdots, r.$$

Hence, using the relationships from Example 3, the problem is of the form: minimize

$$\lim_{T\to\infty} \sum_{k=1}^{K} \phi_{Tik}(i),$$

subject to the constraints

$$\lim_{T\to\infty} \sum_{k=1}^{K} \phi_{Tj_v k}(i) \leqq (1 - c_v) \lim_{T\to\infty} \sum_{k=1}^{K} \phi_{Tik}(i), \qquad v = 1, \cdots, r.$$

Before proceeding it is helpful to classify the possible rules. Let C denote the class of *all* rules and let C' denote the sub-class of rules for which

$$D_k(Y_0, \Delta_0, \cdots, Y_t = i) = D_{ik}, \qquad i = 0, \cdots, L;\ k = 1, \cdots, K;\ t = 0, 1, \cdots,$$

where $\{D_{ik}\}$, $i = 0, \cdots, L$; $k = 1, \cdots, K$, are non-negative numbers such that

$$\sum_{k=1}^{K} D_{ik} = 1, \qquad i = 0, \cdots, L.$$

Let C'' be the sub-class of C' for which $D_{ik} = 0$ or 1.

When $R \in C'$, the stochastic process $\{Y_t\}$, $t = 0, 1, \cdots$, is a Markov chain with stationary transition probabilities

$$P_{ij} = \sum_{k=1}^{K} D_{ik} q_{ij}(k), \qquad i, j = 0, \cdots, L.$$

In seeking optimal rules in problems such as described above, a question presents itself. Namely, is it necessary to consider rules outside the class C' (or sometimes C'')? If not, the problems are more manageable, for then we can bring to bear techniques of dynamic programming, linear programming, and Markov chain theory (see [**6**; **7**; **8**; **10**; **12**; **13**; **15**; **16**; **17**; **18**; **19**; **20**]).

This question has been treated for various separate cases (see [**13**; **4**; **8**; **10**; **11**]). The results presented in the next section are more generally relevant to this question for problems of the above type.

2. **Limit theorems.** Let $H_R(i)$ denote the set of limit points of $\{\Phi_T^R(i)\}$ as $T \to \infty$. If $R \in C'$ it is well known from Markov chain theory (see [**5**, p. 32]) that $\lim_{T\to\infty} \Phi_T^R(i)$ exists. However, in general, $\{\Phi_T^R(i)\}$, $T = 0, 1, \cdots$ may have more than one limit point. Let

$$H(i) = \bigcup_{R\in C} H_R(i),$$

$$H'(i) = \bigcup_{R\in C'} H_R(i),$$

and

$$H''(i) = \bigcup_{R\in C''} H_R(i).$$

In keeping with the terminology employed by Feller [**14**], we call a Markov chain with stationary transition probabilities *irreducible* if for every ordered pair of states (i, j) there exists a t such that

$$P(Y_t = j \mid Y_0 = i) > 0.$$

The following theorem in a slightly different form was proved in [**9**].

THEOREM 1. *If the Markov chain* $\{Y_t\}$ *is irreducible for every* $R \in C''$, *then* $H(i) = H'(i)$.

The proof makes use of the fact that under the irreducibility assumption $H'(i)$ is a closed convex set. This, and the result obtained in [8] that $\lim \sup_{T\to\infty} Q_t(i)$ of Example 1 is minimized over all $R \in C$ by a rule $R^* \in C''$, combine to imply Theorem 1.

Theorem 1 applies to Examples 3 and 4 for most cases of interest. However, the irreducibility assumption is too strong to be of general use in Example 2. The following theorem applies.

THEOREM 2. *Let* $\bar{H}'(i)$ *and* $\bar{H}''(i)$ *denote the convex hulls of* $H'(i)$ *and* $H''(i)$, *then we always have*

$$\bar{H}''(i) = \bar{H}'(i) \supset H(i).$$

The proof of Theorem 2 is suggested, with certain modifications, by that of Theorem 1 and will appear elsewhere.

Since all the extreme points of $\bar{H}''(i)$ are points of $H''(i)$, this theorem asserts that every point of $H(i)$ can be achieved by an appropriate initial random selection of rules $R \in C''$. With respect to Example 2, since $\lim \sup_{T\to\infty} Q_T(i)$ is maximized at the extreme points of $\bar{H}''(i)$ it follows that there exists a rule $R^{**} \in C''$ such that $\lim \sup_{T\to\infty} Q_T(i)$ is maximized over all $R \in C$ at R^{**}.

Though we shall not go into its possible applications we can state a strong law counterpart to Theorem 2. Suppose $P(Y_0 = i) = 1$. Let

$$Z_{tjk} = \begin{cases} 1, & \text{if } Y_t = j,\ \Delta_t = k, \\ 0, & \text{otherwise,} \end{cases}$$

for $j = 0, \cdots, L$; $k = 1, \cdots, K$; $t = 0, 1, \cdots$. For any rule R let $\hat{\Phi}_T^R(i)$ be the vector of components

$$\hat{\phi}_{Tjk}(i) = \frac{1}{T+1} \sum_{t=0}^{T} Z_{tjk}, \qquad j = 0, \cdots, L;\ k = 1, \cdots, K.$$

Let $U^R(i)$ denote the set of limit points of $\{\hat{\Phi}_T^R(i)\}$, as $T \to \infty$; $U^R(i)$ is a random set dependent upon the sample sequence $\{Y_t, \Delta_t\}$, $t = 0, 1, \cdots$ we can state

THEOREM 3. *For each* $R \in C$,

$$P(U^R(i) \subset \bar{H}'') = 1,$$

where $\bar{H}''$ *is the convex hull of* $\bigcup_{j=0}^{L} H''(j)$.

The proof of this theorem will also appear elsewhere.

3. **A counter-example.** A simple counter-example given in [9] shows that the irreducibility condition of Theorem 1 is relevant. Suppose $L = 1$, $K = 2$ with

$$q_{00}(1) = 0, \quad q_{01}(1) = 1, \quad q_{10}(1) = 0, \quad q_{11}(1) = 1, \quad q_{00}(2) = 1,$$
$$q_{00}(2) = 1, \quad q_{01}(2) = 0, \quad q_{10}(2) = 0, \quad q_{11}(2) = 1.$$

In words, decision 2 permits the system to remain in state 0; decision 1 carries the system to state 1 from state 0; once the system enters state 1 it remains there.

From elementary Markov chain considerations, if $R \in C'$ we have that

$$\lim_{T\to\infty} \frac{1}{T} \sum_{t=1}^{T} P(Y_t = 0 \mid Y_0 = 0) = \lim_{T\to\infty} \sum_{k=1}^{2} \phi_{T0k}(0) = 1 \text{ or } 0;$$

no intermediate value is possible. However, consider R as follows:

$$D_2(Y_0, \Delta_0, \cdots, Y_t = 0) = e^{-(1/2)^t}, \qquad t = 0, 1, \cdots,$$

$$D_2(Y_0, \Delta_0, \cdots, Y_t = 1) = 1, \qquad t = 0, 1, \cdots;$$

i.e., make decision 1 with probability $1 - e^{-(1/2)^t}$ if the system is in state 0 at time t and make decision 2 whenever the system is in state 1. Under this rule it follows that

$$P(Y_t = 0 \mid Y_0 = 0) = \exp\left[-\sum_{v=0}^{t-1} (1/2)^v\right], \qquad t = 1, 2, \cdots,$$

from which we have

$$\lim_{T\to\infty} \sum_{k=1}^{2} \phi_{T0k}(0) = e^{-2},$$

a value unattainable with a rule $R \in C'$.

This counter-example makes use of the fact that there is no $R \in C'$ for which the Markov chain $\{Y_t\}$, $t = 0, 1, \cdots$, is irreducible. However, this is not crucial; it is possible to construct a similar example for which there are some rules $R \in C'$ which result in irreducible chains.

With this counter-example we can formulate an optimization problem where the optimal rule is not a member of C'. Namely, find that rule $R \in C$ such that

$$\limsup_{T\to\infty} \sum_{k=1}^{2} \phi_{T0k}(0)$$

is minimized subject to the constraint that

$$\limsup_{T\to\infty} \sum_{k=1}^{2} \phi_{T0k}(0) \geqq e^{-2}.$$

Clearly the above rule is optimal for this problem.

4. **Linear programming formulations.** We can easily show how the linear programming formulation of Manne **[19]** and Dantzig and Wolfe **[6]** applies to the four examples. We assume that the irreducibility condition of Theorem 1 holds. From Theorem 1 it then follows that we can restrict our attention to rules in C' for which we can assert that

$$\lim_{T\to\infty} \phi_{Tjk}(i) = \pi_j D_{jk}, \qquad j = 0, \cdots, L;\ k = 1, \cdots, K,$$

where $\{\pi_j\}$, $j = 0, \cdots, L$ are all positive and uniquely satisfy the system of inequalities and equalities

$$\pi_j \geqq 0 \qquad j = 0, \cdots, L,$$

(*)
$$\pi_j = \sum_{i=0}^{L} \pi_i p_{ij}, \qquad j = 0, \cdots, L,$$

$$\sum_{j=0}^{L} \pi_j = 1.$$

Letting $x_{jk} = \pi_j P_{jk}$, $j = 0, \cdots, L$; $k = 1, \cdots, K$, Example 1 (Example 2) becomes: minimize (maximize)

$$\sum_{j=0}^{L} \sum_{k=1}^{K} w_{jk} x_{jk}$$

subject to

$$x_{jk} \geqq 0, \qquad j = 0, \cdots, L; \; k = 1, \cdots, K,$$

(**)
$$\sum_{k=1}^{K} x_{jk} = \sum_{i=0}^{L} \sum_{k=1}^{K} x_{ik} q_{ij}(k), \qquad j = 0, \cdots, L,$$

$$\sum_{j=0}^{L} \sum_{k=1}^{K} x_{jk} = 1.$$

Example 4 becomes: minimize

$$\sum_{k=1}^{K} x_{ik}$$

subject to (**) and

$$\sum_{k=1}^{K} x_{i_v k} \leqq (1 - C_v) \sum_{k=1}^{K} x_{ik}, \qquad v = 1, \cdots, r.$$

Example 3 is of the form: minimize

$$\frac{\sum_{j=0}^{L} \sum_{k=1}^{K} w_{jk} x_{jk}}{\sum_{k=1}^{K} x_{ik}}$$

subject to (**). The transformation

$$Z_0 = \frac{1}{\sum_{k=1}^{K} x_{ik}}$$

and

$$Z_{jk} = \frac{x_{jk}}{\sum_{k=1}^{K} x_{ik}}, \qquad j = 0, \cdots, L; \; k = 1, \cdots, K,$$

yields the linear programming problem: minimize

$$\sum_{j=0}^{K} \sum_{k=1}^{K} w_{jk} Z_{jk}$$

subject to

$$Z_0 \geqq 0, \quad Z_{jk} \geqq 0, \quad j = 0, \cdots, L; \; k = 1, \cdots, K,$$

$$\sum_{k=1}^{K} Z_{jk} = \sum_{i=0}^{L} \sum_{k=1}^{K} Z_{ik} q_{ij}(k), \qquad j = 0, \cdots, L,$$

$$\sum_{j=0}^{L} \sum_{k=1}^{K} Z_{jk} = Z_0,$$

$$\sum_{k=1}^{K} Z_{ik} = 1.$$

The optimal rule is obtained from the linear programming solutions by setting for each j and k

$$D_{jk} = \frac{x_{jk}}{\sum_{k=1}^{K} x_{jk}} \qquad \text{(for Examples 1, 2, and 4)}$$

$$= \frac{Z_{jk}}{\sum_{k=1}^{K} Z_{jk}} \qquad \text{(for Example 3).}$$

Because of the irreducibility assumption the denominators in the above are always positive so that D_{jk} is well defined for every j and k (see **[8]** for the details).

References

1. R. Bellman, *Dynamic programming*, Princeton Univ. Press, Princeton, N. J., 1957.

2. ———, *A Markovian decision process*, J. Math. Mech. **6** (1957), 679–684.

3. D. Blackwell, *On the functional equation of dynamic programming*, J. Math. Anal. Appl. **2** (1961), 273–276.

4. ———, *Discrete dynamic programming*, Ann. Math. Statist. **33** (1962), 719–726.

5. Kai Lai Chung, *Markov chains with stationary transition probabilities*, Springer, Berlin, 1960.

6. G. Dantzig and P. Wolfe, *Linear programming in a Markov chain*, Operations Res. **10** (1962), 702–710.

7. C. Derman, *On optimal replacement rules when changes of state are Markovian*, Mathematical optimization techniques, pp. 201–210, Univ. of California Press, Berkeley, Calif., 1963.

8. ———, *On sequential decisions and Markov chains*, Management Sci. **9** (1962), 16–24.

9. ———, *Stable sequential control rules and Markov chains*, J. Math. Anal. Appl. **6** (1963), 257–265.

10. ———, *Optimal replacement and maintenance under Markovian deterioration with probability bounds on failure*, Management Sci. **9** (1963), 478–481.

11. C. Derman, M. V. Johns and G. J. Lieberman, *Continuous sampling procedures without control*, Ann. Math. Statist. (1959), 1175–1191.

12. J. H. Eaton and L. A. Zadeh, *Optimal pursuit strategies in discrete state probabilistic systems*, Trans. ASME Ser. D, J. Basic Engrg. **84** (1962), 23–29.

13. F. d'Epenoux, *Sur un problème de production et de stockage dans l'aléatoire*, Revue Française de Recherche Opérationnelle **4** (1960), 13–15.

14. W. Feller, *An introduction to probability theory and its applications*, 2nd ed., Wiley, New York, 1957.

15. M. Freimer, *On solving a Markovian decision problem by linear programming*, Institute for Defense Analyses, Cambridge, Mass., December 1961, rev. March 1962 (unpublished).

16. Guy de Ghellink, *Solving a multistage decision problem*, Carnegie Institute of Technology, Pittsburgh, Pa., November, 1959 (unpublished).

17. R. A. Howard, *Dynamic programming and Markov processes*, Wiley, New York, 1960.

18. M. Klein, *Inspection-maintenance replacement schedules under Markovian deterioration*, Management Sci. **9** (1962), 25–32.

19. A. S. Manne, *Linear programming and sequential decisions*, Management Sci. **6** (1960), 259–267.

20. H. M. Wagner, *On the optimality of pure strategies*, Management Sci. **6** (1960), 304–318.

COLUMBIA UNIVERSITY,
NEW YORK, NEW YORK

THE APPLICATION OF TRUNCATED HIERARCHY TECHNIQUES IN THE SOLUTION OF A STOCHASTIC LINEAR DIFFERENTIAL EQUATION

BY

J. M. RICHARDSON

1. **Introduction.** Stochastic linear operators, in particular linear equations with stochastic coefficients, constitute a topic receiving rapidly increasing attention in recent years.[1] In spite of the work devoted to the general properties of such equations (see, for example, [**1**]) and to the derivation of exact solutions to special problems [**2**], it is the writer's conviction that there exists a paucity of effective approximation techniques of general applicability. It is the purpose of the present paper to present several truncated hierarchy methods that appear to have promise.

In general, a hierarchy of the type considered here is obtained from a stochastic linear differential equation by multiplication by the stochastic coefficient an arbitrary number of times with each factor involving a different value of its argument. The hierarchy can either be averaged or left as is. The hierarchy can be truncated at a given member by a suitable approximation yielding a finite sequence of equations with a consistent number of dependent variables. Truncated hierarchy methods apparently are not well known in most areas of mathematical research although in theoretical physics they are used extensively: for example, the Tamm–Dancoff method [**3**], the Green's function method of Martin and Schwinger [**4**], and the BBGKY hierarchy with various truncation approximations [**5**].

In the succeeding paragraphs we develop three related truncated hierarchy methods and apply them to a very elementary stochastic linear differential equation chosen for purposes of illustration. However, it is clear that the same methods with rather trivial extensions can be applied to equations involving general stochastic linear operators. The present investigation has a somewhat exploratory character and is pursued at a rather low level of rigor.

2. **Formulation of the problem.** Specifically, it is desired to apply various truncated hierarchy methods to the simple stochastic differential equation for the t-dependent scalar variable x:

$$\frac{d^2x}{dt^2} + (1 + u)x = 0, \qquad x(0) = c_1, \qquad \frac{dx}{dt}(0) = c_2, \tag{2.1}$$

[1] See, for example, the contribution of A. T. Bharucha-Reid to this volume.

where c_1 and c_2 are nonrandom constants, and where $u = u(t)$ is a stationary random process with zero mean whose further properties will be defined later. It follows that $x = x(t)$ is also a random process but not necessarily with zero mean. The general problem is then the determination of the statistical properties of $x(t)$ given those of $u(t)$. For the sake of brevity this paper will treat only the problem of determining the mean value of $x(t)$. The considerably more complicated problem of calculating the further statistical properties of $x(t)$ (e.g., the second and higher order correlation functions) will be treated in a future communication.

3. **Perturbation theory.** It is instructive to review the application of the usual perturbation method to the solution of the above problem and the deficiencies of this method will provide a motivation for the search for more sophisticated procedures. By the term perturbation method we simply mean the following. Let u in (2.1) be replaced by εu, where ε is a parameter of smallness, with the result

$$\frac{d^2x}{dt^2} + (1 + \varepsilon u)x = 0 \tag{3.1}$$

where the initial conditions of (2.1) are retained. Now, the solution of (3.1) depends[2] also upon the parameter ε, namely $x = x(t, \varepsilon)$. The solution of the original problem defined by (2.1) is of course regained by setting $\varepsilon = 1$. The perturbation method then consists in expanding x in a power series in ε and then solving the sequence of equations obtained by substituting the series in (3.1) and setting to zero the resultant coefficient of each power of ε.

Before going further into the application of the perturbation method it is illustrative to consider the properties of the exact solution for the special case in which u is independent of t. The exact solution of (3.1) is then

$$x = \operatorname{Re}\{[c_1 - ic_2(1 + \varepsilon u)^{-1/2}] \exp [it(1 + \varepsilon u)^{1/2}]\}. \tag{3.2}$$

Writing x in the power series

$$x(t, \varepsilon) = \sum_{n=0} x_n(t)\varepsilon^n, \tag{3.3}$$

we find that the first two orders of $x_n(t)$ are given by

$$\begin{aligned} x_0(t) &= \operatorname{Re} [(c_1 - ic_2)\, e^{it}], \\ x_1(t) &= \tfrac{1}{2} \operatorname{Re} \{[c_1uit + ic_2u(1 - it)]\, e^{it}\}. \end{aligned} \tag{3.4}$$

It is to be noted that the second of (3.4) oscillates with increasing amplitude as $t \to \infty$ because of the linear dependence on t in the coefficient of e^{it}. This is the well-known secular term difficulty. The expression for $x_n(t)$ will involve a factor

[2] It is tacitly understood that x is also a function of c_1 and c_2, and a functional of $u(t)$.

of t^n multiplied into e^{it}. Thus we see that the asymptotic behavior of x_n (as $t \to \infty$) becomes progressively worse as n increases. Taking the mean values of x_n does not improve the asymptotic behavior (except for $n = 1$, in which case the mean of x_1 vanishes).

Consider now the case in which u is dependent upon t. The perturbation method gives the following sequence of coupled equations

$$\text{(3.5)} \qquad \begin{aligned} &\left(\frac{d^2}{dt^2} + 1\right) x_0 = 0, \\ &\left(\frac{d^2}{dt^2} + 1\right) x_n = -\, u x_{n-1}, \qquad n = 1, 2, \cdots, \end{aligned}$$

with the initial conditions

$$\text{(3.6)} \qquad \begin{aligned} &x_0(0) = c_1, \qquad \frac{dx_0}{dt}(0) = c_2, \\ &x_n(0) = 0, \qquad \frac{dx_n}{dt}(0) = 0, \qquad n = 1, 2, \cdots. \end{aligned}$$

The zeroth order solution is

$$\text{(3.7)} \qquad x_0 = A \sin (t + \theta),$$

where

$$\text{(3.8)} \qquad \begin{aligned} A^2 &= c_1^2 + c_2^2, \\ \tan \theta &= c_1/c_2. \end{aligned}$$

The higher order solutions are obtained by repeated application of the recursion relation

$$\text{(3.9)} \qquad x_n(t) = -\int_0^t dt' \sin (t - t') u(t') x_{n-1}(t')$$

with the results

$$\text{(3.10)} \qquad \begin{aligned} x_1 &= -\int_0^t dt' \sin (t - t') u(t') x_0(t'), \\ x_2 &= \int_0^t dt' \int_0^{t'} dt'' \sin (t - t') \sin (t' - t'') u(t') u(t'') x_0(t''), \text{ etc.} \end{aligned}$$

Denoting the mean value of a quantity by surrounding it by the angular brackets $\langle\ \rangle$, we obtain finally

$$\text{(3.11)} \qquad \begin{aligned} \langle x_0(t) \rangle &= x_0(t) = A \sin (t + \theta), \\ \langle x_1(t) \rangle &= 0, \\ \langle x_2(t) \rangle &= A \int_0^t dt' \int_0^{t'} dt'' \sin (t - t') \sin (t' - t'') C(t' - t'') \sin (t'' + \theta), \text{ etc.,} \end{aligned}$$

where $C(\tau)$ is the second order correlation function of $u(t)$ given by

$$C(\tau) = \langle u(t)u(t+\tau)\rangle. \tag{3.12}$$

Since $u(t)$ is assumed to be a stationary process, the r.h. side of (3.12) is independent of t. Integration of the third equation of (3.11) on t' gives

$$\begin{aligned}\langle x_2(t)\rangle &= \tfrac{1}{4}A\int_0^t ds\, C(s)\sin s \\ &\cdot [\cos(t-s+\theta) - \cos(t-s-\theta) - 2(t-s)\cos(t-s-\theta)].\end{aligned} \tag{3.13}$$

If $C(s)$ is constant, corresponding to the case previously considered in which $u(t)$ is independent of t, the asymptotic behavior of (3.13) is the same as previously stated, namely $\langle x_2(t)\rangle$ oscillates with an amplitude that increases as t^2. Consider now the case in which

$$\int_0^\infty ds\, C(s) \qquad \text{and} \qquad \int_0^\infty ds\, sC(s)$$

exist.[3] Now, it can be shown that $\langle x_2(t)\rangle$ oscillates with an amplitude that increases as t. In either case, the asymptotic behavior of $\langle x_2(t)\rangle$ is an unsatisfactory representation of the actual behavior of $\langle x(t)\rangle$ and one is obliged to consider other methods of solution.

4. **Hierarchies.** In developing various truncated hierarchy methods (at least as applied to (2.1)), we first define several types of hierarchies. The first type is obtained by multiplying (2.1) by factors of the form $u(t_1)$, $u(t_1)u(t_2), \cdots$, $u(t_1)\cdots u(t_n)$ with the result

$$\begin{aligned}\left[\frac{d^2}{dt^2} + 1 + u(t)\right] x(t) &= 0,\\ \left[\frac{d^2}{dt^2} + 1 + u(t)\right] u(t_1)x(t) &= 0,\\ \left[\frac{d^2}{dt^2} + 1 + u(t)\right] u(t_1)u(t_2)x(t) &= 0, \text{ etc.}\end{aligned} \tag{4.1}$$

It should be noted that the number of independent variables increases with the order of the member of the hierarchy. In subsequent work (4.1) will be called the *unaveraged hierarchy*.

[3] We exclude the uninteresting case of "white noise" in which $\int_0^\infty ds\, s^m C(s)$, $m = 1, 2, \cdots$, all vanish.

The second type of hierarchy is obtained from the first type (see (4.1)) by averaging. We obtain

$$
\begin{aligned}
&\left(\frac{d^2}{dt^2}+1\right)\langle x(t)\rangle + \langle u(t)x(t)\rangle = 0,\\
&\left(\frac{d^2}{dt^2}+1\right)\langle u(t_1)x(t)\rangle + \langle u(t_1)u(t)x(t)\rangle = 0, \qquad (4.2)\\
&\left(\frac{d^2}{dt^2}+1\right)\langle u(t_1)u(t_2)x(t)\rangle + \langle u(t_1)u(t_2)u(t)x(t)\rangle = 0, \text{ etc.}
\end{aligned}
$$

The above mean values are well defined by virtue of the fact that $x(t)$ is a functional of $u(t)$ and the statistical properties of $u(t)$ are assumed to be completely specified. It is not in general possible to compute the above mean values directly, but it is possible to infer them indirectly from the hierarchy. (4.2) are subject to the initial conditions

$$
\begin{aligned}
&\langle x(t)\rangle|_{t=0} = c_1, \qquad \frac{d}{dt}\langle x(t)\rangle|_{t=0} = c_2,\\
&\langle u(t_1)x(t)\rangle|_{t=0} = 0, \qquad \frac{d}{dt}\langle u(t_1)x(t)\rangle|_{t=0} = 0, \qquad (4.3)\\
&\langle u(t_1)u(t_2)x(t)\rangle|_{t=0} = \langle u(t_1)u(t_2)\rangle c_1,\\
&\frac{d}{dt}\langle u(t_1)u(t_2)x(t)\rangle = \langle u(t_1)u(t_2)\rangle c_2, \text{ etc.}
\end{aligned}
$$

Henceforth we will call (4.3) the *averaged hierarchy*.

The above equations ((4.2) and (4.3)) suffice for the determination of the mean value of $\langle x(t)\rangle$ and all correlation functions that are linear in $x(t)$. In order to determine $\langle x(t)x(t')\rangle$ and all correlation functions linear in $x(t)x(t')$ (a task beyond the scope of the present paper) it would be necessary to consider a somewhat more complex hierarchy obtained by multiplying every member of (4.1) by $x(t')$ and then averaging. The still higher order statistical properties of $x(t)$ would involve the derivation of still more complex hierarchies by analogous procedures.

5. **Truncation approximations.** Since it is not possible in the general case to solve the averaged hierarchy (4.2) with the associated initial conditions (4.3), it is necessary to terminate or truncate the sequence at some finite (preferably low) order. Clearly, it is possible in principle to achieve an *exact* truncation at the nth order member by solving for $\langle u(t_1)\cdots u(t_n)x(t)\rangle$ in the infinite hierarchy starting with the $(n+1)$st member, thus obtaining a closed set of n equations with a compatible set of dependent variables. This finite hierarchy is precisely equivalent to the original infinite one and unfortunately the difficulties in deriving and solving the former are also precisely equivalent to the difficulties in solving the latter.

Thus, in order to obtain truncated hierarchy methods with any degree of tractability it is necessary to consider various approximate truncation procedures. We discuss three such procedures in the following paragraphs.

(a) *Correlation discard.* In the nth member of the averaged hierarchy we neglect the correlation $\langle u(t_1)\cdots u(t_{n-1})u(t)\,\Delta x(t)\rangle$, where $\Delta x(t) = x(t) - \langle x(t)\rangle$. This is equivalent to the approximation

$$\langle u(t_1)\cdots u(t_{n-1})u(t)x(t)\rangle \cong \langle u(t_1)\cdots u(t_{n-1})u(t)\rangle\langle x(t)\rangle. \tag{5.1}$$

This approximation is equivalent to the "local independence" assumption of Bourret [6], employed in other examples.

(b) *Cumulant discard.* This approximation involves the neglect in the nth member of the averaged hierarchy of the $(n + 1)$st order cumulant (Thiele semi-invariant) corresponding to the variables $u(t), \cdots, u(t_{n-1}), u(t)$, and $\Delta x(t)$. More specifically, we neglect the quantity

$$i^{-n-1}\left[\frac{\partial^{n+1}}{\partial\lambda_1\cdots\partial\lambda_{n-1}\partial\lambda\,\partial\mu}\log\left\langle\exp\left[i\sum_{j=1}^{n-1}\lambda_j u(t_j) + i\lambda u(t) + i\mu\Delta x(t)\right]\right\rangle\right]^0, \tag{5.2}$$

where the super zero on the square bracket implies that the auxiliary variables $\lambda_1, \cdots, \lambda_{n-1}, \lambda$, and μ are all set equal to zero after the differentiations have been performed. The neglect of (5.2) implies that the unwanted quantity in the nth member of the hierarchy, namely $\langle u(t_1)\cdots u(t_{n-1})u(t)x(t)\rangle$, is approximated by a linear combination of the lower order quantities $\langle x(t)\rangle$, $\langle u(t_1)x(t)\rangle, \cdots,$ $\langle u(t_1)\cdots u(t_{n-1})x(t)\rangle$ with coefficients composed of combinations of various correlation functions of the $u(t_1)$, $u(t_2)$, etc. It is known that the second and third order cumulants are equal to the corresponding order correlation functions, namely

$$i^{-2}\left[\frac{\partial^2}{\partial\lambda\,\partial\mu}\log\langle\exp(i\lambda u(t) + i\mu\,\Delta x(t))\rangle\right]^0 = \langle u(t)\,\Delta x(t)\rangle, \tag{5.3a}$$

$$\begin{aligned} &i^{-3}\left[\frac{\partial^3}{\partial\lambda_1\,\partial\lambda\,\partial\mu}\log\langle\exp(i\lambda_1 u(t_1) + i\lambda u(t) + i\mu\,\Delta x(t))\rangle\right]^0 \\ &\qquad = \langle u(t_1)u(t)\,\Delta x(t)\rangle. \end{aligned} \tag{5.3b}$$

Thus, when applied to the first and second members of the hierarchy the correlation discard and cumulant discard procedures are identical. A difference occurs in the next order. Where one considers the fourth order cumulant

$$\begin{aligned} &i^{-4}\left[\frac{\partial^4}{\partial\lambda_1\,\partial\lambda_2\,\partial\lambda\,\partial\mu}\log\langle\exp(i\lambda_1 u(t_1) + i\lambda_2 u(t_2) + i\lambda u(t) + i\mu\,\Delta x(t))\rangle\right]^0 \\ &\quad = \langle u(t_1)u(t_2)u(t)\,\Delta x(t)\rangle - \langle u(t_1)u(t_2)\rangle\langle u(t)\,\Delta x(t)\rangle \\ &\quad\quad - \langle u(t_1)u(t)\rangle\langle u(t_2)\,\Delta x(t)\rangle - \langle u(t_2)u(t)\rangle\langle u(t_1)\,\Delta x(t)\rangle. \end{aligned} \tag{5.3c}$$

(c) *Least mean square error.* The third truncation procedure is substantially different from the previous two in several respects, a basic one being the fact that the procedure is applied to the *unaveraged* hierarchy. We will not define the procedure in a general way to the nth member of the hierarchy. Instead, we will illustrate the procedure for the first member of the unaveraged hierarchy (4.1), namely, the original stochastic differential equation

$$\left(\frac{d^2}{dt^2} + 1\right) x(t) + u(t)x(t) = 0. \tag{5.4}$$

To obtain an approximating linear *nonstochastic* differential equation with x as the dependent variable we proceed as follows. We approximate the troublesome term $u(t)x(t)$ as a linear functional of $x(t')$ in the range $(0, t)$, namely,

$$u(t)x(t) = w(t) + \int_0^t dt' K(t, t')x(t') + \eta(t), \tag{5.5}$$

where functions $w(t)$ and $K(t, t')$ are to be determined in such a way as to minimize the error $\eta(t)$ in the mean square sense. More specifically, we wish to minimize $\langle\eta^2(t)\rangle'$ with respect to the choice of the functions $w(t)$ and $K(t, t')$. The averaging process denoted by $\langle\,\rangle'$ need not be the same as that previously denoted by $\langle\,\rangle$. It should involve the correct statistical properties for $u(t)$ but it may differ otherwise in that $x(t)$ will not necessarily be the correct functional of $u(t)$ and the initial conditions may possibly be temporarily regarded as stochastic. It must be emphasized that the divergence of $\langle\,\rangle'$ from $\langle\,\rangle$ is not critical since if the error is small according to the criterion $\langle\eta^2 t\rangle'$ it is likely to be small also according to $\langle\eta^2(t)\rangle$ as long as $\langle\,\rangle'$ does not involve strong biases compared with $\langle\,\rangle$. In §8 a specific example of $\langle\,\rangle'$ will be treated. The procedure just outlined, when generalized to the nth order, involves the approximation of $u(t_1) \cdots u(t_{n-1})u(t)x(t)$ by a linear combination of linear functionals of $x(t)$, $u(t_1)x(t), \cdots, u(t_1) \cdots u(t_{n-1})x(t)$. These linear functionals involve integrations on t; t_1, t; t_1, t_2, t; etc.

6. **Application of correlation discard truncation.** In the present section we give a detailed discussion of the truncation procedure of paragraph (a) of the last section to the averaged hierarchy (4.2) with the initial conditions (4.3). The explicit expressions for the truncated hierarchies containing one, two, and three members will be discussed. Using the approximation (5.1) with $n = 1, 2$, and 3, respectively, we obtain the following results.

$n = 1$:

$$\left(\frac{d^2}{dt^2} + 1\right) \langle x(t)\rangle = 0, \tag{6.1a}$$

$$\langle x(0)\rangle = c_1, \qquad \frac{d}{dt}\langle x(t)\rangle\big|_{t=0} = c_2; \tag{6.1b}$$

$n = 2$:

(6.2a)
$$\left(\frac{d^2}{dt^2} + 1\right) \langle x(t)\rangle + \langle u(t)x(t)\rangle = 0,$$
$$\left(\frac{d^2}{dt^2} + 1\right) \langle u(t_1)x(t)\rangle + \langle u(t_1)u(t)\rangle\langle x(t)\rangle = 0,$$

(6.2b)
$$\langle x(0)\rangle = c_1, \quad \frac{d}{dt}\langle x(t)\rangle|_{t=0} = c_2,$$
$$\langle u(t_1)x(0)\rangle = 0, \quad \frac{d}{dt}\langle u(t_1)x(t)\rangle|_{t=0} = 0;$$

$n = 3$:

(6.3a)
$$\left(\frac{d^2}{dt^2} + 1\right) \langle x(t)\rangle + \langle u(t)x(t)\rangle = 0,$$
$$\left(\frac{d^2}{dt^2} + 1\right) \langle u(t_1)x(t)\rangle + \langle u(t_1)u(t)x(t)\rangle = 0,$$
$$\left(\frac{d^2}{dt^2} + 1\right) \langle u(t_2)u(t_1)x(t)\rangle + \langle u(t_2)u(t_1)u(t)\rangle\langle x(t)\rangle = 0,$$

(6.3b)
$$\langle x(0)\rangle = c_1, \quad \frac{d}{dt}\langle x(t)\rangle| = c_2,$$
$$\langle u(t_1)x(0)\rangle = 0, \quad \frac{d}{dt}\langle u(t_1)x(t)\rangle|_{t=0} = 0,$$
$$\langle u(t_2)u(t_1)x(0)\rangle = C(t_2 - t_1)c_1,$$
$$\frac{d}{dt}\langle u(t_2)u(t_1)x(t)\rangle|_{t=0} = C(t_2 - t_1)c_2.$$

The case of $n = 1$ is trivial since it corresponds to the neglect of u altogether and it is consequently identical to the zeroth order perturbation equation (the first of (3.5)).

The $n = 2$ case is nontrivial and will be given a detailed discussion. Let us first eliminate $\langle u(t_1)x(t)\rangle$ from the two equations in (6.2a) with the result

$$\left(\frac{d^2}{dt^2} + 1\right) \langle x(t)\rangle = \int_0^t dt' \sin(t - t')C(t - t')\langle x(t')\rangle, \tag{6.4}$$

where $C(t - t')$ is the second order correlation function of $u(t)$ given by (3.12). In lieu of a rigorous analysis of the properties of (6.4), we shall discuss several limiting cases.

If $t << C(0)$, (6.4) can be solved by iteration with the result

(6.5)
$$\langle x(t)\rangle = A \sin(t + \theta)$$
$$+ A\int_0^t dt' \int_0^{t'} dt'' \sin(t - t') \sin(t' - t'')\, C(t' - t'') \sin(t'' + \theta) + \cdots.$$

This expression, as far as it goes, is in exact agreement with the results of the perturbation method presented in (3.11). As might be expected, the higher order terms, not shown in (6.5), will disagree with the corresponding higher order terms of the perturbation series.

Another interesting limiting case is one in which the correlation time is short. More specifically it is assumed that

$$\int_0^\infty ds\, C(s) >> \left| \int_0^\infty s\, ds\, C(s) \right| >> \left| \int_0^\infty s^2\, ds\, C(s) \right| >> \cdots. \tag{6.6}$$

On this basis, we can with acceptable accuracy reduce (6.7) to the form

$$\frac{d^2}{dt^2}\langle x(t)\rangle + C_2 \frac{d}{dt}\langle x(t)\rangle + (1 - C_1)\langle x(t)\rangle = 0, \tag{6.7}$$

where

$$C_n = \int_0^\infty s^n\, ds\, C(s). \tag{6.8}$$

It can thus be seen that if $C_2 \geqq 0$, the solution of (6.7) is bounded for all positive values of t, and in particular if $C_2 > 0$ the solution is dominated by an exponentially damped function. On the other hand if $C_2 < 0$ the solution is dominated by an exponentially growing function.

Another interesting limiting case is the one in which u is a random variable independent of t having zero mean and variance σ. It follows immediately that $C(t) = \sigma^2 =$ constant. (6.4) reduces to

$$\left(\frac{d^2}{dt^2} + 1\right)\langle x(t)\rangle - \sigma^2 \int_0^t dt' \sin(t - t')\langle x(t')\rangle = 0. \tag{6.9}$$

Solution of the above integro-differential equation by use of the Laplace transform gives

$$\begin{aligned} x(t) &= \tfrac{1}{2}c_1[\cos(1+\sigma)^{1/2}t + \cos(1-\sigma)^{1/2}t] \\ &\quad + \tfrac{1}{2}c_2[(1+\sigma)^{-1/2}\sin(1+\sigma)^{1/2}t + (1-\sigma)^{-1/2}\sin(1-\sigma)^{1/2}t], \end{aligned} \tag{6.10}$$

where $c_1 = x(0)$ and $c_2 = (dx/dt)(0)$. This result is remarkable in that it is the *exact* solution for a particular distribution of u, namely, one in which u assumes the two values $\pm\sigma$ with equal probability.

In the case of $n = 3$, the elimination of $\langle u(t_1)x(t)\rangle$ and $\langle u(t_1)u(t_2)x(t)\rangle$ from (6.3a) with the help of the initial conditions (6.3b) yields

$$\left(\frac{d^2}{dt^2} + 1\right)\langle x(t)\rangle - \int_0^t dt'\, L(t, t')\langle x(t')\rangle = g(t), \tag{6.11}$$

where

$$L(t, t') = -\int_{t'}^t dt'' \sin(t - t'')\sin(t'' - t')\langle u(t)u(t')u(t'')\rangle, \tag{6.12}$$

$$g(t) = -\int_0^t dt' \sin(t - t')C(t - t')A\sin(t' + \theta). \tag{6.13}$$

In (6.13) the quantities A and θ are defined by (3.8). It is to be noted that the integro-differential equation of the third approximation (see (6.11)) possesses a substantially different structure from that of the corresponding equation of the second approximation. It also involves rather surprisingly the zeroth order solution (the factor $A \sin(t' + \theta)$ in (6.13)).

7. **Application of cumulant discard truncation.** As stated in paragraph (b) of §5, the cumulant discard truncation procedure is identical to the correlation discard procedure in the first and second orders ($n = 1, 2$). It is only in the next order that significant differences appear. For the case of $n = 3$ we obtain

$$\left(\frac{d^2}{dt^2} + 1\right) \langle x(t)\rangle + \langle u(t)x(t)\rangle = 0,$$

$$\left(\frac{d^2}{dt^2} + 1\right) \langle u(t_1)x(t)\rangle + \langle u(t_1)u(t)x(t)\rangle = 0,$$

$$\begin{aligned}\left(\frac{d^2}{dt^2} + 1\right) &\langle u(t_2)u(t_1)x(t)\rangle + \langle u(t_2)u(t_1)u(t)\rangle\langle x(t)\rangle \\ &+ \langle u(t_1)u(t_2)\rangle\langle u(t)x(t)\rangle + \langle u(t_1)u(t)\rangle\langle u(t_2)x(t)\rangle \\ &+ \langle u(t_2)u(t)\rangle\langle u(t_1)x(t)\rangle = 0\end{aligned} \tag{7.1}$$

subject, of course, to the initial conditions (4.3). The elimination of quantities of the type $\langle u(t')x(t)\rangle$ and $\langle u(t')u(t'')x(t)\rangle$ from the above set of equations is not elementary. It can be seen, however, that the integro-differential equation for $\langle x(t)\rangle$ obtained by the above elimination will be substantially more complicated than (6.11) resulting from the third order correlation discard approximation.

8. **Application of the least mean square error procedure.** In the present section we discuss the use of the procedure of paragraph (c) of §5, in the first order case ($n = 1$). To review our earlier statements, we consider the first member of the unaveraged hierarchy (4.1), namely,

$$\left(\frac{d^2}{dt^2} + 1\right) x(t) + u(t)x(t) = 0 \tag{8.1}$$

and approximate $u(t)x(t)$ by the least mean square error procedure discussed previously. This procedure involves the minimization of the mean square error $\langle \eta^2(t)\rangle'$ where according to (5.5) the error is given by

$$\eta(t) = u(t)x(t) - w(t) - \int_0^t dt'\, K(t, t')x(t'). \tag{8.2}$$

The next step is to minimize $\langle \eta^2(t)\rangle'$ with respect to $w(t)$ and $K(t, t')$. However, before taking this step we must make more precise the meaning of $\langle\,\rangle'$ (as contrasted with $\langle\,\rangle$). As stated before, we will retain the given statistical properties of $u(t)$.

However, we will use the perturbation method of §3 to determine $x(t)$ as an approximate functional of $u(t)$. We will also regard for the moment that c_1 and c_2 are not constant but are random in such a way that A given by (3.8) is a constant and θ also given by (3.8) is random with a uniform probability distribution in the interval $(0, 2\pi)$.

In applying the perturbation method of §3, we proceed in an unrigorous and experimental way. As a first step, let us apply the perturbation method to the zeroth order. We then obtain to this order

$$\eta(t) = u(t)x_0(t) - w(t) - \int_0^t dt' K(t, t')x_0(t'), \tag{8.3}$$

whereupon the minimization of $\langle \eta^2(t) \rangle'$ with respect to $w(t)$ and $K(t, t')$ gives $w(t) = 0$ and $\int_0^t dt' K(t, t') \cos (t' - t'') = 0$. A sufficient condition for the vanishing of the last integral is the vanishing of $K(t, t')$; however, if we require that $K(t, t') = K(t - t')$, the vanishing of $K(t - t')$ is a necessary *and* sufficient condition.

Thus, the use of the zeroth order perturbation approximation for $x(t)$ in $\eta(t)$ leads to trivial results and it is clear that this situation arises from the fact that, due to the low order of approximation, the term $u(t)x(t)$ of $\eta(t)$ disappeared from the variational problem. To rectify this, we replace the zeroth order approximation into $u(t)x(t)$ by the first order obtaining

$$\eta(t) = u(t)[x_0(t) + x_1(t)] - w(t) - \int_0^t dt' K(t, t')x_0(t'), \tag{8.4}$$

where $x_1(t)$ is given by (3.10) of §3. Minimization of $\langle \eta^2(t) \rangle'$ with the new approximate expression yields $w(t) = 0$ and

$$\int_0^t dt' \, [K(t, t') + \sin (t - t')C(t - t')] \cos (t' - t'') = 0. \tag{8.5}$$

It is clear that a sufficient condition for the vanishing of (8.5) is that

$$K(t, t') = -\sin (t - t')C(t - t'). \tag{8.6}$$

However, if we require in advance that $K(t, t') = K(t - t')$ then it follows that (8.6) is a necessary *and* sufficient condition. It is to be remarked that (8.4) does not represent a consistent perturbation expansion of $\eta(t)$. The rationale behind the above procedure is that every term in $\eta(t)$ involving $x(t)$ is treated in the lowest order that will insure its survival in the minimization of the mean square error.

Proceeding further, it follows that according to the last least mean square error procedure, the term $u(t)x(t)$ of (8.1) is to be approximated in the manner below:

$$u(t)x(t) \cong -\int_0^t dt' \sin (t - t')C(t - t')x(t') \tag{8.7}$$

and consequently (8.1) reduces to

$$\left(\frac{d^2}{dt^2} + 1\right) x(t) - \int_0^t dt' \sin (t - t')C(t - t')x(t') = 0. \tag{8.8}$$

The mean value of (8.8) is identical to the second order approximation resulting from both the correlation and cumulant discard procedures.

The same type of truncation procedure could be applied to later members of the unaveraged hierarchy. However, in the interest of brevity such considerations are excluded from this paper.

9. **Discussion.** In the preceding paragraphs, the derivation of the average solution of the simple stochastic differential equation (2.1) was treated by three different methods, all involving truncated hierarchies. Two kinds of hierarchies have been considered: the unaveraged hierarchy (see (4.1)) and the averaged hierarchy (see (4.2)). Two types of truncation procedures have been applied to the averaged hierarchy, namely, correlation discard and cumulant discard. The unaveraged hierarchy has been truncated by means of the least mean square error method. We now compare the results of these procedures.

When the averaged hierarchy is truncated at the first member by either correlation or cumulant discard, the result is identical to the solution of the original differential equation with the stochastic coefficient set equal to zero. In the truncation of the first member of the unaveraged hierarchy by the least mean square error procedure the same result is obtained if in the computation of the mean square error the dependence of $x(t)$ on $u(t)$ is treated by zeroth order perturbation theory.

However, if the averaged hierarchy is truncated at the second member by either correlation or cumulant discard, the result is more interesting. After appropriate eliminations of variables, we find that the average $x(t)$ approximately satisfies the integro-differential equation

$$\left(\frac{d^2}{dt^2} + 1\right) \langle x(t)\rangle - \int_0^t dt' \sin (t - t')C(t - t')\langle x(t')\rangle = 0, \tag{9.1}$$

where $C(t - t') = \langle u(t)u(t')\rangle$. In the case of the *first* member of the unaveraged hierarchy an identical result (after averaging) is obtained if in the least mean square error procedure a higher order perturbation approximation is used for the relation between $x(t)$ and $u(t)$ as discussed at the end of §8. This result suggests that the mean least square error truncation procedure may be basically more powerful since when it is applied with suitable refinement to the first member of the unaveraged hierarchy, it gives (after a final averaging) the same result as the other truncation procedures applied to the second member of the averaged hierarchy.

The integro-differential equation (9.1) was investigated to a limited extent and the following properties were established:

(a) For $t << C(0)$ the solution of (9.1) is identical to the result of the perturbation method carried to the second order.

(b) If u is time-independent, that is, the correlation function $C(t) = \sigma^2 =$ const., the solution of (9.1) is exact if it is assumed to take the values $\pm\sigma$ with equal probability.

(c) The solution of (9.1) is in some cases bounded as $t \to \infty$, depending upon the correlation function $C(t)$, while the perturbation method never yields bounded solutions (except in the trivial case where $\int_0^\infty dt\, t^n C(t) = 0$, $n \geqq 1$).

Although much remains to be done in the study of (9.1), the above properties suggest that truncated hierarchy techniques have promise, at least as applied in the above manner.

The correlation discard and cumulant discard procedures yield quite different results when applied to the third member of the averaged hierarchy. Further study is required in order to understand the nature of these results.

In this paper only one statistical property of $x(t)$, namely $\langle x(t) \rangle$, has been considered. The extension of the methods discussed here to the calculation of higher order statistical properties (e.g., $\langle \Delta x(t)\, \Delta(t') \rangle$, etc.) will be treated in a future communication.

REFERENCES

1. A. T. Bharucha-Reid, *On random operator equations in Banach spaces*, Bull. Acad. Polon. Sci. Sér. Sci. Math. Astronom. Phys. **1** (1959), 561–564.

2. F. Dyson, *The dynamics of a disordered linear chain*, Phys. Rev. **92** (1953), 1331–1338.

3. I. Tamm, *Relativistic interaction of elementary particles*, J. Phys. (USSR) **9** (1945), 449.

4. P. C. Martin and J. Schwinger, *Theory of many-particle systems*. I, Phys. Rev. (2) **115** (1959), 1342–1373.

5. N. N. Bogoliubov, *Problems of a dynamical theory in statistical physics*, Studies in statistical mechanics, de Boer and G. E. Uhlenbeck, eds., pp. 5–116, North-Holland, Amsterdam, 1962.

6. R. C. Bourret, *Stochastically perturbed fields, with applications to wave propagation in random media*, Nuovo Cimento (10) **26** (1962), 1–31.

HUGHES RESEARCH LABORATORIES,
MALIBU, CALIFORNIA

AUTHOR INDEX

Italic numbers refer to pages on which a complete reference to a work by the author is given.

Roman numbers refer to pages on which a reference is made to a work of the author. For example, under Minkowski would be the page on which a statement like the following occurs: "This theorem was proved earlier by Minkowski [7, §2] in the following manner . . ."

Boldface numbers indicate the first page of articles in this volume.

SUBJECT INDEX

BCDEFGHIJ–CM–89876543210